BIOLOGY

THE UNITY AND DIVERSITY OF LIFE

FOURTH EDITION

CECIE STARR
Belmont, California

RALPH TAGGART
Michigan State University

General Advisors and Contributors

John Alcock
Arizona State University
Author of Chapters 48, 49

James Bonner, Professor Emeritus
California Institute of Technology
Advisor on Cell Biology and Genetics

Robert Colwell
University of California, Berkeley
Author of Chapters 43, 44; Advisor on Ecology

George Lefevre, Professor Emeritus
California State University, Northridge
Advisor on Genetics

William Parson
University of Washington
Advisor on Biochemistry

Cleon Ross
Colorado State University
Author of Chapters 20, 22; Advisor on Botany

Samuel Sweet
University of California, Santa Barbara
Author of Chapter 37; Advisor on Evolution

Wadsworth Publishing Company
Belmont, California
A Division of Wadsworth, Inc.

On the cover; Galápagos hawk at sunrise, on the rim of an active fumerole. The diversity of life on the Galápagos Islands profoundly influenced Charles Darwin's perception of the mechanisms underlying evolution, which is a major organizing principle in biology. (Photograph courtesy Tui De Roy)

Biology Editor: Jack C. Carey

Production Manager: Mary Forkner, Publication Alternatives

Editorial Associate: Ev Sims

Art Director: Stephen Rapley

Copy Editor: Toni Haskell

Permissions: Marion Hansen

Production Assistants: Cheryl Carlson, Philippa Webster, Claudia Willrodt

Artists: Lewis Calver, Joan Carol, Raychel Ciemma, Ron Ervin, Laszlo Meszoly, Darwen Hennings, Vally Hennings, Joel Ito, Julie Leech, Victor Royer

Graphic Artists: Alan Noyes, Susan Breitbard, Pat Rogondino, Jeanne Schreiber, Salinda Tyson

Marketing: Suzanna Brabant, Joy Westberg

Typography: Jonathan Peck Typographers, Ltd.

Color Processing: Image International, Focus 4, Pre-Press Graphics, Star Graphics

Printing: Rand McNally

ISBN 0-534-06924-X

Printed in the United States of America

1 2 3 4 5 6 7 8 9 10—91 90 89 88 87

Library of Congress Cataloging in Publication Data

Starr, Cecie.
 Biology: the unity and diversity of life.

 Includes bibliographies and index.
 1. Biology. I. Taggart, Ralph. II. Title.
QH308.2.K57 1987   574   86-24608
ISBN 0-534-06924-X

PREFACE

Biology, like life itself, has a history of radiating in diverse directions; and in the past few decades alone its pioneers have made truly spectacular discoveries. At the start of each book revision we renew our commitment to do justice to this discipline, which is doing so much to enrich our knowledge of a world we share with more than four million species; for we believe deeply that a biological perspective is one of the greatest gifts a student can receive.

This revision, by far the best, benefited from a gathering of talented individuals who share our commitment to quality in education and who understand there are certain standards governing the content and presentation of introductory biology books. Repeatedly when writing this fourth edition, we had reason to thank the several hundred instructors and reviewers who worked with us early on to identify those standards, which continue to guide our work:

1. State the principles that apply to all fields of biological inquiry, the most pervasive being the principles of evolution and energy flow—and *use* those principles as a framework for presenting the material throughout the book.

2. Distill the main concepts and outline the research trends in all major fields.

3. Give enough examples of problem solving and experiments to provide familiarity with a scientific approach to interpreting the world.

4. Include enough comparative biology to convey a sense of the unity and diversity among living things.

5. Include enough human biology to enhance understanding of our evolution, behavior, and ecology as well as our body structure and functioning.

6. Be selective in developing the vocabulary necessary to comprehend what is being talked about today in each field.

7. Present the material accurately but not at a high level.

8. Write clearly, in a style that is not boring or patronizing.

9. Excise teleological wording.

10. Create easy-to-follow line illustrations and select photographs that are informative as well as beautiful.

11. Keep the writing free of bias, thereby giving students the chance to form their own opinions about the material being presented.

SOME IMPORTANT CONTENT CHANGES

The book remains much the same in its overall organization, which our users consistently find to be effective for their teaching purposes. The revision effort, which was extensive, resulted in the following improvements:

1. The most encompassing change is the extent to which all chapters have been updated and checked for accuracy. Biology is startling in its breadth and depth, and it has continued to grow rapidly during the three years since the publication of our third edition. Today, producing a timely and authoritative survey book requires a strong community effort. We worked closely with more than two hundred highly respected specialists who recognize the need to give students a current, accessible view of their fields. Some were general advisors or consultants. Others wrote original manuscripts, with the understanding that those manuscripts would be rewritten to integrate the style, level, organization, and coverage with the book as a whole.

2. The second most encompassing change is the extent to which we tightened up each chapter to make it easier to follow. We collaborated with instructors and research specialists, some of whom helped us research and interpret a monumental number of new and sometimes conflicting reports and articles. Together we developed a clear picture of what is important in biology today and, consequently, were able to prune each chapter of unnecessary details.

3. We added end-of-chapter summaries to reinforce major concepts.

4. In keeping with our goal to make chapters short enough to be read in one sitting, more of the material was split into smaller chapters. The splits also allow more flexibility than before in adapting the book to different course sequences. For example, mitosis and meiosis are now treated separately but in adjacent chapters.

5. We have essentially new chapters covering three of the most rapidly changing fields in biology: recombinant DNA and genetic engineering, neurophysiology, and immunity. Our reviewers were especially pleased with the clarity, accuracy, and writing style of those chapters.

6. We reorganized some sections of the cell biology and genetics units. The properties of water are now covered in a short chapter on basic chemistry, which precedes a

chapter on the organic compounds found in cells. Mutation is now covered at the end of the chapter on protein synthesis. The chapter on human genetics is now placed directly after the chapters on basic Mendelian genetics.

7. Largely as a result of Thomas Rost's excellent advice, we reorganized and rewrote Chapter 19 on plant tissues. In Chapter 21 the treatment of coevolution of flowering plants and their pollinators has been expanded into a beautifully illustrated case study.

8. Our users have reported so much student interest in human anatomy and physiology that we expanded our coverage of that material. We address numerous topics of social concern from a biological perspective, including AIDS and other sexually transmitted diseases, the effects of smoking (covered in the respiration chapter and in the discussion of human embryonic development), anorexia nervosa, and bulimia, the latter being rampant among college-age women. We included a large number of full-color anatomical paintings, done by some of the country's top medical illustrators.

9. The evolution unit now includes a separate chapter on phylogeny and macroevolution. Processes of speciation, the fossil record, and the history of major radiations and extinctions are discussed authoritatively.

10. The ecology unit underwent major organizational shifts and expansions. Life history patterns are given a modern treatment. Succession is now covered in Chapter 44 (community interactions) which also includes new sections on island, mainland, and marine patterns of species diversity. All biogeochemical cycles are now treated in the ecosystems chapter, which includes coverage of some of the Hubbard Brook studies. The increases in atmospheric concentrations of carbon dioxide are described, as well as the global consequences of the destruction of tropical forests.

OUR APPROACH TO WRITING

Our book has a reputation for its distinctive writing style, and we are grateful for the praise. Even so, we subjected the writing to rigorous scrutiny, not only from more than a hundred instructors who use the book in class but also from Ev Sims, one of the most intelligent editors in publishing. Ev is competent, thorough, and attuned to the needs of student readers, and he put us through the hoops in our mutual desire to improve readability.

We still shun Platte River writing (a mile wide and an inch deep). Such writing encourages students to get quickly to the other side of a topic without getting their feet wet, but it does not really help them learn to self-navigate in deeper waters—which they most assuredly will be encountering throughout their lifetime in this increasingly complex world.

On the other hand, Amazon River writing (very long, often tricky) does introductory students no good either. For this edition, we condensed sections while preserving the basic concepts and the most illuminating examples. This is especially true of our chapters on cell biology and genetics, which are now much more accessible as well as informative.

A NEW ILLUSTRATION PROGRAM

Developing a new art program has been one of the most rewarding and stimulating aspects of the revision. As in past editions, we linked the illustrations intimately with the text—and those 1,300 pieces of art took almost as much research and developmental time as did the text itself. We personally designed the illustration layouts for maximum clarity and pedagogical value, then we laid out the actual book pages so the illustrations would be juxtaposed with the text passages they help clarify.

Reviewers told us that we have made breakthroughs in biological illustration. We created more full-color pieces of art that make hard-to-visualize topics (such as gated channels across a cell membrane) more tangible. We developed numerous full-color anatomical paintings of plants and animals. For the line art, we judiciously selected multiple color combinations that will help students track the information being presented. Many more explanatory diagrams accompany the micrographs.

We greatly expanded the number of original illustrations in which visual and written summaries are combined to make concepts easier to grasp. Wherever possible, we broke down the information into a series of steps, which are far less threatening than one large, complicated diagram. Instructors report that students find this approach exceedingly useful, particularly with ard to the illustrations on mitosis, meiosis, and protein synthesis. It works just as effectively in the new art on such complex topics as antibody-mediated immunity, reflex pathways, and neural functioning.

ADDITIONAL FEATURES

Case Studies, Commentaries, and Perspectives

Our Case Studies show how general concepts apply to specific situations. For example, after the discussion of immunity, a case study shows students how their body would mount an immune response to a bacterial attack (page 413).

The Commentaries, written by ourselves and others, are "outside readings" built into the book. They explore such thought-provoking topics as cancer, death, cardiovascular and lung disorders, and sexually transmitted diseases.

The Perspectives are end-of-chapter sections that encourage students to take a moment to think about the connections between chapters and units. Many also bridge chapter topics and the student's world, inviting reflection on the past and possible futures. See, for example, pages 131 and 747.

Study Aids

Students taking introductory biology will not already know enough about biology to spot all key concepts in page after text page. *Major summary statements* are boxed and printed in boldface for emphasis. Taken together, the statements are an easy-to-identify, *in-context summary* of important concepts. Many are in list form (see, for example, pages 160, 264, and 406). Concepts are further reinforced by *end-of-chapter summaries* as well as *summary illustrations* (see pages 156, 204, 216, and 257 for examples of the latter).

Review questions are keyed to italic and boldfaced sentences as a way of reinforcing the main concepts. Chapters Twelve through Fourteen have *genetics problems*, with *detailed solutions* in Appendix II. The *glossary*, which brings together the text's main definitions, includes pronunciation guides. It also includes origins of words, when such information will make seemingly formidable words less so. The *index* is comprehensive, simply because students may find a door to the text more quickly through finer divisions of topics.

Supplement Package

An extensive package of ten supplements accompanies this text. There are many more *full-color transparencies* of the book's illustrations, all labeled with large, boldface type. The *Instructor's Manual* includes 2,500 new test questions by David Cotter. The questions are also available on IBM and Apple IIe as *Micro pac* (a test generator, editor, and authoring program). TnT (Test and Tutor) *software* for the IBM PC includes simple, easy-to-read menus that assist professors in creating tutorials and interactive tests and that allow them to retrieve student scores.

Jane Taylor wrote a new *Study Guide* in which learning aids are organized by chapter section, allowing students to focus on smaller amounts of material and to skip over unassigned sections. Each chapter section has a detailed summary, list of key terms, learning objectives, and a self-quiz. *Self-instructional computer software* and *self-instructional film strips and slide sets* are available.

Current Readings in Biology, an anthology compiled by John Crane, includes 1985–1986 articles on applications of biological research.

Finally, Jim Perry and David Morton have written a new *Laboratory Manual* to accompany our book. All of its thirty-three experiments and exercises are divided into distinct parts that can be assigned individually, depending on the time available in the laboratory. All have the same format, with objectives, a discussion (introduction, background, and relevance), a list of materials for each part of the experiment, procedure, pre-lab questions, and post-lab questions. The procedure is in list form with each step numbered. The procedures are detailed enough so each exercise is self-explanatory. An *Instructor's Manual* by Joy Perry accompanies the laboratory manual. It covers quantities, procedures for preparing reagents, time requirements for each portion of the exercise, hints to make the lab a success, and vendors of materials, with item numbers.

A COMMUNITY EFFORT

Over the past eleven years, more than five hundred instructors and research specialists have given us the benefit of their insights, and to call our book "the Starr-Taggart book" is something of a misnomer; it is the outcome of a large-scale community effort.

As with previous editions, we extended our interactions with reviewers beyond consideration of the comments on manuscript review sheets. Multiple drafts, multiple and detailed reviews, extended correspondence and phone calls served as the foundation for each chapter, as did feedback from hundreds of instructors who have used the book in class.

How can we convey the closeness of so many of those interactions? Consider Joyce Maxwell, an outstanding reviewer who became so deeply involved with the development of the fourth edition that we experienced withdrawal symptoms when we no longer had problems to solve together. Consider Judith Brown, Stephen Hedman, Kenneth Jones, Heather McKean, John Rickett, and Pat Starr, urging us on to new heights from the beginning until the last paragraph of the last chapter. These and many other biologists transcend the "reviewer" designation; to us they have become part of an extended family, united in a shared commitment to education.

Similarly, the individuals listed as follows in alphabetical order are deserving of credit, not only for their contribution to the book's currency and accuracy, but also for their willingness to accept our strong control over the presentation of its content:

John Alcock, author of an acclaimed book on animal behavior, expanded the excellent material he wrote for the third edition into two chapters that present a balanced picture of the field.

Gary Atchison, one of the best reviewers of the ecology unit, expanded and helped update Chapter 47 with such diligence that he really should be considered its coauthor.

James Bonner, a general advisor since the second edition, helped update both the cell biology and genetics units. Dr. Bonner has given us much of his time and he continues to be our teacher, in the best sense of the word.

Robert Colwell wrote Chapters 43 and 44 and was general advisor on ecology. The entire ecology unit benefited from Dr. Colwell's razor-sharp mind, which has earned him considerable respect in his field. This truly gifted writer endured our manuscript development process with utmost grace.

Jerry Coyne and Janet Dunaif-Hattis revised Chapter 42, moving with a sure hand through the maze of interpretations concerning human evolution. Nancy Dengler helped us rewrite the material on nonvascular and vascular plants in Chapter 40. Stephen Hauschka helped update and reorganize the chapters on molecular genetics.

Eugene Kozloff, one of the country's leading invertebrate zoologists, contributed a resource manuscript for Chapter 41 and, in so doing, will help eliminate some of the fairy tales about animal diversity that abound at the introductory level. Dr. Kozloff is a gracious and meticulous scholar, and a friend of long standing.

George Lefevre was a general advisor for the genetics unit and wrote new genetics problems and solutions. He also was one of our most useful critics; we never will get lazy with him around (we wouldn't dare).

Elizabeth Moore-Landecker, author of a highly respected book on fungi, helped revise our coverage of fungal diversity in Chapter 40. David Morton kindly reviewed Chapter 23, wrote a case study for it, and helped us expand sections in Chapter 28. David Murrish graciously agreed to write the case study for Chapter 33.

William Parson was the major advisor for the chapters on biochemistry. His knowledge of cell metabolism is impressive, and so is his sensitivity to the trepidation with which many students approach the subject. Dr. Parson, coauthor of a leading biochemistry text, did a valuable line-by-line analysis to enhance clarity and verify the accuracy of the material.

Cleon Ross, coauthor of the bestselling book on plant physiology, updated Chapters 20 and 22, which he wrote for the preceding edition. As good friend and general advisor on botany, he accepted calls at all hours to verify the most minute points.

Samuel Sweet, one of the best thinkers in his field, made a major contribution to updating the evolution unit. Besides being general advisor on evolution, he wrote Chapter 37 and worked so intensely on Chapters 36 and 38 that he should be considered their coauthor. Our interaction included spirited conversations and mutual broodings, and we are indebted to him for his good-natured flexibility in developing chapters that introductory students can handle with ease.

Mark Wheelis, a coauthor of the leading microbiology textbook, served as a consultant for Chapter 39, and his suggestions have strengthened our coverage of microbial diversity.

Finally, Peter Armstrong, Joseph Bagnara, George Brengelmann, Fred Delcomyn, Mac Hadley, Douglas Kelly, William Lassiter, Robert Little, Dorothy Luciano, Richard Murphy, Gordon Ross, Gordon Shepherd, Ian Tizard, James Sherman, and Brian Whipp must be given special acknowledgment. Those individuals advised us on the various chapters covering different aspects of animal physiology—which is not one but at least fourteen different and extremely specialized fields. Their exceptionally detailed suggestions helped make the animal anatomy-physiology unit one of the strongest in the book.

A word for Wadsworth. A revision of this magnitude would not have been possible without the unqualified support of Dick Greenberg, Wadsworth's president, who mobilized the resources of his company for us and also fixed a sentence on page forty-four. Kathie Head, Bill Ralph, Stephen Rapley, and Steve Rutter, besides managing their departments, took a personal interest in smoothing the road to the book's publication, as they have done for past editions.

We wish we could keep our production manager and art coordinator, Mary Forkner, under wraps so she will be available for the next edition because we never want to work with anybody else. Mary kept things moving, she never lost her composure or wit, she held our demands in check with integrity and grace, and she exercise skilled judgment in doing what was right for the book.

Stephen Rapley was the art director for the book. This was a challenging project, given the need to fit hundreds of existing illustrations into a new book format even while developing hundreds of additional pieces of multiple-colored line art. Stephen is a rare individual in publishing—a skilled manager *and* a talented designer.

Toni Haskell, Cheryl Carlson, John Prickett, Ruthie Singer, Claudia Willrodt, Philippa Webster, and Judi Walcom, who worked on the editorial aspects of this book, were marvelous in their attention to details.

This time, we told Marion Hansen, the photo research and permissions would be a snap. They weren't, but once started, Marion doesn't quit. She is a tenacious, crackerjack sleuth who tracked down photographs cleverly hidden all over the world.

Once again, Darwen Hennings, Vally Hennings, and Joel Ito took on major portions of the art program, put up with our nitpickings, and created illustrations of a quality for which they are deservedly well known. Lewis Calver, one of the finest medical illustrators in the country, developed many stunning full-color paintings, as did Ron Ervin. Victor Royer took on most of all. We marvel at his skill and intuitive sense of knowing what we want the first time around, thereby protecting time and nerves.

So much outstanding art was developed for this edition that we also must acknowledge the skilled contributions of Susan Breitbard, Joan Carol, Raychel Ciemma, Julie Leech, Laszlo Meszoly, Alan Noyes, Pat Rogondino, Jeanne Schreiber, and Salinda Tyson. We acknowledge, too, the extraordinary skill and speed of Pete Shanks, Merry Bilgere, Laurie Osterhoudt, and the staff at Jonathan Peck Typographers.

Finally, the revision never would have been finished this soon if it were not for our editor, Jack Carey, who worked just as hard as we did on it. His persistence is legendary. One professor told us she asked herself, on the way to the post office late at night on a holiday weekend to mail a review, Why in the world am I doing this?? Probably for the same reason we do: Jack somehow makes it seem like all things worth doing will come to an end if we don't. And he couldn't make us believe it unless he believed it himself. Jack is the catalyst who makes things happen as they should, and faster than they otherwise would. If this book turns out to be as useful as we think it will be in the classroom, he deserves a significant share of the credit.

FOURTH EDITION REVIEWERS

Mike Adams, Eastern Connecticut State University
John Alcock, Arizona State University
Peter Armstrong, University of California, Davis
Gary Atchison, Iowa State University
Louis Avosso, Nassau Community College
Joseph Bagnara, University of Arizona College of Medicine
Aimee Bakken, University of Washington
Mary Barkworth, Utah State University
William Barnes, Clarion University of Pennsylvania
Barry Batzing, State University of New York, Cortland
Edward Barrows, Georgetown University
Wayne Becker, University of Wisconsin, Madison
William Bessler, Mankato State University
Louis Best, Iowa State University
Samuel Bieber, Old Dominion University
Carl Biehl, University of California, San Diego
Paul Biersuck, Nassau Community College
William Birky, Ohio State University
William Bischoff, University of Toledo
Antonie Blackler, Cornell University
John Blamire, Brooklyn College
David Bohr, University of Michigan Medical School
Joe Bonfiglio, West Valley College
James Bonner, California Institute of Technology
Robert Brasted, University of Minnesota
George Brengelmann, University of Washington School of Medicine
Edmund Brodie, University of Texas
Virginia Brooks, Oregon Health Sciences University
Leon Browder, University of Calgary
Arthur Brown, Oregon Health Sciences University
Judith Brown, California State University, Stanislaus
David Brumagen, Moorhead State University
Ann Burgess, University of Wisconsin, Madison
Robert Burnes, University of Arkansas for Medical Sciences
John Cabot, State University of New York, Stony Brook
Clyde Calvin, Portland State University
Donald Campbell, St. John's University, New York
Bruce Carlson, University of Michigan
Christine Case, Skyline College
David Chapman, University of California, Los Angeles
Neil S. Cherniack, University Hospital, Cleveland
Rubin Cherniack, National Jewish Center of Immunology and
    Respiratory Medicine
A. Kent Christensen, University of Michigan Medical School
Donald Christian, University of Minnesota
Paul Churchill, Wayne State University School of Medicine
William Clark, University of California, Los Angeles
Richard Coles, Washington University, Missouri
Robert Colwell, University of California, Berkeley
John Conway, University of Scranton
Joyce Corban, Wright State University
David Cormack, University of Toronto, Faculty of Medicine
Jerry Coyne, University of Maryland
George Cox, San Diego State University
David Crews, University of Texas
William Crumpton, Iowa State University
Fred Delcomyn, University of Illinois, Urbana
Darleen DeMason, University of California, Riverside

Jerry Dempsey, Medical School, University of Wisconsin, Madison
Nancy Dengler, University of Toronto
Megan Dethier, Friday Harbor Laboratory
Alfred Diboll, Macon Junion College
Stanley Duke, University of Wisconsin, Madison
Diane Eardley, University of California, Santa Barbara
Gerald Eck, University of Washington
John Endler, University of California, Santa Barbara
Larry Epstein, University of Pittsburgh
James Estes, University of Oklahoma
Craig Evinger, State University of New York, Stony Brook
Richard Falk, University of California, Davis
Donald Fisher, Washington State University
Karl Flessa, University of Arizona
Joe Fondacaro, Smith Kline French Laboratories
Douglas Futuyma, State University of New York, Stony Brook
William Ganong, University of California, San Francisco
Michael Ghiselin, California Academy of Sciences
James Giesel, University of Florida
Edward Golub, Purdue University
Anne Good, University of California, Berkeley
H. Maurice Goodman, University of Massachusetts
    Medical School, Worcester
Barbara Gordon, Indiana University
James Grendell, University of California, San Francisco
Stanley Gunstream, Pasadena Community College
William B. Gurley, University of Florida
Sam Ha, Millersville University
Mac Hadley, University of Arizona Medical School
Jim Hanken, University of Colorado
Keith Hartberg, Georgia Southern College
Aslam Hassan, University of Illinois, Urbana
Janet Dunaif-Hattis, Northwestern University
Stephen Hauschka, University of Washington
Stephen Hedman, University of Minnesota
Steven Heidemann, Michigan State University
Lois Jane Heller, School of Medicine, University of Minnesota, Duluth
Thomas Hemmerly, Middle Tennessee State University
Wilford Hess, Brigham Young University
Bruce Holmes, Western Illinois University
Kenneth Holmes, University of Illinois, Urbana
Erik Holtzman, Columbia University
Alfred Hopwood, St. Cloud State University
Edwin House, Idaho State University
Nicki Huls, West Valley College
Robert Huskey, University of Virginia
John Jackson, North Hennepin Community College
Meyer Jackson, University of California, Los Angeles
Ray Jackson, Texas Tech University
Duane Jeffery, Brigham Young University
Sam Jones, University of Georgia
Kenneth Jones, California State University, Northridge
George Karleskint, St. Louis Community College at Meramec
Gerald Karp, formerly of University of Florida
Douglas Kelly, University of Southern California School of Medicine
Steven Kelsen, University Hospital, Cleveland
Bryce Kendrick, University of Waterloo, Ontario
Elaine Kent, California State University, Sacramento

Jack Keyes, University of Oregon
Arnold Kluge, University of Michigan
Richard Korf, Cornell University
Eugene Kozloff, Friday Harbor Laboratory
Armand Kuris, University of California, Santa Barbara
Howard Kutchai, University of Virginia Medical School
Clifford LaMotte, Iowa State University
Debbie Langsam, University of North Carolina, Charlotte
William Lassiter, School of Medicine, University of North Carolina, Chapel Hill
George Lauder, University of Chicago
Elias Lazarides, California Institute of Technology
Thomas Leeson, University of Alberta Faculty of Medicine
George Lefevre, California State University, Northridge
Georgia Lesh-Laurie, Cleveland State University
Matthew Levy, Mt. Sinai School of Medicine, City University of New York
David Lindsay, University of Georgia
Mary Lindstrom, Center for Great Lakes Studies
Robert Little, Medical College of Georgia
Elizabeth Lord, University of California, Riverside
John Lott, McMaster University
Jane Lubchenco, Oregon State University
Dorothy Luciano, formerly of University of Michigan
David Lygre, Central Washington University
Richard R. MacMillen, University of California, Irvine
Michael Madigan, Southern Illinois University
David Mallach, University of Toronto
Phil Mathis, Middle Tennessee State University
Gary Matthews, State University of New York, Stony Brook
Joyce Maxwell, California State University, Northridge
Jerry McClure, Miami University
David McGrane, Morehead State University
Heather McKean, Eastern Washington State University
Robert McKinnell, University of Minnesota
Charles McMullen, South Dakota State University
John McReynolds, University of Michigan Medical School
G. Tyler Miller, St. Andrew's Presbyterian College
Alan Mines, University of California Medical School, San Francisco
David Mohrman, School of Medicine, University of Minnesota, Duluth
Elizabeth Moore-Landecker, Glassboro State College
David Morton, Frostburg State College
August Mueller, State University of New York, Binghamton
Richard Murphy, University of Virginia Medical School
David Murrish, State University of New York, Binghamton
Peter Narins, University of California, Los Angeles
John Nelson, Oral Roberts University
Eugene Nester, University of Washington
Karl Niklas, Cornell University
John O'Connor, University of California, Los Angeles
Robin Panza, University of California, Santa Barbara
Jal Parakh, Western Washington University
Gordon Parker, University of Toledo
William Parson, University of Washington
Thomas Parsons, University of Toronto
Vicki Pearse, University of California, Santa Cruz
Jan Pechenik, Tufts University
Michael Pelczar, Jr., formerly of University of Maryland
James Perry, Frostburg State College
Jane Phillips, University of Wisconsin, Madison
Carl Pike, Franklin and Marshall College
John Pleasants, Iowa State University
Karen Porter, University of Georgia
C. Ladd Prosser, University of Illinois, Urbana
Alfred Rampone, Oregon Health Sciences University
Oscar Ratnoff, Case Western Reserve University
Marian Reeve, Merritt College

Doug Reynolds, Eastern Kentucky University
John Rickett, University of Arkansas
Robert Robbins, Michigan State University
Michael Rose, Dalhousie University
Marvin Rosenberg, California State University, Fullerton
Cleon Ross, Colorado State University
Gordon Ross, University of California Medical Center, Los Angeles
Michael Ross, College of Medicine, University of Florida, Gainesville
Thomas Rost, University of California, Davis
Joan Ruderman, Duke University
Rudolfo Ruibal, University of California, Riverside
Frank Salisbury, Utah State University
Joseph Sanger, University of Pennsylvania School of Medicine
Stephen Scheckler, Virginia Polytechnic Institute and State University
William Schlesinger, Duke University
Rudi Schmid, University of California, Berkeley
Richard Searles, Duke University
Duane Sears, University of California, Santa Barbara
Bradley Shaffer, University of California, Irvine
Gordon Shepherd, Yale University Medical School
James Sherman, University of Michigan Medical School
Steven Shimada, Yale University Medical School
Walter Shropshire, Jr., George Washington University
David Smith, Valencia Community College
Ralph Smith, University of California, Berkeley
Robert Smith, West Virginia University
Travis Solomon, Truman Veterans Administration Hospital, University of Missouri
John Southin
Gerald Spangrude, Stanford University School of Medicine
Pat Starr, Mount Hood Community College
Burton Staugaard, University of New Haven
Taylor Steeves, University of Saskatchewan
Janet Stein, formerly of the University of British Columbia
Eugene Strobel, Middle Tennessee State University
Lawrence Sullivan, University of Kansas Medical Center
William Surver, Clemson University
Lee Swedberg, Eastern Washington State University
Samuel Sweet, University of California, Santa Barbara
Paula Timiras, University of California, Berkeley
Ian Tizard, Texas A&M University
Kenneth Todar, University of Wisconsin, Madison
Jane Underwood, University of Arizona
James Valentine, University of California, Santa Barbara
Paul Van Houtte, Mayo Clinic
Richard Van Norman, University of Utah
Robert Waaland, University of Washington
Michael Wade, University of Chicago
Betty Ward, Mount San Antonio College
Robert D. Warmbrodt, University of Maryland
George Welkie, Utah State University
David West, Virginia Polytechnic Institute and State University
John West, University of California, Berkeley
Mark Wheelis, University of California, Davis
Brian Whipp, Harbor University of California, Los Angeles Medical Center
Bernadine Wisnieski, University of California, Los Angeles
Dana Wrensch, Ohio State University
Raphael Zidovetski, University of California, Riverside

CONTENTS IN BRIEF

CONTENTS

UNIT THREE / THE ONGOING FLOW OF LIFE

UNIT ONE

UNIFYING CONCEPTS IN BIOLOGY

1

ON THE UNITY AND DIVERSITY OF LIFE

Buried somewhere in that mass of nerve tissue just above and behind your eyes are memories of first encounters with the living world. Still in residence are sensations of discovering your own two hands and feet, your family, the change of seasons, the smell of rain-drenched earth and grass. In that brain are traces of early introductions to a great disorganized parade of insects, flowers, frogs, and furred things, mostly living, sometimes dead. There, too, are memories of questions—*"What is life?"* and, inevitably, *"What is death?"* There are memories of answers, some satisfying, others less so.

Observing, asking questions, accumulating answers—in this manner you have acquired a store of approximate knowledge about the world of life. During the journey to maturity, experience and education have been refining your questions, and no doubt the answers are more difficult to come by. What *is* life? What characterizes the living state? The answers you get may vary, depending, for example, on whether they come from a physician, a clergyman, or a parent of a severely injured person who must be maintained by mechanical life support systems because the brain no longer functions at all.

Yet despite the changing character of the questions, the world of living things persists much as it did before. Leaves still unfurl during the spring rains. Animals are born, they grow, reproduce, and die even as new individuals of their kind replace them. *The most important difference is in the degree of insight you now bring to observations, questions, and answers about such events.*

It is scarcely appropriate, then, for a book to proclaim that it is your introduction to biology—"the study of life"—when you have been studying life ever since awareness of the world began penetrating your brain. The subject is the same familiar world that you have already thought about to no small extent. That is why this book proclaims only to be biology *revisited*, in ways that may help carry your thoughts about life to deeper, more organized levels of understanding.

Let us return at the outset to the question, What is life? It happens that the answer has yet to be reduced to a simple definition. The word embodies a story that has been unfolding in myriad directions for several billion years! To biologists, "life" is what it is by virtue of its ancient molecular origins and its degree of organization. "Life" is a way of capturing and systematically using energy and materials. "Life" is a commitment to some specific program of growth and development; it is a capacity for reproduction. "Life" is adjustment to changing conditions—it is *adaptive* to environmental change, both in the short term and through successive generations. Clearly, a short list of definitions only hints at all that the word conveys. Deeper insight into its meaning comes with wide-ranging study of its characteristics, *for life cannot be understood in isolation from its history and its adaptive potential.*

Throughout this book you will encounter examples of living things—how they are constructed, how they function, where they live, what they do. You also will come across statements about those examples and the generalizations that can be drawn from them. Such statements are highlighted with dark type and separated by lines from the text. All the statements, taken together, will give you a sense of what "life" is.

With this in mind, let us turn to a few examples that can illustrate the most general concepts of all. Although the concepts are explored more fully in later chapters, they are summarized here to provide perspective on things to come. You may also find it useful to refer to them later on, as a way of reinforcing your grasp of details.

ORIGINS AND ORGANIZATION

Suppose someone asks you to point out the difference between a frog and a rock. The frog, you might say, has a body of truly complex organization. Its hundreds of

thousands of individual cells are organized into tissues. Its tissues are arranged into organs such as a heart and a stomach. The frog can move about on its own. And sooner or later (given a receptive member of the opposite sex), it can reproduce. A rock shows no such ordered complexity, it cannot move by itself, and it certainly cannot reproduce either on its own or in the company of another rock. If you deduce from this that a living organism has complex regional organization, the capacity to move, and the capacity to reproduce, then the frog is alive and the rock is not.

Now suppose someone asks you to point out the difference between a bacterium and a rock. A bacterium is one of the simplest organisms, no more than a single cell. Yet microscopic examination shows that a bacterial body is regionally organized in complex ways. All but a few bacteria have an outer wall. All have a plasma membrane (a saclike outer membrane that helps control the kinds of substances moving into and out of the cell). The membrane encloses a semifluid substance in which specific structures are embedded (Figure 1.2). All bacteria can divide and reproduce. Some can move on their own

Figure 1.1 A common egret on the wing, against a background of bald cypress and Spanish moss. These three kinds of organisms are diverse in appearance, yet they have much in common. They illustrate the unity and diversity inherent in *all* of life, which is the subject of this chapter.

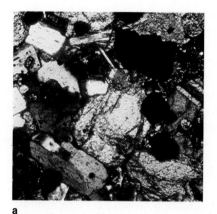

a

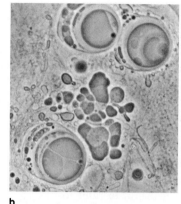

b

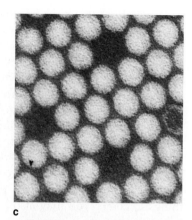

c

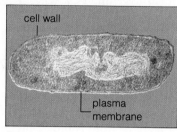

d

Figure 1.2 A hierarchy of structural organization between a frog and a rock. The sizes of the subjects in these photographs are not to the same scale relative to one another. (**a**) Basalt rock, thin-section, from Marianas Trench. (**b**) Liposomes, artificial lipid spheres with membrane properties that mimic living cell membranes. (**c**) Virus particles, with outer layers enclosing hereditary instructions. (**d**) Bacterial cell, sliced to show the inside. (**e**) Some of the complex structures within a single cell in a multicellular plant. (**f**) The many-celled frog body.

through the surroundings; others cannot. The "movement" criterion is becoming a little fuzzy now. However, by the other two standards (complex organization and the capacity to reproduce), a bacterium is alive.

Suppose you are now asked to compare a virus with a rock. A virus is a peculiar particle in the shadowy world between the living and the nonliving. Viruses do have a distinctive organization. Some, for instance, have a "head" end, a sheathlike midsection, and a "tail" end. A virus has no means whatsoever of moving on its own. It cannot reproduce on its own. Yet all viruses contain instructions for producing copies of themselves. To be carried out, the instructions must become incorporated into a living host cell. In effect, a virus takes over the cellular machinery, to the extent that its host starts following *viral* instructions. The machinery starts churning out parts that cannot be used by the cell, but that can be used for building new viruses! By your initial criteria—complex organization, movement, reproduction—is a virus alive? In some respects yes, in others no.

Somewhere below the boundary to the living world are microscopically small, water-filled sacs called liposomes (Figure 1.2b). Under the right conditions, they assemble spontaneously from simple lipid molecules. (You can observe this kind of assembly by stirring some oil into a glass of water. Oil molecules will not mix with water molecules; they cluster into droplets.) Intriguingly, lipids also are the main structural component of cell membranes—and many properties of liposomes correspond to properties of cell membranes. Both self-assemble, and both have the capacity for self-repair (for

example, they seal themselves when punctured). Both let water molecules enter freely but keep certain ions out. Thus liposomes show at least some *potential* for organization.

What about reproduction? When lipid molecules are added to them, liposomes grow in size. They can even grow to the extent that parts break off and form new spheres, which grow in turn. But this is not reproduction; this is only random chemical growth and fragmentation into (generally) nonidentical parts. So you are left to conclude that liposomes are not alive. Still, the way they are organized seems more intriguing than the organization of a rock.

And what about that rock? Although it seems so different from a bacterium, at the levels of viruses and liposomes the difference begins to blur. At a still deeper level, the difference becomes nonexistent. Frog, bacterium, virus, liposome, rock—all turn out to be composed of the same raw materials (particles called protons, electrons, and neutrons). And those materials become organized relative to one another according to the same physical laws. At the heart of those laws is something called **energy**—a capacity for interaction between particles, a capacity to make things happen, to do work. Energetic interactions join particles together, in predictable ways, and form atoms. They bind atom to atom in predictable patterns, thereby giving rise to molecules that form (for example) all frogs and rocks. Energetic interactions hold a frog and a rock together; the flow of energy organizes and holds entire communities of organisms together. Thus we have a profound concept:

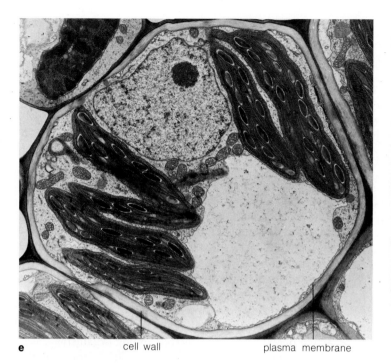

e   cell wall   plasma membrane   f

The structure and organization of the nonliving and the living world arise from the same fundamental properties of matter and energy.

This concept is even used to explain the origin of life. It now appears that the first cells emerged through the evolution of complex systems of molecules. Look carefully at Figure 1.3, which outlines the levels of organization in nature. Then consider the idea that these levels echo successive stages in the history of life. In the beginning, atoms and molecules interacted under certain conditions and formed special large molecules that are now found in all living things. Increasing molecular organization led to spheres much like liposomes. Within those spheres, molecules became organized in ways that allowed them to *duplicate* themselves—and thereby to lay the foundation for reproduction. This capacity led to the *cell*—the smallest unit of life that still retains the properties of life.

What we are unfolding here is a picture of increasingly ordered patterns in the use of materials and energy. It accounts, as any speculation about the origin of life must do, for this apparent fact:

The "difference" between the living state and the nonliving state lies in the degree to which energy is used and materials are organized.

UNITY IN BASIC LIFE PROCESSES

Metabolism

So far, we have touched on the substances that are assembled in orderly ways to form each living thing. Underlying the assembly of those substances, and underlying all activities of the organisms in which they are found, are *energy transfers*. All substances have energy stores, and some of these stores are more readily tapped than others to do work, such as building cell parts. However, there are limits to the amount of useful energy available to do work. "New" energy cannot be created from nothing; to stay alive, an organism must acquire energy from someplace else. The organism must tap an *existing* energy source from its surroundings, then transform the energy into forms it can use.

For example, plant cells acquire energy from sunlight in a process called photosynthesis. Sunlight travels in packets of energy, which plants can absorb. Some of the energy is transferred to molecules such as adenosine triphosphate (ATP), which can readily give up its energy cargo to drive cell activities. As another example, energy stored in carbon-containing molecules can be released during programs for building more ATP. The most pervasive of these energy-releasing programs is aerobic respiration.

All forms of life extract and transform energy from their surroundings, and they use it for manipulating

Biosphere
Entire zone of earth, water, and atmosphere
in which life can exist on our planet's surface

↑ ↓

Ecosystem
All the energetic interactions and materials cycling
that link organisms in a community with one another
and with their environment

↑ ↓

Community
Two or more populations of different organisms that occupy
and are adapted to a given environment

↑ ↓

Population
Group of individuals of the same kind, occupying
a given area at a given time

↑ ↓

Multicellular Organism
Individual composed of specialized, interdependent cells
arrayed in tissues, organs, and often organ systems

↑ ↓

Organ System
Two or more organs whose separate functions
are integrated in the performance of a specific task

↑ ↓

Organ
One or more types of tissues
interacting as a structural, functional unit

↑ ↓

Tissue
One or more types of cells
interacting as a structural, functional unit

↑ ↓

Cell
Smallest *living* unit; may live independently
or may be part of a multicellular organism

↑ ↓

Organelle
Structure inside a cell whose molecular
organization enhances specific cell activities

↑ ↓

Molecule
Two or more identical or different kinds of atoms bonded together

↑ ↓

Atom
Smallest unit of a pure substance that
retains properties of that substance

↑ ↓

Subatomic Particle
Unit of energy and/or mass; electron, proton, neutron

Figure 1.3 Levels of organization in nature.

materials in ways that assure maintenance, growth, and reproduction. More briefly, they show what is called "metabolic activity."

All forms of life can acquire and use energy to stockpile, tear down, build, and eliminate materials in ways that assure survival and reproduction. This capacity is called metabolism.

Growth, Development, and Reproduction

Through metabolic events, living things come into the world, grow and develop, and reproduce. Most then move on through decline and death according to a timetable for their kind. Even as individual organisms die, reproduction assures that the form and function of the living state are perpetuated.

Yet "an organism" is much more than a single organized form having a single set of functions during its lifetime. One example will make the point, even though actual details vary considerably among organisms.

A tiny egg deposited on a branch by a female moth is a compact transitional form (Figure 1.4). It contains all the instructions necessary to become an adult moth. Before becoming a moth, developmental events inside the egg lead to an entirely different form: a wingless, many-legged larva called a caterpillar.

The caterpillar hatches during a warm season when tender new leaves unfold. Not coincidentally, the caterpillar is a streamlined "eating machine" able to tear and chew tender plant tissues. It has a capacity for extremely rapid growth. It eats and grows until some internal alarm clock goes off, setting in motion events that lead to profound changes in form. Some cells are disassembled, other cells multiply and are assembled in entirely different patterns. Tissues, too, are moved about during this wholesale remodeling, the so-called pupal stage. From the pupa, the adult moth emerges.

The moth is the "reproductive machine" stage. Its head has a tubelike extension (a proboscis) that draws nectar from flowers. From the nectar comes energy that powers free-wheeling flights. For this insect, wings are emblazoned with colors and move at a frequency that can attract a potential mate. The moth has organs in which egg or sperm develop, and which enhance fertilization of an egg.

None of these stages is "the insect." "The insect" is a series of stages in organization, with different adaptive properties emerging at each stage.

Each living thing undergoes development: it proceeds through a series of changes in form and behavior. These stages unfold at about the same rate and in the same way for all organisms of a given type.

Homeostasis

Any attempt to define the nature of life cannot focus only on the organism, for the organism cannot exist apart from its surroundings. The living state happens to be maintained within rather narrow limits. Concentrations of substances such as carbon dioxide and oxygen must not rise above or fall below certain levels. Toxic substances must be avoided or eliminated. Certain foods must be available, in certain amounts. Water, oxygen, carbon dioxide, light, temperature—such environmental factors dictate the terms of survival. *And such terms are subject to change.*

How do living things respond to changes in the environment? They respond in two ways. First, all organisms have built-in means of making internal adjustments to outside changes. The adjustments help maintain operating conditions within some tolerable range. This capacity for maintaining the "internal environment" is known as **homeostasis**. Individual cells have homeostatic controls. (For instance, they have mechanisms for bringing in substances that are in short supply and for eliminating other substances.) Multicelled organisms also have homeostatic controls. (Birds, for instance, have sensors that signal the brain when the outside temperature drops. The brain may send signals to cells that control feather movements. Special movements lead to feather fluffing, a behavior that retains heat and thereby helps maintain body temperature.)

Homeostasis implies constancy, a sort of perpetual bouncing back to some limited set of operating conditions. In some respects, constancy is indeed vital. Your red blood cells will not function unless they are bathed in water that contains fairly exact amounts of dissolved components. Your body works so that the bathwater, so to speak, is always much the same.

Yet living things also respond in a second way to changing conditions. All organisms adjust to certain *directional* changes in the internal and external environments. We might call this **dynamic homeostasis**, for the living state is maintained through adjustments that shift the form and function of the organism over time.

A simple example will do here. In humans, irreversible chemical changes trigger puberty, the age at which sexual reproductive structures mature and become functional. At puberty, the body steps up its secretions of such hormones as androgens (in males) and estrogens (in females). The increased secretions are necessary for sexual maturation. They call for entirely new events such as the menstrual cycle. This cycle includes a rhythmic accumulation of substances that prepare the female body for pregnancy, followed by disposal of substances when pregnancy does not occur. It is not that homeostasis no longer operates. It is that developmental events now demand new kinds of adjustments in the internal state.

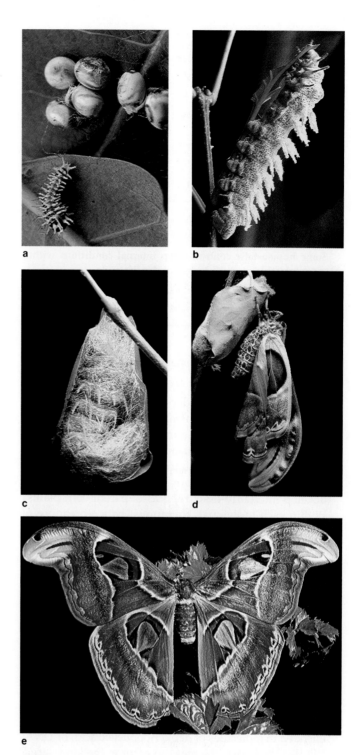

Figure 1.4 "The insect"—a continuum of developmental stages, with new adaptive properties emerging at each stage. Shown here: the development of a giant moth, from egg (**a**) to larval stage (**b**), to pupal form (**c**), to emergence of the resplendent moth form (**d, e**).

Figure 1.5 Underwater tropical reef off the Florida Keys. Elkhorn coral rises above the forestlike growth of staghorn coral.

All forms of life depend on homeostatic controls, which maintain the living state as internal and external conditions change.

Some homeostatic controls keep internal conditions within some tolerable range throughout the life cycle.

Some homeostatic controls govern new adjustments in the internal state as the life cycle unfolds.

DNA: Storehouse of Constancy and Change

Upon thinking about the preceding examples of development, you might wonder what could be responsible for **inheritance**—the transmission, from parent to offspring, of structural and functional patterns characteristic of each kind of organism. How is it that a bacterium can divide and develop into two fairly exact copies of itself? How is it that corn seeds can germinate and develop into fairly exact replicas of parent corn plants? Within each individual, there must be a storehouse of hereditary information.

This storehouse of information has a remarkable characteristic. Although offspring resemble their parents in form and behavior, *variations* can exist on the basic plan. A newly produced bacterium might not be able to assemble (as it is supposed to) some molecule that is vital to its functioning. Some humans are born with six digits on each hand instead of five. *Overall, hereditary instructions must remain intact to assure fidelity in the transmission of traits—yet they also must be subject to change in some details!*

As you have probably learned by now, we know where the instructions reside. In all living cells, they are encoded in molecules of deoxyribonucleic acid, or DNA. We also know that changes can occur in the kind, structure, sequence, or number of component parts of DNA. These changes are **mutations**. Most mutations are harmful, for the DNA of each kind of organism is a package of information that is finely tuned to a given environment. In addition, its separate bits of information are part of a coordinated whole. When one crucial part changes, the whole living system may be thrown off balance. Such is the fate of the bacterium mentioned in the preceding paragraph, for the change probably means that it is doomed.

However, sometimes a mutation may prove to be harmless, even beneficial, under prevailing conditions. For example, mutation produces a dark-colored form of a moth (*Biston betularia*) that otherwise is light-colored. When the mutant moth rests on soot-covered trees, bird predators simply do not see it. In places where there happen to be lots of soot-covered trees (as in industrial regions), the mutant stands a better chance of not being eaten—hence surviving and reproducing—than its light-colored kin.

DNA is a storehouse of patterns for all heritable traits.

Mutations introduce variations in the patterns.

The environment—internal and external—is the testing ground for the combination of patterns that come to be expressed in each individual.

DIVERSITY IN FORM AND FUNCTION

Until now, we have focused on the unity of life—on characteristics shared by all organisms. We have suggested that the structure and organization of all living things arise from basic properties of matter and energy, that all organisms rely on metabolic and homeostatic processes, and that all organisms have the same molecular basis of inheritance. At one time people had no idea that living things hold these characteristics in common. What *was* apparent, and difficult to explain, was the tremendous *diversity* of life. Why is it that almost every environment is host to an astonishing array of different organisms? Before trying to answer this question, let's briefly consider some aspects of diversity in two different settings.

The Tropical Reef

Imagine yourself exploring a tropical reef of the sort shown in Figures 1.5 and 1.6. Long ago, small animals called corals began to grow and reproduce in the warm, nearshore waters of tropical seas (Figure 1.7). The skeletons they left behind served as a foundation for more corals to build upon. As skeletons and residues accumulated, the reef grew. All the while, tides and currents carved ledges and caverns into it. Today, a reef's spine can be decked out with any number of *750 different kinds* of animals called corals.

Plants called red algae typically encrust the coral foundation. In shallow waters behind the reef, red algae give way to blue-green forms. Many small, transparent animals feed on algae and other plants. These animals in turn are food for still larger animals, including some of the world's *20,000 different kinds* of fishes. Squatting

Figure 1.6 Who eats whom on coral reefs. (**a**) Crown-of-thorns, a sea star that feasts on tiny corals. (**b**) Sea anemone, an animal with weapon-studded tentacles, which ensnare tiny animals floating past. (**c**) Sponges, with pores opened toward the oncoming food-laden currents. (**d**) Clownfish, curiously at home above the mouth of a sea anemone—a mouth through which other kinds of edible fish quickly disappear. (**e**) Green algae, plants that are food for various reef organisms. (**f**) Red algae, food for various animals (but not for this chambered nautilus, a shelled animal that swims expertly after shrimp and other prey).

(**g**) A school of goatfish, which feed on small, spineless animals on the sea floor. Goatfish are tasty to humans, also to large fish. (**h**) Some fish are not on the general menu. Here, a blue wrasse safely picks off and dines on parasites that live on this large predatory fish. (**i**) Stone crab. Depending on the species, crabs eat plants, animals, and organic remains. The moray (**l**) prefers meat. (**j**) Lion fish, with its fanned, poison-tipped spines warning away intruders. (**k**) Find the scorpion fish—a dangerous animal that lies camouflaged and motionless on the sea bottom, the better to surprise unsuspecting prey.

a   b   c

Figure 1.7 A few master builders of coral reefs: (**a**) pillar coral, (**b**) daisy coral, and (**c**) green tube coral.

Figure 1.8 (right) The East African Rift Valley, some 6,400 kilometers (4,000 miles) long. The sparsely wooded grasslands in this valley are home for a diverse array of animal life and were the probable birthplace of the human species.

on the reef are predatory sea anemones, each with a mouth fringed with tentacles that capture smaller fish. Yet, hovering above the tentacles *is* a certain kind of fish! It is as edible as most others, but somehow it is not recognized as prey. The fish moves away, captures food, and returns to the anemone's tentacles—which give it protection from other predators. The anemone eats food scraps that fall from the mouth of the fish. In effect, the animals are allies: one receives protection, the other receives food (Figure 1.6d).

The reef is also home for different kinds of sea stars. When feeding, a sea star extends its stomach from its body and puts it into coral chambers. Each chamber resident is digested in place before the stomach is pulled out. When sea stars reproduce, millions of larvae emerge and feed on microscopic algae. Then, as the larvae grow, they become food for the meat-eating corals! It is the corals that now grow and reproduce. In time they repopulate the reef regions that the earlier generation of sea stars had stripped clean. Sea star larvae that do escape grow to become diner instead of dinner, and thereby initiate a new cycle of death and life.

Before speculating on what could account for the diversity of reef organisms, imagine yourself in another setting to see whether a comparison yields any similarities or differences that might provide added insight.

The Savanna

In the shadow of Kilimanjaro, a volcanic peak rising above the edge of the East African Rift Valley, grasslands sweep out to the northeast. This is the African *savanna*, a region of warm grasslands punctuated by scattered stands of shrubs and trees (Figure 1.8). More large ungulates (hoofed, plant-eating animals) live here than anywhere else. One form, the giraffe, browses on leaves beyond the reach of other ungulates. Another form is the Cape buffalo (Figure 1.9). An adult male can weigh a ton, it has formidable horns, and its behavior is unpredictable. It is rarely troubled by predators. Other forms include zebra and impala, which are smaller, more vulnerable, and more abundant than Cape buffalo. They are constantly troubled by such predators as lions and cheetahs. Their remains (as well as the remains of lions and cheetahs) are picked over by scavengers—hyenas, jackals, vultures, and marabou storks (Figure 1.10).

Buffalo, zebra, impala, rhinoceros, giraffe—these and *eighty-five other kinds* of large, plant-eating animals live in the immense valley, as do different kinds of animals that feed on them. They exist side by side in time, moving westward, southward, and back again as dry seasons follow rains, as scorched earth gives way to a resurgence of plant growth.

A Definition of Diversity

When you compare the diversity of the reef with that of the savanna, what conclusions can be drawn? One thing their diverse occupants have in common is some specialization in "who eats whom," beginning with plants and proceeding through different forms of animals that eat the plants and one another. You could spend years observing organisms in any setting, at microscopic as well as macroscopic levels, and you would find that they speak eloquently of the same challenge. *All organisms must be equipped to obtain a share of available resources.* In large part, diversity in form, function, and behavior represents specialized ways to get and use resources—and to avoid becoming a resource for some other organism. In light of this observation, let us now address the question of how such diversity could have come about.

Since the time of origin, living things have required a constant supply of energy and materials. Yet think about the times when you yourself have encountered shortages (for example, of water, gasoline, electricity, or lettuce). In the past, as today, resources were not always abundant. More often than not, *members of every group of organisms must have been demanding a share of limited resources.*

Imagine, next, that *variant* members occasionally appeared (perhaps through DNA mutations). Some variant forms might have been better equipped for securing resources. Some might have been better equipped for responding to predators, prey, or inadvertent allies around them. Accordingly, they would have tended to be the ones that survived and reproduced. Through reproduction, the heritable basis for the variation would have been transmitted to offspring. Given that such offspring were more likely to survive, they would have occurred with increasing frequency in successive generations. There would have been *natural selection,* within the group, of those individuals better adapted to prevailing conditions.

This line of thought amounts to one view of a main road to diversity. Other roads opened up also, but whatever the pathways taken, a strong argument can be made that the following is true:

Diversity is the sum total of variations in form, functioning, and behavior that have accumulated in different lines of organisms.

Such variations tend to be adaptive (or at least not harmful) under prevailing conditions.

Figure 1.9 A sampling of the ninety kinds of large plant-eating animals that live in the savanna—a clear example of diversity in a single environment. (**a**) A herd of Cape buffalo. Imagine yourself a predator this close to the herd and you get an idea of one of the benefits of group living. (**b**) Zebra mother and offspring. (**c**) Male and female impala on the alert, ready to take cover in the nearby woods. (**d**) The rhinoceros, another formidably decked-out plant eater. (**e**) The giraffe, browsing on vegetation high up.

ENERGY FLOW AND THE CYCLING OF RESOURCES

Let's now put this view of diversity in the context of ecological interactions. The geologic record suggests that the first forms of life on earth lived independently of one another, perhaps in tidal flats along the margins of ancient seas. They must have fed on already existing substances, such as simple carbon-containing compounds that had accumulated through volcanic eruptions and other geologic processes. Eventually, perhaps when food supplies began to dwindle, those organisms

d

e

began relying more *on each other* as sources of energy and materials. Thus, by necessity, community interactions such as predation began and have continued in ever richer diversity. Through these interactions, few existing energy sources are unexploited. One example will make the point, even though the cast of characters seems of a most improbable sort.

First we have the adult male elephant of the African savanna (Figure 1.11). It stands almost two stories high at the shoulder and weighs more than eight tons. This grazing animal eats quantities of plants, the remains of which leave its body as droppings of considerable size. Appearances to the contrary, locked in the droppings are substantial stores of unused nutrients. With resource availability being what it is, even waste products from one kind of organism are food for another.

And so we next have little dung beetles rushing to the scene almost simultaneously with the uplifting of the

a

b

Figure 1.10 Predators and scavengers of the savanna. (**a**) An adult lioness standing over a fresh kill. These large cats stalk the herds at dusk or afterward, typically concealing themselves in dense or low-lying vegetation. (**b**) Vultures, together with marabou storks, feed on locusts, small birds, and small mammals—but they also clean up whatever carrion becomes available to them. In this dual predator-scavenger role, they are like other diverse animals of the savanna, including hyenas and jackals.

Figure 1.11 An ecological interdependency of a most improbable sort, beginning with the plants that feed the elephants (**a**), the dung that leaves the elephant (**b**), the beetles that roll dung balls away and bury them (**c**), ending with the beetle larva (**d**) that hatches in the dung—and the remains of the dung itself, enriching the soil in which plants grow, eventually to feed the elephants.

elephant tail. With great precision they carve out fragments of the dung into round balls. The dung balls are rolled off and buried underground in burrows, where they serve as compact food supplies. In these balls the beetles lay eggs, a reproductive behavior that assures the forthcoming offspring of a food supply. Also assured is an uncluttered environment. If the dung were to remain aboveground, it would dry out and pile up beneath the hot African sun. Instead, the surface of the land is tidied up, the beetle has its resource, and the remains of the dung are left to decay in burrows—there to enrich the soil that nourishes the plants that sustain (among others) the elephants.

Such interactions of organisms with their environment and with one another are the focus of **ecology**. Everywhere you look you will find different organisms locked in ecological interdependency. Some interdependencies are simple and others complex, and some seem outlandish—yet the underlying principle is the same:

Most forms of life depend directly or indirectly on one another for materials and energy.

At its most inclusive level, ecological interdependency encompasses the biosphere (the entire zone of earth, water, and atmosphere in which life can exist). With few exceptions, living organisms are linked together by a one-way flow of energy from the sun and by a cycling of materials (such as carbon dioxide and oxygen) on a global scale. Only plants (and some photosynthetic microorganisms) can harness sunlight energy. They use this energy in constructing and maintaining the plant body. Directly or indirectly, other organisms feed on energy stored in plant parts. Microscopic decomposers obtain energy by breaking apart molecules of plant and animal remains. Through their activity, they make available many raw materials for new generations of life (Figure 1.12).

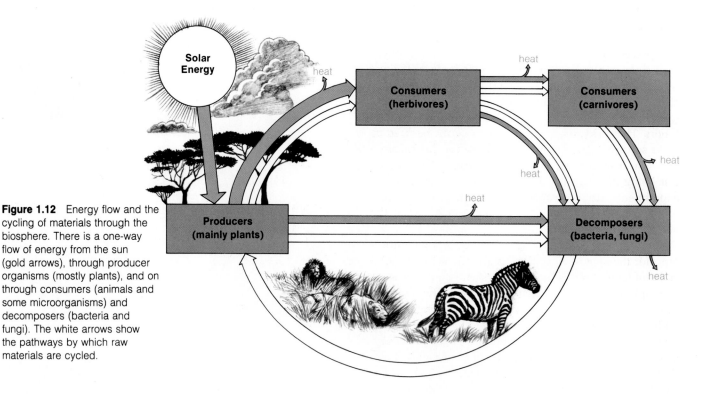

Figure 1.12 Energy flow and the cycling of materials through the biosphere. There is a one-way flow of energy from the sun (gold arrows), through producer organisms (mostly plants), and on through consumers (animals and some microorganisms) and decomposers (bacteria and fungi). The white arrows show the pathways by which raw materials are cycled.

PERSPECTIVE

This chapter has touched on two fundamental aspects of life: its unity and diversity. All organisms are *alike* in sharing common origins, in adhering to the same rules that govern the organization of matter and energy, in relying on metabolic and homeostatic processes, and in having the same molecular basis of inheritance. They are also dramatically *unalike* in appearance and behavior.

There are *many millions* of different organisms, and only a fraction have been catalogued and studied. Many millions more once existed and became extinct. Explaining how such immense diversity arose, and at the same time accounting for its underlying unity, would be quite an accomplishment. In biology, such an explanation was formulated. It is called the principle of evolution by means of natural selection. It is, in fact, the formal statement of the informal picture of natural selection presented earlier in the chapter.

A principle is an idea whose validity holds up even when many different observations and experiments are used to test it. The one we are talking about here will be described in the next chapter, together with examples of observations that led to its formulation. This powerful, integrative concept will help you interpret many of the observations and experimental results you will encounter throughout this book. In turn, you will see how observations and experiments from many lines of inquiry lend support to the principle. It is because of this depth and range of substantiating studies that biol-ogists in general view the principle as the most logical explanation of the seeming contradiction inherent in life: its unity *and* diversity.

Review Questions

1. For this and subsequent chapters, make a list of the **boldface** terms that occurred in the text. Write a definition next to each, then check it against the one in the text. (You will be using these terms later on.)

2. Why is it difficult to give a simple definition of life?

3. What is meant by "adaptive"? Give some examples of environmental conditions to which plants and animals must be adapted in order to stay alive.

4. If the structure and organization of all things arise from the basic nature of matter and energy, then what is the essential difference between living and nonliving things?

5. Study Figure 1.3. Then, on your own, arrange and define the levels of biological organization. What key concept ties this organization to the history of life, from the time of origin to the present?

6. What is metabolic activity?

7. What aspect of life is being overlooked when you talk about "the animal" called a frog? (Hint: What's a tadpole?)

8. What is DNA? What is a mutation? Why are most mutations likely to be harmful?

9. Outline the one-way flow of energy and the cycling of materials through the biosphere.

ON SCIENTIFIC PRINCIPLES

The preceding chapter claimed that this book can help carry your thinking about life to deeper, more organized levels of understanding. By itself that is a presumptuous claim, for success depends partly on how open you are to thinking about things in light of the principle of evolution. This principle is so powerful it can be used to make sense of observations at all levels of biological organization, from molecules to the biosphere. That is why it is being introduced here. Details of observations and experiments are a necessary part of the chapters that follow, for they represent evidence in support of the statements being made.

What, exactly, is a "principle"? The question is important, for the answer provides insight into why the principle of evolution is used with such confidence. A **principle** is a way of explaining a major phenomenon of nature, one that has been synthesized from a large body of information. Thus the idea of evolution developed over centuries, as naturalists and travelers observed and collected specimens of living and extinct forms, then asked questions about the remarkable diversity those specimens represented. It became clear that almost all organisms alive today are very different from organisms of the remote past. Eventually there was overwhelming evidence that the difference was a consequence of evolution—of changes in lines of descent that have accrued since life began.

If we were to idealize the route from a question about such a major aspect of the world to a fundamental explanation for it, we might end up with a list like this:

1. Ask a question (or identify a problem).

2. Make one or more **hypotheses**, or educated guesses, about what the answer (or solution) might be. This means using the process of **induction**: sorting through clues, hunches, and observations, then combining bits of information and logic to produce a general statement (the hypothesis).

3. Predict what the consequences might be if a hypothesis is valid. This process of reasoning from a general statement to predicting consequences is called **deduction** (and sometimes the "if-then" process).

4. Devise ways to *test* those deductions by making observations, developing models, or performing experiments.

5. Repeat the tests as often as necessary to determine whether results will be consistent and as predicted.

6. Report objectively on the tests and on conclusions drawn from them.

7. Examine alternative hypotheses in the same manner.

2

METHODS AND ORGANIZING CONCEPTS IN BIOLOGY

Figure 2.1 Galápagos tortoises, one of many diverse species that Charles Darwin encountered on the Galápagos Islands.

This route represents what might be called a scientific approach to interpreting the natural world. Hypotheses are proposed, then deductions are made and tested (see *Commentary*). There is no such thing as any one "scientific method" of doing this. In practice, insights arise from accident and intuition as well as from methodical search. Some individuals adhere to existing procedures, others improvise as they go. No matter what the individual "method," however, the bottom line in science is this: *Hypotheses must be testable—and the tests must not be so loosely conceived that they cannot be duplicated or verified.*

No scientist can put forward an idea and demand that it be believed as true, no questions asked. In science, there are no absolute truths. There are only high probabilities that an idea is correct in the context of observations and tests made so far. Instead of absolutes, there is **suspended judgment**. This means a hypothesis is tentatively said to be valid if it is consistent with observations at hand. You won't (or shouldn't) hear a scientist say, "There is no other explanation!" More likely you will hear, "Based on present knowledge, this explanation is our best judgment at the moment."

Often the weight of evidence is so convincing that the hypothesis becomes accepted as a **theory**: a coherent set of ideas that form a general frame of reference for further studies. In science, the word "theory" is not used lightly. It is bestowed only on hypotheses that can be relied upon with a very high degree of confidence.

Sometimes substantiating evidence seems so overwhelming that a theory is elevated to the status of principle and serves as a doctrine from which other concepts are drawn. Even a principle is not beyond scrutiny, for new observations and test results may call for its modification or replacement. Far from being a disaster, this activity stimulates the development of even more adequate explanations.

Obviously, individual scientists would rather have their name associated with useful explanations than with useless ones. But they must be objective and keep asking themselves: "Is my thinking consistent with available observations and tests of what I hope to explain?" This does not mean all scientists are objective all of the time or even most of the time; no one can lay claim to that. It means only that they are expected as individuals to forsake pride and prejudice by testing their beliefs, even in ways that might prove them wrong. Even if an individual scientist doesn't, or won't, *others will*—for science proceeds as a community that is both cooperative and competitive. Ideas are shared, with the understanding that it is just as important to expose errors as it is to applaud insights.

This call for objectivity strengthens the theories and principles that do emerge from scientific studies. Yet it also puts limits on the kinds of studies that can be carried out. Beyond the realm of what can be analyzed with the methods and technology available, certain events remain unexplained. Why do we exist, for what purpose? Why does any one of us have to die at a particular moment and not another? Why do we sense beauty in some things and recoil in horror from others?

Answers to such questions are *subjective*; they come from within, as a consequence of all those factors shaping the consciousness of each individual. Because these factors can be infinitely variable, they do not readily lend themselves to scientific analysis.

This is not to say that subjective answers are without value. No human society can function without a shared commitment to standards for making judgments, however subjective those judgments might be. Moral, aesthetic, economic, and philosophical standards vary from one society to the next. But all guide their members in deciding what is important and good, and what is not. All attempt to give meaning to what we do.

Every so often, scientists stir up controversy when they explain part of the world that was previously considered beyond natural explanation, or belonging to the *supernatural*. This is sometimes true when moral codes are interwoven with religious narratives, which grew out of observations by ancestors. Questioning some long-standing view of the world may be misinterpreted as questioning morality, even though the two are not remotely synonymous.

For example, centuries ago Nicolaus Copernicus studied the movements of planets and stated that the earth circles the sun. Today the statement seems obvious. Back then, it was heresy. The prevailing belief was that the Creator had made the earth (and, by extension, humankind) the immovable center of the universe! Not long afterward a respected professor, Galileo Galilei, studied the Copernican model of the solar system. He thought it was a good one and said so. He was forced to retract his statement publicly, on his knees, and to put the earth back as the fixed center of things. (Word has it that when he stood up he muttered, "But it moves nevertheless.")

Today, as then, society has its sets of standards. Today, as then, those standards may be called into question when a new, natural explanation runs counter to a supernatural belief. When this happens it doesn't mean that scientists as a group are less moral, less lawful, less sensitive, or less caring than any other group. It means only that their individual and collective work has been guided by one additional standard: *The external world, not internal conviction, must be the testing ground for scientific beliefs.*

Systematic observations, hypotheses, predictions, tests—in all these ways, science differs from systems of belief that are based on faith, force, authority, or simple

Testing the Hypothesis Through Experiments

William H. Leonard, Clemson University

How is it that scientists probe so skillfully into the monument of life and discover so much about its foundations? What is it about their manner of thinking that yields such precise results? Simply put, scientific inquiry routinely depends on systematic observation and test.

Observations can be made directly, through systems of vision, hearing, taste, olfaction, and touch. They can be made indirectly, through use of special equipment (such as a microscope) that extends the range of perception. With practice, we can become skilled at *making systematic observations*. This means focusing one or more senses on a particular object or event in the environment, and screening out the "background noise" of information that probably has no bearing on our focus.

Hypothesizing means putting together a tentative explanation to account for an observation. When a hypothesis is scientific, it is *testable* through experiments. Experiments are devised to test whether predictions that can be derived from the hypothesis are correct. Thus the hypothesis must be constructed so that it provides a framework for stating the results of an experiment. Its content must be more specific than a problem statement, and often it is worded in the negative. Why is this so? Scientists tend to accept tentatively a plausible idea until it is shown to be false. It is difficult to prove experimentally that a hypothesis is true, because its validity would have to be demonstrated for all possible cases and under all possible conditions. That, obviously, is impossible. Scientists therefore continue to test hypotheses by devising experiments that might show them to be false. If they succeed, then the hypothesis must be modified or discarded. That is why hypotheses are expressed in the negative. For example, "DDT concentrations of 0.0001 percent by weight in the food of laboratory rats will not have harmful effects on the maintenance of the rat population over five years." If experiments reveal harmful effects at that dosage, then the hypothesis is not correct, and support is given to the idea that DDT is harmful.

Testing the hypothesis through experiments is at the heart of scientific inquiry. The goal is to control all variables except the one under study. Variables are events or conditions subject to change. For example, variables that are common to many biological experiments are the amount of light, temperature, and moisture. Others are concentrations of substances and numbers of organisms (population density) in a defined space. There are three general categories of variables:

independent variables	*the condition or event under study*
dependent variables	*ones that can possibly change because of the presence of, or change in, an independent variable*
controlled variables	*conditions that could affect the outcome of an experiment but that do not, because they are held constant*

An experimenter observes or manipulates one independent variable at a time, to identify any effects it has on dependent variables. If more than one independent variable were studied simultaneously, it would not be clear which one was responsible for the observed experimental results.

In one classic experimental design, a population of organisms is divided into two groups. The experimental group is the one subjected to the independent variable; the control group is not. All other variables are held the same in both groups. Thus, any differences that show up in test results for the two groups can be attributed to the independent variable. The illustration on the next page is an example of the use of experimental and control groups. This experiment has been used to test the hypothesis that laboratory rats ingesting DDT with normal food will lose weight, show less resistance to disease, and have a lower reproductive rate than rats not ingesting DDT. Notice

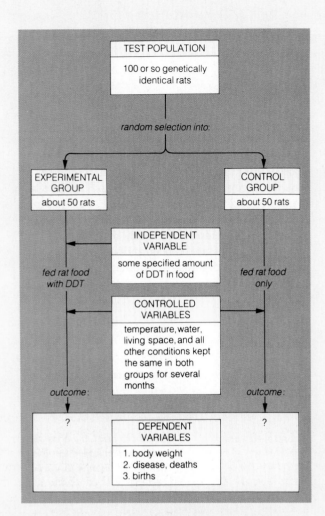

TEST POPULATION

100 or so genetically
identical rats

random selection into:

EXPERIMENTAL
GROUP

about 50 rats

CONTROL
GROUP

about 50 rats

fed rat food
with DDT

INDEPENDENT
VARIABLE

some specified amount
of DDT in food

fed rat food
only

CONTROLLED
VARIABLES

temperature, water,
living space, and all
other conditions kept
the same in both
groups for several
months

outcome:

outcome:

?

DEPENDENT
VARIABLES

1. body weight
2. disease, deaths
3. births

?

An example of a classic experimental design in biology.
The experiment is designed to test the hypothesis that DDT
ingested with food will not have harmful effects on laboratory
rats over a period of time. With all other variables held
constant, test results should refute or support the hypothesis.

that the rats were randomly assorted into either the experimental group or the control group. *Randomization* ensures that both groups are representative (or equivalent) samples of the original population. When any test group is not equivalent to a natural population, *sampling error* is introduced into the experiment. Then one could argue that any experimental results were due to differences in the composition of the different test groups, instead of a result of the independent variable.

Collecting and organizing test results is a necessary process in biological experiments. Data tables or graphs are used to organize and display information for analysis. Graphs are especially useful in illustrating trends or patterns. Data analysis is less mechanical and more conceptual than collecting and organizing the information. Often, statistical tests are used to determine if differences between experimental data and control data are *significant* or are likely due only to chance. If it can be argued that the differences are due to chance only, then it can also be argued that the independent variable had no effect. For example, say that at the end of an experiment on DDT effects, the average adult rat weight was 187.4 grams in the DDT-fed group, and 206.7 grams in the control group. Is this difference significant enough to suggest that there was an actual effect? The use of mathematical tools characteristic of statistical analysis could help in finding an answer.

Generalizing from test results requires careful and objective analysis of the data gathered. Usually, the hypothesis under test is accepted or rejected on the basis of conclusions drawn. A statement is written about what new insights (if any) have been gained into the original problem. Apparent trends are noted when the same data appear in test results gathered over a period of time. Often, further questions and hypotheses are posed in an attempt to guide additional studies of the problem.

consensus. It is not any "law" that is the focus of science. Rather the focus is on the observations that the "law" attempts to explain. A "law" can be invalidated by new evidence, gathered through ongoing tests and clarification of what those observations really mean.

There are, in the history of science, a few individuals who challenged beliefs held not only by society but by the scientific community within it. In biology, Charles

Darwin is among them. More than a century ago, he put together a theory and started lines of investigation that are still flourishing. Tracing Darwin's story and its antecedents will show that he relied on the scientific approach. In his words, "The line of argument pursued through my theory is to establish a point as a probability by deduction and to apply it as hypothesis to other points to see whether it will solve them."

EMERGENCE OF EVOLUTIONARY THOUGHT

More than two thousand years ago, the seeds of biological inquiry were taking hold among the ancient Greeks. This was a time when popular belief held that supernatural beings intervened directly in human affairs. For example, the gods were said to cause a common ailment known as the sacred disease. Yet from a physician of the school of Hippocrates, these thoughts come down to us:

It seems to me that the disease called sacred . . . has a natural cause, just as other diseases have. Men think it divine merely because they do not understand it. But if they called everything divine that they did not understand, there would be no end of divine things! . . . If you watch these fellows treating the disease, you see them use all kinds of incantations and magic—but they are also very careful in regulating diet. Now if food makes the disease better or worse, how can they say it is the gods who do this? . . . It does not really matter whether you call such things divine or not. In Nature, all things are alike in this, in that they can be traced to preceding causes.

—On the Sacred Disease (400 B.C.)

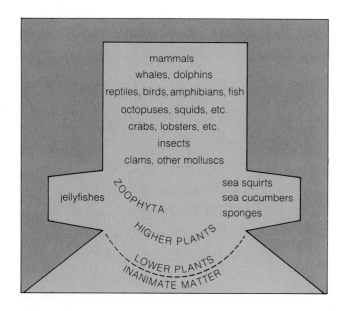

Figure 2.2 Scala Naturae—Aristotle's "ladder of life," the prototype of modern classification schemes.

Such passages reflect the start of a commitment to finding natural explanations for observable events.

Aristotle was foremost among the early naturalists and described the world around him in excellent detail. He had no reference books or instruments to guide him, for biological science in the Western world *began* with the great thinkers of this age. Yet here was a man who was no mere collector of random bits of information. In his descriptions is evidence of a mind perceiving connections between observations and making hypotheses to explain the order of things.

When Aristotle began his studies, he believed (as did others) that each kind of organism was distinct from all the rest. Later he wondered about bizarre forms that could not be readily classified. In structure or function, they so resembled other forms that their place in nature seemed blurred. (For example, to Aristotle some sponges looked like plants but in their feeding habits they were animals.) He came to view nature as organized gradually from lifeless matter through complex forms of plant and animal life. This view is reflected in his model of biological organization (Figure 2.2), the first theoretical framework in the history of biology.

By the fourteenth century, this line of thought had become transformed into a rigid view of life. A great Chain of Being was seen to extend from the lowest forms to humans and on to spiritual beings. Each kind of being,

or **species**, as it was called, was seen to have a fixed, separate place in the divine order of things. Each had remained unchanged since the time of creation, a permanent link in the chain. Scholars thought that once they had discovered, named, and described all the links, the meaning of life would be revealed to them. Contradictory views were not encouraged; scientific inquiry was channeled into an encyclopedic recording of the links.

As long as the world of living things meant mostly those forms in Europe, the task seemed manageable. With the global explorations of the sixteenth century, however, "the world" expanded enormously. Naturalists were soon overwhelmed by descriptions of thousands of plants and animals discovered in Asia, Africa, the Pacific islands, and the New World. Some specimens appeared to be quite similar to common European forms, but some were unique to different lands. *How could those extraordinarily diverse organisms be classified?*

The naturalist Thomas Moufet, in attempting to sort through the bewildering array, simply gave up and recorded such gems as this description of grasshoppers and locusts: "Some are green, some black, some blue. Some fly with one pair of wings; others with more; those that have no wings they leap; those that cannot fly or leap they walk; some have long shanks, some shorter. Some there are that sing, others are silent. . . ." It was not exactly a time of subtle distinctions.

Table 2.1	Classification of Three Organisms According to a Linnean Scheme		
Category (taxon)	Corn	Housefly	Human
Kingdom	Plantae	Animalia	Animalia
Phylum (or division, in botanical schemes)	Anthophyta	Arthropoda	Chordata
Class	Monocotyledonae	Insecta	Mammalia
Order	Commelinales	Diptera	Primates
Family	Poaceae	Muscidae	Hominidae
Genus	Zea	Musca	Homo
Species	mays	domestica	sapiens

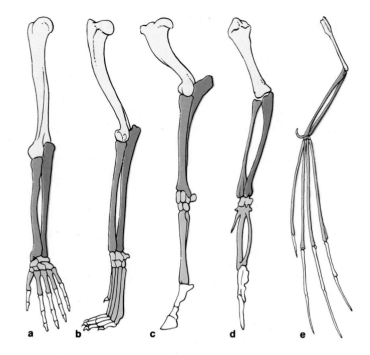

Figure 2.3 Similarities in forelimb structure among a few vertebrates: (**a**) human, (**b**) dog, (**c**) horse, (**d**) bird, and (**e**) bat. The drawings are not to scale relative to one another. Similar structures are shaded the same way from one animal to the next.

Linnean System of Classification of Organisms

The first widely accepted method of classification is attributed to Carl von Linné, now known by his latinized name, Linnaeus. This man was an eighteenth-century naturalist whose enthusiasm knew no bounds. He sent ill-prepared students around the world to gather specimens of plants and animals, and is said to have lost a third of his collectors to the rigors of their expeditions. Although perhaps not very commendable as a student adviser, Linnaeus did go on to develop the **binomial system of nomenclature**. With this system, each organism could be classified by assigning it a Latin name consisting of two parts.

For instance, *Ursus maritimus* is the scientific name for the polar bear. The first name refers to the **genus** (plural, genera), and the first letter of that name is capitalized. Distinct but obviously similar species are grouped in the same genus. Thus other bears are *Ursus arctos*, the Alaskan brown bear; and *Ursus americanus*, the black bear. The second, uncapitalized name is the **specific epithet**. The specific epithet is never used without the full or abbreviated generic name preceding it, for it also can be the second name of a species found in

an entirely different genus. For instance, the Atlantic lobster is called *Homarus americanus*; the American toad, *Bufo americanus*. (Hence one would not order *americanus* for dinner unless one is willing to take what one gets.)

The binomial system was at the heart of a scheme that was thought to mirror the patterns of links in the great Chain of Being. This classification scheme was based on perceived similarities or differences in physical features (coloration, number of legs, body size, and so forth). Eventually it became structured into a system with more inclusive levels of organization. For example, a **family** was said to include all genera that resemble one another more than they do the genera of other families; an **order** includes all families that resemble one another; and so on (Table 2.1).

In retrospect, we can say that the Linnean system was the basis of the first widely accepted, shared language for naming and classifying organisms. It came at a time when ordering was desperately needed. Yet we must also say that the Linnean system reinforced the prevailing view—that species were unique and *unchanging* kinds of organisms, each locked in place in the Chain of Being. To this day, the use of rigid categories for classifying organisms works in subtle ways on our perceptions of the diversity of living things.

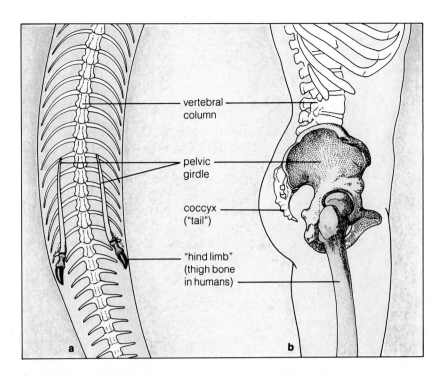

vertebral
column

pelvic
girdle

coccyx
("tail")

"hind limb"
(thigh bone
in humans)

a

b

Figure 2.4 (**a**) Bony structures in a python that correspond to the pelvic girdle of vertebrates. Small "hind limbs" protrude through the skin on the underside (ventral surface). (**b**) Pelvic girdle in a human.

Challenges to the Theory of Unchanging Life

By the late eighteenth and early nineteenth centuries, the somewhat passive cataloging of life was disrupted. Puzzling information was emerging from comparative anatomy (the dissection and comparison of body structure and patterning in major groups of animals). For example, most mammals have two forelimbs. Whales have two flippers, humans have two arms, bats have two wings, and so forth (Figure 2.3). These structures vary in size, shape, and how they are put to use (as in flying versus swimming). Yet when early anatomists analyzed forelimbs of different mammals, they found them to be similar in general body position, in component materials, and in how they developed in the embryo. What meaning could be assigned to the unmistakable resemblances and variations in body plan?

According to one explanation, at the time of creation there was no need to come up with completely new body patterns for each organism, because some patterns were perfect and worked so well. Yet if that were true, then what could explain another discovery—that some body parts have no apparent function at all?

For example, snakes and humans belong to the major group called vertebrates (animals with backbones).

Unlike humans, snakes have no limbs. However, if snakes had been created in a state of limbless perfection, then why do some have bony parts that correspond to a pelvic girdle? A pelvic girdle (Figure 2.4) is a set of bones to which *hind limbs* attach! As another example, at the tip of the human backbone, anatomists discovered bony parts that looked exactly like bones in an animal tail. What would the remnants of a tail be doing in a perfectly constructed human body? It was not a time of easy answers.

Also puzzling was the world distribution of plants and animals. For instance, naturalists found that marsupials (pouched mammals such as kangaroos) were not at all common to most places but abounded in Australia. Cactus plants were observed in North and South American deserts yet were nowhere to be seen in Australian and Asian deserts. If all species had been created at the same time in the same place, as most scholars then believed, *how could so many be restricted to one part of the world or another?*

In the late eighteenth century, two hypotheses were advanced to explain these observations. Both are credited to the zoologist Georges-Louis Leclerc de Buffon. If there had been only a single center of creation, thought Buffon, then species spreading out from it would have been stopped sooner or later by mountain barriers and oceans. But what if there had been *several* "centers of creation"? *Perhaps the creation of species had been "spread out" in space.*

Buffon's work in zoology also led him to suggest that species might not have been created in a perfect state. For instance, if the pig had been so perfectly constructed, then why did it have lateral toes—which are too high on the leg to reach the ground? *Perhaps species became modified over time.*

Evidence favoring these two hypotheses came from fossils: the remains or impressions of once-living organisms entombed in the earth. Fossils of different types were buried in a progression of distinct layers under surface rocks and soil. In underlying (earliest deposited) rock layers, fossils of marine organisms were somewhat simple in structure. In layers above them, fossils of similar structure showed more complexity. Finally, in the uppermost (most recently deposited) layers, they closely resembled living marine organisms. Some naturalists thought these patterns were a record of successive changes in various forms of life. The reasons for change and the time required for it were not known—*but the very concept of change ran counter to the concept of the fixity of species.*

Many naturalists now tried to reconcile the new observations of changing fossil patterns with a traditional conceptual framework that did not allow for change. The nineteenth-century anatomist Georges Cuvier had spent twenty-five years comparing fossils with living organisms. He perceived that the fossil record changed abruptly in certain rock layers. The layers were so distinct from ones that had been deposited before them that they seemed to be boundaries for dramatic change in ancient environments. Four broad eras were recognized, each with a distinctive array of fossils. (These eras are now called the Proterozoic, Paleozoic, Mesozoic, and Cenozoic.)

Cuvier actually had made some astute inferences about past episodes of catastrophic change. He attempted to explain these changes with his concept of **catastrophism**. There was only one time of creation, said Cuvier, which had populated the world with all species. Many species had been destroyed in a global catastrophe. The few survivors repopulated the world. It was not that the survivors were *new* species. Naturalists simply hadn't got around to discovering earlier fossils of them, fossils that *would* date to the time of creation. Another catastrophe wiped out more species and led to repopulation by the survivors, and so on through various catastrophes.

Investigations never have turned up the fossils needed to support Cuvier's concept. Rather, they have turned up considerable fossil evidence against it (Chap-

ter Thirty-Seven). The concept is illuminating, however, *for it shows how prevailing beliefs may influence explanations of what is being observed.*

Lamarck's Theory of Evolution

One of Cuvier's contemporaries viewed the fossil record differently. Jean-Baptiste Lamarck believed that life had been created long ago in a simple state. He believed further that it gradually improved and changed into complex levels of organization. The force for change was a built-in drive for perfection, up the Chain of Being. The drive was centered in nerve fibers, which directed "fluida" (vaguely defined substances) to body parts in need of change (in a manner unspecified).

For instance, suppose the ancestor of the modern giraffe was a short-necked animal. Pressed by the need to find food, this animal constantly stretched its neck to browse on leaves beyond the reach of other animals. Stretching directed fluida to its neck, making the neck permanently longer. The slightly stretched neck was bestowed on offspring, which stretched their necks also. Thus generations of animals desiring to reach higher leaves led to the modern giraffe. Conversely, a vestigial structure was an organ no longer being exercised enough. It was withering away from disuse, and each newly withered form was passed on to offspring.

Such was the Lamarckian hypothesis of **inheritance of acquired characteristics**—the notion that changes acquired during an individual's life are brought about by environmental pressure and internal "desires," and that offspring inherit the desired changes.

Lamarck's contemporaries considered the hypothesis a wretched piece of science, largely because Lamarck habitually made sweeping assertions but saw no need to support them with observations and tests. In retrospect, perhaps we can find kinder words for the man. His work in zoology was respected. And he did indeed piece together a foundation for an evolutionary theory: *Species change over time, and the environment is a factor in that change.* It was his misfortune that he made some crucial observations but came up with a hypothesis that has never been supportable by tests.

EMERGENCE OF THE PRINCIPLE OF EVOLUTION

Naturalist Inclinations of the Young Darwin

Charles Darwin (Figure 2.5) was to develop an evolutionary theory that has had repercussions throughout Western civilization. His early environment may have influenced that destiny. His grandfather, a physician and naturalist, was one of the first to propose that all organisms are related by descent. Being from a wealthy family, Darwin had the means to indulge his interests. When he was eight years old, he was an enthusiastic but haphazard shell collector. At ten, he focused on the habits of insects and birds. At fifteen, he found schoolwork boring compared to the pursuit of hunting, fishing, and observing the natural world.

At college, Darwin attempted to study medicine. He abandoned the study after realizing he never could practice surgery on his fellow humans, given the crude and painful procedures available. For a while he followed his own inclinations toward natural history. Then his father suggested that a career as a clergyman might be more to his liking, so Darwin packed for Cambridge. His grades were good enough to earn him a degree. But most of his time was spent among faculty members with leanings toward natural history. It was the botanist John Henslow who perceived and respected Darwin's real interests. Henslow arranged for him to take part in a training expedition led by an eminent geologist. At the pivotal moment when Darwin had to decide on a career, Henslow arranged that he be offered the position of ship's naturalist aboard H.M.S. *Beagle*.

Figure 2.5 Charles Darwin, at about the time he accepted the position of ship's naturalist aboard H.M.S. *Beagle*.

Voyage of the *Beagle*

The *Beagle* was about to sail for South America to complete earlier work on mapping the coastline. Prolonged stops at islands, near mountain ranges, and along rivers would give Darwin a chance to study many diverse forms of life. Almost from the start of the voyage, the young man who had hated work suddenly began to work with enthusiasm, despite lack of formal training. Throughout the journey to South America, he collected and examined marine life. And he read Henslow's parting gift, the first volume of Charles Lyell's *Principles of Geology*.

Amplifying earlier ideas of the geologist James Hutton, Lyell argued that processes now molding the earth's surface—volcanic activity, the gradual uplifting of mountain ranges, erosion by wind and water—had also been at work in the past. This concept is called **uniformitarianism**. It called into question prevailing ideas about the age of the earth. (For example, Jewish calendar-years were based on the concept that the earth was less than 6,000 years old.) Given the rates at which known geologic processes proceed, Lyell had reckoned that it would have taken not a few thousand years but millions of years to mold the land into its current configurations.

The implications of this concept were staggering. One of the reasons that evolution initially seemed so implausible was that it was hard to imagine all of life's exuberant diversity evolving in the space of a few milenia. If Lyell were interpreting the geologic record correctly, *then there had indeed been time enough for evolution.*

Darwin and Wallace: The Theory Takes Form

Clues From the Voyage. Darwin returned to England in 1836, after nearly five years at sea. In the years to follow, his writings established him as a respected figure in natural history. All the while, his consuming interest was the "species problem." *What could explain the remarkable diversity among organisms?* As it turned out, field observations he had made during his voyage enabled him later to recognize two clues that pointed to the answer.

First, while the Argentine coast was being mapped, Darwin repeatedly got off the boat (he was prone to seasickness). He made many excursions inland, where he made detailed field observations and collected various

Figure 2.6 The armadillo. This living animal, along with fossils of animals that were very similar to it (fossils of glyptodont, shown in the reconstruction at left), provided Darwin with a clue that helped him develop his theory of evolution.

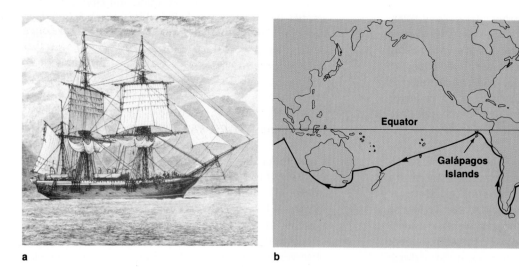

a   b

Figure 2.7 (**a**) H.M.S. *Beagle,* shown in the Straits of Magellan. (**b**) The route of the *Beagle*'s five-year voyage around the world. The Galápagos Islands, about 1,000 kilometers (600 miles) off the coast of Ecuador, support a number of unique plant and animal species. The diversity of life on these isolated islands profoundly influenced Darwin's thinking about the evolution of species.

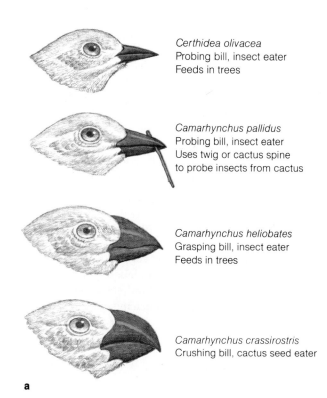

Certhidea olivacea
Probing bill, insect eater
Feeds in trees

Camarhynchus pallidus
Probing bill, insect eater
Uses twig or cactus spine
to probe insects from cactus

Camarhynchus heliobates
Grasping bill, insect eater
Feeds in trees

Camarhynchus crassirostris
Crushing bill, cactus seed eater

a

b

Figure 2.8 (**a**) Examples of variation in beak shape among different species of Darwin's finches, as correlated with feeding habits. (**b**) Woodpecker finch (*C. pallidus*), using a cactus spine to pry out wood-boring insects.

fossils. He saw for the first time many unusual species, including the armadillo (Figure 2.6). Among the fossils were remains of glyptodonts: extinct animals that looked suspiciously like living armadillos. Here were two bizarre but very similar animals, separated in time but confined to the same part of the world. If both had been created separately at the same time, then why were living armadillos lumbering about the very spot where fossil remains of others like them lay buried? *Wouldn't it make more sense to assume that one form evolved from the other?*

Second, Darwin had observed local populations of the same species at many different places along the east coast of South America. It later struck him that individuals within those populations varied in such traits as size, form, coloration, and behavior. When two populations were geographically close, the differences often didn't amount to much. But when populations were separated by some distance, the differences were so pronounced that they might almost be judged as separate species. Darwin had seen nothing like this geographic variation in England—a small island where environmental conditions do not vary much.

The phenomenon of geographic variation was pronounced in the Galápagos Islands, about 1,000 kilometers off the coast of Ecuador (Figure 2.7). These tiny islands arose long ago as volcanoes from the sea floor, hence originally had been devoid of life. Every island or island cluster housed diverse organisms, including its own species of finch. Each finch species had a distinct beak shape, which seemed to be related to particular types of food in the local environment (Figure 2.8). Now, if all those species had not changed since the moment of their creation, then a slightly different finch must have been created for each speck of land in the Galápagos group! *Would it not be simpler to assume that geographic variants had evolved from a single ancestral species of finch?*

Darwin's Deductions. If Darwin were correct in his hypothesis that species evolve, then what mechanism could account for their evolution? No one had ever seen one species change into another. Darwin thought no one would because he believed evolution was exceedingly slow. Even so, he did come up with a plausible mech-

Figure 2.9 A few examples of the more than 300 varieties of domesticated pigeons. Such forms have been derived, by selective breeding, from the wild rock dove (**a**).

anism after reading *Essay on the Principle of Population* by Thomas Malthus—a clergyman and economist.

Malthus stated that "nature has scattered the seeds of life abroad with the most profuse and liberal hand [but] has been comparatively sparing in the room and the nourishment necessary to rear them." Thus, any population tends to outgrow its resources, and its individual members must compete for what is available.

When Darwin reflected on his observations of the natural world, it dawned on him that the normal variation in local populations of a species could include differences in the ability to acquire resources. If there were indeed a struggle for existence, then some *variant* individuals might have a competitive edge in surviving and reproducing. Nature would *select* certain traits and eliminate others—and slowly the population could change! Thus Darwin deduced that "natural selection" among variant individuals could be a mechanism of evolution.

The Theory of Natural Selection. Now the problem became one of compiling evidence of natural selection. Darwin knew he had to find a way to convince the scientific community that the normal variation within species could have occurred without divine creation. He did this by turning to a familiar practice among plant and animal breeders of the day: the development of new varieties within a species.

A compelling example was the flamboyant variation among domesticated pigeons (Figure 2.9). Then, as now, breeders decided which traits were desired, such as a black tail, and they selected individual variants having the most black coloration in their tail feathers. By permitting only those birds to mate, they fostered the trait and eliminated others from the population.

Thus Darwin used *artificial* selection as a model for *natural* selection. He also put together ample examples of variation in natural populations and thereby provided compelling evidence for one of his key deductions: *If evolution occurs by natural selection, then there must be variation in populations.*

Now, for natural selection to operate, more offspring must be produced than can survive; otherwise there would be no competition among variant individuals, hence no requirement for selection among them. Darwin argued that this deduction is probably valid, given that populations typically remain much the same in size. (To give a simple example, a single sea star can release 2,500,000 eggs every year, but the oceans obviously do not fill with sea stars.)

On the basis of such reasoning, Darwin put together his **theory of natural selection**, which is expressed here in modern form:

Figure 2.10 Alfred Wallace. Although Darwin and Wallace had worked independently, they both arrived at the same concept of natural selection. Darwin tried to insist that Wallace be credited as originator of the theory, being the first to circulate a report of his work. Wallace refused; he would not ignore the decades of work Darwin had invested in accumulating supporting evidence.

1. In any population, more offspring tend to be produced than can survive to reproductive age.

2. Members of a population vary in form and behavior. Much of the variation is heritable.

3. Some varieties of heritable traits improve chances of surviving and reproducing under prevailing conditions.

4. Because bearers of such traits are more likely to reproduce, their offspring tend to make up more of the reproductive base for each new generation. This tendency is called *differential reproduction.*

5. *Natural selection is the result of differential reproduction.* Some traits show up (are selected for) with increased frequency because their bearers contribute proportionally more offspring to succeeding generations.

Having formulated his concept of natural selection, Darwin continued his research and sifted his data for flaws in his reasoning. Then, in 1858, he received a paper from Alfred Wallace outlining the same concept! Like Darwin, Wallace was a respected naturalist, with thirteen years of research in South America and the Malay Archipelago to his credit. However, Darwin had only circulated a letter outlining the main points of his ideas even though he had been working for more than twenty years on the problem; the same ideas flashed into Wallace's mind and he had written out his paper in two days (Figure 2.10).

Despite the shock, Darwin sent Wallace's paper to colleagues and suggested that it be published. His colleagues prevailed on him to gather his own notes into a paper that could be presented along with Wallace's. In 1858, both papers were presented to the Linnean Society. The next year Darwin published his monumental book, *On the Origin of Species by Means of Natural Selection.*

On the Origin of Species. Darwin's theory still faced a crucial test. *If evolution occurs, then there should be evidence of one kind of organism changing into a different kind.* The fossil record seemed to contain no transitional forms, the so-called "missing links" between one major type of organism and another. Oddly enough, two years after Darwin's book was published such a fossil did turn up—and no one paid much attention to it. The fossil speci-

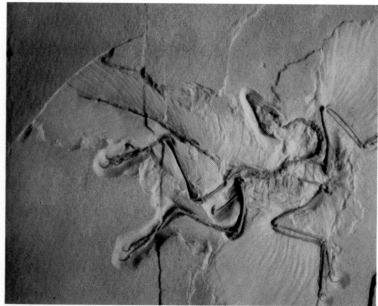

Figure 2.11 *Archaeopteryx*, a link between reptiles and birds. To the left, a restoration based on the fossil shown above.

men, named *Archaeopteryx*, had traits reminiscent of small, two-legged reptiles and modern birds (Figure 2.11). Like reptiles, it had teeth and a long, bony tail. Yet its entire body was covered with feathers!

No other evidence was forthcoming in Darwin's time. Almost seventy years would pass before solid evidence would show that his idea of evolution holds under many different tests, on many different levels of biological organization. Ironically, even though the idea itself had at last gained respectability, it would be that long before most of the scientific community would agree with Darwin and Wallace's remarkable insight—that *natural selection* is a major mechanism by which evolution occurs. In the meantime, their names would be associated mostly with the concept that life evolves—something that others had proposed before them.

AN EVOLUTIONARY VIEW OF DIVERSITY

With widespread acceptance of evolutionary thought, fresh winds began blowing through the rigid framework of biological classification. It was not that classification schemes were blown away; they were and continue to be useful ways of storing and retrieving information.

Rather, the emphasis shifted to identifying evolutionary links between species, including those already categorized.

Today, particular characteristics are viewed as being indicative of different lines of descent. A *genus* includes only those species related by descent from a fairly recent, common ancestral form. A *family* includes all genera related by descent from a more remote common ancestor, and so on up to the highest (and most inclusive) levels of classification: *phylum* and *kingdom*. A scheme that takes into account the evolution of major lines of descent is known as a natural system, or **phylogenetic system of classification**.

In this book we use a modified version of Robert Whittaker's phylogenetic system of classification. In this scheme, all organisms are assigned to one of five kingdoms: Monera, Protista, Fungi, Plantae, and Animalia (Table 2.2).

Regardless of its strengths, no classification system should be viewed as *the* system. As long as observations continue to be made, different people will interpret relationships among organisms in different ways. (For example, some researchers group all forms of life into two kingdoms; others group them into as many as twenty.) The system used here helps summarize knowledge about the diversity of life but it, too, is subject to modification as new evidence turns up.

PERSPECTIVE

When Darwin disembarked from the *Beagle* with his observations and thoughts on diversity, he set in motion a chain of events that made the study of life simpler—and, at the same time, more complex. As you will discover in chapters to follow, the concept of evolving life provides a clear path through the seeming maze of species diversity. However, even though species are no longer generally regarded as unchanging, major questions have been raised concerning exactly *how* they change. *As in all of science, Darwin and Wallace's principle remains open to test, open to revision.* Is evolution always as extremely gradual as Darwin envisioned it? Or do some evolutionary changes occur rapidly (as some current investigators believe)? Is natural selection the only evolutionary process? What, exactly, causes the variation that selective agents act upon? Do species change only in directions corresponding to environmental pressures, or are there also built-in limits on which ways they *can* go? We will be returning to the more recent questions, hypotheses, and tests later in the book.

SUMMARY

1. A scientific *principle* is a way of explaining a major aspect of nature, put together from a large body of information. An example is the principle of evolution, which was developed from information gathered over centuries.

2. Although there is no such thing as "the scientific method" of gathering information, scientists do use these processes:

a. *Hypothesizing* about the meaning of some aspect of the world (for example, proposing that the tremendous diversity among organisms, both living and extinct, came about by evolution.)

b. *Predicting* what the consequences will be if a hypothesis is valid (the "if-then" process).

c. *Testing* the hypothesis in ways that can be duplicated or verified by others, as through systematic observation and through experimentation.

3. With enough substantiating tests, a hypothesis may become a *theory* (a coherent set of ideas that form a general frame of reference for further studies and that can be used with a high degree of confidence). When overwhelming evidence favors it, a theory may become a principle. In science, theories and principles are always open to further test and modification.

Table 2.2	Classification Scheme Used in This Book
Kingdom	General Characteristics
Monera	Single cells. Some are autotrophs (able to build own food from simple raw materials, such as carbon dioxide and water, using sunlight or other environmental energy source). Some are heterotrophs (depend on tissues, remains, or wastes of other living things for food). Cell body is prokaryotic: it has no true nucleus or other membrane-bound internal compartment. *Representatives:* archaebacteria, eubacteria (including the cyanobacteria, or blue-green algae)
Protista	Single cells. Includes autotrophs and heterotrophs. Cell body is eukaryotic: it has a true nucleus and other membrane-bound internal compartments. *Representatives:* golden algae, diatoms, amoebas, sporozoans, ciliates
Fungi	Multicelled. Heterotrophs. Most rely on digestion outside the fungal body, then absorption (they first secrete substances that break down food; then breakdown products are absorbed across the fungal cell wall). All eukaryotic. *Representatives:* slime molds, true fungi
Plantae	Multicelled. Autotrophs. With few exceptions, plants build all of their own food through photosynthesis. All eukaryotic. *Representatives:* red algae, brown algae, green algae, mosses, horsetails, lycopods, ferns, seed plants (such as cycads, ginkgo, conifers, flowering plants)
Animalia	Multicelled. Heterotrophs of varied sorts, including plant eaters, meat eaters, parasites. All eukaryotic. *Representatives:* sponges, jellyfishes, flatworms, roundworms, segmented worms, mollusks, arthropods (such as insects and lobsters), echinoderms (such as sea stars), chordates (fishes, amphibians, reptiles, birds, mammals)

4. Attempts to find meaning in the diversity of organisms led to classification schemes based on perceived similarities and differences in physical traits. Thus each distinct kind of organism was called a *species*, distinct species resembling one another more than they resembled other species were grouped into the same *genus*, and so on with increasingly inclusive groupings into *family, order, class, phylum,* and *kingdom*.

5. The idea that species *evolve* (change over time) emerged through:

   a. Comparisons of the body structure and patterning among major groups of animals.

   b. Questions about the world distribution of plants and animals.

   c. Observations of fossils of different types (structurally simple to complex) buried in a series of distinct layers of the earth (most ancient layers to more recent layers).

6. Darwin and then Wallace observed that individuals in local populations of a species vary in size, form, and other traits that might influence their ability to acquire resources—hence to survive and reproduce. If those traits had a heritable basis, then nature would "select" certain varieties of traits and eliminate others through the generations, and the population would change (evolve). Darwin further observed that for evolution to occur, more offspring must be produced than can survive to reproductive age; otherwise there would be no need for selection among them. These concepts are central to the *theory of evolution by natural selection*.

7. One "test" of Darwin's theory would be evidence of one major kind of organism changing into another kind. *Archaeopteryx* provided early evidence; later work at many levels of biological organization has provided a large body of substantiating evidence.

8. Classification schemes are now constructed to indicate possible evolutionary links between organisms. In one current scheme, the most inclusive evolutionary categories are the kingdoms Monera, Protista, Fungi, Plantae, and Animalia.

Review Questions

1. Define inductive reasoning and deductive reasoning.

2. How do beliefs derived from a scientific approach differ from beliefs based on faith, force, authority, or simple consensus?

3. The following terms are important in scientific testing: control group, experimental group, independent variables, dependent variables, controlled variables, randomization, sampling error. Can you define them?

4. Design a test to support or refute the following hypothesis: The body fat in rabbits appears yellow in certain mutant individuals—but only when those mutants also eat leafy plants containing a yellow pigment molecule called xanthophyll.

5. Witnesses in a court of law are asked to "swear to tell the truth, the whole truth, and nothing but the truth." What are some of the problems inherent in the question? Can you think of a better alternative?

6. Spend some time watching television (commercials, news broadcasts, documentaries), and write down examples of statements presented as facts. Also write down why you think each statement is either plausible or nonsense. (For example, when "leading doctors" are said to recommend something, what kind of doctors are they, how many are they, are they doctors everywhere or in a village in Samoa, and what, exactly, are they leading in?)

7. List and define the main categories in the Linnean system of classification. How do you suppose this system influences our perceptions of the diversity of living things?

8. What is a phylogenetic system of classification? How are organisms grouped in the Whittaker system?

9. State the key points of the theory of natural selection. What is meant by differential reproduction?

Readings

Darwin, C. 1957. *Voyage of the Beagle.* New York: Dutton. In his own words, what Darwin saw and thought about during his global voyage.

Dobzhansky, T. 1973. "Nothing in Biology Makes Sense Except in the Light of Evolution." *The American Biology Teacher* 35(3):125–129. Personal views of one of the world's leading geneticists, who argues that the principle of evolution does not clash with religious faith.

Gould, S. 1982. "The Importance of Trifles." *Natural History* 91(4):16–23. Gould discusses the principles of reasoning that are evident in one of Darwin's last books (on worms). He argues persuasively that Darwin was indeed one of the great thinkers and not "a great assembler of facts and a poor joiner of ideas," as a detractor once wrote.

Mayr, E. 1976. *Evolution and the Diversity of Life.* Cambridge, Massachusetts: Belknap Press. Insights into a mind searching for ways to untangle the knot of biological diversity.

Moore, J. 1984 "Science as a Way of Knowing—Evolutionary Biology." *American Zoologist* 24:467. Recommended for those who would like more evidence for the line of thought presented in this chapter. An excellent essay.

Moorhead, A. 1969. *Darwin and the Beagle.* New York: Harper and Row. Well-illustrated account of the places Darwin visited, what he observed, and the home to which he returned.

Singer, C. 1962. *A History of Biology to About the Year 1900.* New York: Abelard-Schuman. Out of date, but contains absorbing portrayals of the men and women who led the way in developing basic biological concepts.

UNIT TWO

THE CELLULAR BASIS OF LIFE

3

CHEMICAL FOUNDATIONS FOR CELLS

Sunlight energy enters a photosynthetic cell in a blade of grass and sets up a commotion among some of its molecules. Because of the way those molecules are arranged relative to one another, the commotion is highly channeled: energy is absorbed, converted, transferred to other molecules, and used in building sugar. A bacterium present in your gut depends on energy stored in the sugar present in milk. It can extract that energy because some of the enzymes it produces can dismantle milk-sugar molecules quickly and specifically. What happens if you stop drinking milk? If the bacterium does not have other enzymes able to handle other foods, that is the end of the bacterium.

No matter what examples come to mind, all events in the living world begin with the organization and behavior of atoms and molecules. It is here that energy transfer within and between living things originates. It is here that the shape and function of cells, and the multicelled body, are determined. What is it about carbohydrates, lipids, and proteins that makes them such suitable building blocks in cell architecture? How can some molecules change shape or be zipped open so that vital reactions can be played out on their surfaces? Why is it that some molecules can be broken apart more easily than others? Answers to such questions are found in rules governing (1) the internal organization of atoms and (2) how atoms and molecules behave relative to one another.

ORGANIZATION OF MATTER

All the diverse substances that occur naturally on earth are alike in two respects: they occupy space and have mass. All contain one or more types of about ninety naturally occurring **elements**, which are materials that cannot be decomposed into substances with different properties.

By international agreement, a one- or two-letter chemical symbol stands for each element, regardless of the element's name in different countries. For example, what we call *nitrogen* is called *azoto* in Italian and *stickstoff* in German. But the symbol for this element is always N. Similarly, the symbol for the element sodium is always Na (from the Latin *natrium*). Table 3.1 lists elements that are most common in living things, along with their chemical symbols.

Different elements combine in proportions that are fixed and unvarying to form **compounds**. For example, the compound water has a fixed proportion of two elements: 11.9 percent hydrogen to 88.1 percent oxygen by mass. Compounds are unlike **mixtures**, in which two or more elements can be present in varying proportions. Seawater is a mixture. It consists of sodium, chlorine,

Table 3.1	Atomic Number and Mass Number of Elements Commonly Found in Living Things			
Element	Symbol	Atomic Number	Most Common Mass Number	Abundance in Human Body* (% Weight)
hydrogen	H	1	1	10.0
carbon	C	6	12	18.0
nitrogen	N	7	14	3.0
oxygen	O	8	16	65.0
sodium	Na	11	23	0.15
magnesium	Mg	12	24	0.05
phosphorus	P	15	31	1.1
sulfur	S	16	32	0.25
chlorine	Cl	17	35	0.15
potassium	K	19	39	0.35
calcium	Ca	20	40	2.0
iron	Fe	26	56	0.004
iodine	I	53	127	0.0004

(hydrogen, carbon, nitrogen, oxygen grouped: 96%)

*Approximate values.

potassium, calcium, sulfur, magnesium, and other substances dissolved in water, but the percentage of each substance varies from place to place. The compound water is 11.9 percent hydrogen and 88.1 percent oxygen no matter where you find it.

Atoms and Ions

In even a small sample of an element such as carbon (C), a gigantic number of atoms are massed together. By definition, an **atom** is the smallest portion of an element that still retains the properties of the element. Atoms contain no more than three major kinds of particles, called *protons, neutrons, and electrons.* Yet if this is true, how is it that each element displays unique properties? We can approach the question through a simple generalization:

Atoms of each element differ from those of all other elements in their number of protons and electrons, and in their electron arrangements.

The number and arrangement of subatomic parts dictate how atoms can combine to form **molecules**, which are units of two or more atoms of the same or different elements bonded together. They dictate what the properties of molecules will be and how (if at all) different molecules will interact. Thus a look at what goes on inside atoms will help explain why substances in living cells behave as they do.

Let's begin with the fact that an atom has one or more protons, which have a positive (+) charge. (Electric charges are positive or negative. Two opposite charges attract each other; identical charges repel each other.) Every atom except a form of hydrogen also has one or more neutrons, which are electrically neutral.

Protons and neutrons form the **atomic nucleus**, the core of the atom that accounts for almost all of its mass. Electrons, which have a negative (−) charge, are attracted to the positively charged nucleus and move rapidly around it. They occupy most of the volume of an atom.

The number of protons in the nucleus is called the **atomic number**, and it is different for each element. For instance, a hydrogen atom has one proton; its atomic number is 1. A carbon atom has six protons; its atomic number is 6.

The total number of protons *and* neutrons in the nucleus is called the **mass number**. For example, the most common form of carbon atom has six protons and six neutrons; its mass number is 12. The masses of different types of atoms relative to one another are also called *atomic weights.* This term is not really precise (mass is not quite the same thing as weight), but it has been entrenched for more than a century and its use continues.

Suppose you managed to isolate one atom from all others. You would find that its number of protons (positive charges) is exactly balanced by its number of electrons (negative charges). Thus an atom as a whole is electrically neutral; it has a *net* charge of zero.

Of course, most atoms in the world do not float about all by themselves. They bump into other atoms all the time. Often they are bombarded by sunlight. Disturbances of this sort sometimes upset the balance between protons and electrons. The number of protons in an atom remains fixed, but one or more *electrons* can be knocked out of it, pulled away from it, or added to it. An atom that loses or gains one or more electrons ends up being positively or negatively charged; and in this state it is called an **ion**.

For example, a sodium (Na) atom has eleven protons and eleven electrons. It can lose an electron, thereby becoming a positively charged sodium ion (Na^+). A chlorine (Cl) atom has seventeen protons and seventeen electrons. It can gain another electron, thereby becoming a negatively charged chloride ion (Cl^-).

An isolated atom has a net electric charge of zero.

An ion is an atom (or a compound) that has gained or lost one or more electrons, hence has acquired an overall positive or negative charge.

Isotopes

All atoms of an element have the same number of protons, but they can vary slightly in the number of neutrons. Most elements have these variant forms, which are called **isotopes**. For example, "a carbon atom" might be carbon 12 (containing six protons, six neutrons), carbon 13 (six protons, seven neutrons), or carbon 14 (six protons, eight neutrons). These can be written as ^{12}C, ^{13}C, and ^{14}C.

Some isotopes are stable, in that they do not change into other atomic forms. **Radioactive isotopes** have unstable nuclei. Over a given period, they spontaneously give off subatomic particles and energy, and they break down (decay) into atoms of different types. For example, carbon 14 has too many neutrons for stability and it decays into nitrogen 14. It takes 5,730 years for half the carbon 14 atoms in a sample of material to do this. Such unvarying rates of decay are used to determine the age of rock samples and fossils (Chapter Thirty-Seven).

Radioactivity can be detected by various methods, hence radioactive isotopes can be used as **tracers**. They can be introduced into some system in order to identify

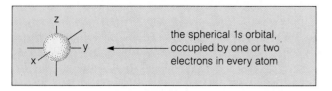

the spherical 1s orbital, occupied by one or two electrons in every atom

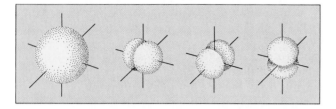

Figure 3.1 Models of electron orbitals. In such models, the nucleus is located where the lines of three axes (x, y, z) intersect. At the first energy level, one or two electrons in every atom occupy the spherelike 1s orbital shown in (**a**). At the second energy level, electrons can occupy the four different orbitals shown in (**b**).

Table 3.2	Electron Distribution Among Energy Levels for a Few Elements				
			Energy Level*		
Element	Chemical Symbol	Atomic Number	First	Second	Third
hydrogen	H	1	1	—	—
helium	He	2	2	—	—
carbon	C	6	2	4	—
nitrogen	N	7	2	5	—
oxygen	O	8	2	6	—
neon	Ne	10	2	8	—
sodium	Na	11	2	8	1
magnesium	Mg	12	2	8	2
phosphorus	P	15	2	8	5
sulfur	S	16	2	8	6
chlorine	Cl	17	2	8	7

*There can be a maximum of two electrons at the first energy level, eight at the second, and eighteen at the third.

the pathways or destination of a particular substance in that system. Such "systems" might be cells, the human body, even whole ecosystems.

For example, plants use carbon as a building block in photosynthesis. By putting plant cells in a medium enriched in carbon 14, researchers were able to identify the exact steps by which plants take up carbon and use it in assembling complex food molecules. As another

example, by injecting a tiny amount of the isotope iodine 131 into a patient's bloodstream, physicians can determine whether the patient has a normal thyroid gland. A normal gland will quickly take up all iodine—including that used as a tracer—from the bloodstream. Isotopes were also used to help determine the location of hereditary instructions inside the cell.

All atoms of an element have the same number of protons but can vary slightly in the number of neutrons. The variant forms are isotopes.

How Electrons Are Arranged in Atoms

Energy Levels for Electrons. No matter how many electrons there are in an atom, each is attracted to the positively charged protons but repelled by other electrons that may be present. How do they spend as much time as possible near the nucleus and, simultaneously, keep far away from each other? They do so by moving in different **orbitals**, which are regions of space around the nucleus in which electrons are likely to be at any instant (Figure 3.1). Each orbital can be occupied by one or, at most, two electrons.

The single electron of a hydrogen atom occupies the 1s orbital (the s stands for spherical), which is the one closest to the nucleus. So do the two electrons of the helium atom. It takes a lot of energy to remove an electron from that orbital, so the electron is said to be at the *lowest energy level.*

What happens in atoms with more than two electrons? Because there are no vacancies in the 1s orbital, other electrons occupy different orbitals that take them farther away, on the average, from the nucleus. It takes less energy to remove an electron that is at a greater distance from the nucleus; hence these electrons are at *higher energy levels.* At the second level, there can be as many as eight electrons in four different orbitals (Table 3.2).

(Another, although not quite accurate way to think about energy levels is to visualize the nucleus surrounded by circles, or shells. The innermost shell would correspond to the lowest energy level and successive shells to higher energy levels, as Figure 3.2 indicates.)

Electrons occupy different orbitals (regions of space around the nucleus in which electrons are likely to be at any instant).

Each orbital can be occupied by one or at most two electrons.

Electrons closest to the nucleus are at the lowest energy level; those that tend to be at a greater distance from it are at higher energy levels.

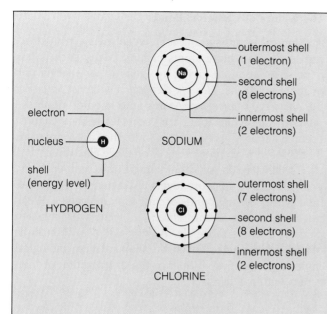

Figure 3.2 Keeping track of electrons: some options.

Atomic structure is most accurately represented by the *orbital model*, in which one or more rapidly moving electrons occupy a volume of space around the nucleus. Electron orbitals of various shapes are associated with each energy level (Figure 3.1 and Table 3.2).

The orbital model is not the most convenient one to use when tracking interactions between atoms. A simplified (if not quite accurate) model is often used instead. In the *shell model*, energy levels that electrons occupy are represented by circles (shells) around the nucleus. Electrons are drawn as dots somewhere on the circles.

An even more stripped-down convention uses only the electron "dots" of the outermost energy level (or shell) with the chemical symbol for the atom. For example, hydrogen and sodium both have only one electron in the outermost shell; chlorine has seven. They would be represented as

$$H\cdot \qquad Na\cdot \qquad :\overset{..}{C}l\cdot$$

The dots can be grouped in pairs to convey information on how the electrons are distributed among the different orbitals of the shell.

Electron Excitation. When an atom (or molecule) is hit with light or heat energy, it can *absorb* some of the energy. When that happens, an electron may move briefly into an orbital at a higher energy level, farther from the nucleus (Figure 3.3). Within a fraction of a second the electron returns to the lowest available energy level, releasing energy when it does. Many metabolic events, including photosynthesis and visual perception, depend on such electron boosts.

To reach a higher energy level, an electron must get energy boosts of specific sizes. It can move to one level, or the next level, or the next, but never in between. It's something like standing on a ladder. You can stand on any of the rungs, and you can move your feet from rung to rung. But you certainly can't stand between rungs and neither, by analogy, can electrons.

For example, electrons in certain pigment (light-absorbing) molecules in photosynthetic cells can be excited to higher energy levels by absorbing wavelengths of sunlight energy. But not any old wavelength will do. Different pigments require *specific* wavelengths to boost their electrons to the energy levels required, (light of different wavelengths varies in how energetic it is).

In looking at Table 3.2, you can see that some atoms have vacancies at the highest occupied energy level. Such atoms tend to form bonds with other elements. This is notably true of hydrogen, carbon, nitrogen, and oxygen—the main building blocks of organisms.

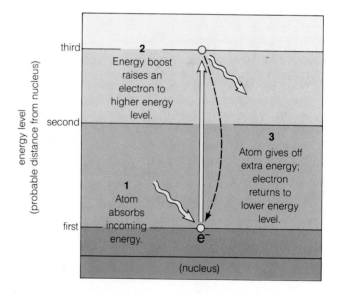

Figure 3.3 Movement of an electron to a higher energy level when an atom (or molecule) absorbs a precise amount of incoming energy, and return of the electron to a lower energy level when the atom releases the extra energy. Photosynthesis and other important biological events are powered by the energy released, in controlled ways, from excited molecules.

Figure 3.4 Chemical bookkeeping.

Symbols for elements are used in writing *formulas*, which identify the composition of compounds. (For example, water has the formula H_2O. The subscript indicates two hydrogen atoms are present for every oxygen atom.) Symbols and formulas are used in *chemical equations*: representations of reactions among atoms and molecules.

In written chemical reactions, an arrow means "yields." Substances entering a reaction (*reactants*) are to the left of the arrow. Products of the reaction are to the right. For example, the overall process of photosynthesis is often written this way:

$$6CO_2 \quad + \quad 6H_2O \quad \longrightarrow \quad C_6H_{12}O_6 \quad + \quad 6O_2$$

6 carbons	12 hydrogens	6 carbons	12 oxygens
12 oxygens	6 oxygens	12 hydrogens	
		6 oxygens	

Notice there are as many atoms of each element to the right of the arrow as there are to the left (even though they are combined in different forms). Atoms taking part in chemical reactions may be rearranged but they are never destroyed. The *law of conservation of mass* states that the total mass of all materials entering a reaction equals the total mass of all the products.

When thinking about cellular reactions, keep in mind that no atoms are lost, so the equations you use to represent them must be balanced in this manner.

Both the reactants and products can be expressed in moles. A "mole" is a certain number of atoms or molecules of any substance, just as "a dozen" can refer to any twelve cats, roses, and so forth. Its weight (in grams) equals the total atomic weight of the atoms that compose the substance.

For example, the atomic weight of carbon is 12; hence one mole of carbon weighs 12 grams. A mole of oxygen (atomic weight 16) weighs 16 grams. Can you show why a mole of water (H_2O) weighs 18 grams, and why a mole of glucose ($C_6H_{12}O_6$) weighs 180 grams?

BONDS BETWEEN ATOMS

So far, we have looked at the basis of reactions among atoms. In case you are not familiar with such reactions, take a moment to review Figure 3.4, which summarizes a few conventions used in describing them.

The Nature of Chemical Bonds

A **chemical bond** is a union between the electron structures of two or more atoms or ions. Many such bonds depend on an atom giving up, gaining, or sharing one or more electrons. *Hence a chemical bond is not an object; it is an energy relationship.* Atoms tend to enter into these relationships when there are vacancies at their highest occupied energy levels. In this state, they are more readily disposed to donate, accept, or share electrons.

Certain bonds predominate at the temperatures, pressures, and moisture levels characteristic of the cellular world. Strong bonds occur *within* molecules. Weak interactions also occur *between* molecules or between parts of a single molecule. As you will see, thousands of such interactions stabilize the shape of many biological molecules, and they influence the organization of molecules within cells.

Ionic Bonding

An ion forms when an atom loses or gains one or more electrons. For this to happen, another atom of the right kind must be nearby to accept or donate the electrons. Since one loses and one gains electrons, *both* become ionized. Depending on the environment in which the transfer is made, the two ions can go their separate ways or remain together as a result of the mutual attraction of their opposite charges. An association of two oppositely charged ions is an **ionic bond**.

For example, only one electron is present at the highest occupied energy level of a sodium atom (Table 3.2). There is only one vacancy left at the highest occupied energy level of a chlorine atom. When these two atoms encounter each other, sodium tends to transfer an electron to chlorine (Figure 3.5). The two atoms thus become oppositely charged ions (Na^+ and Cl^-) and remain in association as NaCl, or sodium chloride, which is our familiar table salt.

In an ionic bond, a positive and a negative ion are linked by the mutual attraction of opposite charges.

Covalent Bonding

Nonpolar and Polar Covalent Bonds. Sometimes an attraction between two atoms is almost but not quite enough for one to pull electrons away from the other. The atoms end up sharing electrons, in what is called a **covalent bond**.

A single covalent bond between two hydrogen atoms may be written as H—H. (Such representations, in which a line signifies a bond, are called structural for-

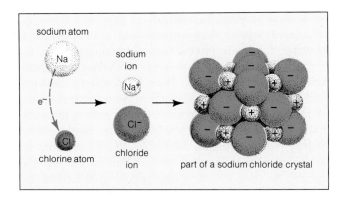

Figure 3.5 Ionic bonding in sodium chloride, or table salt. Strictly speaking, there is no such thing as a "molecule" of NaCl in table salt, for the attraction is not restricted to a single pair of ions. Attractions organize many NaCl units into a crystalline structure, in which units are repeated over and over in three dimensions in a regular, latticelike pattern.

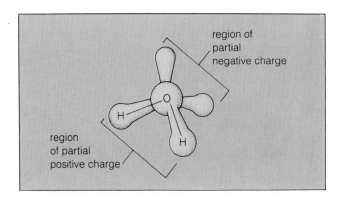

Figure 3.6 Polarity of a water molecule (H_2O). Because of the arrangement of electron orbitals and the bonding angles between oxygen and hydrogen, the molecule as a whole is polar (it carries a slight negative charge at one end and a slight positive charge at the other). Many properties of liquid water—indeed, of organisms—can be traced to this polarity.

mulas.) Each hydrogen atom shares its electron with the other:

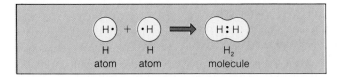

where the dots signify electrons. Similarly, in a double covalent bond, two atoms share two pairs of electrons. An example is the O_2 molecule, or $O{=}O$. In a triple covalent bond (such as $N{\equiv}N$), two atoms share three pairs of electrons.

In *nonpolar covalent bonds*, both atoms exert the same pull on shared electrons. The word nonpolar implies that there is no difference between the two "ends" (or poles) of the bond. An example is the H—H molecule. Each hydrogen atom has only one proton, hence the same positive charge and the same pull on shared electrons.

In *polar covalent bonds*, one atom exerts more of a pull on shared electrons. This happens when atoms of two different elements share electrons. Because unlike elements do not have the same number of protons, they are not equally attractive to electrons—which end up associating with one atom more than with the other. For example, electrons shared in a water molecule (H—O—H) are less attracted to the hydrogens than to the oxygen, which has more protons in its nucleus.

In a polar covalent bond, the atom exerting the greater pull on the electrons ends up carrying a slight negative charge (it is more "electronegative"). But the other atom carries a slight positive charge, so the bonding arrangement as a whole has no *net* charge.

In a nonpolar covalent bond, atoms share electrons equally, hence there is no difference in charge between the two poles of the bond.

In a polar covalent bond, atoms share electrons unequally, hence there is a slight difference in charge between the two poles of the bond.

Polarity of a Water Molecule.   Oxygen has two electrons at the first energy level and four at the second. When oxygen interacts with hydrogen to form a water molecule, the four electrons jockey for position and end up in teardrop-shaped orbitals. In two of the teardrops, oxygen shares an electron with hydrogen. The two unpaired electrons remaining at oxygen's second energy level take up positions on the other side of the oxygen nucleus (Figure 3.6).

With this arrangement, a water molecule carries a slight negative charge near the oxygen relative to the hydrogens. Also, because of repulsive forces among the electrons, the two teardrops associated with the hydrogen atoms tend to be pushed slightly together. Overall, then, it is possible to identify two regions of opposite charge; *the water molecule itself is polar.* Many of the remarkable properties of water arise from this polarity.

To keep things simple in sketches that follow, the polar water molecule will be illustrated in this fashion:

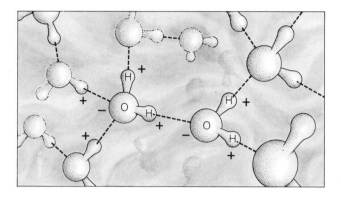

Figure 3.7 Hydrogen bonds between water molecules in liquid water.

Hydrogen Bonding

An atom taking part in a polar covalent bond may also come under the influence of other atoms and molecules. For example, because each hydrogen atom in a water molecule carries a slight positive charge, it can be somewhat attracted to another oxygen atom or to a nitrogen atom in the vicinity. These atoms carry a slight negative charge. Thus we have a **hydrogen bond**, in which an electronegative atom interacts weakly with a hydrogen atom that is already participating in a polar covalent bond:

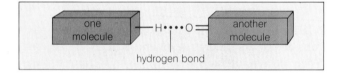

Hydrogen bonds help stabilize the structure of many large biological molecules. They also impart some structure to liquid water. Because a water molecule is polar, each of its hydrogen atoms can form a weak hydrogen bond with the oxygen of a neighboring water molecule. Among neighbors, oppositely charged parts tend to orient toward each other, and parts of like charge away from each other. The bond is strongest when the hydrogen lies in a straight line with two oxygens (Figure 3.7).

In a hydrogen bond, an electronegative atom weakly attracts a hydrogen atom that is covalently bonded to a different atom.

Hydrophobic Interactions

Given that water molecules show polarity, how do other substances act in their presence? If a substance is polar, it is attracted to water molecules and is called **hydro-**philic (water-loving). In contrast, if a substance is wholly or largely nonpolar at its surface, it tends to be repelled by water. It is **hydrophobic** (water-dreading).

Consider what happens when you shake a bottle containing salad oil and water. The oil and water molecules hold little attraction for each other. As water molecules are reunited by hydrogen bonds (which replace the ones that were broken when you shook the bottle), they actually push out the oil. Molecules of oil are forced to cluster together in droplets or in a film on the water's surface. As they do, they expose less of their surface area to water—so the oil interferes less with the attraction that water molecules hold for one another.

In a hydrophobic interaction, nonpolar groups cluster together in water. The clustering is not a true bond; the surrounding water molecules simply push out the nonpolar groups.

Although hydrophobic interactions are not bonds, the clustering effect does influence the shapes of many large molecules, such as proteins and lipid compounds. As such, it also influences the architecture of membranes and other cell structures.

Bond Energies

How much energy is involved in the different kinds of bonds described here? The values have been determined by measuring the amounts of energy needed to break the bonds apart.

Bond energy is measured in terms of kilocalories per mole. A **kilocalorie** is the same thing as a thousand calories—the amount of energy needed to heat 1,000 grams of water from 14.5°C to 15.5°C at standard pressure. (Because energy can be converted from one form to another, bond energies can be expressed in kilocalories even though the energy is used for something other than heating water.)

For covalent bonds between atoms of carbon, nitrogen, oxygen, and hydrogen, bond energies typically range between 80 and 110 kilocalories per mole. Under cellular conditions, ionic bonds are much weaker. Typically the bond energy is about 5 kilocalories per mole. Bond energies average 4 to 6 kilocalories per mole for hydrogen bonds.

Your cells break apart covalent bonds and form new bonds when they degrade the sugar glucose. The overall reaction can yield as much as 686 kilocalories per mole. Even with the energy inherent in sugars (and other substances) stored in your body, it still takes additional energy to keep your body active and functioning. On the average, you take in about 2,000 to 2,800 kilocalories each day to make up the difference.

ACIDS, BASES, AND SALTS

Acids and Bases

Hydrogen takes part in many cellular events, being donated by one molecule and accepted by another when substances are put together and pulled apart. When a hydrogen atom is pulled away from a molecule, it becomes for one fleeting moment a naked proton, stripped of its electron. It is a **hydrogen ion**, or H^+.

Most cellular reactions occur in solution (in some kind of liquid mixture). A substance that releases a hydrogen ion in solution is called an **acid**. A substance that combines with a hydrogen ion in solution is a **base**.

Upon thinking about it, you can deduce that "acid" and "base" can refer to two different states of what is essentially the same molecule. Once stripped of a hydrogen ion, an acid may be free to accept another (serve as a base). Similarly, once a base accepts a hydrogen ion, it may be free to donate it elsewhere (serve as an acid).

Suppose you add hydrochloric acid (HCl) to water. The molecules of this acid dissociate (separate) into ionized parts (H^+ and Cl^-). The hydrogen ion is attracted to any neighboring water molecule, which thereby becomes a hydronium ion:

In this reaction, water acts as a base. However, other basic substances besides water are present in cells, and they tend to be more powerful acceptors of hydrogen ions. Hence the hydronium ion quickly does an about-face and acts as an acid, donating its extra hydrogen to its stronger neighbors.

The most powerful hydrogen ion acceptors are **hydroxide ions**, which can be written as OH^-. Many compounds are sources of hydroxide ions. For example:

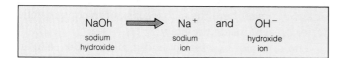

The pH Scale

Depending on how many molecules give up and latch onto hydrogen ions, cells are subject to slight shifts in hydrogen ion concentrations. Changes in acidity can be measured in terms of the **pH scale**, which is based on the concentration of hydrogen ions in solution. The most useful part of the scale ranges from 0 (most acidic) to 14 (most basic). The midpoint of this range, 7, represents a neutral solution in which the H^+ concentration equals

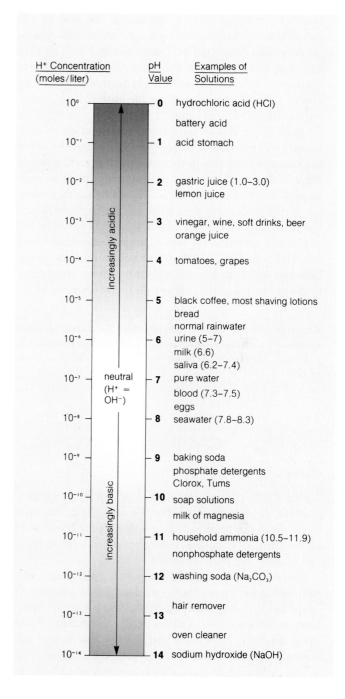

Figure 3.8 The pH scale, in which a fluid is assigned a number according to the number of hydrogen ions present in a liter of that fluid. The most useful part of the scale ranges from 0 (most acidic) to 14 (most basic), with 7 representing the point of neutrality.

A change of only 1 on the pH scale means a tenfold change in hydrogen ion concentration. Thus, for example, the gastric juice in your stomach is ten times more acidic than vinegar, and vinegar is ten times more acidic than tomatoes.

the OH⁻ concentration. A change in one unit in pH means a tenfold change in hydrogen ion concentration.

Pure water has a pH of 7. Any solution with a pH of less than 7 has more H^+ than OH^- ions. The converse is true of any solution with a pH of more than 7. As Figure 3.8 indicates, *the greater the hydrogen ion concentration, the lower the pH value.*

Each living cell is sensitive to pH, and its interior usually will not range far from neutrality. But the pH of the surroundings may be quite different from that inside the cell. For instance, cells of sphagnum mosses grow in peat bogs, where the pH is 3.2 to 4.6 (highly acidic). Some nematodes (a type of worm) thrive in places where the pH is 3.4. For plants and animals living in rivers, the pH of the surrounding water ranges between 6.8 and 8.6. Fluids bathing most cells of your own body range between 6.9 and 7.5. (Exceptions include the highly acidic gastric juices that bathe cells lining your stomach.)

Although the cell interior usually does not range far from neutrality, each cell is adapted to a particular environmental pH range. That range can be quite different for different kinds of cells.

Buffers

Given that hydrogen ions are continually being produced and used in cells, why is it that the cell interior normally does not show drastic shifts in pH? The answer is that most cells have several mechanisms for keeping the internal pH fairly constant.

For example, some substances in the cell act as **buffers**: they combine with and/or release hydrogen ions in response to changes in pH. When some reactions produce an excess of H^+, buffers accept the excess. When other reactions deplete H^+, buffers are able to dole out reserves. Carbonic acid is one of the major buffers. This acid can dissociate into a bicarbonate ion and H^+ in water:

In turn, bicarbonate can combine with hydrogen ions to form carbonic acid, as the reverse arrow indicates in the above equation.

Buffer molecules combine with or release hydrogen ions in response to changes in cellular pH.

Dissolved Salts

A **salt** is an ionic compound formed by the reaction between an acid and a base. Sodium chloride (NaCl) is a salt:

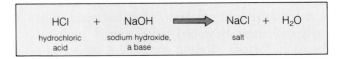

Many salts dissociate into their positively and negatively charged ions when placed in water. In cells, the ions formed from salts include positively charged ions of potassium (K^+), sodium (Na^+), calcium (Ca^{++}), and magnesium (Mg^{++}). They also include the negatively charged chloride ions (Cl^-).

Ions play vital roles in cells. For example, potassium ions help activate many enzymes. In plants, they also seem to be involved in transporting nitrates and phosphates. In animals, signals cannot travel through the nervous system without sodium and potassium ions. Calcium ions take part in cell movements, cell division, nerve functioning, muscle contraction, and blood clotting. In the human body, sodium and chloride ions make up about ninety percent of all ions dissolved in the fluid around cells. The ionic composition of this fluid influences the movement of water and dissolved substances into and out of cells.

Salts are ionic compounds that are formed by the reaction between an acid and a base, and that dissociate into positively and negatively charged ions in water.

Water Molecules and Cell Organization

Water makes up about seventy-five to eighty-five percent of the weight of an active cell, on the average. Water also bathes the cell surface. Surface parts of multicelled plants, animals, and fungi may be exposed to dry soil or air, but even here the active cells are in contact with water present inside the body.

Life does not necessarily end if most of the cellular water is removed. Some plant spores can resume active growth after being dried out for centuries. The water bear also can survive dry spells (Figure 3.9). Water bears, nematodes, rotifers, lichens, mosses, spores, and seeds —all are well matched to environments in which water periodically becomes scarce. Yet with their seeming independence of free-flowing water, the key word in all of this is *active*:

None of the activities associated with the term "living" proceeds without water.

What is it about water that makes it so central to cell functioning? As you will now see, cells depend especially on three properties of water: its internal cohesion, its ability to stabilize temperature, and its capacity to dissolve many substances.

Hydrogen Bonding in Liquid Water

Cohesion. When a substance is overly stretched (placed under too much tension), it can rupture. *Cohesion* is the capacity to resist rupturing under tension. It results from attractions between molecules that make up the substance, such as the hydrogen bonds between water molecules in liquid water.

At interfaces between air and water, hydrogen bonds exert a constant inward pull on molecules at the water's surface and impart a high surface tension to it. The cohesive forces of water resist surface breakthrough. That is why beads of water form and a pond surface resists penetration by small leaves and insects (Figure 3.10). Cohesion, combined with other factors, allows whole columns of water to be pulled through narrow, cellular pipelines to the tops of even the tallest trees.

Numerous hydrogen bonds between water molecules give liquid water high cohesion. The bonds (which individually are weak) resist breaking even when an external force puts the water surface under tension.

Temperature Stabilization. Water helps stabilize temperatures inside and outside the cell because of its high specific heat, high heat of vaporization, and high heat of fusion. All three properties arise largely through hydrogen bonds among water molecules.

Specific heat is the amount of heat energy needed to increase the temperature of one gram of a substance by 1°C. (As you probably know, temperature is a measure of the rate of molecular motion.) Some substances require more energy than others to reach a given level of molecular motion. When liquid water is heated, its individual molecules cannot move faster until the hydrogen bonds among them are broken. Thus water absorbs considerable heat before its temperature increases markedly.

Heat of vaporization is the amount of heat energy that one gram of liquid at its boiling point must absorb before it is converted to gaseous form. It takes 539 calories of heat energy to convert a gram of liquid water into steam at the boiling point of 100°C. (The heat of vaporization for water is twice that for ethanol.)

Figure 3.9 A tardigrade (water bear). This tiny animal survives in ponds and damp soil even when its home freezes over or bakes in the sun. When liquid water becomes scarce, the tardigrade dries out systematically and enters a state of suspended animation. Body parts through which water normally escapes are withdrawn from exposure to air. As the body contracts, its cells produce a compound that replaces much of the water normally surrounding large biological molecules. The compound holds the molecules in place and protects them from mechanical damage. When moist conditions return, the water bear revives and actively goes about its business.

Figure 3.10 A water strider, able to walk on water because of water's high surface tension. This surface tension results from the tenacity with which water molecules cling to one another through hydrogen bonds.

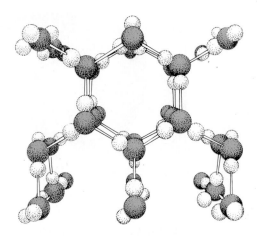

Figure 3.11 Crystal lattice structure of ice. The brown spheres represent oxygen; the white spheres, hydrogen. The "sticks" connecting them represent hydrogen bonds. Below 0°C, each water molecule becomes locked by four hydrogen bonds into a crystal lattice. In this bonding pattern, molecules are spaced farther apart than they would be in liquid water at room temperature (where constant molecular motion usually prevents the maximum number of hydrogen bonds from forming).

Because of the extended bonding pattern, ice is less dense than liquid water and is able to float on it. During winter freezes, sheets of ice that form on surfaces of ponds, lakes, and streams act like a blanket that holds in the water's heat.

Because of hydrogen bonding, water molecules resist separating from each other in liquid water. When a molecule absorbs enough heat energy, however, the bonds can be broken and it can escape from the water's surface. This process is called **evaporation**. Most of the energy needed for the escape is obtained from the surrounding liquid, hence evaporation lowers the surface temperature of water. Oasis plants and many animals cool off through evaporative water loss.

Water also has a high *heat of fusion*: it resists changing from liquid to solid when heat is lost at the freezing point. At room temperature, water is fluid because its relatively weak hydrogen bonds are constantly breaking and rapidly forming again, which allows the molecules some freedom of movement. Only when the temperature drops below 0°C do the molecules become locked in the rigid bonding pattern characteristic of ice (Figure 3.11).

Hydrogen bonds between water molecules give water the capacity to help stabilize temperatures inside and outside the cell.

Solvent Properties of Water

Because of the polar nature of its molecules, water is an excellent solvent for ions and polar molecules. A **solvent** is any fluid in which one or more substances can be dissolved. To understand what dissolving means, consider the following example.

Crystals of table salt (NaCl) separate into Na^+ and Cl^- when they are placed in water. Water molecules tend to cluster around each positively charged ion with their "negative" ends pointing toward it:

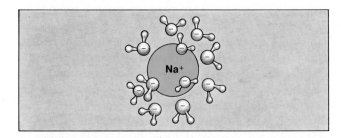

Similarly, water molecules tend to cluster around each negatively charged ion with their "positive" ends pointing toward it:

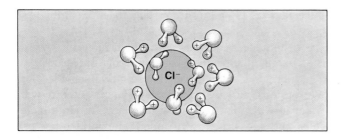

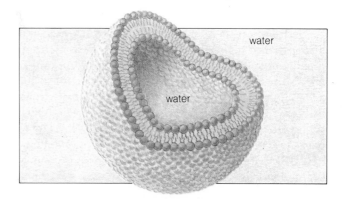

These so-called **spheres of hydration** shield charged ions and keep them from interacting. Thus they force ions to remain dispersed in water.

A charged substance is said to be *dissolved* in water when spheres of hydration form around its individual molecules. The dissolved substances are known as **solutes**. Most ionic substances are at least partially soluble in water. So are most polar molecules, especially those which tend to form hydrogen bonds with water.

Water and the Organization Underlying the Living State

The organization of almost all large biological molecules is influenced by their interaction with water. For example, proteins have many acidic and basic groups on their surfaces. Depending on cellular pH, the protein surface can be positively or negatively charged overall. The charged regions and polar groups attract water molecules. They also attract ions—which in turn attract water. In this manner, an electrically charged "cushion" of ions and water forms around the protein surface:

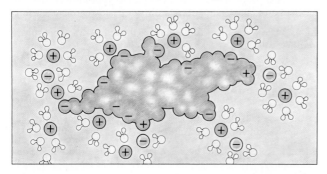

Thus, through interactions with water and ions, proteins can remain dispersed in cellular fluid. (In contrast, particles in a suspension are large enough to settle out.)

As another example, many lipids have hydrophobic tails (which repel water) attached to a hydrophilic head (which, being polar, attracts water). In the presence of water, these lipid molecules tend to cluster, with all of the hydrophilic heads facing the water and all of the hydrophobic tails packed in a sheetlike array that excludes water. Imagine two such sheets arranged into a sphere that has water on the inside and out:

where orange indicates the hydrophilic heads and yellow, the hydrophobic tails. *This "lipid bilayer" arrangement is the framework for all cell membranes.*

Many more examples could be given of interactions between the polar water molecule and other substances characteristic of the cellular world. For now, the point to keep in mind is this:

The properties of water profoundly influence the organization and behavior of substances that make up cells and the cellular environment.

SUMMARY

1. Matter is composed of elements. The atoms of each element have a unique number of protons in the nucleus. Except in the case of a form of hydrogen, the nucleus also contains neutrons. The number of electrons arranged around the nucleus equals the number of protons. The number and arrangement of electrons dictate how atoms can form molecules (units of two or more atoms of the same or different elements bonded together). They dictate the properties of molecules and how (or whether) molecules will interact.

2. In compounds (such as water), certain elements are always present in fixed, unvarying proportions. In mixtures (such as seawater), two or more elements can be present in varying proportions.

3. An isolated atom has a *net* electric charge of zero. An ion is an atom or compound that has gained or lost one or more electrons (it has acquired an overall positive or negative charge).

4. *All* atoms of an element have the same number of protons but they can vary slightly in the number of neutrons. Variant forms of atoms of an element are called isotopes. Radioactive isotopes have unstable nuclei; over a specific period they spontaneously decay into atoms of different types. They are used as tracers in biological systems and are used in determining the age of fossils (remains of once-living organisms).

5. Electrons can be at low energy levels (close to the nucleus) or at higher ones (farther away from it). They occupy orbitals (or shells) at different levels. An orbital may be vacant or may have one or two electrons, but never more than two. Energy inputs can boost electrons to higher energy levels (or out of the atom entirely). When electrons return to a lower energy level, atoms release energy. Photosynthesis and other important biological events are powered by energy released from molecules that have absorbed extra energy.

6. Atoms of hydrogen, carbon, nitrogen, and oxygen (the main structural elements of cells) have vacancies at their highest occupied energy levels. They tend to form bonds with other elements, thereby gaining enough electrons to fill their highest occupied energy level.

7. A chemical bond is a union between the electron structures of two or more atoms or ions:

    a. Ionic bond: a positive and a negative ion remain together by the mutual attraction of opposite charges.

    b. Nonpolar covalent bond: atoms share one or more electrons equally; there is no difference in charge between the two poles of the bond.

    c. Polar covalent bond: atoms share one or more electrons unequally; there is a slight difference in charge between the two poles of the bond (one end is electronegative).

    d. Hydrogen bond: an electronegative atom weakly attracts a hydrogen atom that is already covalently bonded to a different atom.

8. Acids are substances that release hydrogen ions (H^+) in solution; bases are substances that combine with hydrogen ions. Typically, the H^+ concentration in cells equals the hydroxide ion (OH^-) concentration; this represents neutrality on the pH scale. Cells are sensitive to changes in pH.

9. A salt is an ionic compound (such as NaCl) formed by a reaction between an acid and a base. Salts dissociate into positively and negatively charged ions in solution, and these ions play vital roles in cell functions.

10. A water molecule shows polarity; due to its electron arrangement, one end is electronegative. Other polar molecules are attracted to water (they are hydrophilic); nonpolar molecules are repelled by it (they are hydrophobic).

11. The properties of water profoundly influence the organization and behavior of substances that make up cells and cellular environments.

    a. Because of hydrogen bonding between its molecules, water has the capacity to help stabilize the temperatures inside and outside the cell.

    b. Because of water's capacity for hydrogen bonding, its molecules show cohesion (a capacity to resist rupturing when placed under tension, as when water molecules are pulled up from roots to leaves at the tops of trees).

    c. Because of the polar nature of its molecules, water is an excellent solvent for ions and polar molecules.

    d. Because of the polar nature of water molecules, hydrophobic and hydrophilic substances interact with water in ways that influence the shapes of large molecules and the structure of membranes and other cell components.

Review Questions

1. Define the following: element, atom, molecule, compound. What are the six main elements (and their symbols) in most organisms?

2. Define proton, neutron, and electron. How are electrons arranged around the nucleus of an atom?

3. Explain the difference between an atom, an ion, and an isotope.

4. Atoms of chemically nonreactive elements have _____ electrons in the highest occupied orbital. Atoms of chemically reactive elements have _____ electrons in this orbital.

5. Explain the difference between covalent, ionic, and hydrogen bonds.

6. What is the difference between a hydrophilic and a hydrophobic interaction? Is a film of oil on water an outcome of bonding between the molecules making up the oil?

7. Define an acid, a base, and a salt. On a pH scale from 0 to 14, what is the acid range? Why are buffers important in living cells?

8. Cell functioning depends on the properties of water, such as its internal cohesion. What type of chemical bond represents the basis of these properties?

9. Describe how the polarity of the water molecule is the basis of the solvent properties of water. (As part of your answer, describe how spheres of hydration form around a positive or negative ion.)

10. How do water molecules influence the organization of proteins and lipids in a cell?

Readings

Breed, A., et al. 1982. *Through the Molecular Maze.* Los Altos, California: Kaufmann. A short guide to basic chemical concepts.

Frieden, E. July 1972. "The Chemical Elements of Life." *Scientific American* 227 (1):52–64. Good summary of the biological roles of different elements.

Lehninger, A. 1982. *Principles of Biochemistry.* New York: Worth. Classic reference book in the field.

Mertz, W. 1981. "The Essential Elements." *Science* 213: 1332–1338.

Miller, G. 1987. *Chemistry: A Basic Introduction.* Fourth edition. Belmont, California: Wadsworth. Well-written, accessible introduction to chemical principles.

The preceding chapter slipped in a rather odd statement, that living cells are between seventy-five and eighty-five percent water. Whether or not you accepted the statement without giving it a second thought may be a test of how critically you are evaluating what you read. For example, given the many trillions of cells comprising your body, might you not wonder why you do not ooze forward when you walk or quiver like Jell-O, which also is mostly water? Obviously there must be more than free water molecules in your cells.

By far, the three most abundant elements in your body are oxygen, hydrogen, and carbon. (As Table 3.1 indicated, they represent ninety-three percent of its weight. Nitrogen, calcium, and phosphorus account for a little over six percent, and very small amounts of other elements make up the rest.) Much of the oxygen and hydrogen is indeed linked together in the form of water. But these two elements also are linked in significant amounts to carbon—the most important structural element in the body.

THE ROLE OF CARBON IN CELL STRUCTURE AND FUNCTION

A carbon atom can form as many as four covalent bonds with other carbon atoms as well as with other elements. In cells, carbon atoms linked one after another in chains or rings form the backbones (or skeletons) for diverse compounds. These backbones occur in strandlike, globular, and sheetlike molecules, some of which contain thousands, even millions, of atoms. The carbon compounds assembled in cells are *organic* molecules. (The term distinguishes them from the simple *inorganic* compounds, such as water and carbon dioxide, which have no carbon chains or rings.)

Families of Small Organic Molecules

Compounds having no more than twenty or so carbon atoms are considered to be small organic molecules. The four main families of these small molecules are called simple sugars, fatty acids, amino acids, and nucleotides. Usually they form pools of dissolved materials that the cell can tap for energy or for building blocks in the synthesis of large molecules (macromolecules). The main macromolecules in cells are carbohydrates, lipids, proteins, and nucleic acids.

Four families of small carbon compounds—simple sugars, fatty acids, amino acids, and nucleotides—serve as energy sources and as building blocks for the macromolecules present in cells.

4

CARBON COMPOUNDS IN CELLS

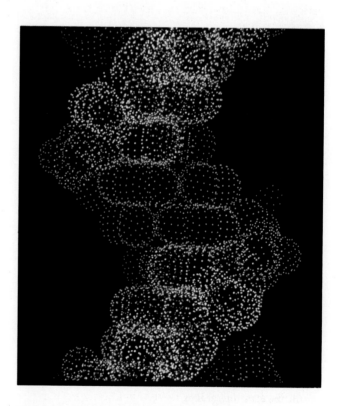

Figure 4.1 Model of a segment of DNA, a molecule that is central to maintaining and reproducing the cell. Carbon atoms are in green. Other atoms represented are nitrogen (blue), oxygen (red), and phosphorus (yellow). Most of the carbon atoms are bonded in ring structures that interconnect to form the carbon skeleton of DNA.

Table 4.1 Some Classes of Organic Compounds

Class	Some Characteristic Functional Groups		Occur in:
Hydrocarbons	methyl (—CH_3):	H | —C—H | H	fats, oils, waxes
Alcohols	hydroxyl:	—OH	sugars
Carbonyls	aldehyde (—CHO):	O || —C—H	sugars
	ketone ($>$C=O):	O || —C—C—C	sugars
	carboxyl (—COOH):	O || —C—OH	sugars, fats, amino acids
Amines	amino (—N, —NH, —NH_2):	H | —N—H	amino acids, proteins
Phosphate compounds	phosphate (—P):	O || —P—O | O	energy carriers (such as ATP)

Properties Conferred by Functional Groups

All organic molecules have a carbon backbone, but they differ in what is attached to it. Atoms or groups of atoms covalently bonded to a carbon backbone are called **functional groups** (Table 4.1). Organic molecules are distinctive in their structure and properties because of their characteristic functional groups, as a few examples will illustrate.

Carbon-Hydrogen Compounds. There is a TinkerToy quality to carbon compounds, in that a single carbon atom can be the start of truly diverse molecules assembled from "straight-stick" covalent bonds. Consider the *hydrocarbons*, which consist only of hydrogen and carbon. Methane (CH_4) is the simplest hydrocarbon. If you were to strip one hydrogen from methane, the result would be a *methyl group*:

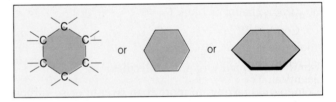

Now imagine that two methane molecules are each stripped of a hydrogen atom and bonded together. If the resulting structure were to lose a hydrogen atom, you would end up with an *ethyl group*:

In this mental exercise, you could go on building a continuous chain, with all the carbon atoms arranged in a line:

To such linear carbon chains, you could add branches of this sort:

You might even have chains coiled back on themselves into rings:

All of the structures illustrated so far have nonpolar bonds and will not dissolve in water; they are hydrophobic. These are properties of fats, oils, and waxes—which are largely hydrocarbon.

Carbon-Oxygen Compounds. The compounds called *alcohols* are mostly carbon and hydrogen, but they also incorporate one or more *hydroxyl groups* (—OH). Methyl alcohol can cause blindness and death, even in small amounts. Yet ethyl alcohol occurs in alcoholic beverages. Glycerol, one of the building blocks of fats and oils, is an alcohol. So are all sugars. All of these compounds will dissolve in water because of hydrogen bonding at their —OH groups.

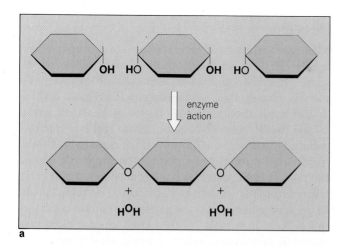 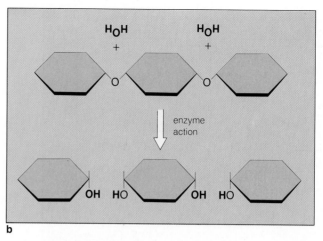

Figure 4.2 (**a**) Condensation of three subunits into a larger molecule. Water can be formed during the reaction. (**b**) Hydrolysis of a molecule into three subunits.

Condensation and Hydrolysis

How do small organic compounds combine to form macromolecules? They don't just get together on their own. Their union depends on the action of enzymes, which are a special class of proteins that speed up reactions between specific substances.

For example, in **condensation**, enzymes speed up the covalent linkage of small molecules (such as short carbon-based chains and rings) in a reaction in which water also may form. Essentially, one molecule is stripped of a hydrogen ion and another is stripped of an —OH group, then the two molecules become joined. At the same time, the H^+ and OH^- that are released can combine to form a water molecule (Figure 4.2).

Condensation reactions can produce a **polymer**: a molecule composed of anywhere from three to millions of relatively small subunits, which may or may not be identical. The individual subunits incorporated in polymers are called **monomers**.

Often the substances used in assembling macromolecules are obtained by breaking down other molecules. A common breakdown process, called **hydrolysis**, is like condensation in reverse. Covalent bonds between parts of molecules are broken and an H^+ ion and an OH^- group derived from water become attached to the fragments (Figure 4.2).

Condensation is the covalent linkage of small molecules in a reaction that can also involve the formation of water.

Hydrolysis is the cleavage of a molecule into two or more parts by reaction with water.

CARBOHYDRATES

Almost all eukaryotic cells use carbohydrates directly or indirectly for energy. Bacteria, plants, and fungi also use carbohydrates as major structural materials. **Carbohydrates** are monomers or polymers of a sugar. A *sugar* is a compound in which carbon, hydrogen, and oxygen atoms are combined in about a 1:2:1 ratio. This means that for every carbon atom present, there typically are two hydrogen atoms and one oxygen atom. Thus the composition of a sugar is $(CH_2O)_n$.

Monosaccharides

The simplest carbohydrates are sugar monomers, or **monosaccharides**. ("Saccharide" comes from a Greek word meaning sugar.) Typically, monosaccharides have a backbone of three to seven carbon atoms. They also have an aldehyde group or a ketone group and two or more hydroxyl groups (Table 4.1).

Of the more than 200 monosaccharides known, the most common have a backbone of three, five, or six carbon atoms. Those with five or more tend to form ring structures when dissolved in cellular fluids. *Ribose* and *deoxyribose* (each with five carbon atoms) are key components of nucleic acids. *Glucose* and *fructose* (each with six carbon atoms) are two of the most abundant sugar monomers.

Both glucose and fructose have the same molecular formula ($C_6H_{12}O_6$), but they differ slightly in their structure and properties. These sugars are structural isomers: they have the same numbers of the same kinds of atoms, but the atoms are not attached to one another in the same ways (Figure 4.3).

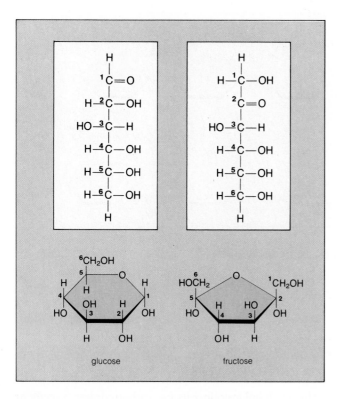

Figure 4.3 Straight-chain and ring forms of two monosaccharides: glucose and fructose. (For reference purposes, sometimes the carbon atoms of sugars are numbered in sequence, starting at the end of the molecule closest to the aldehyde or ketone group.)

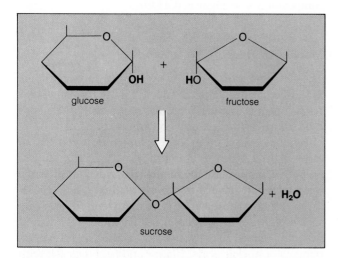

Figure 4.4 Condensation of two monosaccharides (glucose and fructose) into a disaccharide (sucrose).

Disaccharides

Monosaccharides have several free —OH groups, where bonds can be made with another sugar monomer or some other compound. A **disaccharide** results when two monosaccharides are covalently bonded (Figure 4.4).

The most abundant of all sugars is the disaccharide *sucrose*, which consists of one glucose and one fructose subunit. Sucrose is the form in which carbohydrates are transported through leafy plants. The sugar you buy at the market is sucrose that has been extracted and crystallized from plants such as sugarcane. The disaccharide *lactose* (one glucose and one galactose subunit) occurs in the milk of mammals. *Maltose* (two glucose subunits) occurs in germinating seeds. Among other things, maltose is an ingredient in beer production.

Polysaccharides

When more than two monosaccharides are bonded covalently, the result is a **polysaccharide**. In some polysaccharides the sugar subunits are all the same kind; others incorporate two or more different kinds.

Land plants have an abundance of the polysaccharide *starch* (a storage form for sugar). They also may have an abundance of the polysaccharide *cellulose* (a structural material in their cell walls). Although starch and cellulose are both assembled from glucose units, they have different properties. In plant cells, starch molecules can cluster in large granules called starch grains, which can be quickly hydrolyzed into sugar subunits used in metabolism. In contrast, cellulose is tough, fibrous, and insoluble in water.

The differences between starch and cellulose arise from differences in the bonding alignments between their glucose subunits. The oxygen atoms that link the adjacent glucose subunits are not oriented the same way in the two polysaccharides. In starch, the alignment permits the glucose chain to twist into a coiled structure that favors granule formation (Figure 4.5). In cellulose, the glucose chain is extended and many chains lie side by side, hydrogen-bonded to each other at —OH groups. The hydrogen bonds stabilize the glucose chains in tight bundles (Figure 4.6). This bonding arrangement resists most digestive enzymes except those present in termites, wood-rot fungi, and certain bacteria.

Glycogen is a highly branched polysaccharide (Figure 4.7). It is a form in which sugar is stored in fungi and certain animal tissues (such as liver and muscle tissues). *Chitin* incorporates nitrogen and is called a modified polysaccharide. Many animals and fungi have chitin-secreting cells. The chitin secretions are the main structural material in external skeletons and other hard parts of many insects and crustaceans (including crabs). It imparts firmness to the cell walls of most fungi.

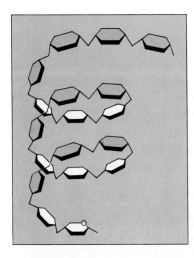

Figure 4.5 Oxygen bridges between the glucose subunits of amylose, a form of starch. The boxed inset depicts the coiling of an amylose molecule, which is stabilized by hydrogen bonds.

Figure 4.6 Structure of cellulose, which is composed of glucose subunits. Neighboring cellulose molecules link together at —OH groups to form a fine strand (microfibril). These strands can be twisted into a threadlike fibril. In some cells, fibrils are coiled into a macrofibril, which is as strong as a steel thread of the same thickness.

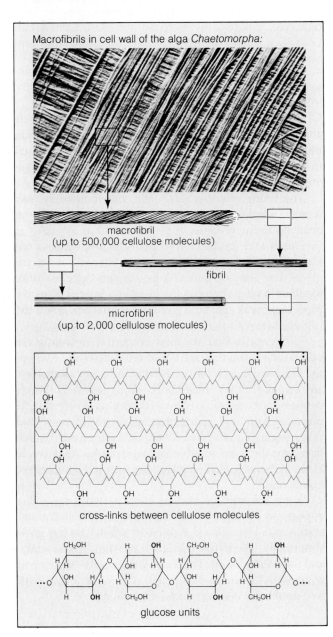

Macrofibrils in cell wall of the alga *Chaetomorpha:*

macrofibril
(up to 500,000 cellulose molecules)

fibril

microfibril
(up to 2,000 cellulose molecules)

cross-links between cellulose molecules

glucose units

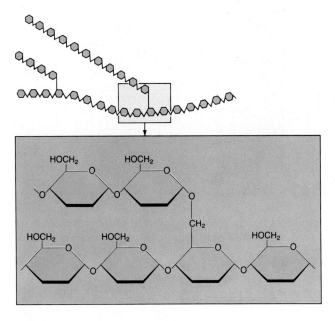

Figure 4.7 Branched structure of glycogen, a form in which sugars are stored in some animal tissues.

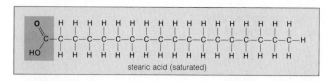

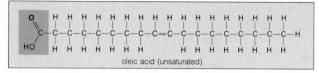

Figure 4.8 Structural formulas for a saturated and an unsaturated fatty acid.

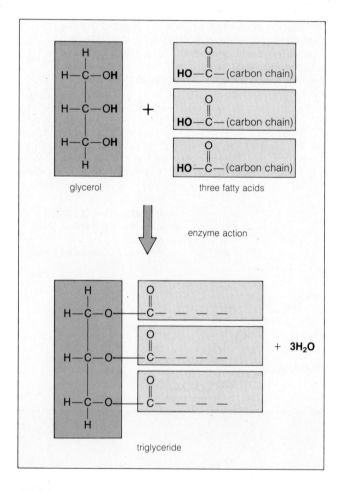

Figure 4.9 Formation of a triglyceride (a fat).

LIPIDS

Lipids are largely hydrocarbon and show little tendency to dissolve in water, but they do dissolve in nonpolar solvents (such as ether). Some lipids function in the storage and transport of energy; others are key components of membranes, protective coats, and other cell structures. Here we shall consider two subcategories of lipids: those with and those without fatty acid components.

Lipids With Fatty Acid Components

Many lipids such as butter and vegetable oils have fatty acid components. A **fatty acid** is a long, unbranched hydrocarbon with a —COOH group at the end (Figure 4.8). When a fatty acid is part of a more complex lipid molecule, it is usually stretched out like a flexible tail. Let's look briefly at three common lipids having fatty acid tails: the glycerides, phospholipids, and waxes.

Glycerides. Glycerides are the most abundant lipids and the richest source of energy in the body. A **glyceride** molecule has one, two, or three fatty acid tails attached to an alcohol backbone. In this case, the alcohol is glycerol (see Figure 4.9).

The terms *monoglyceride, diglyceride,* and *triglyceride* refer to whether one, two, or three fatty acid tails are attached to the glycerol.

Plants and animals store lipids in the form of triglycerides (Figure 4.9). Most of the fats we eat are triglycerides. They are insoluble in water and tend to clump together in fat globules in the intestine. Special digestive processes break apart the globules, then break down the triglycerides into parts that the body can absorb.

Triglycerides that are solid at room temperature are called **fats**; those that are liquid are called **oils**. Most of the fatty acid tails in butter, bacon, and other animal fats are *saturated*: they have only single covalent bonds between carbon atoms. Saturated fatty acids can snuggle up to one another in parallel arrays. Soybean, corn, and other vegetable oils have *unsaturated* fatty acids, which have one or more double covalent bonds that create kinks in the tails. Tight packing is disrupted where these bonds occur.

Some amount of certain unsaturated fatty acids appears to be important in nutrition. In one study, immature rats were placed on a fat-free diet. The rats grew abnormally, their hair fell off, their skin turned scaly, and they died young. These conditions never developed when small amounts of linoleic acid (a fatty acid with two double bonds) were added to the diet.

Phospholipids. Membranes of plant and animal cells have abundant phospholipids. A **phospholipid** molecule had a glycerol backbone, two fatty acid tails, and a phosphate group to which an alcohol is attached. Figure 6.1 shows one of the most common phospholipids in membranes.

Waxes. In **waxes**, long-chain fatty acids are linked to long-chain alcohols or to carbon rings. Wax secretions help form coats on leaves, fruits, animal skin, feathers, and fur. Beeswax is a structural material used to construct the honeycomb of beehives. In some plants, waxes are important components of *cutin*, a secretion that covers and waterproofs stem, leaf, and fruit surfaces and that may impart some resistance to disease-causing organisms. Figure 4.10 shows the waxy surface coat of a eucalyptus leaf.

Lipids Without Fatty Acid Components

Lipids that have no fatty acid tails are less abundant than the ones described so far, but many play important roles in cell membranes and in the regulation of metabolism. Some (such as the terpenes) are long, water-insoluble chains; others (such as the steroids) have ring structures.

For example, all **steroids** start out with the same backbone of four carbon rings:

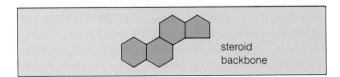

steroid backbone

However, they differ in the number and location of double bonds in the backbone, and in the number, position, and type of functional groups attached to it.

Cholesterol, the most common steroid in animal tissues, is a component of cell membranes. A cholesterol molecule can undergo rearrangements that lead to the formation of such substances as sex hormones and bile acids (which function in digestion). Plant tissues do not contain cholesterol; their steroids are called phytosterols.

PROTEINS

In addition to carbon, hydrogen, and oxygen, the molecules called **proteins** also contain nitrogen and (usually) sulfur. Of all biological molecules, proteins are the most diverse. In this class of molecules are thousands of dif-

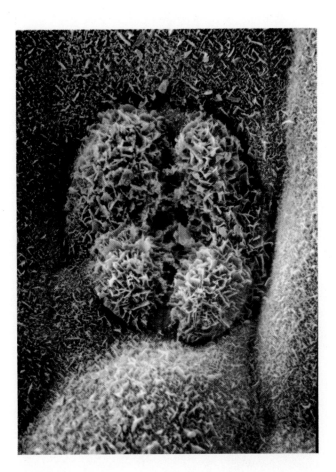

Figure 4.10 Scanning electron micrograph of wax deposits on cells making up the upper epidermal surface of a blade of grass. Here, the shredlike wax depositions surround a stoma, one of the openings that help control movements of gases and water across the leaf epidermis.

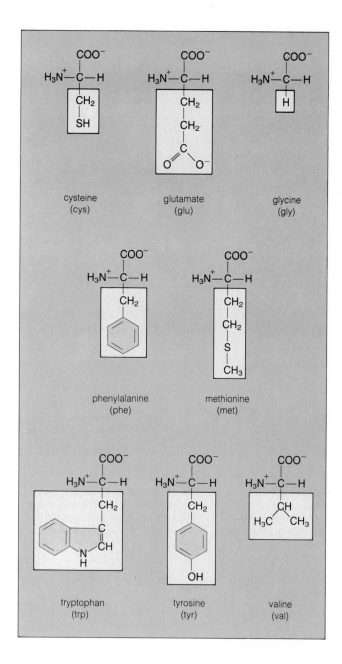

Figure 4.11 Structural formulas for eight of the twenty common amino acids. The R groups are shaded in lavender.

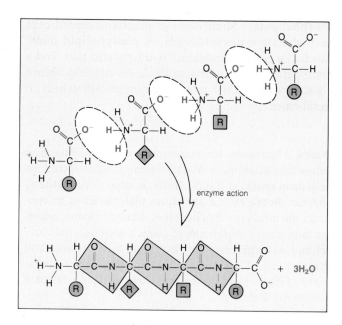

Figure 4.12 Condensation of a polypeptide chain from four amino acid units.

ferent enzymes, which make specific reactions proceed faster than they would on their own. In this class are substances concerned with cellular movements. Here, too, are storage molecules and transport molecules such as hemoglobin (which carries oxygen in blood). Some proteins are hormones; some help defend the vertebrate body against damage or attack. Other proteins are structural materials of the first rank, the stuff of cell walls and membranes, bone and cartilage, hoof and claw.

Primary Structure: A String of Amino Acids

Despite their diversity, proteins typically contain only twenty different kinds of amino acids. An **amino acid** has four parts, all bonded covalently to one carbon atom. The parts are an amino group (—NH₂), a carboxyl group (—COOH), a hydrogen atom, and some distinct atom or cluster of atoms designated the *R group* (Figure 4.11). Under cellular conditions, the amino and the carboxyl parts are ionized as shown here:

How do amino acids become linked to form proteins? A covalent bond forms between the amino group of one amino acid and the carboxyl group of another (Figure 4.12). This covalent linkage, called a **peptide bond**, results in a molecule called a dipeptide. Three or more amino acids linked together form a **polypeptide chain**.

Which kind of amino acid follows another in the chain is always the same for all proteins of a given type. For example, the two chains making up the protein insulin always have the sequences shown in Figure 4.13. The specific sequence of amino acids in a polypeptide chain constitutes the **primary structure** of a protein.

Spatial Patterns of Protein Structure

The sequence of amino acids influences the shape that a protein can assume, what its function will be, and how it will interact with other substances. It does so in two major ways. First, it influences the patterns of hydrogen bonding along the chain. Second, the R groups in the sequence interact and determine the way the chain can bend and twist into its three-dimensional shape.

In Figure 4.12, notice the so-called peptide groups, which are indicated by the tan-shaded squares. Because of the way electrons are shared in each square, the atoms linked by the red bonds tend to be positioned rigidly in the same plane (the square). Only the atoms outside the squares have some freedom in how they become oriented.

These bonding patterns impose some limits on the protein structures possible. Most often, hydrogen bonds form between every fourth amino acid and hold the chain in a helical coil about its own axis (Figure 4.14). Coils of this sort occur in hemoglobin, for example. In other cases, the chain is almost fully extended, and hydrogen bonds form between *different* chains (Figure 4.14). These bonds hold many chains side by side in a sheetlike structure, as they do in the protein that makes

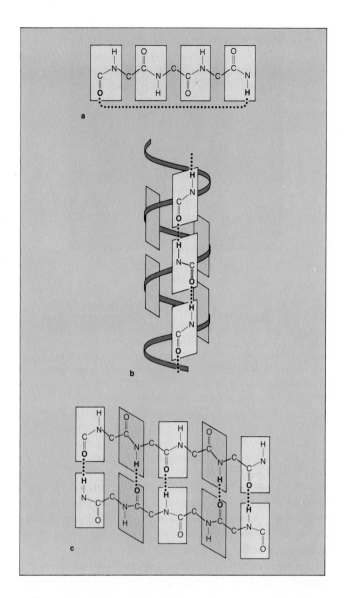

Figure 4.14 (**a**) Hydrogen bonds (dotted lines) in a polypeptide chain, which can produce a coiled chain (**b**) or a sheetlike array of chains (**c**).

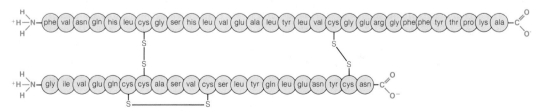

Figure 4.13 Linear sequence of amino acids in bovine (cattle) insulin, as determined by Frederick Sanger in 1953. This protein is composed of two polypeptide chains, linked together by disulfide bridges (—S—S—).

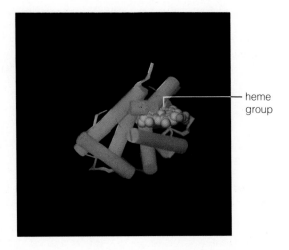

— heme group

Figure 4.15 Model of the three-dimensional structure of one of the four polypeptide chains in a hemoglobin molecule. Red cylinders represent parts of the chain that are helically coiled. Irregular loops that disrupt the regular coiling are shown as ribbons. The amino acid proline causes such disruptions at four different points in the primary sequence of amino acids. When oxygen is transported in blood, it binds with an iron-containing component (heme group) of the molecule.

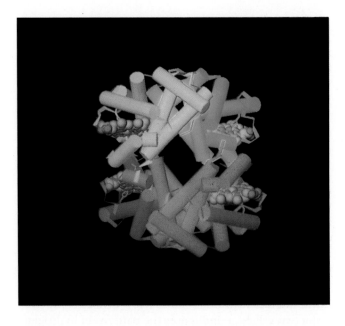

Figure 4.16 Quaternary structure of the protein hemoglobin in human blood. Hemoglobin is a red pigment circulating in animal blood and carrying vital oxygen to tissues. The hemoglobin molecule consists of four polypeptide chains, held tightly together by numerous weak bonds. (Compare Figure 4.15.)

up silk. The term **secondary structure** refers to the helical or extended pattern brought about by hydrogen bonds at regular intervals along a polypeptide chain. Some amino acids tend to favor helical patterns, others tend to favor sheetlike patterns.

Protein structure is also affected by interactions among R groups. Most helically coiled chains become further folded into some characteristic shape when one R group interacts with another R group some distance away, with the backbone itself, or with other substances present in the cell. The term **tertiary structure** refers to the folding that arises through interactions among R groups of a polypeptide chain. Figure 4.15 shows one of the diverse shapes achieved through such interactions.

Quaternary structure, the fourth level of protein architecture, results from interactions between two or more polypeptide chains in some proteins. The resulting protein can be globular, fiberlike, or some combination of the two shapes. For example, hemoglobin is globular, overall (Figure 4.16). Collagen, the most common animal protein, is fibrous (Figure 4.17). Skin, bone, tendons, cartilage, blood vessels, heart valves, corneas—these and other structures depend on the strength inherent in collagen.

Protein Denaturation

Many experiments support the idea that primary structure dictates the shape that a protein maintains under normal conditions. Among these are studies of proteins that have undergone **denaturation**: a disruption of the interactions holding a molecule in its three-dimensional form. When denatured, the polypeptide chain unwinds or changes shape.

Denaturation can be brought about by exposure to high temperatures (typically above 60°C) or to chemical agents that disrupt the bonds on which secondary, tertiary, and quaternary structure are based. Following the drastic structural changes brought about by denaturation, the protein no longer can perform its biological functions.

For example, the white portion of an uncooked chicken egg is a concentrated solution of the protein albumin. When you cook an egg, the heat does not affect the strong covalent bonds of albumin's primary structure, but it destroys the weaker bonds that maintain secondary and tertiary structure. Although denaturation can be reversed for some kinds of proteins when normal conditions are restored, albumin isn't one of them. There is no way to uncook a cooked egg.

In itself, a stretched-out polypeptide chain plays no functional role in a cell. However, its amino acid sequence dictates the final three-dimensional structure of a protein—and this structure dictates how the protein will interact with other cell substances.

NUCLEOTIDES AND NUCLEIC ACIDS

Nucleotides are vital to the operation and the reproduction of cells. Each **nucleotide** contains three different kinds of components: a five-carbon sugar (ribose or deoxyribose), a nitrogen-containing base (either a single-ringed pyrimidine or a double-ringed purine), and a phosphate group. For example, one nucleotide has the three components hooked together in this way:

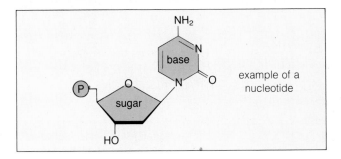

Three kinds of nucleotides or nucleotide-based molecules are the adenosine phosphates, the nucleotide coenzymes, and the nucleic acids. Later chapters will explore the structure and function of these molecules. Here, we will simply summarize their functions.

Adenosine phosphates are relatively small molecules that function as chemical messengers within and between cells, and as energy carriers. Cyclic adenosine monophosphate (cAMP) is a chemical messenger. Adenosine triphosphate (ATP) is a nucleotide that serves as an energy carrier.

Nucleotide coenzymes transport the hydrogen atoms and electrons necessary in metabolism. Nicotinamide adenine dinucleotide (NAD⁺) and flavin adenine dinucleotide (FAD) are two of these coenzymes.

Nucleic acids, which are nucleotide-based molecules, are single- or double-stranded. A single strand consists of nucleotide units strung into long chains, with a phosphate bridge connecting sugars and with bases sticking out to the side (Figure 4.18). The sequence in which the four kinds of bases follow one another varies among nucleic acids.

Deoxyribonucleic acid (DNA) and the ribonucleic acids (RNAs) are built according to the plan just outlined. DNA is usually a double-stranded molecule that twists

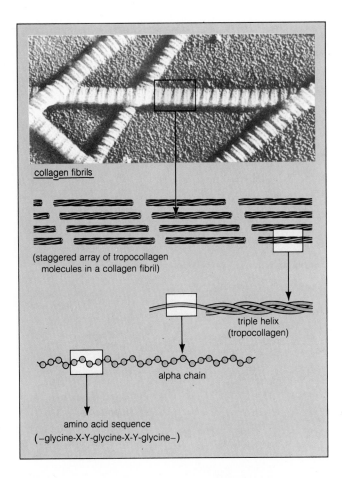

Figure 4.17 Structure of collagen, from fibrils down to its amino acid sequence. The X designates proline; the Y designates a modified proline (it has an —OH group attached).

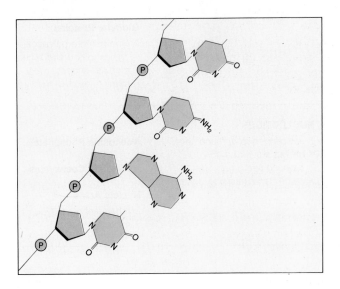

Figure 4.18 Example of bonds between nucleotides in a nucleic acid molecule.

Table 4.2 Summary of the Main Biological Molecules

Category	Main Subcategories		Some Examples and Their Functions
CARBOHYDRATES *contain an aldehyde or a ketone group, one or more hydroxyl groups*	**Monosaccharides**	Glucose	Energy reservoir for metabolism, structural unit for biosynthesis
	Disaccharides	Sucrose	Sugar transport in complex plants
	Polysaccharides	Starch	Food storage
		Cellulose	Structural roles
LIPIDS *are largely hydrocarbon, generally do not dissolve in water but dissolve in nonpolar substances*	**Lipids With Fatty Acid Components:**		
	Glycerides: one, two, or three fatty acid tails attached to glycerol backbone	Fats (e.g., butter) Oils (e.g., corn oil)	Forms in which energy is stored
	Phospholipids: phosphate group and (often) two fatty acids attached to glycerol backbone	Phosphatidylcholine	Key component of cell membranes
	Waxes: long-chain fatty acid tails attached to alcohol	Earwax Waxes in cutin	Protective barrier to inner ear Help restrict water loss from some plant parts
	Lipids With No Fatty Acid Components:		
	Steroids: four carbon rings; the number, position, and type of functional groups vary	Cholesterol	Component of animal cell membranes; can be rearranged into other steroids (e.g., bile acids, male and female sex hormones)
PROTEINS *are polypeptides (up to several thousand amino acids, covalently linked)*	**Fibrous Proteins:** Individual polypeptide chains, often linked into tough, water-insoluble molecules	Keratin Collagen	Structural element of hair, nails Structural element of bones
	Globular Proteins: One or more polypeptide chains folded and linked into globular shapes; water-soluble, many roles in cell activities	Polymerases Hemoglobin Insulin Antibodies	Serve as enzymes Transport of oxygen Hormone that influences metabolism Tissue defense
NUCLEOTIDES *are derived from monomers having a five-carbon sugar, a phosphate group, and a nitrogen-containing base*	**Adenosine Phosphates**	ATP	Energy carrier
	Nucleotide Coenzymes	NAD$^+$	Transports hydrogen and electrons in cells
	Nucleic Acids (chains of nucleotide monomers)	DNA, RNAs	Storage, transmission, translation of genetic information

helically about its own axis (Figure 4.1). The bases of one strand are connected by hydrogen bonds to bases of the other strand. You will be reading more about these molecules in chapters to come. For now, it is enough to know the following: (1) Genetic instructions are encoded in the sequence of bases in DNA, and (2) RNA molecules function in the processes by which genetic instructions are used in building proteins.

SUMMARY OF THE MAIN BIOLOGICAL MOLECULES

Table 4.2 summarizes the main categories of biological molecules that have been described in this chapter. Included in this table are the most common classes of molecules within each category. We will have occasion to return to their nature and roles in diverse life processes.

Review Questions

1. What are the four main families of small organic molecules used in cells for the assembly of carbohydrates, lipids, proteins, and nucleic acids (the large biological molecules)?

2. Identify which of the following is the carbohydrate, fatty acid, amino acid, and polypeptide:

   a. $^+NH_3$—CHR—COO$^-$
   b. $C_6H_{12}O_6$
   c. (glycine)$_{20}$
   d. $CH_3(CH_2)_{16}COO^-$

3. Amylose (a form of starch) and cellulose are both assembled from glucose subunits, but they have very different properties. Can you describe these properties and explain how they arise?

4. Explain the difference between saturated and unsaturated fats in terms of the carbon skeletons of their fatty acid tails. Is butter a saturated or unsaturated fat?

5. Is this statement true or false? Not all proteins are enzymes, but all enzymes are proteins.

6. Describe the four levels of protein structure. How do the side groups of a protein molecule influence its interactions with other substances? Give an example of what happens when the bonds holding a protein together are disrupted.

7. Distinguish between the following:

   a. monosaccharide, polysaccharide
   b. peptide, polypeptide
   c. glycerol, fatty acid
   d. nucleotide, nucleic acid

8. Define the general structure and function of three kinds of nucleotides or nucleotide-based molecules that are important to cell functioning and reproduction.

Readings

Alberts, B., et al. 1983. *Molecular Biology of the Cell.* New York: Garland Publishing. Extraordinarily clear descriptions of the biological molecules; good illustrations.

Dickerson, R., and I. Geis. 1976. *Chemistry, Matter, and the Universe.* Menlo Park, California: Benjamin/Cummings. Introductory text with a biological point of view.

Eyre, D. 1980. "Collagen: Molecular Diversity in the Body's Protein Scaffold." *Science* 207:1315–1322.

Karplus, Martin, and J. Andrew McCammon. April 1986. "The Dynamics of Proteins." *Scientific American* 254(4):42–51.

"The Molecules of Life." October 1985. *Scientific American.* This entire issue is devoted to articles on current insights into DNA, proteins, and other biological molecules. Excellent illustrations.

Sharon, N. November 1980. "Carbohydrates." *Scientific American* 243(15):90–116. Describes carbohydrates and the roles they play in organisms.

Weinberg, R. October 1985. "The Molecules of Life." *Scientific American* 253:48–57. Survey of the new techniques and discoveries of molecular biology.

5

CELL STRUCTURE
AND FUNCTION:
AN OVERVIEW

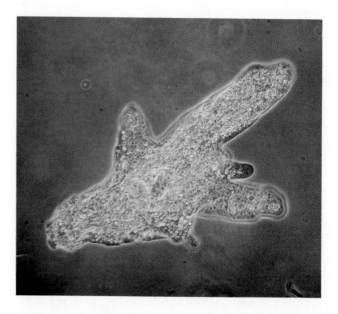

Figure 5.1 A "cell"—a vaguely dreary name more suggestive of carton boxes than of things seething with life. Shown here, *Amoeba proteus* on the move through a water droplet.

GENERALIZED PICTURE OF THE CELL

Emergence of the Cell Theory

Early in the seventeenth century, Galileo Galilei arranged two glass lenses in a cylinder. With this instrument he happened to look at an insect and thereby came to describe the stunning geometric patterns of its tiny eyes. Thus, Galileo, who was not a biologist, was the first to record a biological observation made through a microscope. The study of the cellular basis of life was about to begin. First in Italy, then in France and England, biologists began to explore a world whose existence had not even been suspected.

At mid-century Robert Hooke, "Curator of Instruments" for the Royal Society of England, was at the forefront of these studies. When Hooke first turned one of his microscopes to a thinly sliced piece of cork from a mature tree, he observed tiny, empty compartments. He gave them the Latin name *cellulae* (meaning small rooms); hence the origin of the biological term "cell." They were actually walls of dead cells, which is what cork is made of, although Hooke did not think of them as being dead because he did not know that cells could be alive. He also noted that cells in other plant materials contained "juices." He did not speculate on what the juice-filled structures might represent.

Given the simplicity of their instruments, it is amazing that these pioneers saw as much as they did. Antony van Leeuwenhoek, a shopkeeper, had the greatest skill in constructing lenses and, possibly, the keenest vision. He even observed a single bacterium—a type of organism so small it would not be seen again for another two centuries! Yet this was mostly an age of exploration, not of interpretation. Once the limits of those simple instruments had been reached, biologists gave up interest in cell structure without having been able to explain what they had seen.

Then, in the 1820s, improvements in lens design permitted closer views of the cell. The botanist Robert Brown observed a spherelike structure in every plant cell he examined; he called the structure a "nucleus." By 1839, the zoologist Theodor Schwann reported the presence of cells in animal tissues. He began working with Matthias Schleiden, a botanist who had concluded that cells are present in all plant tissues and that the nucleus is somehow paramount in cell reproduction. Both investigators proposed that each living cell has the potential to exist independently—in other words, in the absence of other cells. It was Schwann who distilled the meaning of these new observations in what came to be known as the first two principles of the cell theory:

Table 5.1 Units of Measure Used in Microscopy			
Unit	Equivalence in Millimeters	Equivalence in Micrometers*	Equivalence in Nanometers**
Millimeter (mm)	1	1,000	1,000,000
Micrometer (μm)	0.001	1	1,000
Nanometer (nm)	0.000001	0.001	1

*The micrometer is used in describing whole cells or large cell structures, such as the nucleus.

**The nanometer is used in describing cell ultrastructures (those ranging from, say, mitochondria downward), even of macromolecules. Lipid molecules, for example, may be about 2 nanometers long; cell membranes may be 7 to 10 nanometers thick; ribosomes are about 25 nanometers across.

All organisms are composed of one or more cells.

The cell is the basic living unit of organization.

Yet a question remained: Where do cells come from? A decade passed before the physiologist Rudolf Virchow completed studies of cells growing and reproducing (dividing into two new cells). He reached this conclusion, which became the third principle of the cell theory:

All cells arise from preexisting cells.

Not only was a cell viewed as the smallest living unit, the continuity of life was now seen to arise directly from the division and growth of single cells. Within each tiny cell, events were going on that had profound implications for all levels of biological organization!

Basic Aspects of Cell Structure and Function

Describing the cell might seem to be impossible, because cells vary enormously in size, shape, and complexity. However, we can begin with three features that all cells have in common: a *nucleus* (or nucleoid), surrounded by *cytoplasm*, which is surrounded by a *plasma membrane*:

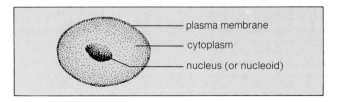

- plasma membrane
- cytoplasm
- nucleus (or nucleoid)

These features, which are the structural basis of all cellular events, can be defined in this way:

---

1. Plasma membrane. This membrane is the boundary between events in the cell (which are organized and controlled) and in the environment (which the cell cannot control). Many events occur at the membrane itself. Nutrients, ions, and wastes cross the membrane at built-in channels and shuttles. Some membrane components send signals about changing conditions into the cell; others serve as docks for specific substances.

2. Nucleus (or nucleoid). Most cells have a nucleus: a membrane-bound compartment which contains hereditary instructions (DNA) and other molecules that function in how the instructions are read, modified, and dispersed. In bacterial cells only, DNA is concentrated in a region called the nucleoid; there is no surrounding membrane.

3. Cytoplasm. The cytoplasm includes everything except the plasma membrane and nucleus. It consists of internal membranes, particles, and strands bathed in a semifluid matrix (the cytosol). The cytoplasm serves as a warehouse for substances moving into and out of the cell. It has pathways for channeling the movement of materials in different directions, and regions where materials are synthesized and broken apart. It also contains housekeeping equipment, machinery for cell movements, and support elements that give cells their shape.

Cell Size and Cell Membranes

Bird eggs are exceptionally large single cells, which you can observe with the unaided eye. So are cells in the red part of a watermelon. Parts of nerve cells running through a giraffe leg can be three meters long and are certainly what you would call enormous. However, most cells cannot be seen without the aid of a microscope (Table 5.1).

In massive multicelled bodies, most cells are between 5 and 20 micrometers in diameter. (Your red blood cells are about 6 to 8 micrometers across; a string of about 2,000 would only be as long as your thumbnail is wide.) The finest detail the light microscope can resolve is about 0.2 micrometer across. However, recent biochemical studies and advances in electron microscopy have enabled us to visualize cell structures much smaller than this (Figures 5.2 and 5.3).

Why are most cells so small? In large part, cell size is governed by how efficiently materials cross the plasma membrane and become distributed through the cytoplasm. A small single cell has enough surface area to

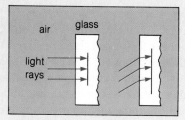

air   glass

light
rays

a *Refraction of light rays
(The angle of entry and the
molecular structure of the
glass determine how much
the rays will bend)*

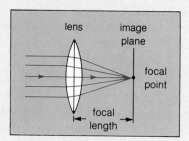

lens   image
        plane

        focal
        point

        focal
        length

b *Focusing of light rays*

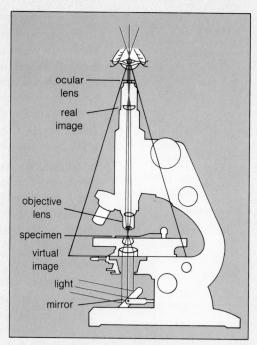

ocular
lens

real
image

objective
lens

specimen

virtual
image

light

mirror

c *Compound light microscope*

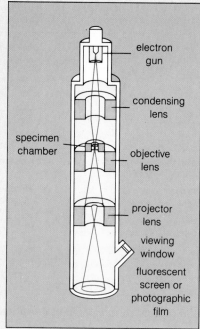

electron
gun

condensing
lens

specimen
chamber

objective
lens

projector
lens

viewing
window

fluorescent
screen or
photographic
film

d *Transmission electron microscope*

Figure 5.2 Microscopes—gateways to the cell.

Light Microscopes (a) Light microscopy relies on the bending, or refraction, of light rays. Light rays pass straight through the center of a curved lens. The farther they are from the center, the more they bend. (**b, c**) The *compound light microscope* is a two-lens system. All rays coming from the object being viewed are channeled to a single place behind the curved lens system.

If you wish to observe a *living* cell, it must be small or thin enough for light to pass through. Also, structures inside cells can be seen only if they differ in color and density from their surroundings—but most are almost colorless and optically uniform in density. Specimens can be stained (exposed to dyes that react with some cell structures but not others), but staining usually alters the structures and kills the cells. Finally, dead cells begin to break down at once, so they must be pickled or preserved before staining. Most observations have been made of dead, pickled, or stained cells. Largely transparent living cells can be observed through the *phase contrast microscope*. Here, small differences in the way different structures refract light are converted to larger variations in brightness.

No matter how good a glass lens system may be, when magnification exceeds 2,000× (when the image diameter is 2,000 times as large as the object's diameter), cell structures appear large but are not clearer. By analogy, when you hold a magnifying glass close to a newspaper photograph, you see only black dots. You cannot see a

detail as small as or smaller than a dot; the dot would cover it up. In microscopy, something like dot size intervenes to limit *resolution* (the property that dictates whether small objects close together can be seen as separate things). That limiting factor is the physical size of wavelengths of visible light.

Light comes in different wavelengths, or colors. Red wavelengths are about 750 nanometers and violet wavelengths are about 400 nanometers; all other colors fall in between. If an object is smaller than about one-half the wavelength, light rays passing by it will overlap so much that the object won't be visible. The best light microscopes resolve detail only to about 200 nanometers.

Transmission Electron Microscopes Electrons are usually thought of as particles, but they also behave like waves. Electron wavelengths are about 0.005 nanometer—about 100,000 times shorter than those of visible light! Ordinary lenses cannot be used to focus such accelerated streams of electrons, because glass scatters them. But each electron carries an electric charge, which responds to magnetic force. A magnetic field can divert electrons along defined paths and channel them to a focal point. Magnetic lenses are used in *transmission electron microscopes* (**d**).

Electrons must travel in a vacuum, otherwise they would be randomly scattered by molecules in the air. Cells can't live in a vacuum, so living cells cannot be observed at this higher magnification. In addition, specimens must be

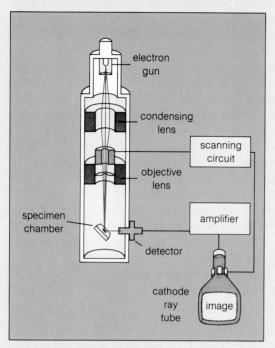

e *Scanning electron microscope*

sliced extremely thin so that electron scattering corresponds to the density of different structures. (The more dense the structure, the greater the scattering and the darker the area in the final image formed.) Specimen fixation is crucial. Fine cell structures are the first to fall apart when cells die, and artifacts (structures that do not really exist in cells) may result. Because most cell materials are somewhat transparent to electrons, they must be stained with heavy metal "dyes," which can create more artifacts.

With *high-voltage electron microscopes*, electrons can be made ten times more energetic than with the standard electron microscope. With the energy boost, intact cells several micrometers thick can be penetrated. The image produced is something like an x-ray plate and reveals the three-dimensional internal organization of cells (see, for example, Figure 5.20).

Scanning Electron Microscopes **(e)** With a *scanning electron microscope*, a narrow electron beam is played back and forth across a specimen's surface, which has been coated with a thin metal layer. Electron energy triggers the emission of secondary electrons in the metal. Equipment similar to a television camera detects the emission patterns, and an image is formed. Scanning electron microscopy is limited to surface views, and it does not approach the high resolution of transmission instruments. However, its images have fantastic depth.

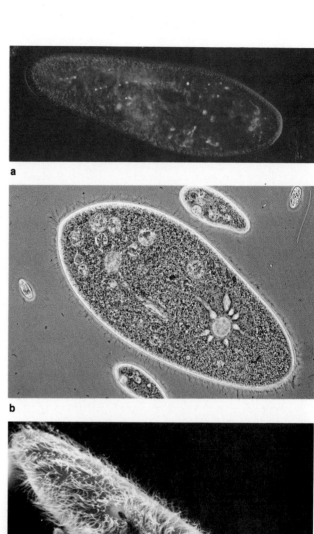

a

b

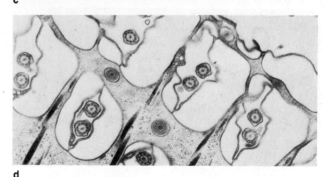

c

d

Figure 5.3 Comparison of image-forming abilities of different microscopes. The specimens are different species of *Paramecium*. **(a)** Conventional light microscope, bright-field, 2900×. **(b)** Phase contrast microscope, 2,000×. **(c)** Scanning electron microscope, 500×. Notice hairlike cilia (motile structures) on surface. **(d)** Transmission electron micrograph, glancing section through body surface, 17,800×. The cilia (circles) have been cut, like whiskers with a razor.

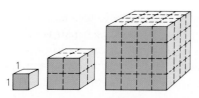

surface area (cm²):	6	24	96
volume (cm³):	1	8	64
surface-to-volume ratio:	6:1	3:1	1.5:1

Figure 5.4 Simple portrayal of the relationship between surface area and volume when a cube is enlarged. The cube to the left measures one centimeter on a side. As its linear dimensions increase, its surface area does not increase at the same rate as its volume.

This relationship puts constraints on the nature of cell enlargement. Past a certain point, a volume of cytoplasm enlarging in all directions would outstrip the amount of plasma membrane surface available to serve it.

accommodate exchanges between the environment and the cell interior. And diffusion (the random motion of molecules) easily distributes substances through the small volume of cytoplasm.

But suppose a small round cell did nothing but expand in volume. Its surface area would not increase at the same rate. (As Figure 5.4 indicates, volume increases with the cube of the diameter, but surface area increases only with the square.) Thus the cell would have an inefficient *surface-to-volume ratio*.

If that cell enlarged four times in diameter, its volume would increase (4 × 4 × 4) or sixty-four times. But its *surface area* would increase by only sixteen times—so each unit of plasma membrane would have to serve four times as much cytoplasm as before! Past a certain point, the inward flow of nutrients and outward flow of wastes (some toxic) would not be fast enough, and the cell simply would die.

Body plans of multicelled organisms are also influenced by surface-to-volume constraints. Thus we see algal bodies that are strings of cells, with each cell in

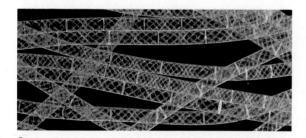

Figure 5.5 Multicellular strategies for getting around the constraints of the surface-to-volume ratio. (**a**) One-dimensional growth in a cyanobacterium. (**b**) Two-dimensional growth in a red alga. (**c**) Three-dimensional growth in a whale. Here, membrane specializations in internal transport systems assure even the internal cells of efficient exchange of materials.

direct contact with the environment (Figure 5.5). Other algae and a few protistans have sheetlike bodies in which cells are kept at or near the surface. Photosynthetic cells near the surface of flat plant leaves are in a good position to intercept sunlight energy and carbon dioxide from the surrounding air.

In massive multicelled bodies, internal transport systems (such as a circulatory system) rapidly move materials to and from cells in tissue regions deep in the body. Here also, we see cell membrane specializations that provide greater surface area for reacting to the materials flowing past. For example, slender extensions of cells lining the small intestine increase the surface area exposed to ingested nutrients, which the cells absorb.

PROKARYOTIC CELLS

Bacteria are typically about ten times smaller than most other cells, and they function efficiently without a lot of complex structures. The plasma membrane itself houses enzymes and other molecules used in cell activities. All **ribosomes**, which are the construction sites for protein synthesis, are simply dispersed in the small volume of cytoplasm. The DNA is more or less concentrated in the irregularly shaped region called the nucleoid (Figure 5.6).

Because bacteria do not have a nucleus, they are called **prokaryotic** (which means "before the nucleus"). The word implies that ancestors of these cells were present before the evolution of nucleated cells. All living bacteria are prokaryotic (including cyanobacteria, which formerly were called blue-green algae).

Most prokaryotic cells have another distinctive feature. They have a rigid or semirigid **cell wall** outside the plasma membrane that supports the cell and imparts shape to it (Figure 5.6). The wall is composed of secretions from the cell itself, and sometimes the wall has a protective coating.

EUKARYOTIC CELLS

Function of Organelles

All cells except bacteria are called **eukaryotic**. They contain many **organelles**, which are membranous sacs, envelopes, and other compartmented portions of the cytoplasm. The most conspicuous organelle is the nucleus (hence the name eukaryotic, which means "true nucleus").

No chemical apparatus in the world can match the eukaryotic cell for the sheer number of chemical reactions that proceed in so small a space. Many of these reactions are totally incompatible, yet they proceed smoothly. Why? There are two reasons:

Organelles physically separate different chemical reactions (many of which are incompatible) in the space of the cytoplasm.

Organelles also separate different reactions in time, for molecules being used and produced are passed from one organelle to another in specific reaction sequences.

Keep these two points in mind as you read through the rest of this unit, for they give insight into why many metabolic activities proceed as they do.

Types of Organelles

Eukaryotic cells generally contain the following organelles, each with specific functions:

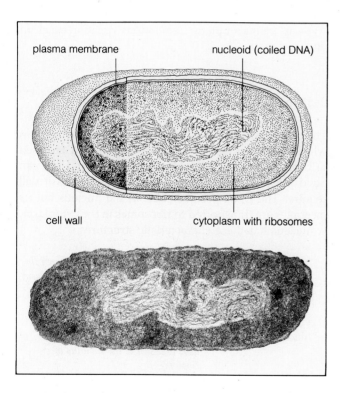

plasma membrane nucleoid (coiled DNA)

cell wall cytoplasm with ribosomes

Figure 5.6 Body plan of the bacterium *Escherichia coli*. The transmission electron micrograph shows this prokaryotic cell in longitudinal section, 36,500×.

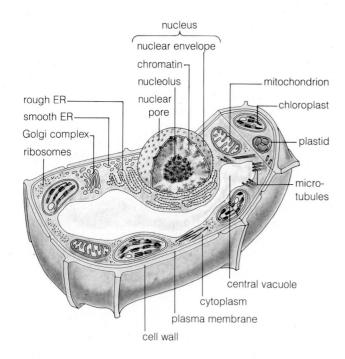

Figure 5.7 Generalized sketch of a plant cell, showing the types of organelles that may be present.

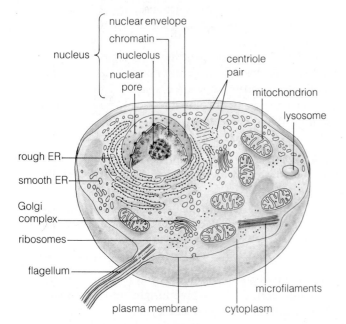

Figure 5.8 Generalized sketch of an animal cell, showing the types of organelles that may be present.

nucleus	storage, control of hereditary instructions
endoplasmic reticulum	manufacture of protein, lipid components of membranes; modification of proteins destined for secretion
Golgi bodies	transport of proteins and other materials
lysosomes	degradation, recycling of materials
transient vesicles	transport of materials
mitochondria	extraction of energy from carbohydrates; ATP formation

Other organelles are found in some eukaryotic cells but not in others. For example, all photosynthetic cells have one or more *plastids*, which function in food production and in storage. Cells of many fungi and plants have one or more fluid-filled sacs called *central vacuoles*. Cells of many protistans, fungi, and plants have a *cell wall*.

Eukaryotic cells also have a *cytoskeleton*, which serves as an internal framework for organizing organelles and ribosomes. Some of its components move internal parts and often the entire cell body. Eukaryotic cells also have *ribosomes*, the sites of protein synthesis.

Figures 5.7 through 5.10 show how some of these organelles and structures might be arranged in a typical plant and animal cell. Keep in mind that calling them "typical" is like calling a squid or a watermelon plant a "typical" animal or plant; tremendous variation exists on the basic plan.

THE NUCLEUS

The **nucleus** is a membrane-bound compartment in which hereditary instructions (DNA) are stored and used in controlled ways. With these instructions, a cell can build most of the proteins and RNA molecules it will require. The DNA itself remains in the nucleus but its messages are transported to ribosomes in the cytoplasm. A nucleus has these characteristic structures:

nuclear envelope	a two-membrane boundary between the inside of the nucleus and the cytoplasm
nucleoplasm	the fluid portion of the nucleus
chromosomes	DNA strands (and associated proteins)
nucleolus	dense cluster of DNA and materials used to assemble the subunits of ribosomes

Electron micrographs reveal that the nucleus also has a "skeleton" of protein-containing strands that organize the DNA.

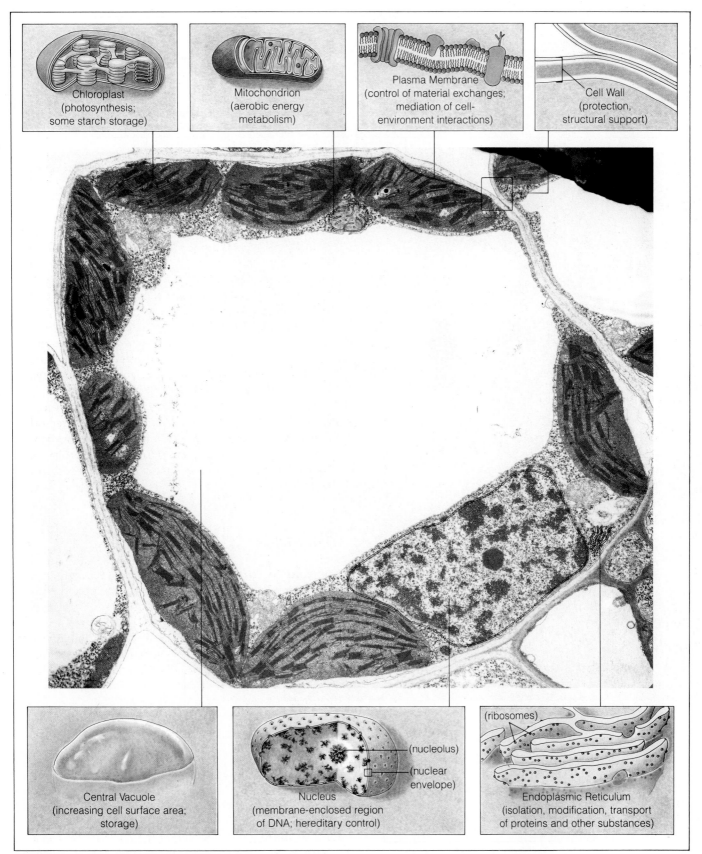

Chloroplast (photosynthesis; some starch storage)

Mitochondrion (aerobic energy metabolism)

Plasma Membrane (control of material exchanges; mediation of cell-environment interactions)

Cell Wall (protection, structural support)

Central Vacuole (increasing cell surface area; storage)

Nucleus (membrane-enclosed region of DNA; hereditary control)

(nucleolus)

(nuclear envelope)

Endoplasmic Reticulum (isolation, modification, transport of proteins and other substances)

(ribosomes)

Figure 5.9 Transmission electron micrograph of a plant cell from a blade of timothy grass, cross-section, 11,300×.

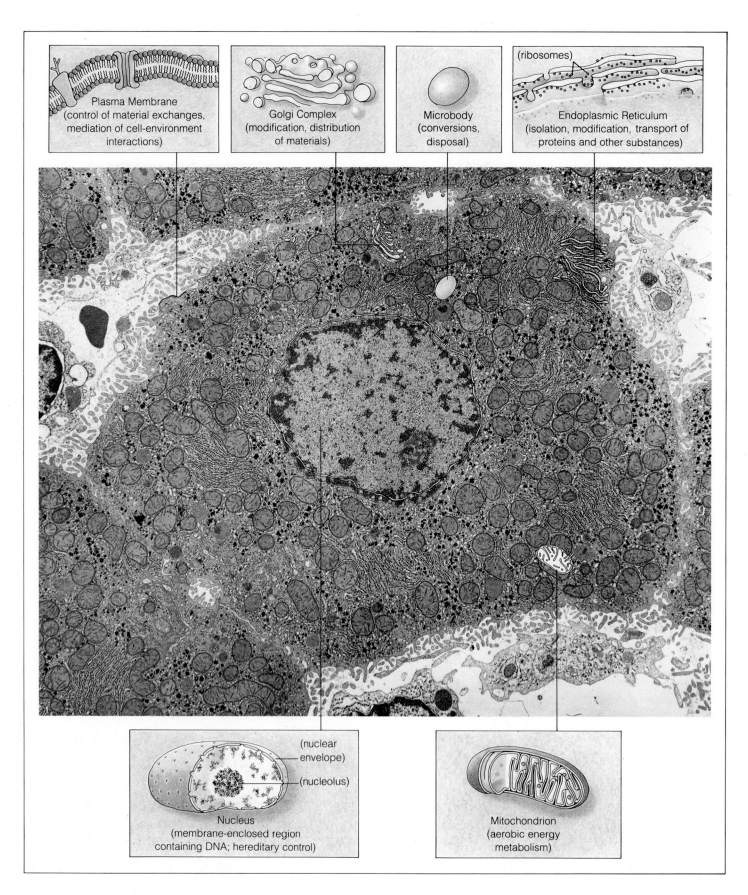

Plasma Membrane
(control of material exchanges,
mediation of cell-environment
interactions)

Golgi Complex
(modification, distribution
of materials)

Microbody
(conversions,
disposal)

(ribosomes)

Endoplasmic Reticulum
(isolation, modification, transport of
proteins and other substances)

(nuclear
envelope)

(nucleolus)

Nucleus
(membrane-enclosed region
containing DNA; hereditary control)

Mitochondrion
(aerobic energy
metabolism)

Figure 5.10   Transmission electron micrograph of an animal cell from a rat liver, cross-section, 15,000×.

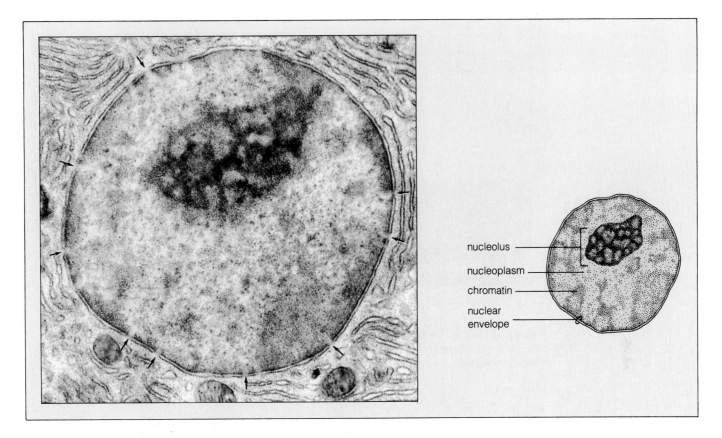

nucleolus

nucleoplasm

chromatin

nuclear
envelope

Figure 5.11 Electron micrograph of the nucleus (in cross-section) from a bat pancreatic cell, 12,600×.
Arrows point to pores in the nuclear envelope.

Nuclear Envelope

Figures 5.11 and 5.12 show details of the **nuclear envelope**. This two-membrane envelope is the outermost
component of the nucleus. Ribosomes are attached to its
outer surface, which faces the cytoplasm, and pores
extend across the envelope at regular intervals. The
structure of the nuclear envelope suggests that it is the
boundary for exchanges between the nucleus and the
cytoplasm, with its pores being passageways.

Chromosomes

A cell reproduces itself by dividing in two. When a non-
dividing cell is stained for observation through a light
microscope, the DNA and proteins in its nucleus take
up some of the stain, and they look rather grainy. In this
form they are called **chromatin**, which simply means
colored material (Figure 5.11).

At much higher magnification, the DNA molecules
are seen to be thin strands, with proteins tightly attached
to them like beads on a chain. When the cell is about to
divide, the "beaded chains" fold and twist up into condensed structures that can be observed through light

microscopes. These structures were named **chromosomes** (which means colored bodies), although we now
call the DNA molecule and its associated proteins "a
chromosome" regardless of whether it is in its uncon-
densed or condensed form.

Some of the proteins bound to DNA serve as the
scaffold for organizing DNA when it is condensed into
the compact chromosome structure; others may help
control how DNA instructions are used.

Nucleolus

As cells grow and develop, dense masses of irregular
size and shape form in the nucleoplasm. Two or three
masses form in most cells, but there can be a thousand
or more in cells that develop into amphibian eggs. Each
mass is a **nucleolus** (plural, nucleoli); see Figure 5.11.

A nucleolus contains DNA and the materials used to
assemble parts of ribosomes. Each ribosome has two
subunits, composed of proteins and RNA. A complete,
two-part ribosome never is found in the nucleus;
although the subunits themselves are synthesized in the
nucleolus, they are all shipped to the cytoplasm.

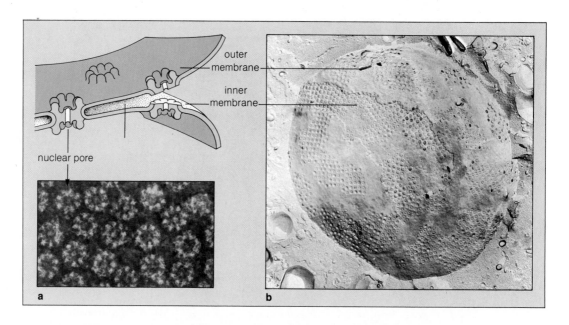

Figure 5.12 Nuclear envelope structure. (**a**) Electron micrograph of the pores that span the envelope. (**b**) Electron micrograph of a freeze-fractured nucleus (page 88), revealing both membranes of the envelope, 10,300×.

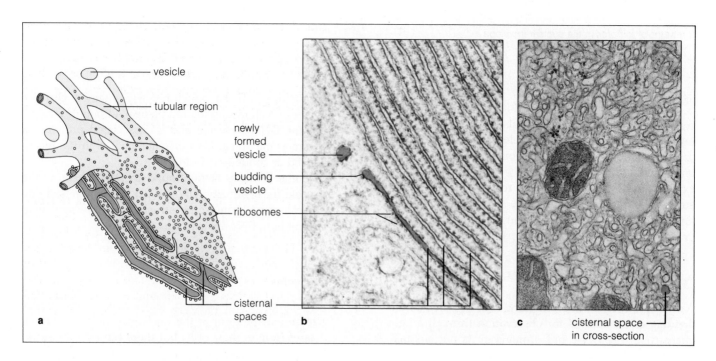

Figure 5.13 (**a, b**) Rough endoplasmic reticulum, showing how the membrane surface facing the cytoplasm is studded with ribosomes, 25,000×. (**c**) Smooth endoplasmic reticulum, in cross-section, 35,000×.

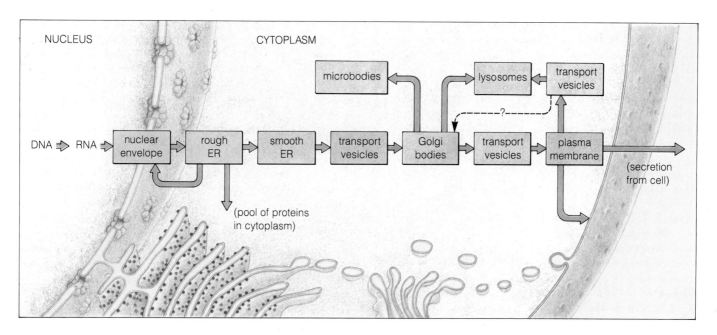

Figure 5.14 Direction of protein traffic through the cytomembrane system of eukaryotic cells. The green boxes indicate the secretory pathway.

CYTOMEMBRANE SYSTEM

Cells are constructed and maintained through the action of proteins called enzymes, and proteins contribute to cell structure. The instructions for building proteins are transported from DNA in the nucleus to ribosomes in the cytoplasm, where protein synthesis takes place.

What happens to the newly formed proteins? Some end up in the pool of proteins dissolved in the cytoplasm. Others enter the **cytomembrane system**, which includes the endoplasmic reticulum (Figure 5.13), Golgi bodies, some enzyme-filled sacs, and assorted transport vesicles. These organelles have roles in the following tasks:

1. Modifying newly formed proteins for use in the cell or for secretion from the cell.

2. Transporting proteins to specific organelles or to the cell surface.

3. Synthesizing lipids.

4. Adding to or replacing the protein and lipid components of organelle membranes.

As Figure 5.14 suggests, protein "traffic" through the cytomembrane system is highly organized. Proteins destined for secretion are segregated from the cytoplasm and are transported through the channels of the system on their way to the cell surface. ("Secretion" means the production and release of a substance from a gland cell.) Other proteins become part of the contents of organelles. Many are incorporated into organelle membranes; and some of these proteins become active as the enzymes that are used to synthesize lipids.

Endoplasmic Reticulum and Ribosomes

The **endoplasmic reticulum** (ER) is a system of membranous tubes, channels, and flattened sacs that form compartments within the cytoplasm (Figure 5.14). Proteins that eventually will be secreted from the cell begin their journey at the ER. Also, the protein and lipid components of most organelles are manufactured here. (ER membranes contain many different enzymes, including the ones that help synthesize lipids and attach carbohydrate groups to proteins.)

The surface appearance of ER membranes varies, depending largely on whether ribosomes have become attached to the side of the membrane facing the cytoplasm. *Rough ER* has ribosomes attached, giving the membranes a grainy, or rough, appearance in electron micrographs. *Smooth ER* has no ribosomes attached. Most cells contain both types, although the relative proportion varies among different cells.

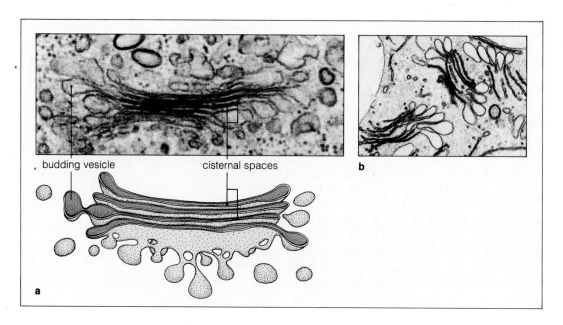

budding vesicle   cisternal spaces

b

a

Figure 5.15 (**a**) Electron micrograph and sketch of a Golgi body from an animal cell, 53,000×. (**b**) Golgi bodies in a root cap cell of corn, 14,000×.

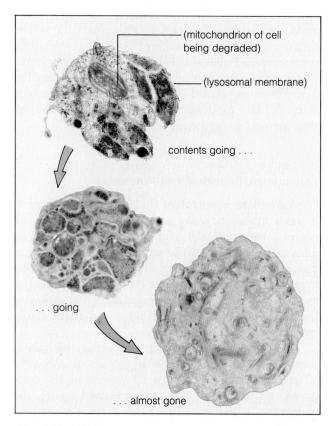

(mitochondrion of cell being degraded)

(lysosomal membrane)

contents going . . .

. . . going

. . . almost gone

Figure 5.16 Digestion of organelles from a destroyed cell, as seen in a lysosome, 11,000×.

Rough ER. According to one hypothesis, the first part of a newly forming protein may or may not have a "mailing address" of about fifteen to twenty specific amino acids. If there is no such address, the protein joins the pool of proteins in the cytoplasm. If an address is present, the protein moves into the rough ER. Proteins inside this compartmented space are modified (for example, complex sugar groups are typically attached to them). Vesicles bud from the membrane and transport the proteins to Golgi bodies (Figure 15.14). Rough ER is especially abundant in cells specialized for protein secretion.

Smooth ER. The interconnected pipelines called smooth ER curve through the cytoplasm. (Often they look like clustered sacs in micrographs, but this is because the "pipes" were sliced crosswise during specimen preparation.) This part of the cytomembrane system has different functions in different cells. In protein-secreting cells, it is continuous with rough ER and buds off to form transport vesicles. Enzymes in some smooth ER break down fat or glycogen molecules. Smooth ER is highly developed in cells that produce an abundance of lipids, such as cells in developing seeds and animal cells that secrete steroid hormones. In skeletal muscle, smooth ER forms the *sarcoplasmic reticulum*, a system for storing and releasing calcium ions (which play a role in muscle contraction).

Golgi Bodies

Golgi bodies, which are stacks of flattened, membranous sacs, receive substances from the ER. (Botanists sometimes refer to Golgi bodies as *dictyosomes*.) In outward appearance, a Golgi body resembles a stack of pancakes—usually eight or less, and usually curled at the edges (Figure 5.15). The topmost pancakes (the ones farthest from the nucleus) bulge at the edges, then the bulges break away as vesicles.

Many molecules travel through a Golgi body on their way to other organelles or to the plasma membrane. Some are modified and others are assembled almost entirely inside the Golgi sacs. Both the ER and Golgi membranes contain enzymes that attach carbohydrate groups to many proteins. Golgi bodies in plant cells may be production sites for cell wall materials (such as pectin) in which cellulose strands become embedded.

Lysosomes

Cells continually disassemble and recycle materials for use in their building programs, and **lysosomes** are central to the process. Enzymes in these organelles can break down virtually every polysaccharide, nucleic acid, and protein, along with some lipids.

For example, cholesterol becomes associated (complexed) with other substances when it is transported in the bloodstream. When the complex is moved into cells, lysosomal enzymes break down the complex and free the cholesterol, which becomes available for cellular use. As another example, lysosomal enzymes degrade organelles that are worn out or that have served their function (Figure 5.16). They also can destroy bacteria and foreign particles.

Lysosomes act on substances that the cell has already wrapped in a membranous sac. When one of these sacs and a lysosome make contact, their membranes fuse and their contents are brought together. The lysosomal enzymes act rapidly, and the molecular leftovers are small enough to be transported across the lysosomal membrane and into the cytoplasm, where they are recycled.

Microbodies

Some proteins and lipids manufactured in the cytomembrane system end up in **microbodies**, which are present in at least some cells of all eukaryotes. These organelles help convert various substances into other substances that are in short supply. The microbodies called *peroxisomes* are present in plant and animal cells. Some of their enzymes use oxygen to strip hydrogen atoms from molecules. One of their enzymes helps convert hydrogen

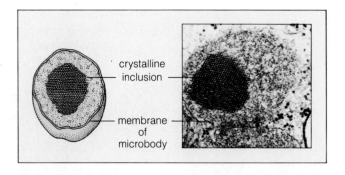

Figure 5.17 Glyoxysome from a castor bean cell, in thin section. This kind of microbody is abundant in lipid-rich cells in germinating seeds.

peroxide to water. (If allowed to accumulate, hydrogen peroxide can kill cells.) The microbodies called *glyoxysomes* occur in lipid-rich plant cells (Figure 5.17). They are abundant in germinating seeds of peanuts, watermelons, cucumbers, and castor beans. Some of their enzymes help convert fats and oils to the sugars necessary for rapid, early growth of the plant.

OTHER CYTOPLASMIC ORGANELLES

Mitochondria

As you have seen, protein synthesis, transport, and recycling are vital aspects of cell function. Another vital aspect is the extraction of energy stored in carbohydrates and the formation of molecules such as ATP—which can give up energy to drive many diverse cellular reactions. The **mitochondrion** (plural, mitochondria) is an organelle specialized for this task. Mitochondria can extract far more energy from carbohydrates than can be done by any other means, and they do so with the help of oxygen atoms. When you breathe in, you are taking in oxygen for your mitochondria.

Mitochondria commonly are about one to five micrometers long, or about the size of some bacteria. They may be round, potato-shaped, tubelike, or threadlike, but these shapes are not rigid. Depending on cel-

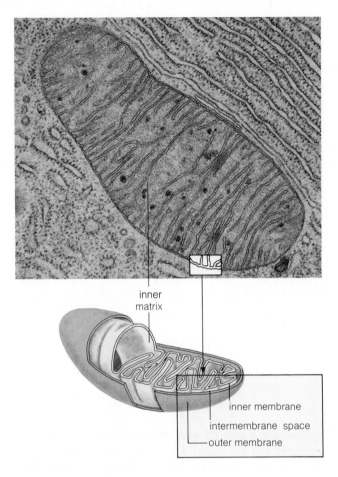

inner
matrix

inner membrane

intermembrane space

outer membrane

Figure 5.18 Micrograph and generalized sketch of a mitochondrial membrane system, 36,500×.

lular conditions, mitochondria can grow and branch ou', even fuse with one another and divide in two.

A mitochondrion has a double-membrane system. The outer membrane faces the cytoplasm; the inner membrane usually is infolded deeply into the mitochondrion (Figure 5.18). The membrane arrangement creates two functional compartments: the intermembrane space (between the two membranes) and the mitochondrial matrix (the innermost space).

Eukaryotic cells typically contain anywhere from a dozen to a thousand mitochondria. Animal cells generally have more mitochondria than plant cells do. Mitochondria are abundant near the surface of cells that specialize in absorbing or secreting substances. They also

are abundant in muscle cells and in parts of nerve cells. All of these cells have high energy demands.

Chloroplasts and Other Plastids

Most plant cells contain **plastids**: organelles specialized for producing food by means of photosynthesis and for storage. There are three main kinds of plastids:

chloroplasts	*with photosynthetic pigments and starch-storing capacity*
chromoplasts	*with pigments that are not functioning in photosynthesis*
amyloplasts	*with starch-storing capacity; no pigments*

Chloroplasts, which specialize in photosynthesis, are often oval or disk-shaped organelles about two to ten micrometers long. Like mitochondria, they have a double-membrane system. In the most common types, the inner membrane is the site where sunlight energy is trapped and ATP is produced. The inner membrane is arranged as a system of stacked disks, called *grana*, which are surrounded by a semifluid matrix called the *stroma* (Figure 5.19). Sucrose, starch, and proteins are assembled in the stroma from small inorganic molecules.

In land plants, chloroplasts appear green; among algae they appear green, yellow-green, or golden brown. Their color depends on the kinds and numbers of light-absorbing pigment molecules embedded in their membranes. For instance, *chlorophyll* appears green because it absorbs most wavelengths (colors) of visible light except those corresponding to green light, which it transmits. Although all chloroplasts contain chlorophyll, its presence may be masked by other photosynthetic pigments, as it is in brown algae.

In land plants, chloroplasts also serve as temporary storage sites for *starch grains*. During daylight, starch grains are assembled in chloroplasts from newly produced sugars. At night (when there is no photosynthetic activity), starch grains are broken down, and the products are transported to longer-term storage sites or to actively growing regions of the plant body.

The plastids called **chromoplasts** store red or brown pigments that are not functioning in photosynthesis. These pigments give petals, fruits, and some roots (such as carrots) their characteristic color. The colorless **amyloplasts** occur in plant parts exposed to little (if any) sunlight. Starch may become concentrated in amyloplasts, out of the way of metabolic activity. These colorless plastids are abundant in cells of potato tubers and many seeds, where they serve as storage sites for starch grains.

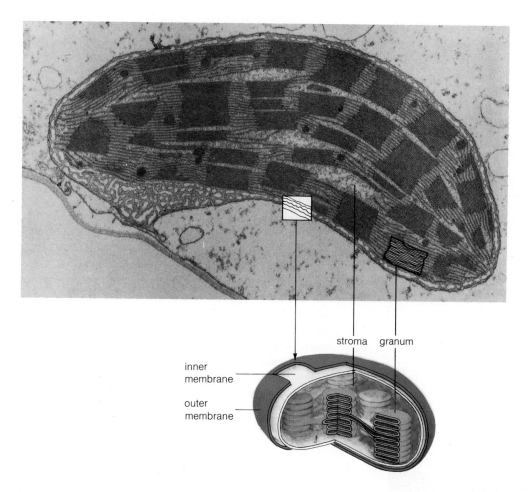

inner
membrane

outer
membrane

stroma   granum

Figure 5.19 Micrograph and generalized sketch of a chloroplast membrane system, 19,800×.

Central Vacuoles in Plant Cells

Mature and still-living plant cells often have a large, fluid-filled **central vacuole** (Figure 5.9). This organelle forms when smaller vesicles fuse during cell growth and development. A central vacuole usually occupies as much as fifty to ninety percent of the cell interior; the cytoplasm is relegated to a narrow zone between the membrane of the vacuole and the cell wall.

A central vacuole can serve as a storage area for metabolic products (such as amino acids and sugars), ions, and toxic compounds. However, its main function may be to increase cell size and surface area. During plant growth, fluid pressure builds up inside the vacuole, which presses against the cell wall. The cell enlarges under this force and, because the cell becomes rigid at maturity, the enlargement is permanent. The increased surface area enhances the rate of absorption of mineral ions, which are vital in plant nutrition.

INTERNAL FRAMEWORK OF EUKARYOTIC CELLS

The micrographs included so far give an idea of the kinds of organelles present in eukaryotic cells. However, they should not be interpreted to mean that organelles remain located precisely this way relative to one another. *Few living cells are internally rigid.* If you were to observe a living cell with the aid of a microscope, you would see constant internal motion.

The fact that cells show internal movements does not mean their structures and organelles float about at random. *If a cell is to function properly, its internal structures and organelles must be highly organized relative to one another.* In the past decade, it has become clear that a cytoskeleton is the basis for this organization. A **cytoskeleton** is composed of several elements, including microfilaments and microtubules, which are organized into a

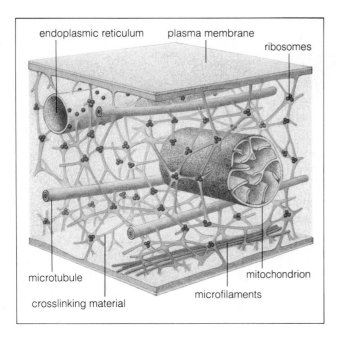

Figure 5.20 Model of the cytoskeleton of an animal cell, showing how some of the microtubules, microfilaments, and other components are arranged relative to the plasma membrane and some organelles.

three-dimensional network through the cytoplasm. Figure 5.20 shows one model of a cytoskeleton, as developed by Keith Porter.

The flexible strands called **microfilaments** are only about five nanometers thick (Figure 5.21). They contain actin, a protein involved in contraction. Microfilaments take part in changing the shape of the cell, moving materials inside it, and moving the cell itself through the environment. For example, when an animal cell divides, microfilaments assembled in a ring around its midsection have contracted and caused the parent cell to pinch in two. In plant cells, microfilaments help move organelles in a constant direction. For example, they move chloroplasts in response to changes in the sun's overhead position (Figure 5.22).

Microtubules are usually about twenty-five nanometers thick (Figure 5.23). During cell division, microtubules help distribute chromosomes to the new cells. Microtubules just beneath the plasma membrane help dictate the shape of many cells and their surface extensions. In growing plant regions, they probably help orient the cellulose strands that are deposited in cell walls. They are components of cilia and flagella, which will be described shortly.

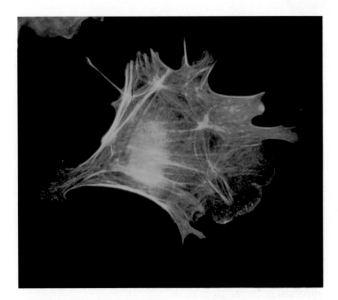

Figure 5.21 Locating structural elements of the cytoskeleton that contain protein. With *indirect immunofluorescence*, the location of a given substance is revealed after a cell takes up molecules labeled with fluorescent dyes. The molecules chosen combine only with specific proteins. When the cell is illuminated with ultraviolet light, it is possible to see the glow from the bound molecules, hence the location of the proteins in question. Here, the protein is one that helps attach microfilaments to membranes.

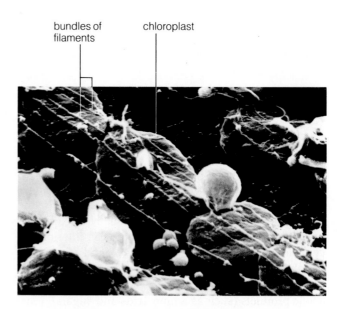

Figure 5.22 Scanning electron micrograph of a plant cell, showing some of its microfilaments. In many plant cells, organelles such as the chloroplasts shown here change position in the cell by moving along well-defined tracks of microfilament bundles.

STRUCTURES AT THE CELL SURFACE

Cell Walls and Other Surface Deposits

For many cells, surface deposits outside the plasma membrane form coats, capsules, sheaths, and walls. **Cell walls** occur among bacteria, protistans, fungi, and plants. (Animal cells do not produce walls, although some secrete products to the surface layer of tissues in which they occur.)

Most cell walls have carbohydrate frameworks, but they are quite diverse in structure. In general, they provide support and resist mechanical pressure. (For example, they keep plant cells from stretching too much when they expand with incoming water during growth.) Even the most solid-looking walls have microscopic spaces that make them porous; hence water and solutes still can move to and from the plasma membrane.

Plant cell walls have cellulose strands embedded in a matrix of materials that resist stretching. In new cells, such strands are bundled together and added to a developing **primary cell wall**. After the main growth phase, many types of plant cells also deposit materials to form an inner, rigid **secondary cell wall** (Figure 5.24).

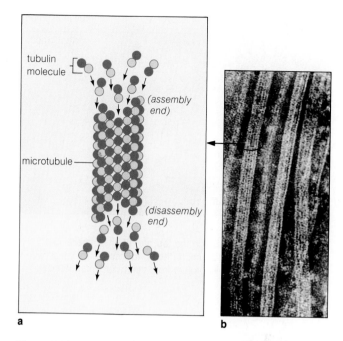

a   b

Figure 5.23 (a) One model of microtubule assembly. Once constructed, microtubules become coated with proteins that stabilize the tubulin walls. (b) Several microtubules, showing the orderly array of their tubulin subunits.

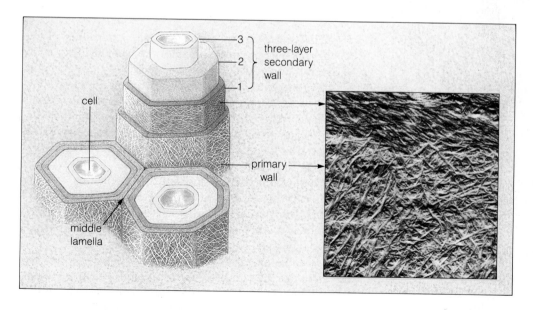

Figure 5.24 Primary and secondary walls of plant cells as they would appear in partial cross-section. The micrograph shows primary and secondary walls of a plant cell. The strands are largely cellulose.

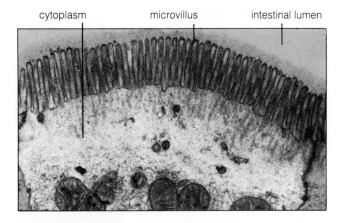

Figure 5.25 A slice through the top part of an epithelial cell, which faces the intestinal lumen (the space inside the intestinal wall). Numerous microvilli project from the surface, 21,000×.

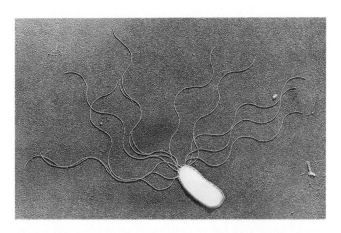

Figure 5.26 *Pseudomonas marginalis,* equipped with profuse bacterial flagella. This motile bacterium causes soft-rot diseases of vegetables, including lettuce and potatoes, 12,200×.

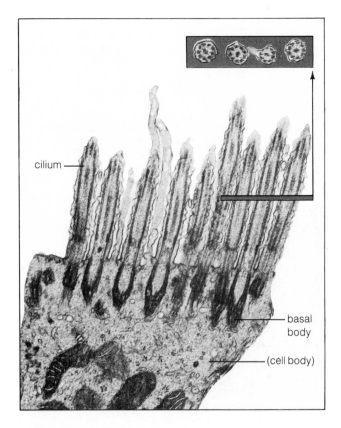

Figure 5.27 Electron micrograph of a ciliated cell from nasal epithelium, long-section, 68,000×. The boxed inset depicts four of the cilia as they would appear in cross-section at the same magnification. Notice the position of the basal body at the base of each motile structure.

Cell Surface Extensions

Microvilli. Among animal cells particularly, extensions of the plasma membrane greatly increase the surface area exposed to the environment. For example, a **microvillus** is a slender, cylindrical extension that functions in absorption or secretion (Figure 5.25).

Cilia and Flagella. Some surface extensions of the plasma membrane are capable of movement. They allow many single-celled organisms to move through liquid environments. Extensions also occur on the surface layers of some animal organs, where their rapid movements propel fluids or mucus in one direction or another.

Many prokaryotic cells have motile extensions called **bacterial flagella** (singular, flagellum). Although a bacterial flagellum looks something like a whip (Figure 5.26), it functions more like a propeller by rotating around the attachment point. The rotation can propel some bacteria a distance that corresponds to thirty-seven of their cell bodies, laid end to end, in a single second. For their size, bacteria move faster than you do.

Motile extensions of eukaryotic cells are either **flagella** or **cilia** (singular, cilium). Single-celled predators and prey living in aquatic environments move by means of cilia or flagella. Flagellated sperm and other male reproductive cells occur among protistans, fungi, some plants, and most animals. Such cells must be able to move through a liquid medium to reach and fertilize eggs. Cilia also project in dense arrays from cells in the nose and the respiratory system, where they help move fluids and dissolved particles (Figure 5.27).

Cilia are shorter and more numerous than flagella, but both are assembled from microtubules and have an outer sheath (which is continuous with the plasma membrane). Most commonly, they have a *9 + 2 array*, with nine pairs of microtubules arranged radially about two central microtubules, as shown in Figure 5.28.

The microtubules in cilia and flagella arise from short, barrel-shaped structures called **centrioles**. After the cilium or flagellum has been formed, the centriole remains attached at its base, just beneath the plasma membrane. Henceforth it is called the **basal body** of the motile structure (compare Figures 5.29b and c).

Cell-to-Cell Junctions

In multicelled plants, fungi, and animals, each living cell must interact with its physical surroundings, but it also must interact with its cellular neighbors. At the surface of tissues, cells must link tightly together so that the interior of the organism (or organ) is not indiscriminately exposed to the outside world. In all tissues, cells of the same type must recognize one another and physically stick together. Finally, in tissues where cells must act in coordinated fashion (as they do in heart muscle tissue), the cells must share channels for exchanging signals, nutrients, or both.

Cell Junctions in Animals.   Three kinds of cell-to-cell junctions are common in animal tissues. *Tight junctions* occur between cells of epithelial tissues (which line the body's outer surface, inner cavities, and organs). At these junctions, sealing strands bar the free passage of molecules across the epithelium. *Adhering junctions* are like welded spots at the plasma membranes of two adjacent cells. One type is profuse in epithelium of the skin, heart, and stomach, which are subject to stretching. At *gap junctions*, small, open channels directly link the cytoplasm of adjacent cells. In heart muscle and smooth muscle, gap junctions function in extremely rapid communication between cells. In liver and other tissues, gap junctions allow small molecules and ions to pass directly from one cell to the next.

Cell Junctions in Plants.   In complex plants, living cells are linked wall-to-wall, not membrane-to-membrane (Figure 5.30). These wall junctions are called **plasmodesmata** (singular, plasmodesma). In a single cell, between 1,000 and 100,000 plasmodesmata can penetrate both the primary and secondary walls. The number of these junctions affects the rate at which nutrients and other substances are transported between adjacent cells.

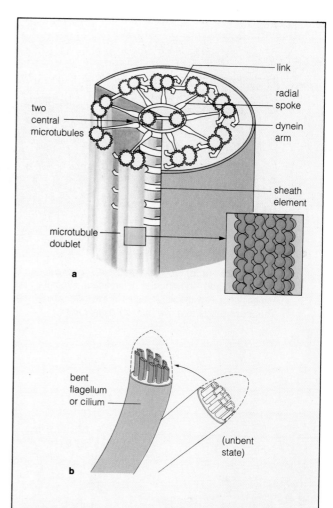

Figure 5.28   Movement of cilia and flagella. (**a**) All radial microtubules in the 9 + 2 array are anchored to the two central microtubules. (**b**) In an unbent cilium or flagellum, all doublets extend the same distance into the tip. With bending, the doublets on the outside of the arc are displaced farthest from the tip.

Numerous clawlike "arms" extend at regular intervals along the length of each microtubule doublet in the outer ring. The arms are composed of dynein, an ATP-hydrolyzing enzyme. All arms protrude in the same direction—toward the next doublet in the ring. When dynein binds and splits an ATP molecule, the angle of the arm changes with respect to the doublet in front of it. The arm bends and is strongly attracted to the doublet in front of it. On contact, the arm "unbends" with great force and causes its neighbor to move. The dynein releases its hold, whereupon it can grab another ATP molecule and attach to a new binding site on the doublet.

Thus dynein arms on one doublet swing back and forth like tiny oars, displacing the neighboring doublet with each oarlike arc. The flagellum or cilium bends toward the side where the displacement is greatest.

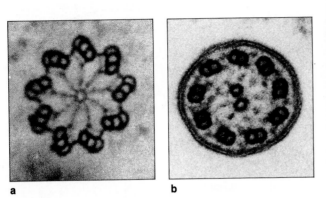 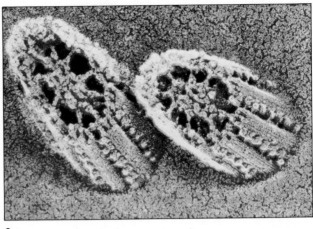

a b c

Figure 5.29 (**a, b**) From the protistan *Trichonympha*, a cross-section of the basal body (left) and the flagellum (right), showing the difference in structural patterning between the two, 140,250×. (**c**) Three-dimensional view of microtubules in two cilia, which project from the surface of the protistan *Tetrahymena*. The arrow points to a row of outer dynein arms.

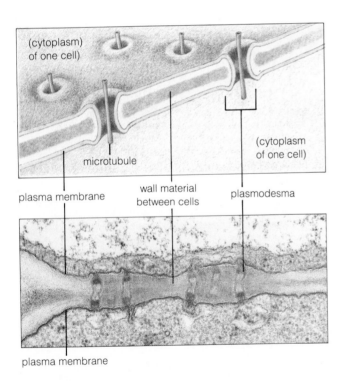

(cytoplasm) of one cell)

microtubule

(cytoplasm of one cell)

plasma membrane

wall material between cells

plasmodesma

plasma membrane

Figure 5.30 Electron micrograph of plasmodesmata crossing the wall material between two corn cells. Notice in the sketch that the plasma membranes of the two cells are continuous. A single microtubule traverses the channel.

We shall be returning to such cell-to-cell interactions. For now, the point to remember is this: *In multicelled organisms, coordinated functioning among individual cells depends on the linkage and communication provided by specialized zones in the plasma membranes of adjacent cells.*

SUMMARY OF CELL STRUCTURES AND THEIR FUNCTIONS

Each living thing is made of one or more cells; each cell is a basic living unit of structure and function; and a new cell arises only from cells that already exist. Beginning with this cell theory, we have looked at the internal structures of this fundamental living unit. We have seen that, at the minimum, all cells have a nucleus (or nucleoid), cytoplasm, and a plasma membrane, which acts as a boundary for exchanges between the internal cell environment and the surroundings.

To this basic plan, eukaryotic cells have additional and diverse embellishments—membranous compartments of varied sorts, concerned with the acquisition and processing of materials and energy in highly controlled, specialized ways. These organelles are summarized in Table 5.2, along with lists of the structures common to both prokaryotic and eukaryotic cells. Details of how they function will occupy our attention in chapters to follow.

Table 5.2 Structures Typically Found in Prokaryotic and Eukaryotic Cells

Cell Structure	Function	Prokaryotic	Eukaryotic			
		Moneran	Protistan	Fungus	Plant	Animal
Cell wall	Protection, structural support	✔*	✔*	✔*	✔	
Plasma membrane	Regulation of substances moving into, out of cell	✔	✔	✔	✔	✔
Nucleoid	Region of DNA (hereditary control)	✔				
Nucleus	Membrane-enclosed region of DNA		✔	✔	✔	✔
DNA molecule	Encoding of hereditary information	✔	✔	✔	✔	✔
RNA molecule	Transcription, translation of DNA messages into specific proteins	✔	✔	✔	✔	✔
Chromosome (DNA + protein)	Packaging of DNA; control of gene expression		✔	✔	✔	✔
Nucleolus	Assembly of ribosomal subunits		✔	✔	✔	✔
Ribosome	Protein synthesis	✔	✔	✔	✔	✔
Endoplasmic reticulum	Isolation, modification, transport of proteins and other substances		✔	✔	✔	✔
Golgi body	Modification, assembly, packaging, secretion of substances		✔	✔	✔	✔
Lysosome	Controlled degradation, material processing		✔	✔*	✔	✔
Microbodies	Isolation of enzymes used in energy and material conversions		✔	✔	✔	✔
Mitochondrion	Aerobic energy metabolism		✔	✔	✔	✔
Photosynthetic pigment	Light-energy conversion	✔*	✔*		✔	
Plastid: Chloroplast	Photosynthesis, some starch storage		✔*		✔	
Chromoplast	Pigment storage				✔	
Amyloplast	Starch storage				✔	
Central vacuole	Increasing cell surface area, storage				✔	
Cytoskeleton	Cytoplasmic organization, support		✔*	✔*	✔*	✔
Microtubule, microfilament, and related structures	Structural support, subcellular and cellular movement		✔	✔	✔	✔
Complex flagellum, cilium	Movement		✔*	✔*	✔*	✔
Centriole	Gives rise to cilia and flagella		✔*	✔*	✔*	✔

*Known to occur in at least some groups.

Review Questions

1. All cells share three structural features: a plasma membrane, a nucleus (or nucleoid), and cytoplasm. Can you describe the functions of each?

2. Why is it highly improbable that you will ever encounter a predatory two-ton living cell on the sidewalk?

3. Are all cells microscopic? Is the micrometer used in describing whole cells or extremely small cell structures? Is the nanometer used in describing whole cells or cell ultrastructure?

4. State the three principles of the cell theory.

5. Suppose you want to observe details of the surface of an insect's compound eye. Would you benefit most from a conventional light microscope, transmission electron microscope, or scanning electron microscope?

6. Eukaryotic cells generally contain these organelles: nucleus, endoplasmic reticulum, Golgi bodies, lysosomes, microbodies, and mitochondria. Can you describe the function of each?

7. Describe the structure and function of chloroplasts and mitochondria. Mention the ways in which they are similar.

8. Is this statement true or false? All chloroplasts are plastids, but not all plastids are chloroplasts.

9. What is a cytoskeleton? How do you suppose it might aid in cell functioning?

10. What are the components of the cytomembrane system? Sketch their general arrangement, from the nuclear envelope to the plasma membrane, and describe the role of each in the flow of materials between these two boundary layers.

11. What are the functions of the central vacuole in mature, living plant cells?

12. Lysosomes dismantle and dispose of malfunctioning organelles and foreign particles. Can you describe how?

13. Name some of the functions of microtubules and microfilaments.

14. Cell walls occur among which organisms: bacteria, protistans, plants, fungi, or animals? Are cell walls solid or porous?

15. In plants, is a secondary cell wall deposited inside or outside the surface of the primary cell wall? Do all plant cells have secondary walls?

16. Describe the difference between a bacterial flagellum, eukaryotic flagellum, and cilium.

17. What gives rise to the microtubular array of cilia and flagella? Distinguish between a centriole and a basal body.

18. In multicelled organisms, coordinated interactions depend on linkages and communication between adjacent cells. What types of junctions occur between adjacent animal cells? Plant cells?

19. With a sheet of paper, cover the Table 5.2 column entitled Function. Can you now name the primary functions of the cell structures listed in this table?

20. Having completed the exercise in item 19 above, can you now write a paragraph describing the difference between prokaryotic and eukaryotic cells?

Readings

Birchmeier, W. April 1984. "Cytoskeletal Structure and Function." *Trends in Biochemical Science,* 192–195.

Bloom, W., and D. Fawcett. 1975. *A Textbook of Histology.* Tenth edition. Philadelphia: Saunders. Outstanding reference book on cell structure.

deDuve, C. 1985. *A Guided Tour of the Living Cell.* New York: Freeman. Beautifully illustrated introduction to the cell; two short volumes.

Fawcett, D. 1981. *The Cell: An Atlas of Fine Structure.* Second edition. Philadelphia: Saunders. Outstanding collection of micrographs.

Kessel, R., and C. Shih. 1974. *Scanning Electron Microscopy in Biology.* New York: Springer-Verlag. Stunning micrographs of cell structures.

Newcomb, E. 1980. "The General Cell." In *The Biochemistry of Plants,* vol. 1 (N. Tolbert, editor). New York: Academic Press. Survey of plant organelles.

Porter, K., and J. Tucker. March 1981. "The Ground Substance of the Living Cell." *Scientific American* 244(3):57–67. Summary of studies on the scaffolding inside the cell.

Rothman, J. September 1985. "The Compartmental Organization of the Golgi Apparatus." *Scientific American* 253(3):74–89. Describes the three specialized compartments of Golgi bodies.

Rothman, J., and J. Lenard. April 1984. "Membrane Traffic in Animal Cells." *Trends in Biochemical Sciences,* 176–178.

Weber, K., and M. Osborn. October 1985. "The Molecules of the Cell Matrix." *Scientific American* 253(4):110–120. Summarizes techniques used to study the cytoskeleton.

Weibe, H. 1978. "The Significance of Plant Vacuoles." *Bioscience* 28:327–331.

As small as it may be, a cell is a *living* thing engaged in the risky business of survival. At any moment its life can be snatched from it if, for example, the concentrations of dissolved ions shoot above or plummet below prescribed levels. For most cells in most places, such concentrations are not constants. Think of the single-celled organisms living in estuaries, where seawater mixes with freshwater currents from rivers and streams. Ion concentrations change with the tides, but ion concentrations *inside* those cells must not change to the same extent. If they did, the cells would die, just as you would die if you were shipwrecked and drank nothing but salt-laden seawater.

So think of the cell for what it is: a tiny compartment in an immense world that is, by comparison, disordered, sometimes benign, and sometimes harsh. No matter what goes on outside, some things have to be kept out and others have to be brought into the cell, where organized structures and activities prevail. For this bit of the world, the bastion against disorganization is the **plasma membrane**—a thin, seemingly flimsy surface layering of little more than lipids and proteins, dotted here and there with guarded passageways and twiglike carbohydrate groups. At this membrane, materials are exchanged between the cytoplasm and the surroundings. Then, within the cytoplasm of eukaryotic cells, exchanges are made across **internal cell membranes**, which form the compartments called organelles. This most fundamental of all cell structures, the membrane, is the focus of this chapter.

6

MEMBRANE STRUCTURE AND FUNCTION

FLUID MEMBRANES IN A LARGELY FLUID WORLD

The Lipid Bilayer

Water bathes the outer surface of living cells and fills most of their interior. You might be thinking that the plasma membrane would have to be a rather solid structure, given that both sides of it are immersed in fluid, yet it happens that the plasma membrane is fluid, too! When you puncture a cell with a fine needle, the cell will not lose cytoplasm when the needle is withdrawn. Rather, the cell surface seems to flow over and seal the puncture. How can a fluid cell membrane remain distinct from fluid surroundings? The answer is found in certain properties of lipid molecules.

Lipid molecules cluster spontaneously when surrounded by water. For example, consider the **phospholipids**, the most abundant type of lipid in cell membranes. A typical phospholipid is shown in Figure 6.1. It has a hydrophilic (water-loving) head and two hydrophobic (water-dreading) tails:

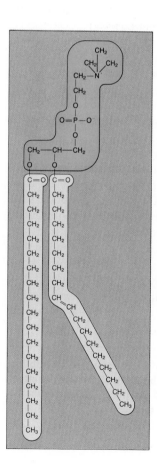

Figure 6.1 Structural formula of a typical phospholipid found in animal cell membranes. The hydrophilic head is shown in orange; the hydrophobic tails in gold. In reality, the sharp bend caused by the double bond in one of the tails is no more than a small kink.

- polar (hydrophilic) head

- nonpolar (hydrophobic) tails

When many phospholipid molecules are surrounded by water, hydrophobic interactions cause their fatty acid tails to cluster together. The clustering may produce a **micelle**, a sphere with all hydrophilic heads at the surface (dissolved in the surrounding water) and all tails pointing inward. Or the clustering may produce a **lipid bilayer**, with the hydrophilic heads at the outer faces of a two-layer sheet and hydrophobic tails sandwiched between them:

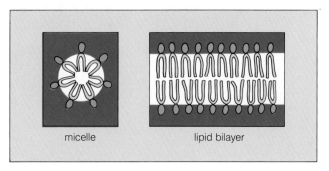

micelle   lipid bilayer

Such bilayers are the structural foundation for the plasma membrane and internal organelles.

Because micelles and lipid bilayers minimize the number of hydrophobic groups exposed to water, the fatty acid tails do not have to spend a lot of energy fighting the water molecules, so to speak. Thus the reason a punctured plasma membrane tends to seal itself is that the puncture is energetically unfavorable (it leaves too many hydrophobic groups exposed).

Ordinarily, of course, few cells are ever punctured with fine needles. But the self-sealing behavior of their membrane lipids is useful for far more than damage control. As you will see, parts of the plasma membrane sink inward and drag extracellular substances in with them. The sinks deepen into spheres that pinch off from the plasma membrane, thereby forming closed vesicles that can move into the cytoplasm. Similarly, buds form from organelle membranes, and when the buds pinch off, the self-sealing behavior of the lipid bilayer maintains the integrity of both the newly formed vesicles and the parent organelle.

Lipid molecules surrounded by water show self-assembling and self-sealing behavior. This behavior maintains the integrity of the plasma membrane and organelles during membrane traffic from one part of the cell to another.

Within the bilayer, individual lipid molecules show quite a bit of movement. They diffuse sideways, they spin about their long axis, and their tails flex back and forth. These movements keep adjacent lipids from packing into a solid layer, hence they impart fluidity to the membrane. Packing is also disrupted by lipids with short tails (which cannot interact as strongly as long-tailed neighbors do) and by unsaturated tails (which have kinks at their double bonds).

Within a bilayer, lipid molecules move rapidly and show variations in their packing tendencies, thereby contributing to membrane fluidity.

Membrane Proteins

The lipid bilayer of cell membranes is not uniformly smooth. Diverse proteins are embedded within the bilayer or weakly bonded to its surface. In 1972, S. J. Singer and G. Nicholson put together the first **fluid mosaic model** of membrane structure. Figure 6.2 shows a more recent version of this model, although keep in mind that actual details vary among membranes. The membrane is "fluid" because of the movements and packing variations of its lipid components. The lipids themselves give the membrane its basic *structure* and its relative *impermeability* to water-soluble molecules. The "mosaic" is the intricate composite of diverse proteins and lipids of the membrane. The protein components are largely responsible for membrane *functions*.

Membrane proteins have been identified in electron micrographs of specimens prepared by freeze-fracture and freeze-etching methods (Figure 6.3). Some of the proteins function as pumps and gated channels through which hydrophilic substances can cross the otherwise hydrophobic membrane. Others are enzymes that take part in different metabolic events. Others intercept chemical signals or serve as binding sites for specific substances.

Membrane Surface Receptors

For all eukaryotic cells, carbohydrate groups project from the surface of the plasma membrane. Most are polysaccharides, bound covalently to membrane proteins and (to some extent) to membrane lipids. The molecules having these groups are called glycoproteins and glycolipids.

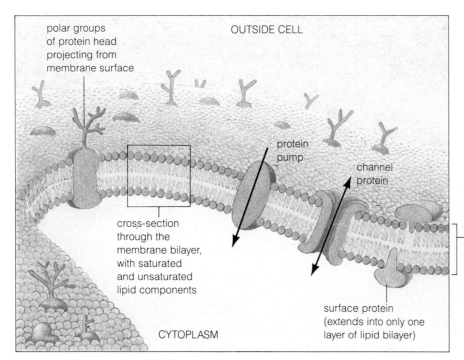

Figure 6.2 Fluid mosaic model of a plasma membrane. The micrograph shows a red blood cell membrane in thin section, at high magnification.

Membrane carbohydrate groups always project outward, never into the cytoplasm. Some preliminary studies indicate these groups function as **surface receptors**: molecular sites on the plasma membrane where cells bind substances and chemically recognize other cells. Activation of some receptors leads to changes in cell metabolism or behavior. Activation of others causes some substance to be brought into the cell.

In multicelled animals, surface receptors also function in cell-to-cell identity. For example, when a human embryo is developing, such receptors help body cells of the same type to come together into tissues. These receptors also function in defense against bacteria and other tissue invaders.

Summary of Membrane Structure and Function

Throughout this book, we shall be looking at different aspects of membrane structure and function. For now, the features that all cell membranes hold in common may be summarized this way:

1. Cell membranes are composed of lipids (phospholipids especially) and proteins.

2. Membrane lipids have hydrophilic heads and hydrophobic tails, and when surrounded by water they assemble spontaneously into a bilayer. All heads are at the two outer faces of a lipid bilayer and all tails are sandwiched between them.

3. The lipid bilayer is the basic *structure* of all cell membranes. It acts as a *hydrophobic barrier* between two fluid regions (either two cytoplasmic regions, or the cytoplasm and the fluid outside the cell).

4. Membrane fluidity arises through rapid movements and packing variations among the individual lipid molecules.

5. Membrane *functions* are carried out largely by proteins embedded in the bilayer or weakly bonded to one or the other membrane surface.

6. Some proteins transport water-soluble substances across the membrane. Some are enzymes. Others are receptors for chemical signals or for specific substances.

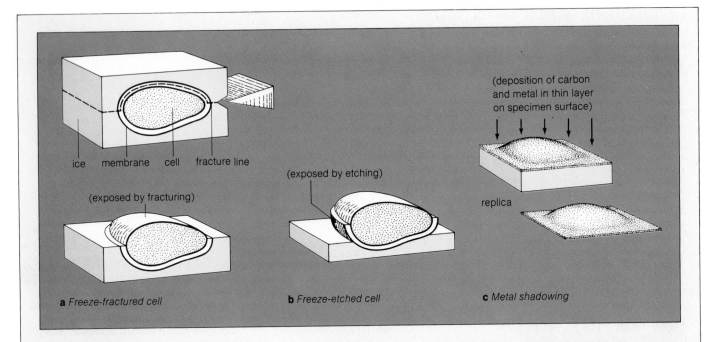

a *Freeze-fractured cell*

(exposed by fracturing)

(exposed by etching)

b *Freeze-etched cell*

(deposition of carbon and metal in thin layer on specimen surface)

replica

c *Metal shadowing*

ice membrane cell fracture line

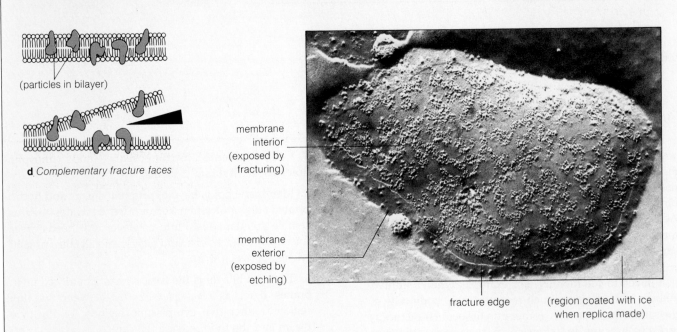

(particles in bilayer)

d *Complementary fracture faces*

membrane interior (exposed by fracturing)

membrane exterior (exposed by etching)

fracture edge

(region coated with ice when replica made)

Figure 6.3 Freeze-fracturing and freeze-etching. **(a)** In the freeze-fracture step, specimens being prepared for electron microscopy are rapidly frozen, then fractured by a sharp blow from the edge of a fine blade. **(c)** In a process called metal shadowing, the fractured surface is coated with a layer of carbon and heavy metal such as platinum. This coating is thin enough to replicate details of the exposed specimen surface. The metal replica, not the specimen itself, is used for preparing electron micrographs.

(b) Sometimes before the metal replica is made, specimens are freeze-etched, which simply means that additional ice is evaporated from the fracture face. In cell membrane studies, such etching exposes the outer membrane surface.

(d) Fractured cell membranes commonly split down the middle of the lipid bilayer, which yields glimpses into internal membrane structure. Typically, one inner surface is studded with particles and depressions, and the other inner surface is a complementary pattern of depressions and particles. The particles are membrane proteins.

The micrograph shows a portion of a replica of a red blood cell, prepared by freeze-fracturing and freeze-etching.

MOVEMENT OF WATER AND SOLUTES

Before exploring the details of membrane function, let's consider the natural, unassisted movements of water and solutes. Much of membrane structure can be interpreted as mechanisms that work with or against these movements.

Gradients Defined

"Concentration" refers to the number of molecules (or ions) of a substance in a given volume of space. In the absence of other forces, molecules of a given type move down their **concentration gradient**, which means they tend to move to a region where they are less concentrated than they were before. They are driven to do so because they are constantly colliding with one another millions of times a second. Random collisions do send the molecules back and forth, but the *net* movement is outward from the region of greater concentration (where more molecules are around to collide).

Similarly, for any defined volume of space, gradients can also exist between two regions that differ in pressure, temperature, or electric charge.

Simple Diffusion

The random movement of like molecules or ions down a concentration gradient is called **simple diffusion**. Simple diffusion of small molecules accounts for the greatest volume of substances moving across cell membranes.

The direction in which a substance diffuses depends on its own concentration gradient, not on any others. In other words, it diffuses *independently* of other substances that may be present. Suppose you put a few drops of red food coloring into one end of a pan filled with water. At first all of the dye molecules remain at that end, but many molecules start careening toward the other end. Even though collisions are also sending some dye molecules back, *more* molecules are bouncing off each other in the concentrated region, and the net movement is down the gradient. For the same reason, water moves in the opposite direction, to the region where water molecules are less concentrated in the pan (Figure 6.4).

Even when dye molecules and water molecules are dispersed evenly in the pan, collisions still occur at random. But now there are no more gradients, hence there is no *net* movement in either direction. The molecules are said to be at dynamic equilibrium.

The *rate* of diffusion depends on several things. For example, the rate is faster when the concentration gradient is steep (molecules collide more frequently in

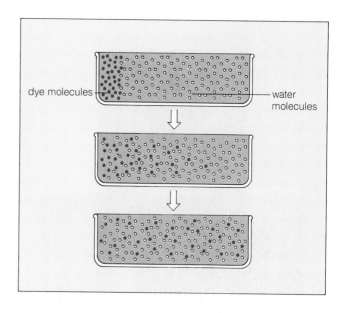

Figure 6.4 Diagram of the simple diffusion of dye molecules in one direction and water molecules in the opposite direction in a pan of water.

regions of greater concentration). It also is faster at higher temperatures, because heat energy causes molecules to move more rapidly (hence to collide more frequently). The rate of diffusion is also affected by molecular size (smaller molecules move faster than large ones do at the same temperature).

Simple diffusion accounts for most of the short-distance transport of substances moving into and out of cells.

Bulk Flow

Diffusion rates are often enhanced by **bulk flow**, which is the tendency of different substances in a fluid to move together in the same direction in response to a pressure gradient.

For example, in complex plants and animals, substances do not diffuse slowly across large tissue masses to reach individual cells. They are moved rapidly by bulk flow through transport tubes, which occur in the respiratory and circulatory systems of animals and in the vascular systems of plants. In effect, bulk flow "shrinks" the diffusion distances between the environment and interior cells. It also shrinks the diffusion distance between one point within the body and another.

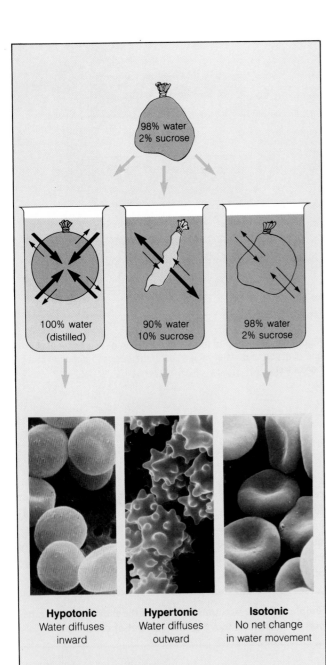

Hypotonic
Water diffuses
inward

Hypertonic
Water diffuses
outward

Isotonic
No net change
in water movement

Figure 6.5 Effects of osmosis in different environments. The sketches show why it is important for cells to be matched to solute levels in their environment. (In each sketched container, arrow width represents the relative amount of water movement.)

The micrographs correspond to the sketches. They show the kinds of shapes that might be seen in red blood cells placed in *hypotonic* solutions (influx of water into the cell), *hypertonic* solutions (outward flow of water from the cell), and *isotonic* solutions (internal and external solute concentrations are matched).

Red blood cells have no special mechanisms for actively taking in or expelling water molecules. Hence they swell or shrivel up, if solute levels in their environment change.

Osmosis

A membrane is "differentially permeable" when it allows some substances but not others to pass through it. **Osmosis** is the movement of water across any differentially permeable membrane in response to solute concentration gradients, a pressure gradient, or both. (A solute, recall, is any dissolved substance.)

Imagine you have a plastic bag that acts like a membrane, in that it is permeable to water molecules but impermeable to larger molecules. You fill the bag with water containing a small amount of table sugar, then you put it in a container of distilled water (which is nearly free of solutes). Because there are fewer water molecules inside the bag than outside, water follows its gradient; there is a net movement of water into the bag (Figure 6.5). However, sugar cannot follow *its* concentration gradient and move out of the bag, for the sugar molecules are too large to pass through membrane. Soon the bag swells with water and eventually it may spring a leak or burst.

Suppose you had put distilled water in that bag and immersed it in water that contained a small amount of dissolved sugar. The net movement of water would have been outward, and the bag would have shriveled (Figure 6.5).

Finally, suppose you had used distilled water inside and outside the bag. Because the solute concentrations would be the same, there would be no water concentration gradient—and there would be no net movement of water in either direction.

Osmosis is the passive movement of water across a differentially permeable membrane in response to solute concentration gradients, a pressure gradient, or both.

The term **tonicity** refers to the relative concentrations of solutes in the fluid inside and outside the cell. When solute concentrations are equal in both fluids, or *isotonic*, there is no net osmotic movement of water in either direction. When the solute concentrations are not equal, one fluid is *hypotonic* (has less solutes) and the other is *hypertonic* (has more solutes). Water molecules tend to move from a hypotonic fluid to a hypertonic one.

If cells did not have mechanisms for adjusting to differences in tonicity, they would shrivel or burst like the plastic bag described above. For example, many solutes are dissolved in your bloodstream and in the cytoplasm of red blood cells. But suppose you immerse a red blood cell in a hypotonic solution. That cell contains many large organic molecules, which cannot cross the plasma membrane. Although those molecules cannot move out, water can move in. Internal pressure builds up, but red blood cells have no mechanism for disposing of the

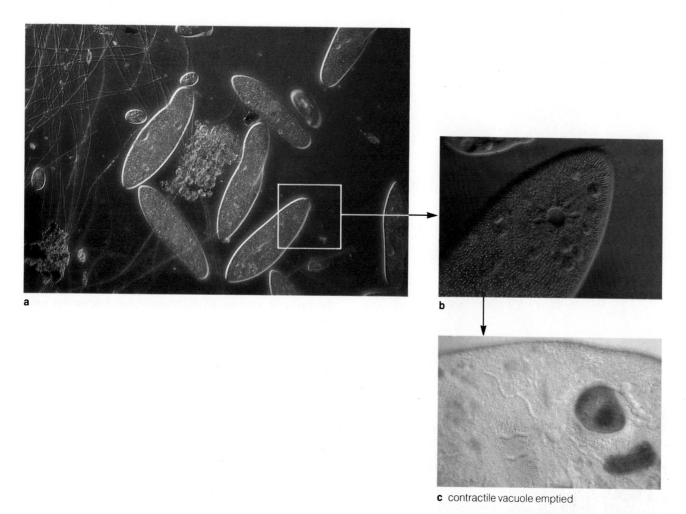

a

b

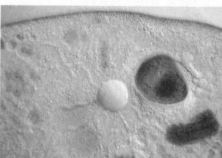

c contractile vacuole emptied

d contractile vacuole filled

excess water. The cell continues to swell until the membrane hemolyzes (it becomes grossly "leaky"), and the cell is destroyed.

If the red blood cell had been placed in a hypertonic solution, water would have moved out of the cytoplasm and the cell would have shriveled. This particular type of cell functions only when solute concentrations stay much the same on both sides of the plasma membrane. However, other types of cells can live in hypotonic or hypertonic environments, as Figure 6.6 illustrates.

Water Potential

The soil in which most land plants grow is hypotonic relative to the cells in those plants. Thus water tends to move by osmosis into the plant body. When individual cells absorb the water, pressure increases against the cell wall, which is fairly rigid. This internal pressure on a cell wall is called *turgor pressure*. Because cell walls are strong enough to resist increases in internal pressure, the excess water is gradually forced out of the cytoplasm and into the surroundings.

Figure 6.6 Osmosis and *Paramecium*, a single-celled protozoan that lives in fresh water (a hypotonic environment). Because its cell interior is hypertonic relative to the surroundings, water tends to move into the cell by osmosis. If the influx were left unchecked, the cell would become bloated and the plasma membrane would rupture. However, the excess is expelled by an energy-requiring transport process that is based on a specialized organelle called the contractile vacuole. Tubelike extensions of this organelle extend through the cytoplasm. Water collects in these extensions and drains into a central vacuolar space. When filled, the vacuole contracts and the water is forced into a small pore that empties to the outside.

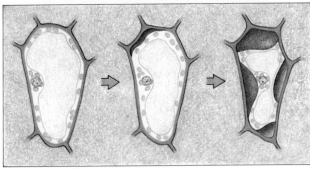

Figure 6.7 Plant wilt resulting from loss of turgor in cells. (**a**) At the start of this experiment, ten grams of salt (NaCl) in about sixty milliliters of water is added to a pot containing tomato plants. (**b**) After about five minutes, wilting is pronounced and the plant is collapsing. (**c**) After twenty-seven minutes, wilting is severe. The corresponding sketches show progressive plasmolysis (a shrinking of cytoplasm away from the cell walls).

Pressure will build up inside any walled cell when water moves inward by osmosis; but water will also be squeezed back out when turgor pressure is great enough to counter the effects of internal solutes. Both forces have the potential to cause the directional movement of water. The sum of these two opposing forces is called the **water potential**.

When the outward flow equals the rate of inward diffusion, turgor pressure is constant. Then, there is enough pressure to keep cell walls from collapsing, and the soft parts of the plant body are maintained in erect positions. However, when the environment is so dry that the movement of water into the plant dwindles or stops, water moves out of the cells and the soft parts of the plant body wilt. Similarly, wilting occurs when solutes reach high concentrations in the environment (Figure 6.7).

MEMBRANE TRANSPORT

What the Membrane Proteins Transport

Oxygen, carbon dioxide, water, and ethanol readily diffuse across the lipid bilayer of cell membranes. These and other small molecules are electrically neutral:

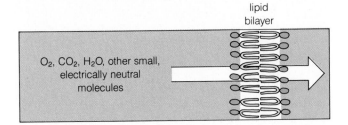

Simple diffusion of such small molecules accounts for the greatest volume of substances moving across cell membranes.

In contrast, glucose and other large molecules that are electrically neutral almost never diffuse across the lipid bilayer. Neither do positively or negatively charged ions, no matter how small they are:

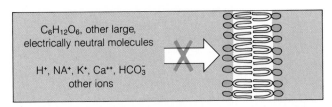

These substances must move through **membrane transport proteins**, which serve as carriers or channels across the lipid bilayer.

Carrier and channel proteins play active or passive roles in moving substances across a membrane. In **active transport**, the protein itself needs an energy boost to move a solute either with or against a concentration gradient. In **passive transport**, the protein needs no energy boost; the solute simply moves through the interior of the protein as it follows its concentration gradient.

Through a combination of simple diffusion, passive transport, and active transport, cells or organelles are supplied with raw materials and they are rid of wastes, at controlled rates. These mechanisms control secretions of cell products. They also help maintain pH and volume inside the cell or organelle within some functional range.

Facilitated Diffusion

Carrier proteins bind and passively transport specific solutes. The solutes are helped across the membrane only in the direction that simple diffusion would take them, hence their protein-assisted crossing is called **facilitated diffusion**.

The carrier proteins span the lipid bilayer. Each type probably has a specific array of hydrophilic groups along its interior surface (Figure 6.8). When water-soluble molecules bind with these groups, the binding apparently triggers a change in the protein shape, and this change permits the solute to move through the hydrophilic interior. While the solute makes its passage, the protein closes in behind it and returns to its original shape.

Active Transport

Some membrane proteins actively transport specific solutes into and out of cells. They undergo changes in shape, somewhat like the changes induced in the carriers responsible for facilitated diffusion. In this case, however, an energy boost leads to a *series* of changes, which cause the protein to pump solutes rapidly across the membrane (Figure 6.9). Most often, ATP donates energy to the protein.

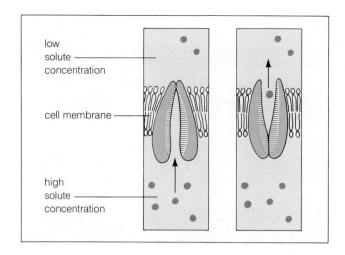

Figure 6.8 Possible mechanism of facilitated diffusion. The sketch shows a carrier protein as if it were sliced down through its midsection. Water-soluble molecules bind with the hydrophilic groups present on the interior surface of the protein (the shaded areas). The binding changes the protein shape in a way that allows a bound molecule to cross the membrane.

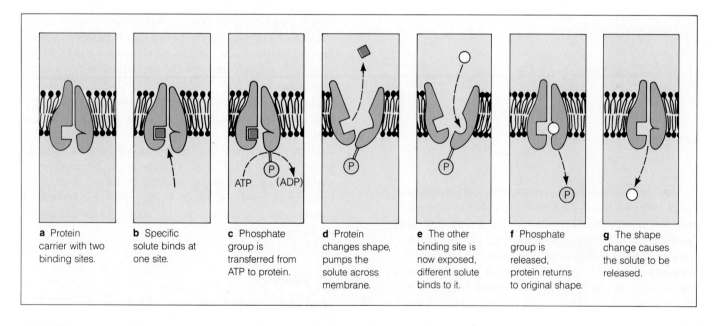

a Protein carrier with two binding sites.

b Specific solute binds at one site.

c Phosphate group is transferred from ATP to protein.

d Protein changes shape, pumps the solute across membrane.

e The other binding site is now exposed, different solute binds to it.

f Phosphate group is released, protein returns to original shape.

g The shape change causes the solute to be released.

Figure 6.9 Simplified picture of active transport. In this example, transport of one kind of solute across the membrane is coupled with transport of another kind in the opposite direction. The transport protein receives an energy boost from ATP and thereby undergoes changes in its shape that are necessary for the transport process.

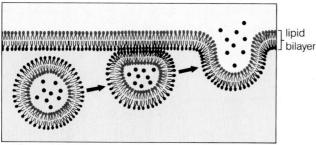

a Exocytosis

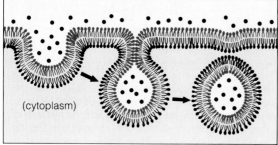

b Endocytosis

Figure 6.10 (**a**) Fusion of a transport vesicle with the plasma membrane during exocytosis, and (**b**) formation of a transport vesicle during endocytosis.

One active transport system, the *sodium-potassium pump*, helps maintain high concentrations of potassium and low concentrations of sodium inside the cell. Another, the *calcium pump*, helps keep calcium concentrations at least a thousand times lower inside the cell than outside. The features that all active transport systems hold in common can be summarized this way:

1. In active transport, small ions, small charged molecules, and large molecules are pumped across a cell membrane, most often against their gradients.

2. Carrier proteins spanning the lipid bilayer are the active transport systems. They are highly selective in the kinds of ions and molecules that they will bind and transport.

3. The proteins act when a specific solute is bound in place and when they receive an energy boost.

Exocytosis and Endocytosis

In some processes, the plasma membrane or organelle membranes pinch off and form transport vesicles around substances. **Exocytosis**, the process by which substances are moved *out* of a cell, calls for fusion of cytoplasmic vesicles with the plasma membrane (Figure 6.10). For example, vesicles pinched off from Golgi membranes travel to and fuse with the plasma membrane, then their contents are secreted to the outside. This process was described in the preceding chapter. In **endocytosis**, a region of the plasma membrane encloses particles at or near the cell surface, then pinches off to form a vesicle that moves into the cytoplasm (Figure 6.10).

The amoeba relies on endocytosis. This single-celled organism is phagocytic (a "cell-eater"). When it encounters a chemically "tasty" particle or cell, it sends out one or two extensions of its body. The lobelike extensions curve back and form a compartment around the particle, and this compartment becomes an endocytic vesicle (Figure 6.11). Most endocytic vesicles fuse with lysosomes and their contents are digested (page 75). Phagocytic white blood cells also rely on endocytosis when they destroy harmful agents such as bacteria.

Endocytosis also transports liquid droplets into animal cells. (Sometimes this process is called pinocytosis, which means "cell drinking.") A depression forms at the surface of the plasma membrane and dimples inward around extracellular fluid. An endocytic vesicle forms and moves inside the cytoplasm, where it fuses with lysosomes.

In *receptor-mediated endocytosis*, specific molecules are brought into the cell through involvement of specialized regions of the plasma membrane that form coated pits. Each pit, or shallow depression, is coated with a dense lattice (Figure 6.12). The pit appears to be lined with surface receptors that are specific (in this example) for lipoprotein particles. When lipoproteins are bound to the receptors, the pit sinks into the cytoplasm and forms an endocytic vesicle.

SUMMARY

Membrane Structure

1. Living cells are bathed in fluid of one sort or another, and their interior is also fluid. Membranes serve as boundaries between the external fluid world and the cytoplasm or between different cytoplasmic regions.

2. The lipid bilayer is the basic structure of cell membranes. There are two layers of lipids (phospholipids especially), with the hydrophobic tails of the molecules sandwiched in between the hydrophilic heads. The membrane is rather fluid because of packing variations and rapid movements among the individual molecules.

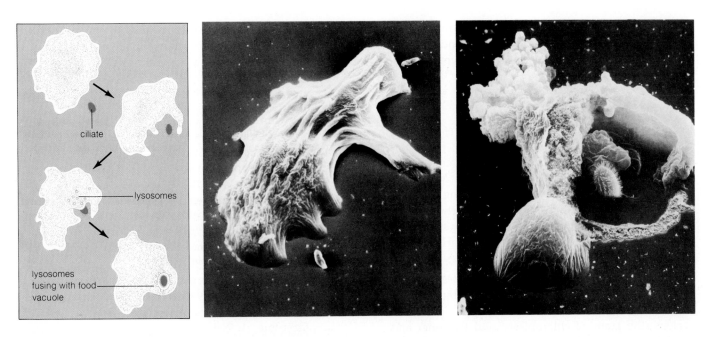

Figure 6.11 Endocytosis as demonstrated by a phagocytic cell. The scanning electron micrographs show *Amoeba* entrapping a ciliated cell. These micrographs correspond to the second and third sketches.

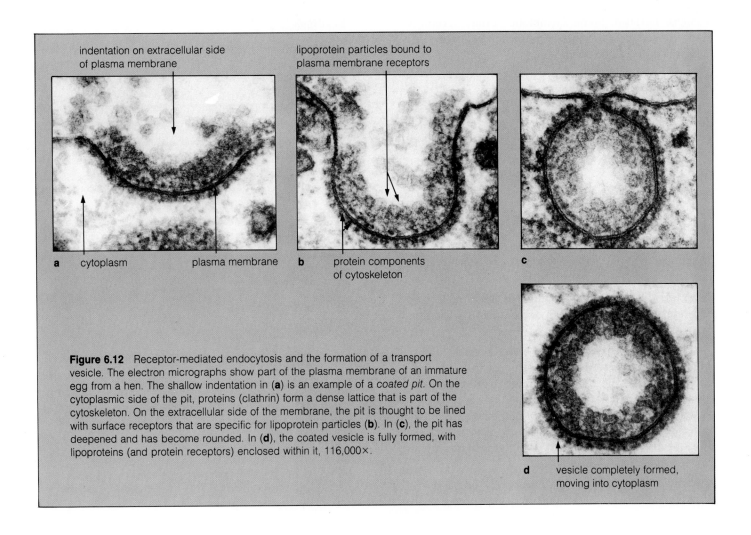

Figure 6.12 Receptor-mediated endocytosis and the formation of a transport vesicle. The electron micrographs show part of the plasma membrane of an immature egg from a hen. The shallow indentation in (**a**) is an example of a *coated pit*. On the cytoplasmic side of the pit, proteins (clathrin) form a dense lattice that is part of the cytoskeleton. On the extracellular side of the membrane, the pit is thought to be lined with surface receptors that are specific for lipoprotein particles (**b**). In (**c**), the pit has deepened and has become rounded. In (**d**), the coated vesicle is fully formed, with lipoproteins (and protein receptors) enclosed within it, 116,000×.

3. Membrane functions are carried out by diverse proteins embedded in the bilayer or weakly bonded to one of its surfaces. These proteins serve as enzymes, transport systems, and receptors.

Membrane Functions

1. *Control of substances moving into and out of cells.* The plasma membrane is a thin, differentially permeable boundary between organized metabolic events in the cytoplasm and events in the outside world. Membrane proteins play passive or active roles in carrying or pumping substances across the lipid bilayer portion of the membrane. The selective movements of these substances affect metabolism, cell volume and cellular pH, nutrient stockpiling, and the removal of harmful substances.

2. *Compartmentalization of internal cellular space.* Within the cytoplasm, membranes form separate compartments (organelles) in which specialized activities occur. The nucleus, the grana of chloroplasts, and the inner and outer mitochondrial membranes are examples. Like the plasma membrane, these internal membranes are differentially permeable, a feature that is vital in the functions of organelles.

3. *Signal reception.* Some membrane surface receptors bind extracellular molecules and function in altering cell metabolism or cell behavior.

4. *Cell-to-cell recognition.* In multicelled organisms, surface receptors also function in identifying cells of like type during tissue formation and, at a later stage of development, during tissue interactions.

5. *Transport of macromolecules and particles.* Through exocytosis and endocytosis, cells take in or eject specific large molecules or particles across the plasma membrane.

Movement of Water and Solutes Across Membranes

1. Exchanges of water and solutes occur across cell membranes, both with and against concentration gradients.

2. Simple diffusion is a natural, unassisted movement of solutes from a region of high concentration to a region of lower concentration.

3. Osmosis is the movement of water across a differentially permeable membrane in response to solute concentration gradients, a pressure gradient, or both. When conditions are isotonic (equal concentrations of solutes across the membrane), there is no osmotic movement in either direction. Water tends to move from hypotonic fluids (with fewer solutes) to hypertonic fluids (with more solutes).

4. Many solutes are transported actively or passively across membranes by channel or carrier proteins. In active transport, the protein requires an energy boost to move a solute. In passive transport, the solute diffuses through the interior of a channel protein.

Review Questions

1. Describe the fluid mosaic model of plasma membranes. What makes the membrane fluid? What parts constitute the mosaic?

2. List the six structural features that all cell membranes have in common.

3. What is meant by differentially permeable? What two general kinds of mechanisms associated with the plasma membrane regulate the movement of substances across a cell membrane?

4. List and define three kinds of passive transport systems, and three kinds of active transport systems.

5. Diffusion accounts for the greatest volume of substances moving into and out of cells. How does diffusion work?

6. What is osmosis, and what causes its occurrence?

7. Under what circumstances would active transport mechanisms come into play?

8. Describe some functions of membrane surface receptors.

Readings

Bretscher, M. October 1985. "The Molecules of the Cell Membrane." *Scientific American* 253(4):100–108. Fairly recent description of the structure and function of the plasma membrane.

Dautry-Varsat, A., and H. Lodish. May 1984. "How Receptors Bring Proteins and Particles Into Cells." *Scientific American* 250(5):52–58. Describes receptor-mediated endocytosis.

Karp, G. 1984. *Cell Biology.* Second edition. New York: McGraw-Hill. Chapter 5 gives a detailed, dynamic picture of membrane function. This is an excellent reference book.

Singer, S., and G. Nicolson. 1972. "The Fluid Mosaic Model of the Structure of Cell Membranes." *Science* 175:720–731.

Unwin, N., and R. Henderson. February 1984. "The Structure of Proteins in Biological Membranes." *Scientific American* 250(2):78–94.

Weissmann, G., and R. Claiborne. 1975. *Cell Membranes.* New York: HP Publishing. Collection of survey articles written by some leading membrane researchers.

When you look at a single living cell with the aid of a microscope, you are watching a form pulsing with activity. Through its movements, it is identifying and taking in raw materials suspended in the water droplet on the slide. To power those tiny movements, the cell is extracting energy from food molecules it had stored away earlier. Even as you observe it, the cell is using materials and energy in building and maintaining its membranes and organelles, its stores of chemical compounds, its information-storage system, its pools of enzymes. It is alive; it is growing; it may reproduce itself. Multiply this activity by *65 trillion cells* and you have an inkling of what is going on in your own body as you sit quietly, observing that single cell!

With this chapter we turn to one of the most fundamental aspects of the living cell, its reliance on **metabolism**—the controlled capacity to acquire and use energy in stockpiling, breaking apart, assembling, and eliminating substances in ways that assure cell growth, maintenance, and reproduction.

7

METABOLISM: GROUND RULES AND OPERATING PRINCIPLES

Figure 7.1 All events large and small, from the birth of stars to the death of a microorganism, are governed by laws of energy. Shown here, eruptions on the sun's surface and, to the right, *Volvox*—colonies of microscopically small single cells that capture sunlight energy necessary to drive their life processes.

THE NATURE OF ENERGY

Two Laws Governing Energy Transformations

Energy is a capacity to make things happen, to cause change, to do work. You use energy to put a hard wax finish on a car. When paste wax is buffed, attractions between molecules in the paste are broken, new ones form, and the molecules become repositioned in ways that produce a hard, glossy finish. A cell, too, uses energy to make things happen, as when it makes large molecules from smaller, simpler ones.

Can you (or the cell) create your own energy from scratch, or must it be obtained from someplace else? You can gain insight into the question by considering two laws of thermodynamics, which deal with the transformation of energy.

The **first law of thermodynamics** deals with the *quantity* of energy available. It may be expressed this way:

The total amount of energy in the universe remains constant. More energy cannot be created. And existing energy cannot be destroyed; it can only undergo conversion from one form to another.

According to the first law, there is a finite amount of energy in the universe. It is distributed among various forms, such as the chemical energy stored in molecular bonds and the radiant energy from the sun. These and other forms of energy are interconvertible. For example, plants absorb sunlight energy and convert it to the chemical energy of sugar. You eat plants and the chemical energy stored in sugar can be converted to mechanical energy that powers your movements. Some energy escapes into the surroundings as heat during these conversions. (Because of ongoing conversions in your cells, your body steadily gives off about the same amount of heat as a hundred-watt light bulb.) However, none of the energy *vanishes*.

More precisely, the total energy content of a system *and its surroundings* remains constant:

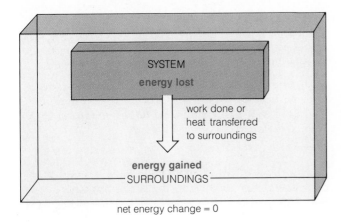

net energy change = 0

As this diagram indicates, the term "system" means all matter within a specified region (a plant, a strand of DNA, a galaxy); "surroundings" can mean the entire universe or some specified region in contact with the system.

The first law means that you cannot increase the universal energy pool by making "new" energy out of nothing, and you cannot destroy any of it either. *All you can do is channel the ways in which energy changes from one form to another, and thereby temporarily hoard or let go of what is already there.*

This brings us to the *quality* of energy available. Energy that is concentrated and organized (for example, in a sugar molecule) is high quality, for it lends itself to conversions. Heat spread out in the atmosphere is an example of low-quality energy; for all practical purposes, it cannot be gathered up and converted to other forms.

According to the **second law of thermodynamics**, the spontaneous direction of energy flow is from high-quality to low-quality forms. "Spontaneous" means the natural or *most probable* direction, regardless of whether it takes a fraction of a second or many billions of years.

Left to itself, any system spontaneously undergoes conversions to less organized forms. Each time that happens, some energy is randomly dispersed in a form (usually low-grade heat) that is not as readily available to do work.

If the first law suggests that we can't get something from nothing, the second law suggests that we don't have a chance of breaking even—because the amount of energy that is unavailable to do work is *increasing!* The reason is that no energy conversion can ever be 100-percent efficient. Each time one system transfers energy to another system, *some* of the energy goes off as low-grade heat.

Entropy is a measure of the degree of disorganization of a system. According to the second law of thermodynamics, the entropy of an isolated system tends to increase. The ultimate destination of the universe is therefore a state of maximum entropy. It has been estimated that some 5 billion years from now, everything will be at the same temperature, and energy conversions as we know them will never happen again.

Living Systems and the Second Law

Can it be that life is one glorious pocket of resistance to the depressing flow toward maximum entropy? After all, with the birth and growth of each new organism, energy becomes more concentrated and organized, not less so! Yet a simple example will show that the second law of thermodynamics does indeed apply to life on earth.

The primary source of energy for almost all organisms is the sun, which is steadily losing energy. Plants intercept some of this energy, then they lose energy to other organisms that feed, directly or indirectly, on plants. At each energy transfer along the way, more heat energy is added to the universal pool. Overall, then, energy is still flowing in one direction. *The world of life maintains a high degree of organization only because it is being resupplied with energy being lost from someplace else.*

There is a steady flow of sunlight energy into the interconnected web of life, and this compensates for the steady flow of energy leaving it.

METABOLIC REACTIONS: THEIR NATURE AND DIRECTION

Energy Changes in Cells

Let's now turn to the energy transformations that occur when cells shuffle materials during metabolic reactions. At the temperature and pressure levels characteristic of the cellular world, the materials present at the end of a reaction (the products) can end up with either less or more energy than the starting materials (the reactants) had.

Reactions showing a net loss in energy are called *exergonic* reactions (the term means "energy out"):

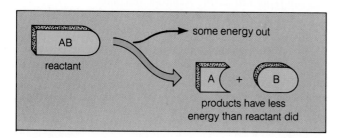

Reactants showing a net gain in energy are called *endergonic* reactions ("energy in"). An endergonic reaction does not violate the second law of thermodynamics. It occurs only when extra energy (lost from some other system) is fed into the reaction:

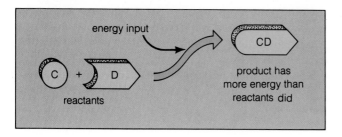

Reversible Reactions

Molecules in solution constantly move about and bump into each other. So when you think about it, molecules concentrated together will be more likely to react spontaneously with one another than if they were less concentrated. The more concentrated they are, the more their random movements will bring them into collision.

Suppose carbon dioxide and water molecules are concentrated together. Given enough time, they will react spontaneously to form carbonic acid, which then dissociates into bicarbonate and H^+ ions:

$$\text{reactants:} \quad CO_2 + H_2O \Longrightarrow \underset{\substack{\text{carbonic} \\ \text{acid}}}{H_2CO_3} \Longrightarrow \underset{\text{bicarbonate}}{HCO_3^- + H^+} \quad \text{products}$$

where the arrows signify that the reaction runs in the "forward" direction. As more and more product molecules form over time, not as many reactant molecules remain. All the while, the concentration of product molecules has been increasing—which means they will collide more frequently than before. Some fraction will now have enough collision energy to be driven in the opposite direction:

$$CO_2 + H_2O \Longleftarrow H_2CO_3 \Longleftarrow HCO_3^- + H^+$$

where the arrows signify this is the "reverse" reaction.

When left to themselves, all reversible reactions approach a state of **dynamic equilibrium**, at which the rates of the forward and reverse reactions are equal (Figure 7.2). Then, there is no *net* change in the concentrations of reactants and products even though molecular movements are still proceeding. (It is like a party with as many people wandering in as wandering out of two adjoining rooms. The total number of people in each room stays the same, even though the mix of people in each room is continually changing.)

Most reactions are reversible. The greater the concentration of reactants, the faster the forward reaction. The greater the concentration of products, the faster the reverse reaction.

All spontaneous reactions tend toward a state of dynamic equilibrium, when the rates of the forward and the reverse reactions are equal. Then, there is no further net change in the concentrations of reactants and products.

Here we are talking about equal *rates*, not equal concentrations. At equilibrium, the concentrations of reactant and product molecules are not necessarily the

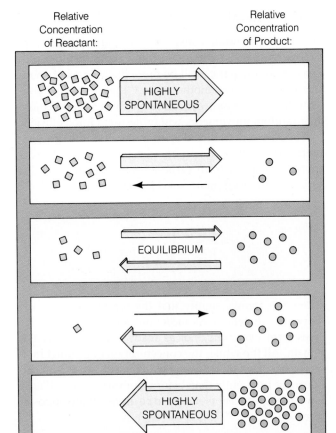

Relative Concentration of Reactant: Relative Concentration of Product:

HIGHLY SPONTANEOUS

EQUILIBRIUM

HIGHLY SPONTANEOUS

Figure 7.2 Chemical equilibrium. With high concentrations of reactant molecules, spontaneous reactions proceed most strongly in the forward direction. With high concentrations of product molecules, they proceed most strongly in reverse. At chemical equilibrium, the *rates* of the forward and reverse reactions are equal. (Note that the *concentrations* are not necessarily the same, as described in the text.)

same. For example, in one metabolic reaction, glucose-1-phosphate is rearranged into glucose-6-phosphate:

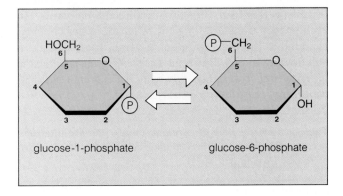

glucose-1-phosphate   glucose-6-phosphate

When the reverse reaction is proceeding as fast as the forward one, there are always 19 molecules of this product for every 1 molecule of reactant, or a ratio of 19:1. Each reaction has its own "equilibrium constant," which is simply the ratio of product and reactant concentrations when the forward and reverse reaction rates are equal.

Metabolic Pathways

Water, again, is central to cell functioning. Yet a cell can hold only so much water, and the water in turn can dissolve only so many atoms and molecules at a given time. *This means there are limits to the number of compounds that can be formed during metabolic reactions.*

Also, many compounds formed during metabolic reactions are able to combine with more than one kind of substance. Some of the resulting combinations would be unsuitable, for their accumulation would impair cell functioning. *Compounds must be produced in concentrations high enough to allow a reaction to run rapidly but low enough to prevent their being funneled into unnecessary and potentially disruptive side reactions.*

How does a cell function, given these constraints? It functions by moving substances through sequential, stepwise reactions called **metabolic pathways**, as illustrated in Figure 7.3.

In *degradative* metabolic pathways, carbohydrates, lipids, and proteins are broken down in stepwise reactions that lead to products of lower energy. In *biosynthetic* metabolic pathways, small molecules (such as sugar subunits) are assembled into large biological molecules (such as carbohydrates). Through interplays between the two kinds of pathways, concentrations of substances are maintained, increased, or decreased in controlled ways.

The participants in metabolic pathways go by these names:

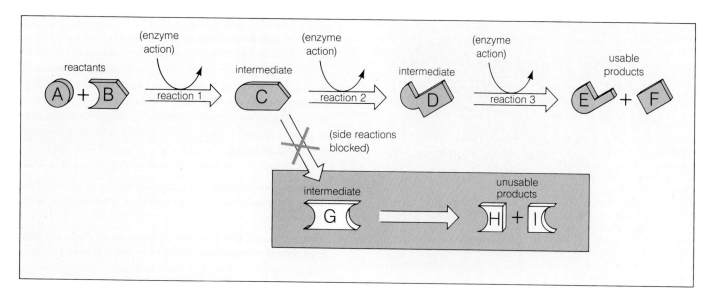

Figure 7.3 Simplified picture of two metabolic pathways. The "product" of the first reaction is not the final compound but an intermediate (C), which can enter into side reactions (a different pathway) as well as the main pathway. Controls over enzyme activity at this branch point help channel the intermediate into the pathway that will lead to the product required at the time.

reactants	organic and inorganic substances able to enter into a reaction; also called substrates or precursors for a reaction
metabolites	compounds being funneled through a metabolic pathway; often, an intermediate form in assembly or breakdown reactions
enzymes	proteins that serve as catalysts (substances that speed up chemical reactions) and in controls over metabolic reactions
coupling agents	compounds (primarily ATP and $NADP^+$) that act as functional links *between* the major degradative and biosynthetic pathways
local electron carriers	compounds (such as NAD^+) that transfer hydrogen and electrons locally, *within* a degradative or biosynthetic pathway
end products	substances formed at the conclusion of a metabolic pathway

Let's take a look at some of these substances and at the roles they play in metabolism.

ENZYMES

If left to themselves, some reactants present in cells would not react until years passed or, in some cases, decades. Yet metabolic reactions typically must occur within a fraction of a second if cells are to function properly. Enzymes account for the difference in reaction rates. **Enzymes** are proteins having enormous catalytic power, which means they greatly enhance the rate at which specific reactions approach equilibrium.

Enzymes do not make anything happen that would not eventually happen on its own. They merely make it happen much more rapidly (usually by at least a million times). And they make it happen again and again; enzyme molecules are not permanently altered or consumed in a reaction.

Also, an enzyme is extremely selective about which reaction it will enhance and in its choice of reactants, called its **substrates**. For example, thrombin (an enzyme involved in blood clotting) will catalyze the breaking of a peptide bond *only* if the bond is between two particular amino acids in this particular order:

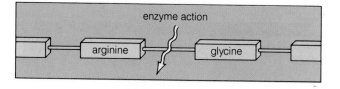

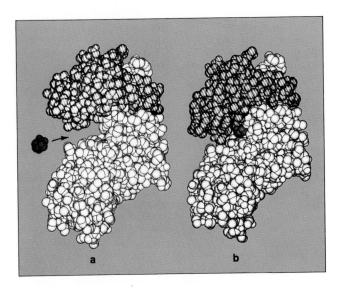

Figure 7.4 Model of the induced fit between an enzyme (hexokinase) and its bound substrate (a glucose molecule, shown here in red).

(a) The cleft into which the glucose is heading is the enzyme's active site. **(b)** In this enzyme-substrate complex, notice how the enzyme shape is altered temporarily: the upper and lower parts now close in around the substrate.

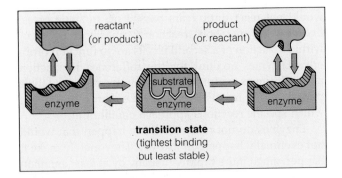

Figure 7.5 Induced-fit model of enzyme-substrate interactions. Only when the substrate is bound in place is the active site complementary to it. The most precise fit occurs during the transition state.

The preference of enzymes for specific substrates is central to metabolism. Cellular concentrations of many reactants must be kept low, to avoid unsuitable side reactions (Figure 7.3). At the same time, concentrations must be kept high enough for required reactions to occur at significant rates. As you will see, metabolism proceeds under these seemingly conflicting conditions because enzymes channel specific substances into and through specific pathways.

Enzymes, which are a class of proteins, serve as catalysts: they greatly enhance the rate at which a reaction approaches equilibrium.

Enzymes assure that the rates of required reactions are high even though overall concentrations of the reactants are low.

Enzyme Structure

Active Sites. Most enzymes are globular (globe-shaped), with at least one surface region folded into a crevice in which a given reaction is catalyzed. The crevice is called an **active site** (Figure 7.4). An **enzyme-substrate complex** forms when substrates become temporarily bound in the crevice. The complex is extremely short-lived, partly because only weak bonds hold it together.

As long ago as 1890, Emil Fischer suggested that the shape of some region of the enzyme's surface matches that of its substrate, much as a lock precisely matches its key. This metaphor is still valid, although it now appears that the match is not a rigid arrangement. As first proposed by Daniel Koshland in 1963, active sites seem to undergo changes during binding.

According to the **induced-fit model**, an active site making contact with its substrate almost but not quite matches it (Figure 7.5). This means that the binding between them is not as strong as it could be. Even so, the interaction is enough to induce structural changes in the active site and to distort the bound substrate, such that the site and the substrate become fully complementary to each other.

Cofactors. Sometimes the chemical groups projecting into an active site are enough to bind substrates tightly or to make them more reactive. However, sometimes the groups require help from nonprotein components called **cofactors.** Some cofactors diffuse freely to and from the enzyme. Others are bound so tightly to it that they are, in effect, part of the enzyme itself (in which case they are called prosthetic groups).

Cofactors include metal ions (such as Fe^{++}) and coenzymes (large organic molecules such as NAD^+, to be described shortly). Water-soluble vitamins such as riboflavin (vitamin B_2) are components of many coenzymes.

Enzyme Function

Activation Energy. How do enzymes enhance reaction rates? We can begin with a simple fact: For any reaction to occur, reactant molecules must collide with each other with some minimum amount of energy. The amount is like a hill over which the molecules must be pushed. For example, H_2 and O_2 molecules are not inclined to react on their own, but when they absorb enough energy (as from an electric spark), they collide with enough force to get to the top of the hill. Then the reaction proceeds spontaneously, just as a boulder pushed up and over the crest of a hill rolls down on its own.

At the crest of an energy hill, reactants are in an activated, intermediate condition called the *transition state*. The height that the hill represents is the **activation energy**: the minimum amount of energy needed to bring all of the molecules in one mole of a particular substance to the top of the hill. As Figure 7.6 suggests, enzymes *lower* the required activation energy.

How do enzymes exert their effects? To give one example, *enzymes orient their substrates in positions and at distances that promote reaction*. Reactants colliding on their own do so from random directions, so their mutually attractive chemical groups may not make contact and reaction may not occur. In contrast, weak but extensive bonding in an active site puts reactants in close and correct alignment.

Effect of Substrate Concentrations. Enzymes lower the crest of the energy hill for both reactant and product molecules. Intermediates at the top of the hill can just as easily roll down one side (to the product form) as to the other (to the reactant form). As you read earlier, the *net direction* in which the reaction proceeds is influenced by the relative concentrations of reactant and product molecules. Enzymes only influence the *rate* at which the forward and reverse reactions reach equilibrium.

Enzymes influence only the rate of a reaction; they do not change what the concentration ratio of reactant to product molecules will be at equilibrium.

Effects of Temperature and pH. With few exceptions, enzymes cannot tolerate high temperatures. The rate of enzyme activity does increase as the environment warms up, until a maximum rate is reached. With further tem-

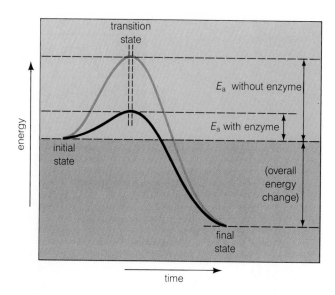

Figure 7.6 Energy hill diagram showing the effect of enzyme action. An enzyme greatly enhances the rate at which a reaction proceeds because it lowers the required activation energy. In other words, not as much collision energy is needed to boost all of the reactant molecules to the crest of the energy hill (the transition state).

perature increases, the reaction rate decreases sharply (Figure 7.7). The increased temperature has disrupted the weak bonds holding the enzyme molecule in its three-dimensional shape, and denaturation has occurred (page 58). Even brief exposure to temperatures above optimum will destroy enzymes. Without enzymes, metabolism grinds to a halt and cells die.

Also, most enzymes are effective at or near pH 7 (Figure 7.7). If the pH rises above or falls below this value, hydrogen bonds and other weak attractions that hold most enzymes in their three-dimensional shape are disrupted and enzyme function is lost. However, there are enzymes that function at pH values other than 7. Pepsin is active in the extremely acid solution in the stomach. Trypsin is active in a more basic medium (about pH 8) in the small intestine.

An enzyme functions only within a limited range of pH and temperature.

Regulation of Enzyme Activity

How much of a product will appear in a given time, if at all? It depends on the number of enzyme molecules available for the reactions leading to its formation. That number can be controlled in several ways. Synthesis of

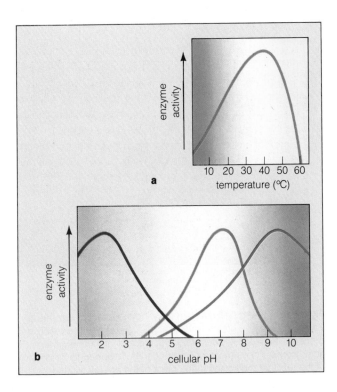

Figure 7.7 (**a**) Effect of increasing temperature on enzyme activity. (**b**) Effect of pH on enzyme activity. The brown line charts the activity of an enzyme that is fully functional in neutral solutions. The red line charts the activity of one that is functional in basic solutions; the purple line, in acidic solutions.

enzyme molecules can be accelerated or slowed down. And the activity of enzymes already formed can be temporarily or permanently shut down.

For example, some *reversible* inhibitors compete with a substrate for an active site, as the example in Figure 7.8 illustrates. When the inhibitor is bound to the site, it inactivates the enzyme. *Irreversible* inhibitors bind with or destroy a key group on the enzyme surface, thereby making catalysis impossible.

Enzyme activity is controlled by several mechanisms, including those which accelerate or inhibit enzyme synthesis or inhibit the activity of enzymes already formed.

Groups of enzymes cooperate in sequence in metabolic pathways. In each group, one or more types of enzymes catalyze the slowest reactions, which (being slowest) set the overall reaction rate for the entire pathway. These enzymes, the so-called pacemakers of metabolism, are known as **regulatory enzymes**.

One type of regulatory enzyme is allosteric. In addition to its active site, an *allosteric* enzyme has one or more regulatory sites at which a specific substance binds and influences enzyme activity (Figure 7.9). Often the substance is the end product of a pathway in which the enzyme takes part. When more product molecules are being made than the cell can use, the substance binds to the enzyme and shuts it down. This is a form of

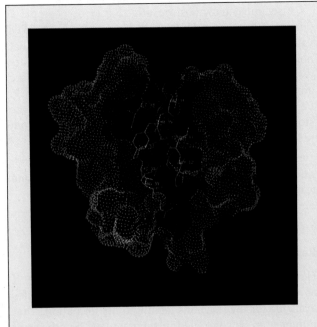

Figure 7.8 Computer-graphics model of the enzyme trypsin (blue) with an inhibitor (green) bound to the active site. The enzyme parts colored red are not exposed to the surroundings when the inhibitor is bound.

This inhibitor is almost exactly complementary in structure to the active site and dissociates slowly from it. The tight, prolonged binding is vital. Trypsin figures in the digestion of proteins in the small intestine. It is so powerful, it is assembled in inactive form in the pancreas then isolated in membranous packages that are not opened until after they are secreted in the small intestine.

Pancreatic trypsin inhibitor serves as a safeguard: it inactivates trypsin molecules that escape packaging in the pancreas. Without this safeguard, trypsin could be unleashed against the protein components of pancreatic tissues and blood vessels. (Acute pancreatitis is a disorder in which this and other pancreatic enzymes are activated prematurely; the results are sometimes fatal.)

feedback inhibition whereby an increase in a substance inhibits the very process leading to its increase. When more product molecules are used up, the substance is released from enzyme molecules, which can function once again.

Through reversible feedback inhibition of regulatory enzymes, rapid adjustments can be made in the rates of biosynthetic pathways.

COUPLING AGENTS

Overall, energy is released in degradative (breakdown) pathways, and considerable energy must be fed into biosynthetic pathways. Notice, in Figure 7.10, that the two types of pathways are linked by two types of compounds: ATP and $NADP^+$. With ATP, energy released when molecules are broken apart can be captured and delivered to reactions by which large biological molecules are built. With $NADP^+$, the hydrogen and electrons released are also delivered to those reactions.

ATP and $NADP^+$ are coupling agents between the main breakdown and biosynthetic pathways.

ATP: The Main Energy Carrier

As Figure 7.11 shows, adenosine triphosphate, or **ATP**, is composed of adenine (a nitrogen-containing compound), ribose (a five-carbon sugar), and three linked phosphate groups (a triphosphate). One or more of these phosphate groups can be transferred to another type of molecule, such as glucose. When ATP gives up a phosphate group, it becomes adenosine diphosphate (ADP). And the molecule to which the phosphate group becomes attached is said to have undergone **phosphorylation**:

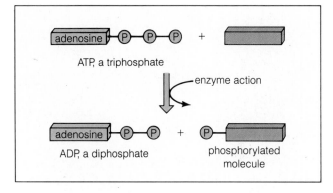

The addition of a phosphate group to a molecule generally increases the molecule's store of energy and thereby primes it to enter specific reactions. So many different molecules can be phosphorylated by ATP that

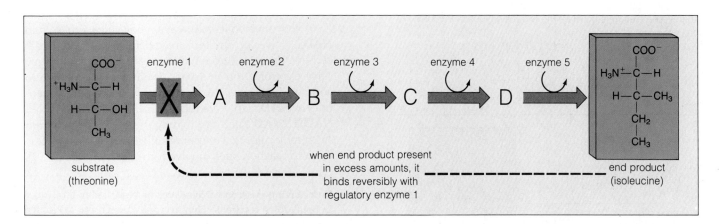

Figure 7.9 Feedback inhibition through an allosteric enzyme. Such enzymes have, in addition to the active site, one or more regulatory sites where a specific metabolite binds and signals the enzyme to shut down its catalytic activity.

Here, the end product (isoleucine) binds reversibly to the first enzyme in the pathway leading to its production. When the isoleucine concentration drops, fewer molecules are around to inhibit the regulatory enzymes present, so isoleucine production can rise again. The rapid reversibility of such forms of feedback inhibition allows cellular concentrations of substances to be adjusted quickly to metabolic needs.

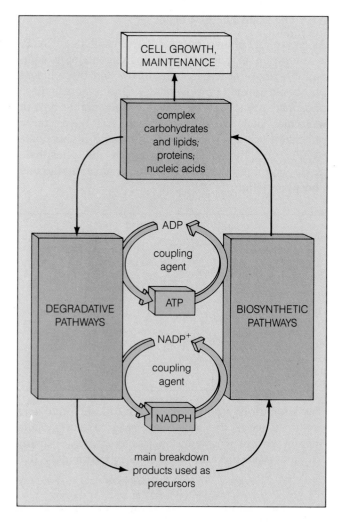

Figure 7.10 Relationships between the degradative and biosynthetic pathways in all cells. Coupling agents carry usable energy and electrons between the two main types of metabolic pathways.

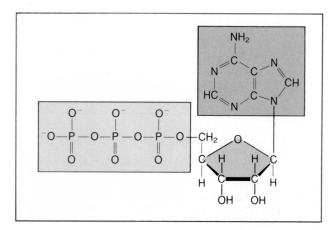

Figure 7.11 Structural formula for adenosine triphosphate, or ATP. The triphosphate group is shaded in yellow, the sugar ribose in pink, and the adenine portion in blue.

it can be likened to a gold coin (rather than, say, a dollar bill or a peso), which is accepted at once as currency in any country in the world. Almost all metabolic pathways directly or indirectly run on ATP energy.

The energy embodied in an ATP molecule is not stored for long at all. Typically an ATP molecule is consumed within sixty seconds of its formation. Even if you were bedridden for twenty-four hours, your cells would turn over about forty kilograms of ATP molecules simply for routine maintenance.

NADP⁺ and Its Hydrogen and Electron Cargo

Another coupling agent between breakdown and biosynthetic pathways is nicotinamide adenine dinucleotide phosphate, or **NADP⁺**. Unlike ATP, this compound keeps its phosphate to itself, as a permanent molecular tag that is recognized by specific enzymes.

NADP⁺ picks up hydrogen atoms and electrons at sites where certain molecules are broken apart, then transfers them to sites where sugars and other hydrogen-containing molecules are assembled. When carrying its cargo, the compound is called **NADPH**.

LOCAL ELECTRON CARRIERS

In metabolic reactions, electrons can be transferred from one substance (the donor) to another (an acceptor). When an atom or molecule gives up one or more electrons, it is said to be *oxidized*. When it accepts one or more electrons, it is *reduced*. The term **oxidation-reduction reaction** refers to an electron transfer.

Within metabolic pathways, molecules other than NADP⁺ take part in electron transfers; they are "local" carriers. Some of these molecules are free-moving; they carry electrons and hydrogen from one reaction site to another. One is called nicotinamide adenine dinucleotide, or **NAD⁺**. In its reduced form, this carrier is called **NADH**. NAD⁺ accepts hydrogen and electrons from sugar molecules that are being split apart, and it transfers them to reaction sites involved in ATP formation.

Free-moving electron carriers such as NAD⁺ play local roles in breakdown reactions. They accept hydrogen and associated electrons from compounds being degraded, then transfer them to reactions related to ATP formation.

Other "local" carriers of electrons are bound in cell membranes, such as the membranes of chloroplasts and mitochondria. Together with enzymes, they are positioned in organized arrays called **electron transport systems**.

In electron transport systems, oxidation-reduction reactions occur in enzyme-mediated steps, one after the other. For example, in some systems the first molecule accepts electrons and hydrogen from NADH. Then the electrons are transferred in stepwise fashion to other membrane-bound carriers that include cytochromes. The last electron carrier in the series gives up the electron to some kind of external acceptor molecule (one that is not part of the transport system). Oxygen is one type of final acceptor molecule.

The point of such oxidation-reduction sequences is the generation of usable forms of energy. By analogy, think of an electron transport system as a staircase. Electrons that have been "raised" (excited) to the top of the staircase have the most energy. They drop down the staircase, one step at a time (they are transferred from one electron carrier to another). With each drop, some of the energy being released is harnessed to do work—for example, to move H^+ in ways that establish pH and electric gradients across membranes. Such gradients, as you will discover, are central in ATP formation.

SUMMARY

1. Cells acquire and use energy to accumulate, break down, synthesize, and eliminate substances in controlled ways. These activities (called metabolism) sustain cell growth, maintenance, and reproduction.

2. Cells use different forms of energy (such as chemical energy and mechanical energy), each of which can be measured in terms of its capacity to do work.

3. Cellular use of energy conforms to two laws of thermodynamics. According to the first law, energy can be converted from one form to another but the total amount in the universe never changes. (No more energy can be created, and none of the existing energy ever vanishes.)

4. According to the second law, with each energy conversion, some of the energy is dispersed in a form (low-grade heat, usually) not as readily available to do work.

5. Cells lose energy during metabolic reactions, but they also replace it with energy being lost from someplace else. In this way they maintain their high degree of organization.

6. Directly or indirectly, some of the energy being lost from the sun is the source of energy replacements for almost all organisms on earth.

7. If left undisturbed, most metabolic reactions approach a state of dynamic equilibrium (in which there is no further net change in the concentrations of reactants and products).

8. A metabolic pathway is a stepwise sequence of reactions in a cell. In biosynthetic pathways, organic molecules are assembled and energy becomes stored in them. In degradative pathways, organic molecules are broken apart and energy is released.

9. The following substances take part in metabolic reactions:

   a. Reactants: the substances that enter the reactions.

   b. Metabolites: intermediate compounds in a series of reactions.

   c. Enzymes: proteins that serve as catalysts (they speed up reactions).

   d. Coupling agents: mainly ATP (which carries energy) and $NADP^+$ (which carries hydrogen atoms and electrons) *between* major metabolic pathways.

   e. Local electron carriers: compounds (such as NAD^+) that transfer hydrogen atoms and electrons *within* a metabolic pathway.

   f. End products: the substances formed at the end of a metabolic pathway.

10. Enzymes do not change what the concentration ratio of reactant to product molecules will be at equilibrium. They only *increase* the reaction rate (by lowering the required activation energy).

11. In oxidation-reduction reactions (electron transfers), energy is released that can be used to do work—for example, to make ATP.

Review Questions

1. State the first and second laws of thermodynamics. Which law deals with the *quality* of available energy, and which deals with the *quantity*? Can you give some examples of high-quality energy?

2. Does the living state violate the second law of thermodynamics? In other words, how does the world of living things maintain a high degree of organization, even though there is a universal trend toward disorganization?

3. In metabolic reactions, does equilibrium imply equal concentrations of reactants and products? Can you think of some cellular events that might keep a reaction from approaching equilibrium?

4. Describe an enzyme and its role in metabolic reactions. How do enzymes affect the proportions of reactants and products that will be present at equilibrium?

5. Define "substrate" and "active site." Why is binding at an active site a readily reversible event?

6. The high temperatures associated with severe fevers can impair cell functioning. Can you explain why?

7. What is an oxidation-reduction reaction? What is its function in cells?

Readings

Atkins, P. 1984. *The Second Law.* New York: Freeman.

Doolittle, R. October 1985. "Proteins." *Scientific American* 253(4):88–99.

Fenn, J. 1982. *Engines, Energy, and Entropy.* New York: Freeman. Deceptively simple introduction to thermodynamics; good analogies. Paperback.

Fersht, A. 1985. *Enzyme Structure and Mechanism.* Second edition. New York: Freeman.

8

ENERGY-ACQUIRING PATHWAYS

Just before dawn in the Midwest the air is dry and motionless; the heat that has scorched the land for weeks still rises from the earth and hangs in the air of a new day. There are no clouds in sight. There is no promise of rain. For hundreds of miles in any direction you care to look, crops stretch out, withered or dead. All the sophisticated agricultural methods in the world can't save them now. In the absence of one vital resource—water—life in each cell of those many thousands of plants has ceased.

In Los Angeles, a student reading the morning newspaper complains that the Midwest drought will probably cause a hike in food prices. In Washington, D.C., economists calculate the crop failures in terms of decreased tonnage available for domestic consumption and export and of what it means to the nation's balance of payments. In Ethiopia, a child with bloated belly and spindly legs waits passively for death. Even if food from the vast agricultural plains of North America were to reach her now, it would be too late. Deprived too long of nutrients, her body's cells will never grow normally again.

You are about to explore the ways in which cells acquire and use energy. You will be considering cellular pathways that might at first seem to be far removed from the world of your interests. Yet the food molecules on which almost all organisms depend—yourself included—cannot be built or used without these pathways and the raw materials required for their operation.

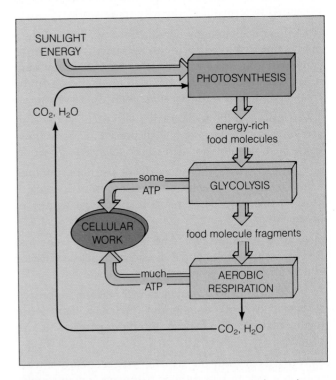

Figure 8.1 Links between three major energy-trapping and energy-releasing pathways.

FROM SUNLIGHT TO CELLULAR WORK: PREVIEW OF THE MAIN PATHWAYS

No matter what the organism, cell structure is based on sugars and other organic compounds (which have backbones of carbon atoms), and cell activities are driven by the chemical energy stored in those compounds. The questions become these:

1. Where does the carbon come from in the first place?

2. Where does the energy come from to drive the linkage of carbon and other atoms into organic compounds?

3. How does the energy inherent in the bonds of those compounds become available to do cellular work?

The answers vary, depending on whether you are talking about autotrophs or heterotrophs. **Autotrophic organisms** get their carbon and energy all by themselves from the physical environment; they are "self-nourishing" (which is what autotroph means). Their carbon source is carbon dioxide (CO_2), a gaseous substance all around us in the air and dissolved in water. The *photosynthetic* autotrophs, which include all plants, some protistans, and some bacteria, harness energy from sunlight for building organic compounds. The *chemosynthetic* autotrophs, which are limited to a few bacteria, extract energy from chemical reactions of inorganic substances (such as sulfur).

In contrast, **heterotrophic organisms** are not self-nourishing; they feed on autotrophs, each other, and organic wastes. Thus they obtain the carbon and energy for their building programs from organic compounds already built by autotrophs. Animals, fungi, many protistans, and most bacteria are heterotrophs.

It follows, from the above, that carbon and energy enter the web of life primarily by way of *photosynthesis*, as carried out by autotrophs. Energy that becomes stored in organic compounds as a result of photosynthesis is almost always released by the pathways called *glycolysis* and *aerobic respiration* (Figure 8.1). These three major pathways are the focus of this chapter and the next.

PHOTOSYNTHESIS

Simplified Picture of Photosynthesis

Photosynthesis consists of two major sets of reactions. In the **light-dependent reactions**, sunlight energy is absorbed and converted to chemical energy, which is stored briefly in ATP and NADPH. In the **light-independent reactions**, sugars and other organic compounds are assembled with the help of ATP and NADPH. Often photosynthesis is summarized this way:

$$\text{sunlight} + 2H_2O + CO_2 \longrightarrow O_2 + (CH_2O) + H_2O$$

Here, hydrogen atoms obtained from water molecules are transferred to CO_2, forming compounds based on some number of (CH_2O) units. For instance, for the reactions leading to glucose formation, you would have to multiply everything by six (to get the six carbons, twelve hydrogens, and six oxygens of the glucose molecule), as shown here:

$$\text{sunlight} + 12H_2O + 6CO_2 \longrightarrow 6O_2 + C_6H_{12}O_6 + 6H_2O$$

The above equation does not provide much insight into what actually goes on. For example, oxygen is shown as a by-product of photosynthesis. Did it come from the H_2O or the CO_2? Also, why is water shown as a by-product as well as a reactant? To answer these questions, we must expand the summary equation with a few more details:

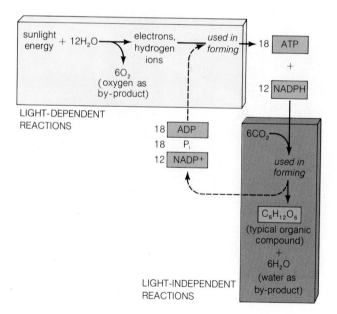

As you can see, the oxygen is obtained early on (when water molecules are split), and "new" water forms later (when the glucose is being synthesized). Notice also how ATP and NADPH bridge the two sets of reactions. The ATP carries energy from one set of reactions to the other, and the NADPH carries electrons and hydrogen atoms (page 105). Both are central to the events of photosynthesis.

Chloroplast Structure and Function

Photosynthesis takes place at specialized cell membranes, such as the **thylakoid membrane** of those chloroplasts described on page 76. The membrane stage on which the reactions are played out is truly tiny; a typical chloroplast is no bigger than one of your red blood cells.

Inside the chloroplast is a semifluid matrix called the **stroma**. Within the matrix, the thylakoid membrane is folded into a system of stacked disks and flattened channels (Figure 8.2c). Apparently the interior spaces of the disks and channels are open to one another. This means the membrane system forms a single compartment separate from the stroma. The compartment is a reservoir for hydrogen ions.

This reservoir is tapped during the light-dependent reactions, in ways that will be described shortly. Actual assembly of sugars and other organic compounds takes place in the stroma.

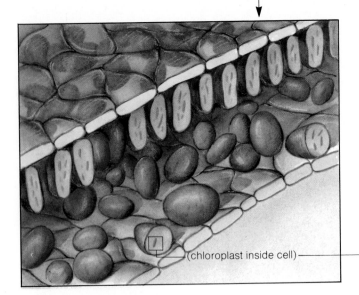

(chloroplast inside cell)

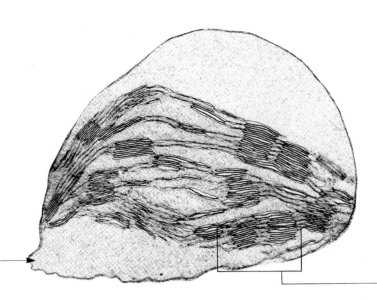

a

b

Figure 8.2 Functional zones of a chloroplast from the leaf of a sunflower plant (*Helianthus*). The light-dependent reactions of photosynthesis occur at thylakoid membranes, and they lead to ATP and NADPH formation. The light-independent reactions occur in the stroma. They lead to production of sugars and other carbon-containing molecules. (**a**) Section through a sunflower leaf, showing chloroplast-containing cells. (**b**) Chloroplast in cross-section, 25,000×. (**c**) Two of the grana, 93,000×. (**d**) Where photosynthetic reactions occur.

LIGHT-DEPENDENT REACTIONS

Three distinct events unfold during the light-dependent reactions. Light energy is absorbed, electron and hydrogen transfers lead to ATP and NADPH formation, and electrons are replaced in the substance that originally gives them up.

Light Absorption in Photosystems

Light-Trapping Pigments. Photosynthetic membranes contain light-trapping proteins called **pigments**. A pigment molecule absorbs photons, which are individual packets of energy from the sun. Some photons have more energy than others, and the difference corresponds to different wavelengths (colors) of light. The more energetic the photon, the shorter the wavelength. Organisms use wavelengths ranging from about 400 to 750 nano-meters for light-requiring processes such as photosynthesis and vision. That is a very small part of the entire electromagnetic spectrum (Figure 8.3).

Each type of pigment can absorb only certain wavelengths. For example, the chlorophylls absorb blue and red wavelengths but cannot absorb green (which they transmit). The carotenoids absorb violet and blue wavelengths but cannot absorb yellow. Wavelengths that *can* be absorbed impart just the right amount of energy to excite one of the electrons in the pigment molecule. Figure 8.3 includes plots (absorption spectra) that show which wavelengths can make this happen in several kinds of pigments.

Photosynthetic organisms use some wavelengths but not others, depending on their assortment of pigments. Figure 8.4 shows an early experiment designed to test which wavelengths are used most effectively by a green alga. In this and other green plants, the main pigments are chlorophyll *a*, chlorophyll *b*, and the carotenoids.

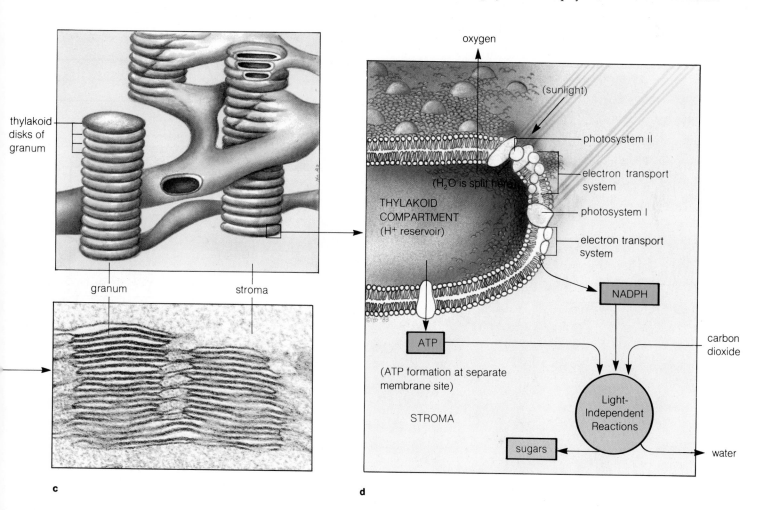

Figure 8.3 (Above) Where wavelengths of visible light occur in the electromagnetic spectrum. For most organisms, the range of photosensitivity is limited largely to wavelengths ranging from about 400 to 750 nanometers. Shorter wavelengths (such as ultraviolet and x rays) are so energetic they can break bonds in organic molecules; hence they can destroy cells. Longer wavelengths (such as infrared) are not energetic enough to power chemical changes in molecules needed to form NADPH, which is used in biosynthesis.

(Below) Ranges of wavelength absorption for photosynthetic pigments. Absorption peaks correspond to the measured amount of energy absorbed and used in photosynthesis. Colors used here correspond to the colors transmitted by each pigment type. (Thus chlorophylls show peak absorption of blue and red wavelengths and transmit wavelengths in between.) Together, different photosynthetic pigments can absorb most of the wavelength energy available in the spectrum of visible light.

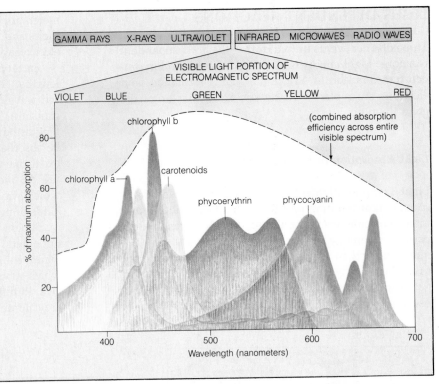

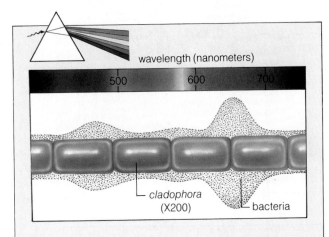

Figure 8.4 T. Englemann's 1882 experiment, which revealed the most effective wavelengths for photosynthesis by *Cladophora*, a filamentous green alga.

Oxygen is a by-product of photosynthesis, and it also is used by many organisms in the energy-releasing pathway called aerobic respiration. Englemann suspected that aerobic (oxygen-using) bacteria living in the same places as the alga would congregate in areas where oxygen was being produced.

Englemann used a crystal prism to cast a tiny spectrum of colors on a microscope slide. Then he positioned an algal filament to run parallel with the spectrum (as shown above). The bacteria did indeed cluster next to the filament where the most oxygen was being released. And those regions corresponded to colors (wavelengths) being absorbed most efficiently—in this case, violet and red.

Photosystems. The light-trapping pigments just described are not scattered at random in the thylakoid membrane of chloroplasts. They are organized into many clusters, each with 200 to 300 molecules. Each cluster is a **photosystem**.

More than ninety percent of the pigments in a photosystem do nothing more than "harvest" light energy. When a light-harvesting pigment absorbs energy from a photon, one of its electrons becomes excited and reaches a higher energy level (page 39). When that electron returns to a lower level, the extra energy is released and is transferred to a neighboring pigment. The energy continues to hop around from one pigment to another and, with each hop, a little energy is usually lost (as heat). Soon the energy remaining corresponds to longer and longer wavelengths, compared to the original photon energy.

Only a few chlorophyll molecules can respond to the longest of those wavelengths. They act like a sink, or trap, for all that energy being harvested by all the other pigments. And they are the only pigments that can *give up* electrons for use in photosynthesis.

When energy flows into one of the energy traps, an electron is excited and is rapidly transferred to an acceptor molecule embedded in the thylakoid membrane. Thus, *the first event of photosynthesis is the light-activated transfer of an electron from a photosystem to an acceptor molecule*. This event is over in less than a billionth of a second.

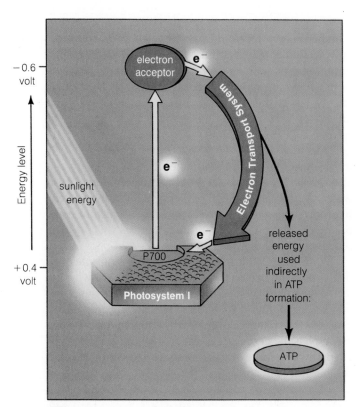

Figure 8.5 Cyclic photophosphorylation, which yields ATP. The vertical scale of this figure indicates the relative tendencies of electron-carrying molecules to release or take up electrons. Electrons are driven uphill by light, and they move back downhill in the transport system.

Two Pathways of Electron Transfer

During their journey, the electrons expelled from a photosystem pass through either one or two **electron transport systems**, which consist of a series of molecules that alternately accept and then donate the electrons to the next molecule in line. Such electron transfers, recall, are called oxidation-reduction reactions. With the energy released at some of the transfers, an ADP molecule is phosphorylated (inorganic phosphate is attached to it) to form ATP.

The Cyclic Pathway. In **cyclic photophosphorylation**, electrons proceed through one transport system and then return to the photosystem that gave them up (Figure 8.5). This so-called **photosystem I** is distinguished by one of its chlorophyll molecules, P700, which absorbs wavelengths of about 700 nanometers. Because electrons travel in a circle (back to the photosystem), the pathway is said to be cyclic.

The cyclic pathway is probably one of the oldest as well as the simplest means of ATP production. Early photosynthetic organisms were no larger than existing bacteria, so their body-building programs could scarcely have been enormous. They could use ATP for building organic compounds even though synthesis reactions based on ATP alone are rather inefficient. However, the energy made available by the cyclic pathway would not have been enough to sustain the evolution of larger photosynthetic organisms, including land plants.

The Noncyclic Pathway. Existing plants rely mainly on **noncyclic photophosphorylation**. In this pathway, electrons are transferred through two photosystems and two electron transport systems, and ADP is still phosphorylated to form ATP. But the electrons do not move in a circle. They end up in NADPH—and hydrogen and electrons can be pulled away directly from this molecule for use in building organic compounds!

The noncyclic pathway of photosynthesis begins at **photosystem II**, which is distinguished by one of its chlorophyll molecules (P680). Absorbed light energy causes P680 to give up an electron to an acceptor molecule. The electron is transferred to a transport system, which delivers it to chlorophyll P700 of photosystem I.

When P700 absorbs energy, an electron is boosted to a higher energy level and passed to a second transport system. Remember that transport systems are like steps on an energy staircase—and the boost starts the electrons at the top of a higher staircase. There is enough energy left at the bottom of this staircase to drive the linkage of electrons and hydrogen to $NADP^+$, the result being NADPH (Figure 8.6).

Thus there is a one-way flow of electrons to NADPH. In the meantime, the P680 molecule that gives up the electrons in the first place is getting replacements—from water. Water molecules are being split into oxygen, hydrogen ions, *and electrons* inside the thylakoid compartment. Photon energy indirectly drives this reaction sequence, which is called **photolysis**.

The oxygen atoms split from water molecules are released as by-products of the noncyclic pathway. In fact, oxygen has been accumulating ever since this pathway emerged, about 1.4 billion years ago. It profoundly changed the character of the earth's atmosphere. And it made possible aerobic respiration, the most efficient of all pathways for extracting energy stored in organic compounds.

Cyclic photophosphorylation leads to the production of ATP alone.

Noncyclic photophosphorylation leads to ATP and NADPH production.

Oxygen, an end product of the noncyclic pathway, has profoundly influenced the character of the atmosphere, and it made possible the energy-yielding pathway called aerobic respiration.

A Closer Look at ATP Formation

So far, you have seen what happens to the electrons and oxygen released when water molecules are split at the start of the noncyclic pathway. What happens to the hydrogen atoms? They accumulate as hydrogen ions inside the thylakoid compartment:

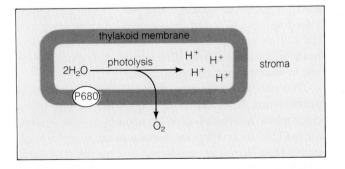

Hydrogen ions also accumulate in the compartment when the transport systems are operating. (This is true of both the cyclic and noncyclic pathway.) When some of the molecules in the transport system accept electrons, they also pick up hydrogen ions from the stroma. But other molecules positioned later in the series take only the electrons, not the hydrogen ions, which end up inside the compartment:

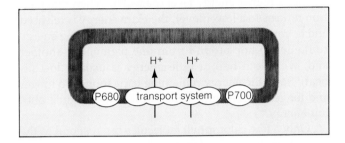

Because of these activities, hydrogen ions become much more concentrated in the compartment than in the stroma. The lopsided distribution of these positively charged ions also creates a difference in electric charge across the membrane. An electric gradient as well as a concentration gradient has become established—*and the energy inherent in those gradients can be tapped to form ATP.*

The combined force of the concentration and electric gradients propels hydrogen ions from the compartment back to the stroma, through channel proteins that span the membrane. The flow of hydrogen ions is linked with enzyme machinery, which causes ADP and inorganic phosphate in the stroma to combine into ATP:

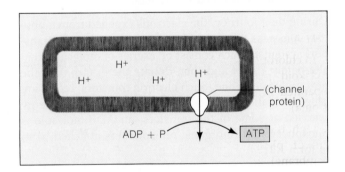

The idea that concentration and electric gradients across a membrane drive ATP formation is known as the **chemiosmotic theory**.

Summary of the Light-Dependent Reactions

So far, we have described the light-dependent reactions as separate events. We can now bring those events together into an integrated picture of the first stage of photosynthesis:

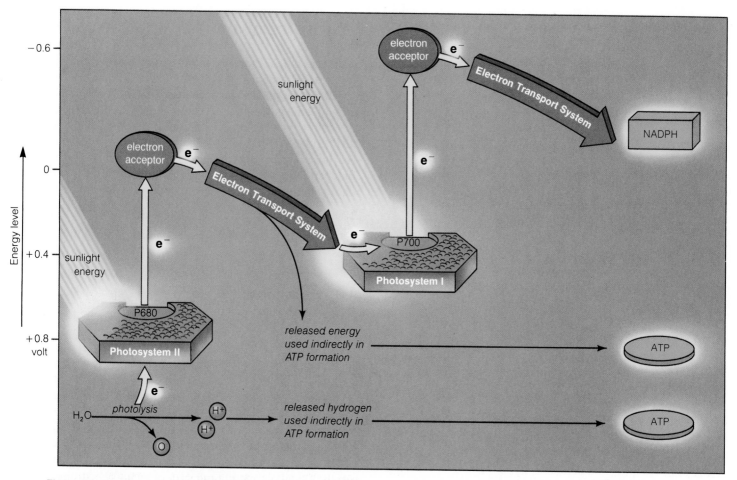

Figure 8.6 Noncyclic photophosphorylation, which yields NADPH as well as ATP. Electrons derived from the splitting of water molecules (photolysis) travel through two transport systems, which work together in boosting the electrons to an energy level high enough to lead to NADPH formation.

Light Absorption

1. In chloroplasts, light absorption occurs at an internal thylakoid membrane, which forms a compartment separated from the surrounding semifluid matrix (the stroma).

2. Light is absorbed by two types of photosystems (clusters of photosynthetic pigments embedded in the membrane).

3. Light absorption activates the transfer of electrons from a special chlorophyll molecule in photosystem I or II to an acceptor molecule, which will donate them to a transport system.

Noncyclic Pathway

1. In noncyclic photophosphorylation, there is a one-way flow of excited electrons from photosystem II, through a transport system, to photosystem I and on through a second transport system.

2. This pathway yields ATP *and* NADPH. Electrons and hydrogen atoms can be stripped from NADPH and used directly in assembling sugars and other organic compounds in the second stage of photosynthesis.

3. Light absorption at the start of the pathway indirectly drives the splitting of water molecules. Electrons released from water replace the electrons being expelled from photosystem II.

4. During electron transfers, hydrogen ions picked up from the stroma are left inside the thylakoid compartment. The hydrogen atoms from water molecules also accumulate here. These two activities establish concentration and electric gradients across the membrane.

5. Hydrogen ions flow down the gradients (through channel proteins that span the membrane), and the flow triggers the attachment of inorganic phosphate to ADP to form ATP.

6. At the end of the second transport system, electrons are donated to $NADP^+$, which combines with H^+ to form NADPH.

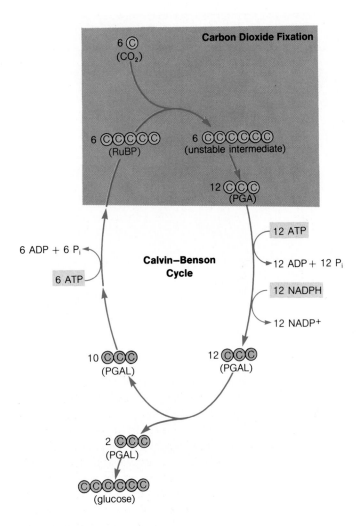

Figure 8.7 Summary of the light-independent reactions of photosynthesis. (Only the carbon atoms of the different molecules are depicted.)

Cyclic Pathway

1. In cyclic photophosphorylation, excited electrons flow from photosystem I, through a transport system, then back to the photosystem. This pathway yields ATP, which can be used in the second stage of photosynthesis.

2. During electron transfers through the transport system, hydrogen ions picked up from the stroma are released inside the thylakoid compartment.

3. The accumulation of hydrogen ions inside the compartment sets up concentration and electric gradients across the membrane. Hydrogen ions flow down the gradients, through channel proteins that span the membrane. The flow triggers the attachment of inorganic phosphate to ADP, thereby forming ATP.

LIGHT-INDEPENDENT REACTIONS

Cells use the ATP and NADPH produced by the light-dependent reactions to build sugars and other carbon-based compounds. The reactions that now occur are called "light-independent" because they do not depend directly on sunlight. (They can proceed as long as ATP and NADPH are available, but these molecules are normally produced during daylight, so the light-independent reactions do not proceed for long in the dark.) The reactions require the following substances:

1. ATP and NADPH from the light-dependent reactions

2. Carbon dioxide from the air around photosynthetic cells

3. Ribulose bisphosphate (RuBP), a five-carbon sugar

4. Enzymes that catalyze each reaction step

Carbon Dioxide Fixation and the Calvin-Benson Cycle

The entire sequence of light-independent reactions is called the **Calvin-Benson cycle** after its discoverers, Melvin Calvin and Andrew Benson. Every carbon atom of every molecule in photosynthetic cells comes from these reactions. In chloroplasts, the reactions take place on the thylakoid membrane surface facing the stroma.

Let's follow the steps whereby six carbon dioxide molecules enter the cycle and are used in forming one glucose molecule. This sequence is illustrated in Figure 8.7.

First, carbon dioxide is attached to RuBP to form a six-carbon intermediate. The intermediate is highly unstable and is broken apart at once into two molecules of phosphoglycerate, or *PGA*, which is a three-carbon compound. This initial sequence, in which carbon dioxide combines with an organic compound, is called **carbon dioxide fixation**.

Next, the PGA receives a phosphate group (donated by ATP) and hydrogen and electrons (donated by NADPH). The result is phosphoglyceraldehyde, or *PGAL*, which is a three-carbon compound.

For every six carbon dioxide molecules fixed at the start of the cycle, twelve PGALs are produced. Ten of the twelve now undergo complex rearrangements into new RuBP molecules—which can be used to fix more carbon dioxide. What happens to the remaining two PGAL molecules? They are combined and rearranged in ways that lead to the six-carbon glucose (Figure 8.7). *Thus the energy of sunlight, which excited electrons in the first stage of photosynthesis, is now stored as chemical energy in glucose.*

The Calvin-Benson cycle yields enough RuBP to replace the six used in carbon dioxide fixation. It also yields glucose. The ADP, phosphate, and NADP$^+$ leftovers are returned to the light-dependent reaction sites—where they are converted once more to NADPH and ATP.

Summary of the Light-Independent Reactions

Figure 8.8 relates the light-independent reactions to the overall events of photosynthesis. The key steps of these reactions can be summarized as follows:

1. Carbon dioxide is fixed to RuBP, making an unstable intermediate that is broken apart into two three-carbon PGA molecules.

2. PGA is phosphorylated (made more reactive) by ATP, and it receives electrons (and hydrogen) from NADPH. The result is PGAL.

3. Through complex reactions, PGAL is rearranged into new RuBP molecules and into glucose.

4. It takes six turns of the Calvin-Benson cycle to produce one molecule of glucose (only one of every six PGAL molecules produced in the cycle is funneled into glucose synthesis).

How Autotrophs Use Intermediates and Products of Photosynthesis

At any moment, you will find little free glucose in photosynthetic cells. Much is used at once as building blocks for compounds that make up the plant body or as fuel to provide energy for cellular work.

Typically, photosynthetic products become converted to easily transportable forms (notably sucrose) or storage forms (such as starch). In some algae, glucose becomes stored as starch within the chloroplast itself. In leaves, stems, and roots of complex plants, starch is the main carbohydrate storage product. For example, sucrose produced in leaves of potato plants is converted and stored as starch in underground stem regions called tubers (the "potatoes"). In sugar beets, onions, and sugarcane, sucrose itself is the main storage form.

Photosynthetic autotrophs use intermediates and products of photosynthesis when they assemble lipids and amino acids. Indeed, some green algae use more than ninety percent of the carbon fixed when they construct proteins and lipids. These plants have a brief life cycle, and they live in places where sunlight and water are plentiful. They put most of the photosynthetic products into growth and reproduction rather than diverting them to storage forms.

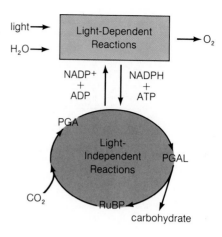

Figure 8.8 Summary of the main reactants, intermediates, and products of photosynthesis.

C4 Plants

Variations in temperature, sunlight intensity, and water availability often affect the rate of photosynthesis. For example, hot, dry conditions put many leafy plants under stress. Typically the leaves of these plants have a waxy surface covering (which retards moisture loss), and any water that does escape from the leaf does so primarily through tiny passages through the surface layers. On hot, dry days, most passages close and water is conserved. But when they are closed, not as much carbon dioxide can wander into the leaf, and oxygen (a by-product of photosynthesis) builds up inside. In other words, *hot, dry conditions promote a high O_2-to-CO_2 ratio inside the leaf.* The stage is set for a rather perverse process, called photorespiration.

In **photorespiration**, oxygen becomes affixed to the RuBP on which the Calvin-Benson cycle turns. It so happens that either dissolved oxygen *or* carbon dioxide can serve as a substrate of the enzyme that attaches carbon dioxide to RuBP. These substances compete with each other for the enzyme's active site, and when there is far more oxygen than carbon dioxide in the leaf, photorespiration wins out:

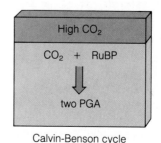

Calvin-Benson cycle predominates

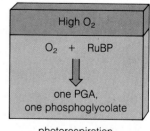

photorespiration predominates

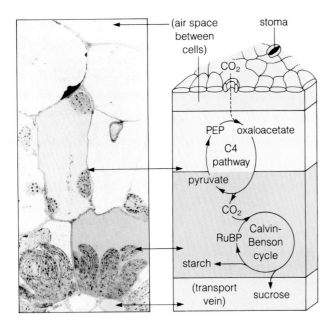

Figure 8.9 Organization of the carbon-fixing system in C4 plants, sometimes referred to as the Hatch-Slack pathway (in honor of M. Hatch and C. Slack, who deduced the entire mechanism). The C4 carbon-fixing system precedes the Calvin-Benson cycle in many plants.

In the photomicrograph, a mesophyll cell is shaded yellow; a bundle-sheath cell is shaded green.

Glucose formation depends on PGA. Photorespiration produces only one PGA (and one phosphoglycolate, which is rapidly broken down into carbon dioxide and water). Thus, when temperatures climb, the rate of photosynthesis drops.

Not all plants are hindered by high oxygen levels. Corn, sugarcane, and crabgrass are among the so-called *C4 plants*, which have a carbon-fixing system that *precedes* the Calvin-Benson cycle. Leaves of C4 plants incorporate two kinds of photosynthetic cells:

mesophyll cells	*preliminary fixation and rapid transport of carbon dioxide to bundle-sheath cells*
bundle-sheath cells	*carbon dioxide fixation through the Calvin-Benson cycle*

As Figure 8.9 shows, carbon dioxide is attached to phosphoenol pyruvate (PEP) in the mesophyll cells, thereby forming oxaloacetate. This first intermediate of carbon dioxide fixation has four carbon atoms, hence the name "C4 plant." (In "C3 plants," three-carbon PGA is the first intermediate.)

The oxaloacetate is transferred to an adjacent bundle-sheath cell, where the carbon dioxide is released and picked up by the Calvin-Benson cycle. Thus carbon dioxide is fixed and then released deeper in the leaf, where it is fixed again. With this system, C4 plants increase the carbon dioxide concentration and thereby produce a more favorable CO_2-to-O_2 ratio in the cells where the Calvin-Benson cycle operates—even under hot, dry conditions.

CHEMOSYNTHESIS

Photosynthesis so dominates the energy-trapping pathways that it is sometimes easy to overlook other, less common routes. The chemosynthetic autotrophs obtain energy from reactions of inorganic substances such as ammonium ions and iron or sulfur compounds, not from sunlight. They use this energy to synthesize organic compounds.

For example, some bacteria present in soil use ammonia (NH_3) molecules as their energy source, stripping them of hydrogen ions and electrons. The electrons and hydrogen ions are transferred to oxygen by way of an electron transport system. Nitrite ions (NO_2^-) and nitrate ions (NO_3^-) are the remnants of these energy-securing routes. Compared with ammonium ions, nitrite and nitrate ions are readily washed out of the soil, so the action of these so-called nitrifying bacteria can lower the soil fertility. Chemicals that inhibit these bacteria are sometimes added to the soil.

We will return later to the environmental effects of chemosynthetic autotrophs. Here, in this unit, we will turn next to the pathways by which the carbon-containing products of autotrophs can be used as energy sources for cellular work.

SUMMARY

1. Carbohydrates and other organic compounds are used in cell architecture and as energy stores that can be tapped to drive cell activities.

2. Autotrophs ("self-feeders") use carbon dioxide from the environment as the carbon source for building carbohydrates. The photosynthetic autotrophs (plants, for example) use sunlight energy for the synthesis reactions. The chemosynthetic autotrophs (some bacteria) use energy from chemical reactions involving sulfer and other inorganic substances.

3. Heterotrophs (animals, fungi, many protistans, and most bacteria) obtain carbon and energy from organic

compounds already synthesized by autotrophs. They do this by feeding on autotrophs, each other, and organic wastes.

4. In land plants, photosynthesis takes place at thylakoid membranes that are organized as stacked disks inside chloroplasts. A semifluid matrix (the stroma) surrounds the stacked disks, which interconnect to form a compartment separate from the stroma.

5. The photosynthetic membranes contain chlorophyll and other light-trapping pigment molecules that are organized into photosystems.

6. Photosynthesis, the main process by which carbon and energy enter the web of life, consists of two sets of reactions (the light-dependent and the light-independent reactions).

7. The light-dependent reactions can be cyclic or noncyclic. In cyclic photophosphorylation, excited electrons from photosystem I flow through a transport system embedded in the thylakoid membrane and then back to the photosystem. ATP is produced.

8. In noncyclic photophosphorylation, there is a one-way flow of excited electrons from photosystem II, through a transport system, to photosystem I, and through a second transport system. The electrons end up in NADPH. Electrons lost from photosystem II are replaced through photolysis (the splitting of water molecules into oxygen, hydrogen ions, and associated electrons). This pathway yields ATP as well as NADPH. Electrons and hydrogen atoms can be stripped from NADPH and used directly in carbohydrate assembly.

9. According to the chemiosmotic theory, ATP is produced during the light-dependent reactions in the following way. Hydrogen ions accumulate inside the thylakoid compartment (as a result of photolysis and electron transfers through transport systems). Thus H^+ concentration and electric gradients are produced between the compartment and the stroma. Hydrogen ions flow down the gradient, through channel proteins that span the thylakoid membrane. Energy associated with the flow drives the coupling of inorganic phosphate and ADP, thereby forming ATP.

10. The light-independent reactions by which carbohydrates are formed take place in the stroma. The reactions are called the Calvin-Benson cycle, and they begin with carbon dioxide fixation.

11. In carbon-dioxide fixation, RuBP combines with carbon dioxide from the air, and then splits into two PGA molecules. The PGA is phosphorylated by ATP and receives electrons and hydrogen atoms from NADPH to form PGAL. Five of every six PGAL molecules are used to regenerate RuBP for the cycle; one can be used in the formation of a glucose molecule.

Review Questions

1. Define the difference between autotrophs and heterotrophs, and give examples of each. In what category do photosynthesizers fall?

2. Summarize the photosynthesis reactions in words, then as an expanded equation. Distinguish between the light-dependent and the light-independent stage of these reactions. Be sure to show the electron and hydrogen "bridges" between these two stages.

3. Oxygen is a product of photolysis. What is photolysis, and does it occur during the first or second stage of photosynthesis? Is water a by-product of the first or second stage?

4. Describe where the light-dependent reactions occur in the chloroplast, and name the molecules formed there. Do the same for the light-independent reactions.

5. A thylakoid compartment is a reservoir for which of the following substances: glucose, photosynthetic pigments, hydrogen ions, fatty acids?

6. Sketch the reaction steps of noncyclic photophosphorylation, showing where the excited electrons eventually end up. Do the same for the cyclic pathway. Which pathway has the greater energy yield?

7. Is oxygen an end product of cyclic or noncyclic photophosphorylation?

8. Describe the chemiosmotic theory of ATP formation in chloroplasts. Use sketches to show the proposed movements of hydrogen ions across thylakoid membranes.

9. Which of the following substances are not required for the light-independent reactions: ATP, NADH, RuBP, carotenoids, free oxygen, carbon dioxide, enzymes?

10. Suppose a plant carrying out photosynthesis were exposed to carbon dioxide molecules that contain radioactively labeled carbon atoms ($^{14}CO_2$). In which of the following compounds will the labeled carbon first appear?

    a. NADPH   c. PGAL
    b. pyruvate   d. PGA

11. How many turns of the Calvin-Benson cycle are necessary to produce one glucose molecule? Why?

12. Give examples of how different autotrophs use the intermediates and products of photosynthesis.

Readings

Clayton, R. 1981. *Photosynthesis: Physical Mechanisms and Chemical Patterns.* New York: Cambridge University Press.

Hinckle, P., and R. McCarty. March 1978. "How Cells Make ATP." *Scientific American* 238(23):104–123. How the chemiosmotic theory explains ATP formation in both chloroplasts and mitochondria.

Miller, K. 1982. "Three-Dimensional Structure of a Photosynthetic Membrane." *Nature* 300:5887.

Moore, P. 1981. "The Varied Ways Plants Tap the Sun." *New Scientist.* 12 February, pp. 394–397. Clear, simple introduction to the C4 plants.

Zubay, G. 1983. *Biochemistry.* Menlo Park, California: Addison-Wesley. Advanced but authoritative introduction to photosynthesis.

9

ENERGY-RELEASING PATHWAYS

It is one of the quirks of the human mind that plants just aren't thought about very often as *living* organisms. Even vegetarians who become nauseated at the thought of eating the flesh of an animal can relish the flesh of a peach. Perhaps it is understandable. Lacking autotrophic equipment of our own, we have to depend on something to produce energy for us; and if we carried a concern for the sanctity of life too far, we'd all starve to death.

Yet there is an undeniable unity among organisms that becomes evident at the biochemical level. Animals, plants, fungi, protistans, and most bacteria use carbohydrates and other molecules for energy and raw materials in much the same way you do. They break apart those molecules to help produce ATP—the prime energy carrier for life in all of its forms. There is, in short, a remarkable similarity in the ebb and flow of energy through the individual—hence through the biosphere. We will return to this idea at the chapter's end.

OVERVIEW OF THE MAIN ENERGY-RELEASING PATHWAYS

When molecules are broken apart, heat and chemical energy are released. Cells cannot use the heat to drive their activities. And they cannot *directly* use the chemical energy; it must become stored first in the phosphate bonds of ATP, which gives up energy easily and directly to a great variety of metabolic reactions (page 105). In the so-called **ATP/ADP cycle**, an input of chemical energy drives the linkage of ADP and a phosphate group (or inorganic phosphate) into ATP; then the ATP donates the phosphate group elsewhere and becomes ADP:

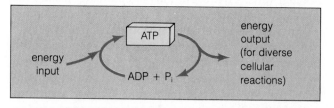

As we saw in the preceding chapter, the energy input necessary for ATP formation can come from sunlight (in photosynthesis) or from small inorganic compounds (in chemosynthesis). It also can come from the breakdown of carbohydrates, lipids, and proteins.

Of the foods we eat, carbohydrates—glucose molecules especially—are the main source of energy. Our cells use glucose first for as long as it is available, then lipids, and then proteins as a last resort. This chapter will focus on **carbohydrate metabolism**, in which chemical energy is released from various carbohydrates as a result of oxidation-reduction and phosphorylation reactions.

Phosphorylation: You have probably heard the expression, "It takes money to make money." It means investing some money you already have into an activity that will produce more money in return. Cells do essentially the same thing in carbohydrate metabolism. When cells break glucose apart, they make an initial investment of two ATP molecules. Glucose simply will not give up stored energy unless it is phosphorylated twice (each ATP transfers a phosphate group to it). On their original investment, cells show a *net* return of two or thirty-six (and sometimes more) ATP molecules for each glucose molecule degraded.

Oxidation-Reduction: Now, there is quite a difference between two and thirty-six ATP molecules as the net energy harvest from a single molecule of glucose. *The outcome depends on the use made of hydrogen atoms and electrons being stripped from glucose.*

The initial acceptor of those hydrogen atoms and electrons might be either NAD^+ or FAD which, recall, are coenzymes that take part in electron transfers (oxidation-reduction reactions). When these coenzymes are carrying electrons from one reaction site to another, they are called NADH and $FADH_2$.

If you liken electron energy levels to steps of a staircase, then NADH is at the top step in glucose metabolism, and $FADH_2$ is one step down. Each can donate electrons to an acceptor molecule at a lower step. In turn, this molecule can donate the electrons to an acceptor at a still lower step. Obviously, there can be more electron transfers if the initial donor is very high on the staircase and the final acceptor very low! This is important to think about, *for the energy released at several steps of the energy staircase can be used to form ATP.* When the initial electron donor is not at the top step or when the final acceptor is not at the lowest, less ATP is formed.

The main pathways of carbohydrate metabolism differ in the final acceptor of electrons stripped from glucose. Figure 9.2 is an overview of these pathways, which are called aerobic respiration, anaerobic electron transport, alcoholic fermentation, and lactate fermentation.

a b c

Figure 9.1 Three pathways of carbohydrate metabolism, in which electrons stripped from sugar molecules are transferred about in ways that produce ATP. Each pathway is characterized by the final destination of the "spent" electrons.

(a) Bacterial residents of sulfur hot springs use *anaerobic electron transport*, in which an inorganic compound in the environment (in this case, sulfate) is the final electron acceptor. (b) The yeasts making up the dustlike coating on these grapes use *fermentation*, in which the partially dismantled sugar molecule itself takes back the electrons. (c) Gorillas and most other organisms (including the plant upon which the gorilla is dining) use *aerobic respiration*, in which oxygen is the final electron acceptor.

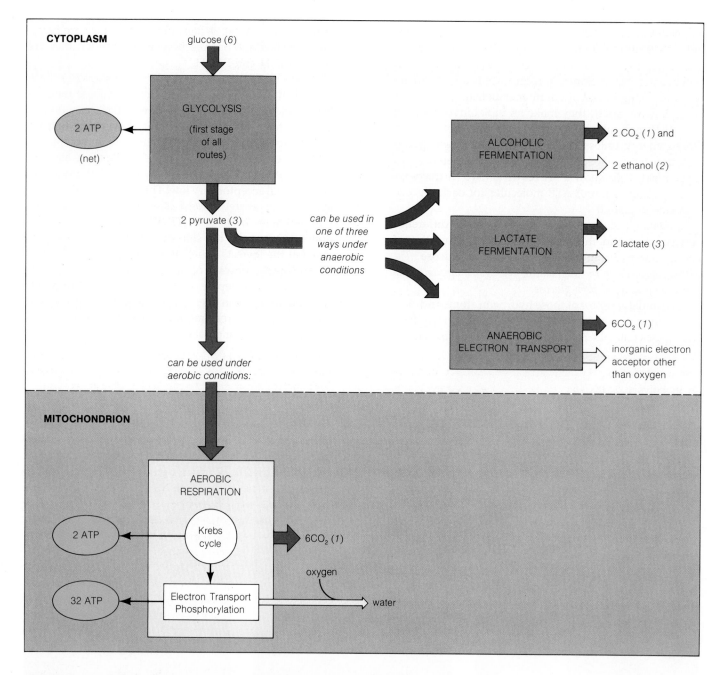

Figure 9.2 Overview of the main pathways of carbohydrate metabolism, with glucose as the starting material. Depending on the organism and on environmental conditions, any one of the four routes shown might be taken.

Yellow arrows indicate the destinations of hydrogen atoms and electrons derived from glucose. Green arrows indicate where its carbon atoms end up. (Numbers in parentheses refer to the number of carbon atoms in each molecule of the compound listed.)

A simple way to remember the *net* energy yield of each route shown is to add up the ATP from start— *in all cases, glycolysis*—to finish. Neither anaerobic electron transport nor the fermentation routes yield more than two net ATP. Aerobic respiration commonly has a net yield of thirty-six ATP for each glucose molecule.

Glycolysis: First Stage of Carbohydrate Metabolism

Glycolysis, the initial stage of all the main pathways of carbohydrate metabolism, takes place in the cytoplasm. The term "glycolysis" refers to the partial breakdown of sugars such as glucose, although other organic compounds can also enter the reactions.

Glucose, recall, has a backbone of six carbon atoms. At the start of glycolysis, two phosphate groups donated by two ATP molecules become attached to the backbone. Then glucose undergoes rearrangements and is split into two molecules of *pyruvate*, each with a three-carbon backbone. One of the intermediate forms produced during the reactions gives up hydrogen and electrons to NAD$^+$ to form NADH.

Energy released in glycolysis is used to produce four ATP. But remember, two ATP were invested to make glucose more reactive, so the *net* yield is two ATP.

At this point, glucose has only been partially degraded (into pyruvate). The pyruvate can travel different routes from here, depending on the type of cell and on environmental conditions. The NADH produced in glycolysis must give up its cargo of hydrogen and electrons along whichever route is taken. If it did not, NAD$^+$ would not be regenerated and glycolysis would grind to a halt.

In glycolysis (the initial stage of all the main pathways of carbohydrate metabolism), glucose is partially broken down into pyruvate, and the energy released leads to a net yield of two ATP.

The pyruvate enters the next stage of carbohydrate metabolism. How it will be put to use depends on which pathway is taken.

Hydrogen atoms and electrons stripped from glucose during glycolysis are transferred to NAD$^+$. The resulting molecule (NADH) gives up the hydrogen and electrons in the subsequent reactions, and NAD$^+$ is thereby regenerated.

Aerobic Respiration

If a cell (or organism) is an *aerobe*, it uses oxygen as the final electron acceptor in carbohydrate metabolism. **Aerobic respiration** is an oxygen-dependent pathway which, in eukaryotes, takes place in mitochondria (page 76). The pyruvate from glycolysis is completely dismantled in this pathway. Its carbon and oxygen atoms end up in carbon dioxide and water molecules. Its hydrogen atoms and electrons are transferred to NAD$^+$ and FAD, which in turn transfer them to an electron transport system. The transport system finally gives up the electrons to oxygen atoms (which combine with hydrogen to form water).

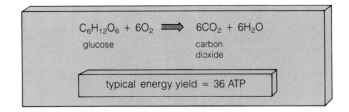

$$C_6H_{12}O_6 + 6O_2 \implies 6CO_2 + 6H_2O$$

glucose   carbon dioxide

typical energy yield = 36 ATP

Figure 9.3 Typical net energy yield of aerobic respiration.

Operation of the transport system leads to the formation of many ATP molecules. From start to finish, the aerobic pathway typically has a net yield of thirty-six ATP (Figure 9.3).

In aerobic respiration, the pyruvate from glycolysis is completely dismantled. Oxygen is the final electron acceptor for the reactions.

From glycolysis (in the cytoplasm) to the final reactions (in the mitochondrion), this pathway commonly yields thirty-six ATP molecules for every glucose molecule degraded.

Not all cells can use oxygen all of the time or even some of the time. Your muscle cells use the aerobic pathway when plenty of oxygen is available, but they can use other final electron acceptors when it is not (as during strenuous exercise). Denitrifying bacteria that live in soil normally use oxygen, but when the soil is saturated with standing water, they can switch to nitrite (NO$_2^-$) as the final acceptor. These and other switch-hitting cells are called "facultative" aerobes.

In contrast, *anaerobes* cannot ever use oxygen as the final electron acceptor. Some are indifferent to the presence or absence of free oxygen, but others are "strict" anaerobes that die on exposure to it. These cells use either anaerobic electron transport or the fermentation pathways.

Anaerobic Electron Transport

In **anaerobic electron transport**, an inorganic compound is the final acceptor of electrons originally stripped from glucose. For example, sulfate-reducing bacteria are strict anaerobes that live in soils, aquatic habitats, and the intestines. They transfer the spent electrons to sulfate (SO$_4^{--}$) in the environment, thereby forming sulfuric acid (H$_2$SO$_4$).

In this pathway, the NADH from glycolysis transfers electrons to a small transport system bound in the plasma membrane, and the NAD$^+$ is thereby regenerated. The reactions do not add any more ATP to the small net yield from glycolysis.

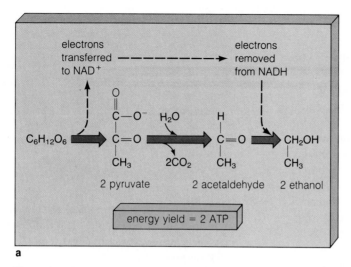
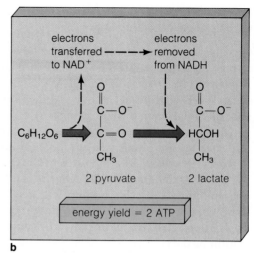

Figure 9.4 Net energy yield of (**a**) alcoholic fermentation and (**b**) lactate fermentation.

Fermentation Pathways

In other anaerobic routes, called **fermentation** pathways, the final electron acceptor is actually an intermediate or product of the glucose molecule being degraded. An "outside" acceptor (such as oxygen or sulfate) is not needed at all.

In *alcoholic fermentation*, pyruvate from glycolysis is broken down to acetaldehyde, which accepts electrons from NADH and thereby becomes ethanol (Figure 9.4a). Yeasts (single-celled fungi) use this pathway. Carbon dioxide, the gaseous product of the reactions, helps yeast dough to rise. Fermentation activities of some yeasts lead to the alcohol portion of beers, distilled spirits, and wines.

In *lactate fermentation*, the pyruvate is converted into lactate, a three-carbon compound (Figure 9.4b). These reactions can occur in some animal cells that normally rely on aerobic respiration. For example, when skeletal muscle cells are contracting vigorously and exceeding the body's capacity to deliver oxygen to them, they can switch to lactate fermentation. One group of bacteria produces lactate exclusively as the fermentation product; milk or cream turned sour is a sign of their activity.

The point to remember about either anaerobic electron transport or fermentation is that the glucose is not completely degraded, so considerable energy still remains in the products. No more ATP molecules are produced, beyond the two formed during glycolysis. Both pathways serve only to regenerate NAD⁺.

Anaerobic electron transport or fermentation has a net yield of two ATP molecules (from glycolysis). The remaining reactions serve only to regenerate NAD⁺.

A CLOSER LOOK AT AEROBIC RESPIRATION

Of the main pathways just outlined, aerobic respiration is the most common. Glycolysis is only the first of three stages in this pathway. The second stage is the **Krebs cycle** and a few reactions immediately preceding it. Here, the two pyruvate molecules from glycolysis are dismantled completely, and two ATP and many electron carriers (NADH especially) are produced. Hydrogen and electrons are stripped from these carriers in the third stage, **electron transport phosphorylation**, and considerable ATP is produced.

Keep in mind that these steps do not proceed all by themselves. *Each step is catalyzed by a specific enzyme, and it produces an intermediate molecule that serves as a substrate for the next enzyme in the series.*

First Stage: Glycolysis

The initial steps of glycolysis are *energy-requiring*; they do not proceed without an energy input from ATP. First glucose receives a phosphate group from ATP, undergoes internal rearrangements, then receives another phosphate group from another ATP (Figure 9.5). The resulting intermediate is split into two different molecules. Because each can be converted easily to the other, we can call them both by the same name: phosphoglyceraldehyde, or PGAL. PGAL formation marks the beginning of the *energy-releasing* steps of glycolysis.

Each PGAL now becomes the substrate of an enzyme that makes it give up hydrogen and electrons to NAD⁺, which leaves the reaction site with its cargo. The inter-

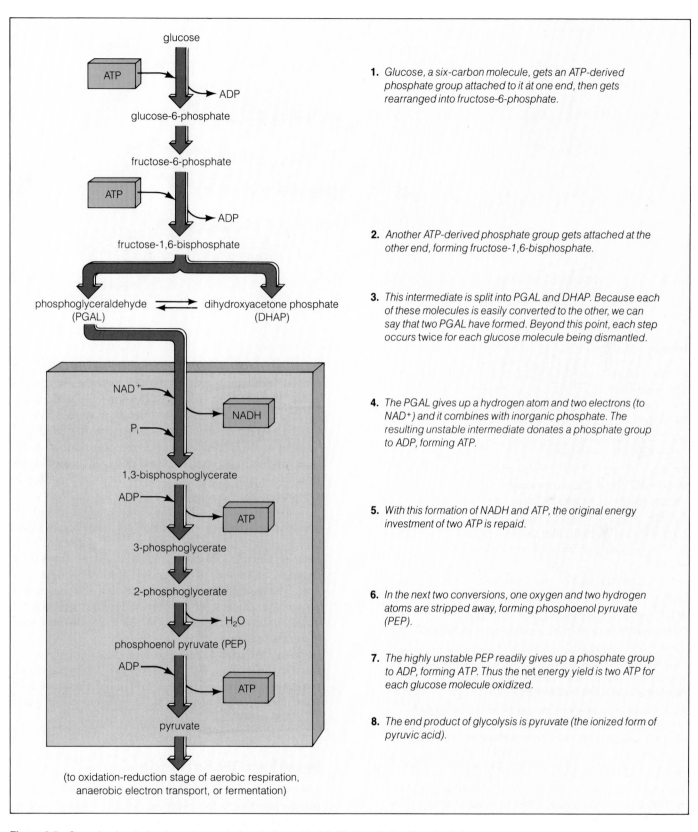

1. *Glucose, a six-carbon molecule, gets an ATP-derived phosphate group attached to it at one end, then gets rearranged into fructose-6-phosphate.*

2. *Another ATP-derived phosphate group gets attached at the other end, forming fructose-1,6-bisphosphate.*

3. *This intermediate is split into PGAL and DHAP. Because each of these molecules is easily converted to the other, we can say that two PGAL have formed. Beyond this point, each step occurs* twice *for each glucose molecule being dismantled.*

4. *The PGAL gives up a hydrogen atom and two electrons (to NAD+) and it combines with inorganic phosphate. The resulting unstable intermediate donates a phosphate group to ADP, forming ATP.*

5. *With this formation of NADH and ATP, the original energy investment of two ATP is repaid.*

6. *In the next two conversions, one oxygen and two hydrogen atoms are stripped away, forming phosphoenol pyruvate (PEP).*

7. *The highly unstable PEP readily gives up a phosphate group to ADP, forming ATP. Thus the* net *energy yield is two ATP for each glucose molecule oxidized.*

8. *The end product of glycolysis is pyruvate (the ionized form of pyruvic acid).*

Figure 9.5 Steps in glycolysis when glucose is the starting material. All steps in the blue-shaded area occur *twice* for each glucose molecule being degraded.

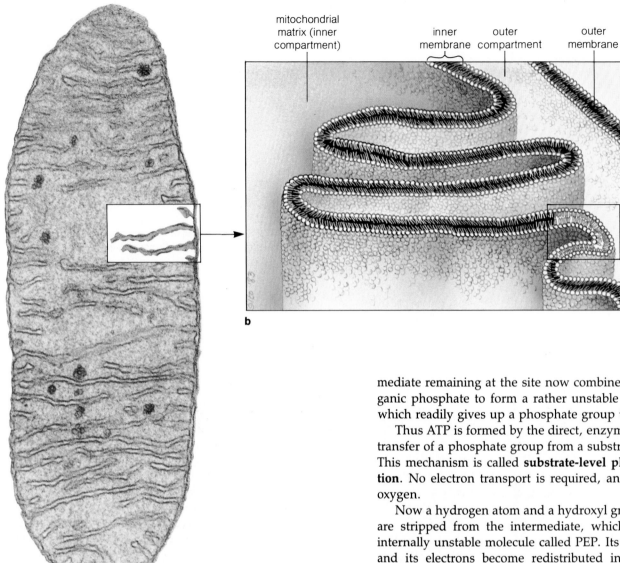

inner membrane / **outer compartment** / **outer membrane** / (cytoplasm)

mitochondrial matrix (inner compartment)

a

b

Figure 9.6 Functional zones of the mitochondrion.
(**a**) Mitochondrion from a bat pancreatic cell, thin section.

(**b**) The inner mitochondrial membrane divides the inside of this organelle into two compartments. No matter how complicated the inner foldings are, the outer compartment extends into all regions between the outer and inner membranes.

(**c**) The inner compartment is the site of the second stage of aerobic respiration, including Krebs cycle activities. Electron carriers (NADH and FADH$_2$) formed during the first two stages are delivered to transport systems embedded in the inner membrane. Operation of these systems during the third stage sets up concentration and electric gradients across the membrane. The gradients are coupled to ATP formation at ATP synthetases (enzyme systems embedded elsewhere in the same inner membrane).

mediate remaining at the site now combines with inorganic phosphate to form a rather unstable molecule—which readily gives up a phosphate group to ADP.

Thus ATP is formed by the direct, enzyme-mediated transfer of a phosphate group from a substrate to ADP. This mechanism is called **substrate-level phosphorylation**. No electron transport is required, and neither is oxygen.

Now a hydrogen atom and a hydroxyl group (−OH) are stripped from the intermediate, which leaves an internally unstable molecule called PEP. Its bonds shift and its electrons become redistributed in ways that increase the amount of energy available for the next reaction. That energy is enough to drive the transfer of a phosphate group from PEP to ADP. After this second substrate-level phosphorylation, the molecule remaining is pyruvate (Figure 9.5).

In all, the partial breakdown of glucose in glycolysis leads to the formation of four ATP, two NADH, and two pyruvate molecules. (But keep in mind that two ATP were needed to start the reactions, so there is a *net* yield of only two ATP.)

Second Stage: The Krebs Cycle

The pyruvate from glycolysis is shuttled from the cytoplasm into the mitochondrion, where an inner membrane forms two compartments (Figure 9.6). The second stage of the aerobic pathway proceeds in this innermost compartment.

Each pyruvate molecule is dismantled completely in the second stage. Its carbon atoms leave the cell in the

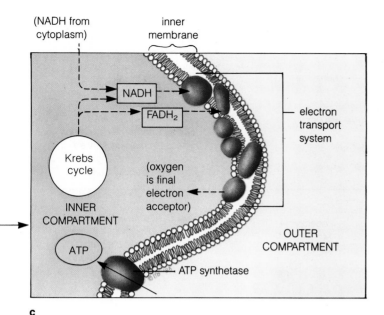

(NADH from cytoplasm)

inner membrane

NADH

FADH$_2$

electron transport system

Krebs cycle

(oxygen is final electron acceptor)

INNER COMPARTMENT

ATP

ATP synthetase

OUTER COMPARTMENT

c

form of carbon dioxide wastes, and its hydrogen atoms and electrons end up being transferred to NAD$^+$ and FAD. Also, enough energy is released at one step to drive a substrate-level phosphorylation that produces ATP.

For all of this to happen, pyruvate must initially be converted to a form that can be attached to a molecule called oxaloacetate. Because a cell has only so many of these molecules, there also must be a way to regenerate the oxaloacetate so the reactions can proceed over and over again. Figure 9.7 outlines the steps by which pyruvate is broken down and oxaloacetate is regenerated. As you can see, they begin with a few preparatory conversions and proceed through the cyclic reaction sequence called the Krebs cycle. (The cycle was named after Hans Krebs, who in the 1930s began working out its details.)

It takes two preparatory reaction sequences and two turns of the Krebs cycle to dismantle the two pyruvate molecules, each with its tiny carbon backbone. Pyruvate breakdown adds only two more ATP molecules to the small net yield from glycolysis. *But it also adds many electron carriers to the number that can be used in the third stage of the aerobic pathway:*

Glycolysis:		2 NADH
Pyruvate conversion preceding Krebs cycle:		2 NADH
Krebs cycle:	2 FADH$_2$	6 NADH
Total electron carriers sent to third stage of aerobic pathway:	2 FADH$_2$ +	10 NADH

Third Stage: Electron Transport Phosphorylation

Electron transport phosphorylation is the most impressive of all ATP-yielding mechanisms. Consider that an adult male uses up about 2,800 kilocalories of energy during a normal day's activities, and that about 150 kilograms of ATP are needed daily to provide this much energy. However, there is only about 0.005 kilogram of ATP in the entire body at any one time! The body makes up the difference by repeatedly phosphorylating ADP, many thousands of times a day. This is an ATP/ADP cycle on a grand scale, and the only mechanism for handling such an enormous turnover is electron transport phosphorylation.

The third stage of the aerobic pathway is carried out by enzymes and other proteins embedded in the inner membrane of the mitochondrion (Figure 9.6). These proteins are organized into two types of reaction sites:

Reaction Site	Function
electron transport system	*establishes concentration and electric gradients across the membrane*
ATP synthetase system	*phosphorylates ADP to form ATP*

The reactions begin in the inner compartment, when NADH and FADH$_2$ give up hydrogen and electrons to an electron transport system (Figure 9.8). At some transfers through this system, the electrons are accepted and passed on, but the hydrogen (in the form of H$^+$) is left behind—in the outer compartment:

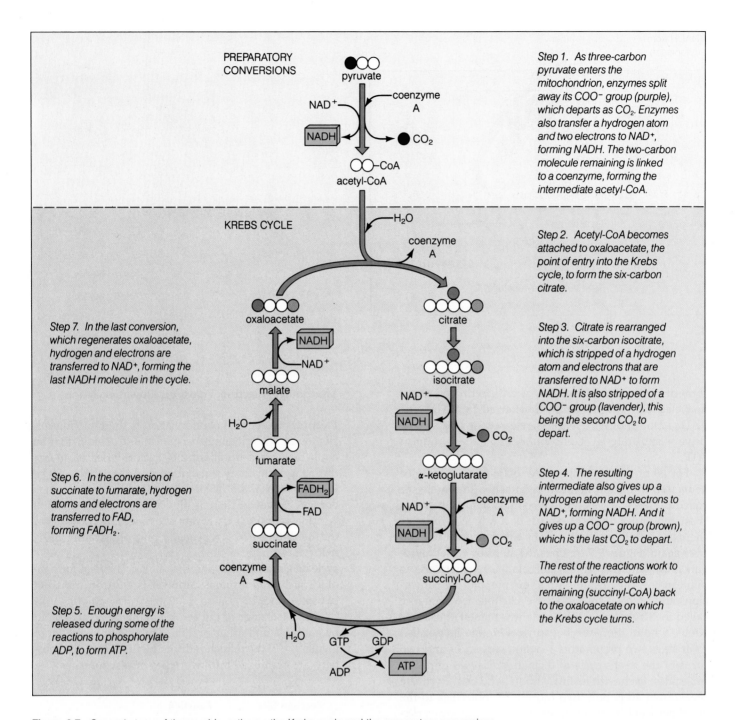

PREPARATORY CONVERSIONS

Step 1. As three-carbon pyruvate enters the mitochondrion, enzymes split away its COO^- group (purple), which departs as CO_2. Enzymes also transfer a hydrogen atom and two electrons to NAD^+, forming NADH. The two-carbon molecule remaining is linked to a coenzyme, forming the intermediate acetyl-CoA.

KREBS CYCLE

Step 2. Acetyl-CoA becomes attached to oxaloacetate, the point of entry into the Krebs cycle, to form the six-carbon citrate.

Step 7. In the last conversion, which regenerates oxaloacetate, hydrogen and electrons are transferred to NAD^+, forming the last NADH molecule in the cycle.

Step 3. Citrate is rearranged into the six-carbon isocitrate, which is stripped of a hydrogen atom and electrons that are transferred to NAD^+ to form NADH. It is also stripped of a COO^- group (lavender), this being the second CO_2 to depart.

Step 6. In the conversion of succinate to fumarate, hydrogen atoms and electrons are transferred to FAD, forming $FADH_2$.

Step 4. The resulting intermediate also gives up a hydrogen atom and electrons to NAD^+, forming NADH. And it gives up a COO^- group (brown), which is the last CO_2 to depart.

The rest of the reactions work to convert the intermediate remaining (succinyl-CoA) back to the oxaloacetate on which the Krebs cycle turns.

Step 5. Enough energy is released during some of the reactions to phosphorylate ADP, to form ATP.

Figure 9.7 Second stage of the aerobic pathway: the Krebs cycle and the preparatory conversions immediately preceding it. This diagram shows what happens to one pyruvate molecule entering the reactions. Because two pyruvate molecules are formed during the breakdown of every glucose molecule in glycolysis, the steps shown occur twice (one for each pyruvate molecule). The circles represent carbon atoms.

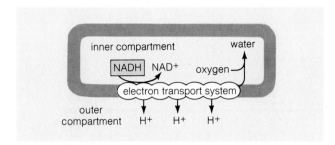

The positively charged hydrogen ions accumulate in the outer compartment, and this sets up concentration and electric gradients across the membrane. The ions follow the gradients and flow inward, through the ATP synthetase system that serves as a channel across the membrane. The energy associated with this flow drives ATP formation:

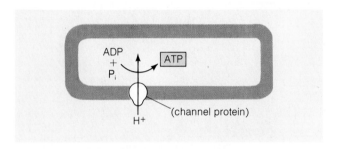

Do these events sound familiar? They should: ATP is formed in much the same way in chloroplasts of photosynthetic cells (page 114). Again, the idea that concentration and electric gradients across a membrane drive the phosphorylation of ADP is called the chemiosmotic theory of ATP formation.

Electron transport phosphorylation in mitochondria involves two events: (1) the transfer of electrons and hydrogen through a membrane-bound transport system, and (2) actual ATP formation.

Figure 9.8 Some components of the system used in electron transport phosphorylation. The system runs on electron energy delivered to it by NADH (from glycolysis and the Krebs cycle) and $FADH_2$ (from the Krebs cycle).

Energy associated with three transfers through this electron transport system drives the formation of three ATP molecules. The NADH can deliver enough energy to form all three. The $FADH_2$ delivers enough energy to form two ATP only.

Glucose Energy Yield

The net energy yield from the three stages of aerobic respiration is commonly thirty-six or thirty-eight ATP molecules for every glucose molecule degraded (refer to Figure 9.9). The actual number varies because each NADH produced in the cytoplasm (by glycolysis) can yield two *or* three ATP, for the following reasons.

In all cells, NADH formed in the mitochondrion (during the second-stage reactions) delivers enough energy to form three ATP molecules. $FADH_2$ does not: it donates electrons at a lower step in the electron transport sequence, hence it delivers only enough energy to form two ATP (Figure 9.8).

It so happens that cytoplasmic NADH only delivers electrons *to* the mitochondrion, not into them. In liver, heart, and kidney cells, a shuttle built into the outer membrane accepts the electrons, then gives them up to an NAD^+ molecule inside. Because the resulting NADH molecule delivers electrons to the top step of the transport system, its energy can produce three ATP in these cells. But in skeletal muscle and brain cells, a different

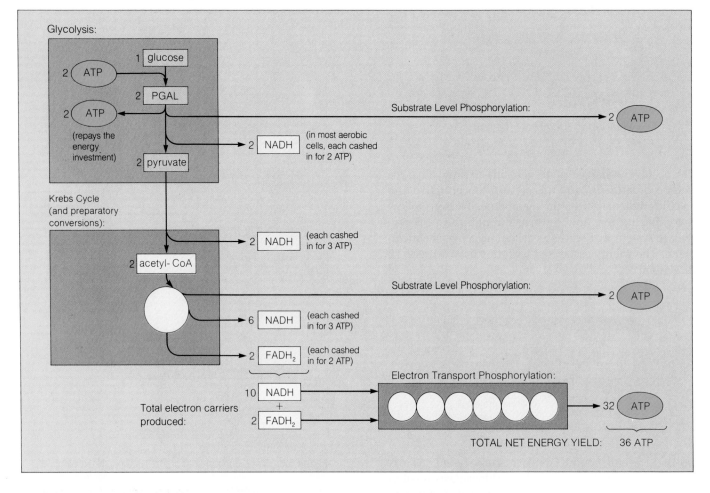

Figure 9.9 Summary of the energy harvest from one glucose molecule sent through the aerobic respiration pathway. Actual ATP yields vary, depending on cellular conditions and on the mechanism used to transfer energy from cytoplasmic NADH into the mitochondrion.

One NADH produced within the mitochondrion can yield three ATP. One cytoplasmic NADH (from glycolysis) may yield two or three ATP, depending on what kind of shuttle mechanism transfers its energy cargo to a membrane-bound transport system in the mitochondrion. In heart, kidney, and liver cells, the cargo is shuttled right to the "top" of the transport chain. Thus three ATP molecules are generated—which means an overall energy harvest of thirty-eight ATP. In other cells, the energy cargo is delivered "lower" in the transport chain, so there is only enough energy to generate two ATP molecules. In this case, the overall energy harvest is thirty-six ATP.

shuttle donates electrons to FAD inside the mitochondrion. Because the $FADH_2$ so formed delivers electrons one step down in the transport system, only two ATP can form.

When glucose is broken down completely to carbon dioxide and water, about 686 kilocalories of energy are released. Of that, about 7.5 kilocalories become conserved for further use in the phosphate bonds of each of the thirty-six ATP molecules typically formed. Thus the energy-releasing efficiency of this pathway is (36)(7.5)/(686), or thirty-nine percent. When the net yield is only two ATP (as in fermentation), then the energy-releasing efficiency of the pathway is only (2)(7.5)/(686), or two percent.

Control of Carbohydrate Metabolism

Glycolysis and the Krebs cycle serve two functions in metabolism. As we have seen, they can dismantle food molecules for energy that can drive ATP formation. But the intermediates formed during the reactions also can be diverted to the synthesis of all lipids, carbohydrates, proteins, and nucleic acids in the cell (Figure 9.10).

What determines whether a given molecule will be broken apart or pulled out of energy-releasing pathways in intermediate forms that can be used as building blocks? The main controls are over enzymes that govern reaction steps in which so much energy is released that the reaction, once carried out, cannot be reversed.

In glycolysis, for example, phosphofructokinase is such an enzyme. It adds a second phosphate group onto fructose-6-phosphate before this molecule is split (Figure 9.5). Once the enzyme catalyzes the reaction, the molecule has become so energized that it is "committed" to entering the glycolytic pathway. When cells are not using much energy, they do not deplete their ATP stores. In high concentration, ATP actually inhibits phosphofructokinase activity, and it thereby slows down glycolysis. When cells use more energy, they tap into their ATP stores and ATP concentrations decrease accordingly. Then the enzyme activity increases—and glucose and other molecules can be broken apart rapidly (to yield more ATP).

By integrating the activity of many different enzymes, cells constantly adjust to the supplies of resources that are currently available and to the demands they are making on those supplies.

Enzymes catalyzing key reactions in glycolysis and the Krebs cycle govern whether molecules are degraded as an ATP-generating energy source or converted to intermediate forms for use in biosynthesis.

PERSPECTIVE

In this unit, you have read about connections that exist in the pathways by which energy is trapped, stored, then released in controlled ways. Trapping environmental energy through photosynthesis, releasing energy from carbon-containing compounds through aerobic respiration—these are central events in the living world.

Later chapters will turn to the greater picture of the origin and evolution of life. However, the following overview of how these pathways came to be linked may give you a sense of how important they are for the continuation of life on this planet.

The first living cells probably obtained energy and materials from environmental sources, such as simple inorganic molecules that contained carbon, hydrogen, oxygen, and nitrogen. Most likely, they produced ATP by pathways similar to glycolysis; and fermentation routes must have predominated because free oxygen was not present in the atmosphere at that time.

Between about $3\frac{1}{2}$ billion and 600 million years ago, these metabolic pathways apparently underwent modification. Through mutations that accumulated in some evolutionary lines, the enzyme systems used to *degrade* carbon compounds were modified and extended, so that the reactions could also run in reverse. In other words, some enzyme systems could now *build* carbon compounds from carbon dioxide and water. Cells could not

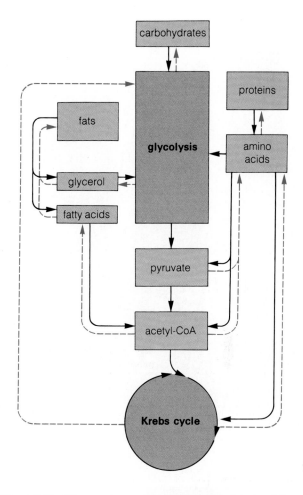

Figure 9.10 Where various substances flow into the glycolytic pathway and the Krebs cycle (black arrows). Different intermediates of glycolysis and the Krebs cycle can be pulled out of the energy-releasing pathways and used in biosynthesis (blue arrows).

have done this without an energy input to drive the reactions in this uphill direction. That energy came from the sun; photosynthesis had emerged.

It was a milestone in the course of evolutionary time. Oxygen, a by-product of photosynthesis, began to accumulate; eventually it would change the character of the earth's atmosphere. In the beginning, at least some photosynthetic cells were opportunistic about the increasing oxygen concentrations. Oxygen, after all, is a ready acceptor for electrons stripped away when glucose is broken apart for its energy. In some cells, oxygen was indeed put to use in an aerobic pathway of carbohydrate metabolism. Once oxygen became abundant, the mutant lines could abandon photosynthetic mechanisms and survive with aerobic machinery alone.

With aerobic respiration, life became self-sustaining. For its final products—carbon dioxide and water—are precisely the materials needed to build sugar in photo-

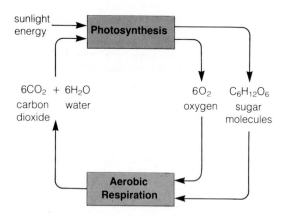

Figure 9.11 The cycle in which molecules of carbon, hydrogen, and oxygen flow through all organisms on earth. In this recycling of matter through time, each birth is affirmation of our ongoing capacity for organization, each death a renewal.

synthesis! Thus the flow of carbon, hydrogen, and oxygen through the energy pathways of living organisms came full circle (Figure 9.11).

Perhaps one of the most difficult connections you are asked to perceive is the link between yourself—a living, intelligent being—and such remote-sounding subjects as energy, metabolic pathways, and the cycling of oxygen, hydrogen, nitrogen, and carbon. Is this really the stuff of humanity? Think back, for a moment, on the discussion of the water molecule. A pair of hydrogen atoms competing with an oxygen atom for a share of the electrons joining them doesn't exactly seem close to our daily lives. But from this simple competition, the polarity of the water molecule arises. As a result of the polarity, hydrogen bonds form between water molecules. And that is a beginning for the organization of lifeless matter which leads, ultimately, to the organization of matter in all living things.

For now you can imagine other kinds of molecules interspersed through some aqueous environment. Many will be nonpolar and will resist interaction with water. Others will be polar and will respond by dissolving in it. Certain molecules, such as phospholipids, contain both water-soluble and water-insoluble regions. They spontaneously form two-layered films in water. Such phospholipid bilayers are the basis for all cell mem-

branes, hence all cells. The cell has, from the beginning, been the fundamental *living* unit.

With membrane organization, the compartment formed by the bilayer is separated from the environment. Through this isolation, chemical reactions can be contained and controlled. The essence of life *is* chemical control. This control is not brought about by some mysterious force. Instead, a class of protein molecules—enzymes—puts molecules into action in precisely regulated ways. It is not some mysterious force that tells enzymes when and what to build, and when and what to tear down. Instead it is a chemical responsiveness to the kinds of molecules present and to energy changes in the environment. And it is not some mysterious force that creates the enzymes themselves. DNA, the slender double strand of heredity, has the chemical structure—*the chemical message*—that allows molecule faithfully to reproduce molecule, one generation after the next. Those DNA strands "tell" the many billions of cells in your body how countless molecules must be built.

So yes, oxygen, hydrogen, nitrogen, and carbon represent the stuff of you, and us, and all of life. But it takes more than molecules to complete the picture. It is because of the way molecules are organized and maintained by a constant flow of energy that you are alive. It takes outside energy from sources such as the sun to drive their formation. Once molecules are assembled into cells, it takes outside energy derived from food, water, and air to sustain their organization. Plants, animals, fungi, protistans, and bacteria are part of a web of energy use and materials cycling that interconnects all levels of biological organization. Should energy fail to reach any part of any one of these levels, life in that region will dwindle and cease.

For energy flows through time in only one direction—from forms rich in energy to forms having less usable stores of it. Only as long as sunlight flows into the web of life—and only as long as there are molecules to recombine, rearrange, and recycle—does life have the potential to continue in all its rich diversity.

Life is, in short, no more *and no less* than a marvelously complex system of prolonging order. Sustained by energy transfusions, it continues because of a capacity for self-reproduction—the handing down of hereditary instructions. With these instructions, it has the means for organizing energy and materials generation after generation. Even with the death of the individual, life is prolonged. With death, molecules are released and can be recycled once more, providing raw materials for new generations. In this flow of matter and energy through time, each birth is affirmation of our ongoing capacity for organization, each death a renewal.

SUMMARY

1. Energy stored in glucose or other organic molecules can be released when electrons and hydrogen atoms are stripped from the glucose and transferred to different molecules. Energy associated with some of the transfers drives ATP formation. (In *substrate-level* phosphorylation, some molecule produced during the reactions simply transfers a phosphate group to ADP to form ATP. In *electron transport* phosphorylation, the operation of an electron transport system causes inorganic phosphate to combine with ADP to produce ATP.)

2. ATP is the main energy carrier in cells. It readily donates a phosphate group to many kinds of enzymes, reactants, and other molecules. The phosphorylation allows the molecules to enter into the various reactions that keep the cell alive. ATP can be formed by photosynthesis, chemosynthesis, and the pathways of carbohydrate metabolism (the main ones being aerobic respiration, alcoholic fermentation, lactate fermentation, and anaerobic electron transport).

3. Glucose itself is phosphorylated by two ATP molecules, and the breakdown of the phosphorylated sugar yields more ATP.

4. Glucose is phosphorylated during glycolysis, the initial stage of *all* the main pathways of carbohydrate metabolism. Glycolysis takes place in the cytoplasm. The glucose is partly broken down (into two pyruvate molecules), and substrate-level phosphorylations provide a *net* yield of two ATP. Also, the electrons and hydrogen stripped from the glucose are transferred to NAD^+ to form two NADH.

5. Until this point, all the main pathways of carbohydrate metabolism are the same. From now on, they differ in the use made of pyruvate and in the destination of the electrons and hydrogen carried by NADH.

6. In *aerobic respiration*, the two pyruvate molecules enter reactions in which six carbon dioxide and six water molecules are produced. These reactions occur only in mitochondria. The electrons and hydrogen stripped from pyruvate are transferred to NAD^+ and FAD, producing eight NADH (in addition to the two from glycolysis) as well as $FADH_2$. Thus a total of ten NADH and two $FADH_2$ deliver their electrons and hydrogen to a transport system for which oxygen is the final electron acceptor. When the transfers are made, NAD^+ and FAD are regenerated. Energy associated with the electron transfers through the transport system commonly yields thirty-six ATP.

7. Specifically, aerobic respiration proceeds through:

   a. *Glycolysis:* breakdown of glucose into 2 pyruvate, formation of 2 NADH; net yield of 2 ATP.

   b. *Krebs cycle:* breakdown of 2 pyruvate to 6 CO_2, formation of 8 NADH and 2 $FADH_2$, formation of 2 ATP.

   c. *Electron transport phosphorylation:* 10 NADH and 2 $FADH_2$ from glycolysis and the Krebs cycle deliver electrons and hydrogen to a transport system embedded in the inner mitochondrial membrane. Operation of the transport system sets up H^+ concentration and electric gradients across the membrane. When H^+ moves down the gradient, energy associated with its flow drives phosphorylation of ADP to produce 32 ATP. The spent electrons are transferred to 6 oxygen to form 6 water molecules.

8. Compared with aerobic respiration, the other main pathways of glucose metabolism produce only the small net yield of 2 ATP from glycolysis. They function only to transfer the electrons away from the NADH formed during glycolysis (thus to regenerate NAD^+).

   a. In *alcoholic fermentation*, the pyruvate from glycolysis is converted to acetaldehyde, which accepts electrons from NADH to form ethanol.

   b. In *lactate fermentation*, the pyruvate accepts electrons from NADH and is converted to lactate.

   c. In *anaerobic electron transport*, electrons from NADH are transferred to inorganic compounds (such as sulfate) in the environment.

Review Questions

1. ATP can be produced by the release of chemical bond energy from carbohydrates. Phosphorylation and oxidation-reduction reactions are needed to do this. Can you describe the roles of these types of reactions in carbohydrate metabolism?

2. Which of the pathways of carbohydrate metabolism occur in the cytoplasm? In the mitochondrion?

3. Is the following statement true? Glycolysis can proceed only in the absence of oxygen.

4. Glycolysis is the first stage of all main pathways of carbohydrate metabolism. If you include the two ATP molecules formed during glycolysis, what is the *net* energy yield from one glucose molecule for each pathway?

5. In anaerobic routes of glucose breakdown, further conversions of pyruvate do not yield any more usable energy. What, then, is the advantage of the conversion?

Readings

Brock, T., B. Smith, and M. Madigan. 1984. *Biology of Microorganisms.* Fourth edition. Englewood Cliffs, New Jersey: Prentice-Hall. Clear descriptions of the energy-releasing pathways of microbes.

Lehninger, A. 1982. *Principles of Biochemistry.* New York: Worth. Clear, accessible introduction to metabolic pathways.

Wolfe, S. 1983. *Introduction to Cell Biology.* Belmont, California: Wadsworth. Concepts-orientated introduction to cell functions.

UNIT THREE

THE ONGOING FLOW OF LIFE

10

CELL REPRODUCTION: MITOSIS

SOME GENERAL ASPECTS OF CELL REPRODUCTION

Reproduction means making copies. Often the word evokes images of identical copies, as might emerge one after another from a Xerox machine. When a bacterial cell divides in two, the two so-called "daughter cells" are almost always identical copies of their parent. When many multicelled organisms reproduce, however, the offspring are not copies down to the last detail. (For example, you resemble your parents in many respects, but you are not an exact copy of either of them.) In the biological sense, **reproduction** means producing offspring that may (or may not) be exact copies of their parents.

Reproduction is part of a **life cycle**, a recurring frame of events in which individuals grow, develop, and reproduce according to a program of instructions encoded in DNA, which they inherit from their parents. Bacterial cells have the simplest life cycles; they grow quickly and reproduce rapidly through cell division. Grow, divide, grow, divide—this "cell cycle" *is* the life cycle. Multicelled plants and animals have more intricate life cycles, with perhaps hundreds or many millions of cells growing, developing, and dividing at different times in the individual. But when it comes time to produce offspring, *a single, dividing cell is still the bridge to the next generation*:

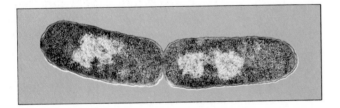

Figure 10.1 Generalized picture of prokaryotic fission. (**a**) A bacterium before its DNA is copied. (**b**) Replication begins at some point on the DNA molecule and proceeds in two directions away from that point. (**c**) The DNA copy is attached at a site close to the attachment site of the parent molecule.
(**d**) Membrane growth occurs between the two attachment sites and moves the two DNA molecules apart. (**e**) New membrane grows through the cell midsection. (**f**) Plasma membrane and cell wall material deposited at the cell midsection divide the cytoplasm in two. The micrograph of the bacterium *Escherichia coli* corresponds to (**e**).

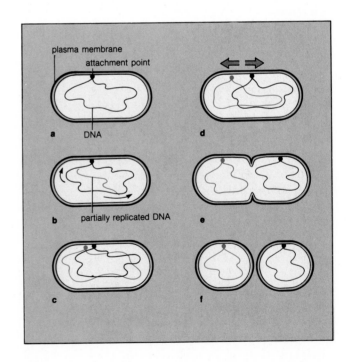

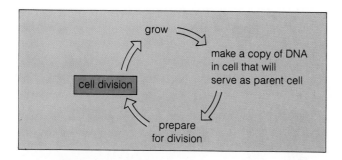

Answers to three questions about this cellular bridge are central to an understanding of reproduction. *First,* what structures and substances are necessary for inheritance? *Second,* how are they divided and distributed into daughter cells? *Third,* what are the division mechanisms themselves? We will require more than one chapter to consider the answers (and best guesses) about the reproductive process. However, the following paragraphs can help you keep the overall picture in focus.

The Ground Rule for Reproduction

The ground rule for reproduction is that *each daughter cell must receive a complete set of DNA and a certain amount of cytoplasm.* DNA, recall, contains the blueprints for making different proteins. Some of the proteins serve as structural materials; others serve as enzymes for reactions by which carbohydrates, lipids, and other substances are constructed. Three substances are used in building the membranes, organelles, and other parts of each new cell:

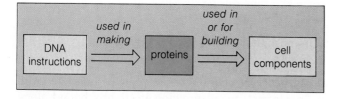

New cells simply cannot grow and they cannot function properly unless they receive a complete set of DNA.

Also, cytoplasm contains organelles, enzymes, and other components that are the operating machinery of the cell. When a parent cell divides, each daughter cell must receive enough of these components as "start-up" machinery for its operation, until it has time to use its inherited DNA for growing and developing on its own.

When cells divide, each daughter cell receives a complete copy of DNA and enough cytoplasmic machinery to start up its own operation.

Kinds of Division Mechanisms in Prokaryotes and Eukaryotes

Bacterial cells are prokaryotic (they have no nucleus), and they use a simple mechanism to assure that their daughter cells end up with a full copy of DNA and enough cytoplasm to start living on their own. In **prokaryotic fission**, growth of the plasma membrane brings about division of the DNA and the cytoplasm at the same time. A simple example will show how this is done.

Suppose you place a bacterium on a nutrient-rich medium in a culture dish. For the next twenty or thirty minutes, the bacterium feeds and grows. During this period, it also copies (replicates) its DNA, which is attached to the plasma membrane. The copy becomes attached near the original DNA molecule, then new membrane is added between the two attachment regions. The two DNA molecules simply are carried away from each other as the membrane grows laterally between them (Figure 10.1). Membrane and wall material now grow across the cell midsection, and the bacterium divides. (This is a "binary" form of fission, for the cell body is partitioned into two equivalent parts.)

In prokaryotic cell reproduction, an original DNA molecule and its copy are both attached to the plasma membrane, and membrane growth between their attachment sites pushes them apart. They end up on opposite sides of the cell midsection before the cell divides in two.

Compared to bacterial DNA, the DNA of eukaryotic (nucleated) cells is more complex and there is much more of it. Depending on the species, there can be two to many hundreds of different DNA molecules, each packaged with proteins into a structure called a **chromosome.** Often there are two *sets* of chromosomes, one inherited from the male and one from the female parent.

As you might imagine, eukaryotic cells use more intricate mechanisms to deliver the DNA and cytoplasmic machinery to daughter cells. The nucleus itself undergoes division by either mitosis or meiosis. Cytoplasmic division, or **cytokinesis,** is a separate event that usually takes place near the end of nuclear division or afterward (Table 10.1).

Mitosis maintains the parental number of chromosome sets in each daughter nucleus. If the parent nucleus has two sets of fourteen chromosomes, so will each daughter nucleus. If it has two sets of twenty-three chromosomes, so will each daughter nucleus. The single-celled eukaryotes called protistans and some fungi use this mechanism when they reproduce. Plants, animals, and some fungi use it as the basis for multicellular growth (by cell divisions).

Table 10.1	Overview of DNA and Cytoplasmic Division Mechanisms		
Mechanism	Outcome	Used by	Functions in
Prokaryotic fission	DNA and cytoplasmic division together	Prokaryotes (bacteria)	Asexual reproduction
Mitosis	Nuclear DNA division only	Protistans, some fungi	Asexual reproduction
		Plants, animals, some fungi	Multicellular growth (by cell divisions)
Meiosis	Nuclear DNA division only	Protistans, fungi, plants, animals	Sexual reproduction (through gamete or meiospore formation)
Cytokinesis	Cytoplasmic division	Protistans, fungi, plants, animals	Distribution of organelles, other start-up machinery to daughter cells

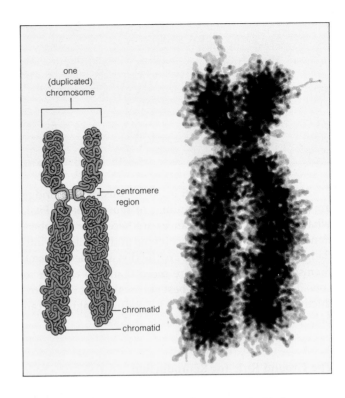

Figure 10.2 A duplicated human chromosome just before mitotic division. The two sister chromatids remain attached for a while at the centromere region, where the chromosome narrows down.

Meiosis changes the parental number of chromosome sets by exactly half in each daughter nucleus. If there are two sets of fourteen to begin with, there will be *one* set of fourteen in each daughter nucleus. This division mechanism is used only in the formation of gametes (sex cells) or meiospores (cells that give rise to gamete-producing bodies in many plants).

As you will see in the next chapter, **sexual reproduction** begins with meiosis, proceeds through gamete formation, and ends at fertilization. In contrast, prokaryotic fission and mitosis are forms of **asexual reproduction**; sex cells and fertilization are not involved.

COMPONENTS OF EUKARYOTIC CELL DIVISION

The capacity to handle large amounts of DNA emerged more than a billion years ago, among the ancestors of modern-day eukaryotes. This capacity was based on two developments: (1) the packaging of DNA and proteins into compact chromosome structures and (2) a spindle apparatus, which can handle more than one chromosome at a time during division.

Chromosomes: Eukaryotic Packaging of DNA

Prokaryotic DNA has only a few proteins associated with it. In contrast, many proteins are tightly bound along the entire length of the eukaryotic DNA molecule; in fact, they make up most of its weight. This molecular complex, which is part nucleic acid and part protein, is called a chromosome.

Chromosome Appearance. In a nondividing nucleus, chromosomes are like extremely thin threads that cannot be distinguished even with the aid of a light microscope. While they are fully extended in their threadlike form, each chromosome is copied. The original and the copy remain attached for a while at the *centromere*, a region where the chromosome narrows down. For as long as they remain attached, they are called the **sister chromatids** of one chromosome.

Just before nuclear division, the threadlike sister chromatids become folded and coiled into a highly condensed form that can be seen with the light microscope (Figure 10.2). This is the form in which the sister chromatids are detached from each other and moved apart during the division process.

Prior to nuclear division, each chromosome is replicated (copied). The original chromosome and the copy remain attached at the centromere, and as long as they are attached, they are called "sister chromatids."

The word "chromosome" can refer to a single or a duplicated chromosome (with its sister chromatids).

Chromosome Number. The cells making up your body are **diploid**: they have *two* sets of chromosomes, one inherited from each of two parents. You inherited one set of twenty-three different chromosomes from your mother and another corresponding set from your father, for a total of forty-six (Figure 10.3). At different times in the life cycle, the chromosomes in a diploid cell may be in the duplicated state (like the one in Figure 10.2) or in the unduplicated state.

Diploid cells of other eukaryotic organisms also have a characteristic number of chromosomes. For example, there are twenty chromosomes in the diploid cells of corn, water fleas, asparagus, and many other species. The fern *Ophioglossum reticulatum* has more than 1,000, one of the highest numbers known (Table 10.2).

In the gametes produced by your body (sperm or eggs), the characteristic diploid number has been reduced by half as a result of meiosis; each has only twenty-three chromosomes. These cells are **haploid**, with "half" the number of chromosome sets. (After a sperm fuses with an egg, the diploid number is restored.)

In forthcoming illustrations on mitosis and meiosis, only a few chromosomes will be shown for the sake of clarity.

Microtubular Spindles

When a nucleus is about to divide, many thin fibers (microtubules) are assembled in parallel array. Together they form a **spindle apparatus**, the actual machinery that organizes and moves the chromosomes for distribution to the two daughter nuclei. The microtubules arc across the nuclear area and their ends come together on opposite sides of it, thereby establishing two poles. A complete set of chromosomes is moved to each pole. Figure 10.4 shows the two sets on their journey to these spindle poles.

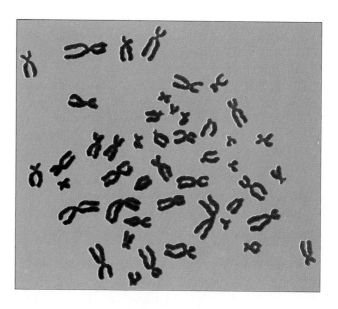

Figure 10.3 The forty-six chromosomes from a diploid cell of a human male, shown here in the duplicated and highly condensed form.

Table 10.2 Number of Chromosomes in the Body Cells of Some Eukaryotes*	
Fungus imperfecti, *Penicillium*	4
Mosquito, *Culex pipiens*	6
Fruit fly, *Drosophila melanogaster*	8
Garden pea, *Pisum sativum*	14
Corn, *Zea mays*	20
Lily, *Lilium*	24
Yellow pine, *Pinus ponderosa*	24
Frog, *Rana pipiens*	26
Earthworm, *Lumbricus terrestris*	36
Rhesus monkey, *Macaca mulatta*	42
Human, *Homo sapiens*	46
Orangutan, *Pongo pygmaeus*	48
Chimpanzee, *Pan troglodytes*	48
Gorilla, *Gorilla gorilla*	48
Potato, *Solanum tuberosum*	48
Amoeba, *Amoeba*	50
Horse, *Equus caballus*	64
Horsetail, *Equisetum*	216
Adder's tongue fern, *Ophioglossum reticulatum*	1,000+

*These examples are a sampling only. Chromosome number for most species falls between ten and fifty.

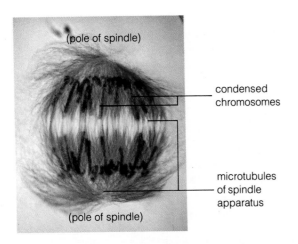

(pole of spindle)

condensed
chromosomes

microtubules
of spindle
apparatus

(pole of spindle)

Figure 10.4 Micrograph of the microtubular spindle in a plant cell undergoing nuclear division.

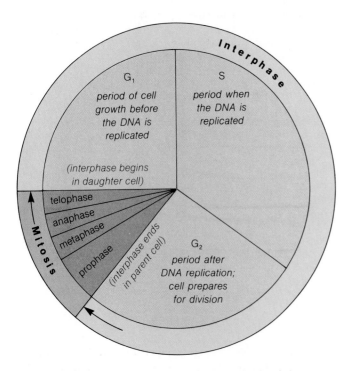

Figure 10.5 Eukaryotic cell cycle. This drawing has been generalized. There is great variation in the length of different stages from one cell to the next.

In eukaryotic cells, nuclear division depends on (1) the formation of poles by a microtubular spindle, and (2) the movement of sister chromatids of each chromosome to opposite ends of that spindle.

WHERE MITOSIS OCCURS IN THE CELL CYCLE

The life cycle of a eukaryotic cell extends from the time that cell is formed until its own division is completed. Mitosis is a very small part of this cycle; it lasts only for a few minutes or an hour or more. Normally, the cell spends about ninety percent of its life in **interphase**, when it increases its mass, approximately doubles the number of its cytoplasmic molecules and organelles, and finally replicates its DNA. As Figure 10.5 and the following list indicate, "interphase" actually consists of three phases of activities that occur between mitotic divisions:

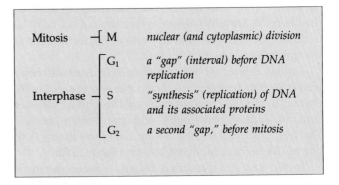

The "gaps" were so named before we knew much about what was going on during those parts of the cell cycle. Today we know that new cell components are synthesized during G_1 and assembled for distribution to daughter nuclei during G_2. Among multicelled organisms, the duration of the cell cycle varies considerably. Most of the variation is associated with the G_1 portion. For example, the mature cells of a complex plant may stay in the G_1 phase for days, even years. In contrast, the cells of an early sea urchin embryo may progress rapidly through numerous mitotic divisions in a relatively brief period.

Cells must have built-in instructions that dictate the duration of a cell cycle, because the duration is fairly consistent for all the members of a given species. But adverse environmental conditions may override heredity and arrest cells in the G_1 phase. For example, this may occur among amoebas and other protistans when they are deprived of an essential nutrient. Normally, if a cell progresses beyond a certain point in G_1 (the so-called restriction point), the cycle will be completed regardless of outside conditions.

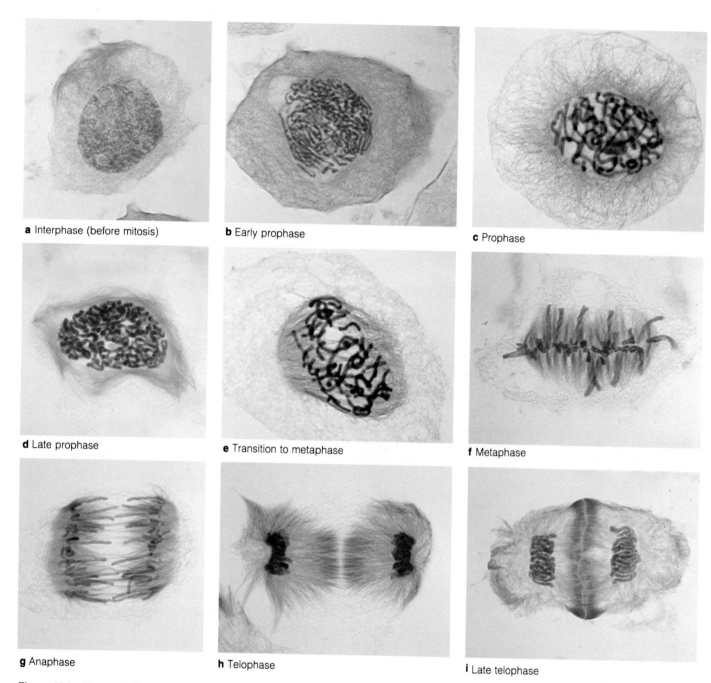

a Interphase (before mitosis)

b Early prophase

c Prophase

d Late prophase

e Transition to metaphase

f Metaphase

g Anaphase

h Telophase

i Late telophase

Figure 10.6 Plant cells (*Haemanthus*) in stages of mitosis.

STAGES OF MITOSIS

When a cell makes the transition from interphase to mitosis, it stops constructing new cell parts. Profound changes now begin to occur in the nucleus. These changes proceed one after another through four sequential stages: prophase, metaphase, anaphase, and telophase (Figure 10.6). The stages do not have distinct boundaries. If you were to watch a film on mitosis, you would see that the photomicrographs in Figure 10.6 are merely single frames of a continuous, changing drama.

Prophase: Mitosis Begins

The first stage of mitosis, called **prophase**, is evident when chromosomes become visible in the light microscope as threadlike structures. (The term mitosis refers to this emergence; it comes from the Greek *mitos*, which

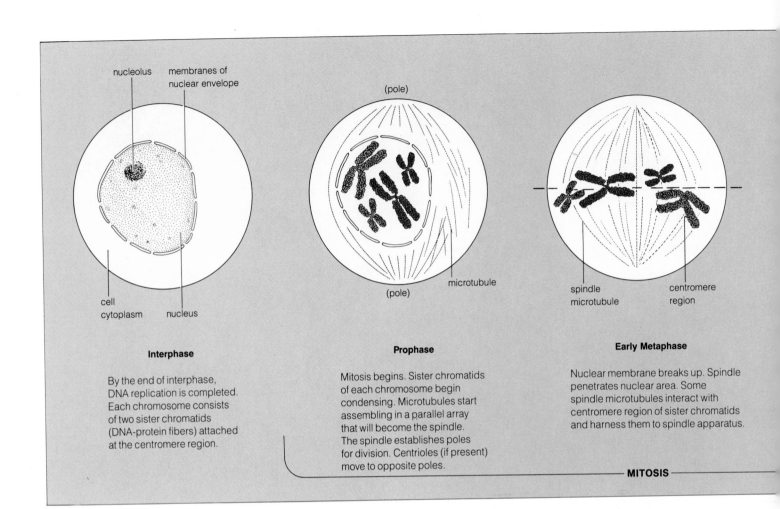

nucleolus — membranes of nuclear envelope

cell cytoplasm — nucleus

(pole)

(pole) — microtubule

spindle microtubule — centromere region

Interphase

By the end of interphase, DNA replication is completed. Each chromosome consists of two sister chromatids (DNA-protein fibers) attached at the centromere region.

Prophase

Mitosis begins. Sister chromatids of each chromosome begin condensing. Microtubules start assembling in a parallel array that will become the spindle. The spindle establishes poles for division. Centrioles (if present) move to opposite poles.

Early Metaphase

Nuclear membrane breaks up. Spindle penetrates nuclear area. Some spindle microtubules interact with centromere region of sister chromatids and harness them to spindle apparatus.

MITOSIS

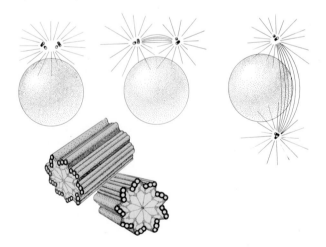

Figure 10.7 Formation of a mitotic spindle in eukaryotic cells having a pair of centrioles in the cytoplasm. Centrioles, like chromosomes, are duplicated prior to division.

means thread.) These chromosomes were already duplicated during interphase, hence each consists of two sister chromatids. By late prophase, the chromosomes become folded and condensed into thicker, rodlike structures.

While the chromosomes are condensing, the spindle begins to form. Figure 10.7 shows how a spindle forms in the nuclear region of a cell that contains a pair of centrioles, the structures that give rise to cilia and flagella (page 81). Like the chromosomes, the centriole pair is duplicated during interphase.

Metaphase

Two key events occur at the onset of **metaphase**. First, the nuclear envelope breaks up and the spindle penetrates the nuclear area (Figure 10.8). Second, some of the spindle microtubules literally capture the chromosomes at the centromere region, thereby harnessing them to the spindle apparatus.

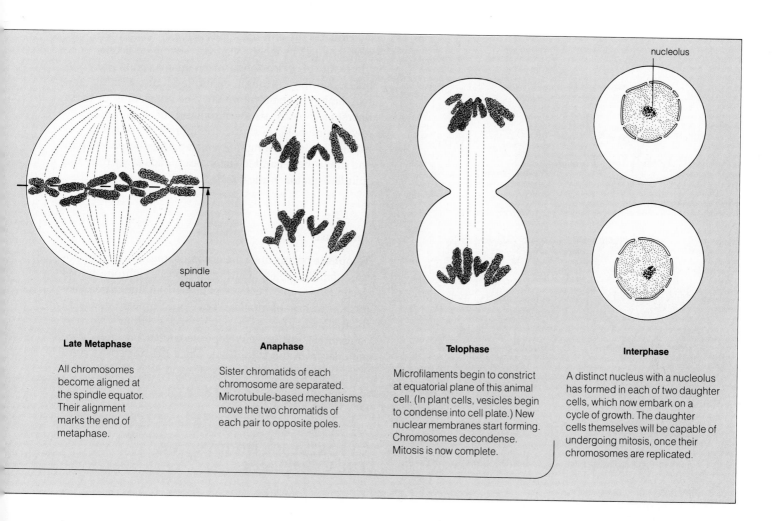

Late Metaphase	**Anaphase**	**Telophase**	**Interphase**
All chromosomes become aligned at the spindle equator. Their alignment marks the end of metaphase.	Sister chromatids of each chromosome are separated. Microtubule-based mechanisms move the two chromatids of each pair to opposite poles.	Microfilaments begin to constrict at equatorial plane of this animal cell. (In plant cells, vesicles begin to condense into cell plate.) New nuclear membranes start forming. Chromosomes decondense. Mitosis is now complete.	A distinct nucleus with a nucleolus has formed in each of two daughter cells, which now embark on a cycle of growth. The daughter cells themselves will be capable of undergoing mitosis, once their chromosomes are replicated.

When you observe living cells at early metaphase, you may wonder why they suddenly make seemingly frenzied movements. This happens when spindle microtubules are making their first random contacts with the centromere regions of chromosomes. The random contacts spin each chromosome around its long axis and yank it toward one pole, then toward the other. Eventually, microtubules extending from one pole connect firmly with one sister chromatid; then microtubules from the opposite pole connect with the other sister chromatid. (Which of the two chromatids become attached to which pole is entirely a matter of chance. When a captured chromosome is forcibly turned around, it just as readily becomes attached to microtubules from the opposite pole.)

Interactions among spindle microtubules orient the centromere regions of sister chromatids toward opposite poles of the spindle. Now, forces associated with the spindle move the chromosomes until the centromere regions are all aligned at the spindle equator, halfway between the two poles (Figure 10.8). This alignment,

Figure 10.8 Mitosis: the nuclear division mechanism by which the number of chromosome sets in a parent nucleus is maintained in daughter nuclei.

The drawing shows a generalized diploid cell, with two sets of chromosomes. (The blue-shaded chromosomes are a set derived from one parent; the red-shaded chromosomes are a set derived from another parent during sexual reproduction, as described on page 139.) To help you in keeping track of the destination of each chromatid, the same kind of chromosome structure is shown through all stages of mitosis proper. But keep in mind that chromosome structure changes considerably through these stages.

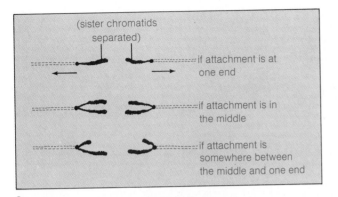

a

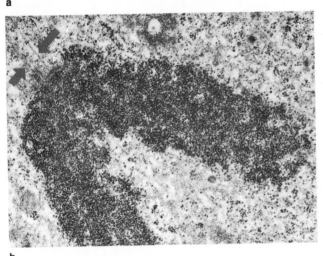

b

Figure 10.9 (a) When chromosomes are being moved to spindle poles, they can appear straight, U-shaped, or J-shaped depending on the location of the centromere region. (b) Micrograph of a chromosome from the alga *Oedogonium* at anaphase, as it is being pulled to a spindle pole by microtubules (red arrows point to two of them).

which is crucial for the separation and movement of sister chromatids to opposite poles, is the dominant event of metaphase. Once alignment is complete, metaphase is over.

Anaphase

Two events characterize **anaphase**: sister chromatids of each chromosome detach from each other, and they move to opposite poles. Once detached, they are thereafter called individual chromosomes. The mechanism underlying their physical separation is not fully understood, but they all separate at about the same time and same rate (one micrometer per second). Although metaphase may last a fairly long time, anaphase begins abruptly and is usually over within a few minutes.

How do the chromosomes actually reach their polar destinations? In most cells, the movement seems to be the result of at least two mechanisms. Some microtubules shorten (through continuous disassembly and reassembly of their subunits). The shortening of these fibers *pulls* each sister chromatid of a chromosome toward one pole, as shown in Figure 10.9. Other microtubules begin to elongate, which *pushes* the poles apart and drags the separated sister chromatids away from the spindle equator in the process.

Telophase

Telophase begins once the chromosomes arrive at opposite spindle poles. During telophase, chromosomes decondense and become extended in the threadlike form. Patches of nuclear envelope appear alongside them. Patch joins with patch, and eventually a new, continuous nuclear envelope separates the hereditary material from the cytoplasm. Once the nucleus is completed, telophase is completed—and so is mitosis.

CYTOKINESIS: DIVIDING UP THE CYTOPLASM

Cytoplasmic division, or **cytokinesis**, usually accompanies nuclear division. In most cases, it coincides with the period from late anaphase through telophase. Exceptions do exist. For example, cytokinesis does not follow mitosis in some insect, plant, and many fungal cells, so the cell body ends up with more than one nucleus.

For most animal cells, the onset of cytokinesis is marked by the appearance of scattered deposits of material around microtubules at the spindle equator. The deposits accumulate until they form a distinct layer across (typically) the cell midsection. Soon a shallow, ringlike depression appears at the plasma membrane in the plane above this layer (Figure 10.10). The depression is a **cleavage furrow**.

At a cleavage furrow, a ring of contractile microfilaments attached to the plasma membrane pulls the membrane inward. (The contractile force generated is strong enough to bend a fine glass needle inserted into dividing cells.) The inward movement eventually divides the cytoplasm in two.

Cytokinesis in the cells of most land plants is foreshadowed by the formation of a layer of material at the spindle equator. However, these cells have fairly rigid walls, which do not lend themselves to the formation of cleavage furrows. Hence plant cells use a different mechanism of cytokinesis, called **cell plate formation**, which is shown in Figure 10.11.

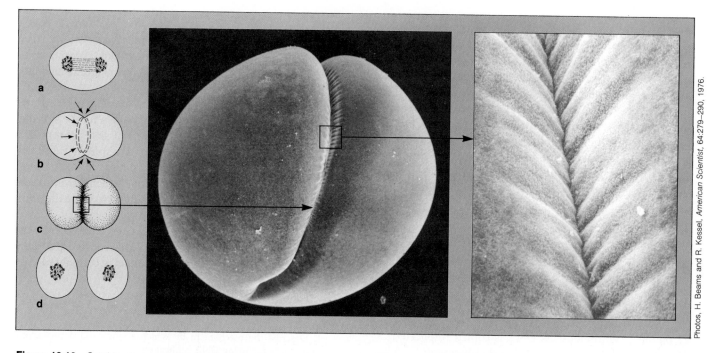

Photos, H. Beams and R. Kessel, *American Scientist*, 64:279–290, 1976.

Figure 10.10 Cytokinesis in an animal cell. The scanning electron micrographs show the furrowing of the plasma membrane, caused by the contraction of a microfilament ring just beneath it. (**a**) Nuclear division is complete; spindle is disassembling. (**b**) Microfilament rings at former spindle equator contract, like a purse string closing. (**c**) Contractions cause furrowing at cell surface. (**d**) Cytoplasm is pinched in two.

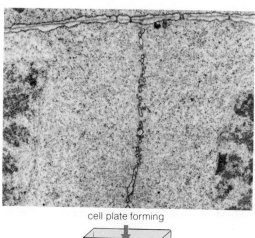

cell plate forming

Figure 10.11 Cytokinesis in a plant cell. The micrograph shows vesicles fusing together at the spindle equator.

a Nuclear division is complete; spindle is disassembling.

b (Top) Vesicles form at spindle equator, around microtubules. (Bottom) Vesicles fuse; their membranes become the plasma membranes of daughter cells, and their contents form the cell plate.

c Wall material is added between the two new membranes; plasmodesmata connect the two new cells.

A cell plate is a cross-wall that forms within the cytoplasm, usually in conjunction with microtubular remnants of the spindle. First the microtubules become arranged in parallel and extend toward both spindle poles (Figure 10.11). Small vesicles filled with wall-building material fuse together at the microtubule array, forming a disklike structure. This structure develops into the cell plate, which marks the boundary between the two new plant cells. Cellulose microfibrils are now deposited inside the plate and the new cell wall is completed.

SUMMARY

1. DNA has all the hereditary instructions necessary to construct a new cell of a given species.

2. Before a cell divides, its DNA is copied (replicated). This is true of both prokaryotes and eukaryotes.

3. When a cell divides, each new cell receives (1) a complete copy of all the DNA present in the parent cell, and (2) a particular array of cytoplasmic substances and organelles that serve as "start-up machinery" for carrying out hereditary instructions.

4. Bacterial cells (prokaryotes) reproduce through prokaryotic fission. The single DNA molecule is replicated, and one whole molecule ends up in each daughter cell when the parent cell divides at its midsection.

5. In a eukaryotic cell, many different proteins are tightly bound to each DNA molecule present, forming the part nucleic acid, part protein structure called a chromosome. When a chromosome is replicated prior to nuclear division, the original and the copy stay attached for a while (as two sister chromatids).

6. Many eukaryotic cells are diploid; they have two sets of chromosomes, inherited from two parents.

7. Mitosis and meiosis are two different nuclear division mechanisms that occur among eukaryotic organisms. Actual cytoplasmic division (cytokinesis) proceeds at a late stage of these divisions or at some point afterward.

8. Mitosis *maintains* the parental number of chromosome sets in each daughter nucleus. It separates the sister chromatids of each (duplicated) chromosome for distribution to daughter nuclei.

9. Mitosis is the basis of asexual reproduction of many single-celled eukaryotes, as well as growth in cell number in multicelled eukaryotes.

10. Interphase represents about ninety percent of the eukaryotic cell cycle. Then, a nondividing cell increases in mass, doubles its number of organelles and molecules, and replicates its DNA.

11. The stages of mitosis are as follows:

a. Prophase: duplicated, threadlike chromosomes condense; a microtubular spindle forms and establishes the poles for nuclear division.

b. Metaphase: the nuclear envelope breaks up and microtubules harness the chromosomes to the spindle, then orient the centromere regions of sister chromatids of each chromosome toward opposite spindle poles. All chromosomes become aligned at the spindle equator.

c. Anaphase: sister chromatids of each chromosome detach from each other and are moved to opposite spindle poles.

d. Telophase: chromosomes decondense to the threadlike form; new nuclear envelope forms around the two parcels of chromosomes; mitosis is completed.

Review Questions

1. Describe the mechanism by which a single-celled prokaryote reproduces following DNA replication.

2. Define the two types of nuclear division mechanisms used by prokaryotes. What is cytokinesis?

3. Table 10.2 indicates that a fern has more than 950 chromosomes in its body cells than we do. What does this tell you about the correlation between the amount of DNA and its information content?

4. What is a chromosome? What is it called in its unduplicated state? In its duplicated state (with two sister chromatids)?

5. Describe a microtubular spindle and its general function in nuclear division processes.

6. Name the four main stages of mitosis, and characterize each stage.

7. When does mitosis take place in the cell cycle?

Readings

Mazia, D. January 1974. "The Cell Cycle." *Scientific American* 230(16):54–64. Discusses the four major stages of cell division.

Mitchison, J. 1972. *Biology of the Cell Cycle.* New York: Cambridge University Press. Basic, comprehensive introduction to cell division.

Prescot, D. 1976. "Reproduction of Eukaryotic Cells." New York: Academic Press.

Sloboda, R. 1980. "The Role of Microtubules in Cell Structure and Cell Division." *American Scientist* 68:290–298. Discussion of how the spindle apparatus is thought to work.

ON ASEXUAL AND SEXUAL REPRODUCTION

When it comes to consistency in reproductive strategies, prokaryotes (bacteria) are without peer. Beginning about 3.5 billion years ago, they have been passing on copies of their DNA to new generations by little more than prokaryotic fission. More than a billion years ago, one line of prokaryotes apparently gave rise to eukaryotes, which have their DNA tucked away in a nucleus. Compared to a bacterium, it takes many more hereditary instructions of a more intricate nature to build and maintain a eukaryote. And it takes special nuclear division processes, called mitosis and meiosis, to pass on a copy of all those instructions to each new generation.

Prokaryotic fission and reproduction of single-celled eukaryotes by way of mitosis were described in the preceding chapter. Both are examples of **asexual reproduction**, whereby a single parent supplies a full copy of hereditary instructions to each of its offspring. Because all the offspring have the same instructions, they end up being practically identical to the parent and to one another.

This consistency in heritable traits seems to have worked well in environments that have not changed much over evolutionary time. Bacteria that are well adapted to living in today's mud flats and hot springs seem to be little changed from bacteria that lived in the mud flats and hot springs of ancient environments.

Most eukaryotes live in environments where the challenges to their existence are not as monotonous as they are in a mud flat. And eukaryotes exhibit sometimes pronounced variation in their inherited traits. Take another look at the pigeons in Figure 2.9 as an example of this variation. Like most other eukaryotes, those pigeons arose through **sexual reproduction**, whereby hereditary instructions end up in offspring as a result of meiosis and fertilization.

No one knows how sexual reproduction emerged or what its initial advantages might have been. Whatever its origins, for our purposes it is enough to know that sexual reproduction has become an extremely important means of shuffling slightly different bits of hereditary information among the individuals of each new generation. The differences give rise to variations in traits, a few of which might make some individuals better or worse off than others when responding to environmental challenges.

For example, if the "challenge" is a chicken breeder who is selecting for chickens that lay exceptionally large eggs, then the individuals that lay the smallest eggs in each generation will probably end up in the stewpot. Only the individuals that can lay the largest eggs will be allowed to live and reproduce, and gradually the character of the chicken population will change.

11

CELL REPRODUCTION: MEIOSIS

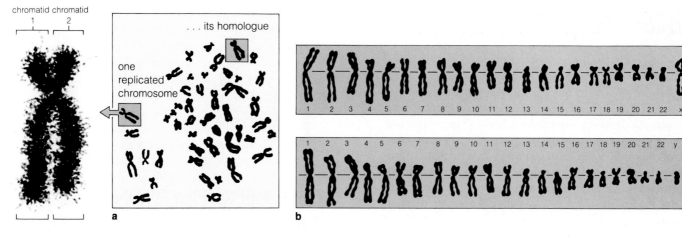

Figure 11.1 (**a**) The forty-six chromosomes in a diploid cell of a human male. Each is in the replicated state; the two sister chromatids remain attached at the centromere region, where the chromosome narrows down. (**b**) Arranging the chromosomes according to length and shape demonstrates that this diploid cell has two *sets* of homologous chromosomes, which resemble each other in length and shape (and in which units of hereditary instructions they carry). Homologous chromosomes pair with each other during meiosis.

Whatever the outcome, sexual reproduction leads to variation in a population—*and variation, remember, is the foundation for evolution.*

MEIOSIS AND THE CHROMOSOME NUMBER

Recall that eukaryotic DNA and many proteins are bound tightly together in structures called chromosomes. Prior to nuclear division, each chromosome is copied, or replicated. The original and the copy remain attached as **sister chromatids** at the centromere, the region where spindle microtubules will attach to the chromosome (Figure 11.1a).

Recall also that each kind of organism has a characteristic number of chromosomes in its cells. For example, body cells of a fruit fly contain eight chromosomes; body cells of a human contain forty-six. Let's focus on the human chromosomes, which are shown in Figure 11.1. As you can see, some of the chromosomes are longer than others and they have different shapes, depending on their centromere location. They also contain different **genes**, or units of hereditary instructions.

On close analysis, you would discover that for each chromosome of a given length and shape, there is another chromosome that has the *same* length and shape. Two chromosomes that resemble each other in length, in shape, and in which genes they carry are called **homologues**, and they pair with each other at meiosis. (Sex chromosomes, such as the X and Y chromosomes in humans, differ in structure but still function as homologues at meiosis.)

By lining up the homologous chromosomes into pairs, you can see that human body cells have two sets of twenty-three chromosomes (forty-six total). Any cell having *two sets* of chromosomes is **diploid**, or 2n. In all but a few cases (for example, in self-fertilizing plants), one of these sets is inherited from a male parent, the other from a female parent.

Diploid cells usually are not used in sexual reproduction, because the chromosome number characteristic of the species cannot be allowed to double with each new generation. (First-generation human offspring would get 46 from one parent and 46 from another, or 92 chromosomes; their offspring would get 92 + 92, or 184 chromosomes, and so on, until cells were stuffed with nothing but chromosomes.)

Some diploid cells in the human body are set aside for reproductive purposes. Meiosis reduces the number of chromosome sets by half in the nucleus of these cells, which are destined to give rise to sperm or eggs. A sperm or an egg is a mature sex cell, or **gamete**. Cells that have only *one set* of chromosomes, as gametes do, are said to be haploid (*n*). The nuclei of two gametes combine at fertilization, thereby restoring the diploid number in the new individual.

Thus sexual reproduction consists of chromosome *reduction* (to *n*) at meiosis and chromosome *doubling* (to 2*n*) at fertilization:

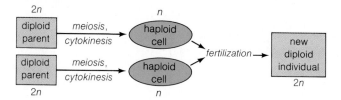

STAGES OF MEIOSIS

Meiosis resembles mitosis in several respects. Each chromosome has already been replicated (during interphase) before nuclear division begins, so each consists of two sister chromatids. In nearly all species, microtubular spindles establish the poles toward which sister chromatids, once detached from each other, are moved. However, unlike mitosis, meiosis consists of *two consecutive divisions*, called meiosis I and II:

First Division	Between Divisions	Second Division
prophase I		prophase II
metaphase I	interkinesis	metaphase II
anaphase I	(*no DNA replication*)	anaphase II
telophase I		telophase II

Briefly, during the first nuclear division the duplicated chromosomes pair with each other, *homologue to homologue*, then each homologue is separated from its partner. Cytokinesis typically follows. Each resulting daughter cell has a single set of chromosomes—but each of those chromosomes still consists of *two* sister chromatids, attached at the centromere.

The transition to the second nuclear division is known as *interkinesis*. Cells proceeding through interkinesis do not replicate their DNA. During the second nuclear division, the two sister chromatids of each chromosome are separated from each other. The cytoplasm is usually divided once again during the last stage of meiosis.

Typically, then, two nuclear divisions and two cytoplasmic divisions produce four haploid cells. Let's now take a closer look at meiosis.

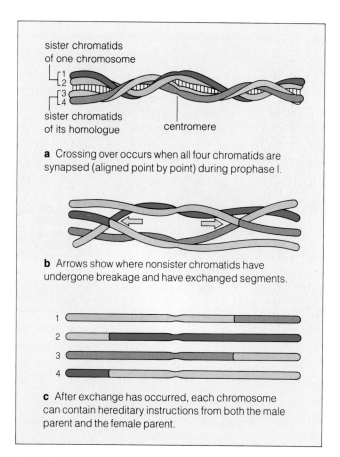

a Crossing over occurs when all four chromatids are synapsed (aligned point by point) during prophase I.

b Arrows show where nonsister chromatids have undergone breakage and have exchanged segments.

c After exchange has occurred, each chromosome can contain hereditary instructions from both the male parent and the female parent.

Figure 11.2 Genetic recombination as a result of crossing over between nonsister chromatids of homologous chromosomes during meiosis I.

Prophase I Activities

At the start of **prophase I**, the first stage of meiosis, each chromosome is like a long, very thin thread. The "thread" is actually two sister chromatids in close alignment with each other. In this extended form, each chromosome begins to pair with its homologue. The homologues are drawn together tightly by a process called **synapsis**. It is as if they become stitched point by point along their entire length, with very little space between them. (Only the two sex chromosomes do not synapse fully at every region.)

Synapsis puts the sister chromatids of one chromosome into parallel alignment with those of the homologous chromosome. The arrangement favors **crossing over**, whereby *nonsister* chromatids undergo breakage at one or more sites along their length and exchange corresponding segments at the breakage points (Figures 11.2 and 11.3).

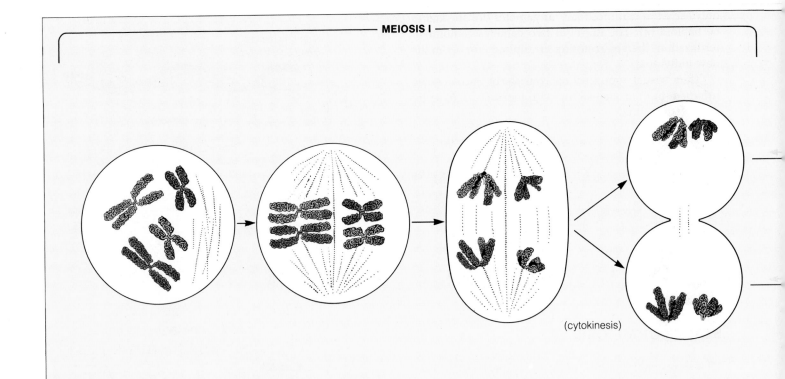

Prophase I

Each chromosome condenses, then becomes paired with its homologue. Crossing over occurs between nonsister chromatids of each homologous pair.

Metaphase I

Microtubular spindle forms during transition to metaphase I. Homologous pairs become aligned at spindle equator, with the two homologues forming attachments to opposite spindle poles.

Anaphase I

Homologous chromosomes are separated; spindle and centromere microtubules move them to opposite poles.

Telophase I

At telophase I, one set of chromosomes is clustered at each of the two spindle poles.

(cytokinesis)

Figure 11.3   Meiosis: a nuclear division mechanism by which the number of chromosome sets in a parent nucleus is reduced by half in each daughter nucleus. Only two pairs of homologous chromosomes are shown here. The blue chromosomes are derived from one parent; the red ones are their homologues from the other parent.

Keep in mind that chromosomes do not really look like this during the different stages of meiosis (at times they look like hopelessly tangled yarn). This diagram simply makes it easier to track the movements of the sister chromatids of each chromosome.

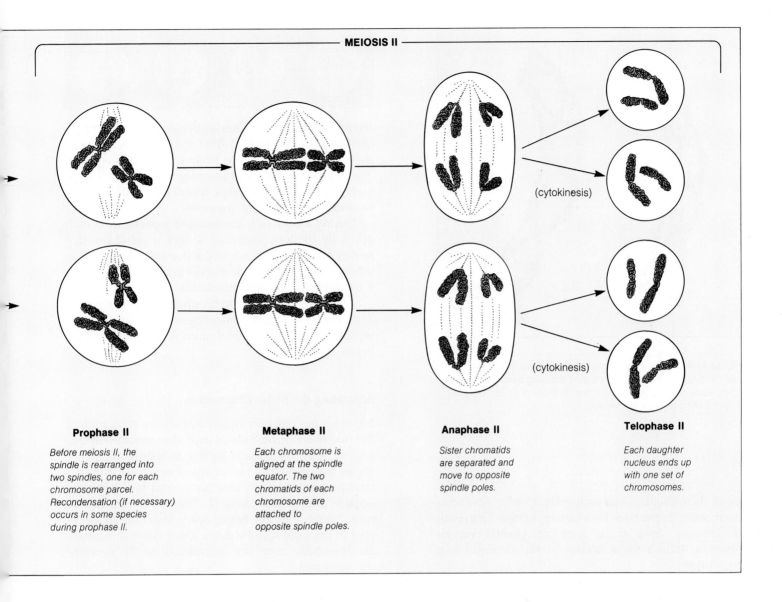

Prophase II

Before meiosis II, the spindle is rearranged into two spindles, one for each chromosome parcel. Recondensation (if necessary) occurs in some species during prophase II.

Metaphase II

Each chromosome is aligned at the spindle equator. The two chromatids of each chromosome are attached to opposite spindle poles.

Anaphase II

Sister chromatids are separated and move to opposite spindle poles.

Telophase II

Each daughter nucleus ends up with one set of chromosomes.

The two homologous chromosomes can move apart after exchanging segments, but they still remain joined at the sites where nonsister chromatids extend across each other (Figure 11.4). Each crossing between two nonsister chromatids is known as a **chiasma** (plural, chiasmata).

A chiasma is evidence that breakage and exchange (a crossover) occurred earlier. It might not indicate precisely where the break occurred because chiasmata are not stationary configurations; they are physically forced toward the chromatid tips as prophase I draws to a close. Apparently, the continued attachment of nonsister chromatids at chiasmata plays a role in aligning homologues at the spindle equator during metaphase I.

The Importance of Crossing Over

Before turning to the events of metaphase I, there is one point that should be addressed. Crossing over is one of the most important events in meiosis. We have said that genes are units of hereditary instructions. It so happens that the instructions for a given trait can vary slightly. Suppose one kind of gene contains the instructions for "flower color." In one chromosome the instructions may specify "white," but in the homologous chromosome they may specify "white," or "yellow," or some other color. Alternative forms of a gene are called **alleles**.

When nonsister chromatids undergo crossing over, they exchange alleles at some number of gene locations

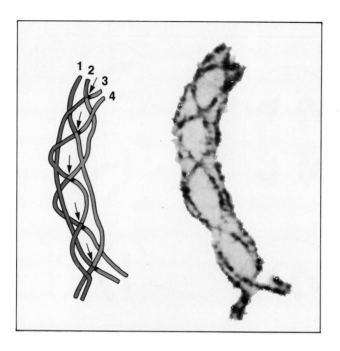

Figure 11.4 Paired homologous chromosomes (grasshopper cell) held together by chiasmata after crossing over. Sister chromatids of one are labeled 1 and 2; those of its homologue, 3 and 4. Arrows show chiasmata.

along their length. Thus each ends up with new combinations of instructions for a variety of traits. This result of crossing over is a form of **genetic recombination**. It is a major source of variation within a population.

Separating the Homologues

The transition from prophase I to **metaphase I** strongly resembles the events in mitosis. A microtubular spindle forms, and chromosomes are lined up at the spindle equator. But there is a major difference: in metaphase I, the lineup is *paired*, homologue with homologue. One chromosome (with its two chromatids) becomes attached to microtubules extending from one spindle pole, and the homologous chromosome becomes attached to spindle microtubules extending from the opposite pole (Figure 11.3).

When chromosomes are lined up at the spindle equator, all of those inherited from one parent do not necessarily line up across from all of those inherited from the other parent. Even if as few as three homologues are involved, any of these alignments might occur at metaphase I:

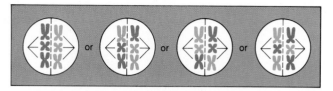

(Here the red chromosomes are derived from one parent, and the blue are derived from the other.) Any one of these four alignments is as likely to occur as any other. The random alignment of homologous chromosomes at metaphase I is another major source of genetic variation in sexually reproducing organisms.

During **anaphase I**, homologues separate from each other. By the time anaphase is over, a set of chromosomes is clustered at each of the two spindle poles (Figure 11.3). Each chromosome is still in the replicated form; it consists of two sister chromatids.

Anaphase I gives way to telophase I and interkinesis. Typically these stages are fleeting, and there is no DNA replication before the final nuclear division (that is, before meiosis II).

Separating the Sister Chromatids

Meiosis II has one overriding function: the separation of the two sister chromatids of each chromosome. In this respect, it is functionally similar to mitosis. Chromosomes may condense into compact structures during a brief prophase II, then they are moved to the spindle equator during **metaphase II**. The two chromatids of each chromosome are moved apart, then they are delivered to opposite spindle poles. Once sister chromatids are detached, they are considered to be separate chromosomes.

At each spindle pole, there are only half the number of chromosomes that were present in the parent nucleus. But that number includes one chromosome of each type; it is a single, complete set of hereditary instructions. Each chromosome set becomes packaged in its own nucleus in telophase II. Meiosis is completed. Fertilization will restore the two sets of hereditary instructions characteristic of the parent diploid cells.

CELL DIVISIONS AND THE LIFE CYCLE

Animal Life Cycles

During the life cycles of most multicelled animals, some diploid cells undergo meiosis and haploid gametes are formed. (This process is called *gametogenesis.*) Two gametes combine to form the first diploid cell of the new

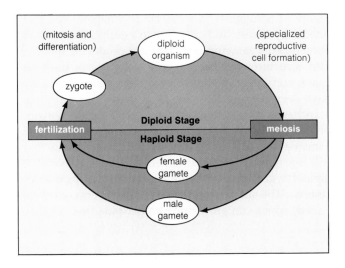

Figure 11.5 Generalized life cycle for animals.

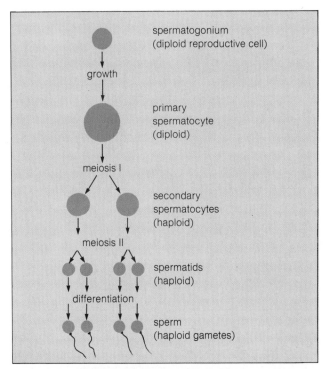

a

individual (the *zygote*). Through mitosis and cytokinesis, the zygote grows and develops into a diploid multicelled organism (Figure 11.5).

In diploid male animals, gamete formation is called **spermatogenesis** (Figure 11.6). A diploid reproductive cell increases in size, thereby becoming a primary spermatocyte. This large, immature cell undergoes meiosis I and gives rise to two cells called secondary spermatocytes. Both of these cells undergo meiosis II. The result is four haploid spermatids, which change in form, develop a tail, and thereby become **sperm** (mature male gametes).

Female gamete formation, or **oogenesis**, follows a similar sequence of events (Figure 11.7). However, cytoplasmic substances accumulate to a much greater extent in a female reproductive cell (primary oocyte) that undergoes meiosis. Also, the cells formed after meiosis are not equal in size or function. After meiosis I, one of the cells (the secondary oocyte) ends up with nearly all the cytoplasm. Following meiosis II, that cell becomes the mature **egg**, or **ovum** (plural, ova). The other cells produced during meiosis I and II (the polar bodies) are extremely small and do not function as gametes.

Plant Life Cycles

There are multicelled diploid bodies and multicelled haploid bodies in the life cycle of most land plants. A pine tree is a multicelled diploid body. It is called a **sporophyte** ("spore-producing plant"), because some of its cells undergo meiosis and give rise to meiospores. As

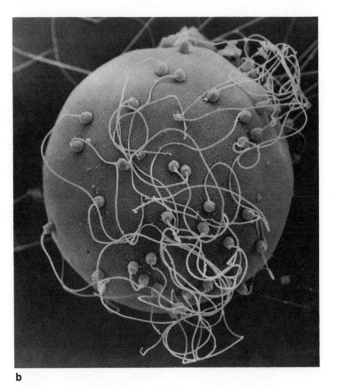

b

Figure 11.6 (a) Generalized picture of spermatogenesis in male animals. (b) This scanning electron micrograph shows how much smaller the sperm are, compared to an egg (these being the sperm and egg of a clam).

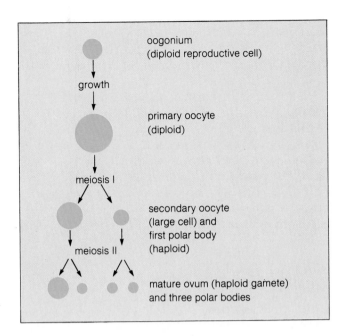

Figure 11.7 Generalized picture of oogenesis in female animals. This sketch is not drawn to the same scale as Figure 11.6. A primary oocyte is *much* larger than a primary spermatocyte, as Figure 11.6b indicates.

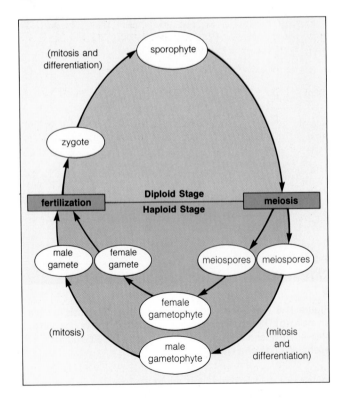

Figure 11.8 Generalized life cycle for complex land plants.

Figure 11.8 suggests, a *meiospore* is a haploid cell that divides by mitosis and gives rise to a multicelled haploid body called a **gametophyte** ("gamete-producing plant"). Tiny structures perched on scales of some of the pine cones are gametophytes. The pollen grains you may have seen drifting like clouds of dust from pine tree branches at certain times of year become male gametophytes. Sooner or later, some cells from both kinds of gametophytes divide by mitosis and produce cells that will function as eggs and sperm. After fertilization, the diploid zygote divides by mitosis and develops into an embryo. The embryo and its protective layers constitute a seed, which can grow into another pine tree.

SUMMARY

The Key Events of Meiosis

1. Asexual reproduction yields offspring that are essentially identical to one another and to the parent. Sexual reproduction gives rise to genetic variation among the offspring and thereby contributes to the variation that is the foundation for evolution.

2. The body cells of sexually reproducing organisms commonly are diploid. Diploid cells have two sets of chromosomes, in which each chromosome of one set has a partner (homologue) in the other set. For most species, one set is inherited from a female parent, the other from a male parent.

3. Homologous chromosomes resemble each other in length, shape, centromere location, and which genes they carry. (Sex chromosomes such as the X and Y chromosome of mammals do not resemble each other in structure but still function as homologues.)

4. Homologous chromosomes carry identical or nonidentical alleles at each gene location. (Genes are units of hereditary information along a chromosome. Alleles are alternative forms of a gene.)

5. Meiosis is a nuclear division process whereby the parental number of chromosome sets is reduced by half in each daughter nucleus. Thus a diploid number of chromosomes ($2n$) is reduced to a haploid number (n) in each daughter nucleus.

6. Prior to meiosis, all of the chromosomes are replicated (so that each consists of two identical sister chromatids, attached at the centromere).

7. Meiosis consists of two consecutive divisions (although the time that passes between the two divisions varies among different species). In meiosis I, homologues separate from each other. In meiosis II, sister chromatids of each chromosome separate from each other.

8. The following events occur during meiosis I:

   a. Prophase I: each homologue undergoes synapsis (pairing) with its partner. Also, crossing over occurs (nonsister chromatids undergo breakage and they exchange segments). Crossing over leads to genetic recombination—new combinations of alleles at different gene locations along the chromosome.

   b. Metaphase I: the microtubular spindle required for chromosome movements completes its formation. The pairs of homologous chromosomes align at random along the spindle equator.

   c. Anaphase I: homologous chromosomes are separated from each other. Typically, two haploid nuclei form.

   d. Telophase I, interkinesis, and prophase II: there is only one set of chromosomes per haploid nucleus, but each chromosome consists of two sister chromatids. No DNA replication occurs.

9. The following events occur during meiosis II:

   a. Metaphase II: a microtubular spindle forms in each daughter nucleus. Spindle fibers attach to the centromere region of each chromosome.

   b. Anaphase II: the two sister chromatids of each chromosome are separated from each other, with each now being called a separate chromosome. The (formerly) sister chromatids move to opposite poles of the spindle.

   c. Telophase II: new nuclear envelope forms around the separated sets of chromosomes. Typically there are now four haploid nuclei.

10. Following cytokinesis (cytoplasmic division), there are four haploid cells, one or all of which may function as gametes.

11. In the life cycle of most multicelled animals, certain diploid cells undergo gametogenesis to form haploid gametes (either sperm in males or eggs in females). Fusion of a sperm and an egg nucleus at fertilization produces a new diploid cell (zygote), which undergoes mitosis and cytokinesis to develop into a multicelled individual.

12. In the life cycle of many land plants, certain diploid cells in the sporophyte stage undergo meiosis to form haploid meiospores. A haploid meiospore divides by mitosis to produce a multicelled haploid gametophyte stage. Some cells of the gametophyte develop into male or female gametes. Fusion of the nuclei of a mature male and female gamete at fertilization produces a diploid zygote. The zygote undergoes mitosis and cytokinesis to develop into a new diploid sporophyte.

Meiosis Compared With Mitosis

So far in this unit, our main focus has been on the division mechanisms that underlie eukaryotic cell division: mitosis as well as meiosis. (Mitosis, recall, is the basis of asexual reproduction of many single-celled eukaryotes and growth of multicelled eukaryotes.) Figure 11.9 summarizes the similarities and differences between the two division mechanisms.

In looking at this illustration, notice that mitotic cell division produces more cells just like the parent cell. Notice also that crossing over and recombination occur *only* during the first division of meiosis. Crossing over and recombination at meiosis can put together new combinations of alleles in chromosomes, then in gametes. *Thus recombination produces variation among individuals of a population.* Such variation is a testing ground for agents of selection, hence for the evolution of populations.

Review Questions

1. Define sexual reproduction. How does it differ from asexual reproduction? What is one of its presumed advantages?

2. Is this statement true: asexual reproductive modes occur only among prokaryotes; eukaryotes rely exclusively on sexual reproduction.

3. The nucleus of diploid cells contains pairs of chromosomes that resemble each other in length, in shape, and in which genes they carry. What are the pairs called?

4. Refer back to Table 10.2, which gives the diploid number of chromosomes in the body cells of a few organisms. What would be the *haploid* number for the gametes of humans? For the garden pea?

5. Suppose the diploid cells of an organism have four pairs of homologous chromosomes, designated AA, BB, CC, and DD. How would its haploid cells be designated?

6. When, and in which type of cells, does meiosis occur?

7. Define meiosis and characterize its main stages. In what respects is meiosis like mitosis? In what respects is it unique?

8. Does crossing over occur during mitosis, meiosis, or both? At what stage of nuclear division does it occur, and what is its significance?

9. Distinguish between:
   a. Gamete and meiospore
   b. Gametophyte and sporophyte

10. Outline the steps involved in spermatogenesis and oogenesis.

Readings

John, B. 1976. "Myths and Mechanisms of Meiosis." *Chromosoma* 54:295–325.

Mitchison, J. 1972. *Biology of the Cell Cycle.* New York: Cambridge University Press.

Raven, P., R. Evert, and S. Eichhorn. 1986. *Biology of Plants.* Fourth edition. New York: Worth. Lovely illustrations of where meiosis occurs in the life cycles of diverse plants.

Strickberger, M. 1985. *Genetics.* Third edition. New York: Macmillan. Contains excellent introduction to chromosomes and meiosis.

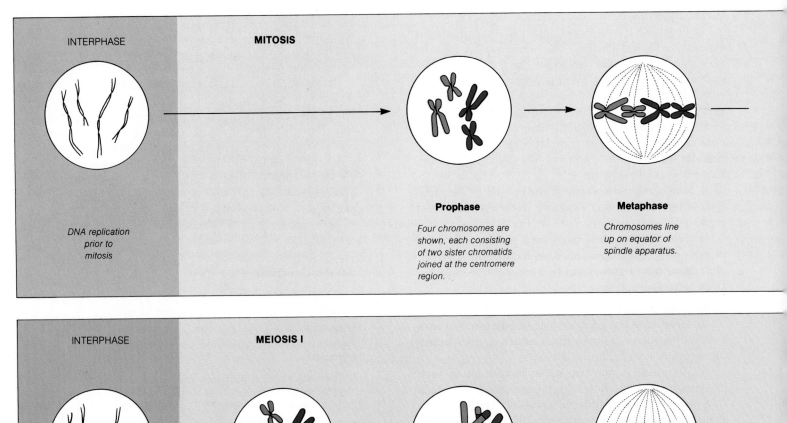

INTERPHASE

MITOSIS

Prophase

Four chromosomes are
shown, each consisting
of two sister chromatids
joined at the centromere
region.

Metaphase

Chromosomes line
up on equator of
spindle apparatus.

DNA replication
prior to
mitosis

INTERPHASE

MEIOSIS I

Prophase I

Four chromosomes are shown. Each double-stranded chromosome
condenses tightly. Each type of chromosome from one parent
pairs with its equivalent (homologue) from the other parent.
Breakage and crossing over occur between nonsister
chromatids in each homologous pair; genetic recombination
will be the eventual result. RNA is synthesized and
stockpiled. Chromosomes condense compactly.

Metaphase I

Homologous pairs of
chromosomes line up on
equator of spindle.

DNA replication
prior to
meiosis

Figure 11.9 Summary of mitosis and meiosis. In these examples, the parent nucleus is diploid, with two
chromosome sets (one red, one blue). The threadlike chromosome structure shown at interphase is not
visible in light micrographs; it can be seen only at the high magnification possible with transmission
electron microscopy.

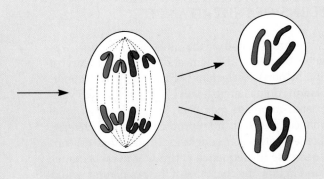

Anaphase

The two sister chromatids of each chromosome are separated and moved to opposite spindle poles.

Telophase

Parental chromosome number is maintained in new nuclei.

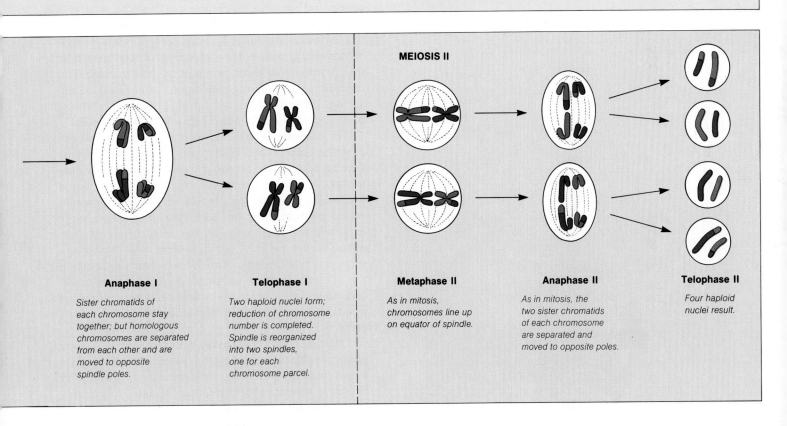

MEIOSIS II

Anaphase I

Sister chromatids of each chromosome stay together; but homologous chromosomes are separated from each other and are moved to opposite spindle poles.

Telophase I

Two haploid nuclei form; reduction of chromosome number is completed. Spindle is reorganized into two spindles, one for each chromosome parcel.

Metaphase II

As in mitosis, chromosomes line up on equator of spindle.

Anaphase II

As in mitosis, the two sister chromatids of each chromosome are separated and moved to opposite poles.

Telophase II

Four haploid nuclei result.

12

OBSERVABLE PATTERNS OF INHERITANCE

Figure 12.1 Gregor Mendel, founder of modern genetics.

MENDEL'S INSIGHTS INTO THE PATTERNS OF INHERITANCE

Biology toward the end of the nineteenth century was dominated by talk of Darwin and Wallace's theory of evolution by natural selection. According to that theory, a population could change (evolve) only if inherited variation existed among its members. Variant forms of traits that enhanced survival and reproduction would increase in frequency with each generation, those that did not would decrease in frequency and might even be eliminated, and in time the character of the population would change (Chapter Two).

Yet many biologists were skeptical of the theory. For one thing, it did not fit with a prevailing notion about the transmission of traits between generations. It was common knowledge that hereditary material from both sperm and eggs had to combine in order to produce a new individual. The widely accepted hypothesis was that the material blended at fertilization, much like cream blending into coffee. This "blending" notion gave little insight into why a distinctive trait, such as freckled skin, can still turn up among generations of nonfreckled offspring. Instead it conjured up pictures of all offspring becoming the homogenized equivalent of *café au lait*.

For example, if blending were the rule, then a herd of white stallions and black mares would produce only gray horses, which thereafter would produce only gray horses. A field of red clover and white clover would eventually give way to pink clover, which thereafter would produce more just like itself. Blending scarcely explained the observable fact that in such populations, not all horses are gray, nor all flowers pink. It was considered a rule anyway. And Darwin had problems on this account, because uniform populations would present no variation whatsoever for selective agents to act upon. That being the case, "evolution" simply could not occur.

But even before the Darwin-Wallace theory was made public, evidence concerning the physical basis of inheritance was accumulating in a monastery garden in Brünn, northeast of Vienna. Gregor Mendel, a scholarly, mathematically oriented monk, was beginning to identify rules governing inheritance.

The monastery of St. Thomas was somewhat removed from the European capitals, which were then the centers of scientific inquiry. Yet Mendel was not a man of parochial interests, who simply stumbled by chance onto principles of great import. Having been raised on a farm, he was well aware of agricultural principles and their application. He kept abreast of breeding experiments and developments described in the available journals. Mendel was a founder of the regional agricultural society. He won several awards for developing improved varieties of fruits and vegetables. After entering the monastery, he spent two years studying mathematics at the University of Vienna.

Shortly after his university training, Mendel began experiments on the nature of plant diversity. Through his combined talents in plant breeding and mathematics, he perceived patterns in the emergence of traits from one generation to the next.

Mendel's Experimental Approach

Mendel experimented with the garden pea plant, *Pisum sativum* (Figure 12.2). This plant fertilizes itself. Male and female gametes (sperm and eggs) develop in different structures of its individual flowers, and fertilization normally occurs between two gametes from the same flower. Some pea plants are **true-breeding**, in that successive generations of offspring always show the same form of one or more traits (some always have white flowers or yellow seeds, for example). Of course, when left to their own devices, true-breeding pea plants show a rather monotonous and uninformative pattern of inheritance.

But pea plants also lend themselves to **cross-fertilization**, whereby sperm from one plant fertilizes eggs from another. To prevent a plant from self-fertilizing, Mendel could open a flower bud and remove the stamens. (Stamens are floral structures that bear pollen grains, in which the sperm develop.) If pollen from another plant were brushed on the "castrated" bud, cross-fertilization could follow.

Why would Mendel bother to manipulate plants this way? He wanted cross-fertilization to occur between two true-breeding plants that showed clearly contrasting traits (for example, one with white flowers and the other with purple flowers). If their offspring ended up with white or purple flowers, he could easily identify one plant or the other as the source of the hereditary material for that trait. Thus, if there *were* patterns in the way hereditary material is transmitted from parents to offspring, the use of contrasting traits might be a way to identify them.

It will be useful to retrace a few of Mendel's experiments. The conclusions he drew from them have turned out to apply, with some modification, to all sexually reproducing organisms.

Some Terms Currently Used in Genetics

Before we begin, let's take a moment to review the following terms, which are important in genetics. Keep in mind that some are our terms, not Mendel's (Figure 12.3). They were coined much later, after the work of explorers who followed him into uncharted territory.

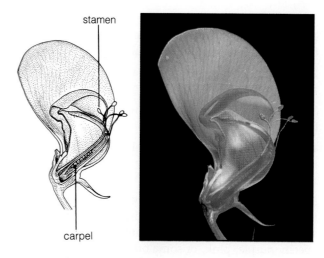

stamen

carpel

Figure 12.2 Flower from a garden pea (*Pisum sativum*), sectioned to show the location of stamens (male reproductive organs) and the carpel (the female reproductive organ).

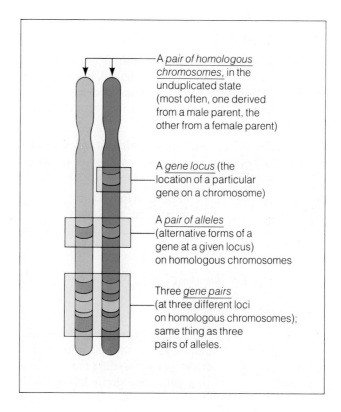

A *pair of homologous chromosomes*, in the unduplicated state (most often, one derived from a male parent, the other from a female parent)

A *gene locus* (the location of a particular gene on a chromosome)

A *pair of alleles* (alternative forms of a gene at a given locus) on homologous chromosomes

Three *gene pairs* (at three different loci on homologous chromosomes); same thing as three pairs of alleles.

Figure 12.3 A few genetic terms illustrated. In Mendel's time, no one knew about meiosis or chromosomes, but it was clear that offspring received hereditary material from parents by way of sperm and eggs. As we now know, the hereditary material (genes) is packaged in homologous chromosomes (one from the male, one from the female parent). Thus at each gene locus along the chromosomes, one allele has come from the male parent and its partner has come from the female parent.

1. Distinct units of heredity—**genes**—are the physical basis for all traits of an individual.

2. Each gene has its own **locus**, a particular location on a chromosome.

3. A gene may occur in one of two or more alternative forms, known as **alleles**.

4. Diploid organisms have two alleles for each trait, on two partner chromosomes (the homologues derived from two parents).

5. The two alleles of a pair may be identical or different. (For instance, both alleles for flower color may specify white. Or one may specify white and the other may specify purple, pink, or some other color.)

6. If the two alleles of a pair are identical, the individual is a **homozygote** for the trait. If they are not identical, the individual is a **heterozygote** for the trait.

7. Often one allele of a pair can mask expression of its partner. It is a **dominant allele**, and the other is a **recessive allele**. Dominant alleles are indicated by capital letters and recessive alleles by lowercase letters (for example, *A* and *a*).

8. A **homozygous dominant** individual has inherited two dominant alleles for the trait (*AA*). A **homozygous recessive** individual has inherited two recessive alleles for the trait (*aa*), and a **heterozygous** individual has two different alleles (*Aa*).

9. The genetic makeup of an individual is called its **genotype**. The term can refer to the sum total of the individual's genes or to one pair of alleles at a time.

10. The term **phenotype** refers to an individual's observable traits (its structure, physiology, and behavior).

The Concept of Segregation

Let's turn to Mendel's first series of experiments, which we would now call "monohybrid crosses." In this type of cross, two parents that are true-breeding for contrasting forms of a single trait give rise to heterozygous offspring (which inherit two different alleles for the trait). Mendel studied seven traits, one at a time (Figure 12.4).

In one monohybrid cross, a purple-flowered plant and a white-flowered plant yielded offspring that had purple flowers *only*. What had happened to the white-flower trait? Had it disappeared? Mendel did not cross these purple-flowered offspring with new plants; he allowed them to self-fertilize and produce seeds. The following season, some of the plants that grew from the seeds had white flowers! The white-flower trait had not been lost after all—but what was going on?

Results were much the same for all of Mendel's monohybrid crosses. One of the contrasting traits seemed to disappear in the hybrid offspring of the first cross (the first-generation plants) only to show up again in some of the second-generation plants.

To explain these results, Mendel assumed that an individual has two "unblended" units of hereditary material for each trait. When it came time to produce sperm or eggs, the two units were segregated from each other and ended up in separate gametes. Mendel reasoned that *both* units had to be either "dominant" or "recessive" in true-breeding parents. Why? Their hybrid offspring, which received one unit from each parent, had *purple flowers*: the unit specifying purple "dominated" the one specifying white.

Let's express Mendel's conclusion in light of what we know about meiosis and fertilization (Chapter Eleven). Pea plants are diploid, with two sets of homologous chromosomes in their cells. If a plant is homozygous dominant (*AA*) for a trait, it has a dominant allele on each homologous chromosome. If the plant is homozygous recessive (*aa*), it has a recessive allele on each one. Homologous chromosomes separate from each other at meiosis, and they end up in separate gametes—and so do the two alleles of the pair:

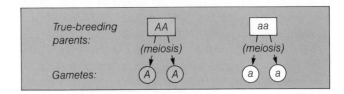

Thus, when gametes from an *AA* plant and an *aa* plant combine at fertilization, there is only one possible outcome. The offspring must be *Aa* (carry alleles for *both* forms of the trait):

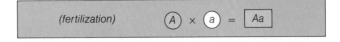

On the basis of such reasoning, a principle was formulated. This principle has been modified somewhat over the years, but it is still a useful starting point for a discussion of the nature of inheritance:

Mendelian principle of segregation. Diploid organisms have a pair of alleles for each trait. The two members of each pair segregate from each other during gamete formation, such that each egg or sperm ends up with one or the other allele, but not both.

Trait	Dominant Form of Trait in One Parent Plant	Recessive Form of Trait in Another Parent Plant	Number of Second-Generation Plants Showing Dominant Form	Number of Second-Generation Plants Showing Recessive Form
seed shape	round	wrinkled	5,474	1,850
seed color	yellow	green	6,022	2,001
pod shape	round, inflated	wrinkled, constricted	882	299
pod color	green	yellow	428	152
flower color	purple	white	705	224
flower position	axial (along stem)	terminal (at tip)	651	207
stem length	tall (6–7 feet)	dwarf (¾–1½ feet)	787	277
	Average ratio for all of the traits tested:		3:1	

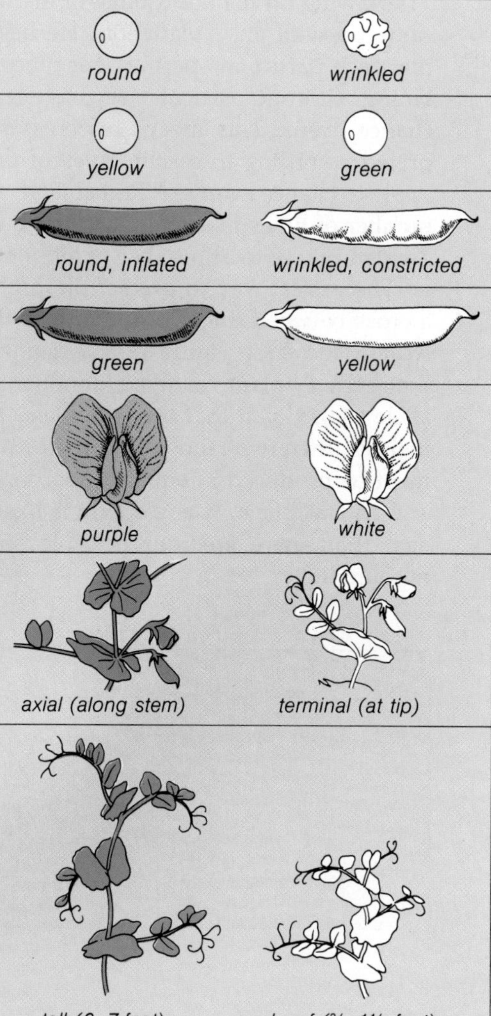

Figure 12.4 Results from Mendel's series of seven monohybrid cross experiments with the garden pea.

Probability: Predicting the Outcome of Crosses

Mendel crossed hundreds of plants and kept track of thousands of offspring, rather than restricting his experiments to a few plants as others had done. Also, he carefully *counted* and *recorded* the number of dominant and recessive offspring for each cross (see, for example, Figure 12.4). Certain numerical ratios emerged with this approach, and they strongly suggested that the hereditary material for each trait does not "blend" at fertilization but rather retains its distinct identity. In describing these ratios, let's make use of a few convenient symbols:

P parental generation

F_1 first generation (the F stands for *filial*, which is taken to mean offspring)

F_2 second generation

An intriguing ratio emerged from the large number of crosses between F_1 plants. On the average, of every

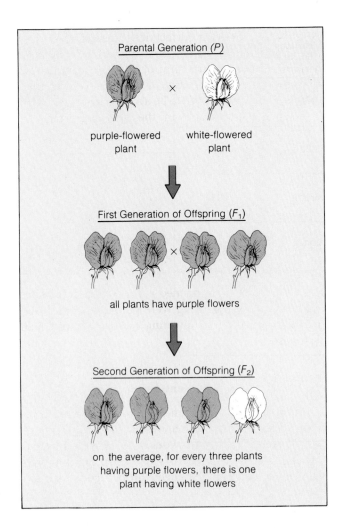

Figure 12.5 Example of a pattern that emerged regularly in Mendel's crosses between true-breeding parents differing only in one contrasting trait (here, flower color). All first-generation plants showed the dominant form of the trait. On the average, though, three of every four second-generation plants showed the dominant form—and one showed the recessive.

four offspring, three showed the dominant form of the trait and one showed the recessive (Figure 12.5). How could this 3:1 phenotypic ratio arise?

Drawing on his knowledge of mathematics, Mendel came up with an explanation. He began by assuming that each particular sperm is not precommitted to combining with one particular egg; fertilization has to be a chance event. This meant his crosses could be interpreted according to certain rules of probability, which apply to chance events. ("Probability" simply means the number of times that one particular outcome will occur divided by the total number of all possible outcomes.)

The easiest way to predict the probable outcome of a cross between the two F_1 plants is the *Punnett-square method*, shown in Figure 12.6. Assume, as Mendel did, that each F_1 plant carried a dominant and a recessive allele (Aa). Given that the two alleles are segregated at meiosis, then two kinds of sperm (or two kinds of eggs) must be produced in equal proportions: half will be A, and half will be a. If any sperm is likely to fertilize any egg, then there are four possible outcomes for each encounter:

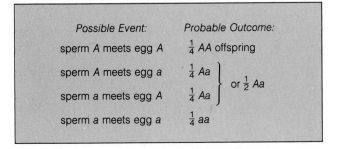

Figure 12.6 Punnett-square method of predicting the probable ratio of traits that will show up in offspring of self-fertilizing individuals known to be heterozygous (Aa) for a trait. The circles represent female gametes, or eggs; circles with "tails" represent male gametes, or sperm. The letters inside gametes represent the dominant or recessive form of the trait being tracked. Each square depicts the genotype of one kind of offspring.

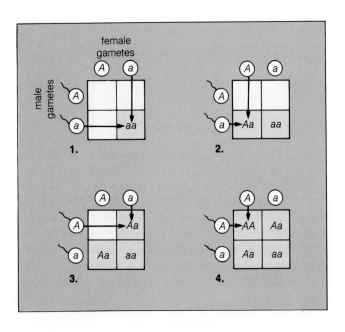

As you can see, there are three chances in four that a new individual will carry one or more dominant alleles. And there is one chance in four that the individual will carry two recessive alleles. Thus the expected phenotypic ratio is 3:1. Figure 12.7 illustrates the outcome for one of Mendel's monohybrid crosses, using the Punnett-square method.

It is important to keep in mind that Mendel's ratios weren't *exactly* 3:1. (See, for example, the numerical results in Figure 12.4.) To understand why, flip a coin a few times. We all know that a coin is just as likely to end up heads as tails. But often it ends up heads, or tails, several times in a row. When the coin is flipped only a few times, the actual ratio may differ considerably from the predicted ratio of 1:1. Only when you flip the coin *many* times will you come close to the predicted ratio. Almost certainly, Mendel's reliance on a large number of crosses and his understanding of probability kept him from being confused by minor deviations from the predicted results.

Testcrosses

Mendel gained support for his concept of segregation through a **testcross**, in which first-generation hybrids are crossed to an individual known to be true-breeding for the same recessive trait as the recessive parent.

For example, purple-flowering F_1 individuals were crossed with true-breeding, white-flowering plants. If Mendel's idea were not correct (if the recessive unit had lost its identity when combined with the dominant unit), then only the dominant form of the trait would show up in the testcross offspring. If his idea were correct, though, there would have to be about as many recessive as dominant individuals in the offspring from the testcross:

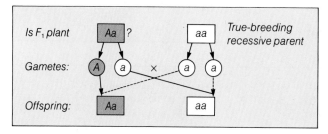

This is exactly what happened in the testcrosses. About half the testcross offspring were purple-flowering (Aa) and half were white-flowering (aa), as Figure 12.8 illustrates.

The Concept of Independent Assortment

In another series of experiments, Mendel crossed true-breeding pea plants that showed contrasting forms of two different traits. We would call these "*di*hybrid crosses," because the heterozygous offspring inherited *two* pairs of nonidentical alleles (at two gene loci; see Figure 12.3).

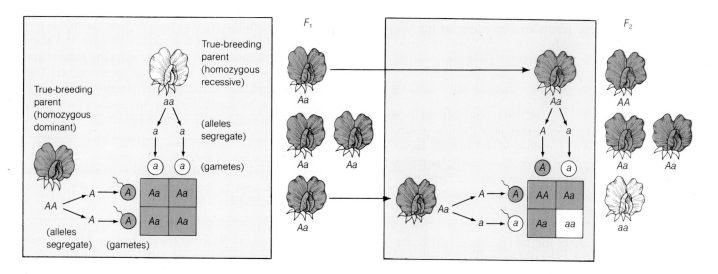

Figure 12.7 Results from one of Mendel's monohybrid crosses, with the Punnett-square diagrams showing the possible allelic combinations in first- and second-generation offspring.

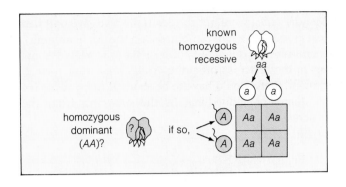

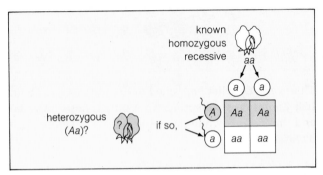

Figure 12.8 Punnett-square method of predicting the outcomes of a testcross between an individual known to be homozygous recessive for a trait (here, white flower color) and an individual that shows the dominant form of the trait. If the individual of unknown genotype is homozygous dominant, all offspring will show the dominant form of the trait. If the individual is heterozygous, about half the offspring will show the recessive form.

In one experiment, plants true-breeding for the following traits were crossed:

purple-flowered,   ×   white-flowered,
tall plant   dwarf plant

Mendel anticipated (correctly) that all of the first generation offspring would be purple-flowered and tall. However, what would happen when these first-generation hybrid plants were crossed? During gamete formation, the hereditary units for flower color and for height would travel either *together* or *independently of each other* into sperm or eggs.

Let *A* and *B* represent the two dominant alleles for flower color and height. Let *a* and *b* represent their recessive alleles. Gametes from a homozygous dominant (*AA BB*) parent would have to be *AB*; those from the

homozygous recessive (*aa bb*) would have to be *ab*. A cross between these plants would produce all *Aa Bb* offspring. Now assume the offspring produce gametes of their own. If the alleles for the two traits travel together as a group, only *two* gene combinations would be possible in those gametes: *AB* (originally from one parent) and *ab* (originally from the other).

If, however, each pair of alleles segregated independently of the other, four combinations would be possible in sperm and in eggs:

$$\tfrac{1}{4}AB \qquad \tfrac{1}{4}Ab \qquad \tfrac{1}{4}aB \qquad \tfrac{1}{4}ab$$

Simple multiplication (four kinds of sperm times four kinds of eggs) shows that sixteen combinations are possible in dihybrid F_2 plants. Figure 12.9 illustrates these possibilities, using the Punnett-square method. When we add up the combinations, we get $\tfrac{9}{16}$ tall purple-flowered, $\tfrac{3}{16}$ dwarf purple-flowered, $\tfrac{3}{16}$ tall white-flowered, and $\tfrac{1}{16}$ dwarf white-flowered plants. That is a phenotypic ratio of 9:3:3:1.

Results were close to this 9:3:3:1 ratio in all of Mendel's dihybrid crosses. That ratio is typical of dihybrid F_2 offspring from a cross involving two different allelic pairs, both of which exhibit dominance.

The variety resulting from independent assortment is staggering. In a simple monohybrid cross, only three genotypes are possible (*AA*, *Aa*, and *aa*). A dihybrid cross (involving two different gene pairs) can yield nine genotypes. When the parents differ in ten gene pairs, almost 60,000 genotypic combinations are possible. When the parents differ in fifteen gene pairs, the possible combinations exceed 14 million!

We now know that the alleles of two or more genes tend to assort independently when those genes are *not* physically located on the same chromosome. During meiosis I, when the pairs of homologous chromosomes line up on both sides of the spindle equator and then separate from each other, the ones from the mother are just as likely to end up at one pole as the other. The same is true of the chromosomes from the father (page 152). Thus, *all* of the alleles from the mother (or father) usually do not end up in the same gamete.

Mendelian principle of independent assortment. The alleles of two (or more) gene pairs that are located on nonhomologous chromosomes tend to be assorted independently of one another into gametes.

On the basis of his experimental results, Mendel was convinced that hereditary material comes in units that retain their physical identity. In 1865 he reported this idea before the Brünn Society for the Study of Natural

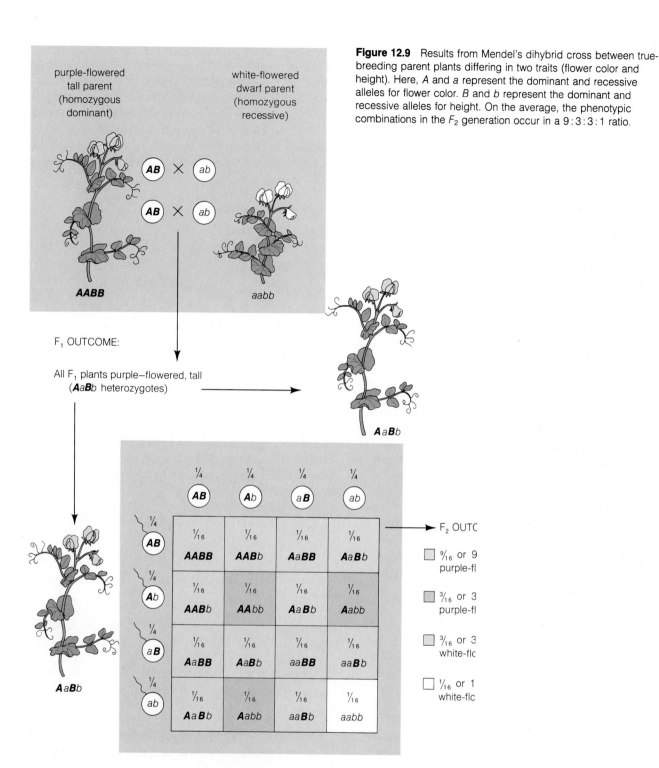

Figure 12.9 Results from Mendel's dihybrid cross between true-breeding parent plants differing in two traits (flower color and height). Here, *A* and *a* represent the dominant and recessive alleles for flower color. *B* and *b* represent the dominant and recessive alleles for height. On the average, the phenotypic combinations in the *F₂* generation occur in a 9:3:3:1 ratio.

purple-flowered tall parent (homozygous dominant)

white-flowered dwarf parent (homozygous recessive)

AABB

aabb

F₁ OUTCOME:

All F₁ plants purple-flowered, tall (**A**a**B**b heterozygotes)

Aa**B**b

Aa**B**b

F₂ OUTC

⁹⁄₁₆ or 9 purple-fl

³⁄₁₆ or 3 purple-fl

³⁄₁₆ or 3 white-flc

¹⁄₁₆ or 1 white-flc

Science. His report made no impact whatsoever. The following year his paper was published. Apparently it was read by few and understood by no one. Remember that Mendel was challenging the well-entrenched blending theory of inheritance. His mathematical analysis of traits probably would not have made sense to anybody

but mathematicians—who probably would not have had the least bit of interest in pea plants.

In 1871, Mendel became an abbot of the monastery, and his experiments gave way to administrative tasks. He died in 1884, never to know his work would be the starting point for the development of modern genetics.

VARIATIONS ON MENDEL'S THEMES

Dominance Relations

It was Mendel's genius to limit his studies to cases of clear-cut dominance. Not all cases are this straightforward. *Some "dominant" alleles of a pair are not fully dominant over their partner, and others are expressed to different degrees (if at all) in various individuals.*

In **incomplete dominance**, one allele of a pair cannot completely mask the expression of its partner. For instance, when homozygous red-flowered and white-flowered snapdragons are crossed, the first-generation plants all have *pink* flowers. Without further tests, this outcome might imply that the hereditary material was blended in offspring. However, cross-fertilization between the first-generation individuals yields these phenotypes: $\frac{1}{4}$ red, $\frac{1}{2}$ pink, and $\frac{1}{4}$ white (Figure 12.10). Apparently a single "red" allele is not enough to form sufficient pigment to make the flowers appear red, as two red alleles can do in homozygous dominant plants.

In **codominance**, the expression of *both* alleles of a pair is discernible in heterozygotes. (In other words, the expression of one does not mask expression of the other.) For example, red blood cells have certain proteins at the cell surface that act like identification markers. (They identify the cell as being of a certain kind.) The gene coding for one of these markers occurs at the so-called ABO blood group locus. It happens that there are more than two forms of alleles for this locus, so they are said to be a *multiple allele system*. Two of the alleles are codominant and code for two forms of the marker (designated A and B). The third allele is recessive, and in homozygous recessive individuals there is an absence of either A or B (a condition called O). Thus four blood types are possible, depending on which two alleles are present in a person's cells. Using the symbols I^A and I^B to represent the codominant alleles and i to represent the recessive allele, then:

Allele Combination:	Blood Type Produced:
$I^A I^A$ (or $I^A i$)	A
$I^B I^B$ (or $I^B i$)	B
$I^A I^B$	AB
ii	O

In **incomplete penetrance**, dominant alleles find expression in some individuals but not in others. For example, identical twins are derived from the same fertilized egg, so they should have identical pairs of alleles for all traits. Yet in one pair of identical twins, one individual developed a genetic defect called a *harelip*, and the other did not (Figure 12.11).

In **variable expressivity**, dominant alleles produce phenotypes that differ, to varying degrees, from one individual to the next. For example, some humans show *polydactyly* (the presence of extra fingers, toes, or both). As the human embryo develops, a dominant allele (*D*) controls how many bone sets will form within the paddlelike appendages destined to become hands and

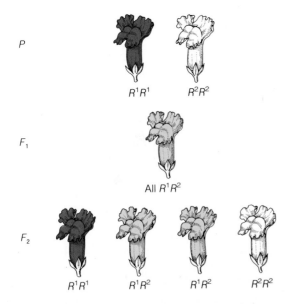

Figure 12.10 Incomplete dominance at one gene locus. Red-flowering and white-flowering homozygous snapdragons produce pink-flowering plants in the first generation. The red allele (R^1) is only partially dominant over the white allele (R^2) in the heterozygous state.

feet. The *Dd* genotype varies in its expressivity. Some carriers of the dominant allele end up with five-fingered hands but six-toed feet; others have five-toed feet but six-fingered hands (Figure 12.12). Still others may end up with five fingers on one hand and six on the other.

Interactions Between Different Gene Pairs

On thinking about the traits considered so far, you might conclude (as Mendel did) that each trait arises from a single gene. However, a single gene generally does not give rise to a trait all by itself. *Genes interact with one another to produce some effect on phenotype.*

For example, two genes that influence comb shape in poultry were identified by W. Bateson and R. Punnett. These two genes cooperate to produce a phenotype that neither gene alone can produce. The most common phenotype, the single comb, is produced by the recessive alleles *r* and *p*. Two dominant alleles (*R* and *P*) can also occur at these loci. Interactions between the two genes can produce single, rose, pea, or walnut combs, depending on the allelic combinations (Figure 12.13).

In one of the more common gene interactions, called *epistasis*, one gene pair masks the expression of another gene pair and some expected phenotypes do not appear at all. For example, some of the genes that produce coat color in mammals help control the type, distribution, and amount of *melanin* (a brownish-black pigment molecule) in a given hair or body region. At one gene locus, the *B* allele for "black coat" is dominant to the *b* allele

for "brown coat." However, the actual phenotype is influenced by a dominant allele (*C*) at an entirely different gene locus. This allele codes for tyrosinase, the first enzyme needed in a series of reactions that produce melanin-containing granules. If an individual is homozygous recessive (*cc*) at this second gene locus, then no melanin can be produced. It makes no difference which

Figure 12.11 Example of incomplete penetrance in identical twins. During embryonic development, certain tissues failed to grow together properly when the upper lip formed in the twin on the left, yet they grew together normally in the twin on the right—even though both twins had identical pairs of alleles for all traits (identical twins arise from the same fertilized egg). Such abnormal closures can be corrected with modern surgery.

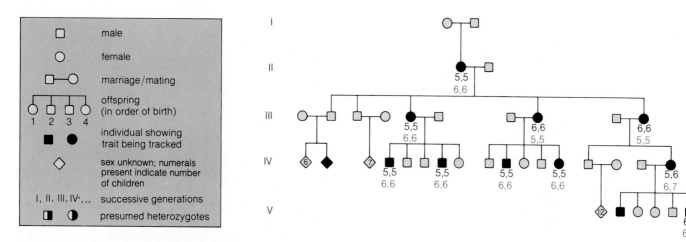

Figure 12.12 Pedigree of polydactyly, showing variable expressivity of the dominant allele for the trait. A "pedigree" is a chart of the genetic relationships of individuals. The phenotype of female I is uncertain, but she probably was polydactylous. The number of digits on each hand is shown in black numerals, the number on each foot is shown in blue. (The boxed inset explains some of the symbols used in constructing pedigree diagrams.)

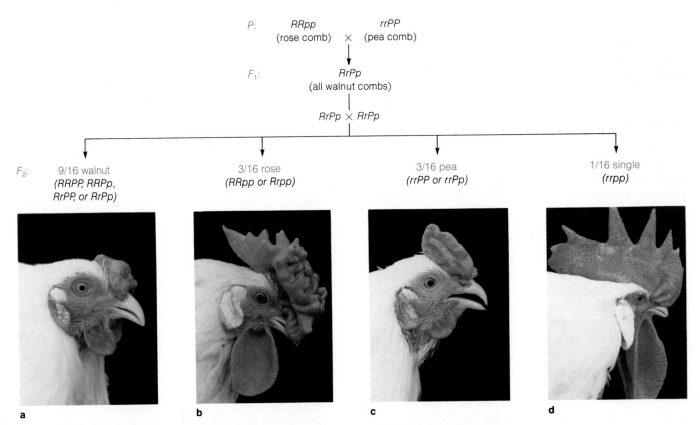

P: RRpp   rrPP
    (rose comb)   ×   (pea comb)

F₁: RrPp
        (all walnut combs)

        RrPp × RrPp

F₂: 9/16 walnut   3/16 rose   3/16 pea   1/16 single
    (RRPP, RRPp,   (RRpp or Rrpp)   (rrPP or rrPp)   (rrpp)
     RrPP, or RrPp)

a   b   c   d

Figure 12.13 Interaction between two genes affecting the same trait in domestic breeds of chickens. The initial cross is between a Wyandotte (with a rose comb on the crest of its head) and a brahma (with a pea comb). With complete dominance at the gene locus for pea comb and at the gene locus for rose comb, the products of these two nonallelic genes interact and give walnut combs (**a**). With complete recessiveness at both loci, the products interact and give rise to single combs (**d**).

combination of *B* and *b* alleles exists at the first gene locus. The animal will be an *albino,* a phenotype arising from the absence of melanin.

Albinism occurs among fishes, amphibians, reptiles, birds, and mammals (Figure 12.14). True albino mammals have white hair, light skin, and red or pink eyes. (With absorptive pigment absent, red light is reflected from blood vessels in the eye.)

When an albino guinea pig with the genotype *cc BB* is mated to a brown guinea pig with the genotype *CC bb,* the F₁ generation will be *Cc Bb* and will have black coats. What will be the phenotypic ratio of the offspring produced when two of these F₁ individuals are mated? (Make a Punnett square to find the answer.)

Many genes modify, interfere with, or prevent the expression of nonallelic genes at different loci.

Multiple Effects of Single Genes

Since Mendel's time, studies have also shown that a single gene can exert effects on seemingly unrelated aspects of an individual's phenotype. This aspect of gene expression is known as **pleiotropy.**

Sickle-cell anemia is actually a group of symptoms produced by an allele that has pleiotropic effects. Hemoglobin A, or HbA, is the red oxygen-carrying pigment in the blood of adult humans, but some individuals carry a variant form known as HbS. The HbS molecules can still carry oxygen, but after giving up the oxygen to other cells in the body, they interlock with one another. They actually stack up like long rigid poles, and the stacking distorts the red blood cells into a "sickle" shape (Figure 12.15).

The deformed cells clump together in capillaries (blood vessels having tiny diameters) and thereby ham-

Figure 12.14 A rare albino rattlesnake, showing the pink eyes and white coloration characteristic of animals that are unable to produce the brownish-black pigment melanin. In birds and mammals, surface coloration is due almost entirely to the color of feathers and fur. In fishes, amphibians, and reptiles, it is due to color-bearing cells located in the skin. Some of the cells contain melanin. Others contain red to yellow pigments. Still others contain crystals that reflect light and alter the effect of other pigments present.

The mutation affecting melanin production in the snake shown here has no effect on the production of yellow-to-red pigments and light-reflecting crystals. Hence the snake's skin appears iridescent yellow as well as white.

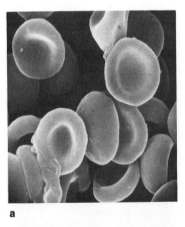

a

b

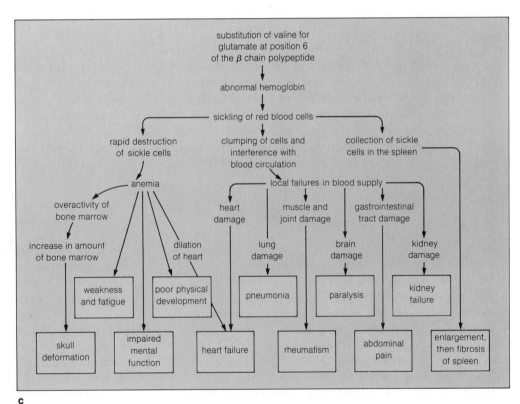

c

Figure 12.15 (a) Normal red blood cells and (b) cells characteristic of the disease sickle-cell anemia. Because of their abnormal, asymmetrical shape, sickle cells do not flow smoothly through fine blood vessels. They pile up in clumps that block blood flow. The tissues served by the blood vessels become starved for oxygen and nutrients even as they become saturated with waste products. (c) Possible pleiotropic effects in sickle-cell homozygotes.

Figure 12.16 Environmental effect on phenotype in the Siamese cat. Fur on the paws and ears is darker than on the rest of the body. The skin temperature in these regions is normally lower than the temperature in the rest of the body. Some of these cats are homozygous recessive for a key gene involved in the formation of the dark pigment melanin. The enzyme produced by this recessive allele is heat-sensitive; it is less active at warmer temperatures. Hence the fur on warmer body parts is lighter in color.

Figure 12.17 Variable expressivity resulting from variation in the external environment. Leaves of the water buttercup (*Ranunculus aquatilis*) show dramatic phenotypic variation, depending on whether they grow underwater or above it. This variation occurs even in the same leaf if it develops half in and half out of water.

per the movement of oxygen to cells and carbon dioxide wastes from them. The impaired gas transfers cause severe damage to many internal organs and tissues. The one normal allele of heterozygotes (HbA/HbS) is fully functional, so heterozygotes show few disease symptoms even though a portion of their blood cells are sickled. It is the homozygous recessive who shows serious phenotypic consequences (Figure 12.15).

Environmental Effects on Phenotype

Interactions of genes or gene products with the environment can lead to variations in phenotypes. For example, temperature variations in different body regions influence the coat color in Siamese cats. In these animals, the main pigment is a brown-black form of melanin. Recall that melanin formation depends on the enzyme tyrosinase. One allele of the gene locus coding for tyrosinase produces a heat-sensitive form of the enzyme; it is less active at warmer temperatures. Thus, in cats homozygous for the mutant allele, warmer body parts have light fur and relatively cool extremities (paws, ears, tail) have dark fur (Figure 12.16).

The water buttercup (*Ranunculus aquatilis*) provides another example of environmental effects on phenotype.

This plant grows in shallow ponds, and some of its leaves develop underwater. The submerged leaves are finely divided, compared with the leaves growing in air. When a leaf-bearing stem is half in and half out of the water, its leaves display both phenotypes (Figure 12.17). Thus the genes responsible for leaf shape produce very different phenotypes when external conditions are different.

Continuous and Discontinuous Variation

The traits Mendel studied would now be called examples of **discontinuous variation**: phenotypes fell into one or another of a few clearly distinguishable classes. Such differences in phenotype were important in Mendel's work, for they were the only markers he could use in identifying and tracking genotypes. You might say that his studies were qualitative in focus, because differences could be established by simple observation, without precise measurements.

However, most of the phenotypic differences in any population are not qualitative. For example, not all humans can be readily classed as tall *or* short, fat *or* thin, and so forth. In most traits, humans and other organisms show **continuous variation**: small degrees of phenotypic

(number of individuals)

| 1 | 0 | 0 | 1 | 5 | 7 | 7 | 22 | 25 | 26 | 27 | 17 | 11 | 17 | 4 | 4 | 1 |

(height, inches)

| 58 | 59 | 60 | 61 | 62 | 63 | 64 | 65 | 66 | 67 | 68 | 69 | 70 | 71 | 72 | 73 | 74 |

a

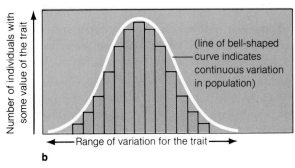

b

Figure 12.18 Example of continuous variation in a population sample: height distribution in a group of 175 United States Army recruits about the turn of the century.

(b) Generalized bell-shaped curve typical of populations showing continuous variation in some trait.

variation occur over a more or less continuous range (Figure 12.18). Measurements of such small differences must be *quantitative*, requiring precise measurements of individuals of the population. The term **quantitative inheritance** refers to the transmission of traits showing continuous variation.

H. Nilsson-Ehle, R. Emerson, and E. East developed the idea that quantitative inheritance arises through the additive influence of three or more gene pairs affecting the same trait. For instance, human skin color ranges through hues of blacks, browns, and whites. As for other mammals, human skin color is influenced by a number of similarly acting gene pairs at different loci. These genes control the type, distribution, and amount of pigment in the skin. Their effect is roughly additive, with the intensity of skin pigmentation being determined by the total number of alleles active at all the different loci.

It is important to understand that quantitative expressions are not a result of "blending," a term that suggests loss of the original identity of hereditary traits. Rather, *all genes retain their physical identity, regardless of the phenotype produced by their combined positive effects and negative effects.*

SUMMARY

1. Through his experiments with the garden pea plant, Mendel demonstrated that the hereditary material (genes) in a sperm does not "blend" at fertilization with the genes in an egg; rather, genes retain their identity from one generation to the next. Thus (in modern terminology), pea plants and other diploid organisms have *two* separate genes for each physical trait, one on the chromosome inherited from the male parent and one on the homologous chromosome from the female.

2. Through his monohybrid crosses (between two true-breeding plants that showed contrasting forms of a single trait), Mendel perceived that there can be alternative forms of each gene (alleles), some of which are dominant over other, recessive forms. Assuming that fertilization is a random event, we find that the possible allelic combinations in F_1 offspring of a monohybrid cross are:

sperm:		egg:		
A	×	A	=	AA (dominant)
A	×	a	=	Aa (dominant)
a	×	A	=	Aa (dominant)
a	×	a	=	aa (recessive)

That is an expected phenotypic ratio of 3:1. (The term phenotype refers to an individual's physical traits.)

3. Analysis of such monohybrid cross results supports the Mendelian principle of segregation: Diploid organisms have a pair of alleles for each trait. The two members of each pair segregate from each other during gamete formation, such that each egg or sperm ends up with one or the other allele, but not both.

4. A diploid individual that inherits two dominant alleles for a trait is homozygous dominant. A homozygous recessive individual has two recessive alleles. A heterozygous individual has two different alleles.

5. Mendel also performed dihybrid crosses (between two true-breeding plants that showed contrasting forms of two different traits). The results from many experiments were close to a $9:3:3:1$ ratio:

9 dominant for both traits (one or two A, one or two B)

3 dominant for A, recessive for b (one or two A, two b)

3 dominant for B, recessive for a (one or two B, two a)

1 recessive for both traits (two a, two b).

5. On the basis of such dihybrid crosses, the Mendelian principle of independent assortment was formulated. In modern terms, the principle is stated this way: The alleles of two (or more) different gene pairs that are located on nonhomologous chromosomes tend to be assorted independently of one another into gametes.

6. Mendel established the foundations for genetic analysis (the study of inheritance patterns). We still rely on his most important insight for analyzing monohybrid and dihybrid crosses—that the two alleles of a gene segregate from each other during gamete formation.

7. As Mendel thought, many genes are indeed assorted independently of one another and give rise to diverse combinations of traits in offspring. However, we now know that other genes do not assort independently. They are physically located on the same chromosome in such a way that they almost always travel together into the same gamete, thereby introducing variations into phenotypic ratios.

8. Mendel's concept of dominance has undergone the greatest modification. In any population, most traits are not clearly contrasting (as in the garden pea); they show continuous variation, which does not lend itself readily to simple analysis. Continuous variation arises through interactions between many different genes and through additive interactions among different alleles. Thus,

   a. Gradations of dominance exist between two alleles for a given gene locus, so that the expression of one or both may be fully dominant, or one may be incompletely dominant over the other.

   b. The activity of a single gene may have major or minor effects on more than one trait.

   c. The activity of a gene in space and time may be influenced by other genes, with the sum total of their positive and negative actions producing some effect on phenotype.

   d. The environment can influence the expression of genes and their contribution to phenotype.

Review Questions

1. State the Mendelian principle of segregation. Does segregation occur during mitosis or meiosis?

2. Distinguish between the following terms:
   a. Gene, allele, and gene pair
   b. Dominant trait and recessive trait
   c. Homozygote and heterozygote
   d. Genotype and phenotype

3. Give an example of a self-fertilizing organism. What is cross-fertilization?

4. Distinguish between monohybrid and dihybrid crosses. What is a testcross, and why is it valuable in genetic analysis?

5. State the Mendelian principle of independent assortment. Does independent assortment occur during mitosis or meiosis? Does independent assortment always occur?

6. How does quantitative inheritance differ from the notion of "blending" of heritable traits?

7. Contrast continuous and discontinuous variation, and outline the genotypic basis for both.

8. Mendel's concept of dominance was based on observations of inheritance patterns in clearly contrasting traits. How has this concept since been modified?

9. List some of the factors influencing the expression of phenotype.

10. Define epistasis. Give a good example.

11. What does codominance mean? Give a good example.

Readings

Dunn, L. 1965. *A Short History of Genetics*. New York: McGraw-Hill.

Goodenough, U. 1984. *Genetics*. Third edition. Philadelphia: Saunders.

Mendel, G. 1959. "Experiments in Plant Hybridization." Translation in J. Peters (editor), *Classic Papers in Genetics*. Englewood Cliffs, New Jersey: Prentice-Hall.

Singer, C. 1962. *A History of Biology to About the Year 1900*. New York: Abelard-Schuman.

Stern, C., and E. Sherwood (editors). 1966. *The Origin of Genetics*. San Francisco: Freeman. Includes a modern translation of Mendel's paper and correspondence.

Strickberger, M. 1985. *Genetics*. Third edition. New York: Macmillan. Still the classic introduction to genetics.

Witkop, C., Jr. 1975. "Albinism." *Natural History* 84(8):48–59. Good account of the various forms of albinism—their biochemical and genetic bases and their phenotypic manifestations.

Genetics Problems (Answers appear in Appendix II)

1. One gene has alleles A and a; another gene has alleles B and b. For each of the following genotypes, what type(s) of gametes will be produced?
   a. *AA BB*
   b. *Aa BB*
   c. *Aa bb*
   d. *Aa Bb*

2. Referring still to the preceding problem, what genotypes will be present in the offspring from the following matings? (Indicate the frequencies of each genotype among the offspring.)
   a. *AA BB* × *aa BB*
   b. *Aa BB* × *AA Bb*
   c. *Aa Bb* × *aa bb*
   d. *Aa Bb* × *Aa Bb*

3. In one experiment, Mendel crossed a true-breeding pea plant having green pods with a true-breeding pea plant having yellow pods. All of the F_1 plants had green pods.
   a. Which trait (green or yellow pods) is recessive? Can you explain how you arrived at your conclusion?
   b. Suppose the F_1 plants are self-pollinated and 135 F_2 plants are produced. What phenotypes should be present in the F_2 generation, and how many of the plants in that generation should show each of those phenotypes?

4. Being able to curl up the sides of your tongue into a U-shape is under the control of a dominant allele at one gene locus. (When there is a recessive allele at this locus, the tongue cannot be rolled.) Having free earlobes is a trait controlled by a dominant allele at a different gene locus. (When there is a recessive allele at this locus, earlobes are attached at the jawline.) The two genes controlling tongue-rolling and free earlobes assort independently. Suppose a woman who has free earlobes and who can roll her tongue marries someone who has attached earlobes and who cannot roll his tongue. Their first child has attached earlobes and cannot roll the tongue.
   a. What are the genotypes of the mother, the father, and the child?
   b. If this same couple has a second child, what is the probability that it will have free earlobes and be unable to roll the tongue?

5. In addition to the two genes mentioned in Problem 1, assume you now study a third gene having alleles *C* and *c*. For each of the following genotypes, indicate what type (or types) of gametes will be produced:
   a. *AA BB CC*
   b. *Aa BB cc*
   c. *Aa BB Cc*
   d. *Aa Bb Cc*

6. A man is homozygous dominant for ten different genes, which assort independently. How many genotypically different types of sperm could he produce? A woman is homozygous recessive for eight of these ten genes, and she is heterozygous for the other two. How many genotypically different types of eggs could she produce? What can you conclude regarding the relationship between the number of different gametes possible and the number of heterozygous and homozygous genes that are present? For a species (or population), what might be the biological benefits, if any, of possessing a large number of heterozygotes?

7. Recall that Mendel crossed a true-breeding tall, purple-flowered pea plant with a true-breeding dwarf, white-flowered plant. All the F_1 plants were tall and purple-flowered. If an F_1 plant is now self-pollinated, what is the probability of obtaining an F_2 plant heterozygous for the genes controlling height and flower color?

8. Assume that a new gene was recently identified in mice. One allele at this gene locus produces a yellow fur color. A second allele produces a brown fur color. Suppose you are asked to determine the dominance relationship between these two alleles. (Is it one of simple dominance, incomplete dominance, or codominance?) What types of crosses would you make to find the answer? On what types of observations would you base your conclusions?

9. The ABO blood system has often been employed to settle cases of disputed paternity. Suppose, as an expert in genetics, you are called to testify in a case where the mother has type A blood, the child has type O blood, and the alleged father has type B blood. How would you respond to the following statements of the attorneys:
   a. "Since the mother has type A blood, the type O blood of the child must have come from the father, and since my client has type B blood, he obviously could not have fathered this child." (*Made by the attorney of the alleged father*)
   b. "Further tests revealed that this man is heterozygous and therefore he must be the father." (*Made by the mother's attorney*)

10. In mice, at one gene locus, the dominant allele (*B*) produces a dark-brown pigment; and the recessive allele (*b*) produces a light-brown, or tan, pigment. An independently assorting gene locus has a dominant allele (*C*) that permits the production of all pigments. Its recessive allele (*c*) makes it impossible to produce any pigment at all. The pigmentless condition is called "albino."
    a. A homozygous *bb cc* albino mouse mates with a homozygous *BB CC* brown mouse. In what ratios would the phenotypes and genotypes be expected in the F_1 and F_2 generations?
    b. If an F_1 mouse from part (a) above were backcrossed to its albino parent, what phenotypic and genotypic ratios would be expected? Diagram the crosses in both parts (a) and (b).

11. In chickens, two different pairs of genes are responsible for producing scales, rather than feathers, on the lower part of the leg (the shank, which is the part just below the "drumstick"). Only when the recessive alleles of both pairs of genes are homozygous will feathers appear on the shanks. Call the two loci *F* and *S*. What phenotypic and genotypic ratios will be found among the progeny of two doubly heterozygous parents?

12. Certain dominant alleles are so important for normal development that the mutant recessive alleles, when homozygous, lead to the death of the organism. However, such recessive alleles can be perpetuated as heterozygotes (*Ll*), which in many cases are not phenotypically different from homozygous (*LL*) normals. (Some *lethal heterozygotes* do have a mutant phenotype.) Consider the mating of two such heterozygotes, *Ll* × *Ll*. Among their surviving progeny, what fraction will be heterozygous?

13. In corn, a series of three independent pairs of gene loci (*A*, *C*, and *R*) affect the production of pigment that leads to kernel color. If any one of the three pairs is in the homozygous recessive state, then no pigment will form in the kernels. However, if at least one dominant allele of each locus is present, then pigment can form in the kernel. Two corn plants with the following genotypes were crossed:

$$Aa\ cc\ Rr \quad \times \quad aa\ Cc\ Rr$$

What fraction of the progeny kernels will be pigmented? (Note: each kernel represents a separate (potential) individual; it will exhibit the pigment phenotype of the plant that can be grown from it.)

13

CHROMOSOMAL THEORY OF INHERITANCE

RETURN OF THE PEA PLANT

The year was 1884, Mendel's paper on pea plant hybridization had been gathering dust in at least a hundred libraries for nearly two decades, and Mendel himself had just passed away. Ironically, separate streams of research were even now moving toward the same principles of inheritance he had stated so carefully in that ignored, forgotten paper. Cytology, the study of cell structure and function, was about to converge with genetic analysis.

Interest in the basis of inheritance had been renewed by new developments in microscopy, and cytologists were beginning to suspect that the hereditary material resides in the cell nucleus. By 1882, Walther Flemming reported seeing threadlike bodies—chromosomes—in nuclei of dividing salamander cells. He called the division process mitosis, after the Greek word for thread. Other researchers had a hunch that the chromosomes were not being divided into daughter cells for no purpose at all. *Could those threadlike chromosomes be the hereditary material?*

Further studies showed that each sperm or egg has *one* set of chromosomes, whereas a fertilized egg has *two*. In 1887, August Weismann proposed that a special division process must reduce the number of chromosomes by half during gamete formation. Sure enough, in that same year the process of meiosis was identified.

Figure 13.1 Overview of some key aspects of chromosome structure and function.

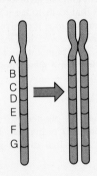 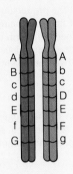 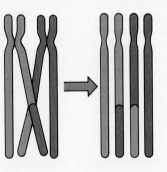 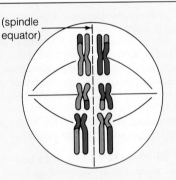

(spindle equator)

1 *Genes are arranged in linear sequence on a chromosome. (Before a cell divides, each chromosome is copied; and the original and the copy remain attached as sister chromatids. The copy has the same gene sequence.)*

2 *Homologous chromosomes can carry identical or nonidentical alleles at corresponding gene locations. (Here, the blue chromosome is from the father; and the red chromosome is its homologue from the mother.)*

3 *During meiosis I, the homologues pair. As a result of crossing over and recombination, a chromosome that ends up in a gamete may include parts of the original maternal <u>and</u> paternal homologues.*

4 *During meiosis I, the chromosomes from the mother randomly align with their homologues from the father on one or both sides of the spindle equator. Thus each pair of homologues assorts independently of the others into gametes.*

Weismann now began to promote his theory of heredity: (1) the chromosome number is halved during meiosis, (2) the original chromosome number is restored when sperm and egg combine at fertilization, and (3) half the hereditary material in offspring is therefore paternal in origin, and half maternal. His views were hotly debated, and the debates drove researchers into devising ways to test the theory. Throughout Europe there was a flurry of quantitative, experimental crosses—just like the ones carried out by Mendel.

Mendel's pioneering work was finally acknowledged in 1900, after three independent researchers came across his paper. They had been checking the literature for reports that might relate to their own hybridization studies—which turned out to reinforce Mendel's concept of segregation. Through these and other studies, it became clear that Flemming's "threads" were probably the carriers of Mendel's "units" of inheritance.

THE CHROMOSOMAL THEORY

This chapter covers some observations, experiments, and hypotheses about hereditary mechanisms that unfolded in the decades after the rediscovery of Mendel's work. Taken together, they lend impressive support to what is now called the **chromosomal theory of inheritance**:

1. The chromosome is the vehicle by which hereditary information is physically transmitted from one generation to the next.

2. In each chromosome, units of hereditary information called *genes* are arranged one after the other, in linear sequence (Figure 13.1).

3. Diploid cells have two sets of *homologous* chromosomes, one maternal and the other paternal in origin. Each homologue resembles its partner in length, shape, and gene sequence.

4. At each gene locus in that sequence, the two homologues may have identical or nonidentical forms of the gene, called *alleles*.

5. Each homologue pairs with its partner at meiosis; then the two separate and end up in separate gametes. Meiosis thus reduces the *diploid* number of chromosomes ($2n$) to the *haploid* number (n) in forthcoming gametes.

6. Which homologue of a pair ends up in which gamete is random, because chromosomes *assort independently*. Thus each gamete may contain some assortment of chromosomes from both parents (Figure 13.1).

7. Different genes on the same chromosome tend to travel together into gametes. But during meiosis, *crossing over* and recombination occur: homologues break and exchange corresponding segments. Thus a chromosome in a gamete may include parts of the original maternal and paternal homologues.

8. Sometimes a chromosome segment is deleted, duplicated, inverted, or moved to a new location. Sometimes chromosomes or whole chromosome sets are improperly distributed into gametes, which end up with an abnormal chromosome number. Such events are *chromosomal aberrations*.

9. Crossing over, independent assortment, and chromosomal aberrations can lead to variations in an individual's genotype (genetic makeup). Thus they all contribute to variations in phenotype (observable traits) upon which selective agents can act. In short, these chromosomal events are sources of diversity and evolutionary change.

CLUES FROM SEX CHROMOSOMES

Enter the Fruit Fly

The growing speculation that genes might be carried on chromosomes was first confirmed by Thomas Hunt Morgan and his students in the early 1900s. Morgan had considered using different kinds of organisms for breeding experiments, and to his mind the fruit fly *Drosophila melanogaster* would be a better choice than the plants and animals that had been used before. These small flies are commonly observed in summer and fall, when they hover over ripe and rotting fruit. They can be grown in bottles on bits of rotting fruit and yeast. Each female can lay several hundred eggs in two or three weeks, and her offspring reach reproductive age in less than two weeks. Thus Morgan could track hereditary traits through nearly thirty generations of thousands of flies in the space of a year. Before long, his laboratory was filled with bottles of busy fruit flies.

Sex Determination

By the time Morgan began his fruit fly experiments, the male and female of many animal species, including fruit flies, were known to be not quite alike in their chromosome sets. All but two of the chromosomes were of the same number and kind in the body cells (somatic cells) of males and females; they were named *autosomes*. The two others were named the *sex chromosomes*, because it is possible to say whether a somatic cell is from a male or female simply by looking at which of those chromosomes it contains.

For example, female grasshopper cells have two of the sex chromosomes designated "X," but male grass-

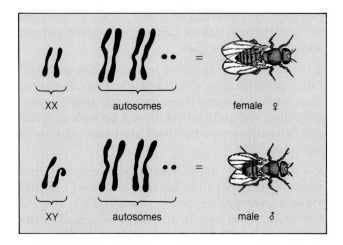

Figure 13.2 Sex chromosomes and autosomes of *Drosophila melanogaster*. Together they represent a diploid number of eight (four chromosome pairs).

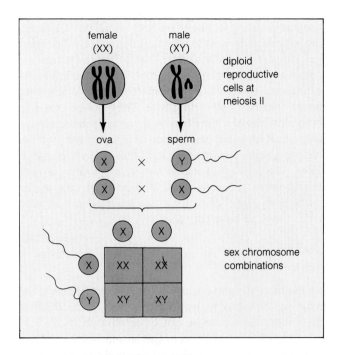

Figure 13.3 Sex determination in humans. This same pattern occurs in many animal species. Only the sex chromosomes, not the autosomes, are shown. Males transmit their Y chromosome to their sons, but not to their daughters. Males receive their X chromosome only from their mother.

hopper cells have only one. Thus the sex chromosome pattern is XX in the female grasshopper and XO in the males (the "O" signifies the absence of a homologue).

More commonly, however, the X chromosome in males has a partner, called the "Y" chromosome, which is often smaller and shaped differently. Thus, for example, female fruit flies are XX, and the males XY (Figure 13.2). The same is true of humans.

The most common sex chromosome pattern is XX (female) and XY (male).

If a female parent has two identical X chromosomes, it follows that each gamete she produces will carry one X chromosome. If a male parent has one X and one Y chromosome (with the X and Y acting as homologues during meiosis), then half the gametes he produces will carry an X and the other half will carry a Y chromosome. When an X-bearing sperm fuses with an X-bearing egg, the resulting zygote develops into an XX female. When a Y-bearing sperm fuses with the egg, the zygote develops into an XY male (Figure 13.3).

This is the meaning of the segregation pattern: Because half of the sperm produced have a Y chromosome and half have an X chromosome, male and female offspring should occur in a 50:50 ratio.

The inheritance of a clear phenotypic difference—maleness versus femaleness—is associated with specific chromosomes.

This association turned out to be important in Morgan's studies of the relationship between genes and chromosomes. Why? If a certain trait—say, eye color—were manifest only in male or in female offspring of an experimental group, *then perhaps the genes coding for those traits could be assigned to specific sex chromosomes.* This would provide strong evidence that genes are indeed located on chromosomes.

Sex-Linked Traits

In Mendel's monohybrid crosses, the phenotypic results in the F_1 and F_2 offspring were the same regardless of which parent carried the recessive allele (page 160). But this turned out *not* to be the case for genes located on the X chromosome. Genes located on the X but not on the Y chromosome are said to be **sex-linked**.

The first detailed study of sex-linked genes took place in Morgan's laboratory. When Morgan began his studies, all the flies he raised were wild-type for eye color; all had brick-red eyes. ("Wild-type" simply refers to the normal or most common allele at a given gene locus.) In 1910, a *white-eyed* male was detected in one of the

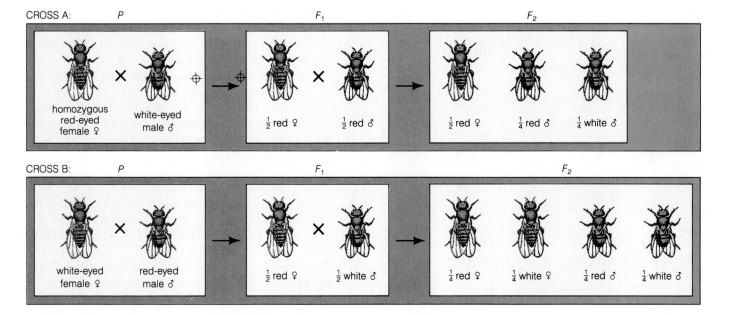

Figure 13.4 Phenotypic results in the F_1 and F_2 generations produced by a reciprocal cross. In the first of these paired crosses, Morgan mated a red-eyed female with a white-eyed male fruit fly. In the second cross, he mated a white-eyed female with a red-eyed male.

laboratory bottles. Apparently, the variant form arose through a spontaneous mutation in a gene controlling eye color.

Morgan established true-breeding strains of white-eyed males and white-eyed females. Then he crossed white-eyed males with true-breeding (homozygous) red-eyed females. Through reciprocal crosses, he discovered that the phenotypic outcome depended on whether the male or female parent exhibited the white-eye trait. A **reciprocal cross** is a pair of crosses. In one cross, one parent exhibits the trait in question; in the second cross, the other parent exhibits the trait.

As Figure 13.4 suggests, Morgan's test results did not seem to fit with Mendel's rules of inheritance. How could they be explained? Morgan had an idea. Female fruit flies (XX) obviously inherit one X chromosome from each parent. Thus the F_1 female offspring of a white-eyed male and a homozygous red-eyed female would have to carry the recessive allele on one of their X chromosomes. Males (XY) inherit their X chromosome only from their maternal parent. *But what if there were no eye-color gene on the Y chromosome?* If only X chromosomes carried the gene, then it would be expressed in males regardless of whether the gene was dominant or recessive!

Figure 13.5 shows the results that can be expected when the idea of a sex-linked gene is combined with Mendel's concept of segregation. By proposing that a specific gene is found only on the X chromosome, Mor-gan was able to explain the seemingly curious dominant-to-recessive ratios of his reciprocal crosses. The actual outcomes matched the predicted outcomes.

Morgan's work confirmed the existence of sex-linked genes, which are located only on an X chromosome and have no alleles on the Y chromosome.

(Before leaving the sex chromosomes, be sure you are clear on this point: Sex chromosomes, which carry major sex-*determining* genes, also carry genes that have nothing to do with sex determination. Similarly, some of the genes on autosomes influence the overall male or female phenotype. In other words, maleness or female-ness is not the exclusive domain of sex chromosomes.)

DISCOVERY OF LINKED GENES

With so many new flies being produced in the *Drosophila* bottles, Morgan and his group were able to collect a number of spontaneously mutated individuals. By 1915 they had isolated more than eighty types of mutants for use in dihybrid cross experiments. (In these crosses, recall, parents true-breeding for different forms of *two* traits produce offspring that are heterozygous at *two* gene loci.)

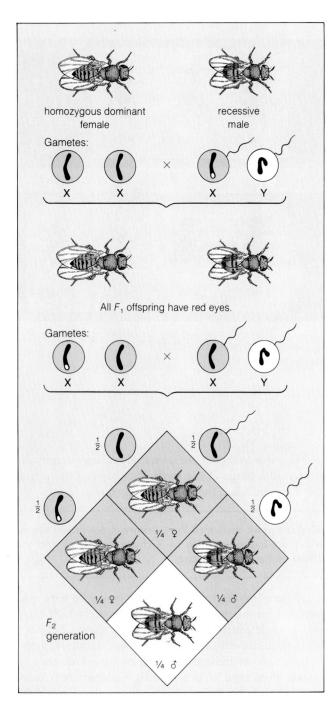

Figure 13.5 Correlation between sex and eye color in *Drosophila*. Given the genetic makeup of the F_2 generation, the recessive allele (depicted here by the white dot) must be carried on the X chromosome only.

This diagram corresponds with Cross A in Figure 13.4. Can you construct a similar diagram to explain the results for Cross B?

In some experiments, flies showing mutant forms of two traits were crossed with wild-type flies, and the results conformed to Mendel's concept of independent assortment. The mutant forms did not show up in the first generation but they reemerged in the second, in a typical 9:3:3:1 ratio (page 165).

However, other dihybrid crosses yielded an unexpected result, for genes coding for the two traits ended up together in the *same* gamete. For instance, a fly mutant for wing shape *and* body color was crossed with a wild-type fly. Wild-type individuals have straight, flat wings (designated *C*) and a gray body (*B*). One mutant form had curved wings (*c*) and a black body (*b*). If the two genes assorted independently, a cross between heterozygous flies (*Cc Bb*) would yield offspring in about a 9:3:3:1 ratio. Yet the ratio seemed to be closer to 3:1. The gene for wing shape obviously was not assorting independently of the gene for body color.

In time, Morgan and his colleagues identified four groups of apparently linked genes in *D. melanogaster*—which happens to have four chromosomes in its gametes. Most likely, each group of genes was located on one of the four chromosomes. The term **linkage** eventually was applied to the tendency of genes physically located on the *same* chromosome to remain together in crosses.

Different genes located on nonhomologous chromosomes can assort independently of each other into gametes.

Different genes physically located on the same chromosome tend to end up together in the same gamete; they do not assort independently.

CROSSING OVER AND LINKAGE MAPPING OF CHROMOSOMES

Even when genes are linked (located on the same chromosome), they do not always end up together in the same gamete. As we have seen, during meiosis the non-sister chromatids of homologous chromosomes undergo breakage, and they exchange segments at corresponding breakage points. This event, called *crossing over*, results in genetic recombination; alleles from both parents end up on a chromosome (Figure 13.6).

Crossing over apparently can disrupt gene linkages at any point along the length of a chromosome. But as Alfred Sturtevant discovered, *the probability of crossing over and recombination occurring at a point somewhere between two genes is proportional to the distance that separates them*. Two genes located very close together on a chromosome

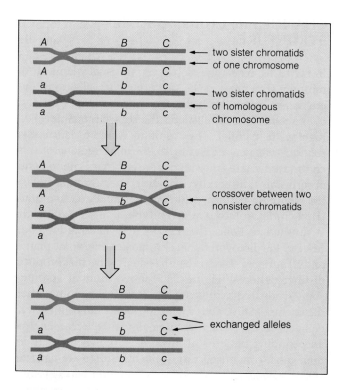

Figure 13.6 Genetic recombination as a result of crossing over between nonsister chromatids of homologous chromosomes during meiosis I.

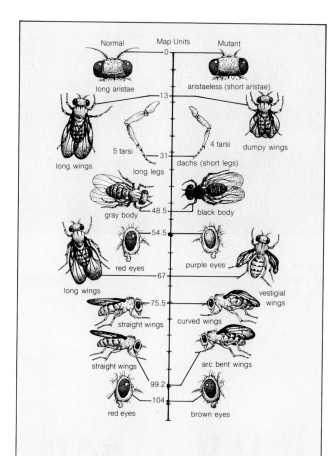

Figure 13.7 Genetic mapping of genes on a segment of chromosome 2 in *D. melanogaster*. Such maps don't show actual physical distances between genes. Rather they show relative distance between gene locations that undergo crossing over and other chromosomal rearrangements. Only if the probability of crossing over were equal along the chromosome's length (which it is not) would it be possible to calculate physical distance exactly.

Here, distances between genes are measured in map units, based on the frequency of recombination between the genes. Thus, if the frequency turns out to be 10 percent, the genes are said to be separated by 10 map units. The amount of recombination to be expected between "vestigial wings" and "curved wings," for instance, would be 8.5 percent (75.5 − 67).

nearly always end up in the same gamete; they are tightly linked. Two genes relatively far apart undergo crossing over and recombination more often than tightly linked genes do; they are loosely linked. Finally, two genes very far apart undergo crossing over and recombination so often that they act as if they assort independently, even though they are located on the same chromosome.

For example, Mendel thought he was observing independent assortment for seven different traits of the garden pea—and this plant has seven chromosomes in its gametes. As it happens, the genes for two of those traits are far apart on the same chromosome—but crossing over and recombination occur so frequently between them that it *seems* as if they are located on separate chromosomes.

The relationship between the linear arrangement of genes on chromosomes and their distribution patterns into gametes is so regular that it can be used to determine the positions of genes relative to one another on a given chromosome. Plotting their positions is called **linkage mapping** of genes. Figure 13.7 is an example of these maps.

Of the several thousand genes in the four chromosomes of *Drosophila* gametes, the positions of about 1,000 have been mapped. There are surely many more genes in human chromosomes, but relatively few of them have been identified and mapped. We still have a long way to go in the mapping of the physical basis of heredity. But work to date shows clearly that genes are carried on chromosomes—*and that they are carried in linear array.*

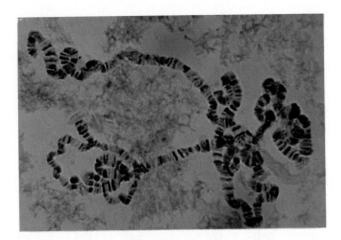

Figure 13.8 Polytene chromosomes from a salivary gland cell of *Drosophila*, an insect that played a key role in the development of the chromosomal theory of inheritance.

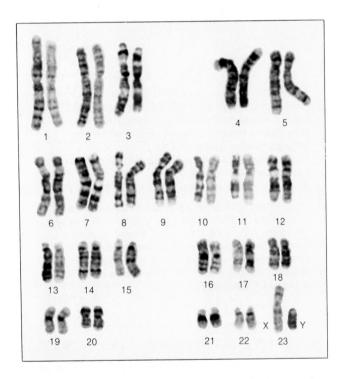

Figure 13.9 Karyotype of a human male. Normally, human cells have a diploid chromosome number of 46. The nucleus contains 22 pairs of autosomes and one pair of sex chromosomes (X and Y). Each chromosome of a given type has already undergone replication.

These chromosomes were prepared for microscopy by G-staining, which provides the banded appearance of chromosomes. The bands are used in karyotype analysis.

CHANGES IN CHROMOSOME STRUCTURE

In the 1930s, researchers observed unusually large chromosome structures in the salivary gland cells from several insect larvae. Each chromosome is copied over and over again in these cells, and the copies remain packed together in parallel. The resulting structure is called a *polytene chromosome* (Figure 13.8). Some regions of this chromosome are more condensed than others and absorb stains differently. The differences produce a distinct banding pattern, which is the same in all nonmutated chromosomes of a given type.

In the late 1960s, thanks to a new staining method, researchers discovered that chromosomes of some plants and all vertebrates also absorb stain in ways that produce distinct patterns. Figure 13.9 shows human chromosomes prepared by this method (called G-banding, after the Giemsa stain). Today, chromosomes can be precisely characterized by such features as their relative lengths and banding patterns at metaphase (when they are in their most condensed form). Such features are used to create a *karyotype*, a visual representation of a chromosome set in which individual chromosomes are arranged in order, from largest to smallest (Figure 13.10).

The distinct banding patterns of chromosomes can be used to establish specific gene locations and to pinpoint where abnormal breaks occurred in altered chromosomes. Four kinds of structural alterations are called duplications, deletions, inversions, and translocations.

In a *duplication*, a repeat of a gene sequence becomes positioned right next to the region that it repeats. Suppose two identical sequences (ABCDABCD) occur in a chromosome. It is possible that homologues will not align properly at this region when they pair at meiosis I. If crossing over affects the improperly aligned region, then one chromosome might end up with three of the sequences and its partner with only one:

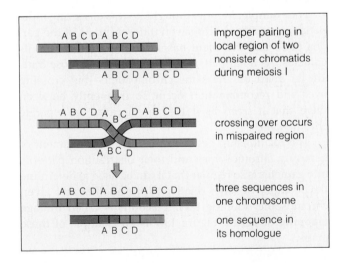

A *deletion* (loss of a chromosome segment) can arise when viral attack, irradiation, or chemical action causes two breaks in a chromosome region. Enzymes can repair the breaks, but sometimes they accidentally leave out the segment in between. There are almost always problems when the altered chromosome is transmitted to offspring, for genes controlling one or more traits may be lost entirely.

An *inversion* is a chromosome segment that has been excised and rejoined at the same place—but in the reverse order, so the position and sequence of its genes are altered:

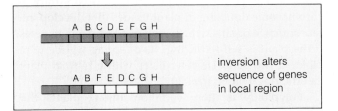

In a *translocation*, a segment from one chromosome is permanently transferred to a nonhomologous chromosome. One translocation is often observed in certain human cells that have become cancerous. In these cells, part of human chromosome 8 has become translocated to chromosome 14. It may be that the translocated segment codes for a growth factor that triggers the cancerous transformation (page 227).

CHANGES IN CHROMOSOME NUMBER

Missing or Extra Chromosomes

On rare occasions, homologous chromosomes fail to separate at meiosis and both end up in the same nucleus. As a result, gametes with the wrong number of chromosomes may be produced. If such a gamete is involved in fertilization, the resulting zygote (hence the new multicelled individual) also will have the wrong number of chromosomes. This condition is called **aneuploidy**.

Suppose improper chromosome separation leads to a gamete having an extra chromosome ($n + 1$). If that gamete combines with a normal one at fertilization, the new individual will have a chromosome number of $2n + 1$. This condition, in which three chromosomes of the same kind are present in the same set, is called "trisomy." Similarly, if the gamete deprived of a chromosome combines with a normal one, the new individual will have a chromosome number of $2n - 1$. This condition is called "monosomy." One form of trisomy, called Down's syndrome, is described in the chapter to follow.

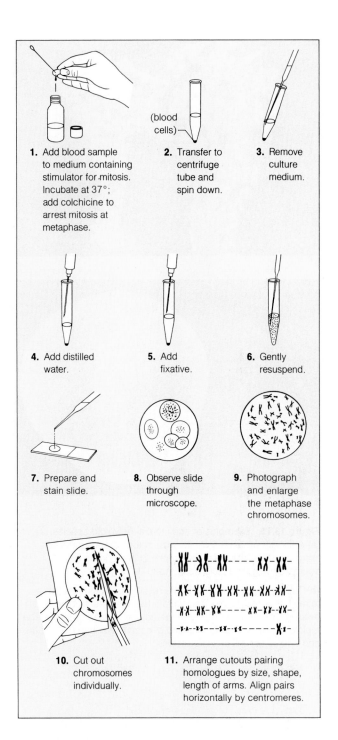

Figure 13.10 Simplified picture of karyotype preparation.

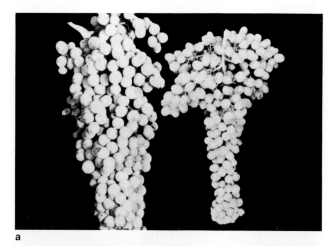

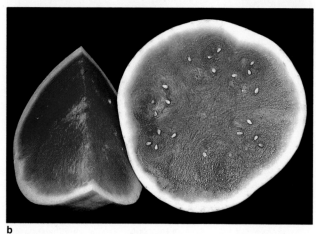

Figure 13.11 Examples of commercial polyploid species.
(**a**) Artificially increased cluster size in a colchicine-induced tetraploid (76n) representative of the Sultana grape (right), compared with its diploid counterpart (left).

(**b**) The seedless watermelon: the result of a cross between a tetraploid and a diploid. The triploid fruit matures, but fertility is reduced; hence most of the seeds do not develop normally. The triploid condition is characteristic of commercial bananas, pears, and apples as well as watermelons.

Three or More Chromosome Sets

Not all eukaryotes are diploid. Some—including about half of all flowering plant species—are **polyploid**, with three or more chromosome sets in their cells.

Polyploidy can arise when improper separation of all the homologues at meiosis leads to diploid gametes. If such a gamete combines with a normal haploid gamete at fertilization, a "triploid" (3n) individual results. If two diploid gametes combine, a "tetraploid" individual results. Usually, triploid individuals are sterile. In fact, any polyploid with an uneven number of chromosome sets (3, 5, 7 . . .) is likely to be sterile, because the extra chromosomes have no homologues with which to pair at meiosis; hence the polyploid cannot produce normal gametes.

Polyploidy also can arise when cells copy their DNA but do not divide. For example, this occurs in human liver cells, which can have two, four, eight, and sometimes sixteen chromosome sets. It also occurs through chromosome doubling in plant tissues that develop into reproductive organs, which give rise to diploid gametes. If the plant is self-fertilizing, its offspring will be tetraploid. If the plant cross-fertilizes with a normal plant, its offspring will be triploid.

Polyploidy is more common among plants than among animals. One reason is that plants generally have no sex chromosomes; for animals, polyploidy can disturb the balance between autosomes and sex chromosomes that is essential for proper development and reproduction. Also, many plants can self-fertilize, whereas most animals cannot. In addition, many sterile plants can still reproduce asexually.

When the number of chromosome sets changes within a species, the change is called autopolyploidy ("self" increase). This change can occur naturally or can be induced artificially. Artificially induced autopolyploids include commercial triploid winesap apples and tetraploid potatoes, peanuts, coffee plants, and grapes (Figure 13.11). All these plants are larger and hardier than their diploid ancestors. The commercial banana (a sterile triploid) does not have the hard, inedible seeds of its diploid parent stock.

Plant breeders often use colchicine to induce autopolyploidy. Colchicine is a chemical derived from the autumn crocus (*Colchicum autumnale*). As one of its effects, colchicine inhibits microtubule assembly. Without microtubules, spindles cannot form during nuclear division, and without a spindle, chromosomes cannot separate. By adjusting colchicine concentrations and the length of exposure to it, plant breeders can produce nuclei with variable numbers of duplicated chromosome sets. When these cells are freed from colchicine exposure, they divide and give rise to polyploid cells. These cells may produce polyploid gametes.

Among plants, polyploids may also arise through multiplication of the number of chromosome sets in hybrids between different species. This process is called allopolyploidy. An example is found in the hybridizations that apparently gave rise to common bread wheat, one of our most important crops (Figure 13.12).

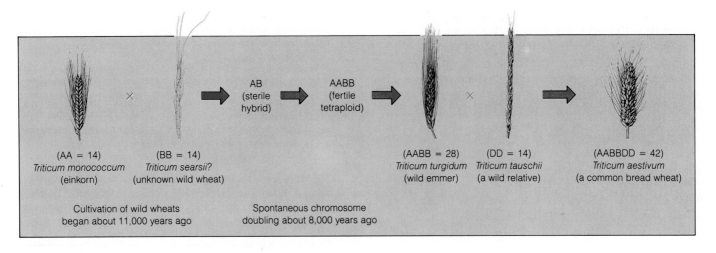

Figure 13.12 Proposed hybridizations in wheat that led to chromosome doublings and polyploidy.

Wheat grains dating from 11,000 B.C. have been found in the Near East. Several species of wild diploid wheat still grow there. They have 14 chromosomes (two sets of 7, designated AA). Also growing in the region is a wild grass with 14 chromosomes, designated BB. They differ from the A chromosomes, judging from their failure to pair with them at meiosis. One tetraploid wheat species has 28 chromosomes; analysis during meiosis shows that they are AABB. The A chromosomes pair with A's, and the B chromosomes pair with B's. A hexaploid wheat has 42 chromosomes (six sets of seven). Its chromosomes are AABBDD, the last set (DD) coming from *Triticum taushchii*, another wild grass.

SUMMARY

1. The chromosomal theory of inheritance was formulated between 1900 (following the rediscovery of Mendel's work) and the 1920s. The key points of this theory were summarized earlier, in the list on page 175.

2. In many animals, all but two pairs of homologous chromosomes in diploid cells resemble each other in length, shape, gene sequence, and other characteristics; they are called autosomes. The remaining two, called sex chromosomes, differ structurally and in their gene content but they still function as homologues at meiosis.

3. Sex chromosomes carry 'major *sex-determining* genes, but they also carry genes that have nothing to do with sex determination (for example, the eye-color gene on the X chromosome in *Drosophila*). Moreover, some genes on autosomes also contribute to the overall male or female phenotype.

4. Some traits are sex-linked; the genes coding for them are located on the X chromosome and have no alleles on the Y. Morgan's work with the sex chromosomes of *Drosophila* provided early evidence that genes are indeed located on chromosomes.

5. The tendency of different genes that are physically located on the same chromosome to remain together in crosses is called linkage. Such genes do not assort independently into gametes. (Different genes located on *nonhomologous* chromosomes do assort independently of each other.)

6. Crossing over disrupts gene linkages along a chromosome. The probability of crossing over occurring at a point somewhere between two gene loci is proportional to the distance separating them. The greater the distance, the more "loosely linked" the genes. Linkage mapping provided clear evidence that genes are carried in linear array on chromosomes.

7. Chromosome structure can be altered in the following ways:

   a. Deletion (loss of a chromosome segment).

   b. Duplication (repeat of a gene sequence right next to the region it repeats).

   c. Inversion (reversal of a gene sequence).

   d. Translocation (permanent transfer of a chromosome segment to a nonhomologous chromosome).

8. In aneuploidy, improper separation of homologous chromosomes at meiosis leads to the wrong number of chromosomes in gametes. If such a gamete is involved in fertilization, the following types of conditions can result:

   a. Trisomy: three homologous chromosomes of the same kind are present in somatic cells ($2n + 1$).

   b. Monosomy: there is no homologue for one kind of chromosome in somatic cells ($2n - 1$).

9. In polyploidy, three or more complete sets of chromosomes are present in somatic cells (3n, 4n, 5n . . .). Polyploidy is more common among plants than among animals. It has played an important role in plant evolution and in the development of commercial crop plants.

Readings

Ayala, F., and J. Kiger. 1984. *Modern Genetics*. Second edition. Menlo Park: Benjamin-Cummings.

Feldman, M., and E. Sears. 1981. "The Wild Gene Resources of Wheat." *Scientific American* 244(1):102–112.

Garber, E. 1972. *Cytogenetics: An Introduction*. New York: McGraw-Hill.

McKusick, V. April 1971. "The Mapping of Human Chromosomes." *Scientific American* 244(4):104–113.

Morgan, T., A. Sturtevant, H. Muller, and C. Bridges. 1915. *The Mechanism of Mendelian Heredity*. New York: Holt. For those who like to browse through original research papers.

Strickberger, M. 1985. *Genetics*. Third edition. New York: McGraw-Hill.

Sturtevant, A. 1965. *A History of Genetics*. New York: Harper & Row. A history of genetics before explosive advances in molecular biology have made it almost impossible to do a historical survey of genetics.

Genetics Problems (Answers appear in Appendix II)

1. Recall that human sex chromosomes are XX for females and XY for males.
   a. Does a male child inherit his X chromosome from his mother or father?
   b. With respect to an X-linked gene, how many different types of gametes can a male produce?
   c. If a female is homozygous for an X-linked gene, how many different types of gametes can she produce with respect to this gene?
   d. If a female is heterozygous for an X-linked gene, how many different types of gametes can she produce with respect to this gene?

2. One human gene, which may be Y-linked, controls the length of hair on men's ears. One allele at this gene locus produces non-hairy ears; another allele form produces rather long hairs (hairy pinnae).
   a. Why would you *not* expect females to have hairy pinnae?
   b. If a man with hairy pinnae has sons, all of them will have hairy pinnae; if he has daughters, none of them will. Explain this statement.

3. Suppose that you have linked genes 1 and 2 with alleles *A,a* and *B,b* respectively. An individual is heterozygous for both genes, as in the following:

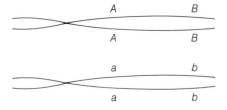

For each of the following crossover frequencies between these two genes, what genotypes would be expected among gametes from this individual, and with what frequencies?
   a. 0 percent
   b. 10 percent
   c. 25 percent

4. Assume the two genes mentioned in Problem 3 are separated by 10 map units. If the double heterozygous individual mates with an individual who is homozygous recessive for these same two genes, what genotypes would you expect among their offspring, and with what frequencies?

5. In *D. melanogaster*, a gene influencing eye color has red (dominant) and purple (recessive) alleles. Linked to this gene is another that determines wing length. A dominant allele at this second gene locus produces long wings; a recessive allele produces vestigial wings. Suppose a completely homozygous dominant female having red eyes and long wings mates with a male having purple eyes and vestigial wings. First-generation females are then crossed with purple-eyed, vestigial-winged males. From this second cross, offspring with the following characteristics are obtained:

252	red eyes, long wings
276	purple eyes, vestigial wings
42	red eyes, vestigial wings
30	purple eyes, long wings

600 offspring total

Based on these data, how many map units separate the two genes?

6. Suppose you cross a homozygous dominant long-winged fruit fly with a homozygous recessive vestigial-winged fly. Shortly after mating, the fertilized eggs are exposed to a level of x-rays known to cause chromosomal deletions. When these fertilized eggs subsequently develop into adults, most of the flies are long-winged and heterozygous. However, a few are vestigial-winged. Provide a possible explanation for the unexpected appearance of these vestigial-winged adults.

7. Individuals afflicted with Down's syndrome typically have an extra chromosome 21, so their cells have a total of 47 chromosomes. However, in a few cases of Down's syndrome, 46 chromosomes are present. Included in this total are two normal-appearing chromosomes 21 and a longer-than-normal chromosome 14. Interpret this observation and indicate how these few individuals can have a normal chromosome number.

8. Refer to Figure 13.12. *Triticum turgidum* should produce gametes having 14 chromosomes (AB), and *T. tauschii* should produce gametes with 7 chromosomes (D). When these gametes combine, offspring having 7 + 14 = 21 chromosomes should be produced. How, then, did *T. aestivum* originate having 42 chromosomes (AABBDD)? If *T. aestivum* is backcrossed to *T. turgidum*, how many chromosomes would be present in the offspring? Would these offspring be fertile?

9. The mugwump, a primitive arboreal tarsier, has a reversed sex-chromosome condition. The male is XX and the female is XY. However, perfectly good sex-linked genes are found to have the same effect as in humans. For example, a recessive allele *c* produces red-green color blindness. If a normal female mugwump mates with a phenotypically normal male mugwump whose mother was color blind, what is the probability that a son from that mating will be color blind? A daughter?

10. Childhood *muscular dystrophy* is a recessive, sex-linked trait in humans. A slowly progressing loss of muscle function leads to death, usually by age twenty or so. Unlike color blindness, this disorder is restricted to males, not ever having been found in a female. Suggest why.

Peas, beans, corn, flies, molds, bacteria—these organisms have been truly accommodating in genetic studies. They grow and reproduce in small quarters, under contrived and controlled conditions. They also reproduce rapidly and in abundance, giving rise to many offspring through successive generations that are far shorter in duration than the life span of the geneticist who observes them. In short, they lend themselves to genetic analysis.

In contrast, humans live in enormously varied and variable environments. They tend not to stay put. Generally they get together and reproduce by preference, not by dictum. Human subjects generally live just as long as the geneticists studying them, so tracking traits through generations is a long, drawn-out process. Besides, humans don't produce families in sizes large enough to make meaningful statistical inferences about the transmission and distribution of traits.

Despite all such obstacles to analysis, human genetics is a burgeoning field. Standardized procedures for constructing pedigrees provide accurate ways of organizing data and tracking family traits through several generations. (A pedigree, recall, is a chart of the genetic relationships of individuals.) Many different families in which the same trait is observed are studied together, as a means of increasing the numerical information on which statistical analyses are based. Data gathered on large populations throughout the world are pooled for comparative analyses.

The information being gathered does more than help satisfy our curiosity about ourselves. It forms a basis for making decisions that affect the well-being of individuals and society at large. Consider that more than 3,000 human disorders have already been traced to genetic origins. Although many can be readily diagnosed and treated, actual treatments are presently limited to *phenotypic* cures, in which certain practices compensate for phenotypic defects. (For instance, in some cases the body is provided with a missing gene product to compensate for a nonfunctional allele.)

Phenotypic cures do alleviate individual suffering. They also circumvent the process of natural selection. Harmful alleles are not only perpetuated, they become represented with increased frequency in a population with the passage of generations. When afflicted persons have children, they may pass on their disorder to the next generation. The "cure," in other words, is illusory.

On the distant horizon are *genotypic* cures, in which defective alleles will be chemically modified or replaced. For instance, if a harmful allele is identified in prospective parents, it theoretically could be isolated, excised, and replaced microsurgically in sperm or eggs by a functional allele obtained with recombinant DNA methods (Chapter Eighteen). Thus a genotypic cure could be brought about before fertilization.

This chapter will focus initially on some human genetic disorders. The examples selected are necessarily

14

HUMAN GENETICS

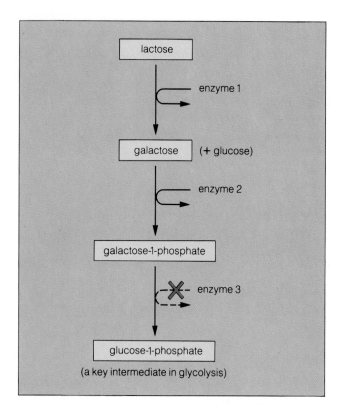

Figure 14.1 Step at which lactose metabolism is blocked in galactosemia, a genetic disorder arising from one mutant allele.

phosphorylated glucose, which can be broken down by glycolysis (Figure 14.1). Individuals who are homozygous recessive for the allele do not have functional enzyme molecules that can act on galactose-1-phosphate, a key intermediate in the reactions. The intermediate accumulates in the blood, and in large concentrations it is toxic.

Galactosemia can be cured phenotypically. Abnormally high amounts of galactose can be detected in urine samples, and if affected infants are detected early enough, they can be placed on a diet that includes milk substitutes. By drinking lactose-free "milk" and avoiding other sources of galactose, they can grow up symptom-free.

Galactosemia is an example of *autosomal recessive inheritance* which, by definition, has the following characteristics:

1. Both males and females can carry the recessive allele with equal probability.

2. The recessive allele is not expressed in heterozygotes of either sex, but it is expressed in homozygotes of both sexes.

3. Two heterozygous parents produce unaffected and affected infants in about a 3:1 ratio. But all the offspring of two homozygous, galactosemic parents are afflicted with the disorder.

SEX-LINKED RECESSIVE INHERITANCE

Hemophilia

Several genetic disorders, including **hemophilia**, lead to abnormalities in the ability to stop bleeding from injury or stress. Several proteins take part in the normal blood-clotting mechanism. Mutations can occur in any of the genes coding for those proteins. In classical hemophilia (called hemophilia A), a gene carried on the X chromosome is affected. It codes for a defective protein (or none at all).

Hemophilia A is an example of *sex-linked recessive inheritance*, for the recessive allele is not expressed in heterozygous females and is almost always expressed in males. A hemophilic male and a heterozygous female produce unaffected and affected infants in about a 1:1 ratio.

In human males, recall, the X chromosome is paired with the Y chromosome. Hence, if a male inherits the mutant allele, it will be the only allele he has for this trait. He will be unable to produce the functional blood-clotting protein. Without medical attention, cuts, bruises, or internal bleeding could lead to death. In heterozygous females, blood-clotting time is more or less

limited, but they will give you an idea of the diversity of genetic problems that we deal with as individuals and as members of society. They will also provide a framework for considering some practical and ethical aspects of genetic screening, counseling, and treatment programs.

AUTOSOMAL RECESSIVE INHERITANCE

Galactosemia is a disorder arising from an inability to metabolize a breakdown product of lactose (milk sugar). It occurs at a frequency of about 1 in 100,000 individuals. Afflicted infants suffer from malnutrition, diarrhea, and severe vomiting. Often the eyes, liver, and brain are damaged. Without treatment, galactosemic individuals usually die.

Galactosemia has been traced to a mutant allele that leads to the absence or nonfunctioning of one enzyme. Normally the enzyme takes part in a metabolic pathway by which lactose is converted first to galactose, then to

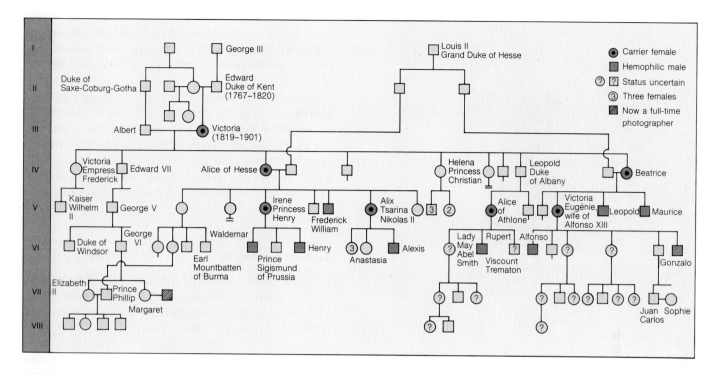

Figure 14.2 Descendants of Queen Victoria, showing carriers and afflicted males that possessed the sex-linked gene conferring the disorder hemophilia A. Many individuals of later generations are not shown in this family pedigree.

normal because the nonmutated allele on one of her X chromosomes codes for enough of the normal protein.

The mutant allele for hemophilia is rare in the human population. (Many more males than females are afflicted by its phenotypic consequences; the mutant allele is so rare that homozygous recessive daughters are not likely to be born unless hemophilic males marry close relatives.) Yet hemophilia occurred frequently among the royal families of Europe during the nineteenth century. Queen Victoria of England was a carrier of the mutant allele. At one time it was calculated that, of her sixty-nine descendants, eighteen were afflicted males or female carriers of the mutant allele. The family inheritance pattern clearly showed that the mutation is sex-linked.

One of Victoria's descendants, Crown Prince Alexis of Russia, was hemophilic (Figure 14.2). His affliction and the cast of characters it indirectly drew together—Czar Nicholas II, Czarina Alexandra (a granddaughter of Victoria and a carrier), the power-hungry monk Rasputin who manipulated the aggrieved family to his political advantage—helped catalyze events that brought an end to dynastic rule in the Western world. The historical novel *Nicholas and Alexandra* is a poignant account of these individuals and the afflicted child.

CHANGES IN CHROMOSOME NUMBER

As we have seen, the number of chromosomes can change as a result of nondisjunction (improper separation of homologous chromosomes at meiosis) and other chromosomal aberrations. Among humans, the phenotypic consequences are usually severe and often lethal. Here we will consider just a few examples of this kind of genetic disorder.

Green Color Blindness

Another example of sex-linked recessive inheritance, called **green weakness**, is a partial color blindness attributable to a reduced amount of retinal pigment that is sensitive to light of green wavelengths. In northern Europe, about five percent of white males show this trait. To them, bright greens appear to be tan colors, olive greens appear to be brown, and reds appear to be reddish browns.

The green-weakness allele (g) occurs on the X chromosome. Hence the trait shows up in homozygous females (gg). In heterozygous (Gg) females, the domi-

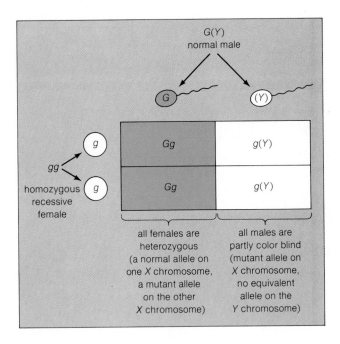

Figure 14.3 Phenotypic outcome among offspring of a female who is homozygous (*gg*) for the recessive green-weakness allele and a male who is normal G(Y). All female children will be heterozygous (carriers), and all males will be affected.

nant allele codes for enough pigment to permit near-normal color vision. The trait shows up in all males with the recessive allele. When an affected male and a female who is heterozygous for the recessive allele have children, the green-weak trait will show up in an average of one child in two (half of the sons, half of the daughters). For a normal male and a homozygous recessive female, all daughters will be normal and carry the recessive allele, whereas all sons will be blind to the color green (Figure 14.3).

Down's Syndrome

Sometimes nondisjunction leads to the presence of two copies of chromosome 21 in a gamete. Fusion of this gamete with a normal one produces an individual with three copies of that chromosome (Figure 14.4a). The condition, called *trisomy 21*, leads to **Down's syndrome**. (The word "syndrome" means a set of symptoms that typically occur together and that characterize some particular disorder.) For about every 750 infants born, one is destined to develop Down's syndrome. In its most

extreme form, the disorder is characterized by severe mental retardation. Most often, afflicted embryos are lost through miscarriage. (They are expelled spontaneously from the uterus before completion of the first six months of pregnancy.)

Women who conceive past age forty run a much greater risk of giving birth to a trisomic 21 infant than do women in their early twenties (see, for example, Figure 14.4b). Apparently, nondisjunction of chromosome 21 in female reproductive cells occurs more frequently with increased age. Recent studies also suggest that the risk also increases when the father is past age fifty-five.

Turner's Syndrome

In about every 5,000 live births, one infant is destined to have **Turner's syndrome**, a *sex chromosome abnormality* that probably arises through nondisjunction at meiosis. Afflicted individuals have only one X and no Y chromosome. The frequency of Turner's syndrome is lower than for other common sex chromosome abnormalities, probably because many affected embryos are miscarried early in pregnancy.

Turner's syndrome leads to a female phenotype, but the phenotypic distortions are pronounced. The females are nearly always sterile. Their ovaries are nonfunctional, and certain secondary sex characteristics fail to develop properly at puberty. At least one-fourth of these females suffer from a defective aorta (a major blood vessel leading away from the heart); they have deformed kidneys, and two ureters instead of one for each kidney. Often these individuals age prematurely and have shortened life expectancies.

Trisomy XXY and XYY

Nondisjunction produces other sex chromosome abnormalities. Some males are trisomic XXY (two X and one Y chromosomes) and show **Klinefelter's syndrome**. Affected males are always sterile, and their testes develop to only about a third of the normal size. Their body hair is sparse, and the deposition of body fat follows female patterns. They may also show some breast enlargement.

About 1 in every 1,000 males born is **trisomic XYY**. Applying the term "syndrome" to this abnormality may be misleading, because there are no critical symptoms. XYY males tend to be tall and may show slight mental retardation, but they are otherwise normal. At one time, these males were thought to be predisposed to become criminals, but the great majority are normal phenotypically and psychologically.

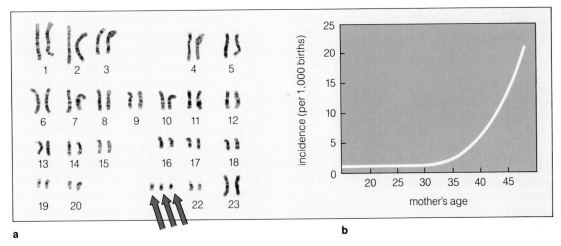

Figure 14.4 (a) Karyotype of a girl with Down's syndrome, with red arrows identifying the trisomy of chromosome 21. (b) Relationship between the frequency of Down's syndrome and the mother's age. The results are from a study of 1,119 children with the disorder who were born in Victoria, Australia, between 1942 and 1957.

PROSPECTS AND PROBLEMS IN HUMAN GENETICS

Chances are, you personally know of someone who suffers the symptoms of a genetic disorder. Of all individuals being born, possibly one percent will be afflicted with pronounced problems arising from a chromosomal alteration. Between one and three percent more will suffer because of mutant genes that either cannot code for vital products or code for defective ones. Of all patients in children's hospitals, between ten and twenty-five percent are being treated for problems arising from genetic abnormalities. As you have seen, the phenotypic consequences of these abnormalities can be severe.

Considerable effort is being devoted to developing diagnostic tools and treatments for genetic disorders. However, some of the work is controversial. Genetically based disorders apparently cannot be approached in the same manner as infectious diseases (such as influenza, measles, and polio). Infectious agents are enemies from the environment, so to speak, that attack without much warning. We have had no qualms about mounting counterattacks through immunizations and antibiotics, which either eliminate these agents or at least bring them under control. With genetic disorders, the problem is inherent in the hereditary material of individual human beings.

How do we attack an "enemy" within? Do we institute regional, national, even global programs to identify afflicted individuals? Do we inform those so identified that they are defective in particular ways, and that they might bestow the defect on one or all of the children they might wish to have? What do we say to a pregnant woman who has been diagnosed as carrying a trisomic 21 fetus that will suffer Down's syndrome? Should society bear the cost of treating genetically based disorders? If so, should society also have some say in whether fetuses diagnosed as having some genetic defect will be born at all, or aborted?

These questions are only the tip of an ethical iceberg. Answers to any one of them have not yet been worked out in a way that can be called universally acceptable. Until such time as permanent, genotypic cures are a reality, however, the questions will remain with us.

The rest of this chapter describes therapeutic and preventive measures currently being used to treat genetically based disorders, as listed in Table 14.1.

Treatments for Phenotypic Defects

Currently, individuals who already suffer from a genetic disorder are treated at the phenotypic level only. In other words, the genetic disorder itself cannot be eliminated, but often its phenotypic expression can be circumvented. Treatments include diet modification, adjustments to the environment, surgery, and chemotherapy.

Diet Modification. The outward symptoms of several genetic disorders can be suppressed or minimized by controlling the diet. Galactosemia, described earlier, can be controlled this way. So can **phenylketonuria**, or PKU. With this disorder, a mutant allele codes for a defective enzyme. The normal enzyme converts phenylalanine (an amino acid) to tyrosine (another amino acid). In homozygous recessive individuals, phenylalanine accumulates in the body. In large amounts, it is diverted to other

Table 14.1 Some Therapeutic and Preventive Measures Used in Treating Genetically Based Diseases

Therapeutic Measures:

Diet modification	Providing substances that the body cannot produce for itself, or restricting intake of substances that the body cannot tolerate
Environmental adjustment	Using corrective devices (such as eyeglasses for visual defects), or avoiding environmental conditions that the body cannot tolerate (for example, low oxygen levels at high altitudes in the case of sickle-cell anemics)
Surgical correction	Surgically repairing deformities (such as harelips, heart defects)
Chemotherapy	Using chemicals (such as drugs) to modify or inhibit gene expression or gene product function

Preventive Measures:

Mutagen reduction	Avoiding or eliminating environmental substances (such as many industrial wastes) that can induce mutations
Genetic screening	Methodically searching through populations and identifying individuals afflicted with (or heterozygous for) a particular genetic disorder
Genetic counseling	Conveying information about genetic disorders, as well as the social and medical options available, to afflicted individuals and their families
Prenatal diagnosis	Detecting chromosomal alterations and metabolic disorders at the embryonic stage
Gene replacement*	Substituting normal alleles for mutant alleles

*Research stage only.

pathways, which produce compounds in high enough concentrations to interfere with normal body functioning. One of the compounds, phenylpyruvic acid, can be detected in urine or blood samples. In children afflicted with PKU, high concentrations of this compound can damage the nervous system and lead to mental retardation.

If PKU is diagnosed early enough, its symptoms can be alleviated. Individuals simply are placed on a diet that provides only as much phenylalanine as required for protein synthesis, so the body is not called upon to dispose of excess amounts. Aside from having a restricted diet, afflicted persons can lead normal lives.

Environmental Adjustments. Some therapeutic treatments for genetic disorders require adjustments to surrounding conditions. Phenotypic defects such as impaired vision or hearing can be circumvented by such devices as eyeglasses or hearing aids. True albinos avoid direct sunlight. Sickle-cell anemics, recall, suffer from impaired oxygen and carbon-dioxide transfers because a mutant allele codes for a variant form of hemoglobin, the oxygen-carrying protein in blood (Figure 12.15). They should avoid strenuous activity in environments where oxygen levels are low (such as high altitudes and unpressurized aircraft cabins). At the extreme are children whose immune system does not function at all. They must be raised in sterile confinement to avoid any encounter with infectious agents.

Surgical Correction. Many phenotypic defects can be corrected or at least minimized by surgical reconstructions. For example, **harelip** is a developmental defect of the upper lip. A vertical fissure is present at the lip midsection and often extends into the palate (the roof of the mouth). Surgery can usually correct the defect in terms of appearance and function (Figure 12.11).

Chemotherapy. Knowledge of the molecular basis of genetic disorders is being used to chemically modify gene products and to compensate for the absence of gene products. For example, **Wilson's disorder** arises from an inability to utilize copper, which is essential for the action of several enzymes. Copper deposits build up in tissues and can lead to brain and liver damage. Convulsions and death are the outcome. Yet by avoiding copper-containing foods (such as chocolate) and taking certain drugs, individuals who are diagnosed early enough can lead relatively normal lives. One drug (a penicillin derivative) binds with copper in the body, and it is flushed from the body by way of the urinary system.

A Holistic View of Human Genetic Disorders

Vernon Tipton, Brigham Young University, Provo, Utah

The sound of carts rolling over cobblestone streets was commonplace in towns and villages of the fifteenth and sixteenth centuries. Ordinarily, it would have served only to break the early morning silence. But the heavy and terrible loads the carts carried lent an ominous tone to the sound and charged the atmosphere with morbid anticipation. Methodically and inexorably, the carts made their rounds to carry away those dead of smallpox and plague.

Samuel Pepys captured the mood of the times in this brief expression of gratitude: "This day I am by the blessing of God, 34 years old, in very good health and mind's content, and in condition of estate much beyond whatever my friends could expect of a child of theirs. This day, 34 years. The Lord's name be praised! And may I be thankful for it." Life expectancy for Pepys and his contemporaries was about one-third of what it is today, primarily because of the ravages of communicable diseases.

Most of our afflictions fit into one of three categories: communicable diseases (such as plague, tuberculosis, and common colds), degenerative disorders, (such as atherosclerosis), and behavioral diseases (including alcoholism). The ten current leading causes of death contrast sharply with those of Pepys's era. Today, few die of plague, tuberculosis, or influenza, and smallpox has been eradicated. Today, most die because of heart diseases, cancer, cerebrovascular disease, accidents, emphysema, and suicide. Thus, *during the past two centuries, there has been a dramatic shift from communicable diseases to degenerative and behavioral diseases, many of which are genetic in origin.*

The causes of the shift are many, but included among them are changes brought about by the industrial revolution, advances in medical technology, mass education, and major changes in life styles. This shift, dramatic and profound, suggests two questions: What is its significance to the individual? What impact does it have on the total population?

The shift from external to internal causation of diseases places a greater burden of responsibility on the individual, who must consider the care of the body as a stewardship that cannot be delegated to others. However, the trend has been in the opposite direction: toward the institutionalization of health care and the abdication of individual responsibility. The trend has now reached immense proportions, as described by Leon Kass: "All kinds of problems now roll to the doctor's door, from sagging anatomies to suicides, from unwanted childlessness to unwanted pregnancies, from marital difficulties to learning difficulties, from genetic counseling to drug addiction, from laziness to crime." It is apparent that an increasing number of attitudes and forms of behavior have become defined as illnesses, the treatment of which is regarded as belonging within the jurisdiction of medicine and its practitioners.

Society has become enormously dependent on medicine, an outgrowth of the "one germ, one disease, one therapy" concept that originated in the latter part of the nineteenth century. There is, however, growing realization that most diseases have multiple causes. John Knowles has suggested that "personal behavior, food, and the nature of the environment around us are prime determinants of health and disease." He further indicates that between 1900 and 1966, life expectancy at age sixty-five increased by 2.7 years, as a result of the efforts of organized medicine. At the same time, it has been possible for the sixty-five year old to extend life expectancy eleven years by practicing good health habits.

The collective impact of the shift from communicable to degenerative and behavioral diseases is more difficult to document, mainly because of the time frame required for genetic changes to become clearly manifest. However, it is reasonable to assume that selective pressures exerted by diseases today are considerably different from those existing two centuries ago. Undoubtedly, the net effect is both positive and negative. The harsh environmental conditions of earlier times produced a high incidence of communicable diseases and an extremely high infant mortality rate that gradually eliminated from the population those most susceptible to disease. Very likely, those suffering genetic abnormalities were among those highly susceptible to communicable diseases. Those who escaped genetic abnormalities and communicable disease became susceptible to chronic degenerative disorders simply by virtue of living longer than those who did not.

During the past thirty years, particularly in Europe and the United States, communicable diseases have not been significant in terms of selective pressure on populations. Thus, many more individuals are reaching reproductive age than in former times. Does this mean that the incidence of genetic abnormalities is increasing? The full significance of the historical shift from communicable to degenerative and behavioral diseases cannot be fully understood without the added perspective that the time factor provides. When selective pressures change over time, the genetic composition of populations also changes.

Some Questions for Debate:

1. Pyloric stenosis, a genetically based disorder, is manifest in some infants. Their pyloric valve is closed, which prevents food from moving from the stomach to the small intestine. Simple surgery corrects the defect; without it, infants would die within a few days. To what extent have such advances changed the genetic composition of populations?

2. Environmental conditions in undeveloped countries are often harsh, and their citizens are exposed to many disease agents which most North Americans seldom (if ever) encounter. In the event of a major disaster requiring prolonged existence in a "nonsanitized" environment, which group would be the more durable? What are some long-term implications of exporting high-technology medicine to undeveloped countries?

3. Why is a cure for cancer (which apparently has a genetic basis) more elusive than the cures for diseases such as polio?

Genetic Screening

The therapeutic measures described so far assist individuals who are already suffering the consequences of genetic disorders. Preventive measures are also being used to allow early detection and treatment before symptoms can develop. Others are used to identify carriers who show no outward symptoms but may give birth to affected children.

Genetic screening usually refers to large-scale programs to detect affected persons in a given population. One of the most extensive programs began in the late 1950s and continues today as a means of detecting PKU, which was described above. Most hospitals in the United States routinely screen all newborns for PKU, so its outward signs are rapidly disappearing from the population.

Genetic Counseling and Prenatal Diagnosis

Sometimes prospective parents are concerned that they will have a severely afflicted child. Their first child may have suffered a hereditary disorder and the parents now wonder if future children will suffer the same defect. How are such individuals counseled? Typically, several different consultants are required, including medical specialists, geneticists, and social workers who can give emotional support to families into which severely afflicted children are born.

Counseling begins with an accurate diagnosis of a particular disorder in the parental genotypes. Biochemical tests can be performed to detect many metabolic disorders. Family pedigrees are constructed as completely and accurately as possible to aid in the diagnosis. For disorders that follow simple Mendelian inheritance patterns, it is possible to predict the likelihood of producing afflicted children—but not all disorders follow Mendelian inheritance patterns. Even those that do can be influenced by other factors, some identifiable, others not. Even when the extent of risk has been determined with some confidence, it is important that the prospective parents understand that the risk is the same for *each* pregnancy. For example, if there is one chance in four that the child will be born with a genetic disorder, there is one chance in four that the next child will be also.

What happens when a female is already pregnant? For example, suppose a woman forty-five years old has just become pregnant. She may want to know whether she will give birth to a child who will suffer Down's syndrome. Through prenatal diagnosis, it is possible to detect this and more than a hundred other disorders in early pregnancy.

A major detection procedure is based on **amniocentesis**, a sampling of the fluid surrounding the fetus in the mother's uterus (Figure 14.5). After the fourteenth to sixteenth week of pregnancy, the thin needle of a syringe is carefully inserted through the mother's abdominal wall and into the fluid-filled sac (amnion) containing the fetus. Floating in this fluid are epidermal cells that have been shed from the fetus. Some of the fluid, with its sample of the fetal cells, is withdrawn by the syringe. These cells can be cultured and can undergo mitosis. Through various tests, including karyotype analysis of metaphase chromosomes (page 180), abnormalities can be diagnosed early in pregnancy. Amniocentesis is also being used to test for biochemical defects that lead to a host of genetic disorders, such as sickle-cell anemia.

A small risk is associated with amniocentesis, for care must be taken not to puncture the fetus or cause infection. A more recent procedure, **chorionic villi analysis**, uses samples drawn from the chorion (a membranous sac surrounding the amnion). This procedure is less risky than amniocentisis and can be used earlier in pregnancy.

Unfortunately, there presently is no cure for disorders arising from abnormalities in chromosome structure or number. If the embryo is diagnosed as having a severe disorder such as Down's syndrome, the parents might elect to request an abortion (an induced expulsion of the embryo from the uterus). Such decisions are bound by ethical considerations. The role of the medical community should be to provide information that the prospective parents need in order to make their own choice. But that choice in turn must be consistent with their own values, within the broad constraints imposed by society.

Gene Therapy

It is not yet possible to replace defective alleles with normal ones in human somatic cells. However, recombinant DNA technology has raised hopes that gene replacement therapy may someday be possible, as will be described in Chapter Eighteen.

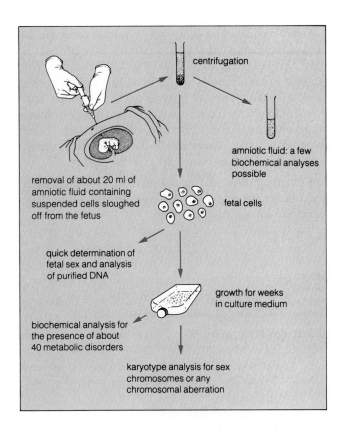

Figure 14.5 Steps in amniocentesis, a procedure used in prenatal diagnosis of many genetic disorders.

SUMMARY

1. Of all humans being born, possibly one percent will be afflicted with pronounced problems arising from chromosomal alterations. Between one and three percent more will have problems because of mutant genes that code for defective proteins (or for none at all). Of all patients in children's hospitals, between ten and twenty-five percent are being treated for problems arising from genetic abnormalities.

2. One type of genetic disorder is called autosomal recessive inheritance. In this case, both males and females can carry the recessive allele with equal probability. The recessive allele is not expressed in heterozygotes of either sex but is expressed in homozygotes of both sexes. Two heterozygous parents produced unaffected and affected infants in about a 3:1 ratio. All infants of two homozygous parents are afflicted with the disorder.

3. Another type of genetic disorder is called sex-linked recessive inheritance. In this case, the recessive allele is not expressed in heterozygous females and is almost always expressed in males. An affected male and a het-

erozygous female produce unaffected and affected infants in about a 1:1 ratio.

4. Nondisjunction is the improper separation of homologous chromosomes at meiosis. It can lead to the following types of disorders:

   a. A trisomic autosomal condition (for example, three copies of human chromosome 21, which results in Down's syndrome).

   b. A monosomic sex chromosome condition (for example, only one X chromosome, which results in Turner's syndrome).

   c. A trisomic sex chromosome condition (for example, two X and one Y chromosome, which results in Klinefelter's syndrome).

5. The four main categories of existing treatment for genetic disorders are diet modification, environmental adjustments, surgical correction, and chemotherapy.

6. Genetic screening refers to programs for detecting individuals affected by a specific genetic disorder and for identifying unaffected carriers (heterozygotes).

7. Genetic counseling refers to programs designed to predict the likelihood that two parents may produce a child affected with some type of genetic disorder. For example, amniocentesis is a form of prenatal diagnosis based on examination of fetal cells, chromosomes, and fluid prior to birth. It can be used to detect a large number of genetic disorders. Chorionic villi analysis (a more recent diagnostic procedure) is less risky than amniocentesis and can be done earlier in pregnancy.

8. Gene therapy (replacing defective alleles with normal ones) is not yet possible, although preliminary experiments are under way.

Readings

Anderson, A., and E. Diacumakos. July 1981. "Genetic Engineering in Mammalian Cells." *Scientific American* 245(1):106–121.

Anderson, W. F. November 1985. "Beating Nature's Odds." *Science 85* 6(9):49–50. Describes prospects for gene therapy.

Fuchs, F. June 1980. "Genetic Amniocentesis." *Scientific American* 242(6):47–53. Excellent summary article of a major diagnostic procedure.

Hartl, D. 1983. *Human Genetics.* New York: Harper & Row.

Lawn, R., and G. Vehar. March 1986. "The Molecular Genetics of Hemophilia." *Scientific American* 254(3):48–54.

Lubs, H., and F. de la Cruz (editors). 1977. *Genetic Counseling.* New York: Raven Press. Includes material on the frequency of genetic disorders and a review of studies of genetic counseling.

Mange, A., and E. Mange. 1980. *Genetics: Human Aspects.* Philadelphia: Saunders. Good introductory textbook.

Genetics Problems (Answers in Appendix II)

1. In hemophilia A, the body's blood-clotting mechanism is defective. This condition has been traced to a recessive allele of an X-linked gene. Refer now to Figure 18.2. Why are only the females shown as carriers of the recessive allele? If Victoria had married Leopold (also in generation IV), what phenotypes would have been expected among their children, and with what probabilities?

2. Huntington's chorea is due to a dominant autosomal allele. Usually this disorder does not manifest itself until after age thirty-five. Individuals having Huntington's chorea are almost always heterozygous. As a genetic counselor, you are visited by a twenty-year-old woman. Her mother has Huntington's chorea but her father is normal. What is the probability that this woman will develop Huntington's chorea as she grows older? Suppose, at her present age, she marries someone with no family history of the disorder. If they have a child, what is the probability that it will have Huntington's chorea?

3. A woman heterozygous for color blindness (*Gg*) marries someone who has normal color vision. What is the probability that their first child will be color blind? Their second child? If they have two children only, what is the probability that both will be color blind?

4. A person afflicted with Turner's syndrome has only a single sex chromosome (X only), yet may survive. In contrast, a person having

a single Y chromosome and no X chromosome cannot survive. What does this tell you about the genetic contents of the X and Y chromosomes?

5. Fertilization of a normal egg by a sperm that has no sex chromosomes (male nondisjunction) can lead to Turner's syndrome. Also, fertilization of an egg that has no sex chromosomes (female nondisjunction) by a sperm carrying one X chromosome can lead to the same disorder. Suppose a hemophilic male and a carrier (heterozygous) female have a child. The child is nonhemophilic and is afflicted with Turner's syndrome. In which parent did nondisjunction occur?

6. The trisomic XXY condition is also called Klinefelter's syndrome. How could this syndrome arise if nondisjunction occurred in the female parent of an afflicted individual? How could it arise if nondisjunction occurred in the male parent?

7. If nondisjunction occurs for the X chromosomes during oogenesis, then some eggs having two X chromosomes and others having no X chromosomes are produced at about equal frequencies. If normal sperm fertilize these two types of eggs, what genotypes are possible?

8. Phenylketonuria (PKU) is an autosomal recessive condition. About 1 of every 50 afflicted individuals is heterozygous for the gene but displays no symptoms of the disorder.
    a. If you select a symptom-free male at random from the population, what is the probability that he will be heterozygous?
    b. If you select a symptom-free female at random from the population, what is the probability that she will be heterozygous?
    c. If you select a symptom-free male and a symptom-free female at random, what is the probability that both will be heterozygous? What is the probability that they could have a child afflicted with PKU?

9. Laws restricting marriage between close relatives (consanguineous matings) are widespread, the rationale being that such marriages generally lead to an increase in the incidence of genetic defects among offspring. Suppose you are a carrier (heterozygous) for PKU. If you pick a potential mate at random from the population, what is the probability that he or she would also be a PKU carrier? If you marry your first cousin, do you think he or she would have the same probability of being a PKU carrier as your randomly selected mate? Why?

DISCOVERY OF DNA FUNCTION

One might have wondered, in the spring of 1868, why Johann Miescher was collecting cells from the pus of open wounds and, later, the sperm of a fish. Miescher wanted to identify the chemical composition of the nucleus, and he was interested in these cells because they are composed mostly of nuclear material, with very little cytoplasm. He succeeded in isolating an acidic substance, which contained an unusually large amount of phosphorus. Miescher called it "nuclein." He had discovered what came to be known as deoxyribonucleic acid, or DNA.

The discovery caused scarcely a ripple through the scientific community. At the time, only a few researchers suspected that the nucleus might have something to do with the whole question of inheritance. In fact, seventy-five years passed before DNA was recognized as a substance of profound biological importance. Many investigations led to that recognition. Some turned out to be blind alleys in terms of their intended purpose—yet they helped lead the way to our current understanding of the molecular basis of inheritance.

Let's look at a few of these investigations, because they reveal how ideas are generated in science. As you will see, ideas typically develop as a community effort, with individuals sharing not only what they can explain but also what they do not understand. There are two points to keep in mind:

1. Even when an experiment "fails," it may turn up information that other researchers can use or lead to questions that others can answer.

2. Unexpected results from an experiment should be reported, for they may be a clue to something important about the natural world.

A Puzzling Transformation

In 1928, Fred Griffith attempted to develop a vaccine against the bacterium *Diplococcus pneumoniae*, which causes the lung disease pneumonia. (Many vaccines are preparations of killed or weakened bacteria which, when introduced into the body, mobilize the body's defenses prior to a real attack.) He never did create a vaccine, but his experiments yielded a bit of information that opened a door to the molecular world of heredity.

Griffith isolated two strains of the bacterium, which he called the *S* and *R* forms. (Colonies of one strain had a *s*mooth surface appearance; colonies of the other had a *r*ough surface appearance.) He used these strains in four different experiments and obtained the following results (Figure 15.1):

1. Laboratory mice were injected with live S cells. The mice contracted pneumonia and died, and blood sam-

15

THE RISE OF MOLECULAR GENETICS

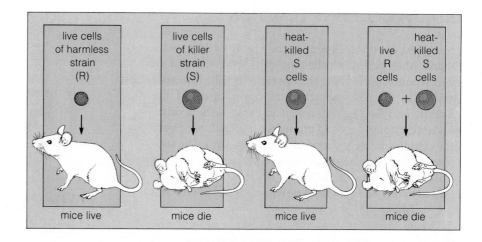

Figure 15.1 Griffith's experiments with harmless (R) strains and disease-causing (S) strains of *Diplococcus pneumoniae*, as described in the text. (You may be wondering why the S form is deadly, and the R form harmless. The killer S form shrouds its cell wall with a protective capsule, which resists attack by the host cell's normal defenses. The R form has no such capsule; hence the host's defense system destroys the R form before it can cause disease.)

ples from them teemed with live S cells. Thus the S form was virulent, or disease-causing.

2. Mice were injected with live R cells. They did not contract the disease; the R form was harmless.

3. S cells were killed by exposure to high temperature. Mice injected with these cells did not develop pneumonia.

4. Live R cells were mixed with heat-killed S cells. Unexpectedly, mice injected with this combination of cells developed pneumonia and died—and blood samples from the mice teemed with *live S cells!*

What was going on in the fourth experiment? Maybe the heat-killed S cells in the mixture were not really dead. But if that were true, the group of mice injected with heat-killed S cells alone should have contracted the disease. Maybe the R cells in the mixture had mutated into the killer form. But if that were true, the group of mice injected with R cells alone should have died. The simplest explanation was that the ability to cause infection had been transferred from the dead S cells to the harmless R cells.

Further experiments made it clear that the "harmless" cells had become permanently transformed—through hundreds of bacterial generations, their progeny also caused infections. *The transformation involved a change in the hereditary system of the bacterium.* This unexpected result had far-reaching implications. Researchers already knew how to subject cells to assays (tests to determine the nature and amount of their component parts). Because Griffith's bacteria were passing on hereditary

instructions in an unconventional manner, perhaps assays of such cells would reveal the identity of the substance carrying those instructions.

News of Griffith's discovery reached Oswald Avery and his colleagues, who worked for the next decade to identify the "transforming" substance. In 1944 they reported that DNA was probably the substance by which hereditary instructions are passed from one generation to the next.

Bacteriophage Studies

Even while work was proceeding in Avery's laboratory, Max Delbrück, Alfred Hershey, and Salvador Luria were forming a core group to study a class of viruses called **bacteriophages**. We now know that all viruses are simple infectious agents, composed only of nucleic acid and protein. Bacteriophages infect specific bacterial cells. The infectious cycle starts when they make contact with a target host cell and bind to it. In some cases, bacteriophages inject their contents into the cell (Figure 15.2). Within sixty seconds, the cell starts making materials necessary to build new bacteriophages. Then the bacterium undergoes lysis: some of the foreign materials degrade its cell wall and the cell bursts, thereby liberating a new infectious generation.

Let's look at a set of bacteriophage experiments conducted in 1952 by Hershey and his colleague Martha Chase. Knowing that a bacteriophage contains only DNA and protein, these workers devised a way to track the two substances through the infectious cycle.

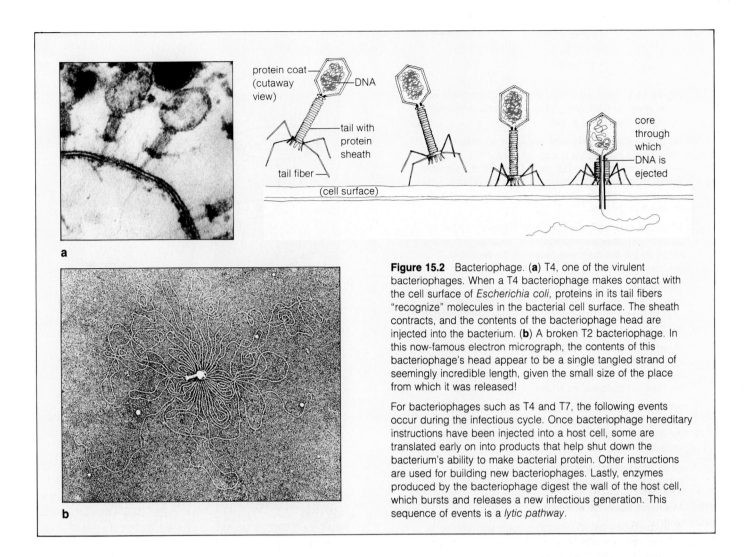

Figure 15.2 Bacteriophage. (**a**) T4, one of the virulent bacteriophages. When a T4 bacteriophage makes contact with the cell surface of *Escherichia coli*, proteins in its tail fibers "recognize" molecules in the bacterial cell surface. The sheath contracts, and the contents of the bacteriophage head are injected into the bacterium. (**b**) A broken T2 bacteriophage. In this now-famous electron micrograph, the contents of this bacteriophage's head appear to be a single tangled strand of seemingly incredible length, given the small size of the place from which it was released!

For bacteriophages such as T4 and T7, the following events occur during the infectious cycle. Once bacteriophage hereditary instructions have been injected into a host cell, some are translated early on into products that help shut down the bacterium's ability to make bacterial protein. Other instructions are used for building new bacteriophages. Lastly, enzymes produced by the bacteriophage digest the wall of the host cell, which bursts and releases a new infectious generation. This sequence of events is a *lytic pathway*.

Bacteriophage proteins contain sulfur but no phosphorus, and DNA contains phosphorus but no sulfur. Both elements have radioactive isotopes (^{35}S and ^{32}P). Cells grown on a culture medium containing ^{35}S would take up this isotope during their protein-building activities and thus would become labeled. If those cells were later infected, the new generation of bacteriophages assembled inside them would also contain the labeled protein. (Why? The new generation could be built only of materials, including radioactive ones, available from their hosts.) Similarly, bacterial cells grown on a culture medium containing ^{32}P would end up with labeled DNA.

As Figure 15.3 shows, Hershey and Chase used bacteriophages labeled in this manner to infect fresh, unlabeled cells. As it turned out, labeled protein was left on the outside of the host cells and labeled DNA ended up on the *inside* of the cells. When the infectious cycle was allowed to run its course, the new bacteriophage generation incorporated labeled DNA. Here was strong evidence that DNA is the hereditary material.

Many different experiments have since been performed on cells of many species from all five kingdoms. For all cells studied to date, the following is true:

Instructions for producing all heritable traits are encoded in DNA.

DNA STRUCTURE

Components of DNA

Even while the studies just described were proceeding, biochemists were working to identify the structure of DNA. By the early 1950s, DNA was known to contain only four kinds of nucleotides, which are the building blocks of nucleic acids (page 59). All four have the same five-carbon sugar (deoxyribose) and a phosphate group.

Figure 15.3 Hershey-Chase experiments to determine the paths taken by DNA and protein from bacteriophages during the infection of *E. coli*.

(a) Some bacterial cells were grown on a medium containing a radioactive isotope of sulfur (^{35}S), an atom usually found in protein but not in DNA. Other cells were grown on a medium containing a radioactive isotope of phosphorus (^{32}P), an atom found in DNA but essentially absent from protein.

Molecules synthesized by the cells incorporated the ^{35}S and ^{32}P, and thus were radioactively labeled. Green and black dots in the diagrams represent the two labels (radioactivity).

(b) Bacteriophages infected the labeled cells. Their infectious progeny became labeled, for they had been assembled from labeled molecules available in the host cells.

(c–d) After lysis, the first-generation (labeled) progeny were allowed to infect fresh, unlabeled bacteria. When they did, part of the bacteriophage body remained attached to the host cell surfaces, as shown in Figure 15.2.

(e) Suspensions of cells bearing their attackers were sheared in a kitchen blender. The shearing force caused the bacteriophage bodies to break away from the cell surface.

When ^{35}S-labeled bacteriophages were used in an experiment, the radioactivity remained in their bodies. When ^{32}P-labeled bacteriophages were used, radioactivity was found inside the host cells.

(f) In other samples, the infectious cycle was allowed to proceed. Second-generation progeny of the ^{35}S-labeled bacteriophage contained very little ^{35}S. But second-generation progeny of the ^{32}P-labeled bacteriophage contained ^{32}P.

Thus the DNA, not protein, was being transmitted through generations of bacteriophages.

first experiment   second experiment

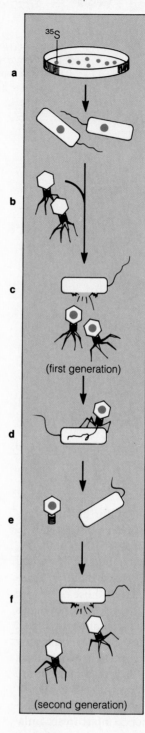

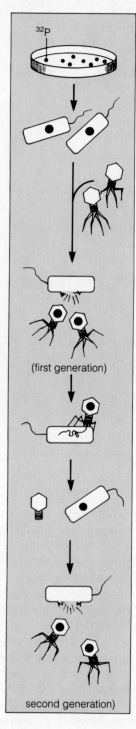

They also have one of four nitrogen-containing bases, as shown in Figure 15.4:

cytosine and **thymine**   (single-ring pyrimidines)

adenine and **guanine**   (double-ring purines)

All four nucleotides show the same general bonding pattern between their component parts (Figure 15.5).

Two more crucial insights about the molecular structure of DNA came from extensive studies by Erwin Chargaff, one of the few who had sensed the importance of Avery's report on DNA:

1. The four bases in DNA vary in relative amounts from one species to the next. Yet the relative amounts are always the same among members of the same species.

2. The amount of adenine present equals the amount of thymine (A = T), and the amount of cytosine equals the amount of guanine (C = G).

But how were the nucleotides arranged in DNA? Maurice Wilkins and his associates used x-ray diffraction methods in their attempt to find the answer. (When a substance is crystallized, its atoms form planes that disperse an x-ray beam. If the planes occur repeatedly in the crystal, the beam will be dispersed in a regular pattern. Such patterns show up as dots and streaks on a piece of film placed behind the crystal. By itself, the dot pattern on the exposed film does *not* reveal molecular structure. But the distances and angles between dots can be used to calculate the position of groups of atoms relative to one another in the crystal.)

The best x-ray diffraction images were obtained by Rosalind Franklin, who became convinced that DNA had the following features. First, DNA had to be long and thin, with a uniform 2-nanometer diameter. Second, its structure had to be highly repetitive: some part of the molecule was repeated every 0.34 nanometer, and a different part was repeated every 3.4 nanometers. Third, DNA could be helical, with an overall shape like a spiral stairway.

Patterns of Base Pairing

In the early 1950s, James Watson (a postdoctoral student from Indiana University) teamed up with Francis Crick (a Cambridge University researcher). By piecing together available data, they deduced the structure of DNA.

According to Franklin's data, DNA has a uniform diameter. According to Chargaff's data, A = T and G = C. Watson and Crick reasoned that the *double*-ringed A and G molecules were probably paired with the *single*-ringed T and C molecules along the entire length of DNA. (Otherwise, DNA would bulge where two double

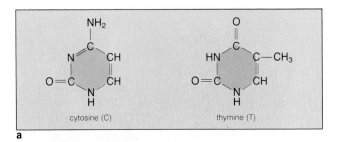

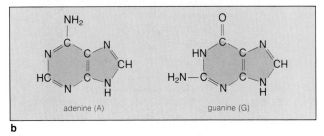

Figure 15.4 Structural formulas for the four nitrogen-containing bases of DNA. (**a**) The single-ring pyrimidines and (**b**) the double-ring purines.

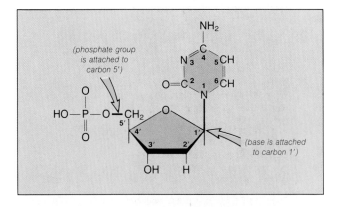

Figure 15.5 Example of one of the four nucleotides of DNA. Each nucleotide has a sugar (ribose), a phosphate group, and one of four bases (shown here, cytosine). The small numerals are used for reference purposes. The carbon atoms of the sugar are numbered 1', 2', and so forth simply to distinguish them from the numerals used for the base.

rings were paired and narrow down where two single rings were paired.)

Watson and Crick shuffled and reshuffled paper cutouts of the nucleotides. They realized that in certain orientations, A and T could become joined by two hydrogen bonds, and G and C could become joined by three. Suppose the nucleotides were strung into *two strands*, with the bases facing each other. Then hydrogen bonds could easily bridge the gap between them, like rungs of a ladder.

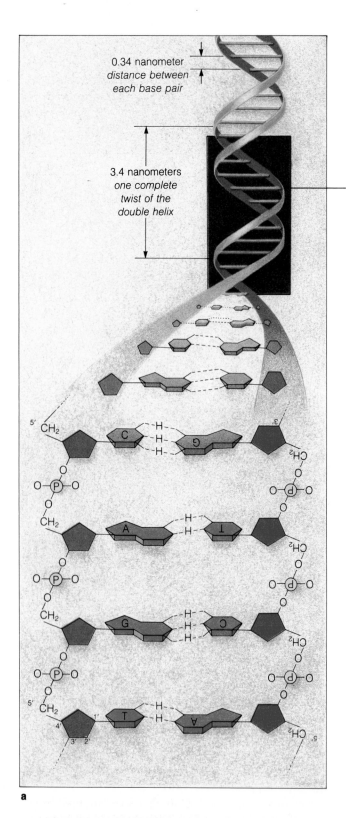

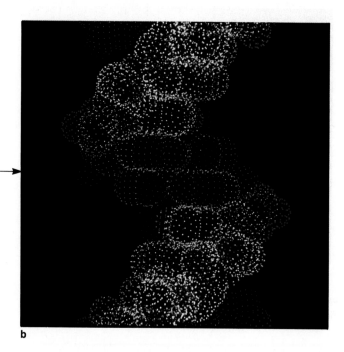

b

a

Figure 15.6 (a) Representation of a DNA double helix. Notice how the two sugar-phosphate backbones run in *opposing directions*. (In other words, the sugar molecules of one strand are upside-down. When you compare the numbers used to identify the carbon atoms of a sugar molecule, you can see that one strand runs in the 5′ → 3′ direction and the other, in the 3′ → 5′ direction.) (b) Computer-graphics model of a section from a DNA molecule, viewed from the side.

Scale models were constructed of how the "ladder" might look, and the only model that fit all available data had A-T and G-C pairs. And those pairs formed the proper hydrogen bonds only when the two DNA strands ran in *opposing directions* and twisted together to form a *double helix* (Figures 15.6 and 15.7).

In the Watson-Crick model, A and T are always present in equal proportions, and the same is true of G and C. Yet the amount of A-T can *vary* relative to the amount of G-C in the DNA of different species because any pair can follow any other in sequence. For example, even in one small region of a DNA double helix, the sequence might be:

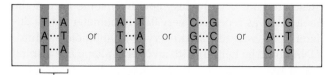

a few base pairs in
the DNA double helix

The number of different sequences possible in a long DNA molecule is staggering.

Thus the molecule of inheritance itself was seen to be a source of the unity and diversity of life:

In all species, there is <u>constancy</u> in base pairing (adenine to thymine, guanine to cytosine) between the two strands of the DNA double helix.

The DNA of different species shows <u>variation</u> in which base follows the next in a strand.

Figure 15.7 James Watson and Francis Crick posing in 1953 by their newly unveiled model of DNA structure.

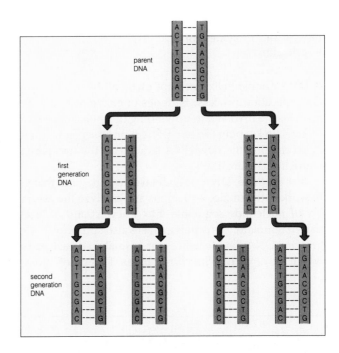

Figure 15.8 Semi-conservative nature of DNA replication, through two generations. The original two-stranded DNA molecule is shown in blue. Two new strands (pink) are assembled on them. The new strands in the second generation are shown in lavender.

DNA REPLICATION

How Copies Are Made of DNA

The discovery of DNA structure was a turning point in studies of inheritance. Until then, no one had any idea of how copies of the hereditary material are made prior to mitotic cell division. Yet the copying process is absolutely crucial; without it, cells simply remain arrested in interphase (page 140).

The Watson-Crick model suggested at once how DNA could be copied (replicated). The two strands of the DNA double helix could be unwound from each other, thereby exposing their bases. Then free nucleotides (drawn from cellular pools) could pair with the exposed bases on both unwound strands. Thus each parent strand would serve as a *template*, or structural pattern, for the assembly of new DNA strands.

If strand assembly followed the principle of base pairing, then the only sequence possible on one DNA template would have to be an exact duplicate of the base sequence of the other template.

Figure 15.8 illustrates this view of DNA replication. As you can see, *each parent strand remains intact, and a new companion strand is assembled on each one.* One parent strand therefore ends up with a new partner, forming a double helix. The other parent strand ends up with a new partner also, forming another helix. In each of the two double helices, one "old" DNA strand has been conserved. This so-called semiconservative mode of DNA replication is now known to be nearly universal among organisms.

Prior to cell division, the double-stranded DNA molecule unwinds and is duplicated through semiconservative replication: each parent strand remains intact—it is conserved—and a new companion strand is assembled on each one.

Thus semiconservative replication produces two "half-old, half-new" DNA molecules.

Semiconservative replication is simple in theory. But stop to think about what is going on even in a relatively simple prokaryote, such as the bacterium *Escherichia coli*. The DNA molecule of *E. coli* is about 1,360 micrometers long, and it turns about 400,000 times on its long axis. Each strand of the double helix has about 4,000,000 nucleotides. For replication to occur, the two strands must be separated and untwisted, so that new nucleotides can base-pair and become joined in precise order on each parent strand. The newly forming molecules must then form two separate helices without getting tangled with each other. All the untwisting, strand separation, assembly of a duplicate strand, and twisting up again can be finished in forty minutes!

Energy and Enzymes for Replication

A DNA double helix does not unzip all by itself during the replication process. Batteries of enzymes and other proteins are involved in unwinding the molecule in local regions, in keeping the two DNA strands separated, and in assembling the new DNA strands on the two parent templates (Figure 15.9).

For example, **DNA polymerases** are enzymes that go to work in the unzipped regions and catalyze the assembly of nucleotide segments into a new strand on each parent template. DNA polymerases also "proofread" the growing DNA double helices for mismatched base pairs, which are replaced with correct bases. This proofreading function is one reason why DNA is copied with such remarkable accuracy. On the average, for every *billion* base pairs formed, only *one* mistake slips through the proofreading net.

Where does the energy come from to drive the replication activities? It so happens that the free nucleotides brought up for strand assembly are not quite in the form shown in Figure 15.5. They are triphosphates (they have three phosphate groups attached, not one). Triphosphates, recall, readily give up energy for cellular reactions (page 105). Energy released when two of the phosphate groups are split away is used by the DNA polymerases for the addition of new nucleotides to a growing chain. ATP provides the energy for the unwinding process.

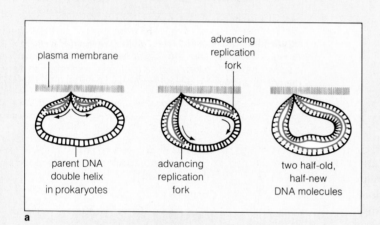

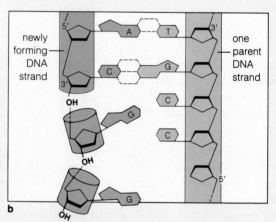

Figure 15.9 A closer look at how a DNA molecule is replicated.

Origin and direction of replication. In prokaryotes, most double-stranded DNA molecules are closed circles. At the start of replication, the two strands separate at an initiation site (a specific base sequence called the "origin"). Replication is *bidirectional*: it proceeds simultaneously in two directions away from an origin, as shown in (**a**). Here, newly forming strands are shaded blue.

In eukaryotes and some viruses, the DNA is linear, not circular. Replication is also bidirectional, but many origins are spaced along the length of the molecule, and replication proceeds in both directions from each one.

Strand assembly. Actual strand assembly proceeds at replication forks. In these limited V-shaped regions, enzymes unzip the parent DNA double helix and assemble new DNA on the exposed regions of both parent strands, which serve as templates.

As discovered by Reiji Okazaki, DNA assembly is *continuous* on one parent template but *discontinuous* on the other. In discontinuous assembly, short stretches of nucleotides are first positioned at intervals on one parent template, then enzymes link the short stretches together into a single chain:

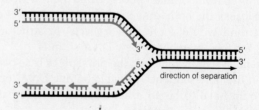

Why are there two assembly mechanisms? As (**b**) suggests, nucleotides can be added to a newly forming DNA strand in the 5′→3′ direction only. Bases projecting from a parent template dictate which kind of nucleotide will follow the next in line. But an exposed —OH group must be available on the growing end of the new strand if enzymes are to catalyze the addition of more nucleotides.

SUMMARY

1. Many different experiments using cells from species of all five kingdoms demonstrate that DNA contains the instructions for producing the heritable traits of organisms.

2. DNA is a macromolecule assembled from small organic molecules called nucleotides. Each nucleotide is composed of a five-carbon sugar (deoxyribose), a phosphate group, and one of the following nitrogenous bases:

cytosine (C)   adenine (A)
thymine (T)   guanine (G)

Cystosine and thymine are single-ring structures; adenine and guanine are double-ring structures.

3. The four nucleotide bases vary in their relative amounts in the DNA molecules of different species (each species has its own characteristic sequence of bases in its DNA). However, the relative amounts of the four bases are the same in all members of the same species.

4. In the Watson-Crick model, DNA is a long molecule of constant diameter. It consists of two nucleotide strands that are twisted together into a double helix. Hydrogen bonds join the bases making up one strand with the bases making up the other, and the two strands run in opposing directions. One of the bases in each hydrogen-bonded pair is a single-ring structure and its partner is a double-ring structure.

5. The Watson-Crick model is based on one of the fundamental principles of genetics—that there is constancy in base-pairing between the two strands of the DNA double helix. Thus adenine always pairs with thymine (A = T), and cytosine always pairs with guanine (C = T).

6. Prior to cell division, the DNA is replicated (copied), so that each of the forthcoming daughter cells will inherit a complete set of hereditary instructions.

7. DNA is replicated in a semiconservative manner. The two strands of the parental DNA double helix unwind and each becomes a template (pattern) for the assembly of a new strand. Following the rule of base pairing, the nucleotide sequence of each new strand is complementary to the sequence on the template strand. Thus two double-stranded molecules result in which one strand is "old" (it is conserved) and the other is "new."

8. Various enzymes (such as DNA polymerases) take part in the replication activities. DNA polymerases carry out the polymerization and also perform proofreading functions with remarkable accuracy. On the average, only one in a billion mistakes are made in the base-pairing process.

9. DNA replication is bidirectional (it proceeds simultaneously in two directions, away from an initiation site).

One new strand is assembled in continuous fashion; the other is assembled in discontinuous fashion. Nucleotides can be added only to the 3' end of a newly forming strand.

Review Questions

1. How did Griffith's use of control groups help him deduce that the transformation of harmless *Diplococcus* strains into deadly ones involved a change in the bacterium's hereditary system?

2. What is a bacteriophage? In the Hershey-Chase experiments, how did bacteriophages become labeled with radioactive sulfur and radioactive phosphorus? Why were these particular elements used instead of, say, carbon or nitrogen?

3. DNA is composed of only four different kinds of nucleotides. Name the three molecular parts of a nucleotide. Name the four different kinds of nitrogen-containing bases that may occur in a nucleotide. What kind of bond holds nucleotides together in a single DNA strand?

4. What kind of bond holds two DNA chains together in a double helix? Which nucleotide base pairs with adenine? Which pairs with guanine? Do the two DNA chains run in the same or opposite directions?

5. The four bases in DNA may *vary* greatly in relative amounts from one species to the next—yet the relative amounts are always the *same* among all members of a single species. How does the concept of base pairing explain these twin properties—the unity *and* diversity—of DNA molecules?

6. When regions of a double helix are unwound during DNA replication, do the two unwound strands join back together again after a new DNA molecule has formed?

Readings

Cairns, J., G. Stent, and J. Watson (editors). 1966. *Phage and the Origins of Molecular Biology.* Cold Spring Harbor, New York: Cold Spring Harbor Laboratories. Collection of essays by the founders of and converts to molecular genetics. Gives a sense of history in the making—the emergence of insights, the wit, the humility, the personalities of the individuals involved.

Felsenfeld, G. October 1985. "DNA." *Scientific American* 253(4):58–67. Describes how the DNA double helix may change its shape during interactions with regulatory proteins.

Taylor, J. (editor). 1965. *Selected Papers on Molecular Genetics.* New York: Academic Press.

Watson, J. 1978. *The Double Helix.* New York: Atheneum. Highly personal view of scientists and their methods, interwoven into an account of how DNA structure was discovered.

Watson, J. 1984. *Molecular Biology of the Gene.* Menlo Park, California: Benjamin-Cummings.

16

THE GENETIC CODE AND PROTEIN SYNTHESIS

DNA is like a book of instructions in each cell. The alphabet used to create the book is simple enough: A, T, G, and C. However, merely knowing the letters does not tell us how they give rise to the language of life, with sentences (genes) evoking precise meanings (proteins). And the letters alone cannot tell us how or why a cell reads certain DNA passages and skips others at any particular moment. The more we learn about DNA, the more we realize it is no simple book. It was put together over a time span of epic proportions, and it is correspondingly complex. It is known to contain passages of great clarity and precision. It contains long stretches of unknown function. It even contains genetic gibberish, arising from the random tamperings of mutations and recombinations.

Even though this evolving, ancient book of inheritance still holds many mysteries, one of the triumphs of our time is that the words themselves have been deciphered! Each word has three letters (some combination of three nucleotides). Genes, the coding regions of DNA, are composed of specific sequences of these triplets, which are strung together one after another. *It is the linear sequence of nucleotide triplets that becomes translated into a linear sequence of amino acids in a polypeptide chain—the basic structural unit of proteins.* How genes are translated into proteins is the subject of this chapter.

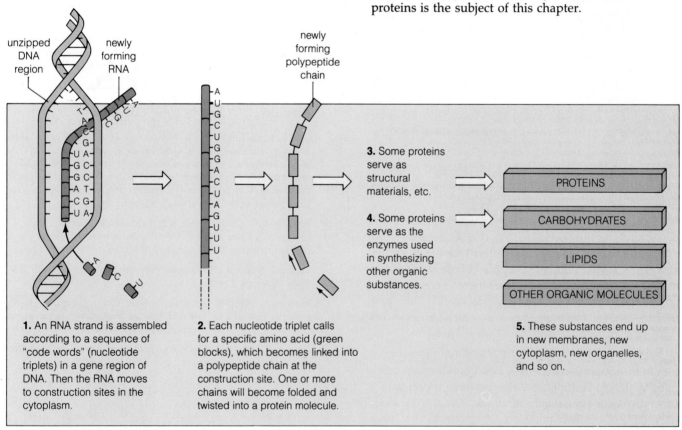

unzipped DNA region

newly forming RNA

newly forming polypeptide chain

1. An RNA strand is assembled according to a sequence of "code words" (nucleotide triplets) in a gene region of DNA. Then the RNA moves to construction sites in the cytoplasm.

2. Each nucleotide triplet calls for a specific amino acid (green blocks), which becomes linked into a polypeptide chain at the construction site. One or more chains will become folded and twisted into a protein molecule.

3. Some proteins serve as structural materials, etc.

4. Some proteins serve as the enzymes used in synthesizing other organic substances.

5. These substances end up in new membranes, new cytoplasm, new organelles, and so on.

PROTEINS

CARBOHYDRATES

LIPIDS

OTHER ORGANIC MOLECULES

Figure 16.1 Simplified overview of how hereditary instructions are translated into the proteins necessary for cell structure and functioning.

THE GENE-PROTEIN CONNECTION

Figure 16.1 is a simplified picture of how the hereditary instructions encoded in genes are carried out. In essence, genes specify all the enzymes and other proteins necessary to construct and maintain the traits of an individual.

The connection between genes and proteins was first suspected in 1908 by Archibald Garrod, who had studied recurring metabolic disorders in some families. Blood or urine samples of afflicted patients revealed a buildup of the intermediate compound produced just before the step that one particular enzyme mediated in a known metabolic pathway (see, for example, page 186). Garrod suspected the heritable disorders arose from a defect in the enzyme, which must have been specified by a defective "unit" of inheritance.

Thirty-three years later, George Beadle and Edward Tatum brought the connection into sharp focus, when they studied the effects of mutations on the metabolic pathways in *Neurospora crassa*, the red bread mold. *N. crassa* will grow on a minimal medium containing only sucrose, mineral salts, and biotin (one of the B vitamins). All other substances (including other vitamins and amino acids) it synthesizes for itself. The synthesis pathways were already known for some of these substances. Thus, if a mutant carried a defective enzyme for one of the steps in a pathway, it would be a *nutritional mutant* (unable to grow on the minimal medium). One mutant strain grew only when vitamin B_6 supplemented the minimal medium. Another grew only when vitamin B_1 was added. Also, analysis of mutant cell extracts revealed a defective enzyme in one strain, and different defective enzymes in other strains. *Each inherited mutation corresponded to a different defective enzyme.*

On the basis of their findings, Beadle and Tatum formulated the "one gene, one enzyme" hypothesis: a single gene codes for a single kind of enzyme.

Studies of sickle-cell anemia provided better understanding of the connection between genes and proteins. As we have seen, this heritable disorder arises from the presence of abnormal hemoglobin (HbS instead of HbA) in the red blood cells of afflicted persons. In 1949 Linus Pauling and Harvey Itano attempted to identify the difference between HbS and HbA through *electrophoresis*, a technique for measuring the rate and direction of movement of organic molecules in response to an electric field (Figure 16.2). Compared to its normal counterpart, HbS moved at a slower rate toward the positive pole of the electric field. HbS, it seemed, had a smaller number of negatively charged amino acids.

A few years later, Vernon Ingram pinpointed the difference. Hemoglobin, recall, is a protein with four polypeptide chains, two designated "alpha" and the other two, "beta." As Figure 16.3 shows, in the beta chains of

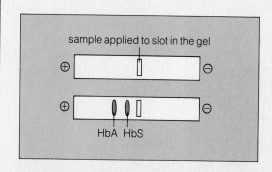

Figure 16.2 Electrophoresis: a technique used to measure the rate and degree of migration of organic molecules in response to an electric field.

Often the electric field is established in a container filled with a gel. One end of the container carries a positive charge, the other a negative charge.

When a sample that has a mixture of different proteins is placed in the gel, each protein molecule will move toward one end of the container or the other. The rate and direction of movement depends partly on a molecule's net surface charge. (They also depend on its size and shape. For example, larger molecules show more resistance than smaller molecules to moving through the gel.)

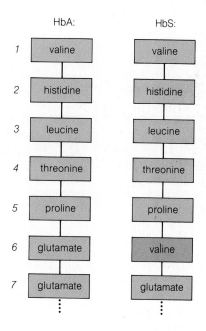

Figure 16.3 A substitution of one amino acid for another in a polypeptide chain. In sickle-cell hemoglobin, valine is substituted for glutamate at position 6 of the beta chain (page 58). This one mutation puts a "sticky" (hydrophobic) patch on the surface of a hemoglobin molecule. When oxygen concentrations in the blood are low, the sticky patches interact and hemoglobin molecules aggregate into rods, distort red blood cells, and cause them to clump in blood vessels (page 169).

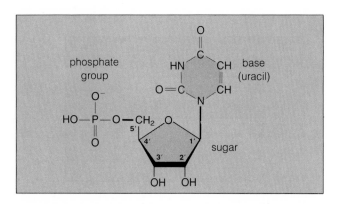

Figure 16.4 Structural formula for one of the four nucleotides in an RNA molecule. In this example, the base is uracil. Compare this illustration with Figure 15.5, which shows one of the DNA nucleotides.

HbS, one amino acid (valine) is substituted for another (glutamate). Unlike glutamate, which carries a negative charge, valine has no net charge—hence the slower rate of movement of HbS toward the positive pole.

Even so, the two alpha chains in the HbS molecule were normal. Clearly, *two* genes coded for hemoglobin (one for each kind of polypeptide chain). Thus the more precise "one gene, one polypeptide" hypothesis emerged:

One gene codes for an amino acid sequence of one polypeptide chain.

A protein molecule consists of one or more polypeptide chains (which may or may not be identical).

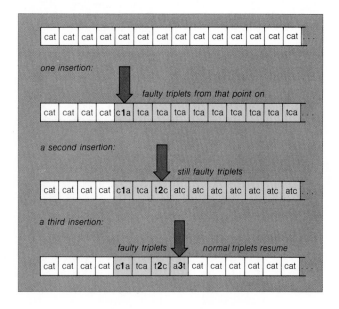

Figure 16.5 Shifts in the "reading frame" for DNA, caused by the insertion of extra nucleotides. If translation of the genetic code depends on reading three adjacent nucleotides at a time, then insertion of one or two extra nucleotides will change the sequence (hence the meaning) of all subsequent base triplets. A third addition will restore much of the normal sequence.

Such insertions into the DNA molecule occur on rare occasions. They are called "frameshift" mutations, and they give rise to defective polypeptide chains. The presence of these mutations often can be detected by the appearance of mutant phenotypes.

RNA STRUCTURE AND THE GENETIC CODE

Instructions for building polypeptide chains are housed in the nucleus (or, in prokaryotes, the nucleoid), but the chains themselves are assembled in the cytoplasm. Molecules of **ribonucleic acid**, or RNA, carry the construction blueprints from the DNA to sites in the cytoplasm.

Unlike the double-stranded DNA molecule, an RNA molecule is a single strand of nucleotides. Each of its nucleotides consists of a sugar (ribose), a phosphate group, and one of four kinds of nitrogen-containing bases. Three bases are the same in DNA and RNA (both have adenine, cytosine, and guanine). However, RNA has the base *uracil* instead of thymine (Figure 16.4). Even so, uracil can base-pair with adenine, just as thymine does. *Hence RNA nucleotides can be strung together and linked into an RNA molecule by using an unwound region of a DNA double helix as the template.* Figure 16.1 is a simple picture of this assembly process.

In itself, the ability of RNA to be assembled on a DNA template suggests it could play a role in carrying instructions from genes in the nucleus to construction sites in the cytoplasm. There is other evidence as well. For example, RNA can be detected in both the nucleus and the cytoplasm. Also, radioactive labeling experiments show that RNA is *produced* in the nucleus and *moves into* the cytoplasm.

Whatever information RNA carries away from DNA, it is still in the form of a linear sequence of nucleotides. So we are still left with a central question: *What are the protein-building "words" encoded in the linear sequence of nucleotides?*

The building blocks of proteins are twenty common amino acids. The "words" along a single strand of DNA (or RNA) must call for some assortment of these amino acids, one after another, to form a polypeptide chain. But neither DNA nor RNA has more than four kinds of nucleotides. Obviously, each nucleotide doesn't code for only one type of amino acid; that would be only four (4^1) choices. A sequence of two nucleotides would yield only sixteen (4^2) choices, not twenty. A sequence of three nucleotides would yield sixty-four (4^3) choices—far more than the twenty required. Could it be that different *triplets* of nucleotides specify the same thing?

Francis Crick and Sidney Brenner discovered the relationship between the nucleotide sequence in DNA (or RNA) and the amino acid sequence in a polypeptide chain—a relationship now called the **genetic code**. *The genetic code consists of nucleotide bases that are read linearly, three at a time, with the sequence of each triplet specifying an amino acid.*

Assume that the nucleotides of a gene are read one after another in a row, and that the "reading frame" proceeds three adjacent nucleotides at a time from the start to the end of a gene. If one extra nucleotide were inserted in a gene, the reading frame would be modified from that point on. And the polypeptide chain would end up with the wrong amino acid sequence. Figure 16.5 shows why this is so. A second insertion would not improve matters. A third insertion, though, would restore the reading frame. Because the region between the first and third extra bases would still be wrong, the corresponding stretch of the amino acid sequence would be defective. However, the sequence coded for by the regions on either side of the insertions would be normal. Hence the protein product could be more or less functional.

Hereditary instructions for building proteins are encoded in linear sequences of nucleotides (in DNA and RNA).

Every three nucleotides (a base triplet) specifies an amino acid, which becomes linked with others into a polypeptide chain.

Through the work of H. Gobind Khorana, Marshall Nirenberg, Robert Holley, and others, the genetic code is now known to consist of sixty-four different nucleotide triplets. Sixty-one actually specify amino acids. The remaining three act like stop signs or termination points, signifying that no more amino acids are to be added to the new polypeptide chain.

In the RNA molecules carrying messages from DNA to the cytoplasm (called, appropriately, "messenger RNA"), each nucleotide triplet that codes for an amino acid or for chain termination is called a **codon**. As Figure 16.6 indicates, only two amino acids (tryptophan and methionine) are specified by one codon only. The other

First Letter	Second Letter				Third Letter
	U	C	A	G	
U	phenylalanine	serine	tyrosine	cysteine	U
	phenylalanine	serine	tyrosine	cysteine	C
	leucine	serine	stop	stop	A
	leucine	serine	stop	tryptophan	G
C	leucine	proline	histidine	arginine	U
	leucine	proline	histidine	arginine	C
	leucine	proline	glutamine	arginine	A
	leucine	proline	glutamine	arginine	G
A	isoleucine	threonine	asparagine	serine	U
	isoleucine	threonine	asparagine	serine	C
	isoleucine	threonine	lysine	arginine	A
	(start) methionine	threonine	lysine	arginine	G
G	valine	alanine	aspartate	glycine	U
	valine	alanine	aspartate	glycine	C
	valine	alanine	glutamate	glycine	A
	valine	alanine	glutamate	glycine	G

Figure 16.6 The genetic code. Each codon consists of three nucleotides. The first nucleotide of any triplet is given in the left column. The second is given in the middle columns; the third, in the right column. Thus we find (for instance) that tryptophan is coded for by the triplet U G G. Phenylalanine is coded for by both U U U and U U C. All such nucleotide triplets (codons) are strung one after another in messenger RNA molecules. Different sequences of triplets code for different protein chains.

eighteen amino acids can be specified by two or more kinds of triplets. For instance, CCU, CCC, CCA, and CCG all specify proline.

The genetic code is nearly universal for all cells. Codons that specify particular amino acids in bacteria specify the same amino acids in almost all the protistans, fungi, plants, and animals studied so far.

Almost all living cells studied to date use the same genetic code as the language of protein synthesis.

Mitochondria: Exceptions to the Rule

In the evolutionary view, the genetic code probably dates back to the most ancient cells from which all existing species arose. First, the code appears to be identical for all living species, from bacterial to human. Second, the absence of variation from one species to the next suggests that the genetic code predates the divergences into separate evolutionary lines. Once the code had been established, variation would have been nearly impossible. (One variant triplet alone could alter the functioning of many different proteins, depending on where and how many times it specified a given product in an amino acid sequence. Most such variations would be harmful or lethal.)

It happens, though, that slightly variant codes exist—in the mitochondrion. A mitochondrion, recall, is an organelle specialized for aerobic energy metabolism. It has its own DNA, which is replicated independently of the nuclear DNA. This DNA governs synthesis of at least some RNA and proteins that the organelle requires for its specialized tasks. Although mitochondrial codes are almost the same as the one used by cells, a *few* codons have different meanings. (For instance, UGA is a termination codon in most genetic messages; for mitochondria, it specifies tryptophan.

With its variant codons, the mitochondrion is genetically isolated from the rest of the eukaryotic cell. What is the significance of this isolation? Lynn Margulis and many other biologists suspect that the original ancestors of mitochondria were free-living cells, similar to some modern bacteria. In some way, these bacteria took up permanent residence inside the cytoplasm of other kinds of cells (page 555). The guest and the host cell became locked in a *symbiotic* relationship, with each gaining some benefit from the activities of the other. Over time, the "guest" lost the means to perform some function that the host cell was performing for it—hence it lost the capacity for independent existence.

If this is what happened, then the mitochondrial code might be a vestige of a primitive code from an entirely separate species, long since vanished. However, it might also be true that the variant code arose only *after* the joint living arrangement became permanent. Mutations leading to a slightly variant code may have been advantageous, for there would henceforth be no accidental role-swapping between nuclear and mitochondrial genes that code for different proteins.

OVERVIEW OF PROTEIN SYNTHESIS

What we have just described is the code for assembling the polypeptide chains of proteins. Let's now take a look at how that code is put to work. There are two distinct stages of protein synthesis, which are called transcription and translation.

In **transcription**, a single strand of RNA is assembled at an unwound portion of a DNA double helix, with one of the DNA strands serving as the template (Figure 16.1). The ribonucleotides are linked one after another into short chains called **RNA transcripts**. The same coding region can be transcribed over and over again during the life of a cell. Often, many thousands of RNA transcripts (hence a large number of protein molecules) are made from the same gene.

In **translation**, three classes of RNA molecules interact in converting the DNA instructions into a particular sequence of amino acids:

ribosomal RNA (rRNA)	"structural" molecule which, with other proteins, forms the *ribosome* (the actual workbench on which the polypeptide chain is assembled)
messenger RNA (mRNA)	the polypeptide "blueprint" (a particular string of codons) that is delivered to the ribosome
transfer RNA (tRNA)	"adaptor" molecule (the means by which free amino acids are brought to the ribosome and matched up to the right codons of the mRNA)

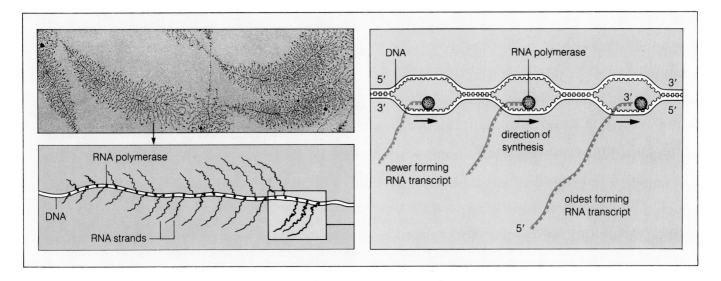

Figure 16.7 Transcription: the transfer of genetic information from DNA to RNA. The micrograph shows genes isolated from a maturing egg of the spotted newt (*Triturus viridescens*). The short RNA strands are in early stages of synthesis; the longer strands are nearing completion.

(By now, you may be wondering where the rRNA and the tRNA molecules come from. They, too, are transcribed from particular gene regions, just like mRNA. But only the mRNA becomes translated into polypeptide chains.)

TRANSCRIPTION

Assembly of an RNA Molecule

The word "transcript" means "copy," although it might help to keep in mind that an RNA transcript is *complementary to* the DNA template strand on which it is formed, not an identical copy of it.

In two respects, the assembly of an RNA strand on the DNA template is like DNA replication (page 202). First, new nucleotides are added to a growing RNA strand according to the rules of base pairing:

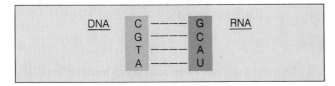

Second, the nucleotides cannot be added any which way; like a DNA strand, the RNA strand grows only in the $5' \rightarrow 3'$ direction (Figure 15.9).

However, RNA strand assembly differs from DNA replication in the following ways:

1. Enzymes called **RNA polymerases** assemble the transcripts.

2. Several RNA polymerases can rapidly move along the same DNA stretch, one after another, thereby assembling identical transcripts one after another.

3. In almost all cases, only *one* (not both) of the two unzipped DNA strands in any given region is transcribed.

Transcription starts at *promoters*, which are specific base sequences on one or the other strand in a DNA double helix. Promoters are the only places where the RNA polymerases can initiate binding to the DNA. As Figure 16.7 shows, each RNA polymerase moves along the template, joining RNA nucleotides one after another into a transcript. When it reaches the end of the gene region, the transcript is released.

Transcript Processing

Not all the nucleotides in a eukaryotic gene code for an amino acid sequence. Between the beginning and end of a gene, there can be some number of noncoding nucleotide sequences called *intervening DNA*, or **introns** for short. The parts that actually do get translated are called **exons**.

All introns and exons that occur between the start signal and termination signal for a gene are transcribed. As a result, a primary transcript contains more than the

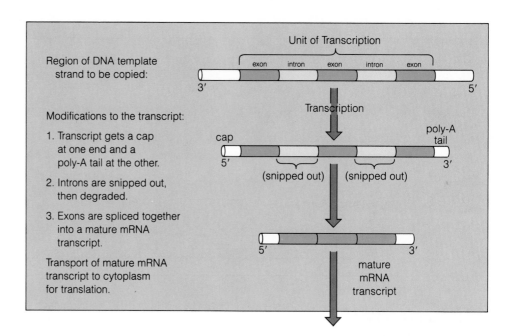

Figure 16.8 Steps in the processing of mRNA transcripts in eukaryotes.

Region of DNA template strand to be copied:

Modifications to the transcript:

1. Transcript gets a cap at one end and a poly-A tail at the other.

2. Introns are snipped out, then degraded.

3. Exons are spliced together into a mature mRNA transcript.

Transport of mature mRNA transcript to cytoplasm for translation.

coding sequences needed to build a particular polypeptide chain. However, as Figure 16.8 indicates, from the time a transcript is released from a gene until it leaves the nucleus it is extensively modified. A "cap" is attached to the start of the transcript and a "poly-A-tail" is attached to its end. (The functions of the cap and of the tail, which consists of about 200 adenine units, have not yet been determined.) Also, the introns are snipped out and the exons are spliced together to form a mature mRNA transcript.

TRANSLATION

In *translation*, the information content of an mRNA molecule is used to build a specific polypeptide chain at the ribosome. The events of translation depend on interactions between mRNA, tRNA, and rRNA.

Codon-Anticodon Interactions

Messenger RNA serves as the template for putting together a particular sequence of amino acids. To keep things simple, we can portray the template this way:

When tRNA molecules bring individual amino acids to the ribosome, they align the amino acids against this template.

Each kind of tRNA contains an **anticodon**: a sequence of three nucleotide bases that can pair with a specific mRNA codon. Each tRNA molecule also has a molecular "hook," an attachment site for an amino acid. Figure 16.9 gives an idea of its molecular structure. For the sake of clarity in later illustrations, the structure of the tRNA molecule can be portrayed in the following simplified way:

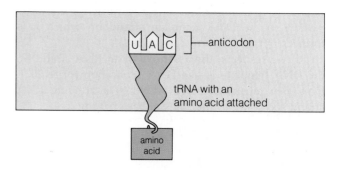

The first two bases of an anticodon must be precisely complementary to an mRNA codon for binding to occur. However, recognition of the *third* codon base is often less precise. (For instance, yeast alanine tRNA can bind to three codons: GCU, GCC, or GCA.) This freedom in codon-anticodon pairing at the third base is called the "wobble" effect.

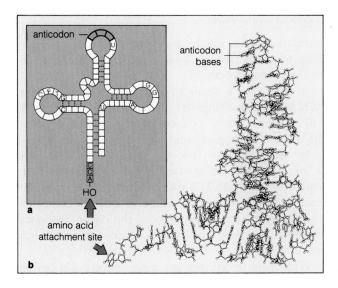

Figure 16.9 (**a**) Structural features that all tRNA molecules hold in common. Notice how the ribonucleotide strand folds back on itself into hairpin loops, which are held in place by hydrogen bonds. (**b**) Three-dimensional structure of one tRNA molecule.

A codon (base triplet of mRNA) is a recognition site for an anticodon (complementary base triplet of tRNA).

Ribosome Function

All codon-anticodon interactions take place on the surface of the **ribosome**, the cytoplasmic structure on which amino acids are linked into polypeptide chains. Ten thousand ribosomes can be present in one prokaryotic cell; a eukaryotic cell may contain many tens of thousands. At its widest dimension, a ribosome is only twenty-five nanometers (about a millionth of an inch).

Each ribosome consists of two parts. The small subunit consists of an rRNA molecule and a number of different proteins; as Figure 16.10 shows, the small subunit has a "platform." Apparently, an mRNA molecule binds to this platform, then a procession of tRNAs binds to the mRNA. The large subunit of the ribosome consists of two or three rRNA molecules and a number of different proteins. It contains the enzyme that catalyzes the linkage between amino acids being delivered by tRNAs to the ribosome.

An mRNA molecule may be translated simultaneously by several ribosomes. The term **polysome** refers to a cluster of ribosomes bound to the same mRNA. Polypeptide chain elongation proceeds independently at each ribosome in the cluster. Figure 16.10 shows polysomes positioned on an mRNA molecule in a bacterium.

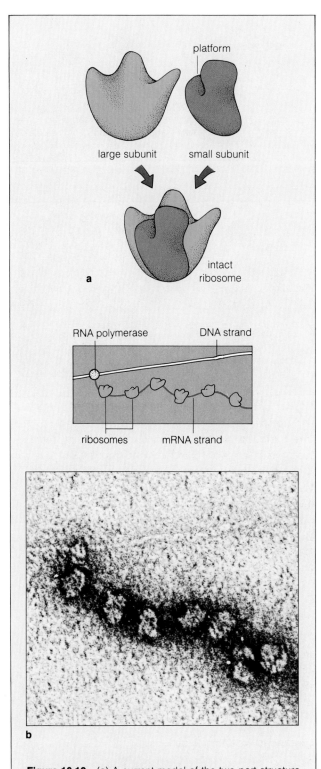

Figure 16.10 (**a**) A current model of the two-part structure of eukaryotic ribosomes. (**b**) A group of ribosomes (polysome) positioned on an mRNA molecule, which was formed from the *E. coli* DNA molecule running from the left to the right edge of the micrograph, 361,200×. In prokaryotes (which have no nucleus), translation occurs even while transcription is still proceeding.

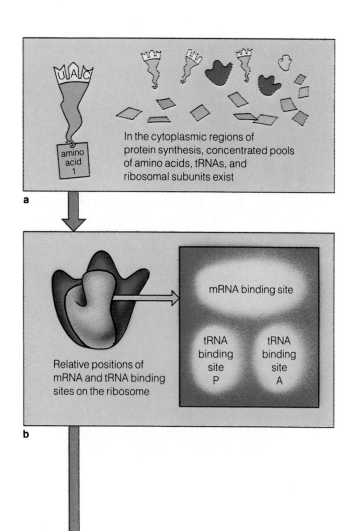

In the cytoplasmic regions of protein synthesis, concentrated pools of amino acids, tRNAs, and ribosomal subunits exist

a

mRNA binding site

tRNA binding site P

tRNA binding site A

Relative positions of mRNA and tRNA binding sites on the ribosome

b

Stages of Translation

Translation of an mRNA transcript into a polypeptide chain proceeds through three stages: initiation, elongation, and termination.

In *initiation*, a small ribosomal unit binds to the mRNA. A special initiator tRNA positions this small unit over the "start" codon on mRNA. The complex binds with the large subunit, and elongation begins.

In *chain elongation*, the start codon on the mRNA defines the reading frame for assembling amino acids in sequence. Figure 16.11 illustrates the manner in which amino acids become positioned and linked together on the ribosomal workbench. Again, the codon sequence of mRNA specifies which amino acids will be bound, one after another, into a polypeptide chain.

With *chain termination*, no more amino acids can be added to the polypeptide chain. The presence of a stop codon (usually UAG, UAA, or UGA) in the mRNA sequence triggers events that allow the polypeptide chain to become detached from the ribosome. In eukaryotes, the chain either joins the pool of free proteins in the cytoplasm or enters the cytomembrane system for further processing, beginning with the endoplasmic reticulum (Figure 5.14).

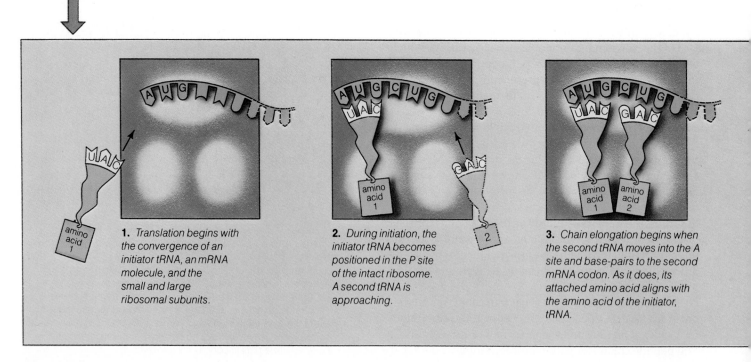

1. *Translation begins with the convergence of an initiator tRNA, an mRNA molecule, and the small and large ribosomal subunits.*

2. *During initiation, the initiator tRNA becomes positioned in the P site of the intact ribosome. A second tRNA is approaching.*

3. *Chain elongation begins when the second tRNA moves into the A site and base-pairs to the second mRNA codon. As it does, its attached amino acid aligns with the amino acid of the initiator, tRNA.*

Figure 16.11 Simplified picture of chain elongation during the translation stage of protein synthesis.

CHANGE IN THE HEREDITARY MATERIAL

Sources of Genetic Variation

As you can deduce from the preceding discussion, the protein-coding messages in DNA must be preserved from generation to generation if cells are to produce functional proteins. Yet changes do occur in the DNA, some of which are reversible and some of which are not. These changes can be grouped into three categories: crossing over and recombination, chromosomal aberrations, and gene mutations (Table 16.1).

At the cytological level, we can detect *crossing over* and *recombination* at meiosis. These events lead to new combinations of alleles at different gene locations in chromosomes. Also detectable are *chromosomal aberrations*, which change the structure or number of chromosomes in a eukaryotic cell (Chapter Thirteen). *Gene mutation* must be described at the molecular level, for it concerns heritable alterations in the amount or sequence of as few as one or two nucleotides. Through many experiments, we know that gene mutation occurs in prokaryotes as well as eukaryotes.

Mutation at the Molecular Level

A change in the nucleotide sequence of DNA is called a gene mutation. Such changes can occur during DNA replication or between replications. Sometimes one kind of base or base pair is replaced by another. At other times, one or more base pairs are added or deleted.

Environmental agents such as viruses, ultraviolet radiation, or compounds foreign to the cell can induce gene mutation. These so-called **mutagens** attack a DNA strand and may cause permanent modifications in its structure.

Other gene mutations are spontaneous rather than induced. They arise mainly from replication errors. For example, the wrong bases may pair during replication (as when adenine pairs with cytosine instead of to thymine). This type of mutation is called a "base-pair substitution." However, more than seventy-five percent of all spontaneous changes are "frameshift mutations" (Figure 16.5). Here, the insertion or deletion of one to several base pairs in the DNA sequence puts the nucleotide sequence out of phase and causes genetic instructions to be followed incorrectly when proteins are being synthesized.

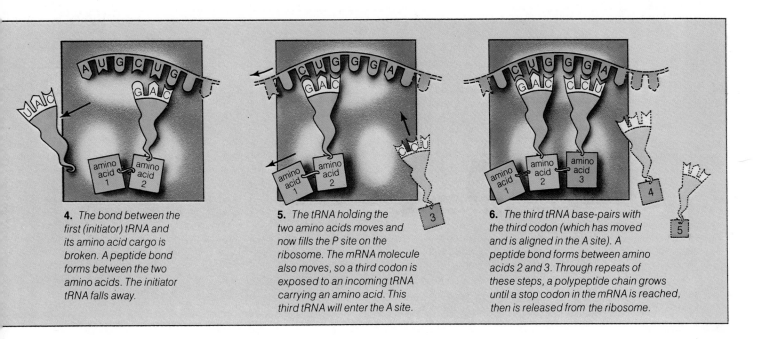

4. The bond between the first (initiator) tRNA and its amino acid cargo is broken. A peptide bond forms between the two amino acids. The initiator tRNA falls away.

5. The tRNA holding the two amino acids moves and now fills the P site on a ribosome. The mRNA molecule also moves, so a third codon is exposed to an incoming tRNA carrying an amino acid. This third tRNA will enter the A site.

6. The third tRNA base-pairs with the third codon (which has moved and is aligned in the A site). A peptide bond forms between amino acids 2 and 3. Through repeats of these steps, a polypeptide chain grows until a stop codon in the mRNA is reached, then is released from the ribosome.

Mutation Rates

The probability that a mutation in a particular gene will occur per cell division is called its **mutation rate** for the gene. On the average, the likelihood of a given gene mutating is only once in a million (10^6) replications.

Genes mutate independently of each other. Thus, to determine the likelihood of any two mutations occurring in the same cell, we would have to multiply the individual mutation rates for the two genes. For example, if the rate for one gene is one in a million per generation and the rate for another is one in a billion (10^9), then there is only one chance in a million billion (10^{15}) that both genes will mutate in the same cell—which is rare to say the least.

The extremely remote likelihood of two gene mutations occurring at once is the reason why two antibiotics are used at the same time in treating certain bacterial diseases, such as tuberculosis. In the rare chance the bacterial invader has a heritable resistance to one antibiotic (say, streptomycin), it is highly unlikely that it will simultaneously be resistant to penicillin.

Mutation and Evolution

In the natural world, gene mutations are rare, chance events. It is impossible to predict exactly when, and in which particular organism, they will appear. They are also inevitable. (For example, as long as sunlight reaches the earth, some ultraviolet wavelengths will disrupt bonds in some DNA.)

A mutation may turn out to be beneficial or harmful. The outcome depends on how the protein specified by the mutated DNA region is received in the environment, and on how it meshes in the coordinated workings of the entire individual. Because of these two variables, most mutations do not bode well for the individual. No matter what the species, that individual generally inherits a combination of many genes already fine-tuned for a given range of operating conditions. A mutant gene is likely to be less functional, not more so, under those conditions. Given this prospect, DNA repair enzymes probably evolved as mechanisms to eliminate replication errors in a package of genes having a history of survival value. The action of those enzymes has protected the

Table 16.1 Sources of Change in the Hereditary Material			
Level of Analysis		Outcome	Occurs in
Crossing Over and Recombination			
Cytological		New allelic combinations of different genes in a chromosome	Viruses, prokaryotes, eukaryotes
Molecular		New base-pair sequence in a stretch of DNA	
Chromosomal Aberration			
Cytological	Depletion	Segment lost from a chromosome	Eukaryotes
	Duplication	Chromosome segment repeated more than once in a chromosome set	
	Inversion	Chromosome segment inverted	
	Translocation	Chromosome segment transferred to a nonhomologous chromosome	
	Aneuploidy	Single chromosome of a given type present once or three or more times in an otherwise diploid set	
	Polyploidy	Three or more chromosome sets present	
Gene Mutation			
Molecular	Base substitution	A purine or a pyrimidine replaces another purine or pyrimidine	Viruses, prokaryotes, eukaryotes
	Frameshift	One to several extra base pairs inserted into the DNA sequence	
		One to several base pairs lost entirely from the DNA sequence	
	Transposition	DNA segments moved to new location in the same or a different chromosome	

overall stability of the vulnerable molecules of inheritance that have been replicated through billions of years.

Yet every so often through that immense time span, mutations appeared and provided their bearers with advantages or did their bearers no harm. Selection processes worked to perpetuate mutations having adaptive consequences. Other mutations, it seems, have produced DNA regions with no currently assignable function, but these so-called "rusting hulks" are still replicated along with everything else. After more than three billion years, molecular descendants of the first strands of DNA are replete with mutations. Each is a patchwork molecule, with variant numbers and kinds of genes stitched into it.

PERSPECTIVE

DNA—deoxyribonucleic acid—is the molecule of inheritance for every living cell. It is the helically coiled, double-stranded blueprint for constructing all the proteins required for proper cell functioning. Every DNA molecule is composed of only three kinds of substances: a sugar, a phosphate group, and nitrogen-containing bases (adenine, thymine, guanine, and cytosine). Every DNA molecule is assembled according to the same rules. When the time comes for new bases from the cellular environment to become paired with old bases sticking out from a DNA sugar-phosphate backbone, adenine is paired only with thymine, and guanine only with cytosine.

What this means is that every living thing on earth shares the same chemical heritage with all others. Your DNA is made of the same kinds of substances, and follows the same base-pairing rules, as the DNA of earthworms in Missouri and grasses on the Mongolian steppes. Your DNA is replicated in much the same way as theirs. In the evolutionary view, the reason you don't look like an earthworm or a flowering plant is largely a result of mutations and recombinations. Mistakes and shufflings in base-pair sequences made their entrance during the past $3\frac{1}{2}$ billion years and they led, in their unique divergent ways, to the three of you. Thus the *sequence* of base pairs along the DNA molecule has come to be different in the three of you.

Dinosaur DNA, too, was presumably assembled from the same chemical stuff as yours. But the mutations and recombinations that gave rise to the unique sequences of base pairs that specified "build dinosaurs" made those creatures unsuitable, when environmental conditions changed, for continuing their journey.

We have, in this chapter, introduced evidence for three concepts of profound importance. *First, DNA is the source of the unity of life. Second, mutations and recombinations in the structure and number of DNA molecules are the source of life's diversity. And finally, the changing environment is the testing ground for the success or failure of the proteins specified by each novel DNA sequence and assortment that appears on the evolutionary scene.*

SUMMARY

1. Hereditary instructions for synthesizing proteins are encoded in the linear sequence of nucleotides in DNA and RNA. Every three nucleotides (a base triplet) specifies either an amino acid, which becomes linked with other amino acids into a polypeptide chain—the basic structural unit of proteins—or a stop signal.

2. The relationship between the sequence of base triplets and the amino acid sequence in proteins is called the genetic code. This code applies to protein synthesis in almost all organisms studied to date. (At the subcellular level, the mitochondrial code is slightly different.)

3. Protein synthesis occurs in two stages. In *transcription*, an RNA strand is synthesized at a given portion of the DNA double helix, which serves as the template. This RNA strand becomes the mRNA transcript. In *translation*, three classes of RNA molecules interact to convert the message encoded in the DNA into the amino acid sequence of a polypeptide chain:

   a. rRNA: a structural molecule which, together with other proteins, forms the ribosome on which a polypeptide chain is assembled.

   b. mRNA: a blueprint for building a polypeptide; it consists of a sequence of specific nucleotide triplets (each triplet in mRNA is a codon).

   c. tRNA: an adaptor molecule that has both an attachment site for an amino acid and an anticodon (a sequence of three nucleotide bases that can pair with a specific codon).

4. Transcription proceeds according to the principle of base-pairing, but in this case uracil (not thymine) is paired with the adenine present in the DNA template strand:

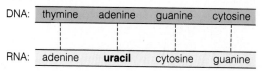

5. In eukaryotes, a primary transcript undergoes processing to form an mRNA transcript. For example, the introns (nucleotide sequences that do not code for parts

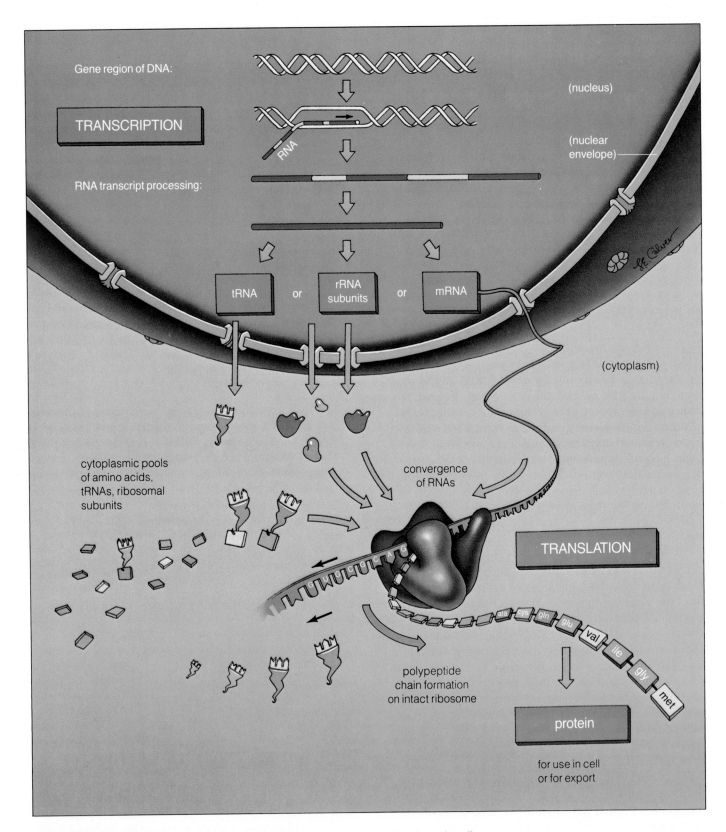

Figure 16.12 Summary of the flow of genetic information in protein synthesis in eukaryotic cells.

of the polypeptide chain) are excised, and the exons (coding sequences) are spliced together.

6. Translation proceeds through three stages. In *initiation*, a small ribosomal subunit binds with an mRNA transcript, and a large ribosomal subunit then becomes bound to the complex. In *chain elongation*, tRNAs deliver amino acids one after another to the ribosome, with the tRNA anticodons matching up to codons on the mRNA; then peptide bonds form between amino acids. In *chain termination*, a stop codon triggers events that cause the polypeptide chain to detach from the ribosome.

7. If cells are to produce functional proteins, the protein-building instructions encoded in DNA must be preserved through the generations. However, the instructions can be altered as a result of three types of events:

a. Crossing over and recombination: introduction of new gene combinations in the chromosome (page 179).

b. Chromosomal aberration (page 180).

c. Gene mutation: for example, insertion or deletion of several bases in the DNA sequence (frameshift mutation); incorrect pairing of bases during replication (base-pair substitution).

8. Most mutations and other changes in the DNA are harmful; an individual generally inherits a combination of many genes already fine-tuned for a given range of operating conditions. However, if conditions change, the products specified by the variant DNA regions may contribute to survival.

9. There is an underlying chemical unity among all organisms. Regardless of the species, DNA is composed of the same substances, it follows the same rule of base-pairing, and it is replicated in much the same way.

10. Mutations and other changes in DNA are the source of diversity among organisms. Whether the changes prove to be harmful or beneficial depends on the environment in which the products of the altered DNA are expressed.

Review Questions

1. Are the products specified by DNA assembled *on* the DNA molecule? If so, state how. If not, tell where they are assembled, and on which molecules.

2. How is a linear sequence of nucleotides in DNA translated into a linear sequence of amino acids in a protein?

3. Name the process by which RNA of three different types is assembled from the parent DNA code. Name the process by which the three different types of RNA cooperatively assemble a sequence of amino acids.

4. If sixty-one triplets actually specify amino acids, and if there are only twenty common amino acids, then more than one nucleotide triplet combination must specify some of the amino acids. How do triplets that code for the same thing usually differ?

5. Is the same basic genetic code used for protein synthesis in all living organisms? What significance is attached to that fact by most biologists?

6. Describe the general structure and function of the three types of RNA. What is a codon? An anticodon? Where is each physically located?

7. Define intron and exon. What happens to introns before an mRNA transcript is shipped from the nucleus to the cytoplasm?

8. If genetic information were transmitted precisely from generation to generation, organisms would never change. What are some of the DNA mutations that give rise to phenotypic diversity?

9. How is your DNA like the DNA of earthworms, grasses, and (presumably) dinosaurs? How is your DNA different?

Readings

Alberts, B. et al. 1983. *Molecular Biology of the Cell.* New York: Garland Publishing. Well-written accounts of the genetic code and protein synthesis.

Bernstein, H., G. Byers, and R. Michod. 1981. "Evolution of Sexual Reproduction: The Importance of DNA Repair, Complementation, and Variation." *The American Naturalist* 117(4):537–549. Summary of arguments that sexual reproduction evolved as a DNA repair process and that an advantage of diploidy is protection against expression of harmful mutations.

Darnell, J. October 1985. "RNA." *Scientific American* 253(4):68–78. Contains some interesting ideas about the original functions of RNA.

Darnell, J., H. Lodish, and D. Baltimore. 1986. *Molecular Cell Biology.* New York: Scientific American Books. Chapters 8 and 9 are comprehensive introductions to RNA synthesis in prokaryotes and eukaryotes.

Nomura, M. January 1984. "The Control of Ribosome Synthesis." *Scientific American* 250(1):102–114.

Rich, A., and S. Kim. January 1978. "The Three-Dimensional Structure of Transfer RNA." *Scientific American* 238(1):52–62.

17

CONTROL OF GENE EXPRESSION

TYPES OF GENE CONTROLS

Each living cell of your body inherits the same genetic information as all the others. Brain cells contain instructions to build skin cell proteins; skin cells contain instructions to build brain cell proteins. Yet each type builds only those proteins it will require throughout its life. Even then, it does not continually produce the same kinds of proteins in unvarying amounts. Certain white blood cells can build proteins called antibodies, which bind and lead to inactivation of bacterial invaders of your body—but they produce antibodies only when they have been stimulated by specific bacterial targets. *These cells and all others control which gene products appear, at what times, at what rates, and in what amounts.*

As we have seen, the genetic information encoded in DNA first becomes transcribed (as mRNA transcripts), then becomes translated (with the help of tRNA and rRNA) into polypeptide chains. Afterward, one or more chains become folded into a protein. Along this journey from genes to proteins, different controls are exerted:

1. *Transcriptional controls* influence which genes will be transcribed (if at all). They also affect the number of RNA molecules transcribed on a given gene at a given time.

2. In eukaryotes, *transcript processing controls* work to modify transcripts after they peel off the genes.

3. *Translational controls* influence the timing or rate of protein production. For example, through various control mechanisms, translation of mRNA transcripts already present in the cytoplasm can be postponed or hastened.

4. *Post-translational controls* govern modifications of the polypeptide chains along the route to a final protein product. For example, some proteins do not function until specific sugar groups or a phosphate group becomes attached to them.

In both prokaryotes and eukaryotes, controls over gene expression are exerted through interactions among molecules. Enzymes and other proteins interact with one another and with the DNA to trigger, enhance, or inhibit the expression of various genes. Extracellular signals (as delivered by hormones), changing pH levels, and other events influence the amount and activity of the proteins being synthesized. The crucial nature of gene control becomes dramatically clear when certain controls are lost and cells become cancerous, as described in the *Commentary* at the chapter's end.

Both prokaryotes and eukaryotes depend on gene controls that are responses to short-term shifts in cellular or extracellular conditions. Multicelled eukaryotes alone also depend on long-term gene controls that govern the development of specialized cell types.

GENE ACTIVITY IN PROKARYOTES

DNA Sequences in Prokaryotes

Among prokaryotes, the most common controls over gene expression deal with the rate of transcription. In these single-celled organisms, each gene is an uninterrupted stretch of nucleotides coding for a polypeptide chain (or a tRNA or rRNA molecule). However, other nucleotide sequences in prokaryotic DNA do not code for particular products. Rather they serve as binding sites for molecules that can initiate or shut down transcription of specific genes. Let's look at two examples of how these "noncoding" stretches are used to control transcription in *Escherichia coli*, a bacterial resident of the mammalian gut.

Operon Controls

Lactose Operon. In newborn mammals, *E. coli* is destined to encounter an abrupt change in diet. For a few weeks or months, the host in which *E. coli* resides takes in nothing but milk, which contains large amounts of the sugar lactose. Once the weaning period is over, most mammalian hosts never take in milk again. (The only exceptions are humans and a few of their pets.) During the first part of the mammalian life cycle, then, *E. coli* present in the gut must be able to metabolize a sugar that subsequent *E. coli* generations never will encounter.

André Lwoff, François Jacob, and Jacques Monod looked into this intriguing aspect of *E. coli* nutrition. When Jacob and Monod grew bacteria on a culture medium in which no lactose was present, the bacteria produced no lactose-degrading enzymes. When the bacteria were switched to a medium in which lactose was the only energy source, they produced thousands of molecules of lactose-degrading enzymes within minutes! Somehow, the presence of lactose *induced* the synthesis of those enzymes used to degrade lactose.

As was revealed by genetic studies of mutant bacteria, three adjacent genes in *E. coli* DNA code for a set of lactose-degrading enzymes. Transcription of those genes is controlled by activity at neighboring nucleotide sequences that go by these names:

regulator gene	*a DNA sequence coding for a "repressor" molecule*
promoter	*a short, noncoding DNA sequence to which RNA polymerase must bind in order to initiate transcription*
operator	*a short, noncoding DNA sequence to which a repressor molecule can bind and thereby block transcription of a set of related genes*

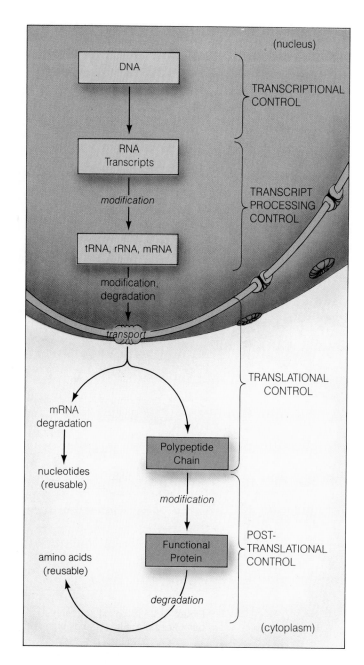

Figure 17.1 Control of eukaryotic gene expression: steps at which regulatory mechanisms can be brought into play. (Here, the steps are superimposed on a sketch of nuclear and cytoplasmic regions of a eukaryotic cell.)

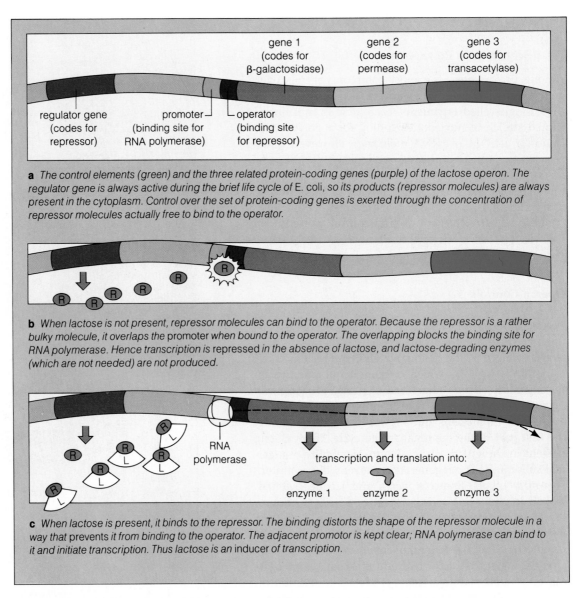

a *The control elements (green) and the three related protein-coding genes (purple) of the lactose operon. The regulator gene is always active during the brief life cycle of E. coli, so its products (repressor molecules) are always present in the cytoplasm. Control over the set of protein-coding genes is exerted through the concentration of repressor molecules actually free to bind to the operator.*

b *When lactose is not present, repressor molecules can bind to the operator. Because the repressor is a rather bulky molecule, it overlaps the promoter when bound to the operator. The overlapping blocks the binding site for RNA polymerase. Hence transcription is repressed in the absence of lactose, and lactose-degrading enzymes (which are not needed) are not produced.*

c *When lactose is present, it binds to the repressor. The binding distorts the shape of the repressor molecule in a way that prevents it from binding to the operator. The adjacent promotor is kept clear; RNA polymerase can bind to it and initiate transcription. Thus lactose is an inducer of transcription.*

Figure 17.2 Induction and repression of gene transcription for the lactose operon. The diagrams show part of one of the two strands of the DNA double helix of *E. coli*.

The term **operon** was coined to signify a *set* of related protein-coding genes transcribed as a coordinated unit under the direction of a single set of controls built into the DNA.

Figure 17.2 shows the relationship between the components of the lactose operon (in other words, the control elements and the set of protein-coding genes). Here, the transcription rate depends on the ratio of repressor to lactose molecules. Repressor molecules are produced all the time, and when the lactose concentration is low, nothing stops them from binding with the operator and inhibiting transcription. However, lactose can bind with a *repressor* and cause it to change in shape—and in its altered state the repressor cannot bind with the operator (Figure 17.2c). At increased lactose concentrations, most of the repressor molecules are inactivated and lactose-degrading enzymes can be rapidly produced.

Tryptophan Operon. The enzymes produced through transcription of the lactose operon are part of a degradative pathway. Enzymes produced through transcription of other operons are part of biosynthetic pathways. For example, *E. coli* produces a set of enzymes necessary to build the amino acid tryptophan. The tryptophan operon is controlled by a mechanism called corepression. Here also, a regulator gene produces repressor molecules. But this repressor cannot bind to an operator *until* a tryptophan molecule binds with it and alters its shape. The repressor-tryptophan complex alone can bind to the operator and shut down transcription. These complexes form only when the tryptophan concentration is high.

In the lactose operon, lactose molecules bind with and inactivate the control element that otherwise represses transcription of genes coding for lactose-degrading enzymes. Thus lactose induces the transcription of genes that allow the cell to use lactose as an energy source.

In the tryptophan operon, tryptophan molecules bind with and activate the control element necessary to repress transcription of genes coding for the enzymes necessary to build tryptophan. Thus, in excess concentration, tryptophan itself shuts down its further synthesis.

GENE ACTIVITY IN EUKARYOTES

DNA Sequences in Eukaryotes

In contrast to prokaryotes, it seems likely that only ten to twenty percent of the nucleotide sequences in the DNA of complex eukaryotes becomes translated into proteins. (For example, mammalian DNA contains about 3 billion nucleotides which, if they did nothing but code for proteins, could provide the cell with 3 million different proteins of average length, and that is far more than has ever been detected in any cell.) The fraction of eukaryotic DNA coding for proteins (and for the tRNA and rRNA necessary for protein synthesis) is called coding DNA. The function of much of the rest is not known.

Chromosome Organization and Gene Activity

Folding of DNA in Chromosomes. Eukaryotic DNA is an enormously long molecule (Figure 17.3). In some chromosomes, the DNA molecules are probably more than a meter long! You might well wonder how regulatory proteins and other agents of transcription even manage to make contact with specific gene regions.

Throughout the cell cycle, eukaryotic DNA is tightly bound with proteins, collectively called **histones**, which are about equal in mass to the DNA. Some histones form a "spool" for winding up small stretches of DNA. Each of these histone-DNA units is a **nucleosome**. Even during interphase, when chromosomes are extended in thin, threadlike form, the nucleosome packing arrangement is maintained. Thus, at high magnification, the interphase chromosome looks like a beaded chain (Figure 17.4).

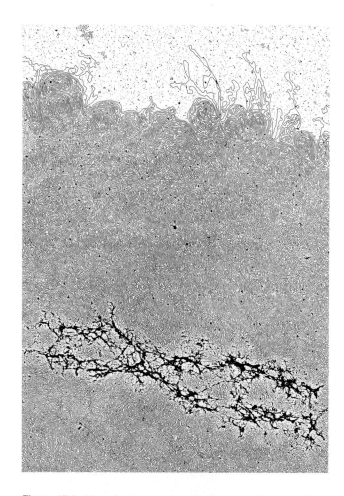

Figure 17.3 Human chromosome at metaphase, stripped of the proteins that normally hold it in its highly condensed form.

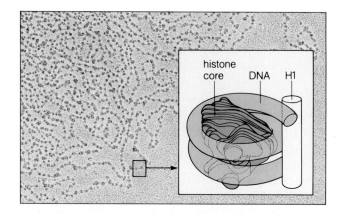

Figure 17.4 Photomicrograph of the nucleosome "beads" on a stretched-out *Drosophila* chromosome. The inset shows one model of the nucleosome. Each nucleosome consists of a double loop of DNA around a core of eight histone molecules. Another histone molecule (H1) binds to the DNA between nucleosomes.

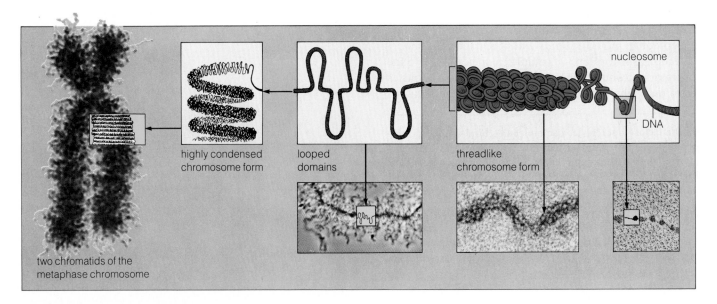

Figure 17.5 A current model of chromosome structure. Each nucleosome consists of DNA wound as a double-stranded thread on an individual spool of histone molecules. The DNA loops twice around each spool. In an unknown way, further levels of folding lead to the condensed form of the metaphase chromosome.

Histone molecules interact and fold up the DNA in orderly ways in chromosomes. The entire DNA molecule apparently is organized as a series of **looped domains**, with each loop containing one or at most a few protein-coding sequences:

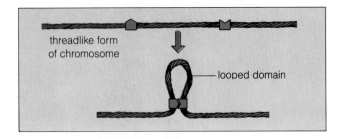

DNA loops and folds in different ways along the length of the chromosome (Figure 17.5). As you will see, this organization gives rise to packing variations that may influence the accessibility and activity of genes in different regions.

Selective Gene Expression During Development

The cells of a multicelled organism arise through mitosis, beginning with a single cell (such as the zygote produced at fertilization). Thus they all inherit the same genetic instructions. Somewhere along the way, however, the cells **differentiate**: they become specialized in appearance, composition, function, and often in their position in a given tissue.

The total set of genetic instructions is still present in differentiated cells—but the cells selectively use only parts of it. The patterns of selective gene expression in differentiated cells are remarkably varied. Some genes might be turned on only once, or left on all the time, or never activated at all. Others are switched on and off throughout the individual's life. Later chapters will describe the developmental outcomes of variable gene activity in specific plants and animals. For now, Figure 17.6 will give you an idea of how variability can arise through controls operating within cells, between cells, and between cells and the environment.

Differentiation among cells that carry the same set of hereditary instructions arises from selective gene expression (activation of some parts of the DNA and suppression of other parts in a given cell type).

Selective gene expression is controlled by agents that operate within cells, between cells, and between cells and the environment.

Figure 17.6 The not-so-simple slime molds: an example of how gene activity is influenced by controls operating within cells, between cells, and between cells and the environment.

At one stage of the life cycle, *Dictyostelium discoideum*, one of the slime molds, produces spores. Each spore gives rise to a single-celled amoeba. The amoebas feed on bacteria, and they grow and divide as long as bacteria are available.

When their food dwindles, amoebas begin streaming toward one another (**a**). A dozen to more than a hundred thousand cells may congregate into the shape of a slug, which crawls around as a coordinated body (**b**).

The dwindling food supply is an environmental signal for metabolic change, for when amoebas are starving, their cell surface becomes quite sticky. Some cells also secrete a chemical signal (cyclic AMP) at intervals of about five minutes. The dispersed amoebas move in the direction where the intervals are shortest, hence where secretions become most concentrated.

The sticky amoebas adhere to one another, and the contact seems to activate membrane surface receptors. Studies of amoebas raised in isolation suggest that receptor signals are transmitted to genes controlling differentiation. Unless amoebas receive the signals, they continue to synthesize only those proteins required for growth and division. And the isolated cells never do differentiate.

(**c–e**) Once amoebas stick together, they differentiate according to where each cell ends up in the slug. Some will form a vertical stalk, strengthened with cellulose. Others will give rise to new spores. The cells at the leading end (which produce the most cyclic AMP) become prestalk cells. The rest become prespore cells. The developmental signal seems to be the cyclic-AMP gradient between the leading end and the tail end of the migrating slug. A slug can be cut in two, each half can be cut in two again—and a gradient still forms, provided the fragments are given enough time to migrate.

Prestalk and prespore cells differ structurally (for example, only prespore cells develop a cytoplasmic vacuole). They also differ in protein composition—which suggests that they are transcribing different genes.

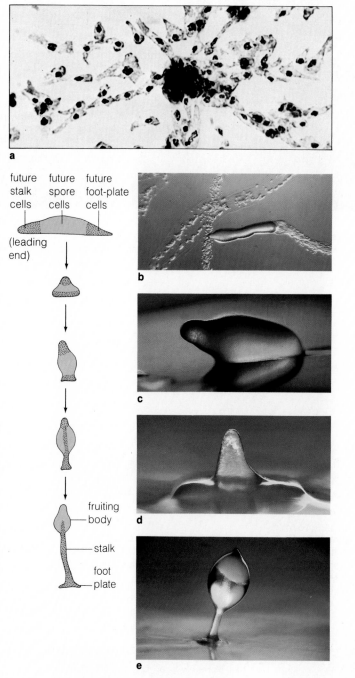

Transcription and Variations in Chromosome Structure

Lampbrush Chromosomes. In some instances, variations in the levels of transcription can be directly correlated with variations in chromosome structure. For example, during prophase I of meiosis, **lampbrush chromosomes** emerge in cells destined to become the eggs of amphibians and other animals (Figure 17.7). These chromosomes have decondensed, and hundreds or thousands of looped domains have uncoiled. When viewed with the light microscope, the decondensed chromosomes resemble bristle brushes that were once used to clean oil lamps (hence the name). The appearance of the looped domains heralds intense transcription. Some mRNA transcripts formed at this time will

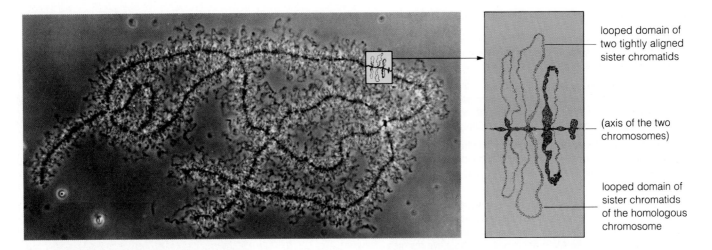

looped domain of two tightly aligned sister chromatids

(axis of the two chromosomes)

looped domain of sister chromatids of the homologous chromosome

Figure 17.7 Micrograph of a pair of lampbrush chromosomes at prophase I of meiosis. These homologous chromosomes were taken from an immature egg of a female newt (*Triturus viridescens*).

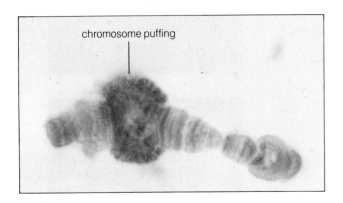

chromosome puffing

Figure 17.8 Chromosome puffing in a polytene chromosome from the salivary gland of a midge. The red-violet stain reveals the regions that are transcriptionally active; the blue stain indicates inactive regions.

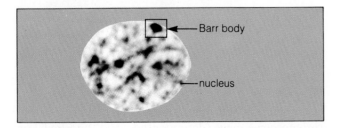

Barr body

nucleus

Figure 17.9 Barr body from a cell of a mammalian female.

be used to direct the initial development of the animal embryo.

Chromosome Puffs. Intense transcription also can be observed during the larval stage of many flies. In secretory cells of larval salivary glands, DNA replication occurs repeatedly, producing the polytene chromosomes described on page 180. During different stages of development, regions of polytene chromosomes open up and extend outward, forming **chromosome puffs** (Figure 17.8). The amount of transcription in cells containing polytene chromosomes has been correlated with how large and diffuse these puffs become.

X Chromosome Inactivation. A dramatic correlation between transcription and chromosome structure comes from studies of cells of mammalian females, which have two X chromosomes. Transcription of genes on both chromosomes is necessary for the normal development of immature eggs (hence for female fertility). However, oocytes are the *only* cells in an adult female mammal that have two active X chromosomes. In all other cells one is almost entirely inactivated and stays almost as tightly condensed as a metaphase chromosome. This condensed chromosome is called a **Barr body** after its discoverer, Murray Barr (Figure 17.9).

During the development of a female, one of the two X chromosomes is randomly inactivated in different cell lineages. However, in all descendant cells of each lineage, genes on that same chromosome remain inactive. This means that in some female cells, the alleles on the chromosome inherited from the father are expressed; in others, the alleles on the chromosome inherited from the mother are expressed. Thus every mammalian female is a "mosaic" of sex-linked traits for which she is heterozygous. The mosaic tissue effect arising from random X chromosome inactivation is called **Lyonization** (after its discoverer, Mary Lyon).

The mosaic effect is most evident in calico cats. These animals are heterozygous for black and yellow coat-color alleles, which reside on the X chromosome. Coat color in a given body region depends on which of the two X chromosomes is functioning, with its particular alleles available for transcription (Figure 17.10).

An example of the mosaic effect in human females is *anhidrotic ectodermal displasia*. Here, a mutant allele on one of the X chromosomes gives rise to a skin disorder characterized by a regional absence of sweat glands. Such females have patches of defective skin, the cells of which express the X chromosome bearing the mutant allele.

Evidence of Translational Controls

We have been focusing on evidence of controls that influence whether a gene will be transcribed into RNA. These are the most common controls over gene expression. But keep in mind that other controls exist, as the following example will illustrate.

In eukaryotic cells, mRNA transcripts may be stored for hours, days, or weeks before they are translated. The existence of such stockpiled genetic information was first suspected in the 1930s, when Joachim Hämmerling experimented with *Acetabularia*, a relatively enormous single-celled alga some three to five centimeters in diameter (Figure 17.11). Each algal cell starts out as a blob of cytoplasm with a nucleus. First a rhizoid develops from the blob and serves to anchor the cell to the seafloor. Then a stalk develops. Eventually a caplike structure

Figure 17.10 Random inactivation of X chromosomes in the female calico cat. One X chromosome carries an allele coding for black coat color and the other carries an allele coding for yellow. In different body regions, random patches of these two colors occur, depending on which of the two chromosomes is activated. (The white patches result from interaction with another gene locus—the so-called spotting gene—that determines whether any color appears at all.)

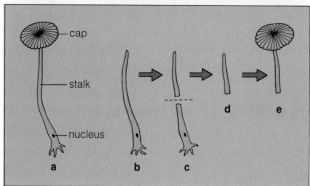

Figure 17.11 *Acetabularia* studies indicating that mRNA can be stored in the cytoplasm for an extended period before being translated into protein. (**a**) A mature *Acetabularia* cell has a base (a rootlike structure where the nucleus is housed), a stalk, and a cap in which gametes eventually form. In (**b**), a cell is developing and a cap has not yet formed. (**c–d**) If the stalk is amputated from the base, it will show metabolic activity for several weeks afterward, even though there is no nucleus to direct its activities. (**e**) After many weeks, the stalk may even regenerate a cap—a process that requires new protein synthesis. Chemical evidence also supports the idea that stable mRNA exists in the cytoplasm: high molecular weight RNA has been shown to accumulate at the stalk tip.

Altered Regulatory Genes and Cancer

Every time the clock ticks off one second, millions of cells in your body divide and thereby replace their worn-out, dead, and dying predecessors. They do not all divide at the same rate. Some have long-term roles and are arrested in interphase; nerve cells are like this. Others have short-term roles and divide rapidly, as do cells that give rise to the protective epithelium between gastric juices and the stomach wall.

No one knows exactly what controls the genes responsible for cell growth and division, but controls clearly exist. When most of a rat's liver is surgically removed, cell divisions accelerate to a phenomenal degree: four *billion* replacement cells are called up in four days. Then brakes are applied and the division rate slows, and by the seventh day, most of the missing tissue has been replaced.

Characteristics of Cancer Cells

On rare occasions, genetic controls over cell division become altered permanently. A cell divides again and again, until its offspring begin to crowd surrounding cells and interfere with tissue functions. The alteration has spawned a **tumor**: a clonal population of abnormally dividing cells.

It is not that tumor cells divide at a horrendous rate; normal cells that replace a surgically removed portion of the liver divide much faster. Rather, tumor cells have lost the controls telling them when to stop. They will not stop as long as conditions for growth remain favorable. We know this from studies of "immortal" tumor cells, which have been dividing for decades under controlled laboratory conditions and still show no sign of dying off.

When tumor cells simply divide more than they should in one place in the body, the tissue mass is a **benign tumor**. When it is surgically removed, its threat to the individual ceases. Sometimes, however, a cell loses not only the controls over division but also the controls over its position in a specific tissue. When that happens, the cell has become malignant, or cancerous. By definition, a **cancer cell** has (at the minimum) the following characteristics:

1. *Profound abnormalities in the plasma membrane*. Membrane transport and permeability are amplified. Some proteins at the surface are lost or altered, and new ones appear.

2. *Profound changes in the cytoplasm*. The cytoskeleton shrinks, becomes disorganized, or both. Enzyme activity shifts (as in a greatly increased reliance on glycolysis, even when oxygen is available).

3. *Abnormal growth and division*. Inhibitors of overcrowding in tissues are lost. Cell populations increase to unusually high densities; proteins are formed that trigger the abnormal proliferation of blood capillaries that service the growing cell mass.

4. *Diminished capacity for adhesion to substrates*. Secretions needed for adhesion dwindle; cells cannot become properly anchored in the parent tissue.

Normally, membrane surface factors enable cells of like type to recognize each other and bind together to form tissues and organs. When the genes coding for some of those factors are altered or suppressed, a cell can lose its identity and ignore the territorial restraints that bar indiscriminate wanderings through the body. The cell can leave its proper place and invade other tissues, a process called **metastasis**.

Whatever their site of origin, cancer cells of a primary tumor can invade other tissues and multiply into secondary tumors. Medical treatment is very difficult in such cases, for even if a malignant tumor can be removed from one site, another may appear elsewhere.

Oncogenes

Until a decade ago, the possibility of unraveling the genetic basis of cancer seemed remote, because cancer cells differ greatly in form, behavior, metabolic requirements, surface properties, and growth rates. However, *it now appears that a small number of altered regulatory genes may give rise to all the diverse kinds of cancer*.

Several viruses can cause cancerous transformations in vertebrate cells. They include the sarcoma virus, or RSV (which causes cancer in chickens), papovaviruses (some species cause warts, others cause tumors), adenoviruses (which cause lung infections as well as tumors), and herpesviruses (different species cause fever blisters, chickenpox, genital infections, and cancer). In all cases, the viral genetic instructions become integrated into the DNA of a host cell and are subsequently expressed in all offspring of that cell.

In the early 1970s, researchers identified RSV mutants that could infect host cells but not render them cancerous. Through DNA cloning and sequencing methods (page 234), it became evident that the mutants were missing a segment of their hereditary material. Because the missing segment coded for a protein that is present in cancerous chicken cells, the conclusion was that this segment, called the *src* gene, could give rise to cancer. Any gene having the potential to induce cancerous transformations is now known as an **oncogene**.

Even with these insights, a general theory of cancer seemed elusive. The problem was that most types of cancer arise with no help at all from viruses or viral genes! However, in 1975, J. Michael Bishop and Harold Varmus found that a gene almost identical to the *src* gene exists in the host DNA—*and it codes for a protein that takes part in normal cell activities!*

More than thirty viral oncogenes have now been isolated. In each case, a nearly identical gene sequence has been discovered in the normal DNA of the host animal—and the cells carrying them rarely become cancerous. The cellular counterparts of viral oncogenes are highly conserved in diverse species, ranging from humans to fruit flies to the enduring yeasts, which have not changed much in over a billion years.

We can assume that selective agents have not removed these genes from cells because their *normal* expression is vital—even though their abnormal expression is lethal. In normal cells they are not oncogenes but **proto-oncogenes**, which can induce cancer only if they are affected by specific mutations. *Proto-oncogenes are inherent parts of vertebrate DNA, and they code for proteins necessary in normal cell functioning. They become cancer-causing genes only on those rare occasions when specific mutations alter their structure or their expression.*

How Oncogene-Encoded Proteins Induce Cancer

Cancer can arise when viral DNA is inserted at a vulnerable regulatory region in the cellular DNA. For example, insertion of a viral oncogene adjacent to an active promoter can cause a high level of transcription from the oncogene. Cancer also can arise by the action of a **carcinogen**, which is an agent that causes gene mutations leading to cancer. Carcinogens include numerous natural and synthetic compounds (such as asbestos and certain components of cigarette smoke), x-rays, gamma rays, and ultraviolet radiation.

Yet cancer seems to be a multistep process, and it may be that two or more oncogenes are activated in sequence or cooperate in other ways to bring it about.

In normal cells, extracellular signals trigger cell division. At the very least, signaling mechanisms of the pathways controlling cell division must include the following:

1. Growth factors (molecules that carry growth signals from one cell to another).

2. Signal receptors at the membrane surface of the receiving cell.

3. Molecules, structures, or both that transmit signals from the cell surface to specific targets inside.

No one knows where oncogene proteins penetrate the pathways controlling division. But this much is clear: groups of oncogene proteins mimic the action of their normal counterparts, and they do so in a grossly amplified manner.

Of the known oncogenes, one group codes for proteins that can enhance transcription of a number of genes, which may be central in cell growth and division. Because the protein products of these genes become concentrated in the cell nucleus, there is speculation that they operate on the agents of transcription or on actual target genes. Intriguingly, the normal counterpart of one oncogene protein appears in interphase cells just before the DNA is replicated—*an event that generally foreshadows cell division.*

forms at the end of the stalk. With cap formation, the nucleus starts dividing, and the daughter nuclei migrate upward through the stalk and into the cap.

Hämmerling discovered that when a developing algal stalk is amputated from the rhizoid before cap formation, the stalk will continue to show photosynthetic and metabolic activity, including protein synthesis. It does so even though it has no nuclear means of directing the activity (Figure 17.11). The stalk even may regenerate a cap, although the cap will not contain gametes. Hämmerling wrote that some "morphogenetic substance" must be transported from the nucleus into a maturing stalk, and that this substance must be present in the stalk for the cap to form. The "substance" almost certainly is a supply of mRNA transcripts, stored in the stalk tip for later translation into the proteins used in cap development.

SUMMARY

We have barely touched on the controls that govern gene expression. However, the examples given are enough to support the following points:

1. All cells have interacting control elements (such as enzymes, regulatory proteins, and noncoding DNA stretches) that govern which gene products appear, at what times, and in what amounts.

2. The most common gene controls are transcriptional. In prokaryotes, operon controls influence the rate of transcription. In eukaryotes, the timing and rate of transcription is influenced by chromosome organization.

3. Prokaryotic and eukaryotic cells depend on controls that are responses to short-term shifts in cellular or extracellular conditions.

4. Eukaryotic cells only also depend on long-term controls over development and differentiation.

5. In complex eukaryotes, cells differentiate (they become different in appearance, composition, function, and often position).

6. Cell differentiation arises through *selective* gene expression. In a given cell type, some genes are activated and others are repressed.

7. Selective gene expression is controlled by factors that operate within cells, between cells, and between cells and the environment.

Review Questions

1. Cells depend on controls over which gene products are synthesized, at what times, at what rates, and in what amounts. List four general kinds of the controls involved, then give an example of how one kind works.

2. What is a regulator gene? A promoter? An operator?

3. Explain how gene transcription is induced for the lactose operon in *E. coli*. Under what conditions is it repressed?

4. Histones and nucleosomes are key structures in chromosome organization. Think about how DNA is spooled around histone cores. Can you speculate on how the arrangement might influence the accessibility of genes to enzymes and other agents of transcription?

5. Somatic cells of human females have two X chromosomes. During what developmental stage are genes on *both* chromosomes active? Can you explain what happens to one of those chromosomes after that stage?

6. Define development and differentiation.

7. A plant, fungus, or animal is composed of diverse cell types. How might this diversity arise, given that all of the body cells in each organism inherit the *same* set of hereditary instructions?

8. What are the characteristics of cancer? Explain why a benign tumor is less serious than one exhibiting metastasis.

Readings

Brown, D. 1981. "Gene Expression in Eukaryotes." *Science* 211:667–674.

Croce, C., and G. Klein. March 1985. "Chromosome Translocations and Human Cancer." *Scientific American* 252(3):54–60.

Gilbert, W. 1981. "DNA Sequencing and Gene Structure." *Science* 214:1305–1313.

Goldman, M., G. Holmquist et al. May 1984. "Replication Timing of Genes and Middle Repetitive Sequences." *Science* 224:686–692. Example of the kinds of experiments that are yielding insights into the mechanisms of gene regulation.

Klug, A. et al. 1981. "A Low-Resolution Structure of the Histone Core of the Nucleosome." *Nature* 287:509–515.

Sachs, L. January 1986. "Growth, Differentiation, and the Reversal of Malignancy." *Scientific American* 254(1):40–47. Describes the isolation of proteins that induce differentiation and that halt the growth of some cancer cells.

Snyder, L., D. Freifelder, and D. Hartl. 1985. *General Genetics.* Boston: Jones and Bartlett.

Weinberg, R. 1984. "Cellular Oncogenes." *Trends in Biochemical Sciences*. Brief summary of insights into the mechanisms of cancer.

It is doubtful that a DNA molecule exists anywhere in the natural world that is not already a patchwork molecule. As we have seen, gene mutations are rare in terms of a human lifetime—yet think about how many must have occurred in the DNA of millions of species during the *billions* of years since life began. Novel gene combinations also arose during the evolution of sexually reproducing species, as a result of independent assortment and crossing over. Moreover, all viruses, bacteria, fungi, plants, and animals are subject to DNA shufflings, not only within but often *between* species. Viral DNA routinely invades the DNA of bacterial, plant, and animal hosts. Many thousands of flowering plants have hybridized in the wild, with the DNA of two or more species becoming integrated during gamete formation.

DNA, in short, is not entirely static in its information content; *naturally occurring mechanisms inevitably change the number, kind, and sequence of its genetic messages.* Today, with recombinant DNA technology, researchers have the capability to alter the information content of DNA in a highly focused way. What does their work hold for the future? Let's lay the groundwork for addressing that question by first considering what is known about the molecular basis of natural recombination. Then we will turn to recombinant DNA technology and to some of its potential applications.

18

RECOMBINANT DNA AND GENETIC ENGINEERING

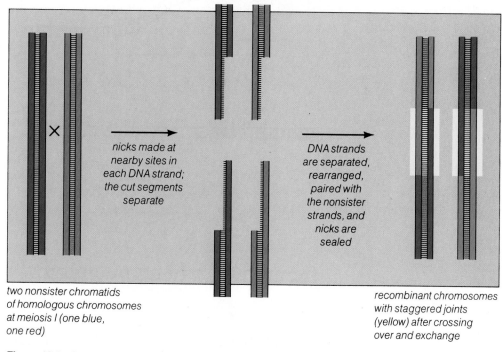

two nonsister chromatids of homologous chromosomes at meiosis I (one blue, one red)

nicks made at nearby sites in each DNA strand; the cut segments separate

DNA strands are separated, rearranged, paired with the nonsister strands, and nicks are sealed

recombinant chromosomes with staggered joints (yellow) after crossing over and exchange

Figure 18.1 Simplified picture of the essential features of homologous recombination, as described in the text.

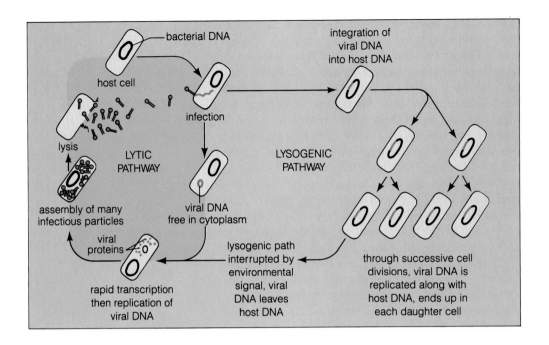

Figure 18.2 (Above) Life cycle of lambda bacteriophage. Depending on environmental factors, infection proceeds by way of either the lytic pathway or the lysogenic pathway.

In the lysogenic pathway, the bacteriophage enters a latent state in which the viral DNA becomes integrated into the host DNA, then remains silent during successive DNA replications and cell divisions. Specific environmental agents (such as ultraviolet radiation, which damages the bacterial DNA and leads to cell death) activate the viral DNA and cause it to leave the host DNA molecule. When it does, the lytic pathway is followed.

Figure 18.3 (Right) Site-specific recombination between the DNA of lambda bacteriophage and *Escherichia coli*. The micrograph shows several bacteriophages.

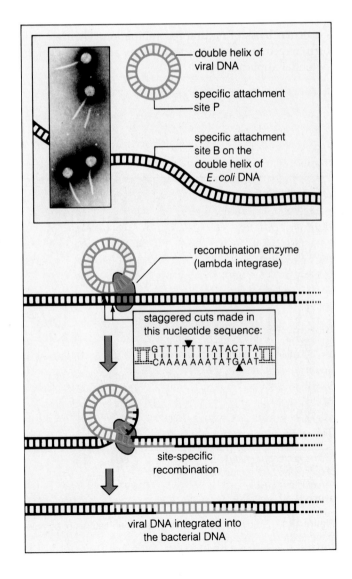

Figure 18.4 Kernels of maize (Indian corn) bearing unstable mutations, as studied by Barbara McClintock. Each colored sector is a clonal population of cells for which a transposable element has affected genes that control the production of color pigments.

NATURAL RECOMBINATION MECHANISMS

Recombination in Homologous Chromosomes

As we described earlier, genetic recombination occurs when homologous chromosomes cross over and exchange segments during meiosis I. This process, called **homologous recombination**, has three characteristic features. First, the exchange may occur almost anywhere along the chromosomes. Second, the exchange is reciprocal (each chromosome donates and receives the same number of base pairs as the other). Third, a fairly long, "staggered joint" must be created between the interacting DNA strands, because the exchange process requires extensive base-pairing (Figure 18.1).

Site-Specific Recombination

In another naturally occurring process, called **site-specific recombination**, exchange occurs only within a small and specific base sequence that is identical in both of the participating DNA molecules; extensive homology is not required. A single enzyme mediates the exchange. The enzyme recognizes the sequence in the two molecules, it makes identical breaks in both, and then the broken ends are exchanged and rejoined.

Site-specific recombination was discovered through studies of lambda bacteriophage, which infects *Escherichia coli*. On rare occasions, the DNA of this virus becomes integrated into the host DNA by this recombination process. The enzyme responsible for the integration is encoded by one of the viral genes (Figures 18.2 and 18.3).

Site-specific recombination is known to occur in several types of bacteria, yeast, and mammals. Certain mammalian cells produce antibodies (proteins that recognize foreign substances or cells in the body and that trigger immune reactions against them). As these cells mature, different protein-coding segments of their DNA undergo site-specific relocation. Thus different patterns of shuffling of a relatively limited number of DNA segments can provide the coding sequences for perhaps a million or more different antibodies!

Transposition

By now, you might be thinking that recombination requires a matching-up of at least a short region of homologous base sequences. However, in **transposition**, recombination occurs between DNA sequences that are *not* homologous. Transposition takes place in both prokaryotic and eukaryotic cells. The breakage and rejoining of DNA sequences are brought about by *transposable elements* (also called jumping genes), which move from one region to another in the same DNA molecule or to another one.

Transposable elements were discovered by Barbara McClintock (who also was the first to recognize that chromosomes break and rejoin during meiosis). McClintock deduced that a transposable element moves back and forth between an inactive site and a specific site adjacent to a gene that regulates kernel color in maize, or Indian corn (Figure 18.4). In *Drosophila*, transposable elements

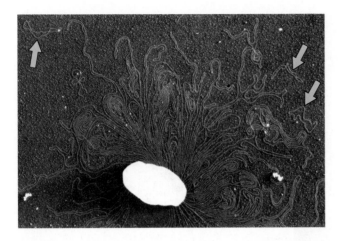

Figure 18.5 An *E. coli* cell that has ruptured, thereby releasing a portion of its main DNA molecule and several plasmids (blue arrows).

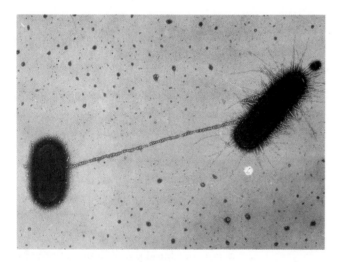

Figure 18.6 Early stage of conjugation between a recipient (F⁻) cell, shown to the right, and a donor (F⁺) cell of *E. coli*. The filamentous structure between the two bacteria is a pilus; it will retract into the F⁺ cell and bring the two participants into close contact.

occasionally insert themselves at random into a gene sequence and thereby inactivate the gene. One such transposition leads to a white-eyed instead of a red-eyed (wild-type) fruit fly.

PLASMIDS

A bacterial cell contains one large, circular molecule of DNA. In addition, many types of bacterial cells also contain small circles of DNA called **plasmids** (Figure 18.5), which replicate somewhat independently of the large molecule. Some plasmids contain transposable elements. If the large DNA molecule has an identical transposable element, the plasmid can become inserted into it through homologous recombination.

F Plasmids

Plasmids were discovered during the course of studies of conjugating *E. coli* cells. In **bacterial conjugation**, DNA is transferred from a donor cell to a recipient cell. One such transfer is directed by the F ("fertility") plasmid. Cells designated F⁺ have a free plasmid, F⁻ cells do not, and Hfr cells have it integrated into the main DNA molecule:

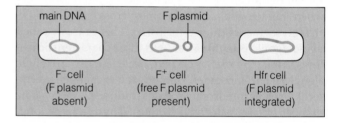

The F⁺ cells have a pilus, a filamentous projection from their surface (Figure 18.6). When the pilus makes firm contact with a potential recipient, it retracts into its owner's cytoplasm, drawing the two cells into intimate contact, whereupon a conjugation bridge is formed. Now the plasmid is nicked. DNA replication begins at the nick, causing one of the DNA strands to peel away from the circular molecule (Figure 18.7). DNA replication proceeds on the newly exposed bases of the plasmid, and the nicked single strand invades the recipient cell. There, DNA replication converts it into a double-stranded molecule, and the ends join to form a circular plasmid. Because the recipient cell now contains the F plasmid, it has become a potential donor, ready to conjugate with another recipient.

How does gene transfer occur from an Hfr cell? Here, the integrated plasmid DNA is nicked and a single strand invades a recipient cell—but in this case it drags one strand of the main DNA molecule with it. The entire transfer takes 100 minutes, and usually the whole strand doesn't make it. *E. coli* live in fluid environments, and random bombardments by water molecules usually knock conjugating cells apart before the 100 minutes are up. However, the presence of the transferred fragment in the recipient cell triggers homologous recombination. Soon after the cells are knocked apart, the fragment and the main DNA molecule in the recipient undergo crossing over. If the donor and recipient genotypes are not the same, recombinants may result.

R Plasmids

One type of plasmid, the *R plasmids*, confers resistance to antibacterial drugs. (The "R" stands for resistance.) One region of an R plasmid is concerned with transfer and replication of the plasmid itself; other regions code for products that impart resistance to as many as ten different antibacterial drugs. For example, the antibiotics streptomycin, tetracycline, sulfanilamide, and chloramphenicol may not work at all against a bacterial strain that carries R plasmids.

R plasmids can transfer between bacteria of different genera. Through conjugation, drug resistance spreads quickly through a bacterial population—for example, from the normally harmless *Escherichia* to *Shigella* (a type of bacterium that causes a dangerous form of dysentery). Conjugation is giving rise to a disturbing number of pathogenic bacteria with enhanced drug resistance. Among them are the pathogens responsible for intestinal tract disorders, gonorrhea, typhoid, and meningitis.

Antibacterial drugs act as selective agents, for they remove susceptible phenotypes from target populations. However, because of the rapid transfer of R factors, the frequency of drug-resistant genotypes rapidly increases in the population. Different antibiotics are then used which, in effect, remove all but the most resistant strains; and the rise of more resistant strains prompts research for more effective antibiotics. One outcome is that in parts of the world where antibiotics are used routinely, genes carried on R plasmids are becoming widespread.

RECOMBINANT DNA TECHNOLOGY

Plasmids are a major research tool in recombinant DNA technology. This technology is based on the ability of certain enzymes to cut DNA into fragments that can be rejoined in particular ways that produce novel gene combinations. Fragments cut from the DNA of any organism can be incorporated into plasmids, which can then be introduced into bacterial cells. In very little time, these

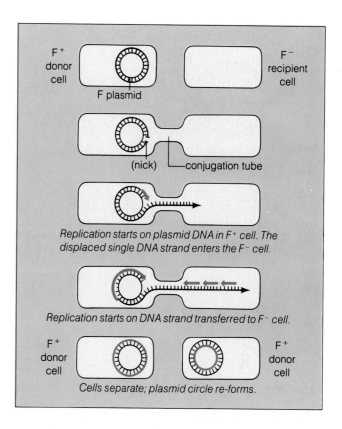

Figure 18.7 Transfer of an F plasmid between two *E. coli* cells during conjugation. For clarity, the main DNA molecule is not shown.

rapidly reproducing cells give rise to millions of genetically identical progeny—a **clonal population**—in which each cell contains the recombinant DNA. This cloning technique can produce cells with novel properties. Moreover, the cells can serve as a source of new gene products or even the genes themselves in pure form. These genes and gene products are used in genetic research; many have proved to be useful in medicine and agriculture.

Cloning Methods

The first step in DNA cloning is to extract and purify the DNA from the cells of interest. The next step is to cut up the DNA into gene-sized pieces, using one of more than 500 known **restriction enzymes**. These enzymes are like highly specialized scissors that cut a DNA molecule only in certain base sequences showing a certain type of symmetry (Figure 18.8). Because of the

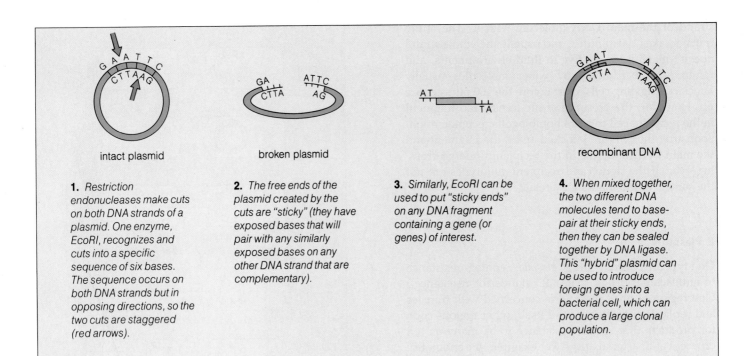

intact plasmid	broken plasmid		recombinant DNA

1. *Restriction endonucleases make cuts on both DNA strands of a plasmid. One enzyme, EcoRI, recognizes and cuts into a specific sequence of six bases. The sequence occurs on both DNA strands but in opposing directions, so the two cuts are staggered (red arrows).*

2. *The free ends of the plasmid created by the cuts are "sticky" (they have exposed bases that will pair with any similarly exposed bases on any other DNA strand that are complementary).*

3. *Similarly, EcoRI can be used to put "sticky ends" on any DNA fragment containing a gene (or genes) of interest.*

4. *When mixed together, the two different DNA molecules tend to base-pair at their sticky ends, then they can be sealed together by DNA ligase. This "hybrid" plasmid can be used to introduce foreign genes into a bacterial cell, which can produce a large clonal population.*

Figure 18.8 Cleavage and splicing methods used in DNA cloning.

symmetry, the resulting fragments have "sticky ends"— *complementary ends that can base-pair with any other fragment cut by the same enzyme.* This is the basis of the cloning method.

The third step in DNA cloning is to cut a plasmid at a single position with the *same* restriction enzymes used to cut the DNA of interest. (Plasmids to be used for cloning are called "cloning vectors.") The sticky ends on the (now) linear plasmid are the same as the sticky ends on all the DNA fragments. When the two ends of a fragment base-pair with the two ends of a cut plasmid, the result is a circular, recombinant plasmid that can be used in cloning.

The fourth step in DNA cloning is to introduce the recombinant plasmid into a population of cells that can serve as hosts, then to isolate the cell that has actually taken up the plasmid. How is this done? Usually, a plasmid has one or more phenotypic traits (such as resistance to a particular drug) that can be used to detect its presence in recipient cells. For example, suppose a DNA fragment is inserted into a plasmid containing two genes that code for resistance to two antibiotics. One protein product degrades ampicillin, the other degrades tetracycline. The DNA is inserted into the plasmid by a restriction enzyme that recognizes the sequence CTGCAG. This sequence happens to be in the middle of the coding sequence for the ampicillinase gene—and

if this sequence is interrupted by an inserted bit of DNA, the outcome will be a nonfunctional ampicillinase protein.

After host cells take up the plasmids, the next step is to spread the cells onto nutrient medium containing tetracycline. Cells that did not take up a plasmid with the gene for tetracycline resistance will die; the ones that did will survive and produce colonies. A sample of each surviving colony can now be transferred to a medium containing ampicillin. The ones unable to grow are likely to have a DNA fragment inserted into the ampicillin gene. Thus the bacterial colonies containing the DNA of interest have been located.

The final step in DNA cloning is to allow colonies that house the desired DNA fragment to multiply. These clones can be put to work in synthesizing useful products. For example, a cloned gene coding for the protein *insulin* is now being mass produced. Insulin is crucial for glucose metabolism but is absent or deficient in diabetics. Afflicted persons can lead normal lives by receiving daily insulin injections. However, before recombinant DNA technology, human insulin could not be obtained. Instead, insulin was extracted from pancreatic tissues of an enormous number of pig cadavers. Also, insulin from pigs is not exactly the same as insulin from humans, so it sometimes caused severe allergic reactions in some diabetics.

Probes for Cloned Genes

When restriction enzymes make their cuts in a sample of DNA molecules, they produce thousands of fragments. How is it possible to detect the one sequence coding for the desired protein among them? During the cloning process, each bacterial cell takes up only one fragment. When individual cells later give rise to colonies, samples of each colony are examined to locate the cells containing the right fragment. One method of doing this is founded on the principle of base pairing.

When DNA in solution is heated, its hydrogen bonds break and the two strands of the double helix separate. (This is called denaturation.) Under specific conditions, single DNA strands will base-pair with any other strands having complementary sequences. Thus, if single-stranded DNA molecules from two different sources are mixed together in solution, they can form a "hybrid" DNA molecule.

When cells of clones are broken apart, the released DNA can be converted to single strands. The cellular contents can be treated with a *probe*—in this case, radioactively labeled DNA known to be complementary to the desired gene sequence. The probe will pair with (and thereby tag) only the DNA sequence being sought.

Figure 18.9 Crown gall tumor on a sunflower plant, caused by a plasmid from *A. tumefaciens*.

Gene Transfer Into Mammalian Cells

Mammalian cells also can be made to take up foreign DNA, although research in this area is still in its infancy. In this case, certain viruses serve as the cloning vectors. First, several of the viral genes are removed, including those coding for the viral coat proteins. (Without the coat proteins, intact viral particles cannot be assembled in host cells, so they cannot infect the cells.) Next, the DNA under study is joined to the whittled-down viral DNA, near some sequence that can control its transcription. The resulting recombinant DNA molecule is inserted into host cells. In some cells, the viral DNA becomes integrated into the host chromosome, and in doing so it carries in a cloned DNA fragment.

Gene Transfer Into Plant Cells

Nature has provided excellent cloning vectors for transferring genes into plant cells. For example, in the early 1980s, researchers succeeded in inserting DNA fragments into a plasmid from a bacterium that infects many flowering plants. The tumor-inducing plasmid from *Agrobacterium tumefaciens* carries genes responsible for crown gall tumors (Figure 18.9). These genes are integrated into the host DNA. When the plasmid is used as a vector, the tumor genes are removed and the desired genes substituted. Plants are then propagated from sin-

gle cells containing the new DNA molecule. In some cases, the foreign genes are usually expressed in the plant tissues.

PROSPECTS FOR GENETIC ENGINEERING

Genetic Engineering of Microbes. The first successful insertion of viral DNA into a plasmid was made by Paul Berg and his colleagues during the mid-1970s. It then became clear that natural restrictions on the transfer of genes between different species could be bypassed. Could such transfers produce bacteria that would be dangerous to humans or to the environment? Exhaustive investigations carried out since that time indicate that the risks probably are small. The *E. coli* strains used in recombination experiments are not pathogenic and have been selected carefully for metabolic and functional defects, which prevent their survival outside the laboratory.

But other microbial species can adapt to new environments. Many can infect different plants or animals and some produce substances that may be toxic to new host cells. (For example, some bacterial species produce a natural antibiotic against other bacteria and fungi.) If researchers are to work with potentially harmful species, they must follow stringent containment procedures, similar to ones that medical microbiologists have used successfully for many decades.

Also, what might happen when remodeled bacteria are reintroduced into the environment? For example, suppose a wild-type bacterium is genetically engineered to produce a potent protein that can work against a fungal parasite of wheat. If the bacterial gene could be transferred to wheat species, would the effect of the foreign protein be limited only to the target? Or might there be repercussions for humans and other animals that consume the wheat? There is some concern about the unknown side effects of field applications. Stringent safety tests have been developed that must be used before genetically engineered species are introduced into the environment.

Genetic Engineering of Plants. Plant genetic engineering holds promise for increasing global food production. A sense of urgency surrounds research in this area, for millions die each year from starvation. In the long run, simply increasing global food production to keep pace with the burgeoning human population is not a solution to the problem, but it is one of the few short-term options available to us.

Consider the research interest in developing salt-tolerant crop plants alone. *Halophytes* are plants, including some species of barley and sugar beets, that are adapted to moderately salty environments. Although they do not produce the high yields of conventional crop plants, they can grow in soils far saltier than *glycophytes* ("sweet-water" plants) can tolerate. Almost all conventional crops are salt sensitive. Can the genes of halophytes and glycophytes be recombined to produce salt-resistant, high-yield strains? The question is not trivial. Much of the world's agricultural regions must be irrigated to be productive. Enormous amounts of salt are brought in with the irrigation water, and typically they cannot be flushed from the soil because of the scarcity of rain in those regions. All over the world, croplands are becoming "salted out," which means the number of acres able to support existing crop species is declining rapidly. Thus in India alone, 250 acres a day are being removed from agriculture.

Genetic Engineering of Animals. In 1982, Ralph Brinster, Richard Palmiter, and their colleagues succeeded in introducing a growth hormone (GH) gene from a rat into fertilized mouse eggs. When the mice grew, it was clear that the foreign gene coding for GH

was integrated into the mouse DNA and was being used, as Figure 18.10 dramatically shows. Mice with the foreign GH gene grew two to three times larger than their normal littermates. Their cells had up to thirty-five copies of this gene, and the level of GH circulating in the blood was several hundred times higher than the normal values.

The success of such experiments depends largely upon combining the genes of interest with regulatory elements that will assure their expression in the proper cell types. Usually, this is the major stumbling block in experiments of this type. More distant prospects are methods of inserting foreign genes into specific chromosomal locations, so that they may be exposed to any region-specific developmental modifications that may be necessary for their normal expression (Chapter Thirty-Four).

Can such work be extended to humans? For example, could a *gene* coding for human insulin be provided to diabetics? Given that understanding of the interrelated functions of human genes may be many years away, any attempt to answer that question would be premature. Even so, some limited experiments are under way. The goal of these experiments is to alleviate suffering among individuals who are already afflicted with lethal genetic disorders, who have no prospect of recovery, and who are willing to risk participating in such research in their hope for a better life.

For example, humans who suffer from a rare genetic disorder, the *Lesch-Nyhan syndrome*, do not have a gene that codes for an enzyme necessary to degrade pyrimidines. Its absence leads to a buildup of uric acid wastes that can cause kidney damage, gout, cerebral palsy, and gross behavioral modifications. Unless afflicted persons are mechanically constrained, they gnaw uncontrollably at their lips and fingers, spit, curse, and pound their head against the wall. They cannot expect to live past their twenties.

Cells taken from afflicted children have been grown in culture and infected with recombinant viruses carrying a gene (obtained from a bacterium) that provides the missing enzyme. The cultured cells were converted to cells that apparently are normal in their biochemistry. After extensive investigations of potential hazards, it may be possible to test the recombinant virus on afflicted children.

Until the genetic makeup of humans is better understood, there is a risk associated with inserting foreign genes into human DNA. (For example, an insertion might put a foreign gene *within* a gene sequence and cause a mutation.) Is the risk worth taking? Many individuals afflicted with severe genetic disorders have expressed a willingness to try anything that might be their passport to a less tortured existence.

Some Ethical Considerations. For nearly ten thousand years, humans have been manipulating the genetic character of diverse species. One need only compare the modern strains of domesticated wheat and corn, or compare the varied breeds of cattle, poultry, dogs, and cats with their wild ancestral stocks to know this is true. Artificial selection has produced chickens and turkeys with broader, meatier breasts; and larger, sweeter oranges, seedless watermelons, and flamboyant ornamental plants. Hybridizations have given us the tangelo (tangerine crossed with grapefruit) and the mule (donkey crossed with horse). In an indirect way, medical practices (such as surgery, transfusions, vaccinations, and drug therapy) are also forms of genetic manipulation, for they preserve genotypes that might otherwise be selected against and lost from the population. When viewed from this perspective, genetic engineering is novel not in concept but rather in the *magnitude* of its potential to cause change.

Some say that the integrity of the DNA of each species is inviolate and should not be tampered with. But as the earlier discussion should make clear, nature itself alters DNA much of the time. The real argument, of course, is whether we as a species have the wisdom to bring about beneficial changes without causing unforeseen harm to ourselves or to the environment.

One is reminded of an old saying: "If God had wanted us to fly, he would have given us wings." Certainly when a plane crashes and everyone on board is killed, we wonder what we are doing up in the air in the first place. But something about the human experience gave us the *capacity* to imagine wings of our own making—and imagination has carried us to the frontiers of space.

Where are we going from here with recombinant DNA technology, this new product of our imagination? To gain perspective on the question, spend some time reading the history of our species. It is a history of survival in the face of all manner of threats, of expansions, of bumblings, and sometimes of disasters on a grand scale. It is also a story of progressively intimate interactions with the environment and with one another. The questions confronting you today are these: Should we be more cautious, believing that one day the risk takers may go too far? And what do we as a species stand to lose if the risks are *not* taken?

Figure 18.10 Ten-week-old mouse littermates, the one on the left weighing twenty-nine grams, and the one on the right forty-four grams. The larger mouse received injections of the rat growth hormone gene and grew much faster under laboratory conditions that triggered the expression of this gene.

SUMMARY

1. DNA undergoes natural recombination through the following three processes:

a. *Homologous recombination:* crossing over and exchange between homologous chromosomes at meiosis I or between DNA molecules showing extensive homology. The exchange is reciprocal, it can occur at any site along the chromosome, and a long staggered joint is created between the interacting DNA strands of the non-sister chromatids.

b. *Site-specific recombination:* exchange only within a short, specific base sequence that is identical on both of two interacting DNA molecules; a single enzyme mediates the exchange.

c. *Transposition:* recombination involving nonhomologous base sequences in which transposable elements move from one site in the DNA to another.

2. Besides the large, circular DNA molecule, many types of bacterial cells also contain small DNA circles called plasmids. Plasmids can be transferred between a donor and a recipient cell. An integrated plasmid can cause the transfer of part of the main DNA molecule to a recipient cell.

3. Recombinant DNA technology uses plasmids and viruses as cloning vectors: DNA fragments obtained from any organism can be integrated into them, and they become the vehicles by which the DNA is transferred into a new cell.

4. Recombinant DNA technology is based on the use of many different restriction enzymes, each of which cuts DNA only in certain base sequences showing a particular kind of symmetry. Because of the symmetry, the fragments produced have sticky (complementary) ends that can base-pair with any other fragment cut by the same enzyme to form a recombinant DNA molecule.

5. Recombinant plasmids or viruses can be introduced into growing bacterial cells, which can yield a large population of genetically identical cells (a clonal population), each containing a copy of the recombinant DNA molecule.

6. DNA cloning has uses in research and extensive practical applications. It provides molecular biologists with a source of purified genes for detailed study. The bacteria containing the recombinant DNA also serve as a source of large quantities of proteins (such as insulin) that are normally difficult to obtain.

Review Questions

1. What is the difference between homologous recombination and site-specific recombination? Are these both naturally occurring processes, or does one (or both) occur only in the laboratory?

2. What is a plasmid? How is it transferred between conjugating *E. coli* cells?

3. What is a restriction enzyme? Describe how restriction enzymes and plasmids are used in recombinant DNA research.

4. Having read through some of the examples of genetic engineering in this chapter, can you think of some additional potential benefits of this technology? Can you visualize some potential problems?

5. At this writing, the AIDS virus is spreading and there is no vaccine and no cure in sight; once the disease has been contracted, death is inevitable. Discuss under what circumstances, if any, it might be all right to "tamper with" the DNA or RNA of viruses and bacteria.

Readings

Anderson, W. F. 1984. "Prospects for Human Gene Therapy." *Science* 226:401–409.

Anderson W. F., and E. G. Diacumakos. July 1981. "Genetic Engineering in Mammalian Cells." *Scientific American* 245:106–121.

Chilton, M. June 1983. "A Vector for Introducing New Genes into Plants." *Scientific American* 248:51–59.

Clowes, R. C. April 1973. "The Molecule of Infectious Drug Resistance." *Scientific American* 228:18–27.

Cohen, S. N., and J. A. Shapiro. February 1980. "Transposable Genetic Elements." *Scientific American* 242:40–49.

Drlica, K. 1984. *Understanding DNA and Gene Cloning: A Guide for the Curious.* (Introduction by Arthur Kornberg.) New York: Wiley. DNA cloning made accessible for those with little or no background in chemistry. Excellent analogies in this little paperback.

Nester, E., et al. 1983. *Microbiology.* Third edition. Philadelphia: Saunders. Contains good introductions to microbial growth in the laboratory (chapter 4) and microbial genetics (chapter 8).

Novick, R. P. December 1980. "Plasmids." *Scientific American* 243:103–127.

Olson, S. 1986. *Biotechnology Comes of Age.* Washington, D.C.: National Academy of Sciences.

Palmiter, R., et al. 1983. "Metallothionein-Human GH Fusion Genes Stimulate Growth of Mice." *Science* 222:809–814. Report on landmark experiments in mammalian gene transfers.

Ream, L., and M. Gordon. 1982. "Crown Gall Disease and Prospects for Genetic Manipulation of Plants." *Science* 218:854–858.

Snyder, L., D. Freifelder, and D. Hartl. 1985. *General Genetics.* Boston: Jones and Bartlett. Clear introduction to DNA recombination mechanisms.

UNIT FOUR

PLANT SYSTEMS AND THEIR CONTROL

19

PLANT CELLS, TISSUES, AND SYSTEMS

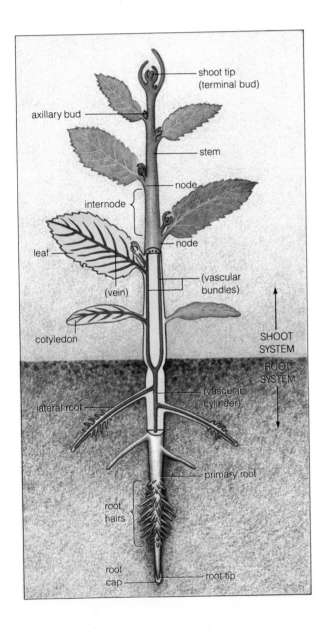

On a summer morning in 1883, a cataclysmic explosion blew apart the South Pacific island of Krakatoa, and life there ended abruptly. All that remained beneath the hot ashes and pumice was the jagged peak of a volcanic cone, which continued to exude lava. For about a year this smoldering remnant of the island was essentially sterile. Yet even then, winds and water were carrying seeds and spores to it from nearby islands—and half a century after the explosion, a dense forest cloaked the "new" Krakatoa.

In the spring of 1980, Mount St. Helens in southwestern Washington exploded violently, and within minutes, hundreds of thousands of mature trees near the volcano's northern flank were blown down like matchsticks. Thick ashes and pumice turned the previously forested region into a barren sweep of land.

Events of this magnitude dramatize what the world would be like without plants, reminding us that we could no more survive without them here than on the rock-strewn surface of the moon. What characterizes these remarkable organisms, which directly or indirectly nourish all others and make the land habitable? Can we identify patterns of structural organization among them? Do plants, like animals, have intricate systems for circulation, gas exchange, and nutrition? How do plants reproduce? What governs their growth and development? These are questions addressed in this unit.

THE PLANT BODY: AN OVERVIEW

There are more than 275,000 species of plants, and no one species can be used as a "typical example" of their body plans. Plants live in fresh water, in seawater, on land, even high above a forest floor (attached to other plants), and their features are correspondingly diverse. In size alone, plants range from microscopic algae to giant redwoods. Which plants, then, should be our focus? Botanists divide the plant kingdom into two broad categories:

vascular plants   *with well-developed conducting tissues through which water and solutes are transported to different body regions*

nonvascular plants   *no internal transport systems (or very simple ones)*

Figure 19.1 Generalized body plan of a dicot, one of the two major classes of flowering plants. The cutaway regions (shaded yellow) show parts of the vascular system by which water, minerals, and nutrients move through the plant body.

Because fewer than 30,000 species (the red, brown, and green algae and the bryophytes) are nonvascular, it is clear that the vascular pattern is more typical. The most familiar vascular plants are the seed-producing gymnosperms and angiosperms:

gymnosperms   *chiefly conifers, such as pines, junipers, and redwoods*

angiosperms   *flowering plants, such as corn, roses, and cherry trees*

The overwhelming majority of vascular plants (about 235,000 species) are flowering, and they will be our focus here.

There are two classes of flowering plants, referred to informally as **monocots** and **dicots**. They both have a shoot system, consisting of stems and leaves, and a root system (Figure 19.1). Both produce flowers, which serve as the reproductive system. However, one of the main differences between the two classes is that monocot seeds have one cotyledon, and dicot seeds have two. (A *cotyledon* is a leaflike structure originating within the seed as part of the plant embryo. After the seed germinates, or starts to grow, cotyledons unfurl somewhere along the length of the tiny seedling.) Other differences between monocots and dicots are shown in Figure 19.2.

PLANT TISSUES AND THEIR COMPONENT CELLS

A *tissue* is a group of cells and intercellular substances functioning together in some specialized activity, such as water conduction. All vascular plants have primary tissues originating at root and shoot tips, and as these tissues grow and develop, the roots, stems, and leaves increase in length. Collectively, these tissues are called **primary growth**. Many species also show **secondary growth**, which results in increases in diameter (girth) at older root and stem regions. For example, secondary growth adds to the woody parts of a maple tree; primary growth gives rise to its new shoots each spring.

Three types of *ground tissues*, called parenchyma, collenchyma, and sclerenchyma, make up the bulk of the primary plant body. *Vascular tissues* (xylem and phloem) thread through the ground tissue system, and *dermal tissues* form a protective covering for the plant.

Ground Tissues

Parenchyma.   By far, **parenchyma** is the most common of the ground tissues (Figure 19.3). Its cells are massed together continuously in stems, roots, leaves, and the flesh of fruits. Most parenchyma cells have thin primary

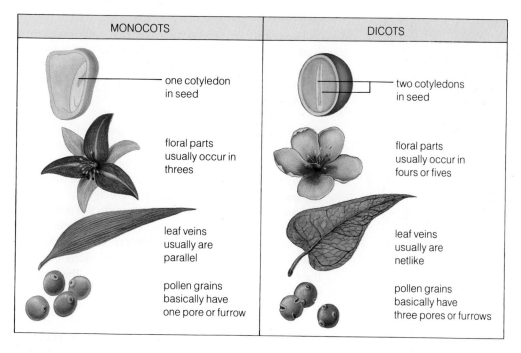

MONOCOTS	DICOTS
one cotyledon in seed	two cotyledons in seed
floral parts usually occur in threes	floral parts usually occur in fours or fives
leaf veins usually are parallel	leaf veins usually are netlike
pollen grains basically have one pore or furrow	pollen grains basically have three pores or furrows

Figure 19.2 Main differences between monocots and dicots. Monocots also have more primary vascular bundles (page 247), which are distributed through the stem tissue; dicots commonly have them positioned in a ring. Most monocots show little or no secondary growth; many dicots show secondary growth.

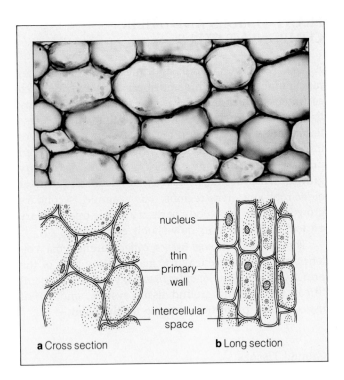

Figure 19.3 Parenchyma from a sunflower stem. The thin-walled cells of this tissue are alive at maturity and function in photosynthesis, storage, and other activities.

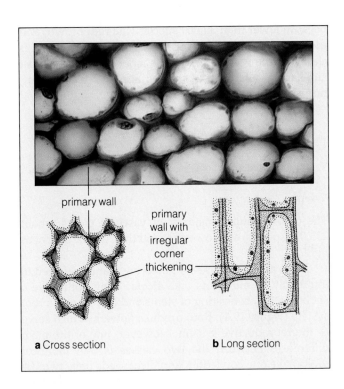

Figure 19.4 Collenchyma from the stem of a sunflower (*Helianthus*). Notice the irregular thickening at the corners where cells meet.

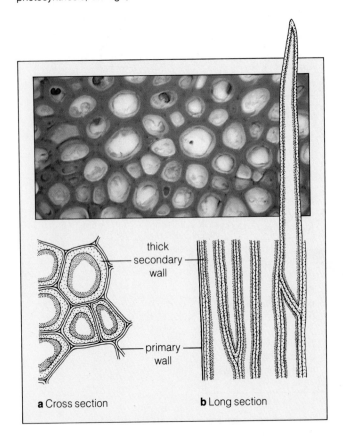

Figure 19.5 Sclerenchyma from a sunflower stem. Some sclerenchyma cells have tapered ends. All have somewhat evenly thickened secondary walls.

walls only, although some also have secondary walls (page 79).

Even after parenchyma cells have matured, they retain the ability to divide. Thus they take part in healing wounds and in regenerating plant parts. As you will see, parenchyma cells also take part in such tasks as photosynthesis, storage, and secretion.

Collenchyma. If you have ever chewed on a stalk of celery, you know that the "strings" near its outer surface are strong yet pliable. These properties, which are typical of many leaf stalks and stems, are attributable to **collenchyma**. Cells of this ground tissue are alive at maturity and are joined into long strands beneath the dermal tissue. Their primary walls become thickened with cellulose and pectin, and the pliability characteristic of collenchyma arises from bonding interactions between the two substances. As Figure 19.4 shows, the primary walls of collenchyma cells are often thickened unevenly at their corners.

Sclerenchyma. The ground tissue called **sclerenchyma** strengthens mature plant parts. Sclerenchyma cells have thick secondary walls, sometimes impregnated with lignin (Figure 19.5). Most often the cells form continuous strands or sheets, but they also may be scattered among other cell types.

Some sclerenchyma cells, called *sclereids*, are the structural element of seed coats and nut shells; they also give pears their gritty texture (Figure 19.6). Others, called *fibers*, are long, tapered cells. Strands of hemp and flax fibers are used in manufacturing paper, textiles, thread, and rope.

Vascular Tissues

Xylem. Both xylem and phloem form a continuous system of vascular tissue through the plant body. **Xylem** conducts water and dissolved mineral salts absorbed from soil, and it also lends some mechanical support to the plant.

The main water-conducting cells of xylem are called *tracheids* and *vessel members*. Both have strong, thick walls, with crisscrossed layers of cellulose impregnated with lignin and other substances. Both cell types are dead at maturity, and all that is left are the walls. In these walls are numerous recesses, or "pits," where secondary wall material has not been deposited over the primary wall. Typically, a pit in one cell matches up with a pit in an adjacent cell:

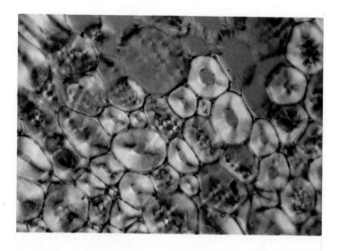

Figure 19.6 From the flesh of a pear, some thick, lignin-impregnated stone cells (a type of sclereid).

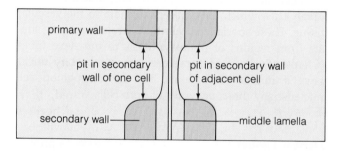

Tracheids are long cells with tapered ends, and water moves from one of these cells into another through pit-pairs that are concentrated at their overlapping regions (Figure 19.7). In contrast, vessel members are somewhat shorter cells joined end to end in a continuous tube called a *vessel*. In some vessels, the end walls are completely digested away, the result being an uninterrupted passageway between cells. In other vessels, the primary portion of the end walls of the vessel members is digested away, the result being a "perforation plate" that affords an uninterrupted passageway between cells (Figure 19.8).

Phloem. The food-conducting tissue in vascular plants is called **phloem**. Its main food-conducting cells are *sieve tube members* (in flowering plants) or *sieve cells* (in gymnosperms). Both are living cells through which sugars and other solutes are rapidly transported. In their walls are clusters of enlarged plasmodesmata, the channels connecting the cytoplasm of adjacent cells (page 82). In

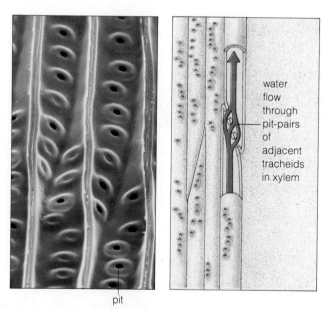

Figure 19.7 Scanning electron micrograph of some tracheids, sectioned lengthwise; these come from a pine tree.

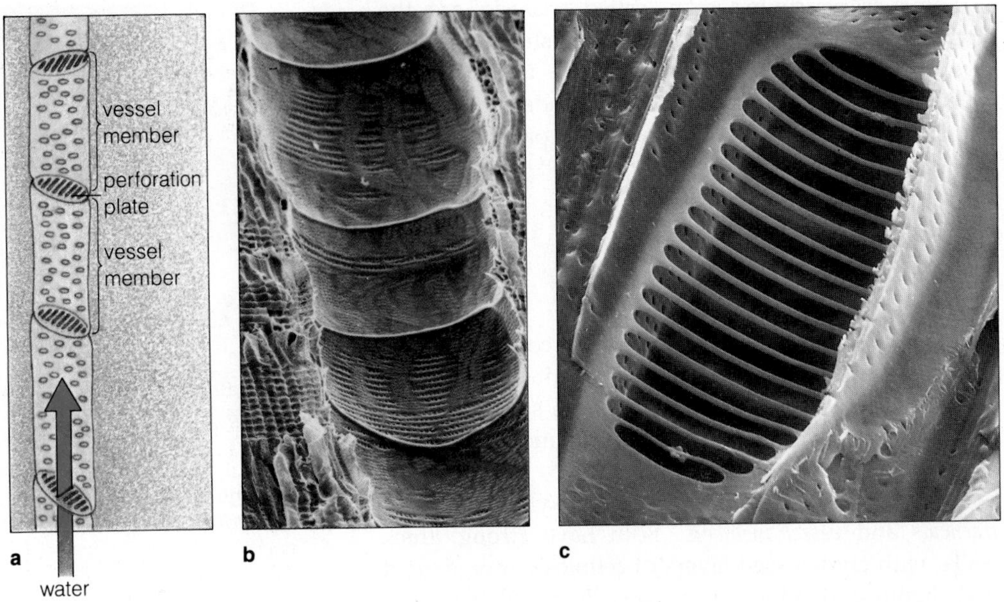

Figure 19.8 (a) How individual members are arranged to form the continuous water-conducting tubes called vessels in xylem. (b) Vessel members from the red oak (*Quercus rubra*). (c) A perforation plate at the end wall of a vessel member from the red alder (*Alnus rubra*).

vessel member

perforation plate

vessel member

a

water

b

c

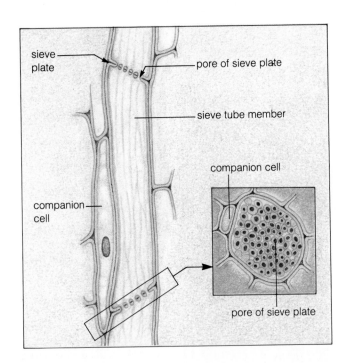

sieve plate

pore of sieve plate

sieve tube member

companion cell

companion cell

pore of sieve plate

Figure 19.9 Mature sieve tube member and adjacent companion cell. The inset shows one type of sieve plate.

sieve tube members alone, the clusters form large pores through the end walls, which are called "sieve plates" (Figure 19.9). These cells join end to end, with plates abutting, to form sieve tubes.

Sieve tube members are functionally linked with *companion cells*, which are adjacent to them in the vascular tissue. These specialized cells have a role in moving sugars from regions of photosynthesis to the sieve tubes (Chapter Twenty). Mature sieve tube members no longer have a functional nucleus, but the companion cell nucleus may direct activities of both cells.

Dermal Tissues

Epidermis. A continuous layer of tightly packed cells, the **epidermis**, usually covers the whole primary plant body (Figure 19.10). On aboveground parts, waxes and cutin impregnate the outer walls of epidermal cells and also form a noncellular surface coating called a *cuticle*. A cuticle restricts water loss and may confer some resistance to microbial attack.

Epidermis often contains highly specialized cells. For example, some thin-walled root epidermal cells have long protuberances. These so-called **root hairs** increase the cell surface; hence they enhance absorption of water and nutrients from the soil. As other examples, leaf epidermis in mint, lavender, and peppermint plants contains oil-secreting structures. Many floral parts, stems, and leaves contain *nectaries*: tissues or glands that secrete sugar-rich fluid derived mainly from phloem.

Periderm. A protective cover replaces the epidermis when roots and stems undergo secondary growth. This cover is the **periderm**. Its outermost cells are not alive at maturity, but the cell walls form a tissue called *cork*. The walls are impregnated with suberin, a waxy substance that functions in waterproofing.

How Plant Tissues Arise: The Meristems

As in animals, cell divisions, enlargements, and differentiation give rise to specialized tissues of the plant body. In animals, however, almost all cells are already committed before birth to being one type of cell only. In contrast, plants have many perpetually young tissue regions where some cells retain the capacity to divide again and again. The shoot tips of fruit trees that give rise to new blossoms and leaves each spring are like this. The term "meristem" refers to a mass of self-perpetuating cells, which are not (or not yet) committed to developing into a specialized cell type.

Meristems at shoot and root tips are responsible for the elongation (primary growth) of plant parts. Each tip has a dome-shaped **apical meristem**. Some cells in this dome-shaped region retain the capacity to divide repeatedly. Others produce daughter cells that become committed to giving rise to three kinds of **primary meristems**, the cells of which produce the mature tissues:

protoderm	*gives rise to the epidermis*
ground meristem	*gives rise to the ground tissue*
procambium	*gives rise to primary vascular tissues*

Two types of **lateral meristems** are responsible for secondary growth (increases in diameter):

vascular cambium	*gives rise to new vascular tissues after the primary plant body has formed*
cork cambium	*gives rise to periderm (the protective covering that replaces epidermis)*

The tissue organization of stems, leaves, and roots will be described next. When looking at the photographs accompanying the text, keep in mind the following terms, which identify the way a given tissue specimen was cut from the plant:

 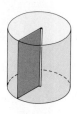 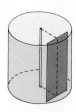

transverse	**radial**	**tangential**
(cross-section cut perpendicular to long axis of root or stem)	(longitudinal section cut parallel with the radius of root or stem)	(longitudinal section cut at right angles to radius of root or stem)

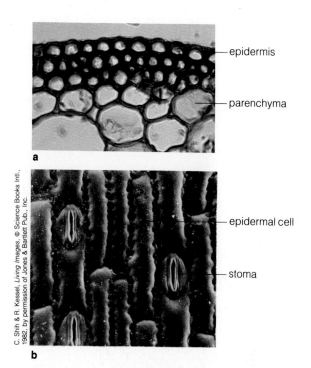

Figure 19.10 The epidermal layer of corn (*Zea mays*). (**a**) Transverse section through the stem; (**b**) surface of the lower epidermis of the leaf.

THE PRIMARY SHOOT SYSTEM

Functions of Stems and Leaves

The cells and tissues described so far do not function in isolation. Rather, they work together in ways that assure survival of the whole plant. In shoot systems, cells and tissues interact in carrying out the following tasks:

1. Providing a structural framework for upright growth, which gives photosynthetic tissues in leaves favorable exposure to light.

2. Transporting water, dissolved minerals, and organic molecules between roots, leaves, and other plant parts.

3. Storing food (in the parenchyma of various regions in the stem).

4. Displaying reproductive structures.

Stem Primary Structure

Figure 19.11 shows the primary structure of a dicot stem. All of its cells are derived originally from apical meristem at its tip. Some cells give rise to the three primary meristems, which in turn produce the specialized cells of the ground, vascular, and dermal tissues.

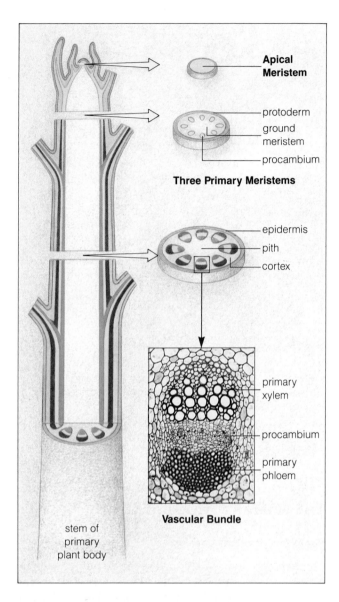

Apical
Meristem

protoderm
ground
meristem
procambium

Three Primary Meristems

epidermis
pith
cortex

primary
xylem

procambium

primary
phloem

Vascular Bundle

stem of
primary
plant body

Figure 19.11 Stem primary structure for one type of dicot. Notice the location of the apical meristem and primary meristems.

Lateral outgrowths of apical meristem called **leaf primordia** (singular, primordium) develop into the mature leaves of the stem. These outgrowths form so rapidly that at first there is no space at all between successive levels of leaves (Figure 19.12). But gradually the stem lengthens between the *nodes* (the parts of the stem where one or more leaves are attached). Each stem region between two successive nodes is an *internode*.

Meristematic cells also give rise to **bud primordia**, the beginnings of either lateral branchings of the stem or its terminal bud (Figure 19.13). Buds form only at nodes, in the upper angles (axils) where leaves are attached to the stem.

How are the primary tissues organized within the stem itself? Two patterns are most common. These patterns are distinctive because of the distribution of **vascular bundles**, which are clusters of strands of xylem and phloem in the ground tissue:

---

1. In conifers and some dicots: discrete cylinders of vascular tissues are arranged as a ring that divides the ground tissue into *cortex* and *pith*. (This arrangement is shown in Figure 19.14a.)

2. In most monocots and some dicots: many bundles of vascular tissues form a system of strands through the ground tissue (Figure 19.14b). Often there is no cortex/pith distinction.

Most monocot stems have only primary growth, but many dicot stems undergo secondary growth and their tissue organization becomes more intricate, as you will see shortly.

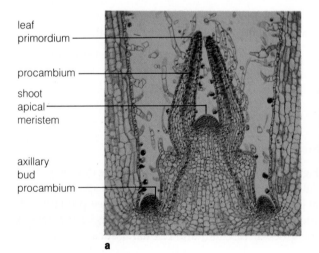

leaf
primordium

procambium

shoot
apical
meristem

axillary
bud
procambium

a

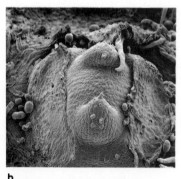

b

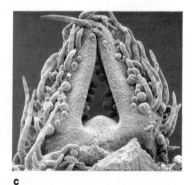

c

Figure 19.12 (**a**) Leaf primordia in the shoot tip of *Coleus*, longitudinal section. (**b,c**) Scanning electron micrographs of leaf development in the same plant.

Figure 19.13 Terminal bud of a dogwood tree showing its development into leaves and the dogwood "flower."

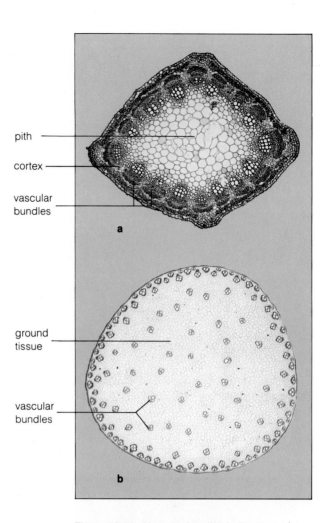

Figure 19.14 Vascular bundles in the stem of (**a**) alfalfa, a dicot; and (**b**) corn, a monocot.

Leaf Structure

Monocot and Dicot Leaves. Each node of a stem may have one or more **leaves**, which usually are sites of photosynthesis. The most common dicot leaf has two parts: a broad *blade* and a stalklike *petiole* attached to the stem (Figure 19.15a). A "simple" leaf has only one blade, although in some species it is deeply lobed. In a "compound" leaf the blade is divided into smaller leaflets, each attached to the petiole by a small, stalklike structure (Figure 19.16).

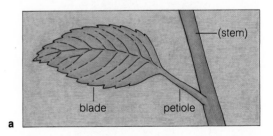

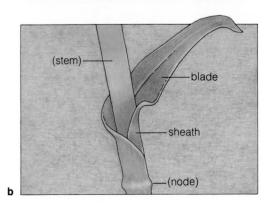

Figure 19.15 (Right) Leaf structure characteristic of dicots (**a**) and monocots (**b**).

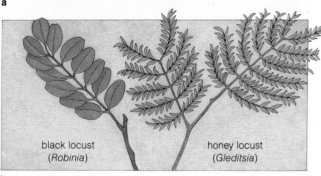

Figure 19.16 Examples of simple leaves (**a**) and compound leaves (**b**).

Most monocot leaves have no petiole; the base of the blade simply encircles the stem, forming a sheath. Corn is an example (Figure 19.15b).

There are numerous variations on these basic leaf plans. For example, some leaves have hairs and scales, others have hooks that impale predators. The two-lobed leaf of the Venus flytrap even turns the table on plant-eating animals, so to speak (Figure 19.17).

Regardless of the diversity, most leaves are alike in being short-lived. "Deciduous" species such as birches are periodically devoid of leaves, which drop away from the stem as winter approaches. Species such as camellias also drop leaves, but they appear "evergreen" because the leaves do not all drop at the same time.

Leaf Internal Structure. The leaves just described have a large external surface area exposed to sunlight and carbon dioxide in the air. Inside the leaf, the membranes of photosynthetic cells collectively represent an enormous surface area for sunlight reception and gas exchange. A network of *veins* (vascular bundles) through the leaf moves water and solutes to photosynthetic cells and carries products from them (Figures 19.18, 19.19).

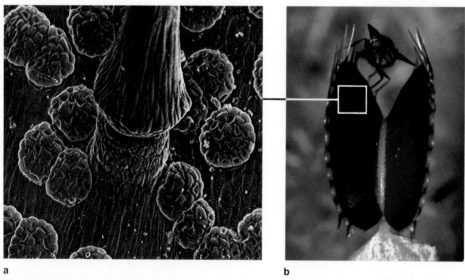

Figure 19.17 Specialized leaves of the Venus flytrap, a carnivorous plant. This plant grows in nitrogen-poor soil. Its two-lobed leaves open and close like a clamshell; the spines fringing the leaf margins intermesh when the lobes close.

Suppose an insect lands on the leaf and moves against one of its long epidermal hairs (the base of one is shown in **a**). The movement triggers cellular changes at the leaf midrib, and the leaf closes (**b,c**). Glandlike epidermal cells (the pincushion-like structures in the micrograph) secrete enzymes that digest proteins of the trapped insect body. Nitrogen is released from the proteins so kindly provided.

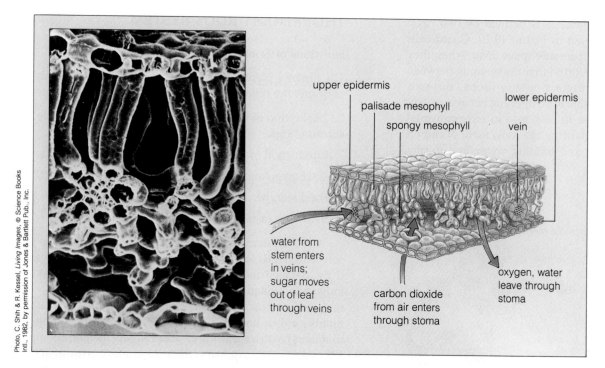

Photo, C. Shih & R. Kessel, *Living Images*, © Science Books Intl., 1982, by permission of Jones & Bartlett Pub., Inc.

Figure 19.18 Example of leaf internal structure. The scanning electron micrograph shows a mature broadbean leaf, transverse section. The sketch identifies the different leaf cells.

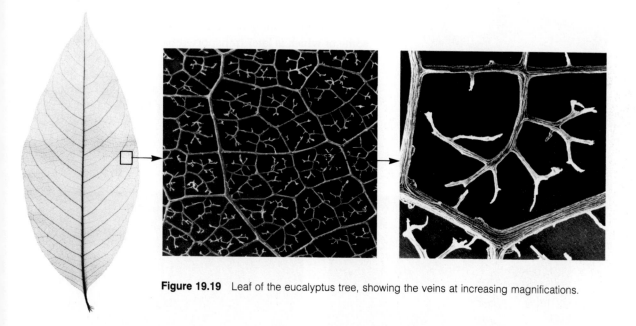

Figure 19.19 Leaf of the eucalyptus tree, showing the veins at increasing magnifications.

Figure 19.18 shows the tissue layers common to many leaves. Uppermost is a protective epidermis, with cuticle covering its outer surface. Next comes the *palisade mesophyll*, a loosely packed tissue of parenchyma cells that are capable of photosynthesis. Below the palisade tissue is the *spongy mesophyll*: even more loosely packed parenchyma cells that also are photosynthetic. Between thirty and fifty percent of a leaf consists of air spaces around the spongy mesophyll and around most of each palisade cell wall. Below the spongy mesophyll is another cuticle-covered epidermal layer.

The lower epidermal layer usually contains most of the tiny openings through which water vapor moves out of leaves and carbon dioxide enters them. Each opening is a *stoma* (plural, stomata). It is defined by two *guard cells*, which are illustrated in Figure 19.20. Guard cells can swell under turgor pressure (page 264). When they do, their shape is so distorted that the opening between them widens. When turgor pressure drops, the cells become flaccid; then, there is no opening between them. Stomata open and close in response to environmental conditions, thereby enhancing the movement of carbon dioxide into the plant and limiting water loss during drought or at night.

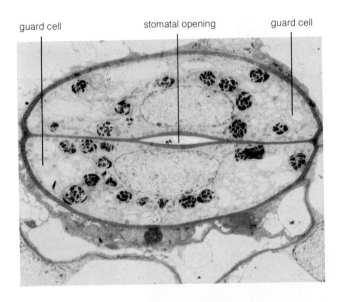

guard cell   stomatal opening   guard cell

Figure 19.20 Paired guard cells of a stoma on the stem of a beavertail cactus (*Opuntia*).

THE PRIMARY ROOT SYSTEM

Functions of Roots

The life of a plant depends on how well a root system performs the following tasks:

1. Absorption of water and dissolved minerals from the surroundings.
2. Conduction of water and solutes to aerial plant parts.
3. Anchorage and, in some species, structural support.
4. Food storage.
5. Synthesis of certain hormones.

A root system must absorb enough water and dissolved minerals to sustain plant growth and metabolism. If the detailed measurements of one rye plant are any indication, adequate absorption requires a tremendous root surface area. By the time that rye plant was four months old, it had developed a root system (including root hairs) with a surface area of 639 square meters, about 130 times more extensive than the surface area of its shoot system!

The root system penetrates downward and spreads out laterally, anchoring the aboveground parts. Most roots also function as storage sites for photosynthetically produced food, some of which is used by root cells and some of which is later transported back to aboveground parts. For example, carrots and sugar beets require two growing seasons to complete the life cycle. During the first growing season, food is stockpiled in roots. During the second, stored food is tapped for the formation of flowers, fruits, and seeds.

a   b

Figure 19.21 (**a**) Taproot system of a dandelion. (**b**) Fibrous root system of a grass plant.

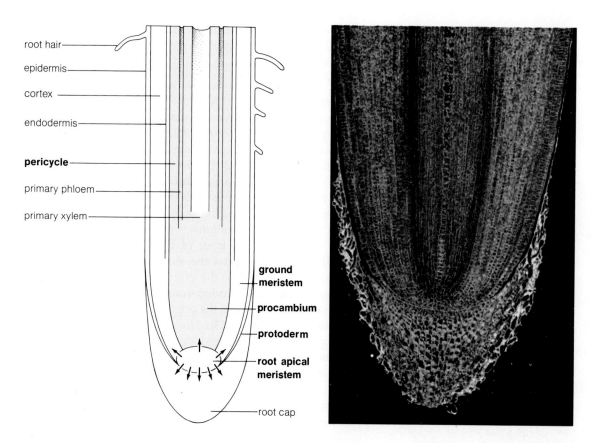

root hair
epidermis
cortex
endodermis
pericycle
primary phloem
primary xylem

ground meristem
procambium
protoderm
root apical meristem
root cap

Figure 19.22 Generalized root tip, sliced lengthwise. The micrograph shows a corn root tip.

Taproot and Fibrous Root Systems

The first plant root, the **primary root**, forms in the seed as part of the embryo. In most dicots and gymnosperm seedlings, the primary root increases in diameter and grows downward, and younger branchings called **lateral roots** emerge sideways along its length (Figure 19.21). The youngest branchings are near the root tip. A primary root and its lateral branchings are a *taproot system*.

A carrot has a taproot system. So does a pine tree, the roots of which can penetrate the soil to depths of six meters and more. In these plants, the taproot is quite distinct from its branchings. In other species, the branchings enlarge and the primary root stops growing, so in time the roots superficially look alike.

Generally, the primary root is short-lived in monocots such as corn. In its place, numerous **adventitious roots** arise from the stem of the young plant. (The term "adventitious" refers to any structure arising at an unusual location, such as roots that grow from stems or leaves.) Adventitious roots and their branchings are all somewhat alike in length and diameter, and they form a *fibrous root system* (Figure 19.21). Generally, fibrous root systems do not penetrate the soil as deeply as do taproots.

Root Primary Structure

The cells of primary roots divide, elongate, and mature in different zones. Cell divisions occur at the apical meristem and in a limited region behind it, where primary tissues start to differentiate. The cells elongate in the next few millimeters behind that region. Cells of older tissues past that region may mature further, but they do not grow longer.

Root Cap. At the root tip is a dome-shaped cell mass, the **root cap** (Figure 19.22). The root apical meristem produces the root cap and in turn is protected by it. As the root elongates, the root cap is pushed forward and some of its cells are torn loose. The slippery remnants lubricate the cap and enhance its penetration of the soil.

Root Epidermis. The epidermis, ground tissue, and vascular tissues form behind the root cap and the root elongates. **Root epidermis** is the absorptive interface with the environment. In the region behind the elongating portion of a root, some epidermal cells send out the long protuberances called root hairs (Figure 19.22). Root hairs greatly increase the surface available for taking up water and solutes. That is why you should never yank a plant out of the ground when transplanting it; doing so would tear off too much of this fragile absorptive surface.

Vascular Column. Most often, root vascular tissues form a central cylinder, or **vascular column**, surrounded by ground tissue (Figure 19.23). In some species such as corn, a ring of vascular tissues divides the ground tissue into pith and cortex (Figure 19.24). The ground tissue itself always has an abundance of air spaces. These spaces are continuous with the shoot system and allow oxygen to reach the living root cells, which depend on it for cellular respiration.

The innermost layer of the root cortex, the **endodermis**, helps control the movement of water and dissolved minerals into the vascular column. Think of each walled cell of the endodermis as having six sides. One side faces the cortex, the opposite side faces the vascular column (Figure 19.23). The remaining four sides press against neighboring endodermal cell walls—and embedded in the walls of these abutting sides is a thin strip of waxy suberin deposits, called the *Casparian strip*.

The Casparian strip prevents water and minerals from moving indiscriminately into the vascular column, because water cannot penetrate through adjoining walls of the endodermis. To get into the vascular column, water and minerals actually must pass through the cytoplasm of endodermal cells (Figure 19.23). This means they must cross the plasma membrane, with all of its

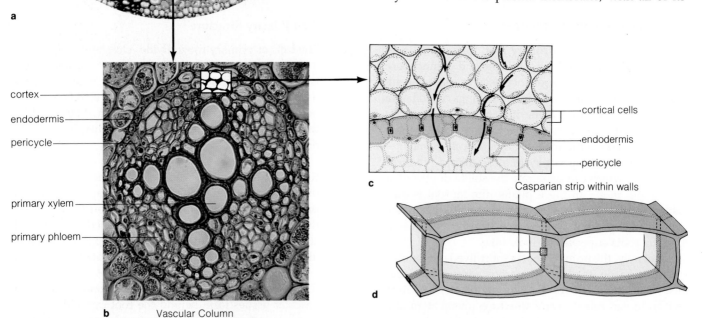

cortex
endodermis
pericycle
primary xylem
primary phloem
b Vascular Column

cortical cells
endodermis
pericycle
Casparian strip within walls
c
d

Figure 19.23 (**a**) Young root from a buttercup (*Ranunculus*), transverse section. (**b**) Closer view of the vascular column. (**c**) Water moving into the root travels along the cell walls and in the spaces between cells of the cortex. Water can move into the vascular column only through the cytoplasm of endodermal cells; it cannot penetrate the walls of these cells at the Casparian strip. (**d**) Cells of the endodermis are walled on all sides and contain cytoplasm; for clarity, only the walls containing the Casparian strip are shown here.

built-in transport mechanisms. *Membrane transport mechanisms enable the endodermis to function as a control point, where different substances can be selectively transported into the vascular column or barred from it.*

Just inside the endodermis is the **pericycle**. This part of the vascular column consists of one or more layers of parenchyma cells that maintain a high degree of meristematic potential. For example, the pericycle gives rise to lateral roots, which grow outward through the cortex and epidermis (Figure 19.25). In species showing secondary growth, the pericycle also contributes to the formation of the vascular and cork cambia, the meristems responsible for increasing the diameter of stems and roots.

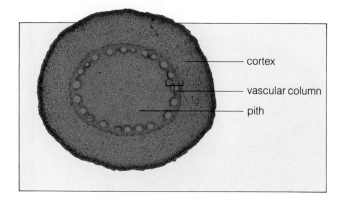

Figure 19.24   Root from a corn plant (*Zea mays*), transverse section.

SECONDARY GROWTH

Seasonal Growth Cycles

The life cycle of flowering plants extends from seed germination to seed formation, then eventual death. Most monocots and some dicots are called nonwoody, or *herbaceous*, plants because they show little or no secondary growth during the life cycle. In contrast, many dicots and all gymnosperms are called *woody* plants because they show secondary growth during two or more growing seasons. Herbaceous and woody plants are characterized as follows:

annuals	*life cycle completed in one growing season; little (if any) secondary growth. Examples: snap beans, corn, marigolds.*
biennials	*life cycle completed in two growing seasons (root, stem, leaf formation the first season; flowering, seed formation, death the second). Examples: carrots, dandelions, foxgloves.*
perennials	*seed germination, formation continue year after year. Some have secondary tissues, others do not. Examples: the herbaceous cacti, and woody shrubs (roses), vines (ivy, grape), and trees (apples, elms, magnolias).*

Vascular Cambium Activity

In species showing secondary growth, stems and roots increase in diameter when lateral meristems (vascular cambium and cork cambium) become active. When fully developed, the vascular cambium is like a cylinder, one or a few cells thick. Its meristematic cells (called "initials") give rise to secondary xylem and phloem that conduct water vertically and horizontally through the enlarging stem or root (Figure 19.26).

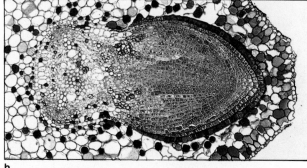

a

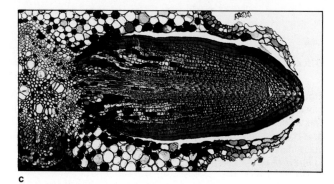

b

c

Figure 19.25   Lateral root formation in a willow (*Salix*).

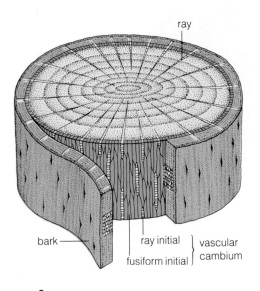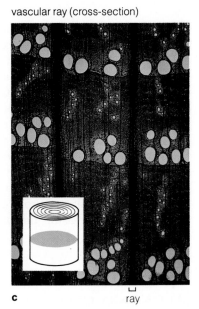

ray

vascular ray (tangential cut)

vascular ray (cross-section)

bark

ray initial
fusiform initial } vascular cambium

a

b   ray

c   ray

Figure 19.26   (**a**) Location of vascular cambium in older roots and stems showing secondary growth. The meristematic cells called fusiform initials produce the "vertical system" of secondary xylem and phloem, which conducts water and food up and down the root or stem. The ray initials produce the "ray system" of (mostly) parenchyma cells that act as horizontal conduits for water and as food storage centers in wood. (**b–c**) Appearance of rays in the hardwood of a red oak (*Quercus rubra*), in tangential and transverse sections.

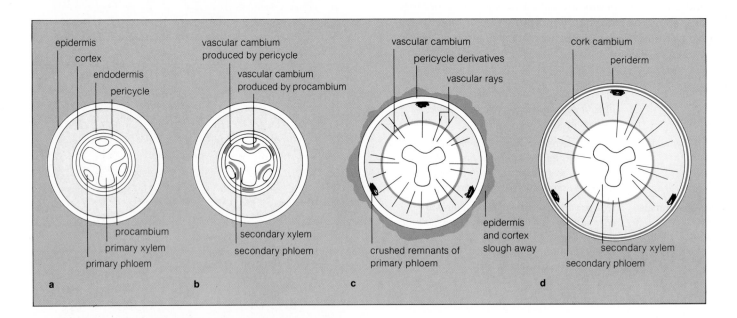

epidermis
cortex
endodermis
pericycle

vascular cambium
produced by pericycle
vascular cambium
produced by procambium

vascular cambium
pericycle derivatives
vascular rays

cork cambium
periderm

procambium
primary xylem
primary phloem

secondary xylem
secondary phloem

crushed remnants of
primary phloem

epidermis
and cortex
slough away

secondary xylem
secondary phloem

a   b   c   d

Figure 19.27   Secondary growth in a root, transverse section. (**a**) Arrangement of tissues at the end of primary growth. (**b–c**) Formation of a complete ring of vascular cambium from two types of meristems. The vascular cambium gives rise to secondary xylem and phloem. The root increases in diameter as cell divisions proceed parallel and perpendicular to the vascular cambium. The cortex ruptures as the tissue mass increases. (**d**) Epidermis is replaced by periderm, which arises from cork cambium.

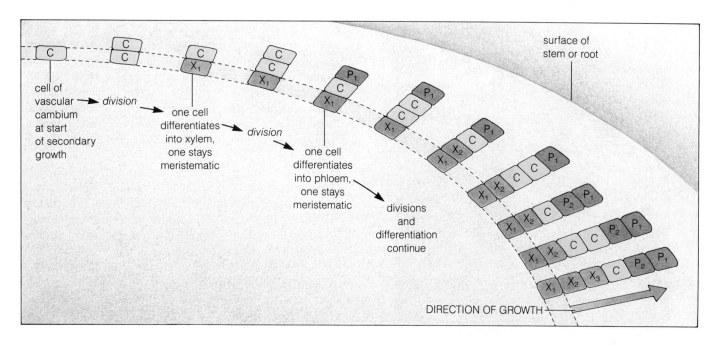

Figure 19.28 Relationship between the vascular cambium and its derivative cells (secondary xylem and phloem). Notice how the ongoing divisions displace the cambial cells, moving them steadily outward even as the core of xylem increases the stem or root thickness.

Xylem forms on the inner face of the vascular cambium, and phloem forms on the outer face (Figures 19.27 and 19.28). As the xylem increases in mass, the vascular cambium is displaced outward. Cells of the vascular cambium do not divide in one direction only, in line with the radius of the root or stem. They also divide in a direction parallel with the root or stem axis, thereby adding new initials that contribute to increases in the circumference of the vascular cambium.

In most species, the number of secondary xylem cells formed is far greater than the number of phloem cells. As the mass of xylem increases season after season, it usually crushes the thin-walled phloem cells from the preceding growth period, leaving only the thick-walled cells. In order to sustain plant growth and development, new rings of phloem cells must be produced each year, outside the growing inner core of xylem.

Early and Late Wood

In regions having prolonged dry spells or cool winters, the vascular cambium of stems and roots becomes inactive during parts of the year. The first xylem cells produced at the start of the growing season tend to have large diameters and thin walls; they represent **early wood** (Figure 19.29). As the season progresses, the cell diameters become smaller and the walls thicker; these cells represent **late wood**.

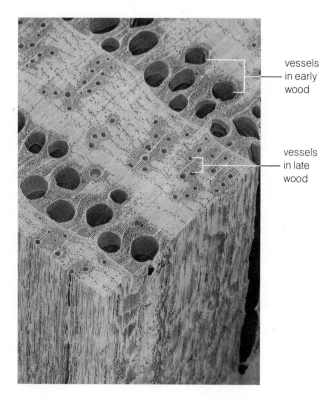

vessels in early wood

vessels in late wood

Figure 19.29 Scanning electron micrograph of red oak (*Quercus rubra*), transverse section.

The last-formed, small-diameter cells of late wood will end up next to the first-formed, large-diameter cells of the next season's growth. Although we don't see individual cells with the naked eye, there is enough difference in light reflection from a stem cross-section to reveal the alternating light bands (early wood) and dark bands (late wood). These alternating bands represent annual growth layers, which often are called "tree rings" (Figure 19.30).

In some wet, tropical regions, the growing season is continuous. Conditions in other tropical regions allow several spurts of growth during one year. The growth layers of many tropical woody plants are faint, nonexistent, or do not correspond to a single year's growth.

Cork Cambium Activity

In time, increases in the stem or root diameter cause the cortex and outer phloem to rupture. Parts of the cortex split away and carry epidermis with them (Figure 19.27). But pericycle cells continue to divide. Their derivative cells—including cork cambium—produce periderm, the corky covering that replaces epidermis. ("Cork" is not the same as "bark," a nontechnical term that refers to all living and nonliving tissues between the vascular cambium and the stem or root surface.) In some plants, secondary growth of this sort occurs year after year and produces massive, woody structures.

As you can see from Figure 19.30, living phloem in older trees is confined to a thin zone beneath the periderm. If this narrow band of phloem is stripped all the way around a tree's circumference, the tree will die. When the phloem cells in this region are stripped away, there is no way to transport photosynthetically derived food down to the roots, which will die.

SUMMARY

1. Of the more than 300,000 known species of plants, most have a vascular system (which conducts food and water from one part of the plant to another) and produce flowers. The flowering plants are classified as monocots and dicots.

2. The roots, stems, and leaves of flowering plants are composed of three kinds of tissues—the ground, vascular, and dermal tissues:

    a. Ground tissues are parenchyma, collenchyma, and sclerenchyma. Parenchyma cells have thin walls and they make up the bulk of fleshy plant parts. Collenchyma cells are strong and pliable. Sclerenchyma cells are rigid and capable of providing support.

    b. Vascular tissues include the xylem and phloem. In the xylem, vessel members are connected end-to-end to

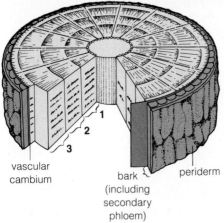

vascular cambium   bark (including secondary phloem)   periderm

Figure 19.30 Growth rings of a pine tree, transverse section.

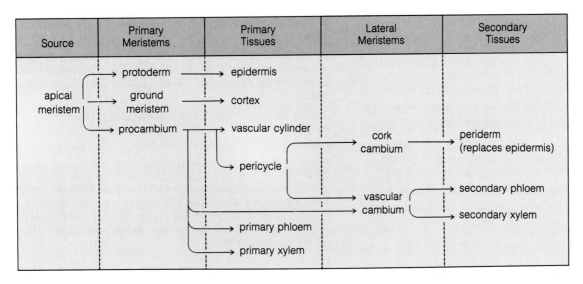

Figure 19.31 Summary of primary and secondary growth during the development of a root from a vascular plant.

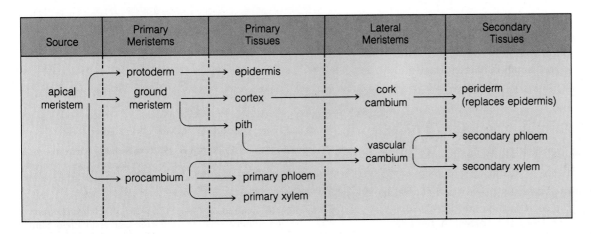

Figure 19.32 Summary of primary and secondary growth during the development of a stem from a vascular plant.

form a continuous vessel through which water and minerals can flow. In the phloem of flowering plants, sieve tube members are joined end-to-end to form a continuous tube through which nutrients are transported.

c. Dermal tissues are the epidermis (which covers and protects the aboveground parts of young plants) and periderm (which replaces the epidermis on plants that undergo secondary growth).

3. Apical meristems are masses of cells at shoot and root tips that are responsible for the elongation (primary growth) of plant parts. Some of the cells retain the capacity to divide repeatedly; others produce daughter cells that give rise to primary meristems (Figure 19.31 and 19.32).

4. Primary meristems divide and differentiate into protoderm (which develops into epidermis), ground meristem (which develops into the ground tissue), and procambium (which develops into the primary vascular tissues).

5. Lateral meristems are responsible for secondary growth, or increases in the diameter of stems and roots. They include vascular cambium (which gives rise to secondary vascular tissues) and cork cambium (which gives rise to periderm).

6. Shoot systems include stems and leaves (organs of photosynthesis). The stems provide structural frameworks for upright growth, giving the photosynthetic tissues favorable exposure to light, and they display the

reproductive structures. The vascular tissues of stems transport substances to and from roots, leaves, and other plant parts.

7. In monocot stems, vascular bundles are scattered through the ground tissue. Dicot and conifer stems have the vascular bundles arranged as a cylinder that separates the ground tissue into an outermost cortex and a central pith.

8. Inside most leaves, the plasma membranes of photosynthetic cells collectively represent an enormous surface area for sunlight reception and for gas exchange. Veins (vascular bundles) move water and solutes to these cells and carry products away from them.

9. Stomata are tiny openings, usually on the lower epidermis of the leaf, through which water vapor moves out and carbon dioxide moves in. A stoma is defined by two guard cells, which can swell under turgor pressure (and thereby widen the opening between them). Stomata open and close in response to environmental conditions.

10. Roots function in absorbing water and dissolved minerals from the surroundings and in conducting them to aerial plant parts; they also function in anchorage and often structural support, food storage, and synthesis of certain hormones.

11. The apical meristem at the root tip is protected by a root cap. Root hairs occur as extensions of the epidermis and greatly enhance the absorption of water and dissolved minerals.

12. Inside the root, the vascular tissues form a central column surrounded by the cortex. The endodermis separates the vascular tissues from the cortex. A beltlike Casparian strip surrounds the adjoining regions of each cell of the endodermis, sealing the endodermis in such a way that substances can reach the vascular tissue only by passing through the endodermal cells. Their passage can be regulated through controls over transport systems built into the cell membranes.

13. The pericycle lies just inside the endodermis. It gives rise to lateral roots and makes additional lateral growth of the root possible.

14. Unlike herbaceous annual plants (which live only one season), perennial plants live for many seasons and show secondary growth.

Review Questions

1. Define meristem. Which meristem regions produce the primary plant body? Which two kinds of active cambium give rise to layers of secondary xylem and phloem?

2. Distinguish between the following:
   a. xylem and phloem
   b. tracheid and vessel member
   c. epidermis and endodermis

3. What is the functional relationship between endodermis and the Casparian strip?

4. Sketch the stem of a monocot and a dicot in cross-section. Can you label the main tissue regions of each, and describe the functions of their cellular components? Do the same for a cross-section of a root.

5. How are annual growth layers formed in woody stems? If you were to strip away a narrow band of phloem from a tree's circumference, what would happen to the tree?

6. With a sheet of paper, cover the last three columns in Figure 19.32. Can you now state which primary tissues of roots arise from protoderm, ground meristem, and procambium? Move the sheet to the right so that only the last two columns are covered. Can you state which lateral meristems and secondary tissues develop from various primary tissues? Do the same exercise with Figure 19.33 for the tissues of stems.

7. Label the following stem regions:

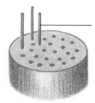

Readings

Bold, H., C. Alexopoulos, and T. Delevoryas. 1980. *Morphology of Plants and Fungi*. Fourth edition. New York: Harper & Row.

Bracegirdle, B., and P. H. Miles. 1971. *An Atlas of Plant Structure* (two volumes). London: Heinemann Educational Books. Excellent micrographs accompanied by diagrams that identify component structures.

Core, H., W. Côté, and A. Day. 1979. *Wood Structure and Identification*. Second edition. New York: Syracuse University Press. Stunning scanning electron micrographs of softwood and hardwood structure.

Esau, K. 1977. *Anatomy of Seed Plants*. Second edition. New York: Wiley. A classic reference for seed plant structure.

Raven, P., R. Evert, and S. Eichhorn. 1986. *Biology of Plants*. Fourth edition. New York: Worth. Exquisite color micrographs and illustrations of plant cells and tissues.

It took you eighteen years or so to grow to your present height. A corn plant can exceed that in three months! Yet how often do we stop to think that plants actually do anything at all impressive? Being endowed with great mobility, intelligence, and varied emotions, we tend to be endlessly fascinated with ourselves and somewhat indifferent to the immobile, expressionless plants around us. Besides, from experience and educational biases, most of us simply have acquired more knowledge about animals than we have about plants.

And yet, think about what it must take for plants such as an elm tree or a dandelion simply to survive. They require only sunlight, water, carbon dioxide, and assorted minerals. But most soils are frequently rather dry, and what does the plant do then? In a given volume of air, the carbon dioxide concentration averages only about 340 parts per million, so how can enough carbon dioxide be taken up to sustain photosynthesis? Minerals dissolved in soil water are scarce, so how does the plant accumulate them against concentration gradients?

Now think about the way those plants are put together. Which aspects of their structure and function might be responses to the low concentrations of environmental resources? For one thing, in most of the living cells, central vacuoles increase the cell volume—hence the volume of the plant as a whole—and thereby increase the surface area for absorbing scarce materials. (As we have seen, a central vacuole is mostly water, and fluid pressure that builds up inside it exerts force on the walls of growing cells and leads to their enlargement.)

20

WATER, SOLUTES, AND PLANT FUNCTIONING

Figure 20.1 Sunlight filtering through a grove of California coast redwoods. How are water and essential nutrients transported to the tops of such giant trees? How are photosynthetically derived organic molecules distributed from leaves to all parts of the massive plant body, even down to the roots? These are questions that will be addressed in this chapter.

Table 20.1	Essential Elements for Most Complex Land Plants			
Element	Symbol	Form Available to Plants	Percent by Weight in Dry Tissue	
Carbon	C	CO_2	45	96% of total dry weight
Oxygen	O	O_2, H_2O, CO_2	45	
Hydrogen	H	H_2O	6	
Nitrogen	N	NO_3^-, NH_4^+	1.5	
Potassium	K	K^+	1.0	
Calcium	Ca	Ca^{++}	0.5	
Magnesium	Mg	Mg^{++}	0.2	
Phosphorus	P	$H_2PO_4^-$, HPO_4^{2-}	0.2	
Sulfur	S	SO_4^{2-}	0.1	
Chlorine	Cl	Cl^-	0.010	
Iron	Fe	Fe^{++}, Fe^{+++}	0.010	
Boron	B	H_3BO_3	0.002	
Manganese	Mn	Mn^{++}	0.0050	
Zinc	Z	Zn^{++}	0.0020	
Copper	Cu	Cu^+, Cu^{++}	0.006	
Molybdenum	Mo	MoO_4^-	0.00001	

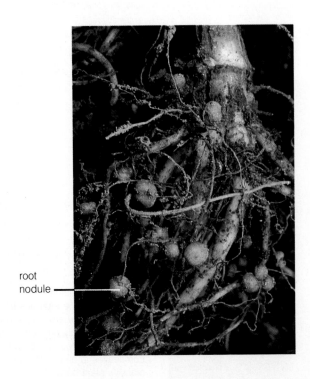

root nodule —

Figure 20.2 Root nodules, where symbiotic nitrogen-fixing bacteria live.

For another thing, their leaves are thin and broad, with a large surface area for absorbing sunlight and carbon dioxide.

Obviously those plants cannot have broad, leaflike roots (imagine trying to push a narrow strip of paper through soil). But they have roots shaped like thin cylinders, which collectively represent a large surface area and individually overcome the soil's resistance to penetration. Also, their root systems grow outward in many directions, thereby exposing the root absorptive surface areas to a large volume of moist soil. Finally, even though shoots move farther and farther away from roots during growth, the plants have vascular systems that conduct water, minerals, and organic molecules from one region to another.

Many aspects of plant structure and function are concerned with exposing a large surface area to a large volume of environmental space and thereby enhancing the uptake of dilute raw materials.

How plant systems function in response to the environment is the focus of *plant physiology*. We have already considered how plants acquire energy (by photosynthesis), then produce and use their own food (by aerobic respiration). In this chapter, we will see how plants acquire and distribute materials used directly or indirectly in these and other activities.

ESSENTIAL ELEMENTS AND THEIR FUNCTIONS

Oxygen, Carbon, and Hydrogen

As indicated in Table 20.1, plants generally require sixteen essential elements to grow and reproduce, although most studies have been carried out with crop plants only. Oxygen, carbon, and hydrogen are building blocks for all organic compounds (carbohydrates, lipids, proteins, and nucleic acids). Together, these three elements account for about ninety-six percent of the plant's dry weight. The oxygen comes from water, gaseous oxygen (O_2), and carbon dioxide (CO_2) in the air; the carbon dioxide also is the source of carbon. The hydrogen comes from water molecules.

Mineral Elements

Thirteen of the essential elements are minerals. These naturally occurring, inorganic substances become available to plants in ionized form; they are "mineral ions." Table 20.2 lists their main functions, along with the symptoms associated with mineral deficiencies. Six of these minerals are called *macronutrients*; each makes up

Table 20.2 Role of Mineral Elements in Plant Function

Element	Some Known Functions	Deficiency Symptoms
Macronutrients:		
Nitrogen	Component of amino acids, proteins, chlorophyll, nucleic acids, coenzymes	Stunted growth, delayed maturity, light green older leaves; lower leaves turn yellow and die
Potassium	Activates enzymes used in protein, sugar, starch synthesis, helps maintain turgor pressure*	Reduced yields; mottled, spotted or curled older leaves; marginal burning of leaves; weak root system, weak stalks
Calcium	Part of middle lamella (helps cement cell walls together); necessary for spindle formation in mitosis, meiosis	Deformed terminal leaves, reduced root growth. Dead spots in dicot leaves. Terminal buds die
Magnesium	Component of chlorophyll; activates many enzymes used in photosynthesis, respiration, protein synthesis	Plants usually chlorotic (interveinal yellowing of older leaves); leaves may droop
Phosphorus	Component of some amino acids, proteins, nucleic acids, ADP and ATP, phospholipids	Purplish veins in older leaves, stems, and branches often turn dark green; reduced yields of seeds and fruits, stunted growth
Sulfur	Component of some amino acids, two vitamins, most proteins	Light green or yellow leaves, including veins; reduced growth. Weak stems. Similar to nitrogen deficiency
Micronutrients:		
Chlorine	Aids in root and shoot growth. Aids in photolysis in noncyclic pathway of photosynthesis	Plants wilt. Chlorotic leaves. Some leaf necrosis. Bronzing in leaves
Iron	Helps synthesize chlorophyll, component of electron transport systems of photosynthesis and respiration	Paling or yellowing of leaves (chlorosis) between veins at first. Grasses develop alternate rows of yellowing and green stripes (veins) in leaves
Boron	Affects flowering, pollen germination, fruiting, cell division, nitrogen metabolism, water relations, hormone movement	Terminal buds die, lateral branches begin to grow, then die. Leaves thicken, curl, and become brittle
Manganese	Chlorophyll synthesis; acts as coenzyme for many enzymes	Network of major green veins on light green background. Leaves later become white and fall off
Zinc	Used in formation of auxins, chloroplasts, and starch; component of several enzymes	Abnormal roots; mottled bronzed or rosetted leaves. Interveinal chlorosis
Copper	Component of enzymes used in carbohydrate and protein metabolism	Terminal leaf buds die. Leaves have chlorotic or dead spots. Stunted growth. Terminal leaves die
Molybdenum	Essential in enzyme-mediated reaction that reduces nitrate; component of enzyme used in nitrogen fixation	Plants may become nitrogen deficient. Pale green, rolled or cupped leaves, with yellow spots

Data from Hartmann et al. *Plant Science*, 1981, and others.
*All solutes contribute to the osmotic solute and ion balance.

at least a tenth of a percent of the total dry weight of the plant. The rest are *micronutrients*, or trace elements, which represent only a few parts per million of the plant's dry weight.

Nitrogen is one of the macronutrients. It is not exactly scarce; N_2 molecules make up seventy-eight percent of the air. But plants do not have the metabolic machinery for breaking apart the three covalent bonds ($N\equiv N$) of these molecules. However, some microbes living independently of plants can break the bonds and reduce each nitrogen atom to ammonium (NH_4^+); and other microbes living in root nodules can do the same thing. **Nodules** are localized swellings on the roots of legumes (for example, peas) and other plants where symbiotic, nitrogen-fixing bacteria live (Figure 20.2). The nodule residents use up some of the plant's organic molecules. However, the ammonium they produce is used in assembling their own amino acids and nucleic acids—and most of the ammonium and amino acids remain in the plants, which thereby are provided with a nitrogen source.

Most of the minerals listed in Table 20.2 function in activating the enzymes of protein synthesis, starch synthesis, photosynthesis, and aerobic respiration. They also function in setting up solute concentration gradients across plasma membranes. Because of these gradients, water moves by osmosis into cells and their vacuoles, and turgor pressure increases. (Turgor pressure, recall, is the internal pressure applied to walls as water is

Figure 20.3 Mycorrhiza (fungus-root) of a lodgepole pine tree. White threads are fungal strands.

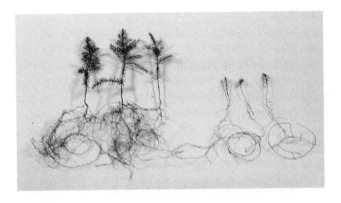

Figure 20.4 Effect of mycorrhizal fungi on plant growth. The six-month-old juniper seedlings on the left were grown in sterilized low-phosphorous soil inoculated with a mycorrhizal fungus. The seedlings on the right were grown without the fungus.

Figure 20.5 Scanning electron micrograph of root hairs, 46×.

absorbed into cells.) This pressure is necessary for cell enlargement during growth. Also, in mature cells with walls that no longer stretch, turgor pressure keeps the plant from wilting (Figure 6.7).

Mineral ions function in many metabolic activities. They also help set up solute concentration gradients; and water necessary for growth and for maintaining plant shape thereby moves by osmosis into cells.

WATER UPTAKE, TRANSPORT, AND LOSS

Water Absorption by Roots

Generally, annual grasses (including corn) have a highly branched, fibrous root system that spreads out near the soil surface. Most dicots have a taproot system that penetrates more deeply into the soil. However, the environment influences the extent to which any root system develops. Millions, sometimes billions of root hairs might develop in a single system. Roots branch out in some locations, then they are replaced by roots that branch into different locations as conditions change. It is not that the roots "explore" the soil in search of minerals or water. Rather, outward root growth (hence growth of the entire plant) is stimulated in regions where water and dissolved mineral ions happen to be more concentrated.

In nearly all vascular plants, water (and solute) absorption apparently is enhanced by mycorrhizae, or "fungus-roots." A **mycorrhiza** is a mutually beneficial association between a fungus and a young root (Figure 20.3). The symbiotic fungi form an extensive mat of very thin filaments. Collectively, the filaments have a tremendous surface area for absorbing mineral ions from a large volume of the surrounding soil. The root uses some of the ions during growth (Figure 20.4); the fungus absorbs some sugars and nitrogen-containing compounds in the root.

Take a moment to look at Figure 20.5, which shows the hair cells of a typical root. Especially in regions of root hairs, water from the soil diffuses inward through epidermal cells. Once inside the root cortex, much of the water apparently is conducted along the porous walls of cortical cells until it reaches the endodermis surrounding the vascular column. There it encounters the waxy Casparian strip, which prevents water from entering the vascular column except by way of the cytoplasm of endodermal cells (Figure 19.23c). Inside the vascular column, water enters the conducting cells of xylem, where its movement is nearly always governed by forces of the sort to be described next.

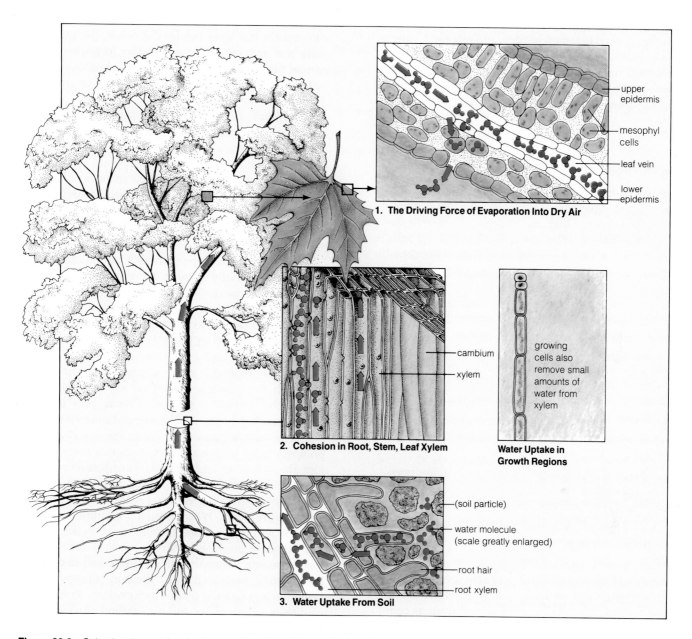

Figure 20.6 Cohesion theory of water transport. Tensions in water in the xylem extend from leaf to root. These tensions are caused mostly by transpiration. As a result of the tensions, columns of water molecules that are hydrogen-bonded to one another are pulled upward.

Transpiration and Water Conduction

Water moves from roots to stems, then into leaves. A small part is used for growth (through cell enlargement) and in metabolism, but most evaporates into the air. Water evaporation from plant parts (stems and leaves especially) is called **transpiration**. How does water get to those parts which, in the tallest trees, are 100 meters aboveground? *Water is pulled upward by continuous negative pressures (tensions) that extend downward from the leaf to the root* (Figure 20.6).

The question becomes this: What causes the tension in the xylem cells? *First*, the air around a plant nearly always causes evaporation from the walls of mesophyll cells inside the leaf. As some water molecules escape, others diffuse out of the cell cytoplasm and into the walls as replacements. When they do, water molecules from the xylem in leaf veins move into the mesophyll cells. *Second*, when water moves out of the veins, replacements are pulled in from the xylem cells leading into the veins from the stem. Because of the pulling action, water inside all of those conducting cells is in a state of tension. *Third,* replacement water molecules move into the root xylem even when the soil is somewhat dry, and more soil water is drawn into the plant, following its osmotic gradient. This inward movement will continue until the soil becomes so dry that an osmotic gradient no longer exists. Figure 20.6 shows how water is pulled uphill, from roots to leaves.

Cohesion Theory of Water Transport

When water moves as a continuous, fluid column through xylem pipelines, why doesn't the "stretching" cause the molecules to snap away from each other? Some time ago the Irish botanist Henry Dixon came up with an explanation, which has since been named the **cohesion theory of water transport**. According to this theory, hydrogen bonds are strong enough to keep water molecules from separating from one another as they are pulled up through the plant body. Dixon had no way of measuring how much tension exists in xylem (and therefore of convincing skeptics that it really exists). Confirmation of his explanation came much later. In any event, the points to remember are these:

1. The drying power of air causes transpiration (evaporation of water from plant parts exposed to air, especially leaves).

2. Transpiration causes a state of tension (negative pressure) in xylem that is continuous from leaves, down through the stems, to roots.

3. As long as water molecules continue to vacate transpiration sites, replacement molecules are pulled under tension.

4. Because of the cumulative strength of hydrogen bonds between water molecules that are confined in the narrow, tubular xylem cells, water is pulled up as continuous columns to transpiration sites.

5. Although hydrogen bonds are strong enough for water molecules to cohere in the xylem, they are not strong enough to prevent them from breaking away from each other during transpiration.

The Dilemma in Water and Carbon Dioxide Movements

Of the water moving into a leaf, more than ninety percent is usually lost through transpiration. About two percent is used in photosynthesis, membrane functions, and other activities. However, when water loss by transpiration exceeds water uptake by roots, the resulting dehydration of plant tissues will interfere with these water-requiring activities.

Even under mild conditions, plants would rapidly wilt and die if it were not for the *cuticle*, the waxy covering that reduces the rate of water loss from above-ground plant parts (page 244). The cuticle does conserve water, but it also limits the rate of diffusion of carbon dioxide into the leaf.

Transpiration and carbon dioxide uptake occur largely at numerous *stomata*, the small openings in the epidermis of leaves and stems (page 250). Two guard cells flank each opening. When the cells are swollen with water, turgor pressure distorts their shape in such a way that they move apart (Figure 20.7). Their separation produces a gap between them (the true stoma). When the water content of the guard cells dwindles, turgor pressure drops and the stoma closes.

When stomata are open, enough carbon dioxide can be absorbed rapidly enough for photosynthesis—but when they are open, water nearly always moves out! In itself, carbon dioxide uptake and water loss at the same sites would present something of a dilemma for the plant, if it were not for controls over stomatal opening and closure.

Stomata typically open during daylight hours. Although the processes involved are not completely understood, this much is clear: *A stoma opens and closes mainly according to how much water and carbon dioxide are present in the two guard cells.* When the sun comes up, photosynthesis begins and carbon dioxide is used in the formation of starch and sucrose. As carbon dioxide concentrations dwindle in guard cells, potassium ions are actively pumped into them from surrounding epidermal cells (Figure 20.8). When the ions accumulate inside, water moves in by osmosis from the surrounding cells. The resulting increase in turgor pressure leads to stomatal opening. Transpiration proceeds, and so does carbon dioxide movement into the leaf. Photosynthesis keeps carbon dioxide concentrations low, so the plant continues to lose water and gain carbon dioxide during the day.

Transpiration nearly always occurs when carbon dioxide, which is present only in dilute concentrations in air, is being absorbed. When stomata are open and carbon dioxide is moving into the plant, water tends to move out.

When the sun goes down, photosynthesis stops. Carbon dioxide is no longer used, but it is still being released during aerobic respiration. As a result, carbon dioxide accumulates in all cells. Much of the potassium inside guard cells now moves out, and water follows it osmotically. Turgor pressure decreases, stomata close, transpiration is greatly reduced, and water is conserved.

As long as the soil is moist, stomata can remain open during daylight. When the soil is dry and the air is also

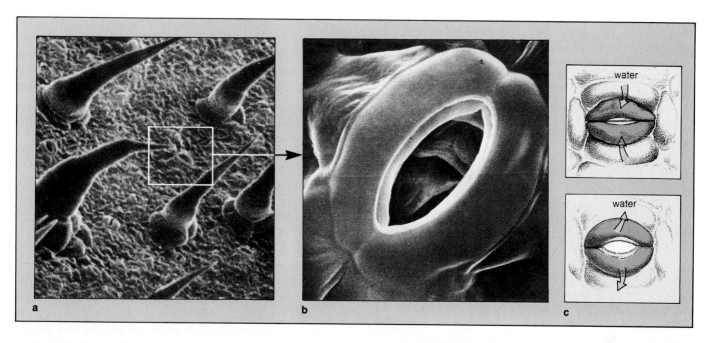

Figure 20.7 Where stomata occur on a typical dicot leaf, and how they function. (**a**) Stomata are found among hairlike structures on a cucumber leaf's lower epidermis. The box identifies one of the stomata. (**b**) Closer look at stomatal structure. Here, we can peer through the gap between the stomatal guard cells and view parts of mesophyll cells inside the leaf. (**c**) Sketches of a closed stoma and one opened as guard cells swell.

dry and hot, the stomata of land plants close or do not open as much, so little water is absorbed and transpired. Although photosynthesis (and growth) slows as a consequence, the plants still survive short drought periods. They can do so repeatedly. Briefly, such stressful conditions trigger the production of a plant hormone called **abscisic acid**. The hormone is synthesized faster when a leaf is water stressed. When abscisic acid accumulates in a leaf, it somehow causes guard cells to give up potassium ions, hence water, so the stomata close.

UPTAKE AND ACCUMULATION OF MINERALS

Active Transport of Mineral Ions

Water uptake depends only on an osmotic gradient. Yet if a cell in a root (or in any other part of the plant) is to retain water, it must maintain a high solute concentration so that an osmotic gradient can be produced. This means that a cell must expend energy to actively accumulate solutes, particularly dissolved mineral ions. Without energy outlays, diffusion would equalize the solute concentrations on both sides of a plasma membrane. Energy from ATP drives the membrane pumps involved in active transport. These pumps are mem-

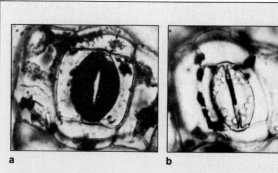

Figure 20.8 Evidence for potassium accumulation in stomatal guard cells undergoing expansion. Strips from the leaf epidermis of a dayflower (*Commelina communis*) were immersed in solutions containing dark-staining substances that bind preferentially with potassium ions. (**a**) In leaf samples having opened stomata, most of the potassium was concentrated in the guard cells. (**b**) In leaf samples having closed stomata, very little potassium was in guard cells; most was present in normal epidermal cells.

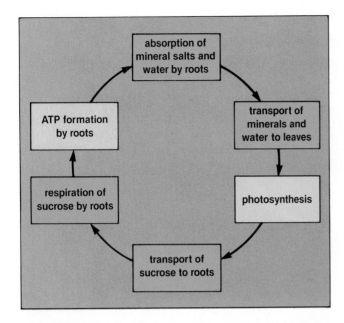

Figure 20.9 Interrelated processes that influence the coordinated growth of roots, stems, and leaves. When one process is rapid, the others also speed up. Any environmental factor limiting one process eventually slows growth of all plant parts.

brane-bound proteins that move substances into the cell even against a concentration gradient (page 93).

In photosynthetic cells, ATP necessary for the membrane pump operation is formed during both photosynthesis and aerobic respiration. What about nonphotosynthetic cells, such as those in roots? How do they get all the ATP necessary for active transport? Here, ATP is formed almost entirely through aerobic respiration in the mitochondria of individual cells.

Controls Over Ion Absorption

Solute absorption and accumulation must be coordinated throughout the plant. For example, root cells receive sugars (commonly sucrose) from leaves, especially when photosynthesis is rapid during the daytime. These cells absorb oxygen from air in the soil (unless the soil is waterlogged). When the soil is moist enough, dissolved ions move rapidly to roots, where some ions are actively transported from cell to cell into the xylem. When the soil is dry, more air is present and more oxygen can be absorbed. But insufficient water limits growth and ion absorption no matter how much oxygen is present. Also, dry soil causes stomata to close partly or completely. Thus, leaves absorb less carbon dioxide when water and ions are in short supply, and photosynthesis and growth slow down. Then, leaf cells cannot send as much sucrose

to roots. Without enough sucrose, aerobic respiration slows down in roots, and so does ion absorption.

Figure 20.9 summarizes part of what is known about coordinated growth between different plant regions.

Mineral ion absorption and accumulation are coordinated throughout the plant body in ways that have profound influences on growth.

TRANSPORT OF ORGANIC SUBSTANCES IN PHLOEM

So far, we have considered some ways in which plants acquire and distribute raw materials. Let's now consider how organic molecules are distributed through the plant body.

Most of the plant is made of sucrose and related organic compounds formed in the leaves. Certain organic molecules not used by the leaf cells themselves are transported to buds, roots, stems, flowers, and fruits, which also require these molecules for growth. Many of the organic molecules are also stockpiled. Starch is the main storage form of carbohydrates in most plants. Fat stores are especially prevalent in many seeds and some fruits, such as the avocado. Proteins are stored in granules in many seeds and grains.

Now, starch molecules are too large to cross cell membranes; they cannot leave the cells in which they are formed. Even if they could, they are too insoluble for transport in water to other regions of the plant body. Fats are largely insoluble in water, and they cannot be transported out of their storage sites. Storage proteins do not lend themselves to transport, either.

The energy and building blocks inherent in starch, fats, and proteins are made available through different chemical reactions. For example, starch molecules are first hydrolyzed, and in most plants the glucose units released are combined with fructose. The resulting molecule is sucrose, which is soluble and transportable. Similarly, proteins are converted to soluble amino acids and amides.

Transport of organic substances through the plant requires that storage starch, fats, and proteins be converted to smaller subunits that are soluble and transportable.

Translocation

How do soluble organic molecules travel from photosynthesis sites or storage sites to organs that require them? Here we must turn to a process known as **trans-**

location. In botany, the word can mean the relatively long-distance transport of water or solutes, but most often it is used to signify the transport of sucrose and other compounds through phloem. As we have seen, sieve tube members of the phloem are joined into long, interconnecting pipelines (Figure 19.9). The pipelines lie side by side in overlapping array within vascular bundles, and they extend from leaf to root. Unlike water-conducting xylem cells, sieve tube members are alive at maturity. Water and organic molecules are transported rapidly through their numerous wall perforations at rates up to 100 centimeters an hour.

Interestingly, the feeding habits of small insects called aphids tell us something about translocation. An aphid feeds on leaves and stems. It forces a mouthpart (stylet) into sieve tube cells, which contain dissolved sugars and other organic compounds. The contents of the cells are under high pressure, often five times as much as in an automobile tire. This pressure seems to force the fluid right through the aphid gut and clear out the other end as "honeydew" (Figure 20.10). Park your car under trees being attacked by aphids and it might get a spattering of sticky honeydew droplets, thanks to sieve-tube pressures.

In some experiments, feeding aphids were anesthetized with carbon dioxide. Then their bodies were severed from their stylets, which were left embedded in the sieve tubes that the aphids had been attacking. Analysis of the fluid being forced out of the tubes verified that sucrose is the main carbohydrate being transported through the plant body in most species.

Pressure Flow Theory

The movement of organic molecules in phloem follows a "source-to-sink" pattern. The main *source regions* are the sites of photosynthesis in leaves. A *sink region* is any plant part that depends on inputs of organic molecules to meet its nutritional needs or that stockpiles organic molecules for later use. Growing leaves, fruits, seeds, and roots are examples of sink regions. (Although storage tissues are sinks when they receive organic molecules, they also serve as sources when the stores are tapped.)

Obviously, a plant doesn't "know" which way it should be translocating organic molecules through the phloem. But what causes the directional movement? According to the **pressure flow theory**, translocation through the phloem depends on pressure gradients between source and sink regions.

To understand the points of the pressure flow theory, let's begin with what happens when sucrose and other organic molecules move into a sieve tube in a leaf. As solute concentrations increase in this source region, the water potential decreases. Water potential, recall, is the

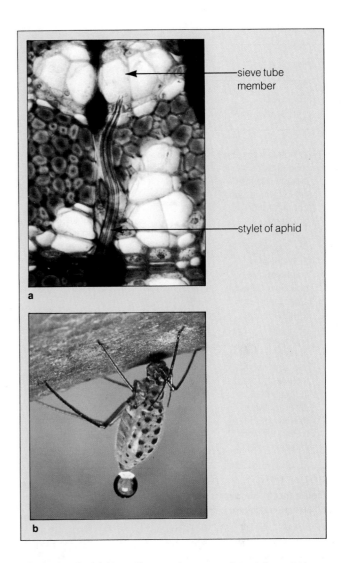

Figure 20.10 (**a**) Naturally occurring research tool: the aphid stylet, here penetrating a sieve tube member. (**b**) Honeydew droplet at the tail end of a well-fed aphid.

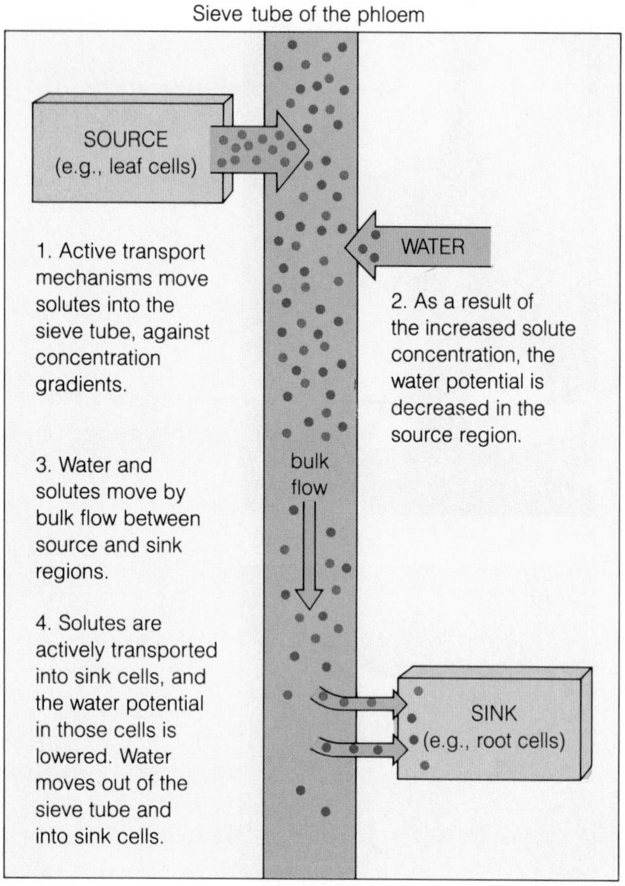

Sieve tube of the phloem

SOURCE (e.g., leaf cells)

1. Active transport mechanisms move solutes into the sieve tube, against concentration gradients.

WATER

2. As a result of the increased solute concentration, the water potential is decreased in the source region.

bulk flow

3. Water and solutes move by bulk flow between source and sink regions.

4. Solutes are actively transported into sink cells, and the water potential in those cells is lowered. Water moves out of the sieve tube and into sink cells.

SINK (e.g., root cells)

Figure 20.11 Proposed mechanism of pressure flow in the phloem of vascular plants.

sum of two opposing forces: water movement into cells by osmosis (when internal solute concentrations increase), and an outward-directed movement of water when turgor pressure increases. The resulting increase in local turgor pressure causes water to flow to regions in the tube system where pressure is lower (Figure 20.11).

Sieve tubes play a passive role in the bulk transport of organic molecules from source to sink regions. These molecules are loaded into the tubes by active transport mechanisms. The energy required for active transport is expended by companion cells alongside the sieve tube members or by neighboring parenchyma cells (page 244).

What maintains the low pressure at sink regions? There, organic molecules are unloaded from the tubes and actively transported into sink cells (which use them in wall building, starch formation, aerobic respiration, and other activities). The water potential in the sink cells is lowered when the solutes move in, so water also moves in by osmosis.

In growing tissues where cells are still enlarging, the expanding cell walls prevent an increase in turgor pressure (which would counter the inward water movement). As water continues to enter the sink cells, the dilution of solutes allows even more solutes to enter rapidly—and the cells continue to grow.

SUMMARY

1. Plants have a large surface-to-volume ratio that favors the uptake of water and nutrients, which often are present in relatively low concentrations in their surroundings.

   a. A central vacuole increases the volume of each cell (hence of the plant), thereby increasing the absorptive surface area.

   b. Leaves are generally thin and broad, with a large surface area for sunlight interception and carbon dioxide uptake.

   c. Root structure and patterns of growth enhance absorption of water and dissolved minerals from a large volume of soil.

   d. Vascular systems conduct water, minerals, and organic molecules through the rather long distances between shoots and roots.

2. Plants build their own organic compounds with oxygen, carbon, and hydrogen (which together make up about ninety-six percent of the plant's dry weight) and with thirteen mineral ions. Water provides oxygen and hydrogen; the air provides gaseous oxygen and carbon dioxide.

3. The mineral ions serve as macronutrients (each being at least 0.1 percent of the total dry weight) or micronutrients (present in trace amounts). They are ionized forms of nitrogen, potassium, calcium, and so on that function in metabolic activities (including enzyme activation) and in establishing solute concentration gradients across the plasma membranes of the plant cells. Such gradients are the basis for water movement into the cells, hence for maintaining cell shape and growth.

4. Water and solute absorption is enhanced by mycorrhizae, mutually beneficial associations between young roots and fungi (which form an extensive mat of thin absorptive filaments around the root). The fungus provides the plant with absorbed mineral ions; the root provides the fungus with some sugars and nitrogen-containing compounds.

5. In the water-conducting system (xylem) of plants, continuous negative pressures (tensions) extend down from leaves to roots. The tension is caused by transpi-

ration (the evaporation of water from leaves and other plant parts exposed to air).

a. When water molecules vacate transpiration sites, replacements are pulled under tension to the site.

b. The cumulative strength of hydrogen bonds between water molecules (which are confined in the narrow, tubular cells of xylem) allows continuous "columns" of water to be pulled up. This concept is the cohesion theory of water transport.

6. A waxy cuticle (which retards water loss) covers most aerial plant parts. Thus transpiration and carbon dioxide uptake occur mostly at stomata, small openings in the epidermis of leaves and stems.

7. Physiological mechanisms cause stomata to open during the day, and this enhances carbon dioxide uptake (hence photosynthesis), although the "cost" is evaporative water loss. However, the mechanisms also cause stomata to close at night, when photosynthesis shuts down, so water loss via transpiration is reduced. The mechanisms involve the movement of potassium ions into and out of the two guard cells flanking each stoma, leading to changes in cell shape (and in the opening between them).

8. Dissolved mineral ions enter cells by active transport, which requires ATP energy. Aerobic respiration in the roots provides the ATP but in turn depends on an adequate supply of sucrose from the leaves.

9. Sucrose and other compounds are translocated to roots and other plant parts through the vascular system called the phloem. These nutrients support growth; some are converted to storage forms (starch, proteins, or fats).

10. Translocation can be explained by the pressure flow theory:

a. Translocation of organic compounds through plants occurs in sieve tube members (living cells in the phloem).

b. Translocation is driven by differences in water pressure between source regions (photosynthesis sites or storage organs) and sink regions (any metabolically active or growing tissue).

c. Organic compounds are actively transported into sieve tube members in source regions. Water potential thereby decreases in those local regions, and the water and solutes move by bulk flow to tube regions of lower pressure.

d. In sink regions, organic compounds are unloaded by active transport mechanisms that move them into individual cells.

e. The movement of organic compounds out of the phloem and into sink cells lowers the water potential in sink cells. Thus water moves out of the phloem and into the sink cells. Cell growth also can lower the pressure

(hence the water potential) in sink cells and cause more water to move out of the sieve tubes. In such ways, low-pressure regions are maintained in sink regions.

Review Questions

1. Give examples of the features that enable land plants to absorb water and nutrients from their surroundings, which have dilute concentrations of these required substances.

2. Which three elements make up most of the dry weight of a land plant? Which six elements are considered macronutrients for land plants?

3. Describe some of the specific roles that mineral ions play in plant functioning. How do solute absorption and accumulation affect plant growth?

4. Define mycorrhiza. Why is it important to include some of the native soil around roots when transplanting a plant from one place to another?

5. Describe transpiration. State how the cohesion theory of water transport helps explain what is going on in this form of water movement.

6. Transpiration competes with other water-requiring cell processes. Can you name some of these processess?

7. Look at Figure 20.9. Then, on your own, diagram the feedback relations that influence the coordinated growth of stems, roots, and leaves.

8. Sucrose transport from one plant organ to another is called translocation. Can you explain how it works in terms of the four key points of the pressure flow theory? How did aphids help show that sucrose is indeed the main substance being transported through the phloem pipelines?

Readings

Apfel, R. 1972. "The Tensile Strength of Liquids." *Scientific American* 227(6):58–71. A fairly simple treatment of theory and experiments providing evidence that liquids can exist under tension as well as pressure.

Epstein, E. 1973. "Roots." *Scientific American* 228(5):48–58.

Galston, A., P. Davies, and R. Satter. 1980. *The Life of a Green Plant.* Englewood Cliffs, New Jersey: Prentice-Hall. A simplified treatment of much of plant physiology.

Hewitt, E., and T. Smith. 1975. *Plant Mineral Nutrition.* New York: Wiley. Techniques and results in studies of plant mineral nutrition.

Mengel, K., and E. Kirkley. 1982. *Principles of Plant Nutrition.* Worblaufer-Bern, Switzerland: International Potash Institute.

Peel, A. 1974. *Transport of Nutrients in Plants.* New York: Wiley. A short, simple treatment of transport processes occurring in xylem and phloem.

Salisbury, F., and C. Ross. 1985. *Plant Physiology.* Third edition. Belmont, California: Wadsworth. Excellent, comprehensive book covering most plant functions.

Torrey, J., and D. Clarkson (editors). 1975. *The Development and Function of Roots.* New York: Academic Press. One of the few modern treatments of root structure and function.

21

PLANT REPRODUCTION AND EMBRYONIC DEVELOPMENT

Although it probably is not something you think about very often, flowering plants engage in sex. They produce sperm and egg cells, as humans do. They, too, have elaborate reproductive systems that protect and nourish sex cells during their formation. As in human females, the female organs of flowering plants house the embryo during its early development. Those exquisite forms called flowers are, in effect, exclusive or open invitations to third parties—pollinators—that function in getting sperm and egg together (Figure 21.1). Long before humans ever thought of it, flowering plants were using tantalizing colors and fragrances in improving the odds for sexual success.

Plants also do something that humans cannot do (at least not yet). They can reproduce asexually. Because asexual reproduction occurs by way of mitosis only, all individuals of the new generation are genetically identical copies of the parent plant. (In contrast, sexual reproduction requires the formation of gametes followed by fertilization. Thus two different sets of genetic instructions, from two gametes, are present in the fertilized egg.) Both sexual and asexual reproduction occur in plant life cycles, and the reproductive details vary from group to group. We will look at these variations in Chapter Forty, which is an evolutionary survey of plant diversity. Here, we will begin with sexual reproduction and embryonic development among angiosperms—the flowering plants.

Figure 21.1 Reproductive shoots called flowers. Many of these exquisite shoots have coevolved with animals, such as this hummingbird, which play indirect but vital roles in the reproductive process.

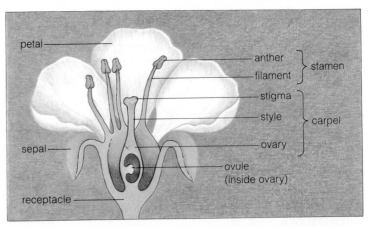

Figure 21.2 Common arrangement of floral appendages. Shown here, a cherry (*Prunus*) flower, with a single carpel.

SEXUAL REPRODUCTION OF FLOWERING PLANTS

Life Cycles of Flowering Plants

What we usually think of as "the plant" is the **sporophyte**, a vegetative body composed of roots, stems, and leaves in which all cells are diploid (or polyploid). Roses, cactus plants, and elm trees are all sporophytes. At some point in the life cycle a sporophyte produces **flowers**, which are reproductive shoots (Figure 21.2). Within the tissues of the flower, some cells divide by meiosis and form the haploid cells called meiospores. (Sporophyte means "spore-producing plant.") Each meiospore divides by mitosis, thereby forming a multicelled, haploid gametophyte. A **gametophyte** is a "gamete-producing body." The ones that produce sperm are called male gametophytes; the ones that produce eggs are called female gametophytes.

Thus the life cycles of flowering plants show *alternation of generations*, in which multicelled diploid bodies alternate with multicelled haploid bodies (Figure 21.3).

Keep in mind that the female gametophytes are not free-living plants. In most species, they are tiny multicelled bodies embedded within floral tissues. Although male gametophytes are released from the flower (as two-celled pollen grains), they are not free-living either. They are more like shipping crates for the sperm-producing cell until they actually land on a female flower part.

The sporophyte, or vegetative body, dominates flowering plant life cycles. For the most part, haploid gametophytes play out their roles inconspicuously, within the specialized sporophyte shoots called flowers.

Floral Structure

A floral shoot has whorls or spirals of specialized structures called carpels, stamens, petals, and sepals. Most or all of the appendages are attached to the *receptacle*, the modified end of the floral shoot.

A **carpel** is a female structure. In this closed vessel, eggs develop, fertilization takes place, and seeds mature. Some flowers have only one carpel; others have several carpels, either separate or fused together.

Figure 21.2 shows the single carpel of a cherry blossom sliced lengthwise. Notice the *ovary*. In this chamber, the structure called an ovule develops into the cherry seed. (Other ovaries contain more than one ovule. For example, a pea pod is a mature ovary in which several ovules have developed into several "peas," or seeds.) Above the ovary is the *stigma*, a landing platform for pollen. In this plant the ovary and the stigma are sep-

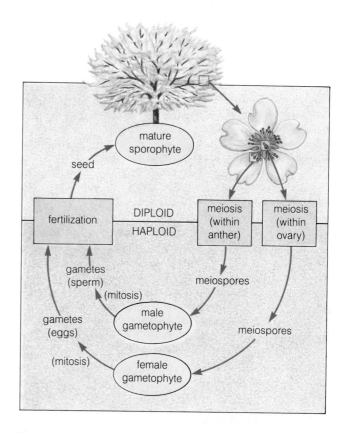

Figure 21.3 Alternation of generations in the life cycle of flowering plants. The sporophyte and the gametophytes are multicelled bodies, produced by way of mitosis.

arated by a *style*, a region in which the carpel narrows into a slender column.

The **stamens** shown in Figure 21.2 are male reproductive structures. Most commonly, a stamen consists of a *filament* (a slender stalk) capped by an anther. An *anther* contains chambers called pollen sacs, in which pollen grains develop.

The flower just described is a "perfect" flower, having stamens and one or more carpels (it produces both sperm and eggs). In contrast, "imperfect" flowers have stamens *or* carpels, but not both. Often they are called male and female flowers. In some species, such as oaks, male and female flowers appear on the same plant. In other species, such as willows and American holly, they occur on separate plants.

Most flowers also have petals and sepals. **Petals** are attached to the receptacle, below the male and female flower parts, and collectively form a "corolla." Often the color, patterning, and shapes of corollas function in attracting bees and other pollinators. The leaflike, often green **sepals** form the "calyx," the outermost whorl of appendages. Commonly, the calyx encloses all other flower parts (as it does in rosebuds) before a bud opens.

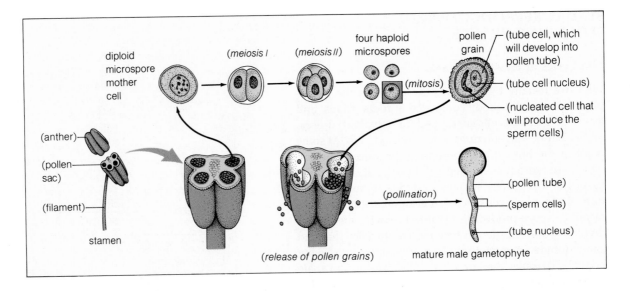

Figure 21.4 Stages in the development of a male gametophyte, beginning with microspore production in the anther. A germinated pollen grain, with its pollen tube and two sperm cells, constitutes the mature male gametophyte.

GAMETE FORMATION

Microspores to Pollen Grains

Let's now turn to reproductive events that typically unfold in a flower. While an anther is growing inside a flower bud, mitotic divisions produce four masses of diploid cells. Each of these so-called mother cells is destined to give rise to four haploid *microspores* (which is the name given to the meiospores produced in anthers). Each cell mass becomes surrounded by several layers of cells that form a **pollen sac**. In this walled chamber, pollen grains develop (Figure 21.4).

Each microspore in a pollen sac divides once, and the two resulting cells stay together as a pollen grain.

One cell will eventually produce the sperm. The other cell will develop into a *pollen tube*, an elongate structure that will grow through carpel tissues and thereby transport sperm to an ovule. Both cells are surrounded by a tough wall (Figure 21.5). This wall will protect the pollen grain from drying out during its journey from the anther to a stigma.

Pollen grains develop from haploid microspores, then later develop into sperm-bearing gametophytes.

Megaspores to Eggs

Meanwhile, in the carpel of the flower, one or more dome-shaped cell masses have been developing on the inner wall of the ovary. In some carpels, only one mass develops; in others, hundreds or thousands form. Each mass is the start of an ovule, which, if all goes well, is destined to become a seed.

As the mass grows, some of its cells form a stalk. The rest develop into an inner *nucellus* and one or two protective layers called *integuments*:

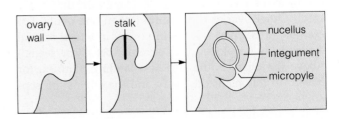

Figure 21.5 Scanning electron micrographs of pollen grains from (**a**) day lily, 260× and (**b**) ragweed, 625×.

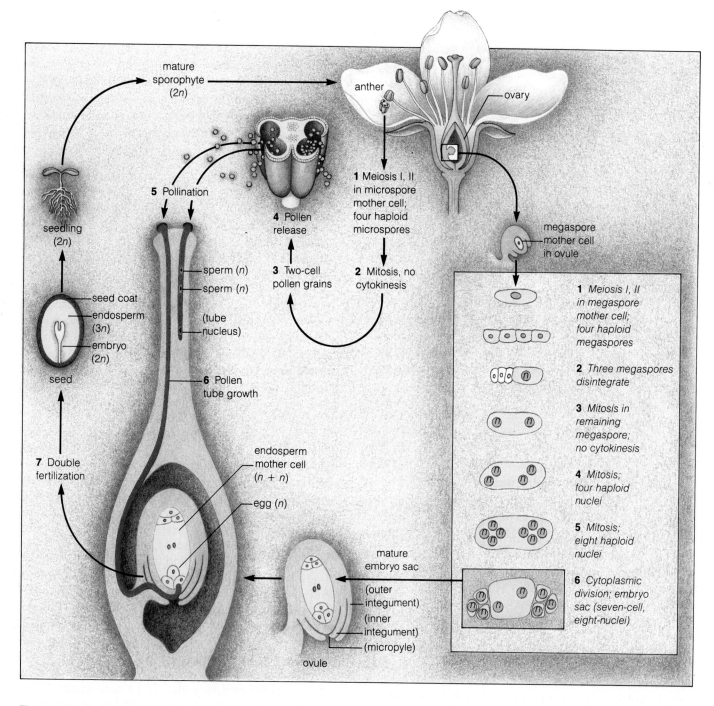

Figure 21.6 Generalized life cycle of a flowering plant, showing details of embryo sac formation within the ovary. Compare this illustration with Figure 21.4.

In the nucellus a diploid mother cell gives rise to haploid *megaspores* (which is the name given to the meiospores produced in carpels). Only a tiny part of the nucellus does not become covered with integuments. Most commonly this tiny gap (the micropyle) is where a pollen tube will penetrate into the ovule.

In most flowering plants, three of the four megaspores disintegrate. The one remaining undergoes mito-

sis three times *without* cytoplasmic division, so at first it is a single cell with eight nuclei (Figure 21.6). Each nucleus migrates to a specific location in the cell, then cytoplasmic division occurs. The result is a female gametophyte, or **embryo sac**, that consists of seven cells. One of the seven cells, the "endosperm mother cell," contains two haploid nuclei. It will help give rise to *endosperm*, a nutritive tissue that will surround the forthcoming embryo. Another cell, which ends up near the micropyle, is the egg.

A female gametophyte (embryo sac) is a seven-celled, eight-nucleate body. One cell of the sac is the egg. Another cell has two nuclei and will help give rise to endosperm, a nutritive tissue.

POLLINATION AND FERTILIZATION

Pollination and Pollen Tube Growth

The word **pollination** refers to the transfer of pollen grains to the surface of a stigma (Figure 21.6). A pollen grain germinates (resumes growth) once it has been deposited on the proper stigma. One cell of the pollen grain starts growing as the pollen tube. The other cell divides, and two sperm cells form before or during tube growth. The pollen tube penetrates the stigma, style, and ovarian tissues on its journey to an ovule. The germinated pollen grain, with its tube nucleus and two sperm, constitutes the mature male gametophyte.

Fertilization and Endosperm Formation

Upon reaching the ovarian chambers, a pollen tube grows toward an ovule. Typically it enters the micropyle and penetrates the embryo sac, then the tip of the pollen tube ruptures and the sperm are released.

"Fertilization" generally means the fusion of one sperm nucleus with one egg nucleus. In flowering plants, **double fertilization** occurs. One of the two sperm nuclei fuses with that of the egg, and a *diploid* zygote (2n) forms. This marks the beginning of a new sporophyte. Meanwhile, the other sperm nucleus and both nuclei of the endosperm mother cell all fuse together, forming a *triploid* (3n) nucleus. This marks the formation of the primary endosperm cell, which will give rise to tissues that will nourish the seedling until leaves form and photosynthesis is under way.

In double fertilization, one sperm nucleus fuses with the egg nucleus, and a diploid zygote results. The other sperm nucleus fuses with the two nuclei of the endosperm mother cell, which will give rise to triploid (3n) nutritive tissue.

Before looking at how the sporophyte initially develops, let's digress for a moment and consider the relationship between floral structure and the pollinators necessary for a new generation to begin. It is one of the most intriguing of all evolutionary stories.

Case Study: Coevolution of Flowering Plants and Pollinators

Origins of Pollination Vectors. Diverse flowering plants live almost everywhere on land, from snow-covered flanks of mountains to low deserts. Many species live in freshwater or saltwater environments. How did this diversity and distribution come about, given that plants (unlike animals) cannot just pick up and move off to new places?

For the answer, we must go back 400 million years or so, to the time when plants began their invasion of the land. When the pioneers took hold, mud-dwelling insects that could scavenge on moist, decaying plant parts and tiny spores were probably right behind them. With the evolution of stronger stems and taller plants, some of the insect scavengers underwent structural modifications that allowed them to leave the moist litter on the ground and withstand exposure to air. The initial absence of competition for edible but aerial plant parts would have favored such modifications. Among other things, sucking, piercing, and chewing mouthparts emerged in some insect lineages. And some of the descendants of those insects developed wings.

The first seed-bearing plants probably made their entrance in the warm, moderately humid coastal forests of the Devonian, some 350 million years ago. These plants gave rise to gymnosperms and, presumably, flowering plants (page 559). Ovules and pollen sacs were borne on modified leaves or scales; often the scales were arranged in conelike structures. The mode of pollination must have been passive indeed, with pollen grains simply drifting on air currents to the ovules.

Now, pollen grains are rich sources of protein. If insects began zeroing in on the pollen at its predictable locations in cones, then they would have begun serving as pollination vectors. (A *vector* is an outside agent that acts like a bridge between two reproductive structures or reproductive stages.) Some of the dustlike pollen would be eaten, but some would cling to the insect body and be transported to ovules. To be sure, insects clambering about reproductive cones would assist pollination in an inadvertent way—but the assistance would be more efficient than that provided by air currents alone. Pollen would be delivered right to the door, so to speak.

The tastier the pollen, the more home deliveries, and the more seeds formed. The greater the number of seeds formed, the greater the reproductive success. What we are describing here is an example of **coevolution**: the joint evolution of two (or more) species interacting in close ecological fashion. When one species evolves, the change affects selection pressures operating between the two species and the other also evolves. Thus, in this case, there was natural selection of variant plants able to attract beneficial insects. Concurrently, there was selection of the vector insects which, because of their ability to recognize a particular food and locate it quickly, would outcompete other foraging insects.

Another, perhaps related change should be mentioned here. Existing pollen-eating beetles have strong mouthparts, and many also chew on ovules that they pollinate. We can speculate that chewing behavior was a selective force contributing to the evolution of floral structure. Instead of being borne naked and vulnerable on cone scales, as they are in gymnosperms, the ovules of today's flowering plants are sequestered inside closed carpels—which offer some protection from hungry insects.

Nectar as an Attractant. Nectaries and their specialized visitors are another example of coevolution. A **nectary** is a glandlike organ that secretes *nectar*, a fluid rich in sugar, amino acids, proteins, and lipids. Before flowering plants emerged, insects were indirectly tapping into nectarlike solutions. The aphid, recall, is a tiny insect that inserts its strawlike mouthpart into phloem (Figure

Table 21.1	Correlation Between Characteristics of Flowering Plants and Their Pollinators
Characteristic of Flowering Plant	Related Characteristic of Pollinator
Floral color Color patterning Floral odor	Sensory apparatus (photoreceptors, olfactory receptors, etc.)
Flower size Flower shape Structure and location of reproductive parts	Size of body Shape of body parts (such as feeding apparatus)
Nectar composition	Diet
Timing of flowering	Foraging behavior (by day, by night, by season)

Figure 21.7 Flowers with red and yellow components, colors that attract bird pollinators. (**a**) Flower of the bird-of-paradise, *Strelitzia reginae*. (**b**) Glorybower, a favorite of hummingbirds. (**c**) Saguaro cactus blossoms, a favorite of gila woodpeckers. (The white component of these blossoms also attracts night-flying bats.)

20.10). The sugary solution transported by phloem is under so much pressure that it courses through the insect body, with the excess exuded from the tail end as a "honeydew" droplet. Ants are so fond of honeydew that they tend "herds" of aphids, even to the extent of protecting them from predators such as wasps. At the same time, some ants protect the plant by attacking insects that chew on it.

Between 100 million and 65 million years ago, the diversity among flowering plants increased dramatically. Fossil ants date from that period. Were those ancient ants tending aphids and protecting plants even then? We might well speculate that nectar coevolved with ben-

eficial insects. Whatever the case, between 60 million and 40 million years ago, nectar became varied in composition and attracted a variety of pollinators with different nutritional needs—insects, then birds and bats.

Existing Flowers and Their Pollinators. Today it is possible to correlate many characteristics of flowering plants with those of their pollination vectors (Table 21.1). As an example of these correlations, consider the glory-bower and similar flowers in which some of the petals form a tube (Figure 21.7). The nectary is deep in the flower, at the base of the tube, and typically it contains copious amounts of nectar. The petals themselves are large and brightly colored, often with red components. Typically these flowers have no odor to speak of.

Could such flowers visually attract insects such as beetles and honeybees? Probably not, because these insects cannot detect red wavelengths. Also, although beetles and honeybees are attracted to strong odors, the flowers offer them no olfactory guides. Even if these insects accidentally scrambled over the flower, there would be no easy way for them to reach the nectary, short of navigating the floral tube, and there is a distinct possibility of drowning in some of the larger nectar cups.

If not insects, then what serves as the pollinators? Birds. Birds have a keen sense of vision, and they can detect red wavelengths—and red petals are like bright flags in the distance for these pollinators. Birds also have a poor sense of smell—so floral odors would be wasted on them. Some birds have long beaks that can penetrate the length of the floral tube, and anthers and stigmas are located where the bird head will brush against them while they drink from the nectary (Figure 21.7). Birds

Figure 21.8 Nectar guide of the marsh marigold (*Caltha palustris*), which is visible to bees. Unlike the human eye, the bee eye can detect ultraviolet light. Although the petals of this flower appear solid yellow to us, its distinctive markings become apparent when the flower is experimentally illuminated with ultraviolet light.

a   b   c

Figure 21.9 Coevolutionary match between a flower and its pollinator. The color of Scotch broom attracts bees. Some of the petals serve as a landing platform that corresponds to the size and shape of the bee body. The pollinator's weight on the platform forces petals apart. The pollen-laden stamens, which are positioned to strike against the bee, are thereby released. The pollen brushes against dense hairs covering the bee body.

Periodically, the bee grooms itself, packing pollen for its trip back to the hive. The orange-colored mass of pollen visible in (**b**) has been packed in a pollen basket, formed by stiff hairs on the outer hind leg. Pollen baskets occur on honeybees, bumblebees, and their kin.

require considerable energy to power their flights, and they depend on more than one cup of nectar to sustain their high metabolic rate. Thus birds visit many plants of the same species—and in so doing they promote cross-pollination.

In contrast, if a tiny beetle were equipped to locate a nectar cup, it probably never would leave the flower, let alone the plant, and cross-pollination would be thwarted. Thus, some flowers pollinated by beetles (and flies) have what we perceive as strong and awful odors, reminiscent of decaying meat, moist dung, and the like. Perhaps olfactory guides of this sort originally resembled the smells of decaying matter in the forest litter, where beetles originally evolved.

Bees, too, are attracted to strong odors, but for them the odors are sweet. Also, bee eyes are sensitive to wavelengths of yellow, blue, purple, and ultraviolet, but not red. Most of the flowers they visit have bright yellow, blue, purple, or ultraviolet components—but not red (Figure 21.8). Most bees are covered with plumelike hairs that retain pollen from the flowers being visited. Some bee-pollinated flowers have landing platforms that actually position the bee body in the most favorable way for brushing against pollen-laden anthers (Figure 21.9).

What about butterflies and moths? Butterflies forage by day and are commonly attracted to flowers that are sweet-smelling, often red, and upright, with a more or less horizontal landing platform. Most moths forage by night. They pollinate flowers with strong, sweet fragrances and white or pale-colored petals, which are more visible in the dark (Figure 21.10). Butterflies and moths both have long, narrow mouthparts that correspond to narrow floral tubes or spurs. The most extreme example is the Madagascar hawkmoth. When uncoiled, its proboscis is twenty-two centimeters long—the exact same length as the floral tube of the orchid *Angraecum sesquipedale!* Like other hawkmoths, this moth does not require a landing platform; it hovers in front of the floral tube.

Winds alone disperse the pollen of some flowers, such as the flowers of diverse grass species. These flowers have little or no color or odor; they have no nectar at all. Many do not even have petals. Pollen of wind-pollinated plants never gets carried much beyond a hundred meters from its release point. These plants are found in temperate regions of North America, Europe, and eastern Asia. Often, many wind-pollinated trees of the same species are crowded together, and many are deciduous. Wind through the bare branches is effective for cross-pollination among them.

Wind-pollinated plants are relatively scarce in wet tropical regions. In these regions, many diverse plants are so crowded that individuals of the same species may be few and far between. Here, cross-pollination depends on birds, bees, and the like, which travel considerable distances and actually search for particular plants.

a

b

c

Figure 21.10 Nectar guides for insect pollinators. (**a**) Bees use the close-range guide of the passion flower (*Passiflora caerula*). (**b**) The vibrant red and red-orange center of *Arctotis acaulis*, a composite, attracts butterflies. (**c**) Stephanotis, like other night-flowering plants, has no distinctive color pattern. Its white petals and strong scent attract moth pollinators (white or pale colors are more visible at night).

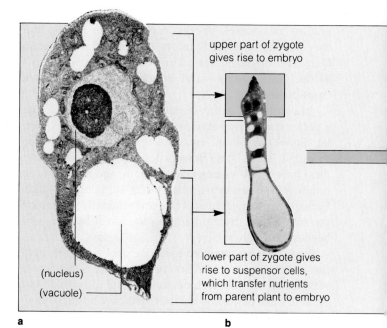

Figure 21.11 A few stages in the development of *Capsella* (shepherd's purse), a dicot. All micrographs are longitudinal section but not to the same scale. (**a**) The zygote, showing the arrangement of cell parts. This arrangement is inherited and leads to differentiation during the first divisions that produce the multicelled embryo, identified by yellow boxes in (**b**) through (**d**) .

upper part of zygote gives rise to embryo

lower part of zygote gives rise to suspensor cells, which transfer nutrients from parent plant to embryo

(nucleus)

(vacuole)

a

b

a

b

Figure 21.12 Two examples of economically important crop plants. (**a**) The seeds of nutmeg (*Myristica fragrans*) are ground to produce one of the most popular spices throughout the world. The spice called mace comes from the fleshy seed coat, which appears as bands of red tissue in this photograph. (**b**) The pods of cacao (*Theobroma cacao*) contain seeds that are processed into chocolate and cocoa.

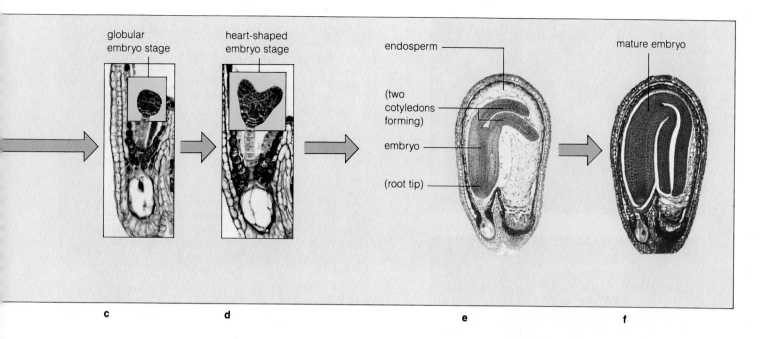

globular embryo stage

heart-shaped embryo stage

endosperm

(two cotyledons forming)

embryo

(root tip)

mature embryo

c   d   e   f

EMBRYONIC DEVELOPMENT

Clearly, the coevolutionary relationships between flowering plants and pollinators are remarkable. As you will see, these plants rely on still other outside agents to assure reproductive success. This part of their story begins with the formation of the zygote, the first cell of the new plant formed through fertilization.

From Zygote to Plant Embryo

When a flowering plant zygote first forms, it is still attached to the parent plant. Even before it begins the mitotic divisions that will produce the multicelled embryo, this single cell already has undergone some development. For example, notice in Figure 21.11 how most organelles, including the nucleus, reside in the top half of a *Capsella* zygote. A vacuole takes up most of the lower half. Once divisions begin, some of the daughter cells give rise only to a simple row of cells (the suspensor) that transfer nutrients from the parent plant to the embryo. Other daughter cells give rise to the mature, multicelled embryo.

Seed and Fruit Formation

The embryo, recall, is housed in an ovule. After double fertilization, the ovule undergoes expansion, integuments harden and thicken, and the endosperm forms. A fully mature ovule is a **seed**; its integuments are the seed coat (Figures 21.12 and 21.13).

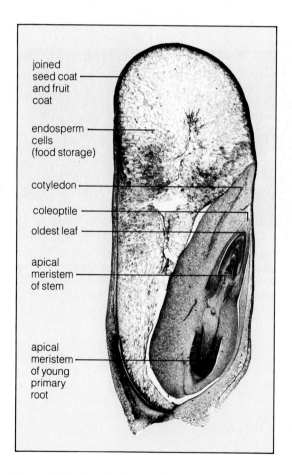

joined seed coat and fruit coat

endosperm cells (food storage)

cotyledon

coleoptile

oldest leaf

apical meristem of stem

apical meristem of young primary root

Figure 21.13 Longitudinal section through a grain of corn (*Zea mays*), a monocot. Most of the volume of this seed is composed of endosperm cells.

Figure 21.14 Fruit formation on an apple tree. (**a**) Petals dropping away from the flower usually signify that fertilization has been successful. After this, the ovary expands (**b**). Sepals and stamens can still be observed on the immature fruit (**c**).

Table 21.2 Kinds of Fruits of Some Flowering Plants

Type	Characteristics	Some Examples
Simple (formed from single carpel, or two or more united carpels of one flower)	1. Fruit wall *dry; split* at maturity	pea, magnolia, tulip, mustard
	2. Fruit wall *dry; intact* at maturity	sunflower, wheat, rice, maple
	3. Fruit wall *fleshy*	grape, banana, lemon
Aggregate (formed from numerous but separate carpels of single flower)	*Aggregate* (cluster) of matured ovaries (fruits), all attached to common receptacle (modified stem end)	strawberry, blackberry, raspberry
Multiple (formed from carpels of several associated flowers)	*Multiple* matured ovaries, grown together into a mass; may include accessory structures (such as receptacle, sepal, and petal bases)	pineapple, fig, mulberry

In some plants, such as *Capsella*, the embryo develops "seed leaves," or cotyledons, which absorb the endosperm and function in food storage (Figure 21.11). In other plants, the cotyledons remain thin but may produce enzymes that aid in transferring stored food from the endosperm to the germinating seedling. For example, endosperm in corn seeds (Figure 21.13) is filled with proteins, fats, oils, and starch.

While an ovule is developing into a seed, the ovary surrounding it is developing into most or all of the structure called a **fruit**. Sometimes the shriveled sepals and stamens are still present on the mature fruit, as they are on apples (Figure 21.14).

We tend to think of fruits as juicy edible structures, as indeed many are. But fruits include any matured or ripened ovary, with or without accessory parts (Table 21.2). Grains and nuts are dry fruits; the fruit wall of a walnut is dry and intact at maturity. Apples and tomatoes are fleshy fruits. A raspberry is an aggregate of many fruits from one flower. Multiple ovaries remain clustered together in a pineapple (Figure 21.15).

All fruits function in seed protection and dispersal in specific environments. Thus, for example, the light-

floral
remnants   ovary

a

seed (in carpel)

wing

b

receptacle

ovary

c

Figure 21.15 (**a**) Pineapple, a multiple fruit; (**b**) maple, a dry fruit; and (**c**) strawberry, an aggregate fruit.

weight, tiny seeds of the orchid fruit can be widely dispersed on air currents when the ovarian walls rupture. A maple fruit has winglike extensions (Figure 21.15). When the fruit drops, the wings cause it to spin sideways. With such spinnings, seeds can be dispersed to new locations, where they will not have to compete with the parent plant for soil water and minerals. Many fruits have hooks, spines, hairs, or sticky surfaces. By such means they adhere to feathers or fur of animals that brush against them—and thereby can be taxied to new locations.

Fleshy fruits such as those of strawberries and cherries are tasty to many animals and are well adapted for moving through the animal gut, which contains powerful digestive enzymes. The enzymes remove just enough of the hard seed coats to enhance the likelihood of successful germination when the indigestible seeds are expelled from the body.

Fruits are specialized structures that protect the seed and enhance its dispersal in specific environments.

ASEXUAL REPRODUCTION OF FLOWERING PLANTS

The sexual reproductive modes just described are the most prevalent among flowering plants. But sporophytes also can be reproduced asexually by several means (Table 21.3). For example, a strawberry plant can send out horizontal aboveground stems, known as *runners*. Along such runners, new roots and shoots develop at every other node. As another example, oranges reproduce every so often by *parthenogenesis*: the development of an embryo from an unfertilized egg. In some plants, parthenogenesis is stimulated when pollen contacts a stigma even though a pollen tube has not grown down the style. Hormones (either formed in the stigma or produced by pollen grains) apparently diffuse down to the unfertilized egg and stimulate embryo formation. The embryos become 2*n* by fusion of products of egg mitosis, or a 2*n* cell outside the gametophyte may be stimulated to form an embryo.

Vegetative reproduction occurs naturally among wounded plants. For example, when a leaf falls or is torn away from a jade plant, a new plant can develop from

Table 21.3 Asexual Reproductive Modes of Flowering Plants

Mechanism	Representative Species	Characteristics
Reproduction on modified stems:		
1. Runner	Strawberry	New plants arise at nodes of an aboveground horizontal stem
2. Rhizome	Bermuda grass	New plants arise at nodes of underground horizontal stem
3. Corm	Gladiolus	New plant arises from axillary bud on short, thick, vertical underground stem
4. Tuber	Potato	New shoots arise from axillary buds on tubers (enlarged tips of slender underground rhizomes)
5. Bulb	Onion, lily	New bulb arises from axillary bud on short underground stem
Parthenogenesis	Orange, rose	Embryo develops without nuclear or cellular fusion (e.g., from unfertilized haploid egg; or adventitiously, from tissue surrounding embryo sac)
Vegetative propagation	Jade plant, African violet	New plant develops from tissue or organ (e.g., a leaf) that drops or is separated from plant
Tissue culture propagation	Carrot, corn, wheat, rice	New plant arises from cell in parent plant that is not irreversibly differentiated; laboratory technique only

meristematic tissue adjacent to vascular bundles just inside the wound. (Meristematic cells, recall, retain the potential for continued division.)

Vegetative propagation also occurs with a little help from humans. For example, a whole orchard of individual pear trees might have been grown from cuttings or buds of a parent tree. Obviously, the "offspring" have the same DNA instructions as the parent. Any organism reproduced asexually in such a way that it is genetically identical with its parent is a **clone**. Many of our food crops, including Bartlett pears, MacIntosh apples, and Thompson seedless grapes, are clonal populations.

In some pioneering experiments, Frederick Steward and his coworkers propagated carrot plants by culturing cells from differentiated carrot tissues. Such induced vegetative propagations are also accomplished with shoot tips or other parts of individual plants. The techniques are particularly useful when an advantageous mutant arises. For example, one such mutant might show resistance to a disease that is particularly crippling to wild-type plants of the same species. Tissue culture propagation can lead to hundreds and even thousands of identical plants from one mutant specimen. This technique is already being used in efforts to improve major food crops, such as corn, wheat, rice, and peas.

SUMMARY

1. In flowering plant life cycles, a multicelled diploid or polyploid stage (the sporophyte, or "spore-producing plant") alternates with a multicelled haploid stage (the gametophyte, or "gamete-producing plant").

2. The sporophyte is a vegetative body with roots, stems, leaves and, at some point, flowers. Certain cells in the floral tissues divide by meiosis and form meiospores (haploid cells that will divide by mitosis to form the gametophytes).

3. A flower consists of one or more carpels (female reproductive structures), stamens (male reproductive structures), petals, and sepals, most or all of which are attached to the receptacle.

   a. Each stamen has a slender stalk (filament) capped by an anther. An anther is a structure containing pollen sacs, or chambers in which pollen grains develop. A pollen grain is a male gametophyte.

   b. Each carpel includes an ovary and a stigma. The carpel is a closed vessel in which eggs develop, fertilization takes place, and seeds mature. The stigma is the landing platform for pollen.

4. "Microspores" are the meiospores produced in pollen sacs of anthers. Each divides once, producing a two-celled pollen grain. One cell produces the sperm; the

other develops into a pollen tube that grows through carpel tissues, transporting the sperm with it.

5. "Megaspores" are the meiospores produced in carpels. They form inside the nucellus, a tissue mass that grows from the inner wall of the ovary and becomes covered with one or two protective layers (integuments). The nucellus, integuments, and the stalk attaching them to the ovarian wall collectively are called an ovule.

6. Of every four megaspores produced by meiosis, all but one usually disintegrate. The one remaining undergoes mitosis three times without cytoplasmic division, the nuclei migrate to specified positions, then cytokinesis produces the female gametophyte (embryo sac). It is a seven-celled, eight-nuclei body, with one cell being the egg. The cell with two nuclei (endosperm mother cell) will help give rise to endosperm, a nutritive tissue.

7. Upon pollination (transfer of pollen grains to an appropriate stigma), the pollen grain germinates, the pollen tube develops and, later, two sperm cells form. The germinated pollen grain (with its two sperm) is the mature male gametophyte.

8. Many flowering plants coevolved with insects, birds, and bats, which serve as pollination vectors (outside agents that act like bridges between male and female reproductive structures of flowers).

  a. Coevolution is the joint evolution of two or more species interacting in close ecological fashion; when one evolves, the change affects selection pressures operating between them and the other species also evolves.

  b. Thus the color, patterning, and odor of flowers can be correlated with the sensory apparatus (such as photoreceptors) of certain pollinators. Floral size and shape, and the structure and location of reproductive parts, can be correlated with the size and shape of the pollinator's body parts (such as mouthparts).

9. At double fertilization, one sperm nucleus fuses with one egg nucleus to form a diploid (2n) zygote. The other sperm nucleus and both nuclei of the endosperm mother cell fuse to form a cell that will give rise to triploid (3n) nutritive tissue in the forthcoming seed.

10. After double fertilization, the ovule expands, the integuments harden and thicken, and endosperm forms. A fully mature ovule is a "seed," with its integuments forming the seed coat.

11. Simultaneously, the ovary undergoes development into the fruit. Mature fruits can be fleshy or dry; they can be simple, aggregate, or multiple structures.

12. Fruits function to protect and disperse the seeds. Fleshy fruits attract animals and other "dispersing" agents; lightweight fruits can be dispersed by winds. Some fruits have hooks and such that attach to animals.

13. The key characteristics of asexual reproduction of flowering plants may be summarized this way:

  a. New plants may arise asexually by mitotic divisions at nodes or buds along modified stems of the parent plant.

  b. Some plants reproduce. asexually by parthenogenesis (development of an embryo from an unfertilized egg).

  c. New plants also may arise by vegetative propagation, either natural or induced.

Review Questions

1. Label the floral parts in the following diagram. Explain floral function by relating some floral structures to events in the life cycle of flowering plants:

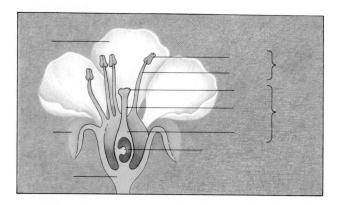

2. Distinguish between these terms:
   a. Megaspore and microspore
   b. Pollination and fertilization
   c. Pollen grain and pollen tube
   d. Ovule and female gametophyte

3. Observe the kinds of flowers growing in the area where you live. On the basis of what you have read about the likely coevolutionary links between flowering plants and their pollinators, can you perceive what kinds of pollinating agents your floral neighbors might depend upon?

4. Describe the steps involved in the formation of an eight-cell, seven-nucleate embryo sac.

5. Describe what happens to the endosperm mother cell and the egg following fertilization.

6. Give some specific examples of adaptations that enhance seed protection and dispersal.

Readings

Proctor, M., and P. Yeo. 1973. *The Pollination of Flowers*. New York: Taplinger. Beautifully illustrated introduction to pollination.

Raven, P., R. Evert, and H. Curtis. 1986. *Biology of Plants*. Fourth edition. New York: Worth. Outstanding illustrations.

Rost, T. et al. 1984. *Botany: A Brief Introduction to Plant Biology*. Second edition. New York: Wiley.

Salisbury, F., and C. Ross. 1985. *Plant Physiology*. Third edition. Belmont, California: Wadsworth.

22

PLANT GROWTH AND DEVELOPMENT

In the preceding chapter, we traced the events by which a zygote of a flowering plant becomes transformed into an embryo, housed inside a protective seed coat. Before and after seed dispersal from the parent plant, embryonic growth idles. But at some point following dispersal, **seed germination** occurs: the embryo absorbs water, resumes growth, and breaks through the seed coat. Now the embryo increases in volume and mass; its cells divide and develop into the tissues and organs of the seedling. For many species, leaves unfurl and, as the life cycle progresses, flowers, fruits, and new seeds develop. Then, in autumn, old leaves drop away from the plant. What internal mechanisms govern such developmental events, and what kinds of environmental signals set them in motion? In this chapter, we will consider some answers (and best guesses) to these questions.

SEED GERMINATION

Water, oxygen, temperature, and usually light are major environmental factors that influence seed germination. Most mature seeds do not contain enough water for cell expansion or metabolism. In many parts of the world, water availability is seasonal and germination coincides with the return of spring rains. In a process called **imbibition**, water molecules move into the seed, being especially attracted to hydrophilic groups of the stored proteins. The seed swells as more and more water molecules move inside and the coat finally ruptures.

Once the seed coat splits, oxygen moves in more easily from the surrounding air. Cells of the embryo switch to aerobic pathways and metabolism moves into high gear. The embryo increases in volume, giving rise to the seedling (Figure 22.1).

Figure 22.1 (a) Seedling of corn, a monocot, and (b) seedling of a bean, a dicot.

a   primary root

b

PATTERNS OF GROWTH

Inside the germinating seeds of nearly all species, cells of the **radicle** (the embryonic root) are the first to grow. Continued cell division and cell elongation in the radicle produce the **primary root** (the first root of the plant). When the primary root visibly protrudes from the seed coat, germination is completed.

As the seed germinates, the embryo within it grows, then embryonic cells and their daughter cells grow in ways that lead to roots, stems, and leaves (Figure 22.2). Often the cell divisions are unequal, with one daughter cell ending up with more cytoplasm than the other. The unequal distribution of cytoplasmic substances gives rise to differences in cell composition and structure. In growing tissues, cells divide in different planes and expand in different directions. These genetically controlled differences lead to tissues and organs with diverse shapes.

Keep in mind that division itself does not constitute growth; the two daughter cells have about the same volume as the parent cell (Figure 22.2). However, on the average, about half the daughter cells enlarge—often by twenty times! Thus, cell division increases the *capacity* for overall growth. The remaining daughter cells stay meristematic (they retain the capacity to divide). They grow no larger than the parent cell, then probably divide again.

Water uptake drives cell enlargement. When water moves into new cells, increased turgor pressure (page 91) forces the cell wall to expand. In some ways, the expansion resembles a balloon being blown up with air. Balloons with soft walls are easy to inflate; cells with soft walls grow rapidly under little turgor pressure. A major difference is that a balloon wall gets thinner as it "grows," but a cell wall does not. New polysaccharides are added to the wall during growth, and more cytoplasm forms between the wall and the central vacuole (Figure 22.2).

The primary walls of young cells are somewhat *elastic* and can be stretched, although they do show some tendency to return to their original shape (much as a filled balloon does when the air is let out). But the walls also are *plastic* and tend to retain much of their stretched shape (as does a blown-up bubble of bubble gum when the air has been let out). These two expansive properties encourage **true growth**, as defined by an increase in cell volume.

PLANT HORMONES

Types of Plant Hormones

Like the growth of other multicelled organisms, plant growth and development are influenced in powerful but poorly understood ways by hormones. A **hormone** is a signaling molecule released from one cell and transported to (typically) nonadjacent target cells. A *target cell* is one with receptor sites for a given signaling molecule. In plants, the receptors have not been identified. They may be proteins at the surface of the plasma membrane or proteins within the cell.

Five hormones (or groups of hormones) are known to exist in most plants, and there is evidence of others. The five are auxins, gibberellins, cytokinins, abscisic acid, and ethylene (Table 22.1).

Auxins are best known for their ability to promote elongation of stem cells. The main auxin is indoleacetic

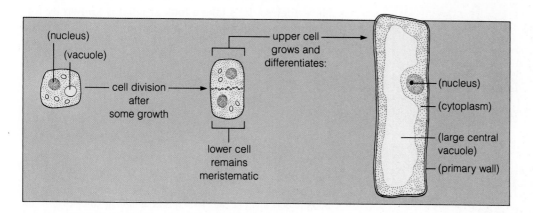

Figure 22.2 The nature and direction of plant growth. Through divisions, meristems provide new cells for growth. At the tip of a young root, small meristematic cells double in size, then divide at right angles to the root axis. The lower cell remains meristematic. The upper cell grows into a mature parenchyma cell of pith or cortex, for example. Tiny vacuoles in young cells absorb water, fuse, and form the central vacuole of mature parenchyma cells.

Table 22.1 Main Plant Hormones and Some Known (or Suspected) Effects

Auxins	*Promote cell elongation in coleoptiles and stems; long thought to be involved in phototropism and gravitropism*
Gibberellins	*Promote stem elongation (especially in dwarf plants); might help break dormancy of seeds and buds*
Cytokinins	*Promote cell division; promote leaf expansion and retard leaf aging*
Abscisic acid	*Promotes stomatal closure; might trigger bud and seed dormancy*
Ethylene	*Promotes fruit ripening; promotes abscission of leaves, flowers, and fruits*
Florigen (?)	*Arbitrary designation for as-yet unidentified hormone (or hormones) thought to cause flowering*

acid (IAA). Several synthetic compounds behave much like natural auxins and are used as *herbicides*, which are selective killers of plant growth. Like other physiologically active compounds, they are toxic when applied in high concentrations to some plant species but not to others. For example, one of the most common herbicides, 2,4-dichlorophenoxyacetic acid (2,4-D), is a potent killer of dicot weeds but has little effect on grasses. This compound is used extensively to kill weeds in fields of cereal grains.

Gibberellins, like auxins, are best known for their promotion of stem elongation. More than sixty kinds have been identified in plants and fungi. The most familiar kind, *gibberellic acid*, is widely used in growth experiments because of its potency and because a certain fungus serves as an inexpensive factory for its production.

Cytokinins were named for their ability to stimulate cell division (the name refers to "cytokinesis"). Leaf cells also grow larger when exposed to these hormones. The most common and abundant cytokinin seems to be *zeatin*, first isolated from immature seeds of *Zea mays* (corn).

Abscisic acid (ABA) is important in promoting stomatal closure. It also promotes seed dormancy, bud dormancy, and resistance to water stress in many species. This hormone was once thought to be important in abscission (the loss of flowers, fruits, and leaves from the plant body), but this effect does not seem to be common.

Among other things, **ethylene** (C_2H_4) stimulates most fruits into ripening. The ancient Chinese knew that a bowl of fruit would ripen faster when incense was burned in the same room, although they didn't know that the ripening was promoted by ethylene present in the smoke.

It is not yet possible to distinguish clearly between hormonal and environmental effects on plant development. One of the reasons is that the environment often controls the amount and distribution of hormones. Nevertheless, let's take a look at what *is* known by following the development of a monocot (corn) and a dicot (soybean).

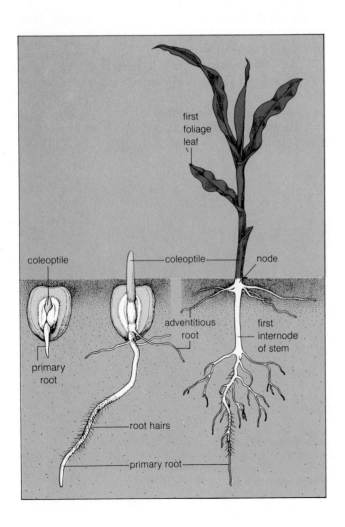

Figure 22.3 A few stages in the development of a corn plant (*Zea mays*), a monocot. The coleoptile encloses the young leaves and protects them as the shoot grows up through the soil. Adventitious roots develop at the first-formed node at the coleoptile base. When a corn seed is planted deeply, the first internode of the stem elongates as shown here. When the seed is close to the soil surface, light inhibits elongation of the first internode; then the primary and adventitious roots look as if they are all originating at the same region, just below the soil surface.

Examples of Hormonal Action

In the seedlings of corn and other grasses, a **coleoptile** is a hollow, cylindrical organ that protects the tender young leaves growing within it (Figure 22.3). Without a coleoptile, the young leaves would be torn apart by soil particles during early growth. While underground, both the coleoptile and the tightly curled oldest leaf grow in a coordinated fashion. Once exposed to sunlight, the coleoptile stops elongating; its task is completed. The leaf now breaks through the coleoptile, uncurls and becomes flattened, engages in photosynthesis, and provides organic molecules for the plant parts below.

The auxin IAA stimulates coleoptile growth (Figure 22.4). IAA moves from the tip of the coleoptile into cells between the tip and the base. Metabolic activities change in those cells and, as a result, cell walls become more plastic—and plasticity encourages elongation.

Gibberellin also can promote growth, especially of stems that never elongate much. For example, dwarf varieties of corn (with short stems) often carry a mutant gene that codes for an enzyme needed in gibberellin production. After application of a gibberellin, the stems of dwarf plants grow longer; yet similar application to stems of a normal variety of corn does not have much of an effect (Figure 22.5).

Most or all of the hormones required by roots and leaves are synthesized by root and leaf cells. Generally, these plant parts do not respond much to hormone applications.

Hormones also promote growth of a soybean stem (Figure 22.6). Like all dicots, soybeans have young, growing cells near the tip of the stem; cells at the base are older and larger. An auxin is synthesized near the tip, especially in very young leaves. When the tip is cut off, the dicot stem stops elongating; when an auxin is applied to the stump, growth resumes. This response suggests that an auxin produced at the tip of a dicot stem promotes growth of stem cells below.

PLANT RESPONSES TO THE ENVIRONMENT

Once an embryonic root breaks through the seed coat and begins its downward growth, the plant cannot move on if conditions are not ideal for growth and development. Yet it still has a genetically based capacity to respond to a range of conditions characteristic of the environment in which that type of plant evolved. *Through interactions with the environment, the plant can adjust its patterns of growth.* It can make adjustments to unique circumstances. (For example, it might sprout alongside a highway, a paper bag tossed out of a passing car may come to rest right on top of it, and the shoot

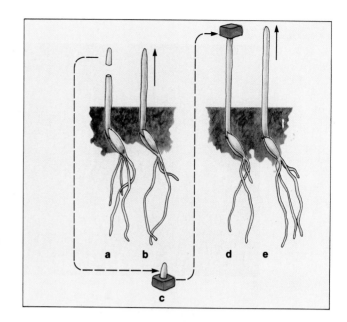

Figure 22.4 Experiment demonstrating that IAA present in the tip of a grass coleoptile promotes elongation of cells below. (**a**) An oat coleoptile tip is cut off. Compared with a normal coleoptile (**b**), the stump does not elongate much. When the excised tip is placed on a tiny block of agar for several hours (**c**), IAA moves into the agar. When the agar is placed on another de-tipped coleoptile (**d**), elongation proceeds about as rapidly as in an intact coleoptile (**e**).

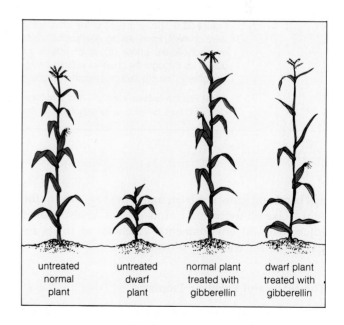

Figure 22.5 Influence of gibberellin on the height of a normal corn plant, compared to its influence on a dwarf corn plant that is mutant for one gene that codes for an enzyme necessary in gibberellin production.

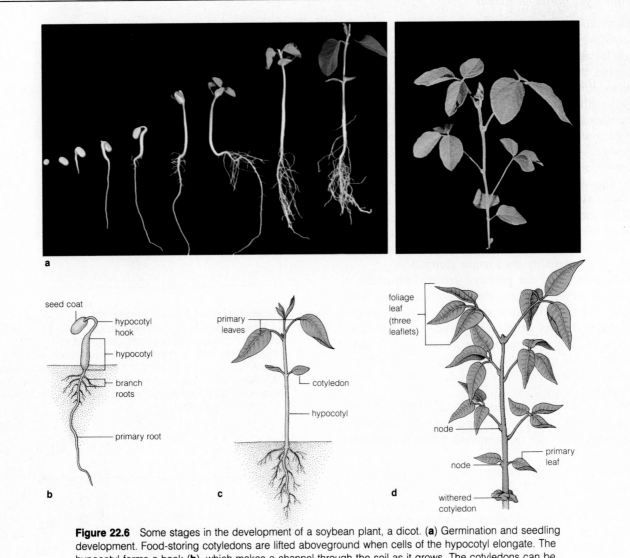

Figure 22.6 Some stages in the development of a soybean plant, a dicot. (**a**) Germination and seedling development. Food-storing cotyledons are lifted aboveground when cells of the hypocotyl elongate. The hypocotyl forms a hook (**b**), which makes a channel through the soil as it grows. The cotyledons can be pulled up through the channel without being torn apart. At the soil surface, light causes the hook to straighten. The cotyledons become photosynthetic for several days, then wither and fall off.

(**c–d**) Foliage leaves, each divided into three leaflets, develop after the first (primary) leaves along the stem. Flowers develop in axillary buds at the four upper nodes shown here.

will bend and grow out from under the bag, toward the light.) Like all individuals of its species, it also makes adjustments to environmental rhythms, as in its responses to the changing seasons.

The Many and Puzzling Tropisms

Often an environmental stimulus is more intense on one side of a stem or some other plant part. The plant responds by making adjustments in the direction and rate of growth of that part. Growth responses of this sort are called "tropisms," and no one knows how they work.

Phototropism. When light strikes one side of a plant more than the other, the stem will curve toward the light (Figure 22.7). Also, the flat surface of its leaf blades will turn until the leaf is perpendicular to the light. This growth response is a form of **phototropism** (growth toward or away from light shining mainly on one side of the organism).

In the late 1800s, Charles Darwin showed that a coleoptile curves more rapidly when light strikes its tip rather than the bending portion below. In the 1920s, Frits Went realized that a growth-promoting substance is present in the tips of coleoptiles. He called the substance

Figure 22.7 Phototropism in bean, pea, and oat seedlings. These plants were grown in darkness, then exposed to light from the right side for a few hours before being photographed.

auxin (after the Greek word meaning "to increase"). Went showed that the substance must move down from the tip, especially into cells that are not as exposed to the light, and then stimulate the cells into elongating.

Today we know that Went's auxin is IAA. We also know that light of blue wavelengths is the main stimulus for phototropism. A large, yellow pigment molecule called *flavoprotein* probably plays a role in phototropic bending, given its capacity to absorb blue wavelengths. We have no idea how that absorption might cause IAA to move horizontally across a coleoptile, to the shaded side.

Gravitropism. A growth response to the earth's gravitational force is called **gravitropism** (or geotropism). You can observe this response after seeds germinate: the primary root always curves down and the coleoptile or stem always curves up. Look at Figure 22.8, which shows what happened when a potted seedling was turned on its side in a dark room. Cell elongation on the upper side of the horizontal stem decreased markedly but increased on the lower side. With the adjusted growth pattern, the stem curved upward, even in the absence of light.

What role, if any, do plant hormones play in gravitropism? At least in bending coleoptiles, some downward movement of IAA has been detected. But the IAA concentration gradient from the tip to the base of the coleoptile just doesn't seem to be great enough or to form fast enough to account for the rapid, extensive differences in growth that lead to bending. Besides, many dicot stems bend in response to gravity even though the IAA gradient seems to be negligible or nonexistent.

Perhaps the main adjustment made by a horizontal stem of a seedling is in the *sensitivity* of cells to IAA already there. Cells on the bottom of the stem, which grow faster, might become more sensitive to IAA; those

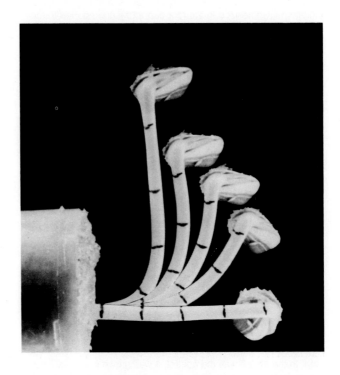

Figure 22.8 Composite time-lapse photograph of gravitropism in a dark-grown sunflower seedling. In this plant, the two cotyledons emerge aboveground, because the stem portion just below the cotyledons elongates. Just before this five-day-old plant was positioned horizontally, it was marked at 0.5-centimeter intervals. After thirty minutes, upward curvature was detectable. The most upright position shown was reached within two hours.

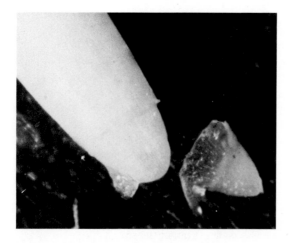

Figure 22.9 Root cap of a young corn plant, shown removed from the root tip. With cap removal, cells just behind the apical meristem elongate faster. With cap replacement, their growth slows. These and other experiments suggest that a growth inhibitor is synthesized in the root cap and is transported to the bottom of a horizontally positioned root. The inhibitor causes the root to curve downward as it continues to grow.

Figure 22.10 Effect of mechanical stress on tomato plants. The plant to the far left was the control; it was grown in a greenhouse, protected from wind and rain. The center plant was mechanically shaken thirty seconds at 280 rpm for twenty-eight consecutive days. The plant to the far right received two such shakings daily for twenty-eight days.

on top might become less sensitive. Perhaps calcium ions or gibberellins figure in the growth response; they, too, are transported from one side to the other of a horizontal coleoptile or stem. These possibilities are just now being investigated, although they were first proposed several decades ago.

What about gravitropism in roots? The root cap plays a role here. A horizontally positioned root will not curve downward if its root cap has been surgically removed, but it will if the cap is put back on (Figure 22.9).

Elongating cells don't stop growing when the root cap is removed; if anything, they grow faster. Could root cap cells contain a growth inhibitor that becomes redistributed in a horizontally oriented root? If gravity somehow causes the inhibitor to move out of the cap and accumulate in cells on the lower side of the horizontally positioned root, those cells would not elongate as much as cells on the upper side—and the root would curve downward. The identity of the growth inhibitor continues to elude us. At one time abscisic acid seemed to be the most likely candidate, but there is now strong evidence that it has nothing to do with root gravitropism.

Thigmotropism. Peas, beans, and many other plants with long, slender stems are climbing vines; they generally do not grow upright without physical support. These plants show **thigmotropism**, or unequal growth resulting from physical contact with solid objects in their surroundings. Suppose one side of the stem grows against a fencepost or the trunk of a tree. Cells stop elongating on the side of the stem making contact, and within minutes the stem starts to curl around the post or trunk. It might do so several times before cells on both sides start elongating at about the same rate once again. How contact affects elongation is a mystery, although auxin and ethylene seem to be involved in the response.

Response to Mechanical Stress

Certain kinds of mechanical stress can inhibit growth of the whole plant, although stem elongation is especially affected. Contact with rain, grazing animals, farm machinery, even winds can inhibit overall growth. Shak-

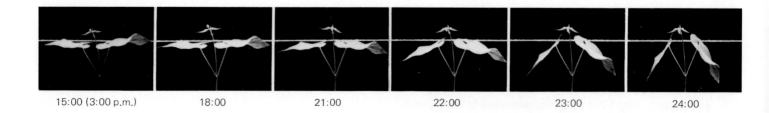

15:00 (3:00 p.m.)   18:00   21:00   22:00   23:00   24:00

ing some plants daily for a brief period can do the same thing (Figure 22.10). Such responses to mechanical stress help explain differences between plants of the same species when some are grown indoors. Often the plants growing outdoors are shorter, have somewhat thicker stems, and are not easily blown over. These characteristics appear to be responses to wind stress, although the response mechanism is not understood.

Biological Clocks in Plants

Plant growth and development—indeed, nearly all plant activities—are influenced by biological clocks. A **biological clock** is an internal time-measuring mechanism that has a biochemical basis. Such mechanisms have not been identified, although there is speculation that components of the plasma membrane are central to their operation.

There is plenty of indirect evidence of biological clocks. Flowers open in the morning and close at dusk. Some flowers secrete nectar at specific times of day, thereby prompting visits only at those times by bees and other pollinators (which have biological clocks of their own). Leaves of many plants, especially beans and other legumes, are horizontal in the daytime but fold closer to the stem at night (Figure 22.11). Remarkably, even if you keep one of those plants in constant light or darkness for a few days, it will fold its leaves into the "sleep" position in the evening anyway! Clearly the plant has a means of measuring time that is independent of light-on (sunrise) and light-off (sunset) signals.

Activities that occur regularly in cycles of about twenty-four hours are called **circadian rhythms** (from the Latin *circa*, meaning about; and *dies*, meaning day).

And yet, environmental conditions in a twenty-four-hour period are not the same in summer as they are in winter. For example, in North America, winter temperatures are cooler and daylength is shorter. Somehow, biological clocks must be reset, for plants make *seasonal adjustments* in their patterns of growth, development, and reproduction. Let's look first at one time-measuring mechanism before we turn to the question of how that mechanism is continually synchronized with changing conditions.

Photoperiodism

A biological response to a change in the relative length of daylight and darkness in a twenty-four-hour cycle is called **photoperiodism**. What is the biochemical basis of this response? **Phytochrome**, a blue-green pigment molecule, is known to be part of a switching mechanism that promotes or inhibits growth of different plant parts. The switch is turned on when phytochrome absorbs light of red wavelengths; it is turned off (or kept off) when phytochrome absorbs light of far-red wavelengths. After sunrise, red wavelengths dominate the sky. Then, phytochrome is converted to its active form (Pfr). In darkness, phytochrome reverts to its inactive form (Pr), and plant responses are curtailed (Figure 22.12).

Pfr takes part in controls over seed germination, stem elongation, leaf expansion, stem branching, and the formation of flowers, fruits, and seeds. When plants that are adapted to full sunlight are grown in darkness, the effects of Pfr deficiency are readily apparent. The plants put more resources into stem elongation and less into leaf expansion or stem branching (Figure 22.13). Shaded plants show the same sort of responses. (Plants growing above them absorb most of the red wavelengths and use the absorbed energy in photosynthesis, hence the shaded plants are deficient in Pfr.)

The biochemical pathways by which Pfr exerts its effects on plant growth are not understood. Phytochrome molecules probably are embedded in plasma membranes, and conversion of Pr to Pfr may be

Figure 22.11 Leaf movements in bean plants. These so-called sleep movements are among the best-studied rhythms that verify the existence of a biological clock. What is their advantage? Perhaps the sleep position prevents bright moonlight from being absorbed by phytochrome in the leaves (moonlight could otherwise interrupt a dark period necessary to induce flowering). Or perhaps it helps slow heat loss from leaves (folded leaves radiate heat to each other rather than to the cold air around them).

The bean plant shown here was kept in constant darkness, and its leaf movements continued independently of sunrise (6 AM) and sunset (6 PM).

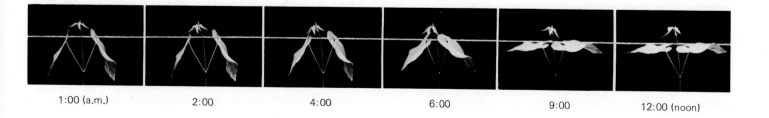

| 1:00 (a.m.) | 2:00 | 4:00 | 6:00 | 9:00 | 12:00 (noon) |

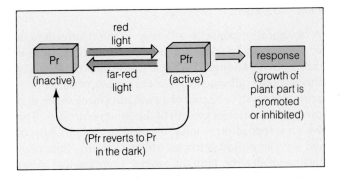

Figure 22.12 Interconversion of phytochrome from the active form (Pfr) to the inactive form (Pr). This pigment is part of a switching mechanism that can promote or inhibit growth of different plant parts.

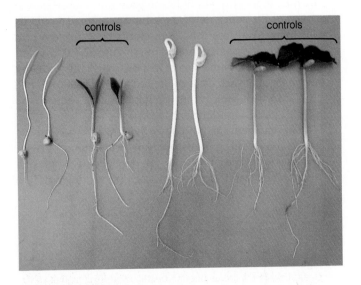

Figure 22.13 Effects of the absence of light on young corn and bean plants. The two plants at the right of each group served as the control; they were grown in a greenhouse. The others were grown in darkness for eight days. Dark-grown plants were yellow: they could form carotenoid pigments but not chlorophyll in darkness. They also had longer stems, smaller leaves, and smaller root systems. (Why do you suppose roots of dark-grown plants grow less, considering that those of light-grown plants aren't exposed to light, either?)

necessary before certain hormones can bind to the membrane or move into the cytoplasm of target cells. Whatever the pathway, there is evidence that some genes but not others are being transcribed when phytochrome is in its active form. This suggests that phytochrome activation helps control which enzymes are being produced in particular cells—and specific enzymes are necessary for specific growth responses.

THE FLOWERING PROCESS

As a flowering plant matures, its activities become directed toward producing flowers, seeds, and fruits. Annuals such as corn, soybeans, and peas begin flowering after only a few months of growth. Perennials such as roses typically flower each successive year or after several years of vegetative growth. Biennials such as carrots and cabbages typically produce only roots, stems, and leaves the first growing season, die back to soil level in autumn, then grow a new flower-forming stem from a bud that remains alive in a protected stem region underground.

The question becomes this: How does a particular plant "know" when to flower? Apparently temperature is one environmental cue. Unless buds of some biennials and perennials are exposed to low winter temperatures, flowers do not form on stems in spring. Low-temperature stimulation of flowering is called **vernalization** (after a Latin term meaning "to make springlike").

In themselves, temperature changes are somewhat unreliable cues about what season is approaching, because weather is often fickle. A better cue is the relative length of day and night, which each year is nearly constant at the same day of the month in a given region. For example, in response to the shorter daylengths of late summer and early fall, many plants form flowers and seeds, develop dormant buds, attain more cold hardiness, and shed their leaves (Figure 22.14). These responses are so predictable that flowering plants can be categorized in terms of daylength:

Long-day plants flower in the spring when daylength becomes longer than some critical value. The critical value can be longer or shorter than twelve hours.

Short-day plants flower in late summer or early autumn when daylength becomes shorter than some critical value. The critical value can be longer or shorter than twelve hours.

Day-neutral plants flower whenever they become mature enough to do so, irrespective of daylength.

To get an idea of the effect of daylength on flowering, look at Figure 22.15, which shows what happens when some long-day plants are grown under short-day conditions and under long-day conditions. These examples illustrate why long-day plants such as spinach never flower and produce seeds in the tropics. In order to flower, spinach needs fourteen hours of light each day for two weeks—and this never happens in the tropics.

The long days of spring promote flowering of many species, including most biennials and winter wheats (which are planted in fall and are able to survive winter conditions). The shorter days and longer nights of late summer promote flowering in other species. At the equator, daylength remains nearly constant, and here you will find day-neutral plants. Many crop plants, including corn, peas, and tomatoes, are also day-neutral. These crop plants have been subject to artificial selection for many generations, and variant forms can flower and produce seeds under a wide range of daylengths.

One short-day plant, the cocklebur, measures time with uncommon sensitivity. Just a single night longer than $8\frac{1}{2}$ hours is enough to promote flowering in most cockleburs. Yet if that dark period is experimentally interrupted with even a minute or two of artificial light, flowering is inhibited! This type of interruption also inhibits flowering in less sensitive short-day plants. Why is this so?

For all short-day plants, red light interrupts the dark period most effectively—and red light is detected by phytochrome. During the day, phytochrome is mostly in the active form (Pfr), which is needed during daylight for flowering to occur. During the night, if red light interrupts the dark period for short-day plants, Pr is then converted to Pfr at night, and flowering is inhibited.

In long-day plants, the extended daylengths provide Pfr for longer periods. As a result, these plants flower more rapidly than do short-day plants, and they produce more flowers.

Again, we do not know how Pfr actually exerts its effects on growth responses such as flowering, but hormonal action is thought to be involved. In anticipation of its discovery, the elusive hormone thought to control flowering was designated **florigen** decades ago. Evidence for the existence of florigen comes from experiments with cocklebur and other plants. Flowers can form only at bud primordia (page 246). Before flowers can form, the bud primordia have to stop producing new cells for leaves and stems. Somehow, meristematic cells in those regions receive signals to start the developmental events that produce flowers.

Experiments suggest that the signal (probably florigen) is produced in leaves and transported to the bud primordia. For example, when all but one leaf is trimmed from a cocklebur plant, and when that leaf is covered with black paper for at least $8\frac{1}{2}$ hours, the plant flowers. But when the leaf is cut off immediately after its dark period, the plant does *not* flower. Apparently some timing mechanism in the leaf detects the length of day and night, and florigen is produced in response. (If the leaf is cut off quickly, the substance remains in the discarded leaf.)

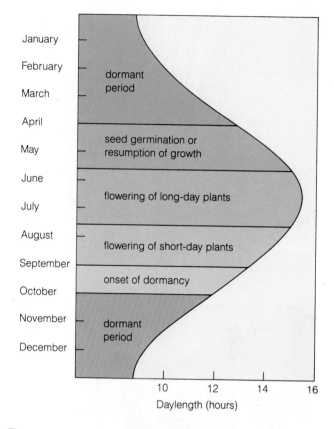

Figure 22.14 Correlation between daylength and plant developmental events in temperate regions of North America. (These regions show seasonal variation in rainfall and temperature.)

Figure 22.15 Effect of daylength on flowering of long-day plants (LDP). In each photograph, the plant on the left was grown under short-day conditions; the plant on the right was grown under long-day conditions. (**a**) *Spinacia* or spinach, (**b,c**) *Sedum spectabile* and *Kalanchoe*.

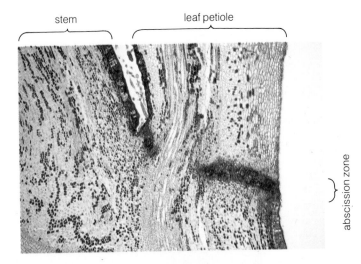

stem   leaf petiole

abscission zone

Figure 22.16 Abscission zone in a maple (*Acer*). This longitudinal section is through the base of the petiole of a leaf.

flowers removed   control group

Figure 22.17 Delay of senescence in soybean plants as a result of the daily removal of flower buds.

SENESCENCE

Plants make a major investment in their reproduction. Nutrients are actually withdrawn from the leaves, roots, and stems and are distributed to newly forming flowers, fruits, and seeds. In older leaves, enzymes synthesized seemingly on demand degrade proteins, chlorophyll, and other organic molecules. The breakdown products then move through the phloem (along with mineral ions) to reproductive structures, where they are converted to fats, starches, and proteins in seeds. Annuals and most perennial plants end up with tan-colored, dead leaves. In deciduous species (which shed leaves at the end of each growing season), nutrients are transported to pa-

renchyma cells in twigs, stems, and roots for storage before the leaves wither and fall off.

The dropping off of leaves (or flowers, fruits, or other plant parts) is called **abscission** (Figure 22.16). Ethylene is formed in cells near the break points and may trigger the process. Studies suggest that ethylene activates enzymes that break down the polysaccharides in cell walls in a narrow zone (the abscission zone).

The sum total of processes leading to the death of a plant or any of its structures is called **senescence**. The drain of nutrients during the growth of reproductive parts might be one stimulus for senescence of leaves, stems, and roots. When the drain of nutrients is stopped (for example, by removing each newly emerging flower from the plant), the leaves and stems stay green and healthy much longer than they otherwise would (Figure 22.17). Gardeners routinely remove flower buds from many plants to maintain vegetative growth.

And yet, other stimuli must be involved in senescence. When a cocklebur is experimentally induced to flower, its leaves turn yellow regardless of whether the nutrient-demanding young flowers are left on or pinched off. It is as if a "death signal" forms during short days and leads to both flowering and senescence. Experiments with soybeans indicate that the signal (whatever it is) counteracts the effects of cytokinins, which delay senescence. Painting the surface of a mature soybean leaf with a cytokinin solution will increase the cytokinin content of the leaf. Leaves painted this way often stay green longer (Figure 22.18).

DORMANCY

As autumn approaches and days grow shorter, growth slows or stops in many evergreen trees, deciduous trees, and nonwoody perennials. Apical meristems in their buds simply stop producing new stem and leaf cells, and they do so even if temperatures are still moderate, the sky is bright, and enough water is available. When a perennial or biennial plant stops growing under conditions that seem (to us) quite suitable for growth, it has entered a state of **dormancy**. Its buds become tolerant of drought and lower temperatures, and ordinarily the buds will not resume growth until early spring.

Short days and long nights represent a strong environmental cue for dormancy. Their effect can be demonstrated with Douglas fir plants (Figure 22.19). When a short period of red light interrupts the long dark period for these plants, the plants respond as if nights are shorter (and the days longer). They continue to grow taller.

In this experiment, conversion of Pr to Pfr by red light during the dark period prevents dormancy. But what happens in nature? Perhaps buds go dormant because less Pfr can form when daylength decreases in late summer. Perhaps also, the decline in Pfr levels causes leaves to synthesize a dormancy-triggering hormone (abscisic acid?) that is transported to buds. Finally, in some unexplained way, cold nights, dry soil, and nitrogen deficiency also promote dormancy.

What breaks dormancy in the spring? Between fall and spring, a major dormancy-breaking process is at work. The process involves exposure to low temperatures for hundreds of hours in winter (Figure 22.20). The actual temperature needed to break dormancy varies greatly among species. For instance, Delicious apples grown in Utah require 1,230 hours near 43°F (6°C); apricots grown there require only 720 hours. Generally, tree varieties growing in the southern United States require less cold exposure than those growing in northern states and Canada. So if you live in Colorado and order a young peach tree from a Georgia nursery, the tree might start spring growth too soon and be killed by a late frost or heavy snow.

Probably both gibberellin and abscisic acid help control dormancy. When gibberellins are applied to dormant buds of several species, the buds grow and dormancy is often broken. Abscisic acid extends dormancy and partially counteracts the effects of gibberellins.

Seeds of most native (noncultivated) species also exhibit dormancy, which provides survival value for them as well as for buds. (In contrast, dormancy is rare in seeds of highly selected agricultural crops.) The mechanisms by which seed dormancy develops and ends are variable. In some species such as honey locust and alfalfa, hard seed coats are formed. These coats prevent absorption of water and oxygen. Dormancy ends when the seed coat is abraded (perhaps as strong winds and rains drive the seed across sand), when it is chemically digested (by bacteria, by fungi, or in the gut of a bird or mammal), perhaps even when fire burns it away.

For many plants, moistened seeds must be exposed to low temperatures for weeks or months before dormancy ends. Such seeds are shed from the plant in summer or autumn, and built-in controls prevent their germination before spring. Without these controls, the seedlings would be killed by frost. Experiments with many plants indicate that gibberellins and abscisic acid help control seed dormancy.

Finally, many species depend on red wavelengths to break dormancy. Once again the light-activated form of phytochrome is involved. At the very least, reliance on this cue helps assure that seedlings will have enough light for photosynthesis and resumed growth. Thus, in the shade of other plants, many types of seeds never do break dormancy.

Figure 22.18 Effect of local cytokinin applications on senescence in a bean plant. These are the primary leaves formed on the stem. Senescence was delayed in one of the primary leaves by covering its upper surface at four-day intervals with a cytokinin solution.

Figure 22.19 Effect of the relative length of day and night on Douglas fir plant growth. The plant at the left was exposed to twelve-hour light and twelve-hour darkness for a year; its buds became dormant because daylength was too short. The plant at the right was exposed to twenty-hour light and four-hour darkness; buds remained active and growth continued. The middle plant was exposed to twelve-hour light, eleven-hour darkness, and one-hour light in the middle of the dark period. This light interruption of an otherwise long dark period also prevented bud dormancy. Such light causes Pfr formation at an especially sensitive time in the normal day-night cycle.

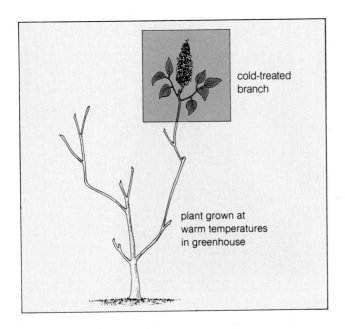

Figure 22.20 Effect of cold temperature on breaking the bud dormancy in many woody plants. In this experiment, one branch (boxed portion) of a lilac plant was positioned so that it protruded from a greenhouse during winter; the rest of the plant remained inside, at warm temperatures. Only the buds on the branch exposed to low outside temperatures grew again in spring. This experiment suggests that low-temperature effects are localized.

Case Study: From Embryogenesis to the Mature Oak

Where the ocean breaks along the central California coast, the land rolls inward as steep and rounded hills. Sixty-five million years ago, these sandstone hills had their genesis on the floor of the Pacific. At that time, violent movements in the earth's crust caused parts of the submerged continental shelf to start crumpling upward into a jagged new coastal range. Since then, the rain and winds of countless winters played across the land, softening the stark contours and sending mineral-laden sediments into the canyons. Grasses came to cloak the inland hills, and their organic remains accumulated and enriched the soil. On these hillsides, in these canyons, the coast live oak (*Quercus agrifolia*) began to evolve more than ten million years ago.

Quercus agrifolia is a long-lived giant; some trees are known to be three hundred years old. They can reach heights of a hundred feet; their evergreen branches may spread even wider. In early spring, clusters of male flowers develop and form pendulous, golden catkins among the leafy branches. Wind carries pollen grains from the catkins to female flowers clustered inconspicuously near the branch tips of the same or neighboring trees. After pollination, a sperm-bearing pollen tube grows through the style, toward the ovary where the eggs reside. After a sperm and egg nuclei unite at fertilization, cells of the newly formed zygote divide repeatedly, giving rise to root and shoot tips and to large cotyledons. Integuments of the ovule form the seed coat, and ovary walls develop into a shell. By early fall, the seed reaches maturity and is shed from the tree as a hard-shelled acorn.

Three centuries ago, long before Gaspar de Portola sent landing parties ashore to found colonies throughout Upper California, oaks were shedding the seeds of a new generation. Suppose it was then that a scrubjay, foraging at the foot of a hillside, came across a worm-free acorn. In storing away food for leaner days, the bird used its beak to scrape a small crater in the soil, dropped in the acorn, and covered the acorn-filled crater with decaying leaves. Although a scrubjay might remember many such hiding places most of the time, this particular acorn lay forgotten. The next spring, it germinated.

From the moment of germination, the oak seed embarked on a journey of dynamic, continued growth. Cells divided repeatedly, grew longer, and increased in diameter. Water pressure forced the enlargement— water taken up osmotically as ions accumulated in the newly forming primary root, and as hormones caused a softening of cell walls that otherwise would have been too strong to allow expansion under pressure.

A root cap formed and protected the primary root during its downward growth through the soil. It produced cortex and epidermis, as well as a vascular column through which water and ions would flow. Lateral roots developed, probably under the influence of growth regulators. As new roots continued to elongate, absorptive surfaces increased. When the primary shoot began its upward surge, separate vascular bundles began forming; eventually they would become consolidated into a continuous cylinder of secondary xylem and phloem.

In parenchyma cells of developing roots, stems, and leaves, large central vacuoles formed. As vacuoles enlarged, they pressed the cytoplasm outward so that it became no more than a thin zone against the cell wall. Cytoplasmic interaction with the environment was thereby enhanced; essential ions, gases, and water could be harvested rapidly in spite of their dilute concentrations in the surrounding air and soil. Mycorrhizae, the symbiotic association of roots and fungal strands, enhanced the absorption process. In leaves, stomata developed and regulated carbon dioxide movements and water loss.

At the whim of a scrubjay, the seed had sprouted in a well-drained, sandy basin at the foot of a canyon, in full sunlight. Rainwater accumulated there each winter, keeping the soil moist enough to encourage luxuriant growth during the spring and through the dry summers.

Figure 22.21 From germination to the mature oak (*Quercus agrifolia*).

Out in the open, red wavelengths of sunlight activated phytochrome pigments in the seedling, and hormonal events were triggered that encouraged stem branching and leaf expansion. All the while, delicate hormone-mediated responses were being made to the winds, the sun, the tug of gravity, the changing seasons. Lignin and cellulose strengthened most of the cell walls in secondary xylem. With this strengthening came resistance to the strong winds racing through the canyon. As the oak matured, phytochrome responded to the subtle shifts in daylength throughout the year and the tree responded to the changing seasons. The shorter days of late summer promoted bud dormancy, hence resistance to autumn drought and winter cold.

As the oak seedling matured into the adult form, roots continued to develop and snake through a tremendous volume of the moist soil. Branches continued to spread beneath the sun. Leaves proliferated—leaves where the oak put together its own food from water, carbon dioxide, the few simple inorganic substances it mined from the soil, and the sunlight energy it harnessed to drive the reactions. Continued activity at meristems increased the girth of roots and branches and the height of the tree.

Thus the oak increased in size every season, year after year, century after century. On their way to the gold fields, prospectors of the California Gold Rush rested in the shade of its immense canopy. The great earthquake of 1906 scarcely disturbed the giant, anchored as it was by a root system that spread out eighty feet through the soil. By chance, the brush fires that periodically sweep through California's coastal canyons did not seriously damage the tree. Fungi that could have rotted its roots never took hold; the soil was too well-drained and the water table too deep. Leaf-chewing insects were kept in check by protective chemicals in the leaves and by the predatory birds abounding in the canyon.

In the 1960s, the human population underwent a tremendous upward surge in California. The land outside the cities began to show the effects of population overflow as the native plants gave way to suburban housing. The developer who turned the tractors on the canyon in which the giant oak had grown was impressed enough with its beauty that the tree was not felled. Death came later. How could the new homeowners, just arrived from the east, know of the ancient, delicate relationships between the giant trees and the land that sustained them? Soil was graded between the trunk and the drip line of the overhanging canopy; flower beds were mounded against the trunk; lawns were planted beneath the branches and sprinklers installed. Overwatering in summer created standing water next to the great trunk—and the oak root fungus (*Armillaria*) that had been so successfully resisted until then became established. With its roots rotting away, the oak began to suffer the effects of massive disruption to the feedback relationships among its roots, stems, and leaves. Eventually it had to be cut down. In their fifth winter, in their red brick fireplace, the homeowners began burning three centuries of firewood.

SUMMARY

1. Following dispersal from the parent, the seeds of flowering plants germinate (the embryo absorbs water in a process called imbibition, it resumes growth, and it breaks through the seed coat). Germination is completed when the primary root (which develops from the radicle, or embryonic root) protrudes from the seed coat.

2. Following germination, the plant increases in volume and mass; tissues and organs of the seedling develop; later, flowers, fruits, and new seeds form, then older leaves drop away from the plant. These developmental events are governed by plant hormones: signaling molecules released from one cell and transported to (typically nonadjacent) target cells, which have receptors for the particular hormone.

3. Five hormones have been identified in most flowering plants:

    a. Auxins (especially IAA) promote elongation of coleoptile and stem cells.

    b. Gibberellins promote stem elongation and may help seeds and buds break dormancy.

    c. Cytokinins stimulate cell division, promote leaf expansion, and retard leaf aging.

    d. Abscisic acid promotes stomatal closure and may trigger seed dormancy and bud dormancy.

    e. Ethylene promotes fruit ripening, abscission.

4. An as-yet unidentified hormone (dubbed "florigen") is thought to cause flowering.

5. Plants adjust their patterns of growth in response to environmental rhythms and to unique environmental circumstances. Among these responses are tropisms (differences in the rate and direction of growth on two sides of an organ such as a stem or root).

6. In phototropism, light is more intense on one side of a coleoptile, stem, or leaf, which responds by bending toward the light. Apparently light of blue wavelengths is absorbed by flavoprotein (a yellow pigment), which causes IAA in shoot tips to move into shaded tissue regions, where it promotes the cell elongation that leads to bending.

7. In gravitropism (a growth response to the earth's gravitational force), cells on the lower side of a horizontally positioned stem grow faster than those on the top, causing the stem to curve up; or cells on the upper side of a horizontally positioned root grow faster than those below, causing the root to curve down.

8. Plants have biological clocks (internal time-measuring mechanisms that have a biochemical basis). They also can "reset" the clocks (make seasonal adjustments in their patterns of growth, development, and reproduction).

9. In photoperiodism, plants respond to a change in the relative length of daylight and darkness in a twenty-four hour period. A switching mechanism involving phytochrome (a blue-green pigment) promotes or inhibits germination, stem elongation, leaf expansion, stem branching, and formation of flowers, fruits, and seeds.

10. Long-day plants flower in spring (when daylength is long relative to the night); short-day plants flower when daylength is relatively short. (Flowering of day-neutral plants is not regulated by light.) The flowering response of all may be controlled by phytochrome, which may regulate florigen activity.

    a. When phytochrome absorbs red wavelengths (after sunrise), it is converted from inactive to active form (Pr to Pfr) and helps control the production of enzymes necessary for specific growth responses.

    b. In darkness (at night or in shade), Pfr reverts to Pr and growth responses are curtailed.

11. Senescence is the sum total of processes leading to the death of a plant or plant structure.

12. Dormancy is a state in which a perennial or biennial stops growing even though conditions appear to be suitable for continued growth. A decrease in Pfr levels may trigger dormancy.

Review Questions

1. List the five known plant hormones (or groups of hormones) and the main functions of each.

2. Which of the following plant cells or organs synthesize most or all of the hormones required in normal growth and development: coleoptiles, dicot stems, root cells, and leaf cells? Describe one experiment that tells us which ones do this, or which ones don't.

3. Explain how sunlight exposure influences leaf expansion, stem elongation, and stem branching during primary growth.

4. What is phytochrome, and what is its role in plant growth?

5. Define plant tropism.

6. Define annual, biennial, and perennial plants. Then describe long-day, short-day, and day-neutral plants.

Readings

Bowley, J. D., and M. Black. 1985. *Seeds: Physiology of Development and Germination.* New York: Plenum Press.

Nickell, L. 1982. *Plant Growth Regulators: Agricultural Uses.* New York: Springer-Verlag. Concise explanations of agricultural practices that include use of growth regulators.

Salisbury, F., and C. Ross. 1985. *Plant Physiology.* Third edition. Belmont, California: Wadsworth. Chapter 20 explains in detail selected examples of time measurement and what we know and don't know about the nature of the biological clock.

Villiers, T. 1975. *Dormancy and the Survival of Plants.* London: Edward Arnold. A short book summarizing major dormancy mechanisms.

Whatley, F. R., and J. M. Whatley. 1980. *Light and Plant Life.* London: Edward Arnold. A short book summarizing effects of light on germination and development with ecological descriptions, too.

UNIT FIVE

ANIMAL SYSTEMS AND THEIR CONTROL

23

SYSTEMS OF CELLS AND HOMEOSTASIS

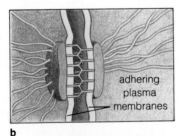

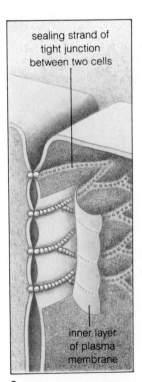

Figure 23.1 What sorts of structures hold animal cells together in tissues? Some examples: (**a**) Tight junction between cells in epithelial tissue. This region of crisscrossed sealing strands prevents leakage between the cells. (**b**) An adhering junction, which acts like a button or rivet between adjacent cells. (**c**) A gap junction, where clustered protein channels permit passage of ions and small molecules between the cytoplasm of adjacent cells.

sealing strand of tight junction between two cells

inner layer of plasma membrane

adhering plasma membranes

matched protein channels bridging the two plasma membranes

a   c   b

SOME CHARACTERISTICS OF ANIMAL CELLS AND TISSUES

In body plan, the water-dwelling animal called the sponge is like a vase riddled with holes. The vaselike body consists of cells positioned rather loosely in a protein gel, and about the only thing that keeps it from collapsing is a "skeleton" of tiny needles, spongy fibers, or both. Unlike most animals, a sponge has no organs, but it does have a thin layer of flat cells at the body surface. If we define a **tissue** as a group of cells and intercellular substances functioning together in a specialized activity, then this sheetlike layer is the closest that sponges come to the tissue level of organization. All other functions (securing food, distributing food, reproduction) are carried out by individual cells, some of which creep about, amoebalike, in the protein gel.

In the 1960s, Tom Humphreys and Aron Moscona thought they might learn something about animal tissues in general by experimenting with the simple sponges. They selected two distinct species (one red, one yellow) so that their component cells could be readily identified. The sponges were forced through a fine sieve to release single cells and tiny clumps of cells, which were then mixed in a swirling flask that contained a nutrient medium. As the cells collided, they adhered to one another—but red cells adhered only to red cells, and yellow cells adhered only to yellow. Moreover, as the cell clusters became longer, cells in the clusters arranged themselves in patterns characteristic of the intact sponge bodies! Such experiments provided the first verification of the following fact:

> Individual cells in animal tissues have properties that promote molecular recognition between similar cells and their adhesion to one another.

Through biochemical and electron microscopic studies, we now know that surface receptors of the plasma membrane promote some of the cell interactions necessary for tissue formation (page 87).

Embryonic Tissues

In complex animals, specialized body tissues cooperate in tasks that help assure survival of the whole organism. Some tissues are the "glue" that binds neighboring cells or tissues; others are coverings, skeletal or muscular elements, absorptive surfaces, and so on. All of these body tissues are composed of **somatic cells**. ("Somatic" comes from the Greek *soma*, meaning body.) The only other cells in animals are **germ cells**, which develop by meiosis into sperm or eggs. Fusion of a sperm and an egg leads to the formation of a zygote, which divides by mitosis to produce the developmental stage called the animal **embryo**.

Soon after an embryo starts to develop, cells become arranged into "primordial" tissue layers. Through cell division and differentiation, these layers give rise to all the specialized tissues of the animal body. In vertebrate embryos, three primordial tissue layers form:

---

1. **Ectoderm** gives rise to the outer layer of skin and to tissues of the nervous system.

2. **Mesoderm** gives rise to tissues of the muscles, skeleton, circulatory system, inner layer of skin, kidney, and reproductive tract.

3. **Endoderm** gives rise to the lining of the gut and to the major organs derived from it.

Cell Junctions in Animal Tissues

Since the time when Humphreys and Moscona squeezed sponges through sieves, advances in microscopy also revealed that cells in many animal tissues are linked at junctions. **Cell junctions** are specialized regions of the plasma membrane that function in intercellular sealing, adhesion, and communication (Figure 23.1).

Sealing junctions form tight seals between the cells of sheetlike tissues, in ways that help prevent the indiscriminate movement of substances across the tissue layer. By sealing off the spaces between adjacent cells, these junctions prevent leakage from one side of the sheet to the other.

Adhering junctions mechanically link adjacent cells and permit them to function collectively as a structural unit. They are found between cells within all tissues and between cells of adjacent tissues. For example, buttonlike junctions form between the plasma membranes of cells of the outer layer of skin, which is subject to stretching. This arrangement helps keep the tissue from being pulled apart under stress.

Communication junctions are regions of numerous channels that influence the passage of ions and small molecules between cells. For example, "gap junctions" are protein channels across the bilayers of adjacent plasma membranes. Through these channels, some substances pass directly from the cytoplasm of one cell into the cytoplasm of its neighbors. Gap junctions help coordinate individual cell activities throughout a tissue.

Definition of Organs

During the development of most animals, tissues become arranged into organs. An **organ** is a structure of definite form and function that is composed of more than one tissue. The individual character of an organ depends on the kinds of tissues present and on the way they are combined and arranged. For instance, organs that move a lot in different directions typically contain layers of muscle tissues oriented at different angles. The stomach is like this.

An **organ system** consists of two or more organs that are interrelated physically, chemically, or both in carrying out specific functions. For example, organs of the human digestive system include the mouth, salivary glands, pharynx, esophagus, stomach, liver, gall bladder, pancreas, small and large intestines, and rectum. All are necessary for a common task: the ingestion and preparation of food for absorption by individual cells and the elimination of food residues.

KINDS OF ANIMAL TISSUES

In chapters to follow, you will be considering the structure and function of organ systems. Because these systems are truly intricate, most of the tissues that compose them and give them their unique properties will be described here, to keep the subsequent chapters as uncluttered as possible.

The components of all organ systems consist mainly of four types of animal tissues, which are called epithelial, connective, muscular, and nervous tissues.

Epithelial Tissue

An **epithelial tissue** is also called *epithelium* (plural, epithelia). Epithelia cover external body surfaces, they line its internal cavities and tubes, and they form the secretory portion of glands. They play many roles, ranging from protection and absorption to secretion. Epithelia are single or multiple layers of cells that are densely packed in continuous sheets, with little space or intercellular material between cells. One surface of the layer is free, the other adheres to an underlying layer called a basement membrane.

Epithelia can be classified by the number of cell layers at the free surface. *Simple epithelium* consists of a single layer of cells. It often functions in diffusion, secretion, absorption, and filtration. (For example, simple epithelium lines air sacs in the lungs, providing a thin surface for gas exchange.) *Stratified epithelium*, which consists of cells stacked into several layers, functions in protection. (Among other things, this tissue forms the outer layer of skin.)

Simple and stratified epithelia can be further classified by the predominant shape of their surface cells (Figure 23.2). These cells are shaped like scales (squamous

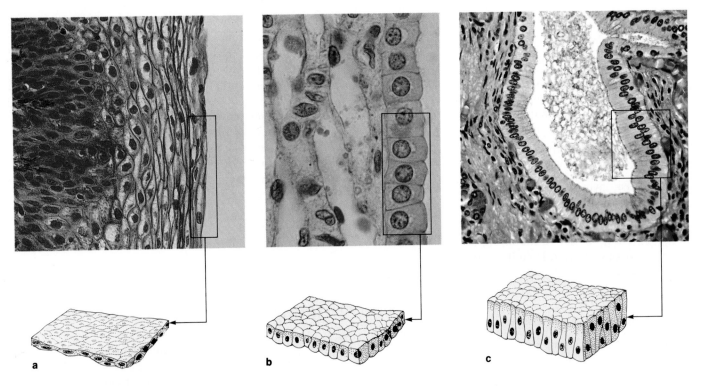

Figure 23.2 Basic cell shapes in epithelial tissues: (**a**) squamous, (**b**) cuboidal, and (**c**) columnar.

Table 23.1 Classification of Some Epithelial Tissues		
Category	Main Functions	Some Common Locations
Simple epithelium:		
Squamous	Filtration, diffusion	Air sacs in lungs, capillary walls, lining of blood and lymph vessels
Cuboidal	Secretion, absorption	Lining of kidney tubules, duct lining for some glands, covering of ovaries
Columnar	Protection, secretion, absorption	Surface layer of lining of stomach, intestines, part of respiratory tract
Stratified epithelium:		
Squamous	Protection	Outer layer of skin, lining of mouth, throat, lining of vagina
Columnar	Protection, secretion, movement of mucus	Lining of parts of respiratory tract

epithelium), cubes (cuboidal epithelium), or columns (columnar epithelium). Table 23.1 summarizes the main varieties of simple and stratified epithelium.

Epithelium is said to be glandular when it contains single-celled or multicelled secretory structures called *glands*. There are two main kinds of glands: endocrine and exocrine. Generally, an **endocrine gland** secretes hormones into the bloodstream. (Animal hormones are signaling molecules, secreted by endocrine glands, that travel by way of the bloodstream and trigger specific reactions in target cells.) In contrast, an **exocrine gland** secretes its products through ducts that empty at the free epithelial surface (Figure 23.3). Exocrine products include mucus, saliva, wax, oil, milk, and digestive enzymes.

Connective Tissue

Of all tissues, **connective tissue** is the most widely distributed through the body. Some types provide support for soft, flexible body parts; others bind different structures together. Compared with epithelia, connective tissues have fewer cells per unit volume but a large amount of extracellular matrix, which consists of fibers and a ground substance. Depending on the tissue, the ground substance is fluid, semifluid, or solid. Often, connective tissues are classified as four groups: connective tissue proper, cartilage, bone, and blood (Table 23.2).

Connective Tissue Proper. In **connective tissue proper**, the ground substance is more or less fluid, with numerous and varied cells scattered through it. The predominant cell type is the fibroblast, which produces the ground substance and tissue fibers. Three common categories of connective tissue proper are called loose, dense, and adipose tissues.

Loose connective tissue supports most epithelia and it surrounds most organs, blood vessels, and nerves. This tissue provides support, strength, and elasticity. These attributes arise from various protein fibers present in the ground substance (Figure 23.4). For example, some are tough yet flexible collagen fibers, which impart strength to the tissue. Thinner fibers containing the protein elastin impart elasticity to it.

In *dense connective tissue*, collagen fibers are bundled together irregularly or in parallel array (Figure 23.4). Both arrangements provide strong connections between different tissues. For example, dense connective tissue with parallel fiber arrays is the main component of tendons (which attach muscle to bone) and ligaments (which connect bones at skeletal joints). In these regions, tension is exerted in one direction only.

Adipose tissue serves as a region of energy reserves and also pads some organs. This tissue contains dense clusters of large cells specialized for fat storage. Fat accumulates in a single vacuole, which increases in volume and pushes the cytoplasm and nucleus to the cell periphery (Figure 23.5).

Cartilage. The connective tissue called **cartilage** cushions body parts and provides a framework for maintaining the shape of some body regions. It consists of a dense network of collagen fibers and elastic fibers, positioned firmly in a jellylike ground substance. Because of these components, cartilage can resist compression while maintaining resiliency.

Cartilage occurs on the ends of bones in many joints and in parts of the nose. It serves as a framework for the nose tip and the external ear. One kind of cartilage forms shock pads (such as intervertebral disks of the backbone). Vertebrate embryos have skeletons of cartilage, but these skeletons are replaced with bone during development.

Bone. Almost all vertebrates have a skeleton composed mostly of a connective tissue called **bone**, or osseous tissue. This tissue forms the flat plates, cylinders, and other structures called bones, which are attached to one another at joints. Bones, and the cartilage in the joints between them, constitute a skeletal system. The bones of this system function in support and protection of softer tissues and organs. Together with muscles, limb bones form a leverlike system that is used for mechanical movements of the body. Bones also function in mineral

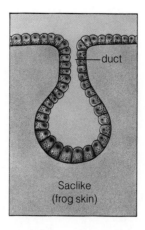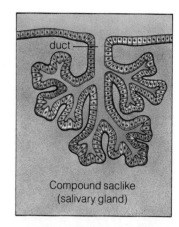

Saclike
(frog skin)

Compound saclike
(salivary gland)

Figure 23.3 Examples of glands found in glandular epithelium, as they would appear in longitudinal section.

Table 23.2	Classification of the Main Types of Connective Tissue	
Category	Main Functions	Some Common Locations
Connective tissue proper:		
Loose connective tissue	Support, elasticity	Beneath most epithelia; subcutaneous layer of skin
Dense connective tissue	Flexible, strong connections between structures; covers some organs	Tendons, around muscles, capsule around kidney, liver
Adipose tissue	Energy reserve, insulation, padding for some organs	Beneath the skin, around the kidneys, on the surface of the heart
Cartilage	Firm, flexible support; maintenance of shape; shock absorption	Nose, ear regions, tracheal rings, intervertebral disks
Bone tissue	Firm support; rigidity necessary for leverage in movements; protection of internal organs	Bones of vertebrate skeleton
Blood	Transport of varied substances to and from cells; transport medium for infection-fighting proteins and cells; stabilization of pH, temperature through body	Inside the heart, inside blood vessels

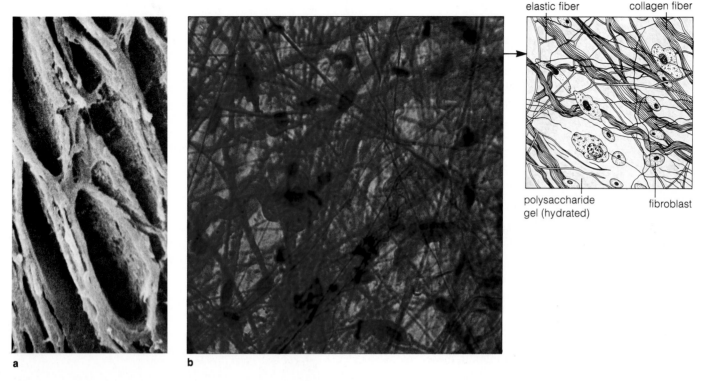

elastic fiber   collagen fiber

polysaccharide   fibroblast
gel (hydrated)

a

b

Figure 23.4 Examples of connective tissue. **(a)** Scanning electron micrograph of dense connective tissue in ligaments. Densely packed fibers give rise to this structural organization. **(b)** Light micrograph of loose connective tissue, showing the weblike scattering of cells and fibers in the semifluid ground substance.

storage, and the marrow of some bones functions in blood cell production.

Cells in bone are scattered in the ground substance, which contains a network of collagen fibers. Unlike other connective tissues, mineral salts become deposited around the fibers during development, such that the tissue becomes ossified (hardened). Calcium phosphate

nucleus

cytoplasm

fat droplet

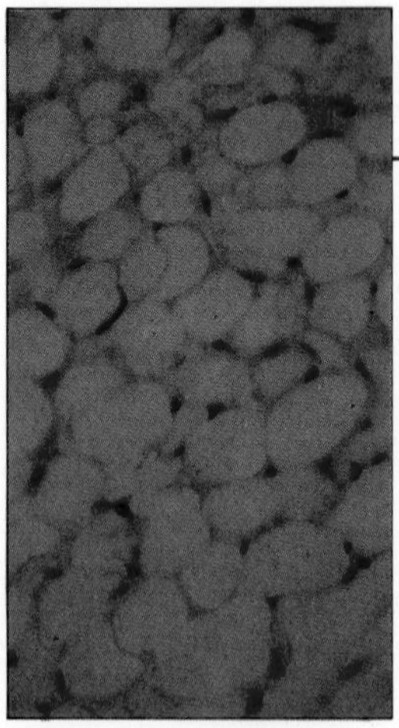

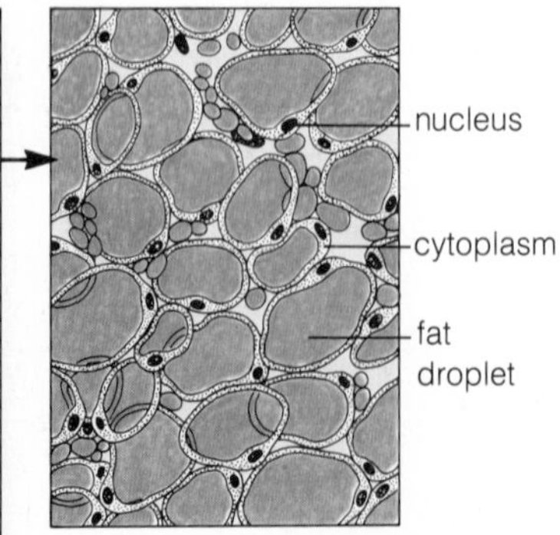

Figure 23.5 Adipose tissue, in which fat droplets fill each cell.

and calcium carbonate deposits make up sixty-seven percent of the weight of mammalian bones.

Bone tissue is not solid. "Canals" of different sizes and lengths thread through the hardened portions, and small spaces called lacunae provide chambers for bone cells. However, some bone regions are less solid than others and appear spongy rather than compact.

Spongy bone tissue is latticelike rather than dense, although it is still firm and quite strong (Figure 23.6). In the spongy tissue found in many bones (such as the breastbone), the spaces are filled with red marrow. *Red marrow* is a major region of blood cell formation. In adults, the interior cavities of most bones are filled with *yellow marrow*, which is a reserve tissue. Whenever a great deal of blood is lost, as it can be during traumatic injury, this tissue assists the red marrow in producing red blood cell replacements.

Compact bone tissue occurs as a dense layer over spongy bone tissue and forms the shaft of long bones. Compact bone helps long bones withstand mechanical stress. In compact bone, the matrix forms thin concentric layers (lamellae). The lamellae are laid down around small channels called Haversian canals. The channels, which are interconnected, run more or less parallel with the bone (Figure 23.6). Haversian canals contain blood vessels (which transport materials to and from living

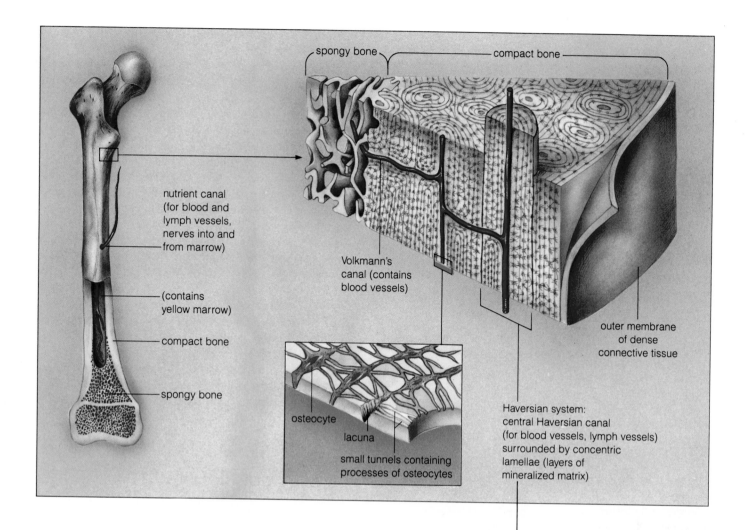

nutrient canal
(for blood and
lymph vessels,
nerves into and
from marrow)

(contains
yellow marrow)

compact bone

spongy bone

spongy bone

compact bone

Volkmann's
canal (contains
blood vessels)

osteocyte

lacuna

small tunnels containing
processes of osteocytes

outer membrane
of dense
connective tissue

Haversian system:
central Haversian canal
(for blood vessels, lymph vessels)
surrounded by concentric
lamellae (layers of
mineralized matrix)

bone cells) and nerve fibers (the signals of which influence blood vessel diameter, hence blood flow to given regions).

Blood.   Some properties of blood are summarized in Table 23.2. Understanding the nature of this tissue and its diverse functions depends on prior understanding of the organ system of which it is part. For that reason, details of blood are postponed until Chapter Twenty-Nine.

Nervous Tissue

The cells of **nervous tissue** specialize in keeping the multicelled body informed of environmental change and in controlling body responses to those changes. **Neurons**, or individual nerve cells, are organized as communication lines that extend through the body. Each can receive and initiate signals in response to specific kinds of change. We will be taking a closer look at the neuron in the next chapter.

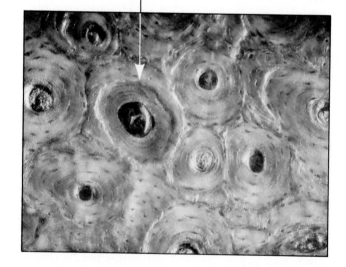

Figure 23.6 Structural organization of the long bones of mammals. The micrograph shows a Haversian system. Through these systems, living bone cells (osteocytes) receive nourishment and integrative signals (through hormones) by way of blood vessels. The osteocytes reside in lacunae (spaces in the bone tissue). Small tunnels connect neighboring lacunae.

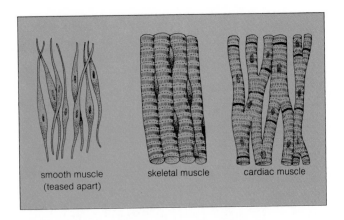

Figure 23.7 Three kinds of muscle tissue.

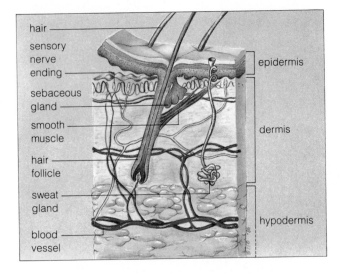

Figure 23.8 A section through vertebrate skin and the underlying hypodermis (or subcutaneous layer). Skin is a two-layered organ, with a thin outer portion (the epidermis) and a thicker, underlying portion (the dermis).

Muscle Tissue

Muscle tissue serves three functions: movement, heat production (which helps maintain body temperature), and maintenance of posture. Tissues containing muscle cells can actively shorten and passively lengthen; in other words, they can *contract* in response to stimulation, then return to their resting state. The three kinds of muscle tissues are called smooth, skeletal, and cardiac. In light micrographs, skeletal and cardiac muscle tissue appear striated (striped), with alternating light and dark bands (Figure 23.7). The bands are not present in smooth muscle.

Smooth muscle tissue consists of spindle-shaped cells held together by connective tissue. This tissue occurs in the walls of hollow internal structures, such as blood vessels, the stomach, the intestines, and some ducts. In vertebrates, smooth muscle is said to be *involuntary*, because the animal usually cannot directly control its contraction.

Skeletal muscle tissue generally attaches by tendons to the vertebrate skeleton and is responsible for voluntarily controlled movements. Skeletal muscle tissue contains many long, cylindrical cells called *muscle fibers*. Some muscle fibers in your body are up to thirty centimeters (twelve inches) long. Typically, a number of skeletal muscle cells are enveloped in connective tissue; together they form a muscle bundle. Several bundles are usually enclosed in a tougher connective tissue sheath and form muscle organs such as the biceps (Figure 28.8).

The contractile tissue of vertebrate hearts is called **cardiac muscle tissue**. Cardiac muscle cells are shorter than most skeletal muscle cells. Also, the membranes of adjacent cells are fused at regions called intercalated disks. Because of communication junctions at these fusion points, the cells do not function on an independent basis. Rather, when one receives a signal to contract, its neighbors are also stimulated into contracting along with it.

OVERVIEW OF ORGAN SYSTEMS AND THEIR FUNCTION

Having reviewed the basic types of animal tissues, we can start to think about how different tissues are combined in organs. Let's use vertebrate skin as an example of the tissue combinations that exist, even in organs that superficially might seem to be rather unremarkable in comparison to others.

Case Study: The Tissues of Skin

Skin is the largest organ of the vertebrate body. It covers the outer surface and is continuous with the mucous membranes found inside the eyelids, nostrils, and other openings of the body. The skin has two main layers, called the *epidermis* and the *dermis* (Figure 23.8). A layer called the *hypodermis* separates the skin from deeper tissues. (The hypodermis is also called the subcutaneous layer.)

Skin structure varies considerably among vertebrates. Some fishes have bony dermal scales; other fishes and amphibians have bare skin covered with mucus. Reptiles have epidermal scales, birds have feathers, and mammals, hairs. Their scales, feathers, and hair, along with beaks, hooves, horns, claws, nails, and quills, are all produced by the epidermis (Figure 23.9). Even the skin covering the human body is not the same in all regions. For example, the epidermis is thicker in areas subject to friction, such as the palms of the hands and the soles of the feet.

Epidermis. The epidermis of mammalian skin is stratified squamous epithelium. As part of a process called *keratinization*, epidermal cells in the deeper layers are transformed into dead bags, so to speak, that contain the protein keratin. Rapid cell divisions in these layers push cells toward the surface of the skin, so that fully keratinized cells pass through the uppermost layer (the stratum corneum) and are shed from the surface. In human skin, their journey takes about forty-five days. The ongoing cell divisions in the deeper layers are responsible for the skin's capacity to repair itself after cuts and abrasions.

Keratin is virtually insoluble. Together with the toughened membrane of fully keratinized cells, it allows skin to act as a barrier that prevents dehydration. The skin's impenetrability makes it a first line of defense against many toxic substances and disease-causing microbes.

Also present in the deeper layers of epidermis are cells called melanocytes, which produce the brown-black pigment melanin. Melanin affords protection from the ultraviolet wavelengths present in sunlight. When light-colored skin "tans" after exposure to sunlight, its melanin concentrations in the epidermis have increased. Other pigments besides melanin (including carotene and oxyhemoglobin) also contribute to skin color.

The surface of human skin is covered with tiny grooves and ridges that form patterns unique to each individual. These patterns are used in the identification procedure called fingerprinting.

Dermis and Hypodermis. The dermis is mostly dense connective tissue that cushions the body from everyday stresses and strains. It is tightly connected to the epidermis by an intervening basement membrane. In the dermis are hair follicles, sebaceous glands, and sweat glands, all of which grow into the dermis from the epidermis during development. Each hair grows from a follicle, and the waxy secretions of sebaceous glands associated with the follicles lubricate both the hairs and the skin surface. Muscles are attached to hair follicles, and when they contract, they "stand hair on end."

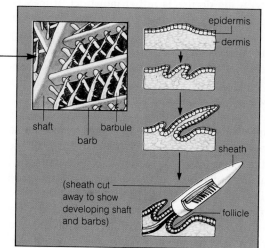

Figure 23.9 Differentiation of a region of epidermal tissue into the organ called a feather. The photograph shows a courtship display of a peacock, which relies on spectacularly specialized features to capture the attention of a peahen.

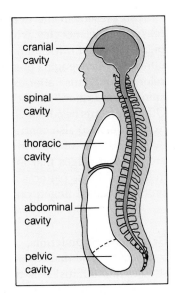

Figure 23.10 Major body cavities in humans. Blue signifies posterior cavities; white, anterior cavities.

Labels in figure:
cranial cavity
spinal cavity
thoracic cavity
abdominal cavity
pelvic cavity

Nerves and the sensory endings of nerve cells are also present in the dermis, where they gather information on touch, pressure, pain, cold, and heat.

The hypodermis is loose connective tissue, and its junction with the dermis is rather indistinct. It may be thin or thickened by the presence of adipose or skeletal muscle tissue. The adipose tissue serves to store energy and provide insulation in cold environments; the skeletal muscle tissue allows skin to move somewhat independently of other underlying tissues. Moreover, the hypodermis also serves as a water-storage tissue.

Blood vessels thread through the dermis and the hypodermis. (They are absent from the epidermis, so oxygen and nutrients from the blood must diffuse through the connective tissues to reach epidermal cells.) The circulation of blood to the skin is regulated, and changes in that flow can affect body temperature.

One final point: Skin is both strong and flexible because of the keratin in its epidermis and the collagen fibers in the dermis (and hypodermis). Young skin stretches easily to accommodate movements by the body. The plasticity of skin is reduced when people grow older and during some aging disorders, and wrinkles form. This is most likely a result of structural changes in the collagen fibers.

Major Organ Systems in Animals

Soon we will be looking at how different organs such as skin as well as complex systems of organs work to keep the animal alive in specific environments. Some assort-

ment of these organs and systems occurs in animals of almost all major phyla. Their main functions may be summarized as follows:

Integumentary system Protection from external stress, protection from loss of internal fluids, body temperature regulation, elimination of some wastes, reception of external stimuli

Nervous system Detection of external and internal stimuli; together with endocrine glands or endocrine system, control over responses to stimuli and integration of body functioning

Endocrine glands or system Internal chemical control; together with the nervous system, integration of body functions

Skeletal system Support, protection, and shaping of some body parts; in many animals, muscle attachment sites, blood cell production sites, calcium and phosphorus storage sites

Muscular system Movement of internal body parts, movement of whole body, maintenance of posture, heat production

Circulatory system Internal transport of materials to and from cells, stabilization of pH and temperature

Defense system Protection against foreign substances and agents, such as those causing infection

Respiratory system Supplying body cells with oxygen, removal of carbon dioxide wastes produced by them, pH regulation

Digestive system Ingestion and preparation of food molecules for absorption; elimination of food residues

Excretory glands or system Disposal of certain metabolic wastes; regulation of salts and fluids in cellular environment

Reproductive system Production of new individuals

Anatomical Planes and Body Cavities

Throughout this unit, we will be referring to some major body cavities to describe the location of organs and other structures. Figure 23.10 shows the location of those cavities in the human body. We also will be using some anatomical terms that apply to animals with bilateral symmetry (Figure 23.11). The body of such animals can be divided into right and left halves that are largely mirror images of each other. (For example, your body is bilaterally symmetrical; so is the body of a cat and a centipede.)

HOMEOSTASIS AND SYSTEMS CONTROL

The Internal Environment

All body cells interact with **extracellular fluid**, a medium through which substances are continuously exchanged between cells. In vertebrates, extracellular fluid falls into two categories. Most is **interstitial fluid**, which fills spaces between cells and tissues. The remainder is **plasma**, the fluid portion of blood. Vertebrate blood constantly exchanges oxygen, nutrients, and metabolic products with interstitial fluid, which then exchanges substances with the cells bathed by this fluid.

In vertebrates, most extracellular fluid is interstitial (occupying spaces between cells and tissues). The remainder is blood plasma (which is contained inside blood vessels and the heart).

Extracellular fluid is the medium through which substances move from cell to cell and from one body region to another.

For relatively simple animals such as marine sponges, the extracellular fluid bathing each cell is the sea. For more complex animals, the "sea" has been internalized. In its composition, extracellular fluid resembles seawater, especially in its high concentration of sodium ions. It also resembles seawater in its concentrations of hydrogen, potassium, calcium, and other ions. Indeed, the resemblance was one of the first clues that the early forms of life evolved in ancient seas.

Homeostatic Control Mechanisms

Through his studies of body functioning, the nineteenth-century physiologist Claude Bernard came to realize that ". . . all the vital mechanisms, varied as they are, have only one object, that of preserving . . . the conditions of life in the internal environment." For instance, he discovered that the liver absorbs many of the nutrients carried to it by blood, then converts some of them to complex storage forms such as glycogen. When nutrient levels fall below a certain point in the blood—hence in extracellular fluid—the liver degrades glycogen. The nutrients so released are distributed through the body and help maintain its cells.

In other studies, Bernard showed that the amount of blood being supplied to different body regions can be regulated by the constriction and dilation of small blood vessels. Blood carries oxygen and nutrients that sustain all body cells. Through controls over blood distribution, blood can be diverted to body regions that require increased amounts of oxygen and nutrients at any particular moment.

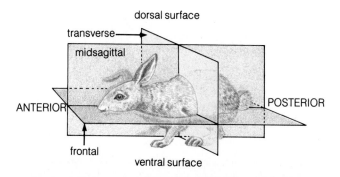

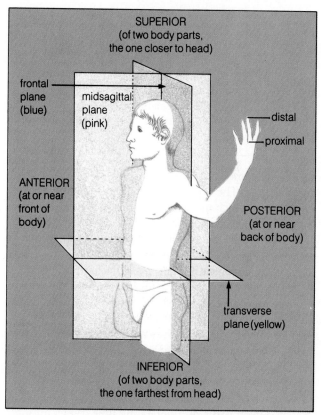

Figure 23.11 (**a**) Directional terms and planes of symmetry for bilateral animals in general. The midsagittal plane divides the body into right and left halves. The transverse plane divides the body into anterior (front) and posterior (back) parts. The frontal plane divides it into dorsal (upper) and ventral (lower) parts. (**b**) Humans move with the main body axis perpendicular to the earth. "Ventral" corresponds to anterior, and "dorsal" corresponds to posterior.

Much later, the physiologist Walter Cannon extended Bernard's line of thinking. Cannon perceived that maintaining internal conditions is possible only through coordinated **homeostatic systems** that operate to keep some physical or chemical aspect of the body within some tolerable range.

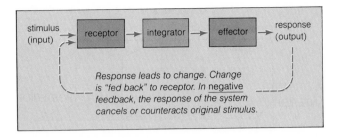

Figure 23.12 Components necessary for negative feedback at the organ level.

In multicelled bodies, homeostatic control systems maintain physical and chemical aspects of the internal environment within some range of tolerance.

Many homeostatic controls are based on feedback mechanisms. The word "feedback" means any circular situation in which information is fed back into a system. The most common is a **negative feedback system**, whereby the initial condition that causes a change is reversed (Figure 23.12). Negative feedback is at work at all levels of biological activity, from biochemical reactions in single cells to ecological interactions in the biosphere.

How does negative feedback work? By analogy, consider how a thermostatically controlled furnace operates. A thermostatic device senses air temperature and activates other devices in a furnace control system when temperature changes from a preset point. When the temperature falls below that point, the thermostat detects the change. Then the thermostat signals an integrating device, which turns on the heating unit. When air temperature has been raised to the prescribed level, the thermostat detects the change and signals the integrating device, which shuts off the heating unit. Similar controls are at work in the human body. For example, feedback mechanisms are geared to maintaining internal body temperature near 37°C (98°F) even during extremely hot or cold weather.

The regulation of organ systems requires three basic components: receptors, integrators, and effectors (Figure 23.12). Sensory cells or tissues act as **receptors** for some particular stimulus. A *stimulus* is a change in some form of energy, such as light or heat, pressure, or chemical energy. Whatever the form, the energy of the stimulus is translated into electrochemical energy by the body's receptors. This form of energy is the "signal" sent to an **integrator**, a control point where different bits of information are pulled together in the selection of a response. (In some animals, the integrating center is little more than some nerve cells clustered at the head end.

In others, the integrating center is a spinal cord and brain.) The integrating center then sends signals to muscles and glands, which are **effectors** in most animals.

Integrating centers receive information that indicates not only how the system *is* operating (the information from sensory receptors), but also how it *should be* operating (information from a "set point," which is sometimes built into the center itself). The difference between these two bits of information influences how the integrating center will respond in increasing or decreasing the activity of effectors and thereby bring the system back to its most effective operating range.

Under some circumstances, **positive feedback mechanisms** operate. These mechanisms set in motion a chain of events that *intensify* the original input. Positive feedback is associated with instability in a system. For example, sexual arousal leads to increased stimulation, which leads to more stimulation, and so on, until an explosive, climax level is reached (Chapter Thirty-Five). As another example, during childbirth, pressure of the fetus on the uterine walls stimulates production and secretion of the hormone oxytocin. Oxytocin causes muscles in the walls to contract, which increases pressure on the fetus, and so on, until the fetus is expelled from the mother's body.

In addition to negative and positive feedback mechanisms, animals also have **feedforward mechanisms**, which generally use parallel inputs to integrating centers. For example, when numerous sensory cells in human skin detect a drop in air temperature, they send signals to the brain that "anticipate" a change in blood temperature. The brain sends signals to metabolic and muscular systems that can function in raising internal body temperature. With feedforward control, corrective measures can sometimes begin even before an outside change significantly alters the internal environment.

What we have been describing here is a general pattern of monitoring and responding to a constant flow of information about the external and internal environments. During this activity, organ systems operate together in coordinated fashion. Throughout this unit, we will be asking the following questions about their operation:

1. What physical or chemical aspect of the internal environment are organ systems working to maintain as conditions change?

2. By what means are organ systems kept informed of change?

3. By what means do they process incoming information?

4. What mechanisms are deployed in response?

With these questions in mind, we will turn next to the two systems under whose dominion all others must fall—those of neural and endocrine control.

SUMMARY

1. A tissue is a group of cells and intercellular substances functioning together in a specialized activity.

2. Individual cells in tissues have properties that promote molecular recognition between similar cells and their adhesion to one another.

3. All body tissues are composed of somatic cells; the only exceptions are the germ cells, which develop into gametes.

4. Cells of the early vertebrate embryo are arranged as three primordial tissue layers (ectoderm, mesoderm, and endoderm); they divide and differentiate to form the specialized tissues of the adult body.

5. Specialized regions of the plasma membrane form cell junctions, which function in intercellular sealing, adhesion, and communication.

6. Tissues become arranged in organs, which are structures of definite form and function, the character of which depends on the tissues of which they are composed.

7. An organ system consists of two or more organs that are interrelated physically, chemically, or both in carrying out specific functions.

8. Epithelial tissue covers external body surfaces, lines its internal cavities and tubes, and forms the secretory portion of glands.

9. Connective tissue provides support for various body tissues or binds them together. Connective tissues can be classified as connective tissue proper, cartilage, bone, or blood.

10. Muscle tissue provides for movement, produces heat, and helps maintain posture. The three kinds of muscle tissue are skeletal, smooth, and cardiac.

11. Skin (the integumentary system of vertebrates) has two main layers: the epidermis and dermis. A layer called the hypodermis separates the skin from deeper tissues. The structures of skin can include scales, feathers, hair, beaks, hooves, horns, nails, and quills.

12. In addition to the integumentary system, other organ systems are the nervous, endocrine, skeletal, muscular, circulatory, defense, respiratory, digestive, excretory, and reproductive systems.

13. All body cells are bathed in extracellular fluid. It is the medium through which substances move from cell to cell and from one body region to another.

14. In complex animals, most extracellular fluid is interstitial; the remainder is blood plasma.

15. Homeostatic control systems maintain physical and chemical aspects of the internal environment within levels tolerable to living cells.

16. Many homeostatic controls are based on feedback, the most common being negative feedback mechanisms in which the response of the system decreases the original disturbance. In positive feedback, the response intensifies the original disturbance. Feedforward mechanisms allow the system to act in anticipation of a disturbance and institute corrective measures before it occurs.

17. The regulation of organ systems requires receptors, integrators, effectors, and ways to transmit information between these components. Sensory cells and tissues act as receptors for a particular stimulus. Integrating centers receive information from the receptors about how the system *is* operating and information from the "set point" about how it *should be* operating; on the basis of this information, an appropriate response is determined.

Review Questions

1. What is an animal tissue? An organ? An organ system?

2. Are there a few tissue types, or a great diversity of tissues? How can a bird wing and a human arm look and function so differently when they are constructed of the same basic tissues?

3. Name some of the functions of epithelium.

4. Describe the structure and function of one of these types of connective tissue: loose connective tissue, dense connective tissue, adipose tissue.

5. Can you describe how cartilage differs from bone?

6. Describe skeletal, cardiac, and smooth muscle tissue.

7. List the major organ systems that occur in animals, along with their main functions.

8. What characteristics of human skin are associated with the protein keratin?

9. Define homeostasis. What are the three components of homeostatic control systems?

10. What is interstitial fluid? What function does it serve in the vertebrate body? How is it functionally related to blood?

11. What are the differences between negative feedback, positive feedback, and feedforward controls?

Readings

Bloom, W., and D. W. Fawcett. 1975. *A Textbook of Histology.* Tenth edition. Philadelphia: Saunders. Outstanding reference text.

Kessel, R., and R. Kardon. 1979. *Tissue and Organs: A Text-Atlas of Scanning Electron Microscopy.* San Francisco: Freeman. Outstanding, unique micrographs.

Staehelin, L., and B. Hull. May 1978. "Junctions Between Living Cells." *Scientific American* 238(5):140–152. Excellent article summarizing research into epithelial cell junctions in animals.

Vander, A., J. Sherman, and D. Luciano. 1985. *Human Physiology: The Mechanisms of Body Function.* Fourth edition. New York: McGraw-Hill. Perhaps the clearest, in-depth introduction to human organ systems and their functioning.

24

INFORMATION FLOW AND THE NEURON

From time to time, the human body has been likened to a city, state, or some other social unit composed of separate but interdependent parts. These analogies are wonderfully optimistic about our capacity for social organization. In truth, our cities and states do not begin to approach the degree of integration of any complex animal. Whether the animal is asleep, relaxed, or alert to danger, diverse body parts are being made to work in coordinated ways. Activities of each part are continually monitored and evaluated, not for their sake alone but for how they are contributing to working patterns of the whole. In this unit, we will look at neural and endocrine systems through which activities are integrated and controlled in the animal body.

NEURONS: FUNCTIONAL UNITS OF NERVOUS SYSTEMS

The functional unit of all nervous systems is the **neuron**, or nerve cell. Neurons are specialized to produce signals that can be communicated rapidly and precisely to other cells. They cannot act alone; they function as part of local and long-distance circuits through the body. It is the *system* of neurons, not the individual neuron, that incessantly coordinates vital functions such as breathing. It is the *system* that senses change, swiftly integrates the sensory inputs, then signals body parts to carry out responses in coordination with all other activities. Parts of the system store and retrieve information about previous experiences, and these activities contribute to memory and learning. Some parts even generate information beyond that based on actual experience and thereby contribute to the highest forms of mental activity—insight and creativity.

Classes of Neurons

The nervous system of snails, spiders, frogs, humans, or any other complex animal is constructed of only three classes of neurons, which are called sensory neurons, interneurons, and motor neurons. We can define these three classes in terms of the roles they play in a control scheme presented earlier, whereby the body monitors and responds to a constant flow of information about change (page 310):

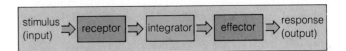

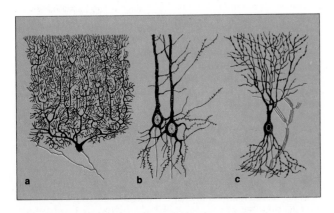

Figure 24.1 Examples of structural diversity in neurons. Those shown occur in the brains of mammals. Axons are shaded in blue. The profuse branches and tufts are dendrites.

Sensory neurons serve as the receptors or are activated by receptors. They are stimulated by changes in the external or internal evironments and respond by relaying signals into integrating centers. In these centers are **interneurons**, which integrate information arriving on incoming sensory lines, then influence other neurons in turn. Your central nervous system (brain and spinal cord) is an example of an integrating center. Motor neurons relay information away from integrating centers, to the body's effectors (muscle and gland cells). Responses to change are initiated when the signals carried by motor neurons alter the activity of effector cells.

Keep in mind that most neurons do not actually touch one another or effector cells. Tiny junctions (chemical synapses) exist between them, and the chemical substances that serve as signals must be transmitted across those junctions.

Structure of Neurons

All neurons have a cell body, and most have slender cytoplasmic extensions called nerve cell "processes." The processes differ in number and in length (Figure 24.1), and they do so to such an extent that there really is no such thing as a "typical" neuron. The one described most often is the motor neuron that *innervates* (carries neural signals to) striated muscles in vertebrates.

The motor neuron shown in Figure 24.2 has many short, slender processes called **dendrites** and a long, cylindrical process called an **axon**. Axons of this sort typically branch, then each branch splits into finer branchings with specialized endings. The endings, called axon terminals, make functional connections with muscle cells. Traditionally, the dendrites (and often the cell body) are viewed as the input zone, where the neuron receives and integrates most information. The conducting zone extends from the start of the axon to its terminals. The axon terminals are output zones.

This picture of functional zones is oversimplified, for a neuron can have many input and output sites. Even so, it is still a good starting point for considering the principles of neural function.

Neuroglia

Nonnervous cells, collectively called **neuroglia**, represent at least half of the volume of vertebrate nervous systems. The term neuroglia ("nerve glue") was coined when it became apparent that such cells help impart structure to the brain, much as connective tissues do for other body regions. Neuroglial cells play other roles, some known, others not yet identified. Some support the neurons metabolically. Others produce sheaths around axons, and these sheaths affect how fast a signal travels over long distances. Still others are scavengers that clean up the debris when neurons are damaged or die.

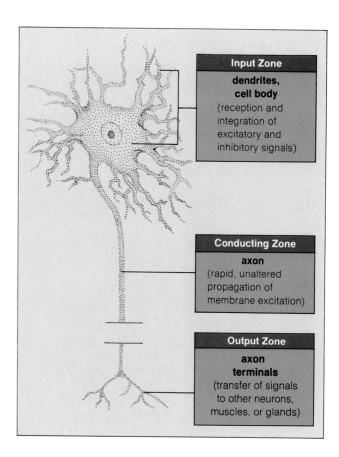

Figure 24.2 Main information-processing zones of a motor neuron present in vertebrates. For neurons other than motor neurons, signals are transferred to other neurons at the axon terminals.

Nerves and Ganglia

Axons are not scattered like tangled spaghetti through all body tissues. To the contrary, axons of sensory or motor neurons are often packed tightly in bundles within connective tissue to form **nerves** (Figure 24.3). These cordlike communication lines connect the brain and spinal cord with the rest of the body. Within the brain and spinal cord, bundles of axons are called nerve *pathways* or *tracts*.

In these communication lines, the cell bodies of the neurons are not strung out here and there, like beads on a chain. Typically the cell bodies are clustered together into distinct structures. Within the brain or spinal cord, such clusters are called *nuclei*; in other body regions they are called *ganglia*.

Much of the human brain consists of enormous numbers of neuron cell bodies. The evolution of the human brain is a fascinating story, one that will make more sense if we begin with the properties of neurons that made it possible.

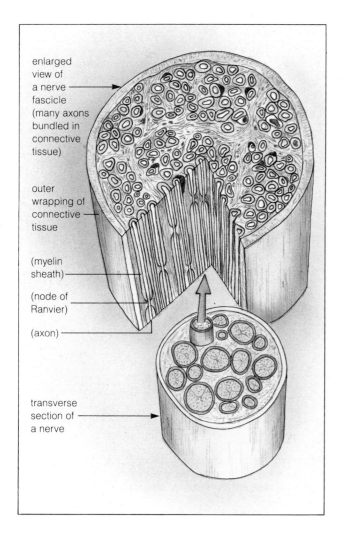

enlarged
view of
a nerve
fascicle
(many axons
bundled in
connective
tissue)

outer
wrapping of
connective
tissue

(myelin
sheath)

(node of
Ranvier)

(axon)

transverse
section of
a nerve

Figure 24.3 Structure of a nerve.

ON MEMBRANE POTENTIALS

To understand how a nervous system works, we can begin with the nature of the signals carried by a neuron. As we have seen, the concentrations of ions and other charged substances inside and outside the plasma membrane of cells are not the same, and as a result of those differences, the cytoplasm is more negatively charged than the fluid outside. When a neuron is stimulated enough, this polarity of charge across the plasma membrane switches for a fleeting moment (the inside briefly becomes more positive than the outside). The reversal is so fleeting that it is like an electrical impulse. Neurons and other cells that produce these impulses are said to show *membrane excitability*.

Membrane excitability depends on two properties. First, being mostly lipid, a plasma membrane tends to limit ion movements across it—yet it is so thin that positively charged ions on one side are attracted to negatively charged ions on the other side. Thus the membrane "stores" ions of opposite charge at its two surfaces. This property is called capacitance.

Second, the plasma membrane has channels through which certain ions can move into and out of the neuron. Some channels are always open, but others have *voltage-sensitive gates*, which open in response to signal-induced changes in the charge across the membrane (Figure 24.4). Some membranes have larger and more numerous channels than others, hence pathways for the flow of ionic current across the membrane can differ. This membrane property (called conductance) is a measure of its ion permeability.

What we have, then, are (1) a separation of ions of opposite charge at the membrane and (2) gated channels for the controlled movement of ions across the membrane. Let's see how these two factors are put to work.

Figure 24.4 Pathways for ions across the plasma membrane of a neuron. These pathways are provided by proteins embedded in the lipid bilayer. Certain ions diffuse through perpetually open protein channels. Other protein channels have gates that only open when the neuron is stimulated; when they are opened, ion movements are more rapid across the membrane. Also, a membrane-bound enzyme system (the sodium-potassium pump) actively transports sodium in one direction and potassium in the opposite direction across the membrane.

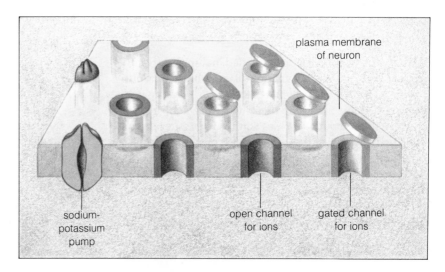

plasma membrane
of neuron

sodium-
potassium
pump

open channel
for ions

gated channel
for ions

The Neuron "At Rest"

Visualize a neuron at rest, when its plasma membrane is not responding to a stimulus. Then, the inside is more negatively charged than the outside, so there is an electric gradient across the membrane (page 89). We say there is a "voltage difference," which in many cells holds steady at about seventy millivolts. (A millivolt is simply a unit for measuring a voltage difference: the amount of potential energy between two differently charged regions.) The steady voltage difference across the plasma membrane is called the **resting membrane potential**.

The electric gradient depends largely on the unequal distribution of sodium ions (Na^+) and potassium ions (K^+) across the membrane. The fluid outside the membrane has far more sodium and far less potassium than the cytoplasm:

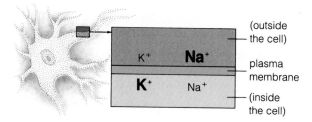

where large letters denote which side of the membrane has the greatest concentration. (As an example of the magnitude of this difference, consider a neuron from a cat. For every 150 potassium ions in a given volume of cytoplasm, there are only 5 in the same volume of fluid outside. For every 15 sodium ions on the inside, there are 150 on the outside.)

What processes establish the ion distributions and the voltage difference across a resting membrane? Two processes are involved:

---

1. A membrane-bound enzyme system, called the **sodium-potassium pump**, actively transports both sodium and potassium across the resting membrane and thereby establishes the ion distributions.

2. Leaks of sodium and potassium ions establish the voltage difference. (*Leaks* refers to passive diffusion of those ions through perpetually open channels across the membrane).

You may be wondering what difference it makes whether sodium and potassium are leaking and being pumped one way or the other across the membrane. After all, both kinds of ions are positively charged—so how can their movements establish a "polarity of charge"? One reason is that the resting membrane is somewhat permeable to potassium, but it is *not* permeable to large, negatively charged ions present in the cyto-

plasm. Thus, when potassium ions diffuse out and leave those negative ions behind, the inside becomes more negative. (Inward leaks of sodium only partly counteract the resulting increase in negative charge, because the resting membrane is much less permeable to sodium than to potassium.)

Under resting conditions, the membrane pumps balance the leaking processes just described and thereby maintain the membrane potential (Figure 24.5). Even while some sodium ions are leaking in, the pumps are sending others back out. Even while some potassium ions are leaking out, the pumps are bringing others back in. (Also, some potassium ions are attracted by the more negative interior and diffuse back in on their own.) What is the overall effect of the pumping and leaking processes? Overall, there is no *net* movement of sodium and potassium ions across the resting membrane.

Changes in Membrane Potential

Suppose our neuron "at rest" is a sensory neuron that detects pressure (mechanical energy) applied to a thumb. When pressure stimulates the neuron, it causes brief electrical disturbances. The disturbances lead to changes in membrane potential, which are called **graded potentials** because they can vary in magnitude, depending on the stimulus energy. (In this case, the stronger the pressure, the greater the electrical disturbance.) Generally, a single graded potential is not enough to excite a neuron into doing anything. But many graded potentials acting at the same time can change the voltage difference across the membrane enough to initiate an **action potential**, or a dramatic reversal in the polarity of charge.

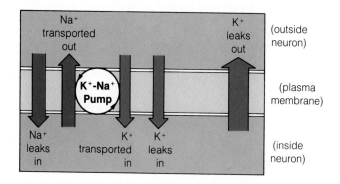

Figure 24.5 Balance between pumping and leaking processes that maintain the distribution of sodium and potassium ions across the plasma membrane of a neuron at rest. The relative widths of the arrows indicate the magnitude of the movements. The inward movement counteracts the outward movement for each kind of ion; hence the ion distributions are maintained.

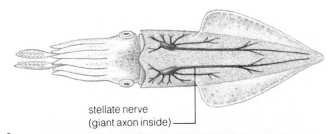

stellate nerve
(giant axon inside)

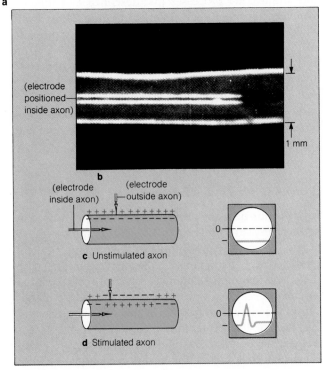

(electrode positioned inside axon)

1 mm

b

(electrode inside axon) (electrode outside axon)

0 —

c Unstimulated axon

0 —

d Stimulated axon

Figure 24.6 (**a**) Approximate location of giant axons that innervate the muscular body wall (the mantle) of the squid *Loligo*. (**b**) The micrograph shows the axon diameter relative to the size of an electrode, a device used in measuring voltage changes. Being large enough to accommodate such devices, the giant axon lent itself to early studies of nerve functioning. (**c**) In a resting neuron, the inside of the axon is negative with respect to the outside, as registered on the screen of an oscilloscope. When the electrodes detect an action potential, a wave form of the sort depicted in (**d**) appears on the screen.

THE ACTION POTENTIAL

An action potential is analogous to a pulse of electrical activity (hence its original name, *nerve impulse*). Its existence can be demonstrated with the use of so-called giant axons from the squid *Loligo*. These axons, which innervate the muscular body wall of the squid, can be up to a millimeter in diameter. This means that an electrode (a device used to measure voltage changes) can be inserted rather easily into one of them. One fine electrode is inserted into the axon; another is positioned in the fluid outside the axonal membrane. Both are connected to a voltmeter and an oscilloscope. With an oscilloscope, voltage changes show up as deflections of a beam traveling across a fluorescent screen (Figure 24.6).

Measurements of the voltage difference across a membrane before, during, and after an action potential reveal this pattern:

1. In a neuron at rest, the membrane is *polarized* (the inside is more negative with respect to the outside).

2. During an action potential, the membrane is *depolarized* (the inside is more positive with respect to the outside).

3. Following an action potential, the membrane is *repolarized* (resting conditions are restored).

Mechanism of Excitation

What sort of ion movements underlie an action potential? The most important are increased movements of *sodium* into the neuron. Sodium ions do not move in significant amounts across the resting membrane. First, there are not many perpetually open channels for sodium, and second, the gated channels for sodium are shut during resting conditions. However, under strong enough stimulation, the gates open to the extent that membrane permeability to sodium increases by several hundredfold. Sodium diffuses rapidly down its steep concentration and electric gradients, into the neuron.

When positively charged sodium ions pass through the gated channels, voltage on the inside becomes less negative. The changing voltage causes more gates to open, which admits more sodium, which increases the positivity inside, and so on until the membrane potential reverses. The escalating flow of sodium is an example of *positive feedback*, whereby an original event is increasingly intensified as a result of its own occurrence.

The minimum voltage change needed to trigger an action potential is called the **threshold** (Figure 24.7). Once threshold is reached, membrane permeability no longer depends on the strength of the stimulus (because the positive feedback cycle that causes the inward rush of sodium is now operating strongly enough to depolarize the cell). Thus, action potentials are **all-or-nothing**

events: If threshold is reached, all of the associated membrane permeability changes will occur; but if threshold is not reached, they will not occur at all.

Once a stimulus causes enough of the gated sodium channels to open (that is, once threshold is reached), the potential energy stored in the concentration and electric gradients for sodium is released automatically. Thus the amount of energy released during the action potential is not related to the strength of the stimulus, any more than the effect of dynamite depends on the size of the match that lit the fuse.

Duration of Action Potentials

Each action potential lasts only a few milliseconds. Several hundred can occur in a neuron in a single second. Certainly billions of action potentials occur in your nervous system during the time it takes you to read about a single one of them.

Why does an action potential end so abruptly? There are two reasons. First, there is *another* gate on the gated channel for sodium. When voltage inside goes from negative to positive during an action potential, this opposing gate slams shut and closes off the sodium flow. Second, about halfway through the action potential, membrane permeability to potassium increases. More potassium now moves out of the neuron (through gated channels). These movements restore the voltage across the membrane to its original value.

Even though the voltage is restored, the sodium and potassium gradients are ever so slightly reduced in size at the end of many action potentials. These tiny reductions in concentration gradients must be countered; otherwise the ability of the neuron to respond to stimuli would gradually disappear. The gradients are restored by the ongoing sodium-potassium pumping process.

Propagation of Action Potentials

Once an action potential has been triggered, it is "self-propagating" along the membrane. In other words, each action potential triggers a *new* one at an adjacent area of the membrane, because the local current flow associated with it changes the membrane permeability. As each new action potential occurs, sodium moves into the neuron and potassium moves out (Figure 24.8). The new disturbance affects the permeability of the next adjacent region, and so it goes, away from the original stimulation point. These alterations in voltage across the membrane constitute the "message" that is propagated along the neuron.

Action potentials always travel in a direction away from the stimulation site. There is no backflow, largely because each action potential is followed by a **refractory period**. This is a time during which the membrane is

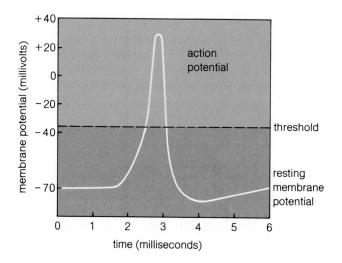

Figure 24.7 Recording of an action potential.

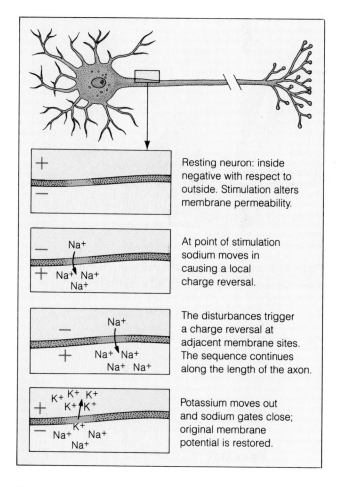

Resting neuron: inside negative with respect to outside. Stimulation alters membrane permeability.

At point of stimulation sodium moves in causing a local charge reversal.

The disturbances trigger a charge reversal at adjacent membrane sites. The sequence continues along the length of the axon.

Potassium moves out and sodium gates close; original membrane potential is restored.

Figure 24.8 Simplified picture of how action potentials are propagated along the axon of a neuron.

insensitive to stimulation. It occurs while sodium gates are shut tight and potassium gates are wide open during restoration of the membrane potential. After the resting membrane potential has been restored, the electrical disturbance has moved away from it, so the sodium gates no longer open in the region of the original action potential.

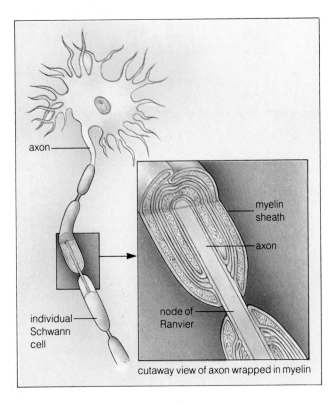

Figure 24.9 Myelinated axon of a motor neuron, formed by wrappings of Schwann cell membranes (shaded in gold).

Saltatory Conduction

Many axons that connect different parts of the nervous system are long and thin, and neuroglial cells form protective sheaths around them. Not only do the sheaths afford protection, some play a passive role in message conduction over long distances.

For example, in peripheral nerves (those outside the central nervous system), modified neuroglial cells called **Schwann cells** have a plasma membrane that becomes wrapped around certain axons like a rolled-up pancake (Figure 24.9). The membrane is so rich in lipids and so tightly wrapped around the axon that it forms a specialized *myelin sheath*. Each sheath is separated from the next by a *node of Ranvier*: a small gap where the axon is exposed to the surrounding fluid. Like the cytoplasm inside the axon, this fluid has low electrical resistance, and current moves readily through it. Each node of Ranvier enhances the movement of current between the extracellular fluid and the cytoplasm. Each node is loaded with gated channels for sodium—some 12,000 per square micrometer! No other axonal membrane comes close to having this channel density.

In contrast, the lipid-rich regions of myelin between nodes of Ranvier have high electrical resistance. This means that current cannot move as readily across the neural membrane where the wrappings occur, so it tends to flow along the outside and inside of the nerve cell process rather than leak across the membrane. In a manner of speaking, the action potential "jumps" from one node to the next in line (Figure 24.10). The rapid, node-to-node hopping via conducting fluids is called **saltatory conduction**, after the Latin word meaning "to jump."

Saltatory conduction affords the best possible conduction speed with the least metabolic effort by the cell. In the largest myelinated axon, signals travel 120 meters per second (270 miles per hour).

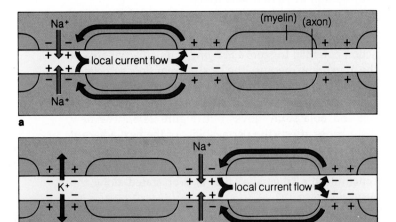

Figure 24.10 Conduction of action potentials in myelinated neurons. (**a**) The influx of sodium ions at the first node triggers an action potential (red arrows), which generates a local current flow to the second node. (**b**) The current flow causes membrane changes that trigger a new action potential, which generates a local current flow to the third node. (Meanwhile, potassium ions leave the first node and restore the resting membrane potential at the first node in line.)

SYNAPTIC POTENTIALS

Chemical Synapses

An action potential being propagated along a neuron is a "message" about some form of stimulation. But the "message" has no meaning for the body as a whole unless the excitation can be conveyed to other neurons or to muscle or gland cells that are necessary for carrying out the response. As it happens, excitation is conveyed from one cell to another at a type of cell junction called the **chemical synapse**. (Cell junctions, recall, are local regions of the plasma membrane where neighboring cells are linked structurally, functionally, or both.) The simplest chemical synapses are sites of near-contact between the terminals of one neuron and some part of another neuron. Only a small extracellular space, the *synaptic cleft*, separates the two cells (Figure 24.11).

At a chemical synapse, information flows *from* the cell that releases molecules of a **transmitter substance** into the synaptic cleft. This substance can change the membrane potential of the other cell or alter its metabolic activity. One transmitter substance is acetylcholine (ACh), which has excitatory or inhibitory effects, depending on the properties of the target cell. Gamma aminobutyrate (GABA) has mostly inhibitory effects.

The neuron that releases transmitter molecules into the cleft is said to be the **presynaptic cell**; the one affected by the molecules is the **postsynaptic cell**. Often it is easy to tell which is which in micrographs, because only one contains numerous vesicles near the synaptic site. These vesicles are thought to be filled with some transmitter substance (Figure 24.11).

Because there is a cleft between the presynaptic and postsynaptic cells, an action potential cannot be propagated directly from one to the other. What happens is this: The arrival of an action potential alters the membrane of the presynaptic cell near the cleft. The membrane becomes more permeable to calcium ions, and when the calcium moves into the cell (down its concentration gradient), vesicles housing the transmitter substance fuse with the membrane. Molecules of the transmitter substance are expelled into the cleft (through exocytosis, page 94), and diffusion carries them to the adjoining cell. These molecules do not enter the postsynaptic cell. Instead they interact briefly with membrane receptors, as will now be described.

Excitatory and Inhibitory Postsynaptic Potentials

Some transmitter substances change the membrane permeability of postsynaptic cells. The membrane responses are graded; they can vary, depending on the amount of transmitter substance reaching receptors and on the state of the recipient cell. Such graded potentials,

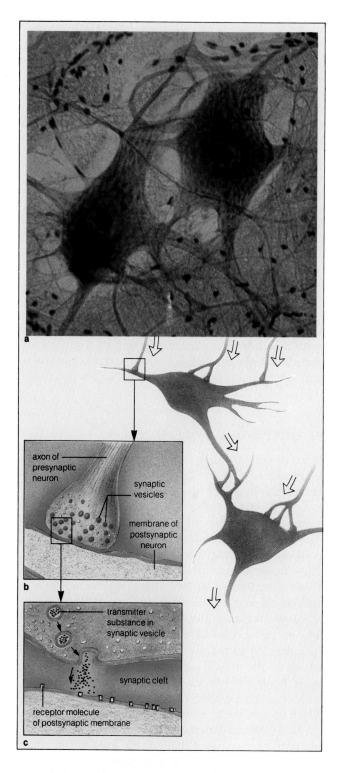

Figure 24.11 Example of a chemical synapse between two neurons. In (**c**), a synaptic vesicle fuses with the presynaptic cell membrane, and its contents are expelled into the synaptic cleft.

arising as they do at synapses, are called **synaptic potentials**.

Like all graded potentials, the synaptic potentials decay a short distance away from the stimulation site. Their function is not to propagate signals over long distances, but rather to influence the membrane so that an action potential *can* be fired off or squelched. Thus a synaptic potential can be either excitatory or inhibitory:

excitatory postsynaptic potential (EPSP)	*brings the postsynaptic membrane closer to threshold (it is depolarizing)*
inhibitory postsynaptic potential (IPSP)	*drives the postsynaptic membrane away from threshold (it is hyperpolarizing) or maintains it at its resting level*

It takes a depolarization of up to twenty-five millivolts to drive a resting membrane to threshold—but a single EPSP provides only about half a millivolt. The input of hundreds of excitatory potentials must be combined before a large postsynaptic cell (such as a motor neuron) will respond to a stimulus with action potentials.

Synaptic Potentials in Muscle Cells

Much of what is known about chemical synapses came from Bernard Katz and his colleagues, who studied **neuromuscular junctions** (the synapses between motor neurons and muscle cells). Where an axonal branching approaches a muscle cell membrane, it is splayed out in finer branchings. The axon terminals in this region are positioned in troughs in the muscle cell membrane. These troughs are called the *motor end plate* (Figure 24.12).

An action potential traveling along the motor axon spreads through all the terminals, which respond by releasing the transmitter substance ACh. Within a mere fifty microseconds, ACh molecules diffuse across the synaptic cleft to the muscle cell. Their arrival is like a whack on the muscle cell membrane. They give rise to many graded responses that are summed to initiate a train of action potentials along the muscle membrane, which in turn leads to contraction (page 383).

ACh interacts with receptors on the muscle cell membrane and causes gated channels for sodium and potassium to open. By some estimates, there are about 10,000 receptors (hence 10,000 gates) in each square micrometer of membrane. While all of those gates are being opened and shut, enzyme molecules on the membrane are inactivating ACh. The enzyme (acetylcholinesterase) gives this particular transmitter molecule no more than 1/500 of a second to act. Having helped open a gate, the ACh molecule is broken down and the receptor site is cleared for the next incoming signal—all of which is for the best,

because in this way the muscle cell is not eternally stimulated.

When foreign substances in the body interfere with synaptic function, the consequences can be deadly. For example, on rare occasions the anaerobic bacterium *Clostridium tetani* is carried accidentally into the body's internal tissues. This can occur when the skin is punctured, as from an animal bite or by a soil-covered nail. If the tissues become necrotic (die off, as happens when the injury prevents further delivery of oxygen to them), the bacterium can multiply. One of its metabolic products functions as a neurotoxin in this setting. It interferes with inhibitory synapses on motor neurons in the brain and spinal cord. The unbalanced excitation causes excessive contraction (as in lockjaw). Muscles cannot be released from the contracted state. The result is *tetanus*: prolonged, spastic paralysis of muscles that can lead to death.

FROM SYNAPSE TO NEURAL CIRCUIT

Synaptic Integration

At each neuron in a nervous system, excitatory and inhibitory signals compete for control of the membrane potential. As Charles Sherrington perceived, the competing signals are *combined* in the neuron, in a process called synaptic integration.

Synaptic integration is the moment-by-moment combining of excitatory and inhibitory signals acting on adjacent parts of a neuron.

Suppose, for example, an excitatory synapse and an inhibitory synapse nearby become active at the same time. The white lines in Figure 24.13 show how each resulting change in membrane potential would register on an oscilloscope screen *if it were occurring alone*. The red line shows what happens when they occur simultaneously. The combined effect is that the excitatory potential is pulled away from the threshold of an action potential.

This simple example should not mislead you into thinking that all synapses have equal influence. The outcome is influenced by the direction and magnitude of the current flows that are actually produced. For instance, some inputs are closer than others to the start of an axon. Because action potentials typically are initiated here, those inputs have a stronger effect. Synaptic potentials are not self-propagating; they decay as they spread from the synapse. The farther they are from the start of the axon, the less effect they have. Also, the faster they are produced in succession, the more effect they can have.

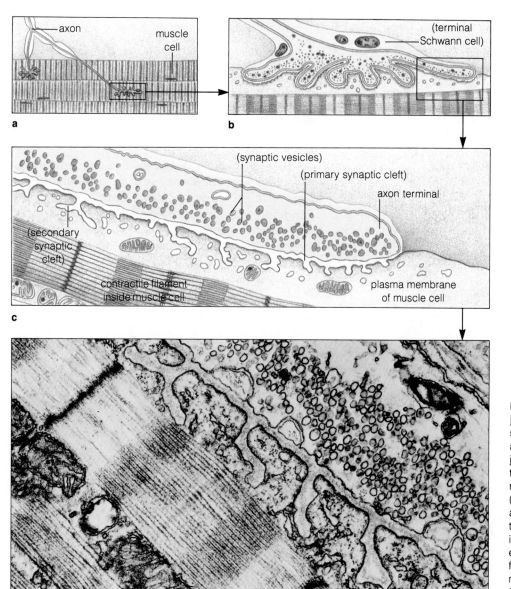

axon

muscle cell

a

(terminal Schwann cell)

b

(synaptic vesicles)

(primary synaptic cleft)

axon terminal

(secondary synaptic cleft)

contractile filament inside muscle cell

plasma membrane of muscle cell

c

d

Figure 24.12 A neuromuscular junction (the region of chemical synapse between a neuron and a muscle cell). (**a**) At this junction, axon terminals act on troughs in the muscle cell membrane (the motor end plate). (**b**) The myelin sheath of the axon stops at the junction, such that the membranes of the two interacting cells are exposed to each other. (**c**) Secondary clefts form deep channels into the muscle cell membrane. (**d**) Transmission electron micrograph of the junction.

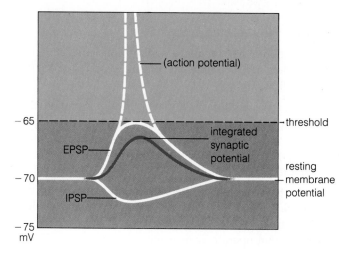

(action potential)

−65 — threshold

EPSP

integrated synaptic potential

−70 — resting membrane potential

IPSP

−75
mV

Figure 24.13 Synaptic integration. In this example, an excitatory synapse and an inhibitory synapse nearby are activated at the same time. The IPSP reduces the magnitude of the EPSP from what it could have been, pulling it away from threshold. The red line represents the integration of these two synaptic potentials. Threshold is not reached in this case; hence an action potential cannot be initiated.

Circuit Organization

Typically, action potentials spread outward along the neural membrane, away from the cell body. The *direction* in which a given message flows through the nervous system depends on the way neighboring neurons and their processes are organized relative to one another, as the arrows in Figure 24.11 suggest. The synapses among neurons form *circuits* for transmitting and processing the electrical and chemical signals. The circuits may involve only a few neurons; they may involve hundreds, thousands, even millions.

When message transfers are confined to a given part of the brain or spinal cord, the interacting neurons form a "local circuit." Local circuits function in processing messages, integrating them with other inputs, and controlling the output to other regions. At a different level of organization are the nerve pathways or tracts. These circuits are formed by connections between two or more regions in the brain or spinal cord.

Reflex Arcs

One of the simplest examples of a nerve pathway is the **reflex arc**. A *reflex* is a simple, stereotyped, and repeatable motor action that is elicited by a sensory stimulus.

Many types of reflex arcs are necessary in maintaining and adjusting body posture. One is the **stretch reflex**, which is the tendency of striated muscle to contract in response to being stretched. Even when you are not aware of the stretch reflex, it is helping you maintain an upright posture despite small shifts in your balance. This reflex also takes part in voluntary movements, such as helping to maintain arm muscles under tension when you are carrying a heavy package that would otherwise pull your arm down.

As Figure 24.14 shows, stretch-sensitive receptors are located within skeletal muscles. These receptors are part of **muscle spindles**, which are made of small muscle cells enclosed in a sheath that runs parallel with the muscle itself. The receptors are the endings of sensory neuron processes.

When a muscle spindle is stretched, action potentials are triggered in the sensory axon. These potentials are conducted rapidly toward the spinal cord, where the axon terminals synapse with (among other things) motor neurons—which have axons leading right back to the muscle containing the spindle that was stretched. Under enough stimulation, action potentials are generated in the motor neurons. They travel to the axon terminals, there to activate the muscle cell membrane and initiate contraction. Thus the stretch reflex may help maintain a degree of tension in the muscle.

In a few reflex arcs, sensory neurons synapse directly on motor neurons. Such direct connections, in which no other neurons intervene to mediate the response, are said to be *monosynaptic pathways*. John Nicholls identified three monosynaptic pathways in the leech. These pathways are involved in the **withdrawal reflex**, a rapid pulling away from an unpleasant or potentially harmful stimulus. By simple reflex action, this invertebrate jerks away from touch, pressure, and noxious chemicals.

In other invertebrates, the withdrawal reflex involves *polysynaptic pathways*. Here, sensory neurons in the pathway make connections with a number of interneurons, whose processes collectively activate or suppress the motor neurons necessary for a coordinated response. The withdrawal reflex is also governed by polysynaptic pathways in vertebrates. If you have ever accidentally touched a hot stove, you know that the withdrawal reflex action can be completed even before you become conscious that it has occurred.

SUMMARY

In this chapter, we have considered several aspects of the neuron. Before we turn to the systems based on this specialized cell, let's put the key points in perspective:

Excitability:

1. All cells show a difference in electric charge across the plasma membrane, in that the inside is more negative than the outside.

2. Neurons and some other cells show excitability: the polarity of electric charge across the membrane can undergo brief, dramatic reversals (action potentials) in response to stimulation.

3. Neurons *collectively* sense environmental change, swiftly integrate sensory inputs, then activate effectors (such as muscles and glands) that can carry out coordinated responses.

The Resting Membrane Potential:

1. A neuron has far more sodium ions and far fewer potassium ions outside the plasma membrane than it has inside.

2. The ion distributions impose concentration and electric forces on the ions themselves. In response to these forces, sodium ions leak in and potassium ions leak out, in ways that create a voltage difference across the membrane.

3. A steady voltage difference is maintained by two opposing activities: (1) leaks of potassium and sodium down their gradients, and (2) the action of membrane pumps that operate against the gradients.

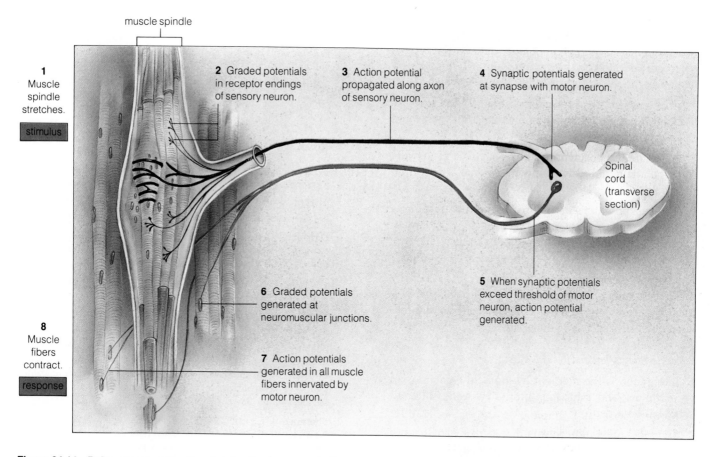

Figure 24.14 Reflex arc governing the stretch reflex in mammals. Sensory axon is shown in purple; motor axon in orange. Other inputs and outputs exist, but they are omitted here for the sake of clarity.

In the figure:

1 Muscle spindle stretches. *stimulus*

2 Graded potentials in receptor endings of sensory neuron.

3 Action potential propagated along axon of sensory neuron.

4 Synaptic potentials generated at synapse with motor neuron.

5 When synaptic potentials exceed threshold of motor neuron, action potential generated.

6 Graded potentials generated at neuromuscular junctions.

7 Action potentials generated in all muscle fibers innervated by motor neuron.

8 Muscle fibers contract. *response*

muscle spindle

Spinal cord (transverse section)

4. Because of the leaking and pumping activities, the inside of an unstimulated neuron stays more negative (commonly by about seventy millivolts) with respect to the outside. A steady voltage difference across the membrane is the resting membrane potential.

The Action Potential:

1. Ions can move through channels across the membrane. Some channels have voltage-sensitive gates, which open in response to changes in membrane potential.

2. When a resting neuron is strongly stimulated, the membrane potential reaches threshold. Voltage-sensitive gates on sodium channels open in an accelerating way (due to positive feedback). The outcome is an action potential: a sudden, dramatic reversal of the polarity of charge across the membrane.

3. An action potential is an all-or-nothing event. If threshold is reached, all of the associated membrane changes occur. If it is not reached, they do not occur.

4. An action potential lasts for only a few milliseconds. Sodium gates shut, potassium gates open, and ion movements across the membrane restore the original voltage difference.

5. Although voltage is restored, the sodium and potassium gradients are slightly reduced. Membrane pumps restore the original gradients.

6. An action potential is self-propagating: it triggers the same permeability changes in the adjacent membrane region, and so on away from the stimulation site.

Synaptic Potentials:

1. Chemical synapses are junctions between two neurons, or between a neuron and a muscle or gland cell. Only a small gap (the synaptic cleft) separates them.

2. At a chemical synapse, only one of the cells releases transmitter substance into the synaptic cleft. Its action identifies it as the *pre*synaptic cell, which relays signals to the other, *post*synaptic cell.

3. Transmitter molecules bind briefly to receptors on the postsynaptic cell membrane. Depending on the type of receptors activated on the postsynaptic cell, the cell may respond with excitatory or inhibitory potentials.

4. An excitatory postsynaptic potential (EPSP) is depolarizing; it brings the membrane closer to the threshold of an action potential.

5. An inhibitory postsynaptic potential (IPSP) is usually hyperpolarizing; it drives the membrane away from threshold.

6. Synaptic potentials are graded membrane responses at chemical synapses. "Graded" means they can vary in magnitude, depending on the stimulus energy. They are not all or nothing and they are not self-propagating.

7. Many excitatory potentials acting together are usually needed to initiate a train of action potentials, and they must be strong enough to override the effects of inhibitory potentials at nearby synapses.

Synaptic Integration and Neural Circuits:

1. Integration is the moment-by-moment combining of all excitatory and inhibitory inputs acting at different synapses on a neuron.

2. Synaptic integration means that signals arriving at a neuron can be reinforced or dampened, sent on or suppressed.

3. The direction of information flow through the body depends on the organization of neurons into interconnecting circuits.

6. Define sensory neuron, interneuron, and motor neuron.

7. What is a synapse? Explain the difference between an excitatory and an inhibitory synapse. Then define neural integration.

8. What is a reflex? Describe the sequence of events in a stretch reflex.

Readings

Berne, R., and M. Levy (editors). 1983. *Physiology*. St. Louis: Mosby. Section II is an authoritative introduction to neural functioning.

"The Brain." September 1979. *Scientific American* 241(3). This entire issue is devoted to brain structure, functioning, and development.

Dunant, Y., and Israël, M. 1985. "The Release of Acetylcholine." *Scientific American* 252(4):58–83. Experiments showing how this major neurotransmitter functions.

Llinás, R. R. October 1982. "Calcium in Synaptic Transmission." *Scientific American* 247(4):56–65.

Morell, P., and W. T. Norton. May 1980. "Myelin." *Scientific American* 242(5):88–116.

Shepherd G. M. February 1978. "Microcircuits in the Nervous System." *Scientific American* 238(2):93–103. For almost a hundred years message conduction has been described largely in terms of axonal pathways. This article describes recent discoveries of inhibitory feedback loops between dendrites only.

Sherrington, C. 1947. *The Integration Action of the Nervous System*. New Haven, Connecticut: Yale University Press. Sherrington was an outstanding neuroscientist, and a poet as well; his writing is near-lyrical, and his insights still rewarding.

Review Questions

1. Label the functional zones of a motor neuron on the following diagram:

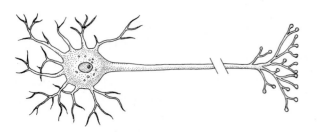

2. What is the difference between a neuron and a nerve?

3. Two major concentration gradients exist across a neural membrane. What are they, and how are they maintained?

4. An electric gradient also exists across a neural membrane. Explain what the electric and concentration gradients together represent. What is a nerve message?

5. Distinguish between an action potential and a graded potential. What is meant by "all-or-none" and "self-propagating" messages?

NEURAL PATTERNING: AN OVERVIEW

Of all organisms on earth, why do animals alone have nervous systems? Plants rely on hormonal integration of activities, and if hormones, why not nerves? Plants also move their leaves in response to environmental stimuli. The Venus flytrap even uses electrical signals in springing its spiny trap—and the response to its insect prey is far more rapid than the response of, say, a sponge to a pinprick. The sponge contracts slowly, in a diffuse sort of way, and the response never extends more than a few millimeters beyond the point of stimulation.

But sponges are just about the simplest members of the animal kingdom. Other animals have specialized means of detecting stimuli and responding swiftly to them. Almost all animals reach out or lunge after food; they pull back, crawl, swim, run, or fly when they are about to become food themselves. And think about what they have to do to find a mate and slow it down or hold its attention. (Think about all the things you have to do.) The more complex the environment and life style, the more elaborate and rapid are the animal modes of sensory reception, signal integration, and response.

There are more than a million known species of animals, so the examples used in this chapter are necessarily limited. Even so, the examples will help reinforce the following points, which together constitute a **theory of neural patterning**:

---

1. Reflexes are simple, stereotyped movements made in response to sensory stimuli. In the simplest reflex pathways, a sensory neuron directly signals a motor neuron, which acts on muscle cells (page 322).

2. Reflex pathways are the basic operating machinery of nervous systems.

3. Nervous systems evolved through accretion: a layering of additional nervous tissues over reflex pathways of more ancient origin.

4. Nervous systems evolved along with sensory organs (such as eyes) and motor structures (such as legs and wings), and together they provided the foundation for more active and intricate life-styles.

5. The oldest parts of the vertebrate brain deal with reflex coordination of sensory inputs and motor outputs beyond that afforded by the spinal cord alone.

6. Even the most recent layerings of nerve tissue deal partly with reflex coordination. But they also deal with storing, comparing, and using experiences to initiate novel, nonstereotyped action. These regions are the basis of memory, learning, and reasoning.

25

NERVOUS SYSTEMS

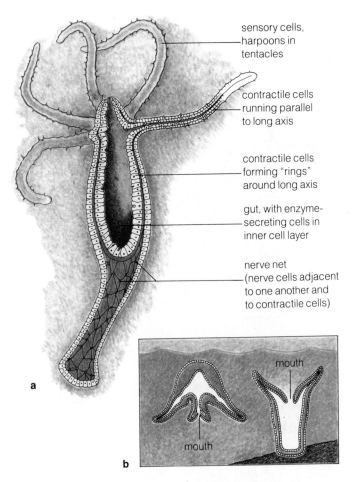

- sensory cells, harpoons in tentacles
- contractile cells running parallel to long axis
- contractile cells forming "rings" around long axis
- gut, with enzyme-secreting cells in inner cell layer
- nerve net (nerve cells adjacent to one another and to contractile cells)

a

b

mouth

mouth

Figure 25.1 (**a**) Nerve net, sensory cells, and contractile cells in the cnidarian *Hydra*. This system is arranged radially about the mouth and gut regardless of whether the cnidarian is free-floating or sedentary (**b**).

INVERTEBRATE NERVOUS SYSTEMS

Nerve Nets

Of all animals, the cnidarians (sea anemones, hydras, and jellyfishes) have the simplest nervous systems. Some forms of cnidarians are attached to the sea floor, others are free-floating, yet they all face similar challenges. For them, food and danger are likely to appear not on water's surface or on the bottom but anywhere in between (Figure 25.1). The systems by which they sense and respond to the environment show **radial symmetry**, with body parts arranged radially about a central axis, much like the spokes of a bike wheel.

The cnidarian nervous system, called a **nerve net**, is based on reflex pathways between epithelial receptor cells and contractile cells (Figure 25.1). For example, in the pathway concerned with feeding behavior, nerve cells extend from receptors in tentacles to contractile cells around the mouth. In jellyfishes, some pathways are involved in slow swimming movements and in keeping the body right-side up.

Cephalization and Bilateral Symmetry

How did complex nervous systems evolve from arrangements as simple as nerve nets? One idea developed through observations of certain cnidarian life cycles, which include a ciliated, cigar-shaped larval stage called a planula. A planula swims or crawls about, then it settles on one end and a mouth forms at the other. It becomes a polyp, a form that looks like the righthand sketch in Figure 25.1b. A polyp survives if it settles where food is plentiful. It does not survive where food is scarce, for a polyp cannot pick up and move on.

Suppose that long ago, mutations in genes governing development prevented the planula from turning into a polyp. And suppose the mutant animal kept right on crawling. It would have looked very much like an existing animal, the flatworm:

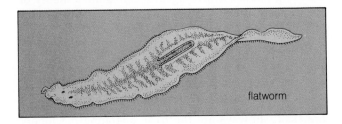

flatworm

Forward-crawling animals could have moved along to places where food was plentiful instead of waiting for it to float past. The forward end would have been the first to encounter stimuli (such as food odors), and responses would have been more rapid and effective with sensory receptors up front. Hence those animals would have been candidates for **cephalization**: an increasing concentration of sensory structures and coordinating centers at the anterior end, or head.

Forward-crawlers also would have been candidates for **bilateral symmetry**, in which the body plan has roughly equivalent right and left halves. For example, motor structures for pulling the body forward would have to develop on both sides of the body, not just one; nerves to control them would have been required on both sides, and so on. The evolutionary importance of cephalization is obvious. What importance can be attached to bilateral symmetry? Perhaps this: *A shift from radial to bilateral symmetry could have led to paired nerves*

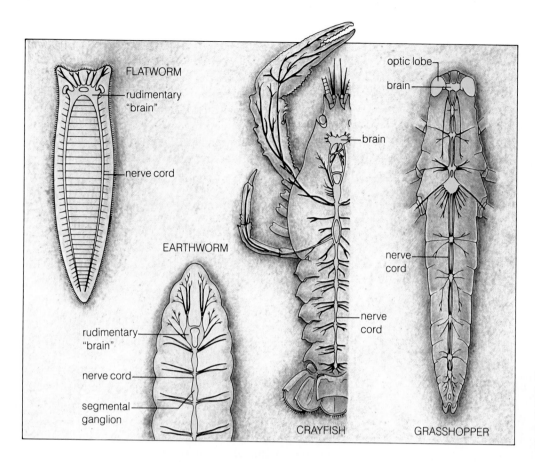

FLATWORM

rudimentary "brain"

nerve cord

EARTHWORM

rudimentary "brain"

nerve cord

segmental ganglion

CRAYFISH

brain

nerve cord

optic lobe

brain

nerve cord

GRASSHOPPER

Figure 25.2 Bilateral symmetry and segmentation evident in the nervous systems of a few invertebrates. The sketches are not to scale relative to one another.

and muscles, paired sensory structures, and paired brain centers.

Intriguingly, simple flatworms have a nerve net, but there are flatworms with *ganglia* (clustered cell bodies of neurons) and with *nerves* and two *nerve cords* (these being bundled-together processes of neurons). The ganglia form a brainlike structure at the head end and coordinate signals from paired sensory organs, including two eye-spots. They also provide some control over the nerve cords, which have nerves branching bilaterally from them and which carry signals to such body parts as the muscles used in swimming and crawling (Figure 25.2).

Segmentation

Like flatworms, the annelids (such as earthworms) and arthropods (such as grasshoppers) have bilateral nerves. These animals also show pronounced **segmentation**: the body is composed of repeating units that are more or less the same in structure and function. Each segment has a pair of nerves and a ganglion, which controls muscles in that segment and (sometimes) in its immediate neighbors. A nerve cord extends through the segments, and at its anterior end, nerve cells are fused into a "brain" (Figure 25.2). The nerve branchings and ganglia of your own nervous system echo this pattern of segmentation.

THE VERTEBRATE PLAN

Evolution of Vertebrate Nervous Systems

It appears that the vertebrate nervous system evolved largely through modification of three key characters: bilateral symmetry, a notochord, and a hollow nerve cord.

A **notochord** is a long rod of stiffened tissue, neither cartilage nor bone, that serves as a supporting structure for the body. It first appeared in marine chordates (which, by definition, have a notochord). Although this rod is present in all vertebrate embryos, it is greatly reduced or absent in adults. In almost all species it is replaced during development by hard, bony segments called **vertebrae**, which are serially arranged in a **vertebral column** (the backbone).

The vertebral column proved to be an ideal skeletal axis against which the force of muscle contraction could be applied. And some bony segments of the skeleton proved to have enormous potential for modification into powerful, bony jaws. The evolution of the notochord into this bony column foreshadowed the evolution of fast-moving, predatory animals.

A related key character in vertebrate evolution was a single, hollow **nerve cord** running dorsally above the notochord. Early on, this hollow structure underwent

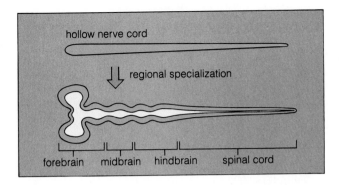

Figure 25.3 Schematic portrayal of the evolution of the anterior end of the vertebrate nerve cord into a spinal cord and specialized brain regions.

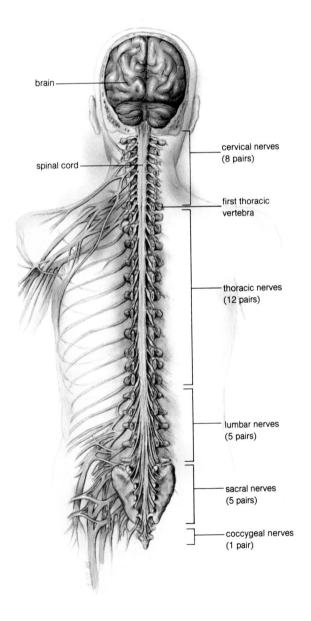

brain

spinal cord

cervical nerves
(8 pairs)

first thoracic
vertebra

thoracic nerves
(12 pairs)

lumbar nerves
(5 pairs)

sacral nerves
(5 pairs)

coccygeal nerves
(1 pair)

expansion and regional modification into a spinal cord and brain. In the changing world of fast-moving vertebrates, sensory receptors became more complex. Paired nasal structures, eyes, and ears gathered information from the outside and fed it into the anterior end of the tube. In time, that end became variably thickened with nervous tissue and functionally divided into three parts: forebrain, midbrain, and hindbrain (Figure 25.3).

Functional Divisions of the Vertebrate Nervous System

The nervous system of all existing vertebrates has two main divisions:

central nervous system	*the spinal cord and brain*
peripheral nervous system	*nerves (leading into and from the spinal cord and brain) and ganglia*

Figure 25.4 shows the relative locations of the thirty-one pairs of nerves that connect with the central nervous system of humans. The sensory and motor axons contained in these nerves make connections with interneurons, the processes and actions of which are confined entirely within the central nervous system. The pathways or tracts running up and down the spinal cord consist primarily of bundled-together long axons of interneurons.

PERIPHERAL NERVOUS SYSTEM

In the peripheral nervous system, the nerves carrying sensory input to the central nervous system are said to be *afferent* (a word meaning "to bring to"). The ones carrying motor output away from the central nervous system to muscles and glands are *efferent* ("to carry outward").

As Figure 25.5 shows, the body's efferent nerves actually form two distinct systems. Efferent nerves leading to skeletal muscles form the **somatic system**. Those leading to the heart, smooth muscles, and glands form

Figure 25.4 Human nervous system. Listed are the paired nerves that connect with the central nervous system (brain and spinal cord). Not visible are twelve pairs of cranial nerves that connect with brain centers.

the **autonomic system**; they service the "visceral" portion of the body (internal organs such as the heart, lungs, and gut).

Nerves Serving Autonomic Functions

The nerves of the autonomic system fall into two categories, called **sympathetic** and **parasympathetic nerves**. They differ in where they connect with the central nervous system and in where their ganglia are located (Figure 25.6).

A superficial look at Figure 25.6 might lead you to believe that both sympathetic and parasympathetic motor nerves service all organs and play off each other, with one type of nerve stimulating an organ and the other slowing down its activity. This is true in *most* but by no means all cases. Some organs have a dual nerve supply; others do not. Also, the nerves of both divisions can have excitatory *or* inhibitory effects. (In other words, one division doesn't turn everything off and the other turn everything on.)

The general functions of the two nerve systems become apparent only when we consider the overall state of the animal body relative to its surroundings. When the animal is not receiving much outside stimulation, parasympathetic nerve action tends to slow down overall body activity and divert energy to basic "housekeeping" tasks, such as digestion. During times of heightened awareness, excitement, or danger, sympathetic nerve action tends to slow down housekeeping tasks and, simultaneously, to increase overall body activities that prepare the animal to fight or flee (or frolic intensely).

For example, under signals from sympathetic nerves, heart rate increases, blood glucose levels rise, and blood circulates faster, distributing oxygen and packets of quick energy (glucose molecules) through the body. Bronchioles dilate and more air enters the lungs, so cells get more oxygen for increased metabolic output.

At all times, both types of autonomic nerves carry signals that bring about minor adjustments in visceral organs. For example, even while the heart may be receiving low levels of sympathetic signals that cause its rate of beating to increase slightly, low levels of parasympathetic signals are opposing this effect. At any moment, the actual rate is the *net* outcome of opposing signals.

One final point should be made here. The word autonomic means "self-managed." It was coined at a time when the reflexes influencing smooth muscles and glands were thought to be self-governed, without inputs from the central nervous system. Today we know that these reflexes are integrated by commands from the brain and spinal cord, as experiments with biofeedback make clear.

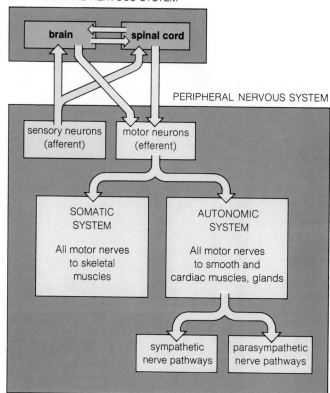

Figure 25.5 Divisions of the vertebrate nervous system.

(*Biofeedback* refers to conscious efforts to enhance or suppress autonomic and other physiological responses. For example, electronic devices can be used to detect changes in heart rate, which are registered on display equipment in the form of auditory or visual cues. An individual uses these cues to recognize and reinforce some desired behavior. Thus, with conscious effort, cardiac muscle contractions detected electronically can be slowed down slightly.)

Cranial Nerves

The peripheral nervous system also includes twelve pairs of **cranial nerves**, which connect directly with the brain. Some cranial nerves contain only sensory axons. (The optic nerves, which carry visual signals from the eyes, are like this.) Others contain sensory and motor axons. For example, the vagus nerves have sensory axons leading into the brain as well as motor axons leading out to muscles in the lungs, gut, and heart.

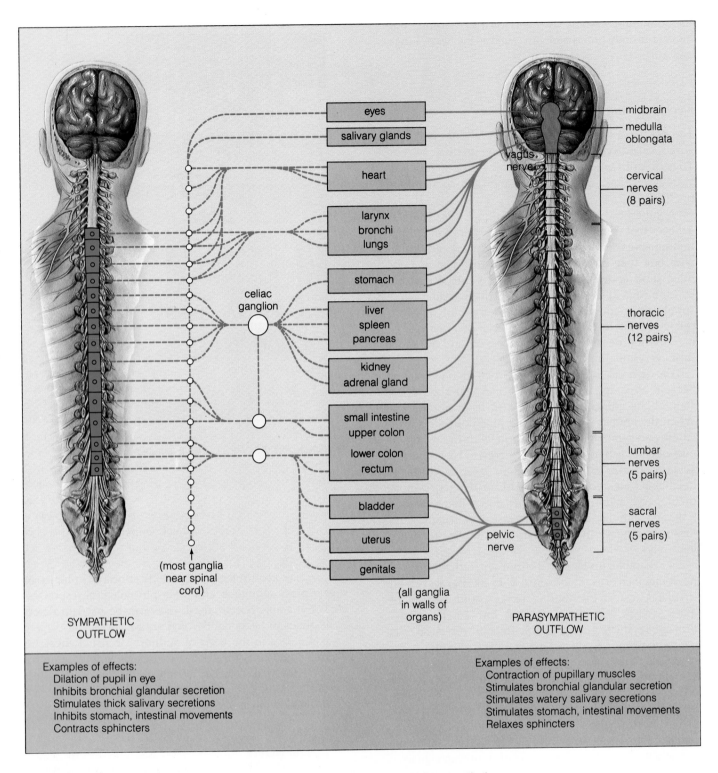

SYMPATHETIC OUTFLOW		PARASYMPATHETIC OUTFLOW

eyes

salivary glands

heart

larynx
bronchi
lungs

stomach

liver
spleen
pancreas

kidney
adrenal gland

small intestine
upper colon

lower colon
rectum

bladder

uterus

genitals

celiac ganglion

(most ganglia near spinal cord)

(all ganglia in walls of organs)

midbrain

medulla oblongata

vagus nerve

cervical nerves (8 pairs)

thoracic nerves (12 pairs)

lumbar nerves (5 pairs)

sacral nerves (5 pairs)

pelvic nerve

SYMPATHETIC OUTFLOW

PARASYMPATHETIC OUTFLOW

Examples of effects:
  Dilation of pupil in eye
  Inhibits bronchial glandular secretion
  Stimulates thick salivary secretions
  Inhibits stomach, intestinal movements
  Contracts sphincters

Examples of effects:
  Contraction of pupillary muscles
  Stimulates bronchial glandular secretion
  Stimulates watery salivary secretions
  Stimulates stomach, intestinal movements
  Relaxes sphincters

Figure 25.6 Autonomic nervous system. Shown here are the main sympathetic and parasympathetic pathways leading out from the central nervous system to some major organs. As the lists of examples suggest, in some cases the sympathetic and parasympathetic nerves operate antagonistically in their effects on the organ. Keep in mind that both systems have *paired* nerves leading out from the brain and spinal cord.

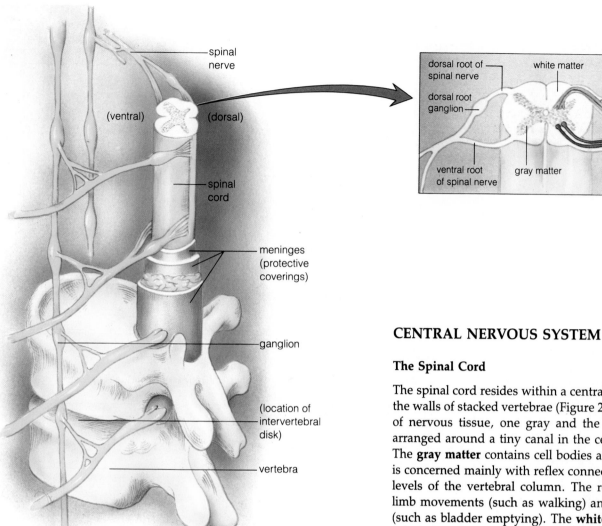

Figure 25.7 Organization of the spinal cord and its relation to the vertebral column.

Labels in figure:
spinal nerve
(ventral) (dorsal)
spinal cord
meninges (protective coverings)
ganglion
(location of intervertebral disk)
vertebra

dorsal root of spinal nerve
dorsal root ganglion
white matter
sensory input lines
ventral root of spinal nerve
gray matter
motor output lines

CENTRAL NERVOUS SYSTEM

The Spinal Cord

The spinal cord resides within a central canal formed by the walls of stacked vertebrae (Figure 25.7). Two regions of nervous tissue, one gray and the other white, are arranged around a tiny canal in the center of the cord. The **gray matter** contains cell bodies and dendrites and is concerned mainly with reflex connections at different levels of the vertebral column. The reflexes deal with limb movements (such as walking) and visceral events (such as bladder emptying). The **white matter** contains mainly axons of interneurons and is so colored because of the glistening myelin sheaths around them. Many of these axons are bundled into sensory nerve tracts that ascend the cord and end at specific brain centers. Others are bundled into motor tracts that run in the opposite direction.

The functional connections made in the spinal cord may be summarized this way:

1. Direct reflex connections between the sensory and motor neurons controlling the limbs and trunk are made in the spinal cord.

2. Most often, interneurons are interposed between sensory input and motor output.

3. Interneurons connect with one another up, down, and laterally in the gray matter of the spinal cord, providing a degree of control over activities within the cord itself.

4. Major nerve tracts (bundles of myelinated axons in the white matter of the cord) ascend into and descend from specific brain centers that provide more refined control over activities.

Divisions	Main Components	Some Functions
FOREBRAIN	Cerebrum	Two cerebral hemispheres. Most complex coordinating center; intersensory association, memory circuits
	Olfactory lobes	Relaying of sensory input to olfactory structures of cerebrum
	Limbic system	Scattered brain centers. With hypothalamus, coordination of skeletal muscle and internal organ activity underlying emotional expression
	Thalamus	Major coordinating center for sensory and motor signals; relay station for sensory impulses to cerebrum
	Hypothalamus	Neural-endocrine coordination of visceral activities (e.g., solute-water balance, temperature control, carbohydrate metabolism)
	Pituitary gland	"Master" endocrine gland (controlled by hypothalamus). Control of growth, metabolism, etc.
	Pineal gland	Control of some circadian rhythms; role in mammalian reproductive physiology
MIDBRAIN	Tectum and other wall regions	Largely reflex coordination of visual, tactile, auditory input; contains nerve tracts ascending to thalamus, descending from cerebrum
HINDBRAIN	Pons	"Bridge" of transverse nerve tracts from cerebrum to both sides of cerebellum. Also contains longitudinal tracts connecting forebrain and spinal cord
	Cerebellum	Coordination of motor activity underlying refined limb movements, maintaining posture, spatial orientation
	Medulla oblongata	Contains tracts extending between pons and spinal cord; reflex centers involved in respiration, cardiovascular function, gastric secretion, etc.

anterior end of ——— spinal cord

Figure 25.8 Summary of the five regional subdivisions of the vertebrate brain. The drawing is highly simplified and flattened. In the developing embryo, the brain flexes forward and the walls develop in complex ways.

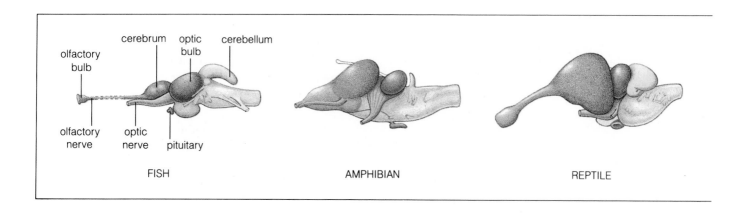

FISH AMPHIBIAN REPTILE

Divisions of the Brain

Anatomically, the brain begins as a continuation of the spinal cord. The protective membranes wrapped around the cord extend up and wrap around the brain also. The central canal of the cord also extends upward, but in the brain it is expanded into fluid-filled chambers (ventricles) and channels (aqueducts). The "walls" around these fluid-filled regions are the actual brain tissue. The cranial cavity (a chamber formed by some bones of the skull) houses the mass of brain tissue.

The brain undergoes its regional expansion during embryonic development. A hollow tube of nervous tissue forms and develops into the primordial hindbrain, midbrain, and forebrain. Further development produces five regional subdivisions, which are more elaborate in some species than in others (Figures 25.8 and 25.9).

Hindbrain

The hindbrain, which is continuous with the spinal cord, includes the **medulla oblongata, cerebellum**, and **pons** (Figure 25.8). The medulla oblongata contains reflex centers for vital functions, including respiration, blood circulation, and vomiting. These centers are part of the **reticular formation**, a loosely organized net of nerve cell bodies that extends from the top of the spinal cord into the forebrain. The formation functions in motor coordination; it helps activate higher brain centers involved in sleeping, dreaming, and arousal; and it coordinates complex reflexes (such as sneezing and coughing).

The cerebellum contains reflex centers for maintaining posture and for spatial orientation. Especially in mammals, it helps coordinate refined limb movements. In the cerebellum, constant integration of sensory signals from the eyes, muscle spindles, tendons, skin, and inner ear provides information on how the body and limbs are positioned, how much particular muscles are contracted or relaxed, and in which direction the body or limbs happen to be moving. When the cerebellum receives commands from higher brain centers, it uses this information to produce the most effective motor response possible at that moment.

The pons is largely a region through which nerve tracts pass on their way from one brain center to another. The word pons (meaning "bridge") refers to prominent bands of axons that extend into each side of the cerebellum.

Midbrain

The midbrain originally was a center for coordinating reflex responses to visual input, but gradually it took on added functions. The roof of the midbrain, the **tectum**, is a thickened region of gray matter where visual and auditory signals are integrated. In fishes and amphibians, the tectum exerts major control over the body. (You can surgically remove the cerebral hemispheres from a frog brain and the frog can still do just about everything it normally does.) The tectum is also important in reptiles and birds. In mammals, sensory information still converges on the tectum but is rapidly sent on to higher centers for further processing.

Forebrain

For the early vertebrates, chemical odors emanating from food must have been a powerful selective agent, for olfactory structures came to dominate the anterior end of the brain. Like their descendants, those vertebrates had two **olfactory bulbs**: centers that receive input from olfactory nerves. They had a primitive **cerebrum**, a brain center that at first was concerned mainly with integrating olfactory input and selecting motor responses to it.

Below the cerebrum, the **thalamus** served as a center for relaying and coordinating sensory signals. Also, some ascending and descending motor pathways con-

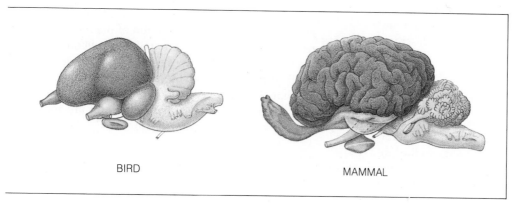

BIRD

MAMMAL

Figure 25.9 Comparison of the brain structure of a fish (codfish), amphibian (frog), reptile (alligator), bird, and mammal (horse). The drawings are not to the same scale.

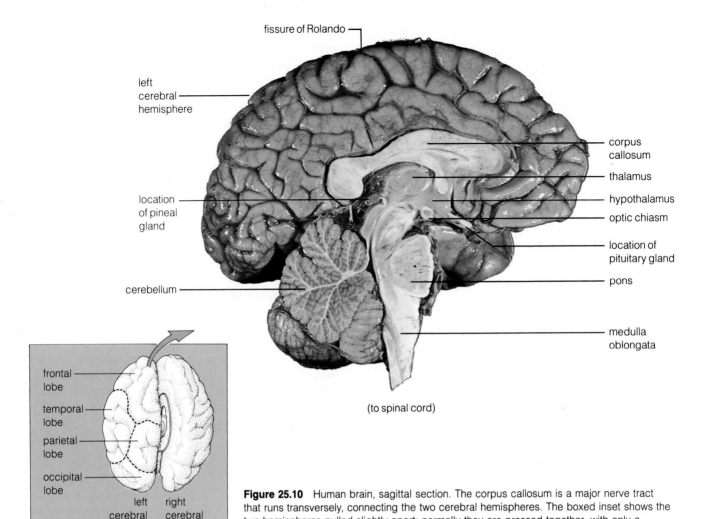

fissure of Rolando

left cerebral hemisphere

corpus callosum

thalamus

location of pineal gland

hypothalamus

optic chiasm

location of pituitary gland

pons

cerebellum

medulla oblongata

(to spinal cord)

frontal lobe

temporal lobe

parietal lobe

occipital lobe

left cerebral hemisphere

right cerebral hemisphere

Figure 25.10 Human brain, sagittal section. The corpus callosum is a major nerve tract that runs transversely, connecting the two cerebral hemispheres. The boxed inset shows the two hemispheres pulled slightly apart; normally they are pressed together, with only a longitudinal fissure separating them.

verged here. The **hypothalamus** monitored visceral activities and influenced forms of behavior related to those activities, such as thirst, hunger, and sex (page 344).

Over evolutionary time, some gray matter came to be positioned as a thin layer over each half of the cerebrum. This **cerebral cortex** provided added control over the body's activities. When mammals first appeared, it expanded spectacularly into information-encoding and information-processing centers, which will be described next.

Parts of the forebrain interconnect to form the **limbic system**, which influences learning and emotional behavior. This system includes parts of the cortex, the thalamus, and the hypothalamus, and it has links with many other parts of the central nervous system. Thus information from diverse receptors may converge in the limbic system, leading (for example) to sweating, laughing, crying, and other responses associated with different emotions.

THE HUMAN BRAIN

The Cerebral Hemispheres

Now that we have considered the vertebrate nervous system in general, we can turn to the intricate cerebrum of the human brain. Much of the gray matter of the cerebrum forms a thin surface layer called the cerebral cortex. This layer folds back on itself to a remarkable extent, suggesting that evolution of the mammalian cerebrum outpaced enlargement of the skullbones housing it. A fissure divides the cerebrum into right and left hemispheres. Although there is some variation from one individual to the next, the folds and fissures of each hemisphere follow certain patterns and are regionally divided into lobes (Figure 25.10).

Nerve Tracts in the Hemispheres. The white matter of the cerebral hemispheres consists of major nerve tracts. One, the *corpus callosum*, is a broad strap of white matter that keeps the two hemispheres in communication with each other (Figure 25.10). Many nerve tracts keep the cerebral cortex in communication with the rest of the body. Although these tracts are rather jammed together in the brainstem, they fan out extensively in the hemispheres. Each hemisphere also has its own set of tracts that provide communication among all of its parts.

Functional Regions of the Cortex. Different regions of the cortex have different functions. Some are *motor centers*, where instructions for motor responses are coordinated (Figure 25.11). For example, parts of the motor cortex connect directly with descending motor tracts. Stimulation of different points on the motor cortex triggers contractions of muscle groups in different body parts. A relatively large area of the motor cortex is devoted to muscles that control thumb and tongue movements (Figure 25.12). The size of this area reflects the control required for intricate hand movements and oral expression.

Other regions are *primary receiving centers* for sensory input, including that from the eyes and ears. Just behind the motor cortex is the somatic sensory cortex, a primary receiving center for sensory input from the skin and joints.

Lying outside the motor and somatic sensory regions, but connected to them by neural pathways, are regions of *association centers*. In these regions, information from memory stores is added to the primary sensory information to give it fuller meaning. It is not known which parts of the brain actually give rise to conscious awareness of the world (Figure 25.13).

Memory

Conscious experience is far removed from simple reflex action. It entails *thinking* about things—recalling objects and events encountered in the past, comparing them with newly encountered ones, and making rational connections based on the comparison of perceptions. Thus conscious experience entails a capacity for **memory**: the storage of individual bits of information somewhere in the brain.

The neural representation of information bits is known as a **memory trace**, although no one knows for sure in what form a memory trace occurs, where it resides, or even if it has a specific regional location. So far, experiments strongly suggest that there are at least two stages involved in its formation. One is a *short-term* formative period, lasting only a few minutes or so; then, information becomes spatially and temporally organized in neural pathways. The other is *long-term* storage; then, information is put in a different neural representation that lasts more or less permanently.

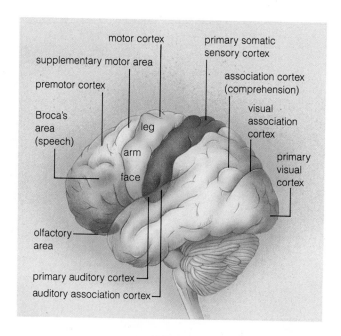

Figure 25.11 Primary receiving and association areas for the human cerebral cortex. Signals from receptors on the body's periphery enter primary cortical areas. Sensory input from different receptors is coordinated and processed in association areas. The text describes the main cortical regions. Also shown here are the *premotor area*, involved in intricate motor activity (as typified by a concert pianist performing); the *supplementary motor area*, which helps coordinate sequential voluntary movements; and *Broca's area*, which coordinates muscles required for speech.

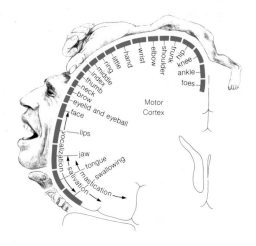

Figure 25.12 Body regions represented in the motor cortex of humans. The size of the body parts in the human figure illustrates the size of the cortical region devoted to controlling that part. The motor cortex runs from the top of the head to just above the ear on the surface of each cerebral hemisphere. The sketch is a cross-section through the right hemisphere of someone facing you.

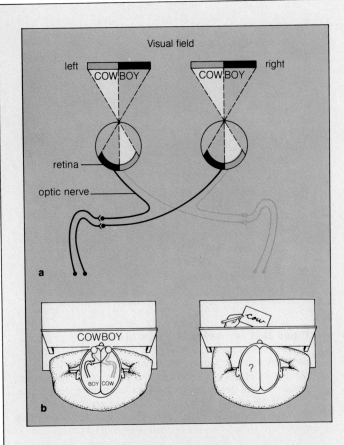

the neural bridge between the two hemispheres put an end to what must have been positive feedback of ever intensified electrical disturbances between them. Beyond this, the "split-brain" individuals were able to lead what seemed, on the surface, entirely normal lives.

But then Sperry devised some elegant experiments to determine whether the conscious experience of these individuals was indeed "normal." After all, the corpus callosum contains no less than 200 million axons; surely *something* was different. Something was. "The surgery," Sperry later reported, "left these people with two separate minds, that is, two spheres of consciousness. What is experienced in the right hemispheres seems to be entirely outside the realm of awareness of the left."

In Sperry's experiments, the left and right hemispheres of split-brain individuals were presented with different stimuli. It was known at the time that visual connections to and from one hemisphere are mainly concerned with the opposite visual field. (Receptors from the left visual field send signals to the right hemisphere; those from the right visual field send signals to the left hemisphere.) Sperry projected words—say, COWBOY—onto a screen. He did this in such a way that COW fell only on the left visual field, and BOY fell on the right (see sketch). The subject reported seeing the word BOY. The left hemisphere, which received the word, controls language. However, when asked to write the perceived word with the left hand—a hand that was deliberately blocked from the subject's view—the subject wrote COW.

The right hemisphere, which "knew" the other half of the word, had directed the left hand's motor response. But it couldn't tell the left hemisphere what was going on because of the severed corpus callosum. The subject knew that a word was being written, but could not say what it was!

The functioning of the cerebral hemispheres has been the focus of many experiments. Taken together, the results have revealed the following information about this functioning:

1. Each cerebral hemisphere can function separately, but it functions mainly in response to signals from the opposite side of the body. Normally, of course, the corpus callosum is intact, and the two hemispheres work together.

2. The main regions responsible for spoken language skills generally reside in the left hemisphere.

3. The main regions responsible for nonverbal skills (music, mathematics, and other abstract abilities) generally reside in the right hemisphere.

Figure 25.13 Sperry's experiments and the association cortex.

Experiments performed by Roger Sperry and his coworkers demonstrated some intriguing differences in perception between the two halves of the cerebrum. The subjects of the experiments were epileptics. Persons with severe epilepsy are wracked with seizures, sometimes as often as every half hour of their lives. The seizures have a neurological basis, analogous to an electrical storm in the brain. What would happen if the corpus callosum of epileptics were cut? Would the electrical storm be confined to one cerebral hemisphere, leaving at least the other to function normally? Earlier studies of animals and of humans whose corpus callosum had been damaged suggested this might be so.

The surgery was performed. And the electrical storms subsided in frequency and intensity. Apparently, cutting

Observations of people suffering from **retrograde amnesia** tell us something about memory. These people can't remember anything that happened during the half hour or so before losing consciousness after a severe head blow. Yet memories of events before that time remain intact! Such disturbances temporarily suppress normal electrical activities in the brain. These observations may mean that whereas short-term memory is a fleeting stage of neural excitation, long-term memory depends on *chemical* or *structural* changes in the brain.

In addition, information seemingly forgotten can be recalled after being unused for decades. This means that individual memory traces must be encoded in a form somewhat immune to degradation. Most molecules and cells in your body are used up, wear out, or age and are constantly being replaced—yet memories can be retrieved in exquisite detail after many years of such wholesale turnovers. Nerve cells, recall, are among the few kinds that are *not* replaced. You are born with billions, and as you grow older some 50,000 die off steadily each day. Those nerve cells formed during embryonic development are the same ones present, whether damaged or otherwise modified, at the time of death.

The part about being "otherwise modified" is tantalizing. *There is evidence that neuron structure is not static, but rather can be modified in several ways.* Most likely, such modifications depend on electrical and chemical interactions with neighboring neurons. Electron micrographs show that some synapses regress as a result of disuse. Such regression weakens or breaks connections between neurons. The visual cortex of mice raised without visual stimulation showed such effects of disuse. Similarly, there is some evidence that intensively stimulated synapses may form stronger connections, grow in size, or sprout buds or spines to form more connections! The chemical and physical transformations that underlie changes in synaptic connections may correspond to memory storage.

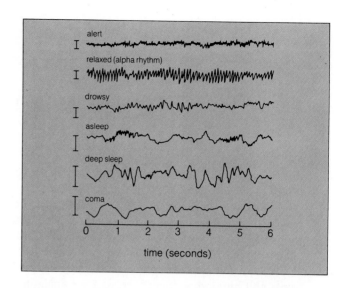

Figure 25.14 Examples of EEG wave patterns. The vertical bars indicate fifty millivolts each, with the irregular horizontal lines indicating the response with time.

States of Consciousness

Between the mindless drift of coma and total alertness are many *levels* of consciousness, known by such names as sleeping, dozing, meditating, and daydreaming. These levels are governed by the central nervous system. They also are subject to alteration by psychoactive drugs (see *Commentary*).

Throughout the spectrum of consciousness, neurons in the brain are constantly chattering among themselves. This neural chatter shows up as wavelike patterns in an **electroencephalogram** (EEG). An EEG is an electrical recording of the frequency and strength of potentials from the brain's surface.

EEG Patterns. Figure 25.14 gives examples of EEG patterns. The prominent wave pattern for someone who is relaxed, with eyes closed, is an *alpha rhythm*. In this relaxed state of wakefulness, EEG waves are recorded in "trains" of one after the other, about ten per second. Alpha waves predominate during the state of meditation. With a transition to sleep, wave trains gradually become larger, slower, and more erratic. This *slow-wave sleep* pattern shows up about eighty percent of the total sleeping time for adults. It occurs when sensory input is low and the mind is more or less idling. Subjects awakened from slow-wave sleep usually report that they were

not dreaming. If anything, they seemed to be mulling over recent, ordinary events.

But slow-wave sleep is punctuated by brief spells of *REM sleep*. *Rapid Eye Movements* accompany this pattern (the eyes jerk beneath closed lids). Also accompanying REM sleep are irregular breathing, faster heartbeat, and twitching fingers. Most people awakened from REM sleep report they were experiencing vivid dreams.

With the transition from sleep (or deep relaxation) into alert wakefulness, EEG recordings show a shift to low-amplitude, higher frequency wave trains. Associated with this accelerated brain activity are increased blood flow and oxygen uptake in the cortex. The transition, called *EEG arousal*, occurs when conscious effort is made to focus on external stimuli or even on one's own thoughts.

The Reticular Activating System. Which brain regions govern changing levels of consciousness? The reticular formation, recall, forms connections with the spinal cord, cerebellum, and cerebrum as well as back with itself. It constantly samples messages flowing through the central nervous system. The flow of signals along

Drug Action on Integration and Control

Classes of Psychoactive Drugs

Broadly speaking, a drug is any chemical substance that has no nutritive value but that is introduced into the body to elicit some effect on physiological processes. A **psychoactive drug** is one that affects those parts of the central nervous system concerned with states of consciousness and behavior.

There are five classes of psychoactive drugs: depressants and hypnotics, stimulants, narcotic analgesics, hallucinogens and psychedelics, and antipsychotics (see Table). Many exert their effects by altering synaptic transmission (for example, by modifying or blocking the synthesis, release, or uptake of transmitter substances).

Depressants, Hypnotics. These drugs lower the activity in certain parts of the brain and nerves, hence they reduce activity throughout the body. Some inhibit synaptic transmission in the reticular activating system and at nerve tracts in the thalamus.

Depending on the dosage, most of these drugs can produce different degrees of general behavioral modification, ranging from relief from anxiety, through sedation, sleep (hypnosis), anesthesia and coma, to death. At low doses, inhibitory synapses are often suppressed slightly more than excitatory synapses, so that initially the person feels excited or euphoric. At increased doses, excitatory synapses are also suppressed and the overall effect is depression. The effects of depressants and hypnotics are additive, in that one will amplify the effect of another. For example, the combined use of alcohol and barbiturates exaggerates behavioral depression.

Alcohol (ethyl alcohol) differs from the drugs just described because it acts not at synapses but directly on the plasma membrane to alter cell function. It is often mistakenly viewed as a stimulant because it produces an initial "high." It is also widely held to be relatively harmless, even though it is one of the most powerful psychoactive drugs and a major cause of death. Short-term use of alcohol in seemingly small doses typically produces disorientation, uncoordinated motor functions, and diminished judgment. Long-term addiction destroys nerve cells and causes permanent brain damage; it can permanently scar and otherwise damage the liver (cirrhosis).

Stimulants. Caffeine, nicotine, amphetamines, and cocaine are examples of stimulants. At first they increase alertness and body activity but inevitably they lead to a period of depression.

Caffeine, one of the most widely used stimulants, is a component of coffee, tea, chocolate, and many soft drinks. Caffeine apparently blocks breakdown of cyclic adenosine monophosphate (cyclic AMP), plays a role in glycogen breakdown into glucose, and influences many receptor functions. At low doses, caffeine stimulates the cerebral cortex first, so its initial effects include increased alertness and restlessness. At higher doses it affects the medulla oblongata and thereby disrupts motor coordination as well as intellectual coherence.

Nicotine, a component of tobacco, has powerful effects on both the central and peripheral nervous systems. It chemically mimics acetylcholine at some receptors and can directly stimulate a number of sensory receptors. Its short-term effects include irritability, increased heart rate and blood pressure, water retention, and gastric upsets. Its long-term effects are serious (pages 429 and 507).

Low doses of amphetamines and cocaine reduce fatigue, elevate mood, and heighten awareness. High doses elicit anxiety, irritability, even psychotic behavior (such as paranoia) and permanent mental abnormalities. Both drugs apparently mimic or enhance the excitatory effect of norepinephrine, which underlies the "fight, flee" responses associated with autonomic nerves. Both drugs can lead to psychological and physiological dependency, with addicts experiencing extreme emotional and physical depression when deprived of them.

Analgesics. When stress leads to physical or emotional pain, the brain produces its own analgesics, or pain relievers. Endorphins and enkephalins are examples. Receptors for brain analgesics have been identified in many parts of the nervous system. When bound to receptors, these naturally occurring substances seem to inhibit neural activity. Endorphins (including enkephalins) have been identified in high concentrations in brain regions concerned with emotions and perception of pain.

The narcotic analgesics sedate the body as well as relieve pain. They include codeine and heroin, both of which are extremely addictive. The body develops a tolerance of these drugs, in that progressively larger doses are needed to produce the same effects. At the same time, the body becomes physically dependent on them. Deprivation following massive doses of heroin leads to cravings for the drug, hyperactivity and anxiety, fever, chills, and painful gastrointestinal disruption (violent vomiting, cramping, and diarrhea).

Psychedelics, Hallucinogens. These drugs alter sensory perception (particularly visual and auditory); hence they have been described as "mind-expanding." Some skew acetylcholine or norepinephrine activity. Others, such as LSD (lysergic acid diethylamide), affect serotonin activity. Even in tiny doses, LSD dramatically warps perceptions.

Marijuana is another hallucinogen. The name refers to the drug made from crushed leaves, flowers, and stems of the plant *Cannabis sativa*. (Hashish, an extract of the same plant, can contain ten times as much of the psychoactive ingredient.) In low doses marijuana is like a depressant; it slows down but does not impair motor activity, it relaxes the body and elicits mild euphoria. Unlike true depressants it cannot, in high doses, lead to coma and death. It can produce disorientation, increased anxiety bordering on panic, delusions (including paranoia), and hallucinations.

Like alcohol, marijuana can affect an individual's ability to perform complex tasks, such as driving a car. In one study, commercial pilots showed a marked deterioration in instrument-flying ability for more than two hours after smoking marijuana. Recent studies point to a link between marijuana smoking and suppression of the immune system.

Antipsychotic Drugs. Emotional states—joy, elation, anxiety, depression, fear, anger—are normal responses to changing conditions in our complex world. Sometimes, through imbalances in the central nervous system, one or another of these states becomes pronounced.

For example, anxiety (a vague sense of uneasiness and tenseness) might give way to emotional instability in the neurotic, who becomes guilt-ridden, anxious, and chronically upset over mundane frustrations. Schizophrenic persons become despairing; they withdraw from society and focus obsessively on themselves. Schizophrenia is an example of psychotic behavior. "Psychosis" means an impairment of a person's ability to communicate and otherwise effectively interact with others, combined with a distorted perception of reality.

Low doses of antipsychotic drugs are tranquilizing, in that they reduce anxiety levels and neurotic behav-

Classes of Psychoactive Drugs	
Class	Examples
Depressants, hypnotics	Barbiturates (e.g., Nembutal, Quaalude)
	Antianxiety drugs (e.g., Valium, alcohol)
Stimulants	Caffeine
	Nicotine
	Amphetamines (e.g., Dexedrine)
	Cocaine
Narcotic analgesics	Codeine
	Opium
	Heroin
Psychedelics, hallucinogens	Lysergic acid diethylamide (LSD)
	Cannabis sativa (hashish, marijuana)
Antipsychotics	Chlorpromazine, lithium

ior. High doses produce tremors, blurred vision, and constipation or diarrhea, all of which may be the reason why you probably will never come across a person addicted to an antipsychotic drug.

A Biological Perspective on Drug Abuse

For the most part, we do not understand much about how any one drug works. Given the complexity of the brain, it could scarcely be otherwise at this early stage of inquiry. Despite our ignorance of the mechanisms of drug action, one of the major problems in the modern world is drug abuse—the self-destructive use of drugs that alter emotional and behavioral states. The consequences show up in unexpected places: in seven-year-old heroin addicts, in the highway wreckage left by individuals whose perceptions were skewed by alcohol or amphetamines, in victims of addicts who steal and sometimes kill to support the drug habit, in suicides on LSD trips who were deluded into believing that they could fly and who "flew" off buildings and bridges.

Each of us possesses a body of great complexity. Its architecture, its functioning are legacies of millions of years of evolution. It is unique in the living world because of its highly developed nervous system—a system that is capable of processing far more than the experience of the individual. One of its most astonishing products is language, the encoding of *shared* experiences of groups of individuals in time and space. Through the evolution of our nervous system, the sense of history was born, and the sense of destiny. Through this system we can ask how we have come to be what we are, and where we are headed from here. Perhaps the sorriest consequence of drug abuse is its implicit denial of this legacy—the denial of self when we cease to ask, and cease to care.

these circuits—and the inhibitory or excitatory chemical changes accompanying them—has a great deal to do with whether you stay awake or drop off to sleep. In fact, damage to some parts of the reticular formation can lead to unconsciousness and coma.

Within this formation are neurons collectively called the **reticular activating system** (RAS). Excitatory pathways connect the RAS to the thalamus (the forebrain's switching station). Messages routed from the RAS arouse the brain and maintain wakefulness.

Also in the reticular formation are *sleep centers*. One center contains neurons that release serotonin, a transmitter substance with an inhibitory effect on RAS neurons (high serotonin levels are linked to drowsiness and sleep). Another sleep center, in part of the reticular formation within the pons, has been linked to REM sleep. Substances released from the second center counteract the effects of serotonin, perhaps enabling the RAS to maintain the waking state.

SUMMARY

1. Nervous systems provide specialized means of detecting stimuli and responding to them swiftly.

2. Reflexes are simple, stereotyped movements made in response to sensory stimuli. In the simplest reflex, activity in sensory neurons directly signals motor neurons, which act on muscle cells. Reflexes are the basic operating machinery of nervous systems.

3. The oldest parts of the vertebrate brain deal with reflex coordination of sensory inputs and motor outputs beyond that afforded by the spinal cord alone. Nervous systems evolved through the gradual layering of additional nervous tissue over reflex pathways of more ancient origin.

4. Increasingly complicated sensory organs and motor structures evolved along with the changing nervous systems.

5. Even the newest layerings of the brain deal partly with reflex coordination. They also deal with storing, comparing, and using experiences to initiate novel, non-stereotyped action. Thus, these regions are the basis of memory, learning, and reasoning.

6. The simplest nervous systems are nerve nets, such as those of sea anemones, hydra, and jellyfishes.

7. More complex nervous systems involve cephalization and bilateral symmetry with paired nerves, muscles, sensory structures, and brain centers.

8. The central nervous system of all vertebrates consists of a brain and spinal cord; the peripheral nervous system consists of the nerves, which pass between the brain and spinal cord and the innervated structures, and ganglia.

9. The peripheral nervous system is further divided into components that are afferent (conduct information to the central nervous system) and efferent (conduct information from the central nervous system to muscles and glands).

10. Efferent nerves are further divided into the somatic nervous system (which innervates skeletal muscles) and the autonomic nervous system (which innervates the heart, smooth muscles, and glands).

11. The autonomic nervous system contains two divisions. Sympathetic nerves function more during times of activity, excitement, and stress, whereas parasympathetic nerves are more active when the body is at rest.

12. The spinal cord has regions of cell bodies (gray matter) and myelinated cell processes (white matter). The white matter contains major nerve tracts. In the spinal cord, connections (often involving interneurons) are made between sensory and motor neurons for reflexes controlling the limbs and trunk.

13. The hindbrain is the part of the brain that connects with the spinal cord. It includes the medulla oblongata, pons, and cerebellum and contains reflex centers for vital functions and muscle coordination. It also contains nerve tracts (passing between the spinal cord and other parts of the brain) and the reticular formation, which is involved in the coordination of sensory and motor functions and arousal.

14. The midbrain functions in the coordination and relay of visual and auditory information.·

15. The forebrain includes the cerebrum, below which are the thalamus (relays sensory information, helps coordinate motor responses) and hypothalamus (monitors visceral activities and coordinates some behavioral, autonomic, and endocrine responses to basic drives such as thirst, hunger, and sex). The limbic system is active in learning and emotional behavior.

16. The cortex, or outer layer of the cerebrum, has regions devoted to specific functions, such as receiving information from the various sense organs, integrating this information with memories of past events, and coordinating motor responses.

17. Memory is thought to occur in two stages: a short-term formative period and long-term storage, which depends on chemical or structural changes in the brain.

18. States of consciousness vary between total alertness and deep coma. The levels are governed by activity in the reticular activating system and sleep centers in the reticular formation. They are subject to the influence of psychoactive drugs. Different EEG patterns are characteristic of the various states of consciousness.

Review Questions

1. Label the parts of the human brain:

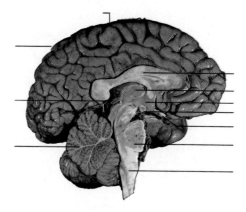

2. Can you list the main points of the theory of neural patterning given at the start of the chapter?

3. Describe three key characters that apparently played central roles in the evolution of the vertebrate nervous system.

4. What constitutes the central nervous system? The peripheral nervous system?

5. Can you distinguish among the following:
   a. neurons and nerves
   b. ganglia and nerve pathways (or tracts)
   c. somatic system and autonomic system
   d. parasympathetic and sympathetic nerves

6. Review Figure 25.8. Then, on your own, describe the main components of the three main subdivisions of the vertebrate brain.

7. What is a psychoactive drug? Can you describe the effects of one such drug on the central nervous system?

Readings

Julien, R. 1981. *A Primer of Drug Action*. Third edition. New York: Freeman. Effectively fills the gap between popularized (and often superficial or misleading) accounts of drug action and the upper-division books in pharmacology. Paperback.

Penfield, W., and T. Rasmussen. 1952. *The Cerebral Cortex of Man*. New York: Macmillan. Fascinating account of early mappings of cortical regions of the brain.

Romer, A., and T. Parsons. 1986. *The Vertebrate Body*. Sixth edition. Philadelphia: Saunders. The classic reference book on vertebrate structure and functioning. Marvelous insights into the evolution of vertebrate nervous systems.

Shepherd, G. 1983. *Neurobiology*. New York: Oxford.

Sperry, R. 1970. "Perception in the Absence of the Neocortical Commissures." *Perception and Its Disorders*. Research Publication of the Association for Research in Nervous and Mental Diseases, vol. 48.

Wurtman, R. April 1982. "Nutrients That Modify Brain Function." *Scientific American* 246(4):50–59.

26

INTEGRATION AND CONTROL: ENDOCRINE SYSTEMS

"THE ENDOCRINE SYSTEM"

The word "hormone" dates back to the early 1900s, when W. Bayliss and E. Starling were trying to figure out what triggers the secretion of pancreatic juices that act on food traveling through the canine gut. At the time, it was known that acids mix with food in the stomach, and that the pancreas secretes an alkaline solution when the acidic mixture has passed into the small intestine. Was the nervous system or something else stimulating the pancreatic response?

To find the answer, Bayliss and Starling cut off the nerve supply to the upper small intestine but left its blood vessels intact. When an acid was introduced into the denervated region, the pancreas still was able to make the secretory response. More telling, extracts of cells taken from the epithelial lining of this intestinal region also induced the response. Glandular cells in the lining had to be the source of a substance that stimulated the pancreas into action.

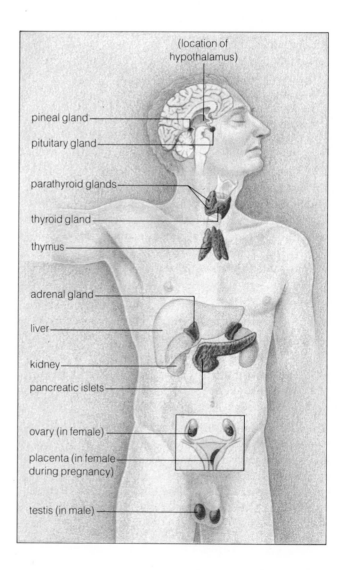

Figure 26.1 Location of endocrine glands in the human body. The liver and kidneys do other things besides secreting hormones, but these organs do have component endocrine cells. Intestinal epithelium also has endocrine cells.

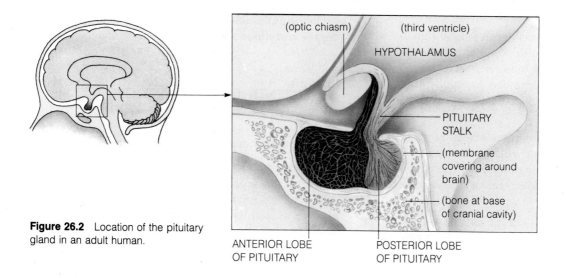

(optic chiasm)   (third ventricle)

HYPOTHALAMUS

PITUITARY STALK

(membrane covering around brain)

(bone at base of cranial cavity)

Figure 26.2 Location of the pituitary gland in an adult human.

ANTERIOR LOBE OF PITUITARY

POSTERIOR LOBE OF PITUITARY

The substance itself came to be called secretin. Demonstration of its existence and its mode of action was the first confirmation of an idea that had been around for centuries: *Internal secretions released into the bloodstream influence the activities of tissues and organs.* Starling coined the word hormone for such internal glandular secretions (the Greek *hormon* means to set in motion).

Later work led to the discovery of many hormones and their sources (Figure 26.1). The sources include but are not limited to the following array, which is characteristic of most vertebrates:

Pituitary gland
Pineal gland
Thyroid gland
Parathyroid glands (four)
Adrenal glands (two)
Gonads (two)
Thymus gland
Pancreatic islets (multiple)
Endocrine cells of gut, liver, kidneys, placenta

In time, these scattered sources of hormones came to be viewed as "the endocrine system," a phrase implying that they formed a separate means of control within the body, apart from the nervous system. (The Greek *endon* means within: *krinein* means to separate.)

As happened in other areas of research, the tidy categorization of control elements into the nervous system *or* endocrine system was shattered by refined biochemical studies and improved techniques in electron micro-scopy. For example, seemingly "separate" glands turned out to be well innervated, some neurons turned out to be secretory, and the substances that certain neurons were secreting were "hormones." The so-called master gland of the endocrine system (the pituitary) turned out to be controlled in major ways by the hypothalamus (part of the brain region shown in Figure 26.2).

At present, it is impossible to say that a natural division exists between the nervous and endocrine systems, given their intricate overlapping. For that reason, we will begin by describing the range of secretions that serve as agents of integration, regardless of their neural or endocrine origin.

HORMONES AND OTHER SIGNALING MOLECULES

All cells have built-in means of responding to chemical changes in their immediate surroundings, including mechanisms for the uptake and release of different substances. In complex animals, the individual responses of thousands, millions, even many billions of cells must be integrated in ways that benefit the whole body.

Integration is accomplished through the action of diverse *signaling molecules*, or chemical secretions that alter the behavior of adjacent or nonadjacent target cells. (By definition, a target cell is one having receptors to which signaling molecules selectively bind.) The cellular or glandular sources of signaling molecules are listed in Table 26.1 and may be described as follows:

1. **Local mediator cells**. Many cells secrete local signaling molecules that alter chemical conditions in their *immediate vicinity*. Most of these molecules act on neighboring cells so swiftly or are degraded so rapidly after their action that they do not reach notable concentrations in the bloodstream.

2. **Synapsing neurons**. Most neurons secrete transmitter substances which, recall, act mainly on an *immediately adjacent target cell*. These signaling molecules reach high concentrations in the synaptic cleft, act swiftly, and are rapidly degraded or recycled.

3. **Neurosecretory cells**. Some neurons (called neurosecretory cells) secrete **neurohormones**, which are transported by the bloodstream to *nonadjacent targets*.

4. **Endocrine glands or cells**. Endocrine glands or cells secrete **hormones**, which also are transported by the bloodstream to *nonadjacent targets*.

5. **Exocrine glands**. Some exocrine glands secrete **pheromones**, which have *targets outside the body*. Pheromones diffuse through water or air and act on cells of other animals of the same species. They integrate activities between animals by serving as sexual attractants, alarm signals, and so on.

To get an idea of how integration is accomplished, let's take a look at the responses of some target cells to signaling molecules and at homeostatic controls over the degree and duration of response.

NEUROENDOCRINE CONTROL CENTER

The Hypothalamus-Pituitary Connection

In vertebrates, integration of many body activities is governed by a **neuroendocrine control center**, which consists of a brain region (the hypothalamus) closely related to a gland (the pituitary). The hypothalamus, recall, monitors internal organs (such as the stomach) and influences forms of behavior related to their activities (such as eating). The pituitary is suspended from a slender stalk that extends downward from the base of the hypothalamus (Figure 26.2).

The pituitary gland is composed of lobes of nervous tissue and glandular tissue. The *posterior lobe* is nervous tissue. It secretes two types of signaling molecules, which are actually produced in the hypothalamus but stored in the lobe. The *anterior lobe* is mostly glandular tissue. It produces and secretes at least six hormones and controls the release of several more from other endocrine glands. The pituitary of many vertebrates (not humans) also has an *intermediate lobe* of mostly glandular tissue. Its secretions induce changes in the coloration of the body's surface of many animals.

Table 26.1 Sources of Signaling Molecules Used in the Integration of Body Activities			
Source	Type of Secretion	Proximity of Target Cell	Examples of Signaling Molecules
Local mediator cells	Local mediator molecules	Neighboring cells in immediate vicinity	Prostaglandins, histamine, nerve growth factor
Synapsing neurons	Transmitter substances	Immediately adjacent, across synaptic cleft	Acetylcholine, dopamine
Neurosecretory cells	Neurohormones	Nonadjacent, reached via bloodstream	Oxytocin, ADH, TRH, dopamine, epinephrine
Endocrine glands or cells	Hormones	Nonadjacent, reached via bloodstream	Glucagon, thyroxine, testosterone
Exocrine glands or cells	Pheromones	Outside body, in other animals of same species	Bombykol, cyclic-AMP

Given its pervasive roles, the pituitary is not as large as might be expected; in adult humans it is about the size of a garden pea.

Posterior Lobe Secretions

Let's first consider the functional links between the hypothalamus and the posterior lobe of the pituitary (Figures 26.3 and 26.4). Two neurohormones, called **antidiuretic hormone** (ADH) and **oxytocin**, are produced in neurosecretory cell bodies clustered together in the hypothalamus. Axons from the cell bodies extend down the pituitary stalk and end next to a capillary bed within the posterior lobe. ADH and oxytocin molecules move down the axons and are stored in the axonal endings.

Capillary beds consist of numerous small-diameter, thin-walled blood vessels (capillaries). Here, substances are exchanged between the blood and interstitial fluid (the fluid that occupies the spaces between cells and tissues). The ADH and oxytocin being stored right next to the capillary bed in the posterior lobe are released in controlled ways into the interstitial fluid. From there, they diffuse into the capillaries and eventually circulate through the body by way of the bloodstream. When they contact cells with receptors specific for them, they exert their effects.

ADH influences the volume and solute concentrations of extracellular fluid. Some of its target cells are in the walls of small urine-conducting tubes and ducts in the kidneys, which are part of the urinary system. When the body must conserve water, ADH makes the tube and duct walls permeable to water. Thus water otherwise destined for excretion moves into interstitial fluid and is retained by the body. ADH also causes smooth muscle in the walls of blood vessels to constrict, which increases blood pressure and the effective blood volume. (These effects account for the hormone's other name: vasopressin.)

Oxytocin plays a role in mammalian reproduction through its effects on smooth muscle walls. It stimulates contraction of the uterus during labor and stimulates the movement of milk into secretory ducts of mammary glands (which provide nourishment for the young).

Both ADH and oxytocin are thought to have evolved from similar ancestral signaling molecules that helped control water loss and, indirectly, solute concentrations. The posterior lobe is notably large in animals that live in arid parts of the world—where water conservation is vital.

Anterior Lobe Secretions

Let's now consider the functional links between the hypothalamus and the anterior lobe of the pituitary (Figures 26.5 and 26.6). The hypothalamus secretes at least

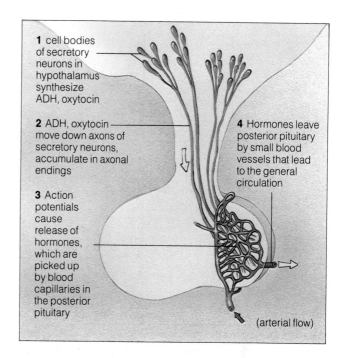

Figure 26.3 Functional links between the hypothalamus and the posterior lobe of the pituitary.

1 cell bodies of secretory neurons in hypothalamus synthesize ADH, oxytocin

2 ADH, oxytocin move down axons of secretory neurons, accumulate in axonal endings

3 Action potentials cause release of hormones, which are picked up by blood capillaries in the posterior pituitary

4 Hormones leave posterior pituitary by small blood vessels that lead to the general circulation

(arterial flow)

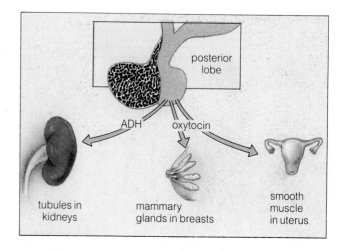

Figure 26.4 Secretions of the posterior lobe of the pituitary and their targets.

posterior lobe

ADH   oxytocin

tubules in kidneys

mammary glands in breasts

smooth muscle in uterus

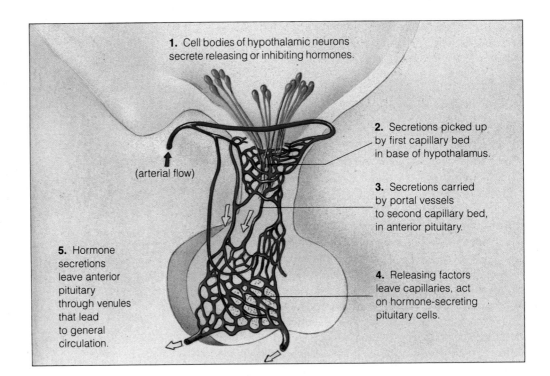

1. Cell bodies of hypothalamic neurons secrete releasing or inhibiting hormones.

(arterial flow)

2. Secretions picked up by first capillary bed in base of hypothalamus.

3. Secretions carried by portal vessels to second capillary bed, in anterior pituitary.

5. Hormone secretions leave anterior pituitary through venules that lead to general circulation.

4. Releasing factors leave capillaries, act on hormone-secreting pituitary cells.

Figure 26.5 Functional links between the hypothalamus and the anterior lobe of the pituitary.

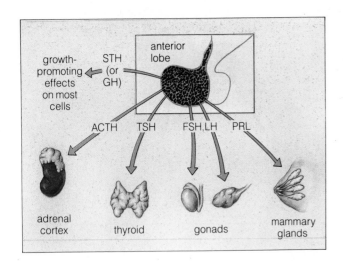

growth-promoting effects on most cells

STH (or GH)

anterior lobe

ACTH   TSH   FSH,LH   PRL

adrenal cortex   thyroid   gonads   mammary glands

Figure 26.6 Secretions of the anterior lobe of the pituitary and their targets.

six different signaling molecules that control the release of hormones from different target cells of the anterior lobe. These molecules are called **releasing** or **inhibiting hormones**. They are produced in neurosecretory cell bodies clustered in the hypothalamus. Then they move down axons that end next to a capillary bed in the base of the hypothalamus. Here, the signaling molecules are released and diffuse into the capillaries.

In this case, the capillaries extend into the pituitary stalk and converge into **portal vessels**, which then diverge into a *second* bed located in the anterior lobe. (A "portal" vessel is one that lies between two capillary beds. In most circulation routes through the human body, blood flowing from and back to the heart passes through only one capillary bed, not two.) Here, the signaling molecules leave the capillaries and stimulate or inhibit the release of hormones. These anterior pituitary hormones move into the capillaries and then circulate through the body.

Releasing and inhibiting hormones were identified by what can only be called monumental research efforts that began in 1955, particularly by Roger Guillemin and Andrew Schally. For example, over one four-year period, Guillemin's team purchased 500 tons of sheep brains from meat processing plants and extracted 7 tons of hypothalamic tissue from it. They eventually ended up with a single milligram of thyrotropin-releasing hormone.

Table 26.2 Hormones Released From the Pituitary Gland

Pituitary Lobe	Tissue Type	Secretions		Known Targets	Primary Actions
Posterior	Nervous tissue (extension of hypothalamus)	Antidiuretic hormone (or vasopressin)	ADH	Kidneys, water storage organs, etc.	Induces fluid movements underlying the control of extracellular fluid volume (and, indirectly, solute concentrations)
		Oxytocin		Smooth muscle in mammary glands	Induces milk movement into secretory ducts
				Uterus	Induces uterine contractions
Anterior	Mostly glandular tissue	Adrenocorticotropic hormone	ACTH	Adrenal cortex	Stimulates release of adrenal hormones involved in stress responses
		Thyroid-stimulating hormone	TSH	Thyroid gland	Stimulates release of thyroid hormones concerned with growth, development, metabolic rate
		Gonadotropins:			
		Follicle-stimulating hormone	FSH	Ovaries, testes	In females, stimulates follicle growth, helps stimulate estrogen secretion and ovulation; in males, promotes spermatogenesis
		Luteinizing hormone	LH	Ovaries, testes	In females, stimulates ovulation, corpus luteum formation; in males, promotes testosterone secretion
		Prolactin	PRL	Mammary glands	Stimulates milk production
		Somatotropin (also called growth hormone)	STH (GH)	Most cells	Generalized growth-promoting effects in young animals; induces protein synthesis and cell division; stimulates release of somatomedins
Intermediate*	Mostly glandular tissue	Melanocyte-stimulating hormone	MSH	Pigment-containing cells in integument	Induces color changes in response to external stimuli; effects on behavior

* Present in many vertebrates (not humans).

Thyrotropin-releasing hormone is just one of several signaling molecules, each of which stimulates the anterior pituitary into releasing one of the following hormones:

Adrenocorticotropic hormone (ACTH)
Thyroid-stimulating hormone (TSH)
Follicle-stimulating hormone (FSH)
Luteinizing hormone (LH)
Prolactin (PRL)
Somatotropin (STH); also called growth hormone (GH)

The first four hormones act only on endocrine glands, as Table 26.2 and Figure 26.6 indicate. In later chapters, those hormones will be described in the context of their roles in the functioning of particular systems. (For example, FSH and LH are central to reproduction.) The last two hormones listed, prolactin and somatotropin, have effects on body tissues rather than on endocrine glands only.

Prolactin regulates a variety of activities among vertebrate species ranging from primitive fishes to humans. One of its functions is to stimulate milk secretion from mammary glands. It works in conjunction with oxytocin, the hormone that increases milk availability for the young. Prolactin exerts its effects only when the tissues of mammary glands have been primed by other hormones.

Somatotropin profoundly influences overall growth, particularly of cartilage and bone. It directly induces protein synthesis in most types of cells by stimulating uptake of amino acids, RNA synthesis, and ribosome activity. It also directly induces the cell divisions necessary for growth. Somatotropin indirectly promotes these same activities by stimulating the release of somatomedins from the liver and other tissues. Somatomedins are growth factors that travel the bloodstream to cartilage, bone, and other target tissues, where they trigger growth-related events.

Table 26.3 Hormone Sources Other Than the Hypothalamus and Pituitary

Source	Its Secretion(s)	Main Targets	Primary Actions
Adrenal cortex	Glucocorticoids (e.g., cortisol)	Most cells	Raise blood sugar level; help control lipid, protein metabolism, mediate responses to stress
	Mineralocorticoids (e.g., aldosterone)	Kidney	Promote sodium reabsorption; salt, water balance
	Sex hormones (e.g., testosterone, estrogens)	General	Influence sexual characteristics, general growth
Adrenal medulla	Epinephrine (adrenalin), norepinephrine	Liver, cardiac muscle	Raises blood sugar level by stimulating glucose production; increases heart rate and force of contraction
		Adipose tissue	Raises blood fatty acid levels
		Smooth muscle in walls of blood vessels	Vasoconstriction or vasodilation
Thyroid	Triiodothyronine, thyroxine	Most cells	Regulates carbohydrate, lipid metabolism; growth, development
	Calcitonin	Bone	Lowers calcium levels in blood by inhibiting calcium reabsorption from bone
Parathyroids	Parathyroid hormone	Bone, kidney, gut	Elevates calcium levels in blood by stimulating calcium reabsorption from bone, kidneys, and absorption from gut
Gonads:			
Testis	Testosterone (an androgen)	General	Essential for spermatogenesis; development of genital tract; maintenance of accessory sex organs and secondary sex traits; influences growth, development
Ovary	Estrogens	General	Essential for oogenesis; stimulate thickening of uterine lining for pregnancy; other actions same as above
	Progesterone	Uterus, breasts	Prepares, maintains uterine lining for pregnancy; stimulates breast development
Pancreatic islets	Insulin	All cells except most neurons in brain and red blood cells	Lowers blood sugar by stimulating glucose uptake by cells; fat storage; protein synthesis
	Glucagon	Liver	Raises blood sugar by stimulating glucose production
Endocrine cells of gastrointestinal epithelium	Include gastrin, cholecystokinin, secretin	Stomach, pancreas, gallbladder	Control of stomach, pancreas, liver, gallbladder
Liver	Somatomedins	Most cells	Stimulates overall growth, development
Kidney	Erythropoietin*	Bone marrow	Stimulates red blood cell production
	Angiotensin*	Adrenal cortex, arterioles	Control of blood pressure and aldosterone secretion
Thymus	Include thymosin	Lymphocytes, plasma cells	Promote development of infection-fighting abilities and lymphocyte function in immune responses
Pineal	Melatonin	Gonads	Influences daily biorhythms, sexual activity and sexual development

*These hormones are not produced in the kidneys but are formed when *enzymes* produced in kidneys activate specific plasma proteins.

Individuals who cannot produce enough somatotropin during childhood suffer *pituitary dwarfism*. Although the adult body is proportionally similar to that of a normal person, it is much reduced in size (Figure 26.7). Excessive production of somatotropin during childhood leads to *gigantism*. The adult body is similar in proportion to that of a normal person but is much larger in size. When somatotropin secretions are excessive during adulthood (when long bones no longer can lengthen), *acromegaly* is the result. Adult bones of the hands, feet, and jaws thicken, as do epithelial tissues of the skin, nose, eyelids, lips, and tongue. Skin thickening is especially pronounced on the forehead and soles of the feet.

Table 26.3 lists major sources of hormones other than the hypothalamus and pituitary. Later we will look closely at the roles these hormones play in digestion, circulation, reproduction, and other activities. For now, we can simply introduce their sources and give a few examples of the kinds of controls governing their secretion.

ADRENAL GLANDS

Adrenal Cortex

Like most endocrine glands, the **adrenal cortex** interacts with the neuroendocrine control center through homeostatic feedback loops. In essence, increased or decreased concentrations of hormones secreted by these glands are detected by the hypothalamus or pituitary (or both), which responds in ways that inhibit or stimulate further secretion of those hormones.

The adrenal cortex is the outer layer of the adrenal gland. Humans have a pair of these glands, positioned above the kidneys (Figure 26.1). Cells in the adrenal cortex produce and secrete glucocorticoids, mineralocorticoids, and sex hormones.

The glucocorticoids, such as cortisol, help regulate carbohydrate, lipid, and protein metabolism. They also function in defense responses to tissue injury or infection. Cortisol secretion increases when the hypothalamus secretes a releasing hormone (CRH), which stimulates the anterior pituitary into secreting adrenocorticotropic hormone (ACTH). This hormone in turn stimulates the adrenal cortex into secreting cortisol, which acts on target cells in the liver, muscles, and elsewhere. Cortisol secretion decreases when both the hypothalamus and pituitary detect the rising blood concentrations of cortisol and respond by reducing CRH and ACTH secretions (Figure 26.8).

The feedback mechanism just described is overridden when the body is stressed, as from a frightful experience, painful injury, or prolonged illness. Then the body is thrown into an emergency state. Unless conditions return to normal, increased secretion of cortisol and other signaling molecules is prolonged.

The adrenal cortex also secretes mineralocorticoids, which help *maintain* concentrations of solutes (such as sodium) in the extracellular fluid when food intake and metabolic activity change the amounts of solutes entering the bloodstream. Aldosterone is important in this

Figure 26.7 Example of pituitary dwarfism. Charles Stratton was proportionally similar to normal adults but was only eighty-four centimeters tall as a result of an underactive anterior pituitary.

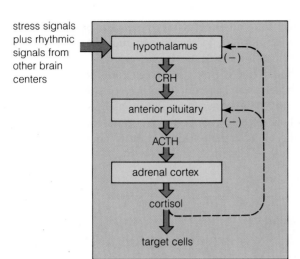

stress signals plus rhythmic signals from other brain centers

hypothalamus

CRH

anterior pituitary

ACTH

adrenal cortex

cortisol

target cells

Figure 26.8 Negative feedback control of cortisol secretion.

respect. By promoting sodium reabsorption in the kidneys, aldosterone also indirectly promotes water reabsorption; hence it influences the volume of extracellular fluid (page 466).

Adrenal Medulla

The inner region of the adrenal gland, the **adrenal medulla**, is under neural control. It contains modified neurons that secrete epinephrine and norepinephrine, both of which help control blood circulation and carbohydrate metabolism. Brain centers including the hypothalamus govern their secretion by way of sympathetic nerves.

During times of excitement or stress, the adrenal medulla contributes to the overall mobilization of the body by the sympathetic nervous system. In response to epinephrine and norepinephrine, the heart beats faster and harder, blood flow increases to heart and muscle cells, airways in the lungs dilate, and more oxygen is delivered to cells throughout the body. These are some of the characteristics of the "fight-flight" response, described earlier on page 329.

THYROID GLAND

The human **thyroid gland** is positioned at the base of the neck, in front of the trachea (Figure 26.9). The major secretions of this gland, triiodothyronine and thyroxine, influence overall growth, development, and metabolic rates.

When thyroid secretions are abnormally high, the outcome can be *hyperthyroidism*. This disorder is char-

acterized by increased heart rate, elevated blood pressure, excessive sweating, weight loss, and intolerance of heat. Typically the affected person shows nervous, agitated behavior and has difficulty sleeping. When thyroid secretions are abnormally low, the outcome is *hypothyroidism*. In adults, this disorder is characterized by lethargy, intolerance of cold, and dried-out skin. If the condition is present in infants and is not diagnosed early enough, permanent mental retardation and dwarfism may result.

A *goiter* is an abnormally enlarged thyroid gland that can result from excessive stimulation of the thyroid. For example, goiters can be caused by a dietary deficiency of iodine, which is necessary for the synthesis of thyroid hormones. When the blood levels of these hormones are low, the anterior pituitary is stimulated to secrete more TSH, which in turn stimulates growth of the thyroid gland, sometimes to gross proportions. Goiter caused by iodine deficiency is no longer common in countries where iodized salt is widely used. Elsewhere, hundreds of thousands of people still suffer from this disorder, which is easily preventable.

Another hormone secreted by the thyroid gland, calcitonin, plays a minor but direct role in controlling extracellular levels of calcium ions (Ca^{++}) by promoting deposition of these ions into bone when the levels rise. Once calcium returns to its homeostatic level, the thyroid cells decrease their secretion of calcitonin.

PARATHYROID GLANDS

As Figure 26.9 shows, four *parathyroid glands* are embedded in the posterior surface tissues of the thyroid gland. Their secretions are directly controlled by the extracellular concentrations of calcium ions. All metabolically active cells require calcium for a variety of tasks. (To give just two examples, calcium concentration profoundly affects gene activation as well as the excitability of nerve and muscle cells.) The extracellular concentration of calcium is commonly maintained within extremely narrow limits by homeostatic feedback. Although the thyroid has a small effect on the calcium concentration, the major feedback loop involves the parathyroid glands.

Parathyroid hormone (PTH) is the main hormone taking part in feedback control of extracellular calcium. Its secretion is not stimulated directly by other hormones or nerves; rather, it responds to the Ca^{++} concentration in extracellular fluid bathing the parathyroid glands. PTH production and release are stimulated when the Ca^{++} level falls. PTH stimulates removal of calcium (and phosphate) from bone and its movement into extracellular fluid. It induces resorption of calcium by the kidney. It also helps activate vitamin D, which enhances the absorption of calcium present in food moving through the gut. (Vitamin D deficiency leads to *rickets*, a disorder arising from the lack of enough calcium for proper bone development.) When calcium levels rise in response to these PTH-induced events, stimulation of the parathyroid glands is stopped.

GONADS

The **gonads** are the primary reproductive organs. They are called testes in males and ovaries in females. Gonads are the sites of gamete production; they also secrete sex hormones that influence reproductive function and secondary sexual traits (such as the distribution of body hair). The hormonal interactions necessary for reproduction are truly intricate, and for that reason we will postpone our consideration of them until Chapter Thirty-Five.

OTHER ENDOCRINE ELEMENTS

Pancreatic Islets

The pancreas is largely an exocrine gland that secretes digestive enzymes. However, about 2 million endocrine cell clusters, called **pancreatic islets**, are also scattered in the pancreas. In response to changing concentrations of glucose and amino acids in the blood, these cells produce and secrete the hormones insulin and glucagon.

Some pancreatic cells secrete *insulin* when glucose levels in the blood are high, as they are after a meal. This hormone promotes the uptake of glucose by the body's cells—including liver cells, where excess glucose can be converted to glycogen (a storage polysaccharide). When glucose levels in the blood are low, other cells of the pancreas secrete *glucagon*, which counterbalances some aspects of insulin action by stimulating the breakdown of glycogen into glucose units, which can be released into the bloodstream (Chapter Thirty-Two).

Several disorders classified under the general term *diabetes* are associated with insulin deficiency or with the ability of cells to take up glucose. In "type 1 diabetes," the person has very little insulin or none at all because the insulin-secreting pancreatic cells have been destroyed. This disorder is the less common form of diabetes. It appears to arise from a convergence of factors, including a heritable susceptibility to the disorder, viral infection, and an autoimmune response mounted against the body's own insulin-secreting cells. Survival for type 1 diabetics is absolutely dependent upon insulin injections. (Insulin cannot be taken orally because it is a protein, hence it would be broken down by protein-degrading enzymes in the digestive tract before it could

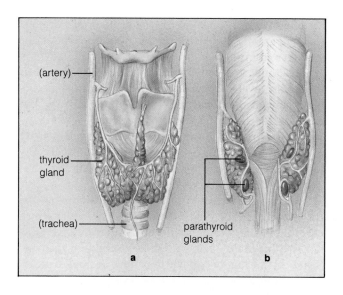

Figure 26.9 (**a**) Anterior view of the human thyroid gland. (**b**) Posterior view, showing the location of the four parathyroid glands embedded in it.

even be absorbed.) Without injections, the person would suffer severe metabolic and urinary disruptions.

In "type 2 diabetes," insulin levels are nearly normal or even above normal—but the target cells cannot respond to the hormone. Either the cells have an insufficient number of insulin receptors, or the receptors themselves are abnormal.

Thymus Gland

The **thymus** is a two-lobed gland located behind the breastbone and between the lungs. This gland is important in the maturation of certain white blood cells (a class of lymphocytes), which defend against infection and form a surveillance system by which damaged or malignant cells are detected and eliminated from the body. The thymus also secretes a group of hormones collectively called thymosins, which may affect the functioning of certain lymphocytes (Chapter Thirty).

Pineal Gland

Until about 240 million years ago, vertebrates commonly were equipped with a third "eye", perched in the middle of the forehead. Lampreys and a few other vertebrates still have one of these photosensitive structures, buried beneath the skin. In mammals, birds, and most reptiles, the "burial" became more pronounced: the structure persists in modified form as an internal **pineal gland**. The pinecone-shaped pineal gland is a small outpouching of the brain (Figure 26.1). It secretes melatonin, a hormone that apparently functions in the development of the gonads and in reproductive cycles.

Melatonin is secreted in the absence of light. Thus melatonin levels vary on a diurnal basis (from day to night) and on a seasonal basis (for example, when winter days are shorter than summer days). Studies of annual reproductive cycles of hamsters show that pineal activity during the long nights of winter suppresses sexual activity. During the long days of summer, pineal activity is suppressed and sexual activity peaks.

In humans, decreased secretion of melatonin may be correlated with the onset of puberty (the age at which reproductive structures mature and start to become functional). Melatonin levels are highest in children aged one to five; by the end of puberty they have decreased steadily to about a fourth of the peak levels. The likely link between the pineal and the timing of puberty was suspected nearly a century ago, when the physician Otto Heubner reported the results of an autopsy on a boy who had undergone puberty prematurely, at age four. Heubner discovered that a brain tumor had destroyed the boy's pineal gland.

LOCAL CHEMICAL MEDIATORS

Within each tissue of the mammalian body, some mediator cells detect changes in the surrounding chemical environment and alter their activity, often in ways that either counteract or amplify the change. These are local responses, confined to the immediate vicinity of the change; most of the molecules secreted are taken up so rapidly that not many are left to enter the general circulation. Among these local mediator molecules are the prostaglandins and growth factors.

Prostaglandins

More than sixteen different prostaglandins have been identified in tissues throughout the body. They are released at a low level on a continuing basis, but the rate of synthesis often increases in response to local chemical changes. The stepped-up secretion can influence neighboring cells as well as the prostaglandin-releasing cells themselves.

At least two prostaglandins help adjust blood flow through local tissues by acting on smooth muscle in the walls of blood vessels. Their production is stimulated by epinephrine and norepinephrine, which cause blood vessel diameters to constrict or dilate. Prostaglandins also act on smooth muscle of airways in the lungs, causing some to constrict and others to dilate. Allergic responses to airborne dust and pollen may be aggravated

by prostaglandins, the massive production of which is stimulated in local tissues.

Prostaglandins have profound effects on reproductive events in the mammalian female. During the menstrual cycle, many women experience painful cramping and excessive bleeding—both of which have been traced to prostaglandin activity. (Anti-prostaglandin drugs can inhibit synthesis of this local mediator and thereby alleviate the discomfort.) Prostaglandins also influence the corpus luteum, a glandular structure that develops from cells which earlier surrounded a developing egg in the ovary. If pregnancy does not follow ovulation (release of the egg from the ovary), the corpus luteum self-destructs by producing copious amounts of prostaglandins that interfere with its own functioning.

Growth Factors

Some signaling molecules, called growth factors, influence growth by regulating the rate at which certain types of cells divide. For example, **nerve growth factor** (NGF), discovered by Rita Levi-Montalcini, promotes the survival and growth of neurons in the developing embryo. (One experiment demonstrated that certain immature neurons survive indefinitely in tissue culture, provided that NGF is in the medium. They die within a few days if it is not.) NGF also may define the direction of growth for the embryonic neurons, laying down the chemical path that leads the elongating processes to target cells. **Epidermal growth factor** (EGF), discovered by Stanley Cohen, plays a role in the growth of many different cell types.

SIGNALING MECHANISMS

Signaling molecules induce remarkably diverse responses in target cells. They cause different proteins to be synthesized, they trigger changes in the rates of synthesis, and they lead to modification of already existing proteins. What determines the nature of the response? First, not all cells can respond to all types of signals. Second, different signals activate different mechanisms inside cells.

In complex animals, each type of cell has a characteristic array of receptors, which means they respond selectively to the many chemical signals diffusing around them. Some receptors occur in nearly all cells, hence the signaling molecules they bind have widespread effects. (The receptor for somatotropin is an example.) Other kinds of receptors occur in only a few cell types, hence the molecules they bind have highly specific effects.

A few hormones are largely water-insoluble. Even so, they are able to travel the bloodstream by riding

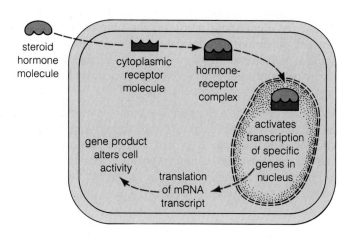

Figure 26.10 Proposed mechanism of steroid hormone action on a target cell.

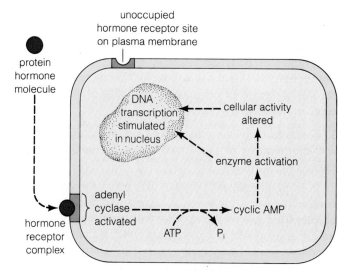

Figure 26.11 Proposed mechanism of protein hormone action on a target cell. In this case the response is mediated by a second messenger inside the cell: cyclic AMP.

piggyback on carrier proteins, which are water-soluble. Upon contact with target cell receptors, they dissociate from the carrier and readily cross the plasma membrane. These hormones are said to have *intracellular receptors*.

Most signaling molecules are water-soluble, which means they are not able to penetrate the plasma membrane of the target cell. Instead they bind with *surface membrane receptors*. These water-soluble molecules include all transmitter substances, protein hormones, and local growth factors.

Intracellular Receptors and Steroid Hormones

Steroid hormones, which are derived from cholesterol, diffuse readily across the plasma membrane and bind to receptors inside their target cells. The hormone-receptor complex then moves into the cell nucleus, where it binds to DNA. In each type of cell, the DNA appears to be packaged in ways that leave only a small number of target areas exposed for the arrival of hormone-receptor complexes. When the receptor is bound to the DNA, certain genes are transcribed (Figure 26.10).

One steroid hormone, called testosterone, is necessary for the development of male sexual traits. In *testicular feminization syndrome*, the receptor to which testosterone binds is defective. Genetically, the affected individuals are males; they have functional testes that secrete testosterone. However, none of the target cell types can respond to the hormone, and the secondary sexual traits that do develop are entirely female or female-like.

Second Messengers

How do protein hormones and other water-soluble signaling molecules induce responses in their target cells, given that they cannot cross the plasma membrane? Some are transported into the cell by receptor-mediated endocytosis, which was illustrated in Figure 6.12. Others bind to a receptor and thereby cause ion channels across the membrane to open. Ion concentrations in the cytoplasm change as a result, and the change affects cell activities. However, most protein hormones activate **second messengers**, which are intracellular mediators of the response to those hormones.

Let's take a look at the pathways by which second messengers are used. Most begin when a signaling molecule binds to a receptor and thereby alters the activity of a membrane-bound enzyme system (Figure 26.11). Typically, adenylate cyclase is activated, and this enzyme catalyzes the transformation of ATP to **cyclic AMP**. (The full name is cyclic adenosine monophosphate.)

Many molecules of adenylate cyclase, not just one, are activated by a hormone-receptor complex. Each of these enzyme molecules increases the rate at which many ATP molecules are converted to cyclic AMP. Each second-messenger molecule so formed then activates many enzymes. Each of the enzymes so activated can convert a very large number of substrate molecules into different enzymes, and so on until the number of molecules representing the final cellular response to the initial signal is enormous. Thus, second messengers are a way of *amplifying* the cellular response to a signaling molecule.

SUMMARY

This chapter concludes our survey of controls over the integration of body activities in multicelled animals. Throughout this unit, we will be looking at specific examples of these controls, so keep the following key concepts in mind:

1. For metabolic activity to proceed smoothly, the chemical environment of a cell must be maintained within fairly narrow limits.

2. In complex animals, thousands to billions of cells are continually removing some substances from the extracellular fluid and secreting other substances into it. The nature and amount of these substances can change with the diet or level of activity; they inevitably change during the course of development.

3. It follows that these myriad withdrawals and secretions must be integrated in ways that assure cell survival through the whole body.

4. Integration is accomplished by signaling molecules: chemical secretions by one cell type that adjust the behavior of another, target cell type. (A target cell is one having receptors to which signaling molecules can bind. It may or may not be adjacent to the signaling cell.)

5. Signaling molecules include local mediators, transmitter substances, neurohormones, hormones, and pheromones.

6. Responses to signaling molecules vary. For example, fast-acting hormones generally come into play when the extracellular concentration of a substance must be homeostatically controlled. Slow-acting hormones have more prolonged, gradual, and often irreversible effects, such as those on development.

7. In all vertebrates, a neuroendocrine control center helps integrate activities for the entire body. This center consists of the hypothalamus and pituitary.

8. At least nine hypothalamic neurohormones (called inhibiting and releasing hormones) control the secretion of hormones from cells of the anterior lobe of the pituitary.

9. Two other hypothalamic neurohormones (ADH and oxytocin) are stored and released from the posterior lobe of the pituitary. ADH influences extracellular fluid volume. Oxytocin influences contractions of the uterus and milk release from mammary glands.

10. Two hormones synthesized and released by the anterior pituitary (prolactin and somatotropin) have effects on body tissues; the remainder (ACTH, TSH, FSH, and LH) act on endocrine glands.

11. Hormone secretion is controlled by neural signals, hormonal signals, and extracellular concentrations of metabolites such as calcium. Secretion of many endocrine glands is controlled through homeostatic feedback inhibition of the hypothalamus, the pituitary, or both.

12. Responses to hormones and other signaling molecules vary, depending partly on the type of receptors present on different cells. (Some cells can respond to some signals but not to others.) Responses also depend partly on which genes or enzymes are activated within any of a number of target cell types.

13. When bound to a receptor, a signaling molecule triggers synthesis of particular enzymes and other proteins, alterations in the rate of synthesis, or modification of existing proteins. The response contributes in some way to maintaining the internal environment or to the development or reproductive program.

Review Questions

1. Name the main endocrine glands and state where each is located in the human body.

2. Define the difference between transmitter substances and neurosecretory hormones.

3. Define hormone. What functions do hormones serve? How do these functions differ from those of transmitter substances?

4. There are two general classes of hormones: steroid and polypeptide. How is each thought to act on a target cell?

5. The hypothalamus and pituitary are considered to be a neuroendocrine control center. Can you describe some of the functional links between these two organs?

6. How does the hypothalamus control secretions of the posterior lobe of the pituitary? The anterior lobe?

7. Name five endocrine glands and a substance that each one secretes. What are the main consequences of their secretion?

8. Which hormone secreted by the anterior pituitary has an effect on most body cells rather than on a specific cell type? What are the clinical consequences of too little or too much secretion of this hormone?

Readings

Alberts, B. et al. 1983. *Molecular Biology of the Cell*. New York: Garland. Chapter 13 is a good introduction to the mechanisms of hormone action.

Berridge, M. October 1985. "The Molecular Basis of Communication Within the Cell." *Scientific American* 253(4):142–152.

Bloom, F. E. October 1981. "Neuropeptides." *Scientific American* 245(4):148–168.

Fellman, B. May 1985. "A Clockwork Gland." *Science 85* 6(4):76–81. Describes some of the known functions of the pineal gland.

Franklin, D. 1984. "Growing Up Short." *Science News* 125:92–94. Describes prospects and problems of using synthetic somatostatin (growth hormone) to boost height of hypopituitary children.

Snyder, S. October 1985. "The Molecular Basis of Communication Between Cells." *Scientific American* 253(4):132–141.

Looking back over the past few chapters, you might be left with the impression that the nervous system is an immense collection of communication lines that are silent until signals are fed into them from the outside, much as telephone lines wait to carry phone calls from all over the country. It is the wrong impression. The crucial feature of a nervous system is not what it is made of but *what it does*—and from its formation in the embryo until death, it never stops doing things, even beyond its responses to signals from the outside.

Many neurons in the brain centers concerned with breathing and other vital functions never do rest. They start firing rhythmically early in development, and although incoming sensory signals may alter the firing frequency, nothing short of damage or death stops their basic activity. Also during development, vast gridworks of neurons are laid down and begin to interact in genetically prescribed patterns. These patterns of activity are inherited **neural programs**. They provide each new individual with a means of responding to situations that members of its species are likely to encounter.

For example, among other things, the python shown in Figure 27.1 eats small mammals that forage at night. At night the temperature difference between its prey and

27

SENSORY SYSTEMS

Figure 27.1 Reticulated python of southern Asia, equipped with infrared-sensitive receptors that enable it to detect warm-blooded prey in the dark. This particular snake has the receptors in thirteen pairs of pits above and below the mouth, as visible here.

Table 27.1   Receptors Associated With Different Sensory Modalities

Sensory Modality	Stimulus	Examples of Receptors	Main Functions
Chemical senses:			
Internal chemical sense	Chemical energy	Carotid body (page 427), free nerve endings	Signal changes in extracellular concentrations of specific ions, molecules
Taste	Chemical energy	Taste receptor cells	Signal presence of nutritious or noxious substances
Smell	Chemical energy	Olfactory receptor cells	Signal presence of specific airborne ions, molecules
Somatic senses:			
Touch, pressure	Mechanical energy	Mechanoreceptors (free nerve endings, Pacinian corpuscles)	Signal changes at body surface (e.g., mechanical distortion)
Temperature	Thermal energy (heat)	Free nerve endings	Signal changes in temperature
Pain	Varied (e.g., mechanical, thermal, chemical energy)	Several, including free nerve endings	Signal damage or threat of damage
Muscle sense and kinesthesia	Mechanical energy (e.g., stretching, rotation at joints)	Mechanoreceptors in joints, tendons, skin	Signal changes in position, movement; contribute to sensations of effort, force, and weight
Balance (equilibrium)	Mechanical energy, gravitational energy	Hair cells; mechanoreceptors in tendons, joints	Signal changes in body position and movements underlying orientation relative to environment
Acoustical sense (hearing)	Varies (compressional waves or distortions in air, water)	Hair cells in lateral line organs, in mammalian inner ear	Signal sound waves and other physical displacements of environmental medium
Photosensitivity, vision	Light energy	Photoreceptors (e.g., rods and cones)	Varies among species (signal changes in light intensity; motion, form, depth, and color in visual field; polarization of light)

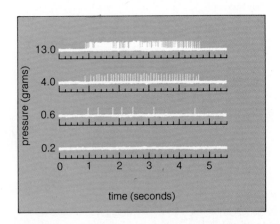

Figure 27.2 Action potentials recorded from a single pressure receptor on the human hand. The recordings correspond to variations in stimulus strength. A thin rod was pressed against the skin with the force indicated. Vertical bars above each thick horizontal line represent individual action potentials. Notice the increases in frequency, which correspond to increases in stimulus strength.

the surrounding air is usually large. This snake has receptors for detecting heat (infrared energy), which is used to pinpoint its prey in the dark. Input from the receptors is fed into a brain center where signals about the location of objects are compared against a neural gridwork. The signals are analyzed with great precision, and the snake directs its strike with great accuracy. (At most, the strike may be only a few degrees off center.) Even so, the same snake might slither past a motionless yet edible frog. Frog skin is cool and typically blends with background colors. The snake has neither receptors nor a neural program for responding to it.

SENSORY PATHWAYS

Receptors Defined

A **stimulus** is any form of energy the body is able to detect by means of its receptors. **Receptors** are finely branched peripheral endings of sensory neurons (or specialized cells adjacent to them) that respond to specific stimuli. Often, receptors, epithelial tissues, and connective tissues are organized into **sensory organs** (such as eyes), which amplify or focus the energy of a stimulus. The major categories of receptors may be described in the following way:

Chemoreceptors detect chemical energy (ions or molecules dissolved in body fluids next to the receptor). They include olfactory and taste receptors.

Mechanoreceptors detect mechanical energy associated with changes in pressure, position, or acceleration. They include receptors for touch, stretch, hearing, and equilibrium.

Photoreceptors detect photon energy of visible and ultraviolet light.

Thermoreceptors detect radiant energy associated with temperature. They include infrared receptors.

Table 27.1 summarizes the common forms of stimuli and lists some of the receptors for them. The kind and number of receptors vary among animals. For example, you do not have receptors for ultraviolet light, as honeybees do; and you do not "see" many flowers the way they do (Figure 21.8). Unlike dogs, you have receptors that contribute to color vision, but dogs have far more olfactory receptors in the nose. Hence humans and dogs sample the environment in different ways and have different perceptions of it.

Principles of Receptor Function

A receptor is a *transducer*: it converts one form of energy into another. When the plasma membrane of a receptor is stimulated, the outcome is a **receptor potential** (a brief change in voltage across the membrane). The change is a graded response. Like the synaptic potentials described in Chapter Twenty-Four, it can vary in magnitude, depending on the intensity and duration of the stimulus. Graded responses acting together lead to action potentials, the means by which messages travel through the nervous system.

As we have seen, action potentials are all alike; they never vary in magnitude. How, then, do they convey information that gives rise to different sensations such as smell, taste, and color? Part of the answer is that some signals from specific receptors always end up in specific parts of the brain. (In the embryo, sensory nerves destined to carry those signals always develop along genetically ordained routes from one part of the body to another.) Thus, when you accidentally poke your eye in the dark, you "see stars." Action potentials from visual receptors are always interpreted as "light," simply because the brain region in which they end up interprets all signals coming in on a particular pathway as "light."

But now the question becomes this: How do action potentials convey information about variations in stimulus intensities? The same receptors detect energy associated with a throaty whisper or a wild screech, and the same sensory pathways carry information about both to the brain—but how? When the strength of a stimulus increases, the magnitude of receptor potentials also increases, so action potentials are fired more frequently. Figure 27.2 shows some frequency variations corresponding to differences in sustained pressure on human skin.

Also, stronger stimuli "recruit" more and more receptors in a larger area of a given tissue. For instance, when you lightly press a finger into skin on your arm, you activate a small number of receptors. When you push the finger firmly into the same region, more receptors are disturbed. The increased disturbance sets off action potentials in many sensory axons at the same time, and the simultaneous activity reflects the increase in stimulus intensity.

A particular sensation is not triggered by action potentials themselves but by their travel along a particular nerve pathway.

Variations in stimulus intensity are encoded in the <u>frequency</u> of action potentials in a single axon.

Variations in stimulus intensity are also encoded in the <u>number</u> of axons recruited into action in a given tissue.

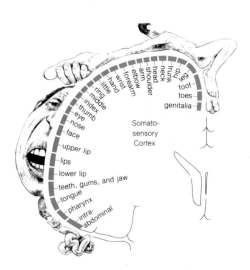

Figure 27.3 Body regions represented in the primary somatic sensory cortex. This region is a strip a little more than an inch wide, running from the top of the head to just above the ear on the surface of each cerebral hemisphere. The diagram is a cross-section through the right hemisphere of someone facing you. Compare this with Figure 25.12.

Primary Sensory Cortex

Sensory nerve pathways from different receptors lead to different parts of the cerebral cortex. For example, signals from receptors in the skin and joints travel to the primary somatic sensory cortex. Cells of this region are laid out like a map corresponding to the body surface. Some map regions are larger than others; they represent areas that are functionally more important and that have more receptors. In humans, a large part of the primary somatic sensory cortex responds to receptors located in the fingers, thumb, and lips (Figure 27.3).

Let's now focus on a few kinds of receptors to illustrate how they detect changes in the surrounding environment.

CHEMICAL SENSES

Animals can distinguish between nutritious and noxious substances with their **taste receptors**. These receptors are present in the mouth of most vertebrates. They also are distributed on snail antennae, insect legs, octopus tentacles, and some fish fins. In many vertebrates, taste receptors are components of sensory organs called **taste buds**, which are distributed mostly on the tongue. A taste bud is composed of modified epithelial cells, includ-

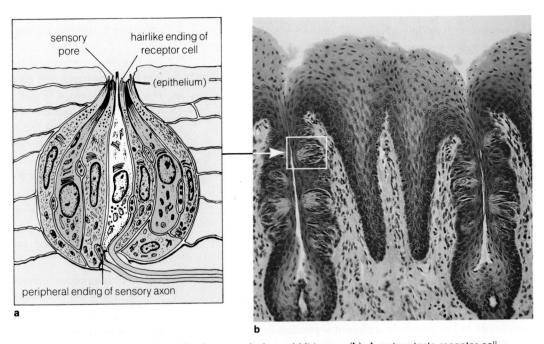

Figure 27.4 Taste bud (**a**) in the walls of narrow pits in a rabbit tongue (**b**). A mature taste receptor cell, which synapses with a sensory axon, is shown in white. An immature taste receptor cell appears in the lower-right region of the taste bud. Cells with supporting functions are shaded orange.

ing the receptor cells (Figure 27.4). Sensory nerves carry signals from the receptors to the brain.

Animals also distinguish among different odors. The detection of odors, or **olfaction**, is one of the most ancient senses. Olfactory receptors still must be important for survival, judging from their sheer abundance. Humans have about 10 million olfactory receptors in the nose; German shepherds have more than 220 million. Olfactory receptors respond to molecules released from food, from predators, and from members of the same species. They all are modified endings of axons that carry signals to the brain.

In many animals, specialized olfactory receptors respond to pheromones (signaling molecules having cellular targets outside the body, in other animals of the same species). Pheromones are social signals. Animals use them to raise an alarm about impending danger, attract a mate, influence sexual behavior, identify individuals of the same social group, disperse a group, or mark territorial boundaries.

Many olfactory receptors are remarkably sensitive to pheromones. Those on the antennae of male silk moths (*Bombyx mori*) can detect one molecule of bombykol in 10^{15} molecules of air! Each receptor, making contact with only one molecule per second, can trigger an action potential. Female silk moths secrete bombykol as a sex attractant, which enables a male to find a female, even in the dark, and even more than a kilometer upwind.

SOMATIC SENSES

The term **somatic senses** refers to sensations of touch, pressure, temperature, and pain near the body surface. Two types of somatic receptors are free nerve endings and Pacinian corpuscles (Figure 27.5). Both are activated when their plasma membrane is mechanically deformed or altered in some other way. Free nerve endings in skin contribute to sensations of light pressure, temperature, and pain; Pacinian corpuscles contribute to sensations of deep, rapid pressure and vibration.

"Pain" is the perception of injury to some region of the body. Specific pain receptors have been identified in invertebrates such as the leech, and it seems likely that they exist in mammals as well. Although invertebrates certainly react to injury, it is not known whether they "feel" pain as we do. The perception of pain seems to have an important emotional component, mediated perhaps by the limbic system (page 334). The invertebrate brain does not have a comparable system.

Responses to some types of pain depend on accurately assessing where the stimulus is located. This can be done for most types of somatic pain and some types of visceral (internal) pain. But most sensations of visceral pain apparently are based on erroneous signals, the outcome being that pain is felt in a tissue remote from the original point of stimulation. (For example, as Figure 27.6 suggests, a heart attack may be felt as pain in the skin

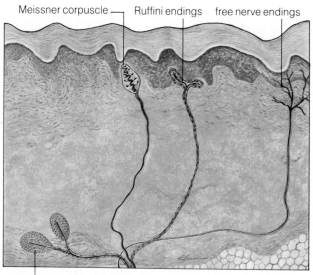

Figure 27.5 Tactile receptors in human skin. Free nerve endings contribute to sensations of temperature, light pressure, and pain. Pacinian corpuscles (see photograph) contribute to sensations of deep, rapid pressure. Meissner corpuscles are stimulated at the onset and end of sustained pressure; the Ruffini endings react continually to ongoing stimuli.

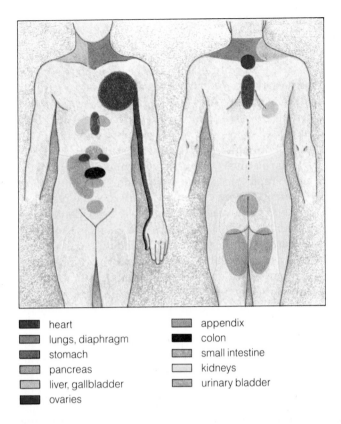

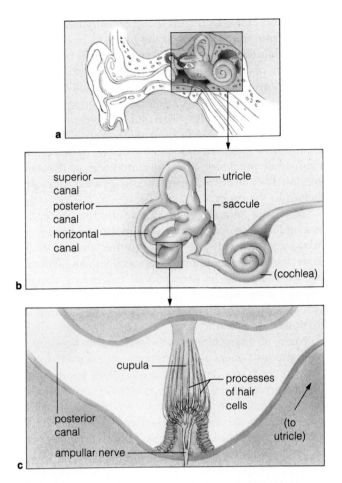

Figure 27.6 Regions of referred pain. Sensations of certain visceral disorders are erroneously localized to the skin areas indicated.

heart
lungs, diaphragm
stomach
pancreas
liver, gallbladder
ovaries
appendix
colon
small intestine
kidneys
urinary bladder

Figure 27.7 (**a**) Location of organs of equilibrium in the human ear. (**b**) The vestibular apparatus, with three semicircular canals and two sacs (utricle and saccule) in which otoliths occur. (**c**) Ampula (bulge) of the posterior canal. The cupula is a gelatinous membrane that bends under fluid pressure. When the head rotates in one direction, fluid in the canal is displaced in the opposite direction (and so are the cupula and the processes of hair cells).

above the heart and along the left shoulder and arm.) This phenomenon is called "referred pain."

What is the basis of referred pain? Nerves from the injured (or malfunctioning) organ and nerves from the skin segment to which the pain is referred usually started out close together when the animal embryo was first developing. By the time they reached their normal position in the adult body, they had become physically separated. Even so, the nerves still enter the same segment of the spinal cord and synapse on the same interneurons. The brain continues to refer the internal pain to the position of the skin segment in the adult.

THE SENSE OF BALANCE

Almost all animals must have a sense of what the "natural" position for their body must be, given the way they return to it after being tilted, turned upside-down, and so on. The baseline against which animals assess displacement from their natural position is called **equilibrium**. At equilibrium, the body (or some part of it, such as a wing) is balanced in relation to gravity, velocity, acceleration, and other forces that influence position and movement.

Organs of equilibrium incorporate mechanoreceptors called **hair cells**. These cells are arranged in an epithelium and have one or more processes ("hairs") projecting from their surface. When the processes bend in response to gravity and other forces, receptor potentials are produced and signals about changing position are sent to the brain.

In vertebrates, hair cells are present in the **vestibular apparatus**, which is a closed system of fluid-filled sacs and canals inside the ear (Figure 27.7). Together with other input from receptors in the eyes, skin, and joints, the vestibular apparatus contributes to the sense of balance. It consists of *semicircular canals* (which detect changing movements) and an *otolith organ* (which detects changes in the head's orientation relative to gravity).

Amphibians, birds, and mammals have three semicircular canals, positioned at angles corresponding to three planes of space (Figure 27.8). Within each canal, hair cells project into a jellylike mass called a cupula. When the body moves, fluid in the canal corresponding to the plane of movement is displaced and pushes against the cupula, causing the hairs to bend. The otolith organ consists of two fluid-filled sacs, each with a jellylike membrane in which calcium carbonate crystals are embedded. The membrane sits on top of a patch of hair cells. When the head tilts, fluid movement causes the heavy membrane to slide over the hair cells and activate them.

Jellyfishes have an organ of equilibrium called a *statocyst* (Figure 27.8). This infolding of epidermis contains hair cells and statoliths, which are dense crystals of sand grains (or mineral salts and organic material). When the body tilts, statoliths also tilt and cause neighboring hair cells to bend. Thus receptor potentials are produced.

THE SENSE OF HEARING

The sense of sound, or **hearing**, also depends on the bending of hair cells under fluid pressure. In this case, vibrations initiate the fluid movements. A "vibration" is a wavelike form of mechanical energy, transmitted outward from a stimulus by a series of compressions and rarefactions of molecules of the surrounding medium. ("Rarefaction" means a low-pressure state.)

For example, clapping your hands produces waves of compressed air. The clapping force drives many molecules together in the air between your hands, creating a high-pressure state in which the molecules collide faster and more often. The collisions send molecules flying outward, where they collide with more distant molecules, and so on away from your hands. Each time molecules are forced outward, a low-pressure state is created in the region they vacated. These pressure vari-

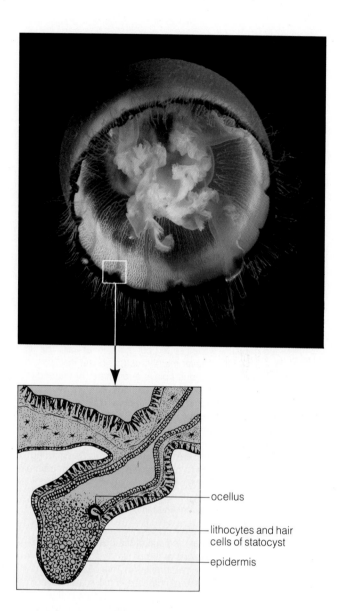

ocellus

lithocytes and hair cells of statocyst

epidermis

Figure 27.8 Statocyst at the bell margin of the jellyfish *Aurelia*. When the bell tilts, lithocytes (statolith-containing cells) press against neighboring hair cells and activate them. Signals from the hair cells lead to vigorous muscle contractions that correct body position.

Figure 27.9 Case study: Receptor activation in the human ear.

When pressure waves reach the outer ear, they are directed inward and funneled through the ear canal. Then they strike the eardrum (*tympanic membrane*), which bows in and out at the same frequency as the waves. The movement activates the middle earbones, a lever system spanning the distance between the eardrum and the *oval window* (an elastic membrane over the entrance to the coiled inner ear; **a** and **b**). The middle earbones transmit the force of pressure waves to the much smaller surface area of the oval window, thereby amplifying the stimulus.

The oval window now bows in and out, producing fluid pressure changes in the inner ear. Pressure waves are propagated through two of the three fluid-filled ducts (the *scala vestibuli* and *scala tympani*). At the end of the second duct, pressure waves reach a membranous *round window*, causing it to bulge. Without this capacity to yield under pressure, fluid would not be able to move inside the inner ear.

How are pressure waves sorted out in the inner ear? The construction of the *basilar membrane* (the basement membrane of the third duct) plays a role here. At the end of the cochlear duct closest to the middle ear, the membrane is narrow and somewhat rigid. It becomes broader and more flexible deep in the coil of the inner ear. High-frequency waves, which carry more energy, cause the greatest displacement of the stiff region. Most of the energy associated with these waves becomes transformed into membrane vibrations here, so the waves die out before traveling deeper into the coil. Low-frequency waves also set up vibrations at the entrance, but the vibrations are lower in amplitude and continue into the more elastic regions.

Displacement of different regions of the basilar membrane stimulates different regions of the *organ of Corti*, which is perched on the basilar membrane. (**c**) This organ contains hair cells. As fluid pressure increases in the scala vestibuli, the vestibular membrane moves down, pressure in the cochlear duct rises, and the basilar membrane is pushed down. When that happens the hair cell processes move in relation to an overhanging flap (the *tectorial membrane*). The movement changes the membrane permeability of the hair cell, leading to receptor potentials that excite associated sensory neurons.

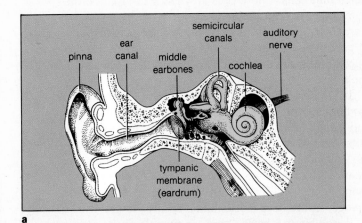

a

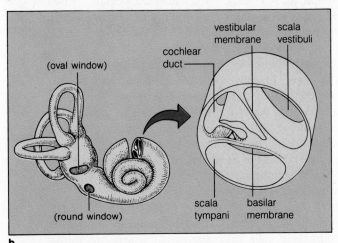

b

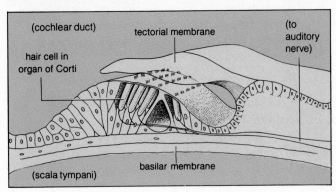

c

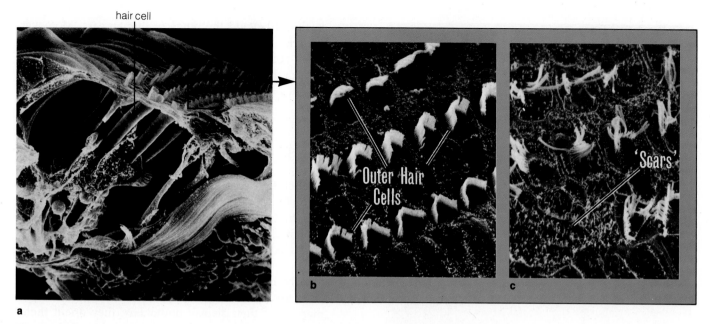

Figure 27.10 Effect of intense sound on the inner ear. Normal organ of Corti from a human (**a**) and from a guinea pig (**b**), showing three rows of outer hair cells. (**c**) Organ of Corti after twenty-four-hour exposure to noise levels approached by loud rock music (2,000 cycles per second at 120 decibels).

ations can be depicted as a wave form in which the peaks represent compressions and the valleys, rarefactions:

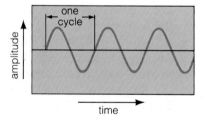

The pressure peaks (amplitude) of sounds are perceived as loudness. The stronger the stimulus, the more compressed the molecules become, and the louder the sound. The frequency of a sound is the number of wave cycles per second. Each "cycle" extends from the start of one peak of the wave to the start of the next. The more cycles per second, the higher the frequency and the higher the perceived pitch of sound.

Which Animals Hear?

Almost all vertebrates have receptors and organs adapted to the higher frequencies associated with hearing. So do many arthropods. (For example, mosquitoes and spiders are attracted to audible, high-frequency vibrations of a tuning fork.) But the most familiar organs of hearing are the paired ears of land-dwelling mammals.

When sound waves spread out through air, they rapidly become weaker with distance. The mammalian **ear**

has three parts in which weak signals are received, amplified, and sorted out. The *outer ear* is a system of external flaps (which collect sound waves) and an internal auditory canal (which channels the waves inward to an eardrum). The *middle ear* is a system of small bones (hammer, anvil, and stirrup) that transfers eardrum vibrations to the inner ear. The *inner ear* contains a coiled tube (cochlea) divided into three fluid-filled ducts, where pressure waves are sorted out and where hair cells are stimulated. Figure 27.9 shows how the three systems operate.

Some sounds are barely perceptible to humans, but others are extremely intense and cause structural damage to inner ear regions. Among these intense sounds are amplified music and the thundering of jet planes taking off. Such recent developments exceed the functional range of the evolutionarily ancient hair cells in the ear. Figure 27.10 shows receptors that have been damaged by intense sound.

Echolocation

In a process called **echolocation**, bats, dolphins, and whales emit high-frequency sound waves, and echoes from the waves bounce back to their ears from objects in the surroundings. By perceiving frequency variations in the echoes, these mammals can pinpoint the direction and distance of predators, prey, and the like. Bats can pinpoint prey even though the echoes are only 1/2,000 the amplitude of their cries. The frequency of those cries

Figure 27.11 An echolocating bat, listening to echoes of self-produced ultrasonic noises as they bounce back from objects in the environment.

is almost beyond the range of human ears. Moreover, the cries have been measured electronically at 110 decibels—about the same intensity as thunder cracking overhead or the rumble of a passing freight train. Figure 27.11 shows a bat echolocating.

SENSE OF VISION

In **photoreception**, pigment molecules embedded in the membranes of receptor cells absorb light energy and thereby trigger events that lead to action potentials in neighboring cells. Photoreception is *not* the same as vision. All organisms, whether they see or not, are sensitive to light. Shine a bright light on a single-celled amoeba moving about and it will stop abruptly. Some small invertebrates have neither photoreceptors nor pigments, yet they display **phototaxis**: they orient themselves toward or away from the direction of incoming light.

What we call "vision" depends on highly developed sensitivity to light. A **visual system** includes (1) structures that focus patterns of light energy onto a dense layer of photoreceptors, and (2) a neural gridwork in the brain that can deal with those patterns. The brain analyzes various aspects of a visual stimulus, including its position, shape, brightness, distance, and movement.

A **lens** is present in most visual systems. This spherical or cone-shaped body of transparent protein fibers channels incoming light to photoreceptor cells located behind it. But a lens alone does not lead to vision. Some invertebrates have eyes equipped with lenses, yet they cannot see as we do. Their lenses channel light either in front of or behind their photoreceptors, the result being a very diffuse kind of stimulation. These invertebrates detect a general change in light intensity, as when another animal passes overhead in the water. But they cannot discern the size or shape of objects.

Vision requires precise light focusing onto a photoreceptor cell layer that is dense enough to sample details of the light stimulus, followed by processing of visual information in the brain.

Invertebrate Photoreception

Many invertebrates have ocelli, or **eyespots**. These organs function in photoreception, but vision does not follow. The eyespots are simply clusters of photosensitive cells, usually arranged in a cuplike depression in epidermis (Figure 27.12). The pigmented membrane of these cells is often folded into tiny, fingerlike projections

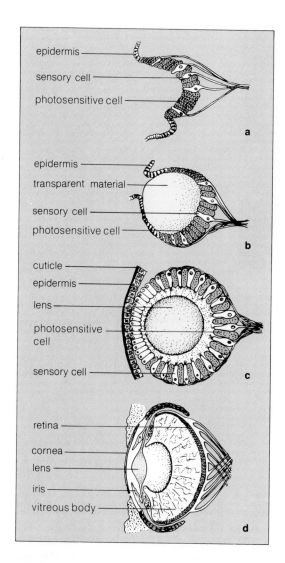

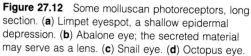

Figure 27.12 Some molluscan photoreceptors, long section. (**a**) Limpet eyespot, a shallow epidermal depression. (**b**) Abalone eye; the secreted material may serve as a lens. (**c**) Snail eye. (**d**) Octopus eye.

Figure 27.13 Examples of the well-developed eyes of mollusks. (**a**) A red-mouthed conch (*Strombus*), peering about the Great Australian Barrier Reef. (**b**) Slit eye of the octopus.

called microvilli. This pattern also occurs in vertebrate photoreceptors.

Mollusks are the simplest animals with **eyes**: well-developed photoreceptor organs that contribute to some degree of image formation. Some molluscan eyes are closed, fluid-filled vesicles (Figure 27.13). They are equipped with a transparent lens, a **cornea** (transparent cover) and a **retina** (a tissue containing densely packed photoreceptors).

Squids, cuttlefishes, and octopuses are fast-moving, predatory mollusks of dimly lit underwater worlds. All have large, paired eyes capable of forming clear images. Both eyes, which are positioned just behind prey-grabbing tentacles, are used in aligning the tentacles at the correct striking distance from prey. Muscles control movements of the eyeball and of the **iris**, an adjustable ring of contractile and connective tissues within the eye. The open center of this contractile ring (the *pupil*) can be varied in size to admit more or less light.

Bright light causes the octopus pupil to shrink into a narrow slit. If you happen to bump into a good-sized octopus while snorkeling or diving, its large slit pupils may flare open suddenly, one or both at a time, in response. The octopus uses its startling stare to secure the attention of a potential mate and possibly to warn away potential enemies.

Insects and crustaceans such as crabs have **compound eyes** that contain closely packed photosensitive

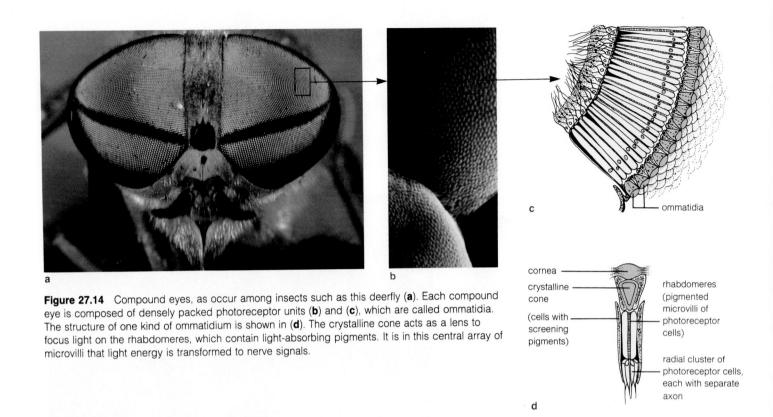

c — ommatidia

cornea —
crystalline cone —
(cells with screening pigments) —

rhabdomeres (pigmented microvilli of photoreceptor cells)

radial cluster of photoreceptor cells, each with separate axon

d

Figure 27.14 Compound eyes, as occur among insects such as this deerfly (**a**). Each compound eye is composed of densely packed photoreceptor units (**b**) and (**c**), which are called ommatidia. The structure of one kind of ommatidium is shown in (**d**). The crystalline cone acts as a lens to focus light on the rhabdomeres, which contain light-absorbing pigments. It is in this central array of microvilli that light energy is transformed to nerve signals.

Figure 27.15 An approximation of light reception in the insect eye. This image of a butterfly was actually formed when a photograph was taken through the outer surface of a compound eye that had been detached from an insect. However, it may not be what the insect actually "sees." Integration of signals sent to the brain from photoreceptors in the eye may produce a more crisply defined image. The representation shown here is useful insofar as it suggests how the overall visual field may be *sampled* by separate ommatidia.

units, of the sort shown in Figure 27.14. Some compound eyes have many thousands of these units, which are called *ommatidia* (singular, ommatidium). We do not know how signals from the photoreceptors in each unit are processed in the brain to form visual images. According to the **mosaic theory** of image formation, each ommatidium samples only a small part of the overall visual field, which varies in light intensities. An image is built up according to signals about different light intensities, with each unit contributing a separate bit to a visual mosaic (Figure 27.15).

Vertebrate Photoreception

Almost all vertebrates have eyes capable of forming clear images. As Figure 27.16 and Table 27.2 indicate, the eyeball has a lens, a **sclera** (a tough outer coat), a **choroid** (a dark-pigmented tissue), and a retina densely packed with receptors. A transparent, light-focusing cornea is continuous with the sclera and covers the front of the eye. Choroid tissue extends inward from the front of the eye to form an iris. The iris is richly endowed with light-screening pigments, and it has radial and circular muscle fibers for controlling the amount of incoming light. A clear fluid (aqueous humor) fills the space between the

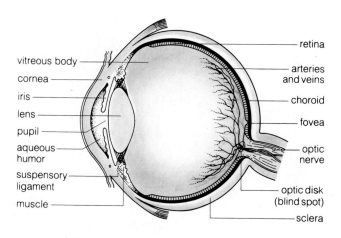

vitreous body

cornea

iris

lens

pupil

aqueous humor

suspensory ligament

muscle

retina

arteries and veins

choroid

fovea

optic nerve

optic disk (blind spot)

sclera

Figure 27.16 Main components of the human eye.

Eye Region	Description	Main Function
Table 27.2 Components of Eyes of Birds and Mammals		
Outer layer:		
Sclera	White fibrous tissue	Protection
Cornea	Curved, transparent tissue	Helps focus light rays entering eyes
Middle layer:		
Choroid	1. Dark-pigmented tissue	Helps prevent light scattering
	2. Iris (extension of choroid tissue having adjustable ring of contractile and connective tissue)	Controls amount of incoming light
	3. Pupil (open center of iris)	Entrance for incoming light
Lens	Spherical or cone-shaped body of transparent protein fibers	Adjustable focusing of light rays onto photoreceptors
Aqueous humor	Clear, alkaline solution between cornea and lens	Light transmission, maintenance of pressure within interior of eye
Vitreous body	Jellylike substance filling chamber behind lens	Light transmission, support for lens and wall of eyeball
Inner layer:		
Retina	Tissue containing densely packed photoreceptors (rods and cones)	Light reception and transduction
Fovea	Funnel-shaped depression near center of retina where cones are concentrated	Area of greatest visual acuity
Beginning of optic nerve	Axons of retinal ganglion cells	Carries signals from photoreceptors to brain

cornea and iris. The lens is positioned behind the iris, and a jellylike substance (vitreous body) fills the chamber behind the lens.

Vertebrates can adjust the way in which light from distant or close-up objects converges on their photoreceptors. Light rays entering the curved cornea are bent toward some focal point. (You might want to review the description of light refraction in Figure 5.2.) If the angle of bending is not enough, the focal point will end up behind the retina. If it is too much, the focal point will end up in front. However, movements of the lens can adjust the path of the light rays. The process by which lens adjustments are made to bring about precise focusing of light onto the retina is called **accommodation**.

Fish use eye muscles that move the entire lens forward or back, thereby adjusting its distance from the retina. Increasing the distance moves the focal point forward; decreasing the distance moves it back. Other vertebrates rely on coordinated stretching and relaxation of eye muscles and fibers, which are attached to the lens (Figure 27.17).

In the vertebrate eye, lens adjustments assure that the focal point for light rays will be on the retina.

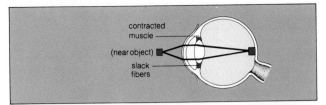

a Accommodation for nearby objects (lens bulges at equator)

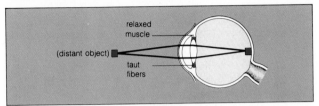

b Accommodation for distant objects (lens flattens out)

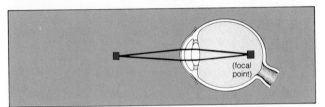

c Focal point in nearsighted vision

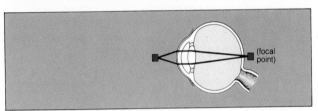

d Focal point in farsighted vision

Figure 27.17 Visual accommodation in the human eye. (**a**) Close objects are brought into focus when eye muscles contract enough to slacken certain fibers interposed between them and the lens, and this causes the lens to thicken at its equator. (**b**) Distant objects are brought into focus when eye muscles relax, thereby putting tension on the fibers and stretching the lens into a flatter shape. (**c**) In the eyes of *nearsighted* people, the retina is too far behind the lens; light from distant objects is focused in front of the retina. (**d**) In the eyes of *farsighted* people, light from nearby objects is focused behind the retina.

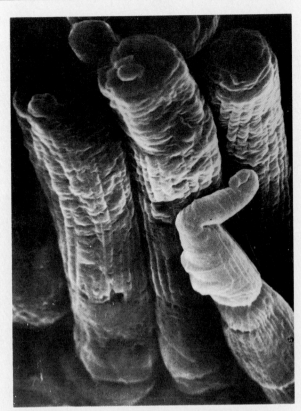

a Scanning electron micrograph of rods and a cone, x 7000

Figure 27.18 Vertebrate rods and cones, and the sensory pathway leading from the retina of the eye to the brain.

Rods and Cones

The retina is well developed in birds and mammals. Its basement layer is composed of pigmented epithelium and covers the choroid. Nerve tissue that contains photoreceptors rests on the basement layer. The pigments of this layer help prevent light scattering and thereby keep the photoreceptors from being stimulated in a diffuse way, which could lead to blurred images.

The photoreceptors are called **rod cells** and **cone cells** because of their shape (Figure 27.18). Rods are sensitive to very dim light; they contribute to coarse perception of changing light intensity caused by movements across the field of vision. Rods are typically abundant in the periphery of the retina.

Cones respond to high-intensity light; they contribute to sharp daytime vision and, usually, color perception. There are three known types of cone cells, which have pigments that are most sensitive to wavelengths corresponding to the colors red, green, and blue. The cones of human eyes are densely packed in the *fovea*, a funnel-shaped depression near the center of the retina,

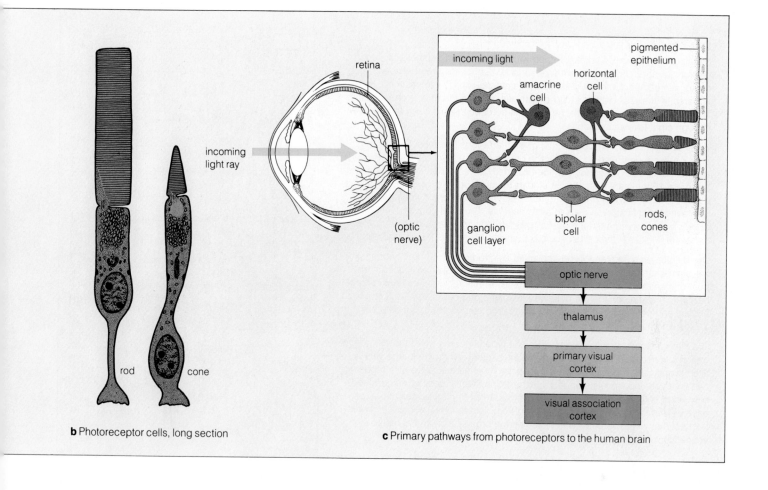

b Photoreceptor cells, long section

c Primary pathways from photoreceptors to the human brain

where nerve tissue is thinner. Although the fovea is only a millimeter across, the cones clustered here contribute the most to visual acuity (precise discrimination between adjacent points in space).

The top of each rod is a dense stack of flattened membrane sacs (Figure 27.18). Embedded in the membranes are **rhodopsin** molecules, each consisting of a protein (opsin) to which a side group (*cis*-retinal) is attached. When the side group absorbs light energy, it is temporarily converted to a slightly different form:

cis-retinal becomes *trans*-retinal

In this altered form, the side group breaks away from the protein. The breakdown of rhodopsin leads to a change in the voltage difference across the membrane in which the protein is embedded. In effect, the change

signals the presence of light to neighboring neurons, which then relay signals to ganglion cells (Figure 27.18c). The axons of these cells lead out from the nerve tissue in the eye and converge to form the optic nerve. From there, signals travel to the thalamus, then on to visual processing centers in the cortex.

Processing Visual Information

Visual information undergoes some processing in the retina before it is conveyed to the brain. Signals converge and diverge among neurons of the retina; they spread out and inhibit activity in adjacent cells. As a result, the frequencies at which signals are transmitted to the brain change in response to changes in the size, location, and intensity of light stimuli.

Our understanding of how visual information is processed in the vertebrate brain comes mainly from experiments with cats and monkeys. Figure 27.19 describes some of the pioneer work done in this area.

Figure 27.19 From signaling to visual perception.

Parts of the cerebral cortex concerned with vision contain neurons stacked in columns at right angles to the brain's surface. Connections run between neurons in each column and between different columns. Each column apparently analyzes only one kind of stimulus, received from only one location. The transformations from signaling to visual perception have been traced through eight levels of synapses.

What do eight synaptic levels tell us, given the *billions* of connections in the brain? A great deal. The neurons in the brain's surface layer fall into a few categories. In each category, the neurons seem to be tripped into action in the same way. For instance, excitatory signals traveling up through the brain's surface layer or traveling parallel to its surface activate certain neurons, which then send out inhibitory signals to other neurons. The excitatory and inhibitory signals between neurons form narrow bands of electrical activity. In fact, *the pattern of excitation through specific columns of neurons is highly focused.*

David Hubel and Torsten Wiesel implanted electrodes in individual neurons in the brain of an anesthetized cat. After the cat woke up, they positioned it in front of a small screen, then projected images of different shapes (including a bar) onto the screen. The cat observed images of the bar tilted at different angles, and changes in electrical activity that corresponded to the different angles were recorded.

The strongest activity was recorded for one type of neuron when the bar image was positioned vertically (numbered 5 in the sketch). When the image was tilted slightly, the signals were less frequent. When the image was tilted past a certain angle, the signals stopped. In other experiments, another neuron fired only when an image of a block was moved from left to right across the screen; another fired when the image was moved from right to left. These experiments suggest that visual perception is based on the organization and synaptic connections between columns of neurons in the brain.

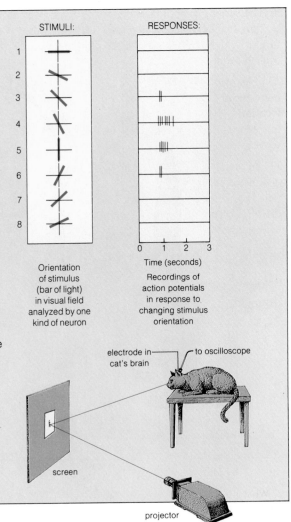

STIMULI:

RESPONSES:

Time (seconds)

Orientation of stimulus (bar of light) in visual field analyzed by one kind of neuron

Recordings of action potentials in response to changing stimulus orientation

electrode in cat's brain — to oscilloscope

screen

projector

SUMMARY

1. Many neurons are active almost continually; their pattern of activity depends on inherited neural programs.

2. A stimulus is any form of energy the body is able to detect by means of its receptors.

3. Receptors are fine peripheral endings of sensory neurons (or specialized cells adjacent to them) that respond to specific stimuli, such as chemical energy, mechanical energy, photon energy, or the radiant energy associated with temperature. Animals can respond to events in the outside world only if they have receptors sensitive to the energy of the particular stimulus.

4. Receptors are transducers which convert one form of energy into another. When they are stimulated, receptors undergo a graded electrical response known as a receptor potential.

5. A particular sensation is not triggered by action potentials themselves but by their travel along a particular nerve pathway. Sensory nerve pathways from different receptors lead to different parts of the cerebral cortex.

6. Variations in stimulus intensity are encoded in the frequency of action potentials in a single neuron and in the number of axons recruited into action in a tissue.

7. Taste receptors are in sensory organs called taste buds. Olfactory receptors respond to molecules released from food and predators and to pheromones released by members of the same species. Taste and olfaction are examples of chemical senses.

8. Somatic senses include the sensations of touch, pressure, temperature, and pain near the body surface.

9. The sense of balance in vertebrates depends on information from hair cell receptors in the vestibular apparatus, which consists of semicircular canals and an otolith organ.

10. Hearing, like the sense of balance, depends on the bending of hair cells by changes in fluid pressure. Echolocation in bats, dolphins, and whales depends on hearing echoes from self-generated, high-frequency sound waves.

11. In vision (photoreception), photopigment molecules are changed by light, which gives rise to receptor potentials in the rod and cone cells of the eyes. Processing of the visual information begins in the retina and continues in the visual cortex.

Review Questions

1. Label the component parts of the human eye:

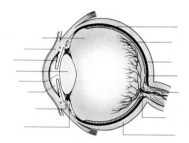

2. What is a stimulus? Receptor cells detect specific kinds of stimuli. When they do, what happens to the stimulus energy?

3. Give some examples of chemoreceptors and mechanoreceptors. What kind of mechanoreceptor occurs repeatedly in sensory organs of different kinds of animals?

4. What is sound? How are amplitude and frequency related to sound? Give some examples of animals that apparently perceive sounds.

5. What is pain? Can you name one of the tactile receptors associated with pain?

6. Which organs contribute to our sense of balance?

7. How does vision differ from photoreception? What sensory apparatus does vision require?

8. How does the vertebrate eye focus the light rays of an image? What is meant by nearsighted and farsighted?

Readings

Eckert, R., and D. Randall. 1983. *Animal Physiology: Mechanisms and Adaptations*. Second edition. New York: Freeman.

Hubel, D. H., and T. N. Wiesel. September 1979. "Brain Mechanisms of Vision." *Scientific American* 241(3):150–162. Describes studies on information processing in the primary visual cortex.

Hudspeth, A. January 1983. "The Hair Cells of the Inner Ear." *Scientific American* 248(1):54–66.

Jacobs, G. 1983. "Colour Vision in Animals." *Endeavor*. 7(3):137–140.

Kandel, E., and J. Schwartz. 1985. *Principles of Neural Science*. Second edition. New York: Elsevier. Advanced reading, but good coverage of sensory perception.

Newman, E. A., and P. H. Hartline. March 1982. "The Infrared 'Vision' of Snakes." *Scientific American* 246(3):116–127.

Parker, D. November 1980. "The Vestibular Apparatus." *Scientific American* 243(5):118–130.

Wu, C. H. November-December 1984. "Electric Fish and the Discovery of Animal Electricity." *American Scientist* 72(6):598–607.

Young, J. 1978. *Programs of the Brain*. New York: Oxford University Press. An extraordinary book, beautifully written.

28

MOTOR SYSTEMS

The sensory systems described in the preceding chapter sample the surroundings with great precision and keep the nervous system informed of change. In turn, the nervous system sends out commands to a motor system so that the body moves in coordinated ways. In all animals, the operation of a motor system is based on units of contraction that can shorten under stimulation and then relax. As you will see, its operation requires the presence of some medium or structural element against which force can be applied.

We have already looked at the "motor systems" of single cells, many of which use cilia and flagella to move through a liquid medium (page 80). Here we will consider a few examples of the main types of motor systems of animals, as listed in Table 28.1.

All motor systems require the presence of some medium or structural element against which force can be applied.

INVERTEBRATE MOTOR SYSTEMS

One of the simplest motor systems is found in sea anemones, and it consists mainly of longitudinal and circular muscles in the body wall. In the longitudinal muscles, the muscle fibers are bundled together by connective tissue and run parallel with the body axis. (A *muscle fiber*, recall, is the same thing as a muscle cell.) When the longitudinal muscles contract, the body shortens. In the circular muscles, the muscle fibers are arranged in rings around the body axis, and when they contract, they cause the body to lengthen (Figure 28.1). Together, these muscles work as an **antagonistic system**, in which the

Figure 28.1 Motor system by which the sea anemone extends its body upward through the water (a feeding behavior) and then relaxes (between feeding periods). Some muscle fibers in the body wall are parallel with the body axis. When they contract, the body shortens (**a**). Other muscle fibers are arranged in rings around the body axis. When they contract, the body is lengthened (**b**).

action of one motor element opposes the action of the other.

Annelids such as earthworms have a different motor system. Their body cavity is divided into a series of segments, each with a flexible wall surrounding a fluid-filled chamber. Each segment has a set of longitudinal and circular muscles. When the circular muscles contract, the force of contraction is applied against the fluid-filled interior. Because fluid resists compression, the fluid interior acts as a **hydrostatic skeleton**. (In all hydrostatic systems, body fluids are used to transmit force.) Thus, instead of being compressed, the fluid is squeezed along the axis of the body, much like toothpaste being squeezed down a tube. As a result, the longitudinal muscles stretch and sets of bristles (setae) that project from the body grip the ground, acting like toes for the stretched-out worm. When the longitudinal muscles contract, their force is applied against the bristles and the body is pulled forward.

There are some intriguing variations on the hydrostatic theme. For example, a rapid surge of blood under high pressure can extend the hind leg spines of spiders. Fluid under pressure in tubes is called hydraulic pressure. Figure 28.2 shows what a jumping spider looks like when it uses blood pressure as a hydraulic source for leaping at prey.

Insects, crabs, and other arthropods also have segmented bodies. Each segment has a *cuticle* (a hardened covering of chitin, protein, and sometimes lipid secretions). Besides covering the segments, the cuticle also covers antagonistic muscles that bridge the gap between segments. The cuticle remains pliable at these gaps and acts like a hinge when muscles move it in different directions. Thus the cuticle forms an external skeleton, or **exoskeleton**.

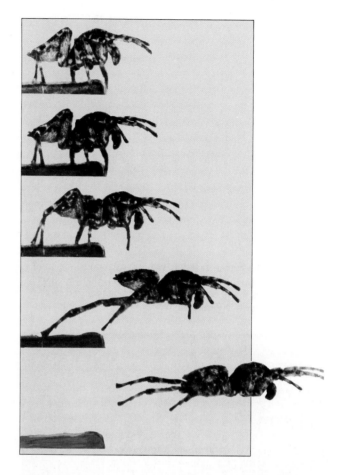

Figure 28.2 Hydraulic leap of the jumping spider (*Sitticus pubescens*), which can soar ten centimeters through the air to pounce on prey. The leap is based on the hydraulic extension of the hind legs when blood surges into them under high pressure.

Table 28.1	Main Categories of Motor Systems	
Type	Representative Animals	
Muscles alone, no skeleton	Sea anemone	
Hydrostatic skeletons:		
1. Body fluid + soft body wall	Earthworm, octopus	
2. Body fluid + rigid body part	Spider (hind legs only)	
Rigid skeletons:		
1. Exoskeleton	Beetle, grasshopper, crab	
2. Endoskeleton	Frog, snake, bird, human	

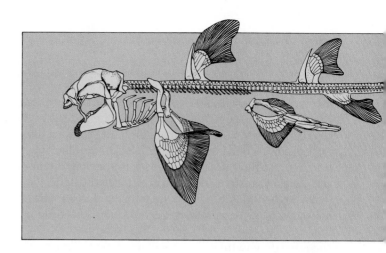

Figure 28.4 The human skeleton, with the axial portion shaded yellow and the appendicular portion shaded tan. Can you identify similar structures in the endoskeletons of the generalized reptile and mammal shown in Figure 28.3?

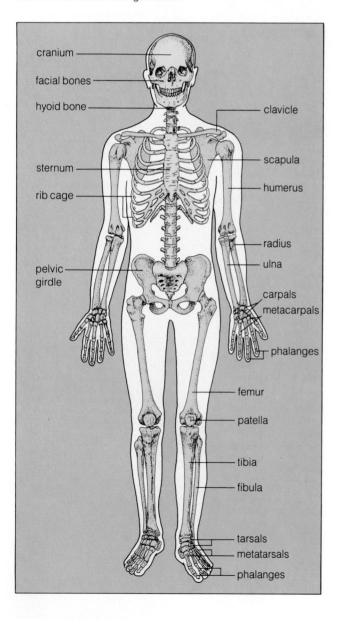

cranium
facial bones
hyoid bone
clavicle
scapula
sternum
humerus
rib cage
radius
ulna
pelvic girdle
carpals
metacarpals
phalanges
femur
patella
tibia
fibula
tarsals
metatarsals
phalanges

VERTEBRATE MOTOR SYSTEMS

Humans and other living vertebrates have an internal skeleton, or **endoskeleton**, of bone and cartilage (or cartilage alone). Some fishes have a flexible skeleton that almost looks like glass; it is composed of an elastic, translucent form of cartilage. Sharks have a skeleton that superficially looks as if it is made of bones, but it is composed of a hardened, opaque form of cartilage in which calcium salts have been deposited (Figure 28.3). For most vertebrates, however, bone is the main skeletal material.

The bones themselves serve as mineral reservoirs, and some are sites of blood cell production. Tough tendons connect skeletal muscles to the bones, and ligaments help hold bones together where they meet at joints. Through their interactions, the bones and skeletal muscles provide support and protection for other parts of the body, they help maintain posture, and they move the body in the external environment.

Human Skeleton

The human skeleton is divided into axial and appendicular portions (Figure 28.4). The *axial skeleton* includes the skull, vertebral column, ribs, and sternum (the breastbone). The *appendicular portion* includes the bones of the arms, hands, legs, feet, pelvic girdle (which is at the hips), and the pectoral girdle (which is at the shoulders).

Axial Skeleton. Of the skull's twenty-eight bones, eight form the cranium, which protects the brain, eyes, and ears. The rest form the facial bones and the middle-ear bones. This collective array of bones balances on the

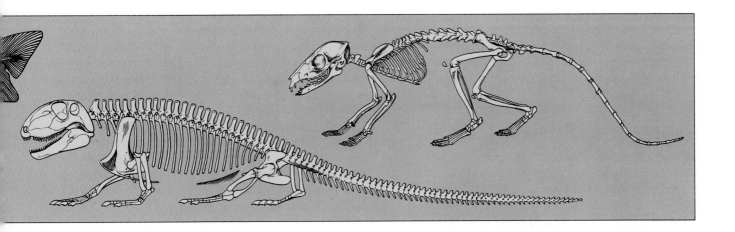

so-called atlas vertebra, and the joint between them allows the head to nod back and forth, as when you nod "yes." In turn, the atlas vertebra rests on the axis vertebra, and the joint between the two allows the head to rotate (as when you shake your head "no").

The atlas and axis vertebrae are the first bones of the vertebral column, the components of which are shown in Figure 28.5. The rib bones (twelve on each side of the chest cavity) form a joint with one or two vertebrae and another joint with the breastbone (or with cartilage that connects to it). Together, the ribs, breastbone, and vertebral column form a cage that protects two of the most vital organs of the body, the heart and lungs. In the entire axial skeleton, only one bone does not form a joint with any other. That is the hyoid bone; it supports the base of the tongue.

Appendicular Skeleton. The appendicular skeleton forms joints with the axial portion at the shoulders and the hips. The pectoral girdle is attached to the breastbone by the clavicle (collarbone), which attaches at its other end to the scapula (part of which is the shoulder blade). The scapula and the humerus (the upper arm bone) meet at the shoulder joint. At its base, the humerus meets with the ulna and radius—the bones of the forearm— to form the elbow joint. The other end of the ulna joins with several carpals at the wrist joint (Figure 28.4). The palm of the hand is supported by metacarpals, which connect with phalanges (the bones of the fingers and the thumb).

The pelvic girdle, which consists of six fused bones, forms joints with the vertebral column and with the bone of each upper leg (the femurs). At its other end, each femur meets the bones of the lower leg to form the knee joint, the front of which is protected by the patella (knee

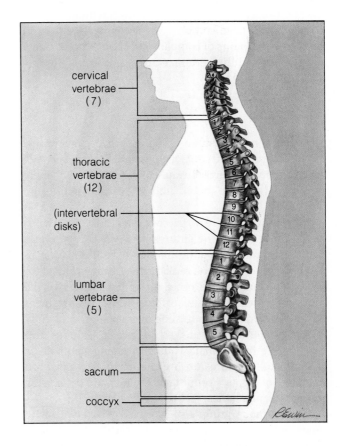

Figure 28.5 Vertebral column of the human skeleton, side view.

cap). At the ankle joint, the bones of the lower leg fit over one of the tarsals (the bones of the back part of the foot). The front of the foot is supported by five metatarsals; the toes are supported by phalanges. The big toe has two phalanges; the rest each have three.

Types of Bones

There are four main categories of bones. *Long bones* are longer than they are wide, and the two ends are usually larger in diameter than the shaft (see, for example, Figure 23.6). If you have ever stopped and looked at a chicken drumstick during dinner, you already know what these bones are shaped like. Similarly structured bones occur in your legs, arms, hands, and feet. *Short bones* are nearly the same in length and width; such bones occur at ankles and wrists. *Flat bones*, such as those of the skull, usually consist of two flat plates of compact bone with a thin layer of spongy bone sandwiched between them. Finally, *irregular bones* have complex shapes, and they vary in the relative amounts of compact and spongy bone. Some facial bones and the vertebrae are examples.

Types of Joints

All changes in the positions of bony parts occur at **joints**, which are points of contact (or near-contact) between bones or between cartilage and bones. Structurally, there are three types of joints: fibrous, cartilaginous, and synovial.

Fibrous joints have no gap between the bones, which hardly move at all. The sutures between the flat bones of the cranium of a young mammal are fibrous joints, but during growth, bone replaces the fibrous connective tissue. The sutures are not tight and the edges of the bones are tapered in human fetuses. The arrangement allows the cranial bones to slide over each other slightly as the full-term fetus makes its rather arduous trip through the birth canal. Thus the arrangement helps prevent skull fractures during childbirth. The newborn still has "soft spots," or fontanels, which are open areas between cranial bones.

Cartilaginous joints also have no gap between the bones, which are held together by cartilage, but there can be some movement at these joints. The intervertebral disks between the bones of the vertebral column are an example (Figure 28.5). These disks form strong joints that allow some movement along the column; they also help absorb vertical shocks, as when you fall out of a tree and land on your feet (but not a very tall tree).

Synovial joints, by far the most common, are freely movable. The bones are separated by a cavity and are held together by a flexible capsule of dense connective tissue (and often by accessory ligaments as well). Cartilage covers the surface of the bone ends in the joint itself. The inner wall of the capsule is lined with a membrane that produces synovial fluid, which lubricates the joint. Unfortunately, freely movable joints sometimes move a little too freely and the structural organization at the joint is disrupted (see *Commentary*).

Freely movable joints obviously are subject to wear and tear over time. In *osteoarthritis*, the cartilage at the bone ends simply wears out. *Rheumatoid arthritis* is a degenerative disorder that has a genetic basis. The synovial membrane thickens and becomes inflamed to the degree that the joint becomes stiff and painful.

Development of Bones

How do human bones develop? Some form directly in connective tissue in the embryo. The flat bones of the cranium are an example. Bone-forming cells called **osteoblasts** secrete material that forms many splinterlike fragments of bone tissue. The osteoblasts become trapped by their own secretions, and when that happens they are known as **osteocytes** (Figure 23.6). Over time, the fragments they produced fuse together; and after a lot of remodeling the fragments develop into bone.

Cartilage models serve as the starting point for other bones in the embryo. For example, when a long bone develops, cartilage inside the shaft of the model calcifies and is invaded by blood vessels, osteoblasts, and osteoclasts (Figure 28.6). The **osteoclasts** are bone-resorbing cells; the enzymes that they release act to dissolve calcified cartilage and bone tissue. The osteoclasts actually excavate cavities in the cartilage model, then osteoblasts start to leave deposits that will form bone tissue along the cavity walls. These activities are repeated in the knobby ends of the long bone. As development continues, cartilage remains only as a covering for the bone in the joints at both ends of the shaft, where it forms epiphyseal plates that extend across the width of the bone (Figure 28.6).

The long bone increases in length as calcified cartilage is removed, as bone tissue is deposited at the inside surfaces of the epiphyseal plates, and as cartilage is replenished. It grows wider as new bone tissue is deposited under the dense connective tissue membrane surrounding the bone. When growth finally ends, the epiphyseal plates disappear.

Marrow fills the cavities in the shafts of long bones and the spaces between the plates of spongy bone almost as soon as they develop. Red marrow, recall, is a major site of blood cell production. In human adults, red marrow is found in the vertebrae, ribs, sternum, and pelvic girdle. Yellow marrow, which is present at other locations, ordinarily is inactive. However, when a great deal

On Runner's Knee

When you run, one foot and then the other is pounding hard against the ground. Each time a foot hits the ground, the knee joint above it must absorb the full force of your body weight. Now, the knee joint can do many things. It allows the leg bones beneath it to swing and, to some degree, to bend and twist. And the joint can absorb a force nearly seven times the body's weight—but there is no guarantee that it can do so repeatedly. Nearly 5 million of the 15 million joggers and runners in the United States alone suffer from "runner's knee," which refers generally to various disruptions of the bone, cartilage, muscle, tendons, and ligaments at the knee joint.

Like most joints, the knee joint permits considerable movement. The two long bones joined here (the femur and tibia) are actually separated by a cavity. They are held together by ligaments, tendons, and a few fibers that form a capsule around them. A membrane that lines the capsule produces a fluid that lubricates the joint, and where the bone ends meet, they are capped with a cushioning layer of cartilage.

Between the femur and tibia are wedges of cartilage that add stability and act like shock absorbers for the weight placed on the joint. Here also are thirteen fluid-filled sacs (bursae) that help cut down the friction between the parts of the joint, tendons, and skin that have to slide over each other.

When the knee joint is hit hard or twisted too much, its cartilage can be torn. Once cartilage is torn, the body often cannot repair the damage. Orthopedic surgeons usually recommend removing most or all of the torn tissue; otherwise it can cause arthritis (a general term for painful disorders that can render the joint practically immovable). Each year, more than 50,000 pieces of torn cartilage are surgically removed from the knees of football players alone. Football players, tennis players, basketball players, weekend joggers—

all are helping to support the burgeoning field of "sports medicine."

The seven ligaments that strap the femur and tibia together are also vulnerable to injury. A ligament is not meant to be stretched too far, and blows to the knee during collision sports (such as football) can tear it apart. A ligament is composed of many connective tissue fibers. If only some of the fibers are torn, it may heal itself. If the ligament is severed, however, it must be surgically repaired. (Edward Percy likens the surgery to sewing two hairbrushes together.) Severed ligaments must be repaired within ten days. The fluid that lubricates the knee joint happens to contain phagocytic cells that remove the debris resulting from day-to-day wear and tear in the joint. The cells will also go to work indiscriminately on torn ligaments and turn the tissue to mush.

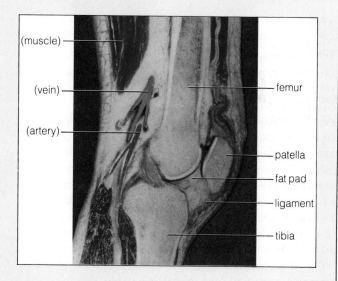

Longitudinal section through the knee joint.

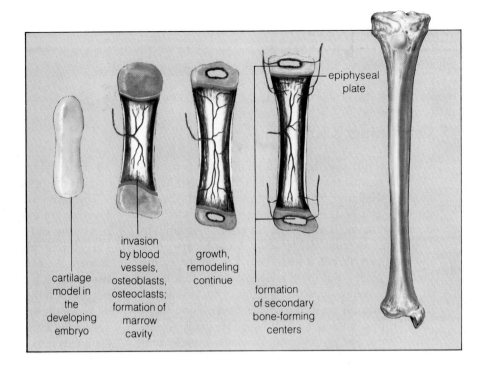

Figure 28.6 Formation of a long bone in mammals.

cartilage model in the developing embryo

invasion by blood vessels, osteoblasts, osteoclasts; formation of marrow cavity

growth, remodeling continue

formation of secondary bone-forming centers

epiphyseal plate

of blood is lost (as during a traumatic injury), this tissue assists the red marrow in producing blood cell replacements.

Bone Tissue Turnover

Bone is constantly being renewed because of continuous deposition and resorption activities. These activities serve two functions: (1) remodeling and (2) maintaining calcium levels for the body as a whole. Osteocytes and osteoclasts are both important in bone resorption. The osteoclasts resorb bone directly at the surfaces during the body's remodeling programs. Their activity is most pronounced during growth but is continuous throughout the individual's life. Osteocytes engage in resorption within the bone tissue; their activity is central to the hormonal control of calcium balance (page 350).

With increasing age, a problem can develop in bone turnover, particularly among older women. The bone mass decreases, especially in the vertebral column, legs, and feet. As a result, the vertebral column can collapse and curve to the extent that ribs drop down and come to rest on the rim of the pelvic girdle. The changing bone positions lead to complications in internal organs as well. This syndrome is called *osteoporosis*. Although the causes of osteoporosis are not known, the suspected factors include decreased activity of the osteoblasts, dwindling levels of sex hormones, calcium deficiency, excessive protein intake, and decreased physical activity.

MOTOR RESPONSES

In vertebrates, only skeletal muscle is concerned with moving the body through the environment. Smooth muscle, which occurs mostly in internal organs, is concerned with propelling or regulating the movement of substances within the body. Cardiac muscle occurs only in walls of the heart, and its action will be described in the next chapter. Here we will focus on skeletal muscle activity.

Skeletal Muscle Action

There are anywhere from a few hundred to many thousands of muscle fibers in a skeletal muscle. Connective tissue holds the fibers together in bundles, and usually tendons attach them to bones. Figure 28.7 shows some major skeletal muscles and their location relative to the skeleton.

The skeleton, together with its muscles, is analogous to a system of levers in which rigid rods (bones) move about at fixed points (the joints). When a typical skeletal muscle contracts, one end (the *origin*) remains stationary; the other end (the *insertion*) moves. Most insertions are close to their joints. This means that the muscle has to contract only a small distance to produce a corresponding large movement of some body part.

A limb can be extended and rotated around a joint because of arrangements between pairs or groups of muscles. Muscles in body limbs are arranged in antagonistic pairs, such as the biceps and triceps shown in Figure 28.8. Notice how these muscles bridge both the elbow and shoulder joints. When one member of this antagonistic pair (the biceps) contracts, the elbow joint flexes (bends). As it relaxes and its partner (the triceps) contracts, the limb extends and straightens out again. **Reciprocal innervation** in the spinal cord contributes to this coordination. Here, inhibitory signals sent to one set of motor neurons prevent one muscle of a pair from contracting while the other muscle is being stimulated.

Thus, when the biceps contracts, inhibitory neurons are acting at the same time on the motor neurons of its partner, the triceps, which relaxes. And when the triceps contracts, inhibitory signals are acting on the motor neurons of the biceps, which relaxes. Reciprocal innervation contributes to the coordinated movements of various skeletal-muscular systems, including the flight muscles of birds.

(Reciprocal innervation can be overridden. For example, you can contract your biceps and triceps at the same time by holding your arm upright, like a stiff pillar.)

Also contributing to the coordination of skeletal muscle contraction are stretch receptors. Within skeletal muscle, muscle cells and stretch-sensitive receptors are enclosed in sheaths to form muscle spindles (Figure 24.14). When the receptors are activated, signals are conducted to motor neurons, which in turn carry signals that cause muscle fibers to contract.

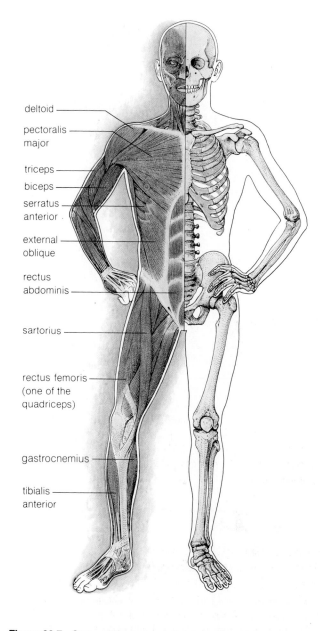

deltoid
pectoralis major
triceps
biceps
serratus anterior
external oblique
rectus abdominis
sartorius
rectus femoris (one of the quadriceps)
gastrocnemius
tibialis anterior

Figure 28.7 Some of the major skeletal muscles of the human skeletal-muscular system.

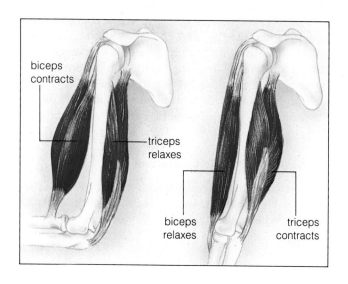

biceps contracts
triceps relaxes
biceps relaxes
triceps contracts

Figure 28.8 Antagonistic muscle pair, showing how two muscles can produce movement in opposite directions.

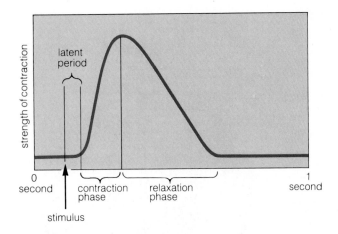

Figure 28.9 Recording of a muscle twitch.

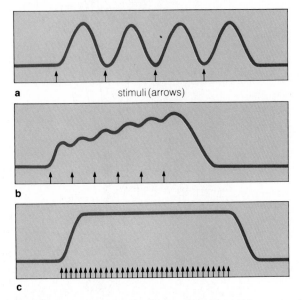

Figure 28.10 Recordings of a series of muscle twitches caused by about two stimulations per second (**a**); recordings of a summation of twitches resulting from about six stimulations per second (**b**); and a tetanic contraction resulting from about twenty stimulations per second (**c**).

Motor Unit Activity

Recall that the axon of a motor neuron has many branched endings called axon terminals. The excitatory signals traveling along the motor neuron cause all muscle fibers at the axon terminals to contract at the same time (Figure 24.11). Together, a motor neuron and all the muscle fibers under its control are called a **motor unit**.

When a single action potential activates the muscle fibers of a motor unit, the resulting contraction is called a **twitch** (Figure 28.9). There is a brief interval (the latent period) between stimulation and the onset of contraction. The contraction builds up to a peak force, after which it diminishes during relaxation.

When a muscle is stimulated again before a twitch response is completed, it contracts again. The strength of the contraction depends on how far the twitch response has proceeded by the time the second signal arrives. A muscle that is stimulated repeatedly does not have time to relax. Instead it is maintained in a state of contraction called **tetanus** (Figure 28.10).

There are low levels of tetanic contraction even in a muscle that appears to be resting. Although most of its motor units are relaxed, some remain in a state of partial contraction. What we call "muscle tone" refers to the sustained, partial contraction of a muscle due to low levels of signals from stretch receptors. Muscle tone is important in maintaining different body positions.

Fine Structure of Skeletal Muscle

A muscle fiber consists of fine, threadlike structures called **myofibrils** (Figure 28.11). Each myofibril, in turn, consists of actin and myosin filaments. An *actin filament* looks like two beaded chains twisted around each other; the "beads" are ball-shaped actin molecules. A *myosin filament* is really 200 to 400 rod-shaped myosin molecules, lying in parallel. As you will see, a globe-shaped head extending from each myosin molecule projects toward a series of binding sites on an actin filament.

The actin and myosin filaments are arranged in alternating bands (the "A" and "I" bands shown in Figure 28.11). This banding pattern gives skeletal muscle its striped appearance in micrographs. The pattern itself is interrupted at regular intervals by Z lines. These lines are fibrous anchors for actin filaments, and they define the **sarcomere**—the fundamental unit of contraction.

Mechanism of Muscle Contraction

The only way that skeletal muscles bring about movement of body parts to which they are attached is to shorten. When a skeletal muscle shortens, its component

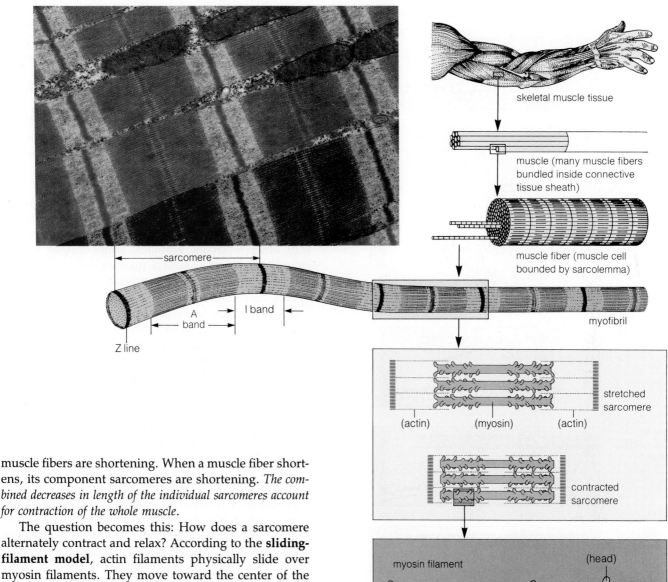

Figure 28.11 Fine structure of skeletal muscle. The micrograph shows a slice through several myofibrils, running diagonally from the lower left to the upper right corner. Interactions between actin and myosin filaments are the basis of skeletal muscle contraction.

muscle fibers are shortening. When a muscle fiber shortens, its component sarcomeres are shortening. *The combined decreases in length of the individual sarcomeres account for contraction of the whole muscle.*

The question becomes this: How does a sarcomere alternately contract and relax? According to the **sliding-filament model**, actin filaments physically slide over myosin filaments. They move toward the center of the sarcomere during contraction and move away from it when the sarcomere is relaxing.

The yellow box in Figure 28.11 shows that each sarcomere has two sets of actin filaments and a set of myosin filaments between them. For the sliding movement to occur, heads on the myosin filaments must first attach to binding sites on the actin molecules. When attached, the myosin heads are *cross-bridges* between the two types of filaments (Figure 28.12). When cross-bridges are activated, the myosin heads tilt inward in a short power stroke, toward the center of the sarcomere. Because the actin filaments are attached to cross-bridges, they move slightly inward, also. The myosin heads then detach, reattach at the next actin binding site in line, and move the actin filaments a little bit more. A single contraction of the sarcomere takes a whole series of these power strokes.

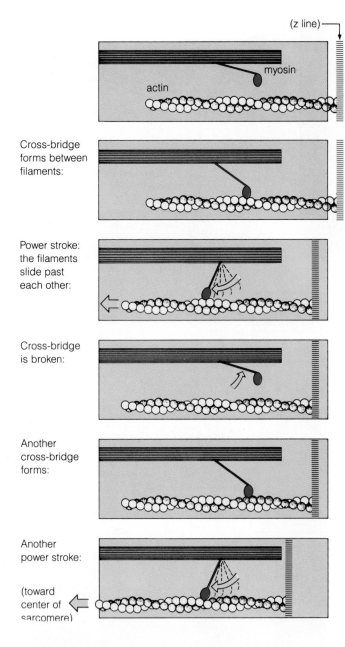

(z line)

myosin

actin

Cross-bridge forms between filaments:

Power stroke: the filaments slide past each other:

Cross-bridge is broken:

Another cross-bridge forms:

Another power stroke:

(toward center of sarcomere)

Figure 28.12 Simplified picture of the sliding filament model, which explains the mechanism of contraction in the sarcomeres of muscle fibers.

The energy needed to change the angle of attachment of cross-bridges during the sliding movements comes from ATP. When a muscle is relaxed, each myosin head has picked up an ATP molecule and has hydrolyzed it into ADP and phosphate. The energy associated with this hydrolysis is used to activate the myosin head, which is now like a loaded, pulled-back spring of a mousetrap. The energy brings about changes in the shape of the myosin head and thereby leads to the power stroke. The myosin head picks up a new ATP molecule, which causes the cross-bridge to detach, and the cycle begins again.

In the absence of ATP, the cross-bridges never do detach. The muscle becomes rigid, a condition known as *rigor*. Following death, ATP production stops along with other metabolic activities. Cross-bridges remain locked in place and all skeletal muscles in the entire body become rigid. This condition, called *rigor mortis*, lasts up to sixty hours after death.

Clearly, ATP is necessary for muscle contraction. As you might expect, many muscle cells are richly endowed with mitochondria, which produce ATP through aerobic respiration. Such cells can contract for extended periods, when oxygen and the substrates for ATP synthesis are supplied continuously by the circulation.

ATP *can* be replaced rapidly by the glycolytic breakdown of glycogen stored in the cells. (Glycogen is the primary fuel supply for cells that consume ATP rapidly.) When oxygen concentrations are low in muscle tissue, as they are during strenuous exercise, glycolysis is critical for ATP synthesis. However, this alternate route cannot be followed for long, for muscle cells fatigue quickly when the glycogen is depleted. In a resting muscle, energy is stored in the form of creatine phosphate. This compound readily gives up phosphate to ADP and thereby helps replace ATP used until metabolic activity increases to match ATP consumption following muscle stimulation.

Control of Muscle Contraction

To understand how muscle contraction is controlled, we have to consider the connections between three types of membranes:

sarcolemma	*the plasma membrane that surrounds the entire muscle fiber*
sarcoplasmic reticulum	*a continuous system of membrane-bound chambers that surround myofibrils within the muscle fiber and that store calcium ions*
transverse tubule system	*a system of tubular membranes that extend from the sarcolemma all the way through the muscle fiber, and that are in intimate contact with the sarcoplasmic reticulum*

Figure 28.13 shows the arrangement of these membranes in a section from a muscle fiber.

The signals that initiate contraction begin at neuromuscular junctions between the sarcolemma and motor neurons. Here, a motor neuron releases acetylcholine (page 320). This transmitter substance interacts with receptors on the sarcolemma, leading to changes that produce an action potential. The action potential travels along the sarcolemma and invades the interior of the muscle by way of the transverse tubule system. There, it increases the permeability of the sarcoplasmic reticulum to calcium ions, which simply diffuse into the cytoplasm. *The calcium ions bind to control sites and thereby clear the way for the formation of cross-bridges between the myosin and the actin filaments.* Figure 28.14 describes the role of calcium ions in the formation of cross-bridges.

When calcium ions are actively taken up after contraction and stored in the sarcoplasmic reticulum, the muscle relaxes. When calcium ions are released from the sarcoplasmic reticulum, the muscle contracts. Now, the nervous system dictates which motor neurons will carry action potentials, and at what frequency. By controlling the action potentials that reach the sarcoplasmic reticulum in the first place, the nervous system controls calcium ion levels in muscle tissue—and thereby exerts control over contraction.

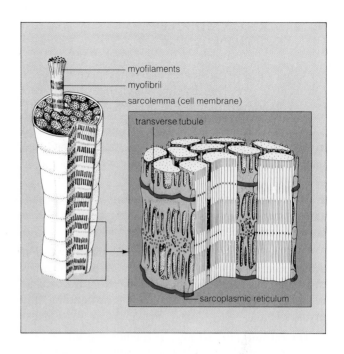

Figure 28.13 Location of sarcoplasmic reticulum, the calcium ion storage site within a muscle fiber.

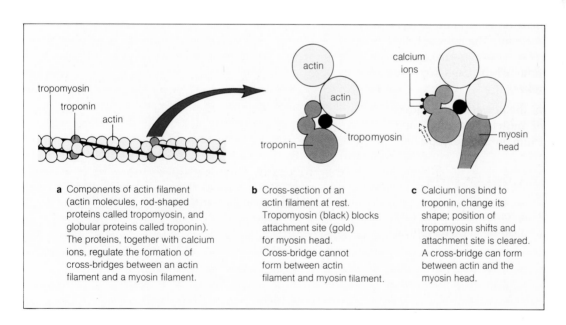

a Components of actin filament (actin molecules, rod-shaped proteins called tropomyosin, and globular proteins called troponin). The proteins, together with calcium ions, regulate the formation of cross-bridges between an actin filament and a myosin filament.

b Cross-section of an actin filament at rest. Tropomyosin (black) blocks attachment site (gold) for myosin head. Cross-bridge cannot form between actin filament and myosin filament.

c Calcium ions bind to troponin, change its shape; position of tropomyosin shifts and attachment site is cleared. A cross-bridge can form between actin and the myosin head.

Figure 28.14 Role of calcium in the formation of cross-bridges between actin and myosin.

SUMMARY OF MUSCLE CONTRACTION

Now that we have considered different aspects of muscle contraction, we can briefly summarize the events involved and how they are controlled:

1. A muscle fiber is composed of fine strands called myofibrils. Each myofibril is partitioned into small segments called sarcomeres, the basic units of muscle contraction.

2. A sarcomere contains actin filaments and myosin filaments, which interact through a sliding mechanism that shortens the sarcomere.

3. Contraction of a sarcomere depends on the arrival of action potentials from motor neurons. Action potentials cause the release of calcium ions from the sarcoplasmic reticulum (membranous chambers surrounding the myofibrils).

4. The calcium ions act to clear the binding sites on the actin filaments to which the heads of myosin molecules can bind.

5. Once a binding site is cleared, a myosin head can attach to the adjacent actin filament. The myosin head is in an energized state (the energy comes from ATP).

6. The energy causes the myosin head to change in shape. The change results in a short power stroke, whereby the actin filament is made to slide past the myosin filament, toward the center of the sarcomere.

7. The deenergized myosin head picks up another ATP and detaches from the actin filament. The myosin head then becomes energized, and reattaches at the next binding site in line. A single contraction of the sarcomere requires a series of power strokes.

8. ATP energy is necessary for the power stroke underlying muscle contraction, for muscle relaxation, and for the accumulation of calcium in the sarcoplasmic reticulum.

9. Calcium ions are necessary to clear the binding site where the power stroke occurs.

10. Because the nervous system controls the release of calcium from the sarcoplasmic reticulum, it exerts control over contraction itself.

Review Questions

1. What are the two basic components of any motor system? Distinguish between a hydrostatic skeleton, exoskeleton, and endoskeleton.

2. Distinguish between the axial and appendicular portions of the vertebrate skeleton.

3. Describe three main types of joints.

4. What are the functions of osteocytes and osteoclasts?

5. What is antagonistic muscle action? Why is reciprocal inhibition of reflexes necessary in producing coordinated contractions?

6. Describe the fine structure of muscle fibers. Explain how the muscle fiber components interact in muscle contraction.

Readings

Alexander, R. M. July-August 1984. "Walking and Running." *American Scientist* 72(4):348–354. The biomechanics of traveling on foot.

Cole, R. 30 April 1982. "Myoglobin Function in Exercising Skeletal Muscle." *Science* 216:523–525.

Eckert, R., and D. Randall. 1983. *Animal Physiology: Mechanisms and Adaptations*. Second edition. New York: Freeman.

Hoyle, G. 1983. *Muscles and Their Neural Control*. New York: Wiley.

Huxley, H. E. "The Mechanism of Muscular Contraction." December 1965. *Scientific American* 213(6):18–27. Old article, great illustrations.

Kandel, E., and J. Schwartz. 1985. *Principles of Neural Science*. Second edition. New York: Elsevier. Advanced reading, but good coverage of motor coordination.

Lester, H. A. February 1977. "The Response to Acetylcholine." *Scientific American* 236(2):106–118.

Luttgens, K., and K. Wells. 1982. *Kinesiology: Scientific Basis of Human Motion*. Seventh edition. Philadelphia: Saunders.

Porter, K. R., and C. Franzini-Armstrong. March 1965. "The Sarcoplasmic Reticulum." *Scientific American* 212(3):72–80.

Shepherd, G. M. *Neurobiology*. 1983. New York: Oxford University Press.

CIRCULATION SYSTEMS:
AN OVERVIEW

A cell survives by exchanging substances with its surroundings, and most of those substances simply diffuse inward and outward across the plasma membrane. (Diffusion, recall, is the random movement of like molecules down their concentration gradients.) The exchanges are not that complicated for the cells of sponges, jellyfishes, and flatworms. These animals do not have massive bodies, and substances simply diffuse through the tissue fluid around individual cells. However, in most invertebrates and all vertebrates, interior cells are too far from the body surface to exchange substances efficiently with the external environment. These animals have a **circulation system**, which consists of the following components:

blood	*a fluid connective tissue composed of water, diverse solutes, and formed elements (for example, blood cells and platelets)*
heart (or heartlike structure)	*a muscular pump that generates the pressure needed to keep blood flowing throughout the body*
blood vessels	*tubes of varying diameter through which blood is transported*

Most animals have a **closed circulation system**, in which the walls of the heart and the blood vessels are continuously connected (Figure 29.1). As you will see, the vertebrate system includes large-diameter blood vessels that function in rapid fluid transport, and an enormous number of small-diameter blood vessels (capillaries) that function as sites of rather leisurely diffusion.

Not all animals have a closed system of fluid transport. Arthropods (including insects and spiders) and most mollusks (snails, clams, and their kin) have an **open circulation system**. Fluid is pumped from the heart into a set of tubes and then is dumped into a space or cavity in the body tissues. There, it mingles with intercellular fluids before moving into open-ended tubes that lead back to the heart (Figure 29.2). In snails and clams, fluid is pumped into a network of spaces in spongy tissue that provides a large surface area for diffusion. In insects, fluid is pumped into a cavity that cannot expand much under the fluid pressure (because of the rigid exoskeleton of the insect body). The fluid has nowhere to go except back to the heart, which pumps it out again.

Even closed circulation systems are not completely sealed off. A slight amount of fluid is always filtering out of the capillaries, and materials are continually passing between the capillaries and the surrounding tissues. A supplementary network of tubes, the **lymph vascular system**, picks up the excess fluid and reclaims proteins from the tissues, returning them to the circulation.

29

CIRCULATION

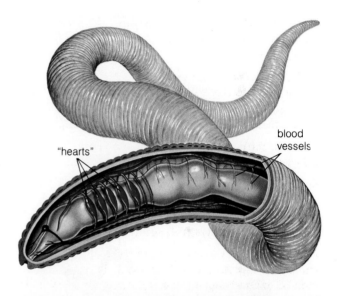

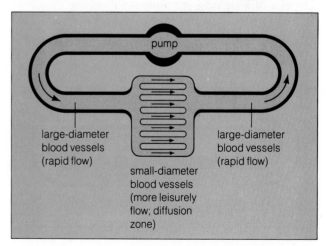

Figure 29.1 Fluid flow through a closed circulation system. The sketch above shows components of the closed system of the earthworm, which has several muscular "hearts" near its anterior end and several blood vessels running lengthwise through its series of body segments. The earthworm also has pairs of transverse blood vessels in each body segment.

Table 29.1 Components of Blood

	Function	Number per Microliter	Volume Percent
Cellular Portion (40%–50% of total volume)			
1. Red blood cells	Oxygen, carbon dioxide transport	4,500,000–5,500,000	
2. White blood cells:			
Neutrophils	Phagocytosis	3,000–6,750	
Lymphocytes	Central to immune response	1,000–2,700	
Monocytes	Phagocytosis	150–720	
Eosinophils	Phagocytosis	100–360	
Basophils	Source of substances that increase capillary permeability and show anticlotting activity	25–90	
3. Platelets	Source of substances that aid in blood clotting	250,000–300,000	
Plasma Portion (50%–60% of total volume)			
1. Water	Serves as solvent		91–92
2. Plasma proteins	Play diverse roles (infection fighting, blood clotting, lipid transport, etc.)		7–8
3. Other solutes (ions, sugars, lipids, amino acids, hormones, vitamins, dissolved gases)	Play diverse roles (maintaining extracellular pH, fluid volume, etc.)		1–2

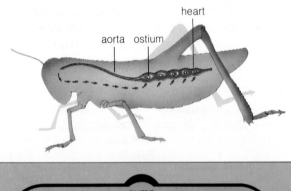

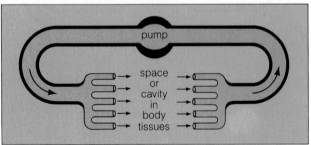

Figure 29.2 Fluid flow through an open circulation system. The sketch above shows the open system of the grasshopper. Like other insects, the grasshopper has a "heart" in the posterior portion of its body that pumps blood through a vessel (aorta) which dumps the blood into tissues at the anterior end of the body. After diffusing through body spaces, blood moves back into the heart through lateral openings (ostia).

CHARACTERISTICS OF BLOOD

Functions of Blood

The simplest animals with a true circulation system are ribbon worms (Figure 41.1), the longest of which is still shorter than your big toe. Their blood functions mainly in carrying wastes away from cells; it also contains phagocytic cells (which, recall, engulf foreign particles). In many invertebrates and all vertebrates, blood not only transports products and wastes from cells but it also transports nutrients and oxygen to them. It contains specialized phagocytic cells that function as scavengers and infection fighters, and it serves as the highway for hormones. By virtue of its composition, blood helps stabilize internal pH. In birds and mammals, blood also helps equalize body temperature by carrying excess heat from regions of high metabolic activity (such as skeletal muscles) to the skin, where it can be dissipated from the body.

In most animals, blood is a transport fluid that carries raw materials to cells, carries products and wastes from them, and helps maintain an internal environment that is favorable for cell activities.

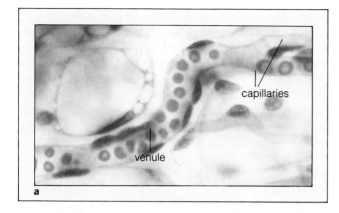

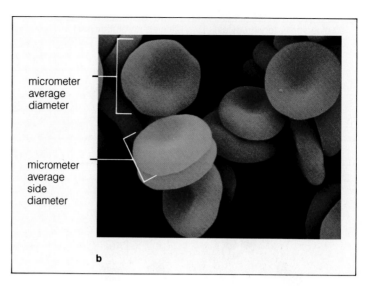

micrometer average diameter

micrometer average side diameter

Figure 29.3 (**a**) Photomicrograph of red blood cells in capillaries and in a venule. (**b**) Scanning electron micrograph showing the biconcave shape of red blood cells.

Blood Volume and Composition

On the average, a fully grown human male who weighs 70 kilograms (about 150 pounds) has a blood volume of about 5 liters, or a little more than 5 quarts. The volume varies, depending on the size of the body and also on changes in the concentration of water and solutes.

In all vertebrates, blood consists of red blood cells, white blood cells, platelets, and plasma (Table 29.1). Normally, the plasma constitutes about fifty to sixty percent of the total blood volume in an adult human.

Plasma. The portion of blood called **plasma** is mostly water, which functions as a solvent. But this straw-colored liquid also contains hundreds of different *plasma proteins*, including albumin, the globulins, and fibrinogen. The concentrations of plasma proteins influence the distribution of water between blood and interstitial fluid. Albumin is most important in this respect, for it represents sixty percent of the total amount of plasma proteins. Some alpha and beta globulins transport lipids and fat-soluble vitamins. Gamma globulins function in immune responses, and fibrinogen is necessary in blood clotting.

Plasma also contains diverse ions, simple sugars such as glucose, amino acids, vitamins, hormones, and dissolved gases (mostly oxygen, carbon dioxide, and nitrogen). The ions help maintain extracellular pH and fluid volume. The lipids present in plasma include fats, phospholipids, and cholesterol. Lipids that are transported from the liver to different body regions are generally bound with proteins to form lipoproteins.

Red Blood Cells. The task of transporting oxygen to cells is mainly the responsibility of erythrocytes, or **red blood cells**. A mammalian red blood cell is a biconcave disk, thicker around the rim than in the center (Figure 29.3). Its red color comes from hemoglobin, an iron-containing protein molecule present in its cytoplasm (page 58). When oxygen from the environment first diffuses into the bloodstream, it quickly binds with the iron and forms *oxyhemoglobin*. Blood rich in oxyhemoglobin is bright red. Blood somewhat depleted of oxygen is darker and appears blue when observed through blood vessel walls (hence the "blue" veins that are visible at your wrists).

In addition to oxygen, hemoglobin also transports some of the carbon dioxide wastes of aerobic metabolism. Most of the carbon dioxide is simply dissolved or combined with water to form bicarbonate (HCO_3^-) in the bloodstream, but about twenty-three percent binds with hemoglobin to form *carbaminohemoglobin*.

Red blood cells originate in red bone marrow (page 304). As each cell matures, its nucleus disappears, but it already has enough enzymes and other proteins to remain functional for its expected life span of about 120 days. The oldest red blood cells are continually removed from the bloodstream, mainly by phagocytic cells in the liver and spleen.

Feedback mechanisms keep the red blood cell count fairly stable. (A *cell count* is the number of cells of a given type in a microliter of blood.) When oxygen levels in tissues are low, the kidneys secrete an enzyme that converts a plasma protein into a key hormone (erythropoietin). The hormone stimulates an increase in red

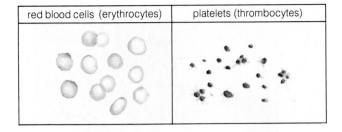

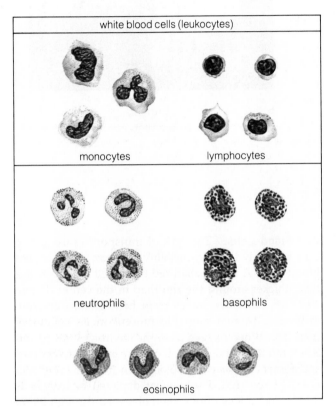

blood cell production in red bone marrow. New oxygen-carrying cells enter the bloodstream, and within a few days there is a rise in oxygen levels in the tissues. Information about the increase is fed back to the kidneys, production of the key hormone dwindles, and the production of red blood cells drops accordingly. With this mechanism, the red blood cell count is maintained, on the average, at 5.4 million in healthy adult males and 4.8 million in females.

White Blood Cells. The day-to-day housekeeping and defense activities that help keep tissues functioning are the responsibility of white blood cells, or **leukocytes**. Some of these cells are scavengers of dead or worn-out cells; others respond to tissue damage and invasion by bacteria, viruses, and other foreign agents. All are derived from immature cells, called *stem cells*, in bone marrow. Blood is like a reservoir of white blood cells, for most of the housekeeping and defense functions of those cells are expressed after they leave the blood capillaries and enter tissues.

Five types of white blood cells can be distinguished on the basis of size, nuclear shape, staining traits, and the presence or absence of dark-staining granules in the cytoplasm (Figure 29.4). The functions of these cell types, which are called lymphocytes, neutrophils, monocytes, eosinophils, and basophils, are summarized in Table 29.1. There are actually two subcategories of lymphocytes, referred to as B cells and T cells. Both are central to immune responses, which are described in the next chapter.

The number of white blood cells in each microliter of human blood varies, depending on whether the body is highly active, in a state of health, or under siege. For example, during bacterial infections, the white blood cell count increases above the levels shown in Table 29.1; during viral infections, it can drop below 5,000.

Platelets. In bone marrow, bits of cytoplasm are pinched off "giant" cells (megakaryocytes). These cell fragments are called **platelets**. As you will see, substances released from platelets aid in blood clotting; hence platelets function in preventing blood loss from damaged blood vessels.

Figure 29.4 Cellular components of vertebrate blood. There are two classes of white blood cells. The "granular" cells (neutrophils, basophils, and eosinophils) have grainy particles in the cytoplasm. The "agranular" cells (monocytes and lymphocytes) do not. These components originate as follows:

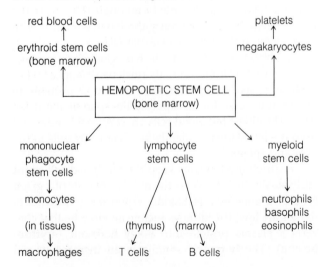

CARDIOVASCULAR SYSTEM OF VERTEBRATES

In all vertebrates, blood pumped out of a muscular heart enters large and then medium-size vessels called arteries. From there it travels into small, muscular arterioles, which branch into tiny vessels called capillaries. Blood travels from capillaries into small vessels called venules.

Finally it flows into larger vessels, the veins, which return it to the heart. The entire system is called a **cardiovascular system** (from the Greek *kardia*, meaning heart; and the Latin *vasculum*, meaning vessel). In humans, it is arranged roughly in this way:

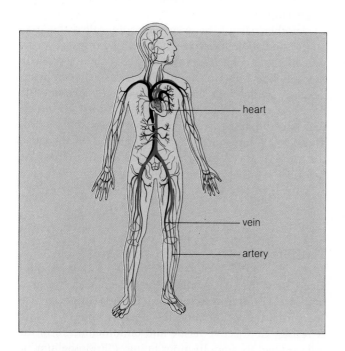

Before we consider the components of this system, let's step back and think about its overall "design." The heart pumps constantly, so the overall volume of flow through each part of the system is also constant. Yet if blood were to flow as rapidly through a capillary as it does through an artery or arteriole, there would not be enough time for adequate diffusion of substances to and from cells. As it happens, blood spreads out into **capillary beds**, each of which contains vast numbers of capillaries that run in parallel. The total cross-sectional area of a capillary bed is much larger than that of the transport tubes leading into it. As a result, the rate of flow through each capillary is decreased, the time available for exchange of materials is increased—and the volume flow for the whole system remains constant:

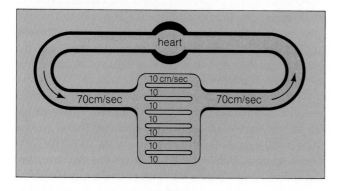

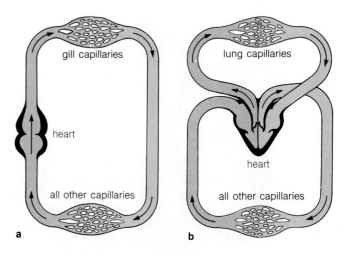

Figure 29.5 Comparison of blood circulation in fishes (**a**) and in mammals (**b**). In fishes, blood must flow through at least two capillary beds on its circuit away from and back to the heart. Hence pressure drops are sharp and blood transport is not as rapid as it is in mammals, where blood most often travels through only one capillary bed during each circuit.

Blood Circulation Routes

In fishes, blood passes through at least two capillary systems during its circuit away from and back to the heart (Figure 29.5a). Blood first encounters the capillary beds of the fish gills, where it picks up oxygen from the surrounding water, then it encounters capillary beds in the rest of the body (page 419).

Whenever blood flows against a blood vessel wall, friction is generated and the pressure forcing blood through the circulation system decreases. Thus, in fishes, there are two sharp drops in pressure during the circulation of blood away from and back to the heart.

In mammals, a given volume of blood being circulated away from and back to the heart generally passes through only *one* major capillary bed. Thus there is less of a drop in blood pressure than there is in fishes (Figure 29.5b). This more efficient circulation pattern is possible because the mammalian heart is divided into two pumps that drive oxygen-poor and oxygen-rich blood along separate routes. In the **pulmonary route**, blood from the right half of the heart is pumped to the lungs, where it picks up oxygen, then it flows to the left half of the heart. In the **systemic route**, the oxygenated blood is pumped through the rest of the body (where oxygen is used), then it flows to the right half of the heart.

As Figure 29.6 indicates, only one part of the systemic route carries blood through two major capillary systems rather than just one. The blood flows through a capillary

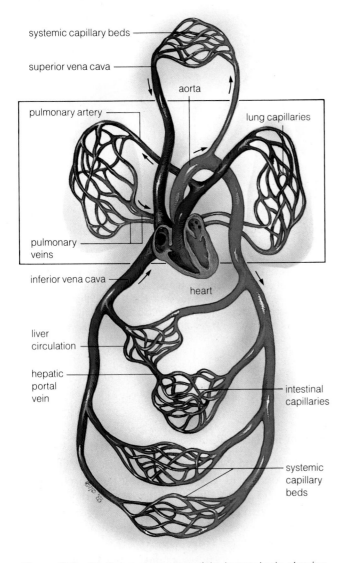

Figure 29.6 Cardiovascular system of the human body, showing the main capillary beds and transport routes. The capillary beds of the pulmonary route are shown in the boxed region.

bed in the intestines, then through the hepatic portal vein, then into a capillary bed in the liver. (A portal vessel, recall, carries blood from one capillary bed to another before the blood flows back to the heart.) Thus the liver, which plays a vital role in nutrition, gets the first crack at food being absorbed from the gut (page 448).

Another part of the systemic route, called the "coronary circulation," services the heart. During a seventy-year life span, the human heart beats some $2\text{-}\frac{1}{2}$ billion times; it rests only during brief intervals between heartbeats. A tremendous amount of oxygen is delivered to the heart muscle cells by way of the coronary arteries, which lead into the heart's own extensive capillary bed. These two arteries are the first to branch off the *aorta*, the artery that carries oxygenated blood away from the heart.

The Human Heart

Heart Structure. The bulk of the human heart, the *myocardium*, is cardiac muscle tissue. Fibrous connective tissue forms the *pericardium*, the heart's protective outer covering. Connective tissue and endothelium form the *endocardium*, its smooth inner lining. ("Endothelium" is a single layer of epithelial cells lining the heart cavities and the space, or lumen, inside blood vessels.)

The left and right halves of the heart are two distinct pumps, each with two chambers. In each half, blood flows first into a thin-walled **atrium** (plural, atria), then into a thick-walled **ventricle** (Figure 29.7). Between the two chambers in each half are flaps of membrane known as an *atrioventricular valve* (or AV valve). Another flap, the *semilunar valve*, spans the exit from each ventricle. The valves open and close in response to fluid pressure changes that are produced when the heart beats. Their passive movements help keep the blood moving in one direction and thereby prevent backflow.

Cardiac Cycle. Each heartbeat is a sequence of muscle contractions and relaxation called the **cardiac cycle**. In each cycle, the four chambers of the heart go through phases of contraction (called *systole*) and relaxation (called *diastole*). The timing of these phases is not the same for all the heart chambers; the atria contract slightly before the ventricles contract.

While the atria are relaxed and filling, the ventricles are also relaxed (Figure 29.8). As pressure rises in the atria, the AV valves are forced open, so the ventricles start to fill even before the atria contract. Then the ventricles contract, causing the AV valves to snap shut, and the pressure inside the ventricles rises sharply above the pressure in the vessels leading out from them. One of

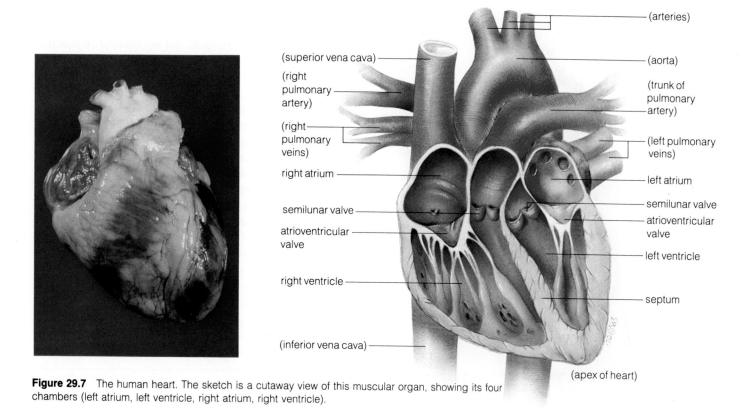

(arteries)

(superior vena cava)

(aorta)

(right pulmonary artery)

(trunk of pulmonary artery)

(right pulmonary veins)

(left pulmonary veins)

right atrium

left atrium

semilunar valve

semilunar valve

atrioventricular valve

atrioventricular valve

left ventricle

right ventricle

septum

(inferior vena cava)

(apex of heart)

Figure 29.7 The human heart. The sketch is a cutaway view of this muscular organ, showing its four chambers (left atrium, left ventricle, right atrium, right ventricle).

those vessels is the aorta; the other is the pulmonary artery. With the increased pressure, the semilunar valves open and blood flows out of the heart. After the blood has been ejected, the ventricles relax and the cardiac cycle starts over.

The blood and heart movements during the cardiac cycle generate vibrations that produce a "lub-dup" sound. The sound can be heard at the chest wall. At each "lub," the AV valves are closing as the ventricles contract. At each "dup," the semilunar valves are closing as the ventricles relax.

The contraction and relaxation phases of the cardiac cycle can be summarized this way:

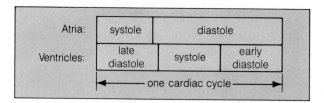

Atria:	systole	diastole	
Ventricles:	late diastole	systole	early diastole
	← one cardiac cycle →		

What can we deduce from this sketch? Clearly the atria contract first and then relax before blood leaves the ventricles. Their contraction simply helps to fill the ventricles; it cannot be the driving force for circulation. *It is the contraction of the ventricles that forces blood through the circulation system.*

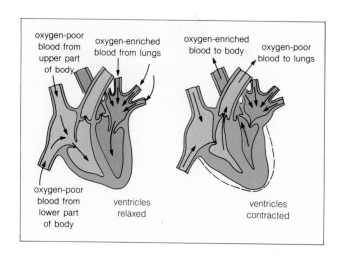

oxygen-poor blood from upper part of body

oxygen-enriched blood from lungs

oxygen-enriched blood to body

oxygen-poor blood to lungs

oxygen-poor blood from lower part of body

ventricles relaxed

ventricles contracted

Figure 29.8 Blood flow through the heart during relaxation (diastole) and contraction (systole) of muscle tissue in the walls of the ventricles.

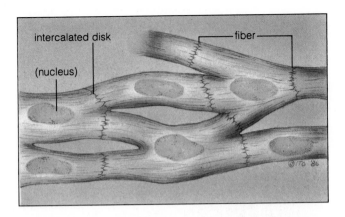

Figure 29.9 Intercalated disks between abutting heart muscle fibers. Each fiber is a separate cell, with its own nucleus, but cell membranes of adjacent cells are fused together at junctions between them. These membrane-to-membrane junctions are intercalated disks.

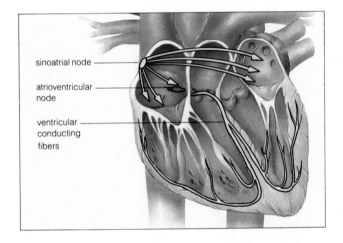

Figure 29.10 Location of conducting muscle fibers that make up the cardiac conducting system.

Heart Muscle Contraction. With each heartbeat, the individual muscle fibers of the heart contract in unison, as if they were a single unit. Yet the units of contraction in cardiac muscle are the same as in skeletal muscle. (Individual sarcomeres are organized one after another along the length of each fiber, as shown in Figure 28.11.) However, unlike skeletal muscle fibers (whose endings are attached to bones), cardiac muscle fibers branch and then abut with one another at the branched endings. Here, the plasma membranes are fused to produce strong cohesion between fibers. Each end-to-end region of membrane is an **intercalated disk** (Figure 29.9).

Intercalated disks contain communication junctions, where signals travel rapidly between cells. Excitatory signals cause ions to flow directly from one muscle fiber to another, changing the membrane properties as they go. The excitation spreads so rapidly that the muscle tissue contracts as a unit in response.

Because of strong cohesion and rapid communication at intercalated disks between individual muscle fibers, cardiac muscle tissue contracts almost like a single unit.

Cardiac Conduction System. As you know, skeletal muscle contracts in response to signals from the nervous system, but cardiac muscle is different. Although the nervous system does control the rate and strength of the heartbeat, the heart will keep on beating even if all autonomic nerves leading to the heart are severed! The heart makes itself contract because some of its muscle fibers are self-excitatory: they initiate and conduct action potentials. These fibers are the basis of the **cardiac conduction system** (Figure 29.10).

In all mammals, the excitation begins in the **sinoatrial node** (or SA node), a region of conducting fibers where major veins enter the right atrium of the heart. Here, membrane properties change repeatedly and action potentials are produced, with no stimulation from the outside. One wave of excitation follows another, seventy or eighty times a minute in the human heart. The rhythmic excitation begins soon after heart muscle cells appear in a developing embryo, and from then on it triggers all cardiac muscle contractions.

Although all cells of the cardiac conduction system are self-excitatory, the SA node fires at the highest frequency and comes to threshold first in each cardiac cycle. Thus the SA node is the *cardiac pacemaker*: its rhythmic firing is the basis for the normal rate of heartbeat.

Each wave of excitation from the pacemaker spreads over both atria, causing them to contract almost at the same time. The wave also passes to the **atrioventricular node** (or AV node), which consists of conducting fibers in the floor of the right atrium. Signals are conducted more slowly through this node. The delay gives the atria enough time to complete their contraction before the ventricles start to contract.

A bundle of conducting fibers leads away from the AV node and branches extensively through the inner walls of both ventricles. This bundle is the *only* conduction pathway between the atria and ventricles, which are otherwise separated from each other by nonconducting tissue. In response to signals carried by this bundle of fibers, all parts of the ventricle walls contract more or less in unison.

The sinoatrial node is the cardiac pacemaker. Its spontaneous, repetitive excitation spreads along a system of conducting muscle fibers that stimulate contractile tissue in the atria, then the ventricles, in a rhythmic cycle.

Blood Pressure in the Vascular System

The fluid pressure that is generated by heart contractions forces blood through the entire circulation system; it is called **blood pressure**. However, blood pressure is not the same along the circuits away from and back to the heart. Normally, pressures are high to begin with and they drop along the way (Figure 29.11). As you will see, the pressure drops result from the loss of energy that is used to overcome resistance to flow as blood moves through the circulation.

Table 29.2 summarizes the differences in blood pressure and volume along the systemic route. Let's now consider some of the reasons for these differences.

Arterial Blood Pressure. The heart ejects blood into **arteries**, the transport tubes that conduct oxygen-poor blood to the lungs and oxygenated blood to the rest of the body (Figure 29.12). The thick, impermeable wall of an artery contains smooth muscle and connective tissue that can be distended under surges of fluid pressure. The wall also recoils elastically, and the recoil forces blood onward (Figure 29.13). Because of these wall properties, the arteries serve as pressure reservoirs that smooth out the changes in blood pressure associated with the cardiac cycle. Pressure is stored here when blood is ejected into the circulation by the contracting ventricles. The *Commentary* on page 396 describes what happens when arterial walls change because of such disorders as arteriosclerosis.

Ordinarily, blood pressure is measured at large arteries of the systemic route, such as arteries in your upper arms. First a systolic reading is taken of the highest pressure during a cardiac cycle (generated by the contracting ventricles). In young adults at rest, systolic pressure is about 120mm Hg. (This means that the measured amount of pressure would make a column of mercury, or Hg, rise a distance of 120 millimeters.) Next,

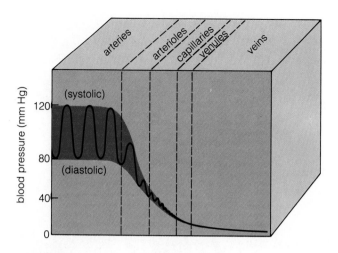

Figure 29.11 Drops in blood pressure in the systemic circulation.

Table 29.2 Blood Volume Distribution Through the Human Cardiovascular System

	Average Blood Volume (percent of total)	Mean Blood Pressure (mm Hg)
Heart chambers:		
Atria	2	0 (right); 6 (left)
Ventricles	10	0–24 (right); 4–120 (left)
Aorta and other arteries:	10	100
Arterioles:	1	60
Capillaries:	5	30
Veins:	54	10
Pulmonary circulation:	18	15
	100 percent	

Data from Robert C. Little, MD.

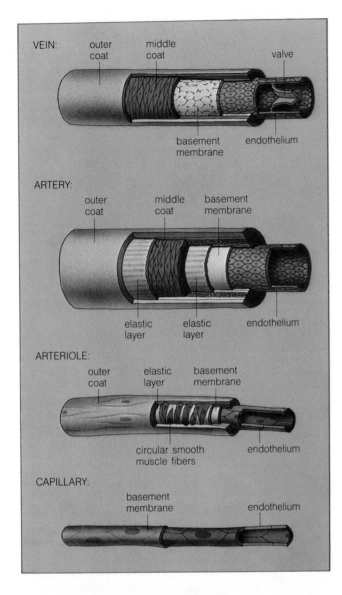

Figure 29.12 Structure of blood vessels. The outer coat consists of loose connective tissue. The middle coat contains elastic fibers and smooth muscle. The basement membrane consists of connective tissue elements including collagen fibers.

a diastolic reading is taken of the lowest arterial pressure at the end of a cardiac cycle, just before blood is pumped out of the heart again. Diastolic pressure is generally about 80mm Hg. Thus the difference between systolic and diastolic readings (the so-called *pulse pressure*) is 120 − 80, or 40mm Hg.

Resistance at Arterioles. Arteries branch into **arterioles**, which are transport tubes of smaller diameter that have rings of smooth muscle fibers in their walls (Figure 29.12). Arterioles are the point at which control is exerted over the relative distribution of blood to different parts of the body.

Arterioles can increase or decrease in diameter and thereby offer variations in resistance to flow. As you will see, arteriole diameter is adjusted by neural and endocrine controls that govern blood pressure for the body as a whole. But the diameter also can be adjusted in response to changes in local chemical conditions of a tissue. For example, when metabolic activity increases in skeletal muscle tissue, the concentration of oxygen decreases, and the concentrations of carbon dioxide, hydrogen ions, potassium ions, and other substances increase. Such chemical changes serve as local signals that act on the smooth muscle fibers in the arteriole wall. The fibers respond by relaxing, the arteriole diameter enlarges, and more blood flows past, delivering more raw materials and carrying away cell products and wastes.

Any enlargement in the diameter of a blood vessel is called **vasodilation**. Any decrease is called **vasoconstriction**. For example, reversal of the local conditions described above could trigger such a decrease. Controlled vasodilation and vasoconstriction of arterioles direct blood to regions of greatest metabolic activity. The more active the cells of a given region, the greater the blood flow to them.

Capillary Function. A **capillary** is a tube about a millimeter long that is specialized for exchanging substances with interstitial fluid. The tube wall is a single layer of

Figure 29.13 Distensibility of an artery. Because of this property, the arterial system serves as a pressure reservoir that is drawn upon to smooth out flow even as fluid pressure changes during the cardiac cycle.

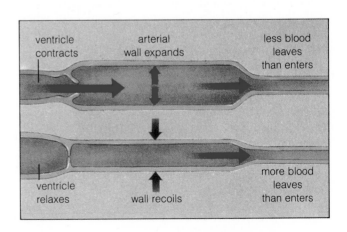

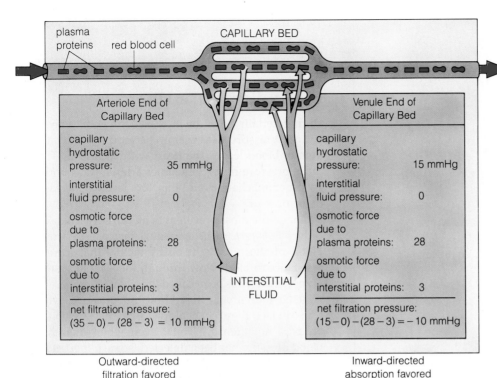

plasma proteins red blood cell CAPILLARY BED

Arteriole End of Capillary Bed	
capillary hydrostatic pressure:	35 mmHg
interstitial fluid pressure:	0
osmotic force due to plasma proteins:	28
osmotic force due to interstitial proteins:	3
net filtration pressure: (35 − 0) − (28 − 3) = 10 mmHg	

Venule End of Capillary Bed	
capillary hydrostatic pressure:	15 mmHg
interstitial fluid pressure:	0
osmotic force due to plasma proteins:	28
osmotic force due to interstitial proteins:	3
net filtration pressure: (15 − 0) − (28 − 3) = − 10 mmHg	

INTERSTITIAL FLUID

Outward-directed filtration favored

Inward-directed absorption favored

Figure 29.14 Example of fluid movements in an idealized capillary bed. Such movements play no significant role in diffusion, but they are important in maintaining the distribution of extracellular fluid between the bloodstream and interstitial fluid.

flat endothelial cells, separated from each other only by narrow clefts. The capillary diameter is so small that red blood cells squeeze through it single file (Figure 29.3). The small diameter resists flow, but the vast numbers of capillaries afford such a huge cross-sectional area that the overall resistance of a given capillary bed is less than that of the arterioles leading into it.

Capillaries thread through nearly every tissue in the body, coming within 0.01 centimeter of every living cell and thereby enhancing diffusion rates. The density of capillaries in a tissue is directly related to metabolic output. For example, a great deal of blood must be circulated through muscle tissue to provide individual cells with oxygen for aerobic respiration. (Only the aerobic pathway can produce enough ATP energy to drive extensive contraction.)

Most of the solutes exchanged with interstitial fluid, including oxygen and carbon dioxide, simply diffuse across the capillary wall. But some proteins are also exchanged here, probably by endocytosis and exocytosis (page 94). Moreover, certain ions and small, water-soluble molecules probably pass through the clefts between cells. The clefts are wider in some capillary beds than in others, and those beds are more "leaky" to solutes.

Some fluid also moves by bulk flow across capillary walls. The movement plays no significant role in the exchange of solutes. Rather, bulk flow helps maintain the balance of extracellular fluid between the bloodstream and the surrounding tissues. This fluid distribution is important, because blood pressure is maintained only when there is an adequate blood volume. Interstitial fluid is a reservoir that is tapped when blood volume drops (as during hemorrhage and other events).

As Figure 29.14 indicates, the fluid movements are determined by two opposing forces. One force is the difference between capillary blood pressure and interstitial fluid pressure. Because of the difference, some plasma (but not the plasma proteins) leaves the capillary. This outward fluid movement is called **filtration**.

The other force is the difference in water concentration between plasma and interstitial fluid. (Plasma has a greater solute concentration, with its protein components, and therefore a lower water concentration.) Because of the difference, some interstitial fluid moves into the capillary. This inward fluid movement is called **absorption**.

As Figure 29.14 suggests, fluid filtration at the arteriole end of a capillary bed tends to be balanced by absorption at the venule end. Normally, there is only a small net filtration of fluid, which is returned to the blood by lymph vessels.

Venous Pressure. Capillaries merge into "little veins," or **venules**. Some diffusion also occurs across the venule wall, which is only a little thicker than that of a capillary. Also, contraction of smooth muscle in the wall can cause fluid pressure to increase in the capillaries and thereby affect filtration.

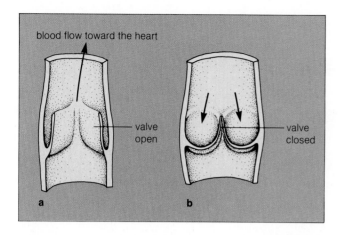

blood flow toward the heart

valve
open

valve
closed

a

b

Figure 29.15 Valve in a vein in the open (**a**) and closed (**b**) positions.

Venules merge into veins, the large-diameter tubes for blood flow back to the heart (Figure 29.12). By the time blood has been circulated into **veins**, fluid pressure has dropped to only 10–15mm Hg. However, fluid pressure in the right atrium of the heart is close to 0mm Hg, and the large diameter of veins presents very low resistance to flow. Thus, even though the pressure gradient between the veins and the heart is rather small, it is enough to allow blood to complete the circuit. Also, many veins have tissue folds into their lumen that serve as valves. When fluid in a vein starts moving backward because of gravity, it pushes the folds into the lumen and thereby prevents backflow (Figure 29.15).

Veins are more than low-resistance transport tubes; they also are blood volume reservoirs. At any given time, veins contain fifty to sixty percent of the total blood volume for the body. This large reservoir can be tapped to increase cardiac output. How? The vein wall is thinner and can distend much more than an arterial wall. When circulation must be stepped up to meet the demands of increased body activities, the nervous system sends signals to smooth muscle in the vein walls. When the muscle fibers contract, the walls become stiffer and cannot distend as much, so pressure in the veins rises. The increased pressure drives more blood from the veins to the heart. With the increase in the cardiac filling pressure, the ventricles contract more forcefully.

The pressure in veins also is increased when limb muscles move. When contracting skeletal muscles "bulge" against an adjacent vein, the diameter of the vein is effectively decreased. Internal pressure rises and forces the blood forward. In addition, breathing plays an indirect role in returning blood to the heart. When

COMMENTARY

On Cardiovascular Disorders

During certain types of increased physical activity, systolic pressure rises mainly because a greater volume of blood leaves the heart with each beat. Thus pulse pressure rises. Sometimes arterial walls undergo structural changes and this, too, leads to increased pulse pressure.

One of the most prevalent of all cardiovascular disorders is *hypertension*, a gradual increase in arterial blood pressure brought on by factors that increase resistance to blood flow through the arterial system. Hypertension has been called the silent killer, because afflicted individuals may show no outward symptoms; in fact, they feel healthy. Often they are neither "hyper" nor "tense." Thus, even when their high blood pressure has been detected, some hypertensive individuals tend to resist exercising, medication, and corrective changes in diet. Of 23 million Americans who are hypertensive, most are not undergoing treatment. About 180,000 will die each year as a result.

In *arteriosclerosis*, calcium salts and fibrous tissue gradually build up within arterial walls, leading to a "hardening of the arteries." As arteries become less and less compliant, they cannot expand under increased blood flow, and pressure builds up in the arterial system. If arterial blood pressure is high enough, small blood vessels supplying brain cells can rupture, which is one cause of a brain *stroke*. If the brain cells deprived of oxygen and nutrients are in a region governing the coordination of body movements, paralysis may follow. If they are in regions controlling the vital function of breathing, death will follow.

In Chapter Four, you read that lipids such as fats and cholesterol are insoluble in water. Such lipids are transported in the bloodstream, where they are bound to protein carriers that keep them suspended in the plasma. In *atherosclerosis* (the major type of arteriosclerotic disease), abnormal smooth muscle cells multiply and connective tissue components increase in arterial walls. Then lipids are deposited within cells and extracellular spaces of the endothelial lining of the wall. Calcium salts are deposited on top of the lipids, and a fibrous net forms over the whole mass. This so-called plaque sticks out into the lumen. Sometimes platelets become caught on rough plaque edges, and are stimulated into secreting some of their chemicals. When they do, they initiate clot formation. As the clot and plaque grow, the artery can become blocked. This shuts off blood flow to tissues that the artery supplies. Heart muscles are supplied by coronary arteries, which branch off the aorta. A *coronary occlusion* may develop suddenly, when a clot forms rapidly on a plaque. Heart attacks may result.

Not all heart attacks are fatal. Survival depends on how much heart tissue is deprived of its blood support system, and on where the damage occurs. Survival also depends on immediate care.

What causes cardiovascular disorders? Cholesterol, obviously, is under suspicion. When transported through the bloodstream, it is bound to one of two kinds of protein carriers: high-density lipoproteins (HDL) and low-density lipoproteins (LDL). Evidence is accumulating that high levels of LDL are related to a tendency toward heart trouble. It appears that LDLs, with their cholesterol cargo, show a greater penchant for infiltrating arterial walls. It may be that HDLs can attract cholesterol out of the walls and transport it to the liver, where it can be metabolized. Atherosclerosis is uncommon in rats; rats have mostly HDL. In monkeys, pigs, and humans, which generally have mostly LDL, atherosclerosis is not uncommon.

Behavior patterns, too, may trigger conditions that lead to cardiovascular disorders. Some individuals are so-called Type A personalities: they tend to be competitive, aggressive, and impatient; they are least stressed when their lives are most organized. Type A individuals tend to become acutely stressed when some aspect of their life—personal or professional—swings out of control. Type B personalities are more relaxed about themselves and about life; they seem to be less prone to cardiovascular troubles.

Intriguingly, Robert Nerem and Fred Cornhill of Ohio State University reported in 1979 on a most unusual effect of their studies of drug action on high-cholesterol diets. Rabbits were the subjects. One group of rabbits was cuddled and played with; the other group received impersonal care. Arterial tissue samples from both groups showed that atherosclerosis was less prevalent by half in the cuddled rabbits. The experiment was repeated; the results were the same. Might we chalk one up for tender loving care?

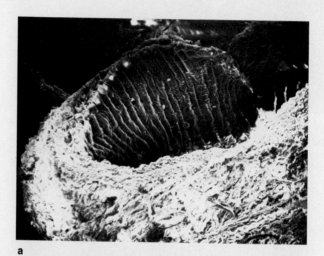

a

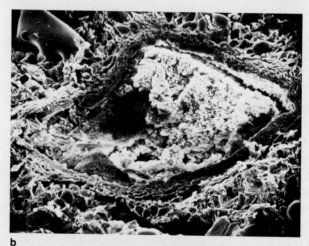

b

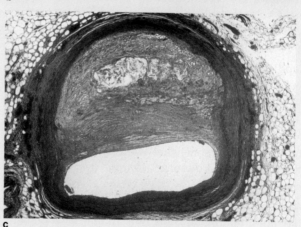

c

(**a**) Normal artery; (**b-c**) artery clogged with plaques.

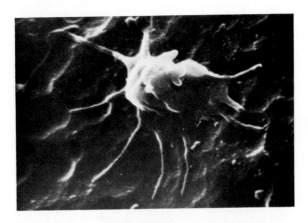

Figure 29.16 Scanning electron micrograph of a spiny blood platelet. The spines are pseudopodia ("false feet"), formed by microtubules.

you inhale, your rib cage expands. Pressure in the chest decreases and the diaphragm pushes down on organs in the abdominal cavity, increasing the pressure there (page 424). The pressure changes help increase the pressure gradient between the heart and the veins that are returning blood from the lower half of the body.

In summary, these are the key points to keep in mind about the differences between the tubes of the cardiovascular system:

1. **Arteries** are pressure reservoirs that keep blood flowing while the ventricles are relaxing. These blood vessels have large diameters that offer low resistance to flow, so there is little drop in blood pressure in the arteries.

2. **Arterioles** are control points for the distribution of different volumes of blood to different diffusion zones. They offer a great deal of resistance to flow, so there is a major drop in pressure in arterioles.

3. **Capillary beds** are diffusion zones for exchanges between blood and interstitial fluid. Because they have a greater cross-sectional area than that of the arterioles leading into the beds, they present less total resistance to flow. There is some drop in pressure here.

4. **Venules** overlap somewhat with capillaries in their function; they afford some control over capillary pressure.

5. **Veins** are highly distensible blood volume reservoirs and are important in adjusting flow volume back to the heart. They offer only low resistance to flow, and blood pressure is very low in veins.

Regulation of Blood Flow

After a large meal, more blood is diverted to your digestive system, which swings into full gear as other systems more or less idle. When your body is exposed to cold wind or snow for an extended time, blood is diverted away from the skin to deeper tissue regions, so that the metabolically generated heat that warmed the blood in the first place can be conserved.

The rate of flow to different body regions is inversely proportional to resistance. When resistance is decreased, the flow steps up. The question becomes this: How is blood pressure raised and lowered through the cardiovascular system as a whole?

The medulla oblongata (a brain region) contains an integrating center that controls blood pressure by coordinating the rate and strength of the heartbeat with changes in the diameter of arterioles (and, to some extent, of veins). This command post integrates information coming in from receptors in certain arteries (such as the carotid arteries in the neck and the aorta) and in cardiac muscle tissue. When blood pressure rises, the receptors signal the medulla, which responds by sending out signals that lead to decreased cardiac output and vasodilation of the arterioles. The combined effect is that blood pressure falls. When it falls too much, signals from the medulla lead to vasoconstriction and, at the same time, to increased heart activity.

Hormones also play a role in the control of blood pressure. The arterioles in different body regions have different receptors that can be activated by such hormones as epinephrine and angiotensin. Depending on which receptors are present, the epinephrine can cause vasoconstriction or vasodilation. Angiotensin causes widespread vasoconstriction.

Hemostasis

Animals can survive all sorts of threats, including the loss of blood when a small blood vessel is ruptured or cut. In many invertebrates and all vertebrates, bleeding can be stopped by mechanisms that include blood vessel spasm, platelet plug formation, and blood coagulation. The collective action of these mechanisms, called **hemostasis**, can be described as follows.

First, smooth muscle in the damaged wall contracts in a reflex response called a spasm. The spasm lasts only a few minutes, but it causes the vessel to constrict and curtail the flow of blood. Second, platelets clump together, temporarily plugging the rupture. The platelets do this by developing spinelike extensions that adhere to exposed collagen fibers in the damaged walls (Figure 29.16). The spiny platelets release substances that attract more platelets. They also release calcium ions (which promote clumping) and a vasoconstrictor (which helps prolong the spasm). Third, blood coagulates (converts to a gel) and forms a clot. Finally, the clot retracts into a compact mass, drawing the ruptured walls of the vessel together.

Blood coagulates mainly through an **intrinsic clotting mechanism** that comes into play when damage exposes the collagen fibers of blood vessels. A plasma protein becomes activated and triggers reactions that lead to the conversion of another plasma protein into thrombin (Figure 29.17). Thrombin is an enzyme that acts on fibrinogen, a rather large, rod-shaped plasma protein. The rods adhere to each other, end to end and side to side. They form long, insoluble threads that stick to exposed collagen on the damaged vessel wall and form a net that entangles blood cells and platelets. The entire mass is a blood clot.

Blood also can coagulate through an **extrinsic clotting mechanism**. The contribution of this mechanism to hemostasis is unclear. However, it is definitely involved in walling off bacteria and preventing the spread of bacterial infection from invaded tissue regions. The damage stimulates tissues into releasing a substance that eventually leads to thrombin formation, and the remaining steps parallel those shown in Figure 29.17.

Blood Typing

All cells in each individual have certain proteins and other molecules that act like identification flags, or markers; they identify the cell as being of a specific type. Each individual also has proteins called **antibodies**, which can recognize markers on *foreign* cells (page 407). When the blood of two people mixes during transfusions, the antibodies will act against any cells bearing the "wrong" marker. They will do the same thing during pregnancy. (Although the circulation systems of the mother and her unborn child remain separate throughout pregnancy, antibodies diffuse from one system to the other.)

ABO Blood Typing. As we have seen, *ABO blood typing* is based on some of the surface markers on red blood cells (page 166). Type A blood has A markers on those cells, type B blood has B markers, type AB has both, and type O has neither one.

Figure 29.18 shows what happens when the blood from different types of donors and recipients is mixed together. Type A blood will agglutinate when mixed with either type B or O. In an **agglutination response**, antibodies act against the "foreign" red blood cells and

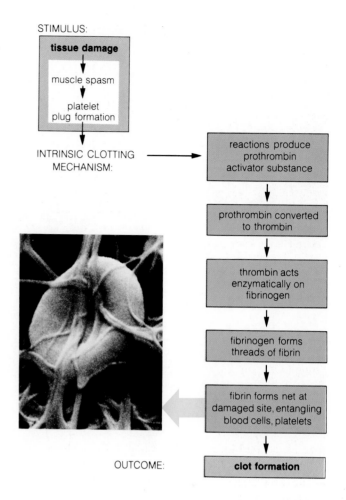

Figure 29.17 Blood coagulation at a cut or ruptured blood vessel tissue. The micrograph shows a red blood cell trapped in a fibrin net.

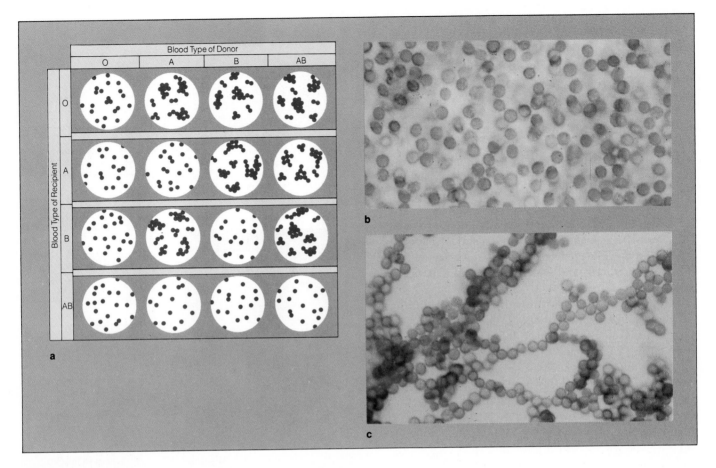

Figure 29.18 (**a**) The agglutination responses in drops of blood of types O, A, B, and AB when mixed with blood samples of the same and different types. (**b**) Micrograph showing the absence of agglutination in a mixture of two different but compatible blood types. (**c**) Micrograph showing agglutination in a mixture of incompatible blood types.

cause them to clump together. Such clumps can clog small blood vessels; they may lead to kidney damage and death. Type B blood will agglutinate when mixed with type A or O, but not with type B or AB. In looking at Figure 29.18, can you determine what the responses will be to type AB blood? To type O blood?

Rh Blood Typing.   Red blood cells also have other surface markers that can cause agglutination responses. For example, the *Rh blood typing* is based on the presence or absence of an Rh marker (so named because it was first identified in the blood of the *Rhesus* monkey). Rh^+ individuals have blood cells with this marker; Rh^- individuals do not. Ordinarily, people do not have antibodies that act against Rh markers. However, if someone who is Rh^- has been given a transfusion of Rh^+ blood, antibodies will be produced against it and will continue circulating in the bloodstream.

If an Rh^- female becomes pregnant by an Rh^+ male, there is a chance that the fetus will be Rh^+. During her pregnancy, some red blood cells of the fetus may leak from the placenta and enter her bloodstream (page 503). If they do, they will stimulate her body into producing antibodies against the Rh markers.

If the woman becomes pregnant again, the Rh antibodies will cross the placenta and enter the bloodstream of the fetus. If this second fetus happens to have Rh^+ blood, the antibodies will cause a hemolytic response. In this response, red blood cells swell and then rupture, releasing hemoglobin into the bloodstream. In extreme cases of this disorder, called *erythroblastosis fetalis*, so many cells are destroyed that the fetus dies before birth. If the child is born alive, all of its blood can be slowly replaced with blood that is free of the Rh antibodies. Currently, known Rh^- females can be treated right after their first pregnancy with a drug that inactivates any Rh antibodies circulating in her bloodstream, thereby protecting the fetus of the next pregnancy.

LYMPHATIC SYSTEM

We conclude this chapter with a brief description of the **lymphatic system**, which supplements the circulation system by returning excess tissue fluid to the bloodstream. But think of this description as a bridge to the next chapter, on immunity, for the lymphatic system is also vital to the body's defenses against injury and attack.

The lymphatic system consists of transport tubes and lymphoid organs. The tubes constitute the lymph vascular system, which supplements pulmonary and systemic circulation. When tissue fluid has moved into these tubes, it is called **lymph**. The lymphoid organs, which take part in defense responses, are structurally and functionally connected with both the blood and lymph vascular systems (Figure 29.19).

Lymph Vascular System

The **lymph vascular system** includes lymph capillaries, lymph vessels, and ducts that drain the processed fluid back into the circulation system. It serves these functions:

---

1. Return of excess filtered fluid to the blood.

2. Return of small amounts of proteins that leave the capillaries.

3. Transport of fats absorbed from the digestive tract.

4. Transport of foreign particles and cellular debris to disposal centers (lymph nodes).

At one end of the lymph vascular system are *lymph capillaries*, which are no larger in diameter than blood capillaries. These vessels occur in the tissues of almost all organs, and they seem to be permeable to all substances dissolved in interstitial fluid. Lymph capillaries are "blind-end" endothelial tubes. They have no entrance at the end residing in interstitial regions; the only opening is one that merges with larger lymph vessels.

Lymph vessels are structurally similar to veins. They have an outer layer of connective tissue, a midlayer of smooth muscle and elastic fibers, and an inner lining of endothelium. Flaplike valves in these vessels prevent backflow. When you breathe, movements of the rib cage and of skeletal muscle adjacent to the lymph vessels assist in moving fluid through them, just as they do for veins. Lymph vessels converge into collecting ducts, which drain into veins in the region of the lower neck (Figure 29.19).

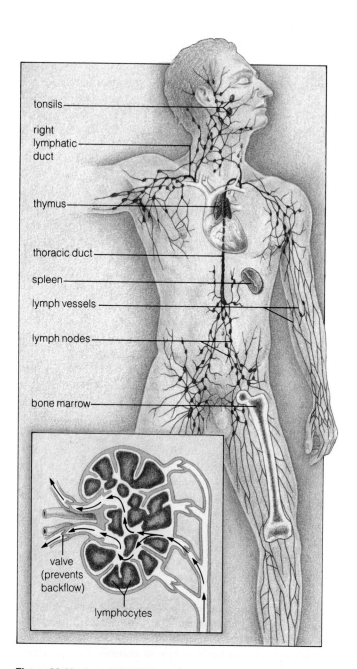

Figure 29.19 Lymphatic system, which includes the lymph vascular network and the lymphoid organs and tissues. Purple dots show some of the major lymph nodes. The inset illustrates the internal structure of a lymph node.

Lymphoid Organs

The **lymphoid organs** include the lymph nodes, spleen, thymus, tonsils, adenoids, and patches of lymphoid tissue in the small intestine and appendix. These organs and tissue patches function as production centers for infection-fighting cells and as sites for some defense responses.

Lymph nodes are located at intervals along lymph vessels (Figure 29.19). All lymph trickles through at least one of these nodes before being delivered to the bloodstream. Each node is internally partitioned into several chambers, which are packed with lymphocytes and plasma cells. Macrophages in the node help clear the lymph of bacteria, cellular debris, and other substances.

The largest lymphoid organ, the *spleen*, is a filtering station for blood and a holding station for lymphocytes. It is located behind the stomach and beneath the diaphragm. The spleen is also partitioned internally into chambers, which are filled with red and white "pulp." The red pulp contains a large store of red blood cells and macrophages. This region is a production site for red blood cells in developing human embryos. In adults, lymphocytes derived from bone marrow differentiate and mature in the white pulp of the spleen and other lymphoid organs.

The *thymus* serves as an endocrine gland, in that it secretes hormones concerned with the activity of lymphocytes. It also is a major organ where certain lymphocytes multiply, differentiate, and mature. The thymus is central to immunity, which is the topic of the chapter to follow.

SUMMARY

1. Cells survive by exchanging substances with the bloodstream, generally by the process of diffusion. A circulatory system allows rapid movement of these substances from one body region to another. Such systems consist of blood, a heart or heartlike structure, and blood vessels.

2. Blood is a transport fluid that carries raw materials to cells, carries products and wastes from them, and helps maintain an internal environment favorable for cell activities. Blood consists of red and white blood cells, platelets, and plasma.

3. Plasma contains water, ions, nutrients, hormones, vitamins, dissolved gases, and the plasma proteins.

4. Red blood cells transport oxygen (bound to hemoglobin) between cells and the lungs. They are also responsible for the transport of some carbon dioxide.

5. Some white blood cells are scavengers of dead or worn-out cells or other debris; others serve in the defense of the body against bacteria, viruses, and other foreign agents.

6. In all vertebrates, the heart pumps blood into arteries. From there it flows into arterioles, capillaries, venules, veins, and back to the heart.

7. The mammalian heart is divided into two pumps that drive oxygen-poor blood to the lungs (pulmonary circulation) and oxygen-rich blood to the other tissues of the body (systemic circulation).

8. Heart contractions generate blood pressure. The pressure is high in the arteries but drops as blood flows through the other vessels.

9. The heart has four chambers (two atria and two ventricles). Valves separate the atrium and ventricle on each side of the heart and span the exit of each ventricle.

10. The cardiac cycle consists of one phase of contraction (systole) and relaxation (diastole),

11. The atria contract before the ventricles. Contractions are initiated by the sinoatrial node. Excitation generated there travels from the atria, to the atrioventricular node, and through a conduction system into the ventricles. Contraction of the ventricles forces blood through the circulation system.

12. Arteries are pressure reservoirs that keep blood flowing while the ventricles are relaxing. The blood pressure normally measured is arterial pressure.

13. Arterioles are control points for the distribution of different volumes of blood to different capillary beds.

14. The exchange of substances between the blood and interstitial fluid occurs in the capillary beds.

16. Venules overlap somewhat with capillaries in their function.

15. Veins are highly distensible blood volume reservoirs and are important in adjusting volume flow back to the heart. Valves in veins prevent backflow of blood to the periphery.

17. Brain centers control blood pressure by varying the rate and strength of the heartbeat and the diameter of arterioles and veins. Hormones also affect blood pressure.

18. Stopping the loss of blood (hemostasis) includes blood vessel spasm, platelet plug formation, and blood coagulation.

19. Some proteins and other organic molecules at the cell surface act as markers that identify a cell as being of a specific type. Antibodies are other proteins that act against foreign cells in the body.

    a. Red blood cells have A or B markers, or both A and B, or neither one (type O).

    b. In an agglutination response, antibodies act

against foreign blood cells (introduced through an improper blood transfusion, for example) and cause them to clump together. The clumping can clog small blood vessels and lead to other complications.

c. Red blood cells also may be Rh⁺ (with an Rh marker) or Rh⁻ (without this marker).

20. The lymph vascular system includes lymph capillaries, lymph vessels, and ducts. This system returns excess, filtered tissue fluid and some proteins to the bloodstream. The fluid being transported in this system is called lymph. It is filtered in lymph nodes and other lymph organs. Lymph vessels transport fat from the digestive tract to the blood.

Review Questions

1. What are some of the functions of blood?

2. Describe the cellular components of blood. Describe the plasma portion of blood.

3. Define the functions of the following:
   a. heart
   b. cardiovascular system
   c. lymph vascular system

4. Distinguish between the following:
   a. open and closed circulation
   b. systemic and pulmonary circulation
   c. lymph vascular system and lymphoid organs
   d. systole and diastole

5. Describe the cardiac cycle in a four-chambered heart.

6. Can you identify the cardiac muscle fibers of the cardiac conduction system and describe how they work?

7. Explain how arteries, arterioles, and capillaries help regulate blood flow to different body regions.

8. State the main function of blood capillaries. What drives solutes out of and into capillaries in capillary beds?

9. State the main function of venules and veins. What forces work together in returning venous blood to the heart?

10. Label the component parts of the human heart:

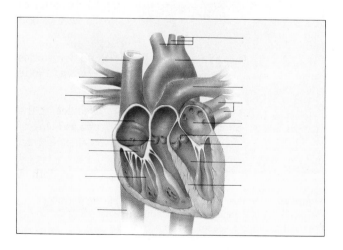

Readings

Berne, R., and M. Levy. 1983. "The Cardiovascular System." In *Physiology*. St. Louis: Mosby. Comprehensive introduction to controls over circulatory functions.

Brown, M. S., and J. L. Goldstein. November 1984. "How LDL Receptors Influence Cholesterol and Atherosclerosis.". *Scientific American* 251(5):58–66.

Doolittle, R. F. December 1981. "Fibrinogen and Fibrin." *Scientific American* 245(6):126–135. Describes clot formation and clot breakdown.

Hickman, C. et al. 1984. *Integrated Principles of Zoology*. Seventh edition. St. Louis: Mosby. Good introduction to invertebrate and vertebrate circulation systems.

Kapff, C. T., and J. H. Jandl. 1981. *Blood: Atlas and Sourcebook of Hematology*. Boston: Little, Brown. Beautiful micrographs of normal and abnormal blood and marrow cells.

Levy, M., and J. Moskowitz. 9 July 1982. "Cardiovascular Research: Decades of Progress, a Decade of Promise." *Science* 217:121–126.

Vander, A., J. Sherman, and D. Luciano. 1985. *Human Physiology*. Fourth edition. New York: McGraw-Hill. Clear exposition of the circulatory system.

Zucker, M. June 1980. "The Functioning of Blood Platelets." *Scientific American* 246(6):86–103. Everything you ever wanted to know about platelets, as of the summer of 1980.

30

IMMUNITY

Until about a century ago, smallpox swept repeatedly through cities around the world. Some outbreaks were so intense that half or more of the people who contracted that contagious disease died. The ones who survived had permanent scars on their face. Intriguingly, the readily identifiable survivors were no longer vulnerable to subsequent attacks; they were immune to smallpox.

No one knew what caused smallpox, but first in Asia, then in Africa and Europe, there were attempts to inoculate against it. ("Inoculation" means intentionally infecting a healthy person with matter taken from someone having a mild case of the disease—in this instance, matter or scar tissue from a lesion of the face.) If all went well, the inoculated person contracted only a mild case of smallpox and thereafter had immunity against it. But inoculation sometimes produced severe cases of the disease—and sometimes it triggered an epidemic.

While this immunological version of Russian roulette was going on, Edward Jenner was growing up in the English countryside. At the time, it was known that a rather mild disease called cowpox could be transmitted from cattle to humans. It also was known that people who contracted cowpox never came down with smallpox. In 1776, America declared her independence, and in a less celebrated event an English farmer, Benjamin

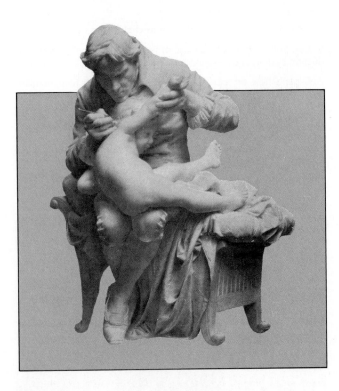

Figure 30.1 Bronze statue commemorating Edward Jenner's development of a vaccination procedure against smallpox, one of the most dreaded diseases in human history.

Jesty, inoculated his family with cowpox matter to protect them against smallpox. But it was not until 1796 that Jenner (by then a physician) scientifically demonstrated the effectiveness of the procedure, and eventually Londoners lined up by the hundreds to be inoculated. The French derided Jenner's procedure, calling it "vaccination" (which translates as "encowment"). Many years later, Louis Pasteur developed similar procedures for other diseases and also called them vaccinations, thereby bestowing respectability on the term.

Today we know that viruses, bacteria, and other microbes cause different diseases, including smallpox. We know also that the body is equipped to defend itself against diverse attacks. With his procedure, Jenner was actually mobilizing the body to make a specific immune response, one of the elegant defenses that are the topic of this chapter. Before turning to the specific responses, let's start out with the body's generalized, nonspecific defenses against invaders.

NONSPECIFIC DEFENSE RESPONSES

Barriers to Invasion

The vertebrate body has a rather impressive array of physical and chemical barriers against invasion. These barriers include the following:

1. Intact skin, which only a few bacteria can penetrate.

2. Ciliated, mucous membranes that line parts of the respiratory tract and that act like sticky brooms to sweep out bacteria and various inhaled particles.

3. Secretions from exocrine glands in surface epithelium (one such secretion is lysozyme, an enzyme that helps degrade the cell walls of many bacteria).

4. Gastric fluid in the stomach, the acids of which destroy many potential invaders.

5. Microbes that normally inhabit the gut (and, in females, the vagina), and that compete effectively with many potential microbial invaders, thereby helping to keep them in check.

Phagocytes

When the barriers listed above are breached, as when the skin is cut or scraped, then invaders encounter phagocytic cells that engulf and destroy them (Figure 6.11). In the adult vertebrate body, all phagocytes are derived from stem cells in bone marrow. (*Stem cells* are immature cells, some of which are "uncommitted" and can develop into one of several mature cell types; others are genetically committed to become one cell type only.)

Phagocytes are strategically distributed through the body. For example, some circulate through the body inside blood vessels, then squeeze out through endothelial cells making up capillary walls when they circulate past a tissue that is damaged or invaded. Some take up residence in lymph nodes and the spleen. (Here you may wish to review Figure 29.19, which shows the tissues and organs of the lymphatic system.) Other phagocytes take up residence in the tissues of the liver, kidneys, lungs, and joints. The brain has its own phagocytic glial cells.

Complement System

During an invasion, the plasma concentrations of a variety of enzymes and other proteins increase dramatically. About twenty of these plasma proteins constitute the **complement system**; through their interactions, they enhance nonspecific *and* specific defense responses.

The complement proteins circulate in inactive form until contact is made with any of a variety of bacterial and fungal cells. Then, the proteins are activated one after another in an amplifying "cascade" of reactions. Every protein molecule switched on at the start of the reaction series switches on many protein molecules at the next step, and so on through the sequence. The reactions have these consequences:

1. Some components of the system act directly on cell membranes, causing the cells to lyse (Figure 30.2).

2. Complement proteins released during the lytic reactions create chemical gradients that attract phagocytes to the scene (this being an example of chemotaxis).

3. Other components of the system attach to the surface of the invading cell, forming a coat that enhances recognition of the invader by the phagocytes.

Inflammation

Activation of the complement system is part of the **inflammatory response**, which consists of a series of events that destroy invaders and restore tissues and intracellular conditions to normal. All of these events occur during both nonspecific and specific defense responses.

Other chemical factors besides the complement proteins contribute to the inflammatory response. For example, mast cells are triggered to release *histamine*, a potent substance that causes blood vessels to dilate (vasodilation), increases capillary permeability, and induces a net outward flow of fluid from the capillaries. (*Mast cells* are derived from stem cells in bone marrow, but they nor-

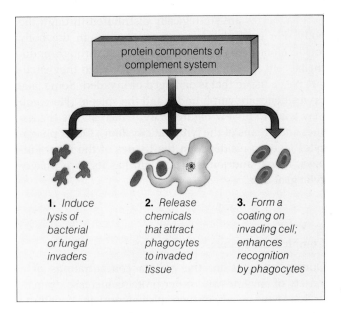

Figure 30.2 Functions of the component proteins of the complement system. These proteins function in specific as well as in nonspecific defense responses.

mally are not found in blood. They are circulated to different parts of the body, where they become lodged in connective tissues. For example, mast cells in nasal membranes are responsible for the sneezing and runny noses of people suffering an allergic reaction.) Other participants in an inflammatory response are the components of clotting mechanisms (page 399). These components work to maintain the integrity of the vascular system and to wall off infected tissues during physical damage or infection.

These and other chemical factors produce the following interrelated events, which are characteristic of the inflammatory response:

1. In response to tissue invasion or injury, small blood vessels dilate, and this increases the blood flow to the local tissue region. The vasodilation causes localized warmth and redness.

2. Local blood vessels become highly permeable to plasma proteins.

3. Net outward flow of fluid in the capillary beds causes local swelling.

4. Phagocytes migrate toward the damaged tissue, then leave the bloodstream and join the battle.

5. The invaders are destroyed (for example, by phagocytosis).

6. The tissue is repaired.

SPECIFIC DEFENSE RESPONSES: THE IMMUNE SYSTEM

The phagocytes summoned to a site of inflammation make a generalized attack response, indiscriminately engulfing cellular debris and anything foreign. But sometimes this general response is not enough to check the spread of an invader, and illness follows. Then, the specific defense responses are activated. These responses are carried out by three types of white blood cells—phagocytes, T lymphocytes, and B lymphocytes—which are the basis of the **vertebrate immune system**. As you will see, this system is characterized by its *specificity* (specific response to a particular invader, rather than generalized attack responses) and by its *memory* (its ability to mount a rapid attack when the same type of invader returns).

The Defenders: An Overview

Of every 100 cells in your body, one is a white blood cell. The ones that are central to immune responses go by these names:

1. **Macrophages.** These phagocytic cells, the "big eaters," can alert helper T cells to the invasion by a specific type of foreign agent.

2. **Helper T cells.** The master switches of the immune system, these cells stimulate the rapid divisions of the lymphocytes which together mount the counterattack.

3. **B cells.** These lymphocytes produce potent chemical weapons (antibodies).

4. **Antibodies.** These Y-shaped proteins bind specific foreign targets and thereby tag them for destruction by phagocytes or by activating the complement system.

5. **Killer T cells.** These lymphocytes directly destroy body cells already infected by viral or fungal parasites, as well as mutant and cancerous cells.

6. **Suppressor T cells.** These lymphocytes slow down or stop the immune response.

7. **Memory cells.** A portion of the B cells and T cells produced during a first-time encounter with a specific invader are held in reserve; they make possible a rapid response to subsequent encounters with that same type of invader.

Recognition of Self and Nonself

Before getting into the details of immunological battles, let's think about an important question: How do the defenders distinguish *nonself* (foreign cells and sub-

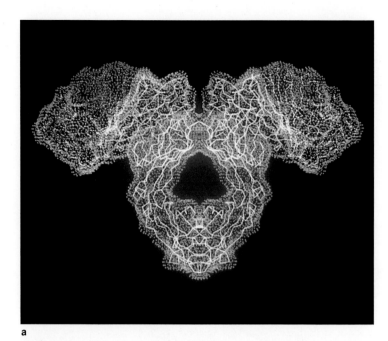

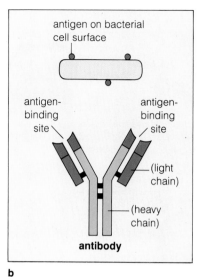

antigen on bacterial cell surface

antigen-binding site

antigen-binding site

(light chain)

(heavy chain)

antibody

b

a

Figure 30.3 Structure of antibodies. An antibody molecule consists of four polypeptide chains of different molecular weight, with disulfide and noncovalent bonds linking the four into a Y-shaped structure. The end of each arm of the Y contains a region where the surface pattern can vary. One antigen molecule can become bound to this region on each arm. The tail of the Y in membrane-bound antibodies differs from that in freely circulating antibodies.

stances) from *self* (the body's own cells)? Such recognition is essential, for the immune reactions unleashed when lymphocytes are activated are extremely destructive. We know this because on rare occasions, the distinction is blurred and lymphocytes turn on the body itself, sometimes with lethal consequences (as in type 1 diabetes, described on page 351).

All cells have "self" markers displayed somewhere on their surface. The markers for cells are usually tiny molecular patterns on the surface of proteins. The main proteins are encoded by genes known as the "major histocompatibility complex" (or MHC); hence the proteins themselves are called **MHC markers**.

Lymphocytes can identify the unique, small patterns of MHC markers. They also can identify **antigens**, which are surface patterns of molecules that are "nonself," or foreign to the body. How do they do it? Lymphocytes bear various recognition proteins at their surface that are complementary to nonself markers. This is really remarkable, given that the body may be exposed to millions (perhaps billions) of different antigens. The recognition proteins are encoded in DNA regions that can undergo recombination, with different DNA segments being randomly shuffled to produce a myriad of different proteins (page 231).

Antibodies are important recognition proteins. They belong to the general class of proteins known as immu-noglobulins (Ig), and they are manufactured only by B cells. Some become bound to the plasma membrane of B cells; others circulate freely in blood and lymph vessels.

Figure 30.3 shows the Y-shaped structure of antibodies. The arms of the Y contain binding sites for specific invaders. The tail of the Y can activate the complement system or bind to receptors on phagocytic cells.

When do the lymphocytes acquire their capacity to distinguish self from nonself? Recall that white blood cells are derived from stem cells which, in adults, are produced in bone marrow. The B cells complete their maturation in the bone marrow, synthesizing the antibodies that will take up positions at the cell surface. After the antibodies are in place, these "virgin" B cells are released into the circulation.

T cells also start their development in bone marrow, but they complete their development in the thymus gland (Figure 29.19). There, they assemble the antigen-specific proteins that will become positioned at their cell surface. There also, they are exposed to MHC markers. Only those T cells able to bind with MHC markers can leave the thymus. Before they are released, their capacity to bind with those self markers is weakened, and this prevents them from attacking the body's own tissues. However, as you will soon see, when the T cell encounters a cell that bears an antigen *and* an MHC marker, its reactivity to the MHC markers is turned on again.

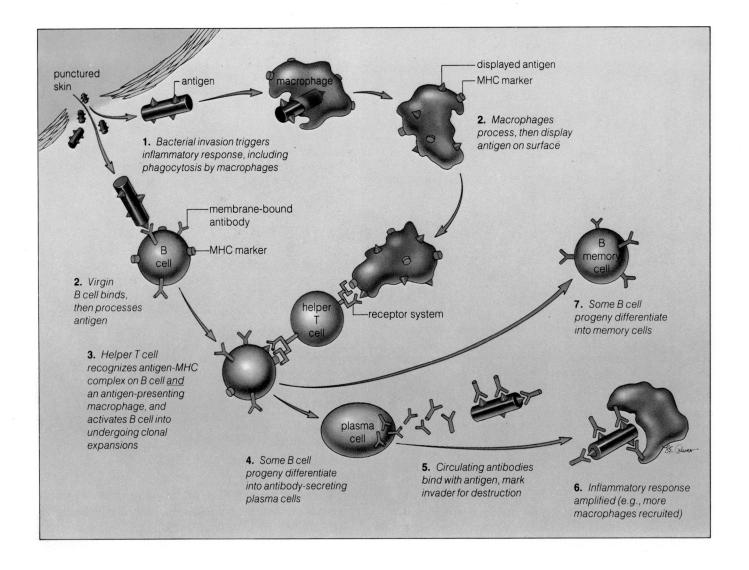

In the figure:

- **punctured skin**
- **antigen**
- **macrophage**
- **1.** *Bacterial invasion triggers inflammatory response, including phagocytosis by macrophages*
- **displayed antigen**
- **MHC marker**
- **2.** *Macrophages process, then display antigen on surface*
- **membrane-bound antibody**
- **B cell**
- **MHC marker**
- **2.** *Virgin B cell binds, then processes antigen*
- **helper T cell**
- **receptor system**
- **3.** *Helper T cell recognizes antigen-MHC complex on B cell <u>and</u> an antigen-presenting macrophage, and activates B cell into undergoing clonal expansions*
- **4.** *Some B cell progeny differentiate into antibody-secreting plasma cells*
- **plasma cell**
- **5.** *Circulating antibodies bind with antigen, mark invader for destruction*
- **B memory cell**
- **7.** *Some B cell progeny differentiate into memory cells*
- **6.** *Inflammatory response amplified (e.g., more macrophages recruited)*

Figure 30.4 Amplification of the inflammatory response by specific immune reactions. This example is of an *antibody-mediated response* to a bacterial invasion. Virgin B cells are induced to manufacture and release antibodies, which circulate and mark invaders for destruction by other defense agents, including more macrophages recruited to the battle scene.

The Primary Immune Response

When a particular antigen turns up in the body for the first time, a *primary* immune response is mounted against it. Suppose bacteria enter the body through a small cut. Inflammation ensues, and a few of the bacterial cells are engulfed by macrophages recruited to the scene. The cells are completely degraded inside the macrophage— but their *antigens* are not. Instead, the antigens are moved through the cytoplasm to the surface of the plasma membrane, where the macrophage "displays" them (Figure 30.4). *What this means is that the macrophage now bears both MHC markers and antigens at its surface.* It is this combination, and this combination only, that is recognized by helper T cells.

When it meets up with a helper T cell, the macrophage is stimulated into secreting interleukin-1, which is one of a class of proteins called **lymphokines**. *These proteins help the cells of the immune system communicate with each other.* Interleukin-1 stimulates helper T cells to

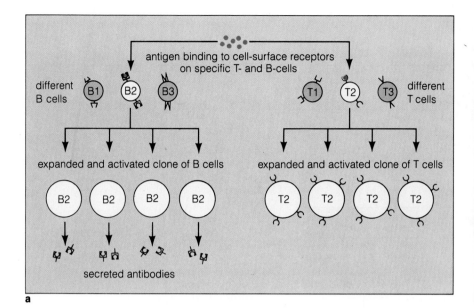

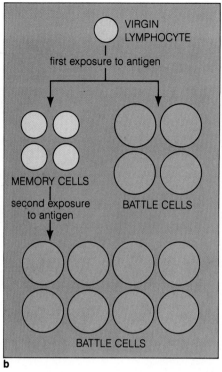

Figure 30.5 (a) Clonal selection of lymphocytes having surface proteins that are able to recognize particular antigens. The recognition proteins are produced through random shufflings of split genes during the maturation of lymphocytes. Only antigen-specific lymphocytes will become activated and will give rise to a population of immunologically identical clones.

(b) Immunological memory. During repeated cell divisions of activated lymphocytes, some clones differentiate at once into battle cells. Others are held in reserve as circulating memory lymphocytes, which will be activated by subsequent invasions by the same antigen.

secrete another lymphokine, which in turn stimulates the rapid growth and division of still more helper T cells. The secretions also cause the rapid division of B cells or killer T cells, depending on the target.

Antibody-Mediated Immune Response. Some secretions of helper T cells act on sensitized B cells. (These B cells are the ones that have recognized and processed the antigen, so that the antigen is displayed on their surface.) The B cells are stimulated to grow, divide, then differentiate into **plasma cells**. A plasma cell never does divide and it dies in less than a week. Before dying off, it secretes large quantities of antibody molecules, which circulate freely in the tissues like time bombs (Figure 30.4). Circulating antibodies cannot destroy an antigen-bearing target directly. Rather, they bind to it and mark it for disposal by macrophages or by the complement system.

The main targets of antibodies are bacteria as well as the *extracellular* phases of viruses, fungal cells, and parasitic cells. In other words, antibodies act against viruses present outside the cells, in fluids bathing the tissues or attached to the surface of a body cell; they cannot act against new virus particles being assembled inside a cell that has already been infected. Similarly, they cannot work against body cells already infected by fungi or parasites.

Cell-Mediated Immune Response. Infected, mutant, or cancerous cells are attacked directly by killer T cells. The lymphokines that are secreted by activated helper T cells also happen to stimulate the rapid growth and division of the few existing killer T cells that bear recognition proteins specific for the invader. In effect, the turned-on killer T cells destroy their targets by "punching" holes in them. Their action also is the main reason

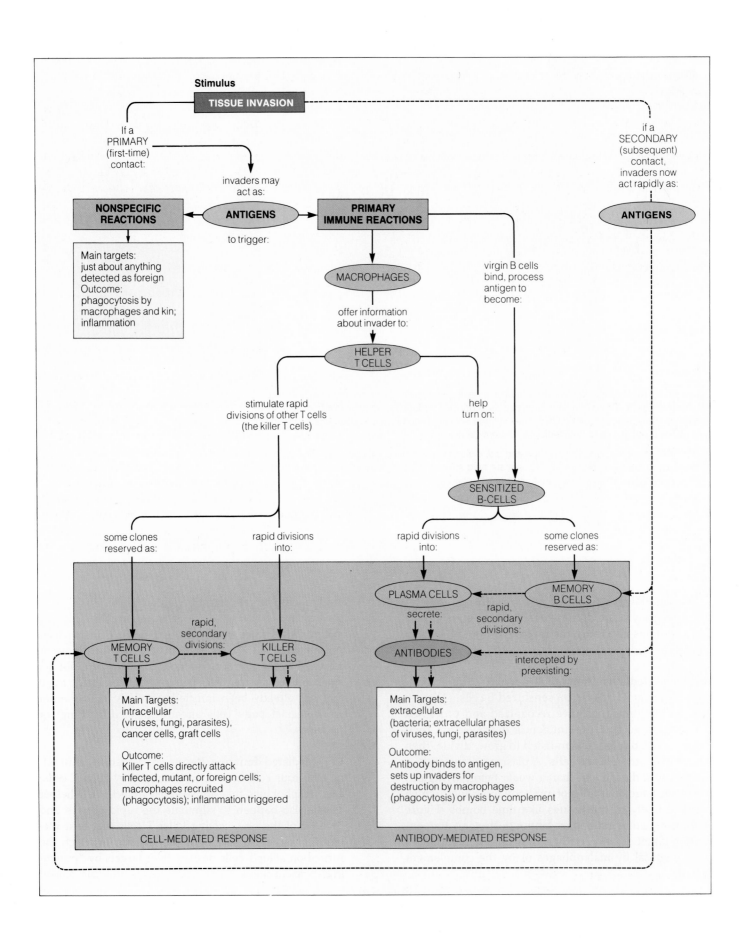

Stimulus

TISSUE INVASION

If a PRIMARY (first-time) contact:

if a SECONDARY (subsequent) contact, invaders now act rapidly as:

invaders may act as:

ANTIGENS

to trigger:

NONSPECIFIC REACTIONS

Main targets:
just about anything detected as foreign
Outcome:
phagocytosis by macrophages and kin; inflammation

PRIMARY IMMUNE REACTIONS

ANTIGENS

MACROPHAGES

virgin B cells bind, process antigen to become:

offer information about invader to:

HELPER T CELLS

stimulate rapid divisions of other T cells (the killer T cells)

help turn on:

SENSITIZED B-CELLS

some clones reserved as:

rapid divisions into:

rapid divisions into:

some clones reserved as:

PLASMA CELLS

MEMORY B CELLS

secrete:

rapid, secondary divisions:

rapid, secondary divisions:

MEMORY T CELLS

KILLER T CELLS

ANTIBODIES

intercepted by preexisting:

Main Targets:
intracellular
(viruses, fungi, parasites),
cancer cells, graft cells

Outcome:
Killer T cells directly attack infected, mutant, or foreign cells; macrophages recruited (phagocytosis); inflammation triggered

Main Targets:
extracellular
(bacteria; extracellular phases of viruses, fungi, parasites)

Outcome:
Antibody binds to antigen, sets up invaders for destruction by macrophages (phagocytosis) or lysis by complement

CELL-MEDIATED RESPONSE

ANTIBODY-MEDIATED RESPONSE

why tissue and organ transplants often fail to take hold. The killer T cells enter the transplanted tissue through its blood vessels, then attach themselves to individual cells and kill them.

The body will normally tolerate tissue and organ transplants between identical twins (who have identical sets of DNA, hence identical MHC regions). But not everyone has an identical twin available and willing to donate replacements. The odds are against transplants between unrelated individuals taking hold. Currently, T cell populations of organ recipients are bombarded with drugs, which kill the transplant-fighting killer T cells. Unfortunately, other kinds of cells are killed along with them. Also, this measure can severely compromise the body's ability to mount counterattacks against other invaders. That is the reason why pneumonia (caused by the bacterium *Diplococcus pneumoniae*) is the leading cause of death among transplant recipients.

Clonal Selection Theory

At the very least, there are millions of different recognition proteins specific for millions of different antigens. They are assembled in each developing lymphocyte through *random* recombination of different DNA segments. Thus it is not that individuals inherit a limited war chest from their ancestors, useful only against types of invaders that were successfully fought off in the past. Even if an entirely new antigen is encountered (as might occur through mutation in *Influenzavirus*), it may be that random recombinations in a maturing lymphocyte produced exactly the protein surface pattern that can lock in on the invader. By happy accident, that particular lymphocyte has the precise weapon needed.

According to the **clonal selection theory**, first proposed by Macfarlane Burnett in 1955, a lymphocyte is activated when an antigen binds with one of its recognition proteins. The lymphocyte multiplies rapidly, and its descendants retain the specificity against that antigen. They constitute a *clone* of cells that are immunologically identical for the antigen that "selected" them (Figure 30.5a).

The clonal selection theory also explains how an individual has "immunological memory," which is the basis of a more rapid response to a subsequent invasion by the same antigen. A *primary* immune response, recall, is the one made upon an individual's first exposure to an antigen. Full mobilization of the body's defense system usually takes five or six days. In contrast, a *secondary* response to the same antigen can occur in a matter of two or three days, and this response is greater and of longer duration.

In the primary response to an antigen, only part of the clonal population of lymphocytes that forms is unleashed. Another part is held in reserve as a battalion of memory lymphocytes. The memory cells are not fully differentiated. They may circulate for years (decades, in some cases). However, if the antigen appears again, they rapidly differentiate and give rise to a large clone of active lymphocytes, thereby amplifying the defenses against it (Figure 30.5b).

Figure 30.6 summarizes the pathways of the primary and secondary immune responses.

IMMUNIZATION

Jenner didn't know why his cowpox vaccine provided immunity against smallpox. Today we know that the viruses that cause the two diseases are related, and they bear the exact same antigen on their surface. Let's express what goes on in modern terms.

Specific immune responses can be triggered by intentionally introducing a particular antigen into the body, the idea being that the body will mount a counterattack and develop an army of memory lymphocytes. Deliberately provoking the production of memory lymphocytes is now called **immunization**. The preparation used to provoke their appearance is known as a **vaccine**. Typically, preparations of some killed or weakened bacterium or virus are injected into the body or taken orally. For instance, Sabin polio vaccine is prepared from live but weakened poliovirus. Other vaccines are prepared from the toxic but inactivated by-products of dangerous organisms, such as the bacteria that cause the disease tetanus (page 320).

A primary immune response occurs after the first injection of a vaccine. A secondary immune response occurs after a second injection (the "booster shot"),

Figure 30.6 Summary of primary and secondary immune pathways: the key events, key participants, and their interactions. Solid lines show primary routes; dashed lines show secondary routes. All events inside the brown-shaded area may also proceed directly upon subsequent encounters with the same kind of invader; hence secondary immune pathways lead more rapidly to the disposal of foreign matter. In the secondary pathways also, macrophages offer information about the invader to helper T cells.

which stimulates the production of antibodies to levels that will impart resistance to the disease.

Through the injection of antibodies themselves, **passive immunity** may be conferred on people who have been exposed to some infectious bacterial diseases (for example, diphtheria, tetanus, and botulism). The antibodies come from individuals who have recovered from the disease or from animals that have been immunized and that are producing the required antibodies. Because the person's own lymphocytes are not producing the antibodies, the effects of passive immunization are not lasting. Still, the injections may help the individual through the immediate attack.

WHEN THE IMMUNE SYSTEM BREAKS DOWN

The immune system is a marvel of surveillance, a restorer of homeostatic balances that have been upset by all manner of threats. The system defends against viruses and bacteria, fungi and parasites. It cleans up bits of dirt and old cell structures, the debris of day-to-day living. It monitors and disposes of cancer cells, which slip from their tissue niche, lodge in improper places, and crowd normal cells to death with their abnormal growth and division. The amazing thing is how *well* the immune system works, given the diversity and unpredictable nature of the tissue invasions to which it responds. Yet the more we learn about the immune system, the more formidable seems the challenge of setting things right when the system breaks down, as the following examples and the *Commentary* will illustrate.

Allergies

Sometimes the immune system can damage the body instead of protecting it. An **allergy** is an altered secondary immune response to a normally harmless substance, and it can cause tissue damage. Some allergic reactions occur explosively within minutes; others are delayed. About fifteen percent of the human population has a genetic predisposition to become sensitized to dust, pollen, insect venom or secretions, drugs, certain foods, and other seemingly innocuous substances. These people produce unusually large amounts of *IgE* antibodies, a class of antibodies that gives rise to many of the symptoms of immediate allergic reactions.

Emotional state, changes in air temperature or pressure, infections, and other conditions can trigger or complicate reactions to dust and other substances that the body recognizes as antigens. With each recurring exposure to the antigen, molecules of IgE antibodies are pro-duced and become attached to cells that release histamine, prostaglandins, and other substances. Histamine causes increased capillary permeability and mucous secretions. Prostaglandins constrict smooth muscle in assorted organs, and prostaglandins and other substances cause platelet clumping. Together, the substances initiate a local inflammatory response. In *asthma* and *hay fever*, the resulting symptoms include a drippy nose, sneezing, congestion, and labored breathing.

In a few hypersensitive individuals, the substances are released in copious amounts, and the inflammatory response can be explosive and life-threatening. There can be massive constriction of air passages leading to the lungs. Circulatory shock can occur when plasma escapes rapidly from capillaries that have become too permeable. That is why people who are hypersensitive to, say, wasp or bee venom can die within minutes following a secondary immune response to a single sting from these insects.

Acquired Immune Deficiency Syndrome

On rare occasions, the body's natural, cell-mediated immunity is weakened, and the individual becomes extremely vulnerable to infections that might not otherwise be life threatening. This is what happens in *acquired immune deficiency syndrome*, or AIDS (Figure 30.7). Individuals with AIDS become vulnerable to so-called opportunistic infections, including Kaposi's sarcoma (a rare type of cancer) and pneumocystosis (a rare lung infection caused by *Pneumococcus carinii*).

The AIDS virus probably emerged in Central Africa within the past two decades, apparently as a mutant form of a harmless virus that occurs in the bloodstream of the green monkey. At some point, the mutant virus was transmitted to human hosts.

Among the more than 22,000 cases of AIDS diagnosed so far in the United States, 73 percent are homosexual or bisexual males. Intravenous drug abusers, some Haitian immigrants, and some hemophiliacs who received contaminated blood transfusions initially represented the remaining afflicted persons, but now AIDS is being spread by heterosexual relations. Female prostitutes constitute an alarming reservoir for infection. In some recent studies, nearly 60 percent of the prostitutes in Nairobi (Kenya), 40 percent in Miami, 19 percent in New York, and 6 percent in Los Angeles were found to carry antibodies to the virus and are presumed to be infectious. The World Health Organization predicts there may be as many as 100 million infected people (including carriers) by 1990.

In the spring of 1984, researchers traced AIDS to a retrovirus, which has been designated "human immu-

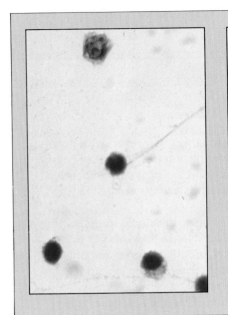
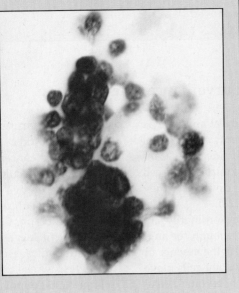

Figure 30.7 (**a**) Normal T cells and (**b**) T cells infected with the virus responsible for AIDS.

a   b

nodeficiency virus" (HIV). It now appears that the virus is transmitted only through blood or through intercourse with an infected person. On the basis of studies to date, AIDS does not appear to be transmitted in food or water, by insect bites, or by casual contact.

The HIV virus leads to alterations in the growth and functioning of lymphocytes called T4 cells. These lymphocytes serve as helper T cells (which turn on antibody-secreting B cells). They also secrete a substance that bolsters the killer T cells; and they produce a form of interferon that prods macrophages into engulfing the virus and into presenting antigen at their surface. Ordinarily, the T4 cells make up between sixty to eighty percent of the circulating T cell population. Obviously they are vital to immunity.

After becoming infected by HIV, the growth of T4 cells slows down. Some infected cells survive and harbor the virus in the latent state. In some way, the virus masks the surface markers that otherwise would inform a macrophage of its presence. In time, the T4 cell population is depleted and the body's ability to mount immune responses against opportunistic infections deteriorates.

Recently, researchers have modified the harmless vaccinia virus (used in smallpox vaccinations) by incorporating an HIV gene and have coaxed it into producing an HIV viral protein. Mice inoculated with the recombinant hereditary material produced antibodies to the HIV protein. Work along these lines may soon produce a vaccine against the virus. For those already infected, however, immune functioning has already been severely compromised.

Case Study: The Silent, Unseen Struggles

Let us conclude this chapter with a case study of how the immune system helps *you* survive attack. Suppose it is a warm spring day and you are walking barefoot across the lawn to class. Abruptly you stop: one of your toes acquired a staple that had been lying prong-upward. You pull out the staple right away, but the next morning the punctured area is red, tender, and slightly swollen. Later on, the swelling begins to subside. Within a few more days your foot is back to normal and you have forgotten the incident.

All that time, your body had been struggling against an unseen enemy. Your toe happened to land on a spot where a diseased bird had fallen the night before and had lain until a scavenging animal carried it off. Your toe came in contact with some bacteria that had parted company from their feathered host—and the prongs of the staple had carried several thousand bacterial cells into the moist, warm flesh beneath your skin.

Within your tissues the bacteria found conditions suitable for growth. And grow they did—they soon doubled in number and were on their way to doubling again. At the same time, they also were releasing metabolic by-products that interfered with your own cell functioning. Left alone, they would have threatened your life.

Yet even as the staple penetrated your foot, events were set in motion that would mobilize your body's defenses. Around the wound, some blood had escaped from ruptured capillaries and it now pooled and clotted. Local tissue cells began releasing histamine and other

Cancer and the Immune Surveillance Theory

Cancer cells might arise in your body at any time as a result of mutations induced by events such as viral attacks, chemical bombardment, or irradiation. "Cancer" refers to cells that have lost controls over cell division. Through their berserk divisions, they destroy surrounding tissues (page 226).

As fast as cancer cells arise, they ordinarily are destroyed before they can multiply enough for anybody to detect them. That, anyway, is the essence of the *immune surveillance theory*. The mutant cells move past circulating lymphocytes, which become sensitized to them. The activated lymphocytes multiply rapidly. Some become killer T cells and others release substances that activate macrophages and enable them to destroy any invaders and mutants—including cancer cells.

Sometimes, though, cancer cells slip through the surveillance net. Maybe the mutation doesn't affect the cell surface markers; or maybe the markers become chemically disguised. Perhaps they are even released from the cell surface and begin circulating through the bloodstream to lead the immune fighters down false trails. Sometimes, too, individuals are not genetically equipped to respond to a particular antigen. The age of the individual and the overall state of health also seem to play a role in resistance to cancer.

At present, surgery, drug treatment (chemotherapy), and irradiation are the only weapons against cancer. Surgery is effective when a tumor is fully accessible and has not spread, but surgery offers little hope when cancer cells have begun wandering (page 226). When used by themselves, chemotherapy and irradiation destroy good cells as well as bad. *Immune therapy* is a promising prospect. The term refers to methods of inducing mobilization of the immune system by deliberately introducing agents that will set off the immune alarm.

Interferons, a group of small proteins, are candidates for immune therapy. Most cells produce and release interferon following a viral attack. The interferon binds to the plasma membrane of other cells in the body and induces resistance to a wide range of viruses. But interferon is species-specific; the interferon from one type of organism cannot normally be used to fight viruses in others. However, with gene-splicing methods, enough interferon is now being produced for cancer research. The goal is to develop interferon molecules with activity tailored against cancer-inducing viruses.

Monoclonal antibodies may also hold promise for immune therapy. Cesar Milstein and Georges Kohler developed a means of producing large amounts of pure antibodies. First they isolated malignant B lymphocytes, whose uncontrolled cell divisions yielded a large clonal population. Then the researchers developed a population of antibody-producing B cells by immunizing a mouse with a specific antigen. Later, B cells were extracted from the mouse spleen and were fused with the malignant lymphocytes. Some of the hybrid cells and their progeny multiplied rapidly, as did the malignant parent, and produced vast quantities of the same type of antibodies as the parent B cells.

Clones of such hybrid cells can be maintained indefinitely and they continue to make antibody. All of the antibody molecules are identical, and all are derived from the same parent cell. Hence the name "monoclonal antibodies."

Monoclonal antibodies are now used in passive immunization against malaria, influenza, and hepatitis B. They also are candidates for cancer therapy. By using scanning machines along with radioactively labeled monoclonal antibodies that are specific for certain types of cancer, it is possible to home in on the exact location of cancer in the body. Such scans indicate whether cancer is present, its location, and its size.

Monoclonal antibodies might also help overcome one of the major drawbacks to drug treatment of cancer. Such treatments have severe side effects because the drugs used are highly toxic and cannot discriminate between normal cells and cancerous ones. A current goal is to hook up drug molecules with a monoclonal antibody. As Milstein and Kohler have speculated, "Once again the antibodies might be expected to home in on the cancer cells—only this time they would be dragging along with them a depth charge of monumental proportions." Such is the prospect of targeted drug therapy.

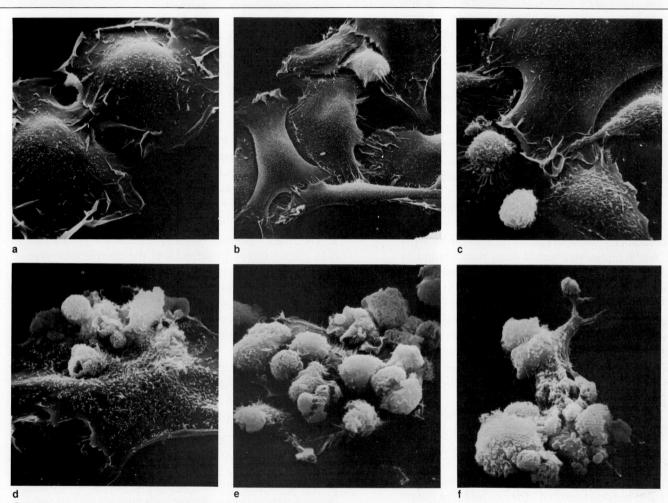

Scanning electron micrographs of the birth and immune-mediated death of cancer cells. (**a**) Cancer cell dividing in a culture dish, magnified about 2,000 times. (**b**) After fifty-six hours, the progeny are crowding each other. (**c**) Even with cancer cells all around them, macrophages (smaller spheres) ignore the invaders; they have not yet been turned on. (**d,e**) Activated macrophages cluster around a tumor cell. Recent work by John Hibbs at the University of Utah suggests that macrophages transfer some substance into the invader. (**f**) One or two days after the transfer, the cancer cell dies—and the macrophages ingest the leftovers.

Therapy for allergy sufferers includes reliance on "blocking antibodies." Once the allergy-producing substance has been identified by tests, it can be injected periodically into the body in increasing amounts. The injections provoke the production of another class of antibodies (the IgG antibodies). Thus, circulating antibodies are available to bind with and inactivate the substance before it interacts with IgE to produce the inflammatory response.

potent chemicals that triggered vasodilation and increased permeability to plasma proteins. The complement system was activated, and phagocytes were summoned to the battleground.

Now phagocytes in the bloodstream attached themselves to the walls of blood vessels and crawled along, amoebalike, until they reached a cleft between endothelial cells. Then they slipped out onto the tissue and crept about between cells and fibers, following the concentration gradient of complement fragments. Like bloodhounds on the trail, they moved in the direction of higher concentration and began engulfing any particle not having an MHC marker that meant, "I'm *self*." Dirt, rust, bits of broken host cells, and bacteria were engulfed indiscriminately.

If bacteria had not entered the wound, or if they had been of a type that could not multiply rapidly in the tissue, then phagocytosis and inflammation would have cleaned things up. As it was, bacterial divisions were outpacing the nonspecific immune responses. It was time to call in the more discriminatory defenders: the lymphocyte lines and their products.

If this had been your first encounter with the bacterial species, few lymphocytes would have been around to respond to the call. A primary immune response would have been mounted, and it would have been five or six days before the lymphocytes divided enough times to produce enough antibody to control the invasion. But when you were a child, your body did fight off the same type of bacterium and it still carries vestiges of the struggle—memory lymphocytes. When the bacterial species showed up again, it encountered a lymphocyte trap ready to spring.

As inflammation progressed, lymphocytes were among the white blood cells creeping out of the bloodstream. Most were specific for other types of antigens and did not take part in the battle. But antigen-specific memory lymphocytes entering the area locked onto the antigens and became activated. They moved into lymph vessels with the foreign cargo, tumbling along until they reached a lymph node and were filtered from the fluid. For the next few days, memory cells with bound antigens steadily accumulated in the node. They divided several times a day, so that the number of lymphocytes and their products increased rapidly.

For the first two days the bacteria appeared to be winning, for they were reproducing faster than phagocytic cells, antibody, and complement were destroying them. By the third day, antibody production reached its peak and the tide of battle turned. For two weeks or more, antibody production will continue in your body until every last bacterium is destroyed. When production finally shuts down, memory lymphocytes will go on circulating, prepared for some future struggle.

SUMMARY

1. The vertebrate body is equipped for the following tasks:

a. *Defense* against a tremendous variety of viruses, bacteria, fungi, and parasites that can invade the body, cause tissue damage, and eventually kill the animal if allowed to multiply unchecked.

b. *Extracellular housekeeping* that helps maintain the tissue environment by eliminating dead or damaged body cells and structures.

c. *Immune surveillance* of tissues, such that malignant cells are selectively eliminated if and when they do appear in the body.

2. "External" lines of defense against invasion include the barrier afforded by intact skin; lysozymes and other secretions from exocrine glands in surface epithelium; gastric fluid; microbial inhabitants of the gut, which outcompete many invaders; and ciliated, mucous membranes that trap and lead to the expulsion of invaders entering the respiratory tract.

3. The initial "internal" lines of defense are phagocytic cells (which engulf invaders) and the complement system. Phagocytes are distributed through the body; many are present in lymph nodes and in the spleen. Proteins of the complement system, which circulate in inactive form, act against bacterial and fungal cells in three ways:

a. Some cause the plasma membrane of the invading cell to lyse.

b. Complement proteins released during the lytic reactions attract phagocytes (by chemotaxis).

c. Some proteins coat the surface of the invading cell, enhancing its recognition by phagocytes.

4. In the *inflammatory response*, the phagocytes, complement proteins, and other chemical factors are mobilized in coordinated ways that destroy invaders, then restore tissues and intracellular conditions to normal. This coordinated activity occurs during both nonspecific and specific responses to invasion.

5. The vertebrate immune system has *specificity* (a specific response is made to a particular invader, rather than being a generalized attack response) and *memory* (a much more rapid attack can be mounted against any subsequent encounter with the same type of invader).

6. Immune responses are carried out by coordinated activities among macrophages and lymphocytes (the helper T cells, antibody-producing B cells, killer T cells, suppressor T cells, and memory cells). These cells communicate with one another by chemical secretions (notably lymphokines), which lead to rapid growth and division of B cells and killer T cells into large armies against particular invaders.

7. MHC markers on the surface of the body's own cells allow the lymphocytes to distinguish "self" from "non-self" (foreign agents).

8. Lymphocytes are precommitted to recognizing a particular antigen (a molecular pattern on the surface of viruses or cells that are foreign to the body).

9. An *antibody-mediated immune response* proceeds as follows against bacteria and extracellular phases of viruses, fungal cells, and parasitic cells:

    a. Macrophages and B cells become sensitized to a particular antigen; they display an antigen as well as an MHC marker at their cell surface.

    b. Helper T cells recognize the antigen-MHC complex on the macrophages and B cells, then stimulate the B cells to grow, divide, and develop into plasma cells, which secrete quantities of antibody molecules.

    c. Antibodies (proteins of the class called immunoglobulins) bind to specific antigens and thereby mark the invader for disposal (by macrophages or by the complement system).

10. A *cell-mediated immune response* proceeds as follows against infected, mutant, or cancerous cells:

    a. Macrophages engulf and destroy a target cell, then display the antigen along with their own MHC marker.

    b. Helper T cells recognize the antigen-MHC complex on the macrophage surface and activate killer T cells precommitted to act against the particular antigen. The killer T cells are stimulated into rapid growth and division.

    c. The killer T cells directly attack and destroy the target cells by "punching" holes in them.

11. Following a primary (first-time) immune response, a portion of the armies of B lymphocytes and T lymphocytes are held in reserve. These cells continue circulating and are available for a rapid, amplified response to subsequent invasion by the same antigen (a secondary immune response).

12. Immunization is the deliberate provocation of memory lymphocytes by a vaccine (typically, a preparation of some killed or weakened bacterium or virus).

Review Questions

1. Sponges aside, all animals have defenses against attack on normal cell functioning. Define the cells and products that constitute the vertebrate immune system of defense against such attacks.

2. Define inflammation. What happens during an inflammatory response?

3. Define antibody, antigen, and memory lymphocyte.

4. What is an immunization? A vaccine?

5. How do the main targets and outcomes of nonspecific and specific immune responses differ?

Readings

Bastian, B. 1986. "AIDS: Education Against Fear." *Carolina Tips.* Burlington, North Carolina: Carolina Biological Supply Company.

Buisseret, P. August 1982. "Allergy." *Scientific American* 247(2):86–95. Describes the biochemical changes associated with an allergic response.

Cooper, E. 1982. *General Immunology.* New York: Permagon Press. Contains an extensive discussion of the evolution of the immune response.

Edelson, R. L., and J. M. Fink. June 1985. "The Immunologic Function of Skin." *Scientific American* 252(6):46–53.

Golub, E. 1981 *The Cellular Basis of the Immune Response.* Second edition. Sunderland, Massachusetts: Sinauer Associates. A readable account of cellular immunology. Paperback.

Krahenbuhl, J., and J. Remington. 1978. "Belligerent Blood Cells: Immunotherapy and Cancer." *Human Nature* 1(1):52–59. Well-written article on immunotherapy as one treatment for cancer.

Leder, P. May 1982. "The Genetics of Antibody Diversity." *Scientific American* 246(5):102–115. Describes how a few hundred units of split genes can be shuffled and recombined to make billions of different antibodies.

Marrack, P., and J. Kappler. February 1986. "The T Cell and Its Receptor." *Scientific American* 254(2):36–45.

Roitt, I., J. Brostoff, and D. Male. 1985. *Immunology.* St. Louis: Mosby. Lavishly illustrated introduction to basic and clinical immunology.

Tizard, I. 1984. *Immunology: An Introduction.* Philadelphia: Saunders. Excellent introduction to immune function.

Tonegawa, S. October 1985. "The Molecules of the Immune System." *Scientific American* 253(4):122–131. Excellent survey article.

31

RESPIRATION

In most animals, energy stored in food becomes available through aerobic metabolism, which requires oxygen. Most aquatic animals get oxygen from the water, land-dwelling animals get it from the air. In both cases, oxygen first crosses the body surface before it reaches individual cells. Carbon dioxide wastes of aerobic metabolism also cross the body surface and are dissipated in the surroundings.

Acquiring oxygen and eliminating carbon dioxide from the body as a whole is called **respiration**. Although the systems of respiration are diverse, all are alike in their reliance on diffusion. To be sure, in many systems the gas molecules are swept along by bulk flow, as when they are sucked into and exhaled from the body or when they hitch rides on the blood circulation highways. But recall that bulk flow has one overriding function in the body: it enhances the *rate of diffusion* between the environment and living cells.

SOME PROPERTIES OF GASES

Diffusion of any substance depends largely on differences in its concentration between two regions. For gases, the differences are expressed in terms of pressure. The more concentrated the molecules of a gas are outside the body, the higher the pressure and the greater the force available to drive individual molecules inside, and vice versa.

Air is a mixture of gases. At sea level, the gases are present in these amounts:

nitrogen	78 percent by volume
oxygen	21
argon, other gases	0.97
carbon dioxide	0.03

Figure 31.1 Countercurrent flow in most fish gills. (**a**) Location of gills beneath the operculum (a bony cover), which has been removed for this sketch. (**b,c**) Each gill filament has an incoming and an outgoing blood vessel. Capillary beds, arranged in thin membrane folds (lamellae), connect the two. (**d**) Blood flow from one vessel to the other in these beds runs counter to the direction of water flow.

water in

water out

a

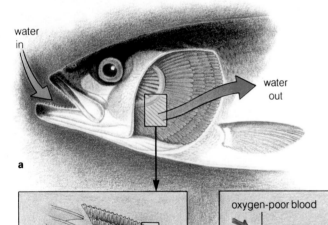

gill arch

blood vessels in gill filament

b

oxygen-poor blood

oxygen-enriched blood

direction of water flow past lamellae

c

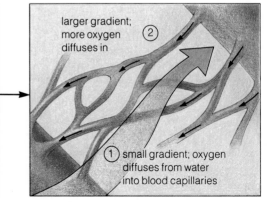

larger gradient; more oxygen diffuses in ②

① small gradient; oxygen diffuses from water into blood capillaries

d

Because individual gas molecules are so spread out in air, we can think of each gas and the pressure it exerts as being independent of all the others. **Partial pressure** is the pressure exerted by one gas in a mixture of gases. Partial pressures can simply be added to give us the total pressure for the gas mixture.

At sea level, atmospheric pressure is about 760mm Hg, as measured by a mercury barometer:

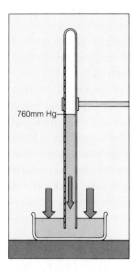

760mm Hg

Atmospheric pressure as measured by a mercury barometer. At sea level, the level of mercury (Hg) in a glass column is about 760 millimeters (29.91 inches). At this level, the pressure exerted by the column of mercury equals atmospheric pressure outside the column.

Oxygen represents about 21 percent of the total atmospheric pressure, so the partial pressure of oxygen is (760 × 21/100), or about 160mm Hg. The partial pressure of carbon dioxide in air is about 0.3mm Hg.

Any gas will diffuse from a region of higher partial pressure to a region of lower partial pressure. For example, oxygen diffuses from the air into surface waters of lakes and seas (where its partial pressure is only about 0.5 percent), and thereby replenishes oxygen supplies in aquatic habitats. Diffusion also figures in the movement of dissolved gases into and out of cells, extracellular fluid, and the bloodstream of animals.

RESPIRATORY SURFACES

In the simplest respiratory system, gases are exchanged across the body surface as a whole. In others, they are exchanged at specialized respiratory surfaces called gills, tracheas, and lungs (Figures 31.1 through 31.4).

Integumentary Exchange

Many animals do not have massive bodies or high metabolic rates. Their demands for gas exchange are not great, and the outer layer of the body itself provides enough of a respiratory surface. These animals rely on **integumentary exchange**, in which gases diffuse through a thin, vascularized layer of moist epidermis at the body surface. In aquatic animals, the layer is kept moist by the surrounding water; in land dwellers, it is kept moist by body secretions.

For example, earthworms secrete mucus that helps moisten the integument, which is a single layer of epidermal cells. Oxygen molecules between soil particles dissolve in the mucus and diffuse across the integument. From there, oxygen diffuses into blood capillaries that project, fingerlike, between the epidermal cells. Pressure generated by muscular contractions of the body wall and by the pumping action of tiny "hearts" causes blood to circulate in the narrow, tubelike body. The bulk flow enhances the diffusion rates for individual cells.

Most annelids, some small arthropods, and mollusks called nudibranchs rely on integumentary exchange. To a large extent, so do amphibians. But the integument of other animals is too thick, is too hardened, or has too few blood vessels to serve as an efficient respiratory surface. Moreover, their integument alone cannot provide enough surface area for adequate gas exchange, and specialized respiratory organs are required.

Gills

The respiratory organ called a **gill** typically has a moist, thin, vascularized layer of epidermis that functions in gas exchange. Larval forms of a few fishes, amphibians, and some insects have *external* gills projecting from the main body mass. Adult fishes have *internal* gills: a series of slits or pockets that originate in the pharynx (a muscular tube at the back of the mouth) and extend to the body surface. Water enters the mouth, moves down the pharynx, and flows out across the gills (Figure 31.1).

Not much oxygen is dissolved in water compared to air, and oxygen takes about 300,000 times longer to diffuse through water, which is more dense and viscous than air. Fish gills can take up adequate oxygen under these conditions. An extensive network of blood vessels services the epidermal surface of the gills. Because of the way gills are oriented, water flows over them in one direction and blood is circulated in the opposite direction. Such movement of fluids in opposing directions is called **countercurrent flow**.

Water passing over a fish gill first encounters the domain of a vessel that transports oxygen-rich blood *into* the body. The partial pressure of oxygen in this region is lower than it is in the water, so some oxygen diffuses in at this time. Then, just before the water moves completely past the gill, it passes over the domain of a vessel carrying blood *from* deep body regions—and this blood has less oxygen than the (by now) oxygen-poor water. With this even greater pressure difference, more oxygen diffuses inward (Figure 31.1d).

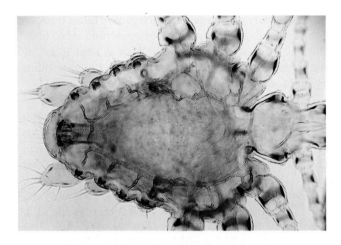

Figure 31.2 Tracheal system of a louse. Notice the spiracles leading inward from the body surface.

Figure 31.3 Closer look at the air-conducting tubes in the louse body.

Figure 31.4 Evolution of vertebrate lungs and swim bladders. The esophagus (a tube leading to the stomach) is shaded gold; the respiratory tissues, pink.

Lungs originated as pockets off the anterior part of the gut; they increased the surface area for gas exchange in oxygen-poor habitats. In some lineages, lung sacs became modified into swim bladders: buoyancy devices that help keep the fish from sinking. Adjusting gas volume in the bladders allows fishes to remain suspended at different depths.

Trout and other less specialized fishes have a duct between the swim bladder and esophagus; they replenish air in the bladder by surfacing and gulping air. Most bony fishes have no such duct; gases in the blood must diffuse into the swim bladder. Their swim bladder has a dense mesh of blood vessels (rete mirabile) in which arteries and veins run in opposite directions. Countercurrent flow through these vessels greatly increases gas concentrations in the bladder. Another region of the bladder allows reabsorption of gases by the body tissues.

Tracheas

Insects, centipedes, some mites, and some spiders use air-conducting tubes called **tracheas** for gas exchange. In most land-dwelling insects, these chitin-reinforced tubes branch finely through the body and provide a rather self-contained system of gas conduction and exchange; assistance by a circulatory system is not required.

Two main tracheal trunks run the length of the insect body (Figures 31.2 and 31.3). In most insects, smaller air-conducting tubes extend from the trunk lines to openings, called spiracles, at the body surface. Many species have a lidlike structure that spans each opening and prevents water loss through evaporation.

Some of the larger and more active insects move their body in ways that enhance air movement through the tracheal system. Have you ever noticed that foraging bees stop every so often and pump the segments of their abdomen back and forth? The segments extend and retract like a telescope, forcing air into and out of the spiracles. The stepped-up oxygen intake and carbon dioxide removal help support the high rate of metabolism required for insect flight.

Lungs

A **lung** is an internal respiratory surface that is generally in the shape of a cavity or sac. Scorpions, trap-door spiders, tarantulas, and a few other invertebrates have lunglike cavities. Fishes that lived more than 450 million

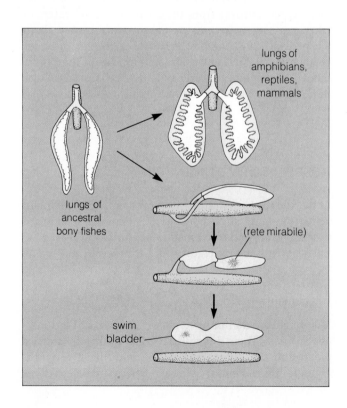

lungs of amphibians, reptiles, mammals

lungs of ancestral bony fishes

(rete mirabile)

swim bladder

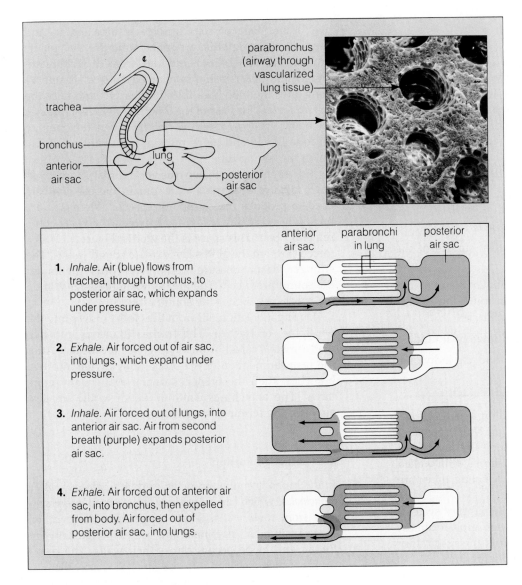

Figure 31.5 Respiratory system of birds. High metabolic rates and efficient gas exchange sustain flight and other activities. The rates are possible because of a unique ventilating system.

Typically, four air sacs are attached to each bird lung, which is somewhat small and inelastic. The sacs are not respiratory surfaces; they are more like bellows.

Inhaled air travels from bronchi into large posterior air sacs. When the bird exhales, this air is forced into bronchial branches that lead to parabronchi: small tubes (open at both ends) present in vascularized lung tissue. This is the respiratory surface (see photograph).

Air is not merely drawn into bird lungs: it is drawn *through* them. Air sacs and intricate lung airways make possible a *continuous* flow of air across the respiratory surface.

Labels in figure:

parabronchus (airway through vascularized lung tissue)

trachea

bronchus

anterior air sac

lung

posterior air sac

anterior air sac — parabronchi in lung — posterior air sac

1. *Inhale.* Air (blue) flows from trachea, through bronchus, to posterior air sac, which expands under pressure.

2. *Exhale.* Air forced out of air sac, into lungs, which expand under pressure.

3. *Inhale.* Air forced out of lungs, into anterior air sac. Air from second breath (purple) expands posterior air sac.

4. *Exhale.* Air forced out of anterior air sac, into bronchus, then expelled from body. Air forced out of posterior air sac, into lungs.

years ago had simple lungs that apparently played a supplementary role in respiration (Figure 31.4). In some of their descendents, the lungs developed into moist, thin-walled swim bladders. (A swim bladder is an organ where gas volume is adjusted to maintain body position and balance; some oxygen is also exchanged with blood and the surrounding tissues.) In other descendents, the lungs became complex respiratory organs.

The evolution of lungs may be reflected in the respiration systems of existing vertebrates. African lungfish have gills, but they also use lungs to supplement respiration. In fact, they will drown if they are kept from gulping air at the water's surface. Integumentary exchange still predominates in amphibians, but they, too, supplement respiration with a pair of small lungs. In all reptiles, birds, and mammals, paired lungs are the major respiratory surfaces.

OVERVIEW OF VERTEBRATE RESPIRATION

The respiratory surface is a boundary layer between the external and internal environments. In all animals with lungs, *airways* carry gas to and from one side of the respiratory surface of the lungs, and *blood vessels* carry gas to and from the other side. Respiration itself involves the following events:

---

1. Air moves by bulk flow into and out of the lungs, and new air is delivered to the respiratory surface. These two events are called *pulmonary ventilation.*

2. Gases diffuse across the respiratory surface of the lungs.

3. Pulmonary circulation (the bulk flow of blood to and from the lung tissues) enhances the diffusion of dissolved gases into and out of lung capillaries.

4. Gases diffuse between blood and interstitial fluid.

5. Gases diffuse between interstitial fluid and individual cells.

Pulmonary ventilation occurs within a series of air-conducting tubes that branch to become shorter, smaller in diameter, and more numerous along the route from the body surface into the lungs. Numerous, thin-walled outpouchings of the smallest air-conducting tubes are the actual sites of gas exchange.

Let's focus now on the human respiratory system, for the principles governing its operation apply to almost all vertebrates. The major exception is the respiratory system of birds, which is shown in Figure 31.5.

HUMAN RESPIRATORY SYSTEM

Air-Conducting Portion

The human respiratory system is shown in Figure 31.6. Normally, air enters and leaves through two narrow channels, called **nasal cavities**, in the nose. (Some air also enters and leaves through the mouth.) Dust and other foreign particles are filtered from the air by hairs at the entrance of the channels and by cilia along the epithelial lining of the channels. Blood vessels embedded in the lining warm the incoming air, and mucous secretions moisten the air before it flows into the lungs.

From the nasal cavities, air moves into the throat, or **pharynx**, which is the entrance to both the respiratory tract and the digestive tract. The throat cavity connects with the **larynx** (which leads to the lungs) and with the esophagus (which leads to the stomach). The larynx consists of muscles and cartilages bound in elastic connec-

tive tissue. Part of the cartilage is attached to and supports the **epiglottis**, a flaplike structure that points upward and allows air to enter the larynx during breathing. When food is being swallowed, muscular contractions force the food back into the pharynx and raise the larynx against the base of the tongue. This action presses the epiglottis down, so that it partly covers the opening into the larynx and helps prevent the food from going down the wrong tube.

The larynx contains two **true vocal cords**. These thickened folds of the larynx wall contain the elastic fibers used to produce the sounds of speech. When you are breathing normally, the space between the vocal cords remains open. This space is the **glottis** (Figure 31.7). Air forced through the glottis gives rise to sound waves. The stronger the air pressure on the vocal cords, the louder the sound produced. The greater the muscle tension on the cords, the higher the sound.

During inhalation, air from the larynx moves into the windpipe, or **trachea**. The trachea branches into two main airways called **bronchi** (singular, bronchus). After the main bronchi enter the lungs, they divide into two branches, then the branches divide again many more times. The branchings continue down to the smallest airways, the terminal bronchioles.

Gas Exchange Portion

In the lungs, terminal bronchioles divide into *respiratory bronchioles*, which can have a few cup-shaped outpouchings from their walls. These outpouchings are called **alveoli** (singular, alveolus), and gas exchange between the air and blood occurs across their walls. The respiratory bronchioles lead into *alveolar ducts*. Here, the outpouchings typically are clustered together to form a larger pouch called an **alveolar sac** (Figure 31.6). Thus the gas exchange portion of the respiratory system extends from the respiratory bronchioles to the alveolar sacs.

Lungs and the Pleural Sac

Human lungs are a pair of elastic, cone-shaped organs separated from each other by the heart and other structures (Figure 31.6). They are positioned in the rib cage above the **diaphragm**, which is a muscular partition between the chest cavity (or thoracic cavity) and the abdominal cavity.

A pair of lungs contains more than 300 million alveoli surrounded by a dense mesh of blood capillaries. Collectively, the alveoli provide a tremendous surface area (about 50 to 100 square meters) for exchanging gases with the bloodstream. If the alveolar epithelium were stretched out as a single layer, it would cover the floor of a racquet ball court!

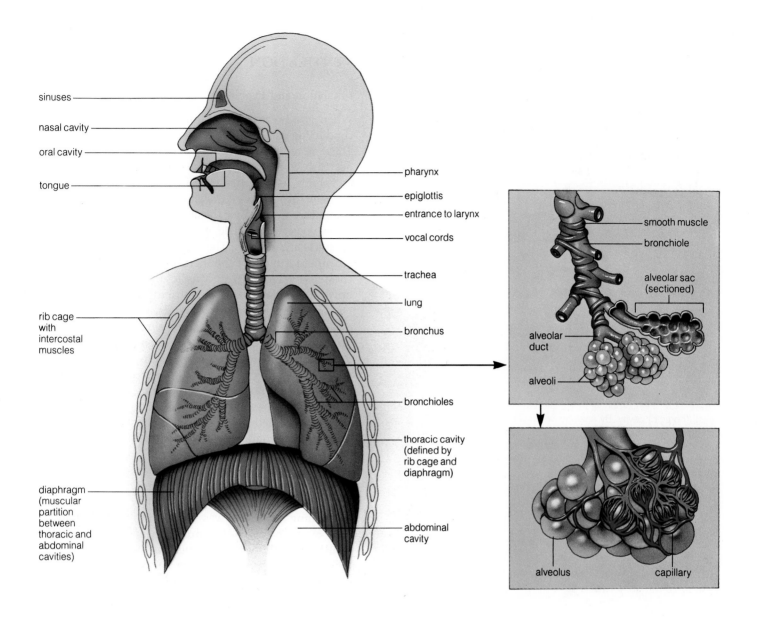

sinuses

nasal cavity

oral cavity

tongue

rib cage
with
intercostal
muscles

diaphragm
(muscular
partition
between
thoracic and
abdominal
cavities)

pharynx

epiglottis

entrance to larynx

vocal cords

trachea

lung

bronchus

bronchioles

thoracic cavity
(defined by
rib cage and
diaphragm)

abdominal
cavity

smooth muscle

bronchiole

alveolar sac
(sectioned)

alveolar
duct

alveoli

alveolus

capillary

Lungs are not attached directly to the wall of the chest cavity. Each lung is positioned within a **pleural sac** (a thin membrane of epithelium and loose connective tissue). By analogy, imagine pushing a closed fist into a fluid-filled balloon (Figure 31.8). A lung occupies the same kind of position as your fist; and the pleural membrane folds back on itself, as does the balloon. Only an extremely narrow *intrapleural space* separates the two facing surfaces of the membrane.

Part of the pleural membrane (the parietal pleura) adheres to the wall of the chest cavity. The other part (the pulmonary pleura) is firmly attached to the lungs. The facing surfaces of the membrane are coated with a thin film of lubricating fluid that prevents friction between them while you breathe. (When the pleural membrane becomes inflamed and swollen, friction does occur and breathing can be painful. This condition is called *pleurisy*.)

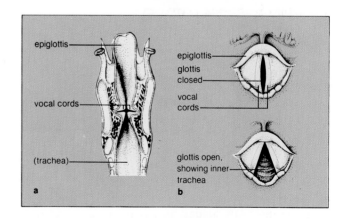

epiglottis

vocal cords

(trachea)

epiglottis

glottis
closed

vocal
cords

glottis open,
showing inner
trachea

a   b

Figure 31.7 Where the sounds necessary for speech originate. (**a**) Front view of the larynx, showing the location of the vocal cords. (**b**) The two vocal cords as viewed from above when the glottis (the space between them) is closed or opened.

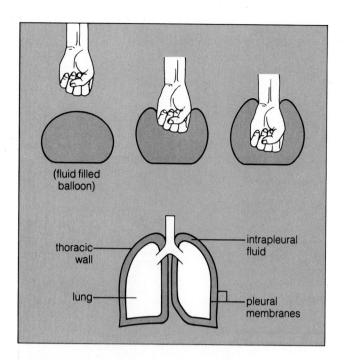

Figure 31.8 Position of the lungs and pleural sac relative to the chest (thoracic) cavity. By analogy, when you push a closed fist into a fluid-filled balloon, the balloon completely surrounds the fist except at your arm. A lung is analogous to the fist; the balloon, to the pleural sac. Here, intrapleural fluid volume is enormously exaggerated for clarity.

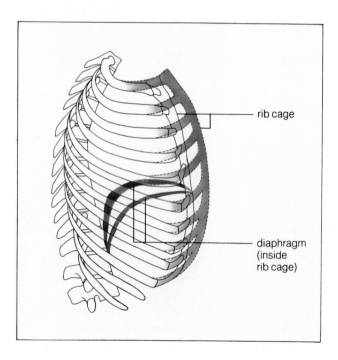

Figure 31.9 Changes in the size of the chest cavity during breathing. Blue lines indicate the position of the diaphragm and the rib cage during inhalation.

VENTILATION

Inhalation and Exhalation

When you breathe, air is inhaled (drawn into the air-conducting tubes), then exhaled (expelled from them). These air movements result from rhythmic increases and decreases in the volume of the chest cavity. The changes in volume lead to reversals in the pressure gradients between the lungs and the atmosphere, and gases in the respiratory tract follow those gradients.

As you start to inhale, your dome-shaped diaphragm contracts and flattens. The action of this muscular sheet accounts for most of the increase in the volume of the chest cavity (Figure 31.9). Muscles that move the rib cage upward and outward also contribute to the expansion.

When the chest cavity expands, the rib cage moves away very slightly from the lung surface, and pressure drops in the intrapleural space. Even between breaths the intrapleural pressure is lower than atmospheric pressure in the lungs, but now the pressure difference becomes large enough to push the lungs outward. Thus incoming air does not expand the lungs; the lungs are *already* expanded. Now fresh atmospheric air flows down the air-conducting tubes, almost to the terminal bronchioles.

As you start to exhale, the muscular contractions that brought about expansion of the chest cavity (and the lungs) have ceased. The muscles now relax, and the elastic lung tissue that was stretched during inspiration now recoils passively toward its resting volume. As a result, the volume of the chest cavity decreases and thereby compresses the air in the alveoli. With this compression, the alveolar pressure becomes greater than the atmospheric pressure, and air follows the gradient and moves out from the lungs.

Lung Volumes

When you are resting, about 500 milliliters of air enter or leave your lungs in a normal breath. This is called the "tidal volume." The maximum volume of air that can move into and out of your lungs in a single, deep breath is called the "vital capacity." You rarely use more than half the total vital capacity, even when you breathe deeply during strenuous exercise. (To do so would exhaust the muscles used in respiration.) Even at the end of your deepest breath and exhalation, your lungs still would not be completely emptied of air; about 1,000 milliliters (the "residual volume") would remain.

How much of the 500 milliliters of inhaled air is actually available for gas exchange? About 150 milliliters remain in the air-conducting tubes between breaths. Thus only (500 − 150), or 350 milliliters of fresh air reach the alveoli with each inhalation. When you breathe, say, ten times a minute, you are supplying your alveoli with (350 × 10) or 3,500 milliliters of fresh air per minute.

GAS EXCHANGE AND TRANSPORT

Gas Exchange in Alveoli

Each alveolus is like a tiny, empty bowl, its rim continuous with the walls of an alveolar duct. Each "bowl" is no more than a single layer of epithelial cells, surrounded by a thin basement membrane. Thus, at most, the gas in alveoli is separated from the blood in pulmonary capillaries by only a thin film of interstitial fluid and the thin capillary and alveolar walls (Figure 31.10). Under normal conditions, gas can diffuse rapidly across this narrow space.

Figure 31.11 shows the partial pressure gradients for oxygen and carbon dioxide through the human respiratory system. Passive diffusion alone is enough to move oxygen across the respiratory surface and into the bloodstream. And it is enough to move carbon dioxide in the reverse direction (from the bloodstream into the spaces inside the alveolar sacs). Carbon dioxide is about twenty times more soluble than oxygen, and it diffuses more rapidly across the respiratory surface.

Driven by its partial pressure gradient, oxygen diffuses from alveolar air spaces, through interstitial fluid, and into the blood capillaries.

Carbon dioxide, driven by its partial pressure gradient, diffuses in the reverse direction.

Gas Transport Between Lungs and Tissues

Diffusion of oxygen and carbon dioxide down their gradients is enough to move adequate amounts of these gases across the respiratory surface in the lungs. But the amounts that can be carried by blood in dissolved form is not enough to meet the needs of the body, and their transport must be enhanced. Oxygen-carrying hemoglobin in red blood cells increases the oxygen transport by seventy times. Carbon dioxide transport is increased seventeen times by a series of reversible reactions, which will be described shortly.

Oxygen Transport. With each new breath, air that is low in carbon dioxide and rich in oxygen enters the lungs and flows down into the alveoli. Within the lung capillaries adjacent to the alveoli, the blood is low in oxygen and rich in carbon dioxide. Oxygen diffuses into the

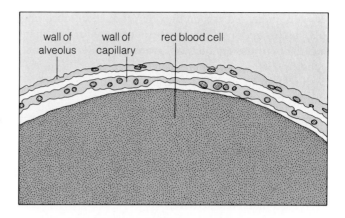

Figure 31.10 Diagram of a section through an alveolus and an adjacent blood capillary. By comparison to the diameter of the red blood cell, the diffusion distance across the capillary wall, the interstitial fluid, and the alveolar wall is exceedingly small.

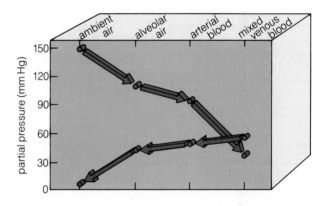

Figure 31.11 Summary of partial pressures for oxygen (blue) and carbon dioxide (red) in different body regions. The graph shows that both gases diffuse down gradients of decreasing partial pressure.

blood plasma, then into red blood cells. When it does, it rapidly forms a weak, reversible bond with hemoglobin. Recall that each hemoglobin molecule can bind four oxygen molecules at the same time. Oxygen combined with hemoglobin is called **oxyhemoglobin** (HbO_2). The amount of oxygen that does combine with hemoglobin depends on the partial pressure of the gas. The higher the pressure, the more oxygen will be picked up (until the hemoglobin binding sites are saturated).

Hemoglobin holds onto oxygen rather weakly and tends to give it up in regions where the partial pressure of oxygen is less than it is in the lungs. It also lets go of oxygen in regions where the blood is warmer and shows a decrease in pH. A decrease in the partial pressure of oxygen, increased carbon dioxide production, declining pH values, and local temperature increases are all characteristics of tissues that are showing greater metabolic activity. That is why more oxygen is released from oxyhemoglobin in highly active tissues (such as muscles during periods of exercise).

---

1. Oxyhemoglobin gives up oxygen when oxygen partial pressure is low, carbon dioxide production is high, temperature is elevated, and pH values are low.

2. These conditions are characteristic of tissues showing increased metabolic activity.

Carbon Dioxide Transport. The partial pressure of carbon dioxide in blood flowing through the capillaries is lower than it is in the surrounding tissues. Thus carbon dioxide diffuses down its pressure gradient, from the tissues into the capillaries. From there, it is transported to the lungs in three forms:

CO_2	(carbon dioxide dissolved in blood plasma)
$HbCO_2$	(carbon dioxide combined with hemoglobin, forming *carbaminohemoglobin*)
HCO_3^-	(bicarbonate ions)

Most of the carbon dioxide produced in the body's tissues is transported to the lungs in the form of bicarbonate. Carbon dioxide combines with water in the plasma portion of blood to form carbonic acid, which then dissociates into bicarbonate and hydrogen ions:

$$CO_2 + H_2O \rightleftharpoons H_2CO_3 \rightleftharpoons HCO_3^- + H^+$$

Although the reaction proceeds slowly in plasma, much of the carbon dioxide diffuses into red blood cells, which contain the enzyme carbonic anhydrase. With this enzyme, the reaction rate increases by 250 times and the concentration of free carbon dioxide in the blood drops rapidly. Its action helps maintain the concentration gradient that keeps carbon dioxide diffusing from interstitial fluid into the bloodstream.

Hemoglobin acts as a buffer for the hydrogen ions that are produced by the dissociation of carbonic acid. (A buffer, recall, is a molecule that combines with or releases hydrogen ions in response to changes in cellular pH.) The bicarbonate ions tend to diffuse out of the red blood cells into the blood plasma. Typically, about 70 percent of the carbon dioxide in blood is transported as bicarbonate. Only about 7 percent remains dissolved in plasma (as carbon dioxide). The remaining 23 percent is transported as carbaminohemoglobin.

In the lungs, where the partial pressure of carbon dioxide in the alveoli is lower than it is in the capillaries, the reactions proceed in the reverse direction. Carbonic acid dissociates to form water and carbon dioxide, which diffuses down its concentration gradient and is exhaled from the body. The blood is now ready for another round trip through the systemic circulation.

---

1. Hemoglobin is central to the transport of oxygen from the lungs to tissues throughout the body. It combines with or releases oxygen in response to shifts in the partial pressure of oxygen, carbon dioxide concentration, pH, and temperature.

2. Hemoglobin helps maintain the partial pressure gradient of oxygen between the lungs and body tissues. It does this by taking up oxygen from the blood plasma in the lungs and releasing oxygen in metabolically active tissues.

3. Carbon dioxide transport is enhanced by reactions that rapidly convert the gas to bicarbonate and hydrogen ions in tissues, then back to carbon dioxide in the lungs. These conversions help maintain the concentration gradient necessary for the gas to diffuse into and out of the bloodstream.

MATCHING AIR FLOW AND BLOOD FLOW DURING VENTILATION

Gas exchange in the alveoli is most efficient when the rate of air flow is matched with the rate of blood flow. Both rates can be adjusted locally (in the lungs) and throughout the body as a whole.

Neural Control Mechanisms

The nervous system controls oxygen and carbon dioxide levels in arterial blood for the entire body. It does this through homeostatic mechanisms that influence the rate and depth of breathing. In general, these are the main elements in respiratory control:

Respiratory centers in the brain

Chemoreceptors in the brain and in the walls of arteries

Respiratory muscles (effectors of change)

Contraction of the diaphragm and muscles that move the rib cage is under the control of cells of the reticular formation (page 333). One cluster of cells, in the medulla oblongata of the brain, is concerned mostly with coordinating the contractions associated with inhalation. Another is concerned with coordinating the signals for

Figure 31.12 The atmosphere contains the same percentage of oxygen at high and low altitudes. But the atmospheric pressure is lower at high altitudes, so the partial pressure exerted by its oxygen component is not as great. Hence less oxygen is available to move from the air into the body. People who are not adapted to high altitudes can experience hypoxia, or cellular oxygen deficiency.

exhalation. Respiratory centers in other parts of the brain can stimulate or inhibit these cell clusters; they work to fine-tune the rhythmic contractions of the muscles concerned with respiration.

Moveover, chemoreceptors in the medulla respond indirectly to rising carbon dioxide levels in the blood. (Such increases affect the level of H$^+$ in the fluid in the brain, and the shift in pH stimulates the receptors.) Other chemoreceptors include the *carotid bodies* (which are located at the branching of the carotid arteries to the brain) and the *aortic bodies* (which are specialized structures in arterial walls near the heart). These receptors detect decreases in the partial pressure of oxygen dissolved in arterial blood, and they can trigger an increase in the rate and depth of respiration (hence more oxygen delivery to tissues).

Local Control Mechanisms

In the lungs themselves, changes in the diameter of bronchioles help control the proportion of air and blood flow to alveoli. When there is not enough air flow and too much blood flow, local levels of carbon dioxide rise. Smooth muscles in the bronchiole walls are sensitive to such increases and they dilate in response—and thereby increase the local air flow. Similarly, a decrease in carbon dioxide levels causes the bronchiole walls to constrict—and thereby decrease the air flow.

Local changes also occur in the diameter of blood vessels that supply different lung regions. If air flow is too large relative to the blood flow, oxygen levels rise in these regions. The increase directly affects smooth muscle in the blood vessel walls, which undergo vasodila-

tion—and thereby increase blood flow to the region. Similarly, if air flow is too small, vasoconstriction leads to a decrease in blood flow (page 394).

Hypoxia

When tissues are not supplied with enough oxygen, *hypoxia*, or cellular oxygen deficiency, is the result. For example, hypoxia can occur at high altitudes, where the partial pressure of oxygen is lower than it is at sea level (Figure 31.12). At an altitude of 2,400 meters (about 8,000 feet), a person attempts to compensate for the oxygen deficiency by hyperventilating (breathing much faster and more deeply than would otherwise be done at a given level of activity). At 3,650 meters (about 12,000 feet), the partial pressure of oxygen is only 100mm Hg, and oxygen deprivation can cause headaches, nausea, and lethargy. At 7,000 meters (23,000 feet), hypoxia can lead to loss of consciousness and death.

Hypoxia also occurs when the partial pressure of oxygen in arterial blood falls because of *carbon monoxide poisoning*. Carbon monoxide (CO) is a colorless, odorless gas that is present in the exhaust fumes from gasoline-powered vehicles, in tobacco smoke, and in the smoke from coal or wood burning. It combines with hemoglobin at least 200 times faster than oxygen does. Even at a partial pressure of only 0.5mm Hg, carbon monoxide will tie up *half* the hemoglobin in the body, forming carboxyhemoglobin (HbCO). When that happens, HbCO levels are high and HbO$_2$ levels are low. This means that the oxygen content of arterial blood is low even though the partial pressure for oxygen remains normal. Thus the chemoreceptors that are supposed to

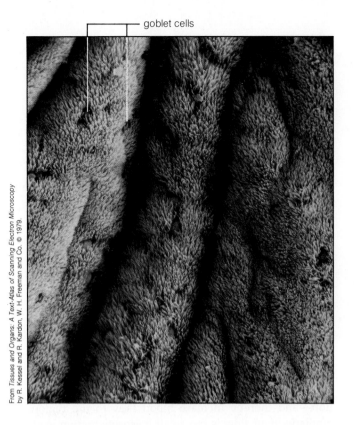

goblet cells

From *Tissues and Organs: A Text-Atlas of Scanning Electron Microscopy* by R. Kessel and R. Kardon, W. H. Freeman and Co. © 1979.

Figure 31.13 Ciliated epithelium of the human trachea, 830×.

detect decreases in oxygen levels are not stimulated, the body does not step up ventilation, and carbon monoxide poisoning occurs. This condition can be reversed in some cases by administering pure oxygen at a partial pressure of 600mm Hg to slowly replace the HbCO.

HOUSEKEEPING AND DEFENSE IN THE RESPIRATORY TRACT

The respiratory surface of the lungs is exposed to what can be a harsh environment. Vapors from acids, airborne particles of dust (which may be carrying bacteria), smoke, ashes, soot, oil, asbestos particles, and other substances may be inhaled daily.

Large airborne particles are filtered out at the nose, and smaller particles are filtered out at the air-conducting tubes, which are lined with ciliated epithelium. Scattered between the hairlike cilia are mucus-secreting goblet cells (Figure 31.13). The small particles stick in the mucus of the epithelium, and the rhythmically upward-beating cilia sweep the debris-laden mucus toward the mouth. The mucus is then swallowed or expelled. In the alveoli, migrating macrophages engulf what they can, and leukocytes eliminate some of the foreign material.

Table 31.1 To Smoke or Not to Smoke: Some Comparisons*

Risks Associated With Smoking	Benefits of Quitting
Shortened Life Expectancy: Nonsmokers live an average of 8.3 years longer than those in mid-twenties who smoke two packs of cigarettes a day	Cumulative reduction of risk; after 10–15 years, the life expectancy of ex-smokers approaches that of nonsmokers
Chronic Bronchitis, Emphysema: Smokers have 4–25 times more risk of dying from these diseases than do nonsmokers	Greater chance of improving lung function and slowing down rate of deterioration
Lung Cancer: Cigarette smoking the major cause of lung cancer	After 10–15 years, risk approaches that of nonsmokers
Cancer of Mouth: 3–10 times greater risk among smokers	After 10–15 years, risk is reduced to that of nonsmokers
Cancer of Larynx: 2.9–17.7 times more frequent among smokers	After 10 years, risk is reduced to that of nonsmokers
Cancer of Esophagus: 2–9 times greater risk of dying from this form of cancer	Risk is proportional to amount smoked, so quitting should reduce risk
Cancer of Pancreas: 2–5 times greater risk of dying from pancreatic cancer	Risk is proportional to amount smoked, so quitting should reduce risk
Cancer of Bladder: 7–10 times greater risk for smokers	Risk decreases gradually over 7 years to that of nonsmokers
Coronary Heart Disease: Cigarette smoking a major contributing factor	Risk drops sharply after a year; after 10 years, risk reduced to that of nonsmokers
Effects on Offspring: Women who smoke during pregnancy have more stillbirths, and weight of liveborns averages less (hence babies are more vulnerable to disease, death)	When smoking stops before fourth month of pregnancy, risk of stillbirth and lower birthweight eliminated
Impaired Immune System Function: Increase in allergic responses, destruction of macrophages in respiratory tract	Avoidable by not smoking

*Based on data published in 1980 by the American Cancer Society, Inc.

Figure 31.14 (a) Normal appearance of human lung tissue and (b) appearance of a lung taken from a person who suffered emphysema.

When Defenses Break Down

In urban environments, in certain occupations, even near a cigarette smoker, airborne particles and certain gases are present in abnormal amounts, and they put an extra workload on the respiratory tract. For example, ciliated epithelium in the air-conducting tubes is extremely sensitive to cigarette smoke, probably because of the chemical nature of the concentrated particles.

Bronchitis. Consider the conditions that lead to a lung disorder called bronchitis. Smoking and other forms of bronchial irritants increase the secretion of mucus while interfering with ciliary action in the air-conducting tubes of the lungs. Mucus and the particles it traps—including bacteria—begin to accumulate in the trachea and bronchi.

Now coughing sets in as the body attempts to clear away the mucus. If the irritation continues, the coughing reflex persists. Coughing aggravates the condition because it further irritates the bronchial walls, which become inflamed and infected. Cilia diminish in numbers, and mucus-secreting cells multiply as the body works to fight against the accumulating debris. With all of this aggravation, fibrous scar tissue forms that can obstruct parts of the respiratory tract.

Emphysema. A person suffering an acute attack of bronchitis who is otherwise in good health responds to medical treatment. But what happens if the irritation persists, particularly in a person who is prone to develop lung infections? As fibrous scar tissue builds up in the respiratory tract, the bronchi become progressively clogged with more mucus. Air becomes trapped in

alveoli, and the alveolar walls break down. The remaining alveoli enlarge and the balance between air flow and blood flow is altered. The outcome is *emphysema*, in which the lungs are distended and gas exchange efficiency is compromised (Figure 31.14). Running, walking, and even exhaling can be difficult.

For some people, early environmental conditions—poor diet, chronic colds, other respiratory ailments—apparently can create a predisposition to emphysema later in life. Also, some who suffer from emphysema have a hereditary deficiency in their ability to form antitrypsin, a substance that inhibits tissue-destroying enzymes produced by bacteria. These individuals may be at a disadvantage in fighting off respiratory infections when they do strike.

Emphysema is insidious. It can develop slowly, over twenty or thirty years. By the time it is detected, the damage to lung tissue is irreparable. About 1,300,000 individuals in the United States alone now suffer from the disorder.

Effects of Cigarette Smoke. Table 31.1 lists some effects of cigarette smoke on the airways. The noxious particles in smoke from one cigarette can prevent the cilia from beating for several hours. The particles also stimulate excessive secretions of mucus, which can eventually clog the airways. "Smoker's cough" is one of the least serious consequences; the coughing can contribute to the development of bronchitis and emphysema, as described above. Cigarette smoke can also destroy phagocytic cells that defend the respiratory epithelium. Extensive lung damage has also been documented in marijuana smokers.

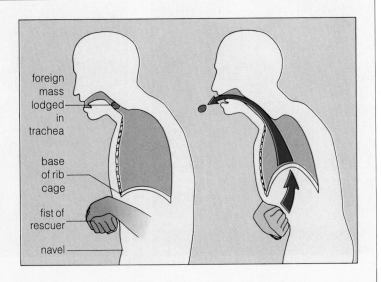

Figure 31.15 The Heimlich maneuver, an emergency procedure used to dislodge foreign matter blocking the respiratory tract.

First, stand behind the victim, make a fist with one hand, then press it thumb-side in against the victim's abdomen. The fist must be slightly above the navel and well below the rib cage.

Second, press the fist into the abdomen with a sudden upward thrust. Repeat the thrust several times if needed. The maneuver can be performed on a victim who is standing, sitting, or lying down.

Once the foreign matter is dislodged, be sure the victim is seen at once by a physician, for an inexperienced rescuer can inadvertently cause internal injuries or crack a rib. It could be argued that the risk is well worth taking, given that the alternative is death.

foreign mass lodged in trachea

base of rib cage

fist of rescuer

navel

Cigarette smoke also contains compounds that can lead to lung cancer. These compounds, including methylcholanthrene, are found in coal tar and cigarette smoke. It appears that they become chemically modified in the body through the action of natural substances, thereby turning into highly reactive intermediates that are the real carcinogens. The carcinogens act on gene expression in the cells of lung tissues in ways that lead to uncontrolled cell division.

Susceptibility to lung cancer is related to the number of cigarettes smoked each day as well as to the extent and depth of inhalation. At least eighty percent of all lung cancer deaths are the legacy of cigarette smoking. In its terminal stage, lung cancer is agonizing. It is a disorder that only ten out of a hundred afflicted individuals will survive, with varying degrees of tissue damage and malfunctioning.

The Heimlich Maneuver

Sometimes large chunks of food become lodged in the respiratory tract. Each year, several thousand people die from strangulation when food enters the trachea instead of the esophagus. Strangulation can occur when the air flow is blocked for as little as four or five minutes.

Such misdirected chunks often can be dislodged by an emergency procedure called the *Heimlich maneuver*. With this procedure, the diaphragm is forcibly elevated, causing a sharp decrease in the volume of the chest cavity and a sudden increase in alveolar pressure. The increased pressure forces air up the trachea, and it is often enough to dislodge the obstruction (Figure 31.15).

SUMMARY

1. Aerobic metabolism requires oxygen and produces carbon dioxide. The process of acquiring oxygen and eliminating carbon dioxide is called respiration.

2. Diffusion of substances, including gases, depends on differences in concentration between two regions. The concentration of gases is measured in terms of partial pressure (the pressure exerted by one gas in a mixture of gases).

3. Respiratory systems make use of the diffusive properties of oxygen and carbon dioxide, which tend to move from a region of higher partial pressure to a region of lower partial pressure.

4. In some respiratory systems, gases are exchanged across the whole body surface (integumentary exchange); in others, they are exchanged at specialized respiratory surfaces known as gills, tracheas, and lungs. All these systems have a moist, thin vascularized layer of epidermis across which gases are exchanged.

5. In vertebrates, airways carry gas to and from one side of the respiratory surface of lungs, and blood vessels carry gas to and from the other side.

6. Air moves by bulk flow into and out of the lungs, and thus new air is delivered to the respiratory surface. This air movement is called pulmonary ventilation. Gases diffuse across the respiratory surface into the blood of the pulmonary circulation.

7. The air-conducting portion of the human respiratory system consists of the nasal cavities, pharynx, larynx, trachea, main bronchi, and bronchioles. Gas exchange

occurs only in the alveoli, which are located at the end of the air-conducting system.

8. Human lungs are paired structures in the thoracic cavity, which is separated from the abdominal cavity by the diaphragm. Each lung is enclosed in a pleural sac and is separated from the thoracic wall by the interpleural space.

9. During inhalation, the chest cavity becomes larger because of contraction of the diaphragm and some of the chest-wall muscles; the pressure in the lungs falls below atmospheric pressure, and air moves by bulk flow down the pressure gradient into the lungs. During normal, quiet exhalation these processes are reversed.

10. There is still air left in the lungs (the residual volume) at the end of each exhalation.

11. Driven by its partial pressure gradient, oxygen diffuses from alveolar air spaces into the blood capillaries. Carbon dioxide diffuses in the reverse direction.

12. Once in the blood, oxygen diffuses into the red blood cells where it combines loosely with hemoglobin to form oxyhemoglobin. In the tissue capillaries, oxyhemoglobin gives up oxygen, which diffuses out of the capillaries and into nearby cells.

13. In the tissues, carbon dioxide diffuses into the capillaries. It is transported in the blood as a dissolved gas, in combination with proteins (carbamino compounds), and as bicarbonate ions. In the lungs, the reactions that occurred in the tissue capillaries are reversed, and the carbon dioxide diffuses from the blood out into the air spaces of the alveoli.

14. Hemoglobin is central to the transport of oxygen from the lungs to the tissues throughout the body. It combines with or releases oxygen in response to shifts in oxygen levels, pH, and temperature.

15. Respiratory control centers in the reticular formation of the brainstem alter the rate and depth of respiration to maintain the levels of oxygen, carbon dioxide, and hydrogen ions within ranges compatible with life.

Review Questions

1. What is the main requirement for gas exchange in animals? What types of systems are used for gas exchange in (a) water-dwelling animals and (b) land-dwelling animals?

2. Explain how a countercurrent flow mechanism works in a fish gill.

3. Define respiration. What events are necessary in human respiration?

4. By what mechanisms do carbon dioxide move out of your body and oxygen into it through the *same* system of branched tubes?

5. What governs the rate and depth of breathing?

6. What force drives oxygen from alveolar air spaces, through interstitial fluid, and across capillary epithelium? What force drives carbon dioxide in the reverse direction?

7. How does hemoglobin help maintain the oxygen partial pressure gradient during gas transport in the body? What reactions enhance the transport of carbon dioxide through the body?

8. Label the component parts of the human respiratory system:

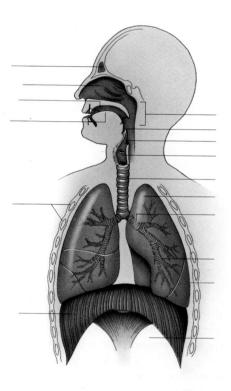

Readings

American Cancer Society. 1980. *Dangers of Smoking; Benefits of Quitting and Relative Risks of Reduced Exposure.* Revised edition. New York: American Cancer Society, Inc.

Baker, P. 1969. "Human Adaptation to High Altitude." *Science* 163:1149.

Ganong, W. 1979. *Review of Medical Physiology.* Ninth edition. Los Altos, California: Lange Publications. Advanced reading, but one of the most authoritative books on respiration.

Hickman, C., et al. 1983. *Integrated Principles of Zoology.* Seventh edition. St. Louis: Mosby.

Vander, A., J. Sherman, and D. Luciano. 1985. *Human Physiology: The Mechanisms of Body Function.* Fourth edition. New York: McGraw-Hill. Clear introduction to the respiration system and its functioning.

West, J. 1985. *Respiratory Physiology: The Essentials.* Third edition. Baltimore: Williams & Wilkins. Excellent, brief introduction to respiratory functions. Paperback.

Wyman, R. 1977. "Neural Generation of the Breathing Mechanism." *Annual Review of Physiology* 39:417.

32

DIGESTION AND ORGANIC METABOLISM

Nutrition—here is a word that has to do with all those processes by which the body takes in, digests, absorbs, and uses food. The word signals that you are about to begin one more educational trek through the animal gut. This time around, however, you will move beyond passive memorization of names for specialized tissues and organs. This time your main concern will be with *systems integration*—with how systems interact in meeting the metabolic needs of the entire body.

Consider the female bear in Figure 32.1 and the destination of that salmon in her mouth. Is it enough, really, to assume the nutritional picture begins and ends in her gut? If nutrients are to reach all living cells, they must cross the gut lining and enter the bloodstream. If cells are to use nutrients, a respiratory system must supply them with oxygen (for aerobic metabolism) and relieve them of carbon dioxide. Because bears do not eat around the clock, there must be ways to store and release nutrients at different times. So "nutrition" in complex animals requires more than food digestion and absorption. It also requires systems of distributing, storing, and using particular substances in controlled ways (Figure 32.2).

This chapter describes the components of digestive systems and their functions. Then it describes how the activities of three systems—digestive, circulatory, and respiratory—are integrated. The main examples are from an organism with which you are already more or less acquainted: yourself.

TYPES OF DIGESTIVE SYSTEMS AND THEIR FUNCTIONS

A **digestive system** is some form of body cavity or tube in which food is first reduced to particles, then to molecules small enough to move into the internal environment. (The internal environment, recall, is the body's *extracellular fluid*. It consists of interstitial fluid that bathes living cells and, in animals with circulatory systems, the plasma portion of blood.) A layer of cells serves as a lining for the body cavity or tube, and nutrients enter the internal environment by crossing that lining.

An **incomplete digestive system** has only one opening, and what goes in but cannot be digested goes out the same way. A flatworm called a planarian has this type of system. A muscular organ (pharynx) opens into a highly branched cavity that serves both digestive and circulatory functions (Figure 32.3). Food is partly digested and transported to cells even while residues are being sent back out through the pharynx. Because of the two-way traffic, this cavity cannot be subdivided into specialized regions for food transport, processing, and storage.

Figure 32.1 Digestion includes those interrelated processes by which food is ingested, prepared for absorption, and moved into the internal environment.

Annelids, mollusks, arthropods, echinoderms, and chordates have a **complete digestive system**. They have an internal tube with an opening at one end for taking in food and an opening at the other end for eliminating unabsorbed residues (Figure 32.3). Between the two openings, food generally moves in one direction through the *lumen* (the space inside the tube). And the tube itself is subdivided into specialized regions for food transport, processing, and storage. For example, one part of the digestive tube of earthworms and birds is modified into a crop (a food storage organ). Another part is modified into a gizzard (a muscular organ in which food is ground into smaller bits).

The specialized regions of complete digestive systems can be correlated with the animal's feeding behavior. Animals with *discontinuous* feeding habits may eat large amounts of food when it is available, then go for long periods without eating at all. Certain organs in these animals store food that is taken in faster than it can be digested and absorbed. Other organs help maintain an adequate distribution of nutrients when food is not being ingested.

Many grazing animals eat almost continuously and face other digestion tasks. For example, they have to digest cellulose, which requires specific enzymes and a rather long processing time for the enzymes to work. **Ruminants**, such as deer, cattle, and goats, have multiple stomach chambers in which the tough cellulose

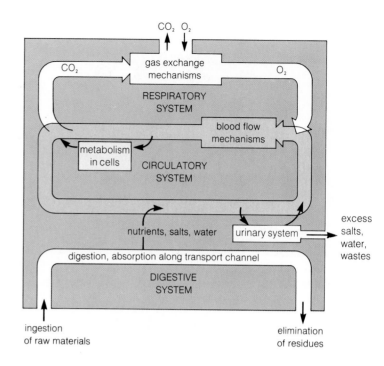

Figure 32.2 Interconnected systems for moving food into the internal environment and for assuring that cells can utilize the energy stored in food once it reaches them. The connections represented here are characteristic of most complex animals. Sensory receptors in each system channel information to the nervous system, which coordinates the interrelated activities. Hormonal controls are also at work.

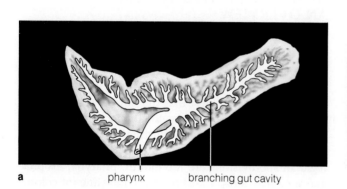

a   pharynx   branching gut cavity

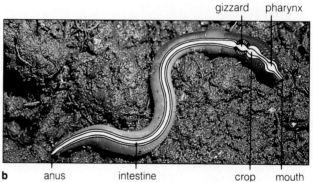

b   anus   intestine   crop   mouth

Figure 32.3 (**a**) Incomplete digestive system of the planarian, a flatworm. The pharynx of this animal is a muscular tube that protrudes from the body during feeding. Complete digestive systems of animals ranging from earthworms (**b**) to birds (**c**) and humans have special food-processing regions and a one-way movement of material.

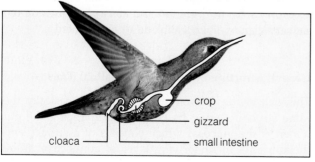

c

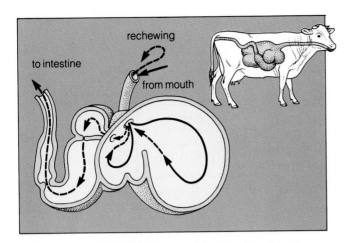

Figure 32.4 Complete digestive system of ruminants, such as cattle. Ruminants swallow partially chewed plant material, which moves into two stomachlike chambers. Then they regurgitate the material, chew it more, and swallow it again. The double chewing time mechanically breaks apart the plant material, which contains tough cellulose fibers. Symbiotic bacteria present in the digestive tract produce enzymes that can digest cellulose. The double chewing gives the enzymes more time and more surface area upon which to act. Altogether, food is processed in four stomachlike chambers before being sent on to the small intestine for absorption.

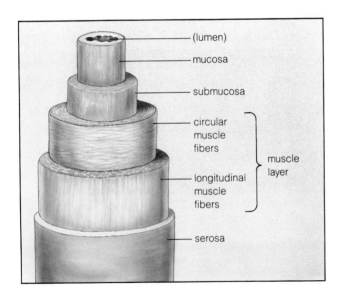

Figure 32.5 Generalized sketch of the wall of the gastrointestinal tract.

fibers are processed gradually (Figure 32.4). The first two chambers house vast microbial populations. Among the microbes are symbiotic bacteria that produce the enzymes capable of degrading cellulose (and other nutrients). Ruminants regurgitate food from the first two chambers and rechew it before swallowing again. Thus the plant material is mixed and pummeled more than once, and more cellulose fibers are exposed to agents of digestion before continuing through the digestive tract.

No matter how regionally specialized they have become, the digestive systems of most animals have four main functions:

Motility: mechanical breakdown and mixing of ingested nutrients, passage of nutrients through the digestive tract, and elimination of undigested or unabsorbed residues from the body.

Secretion: release of enzymes, hormones, and other substances that take part in digestion.

Digestion: chemical reduction of ingested nutrients into particles, then into molecules small enough to cross the lining of the gut and reach the internal environment.

Absorption: passage of digested nutrients from the gut lumen into the blood or lymph, which distributes them through the body.

HUMAN DIGESTIVE SYSTEM: AN OVERVIEW

Components of the Digestive System

Humans have discontinuous feeding habits, and they dine on various and diverse foodstuffs. From this you might deduce (correctly) that the human digestive system is a tube with many regional specializations. The length of the tube, or **gastrointestinal tract** (Figure 32.5), is between 6.5 and 9 meters (21 to 30 feet) in adults. Its regional subdivisions are the mouth, pharynx, esophagus, stomach, small intestine, large intestine (or colon), rectum, and anus (Figure 32.6). Glandular organs having accessory roles in digestion and absorption include the salivary glands, liver, gallbladder, and pancreas.

General Structure of the Gastrointestinal Tract

The wall of the gastrointestinal tract has basically the same structure along its entire length. Facing the gut lumen is the **mucosa**, which consists of a surface epithelium and an underlying layer of connective tissue (Figure 32.5). The mucosa is surrounded by the **submucosa**, a connective tissue layer in which blood and

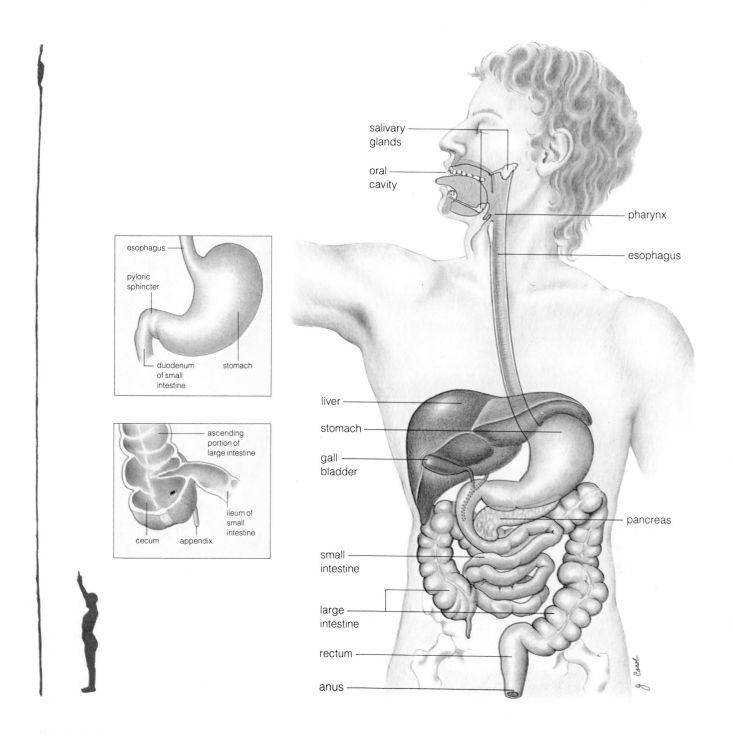

Figure 32.6 Simplified picture of the human digestive system. (Far left: If you have ever wondered how far a stretched-out gastrointestinal tube extends, now you know.)

lymph vessels are meshed. Next is a **muscle layer**, with two sublayers of smooth muscle arranged in longitudinal and circular directions relative to the tube axis. The thin outermost layer of connective tissue is the **serosa**.

Gastrointestinal Motility

Coordinated contractions in the muscle layers of the gastrointestinal tract mix food material with secretions and move it forward. Two common types of movement in the tract are peristalsis and segmentation.

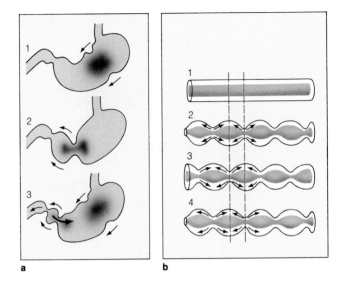

Figure 32.7 (**a**) Peristaltic wave, as it occurs in the stomach. (**b**) Segmentation, or oscillating movement, in the intestines.

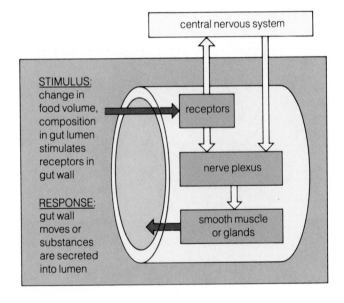

Figure 32.8 Local and long-distance reflex pathways called into action when food is in the digestive tract.

During **peristalsis**, a mass of food material advances through the tract when rings of circular muscles contract behind it and relax in front of it. As the mass moves, it expands the tube wall, the expansion stimulates peristalsis, and so on (Figure 32.7). For example, peristaltic waves move down the stomach walls about three times a minute.

Segmentation occurs only in the intestines. Rings of smooth muscle in the intestinal wall repeatedly contract and relax, creating an oscillating (back-and-forth) movement in the same place. This movement constantly mixes the contents of the lumen and forces them against the absorptive surface of the intestinal wall (Figure 32.7).

Sphincters influence the flow of material from one region to another in the tract, and they prevent backflow. These rings of smooth or striated muscle are located at the beginning and end of specific regions. For example, there is a sphincter between the esophagus and stomach, and another between the stomach and small intestine.

CONTROL OF GASTROINTESTINAL ACTIVITY

Most controls over body activity operate in response to some aspect of the extracellular fluid (such as oxygen concentration). In contrast, controls over the gastrointestinal tract operate in response to the volume and composition of food in the lumen—in other words, to the "external" environment.

Ingested food distends the gut wall, and it varies in solute concentrations and in acidity. During the digestion of carbohydrates, proteins, and fats, the breakdown products as well as certain secretions accumulate in the lumen. Receptors embedded in the gut wall trigger reflex responses to these various stimuli. As Figure 32.8 indicates, a short reflex pathway operates independently of the central nervous system. Signals from the receptors travel through nerve plexuses in the gut wall that can directly influence the wall contractions and secretions into the gut lumen. (A nerve plexus, recall, is a network of nerves outside the central nervous system.) A long-distance reflex pathway connects the receptors and effectors with the central nervous system. One or both pathways can act to control activities of the gastrointestinal tract.

Moreover, several hormones secreted by endocrine cells of the tract help regulate digestion and absorption. These hormones include gastrin, secretin, cholecystokinin (CCK), and gastric-inhibitory peptide (GIP). Table 32.1 lists their sources and functions.

Table 32.1 Functions of Primary and Accessory Organs of Digestion

Organ	Secretions	Main Functions
Mouth	—	Mechanically breaks down food
Salivary glands (accessory organs)	Water	Moistens food
	Mucus	Lubricates and binds food into bolus
	Salivary amylase	Starts breakdown of starch, glycogen
	Bicarbonate	Buffering action neutralizes acidic food in mouth
Stomach	—	Stores, mixes, dissolves food; regulates emptying of chyme into small intestine
Secretory cells in stomach mucosa	Hydrochloric acid	Dissolves food particles; kills many microorganisms
	Pepsinogens	Activated forms (pepsins) split apart peptide bonds in protein chains
	Mucus	Lubricates and protects stomach lining
	Gastrin	Stimulates hydrochloric acid secretion
Small intestine	—	Digestion and absorption of most nutrients; mixes and propels chyme forward
Secretory cells in intestinal mucosa*	Assorted enzymes	Break down major food molecules
	Mucus	Lubricates chyme
	Secretin	Stimulates pancreatic bicarbonate secretion, inhibits gastric acid secretion
	Cholecystokinin	Stimulates gallbladder contraction, pancreatic enzyme secretions; inhibits stomach emptying
	Gastric-inhibitory peptide	Inhibits stomach acid secretion and motility
Pancreas (accessory organ)	Assorted enzymes (e.g., lipase)	Break down all major food molecules
	Bicarbonate	Buffering action neutralizes hydrochloric acid entering small intestine from stomach
Liver (accessory organ)	Bile salts	Hydration of emulsified fat droplets
	Bicarbonate	Buffering action neutralizes hydrochloric acid entering small intestine from stomach
Gallbladder (accessory organ)	—	Stores and concentrates bile from liver
Large intestine (colon)	—	Stores, concentrates undigested matter by absorbing water and salts; mixes and propels material forward
Secretory cells in intestinal mucosa	Mucus	Lubricates undigested residues
Rectum	—	Distension triggers defecation reflex that rids body of undigested and unabsorbed residues

*Most enzymes are embedded in plasma membrane facing the lumen; some are released into lumen when cells are shed and disintegrate.

STRUCTURE AND FUNCTION OF GASTROINTESTINAL ORGANS

Mouth and Salivary Glands

Mechanical reduction of food begins in the **mouth** (oral cavity), as does the digestion of polysaccharides. Most animals have a mouth, but only humans and other mammals *chew* food in the mouth. Adult humans normally have thirty-two teeth (sixteen in the upper jawbone and sixteen in the lower). Each **tooth** consists of an enamel coat (hardened calcium deposits), dentine (a thick bone-like layer), and an inner pulp (which houses nerves and blood vessels). The teeth in the back of the mouth are flat-surfaced *molars*, which grind food. Teeth in the front are chisel-shaped *incisors*, useful in biting off chunks of food. In between are cone-shaped *cuspids*, for grasping and tearing food (see also Figure 42.6).

While the teeth and tongue are mechanically reducing food in the mouth, they are also mixing it with *saliva*, a fluid secreted from several **salivary glands**. Ducts of these exocrine glands empty into different parts of the oral cavity.

Salivary amylase, an enzyme that takes part in the initial breakdown of starch, is a component of saliva. So are bicarbonate ions (HCO_3^-), which act as buffers in keeping salivary pH between 6.5 and 7.5 even when acidic foods are in the mouth. Other components are the *mucins*, glycoproteins that bind bits of food together into a softened, lubricated ball called a bolus.

Pharynx and Esophagus

Once food has been processed into a bolus, voluntary muscle contractions move the tongue toward the roof of the mouth. The movement forces the bolus into the pharynx, where it stimulates mechanoreceptors that call for contraction of the walls of the pharynx and esophagus. That contraction initiates swallowing. A swallow can be *initiated* voluntarily, but thereafter it is a reflex controlled by a brain region.

In humans, the **pharynx** is a muscular tube continuous with the **esophagus**, which leads to the stomach. The pharynx is also continuous with the trachea, which leads to the lungs. The swallowing reflex opens a sphincter at the start of the esophagus, and normally it closes off the trachea (hence prevents choking) while food is moving into the esophagus. Neither the pharynx nor the esophagus contributes to digestion. Their peristaltic movements simply propel food to the stomach.

Stomach

The **stomach** is a muscular, distensible sac having three main functions. First, it stores and mixes food received from the esophagus. Second, it secretes substances that help dissolve and degrade food. Third, it helps control the rate at which food moves into the small intestine.

Components of Gastric Fluid. Among the cells scattered throughout the stomach mucosa are exocrine cells that release hydrochloric acid (HCl), pepsinogens, mucus, and other substances directly into the stomach lumen. Each day they secrete as much as two liters of these substances, which together constitute the gastric fluid. Also, endocrine cells in one part of the stomach mucosa release hormones (such as gastrin), which travel by way of the bloodstream to target cells. The secretions of their target cells also contribute to the gastric fluid.

The HCl secreted into the stomach dissolves bits of food, producing a solution called chyme. It also kills most of the microbes entering the body in food. The H^+ concentration in the lumen increases when the HCl molecules dissociate into H^+ and Cl^-. At times the H^+ level in the stomach becomes three million times higher than it is in arterial blood!

An increase in stomach acidity contributes to protein digestion. First, it alters the ionized groups of proteins and exposes some of their peptide bonds. Second, it converts inactive forms of protein-degrading enzymes (called *pepsinogens*) to active forms (called *pepsins*). These enzymes break certain peptide bonds of proteins and thereby produce peptide fragments.

HCl secretion can be stepped up in several ways. Receptors in the stomach wall trigger reflexes that stimulate secretion whenever food distends the wall. Also,

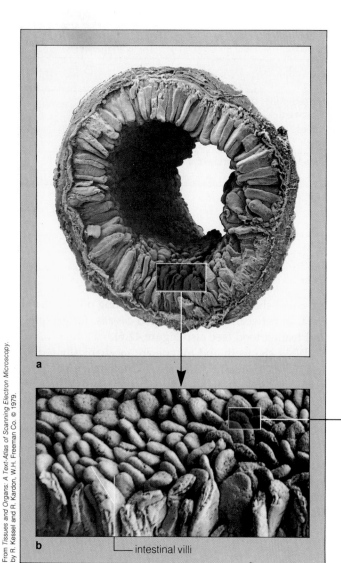

From *Tissues and Organs: A Text-Atlas of Scanning Electron Microscopy.* by R. Kessel and R. Kardon, W.H. Freeman Co. © 1979.

a

b

intestinal villi

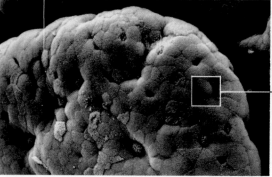

one epithelial cell of villus

c

the peptide fragments of protein breakdown stimulate the release of gastrin, which in turn stimulates the HCl-secreting cells. Other substances, such as the caffeine in coffee, tea, chocolate, and cola drinks also cause an increase in the rate of secretion. Finally, reflexes initiated by food in the mouth as well as by stress can also stimulate HCl secretion by way of the vagus nerve to the stomach (Figure 25.6).

Sometimes part of the stomach mucosa becomes damaged by the digestive action of the gastric fluid. This leads to formation of a *peptic ulcer*. In some way, normal control mechanisms that protect the mucosa are deficient. When the surface of the stomach breaks down, hydrogen ions diffuse into the mucosa and thus trigger the release of histamine. Histamine in turn triggers vasodilation and increased capillary permeability, and it stimulates HCl secretion. Thus a positive feedback loop is set up and leads to tissue damage, which may result in bleeding into the stomach and abdomen.

Stomach Emptying. Peristaltic waves in the stomach mix the chyme. The waves build up force as they approach the pyloric sphincter between the stomach and small intestine (Figure 32.5). Normally the sphincter is relaxed, but with the arrival of a strong peristaltic contraction, it closes. Most of the chyme is squeezed back; only a small amount moves into the duodenum, the first part of the small intestine.

Three factors control how fast the stomach empties. *First,* stomach distension following a meal activates mechanoreceptors in the stomach wall. The larger the meal, the more these mechanoreceptors trigger reflexes that increase the force of contraction (hence stomach emptying). *Second,* increases in acidity, osmotic pressure, and fat content stimulate receptors in the duodenum.

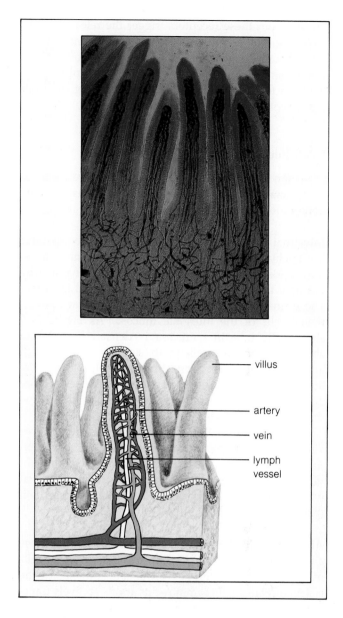

Figure 32.10 Location of blood and lymph vessels within intestinal villi.

microvillus at cell surface

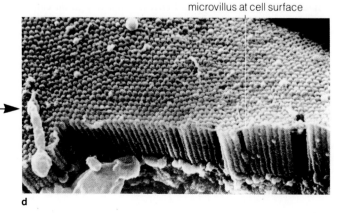

d

Figure 32.9 (**a,b**) Location of villi in the mammalian intestine. (**c**) Tip of a single villus, 825x. (**d**) The dense crown of microvilli at the surface of an epithelial cell. Microvilli face the lumen of the intestine and enhance the absorptive and secretory processes.

Signals from these receptors trigger the release of hormones (such as CCK and GIP) that inhibit stomach motility. Through such slowdowns, food is not moved along faster than it can be processed. *Third*, emotional states (such as fear and depression) can trigger signals from the nervous system that also inhibit motility.

Small Intestine

Digestion is completed and most nutrients are absorbed in the **small intestine**, which has three regions: the duodenum, jejunum, and ileum.

Intestinal Villi. Most proteins, fats, and carbohydrates in the chyme have been broken down to amino acids, fatty acids, monoglycerides, and monosaccharides by the time they are halfway through the small intestine. These small organic molecules readily move across epithelial cells of the intestinal mucosa. As Figure 32.9 shows, the intestinal mucosa is densely folded into absorptive structures called **villi** (singular, villus), which look like tiny tongues. Beneath its one-cell-thick epithelium, each villus houses blood and lymph vessels (Figure 32.10). Contraction of the villus promotes fluid flow through these vessels.

In themselves, villi greatly increase the surface area available for interactions with the chyme. Moreover, epithelial cells of each villus have a surface crown of **microvilli**: threadlike projections of plasma membrane that further increase the surface area available for absorption (page 80 and Figure 32.9).

Absorption Mechanisms. The movement of glucose and certain other monosaccharides into epithelial cells of the villi depends on active transport systems in the cell membranes. Most amino acids are also actively transported; others move passively into the epithelial cells. Free fatty acids and monoglycerides can diffuse across the cells because they are soluble in the lipid bilayer of the plasma membrane.

Once these small molecules have moved across the epithelial cells, they travel one of two routes. Glucose and amino acids diffuse into blood capillaries. Absorbed fatty acids and monoglycerides recombine into triglycerides inside the epithelial cells, and the triglycerides and other fats aggregate into small droplets (called chylomicrons). These droplets leave the cells by exocytosis and then enter lymph vessels, which drain into the circulation system.

Besides absorbing organic molecules, the small intestine absorbs water and dissolved mineral ions (such as Na^+ and Cl^-). Each day, about nine liters of fluid enter the small intestine from the stomach, liver, and pancreas. Of that, all but five percent is absorbed across the intestinal mucosa.

Role of the Pancreas in Digestion. Ducts leading from the pancreas and liver join to form a common duct that empties into the duodenum. Exocrine cells in the **pancreas** secrete enzymes into this duct in response to hormonal and neural signals. The enzymes digest carbohydrates, fats, proteins, and nucleic acids. For example, like pepsin in the stomach, the pancreatic enzymes *trypsin* and *chymotrypsin* digest proteins into peptide fragments. The fragments are then degraded to free amino acids by *carboxypeptidase* (from the pancreas) and by *aminopeptidase* (present on the surface of the intestinal mucosa).

The pancreas also secretes bicarbonate ions, which help neutralize the HCl arriving from the stomach. If there were no such neutralization, pancreatic enzymes could not function.

Other pancreatic secretions do not function in digestion, but they still play a role in nutrition. Certain patches of endocrine cells in the pancreas (the islets of Langerhans) secrete the hormones *insulin* and *glucagon* into the blood. As you will see shortly, these two hormones are important in the feedback control of metabolism.

Role of the Liver in Digestion. The **liver** is the largest glandular organ in the vertebrate body. One of its functions is the secretion of *bile*, a solution containing bile salts, bile pigments, cholesterol, and lecithin. Between meals, bile is stored and concentrated in the **gallbladder**, a small sac that branches off the common duct leading to the duodenum.

Part of a bile salt molecule (its cholesterol part) repels water, but another part dissolves in water. Because of these properties, bile salts enhance the breakdown and absorption of fats. Fats, recall, are insoluble in water. Most of the fats we eat are in the form of triglycerides, which clump together as large fat globules. In the small intestine, fat globules in the presence of bile salts are mechanically broken into smaller droplets by segmentation movements. Once that happens, the water-repelling parts of bile salts dissolve on the surface of the fat droplets. But other parts stick out from the surface and interact with water—which prevents the fat molecules from clumping together again. This suspension of small fat droplets is a type of emulsion. *Through the emulsifying effects of bile salts, pancreatic lipase has access to more triglycerides—hence fat digestion is enhanced.*

Bile salts also combine with the breakdown products of fat digestion (mostly monoglycerides and free fatty acids) to form micelles. A **micelle** is an aggregate of lipid molecules with a surface coat of bile salts. Since they are only three to ten nanometers across, micelles are small enough to move among microvilli of the intestinal epithelium. The monoglycerides and fatty acids diffuse down their concentration gradient—from micelles into the cells of these absorptive structures.

Human Nutrition and Gastrointestinal Disorders

The United States harbors one of the best-fed populations in the world. Yet digestive disorders among its individuals are on the increase.

Along with affluence, it appears we have picked up some bad eating habits. We skip meals, eat too much and too fast when we do sit down at the table, and generally give our gastrointestinal tracts erratic workouts. Worse yet, our diet tends to be rich in sugar, cholesterol, and salt—and low in bulk. (Here, bulk means the volume of fiber and other undigested food materials that cannot be decreased by absorption.) The problem with too little bulk in the diet comes from the longer transit time of feces through the colon. This material has irritating and even potentially carcinogenic effects. The longer the material is in contact with the colon walls, the more damage it can do. Thus, the more steadily the contents of the colon are cleared out by natural processes, the better. Increased bulk produces increased pressure on the colon walls, which stimulates expulsion of the material from the body.

Disorders such as appendicitis and cancer of the colon are practically nonexistent in rural Africa and India, where the inhabitants cannot afford to eat much more than whole grains. Whole grains happen to be high in fiber content. When individuals from those rural areas move to urban centers of the more affluent nations, they tend to become more susceptible to appendicitis and colon cancer. This suggests that diet is a key factor here. In addition, what we eat is known to affect the distribution and diversity of bacterial populations living in the gut. Do these changes somehow contribute to gastrointestinal disorders? That is not known.

Certainly the emotional stress associated with living in complex societies seems to compound the nutritional problem. Urban populations seem to be more susceptible to the irritable colon syndrome (once called colitis). Its symptoms include abdominal pain, diarrhea (excretion of watery feces), and constipation. Diarrhea can be brought on by emotional stress. There seems to be a genetic predisposition to some kinds of ulcers—inflammations of the stomach, the lower end of the esophagus, and the duodenum. But emotional stress apparently is a contributing factor in the development of some ulcers.

Where does this leave us? Short of surgery, there may not be much we can do about many inherited structural disorders of the gastrointestinal tract. Learning to handle stress is one way that we can ease up on the tract, though, and certainly learning how to eat properly is another.

Yet what is "eating properly"? In 1979 the United States Surgeon General released a report representing a medical consensus on how to promote health and avoid such afflictions as high blood pressure, heart disorders, cancer of the colon, and bad teeth. The report advised us to eat "less saturated fat and cholesterol; less salt; less sugar, relatively more complex carbohydrates such as whole grains, cereals, fruits, and vegetables; and relatively more fish, poultry, legumes (for example, peas, beans, and peanuts); and less red meat."

The controversies over what constitutes proper nutrition rage on. In the meantime, it might not be a bad idea to think about your own eating habits and how moderation in some things might help you hedge your bets. Put the question to yourself: Do you look upon a bowl of bran cereal with the same passion as you look upon, say, french fries and ice cream, prime rib and chocolate mousse? Now put the same question to your colon.

Large Intestine

Each day, contractions force the 500 milliliters or so of chyme remaining in the small intestine into the **large intestine**, or **colon**. The colon functions mainly in storing and concentrating *feces*, a mixture of undigested and unabsorbed material, water, and bacteria. This material becomes concentrated when (1) epithelial cells of the colon actively transport sodium ions into the internal environment and (2) water follows passively as a result.

The colon is about 1.2 meters long. The "ascending" part of the colon extends upward on the right side of the abdominal cavity, the "transverse" part cuts across to the other side, then the "descending" part extends down the left side. There it ends in an "S" (sigmoid) shape. The sigmoid colon is continuous with a small tube called the **rectum**. Distension of the rectal wall triggers reflex actions by which material is expelled from the body. This reflex can be overridden by nerves under voluntary control. Those nerves cause contraction of a striated muscle sphincter of the **anus**, the terminal opening of the gut.

The small intestine does not lead into the start of the ascending colon. Because of the location of a sphincter between the two organs, the ascending colon begins as a blind pouch (the cecum). The **appendix** is a small, narrow projection from the cecum (Figure 32.5). Although the appendix has no known digestive functions, it does contain lymphatic tissue, which suggests a role in body defense. The appendix can become infected, a condition called *appendicitis*. If ignored, an infected appendix can rupture. Then, bacteria normally inhabiting the colon can spread into the abdominal cavity and cause serious infections.

Enzymes of Digestion: A Summary

Now that you have completed this tour of the gastrointestinal tract, you may find it helpful to scan Table 32.2. This table summarizes the locations of carbohydrate, protein, fat, and nucleic acid digestion, the enzymes responsible, and the breakdown products at each stage. It should be readily apparent, from this summary, that the small intestine is the major site of digestion and absorption of all foodstuffs.

HUMAN NUTRITIONAL NEEDS

It now appears that the earliest members of the human lineage lived on fruits, seeds, and other plant material. From this nutritional beginning, humans in many parts of the world have moved to diets rich in fats, sugars, and salts—and low in fiber. Some of the suspected consequences of this long-term shift in diet include a predisposition to colon cancer, breast cancer, cardiovascular disorders, kidney stones, as well as obesity (see *Commentary* on page 441).

The body grows and maintains itself in working order when it is kept supplied with energy and materials from foods of certain types, in certain amounts. Let's take a brief look at the energy requirements and nutrients essential for health.

Energy Needs

In nutritional studies, energy is measured in units called "calories," which unfortunately is taken to be equivalent to "kilocalories." (Both units are supposed to mean 1,000 calories of energy.) To avoid confusion, in this book we will express energy requirements in terms of kilocalories. In order to maintain an acceptable weight and keep the body functioning normally, caloric intake must be balanced with energy output. The output varies from one individual to the next, depending on such factors as the extent of physical activity, basic rate of metabolism, age, sex, hormone activity (especially epinephrine and thyroid hormone secretions), and emotional state. Some of these factors have a genetic basis; others are influenced by the social environment of the individual.

In most adults, the energy input and output are so balanced that the body weight remains remarkably constant over long periods. As any dieter knows, it is as if the body has a set point for weight and works to counteract deviations from that set point.

What amount of calories should be taken in daily to maintain what you consider to be an acceptable body weight? You can calculate the amount in two steps. First, multiply the acceptable weight (in pounds) by 10 if you are not very active physically, by 15 if you are moderately active, and by 20 if you are quite active. Then, depending on your age, subtract the following amount from the value obtained from the first step:

Age:	Subtract:
25–34	0
35–44	100
45–54	200
55–64	300
Over 65	400

Thus, for example, if you want to weigh 120 pounds and if you are highly active, 120 × 20 = 2,400 kilocalories. And if you are thirty-five years old, then you should take in a total of (2,400 − 100), or 2,300 kilocalories a day.

Table 32.2 Major Enzymes of Digestion

Enzyme	Source	Where Active	Substrate	Main Breakdown Products*
Carbohydrate digestion:				
Salivary amylase	salivary glands	mouth	polysaccharides	disaccharides
Pancreatic amylase	pancreas	small intestine	polysaccharides	disaccharides
Disaccharidases	small intestine	small intestine	disaccharides	monosaccharides (e.g., glucose)
Protein digestion:				
Pepsins	stomach mucosa	stomach	proteins	peptide fragments
Trypsin and chymotrypsin	pancreas	small intestine	proteins and polypeptides	peptide fragments
Carboxypeptidase	pancreas	small intestine	peptide fragments	amino acids
Aminopeptidase	intestinal mucosa	small intestine	peptide fragments	amino acids
Fat digestion:				
Lipase	pancreas	small intestine	triglycerides	free fatty acids, monoglycerides
Nucleic acid digestion:				
Pancreatic nucleases	pancreas	small intestine	DNA, RNA	nucleotides
Intestinal nucleases	intestinal mucosa	small intestine	nucleotides	nucleotide bases, monosaccharides

*Yellow parts of table identify breakdown products that can be absorbed into the internal environment.

Carbohydrates

The body's main sources of energy are complex carbohydrates. As we have seen, these carbohydrates can be readily broken down to provide the body with glucose—the primary energy source for the brain, muscles, and other body tissues. Many nutritionists recommend that fleshy fruits, cereal grains, legumes, and other fibrous carbohydrates should make up at least fifty-five to fifty-eight percent of the daily caloric intake.

The average American takes in as much as 128 pounds of refined sugar (sucrose) per year. Sucrose adds calories to the diet but does so without the fiber afforded by complex carbohydrates.

Fats

Currently, fats constitute forty percent of the average diet in the United States, and most of the medical community agrees that it should be less than thirty percent. The body can manufacture most fats, including cholesterol, from protein and carbohydrates. (That is exactly what it does when you eat too much protein and carbohydrates.) However, your body needs to be supplied with about one tablespoon a day of polyunsaturated fat (such as corn or safflower oil), which serves as a source of certain essential fatty acids that the body cannot synthesize from other nutrients. Butter and other animal fats are forms of saturated fats, which tend to raise the blood levels of cholesterol. Too much cholesterol is believed to promote atherosclerosis (page 396).

Proteins

In the United States, the typical daily intake of proteins is about 125 grams. The amino acids released through protein digestion are absorbed and used to build the body's own proteins. Of the twenty common amino acids, eight are **essential amino acids**. Our cells cannot build these molecules; they must be obtained from the diet. Those amino acids are phenylalanine (and/or tyrosine), isoleucine, leucine, lysine, threonine, tryptophan, cysteine (and/or methionine), and valine.

Most animal protein contains all of the essential amino acids; plant proteins do not. To get enough protein from plant foods, different plants must be eaten in combination. To compare proteins from different sources, nutritionists use a measure called **net protein utilization** (NPU). NPU values range from 100 (all essential amino acids present in ideal proportions) to 0 (one or more absent; the protein is useless when eaten alone).

Table 32.3	Efficiency of Some Single Protein Sources in Meeting Minimum Daily Requirements				
Source	Protein Content (%)	Net Protein Utilization (NPU)	Amount Needed to Satisfy Minimum Daily Requirement		
			(grams)	(ounces)	
Eggs	11	97	403	14.1	
Milk	4	82	1,311	45.9***	
Fish*	22	80	244	8.5	
Cheese*	27	70	227	7.2	
Meat*	25	68	253	8.8	
Soybean flour	45	60	158**	5.5**	
Soybeans	34	60	210**	7.3**	
Kidney beans	23	40	468**	16.4**	
Corn	10	50	860**	30.0**	

*Average values.
**Dry weight values.
***Equivalent of 6 cups. The figure is somewhat misleading, for most of the volume of milk is water. Milk is actually a rich source of high-quality protein.

Balancing the diet with different proteins can make up for deficiencies.

For much of the world, cereal grains are the main foods. As Table 32.3 suggests, cereal grains such as corn are low in protein content and NPU value. In contrast, beans are high in protein. Although NPU values for beans are no higher than those for cereal grains, beans are deficient in *different* amino acids. When beans are eaten *with* grain, the overall NPU value is raised.

Given the pervasive role of proteins in body structure and function, it is easy to see that protein deficiency has serious consequences. Protein deficiency is most damaging among the young, for rapid brain growth and development occur early in life. Unless enough protein is taken in just before and just after birth, irreversible mental retardation occurs. Even mild protein deprivation can retard growth and affect mental and physical performance.

Vitamins and Minerals

Normal metabolic activity depends on very small amounts of more than a dozen organic substances called **vitamins**. Most plant cells can synthesize all of these substances. In general, animal cells have lost the ability to do so; hence animals must obtain vitamins from food. Human cells need at least fourteen different vitamins (Table 32.4).

In addition to vitamins, cells require inorganic materials known as **minerals**. (Some minerals are called *trace*

elements because they are needed in extremely small amounts.) Most cells require calcium and magnesium in a host of enzyme-mediated reactions. All cells need potassium during protein synthesis, for maintaining osmotic balances, and for muscle and nerve function. All cells require iron in building cytochromes, which are components of electron transport chains (Chapter Seven). Iron is also needed to produce the hemoglobin present in red blood cells (Table 32.5).

The sensible way to supply cells with essential vitamins and minerals is to eat a well-balanced selection of carbohydrates, fats, and proteins. (About 32–42 grams of protein, 250–500 grams of carbohydrates, and 66–83 grams of fat should do the trick.) In recent years, there have been claims that massive doses of certain vitamins and minerals are spectacularly beneficial. To date, there is no clear evidence that vitamin intake exceeding the recommended daily allowance leads to better health. To the contrary, excessive vitamin doses are often merely wasted or even harmful.

For example, the body simply will not hold more vitamin C than it needs for normal functioning. Vitamin C is not fat-soluble and the excess is excreted. Direct chemical analysis shows that any amount above the recommended daily allowance ends up in the urine almost immediately after it is absorbed from the gut. Abnormal intake of at least two other vitamins (A and D) can cause serious disorders. The reason is that, like all fat-soluble vitamins, vitamins A and D can accumulate in the fat deposits of the body (Table 32.4). *Severe shortages and massive excess of vitamins and minerals both can disturb the delicate balances that promote physiological health.*

Objective and Subjective Views of Obesity

By definition, **obesity** is an excess of fat in the body's adipose tissues, caused by imbalances between caloric intake and energy output. Clearly obese persons run a greater risk of developing high blood pressure, atherosclerosis, and diabetes, among other problems. But extremely underweight persons are also risking their health.

What constitutes the "ideal weight" for a person? Many charts have been developed, primarily by insurance companies who want to identify individuals who are overweight and therefore are an insurance risk. Figure 32.11 is an example of this sort of chart. There is widespread agreement that persons who are twenty-five percent heavier than the "ideal" are obese. Some mortality studies suggest that the "ideal" should be ten to fifteen pounds heavier than the charts indicate; adherents to various nutritional programs are convinced the values given in these charts should be less.

Table 32.4 Vitamins Necessary for Normal Cell Functioning

Vitamin	RDA* (milligrams)	Dietary Sources	Major Body Functions	Possible Outcomes of Deficiency	Possible Outcomes of Excess
Water-Soluble					
Vitamin B₁ (thiamine)	1.5	Pork, organ meats, whole grains, legumes	Coenzyme (thiamine pyrophosphate) in the removal of carbon dioxide	Beriberi (peripheral nerve changes, edema, heart failure)	None reported
Vitamin B₂ (riboflavin)	1.8	Widely distributed in foods	Constituent of two flavin nucleotide coenzymes involved in energy metabolism (FAD and FMN)	Reddened lips, cracks at corner of mouth (cheilosis), lesions of eye	None reported
Niacin	20	Liver, lean meats, grains, legumes (can be formed from tryptophan)	Constituent of two coenzymes involved in oxidation-reduction reactions (NAD⁺ and NADP⁺)	Pellagra (skin and gastrointestinal lesions, nervous, mental disorders)	Flushing, burning and tingling around neck, face, and hands
Vitamin B₆ (pyridoxine)	2	Meats, vegetables, whole grain cereals	Coenzyme (pyridoxal phosphate) involved in amino acid metabolism	Irritability, convulsions, muscular twitching, kidney stones	None reported
Pantothenic acid	5–10	Widely distributed in foods	Constituent of coenzyme A, which plays a central role in energy metabolism	Fatigue, sleep disturbances, impaired coordination, nausea (rare in humans)	None reported
Folic acid	0.4	Legumes, green vegetables, whole wheat products	Coenzyme (reduced form) in carbon transfer in nucleic acid and amino acid metabolism	Anemia, gastrointestinal disturbances, diarrhea, red tongue	None reported
Vitamin B₁₂	0.003	Muscle meats, eggs, dairy products	Coenzyme in carbon transfer in nucleic acid metabolism	Pernicious anemia, neurological disorders	None reported
Biotin	Not established. Usual diet provides 0.15–0.3	Legumes, vegetables, meats	Coenzyme in fat synthesis, amino acid metabolism, glycogen formation	Fatigue, depression, nausea, dermatitis, muscular pains	None reported
Choline	Not established. Usual diet provides 500–900	All foods containing phospholipids (egg yolk, liver, grains, legumes)	Constituent of phospholipids. Precursor of neurotransmitter acetylcholine	None reported for humans	None reported
Vitamin C (ascorbic acid)	45	Citrus fruits, tomatoes, green peppers, salad greens	Maintains intercellular matrix of cartilage, bone, and dentine. Important in collagen synthesis	Scurvy (degeneration of skin, teeth, blood vessels, epithelial hemorrhages)	Relatively nontoxic. Possibility of kidney stones
Fat-Soluble					
Vitamin A (retinol)	1	Provitamin A in green vegetables. Retinol in milk, butter, cheese, margarine	Constituent of rhodopsin (visual pigment). Maintenance of epithelial tissues	Xerophthalmia (keratinization of ocular tissue), night blindness, permanent blindness	Headache, vomiting, peeling of skin, anorexia, swelling of long bones, liver damage
Vitamin D	0.01	Cod liver oil, eggs, dairy products, margarine	Promotes bone growth, mineralization. Increases calcium absorption	Rickets (bone deformities) in children. Osteomalacia in adults	Vomiting, diarrhea, weight loss, kidney damage
Vitamin E (tocopherol)	15	Seeds, green leafy vegetables, margarine	Functions as an antitoxidant to prevent cell membrane damage	Possibly anemia; never observed in humans	Relatively nontoxic
Vitamin K (phylloquinone)	0.03	Green leafy vegetables. Small amount in cereals, fruits, and meats	Important in blood clotting (involved in formation of active prothrombin)	Deficiencies associated with severe bleeding, internal hemorrhages	Synthetic forms at high doses may cause jaundice

*Recommended daily allowance, for an adult male in good health.
From "The Requirements of Human Nutrition," by Nevin S. Scrimshaw and Vernon R. Young.

Table 32.5 Minerals Necessary for Normal Cell Functioning

Mineral	Amount in Adult Body (grams)	RDA* (milligrams)	Dietary Sources	Major Body Functions	Possible Outcomes of Deficiency	Possible Outcomes of Excess
Calcium	1,500	800	Milk, cheese, dark-green vegetables, dried legumes	Bone and tooth formation Blood clotting Nerve transmission	Stunted growth Rickets, osteoporosis Convulsions	Not reported for humans
Phosphorus	860	800	Milk, cheese, meat, poultry, grains	Bone and tooth formation Acid-base balance, ATP formation, etc.	Weakness, demineralization of bone, loss of calcium	Erosion of jaw (fossy jaw)
Sulfur	300	(Provided by sulfur amino acids)	Sulfur amino acids (methionine and cystine) in dietary proteins	Constituent of active tissue compounds, cartilage and tendon	Related to intake and deficiency of sulfur amino acids	Excess sulfur amino acid intake leads to poor growth
Potassium	180	2,500	Meats, milk, many fruits	Acid-base balance Body water balance Nerve function	Muscular weakness Paralysis	Muscular weakness Death
Chlorine	74	2,000	Common salt	Formation of gastric juice Acid-base balance	Muscle cramps Mental apathy Reduced appetite	Vomiting
Sodium	64	2,500	Common salt	Acid-base balance Body water balance Nerve function	Muscle cramps Mental apathy Reduced appetite	High blood pressure
Magnesium	25	350	Whole grains, green leafy vegetables	Activates enzymes. Involved in protein synthesis	Growth failure. Behavioral disturbances Weakness, spasms	Diarrhea
Iron	4.5	10	Eggs, lean meats, legumes, whole grains, green leafy vegetables	Constituent of hemoglobin and enzymes involved in energy metabolism	Iron-deficiency anemia (weakness, reduced resistance to infection)	Siderosis (iron deposition in tissues) Cirrhosis of liver
Fluorine	2.6	2	Drinking water, tea, seafood	May be important in maintenance of bone structure	Higher frequency of tooth decay	Mottling of teeth. Increased bone density. Neurological disturbances
Zinc	2	15	Widely distributed in foods	Constituent of enzymes involved in digestion	Growth failure Small sex glands	Fever, nausea, vomiting, diarrhea
Copper	0.1	2	Meats, drinking water	Constituent of enzymes associated with iron metabolism	Anemia, bone changes (rare in humans)	
Iodine	0.011	0.14	Marine fish and shellfish, dairy products	Constituent of thyroid hormones	Goiter (enlarged thyroid)	Very high intakes depress thyroid activity
Cobalt	0.0015	(Required as vitamin B_{12})	Organ and muscle meats, milk	Constituent of vitamin B_{12}	None reported for humans	Industrial exposure: dermatitis and diseases of red blood cells

*Recommended daily allowance, for an adult male in good health.
From "The Requirements of Human Nutrition," by Nevin S. Scrimshaw and Vernon R. Young.
Copyright © 1976 by Scientific American, Inc. All rights reserved.

Man's Height	Size of Frame		
	Small	Medium	Large
5' 2"	128–134	131–141	138–150
5' 3"	130–136	133–143	140–153
5' 4"	132–138	135–145	142–156
5' 5"	134–140	137–148	144–160
5' 6"	136–142	139–151	146–164
5' 7"	138–145	142–154	149–168
5' 8"	140–148	145–157	152–172
5' 9"	142–151	148–160	155–176
5'10"	144–154	151–163	158–180
5'11"	146–157	154–166	161–184
6' 0"	149–160	157–170	164–188
6' 1"	152–164	160–174	168–192
6' 2"	155–168	164–178	172–197
6' 3"	158–172	167–182	176–202
6' 4"	162–176	171–187	181–207

a

Woman's Height	Size of Frame		
	Small	Medium	Large
4'10"	102–111	109–121	118–131
4'11"	103–113	111–123	120–134
5' 0"	104–115	113–126	122–137
5' 1"	106–118	115–129	125–140
5' 2"	108–121	118–132	128–143
5' 3"	111–124	121–135	131–147
5' 4"	114–127	124–138	134–151
5' 5"	117–130	127–141	137–155
5' 6"	120–133	130–144	140–159
5' 7"	123–136	133–147	143–163
5' 8"	126–139	136–150	146–167
5' 9"	129–142	139–153	149–170
5'10"	132–145	142–158	152–173
5'11"	135–148	145–159	155–176
6' 0"	138–151	148–162	158–179

b

Figure 32.11 The "ideal weights" for adult men (**a**) and adult women (**b**), according to one insurance company in 1983. The values shown are for persons twenty-five to fifty-nine years old wearing shoes with one-inch heels and five pounds of clothing (men) or three pounds of clothing (women).

Dieting has become nearly epidemic in the United States especially, where starvation is not the problem that it is in much of the world. Millions are dieting every day. Unfortunately, in a growing number of cases dieting becomes an obsession that leads to a potentially fatal eating disorder called *anorexia nervosa*. The disorder occurs primarily in women in their teens and in their early twenties.

Disturbances in the hypothalamus may trigger some of the weight loss and lead to the skewed perception of body weight that characterizes anorexia nervosa. But it appears that emotional factors contribute more to the disorder. (For example, some individuals fear growing up in general and maturing sexually in particular; others have irrational expectations of what they can accomplish.) Severe cases require psychiatric treatment.

Another eating disorder on the rise is *bulimia* ("an oxlike appetite"). Some surveys show that at least twenty percent of college-age women are now suffering to varying degrees from this disorder. The bulimic person goes on eating binges, taking in enormous amounts of food, then vomits or purges the body with laxatives after each binge. In some cases, the person does this once a month; others go through the binge-purge routine several times a day. Some women start doing this because it seems like a simple way to lose weight. But with repeated vomiting, the stomach acids brought into the mouth can erode teeth to stubs; repeated purgings can severely damage the digestive tract. Psychiatric treatment is also used for severe cases of this disorder.

ORGANIC METABOLISM

So far, we have looked at the routes by which food molecules enter the internal environment. We have also looked at the types and proportions of organic molecules necessary for proper nutrition. Once those molecules are inside the body, some are used as building blocks for structural components of cells. Others are funneled into ATP-producing pathways. Figure 32.12 summarizes the main routes by which organic molecules enter and leave the body.

Figure 32.12 also shows the main routes by which organic molecules are shuffled and reshuffled once they are inside. With few exceptions (such as DNA), most of these molecules are continually being broken down, with some of their component parts picked up and used again in new molecules. At the molecular level, your body undergoes massive and sometimes rapid turnovers.

The Vertebrate Liver

The liver is central to the storage and interconversion of absorbed carbohydrates, fats, and proteins (Figure 32.12). It also helps regulate the concentrations of organic components of blood and removes many toxic substances from blood. Most hormones are inactivated in the liver, then sent to the kidneys for excretion from the body.

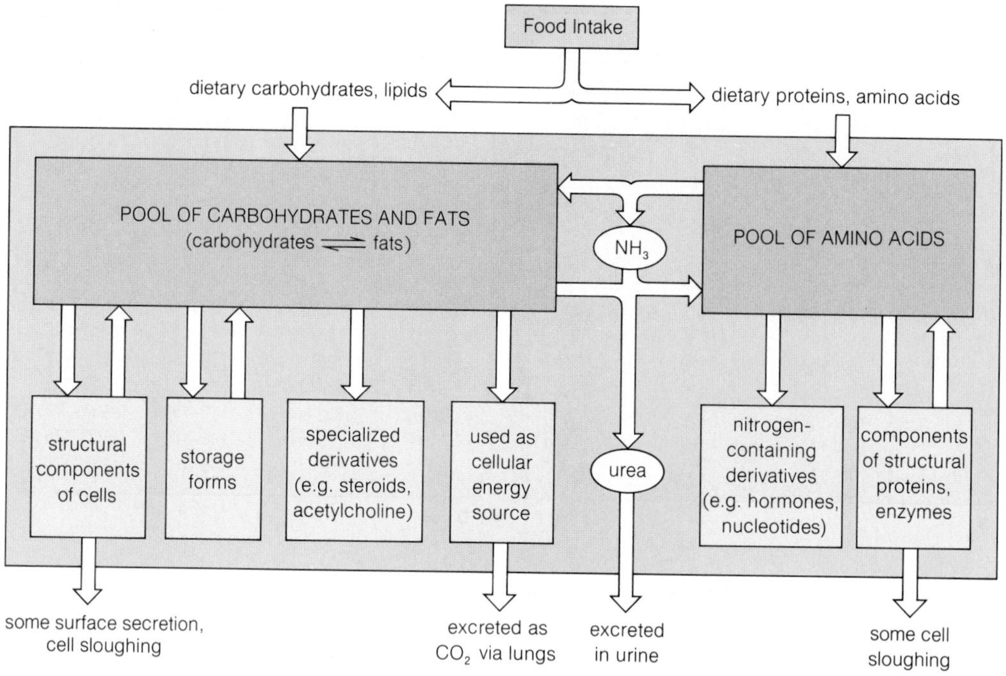

Figure 32.12 Summary of major pathways of organic metabolism. Urea formation occurs primarily in the liver. Carbohydrates, fats, and proteins are continually being broken down and resynthesized.

Table 32.6 Some Activities That Depend on Liver Functioning
1. Carbohydrate metabolism
2. Control over some aspects of plasma protein synthesis
3. Assembly and disassembly of certain proteins
4. Urea formation from nitrogen-containing wastes
5. Assembly and storage of some fats
6. Fat digestion (bile is formed by the liver)
7. Inactivation of many chemicals (such as hormones and some drugs)
8. Detoxification of many poisons
9. Degradation of worn-out red blood cells
10. Immune response (removal of some foreign particles)
11. Red blood cell formation (liver absorbs, stores factors needed for red blood cell maturation)

Glucose and amino acids absorbed across the intestinal wall are transported directly by capillaries to the hepatic portal vein, which leads to the liver capillary bed (Figure 29.6). In the liver, excess glucose is stored as glycogen or converted to fat. Here, too, excess amino acids are converted to forms that can be sent through the Krebs cycle (as an alternate energy source) or converted to fat.

When cells degrade amino acids, ammonia (NH_3) is a reaction product. Ammonia is potentially toxic to cells. However, it travels through the blood to the liver, where it is converted to urea (a much less toxic waste product). Urea is excreted from the body by way of the kidneys.

Table 32.6 summarizes the main functions of the liver.

Absorptive and Post-Absorptive States

In terms of the total nutritional picture, there are two functional states of organic metabolism:

absorptive state	*ingested organic molecules enter the bloodstream from the gastrointestinal tract*
post-absorptive state	*gastrointestinal tract is not supplying nutrients; the body draws from its internal pools of organic molecules*

During the absorptive state, the body builds up its pools of organic molecules. Excess carbohydrates and other dietary molecules are transformed mostly into fats, which are stored in adipose tissue. Some also are

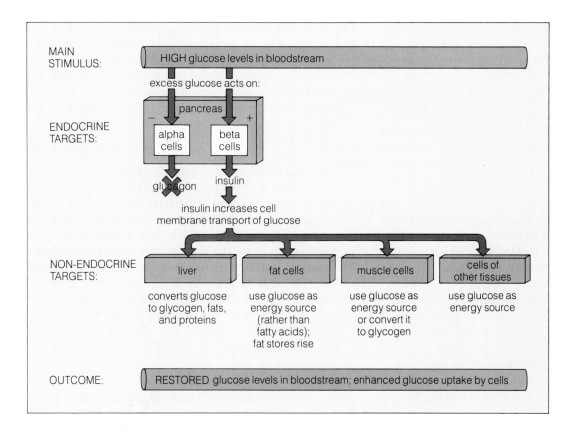

Figure 32.13 Main metabolic routes and the endocrine commands during the absorptive state.

converted to glycogen in the liver and in muscle tissue. Most cells use glucose as the primary energy source at this time; there is no net breakdown of protein in muscle or other tissues.

During the post-absorptive state, there is a notable shift in the type of food molecules used to support cell activities. A key factor in this shift is the need to provide brain cells with glucose, the major nutrient they use for energy.

When glucose is being absorbed, its concentrations in the bloodstream are readily maintained. But how does the body maintain blood glucose concentrations during the post-absorptive state? *First*, glycogen stores (particularly in the liver) are rapidly broken down to glucose, which is released into blood. *Second*, body proteins are broken down to provide amino acids, which are transported by the blood to the liver. There they are converted to glucose, which can be released into blood.

Most cells use fats as the major energy source during the post-absorptive state. Fats stored in adipose tissue are broken down into glycerol and fatty acids, which are released into blood. The glycerol can be converted to glucose in the liver; the circulating fatty acids are used by most cells in ATP production.

During the absorptive state, glucose moves into cells, where it can be used for energy and where the excess can be stored.

During the post-absorptive state, most cells use fat as the main energy source; stored fats are mobilized. Brain cells are kept supplied with glucose mainly by the conversion of amino acids into glucose by the liver.

Controls Over Organic Metabolism

Both endocrine and neural controls govern metabolism during the absorptive and post-absorptive states. The most important control agents are hormones secreted by clusters of endocrine cells in the pancreas (the islets of Langerhans). These clusters include alpha and beta cell types, which function antagonistically. *Beta cells* secrete **insulin**, a hormone that enhances glucose uptake, storage, and use by cells. *Alpha cells* secrete **glucagon**, a hormone that prods liver cells into converting glycogen into glucose and that inhibits glycogen synthesis.

Figure 32.13 illustrates the endocrine controls over organic metabolism during the absorptive state, which we might call times of "feasting." Figure 32.14 illustrates endocrine and neural controls at work during the post-absorptive state, including times of "fasting" (or starvation). A discussion of the control mechanisms themselves would be beyond the scope of this book. However, the following case study will give you a sense of the marvelous nature of their interactions.

Case Study: Feasting, Fasting, and Systems Integration

Suppose, this morning, you are vacationing in the mountains and decide on impulse to follow a forested trail. You fail to notice the wooden trail marker that bears the intriguing name, "Fat Man's Misery." As you walk down the tree-lined corridor, you are enjoying one of the benefits of discontinuous feeding. Having eaten a large breakfast, you have assured your cells of ongoing nourishment; you do not have to forage constantly amongst the ferns as, say, a nematode must do. Food partly digested in the stomach has already entered the small intestine. Right now, amino acids, simple sugars, and fatty acids are moving across the intestinal wall, then into the bloodstream.

With the surge of nutrients, glucose molecules are entering the bloodstream faster than your cells can use them. The level of blood glucose begins to rise slightly. However, your body has a homeostatic program for converting glucose into storage form when it is flooding in, then releasing some of the stores when it is scarce.

With the rise in blood glucose, pancreatic beta cells are called upon to secrete insulin. Blood concentrations of insulin rise—and the hormonal targets (liver, fat, and muscle cells) quickly begin using or storing the glucose molecules (Figure 32.13). At the same time, alpha cells are prevented from secreting glucagon—which slows the liver's conversion of stored glycogen into glucose.

What is the outcome? High levels of glucose that have entered the circulation from your gut move out of the blood and into cells, where it can be burned as fuel or stored for later use.

Even though you are no longer feeding your body, your brain cells have not lessened their high demands for glucose. Neither have your muscle cells, which are getting a strenuous workout. Little by little, blood glucose levels drop. Now endocrine activities shift in the pancreas. With less glucose binding to them, beta cells decrease their insulin output. With less glucose to inhibit them, alpha cells increase their glucagon output. When glucagon reaches your liver, it causes the conversion of glycogen back to glucose—which is returned to your blood. This prevents blood glucose from falling below levels required to maintain brain function.

But the best-laid balance of internal conditions can go astray when external conditions change. In your case, the "miserable" part of the trail has begun. You find yourself scrambling higher and higher on steep inclines. Suddenly you stop, surprised, in pain. You forgot to reckon with the lower oxygen pressure of mountain air, and your leg muscles cramped. Your body has already detected its deficiency of oxygen-carrying red blood cells at this altitude, but it will take days before enough additional red blood cells are available. In the meantime, your muscle cells are not being supplied with enough oxygen for the strenuous climb. They have switched to an anaerobic pathway in which lactate is the end product.

Again, systems interact to return your body to a homeostatic state. The body detects the reduced oxygen pressure and an accompanying increase in hydrogen ion concentrations in the cerebrospinal fluid. Nerve impulses course toward the respiratory center in the medulla. The result: the diaphragm and other muscles associated with inflating and deflating your lungs contract more rapidly. You breathe faster now, and more deeply. In the liver, lactate is converted to glucose—which is returned to the blood.

On checking the sun's position, you see it is well past noon. And guess what: you forgot about lunch. When you start the long walk back, the drop in blood glucose levels triggers new homeostatic mechanisms. Under hypothalamic commands, your adrenal medulla begins secreting epinephrine and norepinephrine. Its main targets: the liver, adipose tissue, and muscles. In the liver, glycogen synthesis stops. In body tissues generally, glucose uptake is blocked. In fat cells, fats are converted to fatty acids, which are routed to the liver, muscles, and other tissues as alternative energy sources (Figure 32.14). For every fatty acid molecule sent down metabolic pathways in those tissues, several glucose molecules are held in reserve for the brain.

You do get back to the start of the trail by sundown. However, your body had enough stored energy to sustain you for many more days, so the situation was never really desperate. The balance of blood sugar and fat is constantly monitored and controlled by the liver and hormones. Glucose levels only drop beyond the set point to stimulate glycogen conversion and fat conversion, and vice versa. It takes several days of fasting before blood sugar levels are markedly reduced.

Even after several days of fasting, your energy supplies would not have run out. Another hypothalamic command would have prodded your anterior pituitary into secreting ACTH (page 349). The ACTH would have signaled adrenal cortex cells to secrete glucocorticoid hormones, which have a potent effect on the synthesis of carbohydrates from proteins and on the further breakdown of fat. Slowly, in muscles and other tissues,

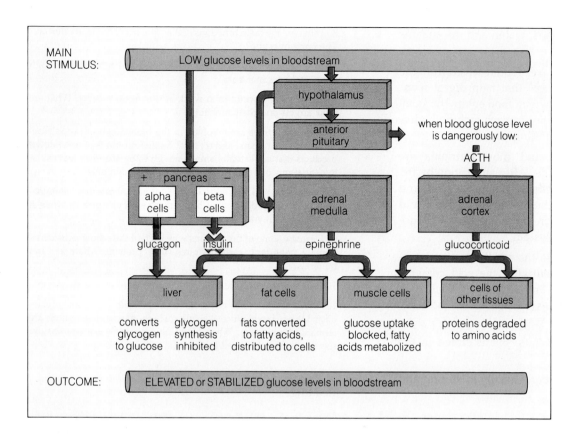

Figure 32.14 Main metabolic routes and the endocrine and neural commands during the post-absorptive state.

your body's proteins would have been disassembled. Amino acids from these structural tissues would have been used in the liver to build new glucose molecules—and once more your brain would have been kept active.

As extreme as this last pathway might be, it would be a small price to pay for keeping your brain functional enough to figure out how to take in more nutrients and bring you back to a homeostatic state.

SUMMARY

1. Nutrition has to do with all the processes by which a body takes in, digests, absorbs, and uses food.

2. A digestive system is a body cavity or tube which has four main functions:

  a. Motility: mechanical breakdown of ingested nutrients and elimination of unabsorbed residues from the body.

  b. Secretion: release of enzymes, hormones, and other substances that take part in digestion.

  c. Digestion: chemical reduction of ingested nutrients into particles, then into molecules small enough to cross the lining of the gut and thereby to reach the internal environment.

  d. Absorption: passage of digested nutrients from the gut lumen into the blood or lymph, which distributes them through the body.

3. The human digestive tract has regional subdivisions: the mouth, pharynx, esophagus, stomach, small intestine, large intestine (colon), rectum, and anus. Glands associated with digestion are the salivary glands, liver, gallbladder, and pancreas.

4. Coordinated contractions of the muscle layer of the gastrointestinal tract mix food with secretions (segmentation) and move it forward (peristalsis). Sphincters control the flow of contents from one region to another.

5. Controls over gastrointestinal activity operate in response to the volume and composition of food in the lumen of the tract. The response can be a change in muscle activity, a change in the secretory rate of hormones or enzymes, or both.

6. Saliva contains mucins, ions, water, and the enzyme salivary amylase.

7. The stomach stores and mixes food received from the esophagus, secretes substances that help digest food, and helps control the rate at which food enters the small intestine. The secretions of the stomach include hydrochloric acid and protein-degrading enzymes.

8. Digestion is completed and most nutrients are absorbed in the small intestine. After absorption by the intestinal mucosa, glucose, other monosaccharides, and most amino acids pass into the intestinal capillaries and then go directly to the liver. Fatty acids and monoglycerides enter the lymph system.

9. The pancreas secretes into the small intestine bicarbonate ions (which help neutralize the acid contents arriving from the stomach) and enzymes (which help digest proteins, carbohydrates, nucleic acids, and fats).

10. The liver secretes bile, a substance essential for the breakdown and digestion of fats. Bile is stored in the gallbladder between meals.

11. The large intestine functions mainly in storing and concentrating feces.

12. Nutritional energy is measured in kilocalories. To maintain weight and health, caloric intake must balance energy output.

13. Complex carbohydrates, which can be broken down to produce glucose, are the body's main energy source. Fats are produced by the body as the storage form of protein and carbohydrate. Dietary protein provides the amino acids needed for the body's synthesis of protein. Eight essential amino acids and a few essential fatty acids must be provided by the diet.

14. Vitamins and inorganic minerals must be supplied by the diet.

15. During the absorptive state, nutrients are being absorbed into the bloodstream by the digestive tract. Secretion of insulin by the pancreas increases and glucose moves into cells, where it can be used for energy and where the excess can be stored.

16. During the post-absorptive state, the gastrointestinal tract is not supplying nutrients and the body draws from its internal pools of organic molecules. Secretion of glucagon by the pancreas increases.

Review Questions

1. Study Figure 32.2. Then, on your own, diagram the connections between metabolism and the digestive, circulatory, and respiratory systems.

2. Explain the difference between digestion and absorption.

3. What are the main functions of the stomach? The small intestine? The large intestine?

4. In what ways are food materials mixed and propelled through the gastrointestinal tract?

5. Name three hormones at work in the digestive tract. What are their targets and their functions?

6. Which enzymes are involved in the breakdown of (a) polysaccharides, (b) proteins, and (c) fats? Name four kinds of breakdown products that are actually small enough to be absorbed across the intestinal mucosa and into the internal environment.

7. A glass of milk contains lactose, protein, butterfat, vitamins, and minerals. Explain what happens to each component when it passes through your digestive tract.

8. Describe some of the reasons why each of the following is nutritionally important: carbohydrates, fats, proteins, vitamins, and minerals.

9. Describe some of the functions of the liver.

10. What are the roles of insulin and glucagon in organic metabolism? When blood glucose levels are high, glucagon secretions are (enhanced/inhibited). When blood glucose levels are low, insulin secretions are (enhanced/inhibited).

Readings

Clemente, C. 1981. *Anatomy: A Regional Atlas of the Human Body*. Second edition. Baltimore: Urban and Schwartzenberg. Stunning, detailed illustrations of human anatomy. Drawings of the gastrointestinal tract are among the best available.

Hamilton, W. 1982. *Nutrition: Concepts and Controversies*. Menlo Park, California: West. Information on digestion, nutrition, diet, and health; evaluates fads and erroneous ideas about nutrition in light of current research.

Kessel, R. G., and R. H. Kardon. 1979. *Tissues and Organs: A Text-Atlas of Scanning Electron Microscopy*. San Francisco: Freeman. Outstanding, unique micrographs, accompanied by well-written descriptions of major tissues and organs.

Krause, M., and L. Mahan. 1984. *Food, Nutrition, and Diet Therapy*. Seventh edition. Philadelphia: Saunders.

Kretchmer, N., and W. van B. Robertson. 1978. *Human Nutrition*. San Francisco: Freeman. Excellent collection of articles from *Scientific American* that consider nutrition at the cellular level and the global level. Paperback.

Vander, A., J. Sherman, and D. Luciano. 1985. *Human Physiology: The Mechanisms of Body Function*. Fourth edition. New York: McGraw-Hill.

Animals, it appears, first evolved in the shallow waters of ancient seas. We can therefore assume that when complex tissues and organ systems developed, they were geared to operating in a salty fluid where temperatures remained relatively stable. In mid-Paleozoic times (about 375 million years ago), some animals began invading the land. In part, they were able to do so by providing their living cells with an "internal environment" that approximated conditions in the seas. But there were new challenges on land, where winds and radiant energy from the sun could dehydrate the animal body, where water might not be available to replenish the body fluids being lost, and where most of the available water was fresh, not salty. In response to these threats to the stability of the internal environment, animals underwent modifications in their body plans, physiology, and behavior.

Today, as then, animals make responses to physical conditions over which they often have little or no control. For example, in cold winters, land-dwelling animals are obligated to give up heat to the surroundings. Yet many of the responses can be adjusted in precisely controlled ways. In essence, *all animals make controlled as well as obligatory exchanges with the external environment to maintain a hospitable internal environment.* Let's look at the obligatory and controlled exchanges necessary to maintain the temperature of the animal body, then at the exchanges necessary to maintain its water and solute balances.

33

TEMPERATURE CONTROL AND FLUID REGULATION

Figure 33.1 After a cold night in their Kalahari Desert dens, meerkats stand seemingly at attention, allowing a large surface area of their bodies to absorb the warm rays of the morning sun. Like many other animals, meerkats rely on such behavioral adaptations to help maintain their body temperature even though the outside temperature changes.

Table 33.1 Temperatures Favorable for Metabolism, Compared With Environmental Temperatures	
Temperatures generally favorable for metabolism:	0°C to 40°C (32°F to 104°F)
Air temperatures above land surfaces:	−70°C to +85°C (−94°F to +185°F)
Surface temperatures of open oceans:	−2°C to +30°C (+28.4°F to +86°F)

CONTROL OF BODY TEMPERATURE

Temperatures Suitable for Life

Enzymes are essential for metabolism, and the enzymes of most organisms usually remain functional within the range between 0°C and 40°C (Table 33.1). At temperatures above 40°C, enzyme molecules cease to function properly as a result of denaturation. (With denaturation, the chemical interactions holding a molecule together in its three-dimensional shape are disrupted.) Also, for every ten-degree drop in temperature, the rate of enzyme activity generally decreases by at least half. Given all the links between enzyme-mediated reactions, it is easy to imagine how the body's metabolic balance can be upset when body temperatures exceed or fall below the proper range.

The question becomes this: How do animals maintain a rather constant body temperature? They do this by balancing heat gains and heat losses.

Heat Gains and Heat Losses

In complex animals, metabolic reactions are proceeding simultaneously in thousands, millions, even many billions of cells. The heat produced as a by-product of this metabolic activity contributes to body temperature. Heat absorbed from the environment also affects what the body temperature will be, and so do heat losses to the environment:

body heat	=	heat produced	+	heat gained	−	heat lost

The body exchanges heat with the surroundings through four processes, called radiation, conduction, convection, and evaporation.

Radiation is a process by which the sun, the animal body, and all other objects having mass emit energy in the form of infrared and other wavelengths. Some of this energy is converted to heat when it is absorbed at the body surface. Thus animals can gain heat by absorbing radiant energy from various objects in the environment as well as from the sun (Figure 33.1).

Conduction refers to the direct transfer of heat energy from one object to another as a result of collisions between each other's atoms and molecules. Because heat moves down thermal gradients, we lose heat by conduction when (for example) we sit on cold ground; and we gain heat when we sit on warm sand at the beach.

Convection is a process of heat transfer by way of moving fluid, such as air or water currents. This process involves both conduction (by which heat moves down the thermal gradient between the body surface and the air or water next to it) and mass transfer (by which currents carry the heat away from or toward the body). For example, if your skin temperature is higher than the air temperature, you will lose heat by convection. Even if there were no breeze at all, your body would still lose heat by creating its own convective current (air rises as it is heated and becomes less dense).

Evaporation, recall, is a process whereby water changes from a liquid to a gaseous state. Humans and a few other mammals have sweat glands that actively move certain solutions through pores to the skin surface. When the skin temperature is high, water at the surface absorbs enough thermal energy from the skin to break the hydrogen bonds holding individual molecules of water together. Some of the molecules depart from the surface, and some body heat is dissipated.

Compared to all other factors contributing to thermal balance, heat loss by way of evaporation has the greatest effect. For example, a horse running in the desert sweats profusely to balance the enormous heat gain resulting from the combination of strenuous muscular activity and the hot environment. As long as the environmental humidity is low enough to permit complete evaporation, sweating can rid the body of the excess heat. Keep in mind that this means of heat loss requires that the water on the surface evaporate. Sweat dripping from the skin has no cooling effect at all.

Animals gain heat by way of metabolism, and they exchange heat with the environment by the processes called radiation, conduction, convection, and evaporation.

Classification of Animals Based on Temperature

Ectotherms. For most animals, body temperature is dictated largely by heat gained from the environment rather than by metabolic heat. Such animals are said to be **ectotherms** (meaning "heat from outside"). Ectotherms have low rates of metabolism and are poorly insulated. They conduct heat rapidly to the environment—and they also absorb heat rapidly from it. These animals maintain a reasonably constant body temperature by *behavioral temperature regulation.*

For example, lizards and other reptiles move about, putting themselves in places where they will minimize heat or cold stress to their body. To warm up, they move out of shade and orient themselves with respect to the sun in order to expose the maximum surface area of the body for heat absorption. They bask on rocks that absorbed heat earlier in the day from the sun. Given the near-absence of insulation, a small lizard can heat up as fast as 1°C a minute.

Of course, the lizard body loses heat just as rapidly when the sun goes down. When its body temperature drops below the range favorable for metabolic activity, the lizard becomes almost immobilized. Before that happens, the lizard usually crawls into a crevice or under a rock, where it is not as vulnerable to predators.

Endotherms and Heterotherms. Most birds and mammals, as well as a few insects and other invertebrates, are **endotherms** ("heat from within"). The body temperature of endotherms is dictated largely by metabolic activity and by precise controls over heat conservation and disposal. Endotherms also respond to temperature stress by making behavioral changes that supplement their precisely controlled physiological activity.

Because most endotherms have a high metabolic rate, they pay a high energy cost for the controls that allow them to maintain a more active life-style. For example, a foraging mouse uses up to thirty times more energy than a foraging lizard of the same weight.

Yet such energy outlays are worth the cost, for they are the main reason why endotherms can be active under a wide range of temperatures. During cold nights or cold parts of the year, mammals and other endotherms can still forage. These animals show sustained stamina when, for example, they are being pursued by a predator or digging a burrow.

Other adaptations of endotherms are related to conserving or dissipating the heat associated with a high rate of metabolism. For example, closely related species of mammals are sometimes distributed in different parts of the world, and the ones in cold regions (such as the arctic hare) are generally more massive than their relatives in warmer regions (such as the jackrabbit). Compared to lightweight or thin-limbed bodies, the massive body has more cells to generate heat and proportionally less surface area for losing heat to the environment. Similarly, mammals living in cool to cold regions have fur and layers of fat. These forms of insulation reduce heat loss.

Like ectotherms, the endotherms also make behavioral responses to heat stress. Some mammals that live in deserts of north temperate regions have much lower body temperatures than the daytime temperature of the air and the ground surface. Temperatures often drop considerably at night, and the air is usually cool or cold in winter. As a result, the soil well below the surface never heats up much, and it is here that most desert rodents and other mammals find refuge from the daytime heat. Typically these animals forage by night and spend the hottest part of the day in burrows or in the shade of bushes or rock outcroppings.

Some birds and mammals fall between the ectothermic and endothermic categories. Part of the time, these so-called **heterotherms** allow their body temperature to fluctuate as ectotherms do—and at other times they control heat exchanges as endotherms do. Tiny hummingbirds have very high metabolic rates, and they devote much of the day to locating and sipping nectar as an energy source for metabolism. Because hummingbirds do not forage at night, they would rapidly run out of energy unless their metabolic rates decreased considerably. At night, these birds enter a sleeplike state and their body temperature approaches that of their cooler surroundings.

Advantages of Ectothermy Versus Endothermy. In general, ectotherms are at an advantage in the warm, humid tropics. They do not have to expend much energy to maintain body temperature, and more energy can be devoted to other tasks, including reproduction. Indeed, in the tropics, reptiles are far more abundant than mammals in terms of the number of species and the number of individuals. However, in moderate to cold environments, endotherms have the advantage and are more abundant. Their high metabolic rates allow some endotherms to occupy even the polar regions, where you would never find a lizard.

The body temperature of different animal groups is determined by their rate of metabolic activity and by anatomical, behavioral, and physiological adaptations.

Temperature Regulation in Mammals

Environmental temperatures can change quickly, and exercise and other activities alter the metabolic rate for the mammalian body. As a result, the normal core temperature increases or decreases slightly. ("Core" refers to the body's internal temperature, as opposed to temperatures near its surface.) Even so, the normal temperature is rapidly restored through feedback control

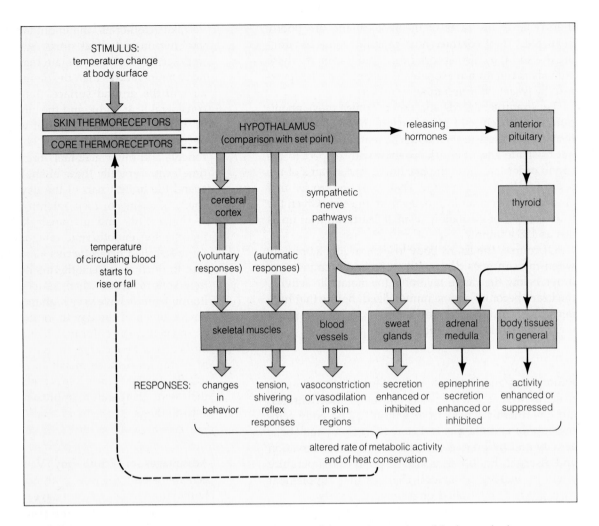

Figure 33.2 Overview of the feedback relationships that control the core temperature of the human body. The dashed line shows how the feedback loop is completed. The purple arrows indicate the main control pathways.

mechanisms. As Figure 33.2 indicates, these mechanisms operate through the brain region called the hypothalamus.

Responses to Cold Stress. When the environmental temperature falls, thermoreceptors in the skin send signals to the hypothalamus, which sends out commands along motor neurons that lead to muscles. Skeletal muscle activity increases and leads to **shivering** (rhythmic tremors in which the muscles contract about ten to twenty times per second). Within seconds or a few minutes, heat production throughout the body increases several times over.

Besides the shivering response, various behavioral responses also combat cold stress. In cold weather, many small mammals such as cats curl up tightly into a ball when they rest; you might stomp your feet or clap your hands.

Responses to cold stress also include changes in the diameter of peripheral blood vessels that carry heat to the body surface. Under signals from the hypothalamus, smooth muscle in the walls of peripheral blood vessels contracts and leads to vasoconstriction. To give you an idea of the effectiveness of this response, all but one percent of the blood flowing to the skin of your fingers is curtailed when they are exposed to the cold.

Severe cold exposure also seems to trigger increased metabolic rates as a result of hormone secretions (Figure 33.2). At least in the initial responses to cold stress, this so-called "nonshivering" heat production is not as important as the changes in muscle activity.

If defenses against cold are inadequate, the animal becomes *hypothermic*; its core temperature falls below normal. In humans, a drop of only a few degrees interferes with brain function and confusion results; further cooling can lead to coma and death (see *Commentary*). Many animals can recover from profound hypothermia. However, cells that have been frozen are destroyed

Falling Overboard and the Odds for Survival

In 1912, the ocean liner *Titanic* set out from Europe on her maiden voyage to America. In that same year, a huge chunk of the leading edge of a Greenland glacier broke off and began floating out to sea. Late at night on April 14, off the coast of Newfoundland, the iceberg and the *Titanic* made their ill-fated rendezvous. Lifeboats and survival drills had been neglected, and only about a fourth of the 2,000 people on board managed to scramble into the lifeboats that could be launched. What happened to the rest of the passengers? Within two hours, rescue ships were on the scene—yet 1,513 bodies were recovered from a calm sea. All were wearing life jackets. None had drowned. Probably every one of those individuals had died from hypothermia—from a drop in body temperature below tolerance levels. The following are responses at the body temperatures indicated:

Temperature	Response
36°–34°C (about 95°F)	Shivering response, increase in respiration. Increase in metabolic heat output. Constriction of peripheral blood vessels; blood is routed to deeper regions. Dizziness and nausea set in.
33°–32°C (about 91°F)	Shivering response stops. Metabolic heat output drops.
31°–30°C (about 86°F)	Capacity for voluntary motion is lost. Eye and tendon reflexes inhibited. Consciousness is lost. Cardiac muscle action becomes irregular.
26°–24°C (about 77°F)	Ventricular fibrillation sets in. Death follows.

(unless they have been frozen by special laboratory procedures). Tissue destruction through localized freezing is known as frostbite.

Responses to Heat Stress. When environmental temperatures rise, peripheral blood vessels dilate. More blood flows from deeper body regions to the skin regions, where the excess heat it carries can be dissipated (Figure 33.3). Heat is also dissipated through evaporation. Moreover, signals from the hypothalamus lead to slowdowns in skeletal muscle activity, hence to a decline in heat production.

In your own body, signals from the hypothalamus stimulate the secretion of sweat from the 2.5 million or so sweat glands in your skin. (With extreme sweating, as might occur in a marathon race, copious amounts of water and an important salt—sodium chloride—are lost from the body. Such losses can change the character of the internal environment to the extent that the individual collapses and faints.) Some mammals sweat very little or not at all, but they make other responses to heat stress. Dogs, for example, are among the mammals that pant. "Panting" refers to shallow, rapid breathing that increases the water loss from the respiratory surface of the lungs. In this response, evaporative cooling occurs

Figure 33.3 A jackrabbit (*Lepus californicus*) cooling off on a hot summer day in the mountains of Arizona. Notice the dilated blood vessels in its large ears. Both the large surface area of the ears and the extensive vascularization are useful for dissipating heat (by way of convection and radiation).

Table 32.2 Summary of Mammalian Responses to Cold Stress and to Heat Stress

Environmental Stimulus	Main Responses	Outcome
Drop in temperature	Vasoconstriction of blood vessels in skin; changes in behavior (e.g., curling up the body to reduce surface area exposed to the environment)	Heat is conserved
	Increased muscle activity; shivering; nonshivering heat production	Heat production increases
Rise in temperature	Vasodilation of blood vessels in skin; sweating; changes in behavior; panting	Heat is dissipated from body
	Decreased muscle activity	Heat production decreases

Table 33.3 Normal Balance Between Water Gain and Water Loss in Humans and in Kangaroo Rats

Organism	Water Gain (milliliters)		Water Loss (milliliters)	
Adult human (measured on daily basis)	Ingested in solids:	1200	Urine:	1500
	Ingested as liquids:	1000	Feces:	100
	Metabolically derived:	350	Evaporation:	950
		2550		2550
Kangaroo rat (measured over four weeks)	Ingested in solids:	6.0	Urine:	13.5
	Ingested as liquids:	0	Feces:	2.6
	Metabolically derived:	54.0	Evaporation:	43.9
		60.0		60.0

in the mouth, on the tongue, and on the lining of the nasal cavity.

Sometimes heat stress is so great that changes in peripheral blood flow and evaporative heat loss are insufficient. Then, the core temperature rises; the animal becomes *hyperthermic*. In humans and other endotherms, an increase of only a few degrees above normal can be dangerous.

Fever. During a *fever*, the hypothalamus actually resets the body's "thermostat" that dictates what the core temperature should be. The same response mechanisms are brought into play, but they are carried out to maintain a higher temperature! At the onset of fever, there is a pronounced decrease in heat loss and a pronounced increase in heat production, leading to an elevated core temperature. At that time the person feels chilled. When the fever breaks, peripheral vasodilation and sweating increase dramatically, and both reduce the core temperature back to normal.

It appears that fever may be an important defense mechanism against infections, and perhaps against cancer, for fever seems to increase the effectiveness of the body's key agents of immunity. Given this possibility, questions are being raised about the widespread practice of administering aspirin and other drugs to suppress modest fevers. (There is no question that such drugs must be used in cases of extreme fevers.)

Table 33.2 summarizes the controlled responses to cold stress and to heat stress in mammals.

CONTROL OF EXTRACELLULAR FLUID

Water Gains and Losses

Just as the mammalian body exchanges heat with its surroundings in obligatory and controlled ways, so does it exchange water and solutes. Ordinarily, water losses are balanced precisely by water gains (Table 33.3). The body *gains* water through two processes:

1. Absorption of water from liquids and solid foods in the gastrointestinal tract.

2. Metabolism (specifically, the breakdown of carbohydrates and other organic molecules in reactions that yield water as a by-product).

The gain of water is influenced by thirst behavior. This behavior is under the control of the hypothalamus, as will be described later in the chapter.

The body *loses* water by several processes, the most important of which are these:

1. Evaporation from the respiratory surface.
2. Evaporation from cells of skin.
3. Sweating.
4. Elimination by way of the gastrointestinal tract.
5. Excretion from the urinary system.

Often, the first two processes listed are called "insensible water losses" because the individual is not aware that they are taking place. Evaporation is controlled by temperature-regulating mechanisms, and very little water leaves the body in feces. The process of greatest importance in controlling water loss is **excretion**: the elimination of excess water as well as excess (or harmful) solutes from the internal environment by way of organs called the kidneys (Figure 33.4).

Solute Gains and Losses

Aside from oxygen (which is absorbed at the respiratory surface), diverse solutes are added to the internal environment by three processes:

1. Absorption from the gastrointestinal tract. The absorbed substances include *nutrients* such as glucose (used as energy sources and as building blocks in synthesis reactions) as well as drugs and food additives. They also include *mineral ions*, such as sodium and potassium ions.
2. Secretion (such as hormone secretions).
3. Metabolism (specifically, the *waste products* of degradative reactions).

Carbon dioxide is the most abundant waste of metabolism, and it is eliminated from the body at the respiratory surface. Aside from carbon dioxide, the main metabolic wastes that must be eliminated from the body are as follows:

1. *Ammonia*, formed in "deamination" reactions whereby amino groups are stripped from amino acids. If allowed to accumulate in the body, ammonia can be highly toxic.
2. *Urea*, produced in the liver in reactions that link two ammonia molecules to carbon dioxide. Urea is the main nitrogen-containing waste product of protein breakdown and is relatively harmless.
3. *Uric acid*, formed in reactions that degrade nucleic acids. If allowed to accumulate, uric acid can crystallize (and sometimes collect in the joints).

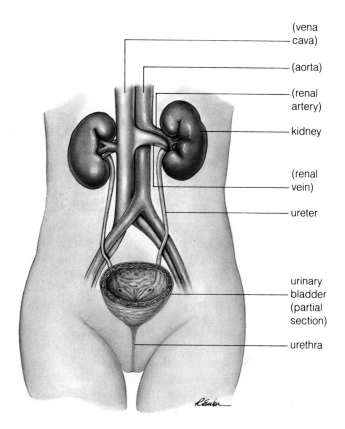

Figure 33.4 Components of the human urinary system.

(vena cava)
(aorta)
(renal artery)
kidney
(renal vein)
ureter
urinary bladder (partial section)
urethra

Figure 33.5 Anatomy of the kidney. (**a**) A human kidney, cutaway view. (**b**) Diagram of a nephron, the functional unit of the kidney. (**c**) Blood vessels associated with the nephron. The glomerular capillaries are shown in detail in Figure 33.10. The peritubular capillaries thread profusely around the tubular parts of the nephron.

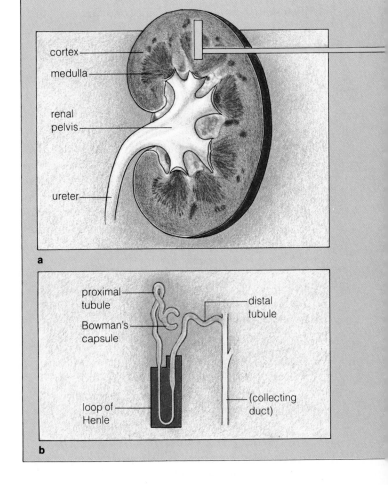

What we have, then, are ongoing inputs of water, nutrients, and ions and the production of metabolic wastes. Let's take a look at the major system for eliminating excess amounts of these substances so that stable conditions prevail in the internal environment.

Urinary System of Mammals

Urine is a fluid composed of the body's excess water and mineral ions, organic wastes, and other substances that have not been metabolized. It is formed by precisely controlled processes in the **kidneys**, a pair of fist-sized organs located behind the back wall of the abdominal cavity, on either side of the aorta (Figure 33.4). The kidneys also take part in endocrine activity. As you will see, they secrete renin and other substances that are components of hormone systems.

Each kidney has a coat of connective tissue, called the renal capsule (from the Latin *renes*, meaning kidney). The *cortex* makes up more than half of the mass of the kidney. It extends from the capsule toward the center of the kidney and surrounds a number of lobes. The inner portion of each renal lobe is called a *medulla*; the outer portion is referred to as its cortical region (Figure 33.5).

Urine formation begins in the cortical region of each lobe, which is composed of a remarkable number of slender tubes (called nephrons) and blood vessels. Water and solutes filter out of the blood and enter these tubes. Most of the filtrate is reabsorbed, but a portion moves into other tubes called the *collecting ducts*. The fluid that leaves the collecting ducts is the final urine. It eventually flows into the *renal pelvis*, which is the kidney's central cavity.

The renal pelvis is continuous with the **ureter**, a tube that carries urine to a storage organ called the **urinary bladder**. Urine from two ureters (one from each kidney) accumulates in this organ. It leaves through a single tube, the **urethra**, which opens to the body surface at the end of the penis (in males) or just in front of the vaginal entrance (in females). The two kidneys, two ureters, urinary bladder, and urethra constitute the **urinary system** of mammals (Figure 33.4).

Controls exist over urine flow from the body. As a urinary bladder fills, tension increases in its strong, smooth-muscled walls. In a reflex response to the increased tension, muscles preventing the flow of urine into the urethra relax; simultaneously, the bladder walls contract and force fluid through the urethra. This reflex response, called *urination*, is basically involuntary. However, it can be consciously inhibited by neural commands. The capacity for voluntary control emerges early in childhood, when the cerebral cortex has developed to the point that it can monitor signals from stretch receptors located in bladder walls. Maturation of this sensory pathway does not occur until children are about two years old.

Nephrons. Urine is formed in long, slender tubes called **nephrons**. There are more than a million of these tubes in each kidney. The tube wall is only a single layer of epithelial cells, but the cells and cell junctions differ

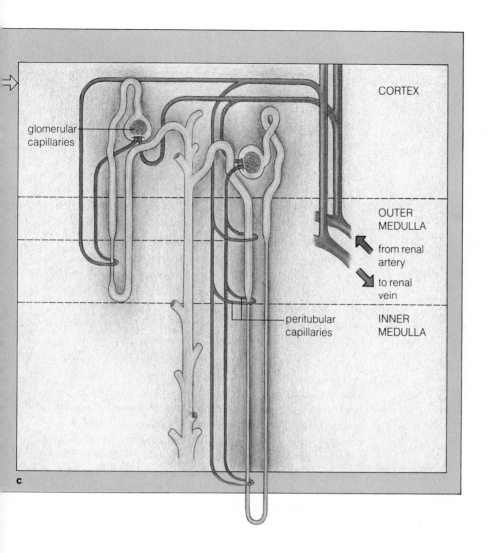

glomerular
capillaries

CORTEX

OUTER
MEDULLA

from renal
artery

to renal
vein

peritubular
capillaries

INNER
MEDULLA

c

in structure and function along the length of the nephron. (For example, some wall regions are highly permeable to water and solutes; other wall regions bar the passage of solutes except at active transport systems, which are built into the cell membranes.)

The start of each nephron is a blood-filtering unit called **Bowman's capsule**, which looks rather like a tennis ball that has been punched in on one side. Here, the nephron wall balloons around a compact cluster of blood capillaries (Figure 33.5c). All of these capsules are located in the cortical region of the renal lobes. Each capsule is the entrance for a tubular section of the nephron, called the **proximal tubule**, which in turn leads into a thin, hairpin-shaped section. This so-called **loop of Henle** descends toward or plunges into the medulla, forms a sharp turn, then thickens again (Figure 33.5b). The last section of the nephron is the **distal tubule** (the part most distant from Bowman's capsule).

The distal portions of neighboring nephrons merge to form one of the preliminary collecting ducts for urine.

In turn, the collecting ducts merge with one another to form larger ducts as they approach the region of the renal pelvis.

Blood Vessels at the Nephron.   Blood is carried to the kidneys by the renal arteries, which branch off the main artery from the heart (the aorta). The renal artery branches into smaller arteries, then into a series of arterioles. Each arteriole is said to be "afferent," for it carries blood *into* a Bowman's capsule. Inside, the arteriole branches into a cluster of tiny blood vessels known as **glomerular capillaries**. A rather unusual vascular connection is made here, for the glomerular capillaries do not merge into veins. Instead they converge to form an "efferent" (outgoing) arteriole, which divides into *another* set of capillaries. These **peritubular capillaries** thread profusely around the tubular portions of the nephron (Figure 33.5c). Eventually they merge to form veins, which carry blood out of the kidney.

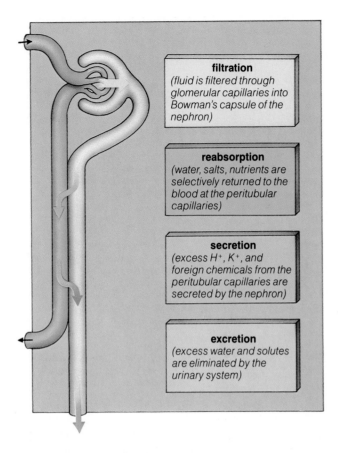

filtration
(fluid is filtered through glomerular capillaries into Bowman's capsule of the nephron)

reabsorption
(water, salts, nutrients are selectively returned to the blood at the peritubular capillaries)

secretion
(excess H⁺, K⁺, and foreign chemicals from the peritubular capillaries are secreted by the nephron)

excretion
(excess water and solutes are eliminated by the urinary system)

Figure 33.6 Overview of the processes involved in the formation and excretion of urine in mammals.

Overview of Urine Formation

Figure 33.6 and the following list provide us with an overview of the processes that lead to urine formation in the nephron:

1. **Filtration** occurs at Bowman's capsule. Blood pressure (generated by contractions of the heart) causes the *bulk flow* of water and solutes out of the glomerular capillaries and into the space inside the capsule. The blood is said to be filtered because blood cells, proteins, and other large solutes are left behind as the smaller solutes (such as glucose, sodium, and urea) are forced out. The filtrate passes into the proximal tubule of the nephron.

2. **Reabsorption** occurs along the tubular sections of the nephron, which are in intimate contact with the peritubular capillaries. With this process, most of the water and usable solutes are reclaimed by the body. They move *out* of the nephron by diffusion or active transport, then back into the bloodstream.

3. **Secretion** occurs in the direction opposite to reabsorption. With this process, hydrogen ions, potassium ions, and a few other substances normally found in the body move *out* of the peritubular capillaries, then are selectively transported across the wall of the nephron, into the urine. The extracellular concentrations of these substances must be regulated precisely. Secretion also rids the body of foreign substances (such as penicillin), uric acid, the products of hemoglobin breakdown, and other wastes.

Through these three processes, urine is formed. The final urine contains waste products as well as water and solutes in excess of the amounts necessary to maintain the extracellular fluid.

A Closer Look at Filtration

Rate of Filtration. As we have seen, capillaries in general are freely permeable to water and small solutes. Yet compared to capillaries outside the kidney, the glomerular capillaries permit filtration to occur on a massive scale. (You may wish to refer to the discussion of filtration on page 395.) Each day, more blood flows through the kidneys than through any other organ except the lungs. That flow amounts to about one-fourth of the cardiac output.

The massive filtration is possible for two reasons. *First,* the afferent arterioles carrying blood to the nephron have a wider diameter (hence less resistance to flow) than most arterioles do. Thus the hydrostatic pressure (caused by heart contractions) does not drop as much when blood flows through them. The hydrostatic

Kidney Failure, Bypass Measures, and Transplants

Sometimes the kidneys can no longer perform their filtration, reabsorption, and secretion tasks. For example, diabetes or immunological reactions can alter the filtering membranes in Bowman's capsule and thereby reduce or stop urine formation. The leading cause of kidney failure is *glomerulonephritis*, an inflammatory disorder in which the glomeruli are damaged. In the United States alone, an estimated 3 million individuals in all age groups suffer from some kidney disorder.

When the kidneys malfunction, solute concentrations in the blood are not regulated properly. Substances such as potassium ions as well as various toxic products of protein metabolism can accumulate in the bloodstream. The build-up may lead to nausea, fatigue, loss of memory and, in advanced cases, death. A *kidney dialysis machine* can be employed to restore the proper solute balances. The machine is often called an artificial kidney, not because it resembles the natural organ but because the end result is the same: concentrations of substances are regulated by their selective removal from (and addition to) the bloodstream.

The artificial kidney is based on dialysis: the separation of substances across a membrane between solutions of differing concentrations. In *hemodialysis*, a patient is plugged into the machine by tubes leading from an artery or a vein. Blood is then pumped through narrow tubes located in a warm-water bath. The bath contains a precisely balanced mix of salts, glucose, acetate, amino acids, and other substances that set up the proper gradients with the blood flowing through the tubes. Thus, substances at too high a concentration in the patient's blood will diffuse into the dialysis fluid.

A similar effect can be obtained in some patients by *peritoneal dialysis*, a process in which a fluid of appropriate composition is instilled into the abdominal cavity and, after a suitable interval, drained out. In this process, the lining of the cavity (the peritoneum) serves as the dialysis "membrane."

On the average, hemodialysis takes about four to five hours. The machine does not approach the kidney's efficiency, and blood must circulate over and over again through the tubes before the solute concentrations of the internal environment are improved. Afflicted persons must be treated three times a week.

For temporary disorders, the artificial kidney is used as a bypass measure until normal kidney function resumes. For chronic kidney disorders, it must be used for the remainder of the patient's life or until a functional kidney is transplanted. With treatment and with controlled diet, many individuals are able to resume normal activity.

Transplant surgery is expensive, but it is far less than the cost of year after year of hemodialysis (which is about 25,000 to 30,000 dollars annually). However, giving up a kidney is something that many healthy people are reluctant to do; donors are scarce. More than half the kidneys donated come from individuals who have just died as a result of accidents. Imagine yourself in the ethical position of being severely afflicted with a kidney disorder, waiting and half-hoping for someone else's death. Besides, about a third of all kidney transplants induce an immune response and are rejected. Is either dialysis treatment or the transplant surgery worth the cost? Would you ask the question if you yourself became one of the stricken, and would your answer be the same?

pressure is further enhanced because the glomerular capillaries do not deliver blood directly to the low-pressure venous system; they deliver it to an efferent arteriole with a high resistance to flow. *Second*, the glomerular capillaries are extremely porous. Their pores are not large enough to allow blood cells, platelets, or proteins to escape, but they are 10 to 100 times more permeable to water and small solutes than other capillaries in the body. Because of the higher hydrostatic pressure and the greater capillary permeability, the kidneys can easily filter an average flow rate of about 180 liters (45 gallons) per day!

Controls Over Filtration. The blood flow through the kidneys (hence its filtration rate) is subject to neural and local controls. The neural controls are concerned more with homeostasis for the body as a whole. For example, while you exercise, your brain sends out commands along sympathetic nerves to smooth muscle fibers in the walls of afferent arterioles. The muscle fibers contract, causing the arterioles to constrict. The constriction cuts down the flow of blood through the kidneys, so more blood can be diverted to the heart and skeletal muscles to sustain your increased body activity. Of course, the resulting drop in blood pressure in the kidneys leads to a decreased rate of urine formation.

The blood flow through the kidneys is also self-regulated. When arterial blood pressure falls, locally produced chemicals stimulate the afferent arterioles into dilating, so more blood flows into the kidneys. When blood pressure rises, the arterioles are stimulated into constricting, so blood flow through the kidneys decreases.

A Closer Look at Reabsorption

A minimum amount of urine must be formed each day in order to rid the body of metabolic wastes. The amount, which is about 400 milliliters, is called the *obligatory water loss*. When water intake exceeds that amount, the urine will increase in volume and it will be less concentrated.

Adjustments in urine composition and volume are made during the reabsorption process. As Table 33.4 indicates, the adjustments involve surprisingly large quantities of water and solutes. If for some reason the kidneys stopped reabsorbing the filtrate, all the water of the bloodstream would be urinated away in less than a half hour!

The question becomes this: How is urine concentration adjusted to maintain optimal water and solute concentrations in the internal environment? For the answer, let's begin with the manner in which sodium ions (Na^+) are reabsorbed in the proximal tubule of the nephron.

Transport Processes at the Proximal Tubule. Epithelial cells of the proximal tubule have sodium pumps, or active transport systems, embedded in their plasma

Table 33.4	Average Daily Reabsorption Values for a Few Substances		
	Filtered	Excreted	Amount Reabsorbed
Water	180 liters	1.8 liters	99%
Glucose	180 grams	none, normally	100%
Sodium ions	630 grams	3.2 grams	99.5%
Urea	54 grams	30 grams	44%

Figure 33.7 Reabsorption of solutes as a result of the active transport of sodium out of the proximal tubule.

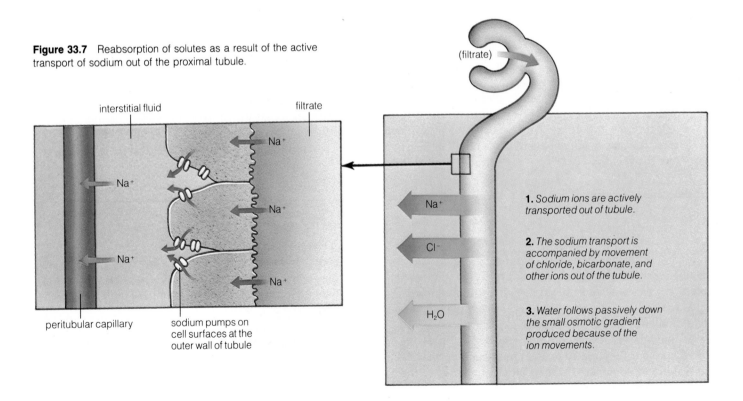

interstitial fluid

filtrate

Na^+

Na^+

Na^+

Na^+

Na^+

Na^+

peritubular capillary

sodium pumps on cell surfaces at the outer wall of tubule

(filtrate)

Na^+

Cl^-

H_2O

1. Sodium ions are actively transported out of tubule.

2. The sodium transport is accompanied by movement of chloride, bicarbonate, and other ions out of the tubule.

3. Water follows passively down the small osmotic gradient produced because of the ion movements.

membrane (Figure 33.7). All the pumps are on the surfaces of cells that make up the wall of the tubule, and they all pump sodium in one direction: out of the filtrate and into the surrounding interstitial fluid.

The active transport of sodium ions is accompanied by the movement of chloride, bicarbonate, and other ions out of the proximal tubule. The solute concentrations inside the tubule drop slightly, and this affects the osmotic gradient for water between the tubule and interstitial fluid. Thus water passively follows its gradient out of the tubule, then it is reabsorbed into the peritubular capillaries.

Countercurrent Multiplication. Water reabsorption is influenced by a solute concentration gradient that is maintained between the nephron and the surrounding interstitial fluid. This gradient, which starts at the boundary between the cortex and the medulla, increases with increasing depth into the medulla. It is established through *countercurrent multiplication*. "Countercurrent" refers to flow in opposing directions through a pair of interacting tubes. As we will now see, such a flow takes place at the loop of Henle.

At the proximal tubule of the nephron, water and solutes are reabsorbed in isotonic proportions. When fluid enters the descending part of the loop of Henle, it has the same total solute concentration as blood. However, the descending part is permeable to water, and as the isotonic fluid inside travels down the loop into the medulla, it encounters a hypertonic environment (one with a higher solute concentration). Water follows its osmotic gradient here and moves out of the tubule, leading to an increased solute concentration inside the loop.

What happens when the fluid rounds the bend of the loop of Henle? The ascending part of the loop is not permeable to water, but cells making up its walls contain transport systems that actively pump out a salt—sodium chloride, or NaCl (Figure 33.8). Because salt is pumped outward, the solute concentration inside decreases progressively as the fluid moves up through the ascending part of the loop. The more salts are pumped out, the saltier the surroundings become, the more water is drawn out of the descending part of the loop, and so on in a positive feedback mechanism that "multiplies" the solute concentration in the surrounding tissue.

Control of Water Reabsorption. The amount of water that actually leaves the body is influenced by antidiuretic hormone, or ADH, which acts on cells of the distal tubule and the collecting duct and makes them more permeable to water. In the absence of ADH, the distal tubule remains impermeable to water, as does the collecting duct. The dilute fluid traveling through these regions eventually reaches the bladder as urine.

When there is need to conserve water, ADH acts to increase water reabsorption from the distal tubule. Sol-

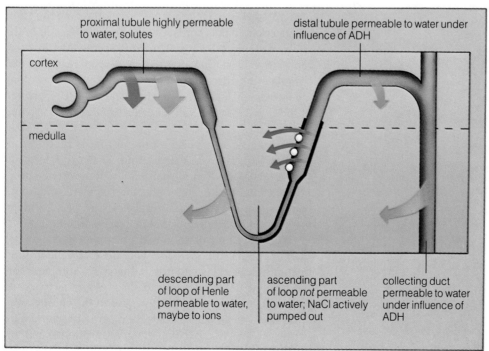

cortex

medulla

proximal tubule highly permeable to water, solutes

distal tubule permeable to water under influence of ADH

isotonic

hypertonic

descending part of loop of Henle permeable to water, maybe to ions

ascending part of loop *not* permeable to water; NaCl actively pumped out

collecting duct permeable to water under influence of ADH

Figure 33.8 Permeability characteristics of the nephron. (Both the distal tubule and the collecting duct have very limited permeability to solutes; most of the solute movements in these regions are related directly or indirectly to active transport mechanisms.)

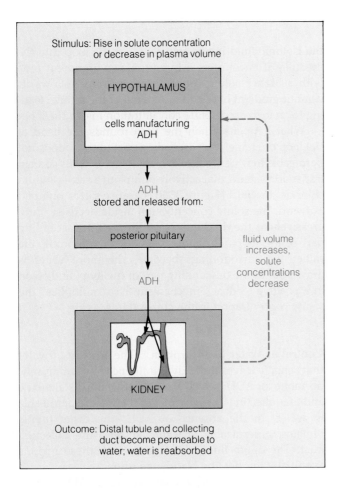

Stimulus: Rise in solute concentration
or decrease in plasma volume

HYPOTHALAMUS

cells manufacturing
ADH

ADH
stored and released from:

posterior pituitary

ADH

fluid volume
increases,
solute
concentrations
decrease

KIDNEY

Outcome: Distal tubule and collecting
duct become permeable to
water; water is reabsorbed

Figure 33.9 Homeostatic control over extracellular fluid volume. The hormone ADH acts on the nephron's distal tubule and the collecting duct walls, making them permeable to water. Water moves into the interstitial fluid and is reabsorbed into the bloodstream. Thus blood volume increases. The change is detected through the associated decrease in solute concentrations. Here, the blue dashed line completes the feedback loop.

utes remain behind, and fluid entering the collecting duct has the same solute concentration as blood. This small volume of isotonic fluid becomes quite concentrated in the collecting duct, where more water is removed from it.

The day-to-day regulation of ADH secretion is mainly under the control of receptors in the hypothalamus that are sensitive to changes in the total solute concentration of the extracellular fluid. Detection of such changes accounts for the delicate balance between variations in water intake and urine output. For example, when you drink a lot of water, the solute concentration of extracellular fluid is diluted slightly, ADH secretion is inhibited, and you excrete a large volume of urine. With water deprivation, the solute concentration of extracellular fluid rises, ADH secretion is stimulated, and the kidney is stimulated into conserving water (Figure 33.9).

Thirst Mechanism. When the body loses more water than it takes in, a thirst mechanism is also activated. The hypothalamus contains a cluster of nerve cells that are also stimulated when ADH secretions are being stepped up. When receptors detect a rise in solute concentrations, thirst center cells are stimulated into sending signals that initiate water-seeking behavior.

Control of Sodium Reabsorption. So far, we have focused on the mechanisms by which water is reabsorbed in the kidneys. But the kidneys also help regulate the reabsorption of sodium, potassium, and other ions.

Sodium retention operates through cells that are part of the *juxtaglomerular apparatus*, a region where the afferent arteriole makes contact with the distal tubule of the nephron (Figure 33.10). The cells secrete the enzyme renin, thereby setting in motion a series of reactions that stimulate the adrenal cortex into secreting the hormone **aldosterone**. This hormone stimulates the reabsorption of sodium across the walls of distal tubules and collecting ducts.

Renin acts on a plasma protein that is produced in the liver and is always circulating, in inactive form, in the bloodstream. Renin catalyzes the removal of a polypeptide from this protein, and the fragment undergoes conversions into the hormone *angiotensin*. This hormone is a powerful stimulator of aldosterone secretion.

When the loss of sodium from the body is greater than the intake, all components of the extracellular fluid (including the plasma) are reduced in volume. This reduction is detected by stretch or pressure receptors in the heart chambers, in blood vessels in the chest cavity, and possibly elsewhere. Detection leads to stimulation of mechanisms that promote the reabsorption of sodium by the kidney, especially the renin-angiotensin-aldo-

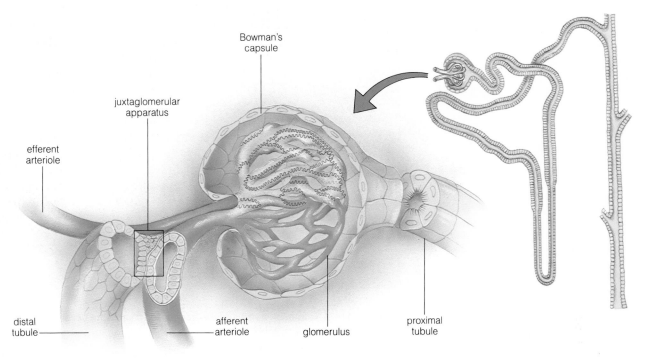

Figure 33.10 Juxtaglomerular apparatus, which plays a role in sodium reabsorption.

sterone system. The reabsorption minimizes any further loss of sodium from the body until the deficit can be corrected. If the sodium deficit (hence the depletion of the extracellular fluid volume) is extreme, ADH secretion may be stimulated, leading to a further increase in water reabsorption by the kidney. This helps to preserve volume, but often at the expense of a reduction in the plasma sodium concentration. If severe enough, such a reduction can have damaging consequences.

Case Study: On Fish, Frogs, and Kangaroo Rats

Let's conclude this chapter with a look at how a few vertebrates maintain their water and solute levels in some entirely different settings—in the seas, in fresh water, and on land.

Compared to the tissues of herring, snapper, and other bony fishes living in it, seawater has about three times more solutes. These fishes continually lose water (by osmosis) to their hypertonic environment, and continual drinking brings in replacements. (If the fishes are experimentally prevented from drinking, they die from dehydration in a few days.) Ingested solutes are excreted against their concentration gradients. Although kidneys are present, they are too small to excrete much water. Instead, most of the excess solutes are pumped out through the membranes of fish gills, the cells of which

actively transport sodium ions out of the blood (and potassium ions into it).

In fresh water (a hypotonic medium), lake trout and other bony fishes tend to gain water and lose solutes. The same is true of amphibians. These animals do not drink water; rather, water moves by osmosis into the body, through the thin gill membranes (or, in the case of adult amphibians such as frogs, through the skin). Excess water leaves by way of well-developed kidneys, which excrete a large volume of dilute urine. Some solutes also are excreted, but the losses are balanced by solutes gained from food and by the active transport of sodium ions across the gills, into the body.

For the desert-dwelling kangaroo rats, water is exceedingly scarce. Moreover, the air is dry and temperatures can approach 45°C, so water losses could be devastating without behavioral and physiological adaptations for conserving water. Like many other desert animals, kangaroo rats spend the day in deep burrows, where the air temperature seldom exceeds 30°C. They forage during the cooler hours of the night. Although some desert rodents eat moist plant parts, the kangaroo rats eat primarily dry seeds, which contain very little water. They do not drink any water at all. They gain most of their water through the metabolic oxidation of carbohydrates, fats, and proteins in the seeds.

Land-dwelling vertebrates in general lose water by way of the skin, respiratory tract, urine, and (to some

Figure 33.11 A kangaroo rat, master of water conservation in the deserts.

extent) feces. Kangaroo rats and other desert rodents reduce all such losses. Their skin has no sweat glands, and it is thick and dry. Their nose is small, with very narrow and convoluted air passages. When these rodents inhale, dry air passing over the moist nasal tissues becomes warmed and saturated with water vapor. Evaporation from the nasal epithelium cools the tissues well below the body temperature. When the rodents exhale, the warm, humid air from the lungs is cooled as it passes over the cooled tissues. Hence water condenses on the nasal epithelium, like it does on the outside of a glass of ice water on a warm day. As a result, considerable respiratory water is recovered, not lost to the environment.

Kangaroo rats are also parsimonious when it comes to giving up water in feces (the water loss is five times less than it is from laboratory rats). Their urine can be twice as concentrated as that of laboratory rats and three times that of humans. This remarkable ability to conserve water loss from the urinary tract is attributed to particularly long loops of Henle in the kidneys. The loops are so long that the renal capsule containing them extends through the renal pelvis, down into the ureter. Through countercurrent multiplication at these long loops, the solute concentration in the surrounding interstitial fluid becomes very high. Thus the osmotic gradient between the interstitial fluid and the urine is so steep that most of the water reaching the (equally long) collecting ducts is reabsorbed; only a small volume of concentrated urine leaves the kangaroo rat body.

SUMMARY

All animals make controlled as well as obligatory exchanges with the external environment and thereby maintain a hospitable internal environment.

Control of Body Temperature

1. Temperatures between 0°C and 40°C are generally favorable for life.

2. In complex animals, body temperature is determined by the balance between heat produced, heat absorbed from the environment, and heat lost to the environment. Animals gain heat by way of metabolism, and they exchange heat with the environment by these processes:

    a. *Radiation* (emission of energy in the form of infrared and other wavelengths which, following absorption by the animal body or some other object, is converted to heat energy).

    b. *Conduction* (the direct transfer of heat energy from one object to an immediately adjacent object).

    c. *Convection* (heat transfer by air or water currents; involves conduction and mass transfer of heat-bearing currents away from or toward the animal body).

    d. *Evaporation* (dissipation of heat energy during the conversion of water from the liquid to the gaseous state).

3. The body temperature of different animals is determined by the rate of metabolic activity and by anatomical, behavioral, and physiological adaptations.

a. Ectotherms are animals whose body temperature is determined more by heat exchange with the environment than by metabolic heat.

b. The body temperature of endotherms is determined largely by metabolic activity and by precise controls over heat produced and heat lost.

c. Heterotherms allow their body temperature to fluctuate at some times, and at other times they control heat balance.

Control of Extracellular Fluid

1. Water and solutes are exchanged with the environment in obligatory and controlled ways.

2. The body gains water through absorption from the gastrointestinal tract and from metabolism. It loses water by evaporation from the respiratory surface and cells of the skin, by sweating, by elimination from the gastrointestinal tract (in feces), and by excretion of urine.

3. Water gain is controlled through thirst. Water loss is regulated by varying the composition and volume of urine.

4. Solutes (nutrients and ions) are gained by absorption from the gastrointestinal tract and by production within the body.

5. The urinary system of mammals includes two kidneys, two ureters, a urinary bladder, and a urethra.

6. Each kidney is composed of a renal capsule, cortex (which represents more than half of its mass), and medulla. Numerous individual tubes (nephrons) produce the urine. The urine drains from the nephrons into collecting ducts.

7. Each nephron consists of a cup-shaped portion (Bowman's capsule), proximal tubule, loop of Henle, and distal tubule.

8. Blood vessels associated with the nephron are an afferent arteriole, glomerular capillaries, an efferent arteriole, and peritubular capillaries.

9. The composition and volume of the final urine is controlled by the selective reabsorption or secretion of substances and involves these processes:

a. *Filtration* of plasma (minus proteins and other large solutes through the glomerular capillaries into Bowman's capsule); blood pressure provides the force for filtration.

b. *Reabsorption*, or the movement of substances from the inside of the nephron, then into the peritubular capillaries.

c. *Secretion*, or the movement of substances from the peritubular capillaries into the space within the nephron.

10. Each day, massive volumes of fluid are filtered at the glomeruli. Most of the water and solutes filtered are reabsorbed.

11. The interstitial fluid of the medulla becomes highly concentrated by a countercurrent multiplier system in the loops of Henle.

12. Antidiuretic hormone (ADH) is formed in the hypothalamus and secreted from the posterior pituitary. It acts on cells in the walls of the distal tubule and collecting ducts, making them permeable to water. ADH is secreted when water must be conserved; its secretion is inhibited when excess water must be excreted.

13. Sodium reabsorption in the distal tubule and collecting ducts is regulated by the hormone aldosterone.

Review Questions

1. Define ectotherm and endotherm. In endotherms, what controls help balance the amount of heat lost and heat gained?

2. In your own body, where are thermoreceptors and the main center of temperature control located?

3. All animals have mechanisms for maintaining body fluid concentration and composition. In your own body, which organs cooperate in these tasks?

4. Describe what happens during (a) filtration, (b) reabsorption, and (c) secretion in the kidney's nephron/capillary unit. What do these three processes influence?

5. Most exchanges between blood capillaries and nephrons are obligatory, but the final urine concentration is under complete homeostatic control. What two interrelated mechanisms act as the controls?

Readings

Bartholomew, G. 1982. "Body Temperature and Energy Metabolism." In *Animal Physiology* by M. Gordon et al. Fourth edition. New York: Macmillan. Good introduction to thermal regulation in animals.

Gottschalk, C., and W. Lassiter. 1980. "The Kidney and Body Fluids." In *Medical Physiology* (V. Mountcastle, editor). Fourteenth edition. Volume 2, part X. St. Louis: Mosby.

Smith, H. 1961. *From Fish to Philosopher*. New York: Doubleday. Available in paperback.

Vander, A., J. Sherman, and D. Luciano. 1985. "Regulation of Water and Electrolyte Balance." In *Human Physiology: Mechanisms of Body Function*. Fourth edition. New York: McGraw-Hill. Excellent introduction to renal functioning.

Vaughan, T. 1986. *Mammalogy*. Third edition. New York: Saunders. Chapter 22 is an excellent introduction to temperature regulation in mammals.

34

PRINCIPLES OF REPRODUCTION AND DEVELOPMENT

With a full-throated croak that only a female of its kind could find seductive, a male frog proclaims the onset of warm spring rains, of ponds, of sex in the night. By August the summer sun will have parched the earth, and his pond dominion will be gone. But tonight is the hour of the frog! Through the dark, a female moves toward the vocal male. They meet, they dally; he clamps his forelegs about her swollen abdomen and gives it a prolonged squeeze. Out streams a ribbon of hundreds of eggs. As the eggs are being released, the male expels a milky cloud of swimming sperm. Each egg joins with a sperm, and soon afterward, their nuclei fuse. With this fusion, fertilization is completed; a zygote has formed.

For the leopard frog *Rana pipiens*, a drama now begins to unfold that has been reenacted each spring, with only minor variations, for many millions of years. Within a few hours after fertilization, the single-celled zygote begins dividing into two, then four, then eight cells, and many more to produce the early embryo. The cells become smaller and smaller with each successive division until there is a ball of tiny cells, no larger than the zygote. It has formed in less than twenty hours.

And now the cells embark on a course of migrations, changes in shape, and cell interactions. A dimple forms on the embryo's surface as some cells sink inward to become the forerunners of internal tissue layers. Other cells lengthen and still others flatten out, causing a groove to form at the surface—a groove destined to become the nervous system. Through interactions between surface cells and interior cells, eyes start to develop. Within the embryo, a heart is forming and will soon beat rhythmically. Fins take shape; a mouth forms. These developments, appearing one after another, are signs of a process going on in *all* the cells that were so recently developed from a single zygote. The cells are becoming different from one another in both appearance and function!

Within twelve days after fertilization, the embryo has become a tadpole, a larval stage that can swim and feed on its own (Figure 34.1). For several months the larva

Figure 34.1 Development of the leopard frog, *Rana pipiens*. (**a**) A male clasping the female in a behavior called amplexus. When the female releases eggs into the water, the male releases sperm over the eggs. (**b**) Frog embryos. (**c**) A tadpole. (**d**) Transitional form between the tadpole and the young adult (**e**).

a

b

c

grows until, in response to hormonal cues, its body starts to change into the adult form. Legs begin to grow from the body; the tail becomes shorter and shorter, then disappears. The small mouth, once suitable for feeding on algae, develops jaws that accommodate insects and worms. Eventually a full-fledged frog leaves the water for life on land. If it is lucky it will avoid hungry predators, disease, and other threats through the months ahead. In time it may even find a pond swollen with the waters of a new season's rains, and the cycle will begin again.

How did that single-celled frog zygote become transformed into all the specialized cells and structures of the adult? With this question we turn to one of life's greatest mysteries—to the incredible orchestration of the processes underlying development, from the time of reproduction to the emergence of the adult form.

THE BEGINNING: REPRODUCTIVE MODES

We have already looked at the cellular basis of **sexual reproduction**, which requires the fusion of nuclei from two (or more) gametes to produce a new individual. We have also looked at the cellular basis of **asexual reproduction**, whereby gametes are not required in the production of offspring. Here we will consider some structural, behavioral, and ecological adaptations that are associated with both reproductive modes. Figure 34.2 hints at the diversity of these adaptations.

Asexual Reproduction

Some of the structurally simple invertebrates can reproduce asexually, most often by fission or budding. In animal **fission**, the entire body of the parent divides into two roughly equivalent parts, with each part then grow-

ing into a whole individual. When certain flatworms reproduce this way, the body may be divided in half in either the transverse or longitudinal direction. In **budding**, the new individual develops as an outgrowth of the parent body. When the bud develops a full set of the parental body structures, it breaks away. The buds of one cnidarian (*Hydra*) protrude from the parent body. The "buds" (gemmules) of sponges are produced internally, and each develops into a separate individual when the parent body disintegrates.

Neither fission nor budding promotes genetic variability; the offspring are *clones*, or genetically identical copies of the parents. Cloning is advantageous only so long as the parents are well adapted to the surroundings, and only so long as the surroundings remain stable. When stability prevails, asexual reproduction is biologically inexpensive, so to speak, because progeny can be produced without much energy outlay. However, most animals live in changing and unpredictable environments. Not surprisingly, most depend primarily on sexual reproduction.

Sexual Reproduction

Usually, sexual reproduction involves the fusion of gametes from a male parent and a female parent. However, some sexually reproducing animals occasionally depart from the basic reproductive mode. **Parthenogenesis** refers to the cleavage and subsequent differentiation of an *unfertilized* egg into an adult. Natural parthenogenesis has been observed in arthropods and in all groups of vertebrates except mammals. For example, every so often between times of sexual reproduction, beetles and aphids produce fatherless offspring. Changes in pH, temperature, salinity, even mechanical stimulation of the egg can trigger parthenogenesis. Such changes can induce parthenogenesis not only in insects but also in frogs, salamanders, turkeys, and lizards.

d

e

liveborn snake in egg sac

a

Figure 34.2 Examples of where the animal embryo develops, how it is nourished, and how (if at all) it is protected. In most mammals, the fertilized egg is retained inside the mother's body and nourished by her tissues until the time of birth. Such animals are called *viviparous* (*viva-*, alive; *-parous*, to produce).

Some fishes, lizards, and a few snakes are *ovoviviparous*. Their fertilized eggs develop within the mother's body. Such eggs are not nourished by the mother's tissues; they are sustained by yolk reserves. (**a**) A copperhead, one of the few ovoviviparous snakes. Her liveborn are still contained in the relics of egg sacs.

(**b**) Snails are *oviparous* (egg-producing) but are not doting parents; their fertilized eggs are left unprotected. (**c**) Monotremes, including the duck-billed platypus, are oviparous, yet they also secrete milk to nourish the juveniles. (**d**) Birds are oviparous animals. Their fertilized eggs, which have large yolk reserves, develop and hatch outside the mother's body.

The opposum (**e**) and kangaroo (**f**), both marsupials, are viviparous. But their young emerge in somewhat unfinished form and undergo further fetal development in a pouch on the ventral surface of the mother's body, where they are nourished from mammary glands. (**g**) Young kangaroo in its mother's pouch.

b

c

Earthworms and parasitic tapeworms also are not bound to straightforward sexual reproduction. These animals are **hermaphrodites** (the individual has both male and female reproductive organs). Two earthworms can cross-fertilize each other by lying head to tail and exchanging sperm. Parasitic tapeworms can fertilize themselves. One need not look askance at tapeworms, when one realizes that these animals can end up living all by themselves somewhere in the tissues of a host animal. Given the bountiful supply of resources and the protected habitat, hermaphroditism does have its advantages for the lone parasite.

The reproductive modes of animals include asexual processes, the development of unfertilized eggs into adults, self-fertilization, and sexual reproduction between separate male and female forms.

Some Strategic Problems in Having Separate Sexes

Complete separation into male and female sexes imposes biologically expensive demands. After all, how is fertilization to be accomplished? Specialized body structures and behavioral activities are necessary.

In behavioral terms, male and female reproductive cycles must be *synchronous*, so that mature gametes are released at the same time. Synchrony requires hormonal control mechanisms and sensory receptors that detect environmental cues, such as changes in daylength from one season to the next. Males and females must be able to recognize each other as being of the same species, so energy outlays are necessary for constructing structural signals (such as bright feathers in birds) and for performing courtship routines.

In structural terms, getting sperm and egg together is not too complicated among certain water-dwelling ani-

d

e

f

g

mals, including sea urchins and frogs. Their gametes are simply released into the same region of water, and the motile sperm can swim to an egg. Such *external* fertilization would be chancy if only one sperm and one egg were released each season. So energy outlays are required for the production of large numbers of sperm and eggs.

In contrast, nearly all land-dwelling animals and even some aquatic forms depend on *internal* fertilization. (Sperm released on dry land obviously do not stand much chance of swimming over to an egg.) The males have internal reproductive organs called **testes**, in which sperm are formed. Many male animals have a **penis**, a copulatory organ by which sperm are deposited into a specialized duct in the female. Many female animals have a **vagina**, a duct between the outside world and the **ovary** (an organ where the immature eggs grow). Thus the reproductive structures of reptiles, birds, and

mammals protect sperm and eggs from harsh external conditions and enhance the probability of successful internal fertilization.

Another problem to consider: How is the embryo to be nourished? Almost all animal eggs contain **yolk** (protein, lipids, and other nutritive substances), but some eggs contain more yolk than others. Fertilized eggs of sea urchins are released from the parent's body and develop to a feeding stage within forty hours. Because these eggs are produced in large numbers, the biochemical investment in yolk for each one is limited. Hence there is a developmental premium on rapid development, with the self-feeding larval stage being reached in the shortest possible time. Bird eggs also are released from the mother's body, but these eggs are mostly a ball of yolk (the cytoplasm and nucleus are positioned as a thin cap on the surface). The large yolk reserves nourish the embryo through a longer period of development,

Figure 34.3 Generalized picture of the stages of animal development. (The same general pattern of gene activity occurs in the development of complex, sexually reproducing plants. However, plant embryos do not undergo gastrulation, and organ formation, growth, and tissue specialization continue through the plant life cycle.)

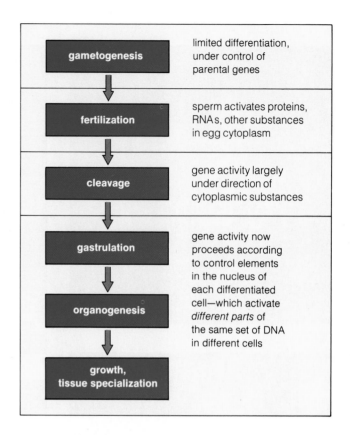

which proceeds inside a hard eggshell; unlike sea urchins, birds cannot feed on their own until they are hatched and fully formed (Figure 34.2). Human eggs have very little yolk—but the developing embryo attaches to the mother's body and receives nourishment by exchanges with her blood.

The point of these few examples is that tremendous diversity exists in reproductive and developmental strategies. However, some patterns are widespread in the animal kingdom, and these patterns will serve as the framework for discussions to follow.

STAGES OF EARLY EMBRYONIC DEVELOPMENT

Figure 34.3 illustrates the stages through which development commonly proceeds. **Gametogenesis** (gamete formation) is the first stage of animal development. In this stage, sperm and egg form and grow within the reproductive systems of the parents (page 153). The second developmental stage, **fertilization**, begins when a sperm penetrates an egg and is completed when the sperm nucleus fuses with the egg nucleus.

In **cleavage**, the third stage, the fertilized egg undergoes mitotic cell divisions that form the many cells of the early embryo. In **gastrulation**, the fourth stage, elaborate patterns of cell migrations in the embryo bring about the formation of two or more primary germ layers. *Germ layers* are regionally positioned layers of cells that will give rise to all organs of the animal body. In many species, the body axis of the embryo is established during gastrulation.

Once the germ layers have formed, cell differentiation and changes in form produce all major organs. This is **organogenesis**, the fifth developmental stage. During **growth and tissue specialization**, organs acquire their specialized structural and chemical properties; this last stage of development continues past the time of birth.

Egg Formation and the Onset of Gene Control

Figure 34.4 shows the appearance of different embryos as they proceed through the five stages of animal development. Let's initially focus on one of those embryos—

the frog embryo—to get an idea of the events taking place at each stage.

To begin, the genes present in the *nucleus* of cells making up the early embryo do not directly control the initial developmental events. Instead, substances that were already present in the egg cytoplasm *before* fertilization direct those events. For example, while an immature egg cell (oocyte) forms inside the female frog, it increases many thousands of times in volume. Ribosomes and Golgi bodies required in protein synthesis are stockpiled, as are mitochondria that will rapidly produce the ATP used to drive synthesis reactions.

Even while the oocyte grows, RNA is transcribed from the DNA (page 216). Many mRNA transcripts are translated at once into enzymes and other proteins, such as histones, which will be used in chromosome replication after fertilization. Other transcripts are sequestered in different cytoplasmic regions. They do not become activated until fertilization; then, their protein products play roles in events such as DNA synthesis.

The components of an oocyte are not randomly distributed. For example, the nucleus takes up position in

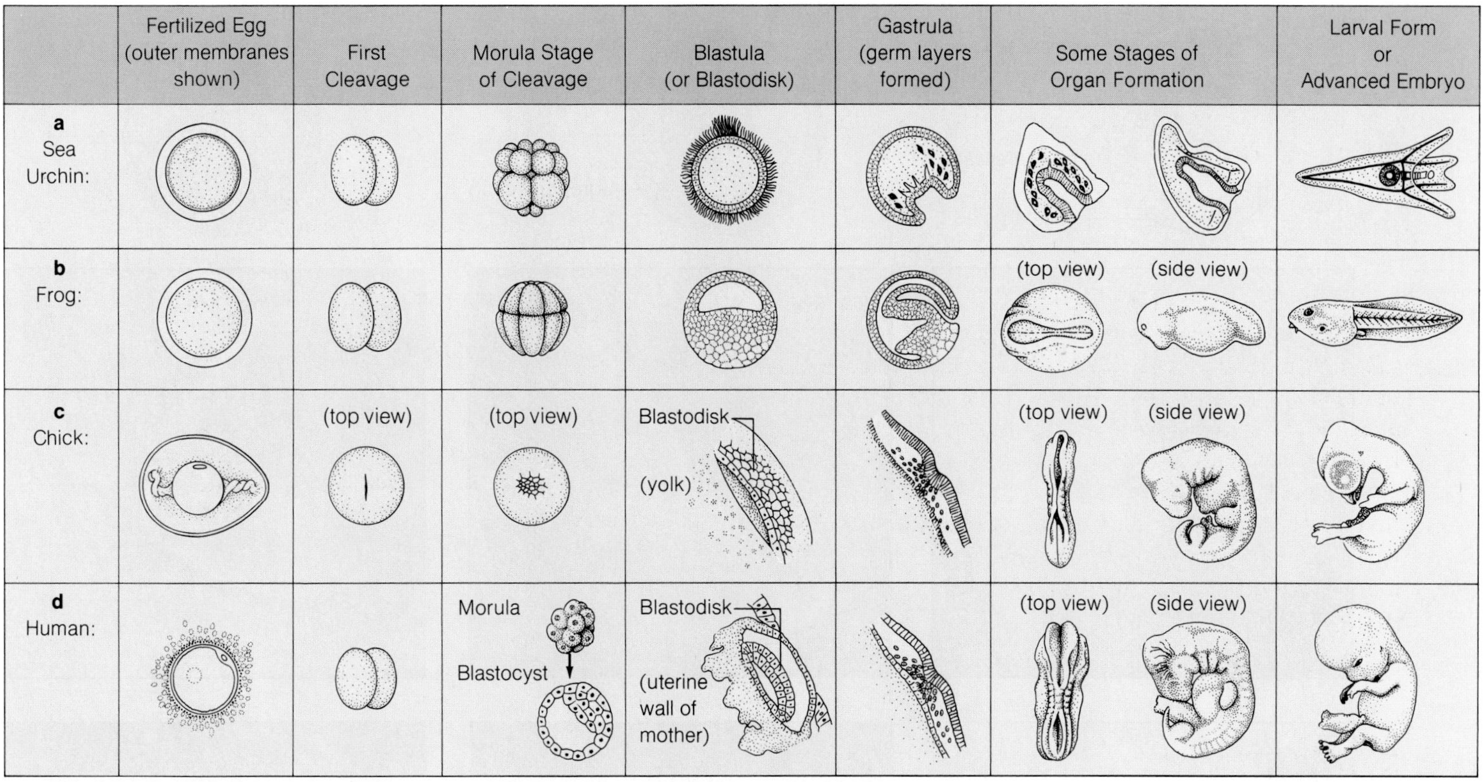

	Fertilized Egg (outer membranes shown)	First Cleavage	Morula Stage of Cleavage	Blastula (or Blastodisk)	Gastrula (germ layers formed)	Some Stages of Organ Formation	Larval Form or Advanced Embryo
a Sea Urchin:							
b Frog:						(top view) (side view)	
c Chick:		(top view)	(top view)	Blastodisk (yolk)		(top view) (side view)	
d Human:			Morula Blastocyst	Blastodisk (uterine wall of mother)		(top view) (side view)	

Figure 34.4 Comparison of embryonic development in four different animals. The drawings are not to the same scale; however, they show the developmental patterns that are common to all four types. For clarity, the membranes surrounding the embryo are not shown from cleavage onward. Blastula and gastrula stages are shown in cross-section; they are described in more detail in this chapter and the next for frogs, birds, and humans.

one region of the cytoplasm and thereby imparts polarity to the oocyte (the oocyte now has two identifiable poles). The *animal pole* is the one closest to the nucleus. Opposite is the *vegetal pole*, where substances such as yolk accumulate. All mature eggs show some degree of polarity, which will influence the structural patterning of the embryo.

Visible Changes in Amphibian Eggs at Fertilization

Sperm penetration into the egg triggers fertilization. It sets in motion morphological and chemical activity within the egg cytoplasm. You can observe indirect signs of this activity in amphibian eggs, which contain granules of dark pigment in their cortex. (The cortex includes the plasma membrane and the cytoplasm just beneath it.) Because the animal pole has a greater concentration of pigment, it is darker than the vegetal pole (Figures 34.5 and 34.6). When a sperm penetrates the egg, the cytoplasm in which the granules are embedded and then

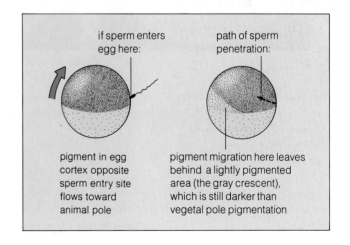

Figure 34.5 Formation of the gray crescent in the eggs of some amphibians, including frogs.

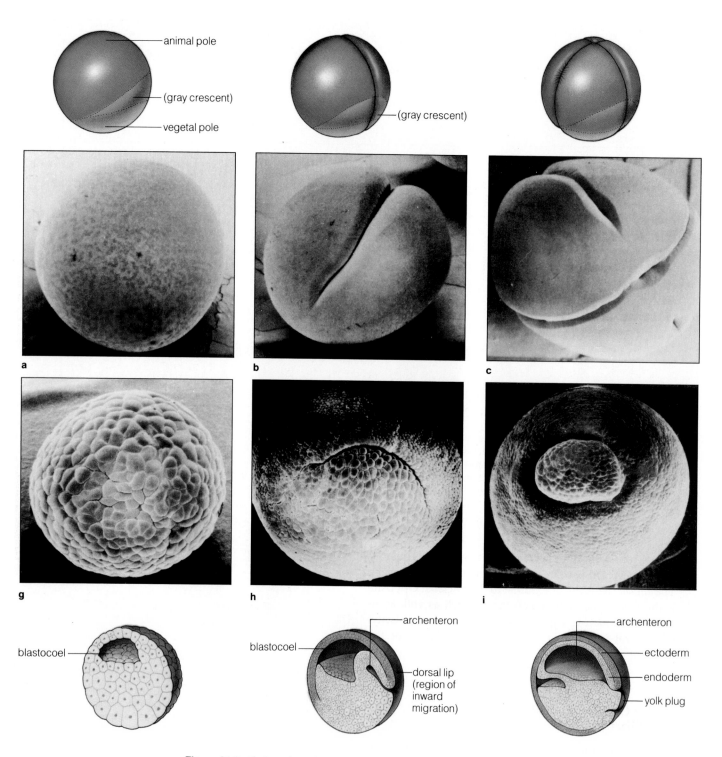

Figure 34.6 Early embryonic development of a frog. For these scanning electron micrographs, the jellylike layer surrounding the egg has been removed. (**a**) Within about an hour after fertilization, a region of differentiated surface cytoplasm (gray crescent) appears opposite the site where the sperm penetrated the egg. (**b–g**) Cleavage leads to a blastula, a ball of cells in which a cavity (blastocoel) has appeared.

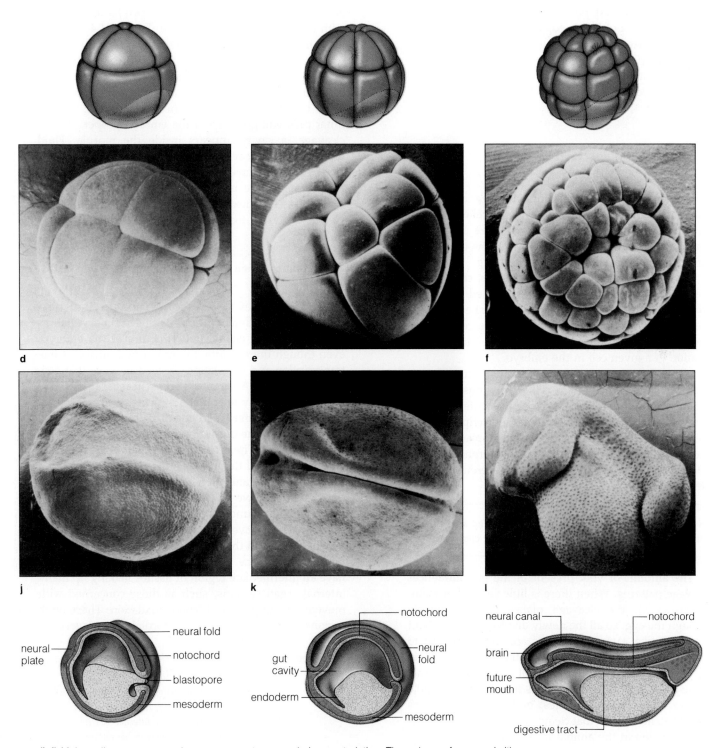

(**h,i**) Major cell movements and rearrangements occur during gastrulation. Tissue layers form; a primitive gut cavity (archenteron) develops. (**j,k**) Neural developments now take place, and the fluid-filled body cavity in which vital organs will be suspended appears. (**l**) Differentiation proceeds, moving the embryo on its way to becoming a functional larval form. (Photos from R. Kessel & C. Shih, *SEM in Biology*, Springer-Verlag, 1976.)

the cortex itself shift position. The outcome is a **gray crescent**, an area of intermediate pigmentation near the equator, on the side of the egg *opposite* the sperm penetration site (Figure 34.5). In this gray crescent, the long axis and bilateral symmetry for the embryo's body will be established and gastrulation will eventually begin.

Cleavage: Distributing the Cytoplasmic Substances

Now the fertilized egg undergoes **cleavage**, a period of cell division that produces the early multicelled embryo. Usually the embryo does not grow in size during this period; the cells produced by the divisions occupy the same total volume as did the fertilized egg.

Even at this early developmental stage, cells vary in their activity. Some cells become flattened and some are rearranged; communication is established between them. Microvilli and ridges appear at the cell surface, and their number and shape vary, depending on the location of a given cell in the embryo.

Most of the genes in the nucleus are inactive during cleavage, so the variations in cell activity must be partly the result of different cells receiving different cytoplasmic substances (such as mRNA transcripts) during the divisions. Thus, the sector of egg cytoplasm "inherited" during cleavage helps establish differences among cells that will help send them down different developmental roads.

The destiny of cell lineages is partly established according to which sector of cytoplasm is inherited during cleavage.

The amount of yolk present in the egg influences cleavage patterns. When there is little yolk (as in mammalian eggs), the cleavage planes pass completely through the egg, so all the newly divided cells are about the same size. Amphibian eggs are about one-half to three-fourths yolk, which is concentrated near the vegetal pole. Cleavage produces small cells in the animal hemisphere and much larger cells near the vegetal pole (Figure 34.6). Eggs of reptiles, birds, and most fishes contain so much yolk that cleavage is restricted to a tiny, caplike region at the animal pole, near the egg surface (Figure 34.4c).

Early in cleavage, newly divided cells are spherical and form a solid mass that looks rather like a mulberry. At this stage, the embryo is called a **morula** (from the Latin word for mulberry). Then cell-to-cell adhesion increases and an epithelium forms. Often the epithelium is in the form of a hollow sphere, called a **blastula**. (The hollow interior is the *blastocoel*.) In yolk-rich eggs, the epithelium has no room to form a sphere. Instead, it sits on the yolk surface as a **blastodisk**: two flattened layers with a thin blastocoel in between (Figure 34.4c).

In mammalian eggs, the outermost cells of the morula eventually form a surface layer that will give rise to extraembryonic membranes. (Those membranes attach the embryo to the mother's uterus and take part in the transfer of nutrients derived from her tissues.) The mass of inner cells will give rise to the embryo proper. By late cleavage, a cavity appears inside the morula. Fluid moves inside the cavity and lifts the inner cell mass away from the surface layer on all but one side (Figure 34.4d). At this stage, the mammalian embryo is known as a **blastocyst**. The next chapter describes the fate of the blastocyst during human embryonic development.

Gastrulation: When Gene Transcription Takes Over

As cleavage comes to a close, the pace of cell division slackens. Now cells and groups of cells begin movements that dramatically change the structure of the embryo, even though there is little (if any) growth in size. Gene transcription and translation proceed in ways that activate different DNA regions in different cells, so that they end up with many new kinds of proteins that were not present in the unfertilized egg. This stage of active cell movement and the onset of new gene activity is known as **gastrulation**. The embryo is now called the **gastrula**.

To understand the point of this activity, think about the structural organization of animals. With few exceptions, animals have an internal region of cells, tissues, and organs that function in digestion and absorption of nutrients. They have a surface tissue region that protects internal parts and that has sensory receptors for detecting changes in the outside world. Almost all animals have an intermediate region of tissues making up many internal organ systems, such as those concerned with movement, support, and blood circulation. These three regions develop from three primordial tissue layers:

endoderm	*inner layer; gives rise to inner lining of gut and organs derived from it*
mesoderm	*intermediate layer; gives rise to muscle, the organs of circulation, reproduction, and excretion, most of the internal skeleton, and connective tissue layers of the gut and body covering*
ectoderm	*surface layer; gives rise to tissues of nervous system and outer layer of body covering*

This three-layered organization, which is typical of most animals, begins with the cell activities that transform the ball- or disk-shaped blastula into the gastrula. For example, some cells at the surface of the blastula migrate to the interior. There, a portion will form the lining of an

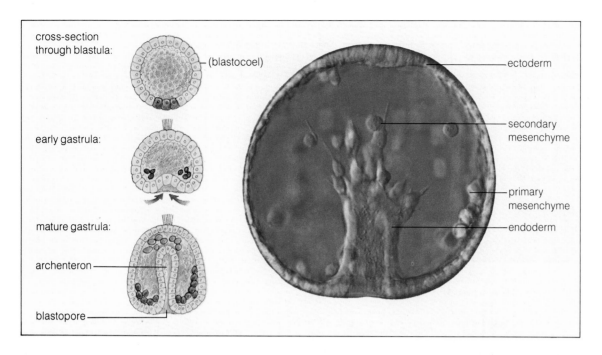

Figure 34.7 Inward migration of surface cells during gastrulation in a sea urchin. The photograph shows the embryo at the mid-gastrula stage. Some of the mesenchyme cells will ultimately develop into a third germ layer, the mesoderm.

internal cavity, the **archenteron**, which will develop into the gastrointestinal tract. Figure 34.7 shows how the archenteron develops in a sea urchin.

Organ Formation

Following their major reorganization in the gastrula, the embryonic cells now embark on a course that will lead to complex organs. Figure 34.8 will give you a sense of how rapidly the organs start to form—in this case, within seventy-two hours after the egg of a chicken has been incubated. (*Incubate* means to induce development, as by the application of heat.) The details in this figure are not meant to be memorized. Rather, they are intended only to show the magnitude of progressive changes that lead to organ formation. The mechanisms by which organs form are not yet fully understood. In the remainder of this chapter, we will consider a few studies that tell us something about what is happening.

MORPHOGENESIS AND GROWTH

It is one thing to ask how a bone cell becomes different from a muscle cell or red blood cell. It is another thing to ask how one bone becomes different from all other bones. All the bones in your hand and arm are practically identical in molecular composition—yet they differ in size, shape, and arrangement. How do such differences arise in different clusters of cells, all of which are committed to making bone? The main processes involved are these:

cell division
cell enlargement
changes in cell shape
local movements of cells
differential growth in local body regions
controlled cell death

These processes are the basis of **morphogenesis**: the growth, shaping, and arrangement of body parts according to genetically predefined patterns. The extent, direction, and rate of morphogenesis depend on genetic as well as environmental controls.

Growth Patterns

During development, the overall growth in size results from an increase in cell number and, often, from cell enlargement. The *capacity* for growth is genetically determined, but the growth actually achieved by an individual is influenced by environmental factors, such as the availability of nutrients. As we have seen, hormones also influence growth and development. Thyroid hormones

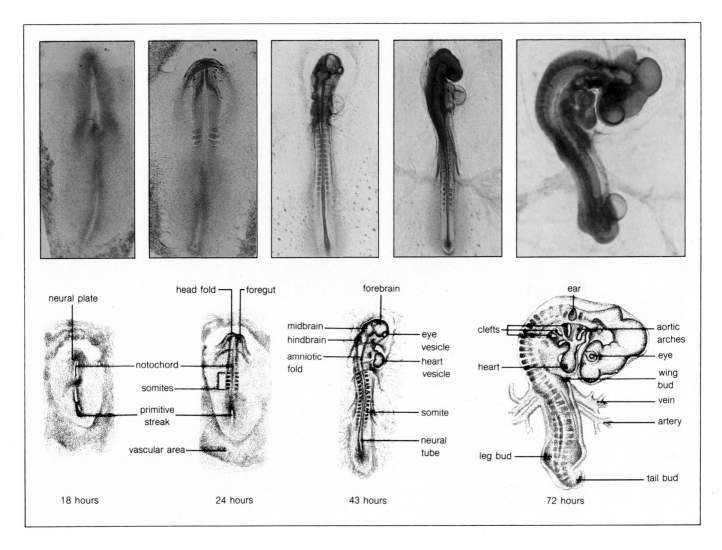

Figure 34.8 Onset of organ formation in a chick embryo during the first seventy-two hours after incubation. The heart begins to beat somewhere between thirty and thirty-six hours.

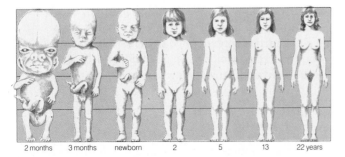

Figure 34.9 Diagram of changes in the proportions of the human body during prenatal and postnatal growth.

and growth hormone especially stimulate the normal growth and maturation of most body regions. When the diet is rich in carbohydrates, the hormone insulin affects overall growth by triggering amino acid transport into cells and thereby contributing to protein synthesis. When the individual does not take in enough nutrients, overall growth will suffer.

Hormones probably start to influence growth patterns early in development. It is quite difficult to isolate hormones from embryos to study their effects because they are present in such tiny amounts. Yet, to give one example, the principal cells of the parathyroid glands differentiate in a human embryo during the fifth week of development. It seems likely that these cells become active in regulating calcium metabolism (hence bone development) even before birth.

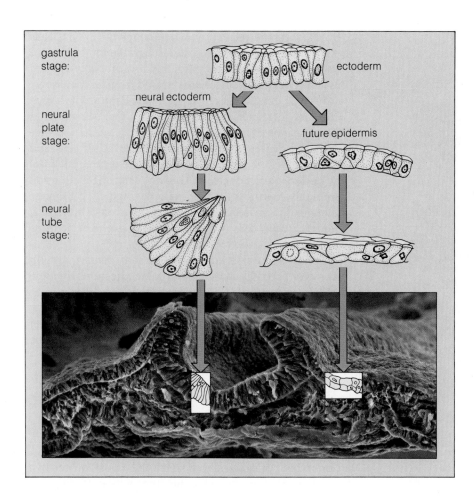

gastrula stage:

ectoderm

neural ectoderm

neural plate stage:

future epidermis

neural tube stage:

Figure 34.10 Example of morphogenetic movements: the shape changes involved in the formation of a neural tube. As gastrulation draws to a close, the ectoderm is a uniform sheet of cells. Some of the cells elongate, forming a neural plate, then they constrict at one end to form the neural tube. Other cells flatten; they will become part of the epidermis. The photograph shows the neural tube forming in a chick embryo.

The growth achieved during development is not uniform; some tissues enlarge more than others, as Figure 34.9 indicates. This differential growth produces changes in the shapes and relative proportions of organs and body parts. How do these changes arise? Let's take a look.

Cell Movements and Changes in Cell Shape

During morphogenesis, cells and entire tissues migrate from one site to another. Three types of morphogenetic movements are common:

    active cell migration
    folding of sheets of cells
    movement of organs

Movements of organs will be described in later chapters. In active cell migration, cells move about by means of pseudopods. (Pseudopods are temporary projections from the main cell body, some rather bulbous, others

fingerlike.) The cells migrate over prescribed pathways, reaching a prescribed destination and establishing contact with cells already there. Those movements must be extremely accurate. Even before you were born, for example, precursors of nerve cells migrated and established the many billions of precise connections that enable your nervous system to function.

Another morphogenetic movement is the inward or outward folding of epithelial sheets, resulting from coordinated changes in the shapes of cells that make up local cell populations. Think about what happens after the three primordial tissue layers form in embryos of amphibians, reptiles, birds, and mammals. At the midline of the embryo, some ectodermal cells elongate and protrude above the surrounding region. They create a *neural plate*, the forerunner of the central nervous system (Figure 34.10). The cells then constrict at one end, causing the edges of the neural plate to fold over and meet at the embryo's midline. In this way a *neural tube* forms. Meanwhile, other ectodermal cells have flattened; they will become the epidermis above the tube.

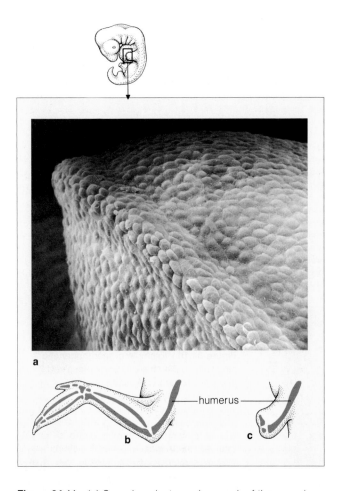

Figure 34.11 (**a**) Scanning electron micrograph of the normal ectodermal ridge of a wing bud in a chick embryo. (**b**) Normal pattern of bone formation. (**c**) Pattern of bone formation when mesodermal cells at the ridge are surgically removed; notice the deficiencies in the wing skeleton.

How do the cells know where to move? In part, cells may move in response to chemical gradients, a behavior called chemotaxis. The gradients are probably created when certain substances are released from target tissues. (This type of movement was shown in Figure 17.6, which focused on the development of the slime mold *Dictyostelium discoideum*.) But cells also move in response to "adhesive cues" on the surfaces of other cells and substances in the extracellular matrix. They migrate from regions of weak adhesion to regions of stronger adhesion in highly specific ways. For example, pigment cells move along blood vessels but not along the axons of neurons; and Schwann cells migrate along the axons of neurons but not along blood vessels.

How do the cells know when to stop? In part, growth slows or becomes channeled in some direction when the cell density increases in a tissue. As we have seen, cells have certain proteins at their surface that recognize the self-markers of adjacent cells. When cells crowd together as a result of divisions and migrations, signals from the recognition proteins inhibit the gene activities responsible for further division. This cell-to-cell interaction is called **contact inhibition**.

Pattern Formation

Embryonic cells somehow sense their position relative to one another and use the information to produce ordered, spatial arrangements of differentiated tissues. The cell activities leading to differentiated tissues and their positioning in space is called **pattern formation**.

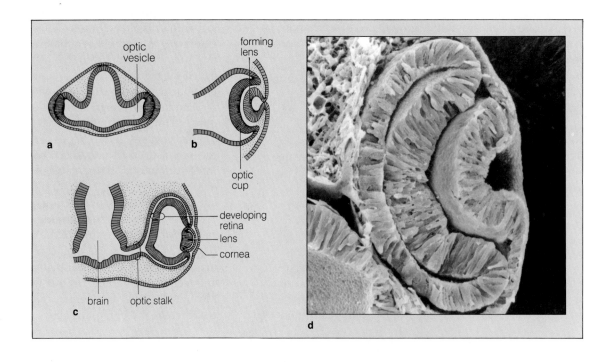

Consider the skeletal pattern that must be followed to produce the wing of a chick. More than a dozen bones must become positioned in precise arrangements relative to one another (Figure 34.11b). When mesodermal cells are removed from a chick embryo, the terminal wing bones never do develop (Figure 34.11c). When those cells are transplanted to another region of the embryo, they may give rise to a wing at that region.

Two cell activities responsible for pattern formation are called ooplasmic localization and embryonic induction. Through **ooplasmic localization**, cells in different regions of the embryo differentiate according to which cytoplasmic substances they inherited during cleavage (page 478). Through **embryonic induction**, one body part differentiates in response to signals from an adjacent body part.

As an example of embryonic induction, consider one of the experiments performed by Hans Spemann, Otto Mangold, and Hilde Mangold that revealed how the vertebrate eye develops. The retina of an eye originates from the forebrain but its lens, which focuses light onto the retina, originates from the ectoderm (Figure 34.12). Otto Mangold experimented with a salamander embryo in which optic cups had already begun to grow out of the forebrain. He surgically removed one of the optic cups and inserted it under the ectoderm of the belly region. Belly ectodermal cells that had come in contact with the transplanted optic cup were induced to form a lens—which fit perfectly into the transplanted part!

What, exactly, acts as an inducer during pattern formation? How is it transmitted from the inducer tissue to the responding tissue? Several experiments indicate that chemical substances diffuse from one tissue to the other. For example, when the two tissues destined to become the epithelium and connective tissue of a pancreas are surgically separated from each other in a mouse embryo, the future epithelial tissue does not differentiate properly. However, suppose the two tissues are grown in a culture medium, separated only by a filter that is permeable to large molecules (but not to cells or cell structures). Then, the responding tissue differentiates properly. Although physical contact with the inducing tissue had been prohibited, signals still reached the responding tissue.

Some inducer substances (hormones, perhaps) act like green traffic lights; they signal target cells that are already committed to a developmental course to continue with their differentiation. Other inducer substances are like flagmen at a road construction site; they instruct target cells on which course to follow.

Experiments with mutant strains of *Drosophila* tell us something about the "instructive" inducers. *Drosophila* larvae contain flattened clusters of cells, each of which will give rise to specific body parts of the adult (Figure 34.13). The clusters are called **imaginal disks**. The fate of the cells in each disk is sealed early in development. Thus, when an "antenna" disk is surgically removed and a "leg" disk inserted in its place, the adult fly will end up with a leg growing out of its head. The same thing happens as a result of homeotic mutations. Apparently, a **homeotic mutation** affects a control gene that activates sets of genes concerned with development. These single-gene mutations lead to alterations in the inducer substance itself or to abnormal responses to it. One

Figure 34.12 (Left) (**a–c**) Formation of the eye in a frog embryo. The retina develops as an outgrowth of the brain; the lens develops as an ingrowth of the ectoderm. (**d**) Scanning electron micrograph of the developing optic cup and lens in a chick embryo.

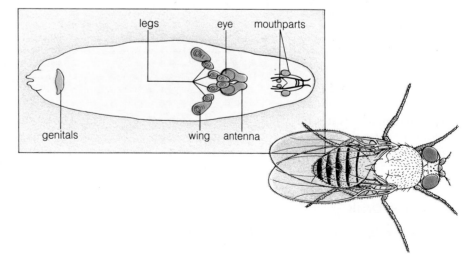

Figure 34.13 Imaginal disks of a *Drosophila* larva. Each disk will give rise to a specific part of the adult fly.

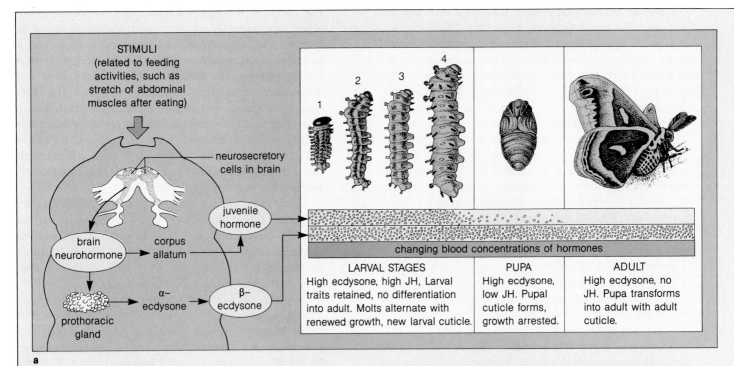

STIMULI
(related to feeding activities, such as stretch of abdominal muscles after eating)

neurosecretory cells in brain

brain neurohormone → corpus allatum

juvenile hormone

prothoracic gland → α–ecdysone → β–ecdysone

changing blood concentrations of hormones

LARVAL STAGES	PUPA	ADULT
High ecdysone, high JH, Larval traits retained, no differentiation into adult. Molts alternate with renewed growth, new larval cuticle.	High ecdysone, low JH. Pupal cuticle forms, growth arrested.	High ecdysone, no JH. Pupa transforms into adult with adult cuticle.

a

Figure 34.14 Neuroendocrine control of the development of a silkworm moth, *Platysamia cecropia*.

A hatched larva eats until it completes five near-doublings in size. Rapid epidermal cell divisions sustain the massive growth—but the cells also secrete chitin to the epidermal surface. Chitin forms a tough, flexible casing (the cuticle) from which the larval legs protrude.

The cuticle sets an upper limit on increases in mass. When the limit is reached, cell division idles and the larva molts (sheds its cuticle). Cell divisions resume, more chitin is secreted, and a larger cuticle is produced. The larva grows and molts five times. Then chitin is deposited over the whole insect, legs and all; this is the pupal stage.

The insect spins a cocoon around itself for three days, then the body undergoes massive cell destruction and tissue reorganization. The contents of degraded cells form a nutrient-rich soup that sustains the growth of imaginal disks. The disks had been growing slowly, but now they rapidly give rise to what will become adult tissues and organs. The pupa lasts eight winter months. In spring, cell death and tissue changes transform the pupa into the adult.

Transformation of a larva into a moth is an example of metamorphosis. The developmental line leading from the egg to the adult is not direct; it proceeds through immature forms that are radically remodeled. The adult "plan" is already laid out in immature tissues, but neural and endocrine controls dictate when the plan will be fulfilled.

In the larval brain, neurosecretory cells release *brain neurohormone*, which acts on paired prothoracic glands. The glands produce and secrete precursors of a hormone, *β-ecdysone*, that activates genes in undifferentiated epidermal cells.

homeotic mutation activates the wrong set of genes in the cells of an "antenna" disk in *Drosophila*, causing those cells to differentiate into a leg.

Controlled Cell Death

The morphogenetic process called **controlled cell death** eliminates tissues and cells that are present for only short periods in the embryo or the adult. The folding of parts, the hollowing out of tubes, the shaping of bones, the separation of one part from another, the opening of eyes, mouth, nostrils, and ears—all involve patterns of cell death.

Perhaps you have noticed that kittens and puppies are born with their eyes sealed shut. From the time of their formation until just after birth, the eyelids consist of an unbroken layer of skin. But then cells stretching in a thin line across each eyelid respond to some internal signal and die on cue. As the dead cells degenerate, a slit forms in the skin, and the upper and lower lids part company.

Cell death also helps to transform four paddlelike appendages in human embryos into hands and feet. Skin

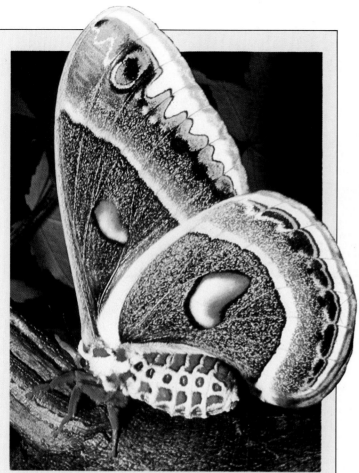

.b Neurohormones also act on two glands (the corpora allata) located behind the brain. The glands secrete *juvenile hormone* (JH) which, at high levels, prolongs the juvenile state. The *absence* of JH triggers cell differentiation into the adult.

High concentrations of both JH and ecdysone promote larval growth and development. But the amount of JH declines steadily and disappears by the fifth larval stage. The pupa forms when ecdysone levels are high and JH levels are low. Then, the developmental restraints are lifted from cells of the imaginal disks.

the mutant skin cells to *respond* to the signal. Controlled cell death in all other parts of the developing embryo proceeded on cue. Apparently the gene controls the response to one signal required for morphogenesis.

METAMORPHOSIS

After being born or hatched, some animals develop directly into the adult form. For them, development of new tissues and organs slows down with increasing age. Some changes do occur, but they are never again as rapid and profound as those at cleavage, gastrulation, and organ formation. Such *direct development* into the adult form is characteristic of reptiles, birds, and mammals.

Other animal groups proceed through *indirect development* into the adult form. The embryo first develops into a larva: a sexually immature, free-living and free-feeding animal that grows and develops into the sexually mature adult form. Morphogenesis is triggered again at some point after the animal has been born or hatched. This process of reactivated development from the larval stage to the adult form is called **metamorphosis**.

Typically, animals that release small and relatively yolkless eggs into water have long larval stages. Animals that lay large, yolky eggs in water often have a short larval stage or none at all. In other animals, such as grasshoppers and caterpillars, the transformation occurs gradually on land. Their larvae grow, then *molt* (shed their too-small skin or cuticle), then grow again. Figure 34.14 shows how hormones control the transformation of a larva into an adult moth. During the pupal stage, most of the larval tissues are destroyed in a massive program of cell death; new structures (such as wings) develop from groups of imaginal disks. In grasshoppers, the larval form (nymph) is quite similar in appearance to the adult (Figure 34.15). In this insect, many larval tissues do not undergo wholesale degeneration and death but are used in the development of the adult.

cells between the lobes of the "paddle" die on cue, leaving separate fingers and toes. Duck embryos also have paddlelike appendages, but cell death normally does not occur in them; that is why ducks have webbed feet instead of separated toes. Now, in some mice and some humans, a certain gene mutation blocks cell death in the paddles, and hands and feet remain webbed. In certain experiments, skin was grafted from normal mice to the mutant mice. In those mice, the toes of the four feet separated normally. Therefore, the results showed that the mutation did not prevent the *generation* of the death signal in the paddles; rather, it changed the capacity of

CELL DIFFERENTIATION

During development, populations of cells having stable phenotypic differences arise from an initial population of cells that all have the same genes (and therefore the same developmental potential). This process is called **cell differentiation**.

At the molecular level, development seems to be a conservative process, one that does usually not involve massive deletions or rearrangements in the genes of each cell. Rather, it appears to involve differences in the activity of genes in different cells. Consider John Gurdon's

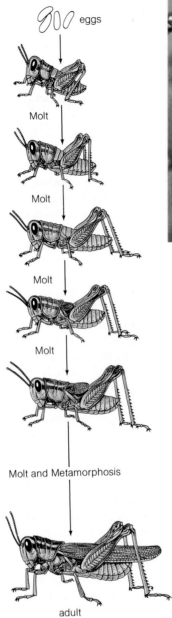

eggs

Molt

Molt

Molt

Molt

Molt and Metamorphosis

adult

Figure 34.15 Metamorphosis leading to the adult grasshopper.

experiments with the South African clawed frog, *Xenopus laevis*. Gurdon and his coworkers removed or destroyed the nucleus of an unfertilized frog egg. At the same time, they isolated intestinal cells from *Xenopus* tadpoles and carefully ruptured the plasma membrane, leaving the nucleus and most of the cytoplasm intact. Then they inserted the nucleus of the ruptured intestinal cell into the denucleated egg. In some cases the transplanted nucleus—which was from a differentiated cell—directed the development of all cell types, leading to a whole frog!

Even at the gastrula stage, all the cells of vertebrate ectoderm still have the potential to differentiate into *any* of the cell types that eventually arise from ectoderm (they are "equipotential"). This is impressive, given that ectoderm cells differentiate into (among other things) neurons, retinal rods and cones, skin glands, lens epithelium, and olfactory epithelium.

It appears that gene controls keep some cells undifferentiated and cause others to follow particular pathways at each stage of development. The controls influence development in space, for they lead to *regional* differences in cell populations. (An adult animal has dozens or hundreds of fully differentiated cell types, such as neurons and muscle cells.) The controls also influence development over time, for the *pace* of events can vary within and between cell populations. (An adult animal also has populations of stem cells that can self-replicate, differentiate, or both. Such populations are the basis of ongoing cell replacements in skin, blood, and the gut mucosa.)

There are two general patterns of cell differentiation. In **regulative patterns** of development, genes that are *not* expressed in differentiated cells still have the potential to be activated. In **mosaic patterns** of development, many genes cannot be activated in fully differentiated cells.

Sea squirts and many other invertebrates show an extreme mosaic pattern in early development and a regulative pattern later. When an early embryo is experimentally split after the first cleavage, each of the two cells gives rise to only half of a sea squirt larva. Yet after each half-larva undergoes metamorphosis, it develops into a whole adult!

Mammals show highly regulative patterns in early development. In a dramatic illustration of this pattern, a fertilized egg was removed from a black Alaska rabbit. After the first cleavage, one of the two newly divided cells was destroyed. Cleavage continued, the embryo was transplanted into a white chinchilla rabbit—and a normal black rabbit developed from the undamaged half of the implanted embryo (Figure 34.16). Thus, the genes in the undamaged half of the embryo had been activated to "cover" the gene activity that normally would have proceeded in the other half.

The capacity for regulation is also observed when a two-celled human embryo splits on its own. The result is not two half-embryos but *identical twins*, or two complete, normal individuals having the same genetic make-up. (In contrast, nonidentical twins occur when two different eggs are fertilized at the same time by two different sperm.)

The word **regeneration** refers to the capacity to replace missing fragments or parts of the body. Regeneration occurs in adults as well as in embryos. For example, sometimes predators tear away arms from sea stars and crabs, or tails from lizards, and the missing arm or tail can be regenerated. Humans can regenerate the distal segments of amputated fingers.

AGING AND DEATH

Following growth and differentiation, the cells of all complex animals gradually deteriorate. Paralleling the deterioration are structural changes and a gradual loss of efficiency in bodily functions, as well as increased sensitivity to environmentally induced stress. Progressive cellular and bodily deterioration is built into the life cycle of all organisms in which differentiated cells show extensive specialization; it is a predictable process called **aging**.

Aging in humans leads to structural changes such as loss of hair and teeth, increased skin wrinkling and fat deposition, and decreased muscle mass. Less obvious are gradual physiological changes. For example, metabolic rates decline in kidney cells, so the body cannot

Figure 34.16 A white chinchilla rabbit that served as a foster mother and her four black Alaska rabbit progeny. Each young rabbit developed from a transplanted two-celled embryo in which one of the two cells had been experimentally destroyed. Why do you suppose the experimenters used a foster mother of a different breed?

respond as effectively as it once did to changes in extracellular fluid volume and composition. As another example, the collagen fibers being produced are structurally changed. Collagen is present in the extracellular spaces of nearly all tissues; it may represent as much as forty percent of the body's proteins. Thus any changes in its structure are bound to have widespread physical effects.

No one knows what causes aging, although researchers have given us some interesting things to think about. More than two decades ago, Paul Moorhead and Leonard Hayflick cultured normal embryonic cells from humans. They discovered that all of the cultured cell lines proceeded to divide about fifty times, then the entire population died off. Hayflick took some of the cultured cells and froze them for a period of years. Afterward, he allowed them to thaw. The cells proceeded to complete the cycle of fifty doublings—whereupon they all died on schedule.

Such experiments suggest that normal cell types have a **limited division potential**, whereby mitosis is genetically programmed to decline at a particular stage of the life cycle. But does the change in mitosis *cause* aging or is it a *result* of the aging process? Consider that neurons, which do not divide at all after an early developmental stage, still deteriorate gradually during the life of an animal. Their predictable deterioration might indicate that similar changes may be occurring in dividing cells throughout the body.

Some researchers believe that cells gradually lose the capacity for DNA self-repair, perhaps as a result of an accumulation of environmental insults. Over time, DNA mutations certainly could thwart the production of enzymes and other proteins required for proper cell functioning. Consider that cells depend on smooth

Death in the Open

Lewis Thomas

(Printed by permission from the New England Journal of Medicine, *January 11, 1973, 288:92–93*

Everything in the world dies, but we only know about it as a kind of abstraction. If you stand in a meadow, at the edge of a hillside, and look around carefully, almost everything you can catch sight of is in the process of dying, and most things will be dead long before you are. If it were not for the constant renewal and replacement going on before your eyes, the whole place would turn to stone and sand under your feet.

There are some creatures that do not seem to die at all; they simply vanish totally into their own progeny. Single cells do this. The cell becomes two, then four, and so on, and after a while the last trace is gone. It cannot be seen as death; barring mutation, the descendants are simply the first cell, living all over again. The cycles of the slime mold have episodes that seem as conclusive as death, but the withered slug, with its stalk and fruiting body, is plainly the transient tissue of a developing organism; the free-swimming amoebocytes use this mode collectively in order to produce more of themselves.

There are said to be a billion billion insects on the earth at any moment, most of them with very short life expectancies by our standards. Someone has estimated that there are 25 million assorted insects hanging in the air over every temperate square mile, in a column extending upward for thousands of feet, drifting through the layers of atmosphere like plankton. They are dying steadily, some by being eaten, some just dropping in their tracks, tons of them around the earth, disintegrating as they die, invisibly.

Who ever sees dead birds, in anything like the huge numbers stipulated by the certainty of the death of all birds? A dead bird is an incongruity, more startling than an unexpected live bird, sure evidence to the human mind that something has gone wrong. Birds do their dying off somewhere, behind things, under things, never on the wing.

Animals seem to have an instinct for performing death alone, hidden. Even the largest, most conspicuous ones find ways to conceal themselves in time. If an elephant missteps and dies in an open place, the herd will not leave him there; the others will pick him up and carry the body from place to place, finally putting it down in some inexplicably suitable location. When elephants encounter the skeleton of an elephant in the open, they methodically take up each of the bones and distribute them, in a ponderous ceremony, over neighboring acres.

It is a natural marvel. All of the life on earth dies, all of the time, in the same volume as the new life that dazzles us each morning, each spring. All we see of this is the odd stump, the fly struggling on the porch floor of the summer house in October, the fragment on the highway. I have lived all my life with an embarrassment of squirrels in my backyard, they are all over the place, all year long, and I have never seen, anywhere, a dead squirrel.

I suppose that is just as well. If the earth were otherwise, and all the dying were done in the open, with the dead there to be looked at, we would never have it out of our minds. We can forget about it much of the time, or think of it as an accident to be avoided, somehow. But it does make the process of dying seem more exceptional than it really is, and harder to engage in at the times when we must ourselves engage.

In our way, we conform as best we can to the rest of nature. The obituary pages tell us of the news that we are dying away, while birth announcements in finer print, off at the side of the page, inform us of our replacements, but we get no grasp from this of the enormity of the scale. There are 4 billion of us on the earth, and all 4 billion must be dead, on a schedule, within this lifetime. The vast mortality, involving something over 50 million each year, takes place in relative secrecy. We can only really know of the deaths in our households, among our friends. These, detached in our minds from all the rest, we take to be unnatural events, anomalies, outrages. We speak of our own dead in low voices; struck down, we say, as though visible death can occur only for cause, by disease or violence, avoidably. We send off for flowers, grieve, make ceremonies, scatter bones, unaware of the rest of the 4 billion on the same schedule. All of that immense mass of flesh and bone and consciousness will disappear by absorption into the earth, without recognition by the transient survivors.

Less than half a century from now, our replacements will have more than doubled in numbers. It is hard to see how we can continue to keep the secret, with such multitudes doing the dying. We will have to give up the notion that death is a catastrophe, or detestable, or avoidable, or even strange. We will need to learn more about the cycling of life in the rest of the system, and about our connection in the process. Everything that comes alive seems to be in trade for everything that dies, cell for cell. There might be some comfort in the recognition of synchrony, in the information that we all go down together, in the best of company.

exchanges of materials between the cytoplasm and the extracellular fluid. And collagen, recall, is present in extracellular spaces throughout the body. If deteriorating regions of DNA code for defective collagen molecules, it is conceivable that the movement of oxygen, nutrients, hormones, and so forth to and from cells could be hampered, with repercussions extending through the entire body.

Or consider what might happen as a result of deterioration of genes coding for membrane proteins that serve as self-markers (page 407). If these identification markers change, does the immune system then perceive the body's own cells as "foreign" and mount an attack on them? According to one theory, such autoimmune responses might intensify over time, thereby producing the increased vulnerability to disease and stress associated with aging.

SUMMARY

1. Asexual reproduction occurs most often by fission or budding. Sexual reproduction requires the fusion of the nuclei of two gametes.

2. The embryos of most animals proceed through five stages of development:

   a. Gametogenesis, during which the egg and sperm mature within the reproductive cycles of the parents.

   b. Fertilization, which begins when a sperm penetrates an egg and is completed when the sperm and egg nuclei fuse. Visible changes occur soon after fertilization.

   c. Cleavage, when the fertilized egg undergoes mitotic cell divisions that form the early multicelled embryo. Most genes are inactive, and the destiny of cell lineages is partly established by the sector of cytoplasm inherited at this time.

   d. Gastrulation, when cell migrations bring about formation of the primary germ layers (the endoderm, mesoderm, and ectoderm).

   e. Organogenesis, when organ systems form through growth and tissue specialization.

3. Morphogenesis is the growth, shaping, and arrangement of body parts according to predefined patterns.

4. During development, growth in size results from an increase in cell number and in the size of individual cells. The capacity for growth is genetically determined, but environmental factors determine how much of this potential is actually reached.

5. During morphogenesis, cells and tissues migrate to prescribed destinations and form functional connections with one another. Coordinated changes in the shapes of cells bring about inward and outward folding in sheets of tissues. The position of organs also may shift.

6. Embryonic cells sense their position relative to each other and use this information to produce ordered, spatial arrangements of differentiated tissues (pattern formation).

7. After being born or hatched, some animals develop directly into the adult form (direct development). Other animals proceed through indirect development, in which one or more larval stages are interposed between the embryo and the adult form. Reactivated development from the larval stage to the adult is called metamorphosis.

8. Each cell of an animal starts out with a complete set of DNA, and all cells have equal developmental potential, but restrictions are placed on that potential and only certain genes are expressed in each cell type. Through these restrictive factors, some cells are kept in their undifferentiated state while others proceed through differentiation and develop into the various specialized cell types.

9. Cells of all complex animals gradually show changes in structure and loss of efficiency of function (aging). Although aging is part of the life cycle of all animals with extensively specialized cell types, its cause is unknown.

Review Questions

1. What are some of the differences between asexual and sexual reproduction? Describe some forms of asexual reproduction. What are some adaptations associated with external and internal fertilization of sexually reproducing animals?

2. Describe what goes on during each of these developmental stages: fertilization, cleavage, gastrulation, organ formation.

3. Discuss the differences between simple cell division by mitosis and the cell divisions of cleavage.

4. Distinguish among endoderm, mesoderm, and ectoderm.

5. Define morphogenesis. Describe some of the cell movements and changes in cell shape involved in morphogenesis.

6. Define two processes of pattern formation. Can you describe an experiment that gave insight into how tissues differentiate and become positioned in specific regions of the embryo?

7. Define indirect and direct development. Give some examples of animals that follow each developmental strategy.

8. During the development of duck embryos and human embryos, the ends of the leg buds differentiate into paddlelike appendages. In ducks, the paddles develop into webbed feet. In humans, the paddles develop separate toes. What morphogenetic process might give rise to the separate digits in humans?

9. Define cell differentiation.

10. During cleavage stages, the embryonic cells form large and coherent masses, but in later stages the tissues and organ rudiments are discrete and the different cell groups do not merge. Suggest mechanisms that form the basis for this phenomenon.

11. Occasionally, the first division in an amphibian egg isolates the gray crescent wholly within one of the two cells formed. If the two cells are separated from each other, only the cell with the gray crescent is able to form an embryo with an axis, notochord, nerve cord, and back musculature. The other cell forms an amorphous mass of gut and blood. What do you think is the explanation of this result?

Readings

Balinsky, B. 1981. *An Introduction to Embryology*. Fifth edition. Philadelphia: Saunders.

Gehring, W. October 1985. "The Molecular Basis of Development." *Scientific American* 253(4):153–162.

Gilbert, S. 1985. *Developmental Biology*. Sunderland, Massachusetts: Sinauer. Excellent introduction to animal development: clear illustrations.

Gordon, R., and A. G. Jacobson. June 1978. "The Shaping of Tissues in Embryos." *Scientific American* 238(6):106–113. Interesting discussion of computer simulations of forces shaping embryonic morphogenesis.

Hayflick, L. January 1980. "The Cell Biology of Human Aging." *Scientific American* 242(1):58–65.

Hopper, A., and N. Hart. 1985. *Foundations of Animal Development*. Second edition. New York: Oxford University Press.

Lewis, J. 1983. "Cellular Mechanisms of Development." In *Molecular Biology of the Cell* (Alberts et al.). New York: Garland.

Patten, B. M. 1971. *Early Embryology of the Chick*. Fifth edition. New York: McGraw-Hill.

Raff, R., and T. Kaufman. 1983. *Embryos, Genes, and Evolution*. New York: Macmillan. Wide-ranging, thoughtful discussion of the relationship between genetics and development and its role in the evolutionary process.

Saunders, J. W. 1982. *Developmental Biology: Patterns, Problems, and Principles*. New York: Macmillan. Excellent introduction to embryology.

In the preceding chapter, we looked at some general principles of animal reproduction and development. Here, we will focus on humans as a way of presenting an integrated picture of the structure and function of reproductive organs, gametogenesis, and the stages of development from fertilization to birth. As part of this picture, we will also consider some of the mechanisms controlling reproduction.

PRIMARY REPRODUCTIVE ORGANS

The human reproductive system consists of a pair of primary reproductive organs, or **gonads**, and accessory glands and ducts. The male gonads are the **testes** (singular, testis); the female gonads are the **ovaries**. Testes and ovaries have the same general functions: they produce gametes and they secrete sex hormones, which influence reproductive functions and the development of secondary sexual traits. The male gametes are **sperm**; the female gametes are "eggs," or (more accurately) **secondary oocytes**. Among the secondary sexual traits are the amount and distribution of body fat, hair, and skeletal muscle. Those traits are associated with maleness and femaleness but play no direct role in reproduction.

Until the seventh week of development, the gonads look the same in all human embryos. Then, activation of genes on the sex chromosomes triggers the differentiation of the early gonads into testes *or* ovaries. Although gonads and accessory organs are already formed at birth, they do not reach their full size and become reproductively functional until twelve to sixteen years later.

35

HUMAN REPRODUCTION AND DEVELOPMENT

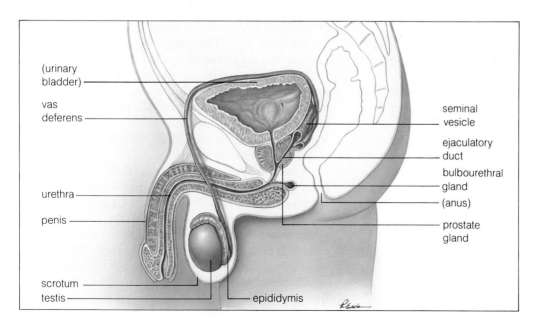

Figure 35.1 Reproductive system of the human male, longitudinal section. There are two each of the following components: testis, epididymis, vas deferens, seminal vesicle, ejaculatory duct, and bulbourethral gland.

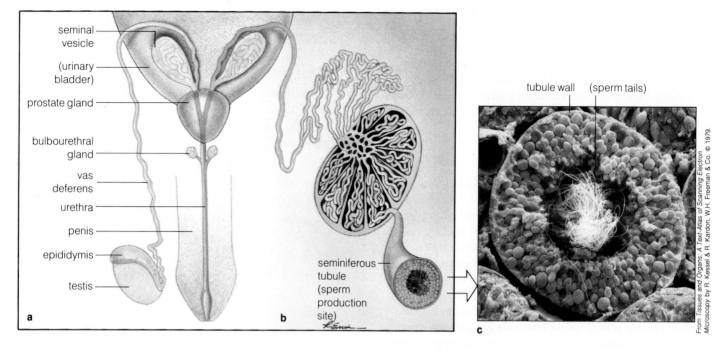

Figure 35.2 (**a**) Posterior view of the male reproductive system. (**b**) A cross-section through a testis. (**c**) Scanning electron micrograph showing the cells inside a seminiferous tubule.

In figure labels:

seminal vesicle
(urinary bladder)
prostate gland
bulbourethral gland
vas deferens
urethra
penis
epididymis
testis

a

seminiferous tubule (sperm production site)

b

tubule wall   (sperm tails)

c

From *Tissues and Organs: A Text-Atlas of Scanning Electron Microscopy* by R. Kessel & R. Kardon, W.H. Freeman & Co. © 1979.

MALE REPRODUCTIVE SYSTEM

Figure 35.1 shows the male reproductive system. The testes initially form in the dorsal wall of the abdominal cavity, and before birth they descend into the *scrotum*, an outpouching of skin below the pelvic region. The interior of the scrotum must be kept a few degrees cooler than the body's core temperature if sperm are to develop properly. A temperature of about 95°F is maintained by controlled contractions of muscles in the tissues of the scrotum. When it is cold outside, contractions draw the pouch closer to the (warmer) body; when it is warm outside, the muscles relax and the pouch is lowered.

The testes themselves are divided internally into 250 to 300 wedge-shaped lobes (Figure 35.2). Each lobe contains two to three **seminiferous tubules** in which sperm form ("seminiferous" means seed-bearing). Although each testis is only about five centimeters long, about 125 meters of these highly coiled tubes are packed into it! The connective tissue between the tubes contains *interstitial cells*, a type of endocrine cell that secretes sex hormones, including testosterone.

Spermatogenesis

Mammals generally reproduce on a seasonal basis, and spermatogenesis (sperm production) generally coincides with the reproductive season. Humans and other pri-

mates are exceptions. From puberty onward, sperm are produced continuously, with many millions of cells in different stages of development on any given day.

Figure 35.3 shows how sperm are produced in a seminiferous tubule. Inside, undifferentiated diploid cells called spermatogonia are closest to the tubule wall. These cells are forced away from the wall by ongoing mitotic divisions and are gradually transformed into primary spermatocytes. They undergo meiosis I, the result being secondary spermatocytes. Although the cells are now haploid, keep in mind that each chromosome is still in the duplicated state; it consists of two sister chromatids (page 152). The sister chromatids of each chromosome are separated from each other during meiosis II. The resulting cells are haploid spermatids, which gradually develop into mature sperm. The entire process takes about nine to ten weeks. All the while, the developing cells receive nourishment and chemical signals from adjacent *Sertoli cells*, which are the only other type of cell present in the tubule.

A mature sperm consists of a head, midpiece, and tail (Figure 35.4). The head contains the nucleus, which is packed with DNA. The *acrosome*, a cap over most of the head, contains lytic enzymes that will assist the sperm in penetrating the outer layer surrounding an oocyte. The tail contains an array of microtubules, which bend in coordinated fashion to produce whiplike movements. Numerous mitochondria in the midpiece supply the energy necessary for these movements.

Sperm Movement Through the Reproductive Tract

After their formation in a testis, sperm move into a long, coiled duct called the **epididymis**. The sperm cannot move on their own yet. Their capacity for movement develops somewhat in the epididymis, but they do not become fully motile until they have entered the female reproductive tract.

Between times of sexual arousal, most sperm are stored in the region of the epididymis closest to a thick-walled tube called the **vas deferens**. The vas deferens functions in rapid sperm transport. It merges with a duct from a gland called the **seminal vesicle** (Figure 35.1). From there, sperm move into a short duct that runs through the tissues of the **prostate gland** and empties into the urethra. (The urethra, recall, terminates at the tip of the penis; it is the channel that serves in the ejaculation of sperm as well as in the excretion of urine.)

As the sperm pass through the reproductive tract, they become mixed with fluids secreted from the prostate gland, seminal vesicles, and bulbourethral glands. This sperm-bearing mixture of fluids is called *semen*. The seminal vesicles secrete a fluid containing fructose (which nourishes sperm) and prostaglandins (which may trigger contractions in the female reproductive tract and thereby assist sperm in their movement). The prostate gland secretes a thin, alkaline fluid that probably helps neutralize acidic secretions in the vagina. (The pH in the vagina is about 3.5–4, but sperm motility and fertility are enhanced when it increases to about 6.) The

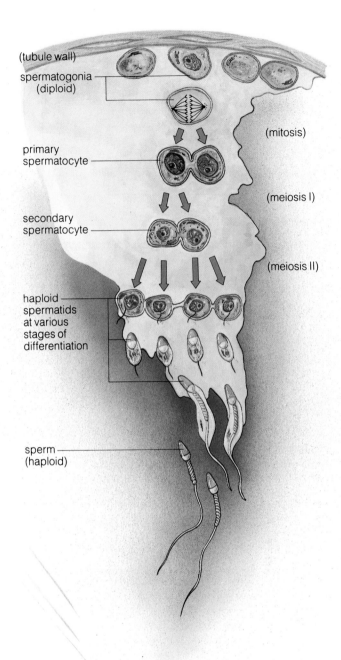

Figure 35.3 Spermatogenesis inside a seminiferous tubule.

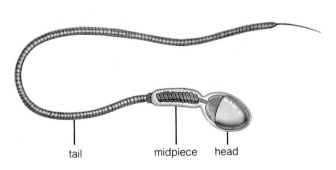

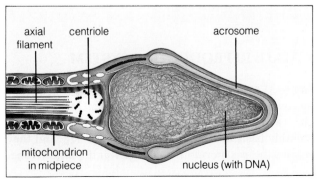

Figure 35.4 Human sperm. The boxed sketch shows the head region in cross-section and in profile view.

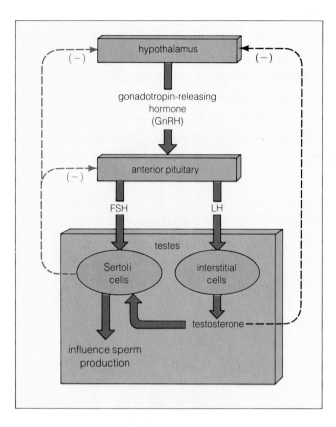

Figure 35.5 Hormonal control of reproductive function in human males. The black dashed line indicates that increased testosterone secretion inhibits LH secretions through its negative effect on hypothalmic GnRH. The blue dashed line indicates that an inhibitory signal from Sertoli cells influences GnRH and FSH secretions.

Table 35.1	**Organs and Accessory Glands of the Male Reproductive Tract**
Organs:	
Testis (two)	Sperm production, sex hormone production
Epididymis (two)	Sperm maturation site, sperm storage
Vas deferens (two)	Rapid transport of sperm
Ejaculatory duct (two)	Conduction of sperm
Penis	Organ of sexual intercourse
Accessory Glands:	
Seminal vesicle (two)	Secretions make up large portion of semen*
Prostate gland	Same as above
Bulbourethral gland (two)	Same as above

*Semen enhances sperm motility in various ways.

bulbourethral glands secrete a small amount of mucus-rich fluid into the urethra during sexual arousal. This fluid lubricates the penis, assisting its penetration into the vagina; it also enhances sperm movement.

Hormonal Control of Male Reproductive Functions

Three major hormones directly or indirectly control the reproductive functions of males. One is **testosterone**, produced by interstitial cells in the testes. The other two are **follicle-stimulating hormone** (FSH) and **luteinizing hormone** (LH). As we have seen, these two gonadotropins are produced by the anterior pituitary. (FSH and LH were first named for their effects in females, but their molecular structure is exactly the same in males.)

Testosterone has multiple functions. First, this hormone stimulates spermatogenesis. Second, it is necessary for the growth, form, and function of all parts of the male reproductive tract. Third, testosterone promotes the normal development and maintenance of sexual behavior. (Aggressive behavior also is somewhat dependent on testosterone.) Fourth, secondary sexual traits such as beard growth and pubic hair growth are testosterone-dependent. Fifth, testosterone has growth-promoting effects on general body tissues.

Testosterone secretion is governed by negative feedback loops involving the hypothalamus, the anterior pituitary, and the testes (Figure 35.5). In response to stimulation, the hypothalamus steps up its secretion of **GnRH**, or gonadotropin-releasing hormone. When GnRH reaches the anterior pituitary, it causes the secretion of LH, which is circulated to the testes. There, LH stimulates interstitial cells to increase their secretion of testosterone, which acts on the Sertoli cells in ways that stimulate sperm production. The GnRH also causes the secretion of FSH, which acts *directly* on the Sertoli cells in the testes.

As the testosterone concentration increases, it exerts an inhibitory effect on the hypothalamus to reduce GnRH secretion. The FSH has a stimulatory effect on the Sertoli cells, which respond by releasing a protein (inhibin) that causes the pituitary to decrease its FSH secretion.

FEMALE REPRODUCTIVE SYSTEM

Figure 35.6 shows the female reproductive system. The two ovaries are positioned against the wall of the abdominal cavity. They produce and release oocytes on a monthly basis in women of reproductive age, and they secrete the sex hormones **estrogen** and **progesterone**.

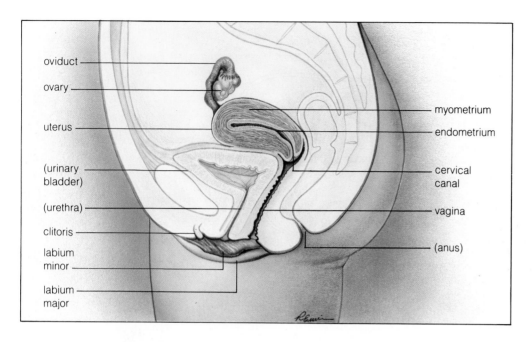

Figure 35.6 Reproductive system of the human female, longitudinal section. There are two ovaries and two oviducts (which lead into the uterus).

Adjacent to each ovary but not connected to it is an **oviduct**, which channels oocytes into the uterus. Ciliated, fingerlike projections from the oviduct extend over one end of the ovary, and their sweeping action moves the oocytes into the entrance of the duct.

The **uterus** houses the developing individual during pregnancy. This hollow organ, which resembles an upside-down pear, is positioned near the floor of the abdominal cavity, above the urinary bladder. The bulk of the uterus is a thick layer of smooth muscle called the *myometrium*. The interior lining of the uterus, the *endometrium*, consists of connective tissue, glands, and blood vessels. The lower portion of the uterus (the narrow part of the "pear") is the *cervix*. The **vagina** is a muscular tube that connects with the cervix at one end and opens to the outside of the body at the other end. It is the organ in which sperm from the male are deposited, and it functions as part of the birth canal (Table 35.2).

At the body surface are the external genitalia (or *vulva*). The outermost structures are a pair of skin folds (the labia majora), which contain adipose tissue. Within the cleft formed by these folds are the labia minora—a smaller pair of skin folds that are highly vascularized but have no fatty tissue. At the anterior end of the vulva, these two interior skin folds partly enclose the *clitoris*, a small organ for sexual stimulation. The opening of the urethra is about midway between the clitoris and the vaginal opening (Figure 35.6).

Oogenesis

Every oocyte that develops in a woman's body was already present, in immature form, before she was born. When she herself was a fetus, certain diploid cells divided by mitosis and enlarged to form primary oocytes, each of which became surrounded by a single layer of *granulosa cells*. A primary oocyte together with its surrounding cell layer is a **primary follicle** (Figure 35.7).

Before the woman was born, her primary oocytes already started meiosis I; that is, the chromosomes in each nucleus paired with their homologues as a prelude to their distribution into haploid gametes (see, for example, Figure 11.7). However, meiosis I was arrested and her primary oocytes remained dormant in her ovaries until puberty.

Table 35.2	Female Reproductive Organs
Ovaries	Oocyte production, sex hormone production
Oviducts	Conduction of oocyte from ovary to uterus
Uterus	Chamber in which new individual develops
Cervix	Secretes mucus that enhances sperm movement into uterus and (after fertilization) reduces the embryo's risk of bacterial infection
Vagina	Organ of sexual intercourse; birth canal

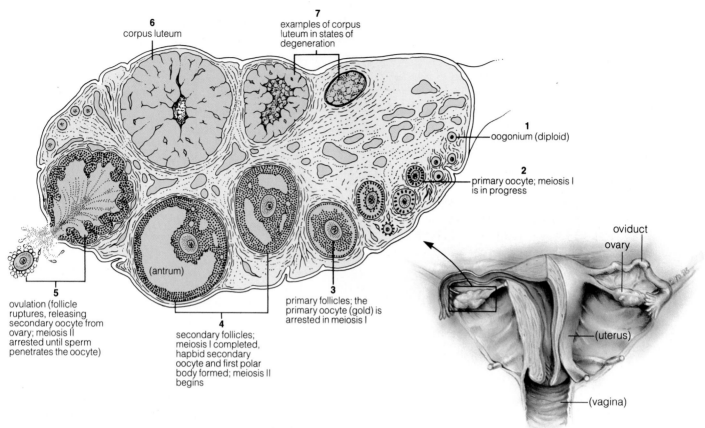

6
corpus luteum

7
examples of corpus luteum in states of degeneration

1
oogonium (diploid)

2
primary oocyte; meiosis I is in progress

oviduct
ovary

(uterus)

(vagina)

5
ovulation (follicle ruptures, releasing secondary oocyte from ovary; meiosis II arrested until sperm penetrates the oocyte)

(antrum)

4
secondary follicles; meiosis I completed, hapbid secondary oocyte and first polar body formed; meiosis II begins

3
primary follicles; the primary oocyte (gold) is arrested in meiosis I

Figure 35.7 A human ovary, drawn as if sliced lengthwise through its midsection. Events in the ovarian cycle proceed from the growth and maturation of primordial follicles, through ovulation (rupturing of a mature follicle with a concurrent release of a secondary oocyte), through the formation and maintenance (or degeneration) of an endocrine element called the corpus luteum. The positions of these structures in the ovary are varied for illustrative purposes only. The maturation of an oocyte occurs at the *same* site, from the beginning of the cycle to ovulation.

Of the perhaps 2 million primary oocytes that were present in her ovaries at the time of birth, less than 300,000 remain at puberty. Of those, about 400 will mature and will be expelled from her body, one at a time on a monthly basis, over the next three decades or so. The longer it takes for the cells to be released from that prolonged meiosis, the greater the risk of chromosome abnormalities (page 188).

When a primary oocyte matures, the granulosa cells deposit a layer of material (the *zona pellucida*) around it. At some point it completes meiosis I and two cells form: a haploid secondary oocyte (which ends up with almost all the cytoplasm) and a tiny, first polar body. The tiny cell functions only as a "dumping ground" for half of the diploid number of chromosomes, so that the gamete will be haploid.

At **ovulation**, a secondary oocyte is released from the ovary and it enters meiosis II. However, this second division is *not* completed unless fertilization occurs.

When meiosis II is completed, one cell again retains most of the cytoplasm; it is the mature oocyte, or **ovum**. The other cell is a tiny, secondary polar body. Polar bodies may eventually complete meiosis like the oocyte, but they degenerate soon afterward.

Menstrual Cycle: An Overview

Unlike males, the reproductive capacity of human females is cyclic and intermittent. In most mammalian females, the reproductive pattern is called the **estrous cycle**. During estrus, the female comes into "heat" (becomes sexually receptive to the male) only at specific times of year. Also during estrus, hormones stimulate the maturation and release of oocytes, and they prime the endometrium for implantation.

In humans and other primates, the cyclic reproductive capacity in females is called the **menstrual cycle**.

Here, too, the reproductive tract and the oocytes maturing within it are primed for fertilization. Unlike estrus, however, there is no correspondence between heat and the time of fertility. All female primates can be physically and behaviorally receptive to the male's overtures at any time. Also unlike estrus, the hormone-primed uterine lining is sloughed off at the end of each cycle. Menstrual cycles begin at about age thirteen and continue until menopause (in the late forties or early fifties).

On the average, it takes about twenty-eight days to complete one menstrual cycle, although the range may be twenty-two to thirty-six days. During this time span, changes take place in the ovary and in the endometrium (Table 35.3). Feedback loops involving the hypothalamus, the anterior pituitary, and the gonads control these changes. In females, as in males, GnRH secreted from the hypothalamus causes the release of LH and FSH from the anterior pituitary. But in females, those pituitary hormones trigger estrogen and progesterone secretions. Estrogen is secreted by cells of the follicles and, after ovulation, by a glandular body called the **corpus luteum**. Progesterone is secreted in very small amounts by the granulosa cells, but the major secretion is from the corpus luteum. Let's look first at how these controls influence ovarian function. Then we will look at the concurrent changes in the endometrial lining of the uterus.

Control of Ovarian Function

Follicular Phase. The follicular phase of each menstrual cycle lasts about thirteen days (Table 35.3). During this phase, several follicles grow but ordinarily only one reaches maturity; the others degenerate. As the follicle develops, estrogen-containing secretions from its granulosa cells begin to accumulate and form a fluid-filled space (the *antrum*) inside the follicle. The antrum eventually becomes so large that the mature follicle balloons outward from the ovary's surface. At ovulation, the "balloon" ruptures and fluid escapes, carrying the secondary oocyte with it (Figure 35.7).

At the start of each menstrual cycle, hormone levels are low (Figure 35.8). Then, FSH and LH levels rise and stimulate estrogen production and secretion in the follicle; the estrogen in turn stimulates the mitotic divisions that underlie follicle growth. The estrogen inhibits the secretion of more LH and FSH early in the follicular phase. However, by about the midpoint of the cycle, a rapid increase in the estrogen level in the blood causes the opposite effect: a brief outpouring of LH from the anterior pituitary. Within thirty-six to forty-eight hours, this **midcycle surge of LH** has triggered the following events:

1. Within the follicle, meiosis I resumes in the primary oocyte, and a secondary oocyte forms. (Thus the parental diploid number of chromosomes has been reduced to the haploid number for the forthcoming gametes.)

2. Estrogen secretion (by cells of the follicle) slows down as the follicle is disrupted.

3. The follicle ruptures, possibly because LH stimulates the production of lytic enzymes that act on the follicle where it balloons out from the ovary. Thus, *high levels of LH trigger ovulation.*

4. The parts of the follicle left behind are transformed into a corpus luteum, which starts secreting quantities of progesterone and estrogen.

Early in the follicular phase, then, estrogen has a *negative* feedback effect on FSH and LH secretion, but later on it has a *positive* feedback effect on LH secretion. These feedback loops are shown in Figure 35.9.

Luteal Phase. The luteal phase of the menstrual cycle follows ovulation. Under the continued influence of LH, the remaining cells of the ruptured follicle become the corpus luteum, which secretes considerable amounts of progesterone and some estrogen. A corpus luteum can persist for about twelve days if fertilization does not follow ovulation.

During that twelve-day span, the high blood concentrations of estrogen and progesterone have a feedback effect on FSH an LH secretions. The FSH and LH levels fall; hence a new follicle is prevented from developing until the menstrual cycle is completed.

Table 35.3 Overview of Events of the Menstrual Cycle		
Phase	Events	Duration* (days)
Follicular phase	Menstruation; endometrium breaks down	1–5
	Follicle matures in ovary; endometrium rebuilds	6–13
Ovulation	Secondary oocyte released from ovary	14
Luteal phase	Corpus luteum forms; endometrium thickens and develops	15–28

*Assuming a 28-day cycle.

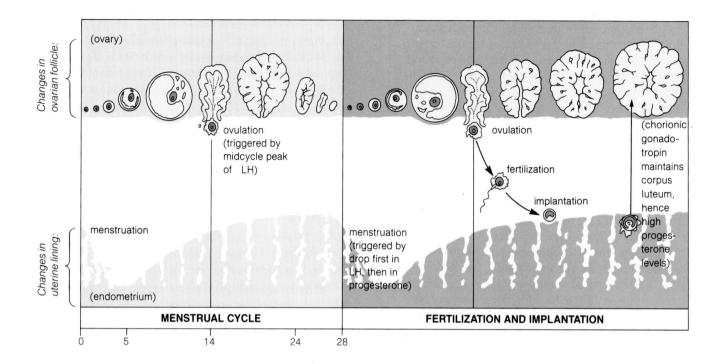

Changes in ovarian follicle:

(ovary)

ovulation (triggered by midcycle peak of LH)

ovulation

fertilization

implantation

(chorionic gonado-tropin maintains corpus luteum, hence high proges-terone levels)

Changes in uterine lining:

menstruation

menstruation (triggered by drop first in LH, then in progesterone)

(endometrium)

MENSTRUAL CYCLE

FERTILIZATION AND IMPLANTATION

0   5   14   24   28

Day of cycle (average)

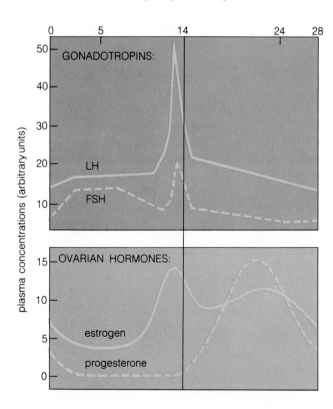

Figure 35.8 Correlation among gonadotropins, ovarian hormones, follicle development, ovulation, and changes in uterine function during the human menstrual cycle.

When fertilization does not occur, the corpus luteum starts to degenerate during the last days of the menstrual cycle. Apparently it self-destructs by secreting large amounts of prostaglandins, which interfere with its functions. Once the corpus luteum is destroyed, estrogen and progesterone levels fall rapidly, so their inhibitory effect on the hypothalamus and pituitary is canceled. FSH and LH secretions rise, other follicles in the ovary are stimulated into developing—and the cycle begins anew.

Control of Uterine Function

The changing estrogen and progesterone levels just described cause profound changes in the uterus. Estrogen stimulates the growth of endometrium and uterine smooth muscle. Under the influence of progesterone, blood vessels proliferate in the thickened endometrium; and glycogen and enzymes are stockpiled in endometrial glands and connective tissues. In such ways, the uterus is prepared for the possibility of pregnancy.

At the time of ovulation, estrogen causes the cervix to secrete a mucus that is thin, clear, and abundant—an ideal medium through which sperm can travel. (Just after ovulation, progesterone secretions from the corpus luteum act on the cervix. The mucus becomes thick and sticky, and it serves as a physical barrier against bacteria that might enter the vagina and endanger a newly conceived zygote.)

In the absence of fertilization, and with the disintegration of the corpus luteum, blood levels of progesterone and estrogen fall. Without hormonal support, the highly developed endometrium starts disintegrating also. Blood vessels in the endometrium constrict and the tissues (deprived of oxygen and nutrients) die off. Arterioles now dilate, and the increased blood flow ruptures the walls of weakened capillaries. Menstrual flow consists of blood and sloughed endometrial tissues. The flow moves out of the body through the vagina, and its appearance marks the first day of a new menstrual cycle. It continues for three to six days, until rising estrogen levels stimulate the repair and growth of the endometrium.

SEXUAL UNION AND FERTILIZATION

A secondary oocyte encounters sperm as a result of sexual union, or **coitus**. Sperm are ejaculated from the penis, which contains three cylinders of spongy tissue arranged around the urethra. At the tip of the penis, the ventral cylinder terminates as a mushroom-shaped structure (the glans penis), which contains abundant mechanoreceptors that are stimulated by friction. During normal circumstances, blood vessels leading into the three cylinders are constricted and the penis is limp. During initial sexual excitation, blood flows into the cylinders faster than it flows out. Blood collects in the spongy tissue, and the organ lengthens and hardens. These changes can occur within seconds of sexual excitation, and they facilitate penetration into the vagina.

During coitus, pelvic thrusts stimulate the penis as well as the vaginal walls and clitoral region of the female body. Mechanical stimulation of the penis triggers rhythmic, muscular contractions that force the contents of seminal vesicles and the prostate into the male urethra. Involuntary contractions of other muscles expel semen into the vagina. (During ejaculation, a sphincter closes off the bladder. As a result, sperm cannot enter the bladder and urine cannot be excreted from it.) The involuntary muscular contractions, ejaculation, and associated sensations of release, warmth, and relaxation constitute the event called *orgasm*.

Female orgasm involves similar physical events, including an intense vaginal awareness, a series of involuntary uterine and vaginal contractions, and sensations of relaxation and warmth. Even if the female does not reach this state of excitation during coitus, she can still become pregnant.

If sperm ejaculation into the vagina coincides with ovulation (if it occurs within a range of about three days before ovulation and three days afterward), pregnancy can result. Within thirty to sixty minutes after ejaculation from the penis, sperm move into the oviduct.

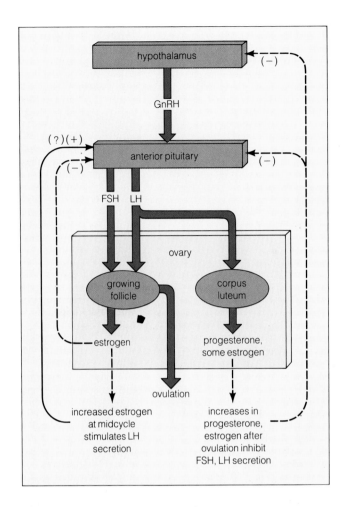

Figure 35.9 Feedback loops among the hypothalamus, anterior pituitary, and ovaries during the menstrual cycle.

Table 35.4 Stages of Human Development: A Summary

Prenatal Period

1.	Zygote	Single cell resulting from the fusion of a sperm nucleus with the nucleus of a secondary oocyte at fertilization; zygote usually forms in the oviduct
2.	Morula	Solid ball of about sixteen cells (called _blastomeres_), produced through cleavages of the zygote (page 478)
3.	Blastocyst	Ball of cells with a surface layer (the _trophoblast_) and an _inner cell mass_. Produced after the morula enters the uterus; then, fluid enters the ball and lifts some cells to form a cavity
4.	Embryo	_Embryo_ refers to all developmental stages from two weeks after fertilization (at which time the inner cell mass forms a two-layer _embryonic disk_) until the end of the eighth week. All major body structures begin forming during the embryonic period
5.	Fetus	_Fetus_ refers to all developmental stages from the ninth week until birth, about thirty-eight weeks after fertilization. The rate of overall growth and structural elaborations increase dramatically during the fetal period

Postnatal Period

6.	Newborn	Individual during the first two weeks after birth (the _neonatal_ period). The physiological transition from life in the uterus to life in the external world requires many gradual changes, as in respiration
7.	Infant	Individual from two weeks to about the first fifteen months after birth. During infancy, body weight triples and height doubles, on the average
8.	Child	Individual from infancy to about twelve or thirteen years of age. Bone formation and overall growth are extensive early in childhood, then slow down; just before puberty, the rate of growth steps up rapidly
9.	Pubescent	Girls between twelve and fifteen years old, and boys between thirteen and sixteen years old; secondary sexual characteristics develop now
10.	Adolescent	Individual from puberty until about three or four years later; a time of physical, mental, and emotional maturation
11.	Adult	During early adulthood (between eighteen and twenty-five years of age), bone formation and growth are finally completed. The maximum human life span is probably 100 years or so, but after early adulthood, developmental changes proceed very slowly

Once a sperm reaches its surface, the secondary oocyte is stimulated into completing meiosis II. Only at this stage is the oocyte referred to as a **mature ovum** (Figure 35.10). Within its cytoplasm, its nucleus fuses with the sperm nucleus and their chromosomes intermingle, restoring the diploid number for the new individual.

On the average, between 150 million and 350 million sperm enter the vagina during ejaculation, but only a few hundred reach the upper region of the oviduct where fertilization usually occurs. Before a sperm can penetrate the secondary oocyte, however, it must undergo a physiological change called _capacitation_, which occurs during its passage through the female reproductive tract. The change somehow primes the sperm membranes so that tiny channels will open up in the wall of the acrosome when the sperm arrives near the secondary oocyte. Digestive enzymes escape through the channels, and their action clears a path for the sperm through the zona pellucida. Although several sperm can penetrate the zona pellucida, usually only one enters the secondary oocyte.

PRENATAL DEVELOPMENT

Table 35.4 summarizes the developmental stages that follow fertilization. Let's briefly consider some key events of the prenatal stages ("prenatal" means before birth).

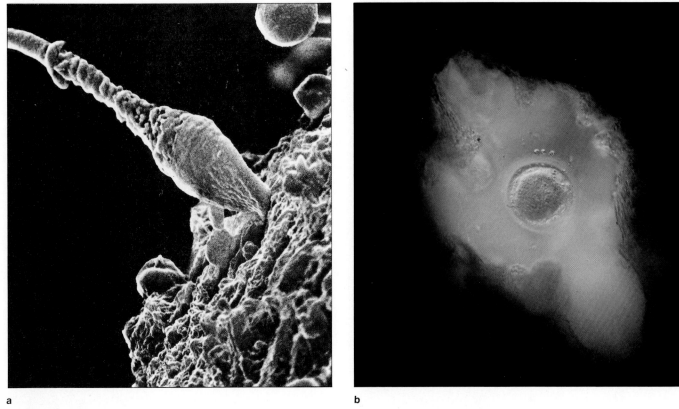

a   b

Figure 35.10 (**a**) Human sperm about to penetrate the zona pellucida around a secondary oocyte. (**b**) A mature ovum. Notice the three polar bodies above it; these products of meiosis will degenerate shortly.

First Week of Development

For the first three or four days after fertilization, the zygote travels down the oviduct, picking up nutrients from maternal secretions and undergoing its first mitotic cell divisions. By the time it reaches the uterus, the cleavages have transformed it into a solid cluster of cells (the *morula*), which develops into a blastocyst. In humans, a **blastocyst** consists of an inner cell mass, and a surface cell layer called the *trophoblast* (Figure 35.11).

Cells of the trophoblast secrete **chorionic gonadotropin**, a hormone that maintains the corpus luteum. Thus the corpus luteum continues to grow and to secrete estrogens and progesterone, which help maintain the uterine lining. (The presence of chorionic gonadotropin in the mother's urine is an early indicator of pregnancy.)

Before the first week ends, the blastocyst contacts and adheres to the uterine lining. Cell divisions in the trophoblast produce two layers, the outer one consisting of a mass of cells that send out fingerlike projections into the uterine lining. Through this invasion, the blastocyst becomes implanted in the mother's tissues (Figure 35.12).

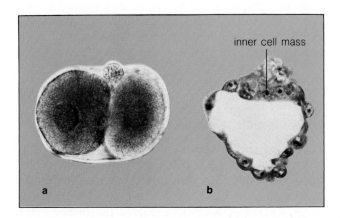

inner cell mass

a   b

Figure 35.11 (**a**) Two-cell stage of cleavage, about thirty-six hours after fertilization. (**b**) Blastocyst in cross-section after four to six days. The layer of surface cells of the blastocyst is called the trophoblast.

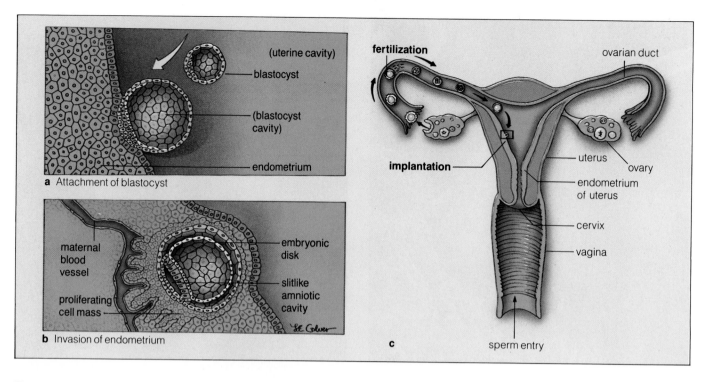

Figure 35.12 Implantation of the blastocyst in the endometrium (the lining of the uterus).

While the invasion proceeds, endoderm starts to form on the inner cell mass of the blastocyst; this is the first primordial tissue layer of the embryo. Eventually there is a two-layered, **embryonic disk** (Figure 35.12). The disk is usually shown in side view, cut down the middle; but keep in mind that it is shaped rather like an oval pancake.

Extraembryonic Membranes

To gain perspective on the events that follow implantation, think about a key character in vertebrate evolution—the shelled egg. Shelled eggs allowed embryonic development to take place on land rather than in the water, where vertebrates first evolved. Inside the shelled eggs of reptiles and birds are several membranes that protect the embryo and function in its nutrition, respiration, and excretion. These membranes are called the **yolk sac**, the **allantois**, the **amnion**, and the **chorion**.

As we have seen, bird eggs have a large store of yolk—but how does the embryo draw nutrients from it? At first, endoderm grows almost completely around the yolk, forming a sac. Then mesoderm spreads over the sac and develops into vessels that convey nutrients to the embryo. Eventually the connection between the yolk sac and the embryo is constricted into a "stalk." Although a human embryo is not housed within a shelled egg, it is still provided with a yolk *sac* (but no yolk) as part of its vertebrate heritage. A membranous sac forms below the embryonic disk *as if* the yolk were still there; parts of this membrane give rise to the digestive tube of the embryo.

As we have seen also, nitrogenous waste products of protein metabolism can be toxic in high concentrations—but how are they disposed of in a hard-shelled egg? In reptiles and birds, they are stored inside the allantois, a bladderlike sac that develops as an extension of the primitive gut. Later, a network of blood vessels develops on the surface of the allantois and helps supply the embryo with oxygen. In humans, the allantois no longer functions in handling nitrogenous wastes but its blood vessels still function in oxygen transport.

In all land vertebrates, the amnion is a membranous, fluid-filled sac, a tiny replica of the aquatic environment in which the embryos of ancient vertebrates developed. In humans, the amnion forms as a slitlike opening in the inner cell mass of the blastocyst about two weeks after fertilization. Eventually it completely surrounds the embryo and keeps it from drying out. The amniotic fluid also acts like a shock absorber. (The "water" that flows

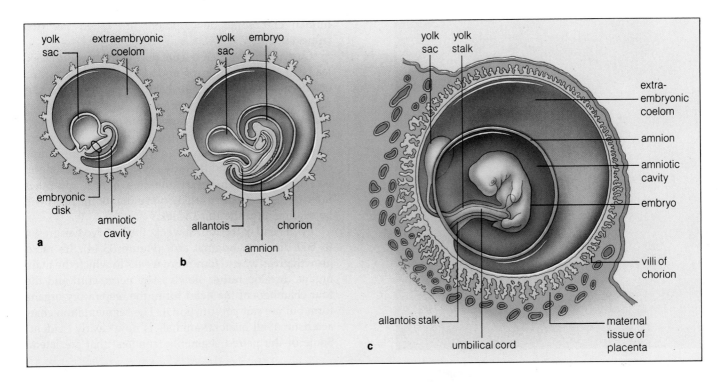

Figure 35.13 Formation of the extraembryonic membranes that protect the embryo and function in its development.

freely from the vagina just before childbirth is amniotic fluid, released when the amnion ruptures.)

As Figure 35.13 shows, the growing human embryo gradually becomes separated from the extraembryonic membranes just described until it is connected to them only by the **umbilical cord**. This cord contains the "stalks" of the yolk sac and allantois, surrounded by the outer layer of the amnion. It is well endowed with blood vessels.

Another membrane, the chorion, develops as a protective membrane around the embryo and the other membrane structures. In humans, it is derived from the trophoblast and reinforcing mesodermal tissues. The chorion becomes a primary component of the placenta.

The Placenta

Three weeks after fertilization, almost a fourth of the inner surface of the uterus has developed into a spongy tissue that is composed of endometrium *and* extraembryonic membranes, the chorion especially. This tissue, the **placenta**, is the means by which the embryo receives nutrients and oxygen from the mother and sends out wastes in return, which are disposed of through the mother's lungs and kidneys.

The tiny, fingerlike projections that were sent out from the blastocyst during implantation develop into numerous chorionic villi, which contain embryonic blood vessels (Figure 35.13). The embryonic blood vessels do not "merge" with those of the mother; the two bloodstreams remain separate throughout pregnancy. Small solutes diffuse both ways across the thin endothelial walls of the two sets of blood vessels, moving to and from the maternal lacunae (which are blood-filled spaces in the uterine lining).

The placenta secretes several hormones, including the chorionic gonadotropin that maintains the corpus luteum. Within about two months after fertilization, it also secretes estrogen and progesterone in such large quantities that pregnancy can be maintained even if the corpus luteum is surgically removed.

Embryonic and Fetal Development

First Trimester. Once the embryonic disk has formed in the blastocyst, human development proceeds along the same general course described in the preceding chapter for all complex animals. The first three months of

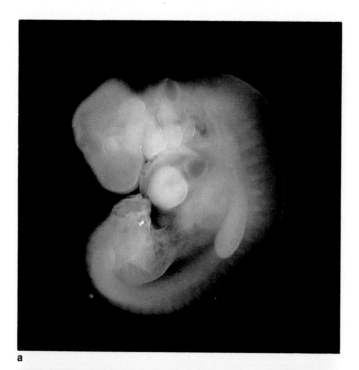

human embryonic development are called the **first trimester**.

Gastrulation, which begins during the second week, leads to the formation of endoderm, mesoderm, and ectoderm—the three primordial tissue layers. Some surface cells of gastrula migrate inward to form the mesoderm. When gastrulation is over, the remaining surface tissue (the ectoderm) will give rise to the nervous system and skin. Inside, the endoderm will form the inner lining of the respiratory and digestive systems; the mesoderm will develop into the heart, muscles, bones, and other internal organs. The tubelike forerunner of the heart will be beating after the third week.

By the end of the fourth week, the embryo has grown to 500 times its original size (Figure 35.14). Its rapid growth gives way to four more weeks in which the main organs develop rather slowly. The nerve cord and the four chambers of the heart form; the respiratory organs form but are not yet functional. The segmentation characteristic of all bilateral animals is now clearly evident. Some of the paired segments (somites) that are lateral

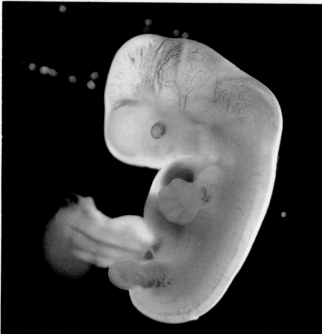

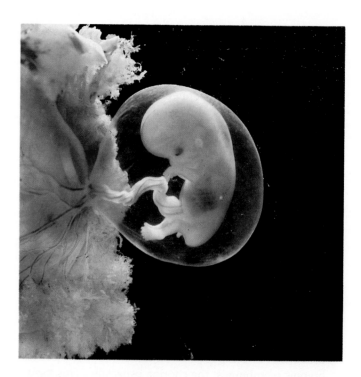

Figure 35.14   (**a**) Embryo at four weeks, about seven millimeters (0.3 inch) long. Notice the tail and the gill arches, which vaguely resemble a double chin. These features emerge during the embryonic development of all vertebrates. Arm and leg buds are also visible now. (**b**) Embryo at the end of five weeks, about twelve millimeters long. The head starts to enlarge and the trunk starts to straighten. Finger rays appear in the paddlelike forelimbs. The umbilical cord has developed.

Figure 35.15   The fetus at nine weeks, floating in fluid inside the amniotic sac. (Here, the chorion, which covers the amniotic sac, has been opened and pulled aside.) Notice the blood vessels in the umbilical cord.

to the main body axis are visible in Figure 34.14. The tissues of these segments develop into connective tissues, bones, and muscles. Arms, legs, fingers, and toes are developing—along with the tail that is characteristic of all vertebrate embryos. After the eighth week, the human tail gradually disappears.

From the start of the ninth week until birth, the developing individual is called a **fetus** (Figure 34.15). Between the ninth and the twelfth week of development, the fetus begins moving its arms and legs. Movements of facial muscles produce what looks like frowns and squints; the sucking reflex also is evident.

Second Trimester. At the start of the **second trimester**, the period extending from the fourth month to the end of the sixth month of development, all the major organs have formed. The fetus is now about 5.7 centimeters, or 3 inches, long. By now, the mother is quite aware of the fetal movements. Bone has already replaced much of the cartilage model of the embryonic skeleton.

When the fetus is five months old, its heart can be heard with the aid of a stethoscope. Soft, fuzzy hair (the lanugo) covers its body. The skin, which is quite wrinkled and rather red, has a thick, cheesy coating that protects it against abrasion. During the sixth month, the upper and lower eyelids separate and eyelashes form; during the seventh month, the eyes open.

Third Trimester. The **third trimester** extends from the seventh month until birth. Not until the middle of the third trimester is the fetus developed enough to survive on its own if born prematurely or if removed surgically from the uterus. By the seventh month, fetal development appears to be relatively complete, but fewer than ten percent of infants born at this stage survive, even with the best medical care. In most cases they are not yet able to breathe normally or maintain a normal core temperature, because the air sacs in the lungs and the parts of the nervous system that govern breathing are not completely developed. By the ninth month, survival chances increase to about ninety-five percent.

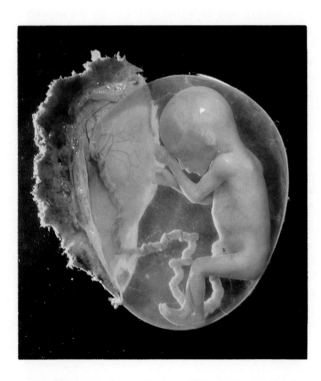

Figure 35.16 The fetus at sixteen weeks (sixteen centimeters, or 6.4 inches, long).

Figure 35.17 Fetus at eighteen weeks, about eighteen centimeters (a little more than seven inches) long. The sucking reflex begins during the earliest fetal stage, as soon as nerves establish functional connections with developing muscles. Legs kick, arms wave, fingers make grasping motions—all reflexes that will be vital skills in the world outside the uterus.

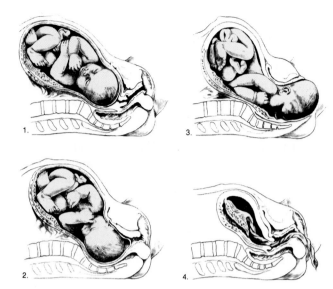

Figure 35.18 Movement of a full-term fetus from the uterus during childbirth. The last frame shows expulsion of the afterbirth (placenta).

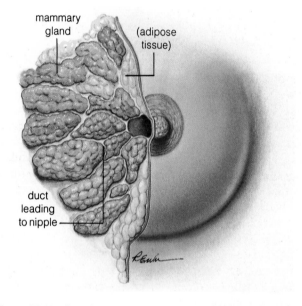

Figure 35.19 Female breast, showing the mammary glands and ducts.

Birth

On the average, the date of birth is approximately 280 days after fertilization. Almost three-fourths of all newborns have arrived within two weeks of that day. The birth process begins with the onset of contractions of the uterus. The contractions build in strength and increase in frequency over a period that usually lasts anywhere from two to eighteen hours. During that period, the cervical canal of the uterus becomes fully dilated and the amniotic sac usually ruptures. Usually within less than an hour, the fetus is expelled. Immediately afterward, uterine contractions cause the expulsion of fluid, blood, and the placenta (Figure 35.18). The umbilical cord—the lifeline to the mother—is now severed, and the newborn embarks on its nurtured existence in the outside world.

Lactation

During pregnancy, estrogen and progesterone levels are high in the bloodstream. Among other things, the hormones stimulate the growth of mammary glands and ducts in the woman's breasts. There are about twenty lobes of glandular tissue in each breast, and a tiny duct leads from each one to the surface of the nipple (Figure 35.19).

For the first few days after giving birth, the woman's mammary glands produce a fluid (colostrum) that is rich in proteins and lactose. Then prolactin secreted by the anterior pituitary stimulates milk production. When the newborn suckles, the pituitary is stimulated into releasing oxytocin as well as prolactin. The hormone oxytocin triggers the release of milk from the mammary glands (pages 345 and 347).

Case Study: Mother as Protector, Provider, Potential Threat

Many safeguards are built into the female reproductive system. The placenta, for example, is a highly selective filter that prevents many noxious substances in the mother's bloodstream from gaining access to the embryo or fetus. Even so, from fertilization to birth, the developing individual is at the mercy of the mother's diet, health habits, and life-style.

Some Nutritional Considerations. During pregnancy, a balanced diet usually provides enough vitamins and minerals for normal development. The mother's vitamin needs are definitely increased, but the developing fetus is more resistant than she is to vitamin and mineral deficiencies (the placenta preferentially absorbs vitamins and minerals from her blood).

A few years ago, it was accepted medical practice for a pregnant woman to keep her total weight gain to ten or fifteen pounds. It is now clear that if the woman restricts her food intake too severely, especially during the last trimester, fetal development will be affected and the newborn will be underweight. Significantly underweight infants face more post-delivery complications than do infants of normal weight; in fact, they represent nearly half of all newborn deaths. They also will suffer a much higher incidence of mental retardation and other handicaps later in life. In most cases, a woman should gain somewhere between twenty and twenty-five pounds during pregnancy.

As birth approaches, the growing fetus demands more and more nutrients from the mother's body. During this last phase of pregnancy, the mother's diet profoundly influences the course of development. Poor nutrition damages most fetal organs—particularly the brain, which undergoes its greatest growth in the weeks just before and after birth.

Risk of Infections. During pregnancy, antibodies transferred across the placenta protect the developing individual from all but the most severe bacterial infections. However, certain viral diseases can have damaging effects if they are contracted during the first six weeks after fertilization, the critical time of organ formation. For example, if the woman contracts German measles during this period, there is a fifty percent chance that her embryo will become malformed. If she contracts the measles virus when the embryo's ears are forming, her newborn may be deaf. (German measles can be avoided by vaccination *before* pregnancy.) The likelihood of damage to the embryo diminishes after the first six weeks. The same disease, contracted during the fourth month or thereafter, has no discernible effect on the development of the fetus.

Effects of Prescription Drugs. During the first trimester, the embryo is highly sensitive to drugs. A shocking example of drug effects came during the first two years after *thalidomide* was introduced in Europe. Women using this prescription tranquilizer during the first trimester gave birth to infants with missing or severely deformed arms and legs. Once the deformities were traced to thalidomide, the drug was withdrawn from the market. However, there is evidence that other tranquilizers (and sedatives and barbiturates) might cause similar, although less severe, damage. Even certain drugs for treating acne increase the risk of facial and cranial deformities. Tetracycline, a commonly prescribed antibiotic, causes yellowed teeth; streptomycin causes hearing problems and may affect the nervous system.

At no stage of development is the embryo impervious to drugs in the maternal bloodstream. Clearly, the woman should take no drugs at all during pregnancy unless prescribed by a knowledgeable physician.

Effects of Alcohol. As the fetus matures, its physiology becomes increasingly like that of the mother's. Alcohol passes freely across the placenta and has the same kind of effect on the fetus as on the woman who drinks it. *Fetal alcohol syndrome* (FAS) is a constellation of deformities that are thought to result from excessive use of alcohol by the mother during pregnancy. FAS is the third most common cause of mental retardation in the United States. It also is characterized by facial deformities, poor coordination and, sometimes, heart defects. Between sixty and seventy percent of alcoholic women give birth to infants with FAS; some researchers now suspect that even three drinks a day during pregnancy may be dangerous for the fetus. Increasingly, physicians are urging total or near-abstention during pregnancy.

Effects of Smoking. Cigarette smoking has an adverse effect on fetal growth and development. Newborns of women who have smoked every day throughout pregnancy have a low birth weight. That is true even when the woman's weight, nutritional status, and all other relevant variables are identical with those of pregnant women who do not smoke. Smoking has other effects as well.

For example, for seven years in Great Britain, records were kept for all births during a particular week. The newborns of women who had smoked were not only smaller, they had a thirty percent greater incidence of death shortly after delivery and a fifty percent greater incidence of heart abnormalities. More startling, at age seven, their average "reading age" was nearly half a year behind that of children born to nonsmokers.

In this last study, the critical period was shown to be the last half of pregnancy. Newborns of women who had stopped smoking by the middle of the second trimester were indistinguishable from those born to women who had never smoked. Although the mechanisms by which smoking exerts its effects on the fetus are not known, its demonstrated effects are further evidence that the placenta—marvelous structure that it is—cannot prevent all the assaults on the fetus that the human mind can dream up.

CONTROL OF HUMAN FERTILITY

Some Ethical Considerations

The remarkable transformation of a single-celled zygote into an intricately detailed adult raises profound questions. *When does development begin?* As we have seen, many key aspects of development emerge even before fertilization. *When does life begin?* During her lifetime, a human female can produce as many as four hundred secondary oocytes, all of which are alive. During one ejaculation, a human male can produce a quarter of a billion sperm, which also are alive. Even before sperm and oocyte merge by chance and establish the genetic constitution of a new individual, they are as much alive as any other form of life. It is scarcely tenable, then, to suggest that "life begins" when they fuse. *Life began billions of years ago; and each gamete, each zygote, and each mature individual is only a fleeting stage in the continuation of that beginning.* This fact cannot diminish the meaning of conception, for it is no small thing to entrust a new individual with the gift of life, wrapped in the unique evolutionary threads of our species and handed down through an immense sweep of time.

Yet how can we reconcile the marvel of individual birth with our growing awareness of the astounding birth rate for the human species as a whole? At the time this book is being written, an average of 2.2 infants are being born each second—132 each minute, 7,920 each hour. By the time you go to bed tonight, there will be 190,080 more people on earth than there were last night at that hour. Within a week, the number will reach 1,330,560—about as many people as there are now in the entire state of Massachusetts. *Within one week.* Our worldwide population growth has outstripped our resources, and each year millions face the horrors of starvation. Living as we do on one of the most productive continents on earth, few of us can know what it means to give birth to a child, to give it the gift of life, and have no food to keep it alive.

And how can we reconcile the marvel of birth with the confusion surrounding unwanted pregnancies? Even in highly developed countries there are not enough educational programs concerning fertility control, and often there is a disinclination to exercise control. Each year in the United States alone there are about 100,000 "shotgun" marriages, about 200,000 unwed teenage mothers, and perhaps 1,500,000 abortions. On the one hand, many parents encourage boy-girl relationships at early ages. On the other hand, they ignore the possibility of premarital intercourse and unplanned pregnancy. Advice is often condensed to a terse, "Don't do it. But if you do it, be careful!"

The motivation to engage in sex has been evolving for more than 500 million years. A few centuries of moral sanctions calling for the suppression of that motivation have not prevented unwanted pregnancies, and complex social factors have contributed to a population growth rate that is out of control. How will we reconcile our biological past and the need for a stabilized cultural present?

Whether human fertility is to be controlled—and how it is to be controlled—is one of the most volatile issues of our time. We will return to this issue in Chapter Forty-Three, in the context of principles governing the growth and stability of populations. Here, we can briefly consider some possible control options.

Possible Means of Birth Control

At present, approaches to birth control fall into three categories:

---

1. *Fertility control* (physical, chemical, surgical, or behavioral interventions that disrupt reproductive function or affect gamete survival).

2. *Implantation control* (physical or chemical interference with the blastocyst's ability to invade the uterine wall).

3. *Abortion* (prevention of embryonic development after implantation has occurred).

Let's consider some behavioral interventions first. The most effective method of preventing conception is complete **abstention**: no sexual intercourse whatsoever. It is unrealistic to expect many people to practice this method. A modified form of abstention is the **rhythm method**, in which intercourse is avoided during the woman's fertile period. The fertile period begins a few days before ovulation and ends a few days after. It is identified and tracked either by keeping records of the length of the woman's menstrual cycle or by taking her temperature each morning when she wakes up. (Just before the fertile period, there is a one-half to one-degree rise in core temperature.) However, ovulation can be irregular, and miscalculations are frequent. Moreover, sperm deposited in the vaginal tract a few days prior to ovulation may survive until ovulation. The method *is* inexpensive (it costs nothing after you buy the thermometer) and it does not require fittings and periodic checkups by a doctor. But its practitioners do run a substantial risk of becoming pregnant (Table 35.5).

Withdrawal is the removal of the penis from the vagina prior to ejaculation. This contraceptive method dates back at least to biblical times. But withdrawal requires extraordinary willpower. In any case, the method may fail: fluid released from the penis just before ejaculation may contain viable sperm.

The practice of **douching**, or rinsing out the vagina with a chemical right after intercourse, is almost useless. Sperm can move past the cervix and out of reach of the douche within ninety seconds after ejaculation.

Other methods involve physical or chemical barriers that prevent sperm from entering the uterus and moving to the ovarian ducts. **Spermicidal foam** or **spermicidal jelly** is packaged in an applicator, which is inserted and emptied into the vagina just before intercourse. These products are toxic to sperm, yet they are not always reliable unless used with another device, such as a diaphragm or condom. A **diaphragm** is a flexible, dome-shaped device that is inserted into the vagina and positioned over the cervix before intercourse. A diaphragm is relatively effective when it is used with foam or jelly, when it has been fitted by a doctor, when it is inserted correctly with each use, and when foam or jelly is reapplied with each sexual contact.

Condoms are thin, tight-fitting sheaths of rubber or animal skin that are worn over the penis during intercourse. They are about eighty-five to ninety-three percent reliable, and they help prevent the spread of sexually transmitted diseases (see *Commentary*). However, condoms can tear and leak, in which case they are rendered useless.

You may have heard about the intrauterine device, or **IUD**. This small plastic or metal device, which must be inserted by a trained specialist, can be left in the uterus for years to prevent implantation. Although the IUD is an effective means of fertility control, controversy surrounds its use. Women who use the device may run a greater risk of becoming infertile and should be aware of the risk if they want to have children later. Infections are sometimes associated with its use; some IUDs were taken off the market because their design actually promoted infection. Also, an IUD may perforate the wall of the uterus, although this is uncommon.

Another method of fertility control is hormonal intervention in the reproductive cycle. Most widely used is **the Pill**, an oral contraceptive of synthetic estrogens and progesterones taken daily by the female. By suppressing the normal release of gonadotropins from the pituitary, these synthetic hormones prevent the cyclic maturation and release of ova. The Pill is a prescription drug. Formulations vary and are selected to match the individual patient's needs. That is why it is not wise for a woman to borrow the Pill from someone else.

If the woman does not forget to take her daily dosage, the Pill is one of the most reliable methods of controlling fertility. It does not interrupt sexual intercourse, and the method is easy to follow. Often the Pill corrects erratic menstrual cycles and decreases menstrual cramping. However, the Pill must be taken for one month (a full cycle) before it can be considered fully effective. In other

Table 35.5 Effectiveness of Some Contraceptive Methods

Method Used	Number of Pregnancies Per 100 Women Per Year
None	115*
Douching	31
Rhythm method	13–24
Spermicidal jelly or foam alone	20
Withdrawal	18
Condom alone	14
Diaphragm with foam or jelly	3–17
IUD	1–1.5**
The Pill	1
The Pill used correctly	0
Vasectomy	0
Tubal ligation	0

*This figure includes women who become pregnant more than once a year, following miscarriage or childbirth.
**Pregnancy rate declines after the first year.
Data from D. Luciano et al., 1978, *Human Function and Structure* and other sources.

Sexually Transmitted Diseases

(This Commentary was written on the basis of information supplied by the Centers for Disease Control, Atlanta, Georgia)

Sexually transmitted diseases (STDs) have reached epidemic proportions, even in countries with the highest medical standards. The disease agents, mostly bacterial and viral, are usually transmitted to uninfected persons during sexual intercourse. In the United States alone, 10 million young adults have reported that they have some form of STD; no one can estimate the number of unreported cases. The economics of this health problem are staggering. By conservative estimates, the cost of treatment is exceeding $2 billion a year. The social consequences are sobering. For example, of every twenty babies born in the United States, one will have a chlamydial infection. Of every 10,000 newborns, as many as three will contact systemic herpes; half may die early on, and a fourth of those surviving will suffer serious neurological defects. Each year, 1 million girls and adult women are stricken with pelvic inflammatory disease, a complication of chlamydial and gonococcal infections. Of these, more than 200,000 are hospitalized, more than 100,000 undergo pelvic surgery that results in permanent sterility, while 900 cannot recover and die. These and other sexually transmitted diseases are expected to increase at alarming proportions during this decade.

Gonorrhea. Gonorrhea ranks first among the reported communicable diseases in the United States. In 1979 alone, an estimated 1.6 to 2 million people became afflicted with the disease. Gonorrhea is caused by *Neisseria gonorrhoeae*, a bacterium that infects the epithelial cells of the genital tract (and the conjunctiva or pharynx if carried there). Males have a greater chance than females do of detecting the disease in early stages. Within a week, yellow pus is discharged from the urethra. Urination becomes more frequent and painful, because the urinary tract becomes inflamed. Females may or may not experi-

ence a burning sensation while urinating. There may or may not be a slight vaginal discharge—and even if there is, it often is not considered abnormal. As a result, the disease often goes untreated. The bacteria may spread into the oviducts, eventually leading to violent cramps, fever, vomiting and, in many cases, sterility.

Complications arising from gonorrheal infection can be avoided with prompt diagnosis and treatment. As a preventive measure, males who engage in sexual activity with more than one partner are being advised to wear condoms to help prevent the spread of infection. Part of the problem is that the initial stages of the disease are so uneventful that the dangers are masked. Also, it is a common belief that once cured of gonorrhea, a person is immune for life, which simply is not true.

Syphilis. Syphilis is caused by a motile spirochaete, *Treponema pallidum*. In its frequency, this reported communicable disease is exceeded only by cases of chickenpox and gonorrhea. Following sexual contact with an infected person, the bacterium penetrates exposed tissues and produces a chancre (localized ulceration), which teems with treponeme progeny. This primary lesion can occur anywhere between one to eight weeks following infection. By the time the chancre is visible, the treponemes have moved into the lymph vascular system and bloodstream. When the second stage of infection occurs, lesions can appear on mucous membranes, the eyes, and bones, and in the central nervous system. After these early lesion stages, an infected person enters a latent stage. During latency, there are no outward symptoms and syphilis can be detected only by means of blood serum tests.

The latent period can last many years. During that time, the immune system works against the bacterium; sometimes the body cures itself, but this is not the usual outcome.

If untreated, syphilis in its tertiary stage can produce lesions on the skin and internal organs, such as the liver, bones, and aorta. Scars form on organs; the walls of the aorta can be weakened. The treponemes also damage the brain and spinal cord in ways that lead to various forms of insanity and paralysis. Women who have been infected typically have miscarriages or stillbirths; if they do give birth, their infants are often sickly and syphilitic.

Chlamydial Infection. Chlamydial infection is one of the most prevalent sexually transmitted diseases. *Chlamydia trachomatis* is the causative agent. This bacterium is an obligate, intracellular parasite. Most often, it infects the genitals and urinary tract. Following infection, the parasites migrate to regional lymph nodes, which become enlarged and tender. The enlargement can obstruct lymph drainage and lead to pronounced tissue swelling in the area. The disease responds to tetracycline and sulfonamides. If untreated, it can lead to serious complications, such as pelvic inflammatory disease.

Genital Herpes. Genital herpes is an extremely contagious viral infection of the genitals. Usually, it occurs through intimate sexual contact with a partner who has active herpes lesions. It is transmitted when any part of a person's body comes in direct contact with active *Herpesviruses* or sores that contain them. Mucous membranes (particularly of the mouth or genital area) are susceptible to invasion, as is broken or damaged skin. The virus probably is not acquired by nonsexual means; it does not survive for long away from the human body.

To the 5 million existing cases of genital herpes already reported by Americans, 300,000 are being added each year. Newborns are among these cases; they can develop the disease when they pass through an infected birth canal or are infected soon after birth.

There are many strains of *Herpesviruses*, which are classed as types I and II. The type I strains infect primarily the lips, tongue, mouth, and eyes. The type II strains cause most of the genital infections. Disease symptoms occur two to ten days after exposure to the virus, although sometimes symptoms are mild or nonexistent. Among infected women, small, painful blisters appear on the vulva, cervix, urethra, or anal area. Among men, the blisters occur on the penis and around the anal area. Within three weeks, the sores crust over and heal without scarring. Once the first infection has occurred, sporadic reactivation can produce new, painful sores at or near the original site of infection. Recurrent infections may be triggered by sexual intercourse, emotional stress, menstruation, and other infections. At present there is no cure for genital herpes. However, a new antiviral drug (acyclovir) decreases the healing time and sometimes decreases the pain and the viral shedding.

AIDS. Acquired immune deficiency syndrome (AIDS) is a constellation of disorders that follow infection by a virus now designated HIV. It appears that the virus is largely transmitted through intercourse with infected persons. AIDS is spreading in the general population. It appears in three forms: asymptomatic AIDS, AIDS-related complex (known as ARC), and full-blown AIDS. The final form is almost always lethal, and at present there is no cure. This sexually transmitted disorder is described on page 412.

words, a woman cannot take one pill the day before intercourse and expect it to work. Also the Pill does have known side effects for a small number of users. In the first month or so of use, it may cause nausea, weight gain, tissue swelling, and minor headaches. Its continued use may lead to blood clotting in the veins of a few women (3 out of 10,000) predisposed to this disorder. There have been some cases of elevated blood pressure among Pill users, and also some abnormalities in fat metabolism that might be linked to a growing number of gallbladder disorders. Longer term effects have been difficult to assess.

Hormonal control of male fertility is a more complicated matter. The hormonal methods for suppressing female fertility are based on the cyclic nature of the woman's reproductive capacity. But sperm production in males is not cyclic, and it is under a more diffuse kind of hormonal control. Although various medications have been developed, they are still experimental. It will probably be several years before they become available, and their effectiveness is not yet assured.

Surgical intervention in the reproductive tract is becoming common. In **vasectomy**, a tiny incision is made in the scrotum and each vas deferens is severed and tied off. The simple operation can be performed in twenty minutes in a doctor's office, with only a local anesthetic. After vasectomy, sperm cannot leave the epididymides, and therefore will not be present in seminal fluid. (However, sperm present in duct regions below the surgical cuts can be present in the ejaculate for several weeks after the operation.) So far there is no firm evidence that vasectomy disrupts the male hormone system, and surveys suggest there is no noticeable difference in sexual activity. Although vasectomies can be surgically reversed, half of the men who have undergone surgery will develop antibodies against sperm and they may not be able to regain fertility.

For females, surgical intervention includes **tubal ligation**, in which the oviducts are cauterized or cut and tied off. Tubal ligation is more complex than vasectomy and is usually performed in a hospital. A small number of women who have had the operation suffer recurring bouts of pain and inflammation of tissues in the pelvic region where the surgery was performed. The operation can be reversed, although major surgery is required and success is not always assured.

Once conception and implantation have occurred, the only way to terminate a pregnancy is **abortion**, in which the implanted embryo is dislodged and removed from the uterus. (In *miscarriages*, or spontaneous abortion, the embryo is dislodged and expelled spontaneously.)

Until recently, abortions were generally forbidden by law in the United States unless the pregnancy endangered the mother's life. Supreme Court rulings have held that the government does not have the power to forbid abortions during the early stages of pregnancy (typically up to five months). Abortion is now legal in this country. Moving the large number of illegal, unsupervised operations to modern medical facilities has reduced the frequency of dangerous, traumatic, and often fatal attempts to abort embryos, either by pregnant women themselves or by quacks. Newer methods have made abortion relatively rapid, painless, and free of complications when performed during the first trimester. Abortions in the second and third trimesters will probably remain extremely controversial unless the mother's life is clearly threatened. For both medical and humanitarian reasons, however, it is generally agreed in this country that a preferable route to birth control is not through abortion but through control of conception in the first place.

In Vitro Fertilization

In the United States, about fifteen percent of all couples are unable to conceive a child, owing to sterility or infertility. For example, hormonal imbalances may prevent ovulation in females; or the male's sperm count may be too low to assure successful fertilization. With **in vitro fertilization**, external conception is possible, provided that the sperm and secondary oocyte obtained from the couple are normal.

A hormone administered to the female prepares her ovaries for ovulation within thirty-three to thirty-four hours. Then a laparoscope allows the physician to locate and remove the preovulatory oocyte with a suction device. (A laparoscope is a long, metal tube containing a light and an optical system; it can be used to identify the ballooning follicle at the ovarian surface.)

Before the oocyte is removed, a sample of the male's sperm is put in a saline solution that simulates the fluid in oviducts. When the suctioned oocyte is placed in the solution with the sperm, fertilization may occur a few hours later. About twelve hours later, the newly dividing embryo is transferred to a solution that will support further development, and about two to four days after that, it is transferred to the female's uterus. If all goes well, implantation may occur.

Since the first successful in vitro fertilization in 1978, clinics specializing in the procedure have opened in several countries. There appears to be no greater risk associated with the procedure than with normally occurring pregnancies. Implantation occurs in only about twenty percent of the attempts, and each attempt costs several thousand dollars. Even so, there are many childless couples who are attempting to have children in this manner.

SUMMARY

1. The human reproductive system consists of a pair of primary reproductive organs (gonads) and accessory ducts and glands. The male gonads are the testes, the female gonads are the ovaries. The gonads produce gametes (sperm in the male, secondary oocytes or "eggs" in the female) and hormones, which influence reproductive functions and the secondary sexual traits.

2. Sperm form in seminiferous tubules within the testes from undifferentiated cells (spermatogonia). Interstitial cells between the tubules secrete sex hormones.

3. Sperm move from the testis to the epididymis, where their capacity for motility is partly developed. (Full motility occurs only in the female reproductive tract.) Sperm pass through the duct system and finally into the urethra. Together with the sperm, secretions from glands along the reproductive tract form the semen.

4. Testosterone, with FSH and LH, stimulates spermatogenesis. It also stimulates the development of the male reproductive tract, promotes male sexual behavior, and supports development of secondary sexual traits.

5. In the female, the ovaries release oocytes and they secrete the sex hormones estrogen and progesterone. An oviduct next to each ovary transports the oocyte to the uterus. The uterus opens by way of the cervix into the vagina, which opens to the exterior.

6. The oocytes develop before birth into primary oocytes which, together with their surrounding cell layer, are called primary follicles. They do not start to mature until puberty, when one at a time on a monthly basis, they develop into secondary oocytes.

7. In the human female, the reproductive periods are menstrual cycles, during which the interplay of FSH, LH, estrogen, and progesterone causes (1) development of the oocyte and (2) preparation of the lining of the uterus to receive the embryo. If fertilization does not occur, the uterine lining is shed (the menstrual flow), the corpus luteum degenerates, and a new cycle begins.

8. At ovulation (caused by a sudden secretory burst of the hormone LH) a secondary oocyte is released from the ovary and begins its final division, which is not completed until fertilization; then, the secondary oocyte becomes a mature ovum. The remaining cells of the follicle develop into a corpus luteum (a glandular structure that secretes progesterone and some estrogen, both of which maintain the endometrium).

9. During sexual intercourse, sperm are ejaculated into the vagina, where they then undergo further maturation. Fertilization normally occurs in the oviduct.

10. Embryonic development depends on the formation of extraembryonic membranes that function in the following ways:

   a. *Yolk sac:* parts give rise to the embryo's digestive tube.

   b. *Allantois:* its blood vessels function in oxygen transport.

   c. *Amnion:* forms a fluid-filled sac that protects the embryo from mechanical shocks and keeps it from drying out.

   d. *Chorion:* forms a protective membrane around the embryo and the other membranes; a primary component of the placenta.

11. The embryo and the mother exchange substances by way of the placenta (a spongy tissue of endometrium and extraembryonic membranes).

12. Human development proceeds with the formation of the three primordial tissues (endoderm, ectoderm, and mesoderm). By the beginning of the second trimester of pregnancy, all major organs have formed.

13. At delivery, contractions of the uterus dilate the cervical canal and expel the infant and afterbirth.

14. Estrogen and progesterone stimulate growth of the mammary glands. After delivery, nursing causes the release of hormones that stimulate milk production and release.

15. During intrauterine life, the placental barrier provides some protection for the fetus, but it may suffer harmful effects from the mother's nutritional deficiencies, infections, intake of prescription drugs, alcohol, and smoking.

16. Control of human fertility raises many important ethical questions. These questions extend to the physical, chemical, surgical, or behavioral interventions used in the control of unwanted pregnancies.

Review Questions

1. Study Table 35.1. Then list the main organs of the human male reproductive tract and identify their functions.

2. What are the accessory glands of the male reproductive tract? What are their functions?

3. Describe spermatogenesis and the route by which sperm leave the seminiferous tubules, then the body.

4. Which hormones have profound influence over male reproductive functioning? Can you diagram how feedback mechanisms link the hypothalamus, anterior pituitary, and interstitial cells in controlling this functioning?

5. Trace the events leading to the formation of the secondary oocyte.

6. Label the component parts of the female reproductive tract:

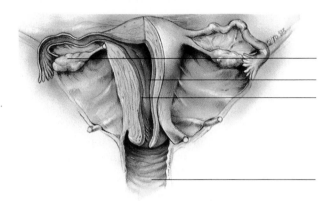

7. What is the menstrual cycle? Which four hormones have profound influence on this cycle? Can you diagram the feedback loops between the hypothalamus, anterior pituitary, and ovary that govern these hormonal secretions? (Review Figure 35.9.)

8. List the four events that are triggered by the surge of LH at the midpoint of the menstrual cycle.

9. What changes occur in the endometrium during the menstrual cycle?

10. Describe these extraembryonic membranes:
    a. amnion
    b. yolk sac
    c. allantois
    d. chorion

11. From what tissues and membranes does the placenta form?

12. With a sheet of paper, cover the summary definitions in Table 35.4. Can you state the definition for each of the stages of human development listed?

Readings

Gilbert, S. 1985. *Developmental Biology*. Sunderland, Massachusetts: Sinauer. Acclaimed reference text.

Greep, R. (editor). 1980. *Reproductive Physiology III* (volume 22 of the *International Review of Physiology*). Baltimore: University Park Press. Distills recent concepts and discoveries in reproductive physiology. The article "Ovarian Follicular and Luteal Physiology" by C. Channing et al. is especially informative.

Nilsson, L. et al. 1977. *A Child Is Born*. New York: Delacorte Press/ Seymour Lawrence. Extraordinary photographs of embryonic development.

Saunders, J. W. 1982. *Developmental Biology: Patterns, Problems, Principles*. New York: Macmillan.

Schatten, G. 1983. "Motility During Fertilization." *Endeavor* 7(4): 173–182.

UNIT SIX

EVOLUTION

36

POPULATION GENETICS, NATURAL SELECTION, AND SPECIATION

a

Where would the mallard duck be now if its destiny had been placed in the hands of that eighteenth-century cataloger of life, Carl von Linné? There, awaiting classification, was a bird with emerald-green head feathers and with wings displaying metallic blue patches. There, in the same ponds and marshes, was a drab little brown-feathered duck bearing no obvious resemblance to the more resplendent waterfowl (Figure 36.1a). Thus did von Linné, on the basis of outward appearance alone, pronounce the male and female mallard duck as separate species. (It goes without saying that the male and female duck, paying no attention whatsoever to his pronouncement, continued to produce more ducks.)

You may be thinking it is unfair to von Linné to dredge up one of his mistakes. Yet anyone might have made the same mistake if he or she had attempted to classify the ducks according to a rigid notion of species, as von Linné had done. Recall, from Chapter Two, that studies of organismic diversity were once funneled through the concept that species do not change. The approach was *typological*: an individual was selected as being the perfect standard, or type, for the species based on a formulation of what the "perfect" physical features were. Then newly encountered individuals were compared against the standard to judge whether they belonged to that species. Small variations among similar individuals were viewed as imperfect renditions of a particular species plan. (Thus, for example, there would be no mistaking any of the individuals shown in Figure

b

36.1b as being anything other than a snow goose.) Dramatic variations in physical appearance were often viewed as evidence in itself of different species—hence von Linné's mistake.

Of course, it took only a little field observation during the breeding season to discover that male and female mallard ducks are simply an extreme case of sexual dimorphism (phenotypic variation between sexes). But adherents of the typological approach fell into other, more serious conceptual traps. *What could be done about the vexing number of obviously similar organisms that nevertheless showed extreme variation in certain traits?* For example, certain snails living in the Caribbean show hundreds of variations in the color and banding pattern of their shells (Figure 36.1c). With the typological approach, each distinctly shelled snail would have to be assigned a separate "species" status—as, indeed, they were—making that taxonomic category rather useless.

The phenotypic variation illustrated in Figure 36.1 only hints at the immense variation among individuals of all the different species, past and present, on earth. How can this variation be explained? How does it arise, and what is its meaning? In this chapter we will cover some possible answers to these questions, using the following premise as our starting point:

Variation is a fundamental attribute of the individuals of a species and the raw material for evolution.

First we will review the sources of variation among the individuals of a species. Then we will consider how variation is maintained and how it can be modified over time.

POPULATION GENETICS

Sources of Variation

As we have seen, variation arises as a result of events at the molecular and cytological levels (Table 16.1). These events in some way alter an individual's *genotype* (genetic makeup), and the alteration may affect the *phenotype* (the individual's structural, physiological, or behavioral traits).

For example, a single amino acid substitution at one gene locus on a human chromosome produces a recessive allele that codes for a variant form of the hemoglobin molecule, designated HbS. (*Alleles*, recall, are alternative forms of a gene at a given locus on a chromosome.) Individuals who are homozygous recessive at this locus will suffer the phenotypic consequences of sickle-cell anemia. As another example, two copies of chromosome 21 may end up in a human gamete as a result of nondisjunction at meiosis (page 188). Fusion of this gamete with a normal one can lead to a variant individual exhibiting Down's syndrome. Independent assortment of homologous chromosomes at meiosis is another source

c

Figure 36.1 Examples of phenotypic variation (and the seeming lack of it) between individuals of the same species. (**a**) Male and female mallard duck, an example of sexual dimorphism (phenotypic variation between sexes). (**b**) Population of snow geese at a wintering ground in New Mexico. Why do all these birds superficially look the same, and why do certain snails—the shells of which are shown in (**c**)—look so different? In more general terms, *what is the meaning of individual variation, and how does it arise?* These are questions addressed in this chapter.

Figure 36.2 Hardy-Weinberg equilibrium. To prove the validity of the Hardy-Weinberg rule stated above, let's follow the course of two alleles, A and a, through succeeding generations.

For all members of the population, the gene locus must be occupied by either A or a. In mathematical terms, the frequencies of A and a must add up to 1. For example, if A occupies half of all the gene loci and a occupies the other half, then $0.5 + 0.5 = 1$. If A occupies ninety percent of all the gene loci, then a must occupy the remaining ten percent ($0.9 - 0.1 = 1$). No matter what the proportions of alleles A and a,

$$p + q = 1$$

You know that during sexual reproduction of diploid organisms, the two alleles at a gene locus segregate and end up in separate gametes. Thus p is also the proportion of gametes carrying the A allele, and q the proportion carrying the a allele. To find the expected frequencies of the three possible genotypes (AA, Aa, and aa) in the next generation, we can construct a Punnett square:

	p Ⓐ	q ⓐ
p Ⓐ	AA (p^2)	Aa (pq)
q ⓐ	Aa (pq)	aa (q^2)

Because the frequency of genotypes must add up to 1,

$$p^2 + 2pq + q^2 = 1$$

To see how these calculations can be applied, let's follow the allele frequencies for a population of 1,000 diploid individuals made up of the following genotypes:

$$450 \ AA$$
$$500 \ Aa$$
$$\underline{\ 50 \ aa}$$
1,000 individuals (or 2,000 alleles)

Theoretically, of every 1,000 gametes produced, the frequency of A will be $450 + \frac{1}{2}(500) = 700$, or $p = 0.7$. The frequency of a will be $\frac{1}{2}(500) + 50 = 300$, or $q = 0.3$. Notice that

$$p + q = 0.7 + 0.3 = 1$$

After one round of random mating, the frequencies of the three genotypes possible in the next generation will be as follows:

$$AA = p^2 \quad = 0.7 \times 0.7 = 0.49$$
$$Aa = 2pq = 2 \times 0.7 \times 0.3 = 0.42$$
$$aa = q^2 \quad = 0.3 \times 0.3 = 0.09$$

and

$$p^2 + 2pq + q^2 = 0.49 + 0.42 + 0.09 = 1$$

Notice that the allele frequencies have not changed:

$$A = \frac{2 \times 490 + 420}{2,000 \text{ alleles}} = \frac{1,400}{2,000} = 0.7 = p$$

$$a = \frac{2 \times 90 + 420}{2,000 \text{ alleles}} = \frac{600}{2,000} = 0.3 = q$$

The genotypic frequencies have changed initially. However, given that the distribution of genotypes fits the equation $p^2 + 2pq + q^2$, the genotypic frequencies will be stable over succeeding generations. You can verify this by calculating the most probable allele frequencies for gametes produced by the second-generation individuals:

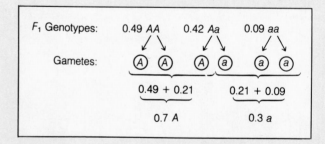

which is back where we started from. Because the allele frequencies are exactly the same as those of the original gametes, they will yield the same frequencies of genotypes as in the second generation.

You could go on with the calculations until you ran out of paper, or patience. As long as the population adheres to the conditions stated in the boxed inset for the Hardy-Weinberg principle, you would end up with the same results. When the frequencies of different alleles and different genotypes remain constant through successive generations, the population is in Hardy-Weinberg equilibrium: it is not evolving.

of phenotypic variation, and so is crossing over. These and other sources of phenotypic variation fall into five broad categories:

1. Gene mutation
2. Chromosomal aberrations
3. Independent assortment of chromosomes at meiosis
4. Crossing over at meiosis
5. Fertilization between genetically varied gametes

Only mutations *create* new alleles. Although individual mutations at any one gene locus are rare events (page 214), the total number of mutations that have occurred in the history of a species has provided the potential for incredible variation. Moreover, the other events listed above shuffle *existing* alleles into new combinations in new individuals. The number of different genotypes made possible by all these events is simply extraordinary.

According to one estimate, more than 10^{600} genetically different human gametes are possible. By comparison, there are fewer than 10^{10} humans alive today, and the estimated number of atoms in the universe is on the order of 10^{70}! Unless you have an identical twin, it is extremely unlikely that another person with your exact genotype has ever lived, or ever will. In short, *far more genetic variation is possible than can ever be expressed in the individuals alive at any one time.*

The Hardy-Weinberg Baseline for Measuring Change

Although it may seem obvious, it is important to keep in mind that individuals don't evolve; *populations* do. Here our focus will be on populations of sexually reproducing, diploid organisms, as defined in the following way:

A population is a group of individuals of the same species for which there are no restrictions to random mating among its members.

The sum total of all the genes of a given population has traditionally been called a gene pool. For sexually reproducing, diploid individuals, it is more instructive to think of it as a pool of alleles. (Such individuals have two sets of chromosomes, hence two alleles for each gene locus.) When you take the whole population into account, you find that for any one locus, some alleles occur more often than others. Thus it is possible to think of variation in terms of **allele frequencies**: the relative abundance of different alleles carried by the individuals in that population.

Early in this century, the mathematician G. Hardy and the physician W. Weinberg independently discovered a principle that can be used as a baseline against which changes in allele frequencies can be measured. They observed the following:

In the absence of disturbing factors, the frequencies of different genotypes in a population will reach an equilibrium and will remain stable from generation to generation.

This observation is based on a mathematical formula called the **Hardy-Weinberg principle**, which is given in Figure 36.2. The principle applies to an idealized population of sexually reproducing organisms that is not evolving. The deduction is that evolution cannot occur under the following conditions:

1. No mutation
2. Infinitely large population
3. Isolation from any other populations of the same species
4. Equal viability, fertility, and mating ability of all genotypes (no selection)

Figure 36.2 gives an example of the genetic results when all of these conditions are met. After one generation of mating, the relative frequencies of the three possible genotypes (*AA, Aa,* and *aa*) reach an equilibrium described by the formula $p^2 + 2pq + q^2$. Those frequencies, and the frequencies p and q of the two alleles (*A* and *a*), stay the same over succeeding generations. Such stability of allelic and genotypic ratios is called **genetic equilibrium**.

As it happens, a natural population never meets all of these conditions, hence never is able to reach genetic equilibrium. Why, then, do we mention the Hardy-Weinberg rule at all? The reason is that genetic equilibrium is useful as a reference point, signifying zero evolution. *Hence the degree to which deviations are measured from this reference point can serve as a measure of the rates of evolutionary change.*

Factors Bringing About Change

When a population is evolving, its allele frequencies are changing through successive generations as a result of one or more of the following factors:

1. **Mutation**. A heritable change in the kind, structure, sequence, or number of the component parts of DNA.

2. **Genetic drift**. A random fluctuation in allele frequencies over time, due to chance occurrences alone. Genetic drift has its greatest effect when population size is small.

3. **Gene flow**. A change in allele frequencies due to immigration (new individuals enter the population), emigration (some individuals leave), or both.

4. **Natural selection**. Differential survival and reproduction of genotypes within a population.

Each of these factors can be important in some circumstances. Overall, genetic drift and natural selection seem to be the most important mechanisms bringing about evolutionary change, with gene flow not far behind in many groups of organisms. Mutations are important in changing the character of a population through long spans of time, but they generally have little effect on a generation-to-generation basis.

As you probably noticed, genetic recombination is not listed as a factor that causes evolution. Recombination does contribute to phenotypic variation, and other forces (such as selective agents) can act on that variation. However, recombination in itself only reshuffles the combinations of existing alleles; it does not change the *frequencies* of those alleles.

Let's now take a look at the effects of mutation, genetic drift, and gene flow before exploring in detail the effects of natural selection.

MUTATION

Mutation is the original source of genetic variation, and mutations have been accumulating for billions of years. In most populations, mutations are random in terms of when they will happen, which locus will be affected and, most importantly, whether they will be harmful or beneficial to the population at the time.

The effects of a mutation depend on how it changes the structure, function, or behavior of the individual in the context of prevailing conditions. For example, a certain gene codes for a protein necessary in cartilage formation. Cartilage is a key component of many pathways in normal animal development. A mutation at this gene locus may lead to such deformities as blocked nostrils, narrowed tracheal passageways, thickened ribs, and the loss of elasticity in lung tissue. Here the new allele produced by mutation is expressed in the context of an intricate developmental program, and it may have lethally disruptive effects.

As another example, suppose an ectothermic animal (which has no physiological controls to maintain body temperature) bears a mutant gene coding for an enzyme that functions at a higher temperature, compared to its normal counterpart. If the animal moves into an environment where temperatures are higher than those typically encountered by its species, enzyme function would not be diminished. Here the mutation might be advantageous in the new environment but not in the old one.

The new alleles that appear by mutation in a given population are neither harmful nor beneficial in themselves; their effects depend on the environment in which their gene products are expressed.

Even though mutations are neither good nor bad in themselves, most mutations with large effects are harmful, even lethal. The reason is that a new allele represents a departure from alleles that have withstood the test of time. Overall, the structural, functional, and behavioral traits of an organism already allow it to function well in a particular environmental context. Any drastic change is more likely to derange things than to enhance them.

Of course, just because a new allele is not beneficial does not mean that it cannot be passed on from one generation to the next. First, its effects might be masked by its allelic partner. Second, the mutation may make no difference whatsoever in a given environmental context, in which case there would be no selection for *or* against the mutation. Third, the new allele might be closely linked in the chromosome to highly adaptive genes. Thus it could hitchhike with the adaptive genes through meiosis, linkage, sexual fusion, and environmental tests.

POPULATION SIZE

Earlier, we defined a population as a group of individuals of the same species, for which there are no restrictions on random mating among its members. Thus, in theory, any male in a population could select any female as a mate (or vice versa), while encounters between males and females of different populations are less likely. (For example, some populations of the same species are separated by areas of unsuitable habitat and they never do make contact.)

The size of a population (its number of individuals) and the area it occupies vary from one species to the next, and both factors influence the mechanisms and rates of evolutionary change. As an example of a truly spread-out population, mallard ducks move around so much from the time of birth to the breeding season, and even between breeding seasons, that all of the individuals in North America could be considered a single population! Toward the other extreme is the fringe-toed sand lizard, the small populations of which are confined to isolated sand dunes in California deserts.

Genetic Drift

Population size influences the rate of genetic drift (the chance increase or decrease in the relative abundance of different alleles). For example, suppose that over several generations in a small population, none of the bearers of an allele designated C reproduces simply by chance. Whether you call it chance or bad luck, they fail to mate, or fall ill, or accidentally die early. Assuming that genetic drift is the only evolutionary factor at work at the time, the C allele would disappear from the population (see Figure 36.3). Such a run of bad luck is much less likely in a large population, where hundreds of individuals might be carrying the C allele; thus there would be a greater chance that at least some of them would reproduce and thereby help maintain the C allele in the population.

By analogy, there are two outcomes when you toss a coin, each with equal probability of occurring. When the coin is tossed ten times, it could by chance end up tails eight times, or two times, or one, rather than the expected 1:1 ratio (half heads, half tails). However, the laws of probability tell you that when the coin is tossed, say, a thousand times, the expected ratio will be more closely approached.

In general, then, genetic drift is the random change in allele frequencies through successive generations of a population, regardless of its size. It can be more rapid in a small population, but even small shifts in each generation of a large population can add up to significant change over time.

One more point can be made here. Some alleles in a population may not be the "best" possible alternatives in contributing to survival and reproduction. Through genetic drift, they may have become established just because other alternatives were lost by chance. Suppose that in a small population of green parakeets and blue parakeets, the allele governing the dominant feather color (green) is lost as a result of genetic drift, and all parakeets of the next generation are blue. This change in feather color has arisen not because the blue color is more advantageous, but because chance events led to a change in genotypes (hence phenotypes).

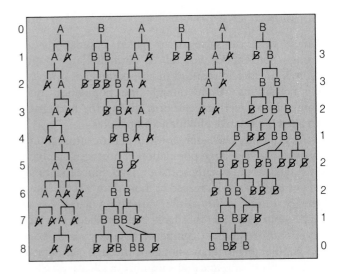

Figure 36.3 Illustration of genetic drift. Each individual of types "A" and "B" produces two identical offspring in each generation. Half the offspring die before reproductive age (population size remains constant) but which ones die is random. The relative abundance of the two types of individuals fluctuates until A no longer is represented in the population and B becomes fixed in the population.

Numbers to the left of the diagram signify the sequential generations; numbers to the right signify the number of type A individuals surviving in each generation. (Which individuals were to die in this example was determined by tossing a coin.)

Founder Effect

Sometimes a small number of dispersed individuals manage to establish a new population. Simply by chance, the allele frequencies at many gene loci are likely to be different in these individuals from what they were in the original population, and the new assortment will dictate the genotypic character of the new population. This extreme case of genetic drift is called the **founder effect**.

The founder effect is important in the colonization of oceanic islands and other isolated locations (such as land-locked, glacier-fed lakes that are "seeded" with a few trout by fishermen). It is also important when a few individuals of a species are accidentally or intentionally introduced into new environments. The American starling is a descendant of a small flock from Europe that was released in New York City in the 1800s. As a result of the limited allelic variation in that small flock of birds, the American and European populations are genetically different—even though there now may be more starlings in America than in all of Europe.

Bottlenecks

Genetic drift can also occur when populations go through **bottlenecks**, whereby a normally large population is drastically reduced in size because of unfavorable conditions. Even though the population may eventually recover, genetic drift during the bottleneck can alter the relative abundances of alleles.

For example, just before the turn of the century, hunters destroyed the large population of northern elephant seals. Only about twenty seals survived. Since that time, the population has increased to more than 30,000. Interestingly, there is no allelic variation whatsoever at twenty-four gene loci that have been studied. The lack of variation is unique, compared to other seal species and populations that have not gone through comparable bottlenecks. It suggests that a number of alleles were lost during the bottleneck.

It may be that cheetahs underwent a similarly severe bottleneck. These cats are so genetically uniform that they will accept skin grafts from unrelated individuals, something that is rarely possible to accomplish even among littermates in other species of mammals. Such extreme genetic uniformity is of concern to conservationists. Without genetic variation the whole population, no matter how large, is extremely susceptible to diseases or environmental changes. Species such as the cheetah and the northern elephant seal could suddenly become extinct.

Founder effects and bottlenecks occur when a population originates or is rebuilt from very few individuals. In both cases, the amount of genetic variation in the population may be severely limited.

GENE FLOW

Probably very few populations are completely isolated from other populations of the same species. Individuals migrate between populations, seeds and pollen grains drift or are transported from one place to another, and ocean currents carry larvae of marine organisms for many hundreds of kilometers.

As a specific example, baboons in Africa generally live in troops. Each troop represents a separate allele pool. Commonly, some of the males wander off or are driven from one troop. They may join another troop some distance away. Assuming that migrant males encounter receptive females in the new population, and assuming they have offspring, the pool of alleles changes, just as it changes in the troop the baboons left behind.

For many plant and animal species, most of the offspring, spores, or gametes end up close to the parent. For example, in 1951, Robert Colwell released and then tracked radioactively labeled pollen grains from individual pine trees. Most of the pollen was dispersed downwind only about three to six meters from the tree that produced it. The amount of pollen then dropped off sharply at a greater distance from the source.

When such a dispersal pattern continues over generations, populations with somewhat distinct genotypes can become established in the same regions. At the same time, however, winds, insects, birds, and other dispersal agents may carry some offspring, spores, or gametes beyond the typical dispersal range for the neighboring populations. A small but probably significant amount of gene flow thereby occurs, the result being a homogenizing effect on the genotypic character of neighboring populations of the same species. Thus gene flow is thought to be a main factor in decreasing the variation between populations—variation that might arise through mutation, genetic drift, and selection.

Gene flow is the physical flow of alleles into and out of populations. It tends to decrease the genetic variation between populations that arises through other evolutionary factors.

NATURAL SELECTION

The simple fact that individuals vary in genotype and phenotype means that they are likely to be equipped in different ways to deal with environmental challenges. Some will be more suitable, in certain contexts, for surviving and reproducing. If their phenotypic differences are genetically based, then alleles associated with the advantageous phenotypes will tend to increase in frequency in the population.

The formal statement of this process is called the principle of natural selection. It was formulated in the nineteenth century by Charles Darwin and Alfred Wallace, as you read in Chapter Two. Here we will take a closer look at this principle.

Darwin, recall, was fascinated with the variation that exists in populations. Much of the variation had to be heritable, for it could be passed on from one generation to the next. At the same time, Darwin knew from his studies of artificial selection that the variation was not necessarily static. For example, selective breeding could lead to diverse breeds of pigeons, as shown in Figure 2.9. In the case of pigeons, at least, humans obviously were doing the selecting—but what could be the basis of selection in *natural* populations?

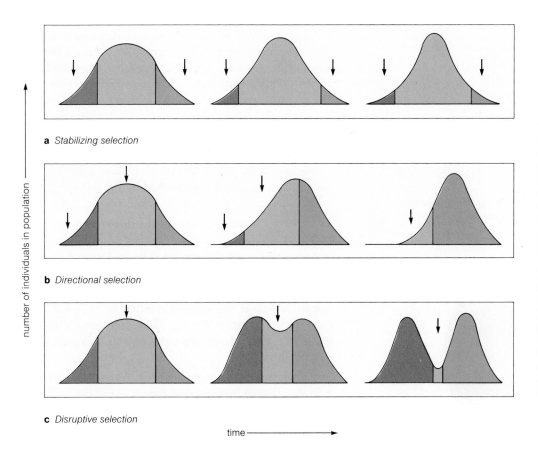

a *Stabilizing selection*

b *Directional selection*

c *Disruptive selection*

time ⟶

number of individuals in population

Figure 36.4 Three modes of natural selection. The blue-shaded and brown-shaded portions of the range of variation encompass individuals that are extreme phenotypes: the tan-shaded region is the range of the most common forms. **(a)** In stabilizing selection, conditions favor the most common forms. The downward-pointing arrows signify that variants are being selected against. **(b)** In directional selection, the phenotypic character of the population shifts as a whole in a consistent direction. **(c)** In disruptive selection, two or more extreme variants are favored and become increasingly represented in the population.

In arriving at the answer, Darwin correlated his observations of inheritance with certain observable characteristics of populations and the environment. First, he knew all populations have enormous reproductive potential. If all the individuals that are born were themselves to reproduce, then population size would burgeon. (The reason is that, with each increase in the number of reproducing members, the potential reproductive base enlarges.) Yet Darwin also knew populations generally do not behave in this fashion. Finally, he knew food supplies and other resources do not increase explosively; in most environments, resources more or less remain within certain limits. Therefore, when the population outstrips the resources necessary to sustain it, there must be *competition* for the resources that are available.

From these and other observations, Darwin put together a concept of natural selection. As noted earlier, observations and experimental work that accumulated since Darwin's time represent such an overwhelming body of evidence in favor of his concept that it is now recognized as one of the most important of all principles.

Because of its importance, the principle of natural selection is restated here:

1. In any population, more offspring can be produced than can survive to reproductive age.

2. Members of the population vary in form and behavior, and much of this variation is heritable.

3. Some heritable traits are more **adaptive** than others—that is, they improve an individual's chances of surviving and reproducing under prevailing environmental conditions.

4. Because the bearers of adaptive traits have a greater chance of reproducing, their offspring tend to make up an increasingly greater proportion of the reproductive base for each new generation. This tendency is called **differential reproduction**.

5. *Natural selection is the result of the differential reproduction.* Adaptive traits increase in frequency in a population because their bearers contribute proportionally more offspring to succeeding generations.

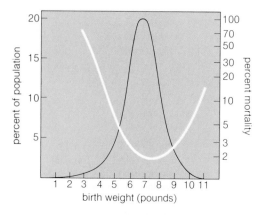

Figure 36.5 Birth weight distribution for 13,730 infants (black curve). The white line is the curve of early mortality in relation to birth weight. The optimum birth weight is between $7\frac{1}{2}$ and 8 pounds.

a

b

Figure 36.6 Two hundred and fifty million years of stabilizing selection? (**a**) Body imprints of a horseshoe crab made that long ago—imprints that could well be made by a modern-day horseshoe crab (**b**), here shown mating and perpetuating the general species form.

It is important to understand that evolution by natural selection is said to be a principle because whenever the conditions necessary for its occurrence (points 1 through 3 above) are met, then its occurrence is inevitable. Hundreds of studies of complex plants, animals, and microorganisms have demonstrated the validity of all the points listed above. Here we will consider just a few examples of natural selection in specific populations.

Modes of Natural Selection

There are three major modes of natural selection, as shown in Figure 36.4 and defined by the following list:

---

1. *Stabilizing selection* favors intermediate forms of a trait and operates against extreme forms; hence the frequencies of alleles representing the extreme forms decrease.

2. *Directional selection* shifts the phenotypic character of the population as a whole, either in response to a directional change in the environment or in response to a new environment; hence the allelic frequencies underlying the range of phenotypes move in a steady, consistent direction.

3. *Disruptive selection* favors extreme forms of a trait and operates against intermediate forms; hence the frequencies of alleles representing the extreme forms increase.

Stabilizing selection tends to counter the effects of mutation, genetic drift, and gene flow by favoring the most common phenotype. In contrast, directional selection tends to favor one phenotype at either extreme of the range of variation. Disruptive selection tends to foster an increase in the frequency of phenotypes at both extremes. The outcome of disruptive selection is **polymorphism** (two or more distinct phenotypes).

Stabilizing Selection

Human Birth Weight.   On the average, human newborns weigh about 7 pounds (3.2 kilograms). Significantly higher or lower birth weight greatly reduces the newborn's chances of survival. The shape of the survival curve in Figure 36.5 suggests that stabilizing selection favors individuals with a birth weight between $7\frac{1}{2}$ and 8 pounds and works against individuals at either extreme. Studies in widely different populations have yielded remarkably similar correlations between birth weight and mortality.

Horseshoe Crabs.   The distinctive horseshoe crab (Figure 36.6) of nearshore North Atlantic waters is not a crab at all. Rather it is representative of a very ancient arthropod lineage. Apparently these animals are very well adapted to scavenging on sandy and muddy bottoms in shallow waters, since the fossil record reveals little morphological change among some members of this group for more than 250 million years. This could be cited as an extreme example of stabilizing selection, in which a single body plan has been conserved in a series of species through time.

Directional Selection

The Peppered Moth. Directional selection has been documented in populations of the peppered moth (*Biston betularia*), which is widely distributed in England. Before the mid-1800s, a speckled light-gray form of this moth was prevalent (Figure 36.7). A dark-gray form also existed but was extremely rare, making up less than 1 percent of the population. Between 1848 and 1898, however, the dark form increased in frequency. Near one industrial city, it came to represent about ninety-eight percent of the population.

Today, two gene loci that influence the wing color and body color of these moths have been identified. Thus the trait has a heritable basis and is subject to selection.

In the 1930s, the geneticist E. B. Ford suggested that natural selection was causing the increased frequency of the dark form in industrial areas. He knew that before the industrial revolution, light-gray speckled lichens grew profusely on tree trunks. He also knew that the peppered moth is active at night but rests during the day on tree trunks, where it is vulnerable to bird predators. Light-gray speckled moths resting on the lichens would be well camouflaged from birds, whereas dark moths would be highly conspicuous (Figure 36.7a).

Ford reasoned that with the spread of factories, soot and other pollutants from smokestacks began to kill the lichens and darken the tree trunks. In this new environmental context, the rare dark form blended in with the blackened trees and the light-gray form was no longer camouflaged. Hence the dark form began to survive and reproduce at a greater rate than its light-colored kin, thereby altering allele frequencies.

In the 1950s, H. Kettlewell used the **mark-release-recapture** method to test Ford's hypothesis. He bred both forms of the moth in captivity, then marked hundreds of them so that they could be identified. He released the moths in two areas, one near the heavily industrialized area around Birmingham and the other in the unpolluted area of Dorset. After a time he recaptured as many moths as he could. Table 36.1 shows the results. He recaptured more dark moths in the polluted area— and more light moths in the pollution-free area. By stationing watchers in blinds near groups of moths tethered

Figure 36.7 An example of variation that is subject to directional selection in changing environments. (**a**) The light- and dark-colored forms of the peppered moth are resting on a lichen-covered tree trunk. (**b**) This is how they appear on a soot-covered tree trunk, which was darkened by industrial air pollution.

Table 36.1	Marked *Biston betularia* Moths Recaptured in a Polluted Area and a Nonpolluted Area			
	Light-Gray Moths		Dark-Gray Moths	
Area	Number Released	Number Recaptured	Number Released	Number Recaptured
Near Birmingham (pollution high)	64	16 (25%)	154	82 (53%)
Near Dorset (pollution low)	393	54 (13.7%)	406	19 (4.7%)

Data after H. B. Kettlewell.

male phenotype
Papilio dardanus

Danaus chrysippus

female morph trophonius
(*P. dardanus*)

Amauris crawshayi

female morph cenea
(*P. dardanus*)

Amauris niavius

female morph hippocoon
(*P. dardanus*)

Inedible Models

. . . and Their Mimics

to tree trunks, Kettlewell also found by direct observations that birds captured more light moths around Birmingham and more dark moths around Dorset.

About a hundred different species of moths underwent the same kind of directional selection in response to pollution in British industrial regions. However, as a result of strict pollution controls (which went into effect in 1952), lichens have become reestablished and tree trunks are largely free of soot. As predicted, the frequency of the dark forms is now declining in the moth populations.

Pesticide Resistance. Insect populations that develop resistance to insecticides provide other examples of directional selection. The first application of an insecticide kills most of the insects. Some individuals may survive, however, because of physiological, behavioral, or morphological differences that enable them to resist the chemical effects of the insecticide. If the resistance has a genetic basis, it can be passed on. As a result of differential survival and reproduction, the next generation will contain more of the resistant individuals.

Then, as the number of resistant forms of the insects increases, heavier and more frequent applications of the insecticide act as a selective agent that favors them even more. The genotypic structure of the insect population shifts rapidly in a consistent direction—toward more resistant phenotypes. Crop damage from insects is now greater than it was before the widespread use of insecticides.

Disruptive Selection

Disruptive selection has been noted in populations of the African swallowtail butterfly (*Papilio dardanus*). Here, selection is associated with **mimicry**, in which one species is deceptively similar in color, form, or behavior to another species that has a selective advantage. The first species is the mimic, the second is the model. For example, if a model has warning coloration that identifies it as inedible to potential predators, then a tasty species that mimics its coloration (as does *P. dardanus*) will have a better chance of being left alone.

The males in all populations of *P. dardanus* have yellow and black wings with "tails" at the tips. In most of tropical Africa, the females of *P. dardanus* are conspicuously different in appearance from the males—and from

Figure 36.8 Effect of disruptive selection on females of African swallowtail butterfly populations. Disruptive selection is apparently favoring several different female forms.

a

b

c

Figure 36.9 Examples of sexual dimorphism in (**a**) the northern sea lion, (**b**) the sage grouse of North America, and (**c**) the lion of Africa.

one another. The wing patterns and coloration of each form of the female mimic those of an inedible species present in the same region (Figure 36.8). In regions where models are absent, there are no mimicking females. Bird predators are apparently the selective agents promoting variation in these populations.

Unlike the females, males of *P. dardanus* are not mimics. Being recognized by females probably has a greater advantage for the males than does the avoidance of predation by mimicry. The fact that all the female variants continue to recognize and accept the same males apparently counterbalances the effect of disruptive selection and keeps the populations from diverging into separate species. Here, sexual selection, which will now be described, may be stronger than other mechanisms of natural selection in guiding the evolution of the male phenotype.

Sexual Selection

Some of the most dramatic examples of selection are found among animals that show **sexual dimorphism**—that is, differences in appearance and behavior between males and females. Sexual dimorphism is especially striking among birds and mammals (Figure 36.9).

Generally, the males are larger, more conspicuous in color and patterning, and more aggressive than females. Of course, such traits make the males of prey species more visible to predators, and the males of predatory species more visible to their prey. So in some respects, the traits are probably selected against. Nevertheless, the traits are used by females to select mates and thus they are associated directly with reproductive success. Such **sexual selection** may be based on any trait that gives the individual a preferential advantage in mating and in producing offspring (page 758).

For example, northern sea lions mate only on small islets and rocky beaches. Males that are large enough to command the rocks enjoy mating privileges with about ten to twenty females. The males that cannot secure a territory do not mate at all, hence they contribute nothing to the allele pool of the next generation. As an evolutionary outcome of sexual selection, males weigh as much as 1,000 kilograms, about twice the weight of the females (Figure 36.9).

As another example, the males of several species of grouse are more strikingly colored than the females, and they engage in behavioral displays. The males congregate in small mating territories, called leks, where they call and display. The drab females, attracted by the fracas, tend to mate with the most dazzling and aggressive males.

Selection and Balanced Polymorphism

Sometimes selection favors a particular allele so strongly that it supplants all other alleles at the gene locus. Often, however, **balanced polymorphism** occurs. Here, two or more alleles of a single gene locus persist at a frequency too high to be maintained by mutation alone; and that frequency, if changed, will return to its former value over several generations.

Sickle-cell polymorphism in humans is one of the best-studied examples of this condition. In West and Central Africa, the HbS allele associated with sickle-cell anemia is maintained at a high frequency relative to the normal HbA allele. Within a given population, HbS/HbS homozygotes make up about 2.5 percent and heterozygotes (HbS/HbA) nearly 30 percent of all genotypes at birth.

HbS/HbS homozygotes often die in their early teens or early twenties, generally before age forty-five. It is the abnormal hemoglobin arising from the mutant allele that puts them at a disadvantage. Whenever the blood oxygen level falls below normal (as it may during overexertion or respiratory illness), HbS molecules crystallize and distort red blood cells into sickle shapes. The sickled cells cannot transport normal amounts of oxygen and tend to clump in the capillaries. As we have seen, the consequences are severe (page 169).

Why doesn't selection remove the HbS allele from the populations? The reason is this: *The survival value of alleles must be weighed in the context of the environment in which they are being expressed.* Wherever the sickle-cell polymorphism persists, malaria is also rampant. About half of all cases of malaria are caused by *Plasmodium falciparum* (page 579), a parasitic protistan that uses only one kind of mosquito to transmit its sporozoites to the human bloodstream. Since these mosquitoes live only in the tropics and subtropics, malaria and the sickle-cell polymorphism occur mostly in those parts of the world.

Many studies suggest that the sickle-cell polymorphism is maintained through (1) *differential mortality* and (2) *differential fertility* between the normal members and the heterozygous members of the populations. (The frequency of the HbS/HbS homozygote among adults is so low that its effect can be ignored here.) Let's look first at the survival advantage of the HbS/HbA heterozygote.

In the absence of malarial infection, the HbA/HbA homozygote has an eighty-five percent greater probability of surviving to reproductive age than does the heterozygote. The presence of malaria alters this probability. In 1954, A. Allison reported that heterozygotes have greater resistance to malaria and are more likely to survive severe infections. After infecting fifteen HbA/HbA and fifteen HbS/HbA volunteers with *P. falciparum*, Allison later found malarial parasites in the blood of fourteen of the normal homozygotes and in only two of the heterozygotes. In another study, severe or fatal infections were found to be twice as high in HbA/HbA children as in HbS/HbA individuals.

Thus, the persistence of an extremely deleterious trait (the HbS allele) becomes a matter of relative evils. On an absolute scale, the survival and reproductive rate of HbS/HbA individuals is poor, but the rate is better than it is for HbA homozygotes in regions where malaria is prevalent. As part of the "cost" of maintaining the more adaptive heterozygote, half the infants born will be either HbA/HbA or HbS/HbS homozygotes—neither of which has much chance of surviving to reproductive age.

EVOLUTION OF SPECIES

All the processes described so far operate at the level of populations. What is commonly overlooked is their potential *reversibility* at this level. Alleles lost through genetic drift or through selection may be introduced again by mutation or gene flow, and changes in allele frequency mediated by selection may be reversed if the environment changes in an appropriate direction. Yet if such reversibility is possible, *then how did the diversity we observe in nature arise in the first place?* What prevents populations from simply cycling through the same events again and again? The answer is that an *irreversible step*, acting like a ratchet, causes genetically isolated populations to branch in different evolutionary directions. This step, "evolution's ratchet," is part of the process by which species originate—a process called **speciation**. We will now consider some characteristics of this process, for which we will need the following definition as our point of reference:

For sexually reproducing organisms, **a species is one or more populations whose members actually (or potentially) interbreed under natural conditions, produce fertile offspring, and are reproductively isolated from other populations.**

Divergence

Within a given population, shifts in allele frequencies either may move the whole population in one direction or another or may produce a balanced polymorphism.

No matter how diverse the individuals become, however, they will remain members of the same species as long as they continue to interbreed successfully and share a common pool of alleles.

Sometimes, however, barriers arise between parts of the population, creating local breeding units. As a result, two or more pools of alleles may arise where there had been only one before. Over time, the absence of gene flow between populations and the action of selection, genetic drift, and mutation can lead to divergence. The term **divergence** refers to a buildup of differences in allele frequencies between reproductively isolated populations. When divergence becomes so great that successful interbreeding is no longer possible under natural conditions, the populations are said to constitute separate species.

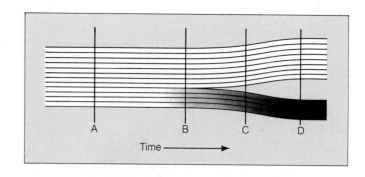

Figure 36.10 Divergence leading to speciation. Because evolution is gradual here, we cannot say at any one point in time that there are now two species rather than one. Each vertical line represents a different population. In A, there is only one species. In D, there are two. In B and C, the divergence has begun but is far from complete.

When Does Speciation Occur?

Except in some unusual cases, which will be described shortly, it is impossible to identify the exact moment at which speciation occurs. As in any gradual process (such as the development of an infant into an adult), species usually do not spring forth at a single moment in time. Figure 36.10 shows the gradual nature of speciation.

Determining whether speciation has occurred is not always a simple matter. Divergence between reproductively isolated populations must have gone so far that interbreeding no longer occurs, even when the opportunity arises. But if the populations are geographically isolated, how can we know for sure that divergence has proceeded to that point? For example, lions and tigers are reproductively isolated in the wild. They certainly look different, and we call them separate species. Yet when they are brought together in captivity, they sometimes interbreed and produce hybrid offspring. Are they, in fact, separate species?

Moreover, if reproductive isolation is one of the criteria used to define a species, how can fossils be assigned to one species or another? We have no idea of whether a now-fossilized individual was once able to interbreed with individuals of similar morphology living in the same region. (Thus, for example, we have different interpretations of what was really going on among the earliest members of the human family, some of which lived in the same regions and were morphologically varied.)

Reproductive Isolating Mechanisms

A **reproductive isolating mechanism** is any aspect of structure, function, or behavior that prevents interbreeding. These mechanisms fall into two categories. First, some take effect before or during fertilization, thereby preventing the formation of hybrid zygotes. Second, some take effect during development of the embryo, resulting in early death, sterility, or even the failure of the hybrid animal to be recognized as an acceptable mate by members of either parental species. All reproductive isolating mechanisms have the same effect: *they prevent the exchange of alleles between populations.*

A reproductive isolating mechanism is any aspect of structure, function, or behavior that prevents successful interbreeding (hence gene flow) between populations.

Mechanical Isolation. Differences in the structure or function of the reproductive organs may prevent individuals of different populations from producing hybrid zygotes. For example, the stigma of one plant species may be so different in size, shape, or length that the pollen tube of another plant species cannot reach its ovary. As another example, two species of sage plants in Southern California differ in the size and arrangement of their floral parts, and their pollinators are so different in size that the pollinator of one species cannot come into contact with the pollen of the other (Figure 36.11).

Isolation of Gametes. Incompatibilies between the sperm of one species and the egg (or the female reproductive system) of another may prevent fertilization. This is the case with certain marine animals that rely on external fertilization. For example, even when the eggs and sperm of two species of sea urchin are released at the same time and in the same place, the gametes are rarely attracted to each other. Probably the female does not release the biochemical substances required to attract

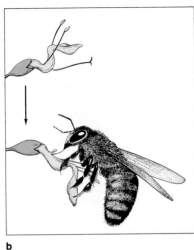

a   b

Figure 36.11 Mechanical isolation between two species of sage (*Salvia mellifera* and *S. apiana*). The first species (**a**) has a small floral landing platform for its small or medium-size pollinators. The second species (**b**) has a large landing platform and long stamens, which extend some distance away from the nectary. Even though small bees can land on this larger platform, they can do so without brushing against the pollen-bearing stamens. It takes larger pollinators to do this. Hence the small pollinators of *S. mellifera* are mostly incapable of spreading pollen to flowers of *S. apiana*; and the large pollinator of *S. apiana* cannot land on and cross-pollinate *S. mellifera*. The plants and their pollinators are all drawn to the same scale.

the sperm to her. Similarly, incompatibilities that exist between the stigma of one plant species and the pollen grains of other species are enough to prevent fertilization.

Isolation in Time.   The timing of reproduction may also serve as an isolating mechanism. For most animals and plants, mating or pollination is a seasonal event of relatively short duration, sometimes less than a day. Even closely related species may be reproductively isolated simply because their times of reproduction do not coincide. Two closely related species of cicadas provide an extreme example. In any one location, one of these insects emerges and mates every thirteen years, the other every seventeen years. The possibility of their meeting arises only once every 221 years!

Behavioral Isolation.   Behavioral isolation is one of the strongest mechanisms at work among related species in the same territory. For example, complex courtship rituals often must precede mating. The song, head-bobbing, wing-spreading, and dancing by a male bird of one species may stimulate a female of the same species—but the female of a related species probably would

not even recognize such behavior as a sexual overture. Table 36.2 reveals the effectiveness of behavioral isolation in some closely related species of *Drosophila*.

Hybrid Inviability and Infertility.   Even when fertilization occurs between the gametes of two species, there may be incompatibilities between the developing embryo and the mother. If the incompatibilities are severe, the embryo will die. Hybrids that do live at first are commonly weak in structure, physiology, or behavior, and their chance of surviving is not good. In a few cases, the hybrid offspring are vigorous but sterile. Thus a cross between a female horse and a male donkey produces a mule, a hybrid that is fully functional *except* in its reproductive capacity.

Finally, even if a first-generation hybrid manages to survive, it may be unable to reproduce. The unusual appearance or behavior of a hybrid animal may preclude its recognition by either species when the time comes for it to mate. Even when such hybrids manage to reproduce, their offspring are typically not vigorous. Crosses between two species of evening primrose produce a partially fertile first generation, but the second-generation plants are slow-growing sterile dwarfs, susceptible to disease.

Modes of Speciation

Allopatric Speciation.   Typically, the populations of a given species are not strung out continuously in space, with one merging into the others and thus open to gene flow. Most often the populations are geographically isolated from one another to varying degrees, with gene flow being more of an intermittent trickle than a continuous stream.

On occasion, even the trickles can be shut off completely through the formation of impassable geographic barriers. Sometimes the barriers can form quickly, as when a flood changes the course of a river and isolates populations of, say, an insect species that cannot swim or fly. Sometimes the barriers form slowly, as when the climate changes over time and causes the breakup of formerly continuous environments. (Thus an extensive forest might give way to grasslands with isolated stands of trees as a result of long-term shifts in the pattern of rainfall.) On a more immense time scale, oceans or mountain ranges can form between populations as a result of exceedingly slow but massive realignments of the continents themselves (Chapter Thirty-Eight).

Once geographic separation is absolute, genetic drift or adaptation to different environments may then bring about divergence. If divergence proceeds far enough, the descendants of the separated populations may become so different that they will not interbreed even if they expand their distribution and make contact at some later time. When new species form in this manner, as a result of geographic isolation, the process is called **allopatric speciation** (from *allos*, meaning "different," and *patria*, "native land").

Allopatric speciation is generally considered to be the major route to diversity. It has been well documented in virtually all groups of sexually reproducing organisms in all types of habitats. Until rather recently some biologists believed it to be the *only* route, but with our increased understanding of the genetics of speciation, that view has become outdated.

Parapatric Speciation. It now appears that speciation also can occur without the formation of absolute geographic barriers. **Parapatric speciation** occurs in populations that live adjacent to one another (*para* means "side by side").

For example, suppose the populations of a species extend across different major environments, as from high mountains into an adjacent desert. Quite probably, selection will favor different phenotypes in the two settings. If the transition between the two environments is abrupt, individuals with intermediate phenotypes may be selected against, for they may not be well adapted to either set of operating conditions. Selection will begin to favor the individuals that choose members of their own population type as mates. In time, reproductive isolating mechanisms will become established and will lead to speciation.

Sympatric Speciation. The word sympatric means "same native land." **Sympatric speciation** is the origin of species as a result of ecological, behavioral, or genetic barriers that arise *within* the boundaries of a single population. Suppose a mutation leading to a shift in food preference or to a shift in the timing of reproduction begins to spread in an insect population. In such cases, individuals with that mutation will be in a position to breed only with one another. Similarly, various types of chromosomal aberrations (such as duplications of some or all of the chromosomes, or changes in chromosome numbers by fusion or fission) can result in instantaneous speciation. The bearers of the altered chromosome complement successfully reproduce only with one another, not with other members of the population. (Offspring that are heterozygous for different numbers of chromosomes die early in development, or are sterile if they do survive.)

Sympatric speciation can occur through *polyploidy*, whereby the original chromosome number is multiplied in a particular zygote. This mechanism has been important in plant evolution; about forty percent of all flowering plants are polyploid. The reason for its widespread occurrence is that plants often are self-fertilizing, and most can reproduce asexually by vegetative propagation. Because the polyploid individual need not wait for a sexual partner with a comparably multiplied chromosome number, speciation is instantaneous. That one individual can give rise to a whole population just like itself.

Polyploidy combined with successful *hybridization* can also produce a new species. Usually, interspecific hybrids are sterile because they inherit chromosomes that differ in number or kind from their different parents; hence the "homologues" probably cannot pair during meiosis. However, if the interspecific hybrid undergoes polyploidy, then the *duplicates* can pair with the originals, meiosis can proceed, and viable gametes can form. Wheat is an example of a successful polyploid hybrid (page 183).

Sympatric speciation is relatively rare in animals. It has been documented in certain flies and grasshoppers, and it may also occur in some mammals.

Table 36.2	Behavioral Isolation Between Three Closely Related *Drosophila* Species			
Contact Limited to the Following Combinations:		Number of Females	Number of Matings	Percent Matings
Females	**Males**			
D. serrata	*D. serrata*	3,841	3,466	90.2
D. serrata	*D. birchii*	1,246	9	0.7
D. serrata	*D. dominicana*	395	5	1.3
D. birchii	*D. birchii*	2,458	1,891	76.9
D. birchii	*D. serrata*	699	7	1.0
D. birchii	*D. dominicana*	250	1	0.4
D. dominicana	*D. dominicana*	43	40	93.0
D. dominicana	*D. serrata*	163	0	0.0
D. dominicana	*D. birchii*	537	20	3.7

Data from F. Ayala. 1965. "Evolution of Fitness in Experimental Populations of *Drosophila serrata*." *Science* 150:903–905.

SUMMARY

1. Within a population, individuals show phenotypic variations which are expressions of gene mutation, chromosomal aberrations, and genetic recombination.

2. Within a population, the frequencies of different alleles change as a result of mutation, genetic drift, gene flow, and natural selection, including sexual selection.

3. A mutation is a heritable change in the kind, structure, sequence, or number of the component parts of DNA. The new alleles produced by mutation arise randomly with respect to timing and location, and without reference to their desirability for the population at the time.

4. Genetic drift is an increase or decrease in the relative abundance of different alleles through successive generations, simply by chance. Its effects are more rapid in small populations.

5. Gene flow is a change in allele frequencies due to immigration or emigration.

6. Natural selection is the result of differential reproduction of genotypes within a population.

7. *Stabilizing* selection tends to counter the effects of mutation, genetic drift, and gene flow by favoring the most common forms of a trait (which are already well adapted to prevailing conditions).

8. *Directional* selection tends to favor phenotypes at one extreme or the other in the range of variation. The frequency distribution of alleles tends to move in a consistent direction, shifting the phenotypic character of the population as a whole.

9. *Disruptive* selection tends to favor an increase in phenotypes at both extremes in the range of variation. The outcome may be polymorphism (in which two or more forms of a trait are common in the population).

10. For sexually reproducing organisms, a species is one or more populations whose members are able to interbreed under natural conditions and produce fertile offspring, and are reproductively isolated from other populations.

11. Over time, one or more populations (or parts within a single population) of a species may become geographically isolated from others. The absence of gene flow between them and the action of selection, genetic drift, and mutation may lead to divergence.

12. Divergence is a buildup of differences in allele frequencies between reproductively isolated populations. Speciation occurs when divergence is so great that successful interbreeding no longer occurs under natural conditions.

13. Any aspect of structure, functioning, or behavior that reproductively isolates different populations may lead to speciation.

14. *Allopatric* speciation may occur when geographic separation prevents gene flow. It is the most common speciation process.

15. *Parapatric* speciation may occur when adjoining populations undergo divergence despite some gene flow.

16. *Sympatric* speciation occurs when populations diverge within the same geographic range. For example, polyploidy has led to sympatric speciation in flowering plants.

Review Questions

1. What is the typological approach to categorizing species? In what fundamental way does the population concept differ from this approach?

2. Define these terms: individual, population, and species.

3. What is genotypic and phenotypic variation? Describe evolution in terms of frequency distributions for a given trait, and in terms of the underlying allele frequencies.

4. What is the Hardy-Weinberg baseline against which changes in allele frequencies may be measured? What is Hardy-Weinberg equilibrium?

5. Changes in allele frequencies may be brought about by mutation, genetic drift, gene flow, and selection pressure. Define these occurrences, then describe the way each one can send allele frequencies out of equilibrium.

6. What implications might the effect off genetic drift hold for an earlier concept of "survival of the fittest?" As part of your answer, define polymorphism and explain how different phenotypes can persist indefinitely in the *same* population.

7. Natural selection is no longer in the realm of pure theory. Can you recount the ongoing sagas of the peppered moth and the pesticide-resistant pests to explain why it is now considered an operating principle of biology?

8. Define stabilizing, directional, and disruptive forms of selection and give a brief example of each.

9. Before labeling a particular genotype as being advantageous or disadvantageous, what must you first consider?

10. Give two examples of reproductive isolating mechanisms, and outline what they accomplish.

11. What is the difference between allopatric and sympatric specialization? Which do you suppose occurs most often in plants?

Readings

Ayala, F. J., and J. W. Valentine. 1979. *Evolving*. Menlo Park, California: Benjamin/Cummings. Short introduction to evolutionary theory. Excellent writing, interesting examples, and well-chosen illustrations.

Cavalli-Sforza, L., and W. Bodmer. 1971. *The Genetics of Human Populations*. New York: Freeman. Objective look at problems of racial differentiation; clear discussion of sickle-cell polymorphism.

Futuyma, D. 1979. *Evolutionary Biology*. Sunderland, Massachusetts: Sinauer. Excellent synthesis of modern evolutionary thought.

Grant, V. 1981. *Plant Speciation*. Second edition. New York: Columbia University Press. Good discussion of speciation in plants.

White, M. J. D. 1978. *Modes of Speciation*. San Francisco: Freeman. Excellent review of speciation mechanisms, particularly those involving chromosomal rearrangements.

The great triumph of evolutionary theory is that it provides a framework for explaining why living organisms are all so similar in their biochemistry and molecular biology yet so different in form and function. Evolution does not create new organisms out of thin air; rather, it proceeds by modifications of the genetic makeup of existing organisms, and this means there is an underlying continuity of relationship among them. Typically this relationship by way of descent is said to resemble a tree, with the living species forming an immense shell-like canopy of leaves that are related by collections of twigs, branches, and major limbs leading to a common ancestral trunk. If we could look back through time, we would see how the failures and successes of vast numbers of previous leaves have uniquely shaped this tree, pruning and nourishing it into its present form. Yet we can directly observe no more than the present leaves and the ones preserved in the fossil record. The rest of the tree must be reconstructed by indirect methods.

It is important to understand that only the *species* are real entities in the tree of descent. The "twigs, branches, and limbs" are not real; they are *categories of relationship* that we superimpose on the species themselves. These categories, called the **higher taxa** (singular, taxon), are listed in Figure 37.1. Obviously, then, we cannot talk about "the evolution" of genera, families, and so on except in terms of the developments that occurred during the history of all the species contained in those increasingly inclusive categories. As we will see, these developments include the origins, persistence, radiations, and extinctions of different groups of species. What we call **macroevolution** refers to the large-scale patterns, trends, and rates of change among groups of *species* since the beginning of life.

37

PHYLOGENY AND MACROEVOLUTION

A QUESTION OF PHYLOGENY

Classification Schemes

Why are species assigned membership in one genus and not another? Why are different genera assigned membership in one family rather than another, and so on? These are questions you should ask when studying any classification system—that is, a system for grouping taxa.

In general, classification schemes are constructed on the basis of phenotypic similarities, relationships through descent, or both. **Phenotype** refers to the morphological, physiological, and behavioral traits of an individual, all of which may be observed or measured. **Phylogeny** refers to evolutionary relationships, starting with the most ancestral species and including all the branchings leading to all of its descendants.

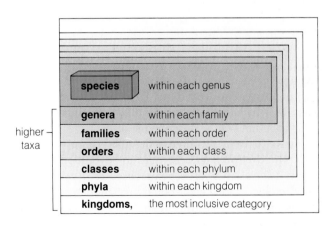

Figure 37.1 The higher taxa—categories used in describing the large-scale patterns, trends, and rates of change among species since the beginning of life.

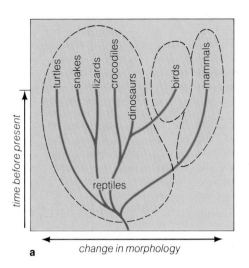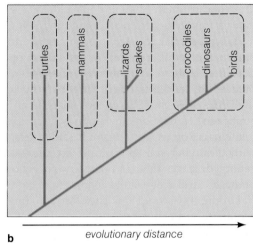

Figure 37.2 How a few major groups of vertebrates are ranked according to two different systems of classification. In evolutionary taxonomy (**a**), the groups are defined by such traditionally accepted characteristics as scales, feathers, or fur. A cladistic classification (**b**) links together the organisms having shared ancestries.

time before present

change in morphology

a

evolutionary distance

b

If you were to construct a tree of descent by taking into account both phenotypic and phylogenetic relationships, each twig or branch would be a single line of descent, or **lineage**. The branch points would represent speciation events, and the angle of each branching would indicate the rate of morphological change in a given time period. Horizontal branchings would mean abrupt change and narrow angles would mean gradual change from the ancestral species form. For example,

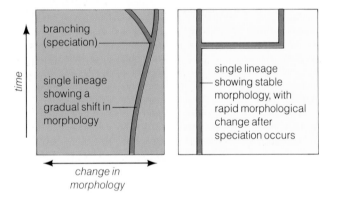

branching (speciation)

time

single lineage showing a gradual shift in morphology

single lineage showing stable morphology, with rapid morphological change after speciation occurs

change in morphology

The time of each branching (how high or low it is in the tree) is now being determined by radioactive dating of rocks in which fossils are found and by biochemical yardsticks that will be described shortly.

Figure 37.2a shows how some major groups of animals would be ranked according to "evolutionary taxonomy," a classification scheme based on a mixture of morphological and evolutionary relationships. Such schemes may be misleading at times, for who is to decide which morphological traits are most important for assigning particular types of animals to one category or another? Here, for example, the dinosaurs are grouped with the reptiles, which are ranked as a class equal to the classes of birds and mammals.

Yet consider how these animals might be ranked in one alternative scheme, called "cladistics." With this approach, relatedness is depicted only in terms of branch points in the lines of descent. (As we will see, branch points are inferred by combining information from the fossil record, comparative morphology, and comparative biochemistry.) What does such a scheme tell us? Notice that there is no class of "reptiles" and that the birds are grouped with their closest relatives, the dinosaurs and crocodiles. With this scheme, we may infer that the dinosaurs may have been much more "birdlike" than "reptilelike" in their lives! Like crocodiles and birds, they may have had a four-chambered heart, territorial behavior and the use of "songs" by males, and complex nesting behavior and parental care of the young. None of these traits characterizes any "reptile" except the crocodiles. The point to keep in mind is this:

The way taxa are categorized affects the ways we think about them and the questions we are led to ask.

Speciation and Morphological Change

Let's now consider a crucial feature of evolutionary trees—specifically, the manner in which they portray the rate of morphological change in an evolving lineage.

Everyday experience tells us there is some order in the diversity among organisms. For example, you know there are several species of cats—lions, tigers, leopards, pumas, bobcats, and so forth—and you could probably identify most of them as cats on sight. You could also identify such members of the dog family as foxes, coyotes, and wolves, and it is unlikely that you would incorrectly identify any member of the dog family as a type of cat. The point is this: *Morphological "gaps" separate even these closely related species, just as increasingly larger "gaps" separate genera, families, and other taxa.*

How does the pattern of clusters and gaps arise—in other words, why are there so few intermediate forms? Let's first address this question at the level of species, then return to it later with reference to higher taxa.

Figure 37.3 shows two ways of interpreting rates of morphological change in an evolving lineage. According to the traditional model, termed **gradualism**, most morphological change occurs *within* species as a result of genetic drift, directional selection, and other processes by which allele frequencies change (Chapter Thirty-Six). For example, continuous layers of deep-sea sediments contain fossils of foraminiferans (a type of shelled protistan) that clearly show gradual changes in form within the same lineage. This fossil sequence, however, is one of only a few well-documented cases of gradualism. Many cases, including the much-repeated example of the evolution of the modern horse, have not held up under close examination.

According to an alternative model, termed **punctuation**, most morphological change takes place rapidly *during* speciation, with relatively little change occurring once a particular species is established. Compared to the average span of existence for a species (often on the order of 2 million to 6 million years), the hundreds or thousands of years that may be required for speciation is a short period. Yet founder effects, bottlenecks, and strong directional selection could accomplish a great deal of genetic change in only a thousand years, and stabilizing selection thereafter could maintain the traits of a well-adapted species within relatively narrow limits.

The punctuational model is consistent with the observation that there are few intermediate forms between closely related species. Moreover, the fossil record provides indirect evidence against gradualism. Although there are cases of gradual change in form, they are greatly outnumbered by cases in which new species were morphologically distinct when they appeared, then changed relatively little thereafter.

RECONSTRUCTING THE PAST

The Fossil Record

Most evidence of extinct species comes from fossilized skeletons, shells, leaves, seeds, and tracks. To be preserved as a fossil, body parts or impressions must be buried before they deteriorate, and the resulting rock layers must remain relatively undisturbed by geologic forces. Some types of organisms and environments are more likely than others to yield fossils. For example, animals with hard, durable shells or skeletons are better represented than are more fragile species, and wholly

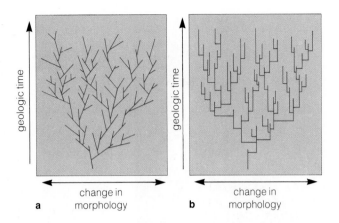

Figure 37.3 Two interpretations of how morphological diversity develops in a lineage. Each vertical line represents a single species. In a *gradual* model (**a**), changes occur more or less steadily but at different rates among species. In a *punctuational* model (**b**), rapid morphological change is associated with speciation, as indicated by the horizontal lines that signify branch points, and the morphologies of established species remain constant through time.

soft-bodied animals such as jellyfishes or worms are preserved only in exceptional environments. The seafloor, flood plains, swamps, and natural traps such as caves and tar pits typically favor fossilization; land surfaces that are undergoing erosion usually do not.

The completeness of the fossil record thus varies as a function of the kinds of organisms involved, the places where they lived, and the geological stability of the region between the time of fossilization and today. Some parts of the record are gone forever, and most of the rest is incomplete. The fossils described so far represent about 250,000 species that lived over the past 600 million years—a very small number when compared to the $4\frac{1}{2}$ million species estimated to be alive today. Still, this is much better than nothing at all. Also, in some cases the fossil record is thought to be nearly complete. For example, in certain shallow marine environments where sediments were deposited continuously, the numbers of shelled invertebrate species known from fossils are closely comparable to the numbers of such species living in the same types of environments today.

The quality of specimens preserved as fossils also varies widely. Because the average fossil is broken, incomplete, and often crushed or deformed (Figure 37.4), painstaking preparation and careful study are needed to gain useful information about the organism it represents. In contrast, a few fossils are spectacular in their completeness and preservation (Figure 37.4), and they tell us a great deal about the structure, function, and even behavior of extinct organisms.

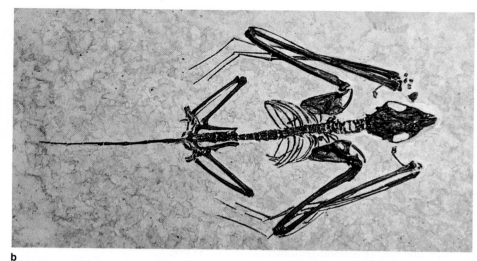

Figure 37.4 Examples of vertebrate fossils. (**a**) This photograph shows a good find—parts of the fossilized skeletons of several individuals of an early ducklike bird. Many hours of very careful preparation and study will be required to establish the identity of the organism and to piece together its morphological characteristics. (**b**) A paleontologist's dream, the complete skeleton of a bat that lived 50 million years ago. Fossils of this quality are extremely rare.

Comparative Morphology

Most hypotheses about evolutionary relationships and most classification schemes are based on **comparative morphology**, the comparison of body form in major groups of animals. The comparative approach can be powerful if certain limitations are recognized, and misleading if they are not.

For example, comparisons show that the early stages of embryonic development have been highly conserved through vertebrate history. Thus gill slits form in all vertebrate embryos, even though adult fish and many amphibian larvae are the only vertebrates with gills. The striking similarities are just one of the reasons why many animals that otherwise might appear to be quite dissimilar are all classified as vertebrates. It is not that, say, a mammalian embryo "passes through a fishlike stage"; evolution is not repeated all over again every time an embryo develops. Rather, the early stages are fundamental steps in major developmental programs in all the lineages. Most mutations affecting these early stages were probably *selected against* during vertebrate history, for they would have been lethal in terms of further development.

Given that development tends to be a conservative process, what accounts for the morphological variation among vertebrates? Most likely, mutations affected certain genes that control the *rate* of growth of organs and other structures.

For example, the series of bones in the vertebral column develop from the tissues of a series of body seg-

ments (somites) in the embryo. A mutation can increase or decrease the rate at which the segments grow and develop. The change can affect the numbers of vertebrae and segmental muscles, which in turn affects the structure of blood vessels and nerves. This sort of modification may help explain how a snake could evolve from a lizard, or a long-necked swan from a short-necked goose. In human embryos, the developmental period in which brain cells rapidly proliferate is longer than in related primates. Many of the differences in skull and facial structure between humans and other primates are the outcome of a developmental program that produces our larger brain.

Divergence and Convergence in Form

Because morphological features often have clearly functional roles, they must be subject to natural selection. However, a structure often serves several different functions, and the outcome of selection may represent a compromise between the most suitable form for one of its roles and the least disturbance to its other roles. Moreover, evolution can only proceed by the modification of *existing* structures, so the results of selection for a particular role may be quite different from the original function. The power of natural selection to bring about change despite these constraints is evident in examples of morphological divergence.

In **morphological divergence**, selection leads to departures from the ancestral form. Figure 37.5 shows

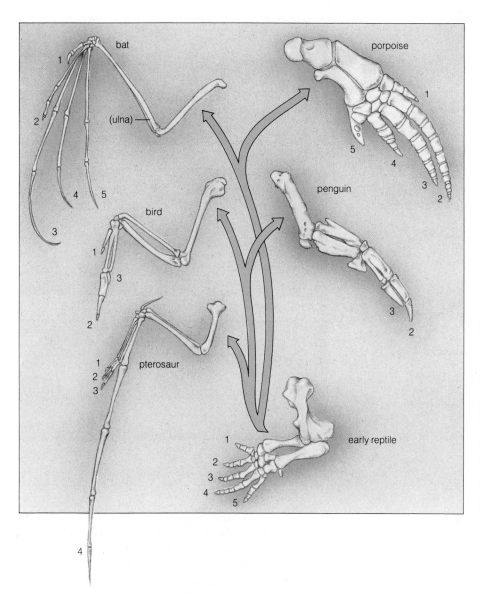

Figure 37.5 Morphological divergence in a structure with a conserved developmental plan— the vertebrate forelimb. From the generalized form of the reptilian limb, structures as diverse as the wings of pterosaurs, birds, and bats, the flippers of penguins and porpoises, and many others have evolved while preserving many similarities in the number and position of bones.

Convergence. The penguin wing has converged on the form of the porpoise flipper, even though the ancestors of penguins had wings used for flight.

examples of divergence in the forelimbs of vertebrates. In each of the three groups of flying vertebrates (pterosaurs, birds, and bats), a different type of wing developed, quite independently, from the five-toed limb characteristic of most terrestrial vertebrates. Even so, all three have the same component parts, as do the flippers of porpoises. We could expand this example to include the long, one-toed limbs of horses, the stout and powerful limbs of moles and other burrowing mammals, and the thick, columnlike limbs of elephants. As another example of morphological divergence, snakes lost all traces of the forelimbs present in their ancestors (Figure 2.4).

In **convergence**, two or more species from dissimilar and only distantly related lineages adopt a similar way of life and come to resemble one another rather closely. For example, penguins evolved from flying birds but now use their wings in swimming. The structure of their wing bones diverged from the ancestral form and converged on the limb form of porpoises (Figure 37.5).

Convergence often results when the physical requirements of a particular life-style are unusually strict. A classic case of convergence in external morphology occurred among sharks, ichthyosaurs, and porpoises (Figure 37.6). The three lineages are among the most distantly related vertebrates, and the ichthyosaurs and porpoises were derived from fully terrestrial stocks. Yet the three are functionally similar in being adapted for rapid swimming, with a common streamlined shape, similar placement and proportions of stabilizing fins, and a powerful tail.

Comparative Biochemistry

All organisms are fundamentally the same in many aspects of their biochemistry and molecular biology; over ninety percent of the known gene products of human cells have readily identifiable counterparts in yeast cells.

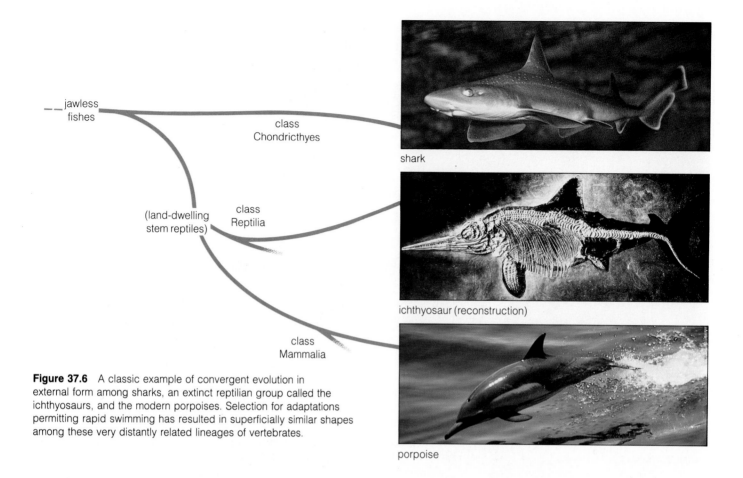

class
Chondricthyes

shark

class
Reptilia

ichthyosaur (reconstruction)

class
Mammalia

porpoise

jawless
fishes

(land-dwelling
stem reptiles)

Figure 37.6 A classic example of convergent evolution in external form among sharks, an extinct reptilian group called the ichthyosaurs, and the modern porpoises. Selection for adaptations permitting rapid swimming has resulted in superficially similar shapes among these very distantly related lineages of vertebrates.

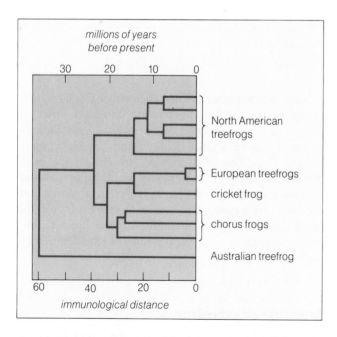

Figure 37.7 Relationships among groups of tree frogs as indicated by immunological data. The immunological distance is simply the number of amino acid substitutions between two taxa.

Yet the genes in the two cell types have been evolutionarily isolated from one another for many hundreds of millions of years.

It is not that the molecular structure of DNA has never been modified in different ways in different lineages. However, we now believe that a significant number of the gene mutations are *neutral* (or nearly so), which means they do not measurably affect the function of the protein being specified. Because neutral mutations have no effect on phenotype, they are not subject to natural selection, and they tend to accumulate through random processes such as genetic drift.

Given the inevitable occurrence of mutations, the probability of neutral mutations becoming permanently incorporated into the DNA of a species increases with time. Through the use of statistical methods, it can be argued that neutral mutations do this at regular rates for major classes of proteins. If that is so, they can be used as a **molecular clock** for dating the divergence of two species from a common ancestor. (Thus, if the sequences of amino acids in a particular protein are quite similar in two species, it can be argued that they shared a recent common ancestor; but if the proteins are quite dissimilar

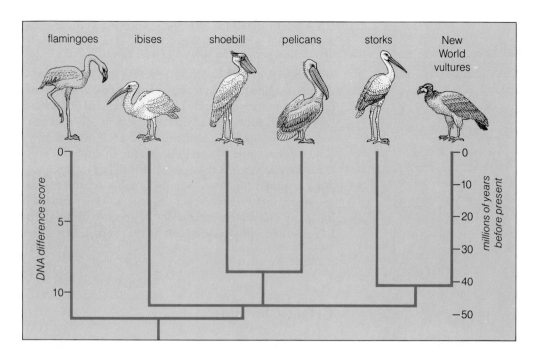

flamingoes   ibises   shoebill   pelicans   storks   New World vultures

DNA difference score

millions of years before present

Figure 37.8 Relationships among the New World vultures and storks, as indicated by DNA hybridization techniques.

as a result of accumulated neutral mutations, the last common ancestor they shared lived in the very remote past.) When such a clock is calibrated against the branching sequences indicated in the fossil record, it may be possible to determine the actual time of each divergence.

Neutral mutations can be estimated by a variety of methods, two of which are described below. In general, it appears that they represent a reliable molecular clock; they have helped to answer some thorny questions about evolutionary relationships.

1. Molecular approaches to phylogenetic reconstruction rely on the premise that neutral mutations accumulate at a regular rate, and thus that distantly related species will differ at more sites than will close relatives.

2. The methods employed can be used to construct a branching sequence and this can be calibrated against the fossil record to give the ages of divergence points.

The fossil record, comparative morphology, and comparative biochemistry each provide information about phylogenetic relationships. Each has its advantages and limitations, and we would be most confident if the evidence from each approach gave the same result. All three approaches have been used for relatively few such comparisons, but in general, the results have been in agreement.

Immunological Comparisons. One method of estimating the differences in the DNA of vertebrate species is based on the antibody-mediated immune response (page 409). As we have seen, the immune system produces antibodies, a class of proteins that recognize and bind with specific foreign substances in the body and mark them for destruction. The behavior of antibodies can be put to use in reconstructing phylogenies. When purified protein obtained from, say, a species of frog is injected into a rabbit, the rabbit will produce antibodies to the foreign protein. The antibodies can be recovered and mixed in a test tube that contains (1) the same protein from a different species of frog and (2) carefully measured quantities of mammalian red blood cells and complement (page 405).

If the antibodies react with the foreign protein, they "turn on" complement, which causes the blood cells to lyse and release hemoglobin. How much hemoglobin is released depends on the amount of complement activated, which in turn depends on the amount of antibodies that recognize the protein as being foreign. (Recognition is less likely when the protein being compared differs from the one initially injected.) The amount of hemoglobin released is a measure of the overall similarity of the two proteins of the two species being compared. The test is repeated for the same protein from many pairs of species, and the resulting data are used to construct a phylogeny such as that shown in Figure 37.7.

Era	Period	Epoch	Age (millions of years)
Cenozoic	Quaternary	Recent	0.01
Cenozoic	Quaternary	Pleistocene	2.5
Cenozoic	Tertiary	Pliocene	7
Cenozoic	Tertiary	Miocene	25
Cenozoic	Tertiary	Oligocene	38
Cenozoic	Tertiary	Eocene	54
Cenozoic	Tertiary	Paleocene	65
Mesozoic	Cretaceous	Late	100
Mesozoic	Cretaceous	Early	135
Mesozoic	Jurassic		195
Mesozoic	Triassic		240
Paleozoic	Permian		285
Paleozoic	Carboniferous		375
Paleozoic	Devonian		420
Paleozoic	Silurian		450
Paleozoic	Ordovician		520
Paleozoic	Cambrian		570
Proterozoic	oxygen (O₂) abundant		2,000
Proterozoic			2,500
Archean	oldest fossils known		3,500
Archean	oldest dated rocks		3,800
Archean	approximate origin of the earth		4,600

Figure 37.9 Absolute geologic time scale, with dates based on radioactive isotopes from rocks of each interval.

DNA Hybridization Studies. Another approach to constructing phylogenies makes use of DNA hybridization techniques (page 235). Here an investigator isolates short segments of DNA from two species, converts them to single-stranded form, and allows them to recombine to form hybrid double helices. The extent of hybridization is directly related to the similarity of base-pair sequences in the DNA of the two species. The similarity can be determined by carefully heating the hybrid DNA until the two strands dissociate. The higher the temperature required for dissociation, the more complete was the match between the two sequences. Data gathered for many pairwise combinations of species can be used to construct a phylogeny such as that shown in Figure 37.8.

GEOLOGIC TIME

Until fairly recently, there was no way to determine the age of the earth or to develop an actual time scale for the events in geologic and evolutionary history. But well over a century ago, geologists and paleontologists established a *relative* time scale. They had observed fossils of different taxa in various rock layers (or strata), which typically lie on top of one another in an orderly sequence. In some cases the fossils and strata could be found in the same association over huge geographic areas, even on different continents.

Because sedimentary rocks form by the gradual deposition of erosion products and by the steady "rain" of skeletons of microorganisms in the sea, it was logical to assume that the now-deepest sedimentary layers were formed first, followed by those overlying them. Layers might be missing (due to erosion or nondeposition in some places), deformed, or even reversed in sequence by geologic upheavals, but the overall pattern through time was clear. By Darwin's time, this pattern was being used as the basis of a relative time scale, which included a series of major and minor subdivisions. These divisions remain in use today.

In what way are these divisions relevant to a discussion of macroevolution? The short answer is that they correspond in time to major evolutionary events. The geologic time scale is in fact a biological one, as evidenced by the names of the largest divisions, or *eras*. From oldest to youngest these are the Proterozoic ("first life"), Paleozoic ("ancient life"), Mesozoic ("middle life"), and the Cenozoic ("modern life"). An even earlier era, the Archean ("ancient times") has now been partitioned away from the Proterozoic, as shown in Figure 37.9.

These divisions of geologic time were initially recognized on the basis of abrupt transitions in the fossil

Figure 37.10 Radioactive dating. For many years, scientists tried to measure the ages of rocks by assuming that erosion, mountain-building, and other geologic processes occurred at a constant rate. Such attempts failed, because there is no constancy. Episodes of uplift and erosion occur irregularly and proceed at highly variable rates.

More recently, radioactivity has proved to be an accurate timekeeper. An atom of a radioactive element has an unstable combination of protons and neutrons in its nucleus, and it breaks down spontaneously. As it decays, radiation is released. Some combinations are inherently less stable than are others, and thus each radioactive isotope has its own characteristic rate of decay. This rate cannot be modified by temperature, pressure, chemical reactions, or any other environmental influence. *The rate of radioactive decay is constant, and differs among elements and their isotopes.*

Each radioactive element has a half-life, which is the time required for half of the atoms in a sample to decay into a lighter element. We can measure the proportions of the original element and its decay product to determine the age of a particular rock. For example, half of the atoms of a radioactive isotope of potassium will decay into the inert gas argon in 1.25 billion years. If a rock contains this isotope of potassium, all of the argon now present in the rock has been derived by radioactive decay. If the ratio of potassium to argon is 1:1, we know that the rock solidified 1.25 billion

Main Radioactive Elements Used in Dating			
Radioactive Isotope	Half-life	Stable Product	Useful Range (years)
Rubidium 87	49 billion years	Strontium 87	>100 million
Thorium 232	14 billion years	Lead 208	>200 million
Uranium 238	4.5 billion years	Lead 206	>100 million
Uranium 235	704 million years	Lead 207	>100 million
Potassium 40	1.25 billion years	Argon 40	>100,000
Carbon 14	5,730 years	Nitrogen 14	0—60,000

years ago. And if the ratio is 1:3, the rock solidified 2.5 billion years ago.

The radioactive dating method is powerful, but not perfect. For example, we must use an isotope whose decay product is otherwise absent in the rock, is chemically inert, and cannot leak away over time. Also, if a rock is melted again, the decay product may escape, resetting the clock to zero. Finally, there is a small percentage error in the measurement of the ratios, such that the more ancient is the date, the greater is its uncertainty. We can often test the dates given by two or more radioactive elements in the same rock sample to minimize the chances of large errors.

sequence, which we now know were the result of mass extinctions. The great Permian extinction brought the Paleozoic Era to a close; the Cretaceous extinction event ushered out many of the Mesozoic forms. Periods and epochs were similarly defined, at least initially, with reference to less dramatic turnovers. We will consider these events shortly.

The actual ages of the divisions of geologic time were established through radioactive dating methods (Figure 37.10), which demonstrate that the history of life began more than 3.5 billion years ago. We are accustomed to thinking of a year as a long time, and few of us will appreciate the duration of a century from personal experience. Yet a million years is ten thousand centuries end-to-end; and this is still only 1/3500, or 0.028 percent, of the history of life. Think about that for a minute.

MACROEVOLUTIONARY PATTERNS

How do lineages come to thrive, become extinct, or undergo sustained trends in their overall pattern of characteristics? Are *microevolutionary* events (changes in the allele frequencies within populations or species) enough to explain what is going on? The answer is probably yes.

However, the term *macroevolution* still can serve as a useful label for certain trends and processes for which a detailed characterization of each species is unnecessary. In this broader sense, the species become *units of evolution*, in the same way that trees are units of a forest. It is a matter of being able to look at the forest without being concerned with the individuality of every tree.

Evolutionary Trends

Evolutionary trends at the level of higher taxa are produced by variations in the *relative rates* of morphological change, speciation, and extinction. Differences in these rates can lead to highly divergent patterns as the history of a lineage develops, as the following examples will illustrate.

Adaptive Radiation

Certain lineages may dribble along, producing a species here, a species there, for tens of millions of years. Others succeed spectacularly, filling the environment with bursts of evolutionary activity in what are termed **adaptive radiations**. Figure 37.11 gives an example of a large adaptive radiation involving mammals.

If we know an adaptive radiation when we see one, can we then specify the conditions under which an adaptive radiation will occur? The answer is no; we can identify conditions under which a radiation *may* occur, but we cannot predict macroevolutionary events.

Nearly half a century ago, George Simpson developed the concept of adaptive zones to help explain the phenomenon of adaptive radiation. An **adaptive zone** is most simply characterized as a way of life, such as "burrowing in the seafloor," or "catching insects in the air at night." It is usually defined in retrospect, with reference to one or more lineages that have succeeded in occupying it, but there is no reason that an unfilled adaptive zone cannot exist as a concept.

According to Simpson, a lineage must have physical, ecological, and evolutionary access to an adaptive zone in order to become a successful occupant. Physical access is a straightforward criterion. For example, it might involve colonization of a new habitat or a new geographic area, such as by the arrival of the ancestor of Darwin's finches on the Galápagos Islands (page 29). Ecological access may be achieved if there are no species already occupying an adaptive zone or if the colonist is competitively superior to the resident species. Thus Darwin's finches reached an unoccupied adaptive zone as well as an unoccupied physical habitat.

The criterion of evolutionary access is less easy to explain. Often it is presented in terms of the acquisition of a **key innovation**, an adaptation that significantly improves some function or allows its possessors to exploit the environment in a novel way. For example, the evolution of wings opened new adaptive zones to the ancestors of insects, pterosaurs, birds, and bats, just as the evolution of limbs opened the way for the radiations of land-dwelling vertebrates.

Adaptive radiations are common features during the first few millions of years following major extinction events. For example, as shown in Figure 37.11, most of the living orders of mammals made their appearance in the first twelve million years of the Cenozoic Era. These included body forms as distinct as those of whales and bats. The mammalian lineage originated more than 140 million years earlier, but mammals were never diverse or numerous during the Mesozoic. It is often suggested that the extinction of the dinosaurs at the close of the Mesozoic opened up an array of ecological opportunities, which the mammals then exploited.

Origin of Higher Taxa

The fossil record repeatedly refuses to show us a gradual and steady development of organic diversity. Instead, organisms as diverse as bats and whales appear quickly and, for the most part, without a succession of intermediate forms. Virtually all phyla whose members have durable shells or skeletons appear "suddenly," in the span of about 30 million years in the early Cambrian. Flowering plants began their adaptive radiation and achieved much of their present diversity in a ten-million-year interval in the early Cretaceous. Darwin termed these events "inexplicable" and "an abominable mystery," and they continue to be cited by opponents of evolution today.

The fossil record is good enough for us to rule out the possibility of long, unrecorded histories for many of these groups. Biologists now face squarely the evidence that group after group originated and reached most of its morphological diversity in less than ten to twelve million years' time. How do we explain the origins of higher taxa and the major changes in form and function that they represent?

For example, the average duration of a mammalian species is comparatively short (about one million years). A lineage of mammals might have made twelve speciational "steps" in a particular direction between the end of the Cretaceous and the early Eocene. Is there a mechanism that could produce a bat, or a whale, from a generalized mammalian ancestor in only twelve steps? No one knows the answer, but mutations that affect developmental pathways are strongly suspected. Probably such mutations are nearly always lethal, but even if one in a million years were not, it would be enough to accomplish the sort of rapid changes in form evident in the fossil record.

Events of this sort are so rare, we cannot reasonably expect to see them happening except by enormous good luck. Suppose an individual able to survive and reproduce despite a major mutation in its developmental program does indeed appear once every million years per species, on the average. This means that somewhere in the world, four or five such mutants occur each year. Would we be likely to notice those mutants among all the individuals making up the estimated $4\frac{1}{2}$ million species, about two-thirds of which have not even been described yet? Probably not.

Still, it is possible to find examples of regulatory gene mutations that result in less pronounced changes in morphology. Some examples were described in the previous chapter. Also, artificial selection has produced the phenotypic diversity among domestic dogs in an "instant" of geologic time, chiefly by inbreeding the lineages in which regulatory gene mutations have occurred. The differences among breeds such as great danes, greyhounds, chihuahuas, and pekingese were produced in a few hundred years at most. Admittedly, all are still dogs—but would they be if the process continued for a million years?

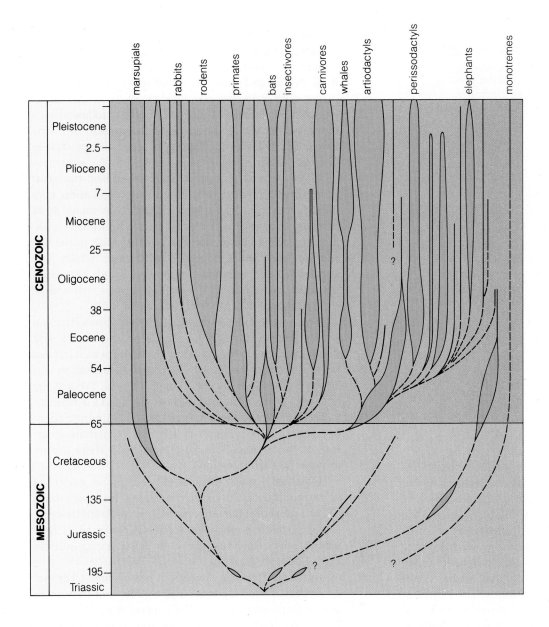

Figure 37.11 Geologic range and phylogenetic relationships of the principal living orders of mammals. In graphs of this sort, the widths of the lines signifying the different lineages indicate the numbers of species through geologic time. Notice the persistent low diversity throughout the Mesozoic, followed by an explosive adaptive radiation of placental mammals in the first 10–12 million years of the Cenozoic Era. Notice that neither the marsupials nor the monotremes (pouched and egg-laying mammals, respectively) participated in this radiation, which is thought to have resulted from the opportunistic invasion of ecological space vacated by the extinction of the dinosaurs in the late Cretaceous.

Extinction and Replacement

Most species that have ever lived are now extinct. Entire lineages that were once dominant have now disappeared or have dwindled in numbers as other radiations have flourished. Often, the extinction of one group has been followed closely by the adaptive radiation of another into the vacated ecological space. In general, it is not that members of the new radiation actively replaced those of the old through competition. Lineages usually declined to extinction millions of years before the radiations of the groups that replaced them began in earnest. Thus, large-scale turnovers are largely opportunistic events. *Radiations may follow extinctions but are rarely the cause of extinctions.*

Table 37.1	Examples of the Estimated Average Durations of Species	
Group		Duration (millions of years)
Protistans:		
	foraminiferans	20–30
	diatoms	25
Plants:		
	bryophytes	20+
	higher plants	8–20+
Animals:		
	bivalves	11–14
	gastropods	10–13.5
	graptolites	2–3
	ammonites	1–2, 6–15
	trilobites	1+
	beetles	2+
	freshwater fishes	3
	snakes	2+
	mammals	1–2+

From Stanley, 1985.

Background Extinction. If competitive interactions have not influenced extinctions at the level of higher taxa, how are such events explained? Careful study of the fossil record in the past decade has begun to provide some answers. Two separate processes, called background extinction and mass extinction, seem to have been at work. **Background extinction** is defined as the steady rate of species turnover that characterizes lineages through most of their histories. The causes of such extinctions seem to be many and varied ecological factors acting at the level of individual species. There is considerable variation in the rate of background extinction from group to group, as shown in Table 37.1. In general, larger and more complex organisms tend to have higher extinction rates. This means, for example, that the rate of species turnover is higher for trilobites, snakes, or mammals than it is for diatoms or clams.

David Raup and John Sepkoski studied background extinction rates for families of marine animals to provide a basis for comparison with mass extinction events. They found that the average extinction rate has declined from about *five* families per million years in the Cambrian to about *two* families per million years in the Tertiary. Standing far above these values are five events in which extinction rates rose to 10–20 families per million years. These global events will now be described.

Mass Extinction. An abrupt increase in extinction rates affecting several higher taxa simultaneously is called a **mass extinction**. Figure 37.12 shows the timing of five major mass extinctions and their effects on marine animals. Notice in Figure 37.12 that diversity rose rapidly in the early Paleozoic and leveled off, then was cut in half at the Paleozoic-Mesozoic boundary, followed by a recovery phase that continues to the present day. The five major mass extinctions appear as abrupt dips in diversity. Four of these each resulted in the extinction of 11–14 percent of marine animal families; the great Permian extinction carried off more than half of the families. According to Raup's calculations, as many as *96 percent of all marine species became extinct* at this time, making the Permian extinction the greatest crisis in the history of life. The Cretaceous extinction, which was substantial for terrestrial organisms, did not affect marine animals as extensively; it appears as only a small drop in Figure 37.12.

These mass extinctions and other, smaller extinction events have had profound effects on the history and diversity of life. Which taxa survived and which did not have repeatedly redefined the course of evolution and affected all aspects of the biological world. What if the ancestors of the dinosaurs had not survived the Permian extinction? Would the mammals have begun their adaptive radiation 100 million years earlier? What if the ancestors of the mammals had not survived? There are no answers to such questions.

Obviously, mass extinctions were global events of catastrophic proportions. In most cases the explanations of their causes remain only partially formulated theories. The explanations advanced for the Permian and Cretaceous mass extinctions are now being tested and are briefly described below.

In the past quarter century, it has become clear that continents move, separate, and collide, and that ocean basins form and close in the course of geologic time. During the latter half of the Paleozoic, widely separated continents began to converge to form a single "world continent" in the late Permian, and sea levels fell drastically. Together, the drainage of shallow continental shelves, loss of coastlines due to the clumping of continents, and substantial climatic changes resulting from altered oceanic current patterns may have caused the late Permian mass extinction. The shallow marine habitats so important to marine organisms virtually disappeared, resulting in the extinction of a large number of groups.

A different mechanism is suspected as the cause of the late Cretaceous mass extinction. In 1980, Luis Alvarez, Walter Alvarez, and other scientists proposed that a comet or an asteroid had collided with the earth. Because the resulting global dust cloud blocked the sun

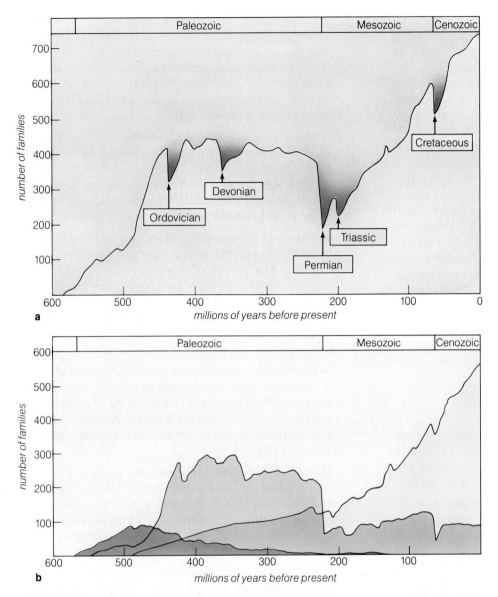

Figure 37.12 Graphs of the family-level diversity of marine animals over the past 600 million years, showing several macroevolutionary patterns. (**a**) All families combined, showing increases in diversity in the early Paleozoic and through Mesozoic and Cenozoic times, separated by a period of stability in the middle and late Paleozoic. Five mass extinction events appear as abrupt decreases in the number of families. The Permian mass extinction was the most severe.

(**b**) The same data as above, partitioned to show the histories of three major assemblages of marine animals. The Cambrian assemblage (pink shading) arose first and declined early, with very few representatives still living today. The Paleozoic assemblage (blue) was catastrophically affected by the Permian mass extinction and has never recovered. In contrast, the Modern assemblage (yellow) suffered little loss at the Permian event and has been dominant ever since.

for a period of months or even years, it drastically reduced temperatures and contributed to or caused widespread extinctions. The evidence for this hypothesis lies in a thin stratum of rock formed precisely at the boundary between the Cretaceous and Tertiary epochs. This stratum contains high concentrations of the element iridium, which is otherwise rare on the earth's surface. Iridium is more common in asteroids, and the thin, worldwide layer of iridium-rich rock may have formed from the dust that settled following the collision.

Such catastrophic, extraterrestrial "bombs" were common in earth's early history, as the ancient craters of the moon testify. Today, a few asteroids remain in earth-crossing orbits, and perhaps there was one more of them in the Cretaceous than there is now. Other theories suggest more exotic forces at work in and around the solar system, for which there is little evidence at present. There is even some evidence that mass extinctions may occur at regular intervals of 26–30 million years, for reasons yet unknown.

SUMMARY

1. The history of life encompasses the origins, persistence, radiations, and extinctions of different groups of species. "Macroevolution" refers to the large-scale patterns, trends, and rates of change among these groups.

2. The premise is that since evolution proceeds by modifications of existing organisms, there is an underlying continuity of relationship among all species, past and present. Evolutionary classification schemes are ways of depicting this relatedness through descent.

3. Classification schemes place species in a hierarchy of increasingly inclusive categories: genera, families, orders, classes, phyla, and kingdoms. Categories above the generic level are often called the higher taxa.

4. In evolutionary taxonomy, the prevailing approach to classification, the higher taxa are ranked on the basis of morphological and evolutionary relationships. Because some morphological traits are given more weight than others, the resulting "family trees" may not accurately reflect relatedness.

5. In cladistics, a newer approach to classification, relatedness is depicted only in terms of branch points in lines of descent. The branch points are inferred by combining information from the fossil record, comparative morphology, and comparative biochemistry.

6. Rates of morphological change in an evolving lineage are interpreted in two ways:

    a. According to the gradualistic model, most morphological change occurs gradually *within* a species.

    b. According to the punctuational model, most morphological change occurs *during the time of speciation*, with little change accruing thereafter.

    c. The fossil record provides indirect evidence against gradualism; in most cases, distinct species appeared abruptly (within only a thousand or a few million years), then changed little after that.

7. The quality and completeness of the fossil record varies as a function of the types of organisms (for example, shelled versus softbodied), the places where they died (natural traps favoring fossilization versus eroding soils), and the geologic stability of the region (tectonically active zones versus undisturbed sedimentary plains).

8. Most hypotheses about evolutionary relationships are based on comparative morphology (comparison of body form in major groups of organisms). For example, gill slits form in the embryos of vertebrates even though they are not present in all adult forms; this is one of many striking similarities used to group diverse organisms together as "vertebrates."

9. Morphological variation within a major lineage may be attributable to mutations in genes that control the rate of growth of organs and other structures. Such mutations would not have been retained if they affected early embryonic stages, which are fundamental steps in major developmental programs for the lineage (hence the conservation of gill slits).

10. In morphological divergence, the process of genetic drift or selection leads to departures from the ancestral form. Because evolution proceeds only through modification of existing structures, the results of selection for a particular role for a given structure may differ from the original role.

11. In convergence, two or more species from dissimilar and only distantly related lineages adopt a similar way of life and come to resemble one another closely. Convergence often results when the physical requirements for a particular life style (for example, the rapid swimming required of certain predatory marine vertebrates) are strict.

12. Molecular approaches to establishing relatedness rely on the premise that neutral mutations (which do not significantly alter gene function and hence have little measurable effect on phenotype) accumulate at a regular rate. Thus distantly related species will differ at more gene loci than will close relatives.

13. Molecular methods (including immunological comparisons and DNA hybridization studies) can be used to construct a branching sequence for a lineage. The sequence can be calibrated against the fossil record to give the ages of divergence points.

14. The geologic time scale is based on succession of major groups of organisms; its eras and most of its

periods are bounded by extinction events that affected numerous taxa simultaneously. Lesser divisions of geologic time are recognized on the basis of smaller "turnovers." The ages of geologic events have been established through radioactive dating.

15. Evolutionary trends result from variations in the relative rates of morphological change, speciation, and extinction of species within a lineage.

16. In one evolutionary trend, called adaptive radiation, certain lineages rapidly fill the environment with bursts of evolutionary activity. Those radiations occur when there are unfilled adaptive zones (ways of life, such as "catching insects in the air at night") to which a lineage has physical, ecological, and evolutionary access.

    a. Physical access may involve colonization of a new habitat or a new geographic area for the species.

    b. Ecological access can occur if an adaptive zone is not yet filled or if the invading species can outcompete the resident species.

    c. Evolutionary access may result when a key innovation appears in a species. A key innovation is an adaptation that significantly improves some function or allows the organism to exploit the environment in a novel way (the chloroplast is an example).

17. Most higher taxa, such as bats and whales, appeared rapidly in evolutionary time and usually without an observed succession of intermediate forms. Mutations that affected developmental pathways are strongly suspected as the basis for their appearance, as defined by major changes in form and function.

18. Extinction is one of the most pervasive trends in the history of life. Through most of their history, lineages show a steady rate of species turnover, probably as a result of ecological factors operating at the species level. This process is called background extinction.

19. Mass extinction refers to an abrupt increase in extinction rates above the background level. These are catastrophic, global events in which several higher taxa disappear simultaneously.

20. In the great Permian extinction, about ninety-six percent of all marine species disappeared. The late Cretaceous extinction may have been triggered by the collision of a comet or asteroid with the earth.

Review Questions

1. Why are species considered to be natural units, but higher taxa not?

2. How does classification affect our thinking about macroevolution?

3. Does morphological change usually accompany speciation or follow it?

4. What factors influence the completeness of the fossil record? On what basis do we subdivide geologic time, and how are actual dates estimated?

5. How do morphological divergence and convergence illustrate constraints inherent in evolutionary processes?

6. What is the basic assumption in employing comparative biochemical techniques for reconstructing phylogenies? How are DNA hybridization and immunological experiments performed to reveal phylogenies?

7. How is an adaptive radiation defined, and what criteria must be met before a radiation can occur? Do extinctions lead to radiations, or do radiations lead to extinctions? Why?

8. What is the biological problem posed by the rapid origin of higher taxa, and how are novel body plans thought to arise in short spans of geologic time?

9. Why is an understanding of background extinction rates necessary for the study of mass extinctions? How have the five major mass extinctions reset evolutionary history?

Readings

Ayala, F. J., and J. W. Valentine. 1979. *Evolving*. Menlo Park, California: Benjamin/Cummings. Short introduction to evolutionary theory.

Gould, S. J. 1982. "Free to Be Extinct." *Natural History* 91:12–16.

Lewin, R. 12 July 1985. "Pattern and Process in Life History." *Science* 229:151–153.

Minkoff, E. 1983. *Evolutionary Biology*. Reading, Massachusetts: Addison-Wesley. Comprehensive introduction to topics covered in this chapter, accessible reading level.

NASA. 1985. *The Evolution of Complex and Higher Organisms*. Washington, D.C. National Aeronautics and Space Administration Publication SP-478. Outstanding, readable review of current topics in macroevolution.

Raup, D., and J. Sepkoski, Jr. 19 March 1982. "Mass Extinctions in the Marine Record." *Science* 215:1501–1502. Landmark report on background and mass extinctions over geologic time.

Stanley, S. M. 1979. *Macroevolution, Pattern and Process*. San Francisco. Freeman. Comprehensive introduction to macroevolutionary theory; well written.

38

ORIGINS AND THE EVOLUTION OF LIFE

By the late 1860s, "the fixity of species" was crumbling as a scientific concept, for the accumulated evidence that species evolve, or change through time, had become overwhelming. By the middle of the twentieth century, "the fixity of continents" met a similar fate. Geologists already suspected that the continents have been buckling upward and eroding downward through time. But now it became clear that continents have formed, split apart, wandered about, and collided with one another, opening and closing ocean basins in the process!

This disconcerting discovery was one more piece of evidence that the earth itself has been evolving on a massive scale. The geologic record is one of stability and instability, involving mountain-building, glaciations, and widespread shifts in climate and in sea level. Although changes in the fossil record were known to be associated with episodes of geologic change, the meaning of the association is only now becoming understood.

Today it is clear that *the evolution of life has been linked, from its origin to the present, to the physical and chemical evolution of the earth.* As yet, there is no comprehensive theory drawing together all of the separate lines of evidence, and interpretations continue to be modified and refined as new evidence comes in. We encourage you at the outset to explore and evaluate this evidence for yourself by continuing with the readings listed at the chap-

Figure 38.1 Representation of the primordial earth, about 4 billion years ago. Within another 500 million years, living cells would be present on the surface. (During its formation, the moon presumably was closer to the earth. Here it looms on the horizon.)

ter's end, for all we can do in these few pages is highlight the discoveries and sketch out some prevailing theories, beginning with our current understanding of how the earth was formed.

ORIGIN OF LIFE

The Early Earth and Its Atmosphere

Billions of years ago stellar explosions ripped through our galaxy, leaving behind a dense cloud of dust and gas that extended trillions of kilometers in space. In time the immense cloud cooled, and countless bits of matter gravitated toward one another. By about 4.6 billion years ago, the cloud had contracted back on itself and was now a flattened disk rotating slowly through space. Temperatures at the dense center of the disk became extremely high, driving the fusion between the nuclei of colliding atoms of hydrogen and, to a lesser extent, helium. Thus thermonuclear chain reactions began that would feed themselves for the next 10 billion years; the sun was born a luminous star.

Farther out from the center of the disk, the earth and other planets were also forming under gravitational forces. At first the earth may have been a cold, homogeneous mass. However, contraction and radioactive

heating would have made its core increasingly dense—and hot. By 3.8 billion years ago, the earth was hurtling through space as a thin-crusted inferno (Figure 38.1).

Long before life appeared in that forbidding place, gases trapped beneath the thin crust or formed during reactions in the earth's molten interior were being forced to the outside. Those emissions were the start of an early atmosphere. Although more than twenty percent of the air we breathe today is oxygen, there probably was very little of this gas on the early earth. (More likely, the early atmosphere resembled those of Venus and Mars. Planetary probes have shown that oxygen accounts for less than one percent of the atmospheres of our two sister planets.) The near-absence of free oxygen was probably a prerequisite for the origin of life. Many of the molecules on which life is based (including amino acids and nucleotides) cannot assemble spontaneously in the presence of free oxygen, which attaches to and chemically alters them.

Another prerequisite for the origin of life was the presence of liquid water; all of the activities associated with life cannot proceed for long without it. Certainly a tremendous amount of water vapor would have been released from the breakdown of rocks during volcanic eruptions. But as fast as water vapor condensed in the early atmosphere, it would have evaporated in the intense heat blanketing the rumbling crust. Most likely liquid water accumulated only after the surface cooled enough for it to rain. When the rain began, so began the stripping of mineral salts from the parched rocks, and salt-laden waters collected where basins had formed in the crust.

If the early earth had settled into an orbit closer to the sun, its surface would have remained so hot that water vapor never would have condensed in liquid form. If its orbit had been more distant, the surface would have become so cold that any water formed would have been locked up as ice. And if the earth had not become as large as it did, it would not have had enough gravitational mass to retain the atmosphere with its water vapor. *Because of its size and its distance from the sun, the earth could retain liquid water on its surface.* Without liquid water, life as we know it never would have originated.

Yet exactly where and when did life originate? There is no fossil record of the event. Most of the rocks from that period have been melted, solidified, and remelted many times over because of ongoing movements in the earth. Some are buried far below more recently formed rocks, where they have been subjected to heat and compression. Thus any clues they might have held about the event would be altered beyond recognition.

Even though there is no direct record of the origin of life on earth, it is still possible to gain insight into the manner in which it occurred. To see how this can be done, we can start with the following questions:

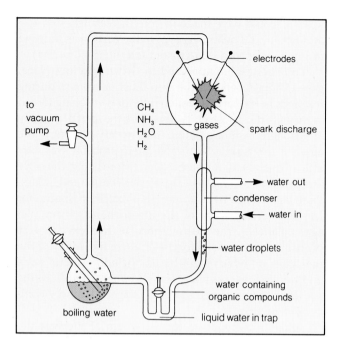

Figure 38.2 Stanley Miller's apparatus used in studying the synthesis of organic compounds under conditions believed to have been present on the early earth. (The condenser cools the circulating steam and causes water to condense into droplets.)

1. What were physical and chemical conditions like at the time of origin?

2. Based on known physical, chemical, and evolutionary principles, could life have originated spontaneously under those conditions?

3. Can we postulate a sequence of events by which the first living systems developed?

4. Can we devise experiments to test whether that sequence could indeed have taken place?

Spontaneous Assembly of Organic Compounds

To begin with, rock samples from meteorites, the earth's moon, and Mars are known to be between 4.5 and 4.6 billion years old. Because meteorites, the moon, and Mars as well as the earth all condensed from the same primordial dust cloud, we can safely assume the rock samples are representative of the earth's early chemical composition. *They all contain the chemical elements found in biological molecules.*

What about energy sources to drive the synthesis reactions? Lightning, hot volcanic ash, even shock waves have enough energy to drive the assembly of carbon-containing compounds. We know this from the pioneering experiments of Stanley Miller and Harold Urey, as

well as from the later work of many other chemists. In 1953, for example, Miller set up a reaction chamber containing a mixture of hydrogen, methane, ammonia, and water (Figure 38.2). For a week he kept the mixture recirculating. All the while, he bombarded it with a continuous spark discharge to simulate lightning as an energy source. By the week's end, he found organic molecules had formed, including many kinds of amino acids.

Such experiments have been repeated many times, with variations in elements, gas mixtures, and energy sources. The results show that all the building blocks in living systems—including lipids, sugars, amino acids, and nucleotides—can form under abiotic conditions. ("Abiotic" means not involving or produced by organisms.) In addition, when inorganic phosphate is present in the starting mixture, ATP will form. That molecule is involved in energy transfers in all living systems.

Speculations on the First Self-Replicating Systems

Given the 300 million years available for it to happen, large quantities of organic material probably accumulated in the shallow waters of the earth. What sequence of events led from this organic "soup" to self-replicating systems? In living cells, these systems include DNA (which contains instructions for building proteins) and RNA (which translates the DNA instructions into actual proteins). The proteins themselves are essential for self-replication. Many are important structural materials, and others are *enzymes*, which greatly increase the rate of chemical reactions. All the reactions needed to maintain and reproduce even the simplest cell could not proceed fast enough without enzymes. So we can rephrase our earlier question: How did the "DNA→RNA→protein" system originate? We do not yet have the answer to this question, although researchers are giving us some interesting things to think about.

Templates for Protein Synthesis. To give you a sense of the kind of work going on, let's look briefly at an idea that the first templates or "patterns" for protein synthesis existed at the bottom of the primordial soup, in the mud of tidal flats and estuaries. G. Cairns-Smith has analyzed microscopic crystals of different clays in which metal ions (such as those of iron and zinc) are embedded. Such crystals can grow by attracting molecules to themselves. When they do, their latticelike organization is repeated again and again, right down to the small imperfections caused by the random inclusions of different metal ions. Now, if you have ever baked meat in a clay or metal pot, you know that bits of protein stick tenaciously to both kinds of heated surfaces. According to Cairns-Smith and J. Bernal, naturally occurring clay crystals might have served as the original templates for assembling free amino acids into protein chains. Without such a template, amino acids might still combine through random collisions in the water, but the template would allow longer chains to form in a shorter period of time and help to prevent them from being broken apart.

Suppose some amino acids brought together on a clay template formed proteins that had weak enzyme activity. Perhaps some of those primordial enzymes hastened the linkages between amino acids. In such ways, some clay templates would have selective advantages over others. *There would have been chemical competition between different types of clay templates for the amino acids available—and selection for the first molecules characteristic of living cells.*

Suppose nucleotides were also attracted by the clay surface or by amino acids stuck to it. Intriguingly, most existing enzymes are assisted by small molecules called coenzymes—*some of which resemble or are identical in structure to RNA nucleotides.*

At this point we run up against an unsolved puzzle. We can show that enzyme-coenzyme systems involving RNA might have developed, and that the more successful combinations would have increased in abundance. We can even hypothesize that in some cases the RNA molecule itself might have replaced the clay crystals as a template for protein assembly. In most living systems, however, RNA plays a secondary role to DNA in mediating protein synthesis. We do not understand how the simple coenzyme relationship between RNA and proteins could have been transformed into a relationship involving DNA as the primary template, coding for RNA as the secondary template, in turn leading to protein synthesis. The problem is that DNA does not work without a suite of enzymes to transcribe it into RNA. It is not at all clear how such enzymes evolved, since both components must be present in order for the system to function. This is the main unsolved problem in devising a scenario for the origin of self-replicating systems, as depicted in Figure 38.3.

Whatever the exact sequence might have been, we should not think of it as a series of purely random chemical reactions. Rather, *there are good reasons to believe the physical and chemical conditions of the early earth made some reactions much more probable than others.* For example, amino acids can exist in two forms that are like mirror images of two hands. Thus one form is called lefthanded, the other righthanded. In almost all living things, amino acids are in the lefthanded form. Was this chemical choice a result of chance? Some recent work by James Lawless suggests it was not. Lawless began by studying meteorites. Samples from meteorites contain disorganized arrays of both lefthanded and righthanded amino acids. Yet Lawless discovered that when such disorganized mixtures are exposed to clay crystals, the clay attracts *only* the lefthanded forms.

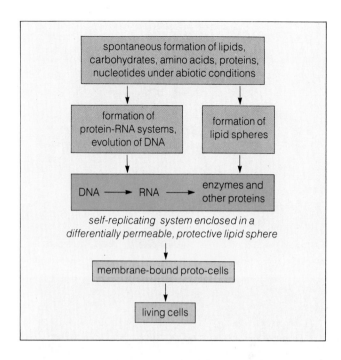

Figure 38.3 One hypothetical sequence of events that might have led to the first self-replicating systems and, later, to the first living cells.

Models for the First Plasma Membranes. All living cells have a surface membrane, or plasma membrane, that is differentially permeable—that is, it lets some substances pass freely into and out of the cell but restricts or controls the passage of other substances. These membranes are composed of lipids and proteins of various sorts. We do not know when membranes emerged in the sequence leading from chemical to organic evolution. What we do know is that metabolism means chemical control—and chemical control is possible only with chemical isolation from the ebb and flow of materials in the environment. If we assert that chemical evolution led to a metabolic system, then we must show that molecular boundaries for a protein-nucleic acid system can arise spontaneously. Here again, experiments suggest some chemical probabilities.

For example, Sidney Fox and his coworkers heated amino acids under dry conditions, which produced a number of long protein chains. The chains were placed in hot water and then allowed to cool. Upon cooling, the proteins assembled into small, stable spheres, or **microspheres** (Figure 38.4a). The microspheres selectively accumulated certain substances in greater concentrations inside themselves than were found on the outside. Moreover, the microspheres tended to pick up lipids from the water—and the outcome was the formation of a lipid-protein film around each sphere.

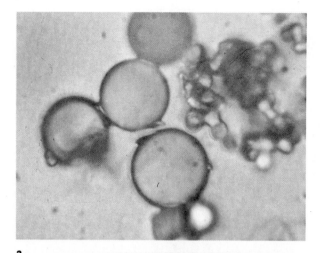

a

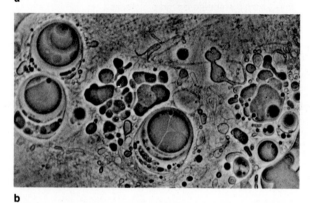

b

Figure 38.4 (**a**) Microspheres of protein chains, as they appear in water solutions. 11,000×. (**b**) Liposomes with multiple internal compartments made from simple, straight-chain lipids.

William Hargreaves and David Deamer have proposed that lipids must have been important in the origin and evolution of biological membranes. Although existing membrane lipids are structurally complex, these workers have experimentally produced membranes from simple lipids having only single hydrocarbon tails and water-soluble heads (page 85). These lipids formed **liposomes** (microscopic, closed sacs filled with water) with the following properties:

1. Self-assembly into spheres

2. Formation of aqueous compartments within the spheres

3. Ion permeability, water permeability

4. Fluidity and elasticity (self-repair)

All of these properties are characteristic of lipids that contribute to cell membrane function.

These and other experiments suggest that membranes as well as the self-replicating systems they enclose could have arisen through spontaneous but inevitable chemical events. If lipids were present in the primordial soup, they may well have formed vesicles, for that is their thermodynamically favored state. Early self-replicating systems could have produced more complex lipids and thereby would have increased membrane stability. Whatever the sequence of chemical events, the first living cells that did appear on earth were probably little more than membrane-bound sacs containing the nucleic acids that served as templates for protein synthesis.

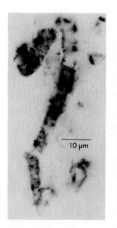

a

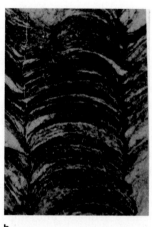

b

c

Figure 38.5 Some of the oldest known fossils. (**a**) From Western Australia, a filamentous form with cross-walls between cells; it is thought to be 3.5 billion years old. (**b**) Section through limestone stromatolite from Rhodesia, about 3.1 billion years old. The stacked organic (dark) and inorganic (light) layering is almost identical with that laid down by modern communities of photosynthetic microbes. (**c**) These stromatolites from Western Australia formed between 1,000 and 2,000 years ago in shallow seawater. Calcium carbonate deposits preserved their structure.

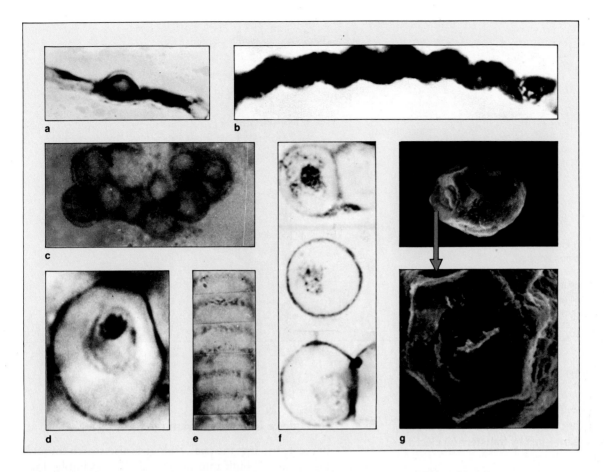

Figure 38.6 (**a**) Fossil cell from South Africa, about 2.25 billion years old; it is similar to modern cyanobacteria. (**b**) From Central Australia, a spiral form 900 million years old. (**c**) Colonial cells found in stromatolites in the USSR, about 650 million years old. (**d**) Filamentous form, about 650 million years old. (**e**) Presumed eukaryotic cell from Central Australia, about 900 million years old; the granules and spots may be remnants of organelles. (**f**) Fossil eukaryotes containing cytoplasmic remnants, about 900 million years old. (**g**) Notice the well-developed mouth of these eukaryotes, 750 million years old.

THE AGE OF PROKARYOTES

Until about 3.7 billion years ago, the earth's crust was highly unstable. Yet rocks that survive from this period contain fossilized cells, which were probably deposited in tidal mud flats. The rocks are dated at about 3.5 billion years, and the cells they contain are already well developed in structure (Figures 38.5 and 38.6).

Prokaryotic Metabolism and a Changing Atmosphere

The first known cells were prokaryotes (cells that do not have a nucleus or other membranous organelles). In their morphology they were like the simple anaerobic bacteria that now live in mud flats, bogs, and pond bottoms. (Anaerobic cells, recall, either cannot use oxygen during metabolism or they die in its presence.) Probably those

early cells were heterotrophs, obtaining energy by consuming organic molecules that had accumulated spontaneously in the environment. The nucleotides such as ATP would have been the most accessible energy source. Most likely, metabolic systems became more and more efficient at using the existing ATP, then later developed the means of producing ATP themselves.

Today, the most common anaerobic pathway for producing ATP is fermentation, and it probably was the first to evolve. (In fermentation, recall, hydrogen atoms and electrons are juggled among intermediates of the reactions but are not transferred anywhere else.)

Possibly in one group, electrons started to be transferred away from fermentation products to inorganic electron acceptors in the environment. Such transfers would generate ATP by anaerobic electron transport (page 123). Today, *Desulfovibrio* and related

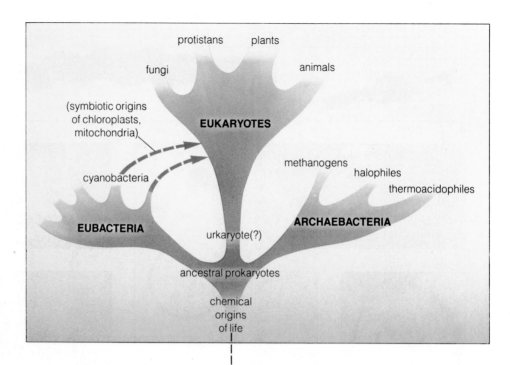

Figure 38.7 Diagram of the proposed relationships among prokaryotic and eukaryotic organisms. The eukaryotes are thought to have gained some of their organelles by entering into a symbiotic relationship with certain prokaryotic eubacteria.

bacteria use this type of pathway. *Desulfovibrio* produces ATP by fermenting sugars—*and* by transferring electrons to sulfate in the surroundings.

Electron transport chains used in such transfers may have been a key character in the evolution of autotrophs ("self-feeders")—including the first photosynthetic cells. In such cells, the transfers leading to ATP formation are driven by a virtually unlimited energy source—sunlight energy (page 109).

When populations of photosynthetic cells expanded —and the fossil record attests to their success—they had to be releasing oxygen as a by-product of their activities. Over time, the accumulation of oxygen in the atmosphere had two irreversible consequences for the origin and evolution of life. First, an oxygen-rich atmosphere meant that life no longer could arise spontaneously; oxygen now prevented further nonbiological synthesis of organic compounds. Second, the oxygen-rich atmosphere was a challenge of global dimensions. Some anaerobic prokaryotes continued to thrive in oxygen-free habitats, and many other lines became extinct. But some cells were able to adapt metabolically in ways that rendered the oxygen harmless. In one such adaptation, oxygen was used as a dump for electrons from transport chains. Hydrogen joined with the oxygen to form water, the toxic element was removed—and aerobic respiration emerged.

Aerobic respiration proved to be highly efficient at generating ATP (page 123). It was a key character in the evolution of larger and structurally complex cells; it was pivotal in the evolution of eukaryotes.

Divergence Into Three Primordial Lineages

In reconstructing the past, biologists also search through the "records" built into living cells. For example, they compare the similarities and differences in the nucleotide sequences of DNA and RNA from diverse cell types. In so doing, they have discovered differences suggesting that the forerunners of eukaryotes—cells that have a nucleus—arose simultaneously with the forerunners of existing prokaryotes. Today, research is pointing to the existence of three primordial lineages that arose from a common prokaryotic ancestor:

eubacteria	*true bacteria, including photosynthesizers such as cyanobacteria*
archaebacteria	*strictly anaerobic, including the methane-producing bacteria*
urkaryotes	*long-since-vanished ancestors postulated for eukaryotes*

The proposed relationship among these lineages is shown in Figure 38.7.

Archaebacteria still exist in many settings, and eubacteria are among the most successful groups of organisms (Chapter Thirty-Nine)—but there is no direct evidence of the urkaryotes. Their supposed descendants—the eukaryotic cells—do not appear in the fossil record until more than 2 billion years *after* the oldest known prokaryotes.

THE RISE OF EUKARYOTES

As Figure 38.7 shows, the protistans, plants, fungi, and animals are all eukaryotes. In structural terms, the simplest ones are the soft-bodied protistans known as amoebas. Because these forms do not have hard parts, they usually disappear quickly when they die. If the ancestors of eukaryotes were no more complex than the simplest of those forms, it is not likely that fossil evidence of them will ever be found. Yet there must have been an interesting history of evolution leading to their descendants.

The hallmark of all eukaryotes is the presence of organelles—especially the nucleus—these being membrane-bound compartments for specialized metabolic activities within the cell body. Given the integrated, coordinated functioning between the nuclear and cytoplasmic machinery, the nucleus probably evolved from infoldings of the plasma membrane around the DNA.

We have an example of such infoldings in some existing bacteria. In most bacteria, replicated DNA simply becomes attached to the plasma membrane, which grows and thereby separates the two molecules during cell division (page 137). In some bacteria, however, the plasma membrane folds elaborately around the replicated DNA in structures called **mesosomes** (Figure 38.8). Mesosomes retain their attachment to the plasma membrane. But perhaps structures similar to mesosomes separated permanently from the outer membrane and became the forerunners of the nuclear envelope.

But how did the eukaryotes acquire such organelles as mitochondria and chloroplasts? Lynn Margulis was the first to propose a symbiotic origin for these organelles. The word **symbiosis**, which means "living together," refers to mutually beneficial interactions in which one species serves as host to another species (the guest).

Origin of Mitochondria

Suppose urkaryotes included predatory, amoebalike forms that were only weakly tolerant of free oxygen. Suppose some of the aerobic bacteria they ingested happened to resist digestion. Oxygen released by their respiring prey could have poisoned most of the predators, yet some of the predators might have been more tolerant of the gas. If that were the case, the two species could benefit from continued association. The guest species would find itself in a protected environment, endowed with energy-rich remnants of the anaerobic host's fermentation activities. The guest cells would divide independently and come to be distributed in the cytoplasm of the host daughter cells. Over time, the guest species might lose its genetic independence, for the host could perform many functions for it. And the host species would develop increased metabolic efficiency as a result of the guest's capacity for aerobic respiration. The extra

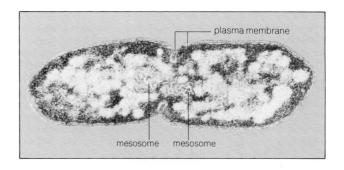

Figure 38.8 Mesosomes in a bacterium that lives in the gut of a rat. These two mesosomes, each an infolding of the plasma membrane that surrounds the DNA, have formed prior to cell division.

energy provided to the host could be channeled into growth in size, increased activity, and the construction of more structures, such as hard body parts. The vestiges of symbiotic bacteria within the increasingly complex host cells would become mitochondria, which are now the basis of aerobic metabolism in eukaryotes.

The scenario is not as farfetched as it might seem. Such partnerships can be observed today between cells of entirely different species, as Figure 38.9 suggests. More to the point, aerobic bacteria that serve as models for such a scenario are uncannily like mitochondria in their size, structure, and biochemistry. Moreover, a mitochondrion has its own set of DNA, separate from the cell in which it is located, as well as a somewhat independent ability to replicate it!

Origin of Chloroplasts

After acquiring a capacity for aerobic metabolism, some urkaryotes may have ingested photosynthetic bacteria, which became a new kind of symbiont. Such guests could have been the ancestors of chloroplasts, foreshadowing the evolution of photosynthetic protistans and plants.

In their metabolism, chloroplasts are like eubacteria. In their DNA they are also like eubacteria. Like the mitochondrion, a chloroplast has its own complement of DNA, and can replicate it in partial independence of the nuclear division of the cell.

If symbiosis was the route by which some urkaryotes acquired the capacity for photosynthesis, the route was followed by a number of ancestral lines. Today, the simplest photosynthetic eukaryotes resemble one another only in the fact that they are photosynthetic. They resemble different species of bacteria that have different body forms and different life-styles. Their chloroplasts are not the same. Their accessory pigments also vary, so they also differ in color.

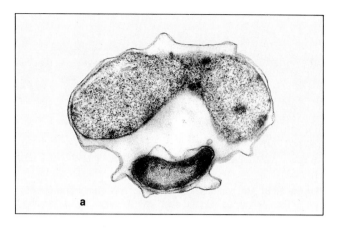

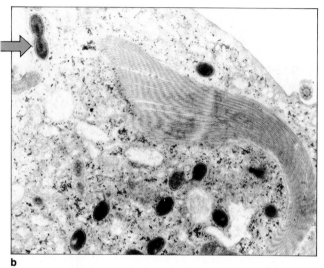

Figure 38.9 Symbiosis in existing prokaryotic and eukaryotic cells. (**a**) A predatory bacterium (small, dark oval) called *Bdellovibrio*, which has ended up living in the space between the cell wall and plasma membrane of a larger bacterium. (**b**) Symbiotic bacteria (small, dark ovals) living in *Pyrsonympha*, a protistan that dwells in the termite hindgut. The bacteria are the same size as mitochondria, they are distributed in the cytoplasm as mitochondria are, and they probably use the breakdown products of sugar molecules as mitochondria do. The arrow points to a bacterium that is dividing inside the cytoplasm (as mitochondria do).

Beginnings of Multicellularity

The evolution of multicelled organisms surely is one of the most important events in the history of life, and it is closely linked to the genetic specializations seen in eukaryotic cells. Unlike prokaryotes, eukaryotes contain organelles such as chloroplasts and mitochondria, and they have genes that regulate organelle activity. Such regulatory genes working *within* cells apparently became adapted to work *between* cells; they began to control the coordinated activities of cells clustered into tissues and, later, organs. In the same way that mitochondria or chlo-

roplasts became dependent on their "host" cell, so did the increasingly specialized cell types of multicelled organisms become dependent on one another to function as a unit.

Multicellularity has many advantages, including larger size and the greater metabolic efficiency that results from the development of specialized organs. Each organ can perform a single function very well, while depending on other tissues or organs to provide for the remainder of its metabolic needs. Thus multicelled organisms benefit from a division of labor, something that is not possible for single cells to do.

While it is easy to picture how multicellularity evolved and how it is advantageous, the fossil record shows that this step took a long time. The first eukaryotes appeared over 1.3 billion years ago. But it is not until about 700 million years ago that we see evidence of multicelled animals (metazoans)—the tracks and burrows on the ocean floor made by soft-bodied animals. Such fossils became widespread by 600 million years ago and were followed quickly by the origin of hard shells and skeletons (Figures 38.10 and 38.11 give examples). This event, 570 million years before the present, marks the beginning of the Paleozoic Era. From that point on, the history of life consists largely of the evolutionary and ecological elaboration of multicelled organisms.

FURTHER EVOLUTION ON A SHIFTING GEOLOGIC STAGE

The events described so far took place in the Archean and Proterozoic eras (Figure 37.8). We have seen how the physical and chemical evolution of the earth influenced the origin of life in that distant time, and how the evolution of photosynthetic pathways led to an oxygen-rich atmosphere. Here we turn to major geologic events during the three subsequent eras—the Paleozoic, Mesozoic, and Cenozoic—and to their effects on the evolution of life.

The continental masses are known to have changed in their position and orientation over the past 540 million years. The *plate tectonic theory*, which is outlined in Figure 38.12, accounts for these changes.

During Paleozoic times, an early continent called **Gondwana** drifted southward from the tropics, across the south polar region, then northward. Other drifting land masses collided to form a tropical continent that is now called **Laurasia**. Then, near the end of the Paleozoic, Gondwana and Laurasia became massed together to form a single world continent, called **Pangea**, that extended from pole to pole. An immense world ocean covered the rest of the earth's surface; to the east of Pangea, it curved around to form the equivalent of a

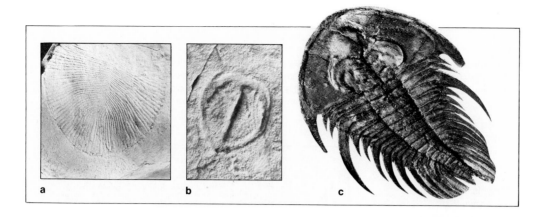

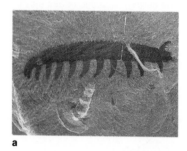

Figure 38.10 (**a, b**) Metazoan fossils, 600 million years old, from South Australia. (**c**) Fossil trilobite from about 570 million years ago.

giant tropical bay, called the **Tethys Sea**. Pangea began to break up in the Mesozoic, and the drifting and collisions of the fragmented land masses continue today.

All such changes in land masses, shorelines, and oceans had a profound effect on the evolution of life. When land masses were widely spread apart, speciation and adaptive radiations proceeded separately, giving rise to diverse arrays of organisms on land and along the shores. When land masses collided, habitats were lost and the overall diversity tended to decline. Thus the mass extinction 240 million years ago at the Paleozoic-Mesozoic boundary reduced the number of species of marine animals by ninety-six percent (page 544). Finally, with each major shift in land masses, the climatic patterns and the direction of warm or cold ocean currents changed, and once again there were evolutionary repercussions for organisms on land and in the seas. The remainder of this chapter can only hint at the outcomes of these and other interactions between organisms and their changing environment.

Life During the Paleozoic

The Paleozoic Era consisted of six periods: the Cambrian, Ordovician, Silurian, Devonian, Carboniferous, and Permian.

The Cambrian. Most organisms of the Cambrian apparently lived on or just beneath the surface of the seafloor, feeding on organic debris or particles suspended in the water. Among the relatively limited array of marine species were the trilobites—crawling and burrowing animals of the sort shown in Figure 38.10c. There was plenty of sand and mud for the trilobites and their neighbors, which included brachiopods (lampshells), mollusks, and echinoderms. Shallow seas covered the low margins of Gondwana and Laurentia and probably represented a rather benign environment; both continents were positioned at tropical latitudes during the Cambrian.

a

b

c

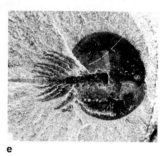

d

e

Figure 38.11 Middle Cambrian soft-bodied animals (**a–c**) and arthropods (**d–e**) from the Burgess Shale in British Columbia. The soft-bodied animals are remarkably well preserved. They died in relatively still, poorly oxygenated water, and were buried gently in a layer of fine silt.

Figure 38.12 Plate tectonic theory. (**a**) The earth's surface is broken up into rigid plates. Today these plates are drifting together or apart, or shifting past one another, in the directions of the arrows. Boundaries between plates are marked by recurring earthquakes and volcanic activity.

(**b**) Seafloor spreading and continental drift. Plate tectonics is based on observations that the seafloor is slowly spreading away from sites called oceanic ridges, and on measured displacements of the continents relative to those ridges. Thermal convection in the mantle is proposed as the mechanism underlying these movements. More heat is being generated deep beneath oceanic ridges than elsewhere. The hotter material slowly wells up and spreads out laterally beneath the crust (much as hot air rises from a stove and spreads out at the ceiling). Oceanic ridges are places where the material has ruptured the crust. As the cooler material moves away from the ridges, it acts like a conveyor belt, carrying older oceanic crust along with it.

Thus plates grow and spread away at the oceanic ridges. As the plates push against a continental margin, they are thrust beneath it. The thrusting causes the crumpling and upheavals that have formed most major mountain ranges (see, for example, Figure 38.13)

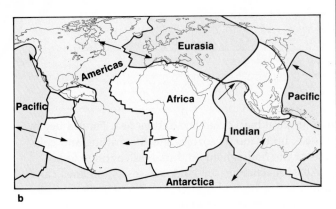

b

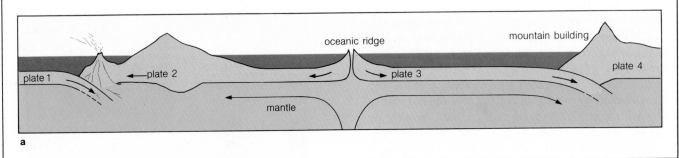

a

Figure 38.13 The view northward across Crater Lake in Oregon, a collapsed volcanic cone aligned with Cascade volcanoes reaching into Washington—and all paralleling the Pacific Coast. These formations are testimony to the violent upheavals that began in Jurassic times.

The Ordovician. Over the next 60 million years, Gondwana drifted southward until it eventually straddled the South Pole. The seas made further transgressions of the land; by the late Ordovician, the inundations were the most widespread ever recorded. As vast marine environments opened up, there were extensive adaptive radiations. During the mid-Ordovician, the mollusks called cephalopods underwent major diversification. The eyes and brain regions of those animals became highly developed, which reflects a trend toward the more active life-style typical of predators. Today their descendants include the squids, octopuses, and nautiluses.

What about the vertebrates? Even before this time, the heavily armored, jawless ancestors of modern vertebrates must have been evolving. From the few fossilized armored plates they left behind, we suspect that they resembled the jawless fishes of later times (page 635).

The Silurian. During the Silurian, Gondwana drifted northward and lesser land masses moved about in the mid-latitudes. This was a time of major evolution for the armor-plated fishes, for by now the vertebrate jaw had developed. That feeding structure, so useful in biting and chewing food, foreshadowed an explosive proliferation of predatory forms.

Even while animal groups were diversifying in impressive ways, a relatively inconspicuous event was taking place that foreshadowed a major evolutionary trend. In the late Silurian, small stalked plants no taller than your little finger were evolving in wet mud at the margin of the land (Figure 38.14).

The Devonian. As the Silurian gave way to the Devonian, several land masses converged and formed Laurasia (Figure 38.15). Overall, the surface area of dry land increased dramatically, and plants adapted to surviving near the land's edge now began their tentative forays onto the land itself. Some plants had leafless, aerial stems extending from horizontal, rootlike structures. Those plants, the Rhyniophyta, had tubelike water-conducting tissue (Figure 38.14b). Lycopods (simple plants with true roots, stems, and leaves) also appeared in tropical land environments during the Devonian, as did fernlike plants, woody trees, and possibly the earliest gymnosperms (a major group of seed-bearing plants).

In the early Devonian, the major groups of fishes came to dominate the seas and freshwater habitats, as their descendants have done ever since. Toward the end of the Devonian, one of the lineages of lobe-finned fishes succeeded in moving onto land (Figure 38.16). The fins of these animals, already thick and muscular, were perhaps adapted for moving slowly through shallow water

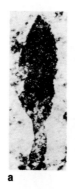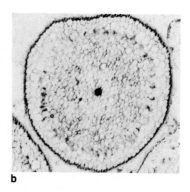
a   b

Figure 38.14 (**a**) Spore capsule of a stalked plant similar to the modern bryophyte *Pellia*. This fossil, from South Wales, dates from the Late Silurian. 12×. (**b**) Cross-section of the stem of *Rhynia*, plants from the Early Devonian. The stem extended from a delicate rhizome, which anchored the plant and absorbed nutrients from wet soil. At the center was a primitive water-conducting tube.

and thick aquatic vegetation; their ancestors had long ago begun to rely on air-breathing, as do many fishes living in stagnant waters today. Although the colonization of wet shorelines and swamps required no major structural or physiological changes, it was a major event foreshadowing the evolution of amphibians and fully terrestrial vertebrates. Forty million years after the pioneering forays of plants, the vertebrate invasion of land was under way.

The Carboniferous. From the beginning to the end of the Carboniferous, land masses were submerged and drained no less than fifty times. Besides these major inundations, there were ever-changing sea levels in between. Immense swamp forests with large, scaly-barked trees became established; over time the seas moved in and buried them in sediments and debris. The organic mess left behind has since compacted into the extensive coal deposits of Britain and of the Appalachians of North America.

Some groups survived—indeed they flourished, for adaptations had appeared among them that allowed movement onto higher (and drier) land. The gymnosperms were one such group. Those plants were not restricted to the water's edge; they could complete their reproductive cycles without free-standing water. The reptiles were another group of survivors. The reptiles, which arose from primitive amphibians, could break away from an aquatic existence because of shelled eggs and internal fertilization. Their adaptations meant that embryonic development could proceed within the eggshell, a moist and (compared to the outside) dependable setting. The Carboniferous also saw the first of the great radiations of insects, which apparently coincided with

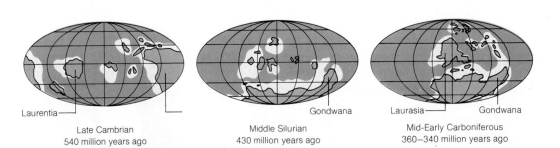

| | Laurentia | | Gondwana | Laurasia | Gondwana |
| Late Cambrian 540 million years ago | | Middle Silurian 430 million years ago | | Mid-Early Carboniferous 360–340 million years ago | |

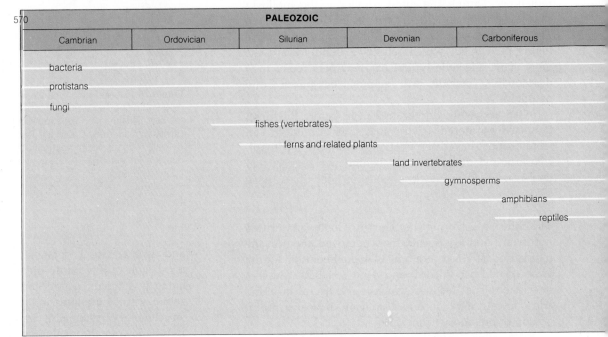

| **PALEOZOIC** | | | | |
| Cambrian | Ordovician | Silurian | Devonian | Carboniferous |

bacteria

protistans

fungi

fishes (vertebrates)

ferns and related plants

land invertebrates

gymnosperms

amphibians

reptiles

Figure 38.15 Correlation between geologic evolution and the rise of major groups of organisms. The light-blue areas on the maps represent known regions of shallow seas. Notice the changing positions of the seas relative to the equator.

Figure 38.16 Reconstruction of lobe-finned fishes venturing onto a Devonian shoreline. Animals such as these were the ancestors of amphibians, reptiles, birds, and mammals.

the evolution of the gymnosperms and, later, the angiosperms, or flowering plants (page 274).

The Permian. As the Carboniferous gave way to the Permian, all land masses had formed one vast continent called Pangea, leaving one immense world ocean. During this transitional period in geologic history, changes in the distribution of water, land area, and land elevation brought pronounced differentiation in world temperature and climate. About ninety percent of all known marine species disappeared at the boundary between the end of the Paleozoic and the beginning of the Mesozoic.

Of the three main lineages of late Permian reptiles, one gave rise to dinosaurs and birds, another to snakes and lizards. The third lineage dominated the Permian animal life. In this lineage were the synapsids, including such animals as the sail-backed pelycosaurs (Figure 38.17) and the diverse therapsids, or mammal-like reptiles (Figure 38.18).

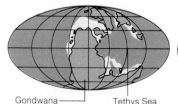
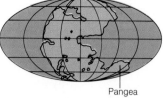
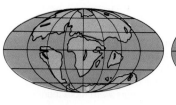

Gondwana — Tethys Sea

Pangea

Early Late Permian
260–250 million years ago

Triassic into Jurassic
240–195 million years ago

40–25 million years ago

Present

240		MESOZOIC		65	CENOZOIC	
Permian	Triassic	Jurassic	Cretaceous		Tertiary	Quaternary

marine reptiles

mammals

birds

flowering plants

marine mammals

Life During the Mesozoic

Once the supercontinent Pangea had formed, the patterns of diversity in the seas changed forever. The corals, brachiopods, cephalopods, fishes, and other lineages comprising the previously stable marine communities had disappeared or were reduced to a few species by the dawn of the Mesozoic. Other lineages, including the bivalves, gastropods, crustaceans, and freshwater fishes, had not been so severely tested. On land, the gymnosperms were becoming the dominant plants, and the therapsid reptiles dwindled to near extinction.

Mesozoic Seas. In the words of James Valentine, the late Paleozoic marine communities were "showy," whereas those of the early Triassic were "grubby." In effect, the most ecologically specialized lineages had been hit hardest during the Permian extinction, particularly tropical reef communities that had been as "showy" as the existing one illustrated in Figure 1.5.

Adaptive radiations of new lineages began in the Mesozoic seas, but this time the increase in diversity did not level off, as it did in the Paleozoic; it has continued to the present day.

Mammals, Dinosaurs, and Flowering Plants. Among the few remaining therapsids were the ancestors of mammals, which made their entrance during the Triassic. Even so, mammals remained small in size and few in number for the next 140 million years, when other vertebrates dominated the land.

As it happened, the reptiles called thecodonts had already begun a modest radiation. Some would give rise to the archosaurs ("ruling reptiles"). Later on in the Triassic, the first vertebrate experiment in flight—the pterosaurs—was under way, and several reptilian lineages had reinvaded the seas (Figure 38.17). On land, two major orders of dinosaurs diversified through the late Jurassic and early Cretaceous, producing the material for fabulous scenes such as shown in Figure 38.17d.

Figure 38.17 Reconstructions of Permian and Mesozoic reptiles and associated plants. (**a**) Triassic thecodont reptiles, ancestral to dinosaurs and birds. (**b**) A pterosaur, representing the first group of winged vertebrates. (**c**) Reptiles that returned to the sea: two ichthyosaurs and a plesiosaur. Some plesiosaurs were more than 15 meters (50 feet) long.

(**d**) A mural depicting four periods of the late Paleozoic and the Mesozoic. On the left, sail-backed synapsid reptiles of the Permian and early vascular plants; at left center, Triassic dinosaurs amid gymnosperms. The radiation of dinosaurs flourished in the Jurassic (right center), with some herbivores exceeding 25 meters (90 feet) in length. Dinosaurs of the late Cretaceous (right side) coexisted with early flowering plants. They included the largest of the carnivorous dinosaurs, such as *Tyrannosaurus*, standing nearly 6 meters (19 feet) tall.

a

d

Most dinosaurs *were* large, many spectacularly so. By reason of their size alone, their body temperatures were likely to have been high and relatively constant. Recent work on fossil trackways and dinosaur skeletal anatomy indicates that many were capable of rapid locomotion and that some of the smaller bipedal carnivores would make a respectable match for modern horses or ostriches in speed. Moreover, at least some of the small, carnivorous forms may have been endotherms, with an insulating coat of feathers. One of the earliest known birds is *Archaeopteryx* (Figure 2.11) from the upper Jurassic. It was so similar to those smaller dinosaurs that only the chance fossilization of feather impressions and the presence of a collarbone distinguish it.

Recently, fossil bones of a crow-sized form that lived 225 million years ago (75 million years before *Archaeopteryx*) were discovered in western Texas by Sankar Chatterjee. With its bony tail, strong hind legs, and pelvis adapted for running, this animal (designated *Protoavis*) was like the small dinosaurs. But the remains indicate that *Protoavis* had ears and eye sockets like birds, as well as small nodes on the bones to which feathers might have been attached. Thus *Protoavis* may be among the earliest transitional forms between dinosaurs and birds, which began their diversification about 100 million years ago.

The dinosaurs disappeared abruptly at the Cretaceous-Tertiary boundary. What happened? Currently, the leading hypothesis is that an asteroid collided with the earth, setting in motion a series of events that led to mass extinctions (page 544).

Dinosaurs so dominate our thinking about this period of evolutionary history that we often overlook one of the most important events of all. In the early Cretaceous, flowering plants began their ongoing radiation. In a span of about 10 million years, they surpassed the already declining gymnosperms in terms of their diversity and distribution.

The Cenozoic: The Past 65 Million Years

If you were to be transported 65 million years back in time to the beginning of the Cenozoic Era, you would probably not see anything that would strike you as being unusual. Marine organisms and the flowering plants would not appear that different from their existing descendants, and the conspicuous insects, reptiles, and birds might evoke only a little surprise as unfamiliar species. On reflection, you might conclude that what you *did not* see was the more unusual. For example, where were all the mammals? If all you could find were some unassuming rodentlike and opossumlike creatures in the shrubbery, you would have done well, because that is all there was.

b

c

Over the next 10–12 million years, however, most of the orders of mammals now living appeared (Figure 38.19). The magnitude of this adaptive radiation into bats, whales, ungulates (hoofed, herbivorous mammals), and so forth is remarkable in comparison to the earlier, rather undistinguished history of the group. For example, in the 140 million years since their origin, the Mesozoic mammals comprised only about 100 genera; in the first 60 million years of the Cenozoic, there were more than 3,000. (Figure 38.20 shows examples.)

The Cenozoic is known as the Age of Mammals, but if sheer *numbers* were the criterion it could be termed the Age of Birds. If the criterion were to be the most *rapid* adaptive radiation, the latter half of the Cenozoic would have to be called the Age of Snakes!

The point is, we live in a period of unparalleled biological diversity. Coinciding with that diversity is a far-ranging dispersal of land masses, which are intersecting the tropics and confining the polar regions. The vast island arcs of the tropical Pacific—remnants and omens

Figure 38.18 Reconstruction of a therapsid. Such reptiles are thought to have been the forerunners of mammals.

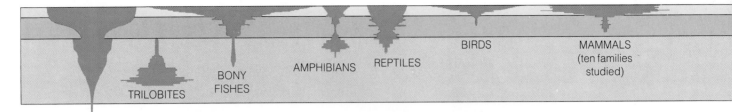

PROTOZOANS

Figure 38.19 A limited representation of the evolutionary histories of organisms. For the sample shown here, notice the diverse patterns of radiation and extinction. Notice also the spectacular current success of the insects. (Here, the yellow band represents the Cenozoic to the present; blue, the Mesozoic; and green, the Paleozoic. The widths of the lineages, all shown in red, represent the approximate numbers of *families* in the case of plants and animals.)

a

b

Figure 38.20 Reconstructions of North American mammals of the Eocene (**a**) and Pliocene (**b**).

of plate tectonic movements—may well be the richest ecosystems ever to appear on the planet, and the extensive tropical forests of South America, Africa, and Southeast Asia are probably not far behind.

PERSPECTIVE

All living things are composed of molecules made of the same chemical elements—primarily carbon, hydrogen, nitrogen, and oxygen. These elements were present in one form or another in the crust and atmosphere of the primordial earth. It has been demonstrated that these elements can combine spontaneously into carbohydrates, lipids, proteins, nucleic acids, even ATP—into the stuff of life—given the right chemical environment and a source of activation energy. It has been demonstrated that, given a particular set of physical and chemical conditions, some of those large molecules can form differentially permeable structures much like simple cell membranes. Even now, research is shedding light on how such structures might have evolved chemically into the first organized, self-reproducing systems—into the first cells.

Whatever the details of such chemical evolution might have been, we suspect that it flowed to the threshold of biological evolution and, eventually, to the explosive diversification of life.

Today you share the earth with perhaps $4\frac{1}{2}$ million species. They, and you, share allegiance to the same principles of energy flow and chemical interactions, and the same molecular and cellular heritage. All these things speak of the underlying unity of life, but they also speak eloquently of its subsequent diversity. For if the environments of the first living things had never changed, if there never had been different abiotic and biotic horizons waiting the first inadvertent explorers, perhaps the world today would hold little more than prokaryotic cells

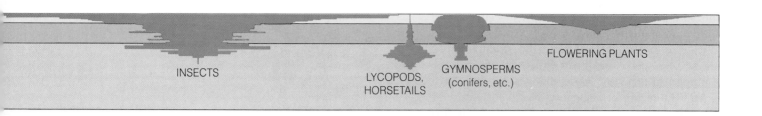

INSECTS   LYCOPODS, HORSETAILS   GYMNOSPERMS (conifers, etc.)   FLOWERING PLANTS

	Millions of Years Before Present	Major Events

Cenozoic Era	Quaternary (1.65–Present)
	  Recent
	  Pleistocene
		North and South America join. Several glacial cycles; extinctions of large mammals; origin of modern humans.

	Tertiary (65–1.65)
	  Pliocene
	  Miocene
	  Oligocene
	  Eocene
	  Paleocene
		Laurasia divides; India, Australia move northward; Alps, Himalayas, Andes, Rockies form. Major radiations of flowering plants; insects, snakes, birds, and mammals. Grasslands arise; succession of mammalian forms; origin of humans.

Mesozoic Era	Cretaceous (135–65)	Pangea continues to separate; broad inland seas; rich marine communities. Major radiation of ray-finned fishes, further radiations of insects and dinosaurs; origin of angiosperms and snakes. Mass extinction removes 11 percent of marine families, and all dinosaurs.
	Jurassic (181–135)	Pangea begins to break up. Marine communities become rich and diverse. Major radiations of dinosaurs; origin of birds. Gymnosperms dominant.
	Triassic (230–181)	Pangea stable. Marine life recovering; gastropods, crustaceans, fishes radiating. Early radiations of thecodont reptiles and dinosaurs; amphibians decline; origin of mammals. Gymnosperms dominant. Mass extinction removes 12 percent of marine families.
Paleozoic Era	Permian (280–230)	Pangea forms, squeezing out shallow tropical seas; now a worldwide ocean. Major radiations of synapsid, therapsid reptiles, and gymnosperms. Mass extinction removes 52 percent of marine families, up to 96 percent of marine species, with lesser effects on land.
	Carboniferous (345–280)	Gondwana moves toward Laurasia, enclosing Tethys Sea. Major radiations of insects and amphibians; origin of reptiles. Spore-bearing plants dominant, gymnosperms present.
	Devonian (405–345)	Gondwana moves north, equatorial land masses converge to form Laurasia. Radiation of fishes continues; origin of amphibians; development of swamp plants. Mass extinction removes 14 percent of known families in sea.
	Silurian (425–405)	Gondwana crosses south pole. Major radiations of jawed fishes; origin of vascular land plants.
	Ordovician (500–425)	Gondwana moves south. Major radiations of corals, crinoids, brachiopods, mollusks, early fishes; swimming arthropods, cephalopods. Marine communities diverse. Mass extinction removes 12 percent of known families.
	Cambrian (570–500)	Continents dispersed near equator. Origin of metazoans with hard parts; 50–100 phyla present; simple marine communities dominated by algae, brachiopods, trilobites.
Archean and Proterozoic Eras	(2,500–570)	Continental drift in progress. Oxygen becomes abundant in atmosphere. Origin of aerobic pathways of metabolism; origin of eukaryotes and multicellular organisms.
	(3,800–2,500)	Formation of continents. Chemical evolution, origin of life; prokaryotes and anaerobic pathways of metabolism.
	(4,600–3,800)	Origin of planets. Cooling, meteorite bombardment; formation of crust, oceans, and atmosphere.

Figure 38.21 Summary of the major features of the evolution of the earth and its life.

of the sort preserved in ancient rocks—cells matted against rocks or suspended in the seas.

Yet the record of earth history tells us that environments *have* changed. Organisms either have been lucky or equipped to adapt to those changes, or they have perished. The record also suggests that the diversity of life has been more than a product of evolution. It also has been an evolutionary force of the first magnitude. "Diversity" not only means adaptations to some combination of temperature, chemical balance, available water, light, dark, and living space. "Diversity" also means adaptations to different kinds of predators, different prey, different competitors after the same resource, and different forms of behavior, coloration, and patterning that help assure reproductive success. *Thus all existing species can be viewed as the evolutionary products of interactions with the environment and one another.*

And therein lies the story of evolution, the story of chemical competition and cooperation leading to the first self-reproducing forms of life, of dinosaurs, of continents on the move. Therein lies the story of simple strategies unchanged since the dawn of life, and of the complex human strategy—as yet unresolved—that can hold a world together or rip it apart. Yet must we predict gloomily that such unresolved activity on our part will end this magnificent story for all time? We doubt it. If the record of earth history tells us anything at all, it is that life in one form or another has survived disruptions of the most cataclysmic sort. That life can evolve tenaciously through tests of flood and fire suggests it has every chance of evolving around and past our transgressions too. *Viva Vida!*

SUMMARY

This chapter has presented a broad historical framework for the evolution of the earth and its organisms. The key points are highlighted in Figure 38.21; details of the evolutionary pathways taken in all five kingdoms are topics of chapters in the unit that follows.

Review Questions

1. Describe the chemical and physical characteristics of the earth 4 billion years ago. How do we know what it was like?

2. Describe the experimental evidence for the spontaneous origin of large organic molecules, the self-assembly of proteins, and the formation of organic membranes and spheres, under conditions similar to those of the early earth.

3. Which steps leading to the origin of living cells remain unexplained, and why are these important?

4. Distinguish between heterotrophic and autotrophic prokaryotes, and describe the origins of photosynthesis and respiration. What are the main biological and environmental consequences of photosynthesis?

5. How might symbiosis have played a role in the evolution of eukaryotes? How do eukaryotes differ from prokaryotes?

6. How does continental drift occur, and in what ways does this process influence changes in biological communities?

7. When did plants, insects, and vertebrates invade the land?

8. The Atlantic Ocean is widening, and the Pacific and Indian oceans are closing. What may be the biological consequences of the forthcoming formation of a second Pangea, and why?

Readings

Bambach, R., C. Scotese, and A. Ziegler. 1980. "Before Pangea: The Geographies of the Paleozoic World." *American Scientist* 68(1):26–38. Excellent summary article, complete with color-coded maps of the changing configurations of continents.

Dott, R., and R. Batten. 1980. *Evolution of the Earth.* Third edition. New York: McGraw-Hill. Findings from diverse lines of research are distilled into a stunning picture of earth and life history. We recommend this book for your personal library.

Eigen, M., W. Gardiner, P. Schuster, and R. Winkler-Oswatitch. April 1981. "The Origin of Genetic Information." *Scientific American* 244(4):88–118. Describes experiments that indicate the chemical nature of the first self-replicating protein-RNA template systems.

Lambert, D. 1983. *A Field Guide to Dinosaurs.* New York: Avon Books. Comprehensive review of the diversity of dinosaurs; well-written and capably illustrated.

Margulis, L. 1982. *Early Life.* Boston: Science Books International. Easy to read introduction to the origin and evolution of prokaryotes and eukaryotes. Paperback.

Schopf, J. 1975. "The Age of Microscopic Life." *Endeavour* 34(122):51–58.

Stebbins, G. 1982. *Darwin to DNA, Molecules to Humanity.* New York: Freeman. Excellent overview of evolutionary theory and of experiments at reconstructing the physical and chemical conditions leading to the origin of life. Paperback.

Valentine, J. (ed.) 1985. *Phanerozoic Diversity Patterns: Profiles in Macroevolution.* Princeton, New Jersey: Princeton University Press. Multi-authored volume on major features of plant and animal histories; technical, but clearly written.

Valentine, J., and C. Campbell. 1975. "Genetic Regulation and the Fossil Record." *American Scientist* 63:673–680. Excellent discussion of the origin of multicellular animals and their rapid adaptive radiation.

Woese, C. R. June 1981. "Archaebacteria." *Scientific American* 244(6):98–125. Presents evidence of three primordial lineages of bacteria.

DIVERSITY: EVOLUTIONARY FORCE, EVOLUTIONARY PRODUCT

39

MONERANS, PROTISTANS, AND VIRUSES

In the preceding unit, we considered the observations, experiments, and concepts being used to discern the family relationships among organisms. With these considerations in mind, let us now turn to the characteristics of the major groups within the five kingdoms—the monerans, protistans, fungi, plants, and animals. The groups we will review in this chapter are the ones with the most ancient ancestry:

Kingdom	Major Groups
monerans	*eubacteria, archaebacteria*
protistans	*euglenids, chryosophytes, dinoflagellates, protozoans*

The organisms in these groups are single-celled *microorganisms*—with few exceptions, they are too small to be seen without a microscope. Viruses and related forms are not cells, but they are microscopic and will also be described here.

Given their small size, the microorganisms do not begin to approach the degree of morphological variation seen in the other kingdoms. And this makes their classification difficult, because often it is necessary to determine what a microorganism *does* in addition to what it looks like. Bacteria, for example, are classified in part according to their ability to perform distinctive chemical

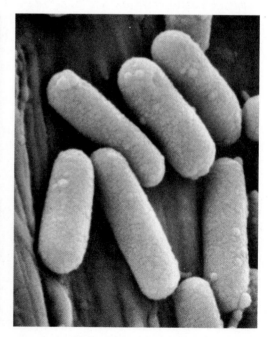

Figure 39.1 How small are bacteria? Shown here, *Bacillus* cells on the tip of a pin at increasing magnification: 85×, 440×, 11,000×.

reactions. Since the range of possible chemical reactions is enormous, an important indication of relationship between one species of bacteria and another is easy to overlook. For such reasons, bacterial classification schemes have changed constantly. The nucleic acid sequencing techniques being used to identify relationships among complex organisms are now being applied to the microbial world, but it will be many years before the full range of bacteria is analyzed in this advanced manner.

Even with the new molecular techniques, classification will not be easy. For most evolutionary studies, a species is defined as a reproductively isolated, interbreeding population. However, microorganisms reproduce by asexual means, for the most part, and so in this chapter we have to be more practical and flexible about our definition. We will call a **microbial species** a collection of strains of a given type of microorganism showing a high degree of similarity to each other and a much lower degree of similarity to other microorganisms. It remains a matter of individual judgment how much similarity justifies the inclusion of microorganisms in the same species.

MONERANS

The simplest and most abundant microorganisms are the bacteria, which are the sole members of the kingdom Monera. Every place in which life is possible, including boiling hot springs, Antarctic deserts, the deepest oceans, as well as soil and water, houses bacteria. A handful of rich soil contains tens of billions of living bacterial cells—several times the number of people on earth. Many bacteria live in or on other organisms. For example, bacterial cells in your intestinal tract and on your skin outnumber the cells making up your body! (Fortunately, animal cells are so much larger than bacterial cells that humans are only a few percent "bacterial" by weight.)

Most bacteria are approximately one cubic micrometer in volume, or about a thousand times smaller than many eukaryotic cells. Because the surface-to-volume ratio of a bacterial cell is high, the cell exchanges nutrients and wastes with the environment at a rapid rate. The rapid exchange rate underlies the rapid growth and division characteristic of bacterial populations, which often exceed a billion cells per milliliter of growth medium. With such rapid growth rates, a considerable amount of genetic variability accumulates through mutations. This genetic variability provides the means for rapid evolution and for survival during episodes of predation or starvation.

Bacterial Form and Function

Most bacterial species have a characteristic cell shape that is maintained by a rigid cell wall. The following terms are often used to designate the most common shapes of bacterial cells:

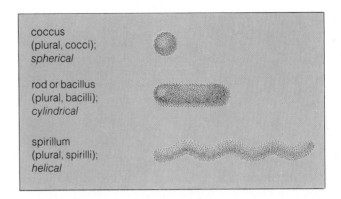

coccus (plural, cocci); *spherical*

rod or bacillus (plural, bacilli); *cylindrical*

spirillum (plural, spirilli); *helical*

Both cocci and rods (and spirilla, much less often) may form chains of cells when the daughter cells fail to separate completely after division. In addition, some bacteria are arranged end-to-end into a filamentlike form, as will be described shortly.

As we have seen, bacterial cells are *prokaryotic*; they have no nucleus or other membranous organelles characteristic of the eukaryotes (page 67). However, their plasma membrane serves many functions that are equivalent to the functions of organelle membranes, including electron transport and ATP synthesis. Typically, several thousand ribosomes are present in the cytoplasm.

Most bacteria have a rigid wall that provides protection against lysis in hypotonic environments. The wall of almost all eubacteria is composed of **peptidoglycan**, a molecule in which long polysaccharide strands are cross-linked to each other through short polypeptides to form a tough molecular mesh around the cell.

Even bacteria that do not have a rigid wall are not completely naked. They have a polysaccharide layer, sometimes quite thick, at the cell surface. Although the layer probably helps to compensate for the absence of a wall, most of these bacteria are sensitive to lysis. These bacteria are parasites in the tissues of plants and animals, where osmotic conditions are moderate.

As we have seen, bacterial cells may have flagella at their surface (Figure 5.26). Other surface structures are external jellylike layers called **capsules** and straight, filamentlike appendages called **pili** (singular, pilus). Capsules provide some protection against phagocytosis. Disease-causing (pathogenic) bacteria may attach to host tissues by means of the capsule and the pili, and thereby resist being washed away (as from the urinary or intestinal tract).

Figure 39.2 Transmission electron micrograph of a magnetotactic bacterium, showing the chain of magnetite crystals that acts like a tiny compass.

Bacterial Metabolism

Bacteria have inhabited the earth for more than 3.5 billion years, and during that time they developed impressively diverse means of acquiring and using energy. Recall that most bacteria are *heterotrophs*; they use preformed organic compounds as a source of the energy and the carbon necessary for biosynthesis. The heterotrophs accumulate wherever organic material is available, including forest soils, the sediments of lakes and continental shelves, and in the tissues of other living organisms. Although a few utilize only a single organic compound, most can utilize between ten and thirty; and collectively they are probably capable of tackling every naturally occurring organic compound.

Some bacteria are *autotrophs*, either photosynthetic forms that absorb light as an energy source, or chemosynthetic forms that use inorganic compounds (page 109). Photosynthetic bacteria include the cyanobacteria (sometimes called "blue-green algae"), the purple bacteria, and the green bacteria. Chemosynthetic bacteria include the methanogens and a variety of eubacteria able to use compounds such as ammonia, iron, hydrogen sulfide, hydrogen gas, or carbon monoxide as an energy source.

One of the main factors dictating whether a bacterial species will grow in a particular place is the presence or absence of oxygen. Aerobic respiration and the noncyclic pathway of photosynthesis predominate in aerobic settings; fermentation, anaerobic electron transport, and the cyclic pathway are used in anaerobic settings (Chapters Eight and Nine). Many species can switch back and forth among some or all of these metabolic routes when conditions change.

Sensory Reception and Response

As small and simple as they are, bacteria have sophisticated ways to sense their environment and to respond appropriately. Photosynthetic bacteria can sense the intensity of light and move toward its source (or away if it is too bright for them). Heterotrophic bacteria can sense and move toward higher concentrations of nutrients. Most bacteria can sense the presence of oxygen and move appropriately (toward for aerobes, away for anaerobes), and many can also move away from several toxic chemical compounds.

The detection of light or chemical compounds normally depends on the presence of membrane receptors that change shape when they absorb light or bind to a chemical compound. The conformational change signals the presence of the stimulus. When a bacterium moves in different directions, variations are introduced in the activity of membrane receptors. The variations are the basis of a fleeting biochemical "memory," allowing the cell to compare present conditions against the immediate past.

Many bacteria simply reverse direction when they sense that conditions are becoming less favorable. Others swim for a period of time, then stop and tumble around for a few seconds, then swim again. Because the tumble orients them in a new and completely *random* way, the new direction of swimming is different from the old. These bacteria modulate the *frequency* of tumbles according to the conditions they sense. When they are moving toward desirable conditions, they tumble rarely, so they tend to keep moving in that direction. When moving away, they tumble frequently and so keep changing direction until they start going the right way.

The "magnetotactic" bacteria respond to magnetism with a novel sensory device. A chain of magnetite crystals in their cytoplasm serves as a tiny compass (Figure 39.2). It happens that the earth's magnetic field is slanted relative to its surface, and the bacteria use their sensing device not only to head north (or south in the Southern Hemisphere) but also down. Thus they swim toward the bottom of a body of water, where the lower oxygen concentrations are more suitable for their growth.

Eubacteria

Most prokaryotic cells are classified as eubacteria, which are further subdivided into **Gram-negative** and **Gram-positive** groups. Originally these two terms referred to differences in the way different cells retain a deep purple stain (devised by Hans Christian Gram). The Gram-positive cells remain colored even when washed with organic solvents; the Gram-negative cells lose the color easily. The staining properties of most eubacteria depend on the structure and composition of the cell wall. The Gram stain is a quick, but not always accurate, test for wall type.

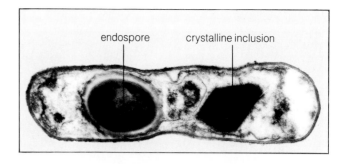

Figure 39.3 Endospore formation in *Bacillus thuringiensis*. The crystalline inclusion is of a kind that may be "harvested" and used as a natural insecticide against certain insect species.

Gram-Positive Eubacteria. Two examples of this group of eubacteria are *Bacillus* (an aerobic heterotroph) and *Clostridium* (an anaerobic heterotroph). Both have the unusual capacity for endospore formation. **Endospores** are dormant cells that are produced when the bacteria run out of nutrients and that convert back to normal cells when nutrients are encountered again. Endospores are resistant to heat, radiation, and toxic chemicals. Most can carry a bacterium through months or years of bad times; some have been converted back to normal cells after several thousand years of dormancy.

During endospore formation, a special division process results in one daughter cell being inside the other. The inner cell becomes the endospore and the outer cell eventually disintegrates. In a number of cases, crystals of protein form inside the endospore (Figure 39.3). Some of the crystalline inclusions are toxic to insects. The highly toxic *B. thuringiensis* inclusion is widely used as an insecticide against the caterpillars of gypsy moths, cabbage worms, and other destructive insects.

Endospore-forming bacteria also include species that cause diseases in animals, and some are particular problems for humans. For example, *C. botulinum* is an obligate anaerobe formed in soil, lake mud, and decaying vegetation. Each year many cattle and birds die after feeding on fermenting grains in which these bacteria are present. *C. botulinum* can produce one of the most dangerous neurotoxins known. According to one estimate, only one milliliter of culture fluid could kill 2 million mice. Sometimes the bacterium can grow and produce toxin in food that has been improperly sterilized and packaged in an anaerobic environment (as in cans and jars). Ingestion of the food can lead to the form of poisoning called **botulism**. (The toxin interferes with normal muscle activity and death can follow as a result of respiratory failure. Individuals who suspect that they have botulism can be given antitoxins; in serious cases, an artificial respirator can be used to prevent breathing failure.)

Exposure to high temperatures for a few minutes is often enough to rid various foods and equipment of most actively growing bacteria. But if the environment is slightly alkaline, endospore-formers can often live through several hours of boiling! That is why hospitals sterilize instruments in **autoclaves**, or "pressure cookers" in which the elevated pressure raises the boiling point of water to a temperature at which even endospores are rapidly killed (121°C).

Gram-Negative Eubacteria. A prokaryote to which we have referred repeatedly in this book, *Escherichia coli*, is a representative Gram-negative eubacterium. *E. coli*, a normal resident of the mammalian gut, produces compounds that enhance the digestion of fats as well as essential vitamins (such as vitamin K). Its activities also help create chemical conditions that do not favor colonization of the gut by disease-causing organisms.

Despite its benign and beneficial roles in healthy people, *E. coli* itself is one of the most serious pathogens in the world. Some *E. coli* carry the genes for a potent toxin and for pili that allow them to colonize the small intestine. When these organisms are ingested along with food, they produce a serious diarrhea. In developed countries, we call this *traveler's diarrhea*, since it often afflicts travelers to the developing countries. However, the disease that it causes in adults is mild compared to what it causes in infants and children. For most of the human population, *E. coli* diarrhea is the leading cause of infant mortality.

The photosynthetic **cyanobacteria** are a structurally diverse group of Gram-negative eubacteria. Some are single cells; others form straight or branching filaments in which cells are held together by a mucous sheath (Figure 39.4). All of these eubacteria have infoldings of the plasma membrane that are studded with light-trapping pigments. Unlike other photosynthetic prokaryotes, they have phycobilins and the same kind of chlorophyll found in plants (chlorophyll *a*). Depending on which accessory pigments are present, each species appears blue, blue-green, red-green, or nearly black.

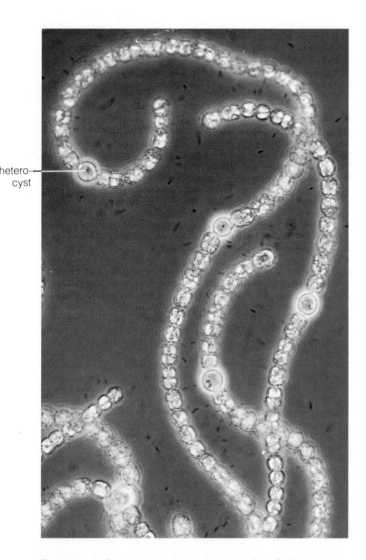

hetero-cyst

Figure 39.4 One species of the cyanobacterium *Anabaena*. Individual cells remain attached following cell division. The resulting filaments become coated in a mucilaginous sheath, formed from cellular secretions.

In many of the filamentous species, rounded cells called **heterocysts** form along the filament or at one end (Figure 39.4). Heterocysts are nitrogen-fixing cells, equipped with the enzymatic machinery needed to reduce gaseous nitrogen in the air to ammonium, which is converted to organic forms during amino acid synthesis. The heterocysts exchange products of nitrogen fixation for products of photosynthesis with their cellular neighbors. The nitrogen-fixing activities of cyanobacteria and other microorganisms are essential for the global cycling of nitrogen, helping to make it available for all other organisms (page 706).

Cyanobacteria live in fresh water, salt water, and land environments. There are species in hot springs, salt lakes, permanent snowfields, and on desert rocks. Two factors underlie their adaptability to such a broad range of environments. First, cyanobacteria have simple requirements for energy and raw materials. Almost any place that has sunlight, carbon dioxide, some water, and simple mineral salts will permit their growth. Second, they reproduce by the simplest means possible. When conditions are favorable, they divide rapidly by bacterial fission. When conditions are unfavorable, some species form a thick wall or a gelatinlike coat that keeps them from drying out. Like prokaryotes generally, the cyanobacteria grow in good times and wait out the bad.

Archaebacteria

A few bacteria were recently recognized as being very different from the eubacteria. The name **archaebacteria** was chosen for the group because many seem to be adapted to conditions that apparently were common on the early earth. Today, archaebacteria are classified as three major ecological subgroups: the methanogens, the halophiles, and the thermoacidophiles.

The **methanogens** inhabit swamps, sediments, and animal intestinal tracts. These strict anaerobes make ATP by converting carbon dioxide and hydrogen gases to methane. Methane (CH_4) is known as "swamp gas" when it bubbles out of marshes and as "natural gas" when it comes out of the ground.

The **halophiles** inhabit brines (water saturated with salt). They and a few animals such as brine shrimp are the only organisms able to endure such extremely hypertonic conditions. These heterotrophic bacteria usually rely on aerobic respiration for their energy. However, one member (*Halobacterium*) uses a primitive photosystem to make ATP under anaerobic conditions. This unique photosystem is composed of a single pigment, *bacteriorhodopsin*. Bacteriorhodopsin is a protein embedded in the plasma membrane, where it acts as a hydrogen ion pump when it absorbs light energy. Hydrogen ions pumped out of the cell reenter through an ATP synthetase system to produce ATP (compare page 129).

The **thermoacidophiles** live in hot springs, highly acidic soils, and even near volcanic vents at the ocean floor. The heterotrophic bacterium *Thermoplasma* has only one known habitat—the waste piles of coal mines! Since this is a habitat of recent origin (only a few hundred years old) there must be habitats we don't know about, where this organism evolved. In the waste piles, eubacteria oxidize sulfur-containing minerals and produce sulfuric acid, which makes the pile very acidic. The heat produced during this metabolism warms the piles to 50°–60°C.

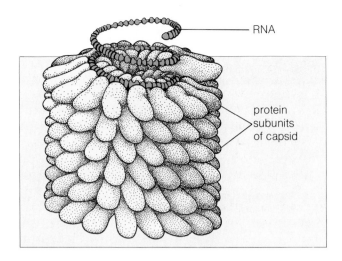

Figure 39.5 Structure of the tobacco mosaic virus. An entire rod-shaped virion (individual virus particle) is much longer than shown here.

RNA

protein subunits of capsid

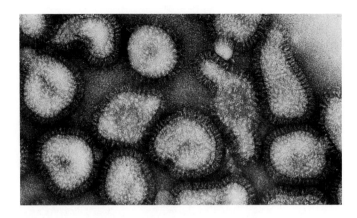

Figure 39.6 Virions of *influenzavirus*, each a package of RNA enclosed in a protein capsid. 250,000×.

Viruses

Characteristics of Viruses

Viruses are agents that infect cells, yet many biologists do not consider them to be alive. Viruses do have their own nucleic acids (single- or double-stranded DNA or RNA). Some contain DNA or RNA polymerase. The nucleic acid is sheathed in a protective protein coat, or *viral capsid*, which ranges from 20 to 250 nanometers across. However, unlike all living cells, viruses are not capable of metabolism; they have no mechanisms for generating their own metabolic energy. Neither are they capable of replicating themselves. Viruses can be replicated only by subverting the biosynthetic machinery of a host cell, which then preferentially follows the instructions encoded in the viral DNA or RNA. Because viruses can be perpetuated only inside the cells of another living organism, they probably developed at some point after cellular life originated.

Even if viruses are not alive, their influence on the biosphere is staggering. They cause many plant and animal diseases, some of which reach epidemic proportions. One of the most thoroughly studied of these disease-causing agents is the **tobacco mosaic virus** (TMV), which causes severe damage to the leaves of tobacco plants. It is a rod-shaped RNA virus with a capsid composed of 2,200 identical protein units (Figure 39.5). The TMV was the first to be purified and crystallized, and when the needle-shaped crystals were subsequently dissolved in water, they infected healthy plants. Other well-studied viruses include the bacteriophages, which were described in Chapter Fifteen.

Animal viruses are extremely destructive, as Table 39.1 suggests. For example, **influenzaviruses** (Figure

Table 39.1   Classification of Animal Viruses

Category	Some Diseases Produced
DNA Viruses:	
Adenoviruses	*Respiratory infections; pinkeye; under some circumstances can cause malignant tumors in hamsters*
Parvoviruses	*Gastroenteritis (diarrhea, vomiting); implicated in hepatitis A in humans*
Papovaviruses	*Warts in humans, rabbits, dogs; cancer in mice, hamsters*
Herpesviruses	*Fever blisters; chickenpox; shingles; genital infections with neurological consequences; some induce cancer; one implicated in infectious mononucleosis*
Poxviruses	*Smallpox; cowpox; formation of fibromas (nodules or benign tumors)*
RNA Viruses:	
Enteroviruses	*Diarrhea; polio; aseptic meningitis*
Rhinoviruses	*Common colds*
Togaviruses	*Yellow fever; German measles; equine encephalitis*
Influenzaviruses	*Influenzas*
Paramyxoviruses	*Mild respiratory disorders: Newcastle disease; measles*
Rhabdoviruses	*Rabies*
Arenaviruses	*Meningitis; hemorrhagic fevers*
Coronaviruses	*Upper respiratory disease*
Retroviruses	*Tumors (sarcomas); leukemia; AIDS*
Reoviruses	*Mild respiratory disorders; severe diarrhea in humans, cattle, mice*

Data from W. Volk, *Essentials of Medical Microbiology*, 1979.

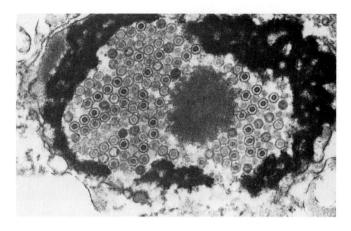

Figure 39.7 Virions of *Herpesvirus* in an infected cell.

Table 39.2 Summary of Protistan Groups
Euglenids
Chrysophytes
Dinoflagellates
Protozoa:
1. Flagellates (trypanosomes, trichosomes)
2. Amoeboids (amoebas, foraminiferans, heliozoans, radiolarians)
3. Sporozoans
4. Ciliates

39.6), which cause the common winter flu, also cause Asian flu, Hong Kong flu, and Spanish flu. These serious diseases occurred in *pandemic* form (as worldwide epidemics). They recur in ten- to forty-year cycles, and there may be localized epidemics every year in between. Between 1918 and 1920, more than 20 million people died under the attack of a Spanish flu virus. That happened not very long ago; your grandparents were alive then. The disease symptoms begin with sensations of chilling, followed by a sudden rise in body temperature to 100°–104°F. Muscles ache severely. The virus attacks the upper respiratory tract. Often, weakened respiratory tissues are susceptible to agents of pneumonia, a complication that accounts for a major part of the fatalities.

New strains of influenzaviruses arise periodically. There is growing evidence that pandemic strains do not arise through mutation, as was previously thought. Rather, new virulent strains appear to be recombinants of human viruses and animal viruses. Whatever the case, the World Health Organization has monitoring stations throughout the world. As soon as new strains are identified, researchers work rapidly to develop modified vaccines.

As another example, **herpesviruses** are infectious agents that are widespread through the animal kingdom. Typically, these viruses produce latent infections that can recur sporadically over months or years, or that can be completely different from the original outbreak. This group of viruses is responsible for fever blisters or cold sores, chickenpox, and shingles. One *herpesvirus* (Epstein-Barr virus) causes infectious mononucleosis, a disease that leads to marked increases in the number of circulating white blood cells. Symptoms include a severe sore throat, fever, enlarged lymph nodes, and overall weakness. Another *herpesvirus* may be involved in certain forms of cancer. These pathogens are also responsible for genital herpes in more than 5 million Americans alone (page 511).

Herpesviruses have double-stranded DNA and a protein capsid. The capsid itself, about 100 nanometers across, acquires a membranous envelope as it buds from the nuclear envelope of the host cell (Figure 39.7). A new cell is infected when the viral envelope fuses with the plasma membrane or when endocytosis transports the virion into the cytoplasm.

Viroids and Prions

Are viruses the most stripped down of all disease agents? It appears not. **Viroids** are infectious nucleic acids that have no protein coat whatsoever. The known viroids are simply linear or closed circles of single-stranded RNA. Under natural conditions, the RNA molecule base-pairs back on itself like a hairpin, so that viroids look like tiny rods.

Viroids are mere snippets of genes, much smaller than the smallest virus. Yet they can wipe out huge fields of seed potatoes and groves of citrus or avocados. Thirty years ago, viroids almost destroyed the chrysanthemum industry in the United States. One viroid has destroyed more than 12 million coconut palms on Philippine plantations, with major economic consequences. There are suspicions that viroids may cause some forms of human cancer, as they are known to do in hamsters.

Prions cause typically slow but fatal diseases of the central nervous system, such as *scrapie* in sheep and *kuru* and *Creutzfeldt-Jacob* disease in humans. Prions act as signals to the host cell to replicate them. Prions are not even nucleic acid! Their extracellular stage appears to consist of protein only.

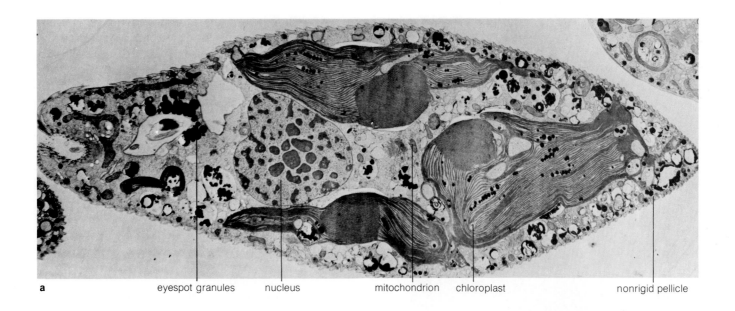

a eyespot granules nucleus mitochondrion chloroplast nonrigid pellicle

PROTISTANS

The single-celled eukaryotes that share membership in the kingdom Protista are listed in Table 39.2. They are almost bewildering in their diversity, and in some cases calling them protistans instead of single-celled plants, fungi, or animals may be rather arbitrary. Photosynthetic forms living at the surface of both fresh water and seawater are collectively called "phytoplankton," and they are probably responsible for more than half the carbon dioxide fixation on earth. Some heterotrophic protistans are enclosed by a cell wall and must absorb dissolved substances from the surroundings. Others are motile, phagocytic predators that ingest bacterial cells and assorted substances, and in this they are like tiny animals. Still others are parasites that absorb nutrients from living hosts. Altogether, protistan diversity may not readily lend itself to classification, but it probably reflects the evolutionary experimentation that is thought to have occurred when eukaryotic lineages first branched from ancestral forms (page 555 and Figure 38.7).

Euglenids

Among the puzzling protistans are the more than 800 known species of **euglenids**, which are flagellated, photosynthetic cells that live in fresh water, especially if it is rich in organic nutrients. Although they do not have a cell wall, euglenids have an outer, elastic *pellicle* composed mainly of protein. They have an *eyespot*, a region in which carotenoid pigments are concentrated in gran-

b

Figure 39.8 Anatomy of a single-celled protistan, *Euglena*. In (**a**), the profusion of internal organelles is evident. 5,200×. In (**b**), notice the flagellum; *Euglena* is a highly motile form. 1,700×.

ules (Figure 39.8). This light-sensitive eyespot is part of a sensory-motor system that is used to position the cell body in regions where light intensity is best for photosynthesis.

Euglenids reproduce by longitudinal fission, a reproductive mode that is common among all flagellated protistans. The cell grows in circumference while all organelles are being duplicated, then the cell divides along its long axis.

Intriguingly, some species of *Euglena* can survive in darkness (in other words, without photosynthesizing) as long as they are provided with nutrients. In fact, given sunlight, a favorable temperature, and a rich source of nutrients, some strains of *Euglena* reproduce faster than they duplicate their chloroplasts and end up as nonphotosynthetic cells. From then on, they are heterotrophs that resemble some flagellate protozoans!

C. Shih, R. Kessel, Living Images, © Science Books, Int'l. 1982, by permission of Jones & Bartlett Publishers, Inc.

Figure 39.9 (a) Scanning electron micrograph of the magnificent silica shell of a diatom. 580×. (b) Closer view of the shell, showing the intricate perforations. 2,340×.

Chrysophytes

The photosynthetic **chrysophytes** are abundant in freshwater and marine environments throughout the world. Among their members are the 500 or so species of Chrysophyceae ("golden algae") and more than 5,000 species of diatoms. All of these single cells contain an accessory pigment (fucoxanthin) that masks the color of their chlorophylls.

Most Chrysophyceae are flagellated and many have silica scales or a silica skeleton. Some are amoebas with chloroplasts and actually ingest bacteria. The diatoms are not flagellated and their walls are thin, double shells of silica that fit together, one on top of the other (Figure 39:9). The diatom shell is perforated with many holes, arranged in a precise, species-specific pattern (Figure

39.9b). These holes permit the underlying cytoplasmic membrane to maintain its close contact with the environment. Some diatoms also have longitudinal slots in the shell, through which the cell can make contact with a solid surface and move in much the same way that an amoeba moves. Other diatoms are immotile.

The silica shells of diatoms began to accumulate 100 million years ago, and extensive deposits of diatomaceous earth are found in different parts of the world. This fine, crumbly substance is used as an abrasive and in filtering or insulating materials. More than 270,000 metric tons are obtained annually from a quarry near Lompoc, California.

Dinoflagellates

The more than 1,000 species of **dinoflagellates** include forms encased in stiff cellulose plates, which have grooves between them. One groove circles the cell body and defines a channel for the movement of a ribbonlike flagellum. Another groove runs perpendicular to it and is a channel for another flagellum (Figure 39.10). When the two flagella beat in their respective channels, they make the cell spin like a top. These spinning species are known as "whirling whips."

Most dinoflagellates are photosynthetic, although a few colorless forms are heterotrophs. The photosynthetic forms contain chlorophylls, but depending on the array of accessory pigments, they may appear yellow-green, brown, or red. They occur as phytoplankton in marine and freshwater settings.

Periodically, red dinoflagellates such as *Gonyaulax* undergo explosive population growth. They actually color the seas red or brown. These so-called **red tides** are extremely devastating, for some of the dinoflagellates produce a powerful neurotoxin. Fish that feed on this plankton are poisoned; sometimes hundreds of thousands are killed and wash up along the coasts (Figure 39.11). Shellfish such as clams, oysters, and mussels are not affected by the neurotoxin. However, the poison builds up in their tissues, and it can be lethal for humans eating the shellfish.

Protozoans

Some single-celled organisms are motile, predatory or parasitic heterotrophs, the equivalent of animals in the microscopic world. Hence the name **protozoans**, meaning "first animals." In one respect the name is somewhat unfortunate, for it implies that these cells are primitive relatives of the multicellular animals. We now know that many are at the pinnacle of single-celled complexity, and their evolutionary relationship to animals is not all that clear.

The protozoans are classified into four groups, according to their form of motility:

Mastigophora	*flagellate protozoans* (flagella)
Rhizopoda	*amoeboid protozoans* (pseudopodia)
Sporozoa	*sporozoans* (gliding or nonmotile)
Ciliophora	*ciliates* (cilia)

Many thousands of protozoan species exist; of these, less than two dozen cause diseases in humans. However, the diseases that these few pathogens do cause are staggering in their extent. Perhaps as much as one-fourth of the entire human population is afflicted at any given time with a protozoan infection!

Flagellate Protozoans. For the most part, the flagellate protozoans are parasites of invertebrates and vertebrates. Characteristically, these cells undergo longitudinal fission in the manner described for *Euglena*.

In this group are the **trypanosomes**, which have a flagellum arising from the base of a small invagination at one end of the cell. During one stage of the life cycle, the flagellum is continuous with an undulating external membrane (Figure 39.12). Some trypanosomes develop in the salivary glands of insect vectors, which infect new hosts by biting. For instance, *Trypanosoma brucei*, which causes African sleeping sickness, is transmitted from host to host by the tsetse fly. Other trypanosomes

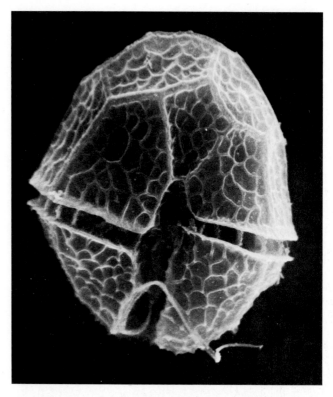

Figure 39.10 One of the "whirling whips"—dinoflagellates with two flagella that beat in opposing grooves in the armor-plated body. One flagellum is visible at the base; another beats in the groove running from left to right of this scanning electron micrograph.

a

Figure 39.11 (**a**) Scanning electron micrograph of *Ptychodiscus brevis*, the dinoflagellate responsible for the outbreaks of red tides along the Florida coast. (**b**) Portion of a fish kill that resulted from a dinoflagellate "bloom."

b

develop in the hindgut of insect vectors, leave in feces, and later penetrate the broken skin or mucous membranes of another host.

Also in this group are **trichosomes** of the sort shown in Figure 39.12. Among humans, *Trichomonas vaginalis* is a worldwide nuisance. It is transferred to new hosts through sexual intercourse. Trichomonal infection can severely damage vaginal membranes if untreated. Among men, infection can occur in the urinary and reproductive tracts.

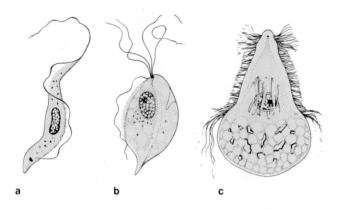

a   b   c

Figure 39.12 A few flagellate protozoans. (**a**) A leaf-shaped trypanosome, with its undulating membrane. (**b**) A trichomonad. These (generally) anaerobic parasites live in vertebrates and invertebrates. (**c**) *Trichonympha*, a flamboyant form that lives in the termite gut.

Amoeboid Protozoans. There are four groups of amoeboid protozoans: the amoebas, foraminiferans, heliozoans, and radiolarians. In all four groups, the adult forms have pseudopods, which are used in capturing prey or in locomotion. Except for the few parasitic species, food for the amoeboids consists of bacteria, algae, diatoms, other protozoans, and even small animals (such as nematodes).

The **amoebas** may or may not have shells. The "naked" species, such as *Amoeba proteus* of biology laboratory fame, constantly change their shape (Figure 5.1). These protozoans can be found in freshwater, seawater, and soil. Although most are microscopic, some species are several millimeters long. Shelled amoebas, which inhabit fresh water, damp soil, and mosses, can extend pseudopods or much of the cell body through a large opening in the shell. Like many other protistans, the freshwater species use organelles called contractile vacuoles, as illustrated in Figure 6.6, to expel excess water from the body.

Some amoeboid protozoans are parasites of humans. For instance, *Entamoeba histolytica* can invade the human gastrointestinal tract. This parasite causes amoebic dysentery, a disease characterized by fever, abdominal cramps, and severe diarrhea. *E. histolytica* proliferates in the gut during part of its life cycle. Then it develops into a thick-walled resting cell (a cyst). The cysts are excreted with feces. Without proper sewage treatment facilities, these cysts can contaminate food and water.

Figure 39.13 Shells of some foraminiferans (the word means, loosely, "bearers of windows").

Figure 39.14 Glass model of *Trypanosphaera regina*, a colonial radiolarian that has a skeleton made of silicon.

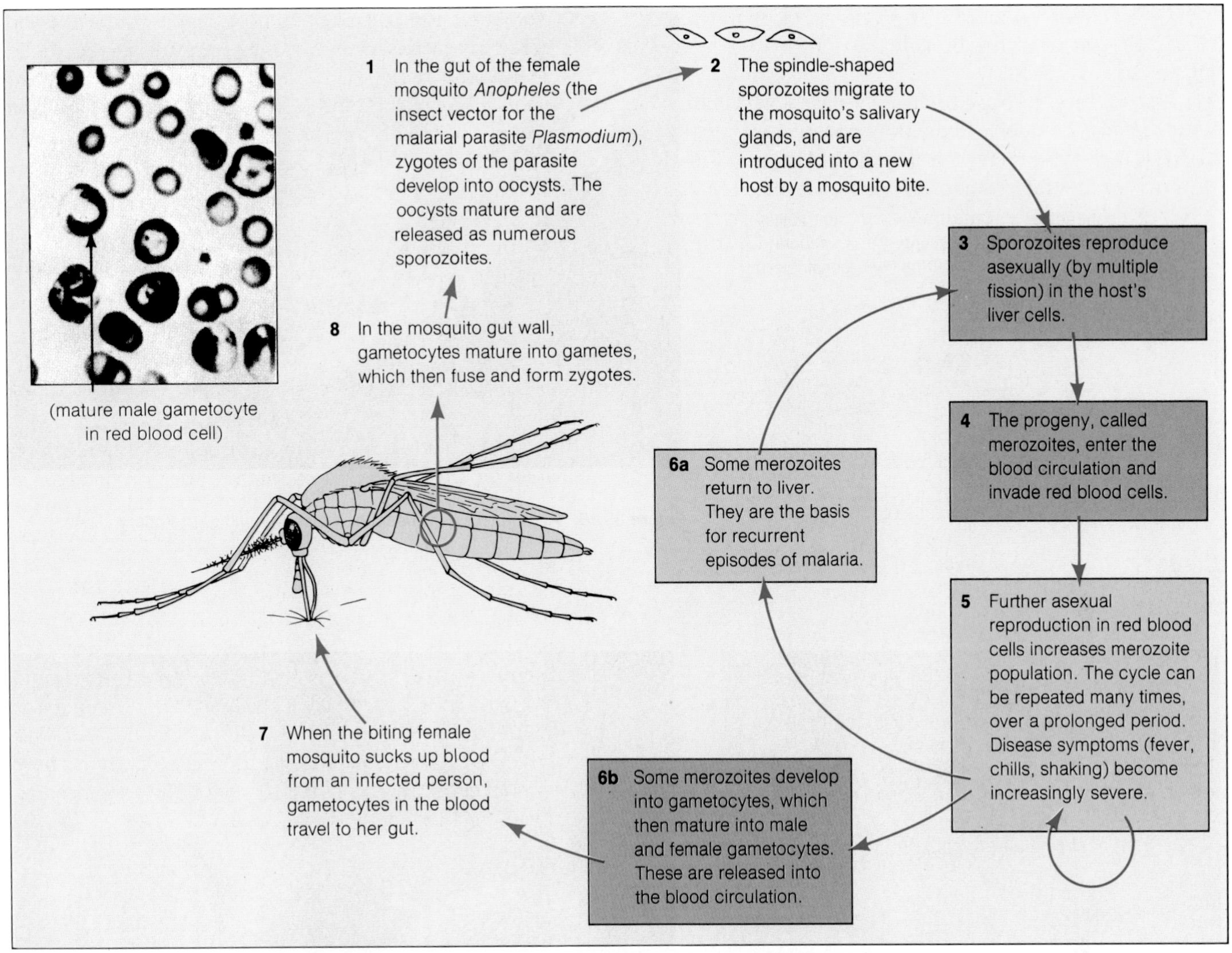

(mature male gametocyte in red blood cell)
1 In the gut of the female mosquito *Anopheles* (the insect vector for the malarial parasite *Plasmodium*), zygotes of the parasite develop into oocysts. The oocysts mature and are released as numerous sporozoites.
2 The spindle-shaped sporozoites migrate to the mosquito's salivary glands, and are introduced into a new host by a mosquito bite.
3 Sporozoites reproduce asexually (by multiple fission) in the host's liver cells.
4 The progeny, called merozoites, enter the blood circulation and invade red blood cells.
5 Further asexual reproduction in red blood cells increases merozoite population. The cycle can be repeated many times, over a prolonged period. Disease symptoms (fever, chills, shaking) become increasingly severe.
6a Some merozoites return to liver. They are the basis for recurrent episodes of malaria.
6b Some merozoites develop into gametocytes, which then mature into male and female gametocytes. These are released into the blood circulation.
7 When the biting female mosquito sucks up blood from an infected person, gametocytes in the blood travel to her gut.
8 In the mosquito gut wall, gametocytes mature into gametes, which then fuse and form zygotes.

Figure 39.15 Life cycle of the sporozoan *Plasmodium*, which causes the disease malaria. The life cycle unfolds in the human body and in an insect vector (the female mosquito *Anopheles*), which transfers the sporozoan to new hosts during bites. The events described in the boxes occur in the human body.

The elaborately shelled **foraminiferans** live mostly in the seas. Their hardened shells are often peppered with hundreds of thousands of tiny holes through which sticky, threadlike pseudopods extend. Figure 39.13 shows some foraminiferan shells minus their inhabitants. Often the shells bear spines, which in some species are long enough that the shell can be seen with the naked eye.

The **heliozoans**, or "sun animals," have fine, needlelike pseudopods that radiate from the body like sun rays. These largely freshwater protozoans are generally floaters or bottom-dwellers. Part of the cytoplasm forms an outer sphere around a core composed of denser cytoplasm and the bases of microtubular rods.

The **radiolarians** are perhaps the most beautiful of all protozoans. They are found mostly in marine plankton. In cytoplasmic structure they resemble the heliozoans, but most also have a skeleton of silica. In addition, there are colonial radiolarians in which many individuals are cemented together (Figure 39.14). The accumulated shells of both radiolarians and foraminiferans are a primary component of many ocean sediments.

Sporozoans. The **sporozoans** are a diverse group of parasites. They have in common the development of an infective, sporelike stage (sporozoites) during the life cycle (see, for example, Figure 39.15). There are both asexual and sexual stages to the life cycle, and often

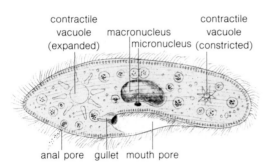

contractile vacuole (expanded) macronucleus contractile vacuole (constricted)

micronucleus

anal pore gullet mouth pore

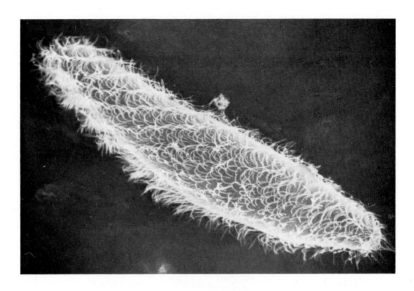

Figure 39.16 Anatomy of *Paramecium*, a fast-swimming, predatory ciliate.

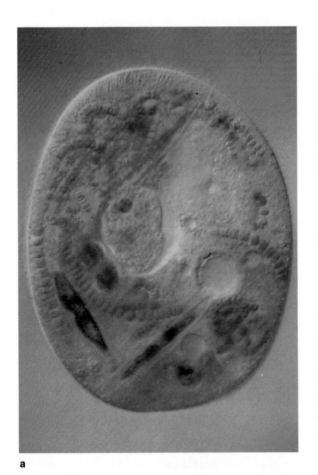

a

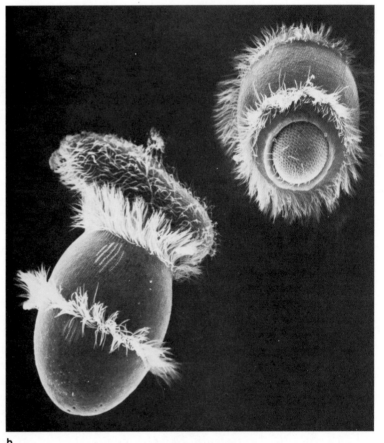

b

Figure 39.17 (**a**) A living ciliate protozoan of unknown affiliation, as viewed with Nomarski optics, a special microscope technique that does not damage the cell. The nucleus (near the center of the cell) is surrounded by food-filled vacuoles in which filamentous algae and diatoms are visible. The "sunken" spots are contractile vacuoles. (**b**) Mealtime for *Didinium*, a ciliate with a big mouth. Dinner in this case is another ciliate, the cucumber-shaped *Paramecium* poised at the mouth (left) and swallowed (upper right).

vectors such as insects transmit the parasites from one host to another.

Probably the most notorious sporozoans are the more than fifty species of *Plasmodium*, four of which cause the disease malaria in humans. The parasite requires an insect vector (mosquitoes) and an animal host (birds, mostly, and humans). As many as 100 million people have been stricken with malaria in a single year.

Ciliate Protozoans. Nowhere in the microbial world is the potential of a single cell expressed more fully than among the **ciliate protozoans**. Some of these cells are covered with thousands of cilia, synchronized for swimming; often the cells bear hundreds of poison-charged, harpoonlike weapons that can be fired at prey and predators.

Perhaps the most widely occurring ciliate is the predatory *Paramecium* (Figure 39.16). Like most ciliates, *Paramecium* has a gullet, a cavity that opens to the outside at the cell surface. Currents created by the beating of rows of specialized cilia carry bacteria and food particles into the gullet. Once inside, the particles become enclosed in food vacuoles, where digestion takes place. Wastes are moved to a region known as the anal pore and are eliminated. Like the amoeboid protozoans, *Paramecium* relies on contractile vacuoles to rid the body of excess water (Figure 6.6).

Paramecium is built for speed. Between 10,000 and 14,000 cilia project like rows of tiny, flexible oars from the cell surface. So efficient is the coordinated beating of the rows that the movements propel some species of *Paramecium* through their surroundings at a remarkable 1,000 micrometers per second. Yet *Paramecium* itself is often outmaneuvered by another free-swimming ciliate, *Didinium* (Figure 39.17). Other ciliates crawl about or simply stay put. *Vorticella*, for example, normally is attached by means of its holdfast on the floors of ponds, lakes, and streams, and turns its mouth to intercept food floating past.

ON THE ROAD TO MULTICELLULARITY

Among the protistans are loose associations of cells that form when cell walls fail to separate after division, and simple colonies cemented together in a glasslike matrix that the cells themselves have secreted. When we think about complex organisms such as redwoods and whales, it seems almost incomprehensible that such simple associations gave rise to multicellular forms of life. Yet the problem may be one of overlooking what must have been an immense evolutionary parade of intermediate forms, some dead-ends, others not.

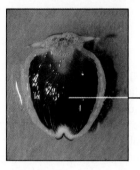

cerata (numerous projections of the dorsal surface) are pushed back here, revealing the extent to which functional chloroplasts become incorporated in tissues

Figure 39.18 One of nature's experiments—*Plakobranchus*, a marine mollusk that feeds on algae, the chloroplasts of which become incorporated in its tissues and continue functioning, providing the animal with oxygen.

As we have seen, chloroplasts and mitochondria may have originated when bacteria ingested by phagocytic cells escaped digestion and took up residence in the host (page 555). An example of this kind of escape is found in an algae-eating mollusk called *Plakobranchus*. The chloroplasts of the ingested algae become incorporated into this animal's tissues and continue to engage in photosynthesis (Figure 39.18). If such jarring examples are observable today, is it too far-fetched to assume that novel forms made their way onto the evolutionary stage in the past?

In its biochemistry, a photosynthetic bacterium called *Prochloron* resembles the chloroplasts of green algae. The simplest green alga, *Chlamydomonas*, has one chloroplast, one eyespot, and two flagella. Cells resembling *Chlamydomonas* occur among the volvocines—colonial organisms that resemble both protistans and plants. The peak of volvocine complexity, *Volvox*, is thought to represent an evolutionary dead-end, but it gives us some interesting things to think about. Depending on the species, *Volvox* is a single-layered, hollow sphere of 500

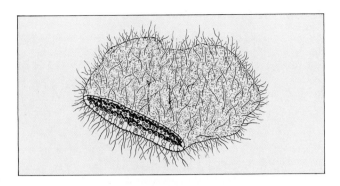

Figure 39.19 *Trichoplax adhaerens*, one of the simplest of all multicelled animals, being little more than a flattened ball of ciliated cells. This tiny animal was discovered crawling about in a seawater aquarium.

to 600,000 cells. When the flagella of those cells beat in unison, the sphere can be propelled forward. Only a few cells reproduce. They divide and give rise to daughter colonies, which float inside the parent sphere (Figure 7.1). When the daughter colonies release certain enzymes, the parent jellylike matrix dissolves and the daughter colonies are on their own. In some *Volvox* species, we also see sexual differentiation (certain cells produce eggs, sperm, or both). Thus *Volvox* represents an experiment in **multicellularity**, with the interdependence and division of labor among specialized cells that the word implies.

Consider, now, a tiny marine animal called *Trichoplax adhaerens*, which is little more than a flattened ball of ciliated cells half a millimeter across (Figure 39.19). It has no right side, left side, front, or back; it simply moves in any direction, amoebalike. It has a few more cilia on its top than on its bottom; but other than this, there is no cellular differentiation to speak of. *Trichoplax* may be reminiscent of simple multicelled animals that made their entrance during the Proterozoic.

Still other living organisms provide us with ideas about the first experiments in multicellularity. Consider the multicelled forms of green algae—straight or branched filaments a mere one cell thick, and simple sheets of cells. As you read in the preceding chapter, the tiny multicelled plants that first invaded the land were not much more complex than this.

Consider, finally, just one of your own multicelled systems—the calcium-containing bones of your endoskeleton. Controls over calcium intake must have existed in the earliest protistans. Calcium, after all, is required in microtubule assembly—and microtubules are required in such essential structures as pseudopods and mitotic spindles. The proteins of existing organisms

preferentially bind and release calcium ions. In ancient protistans, such proteins could have latched onto calcium entering the cell from the environment; they could have been calcium storage centers. Such proteins could have become incorporated into cell surface layers—and into internal structures. In the increasingly elaborate shells and cytoskeletons of protistans, we have hints of the original uses of materials found in our bones—our own skeletal system, and our own calcium reservoirs.

SUMMARY

1. There are two kingdoms of single-celled organisms: the monerans (which are prokaryotic) and the protistans (which are eukaryotic). The monerans are subdivided into eubacteria and archaebacteria. The protistans are subdivided into euglenids, chrysophytes, dinoflagellates, and protozoans.

2. A microbial "species" is a collection of strains of a given type of microorganism showing a high degree of similarity to each other and a lower degree of similarity to other microorganisms.

3. Most bacteria are about 1,000 times smaller than many eukaryotic cells; their surface-to-volume ratio is high and exchanges of nutrients and wastes with the environment are rapid. The rapid exchange rate underlies the rapid growth and division characteristic of bacterial populations.

4. Most often, bacteria are shaped like balls (cocci), rods, and corkscrews (spirilla). Most eubacteria have a rigid wall of peptidoglycan, which forms a tough molecular mesh. One or more flagella, a capsule, and pili may be present. The capsules and pili are structures by which many pathogenic bacteria attach to host tissues.

5. Most bacteria are heterotrophs (they use preformed organic compounds available in various environments and in the tissues of other living organisms). Other bacteria are photosynthetic or chemosynthetic autotrophs (page 109).

6. Different bacteria exhibit phototaxis and chemotaxis; one has a magnetic sense. Such forms of behavior provide a means of moving toward favorable conditions and away from unfavorable ones.

7. The eubacteria include Gram-positive groups (such as *Bacillus* and *Clostridium*) and Gram-negative groups (such as *Escherichia* and the cyanobacteria).

8. The archaebacteria include the methanogens, halophiles, and thermoacidophiles, all of which are adapted to environmental conditions that apparently were present on the early earth.

9. Viruses are infectious agents that have their own nucleic acid (single- or double-stranded DNA or RNA) sheathed in a viral capsid (a protective coat). They are not capable of metabolism or independent reproduction; they rely on the machinery of a host cell.

10. Bacteriophages infect certain bacteria; other viruses infect plants and animals, causing diseases that can reach epidemic proportions.

11. Viroids are infectious agents that do not even have a protein coat; they are folded or closed circles of single-stranded RNA that cause diseases in plants and animals. Prions are simply proteins that apparently act as signals to the host cell to replicate them; they have been linked to slow, fatal diseases of the central nervous system.

12. Protistans are remarkably diverse and may reflect the evolutionary experimentation that is thought to have occurred when eukaryotic lineages first arose from ancestral, prokaryotic forms (page 554).

13. The euglenids are flagellated, photosynthetic cells. Under certain conditions, they reproduce faster than they replicate their chloroplasts and end up as viable heterotrophs (which resemble some flagellated protozoans).

14. Chrysophytes include the photosynthetic golden algae and diatoms, both characterized by an accessory pigment (fucoxanthin) that masks their chlorophylls. Among these cells are forms with flagella and silica scales, skeletons, or shells.

15. Dinoflagellates are largely photosynthetic cells encased in stiff cellulose plates. Some red forms produce a neurotoxin that is responsible for the devastating fish kills associated with "red tides."

16. Protozoans are motile heterotrophs (predators or parasites). They include the flagellate protozoans, amoeboid protozoans (which move by pseudopodia), sporozoans (gliding or nonmotile forms), and ciliates. Of the thousands of species, few infect humans; but those few are infecting about one-fourth of the human population at any given time.

   a. The flagellate protozoans include the trypanosomes that cause African sleeping sickness and the trichosomes that can infect the human reproductive tract.

   b. Amoeboid protozoans include shelled or unshelled amoebas, foraminiferans, heliozoans, and radiolarians, all of which use pseudopods for movement or capturing prey.

   c. Sporozoans have an infective sporelike stage (sporozoites) during the life cycle, and often insects transmit the parasite from one host animal to another.

   d. The ciliates are predatory forms that use numerous cilia for movement and in feeding.

Review Questions

1. Can you describe the main characteristics that distinguish viruses from bacteria? Bacteria from protistans? Protistans from multicelled eukaryotes?

2. Can you explain why some microbes are killed in the presence of oxygen, yet others are not? Why, also, do some microbes get along well whether oxygen is present or not?

3. Can you describe some of the bacteria and protistans that are beneficial to humans, and some that are harmful (or lethal)?

4. What kinds of protistans are photosynthetic? Predatory? Parasitic? Which move by means of pseudopodia?

5. What are the four main categories of protistans? The four main categories of protozoans?

Readings

Brock, T., et al. 1984. *Biology of Microorganisms*. Fourth edition. Englewood Cliffs, New Jersey: Prentice-Hall. One of the best introductions to microbiology.

Diener, T. 1981. "Viroids." *Scientific American* 244(1):66–73. Good summary article on these smallest of all infectious particles.

Jurand, A., and G. Selman. 1969. *The Anatomy of Paramecium Aurelia*. New York: St. Martin's Press. A tribute to the astonishing complexity of a single cell.

Margulis, L. 1982. *Early Life*. Boston: Jones and Bartlett. An exceptional book, written for the general audience. Objective, compelling presentation of evidence for the early evolution of life. Paperback.

Noble, E. R., and G. A. Noble. 1982. *Parasitology: The Biology of Animal Parasites*. Fifth edition. Philadelphia: Lea and Febiger. One of the most authoritative references on parasitic protozoans.

Scagel, R., et al. 1984. *Plants: An Evolutionary Survey*. Belmont, California: Wadsworth. Chapter 6 is an authoritative treatment of the "algae" now classified as protistans.

Stanier, R., et al. 1986. *The Microbial World*. Fifth edition. Englewood Cliffs, New Jersey: Prentice-Hall. Excellent material on bacteria and protistans.

Wiesner, P., and W. Parra. 1982. "Sexually Transmitted Diseases: Meeting the 1990 Objectives—a Challenge for the 1980s." *Prevention* 97(5):409–416.

Woese, C. 1981. "Archaebacteria." *Scientific American*. 244(6):98–125. Summarizes evidence that points to separate origins for archaebacteria, eubacteria, and an ancestral line (urkaryotes) that gave rise to eukaryotes.

40

FUNGI AND PLANTS

With this chapter, we cross the boundary between the protistans and the kingdoms of fungi and plants. Like protistans, some fungi and plants are single cells, but most are complex multicelled forms that live in a wide range of habitats, including the oceans, streams, lakes, forests, prairies, deserts, and tundra. Traditionally, fungi and plants have been the focus of botanists, although fungi are now considered to be as distinct from plants as they are from bacteria, protistans, and animals.

PART I. KINGDOM OF FUNGI

The Fungal Way of Life

Fungi are heterotrophs; they cannot produce their own food. This is the most important way in which fungi differ from the green plants, which are autotrophs.

What fungi can do is decompose just about anything that has organic components—nature's garbage (dead plants and animal wastes, Figure 40.1), your groceries (meat, fruit, vegetables, bread, cheese), and many of your possessions (clothing, paper products, photographic film, shoe leather, and paint). Some fungi can even grow in jet fuel and cause trouble by clogging the fuel lines. Fungi living on the organic materials just mentioned are **saprobes**: they obtain nutrients from nonliving matter. Most fungi are saprobes but some are **parasites**: they obtain nutrients directly from organic matter that is still part of a living host. In both cases, fungal cells secrete enzymes that promote digestion *outside* themselves. Large organic molecules are broken down by the digestive enzymes into smaller components that can be absorbed across the plasma membrane.

Parasitic fungi digest parts of their living host which, as a result, may eventually die. Some of these fungi cause ringworm, athlete's foot, and similar diseases that make human skin redden, crack, or turn scaly. Others cause plant diseases, such as stem rust of wheat or apple scab, which reduce crop yields. Common fungal diseases such as black spot or powdery mildew can take the joy out of growing roses and other ornamental plants. Millions of dollars are lost annually as a consequence of the activities of these and other fungal parasites.

Fungi alter their food source as they feed upon it, and this simple activity makes them extremely important in nature. Think about the annual volume of waste products and remains of insects, birds, and other animals as well as the debris of plants. Each season, one elm tree alone may produce about 400 pounds of leaves! Without the activity of saprophytic fungi and other decomposers, natural communities gradually would be buried in their own garbage, nutrients could not be recycled, and life could not go on.

Figure 40.1 *Pilobilus*, a rather remarkable member of a diverse group of organisms—the fungi. Sporangium-bearing stalks of *Pilobilus* grow on animal feces. (A sporangium is a sac within which asexual spores are produced.) The stalk shows phototaxis: it grows in such a way that incoming rays of sunlight converge at the base of the swollen portion of the stalk (below the brown sporangia). Turgor pressure inside a vacuole in the swollen portion becomes so great that the sporangium can be blasted two meters away—a remarkable feat, considering that the stalk is only five to ten millimeters tall!

In short, some members of the Kingdom Fungi are essential in the web of life, and others are what we consider to be minor nuisances or major problems in terms of their impact on our affairs. Let's take a look at some characteristics of these diverse organisms.

Fungal Body Plans

A complex, multicelled structure, the **mycelium**, develops during the life cycle of most fungi. This mesh of microscopic filaments branches in all directions, spreading over or within the organic matter being used as food. Each filament is a **hypha** (plural, hyphae), a tube with a thin, transparent wall usually reinforced with chitin.

Hyphae are not "multicellular" in the usual sense of the word. Rather, they are tubes with a common, multinucleate cytoplasm. Some hyphae are divided along their length by cross-walls that create a series of compartments; other hyphae are undivided tubes. Even where cross-walls exist, they are sievelike; cytoplasm and nuclei can flow through pores in the cross-walls.

Overview of Reproductive Modes

Asexual reproduction is important in many fungal life cycles. Whenever growing conditions are suitable, asexual reproduction typically includes one of the following events:

1. Formation of asexual spores (reproductive cells that give rise to more of the same kind of hyphae).

2. Fragmentation of the hyphae into parts that grow into new individuals.

3. Binary fission or budding (in single-celled fungi).

Many fungi also reproduce sexually, but the details vary from one group to another. Depending on the species, sexual fusion may occur between haploid gametes, gamete-producing bodies, or hyphae. Sometimes both the cytoplasm and haploid nuclei fuse immediately to produce the diploid zygote. In most species, however, there is a delay between cytoplasmic and nuclear fusion, resulting in a **dikaryotic stage** in which cells contain two separate haploid nuclei, one from each parent (Figure 40.2).

Before considering specific examples of reproductive modes, one more point can be made about sex among the fungi. Sexual reproduction requires the fusion of compatible nuclei. Some fungal species produce sexually compatible gametes on the same mycelium; they are self-fertilizing. Other species require outcrossing between different but sexually compatible mycelia.

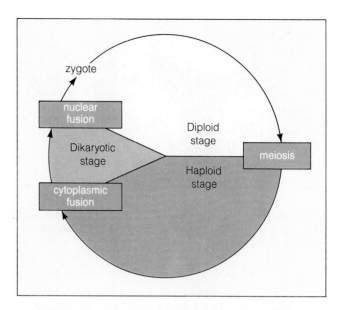

Figure 40.2 Generalized life cycle for fungi showing the alternation of haploid and diploid stages of the life cycle. Some fungal species do not pass through the dikaryotic stage indicated here.

Major Groups of Fungi

The few clues in the fossil record tell us that the fungi are a group with ancient origins (Figure 38.15). Today, 50,000 species of fungi have been catalogued; the actual number of existing species may be more than four times as high. The species may be classified in one of five main divisions (a *division* is equivalent to a phylum in animal classification schemes):

Division	Common Name
Oomycota	*egg fungi*
Zygomycota	*zygospore-forming fungi*
Ascomycota	*sac fungi*
Basidiomycota	*club fungi*
Deuteromycota	*imperfect fungi*

Oomycota. Some egg fungi parasitize land plants, but most are saprobes or parasites on algae or simple animals (such as rotifers) that live in lakes, streams, and other bodies of water. This division contains fungi that produce asexual spores, which swim by beating whiplike flagella. (In most fungal groups, spores are immotile and are dispersed by air currents.)

An extended diploid stage and extensive mycelia are typical of the life cycles for this group. The branching, diploid mycelium produces *gametangia* (multinucleate male and female sexual organs in which haploid "eggs," or gametes, are formed). After fertilization, the resulting zygote becomes a thick-walled resting spore.

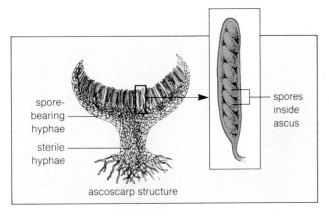

Figure 40.3 Photograph of scarlet cup fungi, of the division Ascomycota. The sketch shows the structure of a cup-shaped ascocarp, composed of tightly interwoven hyphae. Saclike structures (asci) that bear spores occur within ascocarps.

In the sketch: spore-bearing hyphae, sterile hyphae, ascoscarp structure, spores inside ascus

Figure 40.4 Edible morels, rather atypical but delicious sac fungi.

One parasitic member of this group, *Phytophthora infestans*, causes **late blight** (a disease that rots potato and tomato plants). Over a century ago, Irish peasants cultivated potatoes around their cottages and depended upon them as a major source of food. Between 1845 and 1860, there was a succession of cool, damp growing seasons. The parasitic fungus produces more spores in cool weather than in warm weather—spores that swim in the watery film on the plants. Thus the fungus spread rapidly and caused widespread destruction of the potato plants. During this fifteen-year period, a third of Ireland's population either starved to death, died in the outbreak of typhoid fever that followed as a secondary effect, or fled the country.

Zygomycota. This group of fungi has a distinct mode of sexual reproduction. Two similar, club-shaped gametangia fuse to form a zygote. Then a thick wall develops around the zygote, converting it into a *zygospore* (a resting spore).

Many of these fungi are saprobes that live in the soil or on decaying plant matter; some destroy vegetables such as sweet potatoes while they are being stored or transported. Others are parasites on insects such as the common housefly.

The black growth that commonly forms on stale baked products usually is the bread mold *Rhizopus stolonifer*. As is the case for most members of this group, an extensively branched, haploid mycelium forms during the life cycle and produces numerous stalks, each bearing a *sporangium* (a sac within which the asexual spores are formed). So many are formed that the moldy bread looks black. The spores themselves are very light and dry, hence are readily dispersed by air currents. Figure 40.1 describes how another zygomycete, *Pilobilus*, disperses its spores.

Ascomycota. There are more than 15,000 species of sac fungi, which produce haploid spores in saclike cells known as **asci** (singular, ascus). The simplest members are the yeasts. In nature, they occur in the nectar of flowers and on the surfaces of fruits and leaves. One commercially important yeast, *Saccharomyces cerevisiae*, is used to produce the carbon dioxide that leavens bread and the ethanol in wine, beer, and other alcoholic beverages. Yeasts are single-celled organisms that can reproduce asexually by budding or by fission. They also reproduce sexually, most often by the fusion of two gametes that resemble the vegetative cells. Later, meiosis

a

b

c

Figure 40.5 Club fungi. (**a**) A coral fungus. (**b**) Shelf fungi, growing outward from a tree trunk. (**c**) Bird's nest fungi.

occurs in a diploid cell (the ascus). The relative lengths of the haploid and diploid stages vary greatly from one kind of yeast to another, and there is no dikaryotic stage.

The remaining sac fungi are much more complex and numerous than the yeasts. All produce an **ascocarp**, a multicelled structure containing or bearing the asci. Some ascocarps are shaped like globes, others like flasks or open dishes (Figures 40.3 and 40.4). The mycelium grows through soil, wood, and other substrates.

The multicelled sac fungi include some members that are prominent in nature and in human affairs. They include the species that decimated elm trees and chestnut trees in the United States, as well as the truffles and morels that are among the most delicious of the edible fungi (Figure 40.4). Trained pigs and dogs are used to snuffle out truffles, which grow underground; scientists have been frustrated in their efforts to grow truffles commercially. Although the salmon-colored fungus *Neurospora crassa* can become a nuisance in bakeries, it is an important organism in genetic research.

Basidiomycota. You may be familiar with many of the 12,000 or so species of club fungi—the mushrooms, shelf fungi, bird's nest fungi, stinkhorns, and puffballs. Figure

40.5 shows representative species. Most club fungi are saprobes and are especially important decomposers of plant debris. A few (such as the rusts and smuts) cause serious plant diseases. Many species of club fungi are edible; and cultivation of the common mushroom, *Agaricus brunnescens*, is a multimillion-dollar business. However, some are so poisonous that even a mouthful of one can cause painful death; others contain hallucinogenic substances.

The club fungi are distinctive because their haploid spores (called *basidiospores*) are borne on the outside of the cell that produces them. (This is in contrast to the sac fungi, which bear spores inside the ascus.) Usually the spore-bearing cell is club-shaped, giving these fungi their common name.

Typically a short-lived multicelled structure called a *basidiocarp* bears the spore-bearing cells. The part of the fungus that is visible aboveground or on a log (such as the mushroom or a shelf fungus) is actually the basidiocarp; the vegetative mycelium is buried in the soil or decaying wood.

How do the basiodiocarps form during the life cycle? When a haploid basidiospore is dispersed to a suitable habitat and when conditions are favorable, it germinates

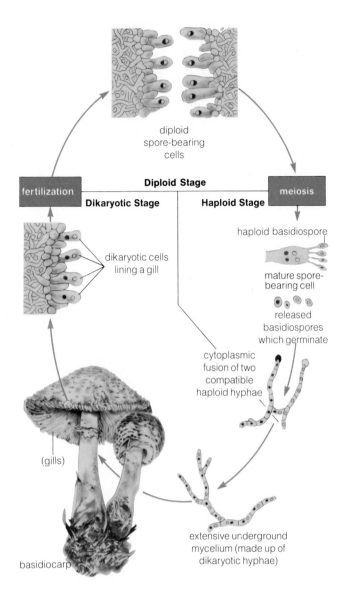

diploid
spore-bearing
cells

Diploid Stage

fertilization

Dikaryotic Stage | **Haploid Stage**

meiosis

haploid basidiospore

dikaryotic cells
lining a gill

mature spore-
bearing cell

released
basidiospores
which germinate

cytoplasmic
fusion of two
compatible
haploid hyphae

(gills)

extensive underground
mycelium (made up of
dikaryotic hyphae)

basidiocarp

Figure 40.6 Life cycle of a typical mushroom. Notice the dikaryotic stage, in which hyphal cells contain two distinct nuclei.

to form a haploid mycelium (Figure 40.6). Often, mycelia of two compatible mating types growing near each other will fuse and exchange nuclei. However, nuclear fusion does not follow immediately and the result is a dikaryotic hypha, the cells of which contain one nucleus of each mating type. The hypha grows into an extensive mycelium that penetrates through the soil or decaying wood, obtaining nutrients. When conditions are favorable, the basiodiocarp forms.

The basiodiocarp, called "the mushroom," consists of a stalk and a cap. The spore-bearing cells are present on the mushroom gills, which are sheets of tissue hanging from the lower surface of the cap (Figures 40.6 and 40.7). At first the spore-bearing cells are dikaryotic, but then the two nuclei fuse to form the zygote. The zygote undergoes meiosis and the haploid spores are produced. Air currents disperse the spores and the mushroom withers. The mycelium, however, can produce a new crop of mushrooms the next season.

Deuteromycota. Of all the fungal species studied, a sexual phase is absent (or undetected) in about 25,000 of them. Often these fungi are called "imperfect fungi" because they apparently lack the sexual, or "perfect" stage. Sometimes a sexual stage is detected later, whereupon the species is reassigned to the sac or club fungi.

Many imperfect fungi cause diseases of plants and animals; in fact, they are responsible for most fungal infections in humans. Some of the saprophytic members grow in stored grain that has become too damp. The toxic by-products of their activities can cause cancer in humans who consume too much of the poisoned grain. Some imperfect fungi are commercially important. For example, *Aspergillus* is used to produce the citric acid that imparts a lemon flavor to candies and soft drinks, and it is used in making soy sauce (the fungus ferments soybeans). Species of *Penicillium* produce the aroma and distinctive flavors of Camembert and Roquefort cheeses. The antibiotic penicillin, used to combat many bacterial

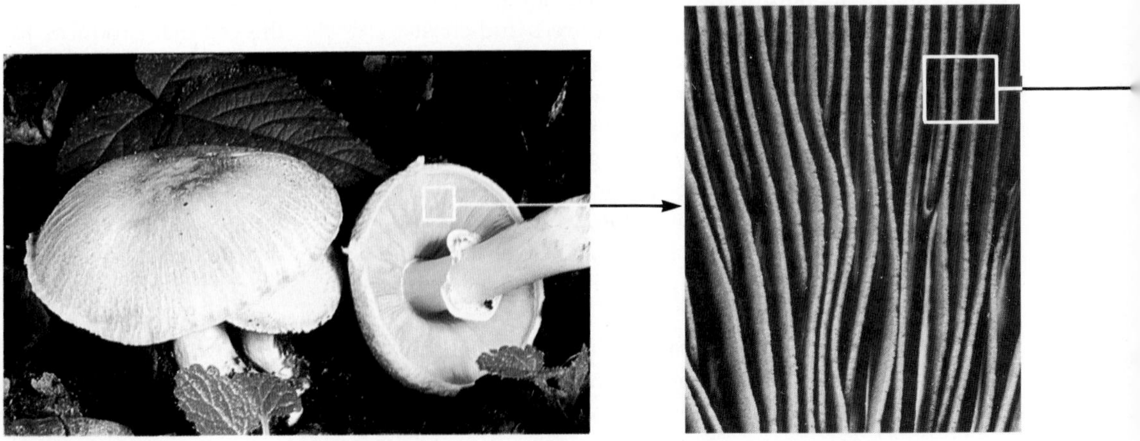

Figure 40.7 A close look at the gills of a basidiocarp from the common field mushroom seen at three magnifications. The highest magnification (far right) shows the spore-bearing cells.

diseases, is also derived from *Penicillium*. Some particularly interesting fungi in this group ensnare nematodes for food (Figure 40.8).

Mycorrhizae

As you read in Chapter Twenty, many complex land plants and fungi enter into *symbiotic relationships*, in which both species benefit from their intimate association. Thus the hyphae of certain club fungi form the dense sheaths around the roots of pine trees and also penetrate between the cells of the superficial layers (Figure 20.4). The combination of roots and hyphae, recall, is a **mycorrhiza**. The fungus actually regulates the flow of mineral ions into the plant, conserving the ions when they are plentiful and making them available to the plant when they are scarce in the soil. The fungus benefits from this partnership by absorbing carbohydrates from the root cells of its host.

Lichens

Some sac fungi (and, to a lesser extent, club fungi) enter into what has been called a symbiotic relationship with a photosynthetic partner, either a cyanobacterium or a green alga. Together they form a unique composite organism, the **lichen**. Figure 40.9 shows the internal structure of one lichen. Figure 40.10 shows two of the 18,000 known species. Lichens live in many seemingly inhospitable or exposed places, including the arctic tundra, bare rocks, and tree trunks. Their acid secretions degrade rock, helping to convert it to soil that can support larger plants. Lichens growing in the arctic tundra, where large plants are scarce, are an important food source for reindeer and other animals. Sometimes air pollution is monitored by observing lichens near cities; lichens cannot grow in badly polluted air.

The relationship often begins when a fungal mycelium contacts a healthy, free-living cyanobacterium or

Figure 40.8 *Arthrobotrys dactyloides*, a predatory "imperfect" fungus with its dinner trapped. The fungal hyphae form nooselike rings that can swell rapidly with incoming water after the hyphal cell walls are stimulated (as when a nematode brushes past). Within a tenth of a second, the increased turgor pressure shrinks the "hole" in the noose and strangles the nematode. Hyphae grow into the animal body and release digestive enzymes.

nematode   noose formed by hypha

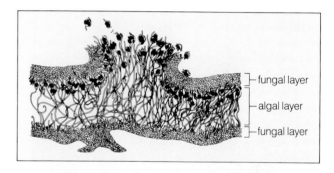

fungal layer
algal layer
fungal layer

Figure 40.9 Cross-section through a complex lichen, *Lobaria verrucosa*. Fungal hyphae with a gelatinlike coat form the upper, protective layer. Just below is a layer of algal cells, which are functionally connected with loosely interwoven, thin-walled hyphae. The lower layer attaches the lichen to the substrate. Notice the fragments containing both hyphae and algal cells; lichens reproduce asexually by such fragmentation.

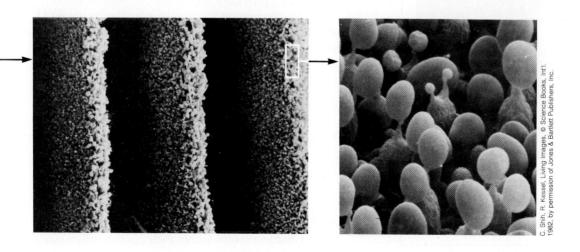

C. Shih, R. Kessel, Living Images, © Science Books, Int'l. 1982, by permission of Jones & Bartlett Publishers, Inc.

a

b

Figure 40.10 Two lichens: (**a**) *Usnea*, known as old man's beard, as it appears on tree limbs, and (**b**) *Cladonia rangiferina*, sometimes called reindeer "moss."

Figure 40.11 Slime molds. Classification of the slime molds is controversial; various members resemble fungi, protistans, and plants. The vegetative body (see *Physarum* above) is a cytoplasmic mass that moves, amoebalike, absorbing dead organic matter, bacteria, and spores. A spore-bearing structure (see *Stemonitis splendens* below) forms during reproduction.

algal cell. The fungus parasitizes the photosynthetic cell, sometimes killing it in the process. If the photosynthetic cell can survive, it multiplies in association with the fungal hyphae. The fungus penetrates cells of its partner with short, specialized hyphae that can absorb as much as eighty percent of the food produced from photosynthesis. Reproduction and growth of the photosynthetic partner suffers from this loss of nutrients, while the fungus depends entirely upon its captive cells for a source of food.

It is difficult to argue that the photosynthetic partner gains much from its enslavement. Of course, by entering into the association, photosynthetic cells can live in places (such as bare rock) that otherwise would be unavailable to them; the lichen can absorb and retain water and the fungal hyphae provide some shading from intense light. Only a few highly specialized green algae truly benefit from the association. These algae, called *Trebouxia*, grow very slowly and cannot on their own compete successfully with other organisms, but they thrive when provided with shelter within the lichen.

Before leaving the fungi, take a look at Figure 40.11, which describes the slime molds. These funguslike organisms are of undetermined affiliation.

PART II. KINGDOM OF PLANTS

Evolutionary Trends Among Plants

Like the fungi, plants are a group with ancient origins; the fossil record indicates that multicelled green algae were well developed 700 million years ago. More than 400 million years ago, simple green shoots were already established on land in moist soil. They had rudimentary water-conducting cells within their photosynthetic aerial

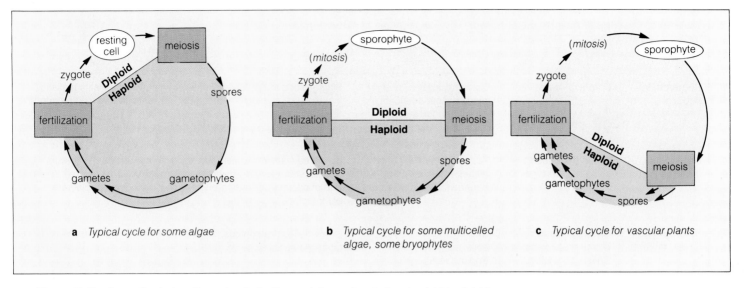

Figure 40.12 Generalized plant life cycles, indicating evolutionary trends from haploid to diploid dominance that occurred among land plants.

stems and they absorbed water and nutrients through mycorrhizal fungi associated with their underground stems. By 345 million years ago, species with large fern-like fronds, tall woody stems, and complex seeds were forming lush green forests.

The fossil record has provided us with the broad outline of plant evolution, but the relationships among the different species are not clearly understood. The best we can do at present is to identify some evolutionary trends that apparently were important, using existing species as our models. Foremost among these was the trend from nonvascular to vascular body plans. Certain reproductive trends, of the sort outlined in Table 40.1, were also among the factors underlying adaptive radiations into diverse environments.

From Nonvascular to Vascular Plants. All plants require water, dissolved mineral ions, and sunlight for photosynthesis, and these requirements undoubtedly influenced the morphological changes that occurred during the evolution of complex land plants. Aquatic plants obviously have no problem with water acquisition; and even such tall aquatic forms as giant kelp grow toward sunlight without specialized support tissues because the surrounding water supports much of their weight. The first land plants were small, and as long as they remained in moist soil near the water's edge, they could simply sprawl out and soak up water from below and sunlight from above.

With the invasion of drier regions, spreading root systems developed that provided the plants with enough water and nutrients, and shoot systems developed that could take advantage of the abundance of sunlight

energy. Increases in size and complexity were sustained by the development of vascular tissues: **xylem** (which transports water and mineral ions through the plant) and **phloem** (which transports sugars and other photosynthetically produced nutrients). Supporting tissues such as sclerenchyma fibers allowed some plants to grow taller and escape the shade created by their neighbors; a cuticle at the surface of aerial parts and numerous stomata in the epidermis permitted some control over water loss in the dry environments. These specialized tissues are characteristic of almost all existing land plants (Chapter Nineteen).

Table 40.1 Summary of Key Reproductive Trends in Plant Life Cycles

	Green Algae	Ferns	Fern Allies	Gymno-sperms	Angio-sperms
Haploid dominance	→ Diploid dominance				→
Homospory (spores of one type)	→ Homospory or heterospory (spores of two types)		→ Heterospory only		→
Isogamy or oogamy	→ Oogamy only				→
Motile gametes			→ Nonmotile gametes → (dispersal agents such as wind, pollinators used)		
No seeds			→ Seeds		→

From James R. Estes, University of Oklahoma.

Table 40.2 Overview of Adaptations of Nonvascular Plants (Algae and Bryophytes)

Division (Phylum)	Some Representatives	Characteristic Habitat	Kinds of Tissue Differentiation
Rhodophyta (red algae)	*Porphyra, Nemalion*	Some freshwater; most marine; more abundant in warmer and tropical seas	Single-celled to branched filaments; some with filaments massed into stemlike and leaflike structures; no vascular (internal transport) tissue
Phaeophyta (brown algae)	*Fucus* (rockweed), kelps, *Sargassum*	Almost all marine, coastal waters especially; most abundant in colder seas	Branched filaments to complex parenchyma bodies; some with leaflike, stemlike (stipe), and anchoring (holdfast) structures; some with ducts for transporting photosynthetic products to different plant regions
Chlorophyta (green algae)	*Ulva* (sea lettuce), *Ulothrix, Spirogyra, Chlamydomonas*	Most freshwater; moist soils; many in shallow tropical seas	Single-celled, filamentous, and simple colonial to simple sheetlike forms; no vascular tissue
Bryophyta (bryophytes)	mosses, hornworts, liverworts	Most land; moist, humid sites; some arid sites; a few submerged sites	Threadlike anchoring structures (rhizoids); leaflike and stemlike structures, often branched; some with simple water- and food-conducting tissues; pores with guard cells in epidermis

Toward a Dominant Sporophyte. The main reproductive mode in plant life cycles is sexual; haploid cells are formed by way of meiosis, then a diploid zygote forms after fertilization. The main variations in the life cycles are in the length of time spent in the haploid and diploid stages and in how large and complex the plant body becomes during each stage.

As we will see, considerable variation exists in the life cycles of algae, which are the simplest plants in terms of their morphology. But the multicelled green algae are of interest here, because they are thought to resemble the forms that gave rise to complex land plants (in particular, the bryophytes and the vascular plants). The life cycles of these algae show a true *alternation of generations*: a multicelled haploid body develops by mitotic divisions, and so does a multicelled diploid body.

The multicelled haploid body is the **gametophyte**; it produces haploid gametes. The multicelled diploid body is the **sporophyte**; it produces haploid spores which, after germinating, develop into the gametophytes (Figure 40.12).

Aquatic environments generally are more stable than terrestrial ones, and perhaps they are more tolerant of variations in the life cycles of water-dwelling algae. On land, where conditions are drier and more hostile, the sporophyte has come to dominate the life cycle. Through the development of extensive root and shoot systems,

the complex sporophytes have the means to gather the raw materials and energy necessary to survive and to produce the next generation.

Isogamy to Oogamy. The gametes produced by some green algae are all motile and of the same size, a condition called **isogamy**. The gametes produced by others are differentiated into motile sperm and immotile eggs. This condition, called **oogamy**, is also characteristic of land plants.

The gametes of water-dwelling species are released into the fluid surroundings, and fertilization occurs when gametes drift together or when one (or both) swims to the other. The bryophytes and a few other land plants are confined to moist environments because gametes cannot get together unless they are released into a liquid medium. As we will see, only among the more specialized land plants do we find reproductive mechanisms that get around this requirement.

The Evolution of Seeds. Some algae and the simplest vascular plants are *homosporous*—that is, all the spores produced by the plant are identical. Many vascular plants are *heterosporous*; they produce two types of spores. Their **megaspores** develop into female gameto-

Light-Harvesting Structures	Gas Exchange Mechanisms	Water Transport and Conservation Mechanisms	Main Reproductive Strategies
Chloroplasts; phycobilin pigments (trap blue-green light in deep water) plus chlorophylls; some species with radial or bilateral branches (nonoverlapping exposure to sunlight)	Diffusion of dissolved gases across individual cell membranes	Direct exchange with surrounding water, across individual cell membranes	Alternation of fertilization and meiosis; sexual reproduction based on oogamy (gametes different in appearance); also vegetative reproduction from plant body fragments or asexual spores
Chloroplasts; xanthophyll pigments plus chlorophylls; leaflike structures			Many with alternation of multicelled generations; sexual reproduction based on isogamy (gametes all identical in appearance), or oogamy; vegetative reproduction from plant body fragments or asexual spores
Well-organized arrays of chloroplasts with membrane stacks; chlorophylls dominant			Alternation of fertilization and meiosis (some have resting spores); sexual reproduction based on isogamy to oogamy; also vegetative reproduction by fragments or spores
Well-organized arrays of chloroplasts with membrane stacks; leaflike structures show radial and bilateral symmetry	Diffusion of dissolved gases across individual cell membranes; some stomatal control	Direct exchange with moisture-laden air; absorption from moist substrate; simple water-conducting cells; waxy covering (cuticle) retards water loss	Alternation of generations; diploid plant body produces homospores (they produce only one type of spore); gametophyte is dominant generation with dependent sporophyte; vegetative reproduction mostly by asexual reproductive bodies (gemmae)

phytes, and their **microspores** develop into male gametophytes. Heterospory was the prelude to the evolution of seeds.

Seed plants, the dominant group of vascular plants, are all heterosporous. The female gametophytes that develop from their megaspores are surrounded by protective tissue and they mature while still attached to the parent plant. Thus the water and food required for their early development are provided by the sporophyte—the stage that is well adapted for obtaining these resources on dry land. The immature male gametophytes, or pollen grains, are released from the parent plant and travel to the female gametophytes by winds, insects, birds, or mammals. The sperm are transported within the protective confines of the pollen grain. Perhaps more than any other factor, the development of pollen grains that carry sperm to the female gametophytes *without* liquid water allowed seed-bearing plants to radiate into diverse environments on land.

Fertilization takes place within the female gametophyte of seed plants, and the zygote develops into a multicelled embryo. Parental tissues that originally surrounded the female gametophyte develop into a protective outer coat around the embryo sporophyte and its food reserves. Together the tissues and the embryo form the seed, the ideal package for surviving dry and otherwise unfavorable environments (page 279).

With these concepts in mind, let's now turn to the spectrum of plant diversity, particularly with respect to the environments in which different species exist. We will begin with the nonvascular plants known as algae and bryophytes. Table 40.2 summarizes the major adaptations of these groups.

Algae

Algae is a term that originally came into use to define simple aquatic "plants." It no longer has any formal significance in classification schemes. The organisms once grouped under the term are now recognized as belonging in three different kingdoms. Thus we have the following divisions:

Division	Kingdom
Cyanobacteria (blue-green algae)	} *Monera*
Chrysophyta (golden algae, diatoms)	
Euglenophyta (photosynthetic flagellates)	} *Protista*
Pyrrophyta (dinoflagellates)	
Rhodophyta (red algae)	
Phaeophyta (brown algae)	} *Plantae*
Chlorophyta (green algae)	

Figure 40.13 (a) Two representative red algae; (b) kelp, a brown alga, as it appears on the water's surface; and (c) a representative green alga.

As you can see, the red, brown, and green algae remain in the plant kingdom. Most aquatic plants belong to these three divisions (Figure 40.13).

Red Algae. The red algae (Rhodophyta) are distinct among algae because of the following combination of features. First, there are no flagellated motile cells in the life cycle. Second, their accessory photosynthetic pigments include phycobilins. Third, sexually reproducing red algae are oogamous, with immotile sperm and specialized, egglike structures.

With the exception of a few freshwater forms, the 4,000 or so species of red algae live in marine environments ranging from shallow intertidal zones to the basements of tropical reefs. Some live as far as 175 meters below the surface when water is clear enough for light penetration and photosynthesis. Most are attached to rocks or other algae; only a few species are free-floating.

Depending on the types and relative amounts of their phycobilins, red algae range in color from green to red, purple, and greenish-black. The phycobilins contribute to photosynthesis, especially in deeper waters. Chlorophyll functions most efficiently in light containing red and blue wavelengths, which do not penetrate far below the surface of water. In deeper waters the available wavelengths are mostly blue-green, which the accessory pigments can absorb.

Patterns of sexual reproduction among red algae are complex. In most species, the zygote develops into a small, diploid sporophyte that is parasitic on the female gametophyte. The diploid spores it produces develop into a larger, independent sporophyte. This second sporophyte finally produces haploid spores that form the gametophyte generation.

Extracts from several red algae are used to make agar (the basis of a culture medium for growing bacteria and fungi in the laboratory) and they are used in the manufacture of ice cream and chocolate milk.

Brown Algae. The brown algae (Phaeophyta) include about 1,500 olive-green, golden, and dark-brown species, many of which are conspicuous seaweeds. They have an abundance of accessory photosynthetic pigments, particularly xanthophylls, that give them their distinct color. Many brown algae live in the intertidal zone along rocky coasts, attached to submerged rocks by *holdfasts* (rootlike structures at the base of the plants). Others, including the kelps, grow profusely offshore but are still anchored to the bottom by holdfasts. The giant kelps sometimes grow fifty meters long, forming underwater forests that sway with the currents. Divers are careful in these forests, for they can become entangled in the dense, waving plants. Some brown algae thrive in the open sea. *Sargassum* floats as tangled masses through the vast Sargasso Sea, which lies between the Azores and the Bahamas.

Many kelps have complex organization. In addition to a holdfast, some species have leaflike *blades* and a *stipe* (a stemlike structure). Sometimes hollow, gas-filled stipe regions called *floats* occur at regular intervals and help hold the plant body upright in water. Surrounded as they are by water, brown algae have no requirements for water-conducting tissues. However, some species have tubelike strands of cells similar to those found in phloem, which transports photosynthetic nutrients.

Like many of their red and green relatives, many brown algae show an alternation of generations. Among simpler, filamentous species, the gametophyte and sporophyte may be much the same in outward appearance. In more complex species, the gametophyte is extremely reduced in size and a large, complex sporophyte dominates. For instance, the kelp shown in Figure 40.13 is the sporophyte; the gametophyte consists of microscopic filaments.

Green Algae. There are at least 7,000 species of green algae (Chlorophyta). They are like complex land plants in terms of their photosynthetic pigments; all have chlorophylls *a* and *b* as well as carotenoids and xanthophylls. Also like land plants, they store excess carbohydrates as starch and have cellulose in their cell walls. Most green algae live in fresh water, but many species grow on snow, soil, patios, and tree trunks.

Green algae reproduce both sexually and asexually, and the details of their reproductive modes are diverse. Some single-celled green algae of the genus *Chlamydomonas* are isogamous; haploid cells of different mating types function as gametes, but the cells are identical in size and structure. There is no multicelled stage; *Chlamydomonas* simply alternates between haploid and diploid cells. Other species of *Chlamydomonas* are oogamous. Their gametes differ in size and motility, much like the motile sperm and immotile eggs of animals.

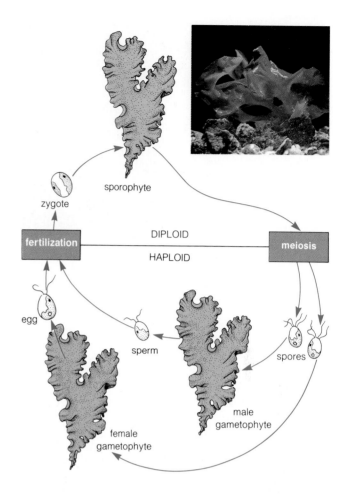

Figure 40.14 Life cycle of sea lettuce (*Ulva*), one of the green algae.

Ulva, or sea lettuce, is an example of a multicelled green alga. As Figure 40.14 shows, it has a broad, leaflike *thallus* (a vegetative body with relatively little cell differentiation). The spores produced by meiosis give rise to "male" and "female" gametophytes that are structurally the same. Motile eggs and sperm produced by these gametophytes fuse to form a new zygote.

The Land Plants

More than 400 million years ago, the first plant pioneers on land had already diverged into two lineages, the bryophytes and the vascular plants. Both lineages share the following features, which must have been adaptive during the transition to land:

1. Some bryophytes and all vascular plants have a cuticle and numerous stomata on aboveground parts.

a

b

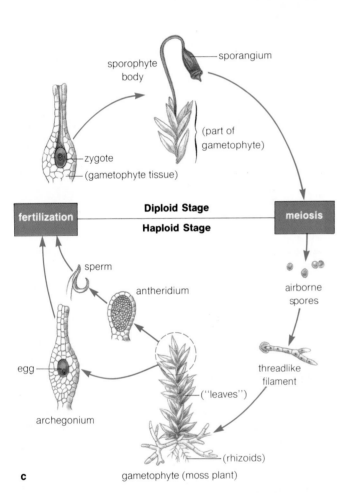

c

Figure 40.15 Bryophytes: (**a**) a moss and (**b**) a liverwort. (**c**) Moss life cycle. (As in the other plant life cycles, the various structures are not drawn to scale.) Notice how the sporophyte body remains attached to the gametophyte (moss plant) body.

2. A protective layer of sterile cells surrounds the reproductive cells, forming a structure that keeps them from drying out. Thus an **archegonium** surrounds the haploid egg in female gametophytes, an **antheridium** surrounds haploid sperm in male gametophytes, and a **sporangium** surrounds haploid spores in the sporophyte.

3. The embryo undergoes early development *within* the female gametophyte. Among most algae, recall, the new individual is pretty much on its own and develops apart from parental cells.

The Bryophytes

The bryophytes include about 16,000 species of mosses, liverworts, and hornworts (Figure 40.15). All are small plants, generally less than twenty centimeters long. Most grow in regions that are moist for at least part of the year. Unless there is a film of water on the plant parts, sperm cannot swim to the egg-containing archegonia and sexual reproduction will not occur.

The bryophytes do not have vascular tissues (xylem and phloem). Without vascular tissue, their leaflike, stemlike, and rootlike organs are not regarded as true leaves, stems, and roots. For example, instead of multicellular roots, most species have *rhizoids* (long single cells or filaments that attach the gametophyte to the substrate and absorb water and mineral ions).

Mosses are the most common bryophytes. When a moss spore germinates, it produces a green, threadlike gametophyte that resembles filamentous green algae. The filament grows by mitotic divisions into the familiar moss plant (Figure 40.15). Often the gametophyte contains a central strand of simple water-conducting cells. A few also contain living food-conducting cells arranged around the central strand. These tissues are not as complex as the conducting cells of vascular plants and are not regarded as true xylem and phloem.

Each haploid gametophyte produces antheridia, archegonia, or both at the shoot tips. In most species, these structures occur on separate plants growing near each other, and sperm must swim through a film of water to reach the eggs. After fertilization, the zygote divides mitotically to form an embryo that develops into a mature sporophyte. The sporophyte consists of a stalk with a sporangium that contains haploid spores (Figures 40.14 and 40.15). Although the sporophyte may be photosynthetic at first, it eventually depends on the gametophyte for all nutrients and water. The bryophytes are distinct from all other land plants in having an independent gametophyte and a dependent sporophyte.

Ferns and Their Allies

Existing vascular plants are listed in Table 40.3. Their ancestors appeared about 400 million years ago, and within a relatively short span of 100 million years, all major lines of vascular plants were established on land. Some forms that once dominated the landscape became extinct (Figure 40.16), but among existing vascular plants are lycopods, horsetails, and ferns—all of which bear strong resemblance to their ancient relatives. Unlike the bryophytes, the sporophyte is free-living and is the most conspicuous stage of the life cycle.

Lycopods. The lycopods (Lycophyta) were highly diverse 350 million years ago, when some forms were even tree-sized. Like all of the existing genera, *Lycopo-*

dium contains miniature descendants, including the club mosses, that typically grow on forest floors. A *Lycopodium* sporophyte has true roots, stems, and small leaves with a single strand of vascular tissue. The sporophyte has *strobili*, which are conelike clusters of leaves bearing the sporangia (Figure 40.17). Haploid spores dispersed from the sporangia germinate to form small, free-living gametophytes. Although most lycopods are homosporous, some are heterosporous. Both types require ample water for sperm to swim to the egg cells.

Figure 40.16 Reconstructions of vascular plants of the Carboniferous. Most of the tall, treelike forms shown here became extinct.

a b c

Figure 40.17 (a) *Lycopodium* sporophytes, showing the cone-like strobili in which spores are produced. (b) Horsetails (*Equisetum*) showing how strobili are carried on stems. (c) Detail of horsetail strobilus showing clusters of sporangia.

Table 40.3 Overview of Adaptations of Vascular Plants

Division (Phylum)	Some Representatives	Characteristic Environment	Kinds of Tissue Differentiation
Ferns and Fern Allies (non-seed-bearing; depend on free water for fertilization)			
Lycophyta (lycopods or club mosses)	*Lycopodium*	Land; wet, shaded sites in tropics and subtropics; some arctic, temperate desert	Sporophyte with true vascular system; roots, stems, leaves; stomata with guard cells in one or both leaf surfaces; vascular system makes possible a higher volume-to-surface ratio than in nonvascular plants. These features occur in all eight plant divisions listed here
Sphenophyta (sphenopsids)	*Equisetum* (horsetails), only living genus	Land; acid soil; sand dunes, swamps, moist woodlands, lake margins, railroad embankments	Sporophyte with true vascular system; extensive underground stem (rhizome) with roots and aerial shoots at nodes; aerial stems jointed, hollow, branched or unbranched
Pterophyta (ferns)	Sword ferns, lady ferns, braken ferns, tree ferns	Land; some epiphytes (attached to but nonparasitic on other plants); most wet, humid sites	True vascular system; conspicuous leaves; most with creeping rhizomes; columnlike stem in tree ferns
Gymnosperms ("naked-seed-bearing"; not dependent on free water for fertilization)			
Cycadophyta (cycads)	*Zamia*	Limited tropical, subtropical land regions	Well-developed, complex vascular system; stems short and bulbous, or columns; palmlike appearance; fernlike leaves; massive cones (reproductive structures bearing pollen or seeds)
Ginkgophyta (ginkgos)	*Ginkgo biloba* (only existing species)	Land; temperate regions	Well-developed, complex vascular system; tall, woody stem (trunk) with long side branches; fan-shaped, deciduous leaves
Coniferophyta (conifers)	pine, spruce, fir, juniper, hemlock, cypress, redwood, larch	Land; widespread through Northern and Southern hemispheres	Well-developed, complex vascular system; some shrubby, others impressively tall, thick-trunked trees; most evergreen; cones
Gnetophyta (gnetophytes)	*Gnetum, Ephedra, Welwitschia*	Land; warm-temperate regions; desert and mountain sand or rocky soil	Well-developed, complex vascular system; some shrubby, branched, with whorled leaves; *Welwitschia* short, bowl-shaped stem, two large, strap-shaped leaves, cones
Angiosperms (flowering, seed-bearing; depend on pollinating agents; not dependent on free water for fertilization)			
Anthophyta Monocotyledonae (monocots)	Grasses, palms, lilies, orchids, onions, pineapple, bamboo	Almost all land zones; some aquatic	Well-developed, complex vascular system; floral structures; one seed leaf (cotyledon); floral parts generally in threes or multiples of threes; parallel-veined leaves common*
Dicotyledonae (dicots)	Most temperate-zone fruit trees; roses, cabbages, melons, beans, potatoes	Almost all land zones; some aquatic	Two seed leaves; floral parts generally in fours, five, or multiples of these; net-veined leaves common*

*See page 241.

Light-Harvesting Structures	Gas Exchange Mechanisms	Water Transport and Conservation Mechanisms	Main Reproductive Strategies
Well-organized arrays of chloroplasts with membrane stacks; most with spirally arranged leaves; some with palisade layer of photosynthetic cells in leaves	Diffusion of dissolved gases across individual cell membranes; regulation through numerous stomata	Well-developed xylem and accessory cells; cuticle (retards water loss)	Alternation of generations; complex, dominant sporophyte independent of gametophyte; some homosporous (they produce only one type of spore), others heterosporous (produce spores of different types, which give rise to male or female gametophytes)
Well-organized chloroplast arrays; upright growth habit; whorled, scalelike leaves but most photosynthesis in stems			Alternation of generations; complex, dominant sporophyte free-living (independent of gametophyte); homosporous
Well-organized chloroplast arrays; most photosynthesis in stems			Alternation of generations; sporophyte free-living; most homosporous
Well-organized chloroplast arrays; complex leaves, arranged in pattern that allows good sunlight exposure	Diffusion of dissolved gases across individual cell membranes; complex stomatal control	Well-developed xylem and accessory cells; cuticle; water storage in roots, stems	Alternation of generations; dominant, woody sporophyte body, heterosporous; well-developed seed- and pollen-bearing cones
			Alternation of generations; dominant, woody sporophyte body; heterosporous; well-developed seeds not borne in cones; separate male and female sporophytes
			Alternation of generations; dominant, woody sporophyte body (e.g., the pine tree); heterosporous; well-developed seed- and pollen-bearing cones
Well-organized chloroplast arrays; in *Ephedra*, most photosynthesis in stems and branches (almost leafless)			Alternation of generations; dominant, woody sporophyte body; heterosporous; seed- and pollen-bearing cones; separate male and female sporophytes
Well-organized chloroplast arrays; complex leaves, arranged in patterns that allow good sunlight exposure	Diffusion of dissolved gases across individual cell membranes; complex stomatal control	Well-developed xylem and accessory cells; cuticle; water storage in roots, stems	Alternation of generations; dominant woody or herbaceous sporophyte body; diverse floral structures adapted to pollinating agents (wind, water, insects, birds, mammals); all are heterosporous; seeds enclosed within ovary; fruits (ripened ovaries) aid in seed dispersal by wind, water, animals

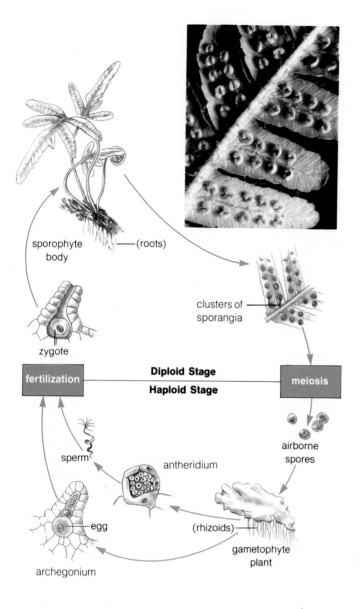

Figure 40.18 Fern life cycle. Clusters of sporangia occur on the underside of the fern leaf.

sporophyte body

(roots)

clusters of sporangia

zygote

fertilization

Diploid Stage

Haploid Stage

meiosis

airborne spores

sperm

antheridium

egg

(rhizoids)

archegonium

gametophyte plant

Figure 40.19 Ferns in a temperate rain forest of Tasmania.

Horsetails. The horsetails (Sphenophyta) are relatively simple vascular plants, but their ancient relatives included treelike forms about fifteen meters tall. A single genus, *Equisetum*, has survived to the present and grows in moist soil throughout much of the world. Often you will see horsetails growing in vacant lots and along railroad tracks.

Horsetail sporophytes typically have underground stems *(rhizomes)*, with roots and aerial branches arising at nodes. Scalelike leaves are arranged in whorls about a distinctive hollow, photosynthetic stem. Pioneers of the American West used horsetails to scrub their cooking pots; the walls of stem cells contain silica, giving the stems a sandpaper quality. All species are homosporous,

with the spores formed in strobili borne at the shoot tips (Figure 40.17). Winds disperse the fragile horsetail spores, which must germinate within a few days to produce haploid gametophytes. The horsetail gametophyte is a free-living plant, about the size of a pea, in which sperm and eggs develop. Here again, the sperm must travel through free water (dew or raindrops) to reach the eggs.

Ferns. The 12,000 or so existing species of ferns (Pterophyta) have well-developed leaves (Figures 40.18 and 40.19). Except for tropical tree ferns, fern stems are underground; only the leaves are aerial. Most species are homosporous and their spores are dispersed through air. The spores of aquatic, heterosporous ferns are dispersed through water. In both cases, spores germinate and develop into small gametophytes, the most common being the green, heart-shaped type shown in Figure 40.18.

Lycopod, horsetail, and fern sporophytes have well-developed vascular systems. Although they are adapted for life on land, they are confined largely to wet, humid regions because their short-lived gametophytes have no

Figure 40.20 Two seed-bearing plants; the cycad (**a**) and the ginkgo (**b**). Cycad cones are shown in (**c**); the seeds of the ginkgo are shown in (**d**).

vascular tissue for water transport. Moreover, the male gametes must have water if they are to reach the eggs. Thus the primitive vascular plants are the "amphibians" of the plant kingdom: *for part of the cycle they are, in essence, aquatic.* Those species that are adapted for survival in extreme environments such as deserts reproduce only when adequate water is available.

Gymnosperms: The Conifers and Their Kin

There are two divisions of seed plants: the gymnosperms and the angiosperms, or flowering plants. In both groups, the gametophytes depend totally on the sporophyte for water and nutrition. The female gametophyte develops inside the seed and the male gametophyte forms the pollen grain.

The young seeds of gymnosperms are borne on surfaces of reproductive structures, without being completely surrounded by protective tissue layers (as they are in angiosperms). The gymnosperms include cycads, ginkgos, gnetophytes, and conifers.

Cycads. During the Mesozoic, the cycads (Cycadophyta) flourished along with the dinosaurs. About ninety species have survived to the present, but they are confined to the tropics and subtropics. At first glance you might mistake a cycad for a small palm tree (Figure 40.20a). Despite having similar leaves and stems, palms and cycads are not closely related. (Palms are flowering

plants, not cone-bearing gymnosperms.) Cycads have massive cones that bear either pollen or seeds (Figure 40.20c). Pollen from male plants is transferred by air currents to the developing seeds on female plants. In tropical Asia, cycad seeds and a starchy flour made from cycad trunks are used as food sources.

Ginkgos. The ginkgos (Ginkgophyta) are even more restricted in native distribution than the cycads. Only a single species has survived (Figure 40.20b,d) despite the diversity of the group during the Mesozoic. It seems that several thousand years ago, trees of this species were planted extensively in cultivated grounds around Buddhist temples. Since then, the natural populations from which the "domesticated" trees were derived must have become extinct. The near-extinction of this living fossil from the age of dinosaurs is puzzling, for ginkgos seem to be hardier than many other trees. They are planted in cities because of their attractive, fan-shaped leaves and their resistance to insects, disease, and even air pollution. Nevertheless, they have survived to the present day only with human protection.

Gnetophytes. The division Gnetophyta contains about seventy species divided into three genera: *Gnetum,* *Ephedra,* and *Welwitschia.* Of all the gymnosperms, *Welwitschia* of arid regions of southwest Africa is probably the most unusual. Most of this seed-producing plant is a deep-reaching taproot; the only exposed part is a woody disk-shaped stem that bears cones and leaves.

a

b

Figure 40.21 (a) *Welwitschia*, a seed-producing gnetophyte. (b) Close-up showing the female cones.

Welwitschia never produces more than two strap-shaped leaves, which split lengthwise repeatedly as the plant grows older and make it look like a pile of straggly leaves (Figure 40.21).

Conifers. Pine, spruce, fir, hemlock, juniper, cypress, and redwood are members of the Coniferophyta, or conifers—the most diverse and widely distributed of all gymnosperms. Most of these cone-bearing, woody trees and shrubs have needlelike or scalelike leaves. Most conifers are evergreens; although old leaves fall from the tree, enough are always attached to distinguish these plants from deciduous species.

Conifer life cycles vary in the duration of events. Here we will focus on the pine life cycle (Figure 40.22). As with all conifers, pine seeds develop on the shelflike scales of a female cone. Each immature seed, or ovule, consists of a tissue layer (the megasporangium) sur-

rounded by protective layers that will become the seed coat. Inside the megasporangium, a reproductive cell undergoes meiosis, and one of the resulting haploid spores (the megaspore) survives to form a female gametophyte.

The male gametophytes originate within the pollen sacs (microsporangia), which are borne on the scales of male cones. Inside the sacs, reproductive cells undergo meiosis and microspores are produced. The microspores develop into pollen grains, which contain the immature male gametophytes (Chapter Twenty-One).

Each spring, a pine tree produces millions of pollen grains. Each floats through the air, having been equipped with two tiny air sacs that impart bouyancy. With such extravagant discharges from the male cones, at least some pollen grains land on the female cones, where they are trapped by a sticky substance at the tip of the ovule. When the sticky fluid dries, it pulls pollen grains inside the immature seeds.

At least a year passes between the time of pollination and fertilization. During this period, the pollen grain produces a pollen tube (a tubelike projection) that starts to grow toward the developing female gametophyte, carrying the nonmotile sperm inside it. Eventually, one cell within each pollen grain divides by mitosis and forms two sperm nuclei (Figure 40.22).

Also during this period, the female gametophyte gradually matures. The gametophyte contains two or more archegonia, each with a single egg (Figure 40.22). When the pollen tube reaches an archegonium, the stage is set for fertilization, zygote formation, and development of the embryo. The whole seed, including the embryo, female gametophyte, and seed coats, is shed from the pine cone. The seed coat protects both the embryo and the gametophyte from drying out and injury. The female gametophyte tissue serves as a food reserve for the embryo during germination.

During the Mesozoic, conifers were the dominant land plants. Their mechanisms of seed production, protection, and dispersal helped assure their radiation into many land environments. Since then, their distribution has been gradually reduced, although they are still the dominant plants in many regions, especially to the north and at high altitudes (Figure 38.19). Conifers are major sources of lumber, paper, and many other commercial and industrial products.

Angiosperms: The Flowering Plants

The word angiosperm means a seed carried within a container; namely, within a mature ovary. This descriptive term applies to flowering plants (the Anthophyta). If numbers are any indication of adaptive success, the flowering plants are indeed successful. There are about

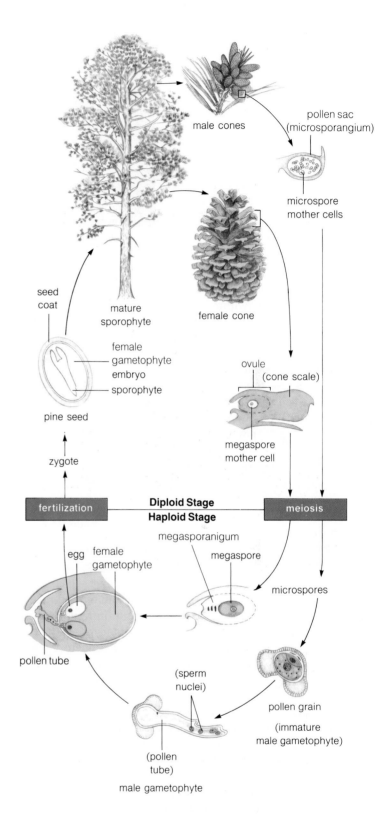

Figure 40.22 Life cycle of a pine. The photograph shows a representative conifer: a ponderosa pine.

250,000 known species (far more than are found in other divisions), and new species are being discovered almost daily in unexplored regions of the tropics. Flowering plants are also the most diverse types of plants. They range in size from tiny duckweeds (about a millimeter long) to *Eucalyptus* trees more than 100 meters tall. Most are free-living and photosynthetic, but some are saprophytic or parasitic. Flowering plant species are found not only on dry land but also on wet land margins, in fresh water, and in seawater.

Flowering plants are grouped into two classes: the monocots and the dicots (Chapter Nineteen). Figure 40.23 shows representatives of both classes. There are at least 200,000 species of dicots, including most of the familiar shrubs and trees (other than conifers), most of the herbs, the cacti, and the water lilies. The monocots include grasses, palms, lilies, and orchids. The main crop plants of the world are wheat, corn, rice, oats, rye, and barley—all monocots, and all domesticated grasses.

Many factors have contributed to the adaptive success of the flowering plants. As with other seed plants, the diploid sporophyte dominates the life cycle. This sporophyte, which retains and nourishes the gametophyte, is well adapted for life on land. Unlike the gym-

Figure 40.23 Floral structure of the wild iris, a representative monocot (**a**); and *Hypericum*, a representative dicot (**b**).

a

b

nosperms, embryos are nourished by a unique tissue, the endosperm, within the seed. Additionally, the seeds of flowering plants are packaged in fruits, which are structures that assure protection and dispersal. Above all, they have a unique reproductive system involving flower formation (page 271). As we have seen, unique and diverse floral structures have in many cases co-evolved with pollinating vectors (insects, bats, birds, and rodents). Although some flowering plants are wind-pollinated, it may have been the effectiveness of pollination vectors that originally led to the gradual displacement of gymnosperms and to the rise of dominance by the flowering plants during the past 100 million years.

PERSPECTIVE

How often, if we think of them at all, do we think of plants as little more than backdrops against which the activities of animals are played out? How often do we look upon fungi as odd or nasty little intruders into human affairs? Yet, without these organisms, without their initial invasion of the land and subsequent interactions, we humans never would have evolved. Until they came to cloak the earth, inch by inch, continent by continent, no other complex forms of life could have appeared on the vacant land. Their evolution, while not smacking of the drama of, say, the rise and fall of dinosaurs, was nevertheless remarkable.

The vegetative body of land plants cannot crawl, leap, run, or fly. Beginning with ancestral forms no more complex than the simple green algae, a long series of mutations and selection processes led to the first simple terrestrial plants. Fossils of those plants tell us they had little more than a few threadlike anchors to sink into muddy margins of pools and lagoons, tiny photosynthetic stalks, and a few scalelike projections to spread out stiffly beneath the sun.

Yet within those simple structures, hollow cellular tubes eventually developed, foreshadowing the trans-

port systems that now run through the roots, stems, and leaves of vascular plants. The diploid stage of the life cycles of certain aquatic plants proved to have remarkable potential. The diploid sporophyte generation, resistant as it was to adverse conditions, gradually became more and more pronounced in the move to higher and drier land. Among some land plants, gametophytes came to be housed inside the parent sporophyte body, which could nurture them with water and dissolved nutrients. In some lineages, seeds took over the dispersal function of spores.

The reproductive structures of plants became attuned to the seasons, to times of rain and winds. Most remarkable of all, some plants coevolved with insects and other animals. They came to depend on animals to carry pollen grains to female reproductive structures and to disperse seeds. Simultaneously, the animal pollinators came to depend on specific plants for food.

Over time, the fungi became part of life on land, as they had been in water—decomposers, recyclers of life-giving nutrients for the animals and plants with which they have been linked in the web of life. Like plants, most fungi rely on wind, splashing rains, and insects for dispersal. Thus, even without their own means of motility, plants and fungi have moved over the barren plains to the mountains, wherever there is land.

SUMMARY

1. Both plants and fungi are mostly multicelled forms in which haploid and diploid stages alternate in the life cycle.

2. Fungi are heterotrophs, either saprobes that obtain nutrients from nonliving organic matter or parasites that obtain nutrients from living organic matter. In both cases, fungal cells secrete digestive enzymes and then absorb nutrients across their plasma membrane.

3. The multicelled fungal body or mycelium is made up of microscopic filaments called hyphae.

4. Asexual reproduction in the fungi may occur by the formation of asexual spores, by the fragmentation of the mycelium, or by binary fission.

5. When fungi reproduce sexually, fertilization results in a diploid stage (a zygote or a multicelled structure). Meiosis results in the haploid phase of the life cycle (gametes only or a multicelled structure). If there is a delay between cytoplasmic and nuclear fusion, a dikaryotic stage occurs in the life cycle.

6. Fungi are involved in two ecologically important associations: those with algae (in lichens) and those with the roots of vascular plants (mycorrhizae).

7. The plant kingdom includes aquatic and terrestrial forms. The land plants are believed to have evolved from multicelled green algae more than 400 million years ago.

8. Some of the important trends in the evolution of plants were (a) a dominant sporophyte stage in the life cycle, (b) a transition from isogamy to oogamy, (c) the development of vascular tissue, and (d) the transition from homospory to heterospory and from unprotected zygotes to the formation of seeds.

9. The algae are grouped in three divisions: the red, brown, and green algae. Species of the red and brown algae are the most common seaweeds. These two groups have chlorophyll, but the green color is masked by accessory pigments. The green algae are similar to the land plants in having chlorophylls *a* and *b* in their plastids, in storing carbohydrates as starch, and in having a cellulose framework within the cell wall.

10. The land plants, which include the bryophytes and vascular plants, typically have a cuticle and stomata that regulate water loss; multicelled reproductive structures on the gametophyte (archegonia and antheridia) and on the sporophyte (sporangia); and a sporophyte that undergoes embryonic development within the gametophyte tissues.

11. The bryophytes, including the mosses and liverworts, are typically found in moist soils: conducting tissues (if any) are simple. Water is required for fertilization of the eggs by the sperm. The bryophytes are distinct from all other land plants in having an independent gametophyte and a dependent sporophyte.

12. The simplest vascular plants are the lycopods, horsetails, and ferns. These groups differ primarily in the size, arrangement, and complexity of their leaves. They all share a similar life cycle, with an alternation between independent gametophyte (haploid) and sporophyte (diploid) generations.

13. The most complex vascular plants are the seed plants (the gymnosperms and flowering plants). The gymnosperms include cycads, ginkgos, gnetophytes, and conifers. The flowering plants are divided into dicots and monocots.

14. In all flowering plants, the gametophyte is wholly dependent on the sporophyte.

15. Sexual reproduction may take two years or more in some conifers such as the pine. In contrast, sexual reproduction in many flowering plants may be completed in weeks. The efficient reproductive structures of the flowering plants, including flowers (which attract pollinators) and fruits, have been largely responsible for the evolutionary success of this group.

Review Questions

1. Name the five main divisions of fungi and describe a representative for each.

2. Which three types of characteristic events are typical of asexual reproduction of a fungus?

3. Can you describe what kind of stage intervenes between the haploid and diploid stages in the life cycle of a common field mushroom?

4. What is the difference between a mycorrhizal mat and a lichen?

5. Distinguish between the megaspores, microspores, and gametes of land plants. As part of your answer, define sporophyte and gametophyte. Which is haploid? Which is diploid?

6. Describe the key evolutionary trends that figured in the invasion of land by plants.

7. What are some similarities and differences between algae and bryophytes? Bryophytes and lycopods?

8. If both gymnosperms and flowering plants are seed-breeding, then how do they differ in reproductive modes?

9. Choose a garden plant, crop plant, or weed that grows in the area where you live. Make a diagram of it, labeling its complement parts. Can you correlate some of those parts with environmental conditions—seasonal variations in temperature, moisture, and so on?

Readings

Christensen, C. 1972. *The Molds and Man*. Third edition. Minneapolis: University of Minnesota Press.

Moore-Landecker, E. 1982. *Fundamentals of the Fungi*. New Jersey: Englewood Cliffs. Very well-written introduction to the kingdom of fungi.

Raven, P., R. Evert, and S. Eichhorn. 1986. *Biology of Plants*. Fourth edition. New York: Worth. One of the most popular of all introductions to botany. Exquisite illustrations; there is nothing else like it.

Scagel, R. et al. 1984. *Plants: An Evolutionary Survey*. Belmont, California: Wadsworth. Excellent reference book.

Stebbins, G., and G. Hill. March 1980. "Did Multicellular Plants Invade the Land?" *The American Naturalist* 115(3):342–353.

Taylor, T. 1981. *Paleobotany*. New York: McGraw-Hill. Everything you ever wanted to know about the ancestors of modern-day plants and fungi.

41

ANIMAL DIVERSITY

It is the last scene of a Western film. The hero has tracked down the rustler, saved the longhorns, fought off a mountain lion, barely missed being bitten by a rattlesnake, and located the rancher's kidnapped daughter by spotting vultures circling in the sky above her. Now, with his faithful dog beside him, he waves goodbye to the girl, mounts his restless steed, splashes across a stream, and rides off into the golden west.

Stories like this one give us a stereotyped idea of what "animals" are. You may have noticed that seven different kinds of animals made their appearance in the film. But did you notice that all were *vertebrates*—animals with backbones? The story is somewhat lopsided, because about ninety-seven percent of all animals do not have backbones; they are *invertebrates*. Although the camera did not record them, there must have been plenty of worms, insects, spiders, scorpions, snails, and clams either on land or in the stream.

The extent of invertebrate diversity is evident in classification schemes, which put all vertebrates into one phylum and the invertebrates into about thirty. In this chapter, we will use examples of a few phyla as models in order to put together a general picture of evolutionary trends that led to a diversity of animals, from sponges and jellyfishes to snakes, horses, and ourselves. Before we begin, let's consider some features that all animals share and some that help us distinguish one group from another.

Figure 41.1 One of the lesser known invertebrates: a ribbon worm, a member of the phylum Nemertea. These animals have a proboscis, a structure that can be thrust out quickly under hydrostatic pressure to capture prey.

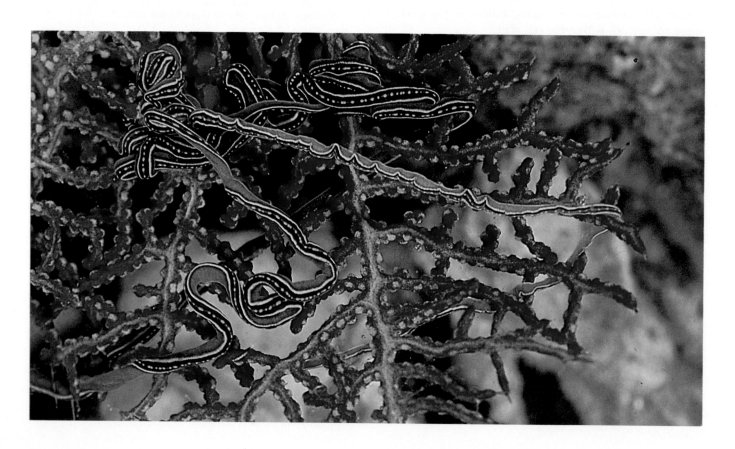

GENERAL CHARACTERISTICS OF ANIMALS

Members of the animal kingdom, **metazoans**, generally have the following characteristics:

1. Animals are multicellular. Except for *Trichoplax* (page 582) and sponges, animals have cells differentiated into tissues, and most have various tissues aggregated into organs.

2. Animals are heterotrophs that usually eat other organisms or the products of other organisms. Some, especially certain parasites, absorb soluble nutrients made available to them by the animal in which they live.

3. Animals are sexually reproducing, diploid organisms. Many also can reproduce asexually.

4. Animal embryos undergo progressive stages of growth and development.

5. Most animals are motile during at least part of the life cycle.

Body Plans

Other features that are important in characterizing animals include the following: body symmetry, type of gut, body cavity (or lack of it), the presence or absence of segmentation, and cephalization.

Body Symmetry. Sponges are among the few animals without a true symmetry; even when they are cup-shaped their microscopic organization is asymmetrical. Almost all animals have either radial or bilateral body plans. Perfect radial symmetry, as seen among the jellyfishes, means the body can be divided into four or more pieces that are equal with respect to the structures they contain. Bilateral symmetry, as in humans, means the body has an anterior and a posterior end, a left and right side, and a dorsal and ventral surface (Figure 23.11). The only plane in which the animal can be divided into equal halves extends from dorsal to ventral along the midline. If certain structures are displaced to one side or the other of the midline, as is often the case, this does not violate the basic bilateral plan.

Type of Gut. Almost all animals have a *gut*, a cavity or tube into which food is taken and within which digestion proceeds. In some cases the gut is a sac with a single opening (the mouth), although the sac may branch extensively through much of the body. No matter how elaborate the sac, undigested residues must be eliminated through the mouth opening.

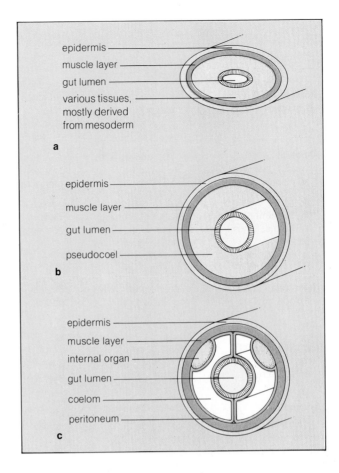

Figure 41.2 Body plans of bilateral animals. (**a**) Acoelomate, or without a body cavity, as in flatworms. (**b**) Pseudocoelomate, without a continuous peritoneal lining; various organs, especially of the reproductive system, occupy the pseudocoel. (**c**) Coelomate, a plan typical of annelids, arthropods, mollusks, echinoderms, and chordates.

In most cases, animals have a *complete* gut (with a mouth at one end of a "tube" and an anus at the other). This is a more efficient arrangement, partly because the mouth region needs to be involved only with taking in food. Moreover, a one-way gut affords more possibilities for regional specialization in such tasks as grinding, digestion, and absorption.

Body Cavities. Although a gut is an internal space, we are concerned here with cavities between the gut and the body wall. For example, the heart, lungs, digestive tract, and other organs of the human body are located in a **coelom**, a type of cavity that is defined by a continuous lining of epithelial cells called the *peritoneum* (Figure 41.2). The peritoneum also lines the outside of the digestive tract and other organs in the coelom.

Table 41.1 Overview of Characteristics of Major Animal Phyla

Phylum (and some representatives)	Typical Environment	Typical Life-Style of Adult Form	Integration	Support and Movement
Porifera (8,000)* sponges	Most marine, some fresh water	Attached filter feeders	No nervous system (cell-to-cell transmission)	Support by spicules, protein fibers, or both; contractile cells change openings at body surfaces
Cnidaria (9,000) hydras, sea anemones, jellyfishes, corals	Most marine, some fresh water	Attached, creeping, swimming, or floating carnivores	Nerve net, cell-to-cell transmission	Hydrostatic support (by fluid in gut) external or internal secreted skeletal elements in many; contractile fibers in epithelial cells
Platyhelminthes (13,000) flatworms, flukes, tapeworms	Marine, fresh water, some terrestrial in moist places; many parasitic in or on other animals	Herbivores, carnivores, scavengers, parasites	Brain, nerve cords	Hydrostatic support (no secreted skeleton); well-developed muscle tissue
Nematoda (12,000) roundworms	Marine, fresh water, terrestrial; many parasitic in animals, plants	Scavengers, carnivores, parasites	Nerve ring, nerve cords	Hydrostatic support (by pseudocoel); tough cuticle; longitudinal muscle in body wall
Annelida (10,000) earthworms, leeches, polychaetes	Marine, fresh water, terrestrial in moist places	Herbivores, carnivores, scavengers, detritus or filter feeders; mostly free-moving	Brain, double ventral nerve cord	Hydrostatic skeleton (using coelom); well-developed musculature in body wall
Arthropoda (1,000,000) crustaceans, spiders, insects	Marine, fresh water, terrestrial	Herbivores, carnivores, scavengers, detritus or filter feeders, parasites; mostly free-moving	Brain, double ventral nerve cord	Exoskeleton (of cuticle); jointed appendages; muscles mostly in bundles
Mollusca (100,000) snails, slugs, clams; squids, octopuses	Marine, fresh water, terrestrial	Herbivores, carnivores, scavengers, detritus or filter feeders; mostly free-moving, some attached	Brain, nerve cords, major ganglia other than brain	Hydrostatic skeleton (using hemocoel) in most; well-developed musculature in foot, mantle, other structures
Echinodermata (6,000) sea stars, brittle stars, sea urchins, sea lilies, sea cucumbers	Strictly marine	Mostly carnivores, detritus feeders, few herbivores; most free-moving, some attached	Radially arranged nervous system	Endoskeleton (of spines, etc.); muscles for body movement, tube feet often used in locomotion
Chordata (39,000) tunicates, lancelets, jawless fishes, jawed fishes, amphibians, reptiles, birds, mammals	Marine, fresh water, terrestrial	Herbivores, carnivores, scavengers, filter feeders; generally free-moving (most tunicates attached as adults)	Well-developed brain, dorsal and tubular nerve cord in most	Notochord or a bony or cartilaginous endoskeleton; well-developed musculature in most

*Number in parentheses indicates approximate number of species.

The body cavity of certain invertebrates lacks a continuous peritoneal lining. Such a body cavity, typical of nematodes, is called a **pseudocoel** (a "false coelom"). Insects, crustaceans, and most mollusks have a **hemocoel**, a body cavity through which blood percolates. This arrangement, recall, is characteristic of open circulation systems (Figure 29.2). A hemocoel is not necessarily obvious when an animal is dissected because it is an aggregate of spaces, most of which may be small.

A coelom, pseudocoel, or hemocoel may function in circulating nutrients, oxygen, and metabolic wastes. It may also function as a hydrostatic skeleton (page 373).

Segmentation. A segmented animal has a series of body units. In earthworms, the segments are externally similar to one another. In insects and crabs, some segments are very different from others and are grouped

Digestive System	Respiratory System	Circulatory System	Mode of Reproduction
No gut; microscopic particles of food captured by individual cells	None; respiration by individual cells	None	Sexual (certain cells become or produce gametes); some asexual budding; production of resistant bodies
Saclike gut (may be branched)	None; respiration by individual cells; gut may distribute oxygen	None, other than via gut	Sexual (usually separate sexes; gonads discharge gametes into gut or to exterior); asexual
Saclike gut (may be branched)	None; gas exchange across body surface	None	Sexual (usually hermaphroditic, with complex reproductive system); some asexual
Complete gut	None; gas exchange across body surface	Pseudocoel	Sexual (sexes separate; reproductive system fairly complex)
Complete gut	Gas exchange across body surface; varied outgrowths of surface in many	Usually closed; coelom also may function in distribution	Sexual (hermaphroditic or sexes separate; reproductive system simple or complex); asexual also in many
Complete gut	Gills; tracheal tubes; book lungs; general body surface in some	Open system, forming a hemocoel	Sexual (usually separate sexes; reproductive system fairly complex)
Complete gut	Ctenidia, other gills; mantle can be modified as lung; gas exchange across body surface	Usually open, forming a hemocoel (closed system in cephalopods)	Sexual (hermaphroditic or separate sexes; reproductive system usually complex)
Usually complete gut (sometimes saclike)	Gas exchange across general body surface or surface or outgrowths of it (such as tube feet)	Coelom around viscera, also water-vascular coelom	Sexual (reproductive system simple; gonads usually discharge gametes directly to exterior); some asexual
Complete gut	Lungs in most vertebrates other than fishes; perforated pharynx; gills; gas exchange across body surface	Closed system in most (open in most tunicates); lymphatic system in many vertebrates	Sexual (sexes usually separate, except in most tunicates); asexual in some tunicates

into body divisions called *tagmata* (singular, tagma). In an insect, for example, the tagmata are the head (with antennae and mouthparts), thorax (with legs and wings), and abdomen (with reproductive openings and organs for copulation).

Cephalization. Jellyfishes and sea stars do not have anything that can be called a head. They do not have a strongly centralized nervous system and there is not a distinct brain. We say these animals are *uncephalized*. Beginning with the simpler bilateral animals such as flatworms, the body has a definite head end with well-developed sensory structures and a rudimentary "brain." Such animals are said to be *cephalized*. When these animals move, the head end generally leads, so it is decidedly advantageous for this end to be equipped to sense and process information.

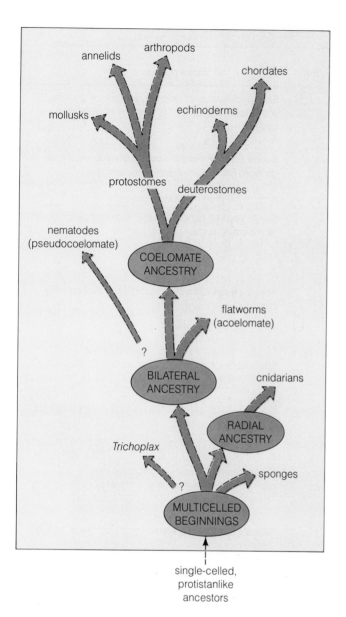

Figure 41.3 Broad phylogenetic framework for considering the relationships among major animal groups. All of the groups shown were established by 570 million years ago.

Major Groups of Animals

Of all the various phyla, let's focus on nine of the larger groups, which reflect several trends in the evolution of complex animals:

Porifera	Nematoda	Arthropoda
Cnidaria	Mollusca	Echinodermata
Platyhelminthes	Annelida	Chordata

As Table 41.1 shows, their numbers range from 6,000 species of echinoderms (sea stars and their kin) to nearly

a million species of arthropods (crustaceans, spiders, insects, and their relatives). Our first subjects are the sponges—anatomically simple animals, but a cut above *Trichoplax*. According to one model (Figure 41.3), sponges are thought to be close to the evolutionary boundary between animals and the ancestral, protistan-like forms from which they arose.

SPONGES

Of the approximately 8,000 known species of sponges (phylum Porifera), only about 100 live in fresh water; the rest are found in the seas. Although all sponges are at the borderline of the tissue level of biological organization (page 300), they are remarkably varied in their shapes and sizes. Some are as small as a fingernail; others are large enough to fill a bathtub. They include encrusting, compact, lobed, tubular, cup-shaped, and vase-shaped species (Figures 41.4 and 41.5).

No matter what a sponge looks like, its cells are organized around a system of pores, canals, and chambers. Water moves *into* the sponge body through thousands or millions of small incurrent pores and *out* through one or more outcurrent openings called oscula (singular, osculum). As Figure 41.6 shows, flagellated **collar cells** line the central cavity of the simplest sponges; they are concentrated in outpocketings of a central cavity or other side chambers in the more complex types. The cells are named for a "collar" of microvilli around the base of the flagellum. As the collective beating of their flagella drives water out through the osculum, water is drawn in through the pores, carrying bacteria and food particles with it—food that becomes trapped by the collars. The collar cells may digest the food completely or transfer it to **amoeboid cells** that further digest food, store it, and distribute it to cells throughout the sponge body.

Flattened cells line the outside of the sponge body and the internal cavities where there are no collar cells. Between the two cell layers is a semifluid matrix through which the amoeboid cells crawl and within which skeletal elements are embedded. The skeletal elements are constructed of substances secreted by the amoeboid cells. Some are spicules, often needlelike, of silica or calcium carbonate; others are fibers of a flexible protein. (The natural bath sponges owe their soft, squeezable texture to their fibrous skeletons.) Many of the spicules project to the outside, forming a protective barrier of sharp points. But the mass of spicules functions mainly to support the sponge body.

Sponges reproduce sexually, with eggs and sperm being derived from collar cells and perhaps from amoeboid cells. Sperm are released into the surrounding water

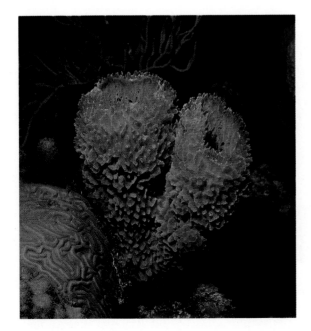

Figure 41.4 Representative sponges in their marine habitat.

Figure 41.5 A sponge (*Verongia archeri*) of the West Indies, releasing sperm to the surrounding water.

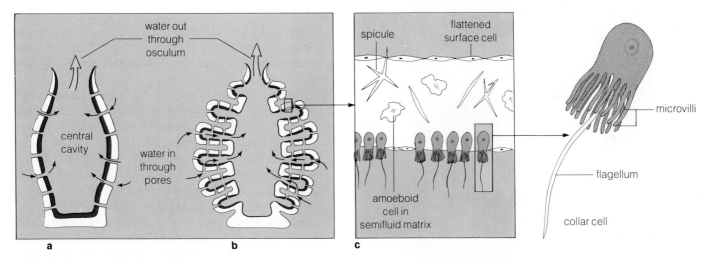

Figure 41.6 Morphology of a simple sponge (**a**) and a more complex form (**b**). Purple shading indicates the location of collar cells, which trap bacteria and small food particles on their "collar" of microvilli. Black arrows indicate the direction of water flow through the sponge, as caused by the collective beating of the flagella of collar cells. (**c**) Small section through a sponge body wall.

(Figure 41.5). After fertilization the eggs may be released or they may spend some time in the canals or chambers of the parent body. Young sponges go through a distinct larval stage.

In some sponges, asexual reproduction occurs when small fragments break off the sponge body, settle down on a substrate, and differentiate into new sponges. The formation of **gemmules** (special resistant bodies) occurs in most freshwater sponges. A group of cells aggregate, and some form a hard covering (with spicules) around the others. The living cells in a gemmule can resist extreme cold or drying out when environmental conditions change. When favorable conditions return, a gemmule germinates and the cells start a new colony.

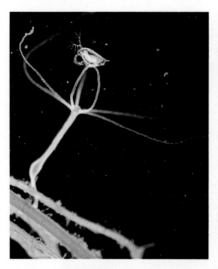
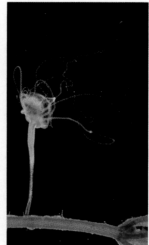

Figure 41.7 A hydra, capturing and swallowing a small crustacean. See also the anatomical sketches of a hydra in Figure 25.1.

Figure 41.8 Sea nettle (*Chrysaora*), one of the large jellyfishes. The frilled structures are oral arms that assist in capturing and ingesting prey. Jellyfishes of this type usually have an abundance of nematocysts which, in the sea nettle and other species, give painful stings.

CNIDARIANS

The phylum Cnidaria, which consists of about 9,000 living species, includes jellyfishes, hydras, sea anemones, corals, and their relatives. Except for the hydras and a few others, nearly all of these animals live in the seas. All cnidarians are carnivores, as the hydra in Figure 41.7 illustrates.

Cnidarians have true epithelial tissues, one covering the outside of the body (epidermis) and another lining the saclike gut (gastrodermis). During development, the epidermis is derived from the germ layer called ectoderm and the gastrodermis, from endoderm (page 478). Some of the cells in these tissue layers are contractile and are involved in the changes in shape necessary for feeding and locomotion (see, for example, Figure 28.1). They are called epitheliomuscular cells. As we have seen, these animals have nerve nets (nerve cells associated with the epidermis and gastrodermis); and the jellyfishes have sensory structures that detect light and changes in orientation. Between the epidermis and gastrodermis is a secreted layer called the **mesoglea** ("middle jelly").

Cnidarians show radial symmetry in their body plans, which are either **medusae** (singular, medusa) or **polyps**. A medusa has the shape of a bell or an upside-down cup. The mouth is usually centered inside the bell, and the downward-pointing rim has tentacles extending from it. Sometimes "oral arms" extend from the region of the mouth and assist in prey capture and ingestion (Figure 41.8). Jellyfish have this form, which is adapted to a floating or swimming life-style. A jellyfish "swims" by contractions of epitheliomuscular cells that deform the mesoglea, causing water to be forced in a jet from

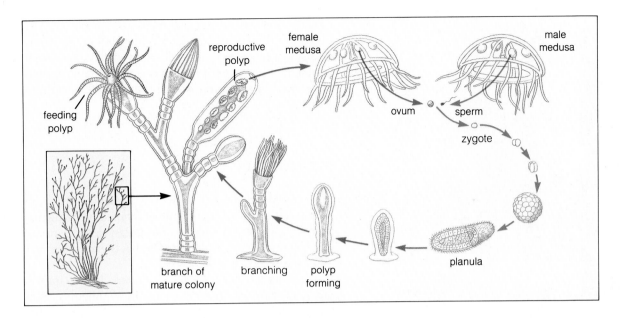

Figure 41.9 Life cycle of the hydra *Obelia*. The inset shows a mature colony, actual size. On its branches are feeding polyps and reproductive polyps. The medusa stage, produced asexually by a reproductive polyp, is free-swimming. Fusion of gametes from male and female medusae leads to a zygote, which develops into a planula. The swimming or crawling planula develops into a polyp, which starts a new colony.

the cavity under the bell-shaped body. After relaxation of the same cells, the bell springs back to its original shape.

Most polyps are tube-shaped, with a tentacle-fringed mouth at one end, and they are attached to the substrate by the opposite end. They have little mesoglea, and it is not especially jellylike. This body form is characteristic of hydras, sea anemones (Figure 1.6), corals, and most other cnidarians. Corals also have a calcium-containing external skeleton. Often polyps live in colonies, with the tissues of one member continuous with those of others (Figure 1.7).

Many cnidarian life cycles proceed through both a medusa stage and a polyp stage. Medusae are produced asexually by budding from a polyp, but they become sexual, producing either eggs or sperm. The zygote that results when an egg is fertilized develops into a swimming or creeping larva called the **planula** (Figure 41.9). The planula settles by one end, a mouth opens up at the other end, and the gut becomes functional. The polyp that results may remain solitary or develop into a colony, but either way it eventually produces medusae.

One of the most characteristic structures of cnidarians is the **nematocyst**, or stinging capsule. The epidermis of a single cnidarian tentacle will have hundreds or thousands of nematocysts. They also may be present elsewhere, even in the gastrodermis. Each nematocyst is within a single cell, and it consists of a capsule with an

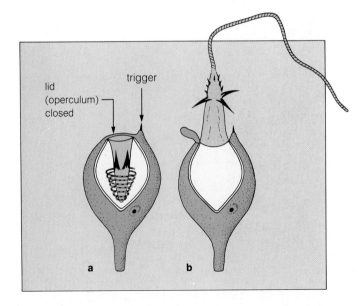

Figure 41.10 A nematocyst, before (**a**) and after (**b**) discharging its barbed thread.

inverted tubular thread inside (Figure 41.10). The cell producing a nematocyst may have a bristlelike trigger; this is a modified flagellum. When the trigger is touched by prospective prey or perhaps by a predator, or when the complex as a whole receives an appropriate chemical stimulus, the permeability of the capsule increases and water diffuses into it. This raises the pressure within the

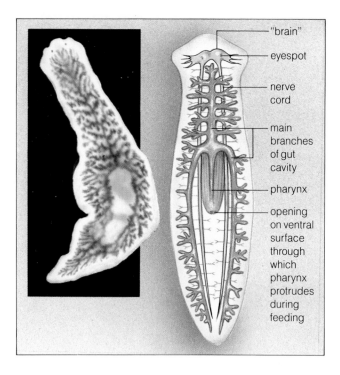

Figure 41.11 A planarian, which illustrates the bilateral symmetry of flatworms. Notice the pharynx and the branching gut.

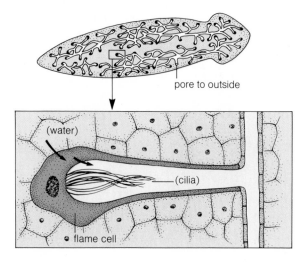

Figure 41.12 In a planarian, a pair of protonephridia, which function in fluid regulation and salt balance.

capsule, and the thread is forced to turn inside-out. The tip of the thread may penetrate the prey or predator, releasing a toxin, or it may adhere by a sticky secretion. Frequently, however, it merely becomes entangled with hairs or other structures on the prey, thereby acting as a lasso.

FLATWORMS

The 12,700 species of flatworms (phylum Platyhelminthes) include free-living predatory or scavenging turbellarians, parasitic flukes, and parasitic tapeworms. Most have more or less flattened bodies (hence the name) and all lack a coelom.

Flatworms differ from the cnidarians in two important respects. First, they are bilaterally symmetrical. Second, besides ectoderm and endoderm, the primordial cell layer called mesoderm forms during the development of their embryos and gives rise to muscles and to reproductive structures (page 478). This primordial cell layer was central to the evolution of complex animals. Besides enabling the contractile cells to develop independently of epidermis and gastrodermis, it eventually became the embryonic source of blood, bones, kidneys, and other internal tissues and organs.

Like cnidarians, the flatworms have a saclike gut, but its arrangement follows a bilateral plan (Figure 41.11). Also, there usually is a muscular **pharynx** by which food is taken into the gut. The flatworm pharynx qualifies as an organ; besides having an epithelial lining, it has muscle tissue and, generally, connective tissue and glands. Other flatworm structures are also at the organ level of complexity.

Turbellarians

Most turbellarians are marine, but some (such as the planarians) live in fresh water. The few terrestrial species are restricted to habitats where there is considerable moisture. As a rule, turbellarians feed on whole small animals or they suck tissues from wounded or dead animals. However, there are herbivores that eat diatoms and multicellular algae.

Especially among the turbellarians living in freshwater (hypotonic) environments, we find a system for fluid regulation and salt balance. The system is based on one or more units called **protonephridia** (singular, protonephridium). Each unit consists of a branched tubule that extends from a pore at the body surface to many cup-shaped *flame cells* in the body tissues (Figure 41.12). Flame cells get their name from a tuft of cilia that "flickers" within their walls. When excess tissue fluid moves

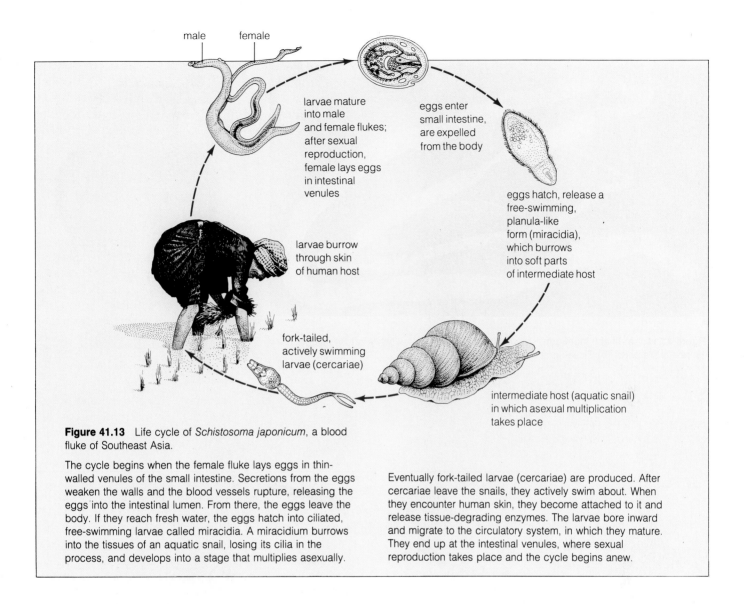

male   female

larvae mature
into male
and female flukes;
after sexual
reproduction,
female lays eggs
in intestinal
venules

eggs enter
small intestine,
are expelled
from the body

eggs hatch, release a
free-swimming,
planula-like
form (miracidia),
which burrows
into soft parts
of intermediate host

larvae burrow
through skin
of human host

fork-tailed,
actively swimming
larvae (cercariae)

intermediate host (aquatic snail)
in which asexual multiplication
takes place

Figure 41.13 Life cycle of *Schistosoma japonicum*, a blood fluke of Southeast Asia.

The cycle begins when the female fluke lays eggs in thin-walled venules of the small intestine. Secretions from the eggs weaken the walls and the blood vessels rupture, releasing the eggs into the intestinal lumen. From there, the eggs leave the body. If they reach fresh water, the eggs hatch into ciliated, free-swimming larvae called miracidia. A miracidium burrows into the tissues of an aquatic snail, losing its cilia in the process, and develops into a stage that multiplies asexually.

Eventually fork-tailed larvae (cercariae) are produced. After cercariae leave the snails, they actively swim about. When they encounter human skin, they become attached to it and release tissue-degrading enzymes. The larvae bore inward and migrate to the circulatory system, in which they mature. They end up at the intestinal venules, where sexual reproduction takes place and the cycle begins anew.

into the cups formed by these cells, the flickering drives it down the tubule system to the outside. Protonephridia are absent or poorly developed in marine turbellarians, which live in isotonic environments.

In cnidarians, simple gonads (primary reproductive organs) remain closely associated with the epidermis or gastrodermis, and they simply rupture and release gametes. In turbellarians and other flatworms, the gonads are nearly always part of a rather complex *her-maphroditic system* (that is, with male and female gonads in the same individual). Usually there is an organ that functions as a penis and a structure for storing sperm received during copulation, as well as glands concerned with producing a protective capsule around one or more eggs after fertilization. Mating between two individuals usually results in the reciprocal transfer of sperm.

Asexual reproduction is common among the turbellarians. Freshwater planarians, for instance, often divide transversely. Both halves must regenerate the parts they

lack at the time they separate. This mode of reproduction is so successful for some planarians that they do not even have a functional reproductive system.

Trematodes

The trematodes, or flukes, are parasites that usually live within vertebrate hosts, especially in the gut, liver, lungs, bladder, or blood vessels. Those living on the body surface feed by sucking in cells, tissue juices, or blood. The vast majority of the internal parasites have complex life cycles, involving asexual multiplication by larval stages (Figure 41.13).

The blood flukes, *Schistosoma*, cause **schistosomiasis**, a serious disease of humans and other animals. At a given time, about 200 million people are afflicted with this disease. An infected host typically makes a cell-mediated immune response to masses of fluke eggs

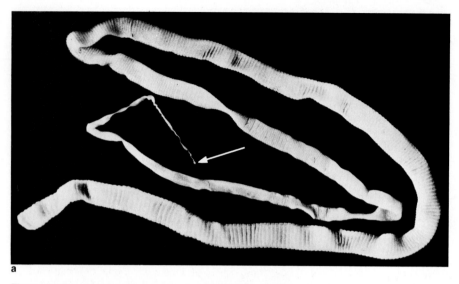

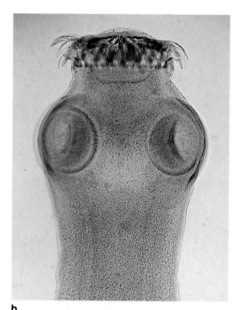

Figure 41.14 (**a**) Sheep tapeworm, showing the size of the scolex (white arrow) relative to the rest of the body. (**b**) Close-up of a tapeworm scolex.

human eats
rare beef

cyst containing
inverted scolex
of future
tapeworm

attachment
to wall
of intestine

scolex
turns
inside-
out

sexually mature
proglottids

embryonated
egg

ripe
proglottid
containing eggs

Figure 41.15 Life cycle of a beef tapeworm (*Taenia saginata*).

being produced within its body (page 409). Leukocytes and other cell types infiltrate and produce granular masses in tissues. The disease leads to deterioration and malfunctioning of the liver, spleen, bladder, and kidneys. Figure 41.13 shows the life cycle of one of the flukes that causes schistosomiasis.

Cestodes

Adult cestodes, or tapeworms, are intestinal parasites of vertebrates. During the course of evolution, they lost their own digestive tract and now feed entirely on soluble nutrients provided by their hosts. A tapeworm attaches to the intestinal wall by a **scolex**, a structure that usually has suckers, hooks, or both (Figure 41.14). Just behind the scolex is a proliferative region from which units called **proglottids** are budded. Each proglottid is almost like an individual, for it has a complete hermaphroditic reproductive system. Proglottids mate and transfer sperm, and the fertilized eggs may accumulate in the older proglottids (the ones farthest from the scolex). Sooner or later these proglottids separate from the chain. They may or may not disintegrate before leaving the body with fecal material, but in either case the eggs reach the outside and thereby become available to infect a new host.

Generally, at least one intermediate host is used in the life cycle of a tapeworm. Pigs are intermediate hosts for one common tapeworm that parasitizes humans; cattle are intermediate hosts for another. Humans become infected when they eat insufficiently cooked pork or beef (Figure 41.15). In another tapeworm of humans and some other animals, the eggs must get into fresh water, where they hatch into ciliated, swimming larvae. When a larva is eaten by a copepod (a type of crustacean), it develops into a stage that will develop further in a fish that eats the copepod. Humans who eat raw, insufficiently cooked, or improperly pickled fish can become hosts for adult tapeworms of the species.

You have probably noticed that the life cycles of the flukes and tapeworms just described take advantage of the life-styles and food habits of various hosts. Sometimes the life cycles are amazingly complex, but they have to be in order to get the parasite through the sequence of hosts in which it reproduces asexually and then sexually.

NEMATODES

Because of their cylindrical bodies, nematodes (phylum Nematoda) are often called roundworms. This phylum has about 12,000 described species but there are probably

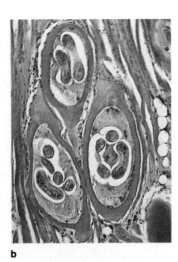

a   b

Figure 41.16 (**a**) Juveniles of *Trichinella spiralis* in muscle tissue. The adults of this small nematode live in the lining of the small intestine of pigs, rats, humans, and some other animals. Female worms release juveniles, and these work their way into blood vessels and travel to muscles, where they become encysted (**b**). The presence of adults and juveniles in humans causes a variety of unpleasant and dangerous symptoms. The usual source of infective juveniles for humans is pork that has not been cooked sufficiently.

at least several times that many. It is mentioned here partly because of the enormous importance of nematodes as parasites and as consumers of various kinds of organic matter. But nematodes also show features that can be viewed as more complex than the flatworm plan. Their origins are not known, and ancestral forms probably did not give rise to any other living groups of animals.

The slender, bilateral nematode body has a **cuticle**, a covering that is tough and not easily stretched, yet flexible. The nematode also has a gut complete with a mouth and an anus. Between the gut and body wall is a pseudocoel (Figure 41.2). Reproductive organs may fill nearly all of the space within the pseudocoel, but the fluid in the cavity nevertheless distributes nutrients and thus serves as a kind of circulatory system. It also serves as a hydrostatic skeleton.

Nematodes live in nearly all habitats where life is possible, including snowfields, deserts, and hot springs. A cupful of rich soil will have thousands of them, and a dead earthworm or a fruit rotting on the ground will have an interesting variety of scavenging types. Nematodes are also immensely successful as parasites of plants and animals. Humans alone are affected by about thirty species, including hookworms and those that cause trichinosis (Figure 41.16), and domesticated animals have their share.

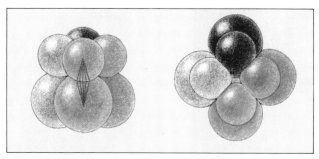

a Radial cleavage, side view (left) and top view

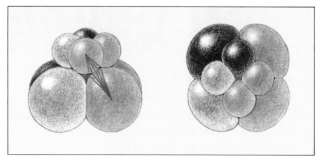

b Spiral cleavage, side view (left) and top view

Figure 41.17 Two major patterns of cleavage in early animal embryos: (**a**) radial, which is characteristic of deuterostomes, and (**b**) spiral, which is characteristic of protostomes.

TWO MAIN EVOLUTIONARY ROADS

After the emergence of bilateral body plans, two key evolutionary roads opened up. The existing travelers of these roads are called protostomes and deuterostomes, and they include the following major phyla:

Protostomes	Deuterostomes
annelids	echinoderms
arthropods	chordates
mollusks	

In general, the animals in both groups have a complete gut and a coelom, but each follows its own patterns of early embryonic development. Recall that at the gastrula stage, an embryo has a single opening (blastopore). In protostomes, this first opening becomes the mouth; the anus forms elsewhere. (Protostome means "first mouth.") In deuterostomes, the first opening becomes the anus and the second opening becomes the mouth.

The fate of the embryonic cells formed during cleavage is fixed in protostomes at the first cell division; each cell can follow only a prescribed developmental pathway. In contrast, the embryonic cells of deuterostomes

do not become committed to one course until later in development. (For example, if the cells of a sea star embryo are separated at the four-cell stage, each cell can develop into a complete larva.)

The cleavage pattern in the two types of animals is also different. As you saw in Figure 34.6, cleavage in frog embryos occurs parallel and at right angles to the axis established by the animal and vegetal poles. The outcome is an array of cells each located directly above or below one another (Figure 41.17a). The pattern is said to be one of *radial cleavage*, and it is characteristic of deuterostomes. In contrast, cleavage proceeds obliquely in almost all protostomes. The outcome is a tiered array of cells, with each cell in a tier snuggled against two cells of adjacent tiers (Figure 41.17b). This pattern is called *spiral cleavage*.

Finally, the coelom itself arises in one of two ways. In protostomes, two masses of mesoderm on either side of the gut split:

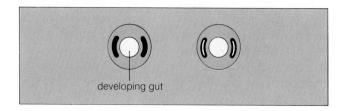

developing gut

The resulting cavities enlarge to form a coelom of the sort shown in Figure 41.2c. In deuterostomes, *pouches* form in the walls of the developing gut cavity (archenteron). The pouches separate to form the coelom:

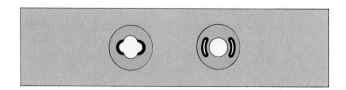

ANNELIDS

Some group of flatwormlike ancestors probably gave rise to the phylum Annelida, the segmented worms. The best known of its approximately 10,000 species are the earthworms and leeches. Less familiar is the larger and far more diversified array of mostly marine worms, the polychaetes.

Oligochaetes

Oligochaetes, including the earthworms, are mostly land-dwelling and aquatic scavengers that burrow in moist soil, mud, and silt. They feed largely on decom-

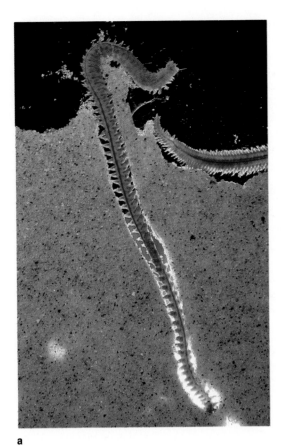

Figure 41.18 Polychaetes. (**a**) This photograph shows a marine polychaete inside its burrow. (**b**) A tube-dwelling polychaete, with featherlike structures used for trapping small food particles.

posing plant material and may also use organic matter that is part of the soil they swallow as they burrow. In soft soil, the earthworm plunges forward and enlarges its highly muscular pharynx, which pushes the soil aside. In compacted soil, it literally eats dirt in order to burrow. Every twenty-four hours, an earthworm can ingest its own weight in soil. Earthworms not only aerate soil (which is beneficial to some plants), they also carry subsoil to the surface, where the nutrients it contains are made available to other organisms.

Leeches are found in fresh water, the sea, and moist regions of the tropics. Most are predators that swallow small invertebrates or kill them and suck out their juices; the leeches people usually have heard about feed on vertebrate blood. Leeches use their sharp jaws to make an incision in their prey, then they plunge a sucking apparatus (a modified pharynx) into the incision. The leech gut has many side branches for storing food taken in during a big meal, this being handy for an animal that may have long waits between meals. The leech body is very muscular and has a posterior sucker for attachment as well as a sucker around the mouth. Using its suckers, a leech can inch forward over such substrates as the surface of an animal about to provide it with dinner.

Polychaetes

Polychaetes include burrowing, crawling, and free-swimming forms (Figure 41.18). Some excavate vertical or U-shaped burrows in soft mud or sand; others live in tubes made of secretions or of sand and shell fragments cemented together. Many polychaetes feed on small invertebrates, some feed on algae, and others use organic matter present in sediments or suspended in the water. The predatory forms have a muscular pharynx equipped with jaws that can be everted to capture prey; herbivorous types use their jaws to grasp and tear algae. Most of the tube-dwelling species have tentacles or featherlike structures originating near the mouth. Their mucous secretions trap bacteria, diatoms, and other small particles. The activity of cilia carries the mixture of food and mucus to the mouth.

Characteristics of the Annelids

Except for leeches and a few other forms, the annelids are characterized by chitin-containing bristles called **setae**. The bristles form from secretions in pockets in the body wall, and there are special muscles for extending

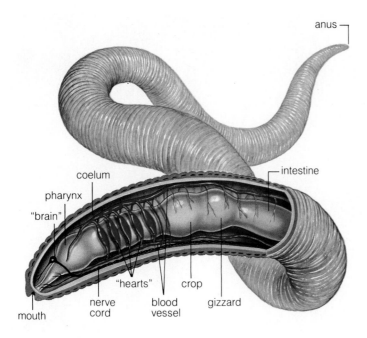

Figure 41.19 Arrangement of nerves, blood vessels, and the digestive system of an earthworm, all of which extend longitudinally through one coelomic chamber after another.

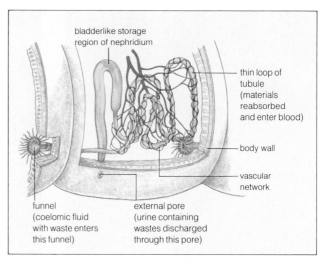

Figure 41.20 Earthworm system of salt and water regulation. The functional unit of the system is the nephridium.

and retracting them. Setae provide the traction required in crawling and burrowing; in species that swim, they are often broadened into paddles. Oligochaetes have only a few setae per segment. Polychaetes usually have many setae, typically concentrated in fleshy lobes (*parapodia*) that project from both sides of most segments (Figure 41.18a). Parapodia function in respiration as well as in locomotion.

Again, the annelids as a group show considerable segmentation. Partitions (septa) separate the segments, which are individual coelomic chambers. However, the gut is continuous and extends through all the chambers, from mouth to anus (Figure 41.19). A "brain" at the head end integrates sensory input for the whole worm. Leading away from the brain is a double ventral nerve cord that extends the length of the body. In each segment the nerve cord is broadened in ganglia that control local activity (Figure 25.2).

The coelomic chambers serve as a hydrostatic skeleton against which circular and longitudinal muscles of the body wall can operate. The circular muscles are mostly segmental, whereas the longitudinal muscles span several segments. The segments become shorter and fatter when the longitudinal muscles contract and the circular ones relax; they become longer and more slender when the pattern reverses. When an earthworm moves forward, the first few segments elongate while

the segments just behind them protrude their setae to hold their position, and the forward tip of the body advances. Then the first few segments contract and protrude their setae, and the segments behind them retract their setae and are pulled forward. The alternation of contraction with elongation takes place throughout the length of the body, the net result being a forward progression of the worm as a whole.

A closed circulation system is typical of animals. The circulating fluid—blood—provides a means of transporting materials through larger bodies than those of flatworms. In the annelid system, forceful contractions in muscularized blood vessels ("hearts") keep blood circulating in one direction. Smaller vessels of the system carry blood to cells of the gut, nerve cord, and body wall (Figure 41.19).

Functionally linked to the closed circulatory system is a system for regulating water and solute levels in the body. In some annelids the components of the system resemble protonephridia, suggesting that the annelids are linked evolutionarily with the flatworms. Most annelids have **nephridia**, which are more complex tubular structures that usually begin as a funnel open to the coelom. As a rule, nephridia are arranged as pairs in the coelomic chambers, with their funnels in the chamber ahead of the one in which the tube opens to the outside (Figure 41.20).

Annelids, in short, have a body plan that reflects some major evolutionary trends. Their coelomate and segmented ancestors were an early branch on the phylogenetic tree, foreshadowing a trend toward increases in size and greater complexity of internal organs. Annelids have a rudimentary brain and segmental ganglia—that is, centralized as well as segmental regions of neural control. They have a complete digestive system with regional specializations; they have a closed circulatory system functionally linked with nephridia—the annelid equivalent of the vertebrate kidneys.

ARTHROPODS

The most diverse living organisms of all are the arthropods, with about a million known species. Evidently, ancestral annelids or animals like them gave rise to four different arthropod groups. (Figure 41.21 shows an existing animal that bears striking resemblances to both annelids and arthropods.) These groups are so distinct that each is now considered to be a subphylum:

Subphylum	Representatives
Trilobita	trilobites (described in Chapter Thirty-Eight); now extinct
Chelicerata	horseshoe crabs, spiders, scorpions, ticks, mites
Crustacea	crustaceans (including copepods, crabs, lobsters, shrimps, barnacles)
Uniramia	centipedes, millipedes, insects

Arthropod Adaptations

The arthropods are more widely distributed through more habitats than any other animal group. Many are herbivores, feeding on algae or land plants; others are carnivores, omnivores, or parasites.

What adaptations have contributed to the success of these diverse animals? Perhaps the most important are these:

1. Hardened exoskeleton
2. Specialization of segments
3. Jointed appendages
4. Specialized respiratory systems
5. Efficient nervous system and sensory organs

The Arthropod Exoskeleton. In annelids, the cuticle covering the body is thin and flexible enough to permit extensive bending as well as integumentary gas

exchange. It is also thin enough to be pierced by predators. From the fossil record, we know that some early annelids developed thicker, hardened cuticles—a pattern that also continued among the arthropods. In arthropods, the cuticle forms an **exoskeleton** (external skeleton). The outer layers of the arthropod exoskeleton contain protein and chitin, a combination that is flexible, lightweight, and protective. In some arthropods, such as crabs and lobsters, it is hardened by substantial deposits of calcium salts, which reduce flexibility but help create "armor" plates. At first, arthropod exoskeletons were probably useful in defense against predators. They also turned out to be useful in other ways when arthropods radiated into land environments. *The arthropod exoskeleton is a superb barrier to evaporative water loss, and it also provides support for an animal body deprived of the buoyancy it would have in water.*

Specialization of Segments. As arthropods evolved, body segments became highly specialized. Groups of segments became arranged into tagmata, such as the head, thorax, and abdomen of an insect, or the cepha-

Figure 41.21 *Peripatus*, a member of the phylum Onychophora. These velvety, "walking" worms resemble both annelids and arthropods. The largest is about fifteen centimeters long. Depending on the species, there are between fourteen and forty-three pairs of stumpy, unjointed legs. Each leg ends in a pair of claws and a rather soft pad.

Onychophorans live primarily in rain forests and in tropical or subtropical forests, and they forage only at night or when the air is extremely humid. Their body plan has not changed much for the past 500 million years.

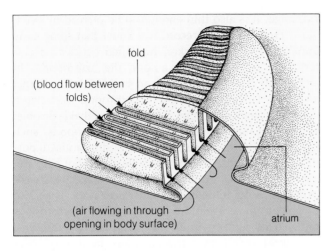

Figure 41.22 A book lung, a type of respiratory organ that occurs among spiders and scorpions.

Figure 41.23 Forebody and hindbody of a spider. Most spiders are harmless to humans and as a group help keep insect pests in check. However, a bite from the brown recluse shown here can be severe to fatal. This North American spider lives under bark and rocks, and in and around buildings. It can be identified by the violin-shaped mark on its forebody.

lothorax and abdomen of a crab or lobster. There were also trends leading to reduction in the number of segments and to fusion of the segments. For instance, the several segments that make up the head of an insect or a crustacean cannot be distinguished, although the appendages that belong to them are evident.

If the arthropod segments were to be enclosed by a firm, unyielding exoskeleton, the muscle layers typical of annelids would be of little use. Arthropod muscles

are concentrated in bundles that bring about specific movements of adjacent segments (and of the parts of appendages). Such movements are somewhat comparable to those you make when you bend your body, raise your head, or throw a ball. Your muscles, however, run from bone to bone, and bones are part of an endoskeleton. In arthropods, muscles run from the exoskeleton of one segment to another. This arrangement limits the extent to which adjacent body segments can be bent or rotated.

Jointed Appendages. Arthropods have jointed appendages (the word arthropod means "jointed foot"). Although their individual appendages are not as versatile as your arms and legs, they are remarkably diverse. Some function in walking or swimming, others in feeding, still others in sensory reception, transfer of sperm, spinning out silk for a web, and so on.

How did jointed appendages originate? We do not know, but it is likely that they were derived from fleshy, segmental outgrowths similar to the parapodia of polychaete annelids.

Respiratory Systems. The main respiratory structures in arthropods are gills, book lungs, and tracheas. *Gills*, described on page 419, are characteristic of marine and freshwater crustaceans. In some cases the gills are an outgrowth from the base of an appendage or from the body proper; in others they are a flattened appendage with only a thin cuticle. When an aquatic arthropod has no recognizable respiratory structures, gas exchange is usually taking place across any body surface where the cuticle is thin.

For the most part, book lungs and tracheas are found in land-dwelling arthropods, although many aquatic insects also have tracheas and fill them with air, at least periodically. *Book lungs*, typical of spiders and scorpions, are deep pockets on the underside of the body (Figure 41.22). The extensively folded walls of the pockets, which resemble the pages of a book, greatly increase the surface area available for gas exchange. The folds are slightly moist, and blood picks up oxygen and gives up carbon dioxide as it passes through them.

The tubes called tracheas are especially characteristic of insects, millipedes, and centipedes. In these animals, the tracheas begin at pores on the surface of the body and branch in such a way that they supply oxygen directly to nearly all tissues (Figure 31.2). A lining of thin cuticle keeps them from collapsing, and they are moist enough to permit gas exchange. The system is efficient enough to support high metabolic rates in the small-bodied arthropods, thus making possible such energy-consuming activities as insect flight.

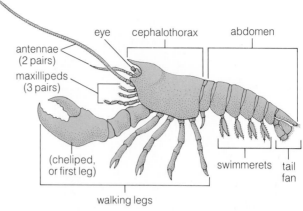

Figure 41.24 Paired appendages of the blue lobster. This and some other crustaceans have a body organized into a cephalothorax (head plus thorax) and an abdomen.

Specialized Sensory Structures. Although we considered many arthropod sensory structures and organs in Unit Five, let's return for a moment to a type that is unique to arthropods, the compound eyes. These sensory organs are typical of present-day insects, millipedes, centipedes, and most crustaceans. Most have a wide angle of vision, permitting the animal to process visual stimuli from many directions. Those of dragonflies are nearly hemispherical and contain more than 30,000 photoreceptor units (ommatidia); they have much to do with the dragonfly's rate of success in capturing other insects in midair.

Chelicerates

The chelicerates originated in the shallow seas of the early Paleozoic. Except for some mites, the only truly marine survivors are the so-called horseshoe crabs and sea spiders. Among the more familiar members of this group are spiders, scorpions, ticks, and mites.

The spider or scorpion body is divided into two regions, which we can call the forebody and hindbody (Figure 41.23). (In ticks and mites, the two regions are indistinct.) The forebody has six pairs of appendages. Four pairs are walking legs; the other two, called chelicerae and pedipalps, are used mostly in subduing prey and food handling. Male spiders also use the pedipalps to transfer sperm to the female, and scorpions use their pincerlike pedipalps to grasp their partner in the mating ritual. Appendages on the hindbody, if any, are extremely modified. Spiders use small appendages of this sort to spin out threads of silk for webs and egg cases.

Spiders and scorpions generally have several eyes on the forebody. The eyes are not developed enough for

Figure 41.25 Specialized appendages of a land crab (a crustacean).

effective image formation, but they are used to detect movements of prospective prey. Body hairs are responsive to air currents and mechanical disturbances, such as the presence of an insect struggling in a web.

Crustaceans

Crustaceans are a largely aquatic group of predators, herbivores, scavengers, and parasites. They include shrimp, crayfish, lobsters, crabs, barnacles, copepods, and water fleas. As you might expect, they show great diversity in the number and organization of body segments and in the specializations of paired appendages (see, for example, Figures 41.24 through 41.27). While a

Figure 41.26 A barnacle, with its body contained in a shell that is hardened with calcium deposits. This crustacean uses featherlike appendages, combing the seawater for microscopic food.

Figure 41.27 A marine copepod, one of the smaller crustaceans (typically no more than a few millimeters long). Some copepods are parasitic, but the ones of greatest ecological importance are the free-swimming herbivores of aquatic communities; they form an important part of the diet of fishes and other carnivores.

crab may tear up seaweed, collect organic debris, or attack another animal, a barnacle uses featherlike appendages for combing microscopic food from the water.

Most crustaceans have fewer than twenty segments, although some have more than thirty. Commonly, as in crabs, some segments are not distinct because the cuticle of the head extends backward, covering several others. For example, the **carapace** (a shieldlike cover) of a crab or lobster covers all of the segments of the head and thorax, forming a cephalothorax. The head appendages of a crustacean usually consist of two pairs of antennae, a pair of mandibles, and two pairs of maxillae. **Mandibles** are jawlike appendages; **maxillae** are food-handling and food-sorting appendages; **antennae** are primarily sensory, although in some crustaceans they are used for swimming or other functions.

Millipedes, Centipedes, and Insects

A distinguishing characteristic of millipedes and centipedes is the presence of numerous pairs of legs of the trunk: a long, segmented region behind the head (Figure 41.28). The trunk is more or less comparable to the thorax-plus-abdomen of an insect. In millipedes, most of the trunk segments are fused in twos; hence what appears to be a single segment has two pairs of legs. Millipedes are mostly scavengers on decaying plant material, whereas centipedes are carnivores, complete with fangs and venom glands to subdue insects and other small animals.

Insects are the most successful invertebrates on land. Many also live in fresh water during at least part of the life cycle; some are adapted for life in the sea or in brackish water. A tracheal system of respiration and a cuticle that retards evaporative water loss were key characters

Figure 41.28 A centipede of Southeast Asia, here dining on a frog. Almost all of its body segments are externally similar, as they are in annelids.

Figure 41.29 One of the most beautiful of the winged insects—the luna moth (*Actias luna*) of North America.

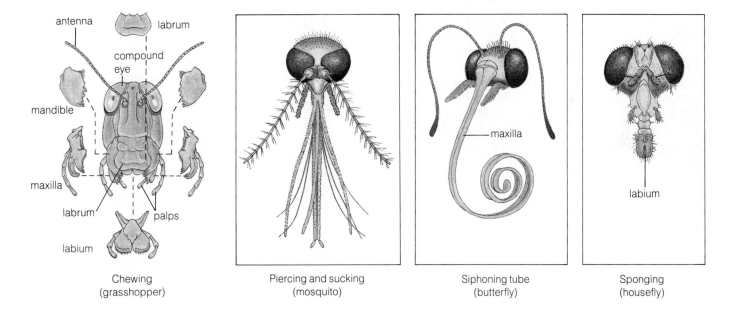

Chewing (grasshopper)

Piercing and sucking (mosquito)

Siphoning tube (butterfly)

Sponging (housefly)

Figure 41.30 Examples of insect headparts, adapted for feeding in specialized ways.

that led to their success on land; but so were the development of wings and the capacity for flight (Figure 41.29). Winged insects disperse themselves rapidly, move into areas where food may be abundant, swoop down on prey or capture other insects in midair, and escape from predators.

The insect body typically consists of a head (with a pair of antennae and three pairs of mouthparts), a thorax (with three pairs of legs and two pairs of wings), and an abdomen. The wings, located dorsal to the legs, are not true appendages; they are mostly expansions of the exoskeleton of the thorax. However, they connect with the thorax in much the same way as appendages, and muscles inserting on their bases provide the force for movement. Insect mouthparts are diverse and can be correlated with specializations in diet (Figure 41.30).

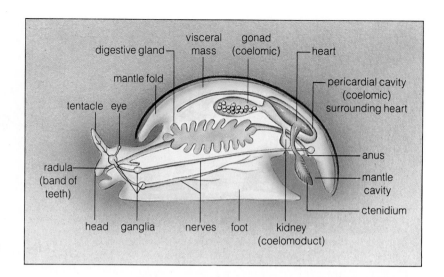

Figure 41.31 Generalized body plan of a mollusk.

MOLLUSKS

The mollusks are a truly ancient group that may have evolved from a flatwormlike ancestor. They are probably closely related to annelids, given that some of the marine species have the same type of larval stage as many marine polychaetes. It is likely that mollusks were first prominent in shallow seas, then later radiated into deeper waters, into brackish and fresh water, and eventually onto land.

Today there are about 100,000 species, most with the same basic body plan. All have a head, a foot, and a visceral mass (Figure 41.31). They usually have a shell composed of calcium carbonate. A **mantle**, a fold of tissue, hangs down like a skirt around some or all of the body. The space between this tissue fold and the rest of the body is the *mantle cavity*.

Mollusks with well-developed heads have obvious sensory structures, such as eyes and tentacles. Most mollusks other than clams have a **radula**, a tonguelike organ with hard teeth (Figure 41.32). The gut runs through the visceral mass. Also in the visceral mass are reproductive organs, the kidneys, and the heart. The kidneys are not always strictly excretory; they sometimes function as ducts through which gametes reach the outside. Both the anus and the kidney openings are in the mantle cavity.

In the mantle cavity of many mollusks is a pair of gills, of a type called **ctenidia**. These outgrowths of the mantle wall probably appeared at an early stage in molluscan evolution. One or both ctenidia are lacking in some mollusks; for them, other outgrowths may function as secondary gills. Except for the cephalopods, mollusks have an open circulatory system. The heart pumps blood into a branching vessel that opens to spaces that form a hemocoel, then oxygenated blood is returned to the heart.

Let's look briefly at three of seven classes of mollusks and see how the basic plan is modified in each of them.

Gastropods

Snails and slugs of varied sorts make up the largest group of mollusks, the gastropods ("belly foots"). They are so named because the foot is spread out under the animal while it is crawling. Many of these mollusks have a spirally coiled or conical shell. Coiling is a way of compacting the visceral organs into a mass that can be balanced above the rest of the body, much as you would balance a backpack full of books.

During the juvenile stage of development, most gastropods undergo a strange internal realignment by a process called **torsion**. In this process, contraction of certain muscles and differential growth cause the posterior mantle cavity to twist around to the right side and then twist even farther forward, until it is above the head:

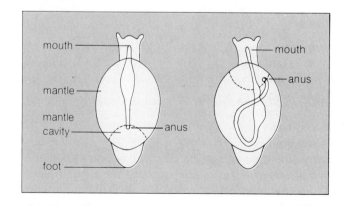

The twisting does provide the head with a cavity in which to withdraw in times of danger. But it also brings the gills, anus, and kidney openings above the head. As

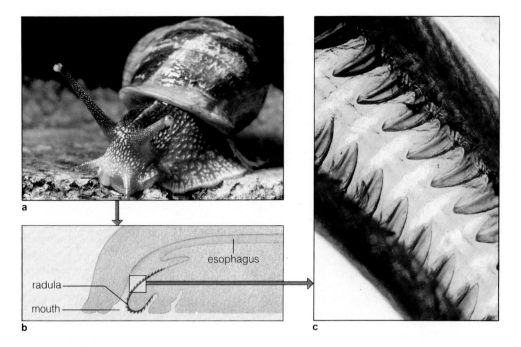

Figure 41.32 (**a**) A land snail. Part of the mantle cavity of this gastropod is modified for air-breathing. (**b**) Diagram of a section through a snail head, showing the radula (a rasping, tonguelike organ). (**c**) Close-up of a radula, showing the teeth that can rasp off pieces of food. The conveyer-belt motion of the radula moves food into the pharynx.

you might imagine, this realignment could create something of a sanitation problem, what with wastes being dumped near the respiratory structures and the mouth. Typically, most gastropods use cilia to create currents that sweep away the wastes.

Certain gastropods, notably sea slugs, have undergone detorsion in the course of their evolution. They have also lost the ctenidia, but most have other outgrowths that function as secondary gills. Many sea slugs are striking in both coloration and their array of these outgrowths (Figure 41.33).

Bivalves

Clams, scallops, oysters, mussels, and shipworms are bivalves (animals with a "two-valved shell"). Most of the bivalves are **filter feeders**. One way or another, they produce water currents that carry food particles to their mouth. These mollusks do not have well-developed heads, and most have a large foot adapted for digging in soft mud and sand. Some are only 1 to 2 millimeters across; the giant clams of the South Pacific may be more than 1 meter across and may weigh in at 225 kilograms (500 pounds).

Most bivalves have two large gills (true ctenidia) that function in respiration and in trapping food brought into the mantle cavity by ciliary currents (see Figure 41.34). Ciliated parts of the gills carry a mixture of food and mucus to fingerlike flaps located beside the mouth.

Figure 41.33 Two of the soft-bodied gastropods called nudibranchs.

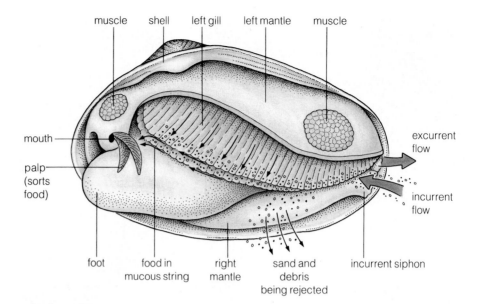

Figure 41.34 Anatomy of a clam. The left valve and mantle have been removed. Food trapped in mucus on the gills is sorted by the palps, and suitable particles are carried by cilia to the mouth.

Labels: muscle, shell, left gill, left mantle, muscle, mouth, palp (sorts food), excurrent flow, incurrent flow, foot, food in mucous string, right mantle, sand and debris being rejected, incurrent siphon

Figure 41.35 A scallop swimming away from a predator (a sea star). Scallops and a few other bivalves clap their two valves together rapidly enough to force a strong jet of water from the mantle cavity.

These flaps are used to sort the food, delivering some particles to the mouth and rejecting others.

Most bivalves have **siphons**, which are extensions of the mantle edges, fused into tubes, that bring water into the mantle cavity and take it out again. Usually the inhalant and exhalant siphons are bound together. When they are separate, the inhalant one is analogous to a hose on a vacuum sweeper, collecting organic debris. Siphons are essential to bivalves that bury themselves well below the surface. The siphons of the giant geoduck (pronounced "gooey duck") of the Pacific Northwest may be more than a meter long.

Scallops and a few other bivalves can swim by clapping their valves together, thus producing a localized jet of water that propels the animal (Figure 41.35). This jet-propulsive behavior is most useful for scallops, which are preyed upon by sea stars, octopuses, and other animals.

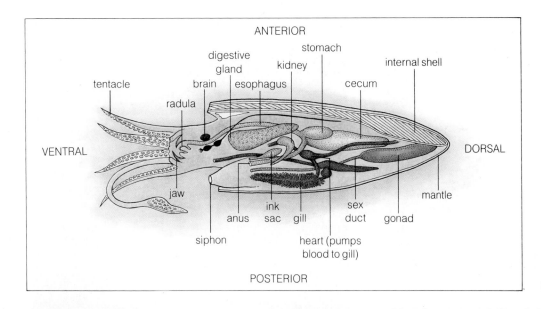

Figure 41.36 Anatomy of a squid.

Figure 41.37 A female and a male cuttlefish (*Sepia*) mating, head end to head end.

Cephalopods

The cephalopods are all sea-dwelling predators. They include the squids, octopuses, nautiluses, and cuttlefish (Figures 41.36 and 41.37). The smallest are only two or three centimeters long; *Architeuthis*, the giant squid, is the largest invertebrate known, with some individuals measuring eighteen meters (about sixty feet) from the tips of their tentacles to the tip of the visceral mass.

Sperm whales prey on these giant squid, and large sucker scars observed on whale bodies indicate there are violent encounters between these animals.

As a group, cephalopods are adapted for swimming rather than for creeping, burrowing, or permanent attachment. The shell is absent or reduced to a small internal structure, the only exception being the nautilus, which has a large shell divided into chambers (Figure 1.6f).

a

b

c

d

e

Figure 41.38 Representative echinoderms. (**a**) Crinoid, with featherlike rays. Both the mouth and anus are on the upper surface. (**b**) Cobalt sea star; this colorful species has six rays (most species have five). (**c**) Brittle star, which moves by rapid, snakelike action of its rays. (**d**) Sea cucumber, with an elongated body and lengthwise rows of tube feet. The branching tentacles of the anterior end are modified tube feet, used in collecting food. (**e**) Sea urchin, with tomato-shaped body and many sharp spines.

The word cephalopod ("head-foot") goes back to the mistaken notion that the tentacles surrounding the head are derived from the foot. They actually are outgrowths of the head itself. The tentacles have suction cups, used in capturing and manipulating prey; octopuses use the suction cups for clinging to firm surfaces (Figure 27.13b). The foot itself is specialized as a siphon, used in swimming (Figure 41.37). When the muscularized mantle wall contracts, water in the mantle cavity is forced out of the siphon in a strong jet that propels the animal, visceral-mass first.

The cephalopod gills are in the mantle cavity, along with the anus, kidney openings, and genital ducts. Closely associated with the rectal portion of the gut is a sac containing an inklike fluid. When the ink is discharged in a rather cohesive cloud, it may delude a pursuer. Perhaps it resembles an object that could be confused with the octopus or squid that produced it, or perhaps it functions as a smoke screen. At least one squid, which lives at depths where no light penetrates, produces luminous ink. Some cephalopod inks have a narcotic effect on sense organs of possible predators.

Unlike that of the other mollusks, the circulatory system of cephalopods is closed; that is, the blood circulates only within arteries, veins, and capillaries. This is a more efficient arrangement for animals that are as active as squids. And squids are remarkably active; freed from a cumbersome shell and equipped with a powerfully muscular mantle, they are the best swimmers in the invertebrate world.

In some ways, the evolution of cephalopods has paralleled the evolution of vertebrates. For instance, most cephalopods have good vision; as Figure 27.12d shows, octopus eyes are structurally very similar to those of vertebrates, although they develop in a different way. Cephalopods also display superb neural control over their movements, not only in swimming and prey capture but also in escaping from predators. The brains of cephalopods are definitely more highly developed than those of any other invertebrates.

ECHINODERMS

Echinoderms (phylum Echinodermata) are another group with ancient origins, but unlike annelids, arthropods, and mollusks, they belong to the deuterostome line. The name of the phylum means "spiny-skinned" and alludes to the calcium-containing spines or plates in the echinoderm body wall.

The members of this phylum are sea stars, sea urchins, sand dollars, brittle stars, sea cucumbers, and crinoids (Figure 41.38). All of the adult forms show radial symmetry, although this is complicated by bilateral characteristics, especially in sand dollars, some sea urchins, and some sea cucumbers.

A unique feature of echinoderms is the **water-vascular system**, which originates from part of the embryonic coelom. It consists of a system of internal canals that connect with external projections called **tube feet** (Figure 41.39). These thin-walled, muscular projections are used in feeding, locomotion, and respiration. In many echinoderms, including sea stars and sea urchins, one of the major canals of the water-vascular system reaches the surface of the body, where there is a porous, skeletal structure called the *sieve plate*. Although the pores admit seawater into the water-vascular system in some cases, the system also contains cells and proteins; it is somewhat similar to the fluid in the main coelom.

The water-vascular system operates on the principle of a hydrostatic skeleton. The lengthening or shortening

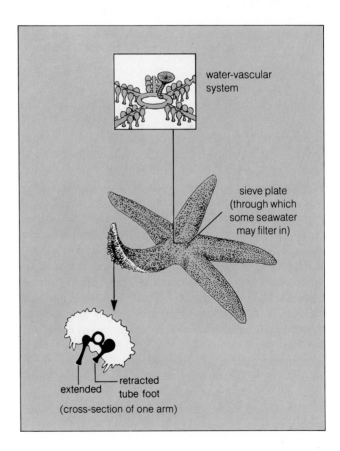

Figure 41.39 Water-vascular system of canals and tube feet, the basis of echinoderm locomotion.

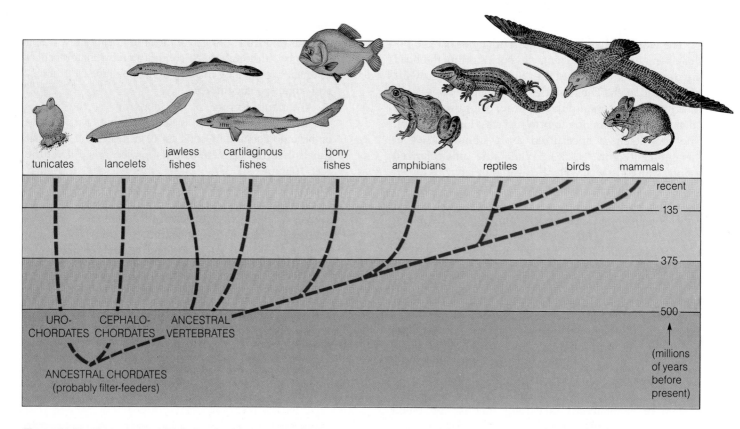

Figure 41.40 One proposed family tree for the vertebrates and their relatives, the urochordates and the cephalochordates. Actual phylogenetic relationships are not known with certainty, hence the dashed lines.

of a tube foot, for instance, depends on contraction of certain muscles of the tube foot and relaxation of others. Notice, in Figure 41.39, that the bulblike part (ampulla) of each foot is connected with a canal of the water-vascular system that runs the length of the ray. If muscles of the ampulla contract and valves in the branch canal leading to it close, fluid will be forced into the external portion of the tube foot, causing it to lengthen. The tube feet of most sea stars have suckers at their tips. When a sucker is touched down to a firm surface, mucus helps it stick, and contraction of certain muscles deepens the cup of the sucker to give it a tight grip on the substrate. The underside of each ray has many tube feet, and the locomotion of a sea star depends on their coordinated activity as they swing forward, lengthen, become attached, swing backward, then become detached. Not all tube feet do the same thing at the same time, but once a sea star starts going in one direction, its locomotion is remarkably efficient, even if rather slow.

The nervous system of echinoderms is not like that of clearly bilateral animals; it is decentralized and there is no brain. The decentralized arrangement is favorable for animals with radial symmetry, for it enables them to respond to stimuli coming from various directions. Any ray of most sea stars may become the leader, and the animal moves in that direction.

Tube feet have functions other than locomotion. Being fluid-filled and relatively thin-walled, they can serve in respiration and probably in elimination of nitrogenous wastes. They are also important in feeding. Some sea stars use tube feet, and the muscles by which they slowly bend their rays, to grasp a whole clam until they get it into the mouth. They also use them to pull apart the valves of a clam in order to insert part of the stomach, turned inside-out. Thus digestion can start without having to swallow the clam.

How did echinoderms originate? We do not know, even though they have left a rich fossil record. Some of the early echinoderms were bilaterally symmetrical, others appear to have been asymmetrical. When the life cycle proceeds through a larval stage, the larva is bilateral, suggesting that some group of ancient bilateral invertebrates gave rise to the ancestral stock of the phylum. When the mineralized endoskeleton developed as a protective measure for the body, it eventually would have imposed a sedentary life-style on the group as a whole.

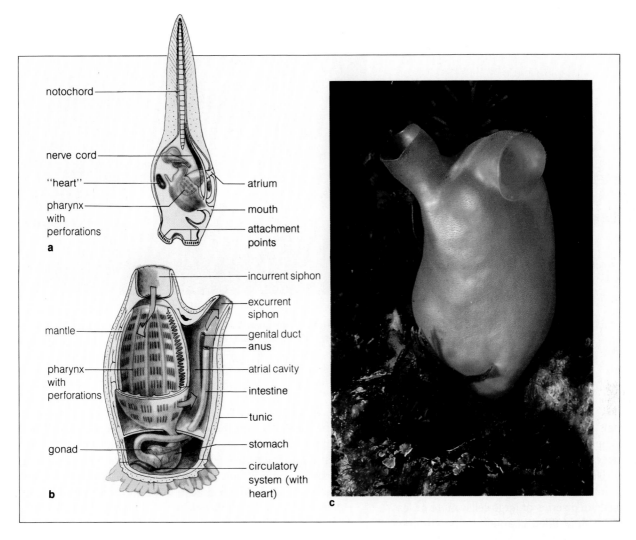

Figure 41.41 One of the tunicates, a "sea-squirt." (**a**) Larval form and (**b**) adult form; these sketches are not drawn to the same scale. (**c**) A typical tunicate, the sea peach.

CHORDATES

Most chordates (phylum Chordata) are vertebrates—animals whose brain and nerve cord are protected by cartilage or bone. The chordates that do not have a backbone have the following features that give them membership in the same phylum with the vertebrates:

1. The nerve cord is dorsal to the gut. It is also tubular, because of the way it originates during embryonic development—as a trough of ectoderm that becomes roofed over and separated from the skin (Figure 34.10).

2. A supporting structure called the *notochord*, located just beneath the nerve cord (page 327).

3. Perforations in the wall of the pharynx, or at least pouches that suggest perforations.

The three features may not persist throughout the life of a chordate, but they are usually evident during the early stages of development. For example, you have a tubular dorsal nerve cord, but your notochord and gill pouches disappeared when you were still an embryo.

Let's now look briefly at the two groups of invertebrate chordates, called **urochordates** and **cephalochordates** (Figure 41.40). All of the members of both groups are confined to the seas, whereas vertebrates are found in marine, freshwater, and terrestrial habitats.

Urochordates

The urochordates are commonly called tunicates, a name alluding to a gelatinous or leathery "tunic" most of these animals secrete around themselves (Figure 41.41). Some also are called sea-squirts, because they expel water through their excurrent siphon when they are touched.

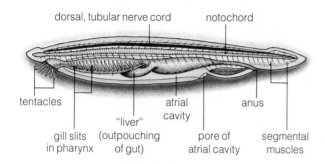

dorsal, tubular nerve cord   notochord

tentacles

gill slits   "liver"   atrial   anus   segmental
in pharynx   (outpouching   cavity   muscles
                 of gut)   pore of
                                atrial cavity

Figure 41.42 Photograph and cutaway view of a lancelet showing the position of its nerve cord and flexible notochord.

These animals do not have a coelom, as all other chordates do. When a notochord is present, it is confined to the tail region of an animal that resembles a tadpole (Figure 41.41a). Usually the "tadpole" form is a larval stage, but sometimes it is characteristic of the adult. The notochord consists of a series of large cells, more or less wrapped around a fluid-filled channel. Being stiffened yet flexible, it functions in much the same way as a torsion bar. After it has been bent by contraction of muscles on one side or the other of the tail, relaxation of these muscles allows the notochord to straighten. As it springs back, it does so forcefully enough to give the tail a strong propulsive force.

As adults, most tunicates are generally attached to solid objects, such as rocks, shells, and wooden pilings. The tadpole larva, shown in Figure 41.41a, is a bilateral animal. The gut at this stage is incomplete and non-functional, but the pharynx has a few perforations on both sides. These perforations open into right and left pockets that are open to the outside.

After attaching itself to a firm substrate, the tadpole undergoes a drastic metamorphosis. It resorbs its tail, thereby losing the notochord, and secretes a gelatinous or leathery tunic around itself. (To a large extent, the tunic consists of a carbohydrate chemically similar to the cellulose of plant cell walls.) The body becomes completely reorganized. The pharynx (into which the mouth opens) enlarges and develops many perforations that are often rather complex. The pharynx is used as a sieve for collecting microscopic forms of food, such as diatoms. The activity of cilia in the pharynx draws water in through the incurrent siphon and drives it through the

perforations into the *atrial cavity* (Figure 41.41b). From there, the water leaves the body through the excurrent siphon. Food trapped in mucus that coats the inside of the pharynx is moved by ciliary activity to the portion of the gut in which digestion takes place. Besides collecting and concentrating food, the pharynx is an efficient respiratory organ, for it is thin-walled and has a good blood supply.

There is no real excretory system in these animals, but relatively insoluble nitrogenous wastes are commonly stored in certain tissues. There is a heart beneath the pharynx. Curiously, it pumps blood for a time in one direction, then in the opposite direction. Blood leaving the heart is at first enclosed by vessels, but it soon escapes into spaces comparable to those of the hemocoel of a mollusk or arthropod.

Are urochordates in the line of evolution of vertebrates? Probably not. The absence of a coelom, the peculiar kind of notochord, and the fact that the perforations of the pharynx are not in a single, linear series suggest there is no direct connection between urochordates and vertebrates. Nevertheless, it is possible that an animal somewhat similar to the swimming larva of a tunicate could have figured in the early evolution of vertebrates.

Cephalochordates

There are only a few species of cephalochordates, which are fishlike animals rarely more than five centimeters long. They are called lancelets because of the way the body is sharply tapered at both ends. "Amphioxus,"

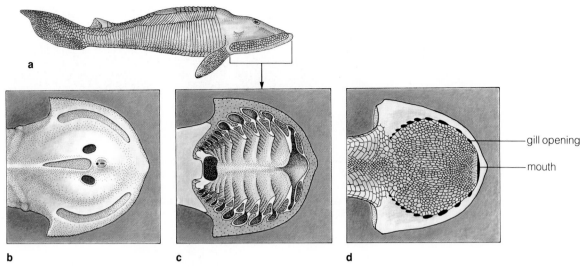

Figure 41.43 (**a**) Reconstruction of an ostracoderm, the oldest and most primitive of the fossil vertebrates. The dorsal view (**b**) shows the openings for three eyes (blue) and a single nostril (red). The ventral view (**c**) shows the small mouth and relatively large, food-straining pharynx; the flattened bones covering all but the gill openings and a slitlike mouth have been removed, but are shown in place in (**d**).

formerly used as a genus name, also refers to this feature (Figure 41.42).

With its well-developed musculature, a lancelet can bury itself quickly in sand or mud. The muscles concerned with the movements are arranged in series on both sides of the notochord, giving the animal a segmented appearance. The notochord itself extends for nearly the full length of the body and consists mostly of disklike muscles. It is decidedly different from the tunicate notochord, as well as from the notochord of living fishes of ancient ancestry.

When a lancelet is buried and in the feeding position, its mouth is at the surface of the sediments. Ciliary activity that draws water into the pharynx also drives it out through numerous gill slits arranged in series on both sides. On passing through the gill slits, the water enters a spacious atrial cavity, then leaves the body by way of a ventral pore. Food sieved out by the pharynx is delivered to the rest of the gut, where digestion, absorption, and compaction of fecal material take place.

Lancelets are almost certainly closer to vertebrates than urochordates are. The larval stage of a lamprey (a jawless fish) is similar to a lancelet, not only in structure but in the method of feeding. The similarity is probably the result of parallel evolution.

Vertebrates

In the fossil record are remains of fishlike animals called **ostracoderms**, a name referring to their external body plates of a material similar to bone (Figure 41.43). The

brain was enclosed within a skull of bone or cartilage, and the notochord was apparently important as a supporting structure. Being without jaws, the mouth was similar to the mouth of a modern lamprey. In other early evolutionary lines of jawless vertebrates, there were distinct vertebral units of cartilage or bone. This arrangement would be much like that in a lamprey or hagfish. In these existing fishes, the notochord is somewhat gelatinlike, but it is surrounded by a sheath of connective tissue and supplemented by cartilaginous elements. Some of these elements support the tissues between successive gill slits (Figure 41.44); others, above the notochord, bear resemblances to the vertebrae of complex vertebrates. There is also a simple skull.

Eventually, well-developed vertebrae evolved in fishes. What are the advantages of having a series of these cartilaginous or bony structures? They fit together in such a way that they allow the body to flex, and they provide attachment sites for muscles concerned with this type of movement. Moreover, they completely surround and protect the nerve cord.

Jaws, which are characteristic of all living vertebrates other than lampreys and hagfishes, evolved from the most forward gill supports (Figure 41.44). It may have taken a long time for this new structural development to become functional in biting and chewing, but it certainly opened up many possibilities for feeding by mechanisms other than the sucking or rasping motions of the mouth of jawless fishes. It must also have intensified competition for prey and increased the prospect that one fish could be eaten by another. Thus there would have been selective advantages for a fish to be able to rec-

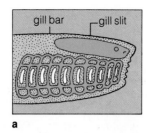 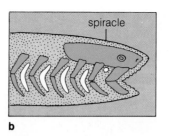 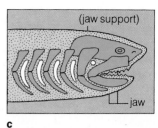

Figure 41.44 Proposed evolution of gill-supporting structures, as found in jawless fishes (**a**) into the hinged vertebrate jaw (**c**). The first gill opening of the mud-dwelling jawless fishes was converted into a spiracle through which water could be drawn. The first gill bars in the series, no longer required as supporting structures, became enlarged and equipped with teeth. The embryonic development of sharks reflects such a sequence.

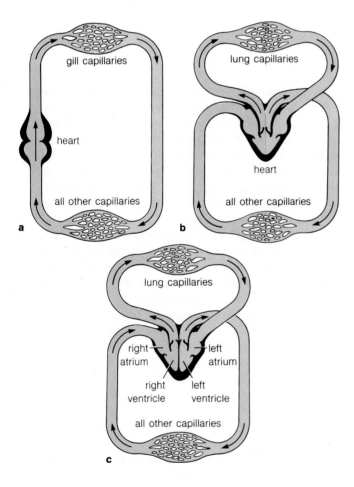

Figure 41.45 Relationship between the heart and blood vessels for (**a**) fishes, (**b**) amphibians, and (**c**) birds and mammals.

ognize food or a predator from a distance. The evolution of more sensitive eyes, olfactory receptors, and sensors that could detect vibrations followed; the brain and motor pathways also became more complex (Chapter Twenty-Five). The evolutionary trends toward more efficient sense organs and nervous systems in the jawed fishes continued in the vertebrates that invaded the land. Thus some of the features that evolved first in the fishes became the legacy for amphibians, reptiles, birds, and mammals.

Vertebrate Gills, Lungs, and Heart

The ancestors of vertebrates were probably filter feeders, and in this respect they were similar to urochordates and cephalochordates. The perforations in their pharynx would have been used for eliminating water, and in this way the food particles that had been retained within the pharynx would have become concentrated. The perforations may have enhanced respiration, too, but this would have been a secondary role. By the time the first jawless fishes evolved, their degree of activity and larger size would have required more efficient ways to absorb and distribute oxygen. When rasping and sucking replaced filter feeding, the gill slits and tissues around them became important in respiration.

In existing lampreys and other fishes, the ventricle of a two-chambered heart initially pumps blood to the gills. Then the oxygenated blood is distributed through the rest of the body, finally returning to the atrium of the heart (Figure 41.45a). Gills are unsuitable respiratory devices for land animals. If its gills and body surface could be kept moist, and if its gill branches did not become so badly stuck together that they could not function, a fish might be able to live temporarily out of water,

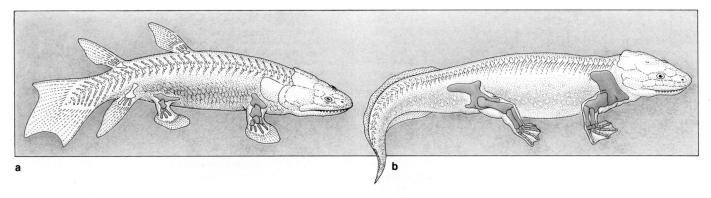

Figure 41.46 Evolution of bony or cartilaginous structures of ancestral lobe-finned fishes (**a**) into the limbs of early amphibians (**b**).

as it would have to do in order to get from a pond that has dried up to one that has water. The first land-venturing animals, which lived during the Devonian period, were fishes that had developed lungs. These were outpocketings of the pharynx, which had already appeared in some fishes and which increased the surface area for aquatic respiration. Lungs continue to function in a few fishes living today in Africa, Australia, and South America, but in most fishes they have been converted into a swim bladder (Figure 31.4). Some of the tissue that lines a swim bladder controls the quantity and proportions of gases within it, and this enables a fish to maintain buoyancy and its position at a particular depth. The swim bladder is not connected to the pharynx except in lungfishes, which periodically go to the surface to gulp air.

Another important development related to the appearance of lungs was the extension of the nostrils to the mouth, as is typical of lungfishes, modern lung-breathing amphibians, and all complex vertebrates. This makes it possible for an animal to breathe while its mouth is closed or full of food.

The pattern of circulation in the first land-dwelling vertebrates was probably the same as it is now in fishes; in other words, blood leaving the ventricle would first be oxygenated in the lungs and gills (if gills were still present), and then would be distributed to the rest of the body. Eventually, however, the atrium became divided, so the heart would have been three-chambered. This is the situation in modern amphibians (Figure 41.45b). In a frog, for instance, blood that has been oxygenated in the lungs is returned to the left atrium, whereas blood from the systemic circulation is returned to the right atrium. From both atria, blood is pumped into the ventricle, where some mixing occurs. Nevertheless, when the ventricle contracts, poorly oxygenated blood, mostly in the anterior part of the ventricle, tends to be forced into the vessels that lead to the lungs. The remainder of the blood, which is moderately well oxygenated, goes to the systemic circulation. The arrangement is adequate for amphibians because they do not have a high rate of metabolism and because the skin and mouth cavity are also generally important in exchange of oxygen and carbon dioxide.

Most reptiles have conserved the amphibian system of circulation, although the ventricle is partly divided into right and left chambers, which reduces the extent to which blood from the right and left atria is mixed. In crocodiles and alligators, in fact, the separation into right and left ventricles is complete, so the pattern of circulation is comparable to that of birds and mammals. Blood from the right ventricle goes to the lungs, that from the left ventricle goes to the systemic circulation. This arrangement is absolutely essential for birds and mammals (Figure 41.45c), which maintain a warm internal body temperature, and which generally have high rates of metabolism.

Evolution of Limbs

The limbs of terrestrial vertebrates can be traced back to fins of certain fishes (Figure 41.46). Paired fins developed in the pectoral and pelvic regions of the body, paving the way for the evolution of *lobed* fins, with internal supports of bone or cartilage. Appendages of this type allowed for more effective locomotion on land, and further evolution led to the four legs of amphibians and eventually to arms and legs of humans, flippers of seals, and wings and legs of bats and birds. Among the few living fishes that have descended from early lobe-finned

Class Osteichthyes: bony fishes. All have mostly bony endoskeletons and overlapping rows of thin, flexible scales. Respiration is through gills with an operculum (lid) on both sides of the body. Many have a swim bladder. Bony fishes include ray-finned fishes (such as sturgeons) and teleosts (such as eels, salmon, minnows, catfish, rockfish, perch, tuna, and deep-sea luminescent fishes). Above, a soldier fish.

Class Chondrichthyes: sharks, rays, skates, and chimaeras. These carnivorous fishes have endoskeletons of cartilage. They have no swim bladder; five to seven gill slits occur in separate clefts along both sides of the pharynx. They have small conelike scales or none at all. Some sharks, fifteen meters long, are among the largest living vertebrates. A blue-spotted reef ray (above) and a gray shark (right).

Class Amphibia: frogs, toads, newts, salamanders, and caecilians. Amphibians have a largely bony endoskeleton and most often, four legs (some have none). Respiration occurs through gills, lungs, skin, and regions of the pharynx. Amphibians have a three-chambered heart. Their moist, thin skin is vulnerable to drying out. Eggs must be laid in water or moist places. Above, a Blue-Ridge Spring salamander.

Class Aves: birds. Most have forelimbs modified for flight. They have a four-chambered heart and lungs with air sacs. They have feathers, leg scales, and horny beaks. Shelled eggs develop outside the body. Strong, lightweight bones are filled with air cavities (the endoskeleton of a frigate bird, with a seven-foot wingspan, weighs only four ounces). Here, an owl on the wing.

Class Mammalia: egg-laying mammals.(Prototheria), pouched mammals (Metatheria), and placental mammals (Eutheria), all descendants of reptiles. Embryonic development is internal; liveborn are nourished by milk-secreting glands. Most mammals are covered with hair; whales are not, and reptilian scales persist on tails of beavers and rats. Adults have a permanent set of teeth (reptiles have successive sets). They have lungs, a four-chambered heart, and a well-developed cerebral cortex. Above, an arctic fox in its winter coat.

Class Reptilia: lizards, turtles, and snakes. Reptiles have a well-developed bony endoskeleton and dry, scaly skin that resists desiccation. Most depend entirely on lungs, with air being sucked in (not forced in by mouth muscles, as in amphibians). Almost all lay shelled eggs on land. Within the shell, the embryo floats in the "aquatic" world of the amniotic sac; it receives nutrients from a yolk sac; it uses the allantois, another membrane, as a surface for gas exchange. The shelled egg was a key character in the vertebrate invasion of land. Above, a green python native to New Guinea and Northeast Australia.

Figure 41.47 Representative vertebrates which, together with the tunicates and lancelets described earlier, are members of the phylum Chordata.

fishes are the coelacanths. They were long thought to be extinct, but since 1938 numerous specimens have been collected in the sea between Africa and Madagascar. Lungfishes are also survivors of the same general group, although their fins are so much modified that they scarcely resemble those of their ancestors.

Vertebrate Nervous System and Sense Organs

In ancient fishes, the brain and cranial nerves gradually became more prominent. By the time amphibians were established on land, they were encountering stimuli to which their aquatic ancestors had not been exposed. The brain underwent dramatic development, particularly in the portions concerned with processing signals related to vision, hearing, and balance.

Eyes already had become well developed in fishes, and they were modified for vision in air when amphibians began their evolution on land. The ear, as a complex of structures concerned with balance, also originated in fishes. These structures—including the semicircular canals and statocysts associated with them—have persisted throughout the diversification of vertebrates.

The cerebral cortex of the brain began to develop markedly during the evolution of reptiles. At first the layer of gray matter was thin, but it nevertheless must have had connections that permitted more complex integration than had been possible in amphibians. It was not until the rise of mammals that the cerebral cortex began to reveal its remarkable potential, gradually expanding to a large mass of information-encoding and information-processing cells (Chapter Twenty-Five). Like other evolutionary developments that have been described in this chapter, the vertebrate brain is the result of many changes taking place over many millions of years.

Today, there are seven classes of vertebrates, representatives of which are illustrated in Figure 41.47:

Agnatha	*jawless fishes*
Chondrichthyes	*cartilaginous fishes*
Osteichthyes	*bony fishes*
Amphibia	*amphibians*
Reptilia	*reptiles*
Aves	*birds*
Mammalia	*mammals*

Details of the anatomy, physiology and life-styles of these diverse animals are covered in other units of the book; a classification scheme indicative of their likely relationships is given in Appendix I.

SUMMARY

1. Multicellular animals, ranging from simple sponges to vertebrates, are assigned to about thirty phyla. The members of each phylum share certain characteristics and are therefore believed to have descended from a single ancestral stock, or at least from closely related stocks.

2. In assigning animals to phyla, many aspects of structure, development, physiology, and the fossil record must be considered. Some of these are mentioned below.

3. Sponges (phylum Porifera) are considered to be the simplest multicellular animals. They have neither a gut nor a nervous system; although they have several kinds of cells, the cells are not organized into tissues. Certain cells trap microscopic food particles from water moving through a system of pores, canals, and chambers.

4. Jellyfishes, sea anemones, and their relatives, which belong to the phylum Cnidaria, have radial symmetry and their cells form tissues. One tissue layer (epidermis) covers the outside of the body; the gastrodermis lines the gut, which is basically saclike because it has no anus. The epidermis and gastrodermis are derived from embryonic ectoderm and endoderm, respectively. They are separated by secreted material, and this may make up much of the total mass of the animal, as it does in a jellyfish.

5. Cnidarians have nerve cells, usually organized into networks closely associated with the epidermis and gastrodermis. A distinctive feature of the phylum is the presence of nematocysts, or stinging capsules, which assist in capturing prey and may also serve in protection.

6. Most animals more complex than cnidarians are bilaterally symmetrical, and during their embryonic development a third germ layer (mesoderm) is formed. Mesoderm gives rise to muscles and various other tissues between the epidermis and gastrodermis. The gut may be saclike as in flatworms (phylum Platyhelminthes), but generally it is complete, with an anus as well as a mouth.

7. Most animals more complex than flatworms have either a coelom or a pseudocoel. These cavities separate the gut from the body wall. A true coelom originates within pouches or masses of mesoderm and is lined by a peritoneum; a pseudocoel is not lined. Some invertebrates (notably of the phyla Mollusca and Arthropoda) have a coelom but it is much reduced. The blood that bathes their tissues, however, flows through the interconnected spaces collectively called a hemocoel. This blood-filled cavity, like a coelom or pseudocoel, may function as a hydrostatic skeleton as well as in distributing oxygen and nutrients.

8. Many invertebrates, mostly of the phyla Annelida and Arthropoda, are segmented (their bodies are partitioned into a series of units). In annelids, many of the segments may be similar, at least externally, but the extent to which the segments are specialized varies greatly. In arthropods, segments tend to be decidedly specialized, and groups of successive segments form distinct body regions such as the head, thorax, and abdomen of an insect. Arthropods also have an exoskeleton and jointed appendages. Both features have contributed to the success of arthropods, of which there are about a million species (ten times as many as in the second largest phylum, Mollusca).

9. Flatworms, the simplest bilaterally symmetrical and cephalized animals, have a brain and two or more longitudinal nerve cords. In annelids and arthropods the brain is more elaborate and there is a double ventral nerve cord. In mollusks, the usual components are ganglia that form a brain and some other ganglia that are linked with it by major nerves. In vertebrates and other chordates, the nerve cord is dorsal.

10. Of the animals having a true coelom, the echinoderms are the only ones that are uncephalized and radially symmetrical. The nervous system is decentralized, there being no brain. This arrangement is favorable for animals with radial symmetry (such as sea stars) because the sensory structures and nerve tracts of any ray can initiate activity in response to a stimulus.

11. Early in the evolution of vertebrates, the brain and dorsal nerve cord became protected by an internal skeleton of cartilage or bone. Other skeletal elements, such as jaws, ribs, and supports for the limbs, also developed.

12. The vertebrae around the spinal cord allow for considerable flexibility, and nearly all skeletal structures are sites to which muscles concerned with chewing, locomotion, and some other functions are attached.

13. During vertebrate evolution, certain fishes became equipped with lungs and fins of a type that enabled them to walk on land, leading to the evolution of amphibians.

14. Reptiles, derived from amphibians, became even more successful on land because of their protective scaly skins, physiological adaptations, and ability to reproduce without going back to water (which most amphibians must do).

15. Birds and mammals evolved from separate groups of reptiles. Their success is related in part to their constant and warm body temperature and to the insulation provided by feathers or hair. These characteristics enable birds and mammals to live in many environments. Good vision, a well-developed brain, appendages specialized for flight, swimming, running, grasping, and other activities are a few more features underlying the success of birds and mammals.

Review Questions

1. Sponges are simple multicellular organisms, but in some situations they colonize much of the available living space. Can you give a few reasons for their success?

2. Taking spiders and insects as examples, explain why certain groups of arthropods have become so successful in terrestrial habitats.

3. How do the structural and developmental characteristics of protostomes differ from those of deuterostomes?

4. Explain the difference between a coelom, pseudocoel, and hemocoel. Using nematodes, annelids, mollusks, arthropods, and vertebrates as examples, discuss the importance of all three of these types of fluid-filled cavities.

5. Why was the appearance of segmentation such an important step in the evolution of certain higher invertebrates? In developing your answer, think about the segmentation of an earthworm or polychaete annelid and the segmentation and presence of jointed appendages of an insect and crustacean.

6. Compare the exoskeleton of arthropods with the endoskeletons of echinoderms and vertebrates. How are these types of skeletons formed, and what advantages do they confer upon the animals that have them?

7. In spite of their great success, insects and spiders do not grow to be as large as lobsters or many crabs. Try to explain this.

8. What were some of the significant steps in the evolution of birds and mammals from early vertebrates that were somewhat comparable to modern lampreys? Be sure to deal with jaws, gill slits, fins and other limbs, lungs, skins, body temperature, and development of eggs and embryos on land versus development in water.

Readings

Barnes, R. 1980. *Invertebrate Zoology*. Fourth edition. Philadelphia: Saunders. Good introduction to the invertebrates.

Hickman, C. P., and L. S. Roberts. 1983. *Integrated Principles of Zoology*. Seventh edition. St. Louis: Mosby.

Hildebrand, M. 1985. *Analysis of Vertebrate Structure*. Third edition. New York: Wiley.

Romer, A. S., and T. S. Parsons. 1986. *The Vertebrate Body*. Sixth edition. Philadelphia: Saunders. Chapter 16 contains detailed pictures of vertebrate nervous systems. Excellent reference book.

42

HUMAN ORIGINS AND EVOLUTION

In the preceding chapter, we covered a tremendous span of animal evolution. We pieced together a picture of adaptations leading to bilateral symmetry and segmentation, foretelling such structures as paddles, fins, and legs. We brought other adaptations into the picture: internal circulation systems for supplying ever-larger bodies with nutrients and for removing metabolic wastes, backbones for strength and flexibility, and complex nervous systems that meet the challenges of increasingly complex environments. In doing so, we watched the sprouting of the first metazoans and their subsequent growth over 2 billion years into the mammalian stem, one of many stems on the animal phylogenetic tree. It is here that we exist, one branch among a diverse array of other mammalian forms.

How did we arrive here? Although the details have not yet been worked out, investigations of fossil hunters over the past ninety years have produced a remarkable array of evidence by which we can follow the development of human forms from ancestral species. In this chapter, we will survey some of the discoveries and hypotheses about human origins from a biological point of view.

GENERAL CHARACTERISTICS OF THE PRIMATES

According to one classification scheme, the order **Primates** has fifty-seven existing genera, which are divided into two suborders:

prosimians	*tree shrews, tarsiers, lemurs, and lorises*
anthropoids	*New World monkeys (ceboids)*
	Old World monkeys (cercopithecoids)
	apes, humans (hominoids)

The primates in Figures 42.1 and 42.2 show some of the distinguishing limb and sensory adaptations that occur among the varied members of these genera. To understand the significance of these adaptations, we must return to the dawn of the Cenozoic, some 65 million years ago—the time of origin for the primates.

Primate Origins

Like all placental mammals, the primates apparently arose from ancestral forms of the order Insectivora. Fossils of those mammals resemble the small, ground-dwelling tree shrews of Southeast Asia (Figure 42.3). As the name implies, the early insectivores ate insects; like shrews, they probably ate seeds, buds, eggs, and other tiny animals as well. Shrews are night-time omnivores—

Figure 42.1 Pygmy chimpanzee walking upright on a bough eighty feet above the ground. Its hands are thus freed to carry objects—in this case, fruits from the boongola tree.

a

b

Figure 42.2 (**a**) The gibbon and (**b**) two tarsiers. Gibbons have limbs and a body adapted for brachiation (swinging arm-over-arm through the trees); tarsiers are vertical climbers and leapers.

highly active, and hungry (they eat the equivalent of their own weight nightly). They have a long snout and a well-developed olfactory sense, useful in snuffling out food and catching the scent of predators, which once included some of the smaller dinosaurs. With the widespread extinction of dinosaurs by the close of the Mesozoic, the shrewlike forms of mammals underwent large-scale adaptive radiations. Some lineages went underground and evolved into moles. Others took to the water and to the air; their descendants include otter shrews and bats. And one lineage took to the trees.

The forest canopy, with its abundance of fruits and insects, and its safety from ground-dwelling predators, was a promising adaptive zone. Yet with all of its advantages, life in the trees also was demanding. It must have been a visually complex world then, as it is now, with dense leaves, dappled sunlight, boughs swaying in the wind, darting insects and other prey, perhaps predatory birds. The mammalian snout would not have been much use in such a world, where air currents disperse the scent of food and predators. However, eyes that could discern color, shape, and movement in a three-dimensional field would have been enormously useful while the animal was leaping and moving among the branches. At the same time, four legs so useful in running would not have been as useful as limbs and digits adapted for climbing and clinging. Later on, modifications of the forelimbs would prove useful for infant survival (offspring could hold on tightly to the mother during foraging activities) and, later still, for making and wielding tools.

Figure 42.3 Common shrew (*Sorex*).

The key characters of primate evolution were (1) body and limbs adapted for tree-climbing and (2) eyes adapted for discerning color, shape, and movement in a three-dimensional field.

Evolutionary Trends

The key characters just described were the basis of the following evolutionary trends that occurred among the primates:

1. *A capacity to move limbs and the head freely in different planes.* Primate arm and leg limbs can swivel freely in their sockets. By swinging the limbs up, down, sideways, backward, or forward, the body can be supported from any direction, not just from the flat ground. Also, primates show considerable head rotation, as the tarsiers in Figure 42.2 illustrate.

2. *Mobile digits capable of grasping.* Among monkeys, apes, and humans, four digits on each hand are complemented by an opposable thumb, which permits detailed manipulation and examination of objects. In humans alone, the thumb is long and powerful enough for strong yet precise gripping of objects.

3. *Upright vertebral column and body posture.* Early in primate evolution, the body's center of gravity shifted until it was well back over the hindlimbs. This development made upright sitting positions possible and freed the forelimbs for manipulation. The development required changes in the proportions and details of the pelvic, leg, and foot bones as well as in the vertebral column. (Interpretation of these changes in fossils is central to attempts to reconstruct primate evolution.)

4. *Enhanced vision.* The early primates had forward-directed eyes, which means they had an overlapping visual field and depth perception. Over time, the retina became increasingly good at detecting light of different intensities (dim to bright) and wavelengths (colors). The eyes themselves became larger and were able to detect more light and detail. As the eyes and their sockets enlarged, the greater reliance on vision was accompanied by a reduction in the snout (and the sense of smell). Gradually the position of the eyes shifted back into the skull and came to be protected by a bony ridge.

5. *A nervous system that integrates and processes diverse stimuli, then coordinates precise, rapid, yet flexible movements in response.* Before walking, swinging, and leaping (especially!) can be undertaken high up in the forest canopy, the brain must assess myriad factors—distance, body weight, wind, the suitability of the destination, how to compensate quickly for miscalculations. The exquisite neural control characteristic of primates began to develop during life in the trees.

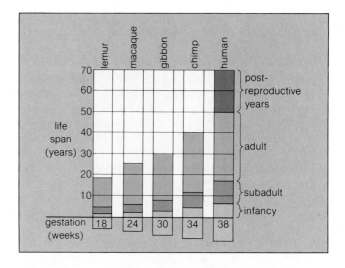

Figure 42.4 Trend toward longer life spans and longer periods of infant dependency among existing primates.

6. *Strong social bonding.* Existing primate groups reflect a trend toward longer life spans, longer periods between pregnancies, single births rather than litters, and longer periods of infant dependency (Figure 42.4). The increased parental investment in fewer offspring has been accompanied by the development of strong social bonds, intense parental care, longer periods of learning, and application of intelligence.

EARLY PRIMATE EVOLUTION

The evolutionary history of the primates is not clear-cut, with one form gradually replacing another. Fossils have not yet been recovered for a few critical time periods. Moreover, there are periods in which closely related forms coexisted for some time, with some destined to leave descendant populations and others to become evolutionary dead-ends. What we will be describing, then, are some of the known branches on a very "bushy" evolutionary tree.

The oldest known primate fossils date from the Paleocene (65 million to 54 million years ago). Again, they were morphologically similar to existing tree shrews, with a relatively small brain and a long snout. Although many of those forms died out, some left descendants that evolved into the true prosimian forms of the Eocene (54 to 38 million years ago).

The Eocene climate was somewhat warmer than the Paleocene, and tropical rain forests flourished—as did the early prosimians. These primates were characterized

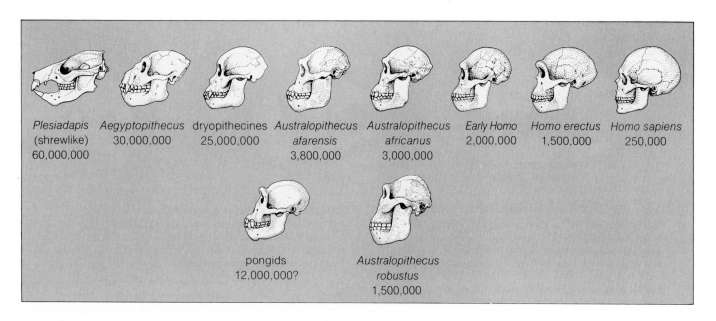

Plesiadapis
(shrewlike)
60,000,000

Aegyptopithecus
30,000,000

dryopithecines
25,000,000

Australopithecus
afarensis
3,800,000

Australopithecus
africanus
3,000,000

Early Homo
2,000,000

Homo erectus
1,500,000

Homo sapiens
250,000

pongids
12,000,000?

Australopithecus
robustus
1,500,000

Figure 42.5 Comparison of the skull shapes and dentition of some anthropoids, living and extinct. (The skulls are not drawn to the same scale; for example, the *Aegyptopithecus* skull was no larger than that of a house cat.) Notice especially the proportion of the cranium relative to the rest of the skull and the variations in teeth. For a little more background on the significance of these dental types, see Figure 42.6.

In general, the australopithecines had small brains and large faces, compared with the larger brained, smaller faced forms of the genus *Homo*.

by an increased brain size, an increased emphasis on vision over smell, and more refined grasping abilities. This was the time of divergences that led, eventually, to the modern-day lemurs, lorises, and tarsiers. It was also the time of divergences that led to the first anthropoids.

The Oligocene (38 to 25 million years ago) was an extremely important time in primate history, but unfortunately our knowledge of what was going on then is largely limited to a small area known as the Fayum Depression, which is now a hot desert southwest of Cairo, Egypt. About 35 million years ago, the land here was cloaked with the lush vegetation of a tropical rain forest. Humid swamps lined the many rivers that ran through the forest and emptied into the Mediterranean Sea. This setting was home to the Oligocene ancestors of the Old World monkeys, apes and, eventually, humans. Those ancestral forms were adapted for life in the trees. Given the altogether nightmarish predatory reptiles that are known to have inhabited the swamps at the time, it is perhaps understandable why the early anthropoids stayed mostly up in the trees.

The fossil record indicates that these small, monkey-like forms had a slightly increased cranial capacity (and perhaps a more complex social behavior), better grasping abilities, and more refined visual abilities than the prosimian forms. An intriguing Fayum primate, *Aegyptopi-*

thecus, dates from about 28 million years ago. This form may be on or close to the line of descent that led to the **hominoids**, which include all ape and human species, living and extinct. *Aegyptopithecus* was about the size of a gibbon. Compared to modern monkeys, its snout was somewhat longer and its cranium not as large—and yet the configuration of one of its molars is seen also in apes and humans (Figures 42.5 and 42.6).

While these hominoids were evolving, major land masses were on the move as crustal plates began to assume their most recent positions (Chapter Thirty-Eight). At the start of the Miocene, some 25 million years ago, the redistribution of land masses led to pronounced shifts in climate. In what is now Africa, Europe, and elsewhere, there began a major cooling trend and a decline in rainfall that would culminate in the Pleistocene. Over time, the tropical and subtropical forests would gradually give way to grasslands. In the interim, there was a mosaic of forests and savannas.

All of this geologic evolution had a profound effect on hominoid evolution. Early populations of forest-dwelling apes spread from Africa across a newly created land bridge that linked the continent with Eurasia. As the climate continued to become cooler and drier, the belts of grasslands spread, and different subpopulations of apes became reproductively isolated within pockets

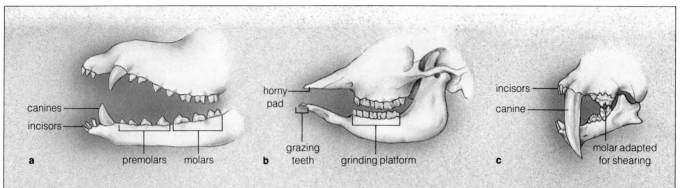

Figure 42.6 (**a**) Generalized representation of the teeth of placental mammals. The individual teeth are highly modified (and some are absent) in different groups. The conelike or chisellike *incisors*, used in nipping and grazing, are retained in most groups; ruminants such as cows retain only the bottom row, which works in opposition to a horny pad on the upper jaw (**b**). Long, pointed *canines*, used in biting and piercing, are pronounced in carnivores and sometimes in nonpredatory groups where they are used in defense and in territorial displays (see, for example, Figure 49.2. Canines were most extreme in the extinct sabertooth cat (**c**).

The *premolars* and *molars* (cheek teeth) form a grinding platform in herbivorous mammals such as cows (**b**). They are reduced in carnivores, which swallow food with little or no chewing. Some of their cheek teeth have sharp crests and are used in shearing tendons and bones.

The modifications on the basic plan are of diagnostic importance to the fossil hunters who are interested in tracing the evolution of early placental mammals into existing genera, including our own.

of forests. There they adapted to local conditions and diverged into a variety of forms. In this manner, a great adaptive radiation of apelike primates took place that resulted in a wide geographical distribution for a large number of species.

The largest group of Miocene apes are collectively known as the **dryopithecines**. There were at least four to seven different species, which came in a variety of forms, sizes, and apparent life-styles. In general, the dryopithecines seemed to differ from the monkeylike primates in terms of diet, as they probably depended more on fruit and the monkeys more on leaves. They were also larger brained and may have been more behaviorally flexible.

Most of the dryopithecines died out. However, evidence suggests that one group was ancestral to the orangutan, while another was the common ancestor of modern African hominoid forms—chimpanzees, gorillas and humans. (The lesser apes, including the gibbons, apparently diverged earlier in the Miocene.) However, the evolutionary interpretation at this point is open to debate. This is especially so with regard to a late Miocene group of apelike hominoids, which are sometimes called the **ramamorphs**. The fossil record for this period is still poor, which is vexing to say the least, since it seems to be the time of divergence leading to the **hominids**—all species on the human evolutionary branch. The ramamorphs also were contemporaries of later species of dryopithecines. They included forms that have been given the names *Ramapithecus* and *Sivapithecus*, although both may belong to a single genus.

The Hominids

The first hominids probably appeared between 8 million and 5 million years ago, during the late Miocene or early Pliocene. The cooling and drying trend that began in the Miocene continued, and vast savannas replaced much of the rain forests in Africa. The evolution of the first hominids coincided with this major shift in climate.

A variety of hominids have roamed the earth. Some were evolutionary side branches; others were on the direct path leading to the only surviving form, *Homo sapiens sapiens*. However, all hominids share a common ancestral form, which was itself descended from some apelike form of the Miocene. They also share the following characteristics:

1. *Bipedalism*: habitually walking and standing on two feet. Bipedalism could not have occurred without many structural modifications to the spinal column, hip bones, legs, and feet.

2. *Omnivorous feeding behavior*: in general, an absence of specialization in jaws and dental patterns that signifies a varied diet.

3. *Brain expansion and reorganization*: as outlined in Chapter Twenty-Five, an increase in the complexity and mass of the cerebral cortex in particular. The expansion of the hominid brain was associated with the evolution of **culture**: the sum total of behavior patterns of a social group, passed between generations by learning.

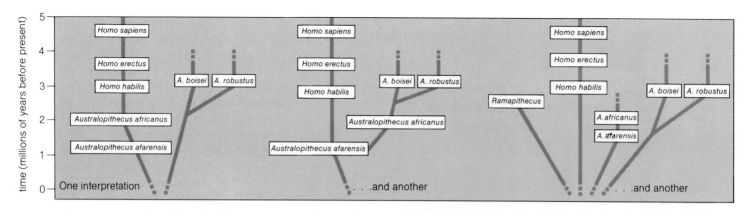

Figure 42.7 Three of the many current interpretations of phylogenetic relationships among fossil hominids. *Homo sapiens* is the only living species.

Why did these characteristics emerge among the hominids? Most researchers believe that they evolved as responses to the transition from the forest habitats to life in the open savannas. The evolutionary details remain elusive, and rather than focus on the numerous hypotheses that have been proposed, we will focus here instead on the plasticity that these characteristics represent. In this context, "plasticity" means the ability to respond to a wide range of demands, especially during the growth and development of an individual. The first hominid niche can be viewed as life in a highly unpredictable and complex world. As humans evolved, they were uniquely equipped—through bipedalism, generalized diet, and embellished brain—to be flexible and to adapt, including culturally, to a broad set of conditions. This represents an extension and embellishment of patterns observed among the other primates, and thus is based in our primate heritage.

Australopithecines

The fossil record of the early hominids suggests that theirs was a "bushy" period of evolution, and the record is not easy to interpret (Figure 42.7). However, since the first remains were discovered in 1924, our knowledge of the earliest undisputed hominids, the **australopithecines**, has increased dramatically. Apparently there were at least four species:

> *Australopithecus afarensis*
> *A. africanus*
> *A. boisei*
> *A. robustus*

and apparently there was some overlapping in time with the earliest members of the genus *Homo*. The australo-

pithecines were smaller overall, and their cranium was much smaller compared to the early forms of *Homo* even accounting for the difference in body size. Unlike the *Homo* fossils, the australopithecine face was rather large in relation to the cranium (Figure 42.5).

Fossils of *A. afarensis*, the most ancestral form, date from 4 million to 3 million years ago. This hominid represents a transition between apelike forerunners and later forms. Its cranial capacity (400 cubic centimeters, on the average) is closer to an ape's. Its teeth have been described as "apish with human tendencies." Unlike the jaws and tooth rows of apes (which tend to be U-shaped), its tooth rows are slightly splayed out in the back of the jaws. (In other australopithecines and in *Homo*, the tooth rows are more strongly splayed.) The canines of *A. afarensis* show human and apelike traits. Most intriguingly, the small back molars of this hominid are closer to the human pattern than they are even to the tooth pattern of the other australopithecines. Overall, the generalized pattern is associated with omnivores. All of this has been interpreted to mean that *A. afarensis* was the most recent common ancestor of the other australopithecines and *Homo*.

These early hominids were fully bipedal. Four-million-year-old fragments of leg bones show muscle insertions typical of bipedal walkers. Some of the strongest evidence comes from the remarkably complete skeleton of a female australopithecine (dubbed "Lucy") who lived three million years ago. As Figure 42.8 shows, Lucy's thighbones were angled from the pelvis toward the center of the body. This position, which allows the body weight to be centered directly beneath the pelvis, is characteristic of two-legged walkers. (Apes have thighbones splayed out from the pelvis and they have a waddling, four-limbed gait.) Dramatic signs of bipedalism are the 3.7-million-year-old fossil footprints made by early australopithecines (Figure 42.9).

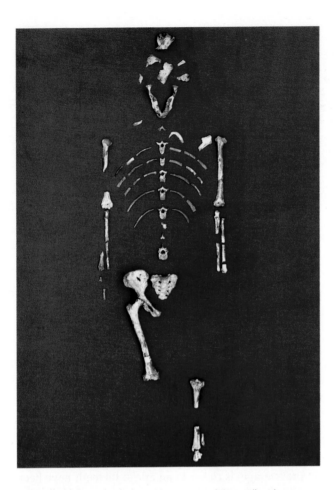

Figure 42.8 Fossil remains of Lucy, one of the earliest known australopithecines. Lucy was only 1.1 meters (3 feet 8 inches) tall and weighed 30 kilograms (about 65 pounds). The density of her limb bones is indicative of very strong muscles. Although her face (with its large, jutting jaws) was apelike, her cranium indicates that the brains of these early hominids were already undergoing some reorganization to the humanlike condition.

A. afarensis fossils more recent than 3 million years ago have not been found. However, numerous remains of *A. africanus* dating from about 3 million to 2.5 million years ago have been recovered, along with remains of *A. robustus* dating from about 2 million to 1 million years ago. The South African specimens indicate that *A. africanus* weighed about sixty pounds, was 4.5 feet tall on the average, and had an average cranial capacity of 450 cubic centimeters. By comparison, *A. robustus* weighed about a hundred pounds, was about 5 feet tall on the average, and had a cranial capacity of 530 cubic centimeters.

The robust australopithecine had a heavy skull, deep jaws, and a bony crest (built-up ridges of cranial bone, which are associated with strong muscle attachments). Like the teeth of *A. africanus*, its cheek teeth formed a grinding platform. Altogether, the fossils indicate that both types of australopithecines lived on rough plant material.

A. robustus and *A. africanus* fossils have been found in southern Africa; fossils of *A. boisei* come from the east. The eastern forms were even taller and more robust than *A. robustus*, although the cranial capacity was about the same. *A. boisei* had an extremely heavy cranial structure with very prominent crests. Its molars were enormous, and those farthest back were heavily wrinkled. Its dental pattern suggests that *A. boisei* was extremely adapted for a diet of rough vegetation, including small, hard objects such as seeds and nuts. The pattern reflects a continuing trend that began in the South African australopithecines. It is possible that an early subpopulation of *A. robustus* moved into eastern Africa, where it became reproductively isolated and increasingly specialized for this sort of diet.

Recently, Alan Walker discovered the skull of a hominid that may have been the most robust form of all. A large crest served to anchor huge muscles that could have been used for chewing tough plant material; the teeth also were specialized for this task. However, in other respects the skull is more primitive—and this specimen may be representative of an intermediate form between *A. africanus* and *A. boisei*.

Early *Homo*

Long before the australopithecines disappear from the fossil record, there is evidence of an omnivorous, larger brained hominid, *Homo*, evolving along with them in time. Fossils of "early *Homo*" (also called *H. habilis*) found in East Africa date from about 2 million years ago. Not only did this early form overlap with its vegetarian cousin, *A. boisei*, there is some tenuous evidence that early *Homo* may have hunted it.

Figure 42.9 Footprints made in soft, damp, volcanic ash 3.7 million years ago at Laetoli, Tanzania, as discovered by Mary Leakey. The arch, big toe, and heel marks are those of upright early hominids.

Although somewhat taller than the australopithecines, early *Homo* differed most strikingly from them in its skull and dental pattern. Its cranial capacity (650 cubic centimeters, on the average) was about thirty percent larger. (It was still small compared to the 1,350 cubic centimeter average for modern humans.) The face was small in proportion to the cranium. One 2-million-year-old find, designated ER 1470, had a thin face, high skull, and a cranial capacity of 775 cubic centimeters (Figure 42.10). Compared to the later australopithecines, early *Homo* had a less specialized dental pattern (Figure 42.5).

Tangible evidence of cultural adaptations, including stone tools and assembled rock foundations for living quarters or hunting camps, provides especially important insights into human evolution. No tools have been indisputably associated with australopithecine remains, although it is possible that they used wood implements and the like, which would not have been preserved over time. However, simple stone tools such as pebble choppers and flakes are clearly associated with early *Homo*. There is also some evidence that these early humans were simple hunters and gatherers. If that is true, then humans have been hunter-gatherers for ninety-nine percent of their cultural history.

Thus the emergence of *Homo* apparently was based on the continuation and embellishment of patterns of hominid *plasticity*, instead of on specializations that would restrict the genus to a particular habitat and lifestyle. It is also a reason why the genus continued to survive, adapt, and spread into a wide range of habitats.

Homo Erectus

A much better known and larger brained human species, *Homo erectus*, appears in the fossil record for the period from about 1.5 million to 300,000 years ago. The Pleistocene (1.8 million to 10,000 years ago), known for its great climatic fluxes embodied in the Ice Ages, was now under way. The extreme changes in climate affected the very shape of the land. Sea levels rose and fell with the comings and goings of the glacial sheets, exposing and then submerging the land bridges and coastal areas. Large sections of land were buried in northern regions

Figure 42.10 Preliminary reconstruction of the skull of one of the early humans, designated ER 1470 and dated at about 2 million years old.

when the glaciers moved south, where they both destroyed and created geographical features as they moved slowly across the land. At the same time, groups of *H. erectus* moved out of Africa and spread into Southeast Asia, China, and Europe. Some sites contain a wealth of cultural materials, including the remains of more advanced forms of stone technology and evidence of the controlled use of fire. There is even the suggestion of ritualized cannibalism.

Although *H. erectus* populations showed regional variation in morphology, the general form is quite distinctive. Compared with *H. sapiens*, the skull was heavy and primitive, with a diagnostic "pentagonal" shape when viewed from behind. The cranial capacity ranged from 775 to 1,225 cubic centimeters, with the average being 1,000 cubic centimeters. Although not yet as large as the brain of contemporary humans, it was significantly larger than the brain of earlier forms. The spectacular expansion of *H. erectus* into diverse environments must have been due in large part to the increased brain size and concurrent evolution of cultural adaptations.

Homo Sapiens

Somewhere between 300,000 and 100,000 years ago, a new species, *Homo sapiens*, evolved out of some population or populations of *H. erectus*, for those earlier forms disappeared from the earth. Their disappearance was not

sudden. The fossil record indicates there was a gradual transition in form, often with a variety of types existing in the world. Transitional forms, with brains larger than *H. erectus* but not yet fully modern, have been found at a number of Old World sites in association with fairly sophisticated tool assemblages.

In general, *H. sapiens* differed from *H. erectus* in a number of ways. The skull was rounder and more vaulted, and the face more delicate in structure. The teeth and jaws were relatively small in size, and a chin (absent in *H. erectus*) was often present. *H. sapiens* was also more slender. Eventually the cranial capacity averaged 1,350 cubic centimeters—the same as modern humans. From that point on, cultural evolution far outstripped biological changes.

It was not until about 40,000 years ago that anatomically modern humans, *H. sapiens sapiens*, emerged. Before then, a number of different archaic groups of *H. sapiens* were found across the Old World, some of which closely resembled modern populations. Another distinctive group has been called the Neandertals.

Neandertal fossils dating from about 125,000 to 35,000 years ago come from Europe and the Near East. The group was morphologically varied, with the "classic" (most extreme) form from European sites. Contrary to common belief, Neandertals really didn't look very different from modern humans. However, when their overall form is taken into account, they were probably unique enough to warrant classification as a subspecies, *H. sapiens neanderthalensis*. The face was heavy, it often included browridges, and there was no chin. The skull shape was long and low rather than round and highly vaulted. The "contents" were similar, however. If anything, Neandertals had a slightly larger cranial capacity than modern humans, ranging from 1,300 to 1,750 cubic centimeters. This may be the result of neural controls for their very strong musculature, rather than reflecting a difference in intellectual capacity.

The culture of the Neandertals was quite rich. They apparently developed diverse tools, and they lived under a variety of conditions, ranging from the arctic conditions of Ice-Age Europe to the temperate climate of the Near East. They lived in caves, rock shelters, and open air camps, and were proficient hunters and gatherers. The remains of the mysterious "cave bear" rituals and ceremonial burial of the dead suggest that the Neandertals had developed complex ideological systems.

About 35,000 to 40,000 years ago, the Neandertals suddenly disappeared from the fossil record. Speculations about why this happened range from extinction due to a massive plague, slaughter by invaders, the evolution of some groups into fully modern humans, to their assimilation into a population of *H. sapiens sapiens* moving into Europe from elsewhere in the Old World. However, no one really knows what transpired.

Human forms were clearly modern by 40,000 years ago. From that point on, human evolution has been primarily cultural, with groups spreading out across the world, rapidly devising cultural means of dealing with a very broad range of environmental conditions. Humans have developed phenomenally rich and varied cultures, and have gone from "stone age" technology to the age of "high tech." Yet, hunters and gatherers still persist in parts of the world, attesting to the great plasticity and breadth of human adaptations.

Our ability to evolve culturally does not mean that we have now stopped evolving physically. To be sure, thanks to dentists, people with poor teeth suffer little reproductive deficit; and natural selection for prowess in hunting and gathering has been largely eliminated by farmers and grocery stores. Yet individuals still die from disease, and modern society imposes new forces of mortality via increased stress and chemical pollution. As long as individuals differ genetically in their ability to withstand such forces, we will continue to evolve.

SUMMARY

1. The primates are a large and diverse group that includes prosimians, monkeys, apes, and humans. They are descended from ancestral forms that evolved during the great mammalian adaptive radiation of the Cenozoic, about 65 million years ago.

2. The following evolutionary trends occurred among the primates:
   a. A capacity to move the head and limbs freely in different planes.
   b. Mobile digits, capable of grasping.
   c. Upright vertebral column and body posture.
   d. Enhanced vision and decreased emphasis on the sense of smell (a sensory adaptation that is pronounced in vertebrates in general).
   e. Strong social bonding, with an emphasis on intense parental care and learned behavior.

3. The oldest known primates, which resemble modern-day tree shrews, date from the Paleocene (65 to 54 million years ago). The first true prosimians appeared during the Eocene (54 to 38 million years ago). Anthropoids were present by the Oligocene (38 to 25 million years ago). Anthropoids now include monkeys, apes, and humans.

4. An adaptive radiation during the Miocene (25 to 7 million years ago) produced a range of ape forms, including the dryopithecines and the ramamorphs. Probably the ancestors of the modern great apes and humans were among those ancient hominoids. Hominoids now include the Old World monkeys, apes, and humans.

5. The first of the hominids (all species on the human evolutionary branch) probably appeared between 8 to 5 million years ago. Their ancestors apparently evolved in response to a major ecological shift from life in the forests to life in the open savanna. Some hominids were evolutionary side branches; others apparently were ancestral to modern-day *Homo sapiens sapiens* (the only existing hominid species).

6. The hominid adaptation, expressed to varying degrees among the different forms, has these components:
   a. Bipedalism (two-legged walking).
   b. A distinctive, generalized array of teeth associated with an omnivorous feeding behavior.
   c. The expansion and reorganization of the brain, with an associated increase in behavioral complexity.

7. The hominid adaptation is unique in its "plasticity"— the ability to respond to a wide range of demands, especially during the growth and development of the individual. This plasticity was a response to the unpredictable and complex environments of the first hominids.

8. *Australopithecus* is generally recognized as the first undisputed hominid. *A. africanus* was a slender South African form; *A. robustus* was the larger of the two species in that region. *A. boisei* (the "super-robust" form) and the ancient *A. afarensis* are two forms that lived in East Africa; both were smaller brained than *Homo*.

9. Fossils of *A. africanus*, *A. robustus*, and *A. boisei* reflect

dietary specializations. *A. afarensis* had a more generalized array of teeth than any of them; it may have been the last common ancestor of both the other australopithecines and the early forms of *Homo*.

10. Early *Homo* may have appeared as long ago as 2 million years. Fossils indicate these hominids were larger brained than the australopithecines, they had a generalized dental pattern (typical of omnivores), and they clearly used simple tools (such as pebble choppers and flakes).

11. *Homo erectus* fossils have been recovered throughout the Old World; they date from about 1.5 million to 300,000 years ago. Diverse and abundant cultural artifacts have been recovered along with the fossils. *H. erectus* was larger brained than early *Homo*, its cultural evolution was more pronounced, and it lived in a wide range of habitats.

12. The larger brained and more culturally complex *H. sapiens* appeared somewhere between 300,000 and 100,000 years ago. The Neandertals, an early form of *H. sapiens*, mysteriously disappeared about 40,000 to 35,000 years ago.

13. By 40,000 years ago, *H. sapiens sapiens* had evolved. From that point on, cultural evolution has greatly outstripped biological evolution of the human form.

Review Questions

1. What are the general primate characteristics and trends? What adaptation was apparently responsible for the establishment of this pattern of "primateness"?

2. What sort of conditions were apparently responsible for the great adaptive radiation of apelike forms during the Miocene?

3. What is the difference between "hominoid" and "hominid"? Are we hominoids, hominids, or both?

4. Describe the three major structural and functional components of the hominid adaptation. How do they relate to the concept of human plasticity?

5. What are the major differences between the australopithecines and the genus *Homo*?

Readings

Cronin, J., N. Boas, C. Stringer, and Y. Rak. 1981. "Tempo and Mode in Hominid Evolution." *Nature* 292:113–122.

Gowlett, J. 1984. *Ascent to Civilization.* New York: Knopf. First-rate introduction to our evolutionary heritage.

Johanson, D., and M. Edey. 1981. *Lucy: The Beginnings of Humankind.* New York: Simon and Schuster. Fascinating account of the search for hominid remains.

Jolly, A. 1985. *The Evolution of Primate Behavior.* Second edition. New York: Macmillan. A thorough, expansive summary of primate adaptations.

Miller, D. 1977. "Evolution of Primate Chromosomes." *Science* 198:1116–1124. Nice description of chromosome banding studies designed to shed light on phylogenetic relationships among the primates.

Nelson, H., and R. Jurmain. 1985. *Introduction to Physical Anthropology.* Third edition. St. Paul: West. A well-executed general text on a wide range of topics, including primate and human evolution.

Rensberger, R. 1981. "Facing the Past." *Science 81* 2(8):41–50. Intriguing look at how artist reconstructions of early hominids can bias our perceptions of how "human" those forms really were. Shows how one of the leading illustrators does his reconstructions, starting from fossil bones and adding skin, and other details.

Weaver, K. 1985. "The Search for Our Early Ancestory." *National Geographic* 168 (5):560–623. Good summary of our current understanding of human origins. Stunning photographs and reconstructions by the artist Jay Matternes, who is in a class by himself.

UNIT EIGHT

ECOLOGY AND BEHAVIOR

43

POPULATION ECOLOGY

The optimist proclaims that we live in the best of all possible worlds; and the pessimist fears this is true—James Branch Cabell

Picture a crowded outdoor marketplace in India, with colorful heaps of fruits and vegetables, a dozen kinds of dried beans and lentils in great baskets, and an endless variety of fragrant herbs and spices. The marketplace links the teeming city with the farms and orchards of the countryside, where nearly every usable plot of land is already in use for producing food. Yet the population of India and many other developing countries throughout the world continues to increase at an alarming rate.

Meanwhile, populations in the developed countries are beginning to stabilize, yet they use resources from around the world at a far greater rate than do the people of less developed nations. Some of these resources, like petroleum, cannot be replaced. Others, like tropical forests, which are often cleared to raise beef and other products for export to the developed nations, will take centuries to replace, if they can be replaced at all. On a global scale, the demand for food and other resources by the increasing human population appears to be on a collision course with a supply of resources that, at best, is scarcely holding its own.

In the natural world, the effects of resource scarcity are often felt more quickly. Shift the scene from the Indian marketplace to a quiet hardwood forest in summer. In an oak tree overhead, a warbler forages for insects among the leaves. Each year, the number of baby warblers hatched depends in part on the supply of insects available. Nearby, a squirrel scuffles noisily in the decaying leaves on the forest floor, adding an acorn to its cache of winter rations. In years of poor acorn crops, many squirrels may not survive the winter.

Nearby, a butterfly alights on a wildflower, briefly takes in nectar, and flies off. When butterflies are scarce, the wildflowers they pollinate will produce few seeds.

Whether we are talking about humans or any other kind of organism, certain principles govern the growth and stability of their populations over time. Over the long term, we must reckon with these principles, which influence the patterns of relationships of organisms with one another and their environment. These patterns, in *all* their varied forms, are the focus of ecology. *As a science, ecology seeks to treat the world of nature—including its human component—with a single set of concepts and principles.*

ECOLOGY DEFINED

The word **ecology** was coined in the last century from the Greek *oikos* (meaning "house") to designate the study of organisms in their natural homes. Specifically, it means the study of the interactions of organisms with one another and with the physical and chemical environment. Although it includes the study of environmental problems such as pollution, the science of ecology also encompasses research on the natural world from many viewpoints, using many techniques. Modern ecology relies heavily on experiments, both in the laboratory and in field settings, and on mathematical models. These techniques have proven helpful in testing ecological theories and in arriving at practical decisions in the management of natural resources.

In this unit of the book, we will be considering ecological interactions at several levels of biological organization, which may be defined in the following way:

---

1. The **population**: a group of individuals of the same species occupying a given area. The place where a population (or an individual) lives is called its **habitat**: tree squirrels live in a "forest habitat," muskrats in a "stream-bank habitat," dandelions in "disturbed habitats."

2. The **community**: the populations of *all* species that occupy a habitat. Ecologists also use the term to refer to certain groups of organisms in a habitat—the "bird community" or the "plant community," for example. Species in a community play different roles:

a. *Producers*: The "self-feeding" (autotrophic) organisms, which include most plants and some microorganisms. The producers synthesize their own organic compounds from simple inorganic substances with the aid of energy from the sun (photosynthetic autotrophs) or from the inorganic substances themselves (chemosynthetic autotrophs).

b. *Consumers*: All organisms that are "not self-feeding" (heterotrophic) and that ingest other (usually) living organisms in whole or in part to obtain organic nutrients. Thus the consumers called *herbivores* eat plants, *carnivores* eat animals, and *parasites* take in blood, sap, and other tissues from living hosts.

c. *Decomposers*: Mostly heterotrophic bacteria and fungi that obtain organic nutrients by breaking down the remains or products of organisms. The activity of decomposers allows simple compounds to be recycled back to the autotrophs.

d. *Detritivores*: Earthworms, nematodes, crabs, and other heterotrophs that feed on particles of organic matter, such as would be produced by the partial decomposition of plant and animal tissues.

3. The **ecosystem**: a community and its physical and chemical environment. An ecosystem has living (*biotic*) and nonliving (*abiotic*) components. Soils, temperatures, rainfall, even organic matter are examples of the abiotic component. The classes of organisms detailed above form the biotic component.

4. The **biosphere**: all living organisms on earth which, together with their interactions with the global physical and chemical environment, maintain a system of energy use and materials cycling. This system runs on energy flowing into it (from the sun) and it gives up energy (primarily as low-grade heat) to space.

Let's focus first on the ecological relationships that influence the size, structure, and distribution of populations. In subsequent chapters we will consider the relationships at the levels of communities, ecosystems and, finally, the biosphere.

POPULATION DENSITY AND DISTRIBUTION

If you have ever collected fossils, or butterflies, or kept a list of backyard birds over a period of time, you will know that for every common species you find, there are several rare ones. In other words, **population density** (the number of individuals per unit area) generally varies widely among species living in the same locality.

Ecological relationships influence the density of a population. To study these influences, we must determine what the density is to begin with so that we can chart any changes in space or over time. For sparse populations, a simple "head count" (or tree count, and so on) can be made in a defined area; for dense populations, counts can be made at random in small sampling areas to arrive at an estimate of overall density.

Distribution in Space

Even when the density of a species in a particular habitat is known, we still don't know how the individuals are distributed (arranged) in space. They might be clumped together in certain parts of the habitat, distributed randomly throughout, or spread out rather uniformly (Figure 43.2).

Clumped Distribution. Most commonly the members of a population are distributed in clumps through their habitat for three reasons. First, physical and chemical conditions suitable for growth typically are "patchy" rather than uniform. In a pasture, for example, certain kinds of plants grow in soil rich in nitrates where individual cowpats fell weeks or months before. Second, some parts of the habitat offer more protection to prey organisms (or, conversely, better hunting to predators). Third, dispersal of seeds, larvae, and other representative forms of each new generation is often limited. For example, the offspring of some intertidal organisms settle near their parents and one another; gypsy moths and other insect pests often spread slowly from the original point of infestation.

Random Distribution. Assuming that environmental conditions are the same throughout a habitat and that members of the population are neither attracting nor repelling each other during a given time period, then distribution may be random. Spiders on a forest floor may be an example of random spacing among individuals. In population studies, a random distribution is the theoretical standard from which clumped or uniform distributions depart in opposite directions (Figure 43.2).

Uniform Distribution. Nearly uniform distribution (even spacing), as typified by orchards, is exceedingly rare in nature. For the few cases in which uniformity is approached, competition among individuals for a limited resource seems to be the cause. An example is the nearly uniform spacing of creosote bushes (*Larrea*) in dry scrub deserts of the sort occurring in the American Southwest. Large, mature plants deplete the soil around them of available water, and seed-eating ants and rodents concentrate their activities near the plants. As a result, seeds and seedlings are unable to survive near mature plants, so clumps do not form (Figure 43.3).

Distribution Over Time

Population distribution also varies with time, often in response to environmental rhythms. Few environments are so productive that they yield abundant resources all year long. For example, temperatures drop and food supplies dwindle during the winter in north temperate regions (such as parts of Canada and Europe), where many migratory birds reproduce. Whole populations move southward to winter feeding grounds in (for example) South America and Africa. With the return of spring, when their breeding grounds are warmer and have

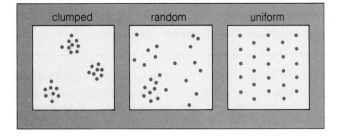

Figure 43.2 Generalized patterns of spatial distribution of individuals in a population.

Figure 43.3 Nearly uniform spacing among creosote bushes (*Larrea*) near Death Valley, California.

abundant new growth of plants and insect populations, they make the return trip north.

Animals may also move locally between habitats with the seasons. In deciduous tropical forests, trees lose their leaves during the dry season—except for broadleaf evergreen species growing as narrow "gallery forests" along watercourses. Many birds and mammals crowd into these narrow habitats during the dry season, greatly increasing the animal density compared to the wet season.

POPULATION DYNAMICS

Variables Affecting Population Size

To understand how populations change in size, we will begin with a simple concept. The size of any population increases or decreases with change in one or more of the following variables:

natality	*births*
mortality	*deaths*
immigration	*individuals from other populations of the same species join the population*
emigration	*individuals leave the population*

Together, these four variables dictate the rate of change in the number of individuals in the population over a given period of time:

$$\begin{pmatrix} \text{population} \\ \text{growth} \\ \text{rate} \end{pmatrix} = \begin{pmatrix} \text{births} \\ + \\ \text{immigration} \end{pmatrix} - \begin{pmatrix} \text{deaths} \\ + \\ \text{emigration} \end{pmatrix}$$

When the birth rate plus immigration is balanced over the long term by the death rate and emigration, population size is stabilized; there is said to be **zero population growth**.

In the next sections, we will assume that there is no immigration or emigration, to simplify things.

Exponential Growth

The variables that affect population size can be measured either for the entire population, as we did in the equation above, or per individual. For example, if 1,000 baby mice per month are born to a population of 2,000 mice living in a cornfield, then the birth rate per individual is 1,000/2,000 = 0.5. Death rates per individual are computed in the same way; if 800 of the original 2,000 mice in the cornfield die during the same month, the death rate per

individual is 800/2,000 = 0.4. Now we can rewrite the relationship for population growth as:

$$\begin{pmatrix} \text{population} \\ \text{growth} \\ \text{rate} \end{pmatrix} = \begin{pmatrix} \text{births} \\ \text{per} \\ \text{individual} \end{pmatrix} - \begin{pmatrix} \text{deaths} \\ \text{per} \\ \text{individual} \end{pmatrix} \times \begin{pmatrix} \text{number} \\ \text{of} \\ \text{individuals} \end{pmatrix}$$

Suppose, for the sake of argument, that the birth and death rates remain constant, regardless of how much or how little the population grows. Then we can lump them together as a single variable, called *r*, the *net reproduction per individual*. (In the mouse example, this would be 0.5 − 0.4 = 0.1, the birth rate minus the death rate.) This simplifies the formula for population growth rate to:

$$\begin{pmatrix} \text{population} \\ \text{growth} \\ \text{rate} \end{pmatrix} = \begin{pmatrix} \text{net} \\ \text{reproduction} \\ \text{per individual} \end{pmatrix} \times \begin{pmatrix} \text{number} \\ \text{of} \\ \text{individuals} \end{pmatrix}$$

In symbols, the same relationship can be represented as:

$$G = rN$$

This equation tells us that, when *r* is held constant, any population will show **exponential growth**. The number of its individuals increases in *doubling* increments—from 2 to 4, then 8, 16, 32, 64 and so on. (This is unlike an arithmetic increase from 1 to 2, 3, and so on.)

We can observe a limited period of exponential growth in the laboratory by putting a single bacterium in a culture flask with a complete supply of nutrients. In thirty minutes the bacterium divides in two; thirty minutes later the two divide into four. If no cells die between divisions, the number will double every thirty minutes. The larger the population base becomes, the more bacteria there are to divide. After only 9-1/2 hours (nineteen doublings), the population size will exceed 500,000; and after 10 hours (twenty doublings), it will soar past 1 million. When size increases are plotted against time, the result is a **J-shaped curve**, which is characteristic of populations undergoing unrestricted, exponential growth (curve *a* in Figure 43.4).

So far, we have assumed none of the bacteria die. To test the effect of mortality on growth rate, let's start over with our bacterium in its nutrient-rich culture flask. This time, assume that twenty-five percent of the population dies between each doubling time. This death rate does slow things down a bit, because now it takes almost two hours instead of thirty minutes to double population size. *But only the time scale changes.* It now takes thirty hours instead of ten to arrive at a million bacteria—but we still have a J-shaped curve (curve *b* in Figure 43.4).

time (hours)	number of individuals for curve a
10	1,048,576
9½	524,288
9	262,144
8½	131,072
8	65,536
7½	32,768
7	16,384
6½	8,912
6	4,096
5½	2,048
5	1,024
4½	512
4	256
3½	128
3	64
2½	32
2	16
1½	8
1	4
½	2
0	1

Figure 43.4 (**a**) Exponential growth for a bacterial population that is dividing by fission every half hour. (**b**) Exponential growth of the population when division occurs every half hour, but when twenty-five percent dies between divisions. Although deaths slow the rate of increase, in themselves they are not enough to stop exponential growth.

As long as the birth rate remains even slightly above the death rate, a population will grow. If the rates remain constant, the population will grow exponentially.

Limits on Population Growth

Clearly there must be factors in nature that limit population growth. (Thus you will never be trampled to death by a billion rabbits when you walk through the woods.) However, those factors are often difficult to determine in nature because of the complex interactions among various populations. For that reason, let's go back to that single bacterium in its culture flask, where we can control the variables.

Assume we enrich the culture medium with glucose and essential elements required for bacterial growth, then allow the bacteria to reproduce for many generations. At first the population grows almost exponentially, then levels off and briefly remains rather stable. Then the population starts to decline rapidly and soon dies out.

What happened? For these bacteria, glucose meant food and energy—but the culture dish held only so much glucose. As the population expanded faster and faster, the glucose was being used up faster and faster. When supplies began to dwindle, so did the basis for growth. *When any essential resource is in short supply, it becomes a limiting factor for population growth.*

Even if we were to supply the bacterial population with all necessary nutrients, it would crash following its initial exponential growth. The increased numbers of bacterial cells would produce increased metabolic wastes which, in high enough concentrations, would drastically alter the environment. Unless we removed the tainted medium every so often from the culture flask and substituted a fresh supply, the bacteria would poison themselves to death.

In sum, the number of limiting factors can be large and their relative effects variable. *For natural populations, the factors that limit population growth can fluctuate considerably over time, with first one and then another setting the upper bound.*

Carrying Capacity and Logistic Growth

As a population increases in size, the same resources must be shared by a greater and greater number of individuals. The decreasing supply of resources may lower the birth rate, increase the death rate, or both—until births and deaths are in balance. At that point, and as long as the resource supply remains constant, the population should stabilize at some equilibrium size, called the **carrying capacity** (K) of the environment. *A sustainable supply of resources (including nutrients, energy, and living space) defines the carrying capacity for a particular population in a particular environment.*

The effect of a limited carrying capacity on population growth is expressed by the **logistic growth equation**:

$$\begin{pmatrix} \text{population} \\ \text{growth} \\ \text{rate} \end{pmatrix} = \begin{pmatrix} \text{maximum net} \\ \text{reproduction} \\ \text{per individual} \end{pmatrix} \times \begin{pmatrix} \text{number} \\ \text{of indi-} \\ \text{viduals} \end{pmatrix} \times \begin{pmatrix} \text{portion of} \\ \text{unexploited} \\ \text{resources} \end{pmatrix}$$

In symbols, this equation becomes:

$$G = r_{max} N\left(\frac{K - N}{K}\right)$$

When a low-density population grows logistically, the growth pattern is slow at first but then steadily accelerates. However, its growth slows more and more after it has passed the halfway mark of the carrying capacity; then it levels off once the capacity is reached. A plot of this growth pattern gives us a sigmoid, or **S-shaped curve**, as shown in Figure 43.5. (Keep in mind that the slope of such curves reflects the rate of population *change*, not population *size*.) Moreover, if a population exceeds the carrying capacity, it declines until it reaches its equilibrium size.

When a low-density population shows a logistic growth pattern, its rate of increase is slow at first, then accelerates and slows again, then levels off as the carrying capacity of the environment is reached.

The logistic growth model is only a baseline for studying population growth. Carrying capacities fluctuate with the seasons and from one year to the next. Moreover, the birth and death rates are not necessarily clearcut (even starving animals can continue to consume resources and bear young). For example, in 1910, four male and twenty-two female reindeer were introduced on one of the Pribilof Islands of Alaska. Within thirty years, the population increased to 2,000—greatly "overshooting" the carrying capacity. The vegetation on which the reindeer grazed almost disappeared, and in 1950 the herd size plummeted to eight members (Figure

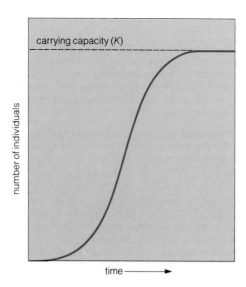

Figure 43.5 S-shaped curve characteristic of logistic growth. Following a rapid growth phase, growth slows and the curve flattens out as the carrying capacity of the environment is approached.

Figure 43.6 Rise and fall of a reindeer herd introduced on one of the Pribilof Islands, Alaska. Rapid population growth led to an overshooting of the carrying capacity of the environment. Growth stopped abruptly, and the population size crashed to eight reindeer—eighteen fewer than were present in the starting population.

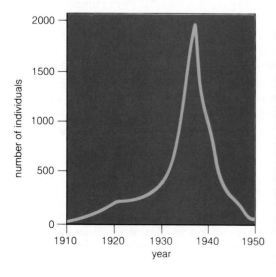

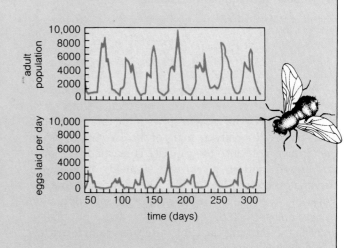

Figure 43.7 Effect of resource availability, time lags, and competition on population density for sheep blowflies (*Lucilia cuprina*). In one laboratory experiment, A. Nicholson fed limited amounts of beef liver to fly larvae. He also fed unlimited amounts of sugar and water to adults that were part of the same experimental group.

When the adult population density was high, so many eggs were laid that the resulting larvae devoured all the food before completing their development, and they all died. Through natural mortality, the size of the existing adult population dwindled and fewer eggs were laid. Fewer eggs meant fewer larvae—and less competition among them. Now some larvae were able to mature into adults. For a while, population density declined because of the time lag between the survival of larvae and the development of egg-laying adults. The delay permitted an increasing number of larvae to survive—and the adult population soared again.

The outcome of this process was a drastic overshoot-undershoot oscillation in population size over time.

43.6). A similar pattern of "overshoot and crashes" is shown in Figure 43.7 for a laboratory culture of flies. These cycles were also caused by time lags in responses to density.

LIFE HISTORY PATTERNS

So far, we have treated populations as if each were made up of identical members. Yet phenotypes obviously differ; and all but the simplest kinds of organisms go through many developmental stages, with each new change in morphology and behavior having its own perils and rewards. In short, *the patterns of birth, death, migrations, and reproduction vary through the life span characteristic of the species.* Let's take a look at a few examples of these **age-specific patterns** before considering some of the environmental variables that might have helped shape them.

Life Tables

Although each species has a characteristic life span, few individuals reach the maximum. The probability of death usually changes with age. Moreover, the reproductive rate and the probability of migration to another population of the same species vary with age in different ways from one species to the next. The study of such age-specific patterns is the subject of *demography*. Although originally developed by the life insurance and health insurance companies, demographic methods are now widely applied by ecologists to populations of plants and animals.

For example, **life tables** are created to summarize the age-specific patterns of birth and death for a particular population in a particular environment. Typically ecologists follow the fate of a group of newborn individuals until the last one dies. (Such a group is called a *cohort*.) Besides recording the age at death of each individual, they keep track of how many offspring (if any) each surviving individual produces during each age interval of its life. The death rate and birth rate for each age interval are then easily computed, producing birth and death "schedules" for the cohort.

The death schedule is usually transformed to its more cheerful reflection, called **survivorship**. A survivorship schedule is produced by writing down, for each age *x*, the number of individuals of the original cohort that *survive* to age *x*.

Table 43.1 presents a life table for a cohort of 996 individuals of an annual plant (*Phlox drummondii*) in Texas. The curve for the death rate, the corresponding survivorship curve, and the "birth" (seed production) pattern are shown in Figure 43.8.

Patterns of Survivorship and Reproduction

Life history patterns are almost bewildering in their diversity, but we can get an idea of the *range* of possibilities by looking at a few extremes.

Table 43.1 Life Table for a Cohort of Annual Plants (*Phlox drummondii*)

Age Interval (days)	Survivorship (number surviving at start of interval)	Number Dying During Interval	Death Rate per Individual During Interval	"Birth" Rate (number of seeds produced per individual) During Interval
0–63	996	328	0.329	0
63–124	668	373	0.558	0
124–184	295	105	0.356	0
184–215	190	14	0.074	0
215–264	176	4	0.023	0
264–278	172	5	0.029	0
278–292	167	8	0.048	0
292–306	159	5	0.031	0.33
306–320	154	7	0.045	3.13
320–334	147	42	0.286	5.42
334–348	105	83	0.790	9.26
348–362	22	22	1.000	4.31
362–	0	0	0	0
		996		

Data from W. J. Leverich and D. A. Levin, *American Naturalist* 1979, 113:881–903.

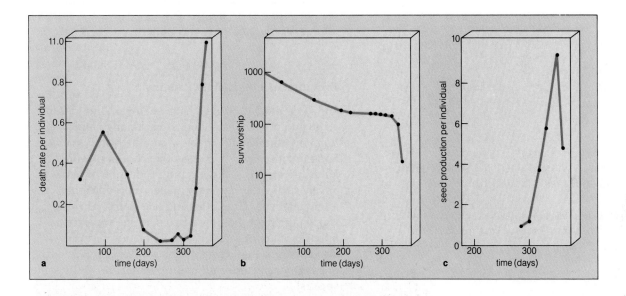

Figure 43.8 Death rate (**a**), corresponding survivorship curve (**b**), and the "birth," or seed production pattern (**c**) for a cohort of 996 individual phlox plants (*Phlox drummondii*), a representative of which is shown in (**d**).

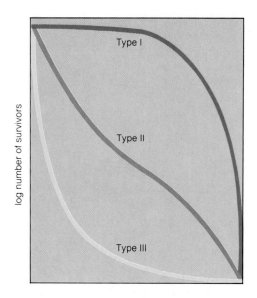

Figure 43.9 (**a**) Three generalized types of survivorship curves. For Type I populations, there is high survivorship until some age, then high mortality. Type II populations show a fairly constant death rate at all ages. For Type III populations, there is low survivorship early in life.

Survivorship Curves. Three "types" of survivorship curves are often noted (types I, II, and III):

I   *survivorship high throughout most of life; most deaths in a short time span after reproduction*

II   *death rate fairly constant at all ages*

III   *high death rate early in life, decreasing with age*

Figure 43.9 provides a generalized picture of these survivorship curves. (Notice that the curve for phlox in Figure 43.8b fits none of them exactly.)

A Type I curve is more or less characteristic of humans in communities with good health care services. Historically, and where health care is poor today, infant mortality puts a sharp drop at the beginning of the curve, followed by a plateau—rather like the phlox curve in Figure 43.8b. In general, large mammals that show extended parental care of offspring approach Type I curves. Survivorship curves approaching a Type II pattern have been reported for some songbirds, lizards, and small mammals, and probably they apply to destruction of seeds prior to germination for many plants. Survivorship patterns approaching the Type III curve are common among marine invertebrates, most insects, and many fishes, amphibians, and reptiles.

Timing of Reproduction. Organisms also vary widely in terms of the age at which they reproduce and in how often they do. Some reproduce once late in life, then die. Mayflies (which spend a single day as adults), annual plants such as the phlox, and certain long-lived plants (such as many bamboos) belong in this category. In contrast, most vertebrates, some insects, and most perennial plants (such as apple trees and roses) reproduce repeatedly throughout maturity.

Number of Offspring. Organisms characterized by a Type I survivorship curve tend to produce only a few offspring, and the offspring are large in relation to the size of the adult. Female elephants, for example, produce only four or five large calves in a lifetime, devoting several years of parental care to each. In contrast, species with a Type III survivorship curve produce vast numbers of offspring that are usually small in relation to the adult size. (Or, to put it more accurately, any species with a high juvenile mortality that did *not* produce large numbers of young is no longer with us.) The small seeds of annual weeds exemplify this pattern.

Evolution of Life History Patterns

For some time, ecologists have been working to understand and predict the relationships among natural selection, environmental variables, and life history patterns. Early on, it seemed that selection processes favored two kinds of patterns—either rapid production of many relatively small offspring early in life, or production of only a few relatively large offspring late in life. However, it is now apparent that these two patterns are extremes at opposite ends of a range of possible life histories. It is also apparent that *both* patterns as well as intermediate ones can characterize different populations of the same species!

Recently, David Reznick and John Endler conducted some elegant studies in Trinidad and, later, in the laboratory to identify which environmental variables were influencing the life history patterns of guppies. These small, live-bearing fish are often purchased for home aquariums (Figure 43.10). Male guppies are smaller than the females, which they attract with brightly colored patterns on the body and with complex courtship displays. Males stop growing once they are sexually mature, but the drab-colored females continue to grow larger as they reproduce.

In the mountains of the island of Trinidad, guppies living in different streams—and even in different parts of the same streams—are subject to different dangers. In some of the streams, a small killifish preys heavily on immature guppies but cannot handle the (larger) adults. In other streams, pike-cichlid and other dangerous pred-

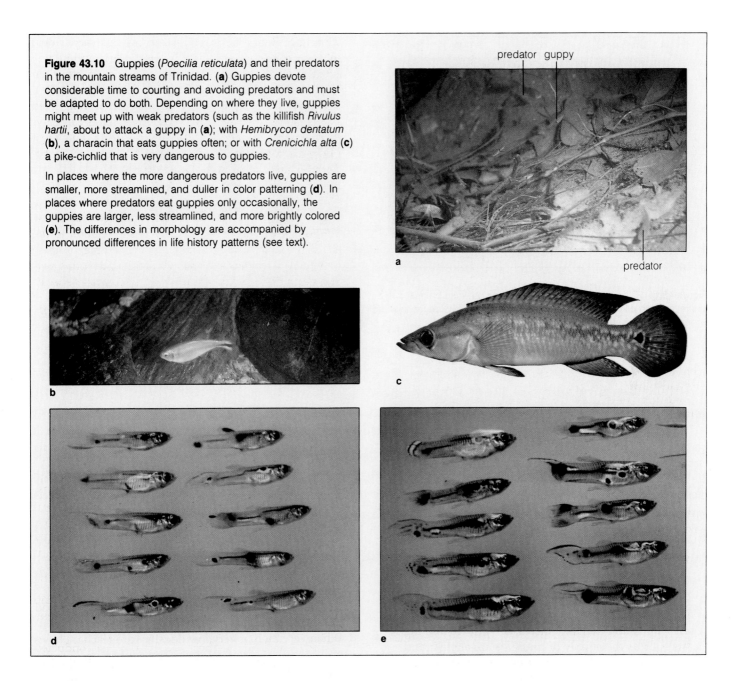

Figure 43.10 Guppies (*Poecilia reticulata*) and their predators in the mountain streams of Trinidad. (**a**) Guppies devote considerable time to courting and avoiding predators and must be adapted to do both. Depending on where they live, guppies might meet up with weak predators (such as the killifish *Rivulus hartii*, about to attack a guppy in (**a**); with *Hemibrycon dentatum* (**b**), a characin that eats guppies often; or with *Crenicichla alta* (**c**) a pike-cichlid that is very dangerous to guppies.

In places where the more dangerous predators live, guppies are smaller, more streamlined, and duller in color patterning (**d**). In places where predators eat guppies only occasionally, the guppies are larger, less streamlined, and more brightly colored (**e**). The differences in morphology are accompanied by pronounced differences in life history patterns (see text).

ators prefer mature (bigger) guppies and tend not to waste time hunting the small ones.

As might be predicted from our understanding of selection processes, the individuals of guppy populations confronted with pike-cichlids and comparable predators (which favor large-bodied dinners) mature sooner, are smaller at maturity, and reproduce at a younger age. Moreover, they produce far more offspring and do so more often than their counterparts in killifish streams (Figure 43.11).

As a check on the possibility that the differences were due to some other, unknown differences between the streams, Reznick and Endler raised guppies from each kind of stream in the laboratory for two generations under identical conditions (with no predation). They discovered that the life-history differences were maintained, indicating a genetic (hence a heritable) basis for the differences.

As a check on the role of predators in the evolution of size differences seen in the field, guppies were grown for many generations in the laboratory—some alone, some with killifish, and some with pike-cichlids. As predicted, the guppy lineage subjected to predation over time by killifish became larger at maturity, whereas the lineage living with pike-cichlids showed a trend toward earlier maturity.

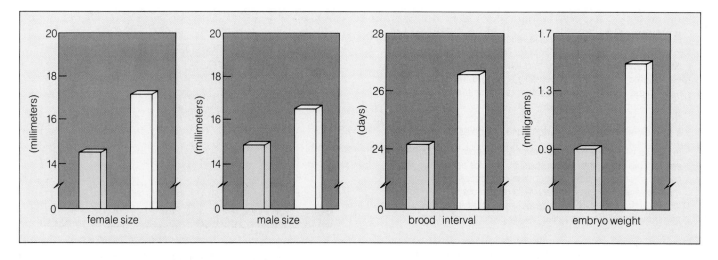

Figure 43.11 Differences in body size, interval between broods, and embryo weights for guppies living in streams with pike-cichlids (lavender) or killifish (yellow) on the island of Trinidad. Pike-cichlids prey on larger guppies, whereas killifish prey on smaller ones, selecting for the differences shown here.

LIMITS ON POPULATION GROWTH

So far in this chapter, we have considered some of the variables shaping the size, structure, distribution, and life histories of populations. Along the way, we have mentioned in passing a few of the factors that act like brakes on population growth (nutrient availability for our flask-bound bacterium, for example). Let's now briefly review the main categories in which these growth-limiting factors are grouped.

Density-Dependent Factors

When population density increases, growth-limiting factors such as the supply of nutrients are altered, and they collectively lead to a reduced birth rate and an increased rate of death, emigration, or both. Such density-dependent factors have a self-adjusting effect on population growth. Once the density decreases, the pressures ease and the net reproductive rate may increase once more.

Density-Independent Factors

Other checks on population growth act more or less independently of population density. Consider that severe changes in temperature or rainfall, wave action in the intertidal zone, and catastrophic events such as hurricanes and volcanic eruptions tend to increase the death rate per individual, *but they do so more or less to the same degree regardless of population density*. A hard freeze in the late spring is just as likely to kill an individual annual plant seedling regardless of whether it is in the middle of a clump of plants or growing alone nearby.

Density-dependent and density-independent factors interact in such complex ways, however, that the effects of one often amplify the effects of the other. For instance, whether rabbits tolerate a sudden freeze depends in part on whether they have enough food and burrows. Food and burrow availability are density-dependent factors, yet here they interact with a change in temperature that would likely kill some rabbits at any density.

For many populations, there seems to be a broad range over which density-dependence is weak or absent. When very high or very low densities are reached, however, density-dependent forces come into play, decreasing or increasing the rate of population growth, respectively. The increase in the net reproductive rate at lower densities is especially important, for if a population were limited only by density-independent factors, there would be nothing to prevent its extinction over the long term.

Competition for Resources

The amount of water, nutrients, and other resources available for each individual often declines as a population increases in density. The decline may temporarily hamper reproductive processes or reproductive behavior without affecting the individual's health over the long term. (For example, in some kangaroos, embryonic development is arrested until there is enough plant food to sustain the mother *and* assure development of the embryo.) However, a decline in resources also can impair health to the extent that the reproductive rate is lowered; this happens to annual plants (such as corn) confronted with drought years.

When individuals of the same species compete for resources, the interaction is called **intraspecific competition**. In general, these interactions take two forms:

exploitation competition	*all individuals have equal access to the resource; only the rate or the efficiency of exploiting it differs*
interference competition	*certain individuals control access to the resource, limiting or preventing use of the resource by others*

The difference between exploitation competition and interference competition becomes clear with a simple analogy. Suppose you and a greedy friend are sharing a milkshake, each with your own straw. If your friend sucks harder than you do, he or she is the superior exploitation competitor. However, if you reach over and pinch your friend's straw, you are the superior interference competitor.

In nature, the most common forms of interference competition among animals are territoriality, social dominance, or simple defense of food sources (Chapter Forty-Nine). Among land plants, larger and taller individuals may shade smaller ones and suppress their growth.

Predation, Parasitism, and Disease

When prey or host populations become increasingly dense, their members face a proportionally greater risk of being dispatched by predators, colonized by parasites, or infected by contagious disease. As the predator, parasite, or pathogen population increases, the abundance of prey or host decreases, and vice versa.

Herbivores—"plant predators"—play an important role in limiting plant populations. For example, foraging rabbits often keep the density of plant populations at depressed levels. Their impact was dramatically illustrated in the English countryside in the 1950s, when the rabbit population was decimated by the disease myxomatosis (itself a good example of population regulation). Shortly afterward, the forage plants grew spectacularly, and formerly rare species became quite common.

Predation, parasitism, and disease are among the most powerful factors shaping the growth of populations. We will look closely at these factors in chapters to follow.

HUMAN POPULATION GROWTH

In the 1986 the human population reached 5 billion, having passed 4 billion only eleven years earlier. Even if that number represented a stabilized population level, we would have to contend with monumental problems. In a given year, between 5 to 20 million people now die of starvation and malnutrition-related diseases. What makes future prospects especially troubling is that our population size is by no means static.

In 1986, the annual population growth rate was higher than 4 percent in some parts of the world—Kenya is an example. Continuing at this rate, Kenya's population would double in seventeen years. In contrast, the population growth rate in some highly developed countries has slowed to nearly zero; the doubling rate for the United Kingdom would now be around 460 years.

Today, for every human death there are about 2.5 births, giving a net increase of 139 people per minute, 212,000 per day, and 77 million per year—that is an annual growth rate of 1.7 percent, down from the historic peak of about 2 percent in 1965. It is estimated that we will reach 6.1 billion by the year 2000, then 7.8 billion by 2020, leveling off (if all goes well) in about the year 2100 at about 10.5 billion—*over twice the present human population*.

With the most intensive effort, we might be able to double food production to keep pace with growth. However, we would succeed in doing little more than maintaining marginal living conditions for most of the world. Under such conditions, deaths from starvation could be 10 to 40 million a year. For a while, it would be like the Red Queen's garden in Lewis Carroll's *Through the Looking Glass*, where one is forced to run as fast as one can to remain in the same place. Can you brush this picture aside as being too far in the future to warrant your concern? It is no farther removed from you than your own sons and daughters.

Where We Began Sidestepping Controls

How did we get into this predicament? For most of our existence as a species, human population growth has been slow. In the past two centuries, there has been an astounding increase in the rate of population growth (Figure 43.12). Why has our growth rate increased so dramatically? There are three possible reasons:

1. We steadily developed the capacity to expand into new habitats and new climatic zones.

2. The carrying capacities of the environments we already occupied were increased.

3. A series of limiting factors was removed.

Let's consider the first possibility. By 50,000 years ago, the human species had radiated through much of the world. For most animal species, such extensive radiation would not have occurred as rapidly. Humans were able to do so with the application of learning and memory—how to build fires, assemble shelters, create clothing and tools, plan a community hunt. Learned experiences were not confined to individuals but spread quickly from one human group to another because of language—our ability for cultural communication. (It took less than seven decades from the time we first ventured into the air until we landed on the moon.) Thus the human population expanded into new environments, and it did so in an extremely short time span compared with the radiations of other organisms.

What about the second possibility? About 10,000 years ago, people began to shift from the hunting and gathering way of life to agriculture—from risky, demanding moves following the game herds to a settled, more dependable basis for existence in more favorable settings. Even in its simplest form, agricultural management of food supplies bypassed the natural carrying capacity of the environment for the old hunter-gatherer way of life. Through the development of irrigation, metallurgy, social stratification (which provided a labor base) and, later, use of fertilizers and pesticides, the limits were expanded and were met again with a resurgence of human population growth. Thus, with the domestication of plants and animals, the carrying capacity has indeed risen abruptly for human populations.

What about the third possibility—the removal of limiting factors? The potential for growth inherent in the development of agriculture began to be realized with the suppression of contagious diseases. Until about 300 years ago, malnutrition, contagious diseases, and poor hygiene kept the death rate relatively high (especially among infants), balancing the birth rate. Contagious diseases are density-dependent factors, and they spread rapidly through crowded settlements and cities. Without proper hygiene and sewage disposal methods, and plagued with such disease carriers as fleas and rats, population size increased only slowly at first. Then plumbing and sewage treatment methods appeared. Bacteria and viruses were recognized as disease agents. Vaccines, antitoxins, and drugs such as antibiotics were developed.

Thus the effect of one after another major limiting factor on human population growth has been reduced. Plague, diphtheria, cholera, measles, malaria—many diseases have been brought under control in developed countries. Smallpox, once a deadly scourge, has apparently been eliminated everywhere. (Some diseases, including malaria, schistosomiasis, and dysentery, are still prevalent in less developed countries and are again on the rise in some areas for lack of adequate funds for treatment and control measures.)

Age Structure and Reproductive Rates

Two important factors influence the degree to which the population growth rate can be slowed. The first is called the **age structure** for a population, this being the relative numbers of its individuals of various ages. Figures 43.13 and 43.14 show examples of age structure diagrams for three human populations growing at different rates. The individuals are assigned to three categories—those before, during, and after reproductive age. (The ages 15 to 44 are used as the average range of childbearing years.)

These diagrams readily show the effects of age structure on population growth. The age structure pyramid for a rapidly growing population has a broad base, as Figure 43.13 suggests. It is filled not only with repro-

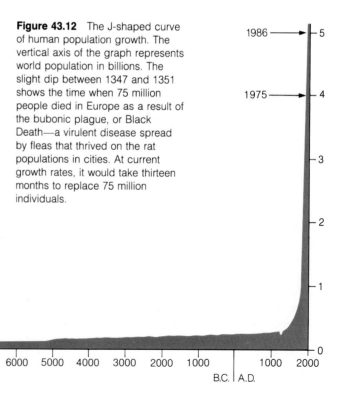

Figure 43.12 The J-shaped curve of human population growth. The vertical axis of the graph represents world population in billions. The slight dip between 1347 and 1351 shows the time when 75 million people died in Europe as a result of the bubonic plague, or Black Death—a virulent disease spread by fleas that thrived on the rat populations in cities. At current growth rates, it would take thirteen months to replace 75 million individuals.

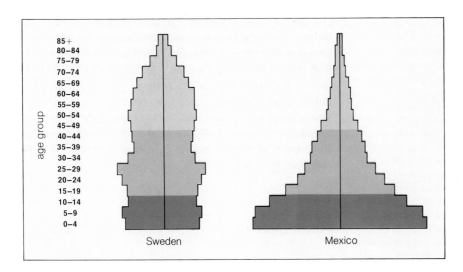

Figure 43.13 Age structure diagrams for two countries in 1977. Dark green indicates the pre-reproductive base. Dark blue indicates reproductive years; light blue, the post-reproductive years. The portion of the population to the left of the vertical axis in each diagram represents males; the portion to the right represents females. Mexico has a very rapid rate of increase. In 1980, Sweden showed zero population growth.

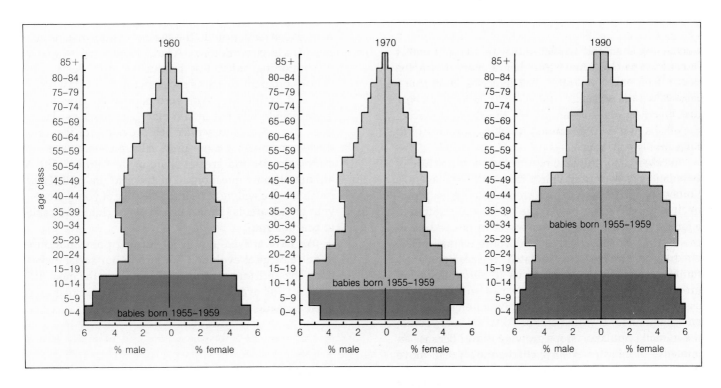

Figure 43.14 Age structure of the U.S. population in 1960, 1970, and 1990 (projected). The population bulge of babies born between 1955 and 1959 will slowly move up.

Table 43.2 Percentage of Population Under Age 15 and Over Age 64 for Ten Geographic Regions in 1986		
Region	Population Under Age 15 (%)	Population Over Age 64 (%)
World	35	6
More developed nations	23	12
All less developed nations	39	4
Less developed nations, excluding China	41	4
Africa	45	3
Asia	37	4
Europe	22	13
Latin America	38	4
North America	22	12
Oceania	29	8

Source: Population Reference Bureau, 1986. *World Population Data Sheet*. Washington, DC: Population Reference Bureau.

ductive-age men and women but with a large number of children who will move into that category during the next fifteen years. As Table 43.2 indicates, *more than a third of the world population now falls in the broad reproductive base.* Such figures give an idea of the challenge facing the efforts to slow population growth by reducing birth rates on a global scale.

Another factor influencing the short-term picture of population growth is the birth rate itself—the average number of infants born to each woman during her reproductive years. Today the average number of children in a family is 2 in the more developed countries and 4.2 in the less developed countries. A world average of 2.5 children per family is the estimated "replacement level" fertility rate that would bring us to zero population growth, given current mortality patterns. But even assuming that a world average of 2.5 children is achieved and maintained, it would still be 70 to 100 years before the human population stops growing. Why? Because an immense number of existing children are yet to move into the reproductive age category.

One way to slow things down is to encourage delayed reproduction—childbearing in the early thirties as opposed to the mid-teens or early twenties. Later childbearing slows population growth by lengthening the generation time and by lowering the average number of children in each family. In Ireland, women customarily marry later in life. In China, the government has raised the age at which marriage is allowed (age twenty for women, twenty-two for men). Moreover, the Chinese government provides financial, educational, housing, and retirement inducements for couples to have only one child, and free birth control measures. It also imposes penalties for having more than two children. The annual birth rate in China dropped from 32 per thousand women in 1970 to only 17 per thousand in 1986; the percentage growth rate is now about 1.0 percent, comparable to that of the United States.

PERSPECTIVE

In this chapter, we began with the premise that all populations have the potential for exponential growth—growth at increasing rates to enormous numbers. For the human population, as for all others, the biological implications of exponential growth are staggering. Yet so are the social implications of achieving and maintaining zero population growth.

For instance, most members of an actively growing population fall in younger age brackets. Under conditions of constant growth, the age distribution means that there is a large work force. A large work force is capable of supporting older, less productive individuals with various programs, such as social security, low-cost housing, and health care. With zero population growth, far more people will fall in the older age brackets. How, then, can goods and services be provided for less productive members if more productive ones are asked to carry a greater and greater share of the burden? These are not abstract questions. Put them to yourself. How much are you willing to bear for the sake of your parents, your grandparents? How much will your children be able to bear for you?

We have arrived at a major turning point, not only in our biological evolution but also in our social evolution. The decisions awaiting us are among the most difficult we will ever have to make, yet it is clear that they must be made, and soon.

SUMMARY

1. Ecology is the study of all organisms, including ourselves, in relation to other species and to the environment. Studies are made of populations of single species, interacting communities of many species, the movement of materials and energy through communities in ecosystems, and global patterns in the biosphere.

2. The number of individuals of a population in a given area determines the population density, and their arrangement in space defines their distribution (which may be clumped, more-or-less random, or fairly uniform).

3. The growth rate of a population (G) depends on the birth rate, the death rate, and the rates of immigration and emigration. If the birth rate per individual exceeds the death rate per individual by a constant amount, the population will grow exponentially (assuming immigration and emigration are zero).

4. If resources are limited, a growing population will level off at a density determined by the carrying capacity of the environment, tracing a pattern of logistic growth.

5. The rates of birth and death among individuals of different ages can be recorded in a life table. Age-specific survivorship and the age-specific pattern of births differ greatly among species, and the variations form one focus of the study of life histories.

6. Some factors that limit population growth in nature (such as starvation) have greater effect with increasing population density; they are density-dependent factors. Other factors (such as freezing weather for plants) tend to limit population growth to the same degree (constant death rate per individual regardless of population density); they are density-independent factors.

7. Currently, human population growth varies from zero in some developed countries to more than 4 percent per year in some developing countries; the world average is about 1.7 percent per year, down from the historic high of 2 percent, but still very high.

8. The present world population of 5 billion is expected to double, at least, before it levels off. The large proportion of humans currently in younger age groups will slow the approach to zero population growth.

9. Rapid growth of the human population during the past 200 years has been made possible primarily by reducing the death rate and expanding our carrying capacity through technology.

Review Questions

1. The following terms will be used repeatedly in this unit. Can you identify them?
   a. population, community, ecosystem, biosphere.
   b. producer, consumer, decomposer, detritivore.
   c. herbivore, carnivore, parasite.

2. Why do populations that are not restricted in some way grow exponentially?

3. If the birth rate equals the death rate, what happens to the growth rate of a population? If the birth rate remains slightly higher than the death rate, what happens?

4. Explain the difference between density-dependent and density-independent factors that influence population growth.

5. What defines the carrying capacity for a particular environment? Can you describe what happens when a low-density population shows a logistic growth pattern?

6. At present growth rates, how many years will elapse before the human population doubles in number?

7. How have human populations developed the means to expand steadily into new environments? How have humans increased the carrying capacity of their environments? How have they avoided some of the limiting factors on population growth?—or is the avoidance illusory?

8. If a third of the world population is now below age fifteen, what effect will this age distribution have on the growth rate of the human population? What sorts of humane recommendations would you make that would encourage this age group to limit the number of children they plan to have?

Readings

Begon, M., J. Harper, and C. Townsend. 1986. *Ecology: Individuals, Populations, and Communities.* Sunderland, Massachusetts: Sinauer Associates.

Bouvier, L. 1984. "Planet Earth 1984–2034: A Demographic Vision." *Population Bulletin.* Vol. 39, no. 1, 1–39.

Krebs, C. 1985. *Ecology.* Third edition. New York: Harper & Row. Chapters 10 through 12 are more advanced treatments of topics covered in this chapter.

Pianka, E. 1983. *Population Ecology.* Third Edition. New York: Harper & Row.

Polgar, S. 1972. "Population History and Population Policies From an Anthropological Perspective." *Current Anthropology* 13(2)203–241. Analyzes often-ignored cultural barriers to programs for population control.

Scientific American. 1974. *The Human Population.* San Francisco: Freeman. Entire issue devoted to world population problems.

44

COMMUNITY INTERACTIONS

Flying through the dense rain forests of New Guinea is an extraordinary pigeon the natives call *gara*. It is about the size of a turkey and has cobalt blue feathers and plumes on its head (Figure 44.1). It flaps so slowly and noisily that its flight has been likened to the sound of an idling truck. Like eight smaller species of pigeon living in the same forest, the *gara* perches on tree branches to eat fruit. Why are there nine variously sized species of fruit-eating pigeons in this forest, and not just one species of some optimal size? Wouldn't competition for fruit eventually leave one species the winner?

In fact, each of these related fruit-eating species has its own role in the forest. Members of the larger species must perch on heavier branches when they feed, and

they eat larger fruit. Members of the smaller species eat fruit hanging from branches too thin to support the weight of a turkey-size pigeon, and they have bills too small to open the larger fruits. The species of trees in the New Guinea forest vary in terms of fruit size and the thickness of fruit-bearing branches, so the bird species end up foraging on different trees. Hence parts of the food supply that are used less by one species are used more by others in the same forest.

All of the pigeons benefit the different tree species by dispersing their seeds. The trees benefit from other animals as well—insects, bats, and other birds that serve as pollinators—and from associations with fungi (such partnerships being the mycorrhizae described on page

262). Meanwhile, many thousands of insect species feed on the leaves, flowers, bark, and roots of those same trees—or eat other insects, decaying vegetation or carcasses, or the blood or droppings of other animals.

This example reminds us that organisms interact as part of **communities**, which are all those populations of all species living together in a given area. In this chapter we turn to what is known about how communities are organized, how they function, and how they change over time.

BASIC CONCEPTS IN COMMUNITY ECOLOGY

Habitat and Niche

As we have seen, a **habitat** is the place where a population (or an individual) of a given species lives—its mailing address, so to speak. The word sometimes refers to a vegetation zone, such as a tropical rain forest, or to a more specific place, such as the height above ground in a forest canopy (some organisms live nowhere else), or a particular depth in the mud at the bottom of a pond (the only place where certain organisms ever will be found).

Each habitat has a characteristic range of physical and chemical conditions, such as the amount of light, typical temperatures, pH of the water, and so on. And each species is adapted to those conditions in terms of its morphology, physiology, and behavior. Moreover, it is adapted to the range of conditions imposed, directly or indirectly, by other organisms in the habitat. (For example, snakes have various adaptations for locating and capturing prey of certain sizes.)

The full range of abiotic and biotic conditions under which a species can live and reproduce is called its **niche**. Within this range, different conditions shift in large and small ways, creating an ever-changing mosaic to which that species responds. Thus a niche is a property of a particular species in a particular setting, including the role it plays in relations with other species.

With enough effort, we could probably identify all the niche characteristics for any species, but we would end up with an unmanageably long list. In attempting to understand how communities work, it is more informative to focus on important *differences* between the niches of species living in the same or similar habitats. For example, given their differences in body size and bill size, the nine species of New Guinea pigeons can eat fruit of different sizes, hence they disperse seeds of different sorts, the differences in dispersal affect where different trees grow—and the tree distribution influences how the entire community is organized.

Figure 44.1 Tropical rain forest of New Guinea—habitat of many diverse species, including the turkey-size *gara* (also known as the Victoria crowned pigeon) and eight species of smaller pigeons.

Within this habitat each species has its own *niche*, an ever-changing mosaic of abiotic and biotic conditions under which it lives and reproduces.

Table 44.1 Types of Interactions Between Two Species

Type of Interaction	Direct Effect of Interaction*	
	Species 1	Species 2
Neutral	0	0
Commensalism	+	0
Mutualism	+	+
Interspecific competition	–	–
Predation	+	–
Parasitism	+	–

*0 indicates no direct effect on population growth,
+ indicates positive effect, – indicates negative effect.

Types of Species Interactions

Even simple communities consist of dozens or hundreds of species, which interact in a bewildering number of ways. To make progress in disentangling this complexity, let's initially focus on the types of interactions between any two species in a community (Table 44.1). Later, we will explore a few of the interactions that affect the structure and function of entire communities.

Most of the interactions between any two species in a community are **neutral**—that is, neither species *directly* affects the other (even though the two may be linked indirectly through a series of interactions with other species). For example, eagles certainly have a neutral effect on any given species of grass. Because rabbits eat grass and eagles eat rabbits, an eagle indirectly benefits the grass by helping to control the rabbit population—and the grass indirectly benefits the eagle by fattening up its prey. Still, by convention, the eagle-grass interaction is regarded as neutral.

Sometimes one species of a pair benefits significantly from the interaction, while the other is *neither helped nor harmed* to any great degree. This kind of interaction is called **commensalism**. For example, by providing many kinds of birds with roosting or nesting sites, a tall tree affords them protection from many kinds of predators. Usually the tree gets little or nothing in return from the birds that do nothing else but roost or nest (and do not, for example, eat insects that damage the tree).

The most important types of community interactions are known as mutualism, competition, predation, and parasitism. In **mutualism**, both members of a pair of species clearly and directly benefit from the interaction.

(Mutually beneficial relationships involving continuous, intimate contact between species are also called *symbiotic* relationships.) In **interspecific competition**, both species are harmed by the interaction. Finally, in **predation** and **parasitism**, one species (the predator or the parasite) benefits directly from the interaction while the other species (the prey or the host) is directly harmed by it. Let's now take a close look at each of these interactions in turn.

MUTUALISM

Mutually beneficial relationships between interacting species were once regarded as quaint biological curiosities. We now realize that the functioning of natural communities depends critically on several main categories of directly mutualistic interactions. One such interaction is the mycorrhiza, an intimate association between a fungus and the young roots of nearly all vascular plants. As we have seen, the extensive fungal filaments absorb mineral ions from the soil, some of which are used by the plant; and the fungus absorbs some sugars and nitrogen-containing compounds from the root (Figure 20.3). The dominant plants of forests and grasslands rely on this interaction.

Natural communities also depend on the mutualistic interactions between flowering plants and their pollinators as well as their agents of seed dispersal. Many of these remarkable interactions are described in Chapter Twenty-One.

These and other examples of mutually beneficial relationships between species are either facultative or obligatory. In **facultative mutualism**, both species benefit from the interaction but each can live without the other, if necessary. Thus some kinds of plants produce seeds by self-pollination if they fail to be visited by their usual pollinators, and the pollinators may do just as well with nectar from other sources.

In **obligate mutualism**, neither one of two interacting species can survive for long without the other. A classic example is the relationship between the yucca moth and the yucca plant (Figure 44.2). This moth obtains pollen only from the yucca plant; even its larval form dines only on yucca seeds. The yucca plant depends exclusively on this one moth pollinator. Hence the moth's private energy source, available throughout its lifetime, helps assure reproductive success. At the same time, the moth helps assure reproductive success for the plant: the pollen is carried exactly where it must go instead of being randomly spread about by a less picky pollinator that visits several plant species.

Figure 44.2 Mutualism in the high desert of Colorado. There are several species of yucca plants (**a**), but each is pollinated exclusively by only one kind of yucca moth species (**b**). The adult stage of the moth life cycle coincides with the blossoming of yucca flowers. Using mouthparts that have become modified for the task, the female moth gathers up the somewhat sticky pollen and rolls it into a ball. Then she flies to another flower and, after piercing the ovary wall, lays her eggs among the ovules. She crawls out the style and shoves the ball of pollen into the opening of the stigma. When the larvae emerge (**c**), they eat a small portion of the yucca seeds. Then they gnaw their way out of the ovary to continue the life cycle. The seeds remaining are enough to give rise to a new yucca generation.

So refined is this mutual dependency that the moth and larva can obtain food from no other plant, and the flower can be pollinated by no other agent.

INTERSPECIFIC COMPETITION

When different species in a community have some requirements or activity in common, their niches potentially overlap. Some resources (such as the oxygen used in aerobic respiration) are unlimited, so the niche overlaps present no problems. In other cases, however, supplies can be limited and the species may end up competing for them. The greater the potential for niche overlap, the greater will be the potential for competition.

Interspecific competition (that is, between species) usually is not as intense as *intraspecific* competition (within a population of a species). Why is this so? Each individual using a resource may deplete or limit it to some extent, so its activity affects the ability of others to use the resource. Because requirements are much the same for all members of a given population, intraspecific competition can be fierce when resources are in short supply. Although requirements for different species might be similar, they commonly differ more than the

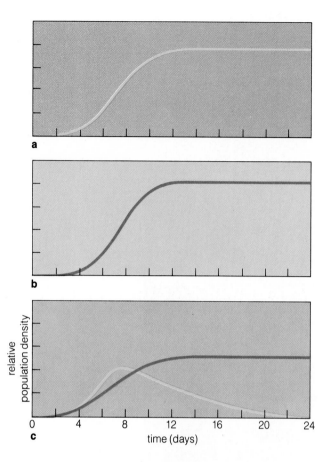

Figure 44.3 Competition between two species of *Paramecium*. When grown separately, *P. caudatum* (**a**) and *P. aurelia* (**b**) established stable populations. (**c**) When grown together, *P. aurelia* (red curve) drove the other species (gold curve) toward extinction.

dant, exploitation competition cannot occur. (Thus, if you and a friend were sharing a milkshake, each drinking from your own straw, you wouldn't care how fast your friend drank if it were a ten-gallon milkshake, replaced daily.)

With interference activities, the picture becomes more complicated, for one species may limit another's access to resources, whatever the level of resource supply. (You could still pinch your friend's straw, even with a ten-gallon milkshake.)

Exploitation Competition

When two species compete for resources in the same habitat, the simplest of all possible outcomes are these: either they will continue to coexist or one will exclude the other from the habitat. Models developed in the 1920s indicated that if the resources required by both species are similar enough, the population growth rate of one species will depress the growth rate of the other and lead to its exclusion from the habitat. (The difference in the rate of increase would be the outcome of some structural, physiological, or behavioral difference between the two species, a difference that would give one the competitive edge.) However, the two species might continue to coexist in spite of some overlap in their use of scarce resources as long as there was enough of a difference in how they use those resources.

In the 1930s, G. F. Gause tested this idea by growing two species of *Paramecium* separately and then together. Because the two species require similar food, Gause reasoned that there would be strong competition between them. His experiments, summarized in Figure 44.3, suggested that the earlier models were correct. Gause reported evidence that complete competitors cannot coexist indefinitely, a concept now called **competitive exclusion**.

In other experiments, Gause used two species of *Paramecium* that did not overlap as much in their use of resources. (When grown together, one species tended to feed on bacteria suspended in the liquid medium in the culture tube, the other tended to feed on yeast cells at the bottom of the tube.) In this case, the growth rate was lowered for both populations—but not enough for either population to exclude the other; they continued to coexist.

The more similar two species are in their use of scarce resources in the same habitat, the less likely they are to coexist.

requirements for individuals of the same species, so competition is usually less intense.

There are two categories of competition, regardless of whether we are talking about individuals of the same or different species. In **exploitation competition**, all individuals have equal access to the resource in question but they differ in how fast or how efficiently they can exploit it. In **interference competition**, certain individuals limit or prevent others from using the resource and thereby control access to it.

Whether competition actually occurs between two or more species that use the same resource depends on two things—the resource supply and the activities of the organisms. In pure exploitation competition, one species detracts from the growth, reproduction, or survival of another only indirectly—by reducing the common supply of resources. Therefore, if shared resources are abun-

alpine
chipmunk

lodgepole
chipmunk

yellow pine
chipmunk

least
chipmunk

Figure 44.4 In the foreground, alpine tundra, with a stand of lodgepole pines in the distance. These are two of the vegetational zones of the eastern slopes of the Sierra Nevada in California. Each zone is the habitat of a different species of chipmunk (*Eutamias*).

Today, ecologists tend to use field experiments to evaluate the effects of exploitation competition. For example, when they suspect competition between two species, they remove one of the species from some test plots, remove the other species from other test plots—and leave still other plots untouched (to serve as controls).

In one field experiment, N. Hairston studied competition between two species of salamanders in the Great Smoky Mountains and in the Balsam Mountains. Although *Plethodon glutinosus* lives at lower elevations than its relative, *P. jordani*, the ranges overlap in certain areas. Hairston did a series of removal experiments in the overlap areas and monitored the results for five years. In the control plots, nothing had changed at the end of the test period—the two species continued to coexist. In plots from which *P. jordani* had been removed, *P. glutinosus* increased in abundance. In the plots from which *P. glutinosus* had been removed, there was no clear change in the abundance of *P. jordani* but the proportion of individuals in younger age classes had increased—a pattern suggesting a growing population.

In another field experiment in Britain, A. Tansley showed that exploitation competition influenced the distribution of two species of bedstraw (plants of the genus *Galium*). One species normally grows on acidic soils and the other, on basic soils. When one species was transplanted to the different type of soil, it could grow—as long as the other species was not there with it! When the two were grown together on either an acidic or a basic soil, the surviving species was the one that normally grew there.

In sum, the niches of two or more species living in the same habitat may overlap in several respects—but the mere existence of overlap is not a sure indicator of competition among them. Birds overlap in their need for oxygen, but they *do not* compete for it. Hairston's salamanders overlap in habitat use, and they *do* compete. Tansley's bedstraws also compete, but they show *no* overlap in habitat use. Finally, of course, most species that share *no* resource do *not* compete (grasses and eagles).

Interference Competition

Nature provides many spectacular examples of interference competition between species. Corals kill neighboring corals of other species by poisoning and growing over them. Hummingbirds chase hummingbirds of other species, even bees, away from defended clumps of flowers. Some limpets slowly but surely "plow" competing species out of defended territories along the seashore.

A strangler fig tree surrounds a victim tree and eventually kills it while growing its own massive canopy of leaves.

In most cases of interference competition, only one species interferes with another species, which survives in or on whatever is left. For example, on the eastern slope of the Sierra Nevada in California, each of four chipmunk species occupies a different habitat (Figure 44.4). The alpine habitat is at the highest elevation, followed by the lodgepole pine habitat, then the piñon pine/sagebrush habitat, and finally, at the base of the mountains, the sagebrush habitat. The so-called least chipmunk of the sagebrush habitat could actually live at higher elevations among the pines (it does so elsewhere). But when it was experimentally introduced into the vegetational zone above it, the aggressive behavior it provoked from the yellow pine chipmunk showed why the least chipmunk is restricted to the sagebrush. On the other hand, because of its dietary requirements, the yellow pine chipmunk could not long survive in the sagebrush habitat.

PREDATION

On "Predator" Versus "Parasite"

Of all community interactions, predation is perhaps the most riveting of our attention (as well as the prey's). Dramatic examples abound, including the confrontation between the leopard and the baboon shown in Figure 44.5. A goat pulling up a thistle plant for breakfast, although less dramatic, is no less a case of predation—the prey is a living organism killed by the predator. But what about grasshoppers or horses, which graze on plants without killing them; or mosquitoes, which take blood from your arm and then fly off? What about ticks and fleas, which live on the host and take blood for long periods but get off to lay their eggs elsewhere? What about internal parasites such as tapeworms, external ones such as lice, or the parasitic plants called mistletoe, which spend their entire lives with the host?

The terminology available for these diverse interactions is extensive and sometimes contradictory. In this section and the next, we will cover all consumer/victim interactions with two broad definitions of *predators* and *parasites*. (To use narrower definitions, we would have to set up half a dozen other categories as well.) **Predators** get their food from other living organisms (their *prey*), but they do not live on or in the prey and may or may not kill it. **Parasites** also get their food from other living organisms (their *hosts*), but they live on or in the host for a good part of their life cycle and may or may not kill it.

Figure 44.5 (Right) Confrontation between predator and prey—the most arresting of all community interactions. As a last resort, the baboon has turned back to face the leopard with threat behavior. The effect has momentarily stopped the predator's advance; under other circumstances it might have meant the difference between capture and escape. Here, in the open, the baboon has run out of alternatives.

Dynamics of Predator-Prey Interactions

Mathematical models predict a great variety of patterns between interacting populations of predator and prey. The patterns range from stable coexistence at steady population levels for both species, to repeated oscillations or cycles of abundance, erratic oscillations, or the extinction of prey through predation. Which outcome is predicted depends on several features of the two populations, including:

1. The carrying capacity of the environment for the prey, in the absence of predation. (The carrying capacity, recall, is the equilibrium size of a population as defined by its sustainable supply of resources.)

2. The reproductive rate of the prey.

3. The **functional response** of predators (the capacity of *individual* predators to respond to increases in the density of prey by eating more prey).

4. The **numerical response** of the predator (its reproductive rate).

Stable coexistence for both populations is likely when predation keeps the prey population in check (it prevents the prey population from overshooting its carrying capacity). Predators can do this when they reproduce quickly relative to the prey and are capable of eating more when there are more prey organisms around.

Oscillations are likely when predators reproduce more slowly than their prey, when they eat only so many prey organisms at a time regardless of how many are around, and when the carrying capacity of the environment is high for their prey.

An idealized cyclic oscillation of predator and prey abundance appears in Figure 44.6. The cycling is caused by time lags in the predator's response to changes in the abundance of prey. (A "time lag" is an interval of time between two related events.) Cyclic oscillations of this sort have been documented in a number of predator-prey systems. For example, one long-term study points to a correlation between population density of the Canadian lynx and the snowshoe hare (Figure 44.7). There is

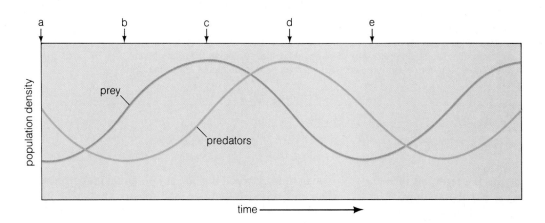

Figure 44.6 Idealized cycles of predator and prey abundance. The cycling is caused by time lags in the predator's response to changes in prey abundance. Starting at time *a*, the prey population is at a low density, so the predator population is hungry and declining. The prey population responds to the decline by increasing, but the predators continue to decline for a while until increased predator reproduction gets under way (time *b*).

Both populations increase until the increasing predation halts further growth of the prey population (time *c*), and it begins to decline. The prey suffer further impact from predation as the predators continue their increase, which slows as they feel the effects of starvation from lower prey density (time *d*). At time (*e*), we are back where we started and a new cycle begins. (The scale on this figure exaggerates predator density—in fact, predators are usually less common than their prey at all points of the cycle.)

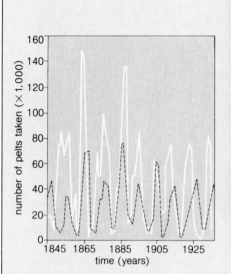

Figure 44.7 Relation between the lynx population (dashed line) and the snowshoe hare population (solid line) in Canada over a ninety-year period. These data are derived not from field observations but from counts of the pelts that trappers sold to Hudson's Bay Co. The curves were long regarded as a general example of the way predation can control the populations of both predator and prey.

This figure is a good test of how willing you are to accept conclusions without questioning their scientific basis. (Remember the discussion of scientific methods in Chapter Two?) For example, what other abiotic and biotic control factors could have been influencing the population levels? Were there also fluctuations in climate over this time span? Were some winters more rigorous, thereby imposing a higher death rate on one or both populations? Although this is called a simple predator-prey system, weren't the hares preying on the vegetation, which may have been overbrowsed in some years but not in others? What about owls, martens, and foxes—which also prey on hares? What if some years there were fewer trappers because of such variables as Indian uprisings? What about fluctuating demands for furs by the fashion industry? What if some years there was more lynx trapping than hare trapping, or vice versa?

indeed a correspondence between the rise and fall of these predator and prey populations, but the assumption that predation alone causes the oscillation is too simplistic.

Time lags certainly contribute to the lynx-hare cycles. When prey population density is at its peak, predators are not usually numerous enough to bring about an immediate decline of the huge hordes of hares. It takes time for the lynx population to increase by way of reproduction and by migration of lynx into areas where prey are abundant. More abundant food can promote survival, more rapid development, even fertility, but there

is a lag time between the birth of potential predators and the point at which they are mature enough to take prey.

In this example, increased predation can make a major contribution to the decline of the prey population, but it probably cannot trigger the decline. What does?

There is some evidence that recurring wildfires, floods, and insect outbreaks have indirect bearing on the hare-lynx cycle. Such environmental disturbances can destroy mature forest canopies that thereby create favorable conditions for different plant species that are adapted to moving in and becoming established in exposed settings. These species flourish until the species typical of the mature forest system become reestablished and gradually succeed them, in ways that will be described shortly.

Now, the so-called early successional species provide ideal forage for the hares; in fact, the peaks in hare population densities correspond to the occurrence of this type of vegetation cover. However, the early successional shrubs and trees include alders, poplar, black spruce, and birch species—all of which produce toxins in especially high concentrations in new shoots. During winter, the hares prefer to feed on other plant parts and suffer no ill effects. When population density is high, though, the hares destroy the harmless parts of their winter forage, and they are forced to begin feeding on the toxic shoots. Experiments show that hares ingesting high concentrations of the plant toxins rapidly lose weight and become severely stressed.

Thus the cyclic fluctuations in population density may be a consequence of hares preying on plants and are simply amplified by lynx preying on hares. The built-in chemical defenses of certain plants prevent even more devastating increases in the numbers of hares (Figure 44.8).

PREY DEFENSES

When you stop to think about it, the fact that the plants described above defend themselves against being eaten by snowshoe hares is rather remarkable. How do such defenses arise? From what is known about evolutionary processes, it is clear that predators and prey exert continual selection pressure on each other. When some new, heritable means of defense spreads through the prey population, only the predators equipped to counter the defense of their prey survive to reproduce. Thus, when the prey evolves, the predator also evolves to some extent because the change affects selection pressures operating between the predator and prey. Again, this type of joint evolution of two (or more) species that are interacting in close ecological fashion is called **coevolution**. Let's take a look at some of the outcomes of predator-prey interactions.

Camouflage

One evolutionary outcome of predation is the capacity of many species of prey (and of predators) to "hide" in the open. **Camouflage** is an adaptation in form, patterning, color, or behavior that enables an organism to blend with its surroundings, the better to escape detection.

Figure 44.8 Snowshoe hare browsing on an early successional shrub. At the peak of the cyclic fluctuation in hare population density, such plants are overbrowsed. The new shoots they put out have high concentrations of toxins, which severely stress the hares. These conditions may last long enough to trigger a decline in the hare population.

Figure 44.9 Camouflage among the rocks. Find the plants (*Lithops*) that have the form, patterning, and color of stones.

Figure 44.10 The fine art of camouflage, as developed in predator-prey interactions. (**a**) One katydid species "hides" in the open from bird predators by looking like leaves—even down to blemishes and chewed-up parts. (**b**) Caterpillars of some moth species tend to look like bird droppings because of their coloration and the body positions that they assume. (**c**) By lurking motionless against its like-colored background, the yellow crab spider is essentially invisible to prey. (**d**) What bird??? With the approach of a potential predator, the least bittern stretches its reed-colored neck and thrusts its beak upward—and even sways gently, like the surrounding reeds in a soft wind.

For example, a desert plant (*Lithops*) resembles small rocks in shape and color (Figure 44.9). Only during the brief rainy season, when other vegetation and water are more plentiful for herbivores, do these "living rocks" put forth brightly colored flowers and draw the attention of pollinators. Figure 44.10 shows examples of camouflage among prey animals. (Camouflage, of course, is not the exclusive domain of prey. Predators that rely on stealth also blend well with the background. Think about polar bears against snow, tigers against tall-stalked and golden grasses, and pastel spiders against pastel flower petals as in Figure 44.10c.)

Moment-of-Truth Defenses

When cornered, some prey species defend themselves with display behavior that may startle or intimidate a predator (Figures 44.5 and 44.11a). Such behavior can create momentary confusion, and a moment may be all it takes for the prey to escape. When attacked, the bombardier beetle raises its abdomen and sprays a noxious chemical at its predator. (It is an effective adaptation in some cases, but grasshopper mice have a behavioral counteradaptation. These mice pick up the beetle, shove its tail end into the earth, and munch the head end.)

The bombardier beetle is one of many animals as well as plant species that release chemicals to repel or otherwise deter a potential attacker. Chemicals serve as warning odors, repellants, alarm substances, and outright poisons. Earwigs, grasshoppers, and skunks can produce awful odors. The foliage and seeds of some plants contain tannins, which taste bitter and which decrease the digestibility of the plant material; the tissues of other plants incorporate terpenes, which can be toxic. Nibbling on a buttercup (*Ranunculus*) leads to highly irritated mucous membranes in the mouth.

Some pheromones serve as alarm substances among prey fish that travel in schools that have neither spines nor other built-in defenses. (Unlike hormones, which have internal targets, a *pheromone* is an exocrine gland secretion with some target outside the animal body. Pheromones can trigger behavioral changes in other animals of the same species by acting as alarm substances, trail markers, sex attractants, and the like.) The pheromones, released when the skin is broken (as during an attack) trigger an escape response in neighboring fish.

Warning Coloration and Mimicry

One consequence of predation is the existence of many prey species that are bad-tasting, toxic, or able to sting or otherwise inflict pain on their attackers. Often, toxic prey species are decked out with conspicuous colors and bold patterns that serve warning to potential predators.

Figure 44.11 Moment-of-truth defensive behavior. (**a**) A cornered short-eared owl spreads its wings in a startling display that must have worked against some of its predators some of the time; it is part of the behavioral repertoire of the species. (**b**) As a last resort, some beetles spray noxious chemicals at their attackers, which works some of the time but not all of the time. (**c**) Grasshopper mice plunge the chemical-secreting tail end into the ground and feast on the head end.

Inexperienced predators might attack a black-and-white striped skunk, a bright orange monarch butterfly, or a yellow-banded wasp. As a result of the experience, they quickly learn to associate the colors and patterning with pain or digestive upsets.

Other prey species *not* equipped with such defenses can still sport warning colors or patterns that *resemble* those of distasteful, toxic, or dangerous species. The resemblance of an edible species to a relatively inedible one is a form of **mimicry** (Figure 44.12).

PARASITISM

True Parasites

Like predator-prey interactions, the interactions between true parasites and their hosts are remarkably diverse. Like predators, a parasite also takes sustenance from other living organisms—but it doesn't eat them outright. Parasites live on or in their host during some part of their life cycle, and the host may or may not die as a consequence of the association. (For example, the parasite can weaken the host so that the host succumbs to secondary infections. The parasitic blood flukes, described on page 615, can do this.)

Generally, parasites cause death only when they infect a host population with no coevolved defenses against them. After all, a host able to live longer may transmit parasites to more new individuals than a vulnerable, rapidly dying host ever could do. Thus, parasite populations also tend to coevolve with their hosts, producing an intermediate level of negative effects on the host itself.

Parasitoids

In contrast to parasites, some insect larvae are **parasitoids**, which kill their host by completely consuming its soft tissues before the host can grow into an adult. Parasitoids sound horrendous, but fortunately their target hosts are not humans but the larvae of other insect species.

Parasitoids are natural controls over other insect populations in a community. Many have been successfully raised and released as an alternative to the use of chemical pesticides. (Many of the worst pest outbreaks have *followed* applications of pesticides. The pest population soon recovers, but parasitoids formerly keeping the pest population under control were also killed and their populations usually fail to recover as quickly.)

What sort of defenses work against attacks of parasitoids? One example was described by Peter Price, who

a A dangerous species that serves as a model...

Figure 44.12 Mimicry. Many animals—especially those bite-sized morsels, the insects—avoid being eaten by having a bad taste, obnoxious secretion, or painful bite or sting. Among predators, knowledge of these traits is usually not inherited. Each young predator learns about them the hard way, by often unpleasant trials.

Among many prey species, dangerous or unpalatable individuals are easily recognized and remembered. If this were not the case, many individuals would be lost as inexperienced predators learned their lessons. Thus repugnant species tend to have distinctive, memorable appearances—bright colors (such as red, which predatory birds see so well), and bold markings (such as stripes, bands, and spots). These flamboyant species make no effort to conceal themselves. Sometimes they even deliberately flash colors with an uplift of the body or the wings. Their coloration and patterning are called "aposematic" (*apo-*, meaning "away," and *sematic*, meaning "signal").

In the natural world, each of the hundreds of dangerous or unpalatable species does not have a distinct warning signal, for too many signals would tax the learning capacity of predators. Instead there are whole groups of related species having almost identical aposematic appearances. Thus, many species benefit from a single taste trial. In turn, many less related and even totally unrelated species avoid predation by mimicking the appearance and behavior of the repugnant or dangerous model species.

Some mimics are as unpalatable as their models; they are called Müllerian mimics (after Fritz Müller, who named the phenomenon). Others may be quite edible yet are still avoided; they are called Batesian mimics (after Henry Bates, who discovered this class of mimics). Most mimicry series comprise both of these types.

There are other types of mimicry. In aggressive mimicry, parasites or predators bear resemblances to their hosts or prey. In speed mimicry, sluggish, easy-to-catch prey species resemble fast-running or fast-flying species that predators have given up trying to catch. (This might well be termed "frustration" mimicry.)

(a) Numerous and pugnacious yellowjackets (here, *Vespula arenaria*) are models for extensive mimicry series. (b) This

b

c

d

...and three of its mimics (above)

e

A speedy model...

f

...and a slow-moving mimic

g An unpalatable model...

h ...and a palatable mimic

masarid wasp is probably a Müllerian mimic of *Vespula*. Other insects, such as beetles (**c**) and flies (**d**) may be Batesian mimics. (**e**) Flesh flies have gray and black bodies, red eyes, and red tail ends. Birds soon give up trying to catch these fast-flying insects. In the American tropics, many sluggish insects, such as the weevil *Zygops rufitorquis* (**f**), closely resemble flesh flies and thus reduce the likelihood of being eaten. This is an example of speed mimicry.

(**g**) Ithomiid butterflies of the New World tropics are frequent models for mimicry. They come in two basic types of coloration: orange with black stripes, and transparent with a white forewing band. Only a specialist can distinguish the many look-alike species (such as *Dismorphia* shown in **h**) and their mimics. In this case, the mimics may be either Müllerian or Batesian. (From Edward S. Ross, California Academy of Sciences)

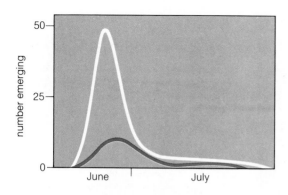

Figure 44.13 Effect of parasitoid wasp activity on sawfly emergence. The white line represents emergence when attack by parasitoids is prevented; the red line represents emergence after parasitoid attack. Notice that most attacks occurred on the would-be early emergers, which did not burrow as deeply into the forest litter.

studied the effects of a parasitoid wasp that lays its eggs on jackpine sawfly cocoons. The sawfly larvae emerge in trees and then drop to the forest floor, where they burrow into the leaf litter and spin their cocoons. The female wasp tends to lay eggs on sawfly cocoons nearest the surface of the litter. From these shallowly buried cocoons, the first adult members of the new generation of sawflies emerge; later in the season, adult sawflies emerge from more deeply buried cocoons (Figure 44.13).

The reproductive activities of the wasps represent a strong selection pressure on the sawfly population by favoring the late-emerging (and deep-burrowing) saw-flies. The sawflies, in turn, exert selection pressure on the wasps, with the advantage going to those individual wasps having the capacity to find sawfly cocoons at ever-increasing depths in the litter. As Price points out, the host stays ahead in this coevolutionary contest. Each time the sawfly larvae burrow deeper, the female wasps have to spend more time searching for them. Consequently, fewer wasp eggs are laid and the wasp population is held in check.

Social Parasitism

Not all parasites feed directly on the tissues of a host. In **social parasitism**, one species depends on the social behavior of another to complete its life cycle. (This case stretches even our broad definition of parasitism, but it has basic parallels with the behavior of true parasites and parasitoids.)

For example, the North American brown-headed cowbird never builds a nest, incubates its eggs, or cares for its offspring. Instead, it removes one egg from the

nest of a bird of another species and lays one of its own eggs as a "replacement." Some birds recognize the foreign egg and shove it out of the nest. Others, including the Kirtland warbler, do not. The warblers hatch the egg and raise the young cowbird. Since it is larger and more aggressive than the warbler nestlings, the young cowbird usually pushes them out of the nest or takes most of the food, leaving the warbler nestlings to starve to death (Figure 48.3).

Increasingly, ecologists have come to appreciate the role of parasites and parasitoids in regulating host populations in nature. The individual insect or marine invertebrate with no parasites at all is rare, and an individual bird or a mammal with neither external nor internal parasites is truly exceptional. Only humans who enjoy the benefits of modern health care are normally free of parasites.

COMMUNITY ORGANIZATION, DEVELOPMENT, AND DIVERSITY

Whatever organization and stability we see in communities is the result of antagonistic forces that have come into balance—sometimes an uneasy balance. The growth rate of a population depends on a balance between births and deaths. The coexistence of predators and prey depends on neither species winning the ecological and evolutionary footrace. Competitors have no sense of fair play whatsoever—straw pinching is common. Even mutualists are really antagonists—the flower gives as little nectar as necessary to attract a pollinator, while the pollinator seeks to get the most reward possible for the least effort. The alga and the fungus in a lichen have different gene pools under different and perhaps conflicting selection pressures. What kinds of community patterns and processes do these conflicting forces produce? Let's take a look.

Resource Partitioning

Through numerous studies, we know that groups of functionally similar species living together in a community do different things. If we assume there is indeed a limit to the degree of ecological similarity that will permit coexistence, then such a community pattern could arise in two ways. First, only species that are dissimilar enough from established ones will succeed in joining an existing community. Second, natural selection and other evolutionary processes may act on the competing populations already established in the community, increasing the differences among them.

However the pattern of coexistence arises, the outcome is **resource partitioning**, whereby groups of functionally similar species share the same resource in different ways, in different areas, or at different times.

N. Weland and F. Bazzaz studied resource partitioning among the plants growing in a field that had been abandoned a year after it was plowed. Three species of annual plants were common in the field. Like all land plants, each required sunlight, water, and mineral ions from the soil. The investigators proposed that each species met those requirements without interfering with the others by exploiting different areas of the habitat space (Figure 44.14).

Bristly foxtail grasses became established in soil where the amount of water available varied greatly from day to day. With their shallow, fibrous root systems, the foxtails could rapidly absorb water after the rains and could recover rapidly from droughts. Mallow plants occupied different areas of the soil. With their taproot systems, those plants could exploit soil depths that were moist early in the growing season but less moist later on. The third species, smartweed, had a taproot system that branched in the topsoil and in the continuously moist soil below the rooting zone of the other two species.

Is interspecific competition usually responsible for such differences among functionally similar species, or are the differences more often due to historical accident? There is no conclusive answer at present.

Effects of Predation on Competition

One day Charles Darwin instructed his gardener to stop cutting a small patch of his lawn. English lawns (unlike most American ones) are composed of many low-growing plant species. If cutting was discontinued, Darwin speculated, then competition among the species might intensify. He was correct. Of the twenty species originally growing in the plot, nine were eventually eliminated by competition with the remaining species once the cutting had stopped.

If we liken the lawnmower to a predator, then predation was minimizing the intensity of competition among the "prey" (plant) species. Many experiments show that predation has this effect in natural communities as well.

For example, for several years, Robert Paine kept predatory sea stars out of experimental plots in a rocky intertidal zone. (He left the sea stars with their fifteen species of invertebrate prey in control plots.) In the absence of the predator, the number of prey species declined to eight—the rest were crowded out by mussels, the main prey of sea stars and yet the strongest competitors in their absence.

Species Introductions

In their field experiments, ecologists are extremely careful about minimizing the disturbance to natural communities (hence Paine's limit on a few experimental plots). In contrast to these planned experiments, great numbers of species have been shuffled among different geographic regions, sometimes intentionally and sometimes inadvertently. Most introductions of species from one continent to another probably fail to become established—although few records are kept of such non-events. What we know the most about are two classes of introductions—agriculturally useful species and pest species.

As an example of a pest, in the 1880s the water hyacinth from South America was put on display at the New Orleans Cotton Exposition. Flower fanciers from Florida and Louisiana carried home clippings of the blue-

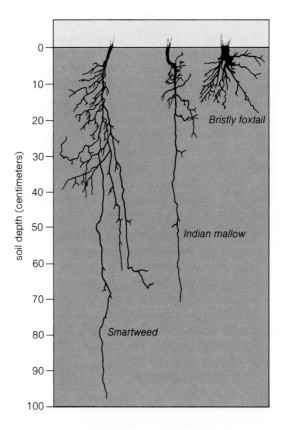

Figure 44.14 Partitioning of a resource (soil, with its nutrients and water) by three annual plant species that became established in a field plowed one year before.

Table 44.2 Effects of Introducing a Few Species into the United States

Species Introduced	Origin	Mode of Introduction	Outcome
Water hyacinth	South America	Intentionally introduced (1884)	Clogged waterways; shading out of other vegetation
Dutch elm disease The fungus *Cerastomella ulmi* (the disease agent)	Europe	Accidentally imported on infected elm timber used for veneers (1930)	Destruction of millions of elms; great disruption of forest ecology
Bark beetle (the disease carrier)		Accidentally imported on unbarked elm timber (1909)	
Chestnut blight fungus	Asia	Accidentally imported on nursery plants (1900)	Destruction of nearly all eastern American chestnuts; disruption of forest ecology
Argentine fire ant	Argentina	In coffee shipments from Brazil? (1891)	Crop damage; destruction of native ant communities
Camphor scale insect	Japan	Accidentally imported on nursery stock (1920s)	Damage to nearly 200 species of plants in Louisiana, Texas, and Alabama
Japanese beetle	Japan	Accidentally imported on irises or azaleas (1911)	Defoliation of more than 250 species of trees and other plants, including commercially important species such as citrus
Carp	Germany	Intentionally released (1887)	Displacement of native fish; uprooting of water plants with loss of waterfowl populations
Sea lamprey	North Atlantic Ocean	Through Welland Canal (1829)	Destruction of lake trout, lake whitefish, and suckers in Great Lakes
European starling	Europe	Released intentionally in New York City (1890)	Competition with native songbirds; crop damage; transmission of swine diseases; airport runway interference; noisy and messy in large flocks
House sparrow	England	Released intentionally (1853)	Crop damage; displacement of native songbirds; transmission of some diseases
European wild boar	Russia	Intentionally imported (1912); escaped captivity	Destruction of habitat by rooting; crop damage
Nutria (large rodent)	Argentina	Intentionally imported (1940); escaped captivity	Alteration of marsh ecology; damage to earth dams and levees; crop destruction

From David W. Ehrenfeld, *Biological Conservation*, 1970, Holt, Rinehart and Winston and *Conserving Life on Earth*, 1972, Oxford University Press.

flowered plants and set them out for ornamental display in ponds and streams. Unchecked by their natural predators and nourished by the nutrient-rich waters of the region, the fast-growing hyacinths spread rapidly, displacing many native species along the way. Eventually they choked off entire ponds and streams, then rivers and canals. They are still thriving—now as far west as San Francisco—and they are still bringing river traffic to a halt in many areas.

Species successfully introduced into established communities do not always lead to such wholesale disasters, but few (if any) are without ecological consequences. Honeybees, mosquitofish, and ring-necked pheasant are among the species that have been absorbed into existing communities in the United States. But the honeybees have displaced native bees in many natural areas. Table 44.2 lists other introduced species and their effects on established communities.

Succession

The potential repercussions of newly introduced species in a community raise an interesting question. How do communities come to exist in the first place? Obviously they do not arise full-blown in the environment. Whether we are talking about an environment initially devoid of life (such as a newly forming volcanic island) or a disturbed patch of a previously inhabited environment (such as an abandoned pasture), *a rather predictable and repeatable pattern of change takes place.*

The first species become established, flourish, and then decline while others are on the rise. Unless some disturbance halts or sets back the process, the changes continue until a more or less constant array of species is reached, called a **climax community**. This relatively self-sustaining assemblage of species tends to be perpetuated until disturbed by some force such as fire, epidemic, or severe weather.

The changes in species composition that lead to a climax community are collectively called **succession**. When these orderly changes occur in an area previously devoid of life, they constitute *primary succession*. When they proceed in a disturbed area that was previously inhabited, the sequential changes are known as *secondary succession*.

Primary Succession.   Primary succession begins with colonization of an uninhabited site by pioneer species, which aid in the development of soil. Pioneer species are able to grow in exposed areas with intense sunlight, wide variations in air temperature, and nutrient-poor soil. Typically the pioneers are small and low-growing. They have short life cycles, and each year they produce abundant small seeds that are quickly dispersed from the parent plants.

At higher latitudes, early successional species must start from scratch with each new season, sprouting from seeds or sending out new shoots from the base of withered stalks. This puts them at a disadvantage when the later successional species begin to appear. Many of the later species are perennials; they live for more than one growing season. Thus they have a head start once they are established and can begin seasonal growth before the pioneer plants get started. They store more biomass and produce relatively few seeds—but the seeds are well endowed with the nutrients necessary for early, secondary growth. Eventually the later species crowd out the pioneers, whose seeds travel as fugitives on the wind or water—destined, perhaps, for a new but equally temporary habitat.

Figure 44.15 shows how primary succession proceeds in a community in the Glacier Bay region of Alaska.

Secondary Succession.   Secondary succession is characteristic of abandoned fields and parts of an established forest where falling trees or other disturbances have opened the canopy of leaves, letting sunlight reach the forest floor. In many respects the pattern of change is similar to all but the first stages of primary succession. In secondary succession, however, many plants grow from seeds or even seedlings that are already present when the process begins.

Different mechanisms account for the rise and fall of species during succession. In some cases, early species apparently set the stage for their own replacement by altering the environment in ways that favor later species. Soil-forming pioneer species in primary successions probably fall in this category. Alternatively, both early and late species may do just as well under existing conditions but the late species simply grow more slowly, eventually excluding the early species through competition. This process seems to be common in secondary successions in communities on land. Finally, early species might actually inhibit the growth of later species, which prevail only if some disturbance removes the early species. Algal succession in the intertidal zone seems to proceed in this way.

Cyclic Replacement.   Even with its array of well-adapted species, a climax community does not perpetuate itself indefinitely. Natural communities that seem most stable are actually a mosaic of successional patches as a result of major and minor disturbances. Winds, fires, insect infestations, and overgrazing all modify and shape the direction of succession by encouraging the proliferation of some species and eliminating others from the community.

One response to such disturbances is **cyclic replacement**. For example, the Sierra Nevada of California supports isolated groves of sequoia trees. Some of the giant trees of this climax community are more than 4,000 years old, which certainly implies long-term stability in the region. Although brush fires sweep through the community every so often, the disturbance they create actually helps maintain the climax configuration. Sequoia seeds germinate only in the absence of smaller, shade-tolerant plant species. If there is extensive litter on the forest floor, no sequoia seeds will germinate. Modest fires eliminate the species of trees and shrubs that compete with the sequoias, but do not damage the sequoias themselves. Mature sequoias have extraordinarily thick bark, which burns poorly and insulates them against heat damage.

Many sequoia groves are protected in national and state parks. Protection traditionally has meant minimizing the incidence of fires—not just accidental fires from

Figure 44.15 Primary succession in the Glacier Bay region of Alaska (**a**), where changes in newly deglaciated regions have been carefully documented. A comparison of maps from 1794 onward shows that ice has been retreating at annual rates ranging from 3 meters (at the glacier's sides) to a phenomenal 600 meters at its tip over bays. (**b**) When a glacier retreats, the constant flow of meltwater tends to leach the newly exposed soil of minerals, including nitrogen. Less than ten years ago, the soil here was still buried below ice. (**c, d**) The first invaders of these nutrient-poor sites are the feathery seeds of mountain avens (*Dryas*), drifting over on the winds. Mountain avens is a pioneer species that benefits from the nitrogen-fixing activities of symbiotic microbes. It grows and spreads rapidly over glacial till.

(**e**) Within twenty years, young alders take hold. These deciduous shrubs also are symbiotic with nitrogen-fixing microbes. Young cottonwood and willows also emerge (**f**) . Eventually the alders form dense thickets (**g**). As the thickets mature, cottonwood and hemlock trees grow rapidly, as do a few evergreen spruce trees. (**h**) By eighty years, the spruce crowd out the mature alders. (**i**) In areas deglaciated for more than a century, dense forests of Sitka spruce and western hemlock dominate. By this time, nitrogen reserves are depleted, and much of the biomass is tied up in peat: excessively moist, compressed organic matter that resists decomposition and that forms a thick mat on the forest floor.

campsites and discarded cigarettes, but also natural fires touched off by lightning. When small, periodic fires are prevented, litter builds up, and other species take hold that are susceptible to fire. Even though these species do not displace the mature sequoias, they prevent the sequoia seeds from germinating and from growing into the replacement trees needed to maintain the climax community.

Moreover, the litter and the undergrowth constitute so much potential fuel that when fires do occur, they are hotter than they otherwise would be—hot enough to damage the giants. Thus fire prevention efforts aimed at preserving the climax community actually have the opposite effect. In recent years, "prescribed burns," set intentionally under carefully controlled conditions, eliminate the underbrush in sequoia groves.

Community stability often requires episodes of instability, which permits cyclic replacements of equilibrium species and thereby maintains the climax community over time.

e

f

g

h

i

Species Diversity: Island Patterns

One day in 1965, about forty kilometers southwest of Iceland, a volcanic eruption began beneath the sea. When it eventually stopped, a new island had been formed and primary succession was under way. Within six months of the island's emergence, the first signs of life had appeared—bacteria, fungi, seeds of several kinds of beach plants, a fly species, and some seabirds. Two years after the eruption, the first vascular plant appeared and, two years later, the first moss. As primary succession and soil formation gradually progressed, the number of plant species continued to increase and is still doing so (Figure 44.16). All these species are colonists from Iceland—none evolved on the island, which is now named Surtsey.

The number of plant and animal species on Surtsey will not increase indefinitely. What will stop it? Detailed studies of community patterns on islands around the world provide us with two ideas. First, the farther an island is from a source of potential colonists, the fewer species it supports (Figure 44.17). This generalization is called the **distance effect**. Islands distant from source areas simply receive fewer colonizing species, at a slower rate, and the ones that do arrive are well adapted for long-distance dispersal. Second, larger islands tend to support more species than smaller islands at equivalent distances from source areas (Figure 44.18). This generalization is called the **area effect**.

The cause of the area effect is not clear-cut. In part, larger islands are physically more complex and often rise higher above sea level, so more kinds of habitat are available. Since many species live only in certain habitats, this effect increases species diversity. To some degree, larger islands probably "intercept" more colonists, being a larger target. However, extinctions may amplify the effect of area on the number of island species. Established colonists on small islands have small populations, which are more vulnerable to extinction from storms, volcanic eruptions, and biotic causes (epidemics, predation, parasitism) than larger populations.

Mainland and Marine Patterns of Species Diversity

In some ways, patterns of species diversity on islands are reflected in communities of the continents and the oceans. The "area effect" also holds for these communities—the larger an area sampled, the more species in the sample. As with islands, the diversity of habitats here is greater in large areas than in small areas. As we enlarge the sampling area, however, we also include more and more rare species with similar habitat requirements yet with only local distributions.

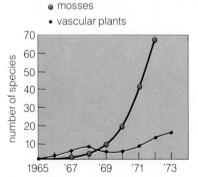

Figure 44.16 The number of species of mosses and vascular plants recorded on the new island of Surtsey from 1965 to 1973.

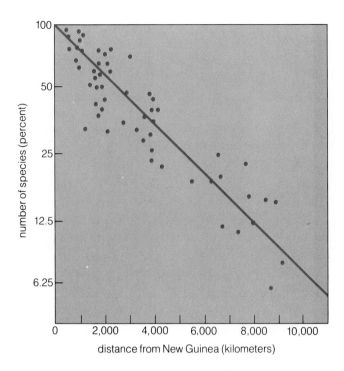

number of species (percent)

distance from New Guinea (kilometers)

Figure 44.17 The "distance effect." The number of species on islands of a given size declines with increasing distance from the source of colonizing species. Here, each point represents the number of species of land birds that live in lowland areas of islands in the South Pacific. The very large island of New Guinea is the source of colonists for these islands, each of which is at least 500 kilometers from this source. To correct for differences in island size (area), the actual number of bird species on each of these distant islands has been expressed as a percentage of the number of bird species on an island of equivalent size close to New Guinea.

The most striking patterns of species diversity correlate with latitude. For most groups, the number of coexisting species rises sharply from the arctic latitudes to temperate zones to the tropics (Figure 44.19). The most obvious difference between tropical regions and higher latitudes is in the annual pattern of sunlight (the tropics receive more direct sunlight spread more evenly throughout the year). Also, where rainfall is high on land in the tropics, communities are extraordinarily productive. In tropical seas, coral reef communities are remarkably productive and diverse.

There are several reasons for the increased diversity at tropical latitudes. First, availability of some resources is more constant than at higher latitudes. For example, although different tree species show different seasonal patterns, the trees as a whole provide new leaves, flowers, and fruit all year long in wet tropical forests and thereby support many resident species of specialized herbivores, nectar foragers, and fruit consumers. In temperate and arctic regions, such specializations would be suicide.

Second, species diversity is self-reinforcing. As soon as more species of plants live together, more species of herbivores emerge, partly because no one herbivore can overcome the chemical defenses of all the plants. (Many wet tropical forests support hundreds of tree species per acre, at least ten times more than temperate forests.) In

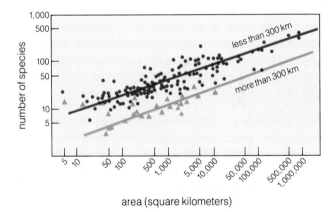

number of species

area (square kilometers)

Figure 44.18 The "area effect." Among islands similarly distant from the source of colonizing species, the larger islands support larger numbers of species. The points in this figure represent the number of bird species on tropical and subtropical islands. The solid circles and the upper line are for islands less than 300 kilometers from their source of colonists, while the triangles and the lower line represent islands more than 300 kilometers from source areas. Thus this figure shows not only the "area effect," but the "distance effect" as well.

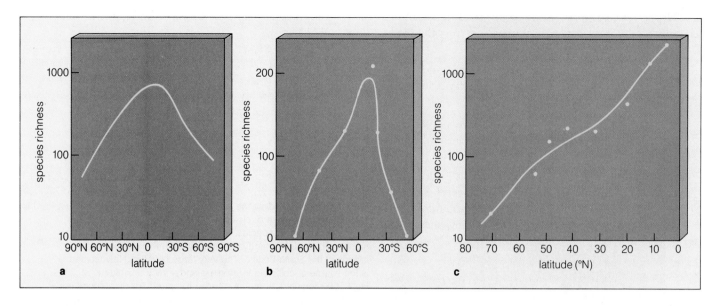

Figure 44.19 Patterns of species diversity corresponding to latitude. (**a**) Marine bivalve mollusks, (**b**) ants, and (**c**) breeding birds of North and Central America.

turn, more kinds of predators and parasites evolve in response to the diversity of prey and hosts. Similarly, the diversity of coral species on a tropical reef supports a pyramid of diverse groups representing many phyla. The reasons for the astonishing diversity of trees in tropical forests and of corals on tropical reefs are still unclear. One hypothesis is that storms and other disturbances, or possibly predation or parasitism, alter community structure in an intermediate way. In other words, they might prevent any one species from becoming dominant, leaving space and resources for many very similar species while causing few extinctions.

Finally, lowland tropical areas have never been glaciated. It may be that the great age of tropical forests itself has contributed to the accumulation of species. In this view, the overall rate of speciation in the tropics has long exceeded the rate of extinction (from natural causes), whereas periodic mass extinctions at higher latitudes have helped to keep diversity low there (Chapter Thirty-Seven).

The next decade will decide the fate of millions of species of organisms in tropical forests, of which ten square miles have been cleared during the past hour. With determination, we may still keep the rate of extinction in the tropical forests low enough to learn the answers to the intriguing questions raised by the patterns of species diversity.

SUMMARY

1. Each component species of a biological community responds to physical factors in its environment as well as to other species in the community. The habitat of the species is defined by the characteristics of the place where it lives. The niche of the species is defined by the conditions it requires for life and by its relations with other species in the community.

2. Many pairs of species in a community derive mutual benefits through their direct interactions, although each partner in such a mutualism has evolved to give as little as necessary to get as much as possible from its partner. Important classes of mutualism include plants and their pollinators, seed dispersal by animals, mycorrhizae, and normal microbial inhabitants of the animal gut.

3. When a resource is in short supply, two species that both require it are likely to compete, either by using up the resource as rapidly or as efficiently as possible (exploitation), or by preventing the other from using it efficiently (interference).

4. Competing species are more likely to coexist in the same community if their niches are rather different than if they are quite similar.

5. Although the population of a predator and the population of its prey may be stable under some conditions,

time lags in the response of the predator population to changes in the prey population may lead to cyclic fluctuations in the abundance of both.

6. The evolutionary race between predators and their prey has produced a rich array of special adaptations for both capture and escape.

7. Like predators, parasites may regulate the population of their hosts. Coevolution tends to favor resistant hosts and only moderately harmful parasites.

8. Related species in a community generally differ in their use of resources. Interspecific competition plays a role in producing this "partitioning" of resources, but the prevalence of competition in producing such patterns is still unclear.

9. Predation can reduce population densities among prey species, reducing competition among them and promoting their coexistence.

10. The introduction of non-native species to communities sometimes produces unexpected and undesirable results.

11. When an unoccupied habitat is colonized, species tend to replace one another in an orderly sequence (succession), eventually leading to a fairly stable "climax" community.

12. The number of species that occupy an area is affected by the size of the area, the rate of colonization by new species, and the rate of extinction of existing species.

Review Questions

1. Define habitat. Why do you suppose it might be difficult to define "the human habitat"?

2. What is the difference between the habitat and the niche of a species?

3. Define mutualism and give a few examples of its occurrence in nature.

4. Define interspecific competition. Explain how this form of behavior is incorporated into the concept of competitive exclusion.

5. Why is it more difficult to observe competitive exclusion in the natural environment, compared to observations of simple two-species interactions (such as bacteria) in a laboratory?

6. How might two species that compete for the same resource coexist? Can you think of some possible examples besides the ones used in this chapter?

7. What effect does predation have on interspecific competition among prey species?

8. Can you explain why predation alone may not account for the long-term oscillations in population growth of the Canadian lynx and snowshoe hare?

9. Define coevolution. How might two species coevolve to the extent that they enter a mutualistic relationship?

10. What is the difference between camouflage and mimicry? Can you give some examples of mimicry among insects? Camouflage among insects?

11. What is the difference between a parasite and a parasitoid? What sort of coevolutionary defenses do host species have against parasitoids?

12. Define primary and secondary succession. What is a climax community?

Readings

Begon, M., J. Harper, and C. Townsend. 1986. *Ecology: Individuals, Populations, and Communities*. Sunderland, Massachusetts: Sinauer Associates. Excellent, current introduction to topics covered in this chapter.

Price, P. 1975. *Insect Ecology*. New York: Wiley. Advanced reading, but excellent examples of ecological interactions.

Smith, R. 1980. *Ecology and Field Biology*. Third Edition. New York: Harper & Row. Particularly good descriptions of succession.

West, D., H. Shugart, and B. Botkin, eds. 1981. *Forest Succession: Concepts and Application*. New York: Springer-Verlag. An excellent overview of the modern approach to succession.

45

ECOSYSTEMS

The regions of the earth's surface are remarkably diverse. In climate, topography, and vegetation cover, deserts differ from mountains, which differ from foothills and arctic tundra, which differ from flat or rolling plains. Oceanic provinces, inland seas, lakes, ponds, and rivers differ in physical and chemical properties as well as in their arrays of organisms. Yet despite the diversity, each region functions as a *system*, in much the same way as the others.

With few exceptions, these systems run on energy from the sun. Photosynthetic autotrophs such as plants—the most common "self-feeders"—capture sunlight energy and convert it to forms they can use in the synthesis of organic compounds from simple inorganic substances.

Autotrophs represent the **energy fixation base** for the entire system. Energy stored in their self-assembled organic compounds is transferred through some array of heterotrophs before being dissipated in the surroundings. The heterotrophs, recall, include *consumers* (herbivores, carnivores, and parasites), *decomposers* (certain bacteria and fungi that break down the remains or products of organisms), and *detritivores* (varied invertebrates

that feed on particles of organic matter, such as would be produced by the partial decomposition of plant and animal tissues).

The autotrophs also represent the **nutrient concentration base** for the system. During growth, they take up water and carbon dioxide (as sources of oxygen, carbon, and hydrogen) along with dissolved minerals, including ionized forms of nitrogen and phosphorus (Table 20.1). With these nutrients they synthesize carbohydrates, lipids, proteins, and nucleic acids—which the heterotrophs use as concentrated food sources. When decomposers and detritivores get their turn at this organic matter, the nutrients are released to the environment. If they are not washed away or otherwise removed from the system, they can be taken up again by the autotrophs.

What we have just described in broad outline is the ecosystem. An **ecosystem** is not a "place." It is a community of organisms functioning together and interacting with their physical environment through (1) a flow of energy and (2) a cycling of materials—both of which have consequences for community structure *and* the environment.

Figure 45.1 Sedges (tufted grasses), mosses, and brightly flowered plants, a few primary producers of the arctic tundra, along with representative consumers—the ptarmigan and lemming (eaters of plant parts) and the snowy owl (eater of lemmings).

As in all ecosystems, these diverse organisms interact with one another and with their physical environment through (1) a flow of energy and (2) a cycling of materials, in ways that will be described in this chapter.

It is important to understand that ecosystems are *open* systems, hence are not self-sustaining. They require an *energy input* (as from the sun) and, often, a *nutrient input* (as from minerals carried by erosion into a lake). Because energy cannot be recycled, all ecosystems have *energy output*, most often as low-grade heat lost to the environment during each energy transfer between organisms. Although nutrients typically are recycled, the process is not 100 percent efficient and some loss occurs from the system (as through soil leaching), so there also is a *nutrient output*.

In this chapter, we will be discussing the inputs and outputs of different ecosystems as well as their internal structure. Keep in mind that the inputs and outputs vary considerably, depending in large part on four factors:

---

1. The size of the ecosystem (the larger it is, the more flexibility there is in accommodating fluctuations in inputs and outputs).

2. The total metabolic rate of its producer organisms (the more they put out, the more energy they must take in).

3. The ratio of producers to consumers (who eats whom, and how fast, influences the rate of energy flow and material passage through the ecosystem).

4. The age of the ecosystem (its requirements during mature stages differ from those of early stages).

STRUCTURE OF ECOSYSTEMS

Trophic Levels

Ecosystems have a layered structure, based on the number of times energy is transferred from one organism to another, away from the initial energy input into the system. Thus all organisms that are the same number of transfer steps away from the energy input are said to be at the same **trophic level** (from *troph*, meaning "nourishment").

For example, the photosynthetic protistans and aquatic plants growing in an estuary represent one trophic level; they are the first step in the transfer of energy from the sun. Herbivorous animals that feed directly on the plants are at another trophic level; primary carnivores (animal-eaters) eating the herbivores, are at still another level. Many organisms obtain energy from more than one source; they are more like a "trophic group" rather than being at a particular level. (For example, humans are "omnivores" that eat plants, herbivores, *and* carnivores.) The categories given in Table 45.1 are useful as a starting point in describing the feeding relationships in an ecosystem.

Table 45.1 Examples of Trophic Levels

Members of Each Level	Energy Source	Representative Organisms
Primary producers:		
photosynthetic autotrophs	sunlight energy	photosynthetic bacteria, plants
chemosynthetic autotrophs	oxidation of inorganic substances	sulfur bacteria
Primary consumers:		
herbivores	primary producers	grasshoppers, deer, periwinkles
Secondary consumers:		
primary carnivores	herbivores	spiders, small squids
Tertiary consumers:		
secondary carnivores	primary carnivores	Emperor penguins

Food Webs

The general sequence of who eats whom is sometimes called a **food chain**. However, the term implies a simple, isolated relationship, which seldom occurs in ecosystems. More typically, the same food resource is part of more than one chain, especially when that resource is at one of the low trophic levels. Thus there are interconnected networks of feeding relationships that take the form of **food webs**, of the sort shown in Figure 45.2.

For example, imagine a fisherman netting some fish that were feeding on algae near the ocean's surface. Come lunchtime, he cooks some of his catch. Should he later lose his footing on the deck and fall into the water, where other carnivores lurk, the "chain" might be portrayed as:

algae → fish → fisherman → shark

Yet this chain would be an oversimplification of feeding relationships, for it would exclude any number of alternatives. Most likely, crustaceans were also grazing on the algae. Small squids and assorted medium-sized fishes might have been feeding on the crustaceans; some larger fishes were probably feeding on smaller ones. Sharks may have been moving in to feed on the large and medium-sized fishes. The fisherman might have cooked his fish in wine and herbs. Thus he would have shifted back and forth between herbivore and carnivore—and would have been even more omnivorous in consuming the alcoholic product of decomposers (yeasts

whose fermentation activities yield wine from crushed grapes).

A food web is a network of many interlinked food chains, encompassing primary producers and an array of consumers and decomposers.

ENERGY FLOW THROUGH ECOSYSTEMS

Primary Productivity

For simplicity, let's now think in terms of ecosystems in which multicelled plants are the primary producers. The *rate* at which the plants fix energy in organic compounds is called the **primary productivity** of an ecosystem. The actual amount of energy stored in a given time interval depends on the balance between energy acquisition by photosynthesis and energy use by the producers (most often by aerobic respiration). It also depends on how much of the stored energy is not utilized by the consumers. Thus,

---

1. The *gross primary productivity* is the total rate of photosynthesis for the ecosystem.

2. The *net primary productivity* is the rate of energy storage in plant tissues in excess of the rate of respiration during the measured time interval.

3. The *net community productivity* is the rate of energy storage in excess of heterotrophic consumption.

In a given ecosystem, net primary productivity is influenced by such variables as the availability of sunlight and nutrients, temperature, rainfall, predation, duration of the growing season, and the occurrence of fires. These variables influence not only the amount but also the distribution of net primary production. For example, the proportion of increased growth in roots and shoots of land plants depends partly on the availability of water and nutrients. The harsher the environment in these respects, the greater the ratio of roots to shoots. The more favorable the environment, the more growth is put into photosynthetic parts (shoots), hence the greater the productivity.

Another variable affecting net primary productivity is the age of the ecosystem. For example, when a forest is young and trees are small, the annual primary productivity is high. The older and larger the trees get, the more the energy captured by photosynthesis is channeled into maintaining the existing plant parts; little is left over for growth.

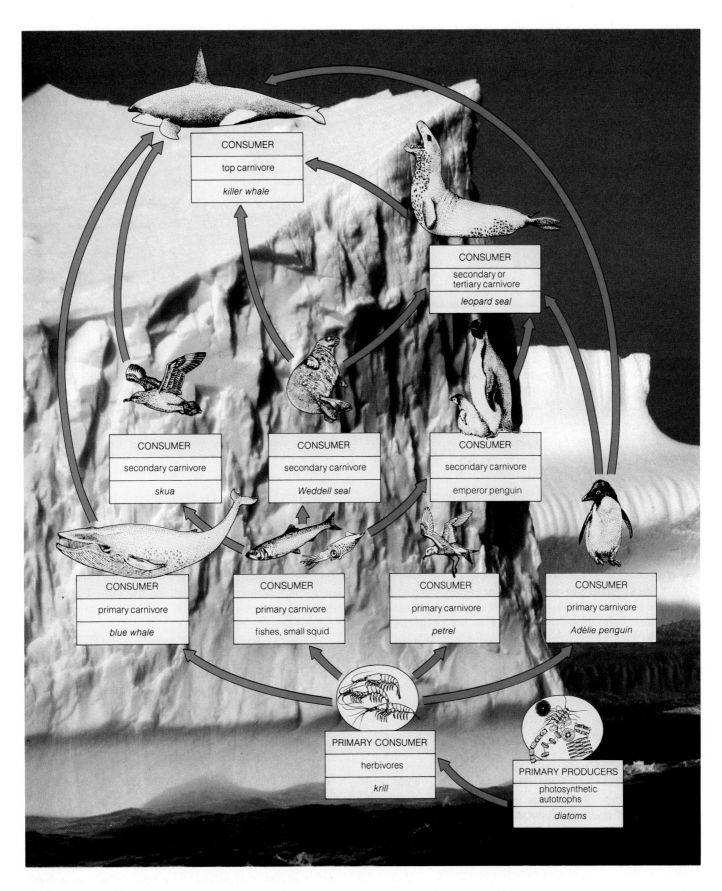

Figure 45.2 Simplified picture of a food web in the Antarctic; there are many more participants, including an array of decomposer organisms.

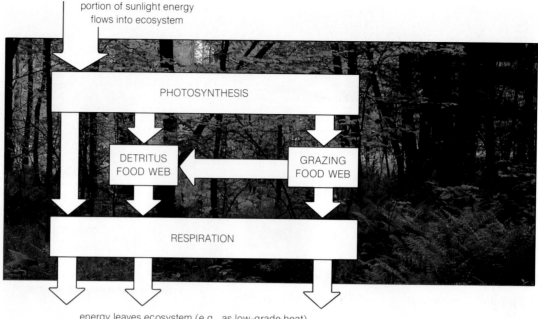

portion of sunlight energy
flows into ecosystem

PHOTOSYNTHESIS

DETRITUS
FOOD WEB

GRAZING
FOOD WEB

RESPIRATION

energy leaves ecosystem (e.g., as low-grade heat)

Figure 45.3 Generalized model of the main pathways of energy flow through an ecosystem. Photosynthesizers capture only a small fraction of solar energy reaching the earth. They use it in assembling organic matter, some of which is incorporated in plant tissues. In a grazing food web, herbivores feed on living plant parts, and carnivores feed on herbivores as well as on each other. In a detrital food web, decomposers and detritivores feed on organic wastes, dead tissues, and partially decayed organic matter. All organisms of the ecosystem release energy to the surroundings (typically as low-grade heat) as a result of energy metabolism. Thus there is a one-way flow of energy into then out of the ecosystem.

Major Pathways of Energy Flow

Figure 45.3 shows the general direction of energy flow through ecosystems. Of all the solar radiation reaching the earth, only about one or two percent becomes stored in organic matter. Then plant respiration reduces (often by half) the energy available for other members of the ecosystem. This tiny fraction becomes available in two forms: as living tissues of the producers and as **detritus** (organic wastes, dead tissues, and partially decayed organic matter). All organisms of the ecosystem lose energy to the surroundings (as low-grade heat, usually) during metabolic activity, hence there is a one-way flow of energy out of the ecosystem.

A **grazing food web** begins with consumption of living plant tissues by herbivores, which are in turn consumed by some array of carnivores. Figure 45.2 illustrates this type of food web. In **detrital food webs**, most of the net primary production is used by decomposers and detrivores (not by herbivores). These organisms break down organic waste products or remains to simple inorganic substances that can be cycled back to the plants.

The fraction of energy that flows through grazing or detrital food webs varies from one ecosystem to the next

(and often throughout the year in the same ecosystem). For example, when cattle feed heavily on plants in a pasture, half or more of the stored energy may flow through the grazing food web. In marshes, consumption of all but about ten percent of the stored energy is delayed until plant parts die and become available for detrital food webs.

All ecosystems have both grazing and detrital food webs, and both are interconnected. For example, cattle do not assimilate all of the energy stored in plants; undigested residues in feces become available for the decomposers and detritivores.

---

1. Energy flows into ecosystems from outside sources such as the sun.

2. Energy flows through ecosystems by way of grazing food webs (based on the consumption of living tissues of photosynthesizers) and detrital food webs (based on use of organic waste products and remains of photosynthesizers and consumers).

3. Energy leaves ecosystems as a result of metabolic activities (aerobic respiration, mostly) of each organism. The energy is lost primarily as low-grade heat.

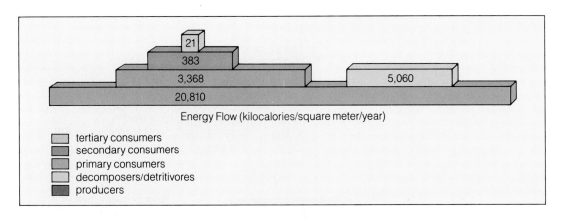

Figure 45.4 Pyramid of energy flow during one year at Silver Springs, Florida.

Pyramids Representing the Trophic Structure

The trophic structure of an ecosystem is influenced by the size of its component organisms, their different metabolic rates, and the loss of energy at each transfer. Often the trophic structure is represented as an ecological pyramid, with the producers forming a base for successive tiers of consumers above them.

Pyramid of Numbers. Sometimes the number of organisms in the ecosystem is depicted in pyramidal form. For example, we might have the following pyramid for a bluegrass field, where the base of the pyramid represents the food production base for higher trophic levels:

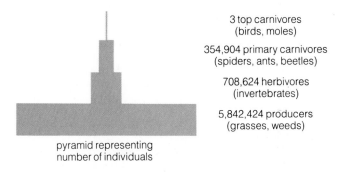

3 top carnivores (birds, moles)

354,904 primary carnivores (spiders, ants, beetles)

708,624 herbivores (invertebrates)

5,842,424 producers (grasses, weeds)

pyramid representing number of individuals

Aside from the monumental patience required to count all the organisms, a pyramid of numbers does not completely define the trophic structure for an ecosystem. For one thing, it defines feeding relationships only at one moment—*and feeding relationships can change cyclically or permanently over time.* For another thing, a pyramid of numbers does not take into account that *the sizes of organisms being counted in each trophic level can vary.* A count in a redwood forest would yield a small number of large producers (the trees) which still manage to support a large number of mostly small herbivores and carnivores (insects)—thus the pyramid would have a small base. One deer would be counted as a single herbivore, as

would a single insect—even though a deer eats far more than an insect ever would.

Pyramid of Biomass. Another approach is to weigh individuals in each trophic level instead of counting them. This would give us a pyramid of **biomass** (the total dry weight of all organisms at a given level). For most ecosystems on land, the pyramids of biomass have a large base of primary production, with ever smaller trophic levels perched on top. In contrast, in some aquatic ecosystems, the producers are tiny phytoplankton that grow and reproduce rapidly. Here, the pyramid of biomass can have a small base, with the consumer biomass at any instant actually exceeding the producer biomass. The phytoplankton are consumed about as fast as they reproduce; it is just that the survivors (few as they may be) are reproducing at a phenomenal rate.

Pyramid of Energy. When we wish to compare the functional roles of the trophic levels in an ecosystem, an energy pyramid is probably the most informative, for the pyramid shape is not skewed by overemphasis on variations in the size and metabolic rates of the individuals. An energy pyramid more accurately reflects the laws of thermodynamics (with energy losses being depicted at each transfer to another trophic level); hence the pyramid is always right-side up, with a large energy base at the bottom.

A pyramid of energy must be based on a determination of the actual amounts of energy that individuals take in, how much they burn up during metabolism, how much remains in their waste products, and how much they store in their body tissues. The energy inputs and outputs are calculated so that energy flow can be expressed per unit of land (or water) *per unit time.* The calculations are difficult to do, but they have been done in a few studies. Figure 45.4 shows an energy pyramid constructed as a result of one study; Figure 45.5 shows the calculations on which it is based.

Figure 45.5 Annual energy flow (measured in kilocalories per square meter per year) for an aquatic ecosystem in Silver Springs, Florida.

The producers are mostly green aquatic plants. The carnivores are insects and small fish, and the top carnivores are larger fish. The energy source (sunlight) is available all year long.

Only 1.2 percent of the incoming solar energy is actually trapped in photosynthesis to generate new plant biomass. And more than 63 percent of the photosynthetic products is metabolized by the plants themselves to meet their own energy needs. Only 16 percent is harvested by herbivores, and the remainder is eventually decomposed by bacteria and fungi. Similarly, most of the herbivore energy is expended in metabolism or goes into the decomposer system: only 11.4 percent is consumed by carnivores. Once again, the carnivores burn up most of the energy they take in and only 5.5 percent is passed on to top carnivores. The decomposers cycle and recycle all the biomass received from all other trophic levels. Eventually all of the 5,060 kilocalories will appear as heat produced during metabolism. (Decomposers, too, are eventually decomposed.)

This diagram has been deliberately oversimplified. No community is completely isolated from all others. Organisms and materials are constantly dropping into the springs. There is a slow but steady loss of other organisms and materials that flow outward in the stream that leaves the community.

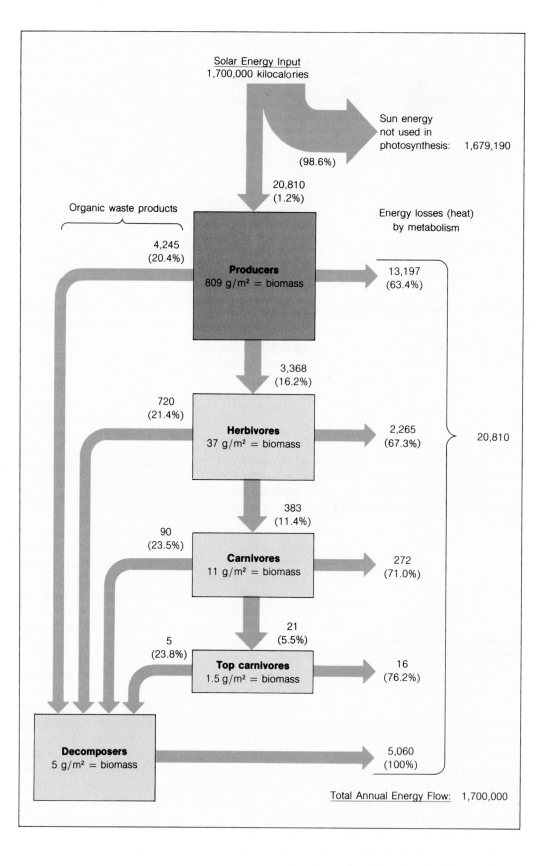

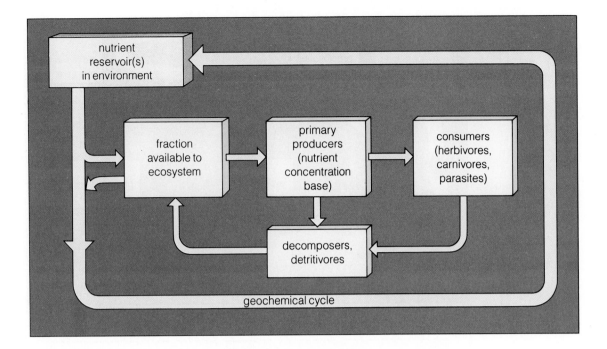

CYCLING OF WATER AND NUTRIENTS

All inorganic substances essential for life tend to move in cycles from the physical environment, to organisms, then back to the environment. These circular movements are called *"biogeochemical cycles."* In general, the physical environment serves as a large reservoir through which the substances move rather slowly, compared to their active, rapid exchange between organisms and the environment.

Figure 45.6 shows a generalized model of nutrient flow into, through, and out from an ecosystem. This model is based on four factors:

1. In general, the nutrients available to the primary producers are in the form of mineral ions (for example, as NH_4^+ and SO_4^{--}).

2. The nutrient reserves in an ecosystem are maintained largely by environmental inputs and by the recycling activities of decomposers and detritivores.

3. In most ecosystems, the amount of a nutrient being cycled within the system is greater than the amount entering or leaving in any given year.

4. Environmental inputs into the ecosystem occur through precipitation, metabolic fixation reactions (as in nitrogen fixation), and the weathering and breakdown of rocks. Ecosystem outputs include losses to the atmosphere and through runoff.

Figure 45.6 Generalized model of nutrient flow through an ecosystem. The overall movement of nutrients from the physical environment, through organisms, and back to the environment constitutes a biogeochemical cycle.

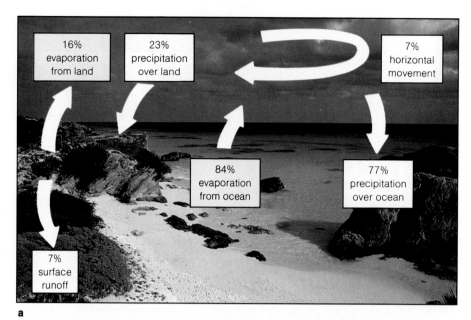

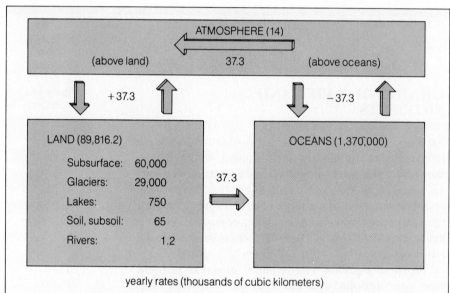

Figure 45.7 (a) Global water budget. The percentages indicate the annual mean distribution of water. (b) Simplified picture of the hydrologic cycle.

Let's now focus on a few examples to gain insight into how these cycles influence the distribution and abundance of organisms.

Hydrologic Cycle

Cold and warm ocean currents, clouds, winds, and rainfall are all part of a cycle, driven by solar energy, that spans the biosphere (page 712). The waters of the earth move slowly and on a vast scale through the atmosphere, on or through the uppermost layers of land masses, to the oceans, and back again. These global water movements represent the **hydrologic cycle**.

The global cycling of water, shown in Figure 45.7, is influenced by the following factors:

evaporation	*release of water as vapor into the atmosphere*
precipitation	*release of water from the atmosphere as rain or snow*
detention	*temporary storage of water on land or in oceans*
transportation	*movement of water on winds or as surface runoff*

Only a small fraction of the water is present in the atmosphere as vapor, clouds, and ice crystals. The amount is

areas of experimental watersheds

Hubbard Brook Valley

Figure 45.8 Hubbard Brook Valley in the White Mountains of New Hampshire. Some of the experimental watersheds are indicated.

greater over oceans than over land, for two reasons. More than ninety percent of all liquid surface water is located in oceans, and the surface of open oceans places less restriction on evaporation.

On the average, a water molecule does not stay aloft for more than about ten days. Thus the turnover rate is rapid for this part of the hydrologic cycle. Water released as precipitation is detained on land for no more than 10 to 120 days on the average, depending on the season and on where it falls. Some evaporates; some is transported by rivers and streams to the seas. There, with large-scale evaporation, the cycle begins again.

In itself, water is essential for life in any ecosystem, but it is also an important medium by which *nutrients* move into and out of ecosystems. In ongoing experiments that started in the 1950s, water movements have been followed through the Hubbard Brook Experimental Forest in the White Mountains of New Hampshire. Here, a stream runs down the middle of a valley characterized by steep slopes and shallow soils (Figure 45.8). Running down the slopes are several smaller streams, each with its own small watershed.

A **watershed** is an area defined by the direction of sloping hills, such that all rain falling on the area is funneled into a single stream. A watershed can be any size. For example, the watershed of the Mississippi River drains roughly one-third of the United States. The watersheds at Hubbard Brook are much smaller, averaging about fifteen hectares (thirty-six acres).

Water enters a watershed mainly as rain or snow, some of which is intercepted by leaves and returned by evaporative water loss to the atmosphere. About eighty to ninety percent of the precipitation passes through the canopy to the forest floor, where it can filter into the soil or run off along the surface (Figure 45.9). In most natural ecosystems, soil infiltration rates are high enough that nearly all water enters the soil. From this shallow soil layer, water can be picked up by plants and moved to the canopy for evaporation back to the atmosphere (transpiration). Water also can filter down through the soil (seepage and percolation) to the water table (groundwater), or it can move laterally to a stream. Water budgets can be constructed for the watershed "ecosystems" by measuring precipitation inputs and stream water and groundwater losses, as well as by estimating evaporation and transpiration outputs.

Long-term measurements have shown that evapotranspiration varies very little from year to year, despite large changes in the amount of precipitation (Figure 45.10). This result is important in studies of the regional water supply; it indicates how much water can be expected for use in cities that are downstream of watersheds of this type.

Further discoveries were made about nutrient inputs and outputs at the watersheds. For example, you might think that the waters draining the watershed would leach away mineral ions, such as calcium ions. However, in young, undisturbed forests at the Hubbard Brook

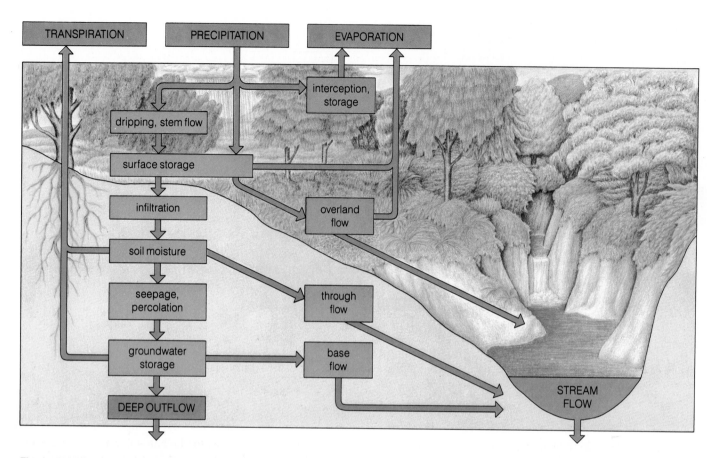

Figure 45.9 Water movements in a watershed.

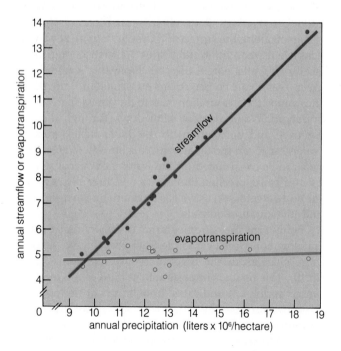

Figure 45.10 Relationship among precipitation, streamflow, and evapotranspiration for the Hubbard Brook Experimental Forest during 1956–1974.

watershed, only about eight *kilograms* of calcium were lost annually from each hectare. Tree roots were "mining" the soil efficiently, moving calcium upward into growing plant parts and making it available for the food webs. Calcium replacements were brought into the watershed by rain and by weathering of the underlying rocks.

In one study, all the trees and shrubs were cut one year at one Hubbard Brook watershed, and herbicides were applied through the next three growing seasons to prevent regrowth. The soil itself was not disturbed; no organic material was removed from the site. Yet the loss of calcium in the stream outflow increased by *six times*, compared to similar but undisturbed watersheds.

Carbon Cycle

Molecules such as carbon dioxide, oxygen, and nitrogen undergo movements that include gaseous forms. They move from reservoirs in the atmosphere and in the oceans, through organisms, then back to the reservoirs (Figure 45.11).

On a global scale, photosynthesizers harness and use a tremendous amount of carbon dioxide. Each year bil-

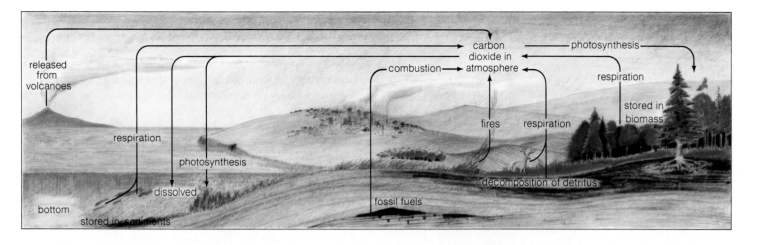

Figure 45.11 Simplified picture of the carbon cycle. Carbon dioxide in the atmosphere is harnessed during the second stage of photosynthesis. Compounds containing carbon and oxygen are assembled into tissues (living biomass). Oxygen is released to the atmosphere during photosynthesis; carbon dioxide is released during aerobic respiration. Conversions to carbon-containing inorganic compounds occur during decomposition. Carbon can also become locked up in peat, coal, oil, and gas, then subsequently released during the combustion of fossil fuels. Some carbon becomes locked up in the bottom sediments of oceans. New carbon enters the atmosphere during volcanic eruptions.

lions of metric tons of carbon become incorporated into organic compounds. Both the photosynthesizers and the consumers that directly or indirectly feed on them release carbon dioxide during respiration. Overall, carbon dioxide fixation during photosynthesis and its release during respiration tend to be more or less in balance, but not exactly so.

The turnover rate for carbon varies, depending on environmental conditions. In tropical forests, decomposition and carbon uptake are rapid, so not much carbon is found in litter on the soil surface. In bogs and marshes, organic compounds are not broken down completely, and much carbon accumulates in forms such as peat. In aquatic food webs, carbon can become bound as carbonate in shells and other hard parts. When the shelled organisms die, they sink and become buried in bottom sediments of different depths. In deep oceans, carbon can remain buried for millions of years, until geologic movements bring sediments once more to the surface. Still more carbon is slowly converted to long-standing reserves of gas, petroleum, and coal deep in the earth. These reserves are used as fossil fuels.

With the burning of fossil fuels, carbon dioxide is rapidly reintroduced into the atmosphere. Through volcanic eruptions, carbon also is released from material present in the earth's crust. About half the total carbon released in the atmosphere is destined for storage in two global sinks: the oceans and accumulated plant biomass. However, there is speculation that increased burning of fossil fuels and increases in volcanic eruptions can intro-

duce more carbon into the atmosphere than can be removed from it by plants and the oceans. Theoretically, increases in atmospheric carbon dioxide could amplify the worldwide "greenhouse effect." (This effect is analogous to heat retention in a greenhouse, which allows penetration of sunlight rays but holds in heat being radiated from objects inside. Thus water vapor and carbon dioxide in the atmosphere absorb heat being radiated away from the earth's surface and reradiate it back toward the earth.) What will happen if the global atmosphere warms even by only a few degrees? There is speculation that global warming could lead to a rise in sea level (because of glacial melting), submergence of coasts, and expansion of the world's great desert regions, to name a few possible consequences.

In the late 1950s a laboratory was established on a mountaintop in the Hawaiian Islands to measure concentrations of different gases in the atmosphere. This remote site was selected because it was free of local contamination and would represent average conditions for the Northern Hemisphere.

Continuous monitoring for many years established two patterns in the concentration of carbon dioxide which have persisted to this day. First, an annual cycle of carbon dioxide concentration follows the stepped-up primary productivity during the summer months and the decline in productivity during the winter months in the Northern Hemisphere (Figure 45.12). The carbon dioxide concentration is lower during summer, when photosynthesis by green plants is most pronounced. But

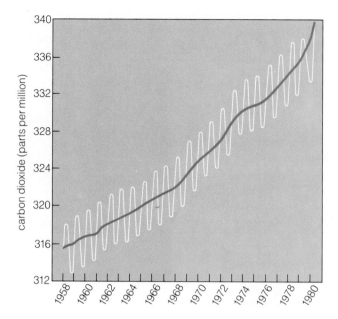

Figure 45.12 Rising atmospheric concentration of carbon dioxide as recorded at Mauna Loa Observatory in Hawaii.

it is higher during winter, when aerobic respiration continues even while photosynthesis levels decline. *For the first time, the integrated effects of the carbon balances of land and water ecosystems of a whole hemisphere were revealed.*

Disturbingly, the peaks and troughs in the carbon cycle showed a continuous increase, providing evidence in support of the idea that the buildup of carbon dioxide may intensify the greenhouse effect over the next century.

The sources and sinks for carbon dioxide in the biosphere have been measured and debated for over a decade. Increasing carbon dioxide levels have been linked both to the burning of fossil fuels and to the clearing of tropical rain forests. These forests, which have large amounts of carbon stored in their tree biomass, are being cleared at a rapid rate for timber harvesting and for the conversion of land to agriculture (see Figure 47.7). Carbon dioxide levels are expected to increase into the middle of the twenty-first century, and global warming by several degrees is expected to occur.

Nitrogen Cycle

Figure 45.13 is a simplified picture of the cycling of nitrogen through organisms and the environment. The atmosphere is the largest reservoir in the cycle; about eighty percent of it is composed of gaseous nitrogen (N_2). However, N_2 molecules are held together by stable, triple covalent bonds ($N\equiv N$), and few organisms have the metabolic equipment for breaking them. A few species of nitrogen-fixing bacteria, volcanic action, and lightning can put nitrogen into forms such as ammonia and nitrates, which can be metabolized by living organisms. These are the only inputs of available nitrogen for ecosystems. Nitrogen is lost to the atmosphere by way of denitrifying bacteria; more nitrogen is lost by leaching to streams, rivers, and oceans.

The Cycling Processes. As Figure 45.13 shows, the movement of nitrogen from the atmospheric reservoir, through organisms, then back to the atmosphere involves six processes: nitrogen fixation, assimilation and biosynthesis, decomposition, ammonification, nitrification, and denitrification.

In **nitrogen fixation**, certain soil bacteria take up nitrogen from the air. They attach electrons (and associated H ions) to the nitrogen through a series of reactions, thereby forming ammonia (NH_3) or ammonium (NH_4^+). They use these compounds in growth, maintenance, and reproduction. When the nitrogen-fixing microbes die, the compounds are released during decay processes. Other bacteria present in soil use the compounds as energy sources.

When ammonia and nitrate dissolve in soil water, they can be assimilated by plant roots and incorporated into organic compounds. These organic compounds are the only nitrogen source for animals, which feed directly or indirectly on plants. Later, in a process called **ammonification**, the nitrogen-containing wastes and remains of plants and animals are decomposed by some species of bacteria and fungi. The decomposers use some nitrogen for their own growth, and they release the excess as ammonia or ammonium. Some of these nitrogen-containing by-products are then reused by plants.

In **nitrification**, soil bacteria strip the ammonia or ammonium of electrons, and nitrite (NO_2^-) is released as a product of the reaction. Still other nitrifying bacteria then use nitrite for energy metabolism, which yields nitrate (NO_3^-) as a product.

Nitrogen is often the limiting nutrient for plant growth. Some plants, however, benefit from increased rates of nitrogen uptake because of a mutually beneficial interaction with free-living or symbiotic microorganisms. Legumes (such as peas and beans) are an example. They harbor nitrogen fixers in nodules in their roots, supplying their guests with energy-rich sugar molecules in exchange for fixed nitrogen (Figure 20.2).

In itself, the continual production of ammonia by innumerable bacterial populations would seem to assure plants of plenty of nitrogen. Yet soil nitrogen is scarce. During crop harvests, of course, some nitrogen leaves the fields with the plants. Because nitrite, nitrate, and ammonia are soluble, some may run off in streams and rainwater. However, the major cause of nitrogen scarcity in wet soils is a bacterial process called **denitrification**: the conversion of nitrate or nitrite to N_2 and nitrous oxide (N_2O). The overwhelming majority of bacteria that take

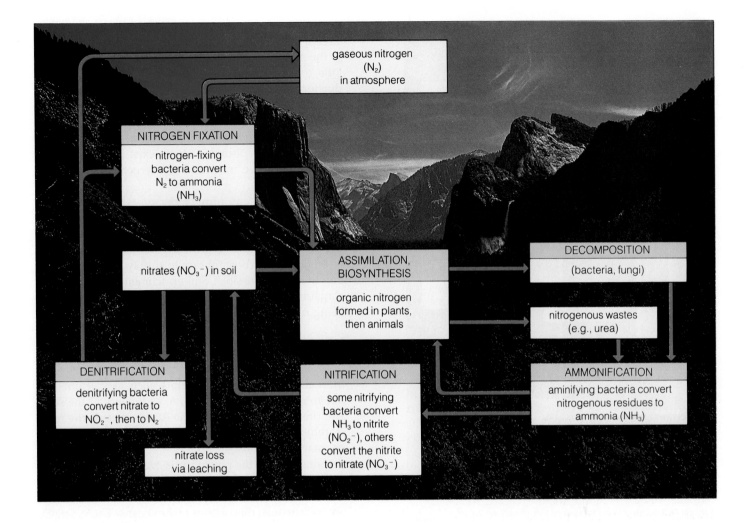

Figure 45.13 Simplified picture of the nitrogen cycle.

part in this process are ordinary species that rely on aerobic respiration. Under some conditions, though, especially when soil is poorly aerated, they switch to anaerobic electron transport. They use nitrate, nitrite, or nitrous oxide instead of oxygen as the terminal electron acceptor.

Nitrogen Scarcity and Modern Agriculture. With any loss of fixed nitrogen, soil fertility (hence plant growth) is reduced. Farms in Europe and in North America have depended on crop rotation to restore the soil. For example, legumes are planted between plantings of wheat or sugar beet crops. This practice has helped maintain soils in stable and productive condition, in some cases for thousands of years.

Modern agriculture, however, depends on nitrogen-rich fertilizers. With plant breeding, fertilization, and pest control, the crop yields per acre have doubled and even quadrupled over the past forty years. With intelligent management, it appears that soil can maintain such high yields indefinitely—as long as water and commerical nitrogen-containing fertilizers are available.

The catch, of course, is that we can't get something for nothing. Enormous amounts of energy are needed

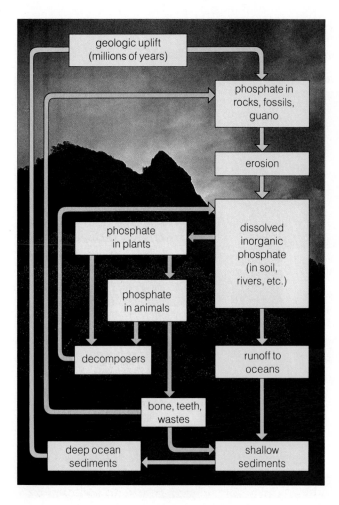

Figure 45.14 Simplified picture of the phosphorus cycle.

to produce fertilizer—not energy from the unending stream of sunlight, but energy from petroleum. As long as the supply of oil was viewed as unending, there was little concern about the energetic cost of fertilizer production. In many cases, we have been pouring more energy into the soil (in the form of fertilizer) than we are getting out of it (in the form of food). Unlike natural ecosystems, in which nutrients such as nitrogen are cycled, our agricultural systems exist only because of constant, massive infusions of energetically expensive fertilizers.

However, as any hungry person will tell you, food calories are more basic to survival than are gasoline calories or perhaps, even, than a car. As long as the human population continues to grow exponentially, farmers will be engaged in a constant race to supply food to as many individuals as possible. Soil enrichment with nitrogen-containing fertilizers is essential in the race, as it is now being run.

Sedimentary Cycles

In sedimentary cycles, minerals such as calcium and phosphorus move from land, to sediments in the seas, and then back to the land. The earth's crust is the main storehouse for mineral nutrients that flow through sedimentary cycles.

For example, the phosphorus cycle is based on phosphate rock formations in the earth. Through weathering and other erosive forces, phosphorus is washed into rivers and streams, then moves to the oceans. There, largely on continental shelves, phosphorus builds up as insoluble deposits. Millions of years pass, and geological events lead to the uplifting and exposure of the seafloor to form land surfaces. Then weathering releases phosphorus from the rocks, whereupon it becomes available for plants. Large phosphate deposits are mined in Florida. (In the United States alone, 2 billion kilograms of phosphorus are mined annually for use in fertilizers.)

There is a biological component to the phosphorus cycle (Figure 45.14). All living organisms require phosphorus, which becomes incorporated in nucleic acids. Phosphorus is released from living tissues in detrital food webs. It may not remain for long in soil and water; photosynthetic organisms can take it up again and thereby recycle it through the ecosystem.

Transfer of Harmful Compounds Through Ecosystems

Through our diverse and varied activities, we humans can have major and minor effects on the functioning of ecosystems. We will consider this idea in detail in Chapter Forty-Seven, but for now a simple example will underscore the point.

In 1955, there was a major campaign to eliminate malaria-transmitting mosquitoes from the island of Borneo (now a state of Indonesia). DDT is a chlorinated hydrocarbon compound that sends insects into convulsions, paralysis, and on to death. It has been instrumental in bringing many pests (including mosquitoes) more or less under control.

DDT is a relatively stable compound; it is nondegradable and insoluble in water. However, it is soluble in fat. It tends to accumulate in the fatty tissues of organisms. For this reason, DDT is a prime candidate for **biological magnification**: the increasing concentration of a nondegradable substance as it moves up through trophic levels. The DDT that becomes concentrated in tissues of herbivores (such as insects) becomes even more concentrated in tissues of carnivores that eat quantities of the DDT-harboring herbivores. The concentration proceeds at each trophic level.

In 1955, the World Health Organization initiated its DDT-spraying program in Borneo. That step was not taken lightly. Nine out of ten people there were afflicted with malaria, an epidemic by anybody's standards. The program worked, insofar as the mosquitoes transmitting this terrible disease were brought almost entirely under control. But DDT is a broad-spectrum insecticide; it kills nontarget as well as target species. Sure enough, the mosquitoes had company: flies and cockroaches that made a nuisance of themselves in the thatch-roofed houses on the island fell dead to the floor. At first there was much applause. Then the small lizards that also lived in the houses and preyed on flies and cockroaches found themselves presented with a veritable feast. Feast they did—and they died, too. So did the house cats that preyed on the lizards. With the house cats dead, the rat population of Borneo was rid of its natural predator, and rats were soon overrunning the island. The fleas on rats were carriers of still another disease, called the sylvatic plague, which can be transmitted to humans. Fortunately, the threat of this new epidemic was averted in time; someone got the inspired idea to parachute DDT-free cats into the remote parts of the island. But on top of everything else, some of the people of Borneo found themselves sitting under caved-in roofs. The thatch in their roofs was made of certain leaves that happen to be the food resource of a certain caterpillar. The DDT did not affect the caterpillar, but it killed the wasps that were its natural predator. When the predator population collapsed, so did the roofs.

This brief example illustrates that disturbances to one part of an ecosystem can have unexpected effects on other, seemingly unrelated parts. A recent approach to predicting such unforeseen effects is through *ecosystem analysis*. Borrowing from the analysis of complex mechanical systems, ecosystem analysis is a method of extracting crucial bits of information about the different components of a system and combining the information through computer programs and models to predict the outcome of the next disturbance. For example, an analysis of which species feed on which other species (as in the food web in Figure 45.2) can be turned into a series of equations describing how much of each species is consumed. The equations can then be used to predict what the effect would be, say, of overharvesting whales. As we attempt to deal with larger and more complex systems, it becomes more difficult and expensive to run the experiments we would like in the natural environment. The temptation is to run them instead on the computer. This is a valid exercise in that the computer memory can contain all the data that we have gathered about a system. The danger is that the most important fact may be one that we do not yet know.

SUMMARY

1. An ecosystem is a community of organisms functioning together and interacting with their physical environment through (1) a flow of energy and (2) a cycling of materials.

2. Photosynthetic autotrophs are the energy fixation base and the nutrient concentration base for almost all ecosystems.

3. Ecosystems are *open* systems, with energy and nutrient inputs and outputs. The following variables affect the degree of input and output: the ecosystem size and age, the overall rate of metabolism of its organisms, and the ratio of producer to consumer organisms.

4. Ecosystems are generally more open for inputs and outputs of water, carbon, and energy, whereas nutrients such as nitrogen are mostly recycled within the ecosystem.

5. Energy fixed by photosynthesis passes through grazing food webs and detrital food webs before being lost (as low-grade heat) through respiration.

6. Biogeochemical cycles include the movement of water, nutrients, and other elements and compounds from the physical environment, to organisms, then back to the environment.

7. Undisturbed ecosystems have predictable rates of water and nutrient losses. The rates are altered when the ecosystem is disturbed, as by fires or clear-cutting.

8. The integrated effect of fossil fuel burning and conversion of natural ecosystems to cropland or grazing land can be measured in terms of an increased carbon dioxide concentration in the atmosphere.

9. Nitrogen availability is often a limiting factor for the total net primary productivity of ecosystems. Although gaseous nitrogen (N_2) is abundant in the atmosphere, it must first be converted to forms such as ammonia and

nitrates before it can be used by the primary producers (mostly plants). A few species of bacteria, volcanic action, and lightning can cause the conversion.

Review Questions

1. Define ecosystem. Autotrophs represent the energy fixation base and nutrient concentration base for ecosystems. Can you explain what these terms mean?

2. Define trophic level. Can you name and give examples of some trophic levels in ecosystems? What is the energy source for each level?

3. Distinguish between a food chain and a food web. Can you imagine an extreme situation whereby you would be a participant in a food chain?

4. Explain the difference between gross primary productivity and net primary productivity. If you were growing a vegetable garden, what variables might affect its net primary production—that is, the amount of energy stored in the organic compounds of plant tissues?

5. There are two major pathways of energy flow through ecosystems: grazing food webs and detrital food webs. Can you characterize each? How does energy leave each one?

6. "Energy budget" refers to the energy entering and leaving an ecosystem in a given time span. Ecologists measure this energy flow as a way of determining how well the ecosystem is functioning in terms of production and energy utilization efficiency. What are some of the ways they do this?

7. What is a biogeochemical cycle? Can you give an example of one and describe the reservoirs and organisms involved?

8. Define these terms: nitrogen fixation, nitrification, ammonification, and denitrification.

Readings

Bormann, F., and G. Likens. 1984. *Pattern and Process in a Forested Ecosystem*. New York: Springer-Verlag. Details of how a forest ecosystem functions from nutrient cycling to succession.

Gibbons, B. September 1984. "Do We Treat Our Soil Like Dirt?" *National Geographic* 166(3):350–388.

Gosz, J., et al. March 1978. "The Flow of Energy in a Forest Ecosystem." *Scientific American* 238(3):93–102. Classic study of the energetics of a forest ecosystem, part of the Hubbard Brook Experimental Forest in the White Mountains of New Hampshire. Quantitative analyses revealed how energy is partitioned among the components of the ecosystem.

Odum, E. 1983. *Basic Ecology*. Philadelphia: Saunders.

Payne, W. 1983. "Bacterial Denitrification: Asset or Defect." *Bioscience* 33(5):319–325. Good summary of the global implications of denitrification.

Pimm, S. 1982. *Food Webs*. London: Chapman-Hall. A new look at food webs and their role in community structure. Somewhat advanced but interesting.

Price, P. 1984. *Insect Ecology*. Second edition. New York: Wiley. Advanced reading but excellent examples of ecological interactions.

Smith, R. 1980. *Ecology and Field Biology*. Third edition. New York: Harper & Row.

At the northern edge of the Kalahari Desert, in the nation of Botswana, the longest river in southern Africa ends in an immense freshwater delta. Here, lush grasses, shrubs, and groves are the primary production base for a magnificent array of wildlife—birds, crocodiles, lions, buffalo, elephants, elands, kudus, wildebeeste, and many more. These large animals are hosts to varieties of trypanosomes—parasites that do not harm the wild animals but can cause the dreaded sleeping sickness in humans and cattle. The parasites spread from one animal to another by way of an insect, the tsetse fly, which thrives in the delta.

Now, it happens that cattle raising is Botswana's economic mainstay. Cattle need water and grazing land, which are becoming scarce. Pressure is on to open up the delta for the cattle industry by eradicating the tsetse fly. That means clearing the woodlands, burning off the brush, draining the swamps—and destroying the wildlife.

Why would people do such a thing? For profit, of course, but also to feed other people. Anyway, how can we demand that they *not* do such a thing? Would Californians so readily turn over the fields and vineyards of their fertile inland valleys to quail and rabbits, coyotes and hawks? Would Texans so readily set aside their open range for the preservation of prairie dogs and rattlesnakes? Would Nebraskans so readily donate their fields of waving grain to support human populations in Africa so that the African wildlife can be left alone?

Questions such as these illustrate that the concerns of any particular group of people often depend on where

46

THE BIOSPHERE

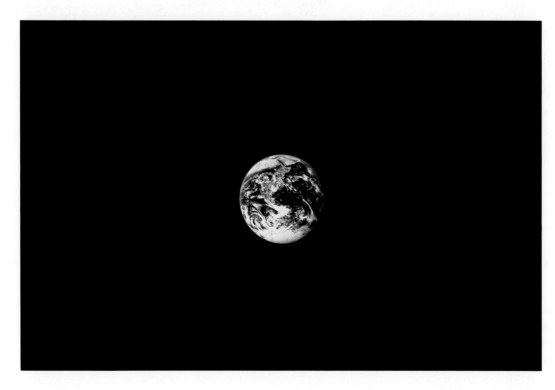

We are no more than sunlight dancing on the stream, and no less.

they happen to live. They also illustrate that what might be a simple answer to one group is no answer at all to another. Through their research, many ecologists have been extending public concerns beyond the immediate horizons, and today there is growing awareness of problems and interdependencies on a global scale. Through their explorations, astronauts have reinforced this awareness. Photographs from space show the earth to be isolated in darkness, beckoning, yet oddly vulnerable. Immense cloud systems appear as thin white traceries across oceans and continents—reduced, from our perspective in space, to mere patches of blues and tans.

Within this thin wrapping of air, water, and surface land are organisms linked by the flow of energy on the stream of time. This is our home, the biosphere; it is all we have.

COMPONENTS OF THE BIOSPHERE

The biosphere is a narrow zone that harbors life; it includes the waters of the earth, a fraction of its crust, and the lower region of the surrounding air. The rocks, soils, and sediments in which organisms occur comprise the surface of the *lithosphere*, the outer portion of a crust divided into rigid plates and extending to depths of 60 to 100 kilometers. Liquid and frozen water on or near the surface of the lithosphere is known as the *hydrosphere*. It includes the oceans and smaller bodies of water, the polar ice caps, and a small amount of airborne water. Enveloping the earth is the *atmosphere*, a region of gases, particulate matter, and water vapor. Most of the molecules making up the atmosphere are distributed within 50 kilometers above the earth's surface.

The biosphere is composed of ecosystems, ranging in size from vast forests and stretches of tundra to small systems that occur, for example, in a pond. All those ecosystems are influenced by the flow of energy and the movements of materials on a global scale. Thus, for example, the distribution of different ecosystems is dictated in large part by the energetics of the earth's climatic system. Their distribution also is influenced profoundly by global fluxes of energy, chemicals, and water.

GLOBAL PATTERNS OF CLIMATE

What determines whether a particular region will be a desert, forest, cold sea, or coral reef? The main determinant is **climate**. The word refers to prevailing weather conditions, including temperature, humidity, wind velocity and direction, degree of cloud cover, and rain-

fall. Climate itself is an outcome of many interacting factors. Here we will consider four of the most important:

1. Variations in the amount of incoming solar radiation

2. The earth's daily rotation and its annual path around the sun

3. Distribution of continents and oceans

4. Elevation of land masses

Through their interactions, these factors produce the prevailing winds and ocean currents in different regions. In turn, the currents influence global weather patterns. Finally, weather affects the composition and character of soils and sediments (through erosion, for example), and these substrates influence the growth and distribution of primary producers (mostly photosynthetic plants), hence the nature of the ecosystem itself.

Mediating Effects of the Atmosphere

The atmosphere has profound effects on the amount and kind of solar radiation reaching the earth's surface. At the outermost layers of the atmosphere, oxygen and ozone (O_3) absorb nearly all of the ultraviolet wavelengths, which are potentially lethal to most forms of life (Figure 46.2). Of the wavelengths that do penetrate the atmosphere, about thirty-two percent is reflected back into space by clouds and particulate matter. Clouds, dust, and water vapor absorb another eighteen percent. Thus, half the solar radiation penetrating the atmosphere is lost through reflection and absorption even before it reaches the earth's surface.

Light of visible wavelengths can penetrate the atmosphere easily enough. But after those wavelengths are absorbed, they are radiated from the earth's surface as infrared wavelengths (heat). Some of these longer wavelengths are absorbed primarily by water vapor and carbon dioxide in the upper atmosphere and reradiated toward the earth—hence, heat is temporarily retained. The effect is analogous to heat retention in a greenhouse, which allows penetration of sunlight rays but holds in the heat radiated from plants and the soil on the inside. *Thus heat indirectly derived from the sun warms the atmosphere through a "greenhouse effect."*

Air Currents

In global terms, the heating effect of solar radiation is not uniform. Light rays strike more directly on the equator than on polar regions (Figure 46.3). Thus solar radiation heats the air more at the equator than at the poles. The warm equatorial air rises and spreads, northward and southward. In the arctic regions the air cools, sinks downward, then flows back toward the equator.

Two factors modify this immense pattern of air circulation. First, heat retention is not the same for land as it is for oceans (the land heats and cools more rapidly), and land masses are not distributed uniformly. These variations disrupt the basic pattern of air movements. Second, a force associated with the earth's rotation deflects air flow in the Northern Hemisphere to the right and, in the Southern Hemisphere, to the left. The outcome is worldwide belts of prevailing east and west air currents (Figure 46.3b).

The prevailing air currents help dictate the distribution of different types of ecosystems. At the equator, for example, hot air gives up moisture as it rises to cooler altitudes, and the rainfall supports luxuriant plant growth. The drier air descends at about 30° latitudes; here, the earth's great deserts occur. The air again becomes warm, picks up moisture, and ascends at 60° latitudes, then travels poleward. Almost no rainfall accompanies its descent in polar regions.

Ocean Currents

The earth's rotation, prevailing air currents, and variations in water temperature give rise to ocean currents and surface drifts that tend to move parallel with the equator. However, land masses intervene and alter the

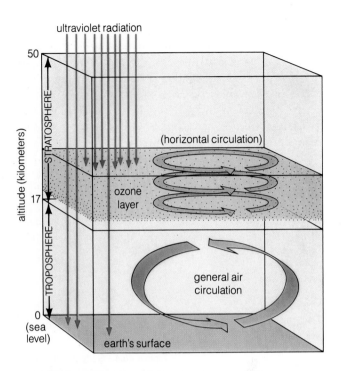

Figure 46.2 Earth's atmosphere. Most of the global air circulation takes place in the troposphere. Circulation patterns are mostly horizontal relative to the earth's surface, and once particulate matter reaches the stratosphere it tends to stay there for a long time. The ozone (O_3) in the stratosphere absorbs most of the ultraviolet radiation from the sun.

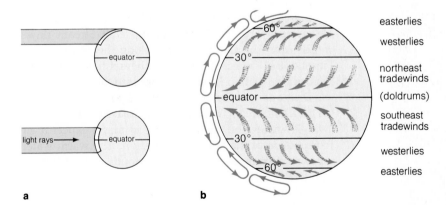

Figure 46.3 Formation of worldwide prevailing air currents. (**a**) During certain times of year, sunlight rays are almost perpendicular (hence more concentrated) at the equator than at the poles, where their angle of incidence spreads them out over a larger surface area. The unequal heating of polar and equatorial air causes air to circulate from the equator, toward the poles, then back to the equator.

(**b**) Because of a force created by the earth's rotation on its axis, this immense pattern of air circulation becomes divided into belts of prevailing east and west winds. Unequal heat retention by oceans and land masses contributes to the formation of these belts, which dictate global patterns of rainfall. In turn, the amount of rainfall helps dictate the distribution of major types of ecosystems throughout the world.

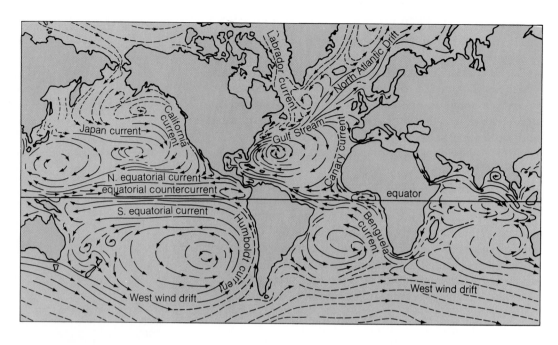

Figure 46.4 Surface drifts and ocean currents in January. The solid arrows indicate warm water movements; dashed arrows indicate cold water movements.

movement of water in the world's great oceans—the Atlantic, Pacific, and Indian oceans. Dominating each ocean are two circular water movements, called **gyres**. In the Northern Hemisphere, currents in the gyre move clockwise; in the Southern Hemisphere, they move counterclockwise. An equatorial countercurrent separates the two gyres and carries water away from the western boundaries of the ocean basins (Figure 46.4).

Gyres move water warmed at the equator northward and southward. For example, the Gulf Stream flows north from the Caribbean and along the southeastern coast of the United States. From there it flows northeast across the Atlantic. Then it divides into the North Atlantic Drift and the Canary Current (Figure 46.4). Because of the Gulf Stream, the climate of Great Britain and Norway is wetter and milder than you might expect if air currents were the only consideration. The warm water spawns moisture-laden fog and clouds, which have a moderating effect on temperature and which lead to abundant rainfall.

Figure 46.5 Monarch butterflies, migrants that gather in trees of California coastal regions and central Mexico. Monarchs typically travel hundreds of kilometers south to these regions, which are cool and humid in winter; if they stayed in their breeding grounds, they would risk being killed by frost.

Gyres also carry cold water from the poles toward the equator, along the western coasts of Africa, North America, and South America. These cold currents have a mediating effect on climate, particularly in coastal regions, which are often shrouded in fog. The mild, wet climate of the Pacific Northwest is largely a result of the California current. As cool air masses from above these waters move inland, they moderate the temperature over even the hottest deserts. In such ways, *ocean currents influence the physical conditions characteristic of diverse ecosystems.*

Seasonal Variations in Climate

Through the year, climates change in both the Northern and Southern hemispheres. Variations in the amount of incoming solar radiation contribute to these changes.

Many biological rhythms correspond with the annual changes in light, temperature, and rainfall. In essence, biological rhythms permit the organisms to anticipate and adjust to seasonal change. In *temperate* regions (which have moderate climates compared, say, to arctic regions), organisms respond more to seasonal changes in light and temperature than to other factors. In tropical regions and deserts, they respond more to changes in amounts of rainfall. Thus there are cycles of leafing out, flowering, fruiting, and leaf drop among many plants. There are cycles of breeding and migration among animals, including caribou, bison, and butterflies (Figure 46.5). In the seas, animals including turtles, seals, and whales also migrate. Such movements correspond with seasonal bursts of primary productivity (page 696) on land and in the seas.

What causes seasonal variations in climate? The circle of illumination that divides the earth into days and nights is always perpendicular to the sun. However, the earth's axis is tilted in relation to its annual path around the sun. In December, the North Pole is most tilted away from the sun. December means winter in the Northern Hemisphere. Temperatures are lower than at other times of year because days are shorter and incoming light rays are more spread out (Figure 46.6). In the Southern Hem-

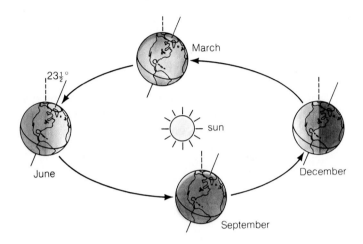

Figure 46.6 Annual variation in the amount of incoming solar radiation. Notice that the northern end of the earth's axis tilts toward the sun in June and away from it in December. Notice also the annual variation in the position of the equator relative to the boundary of illumination between day and night. Such variations in the intensity and duration of daylight lead to seasonal variations in temperature in the two hemispheres and at the poles.

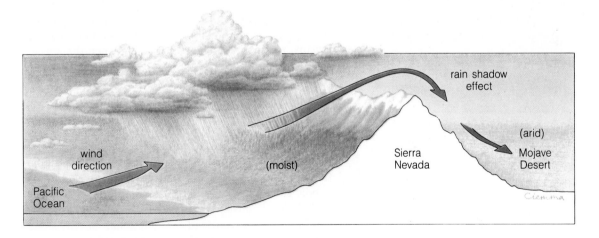

Figure 46.7 Rain shadow effect. As air containing water vapor rises, it cools. Cool air holds less moisture than warm air; hence clouds form on the western side of the Sierra Nevada, and rainfall tends to be heavy on the western flanks of this mountain range. As the air descends on the eastern side it becomes more compact and warm. Warm air holds more moisture than cooler air; hence it picks up moisture from its surroundings rather than releasing it. This rain shadow effect produces desert conditions on the eastern side of the Sierra Nevada.

isphere, Decembers are warmer. Temperatures are higher because days are longer and light rays are not as spread out. The situation is reversed for the two hemispheres in June. Notice, in Figure 46.6, that winter in the north polar regions is a time of near-perpetual darkness, and that daylight is nearly continuous in summer.

Regional Climates

Climate is influenced by more than general patterns of light, temperature, and rainfall. It is influenced also by the presence or absence of bodies of water, mountains, valleys, and vegetation cover.

For example, mountain ranges parallel the west coasts of North and South America. Here, climatic variations are reflected in successive belts of vegetation that change with elevation. At the eastern base of the Sierra Nevada, the belt includes grasslands and shrubs adapted to dry, warm conditions. Above this is the montane belt, with deciduous and coniferous trees that depend on greater moisture and cooler temperatures. Above the montane forests is the subalpine belt, with conifers adapted to the rigors of a still cooler climate. The cold

alpine belt above this cannot support trees; here, grasses and sedges dominate. On the highest peaks, with their cover of snow and ice, a vegetation cover is conspicuously absent.

Such changes in vegetation cover with elevation are due partly to the way that mountains modify patterns of rainfall. When prevailing winds reach a mountain range, the air rises, cools, and loses moisture (cool air holds less moisture than does warm air). Here, the abundant rainfall promotes considerable plant growth. After flowing over mountain crests, the air descends (Figure 46.7). As it does, it becomes warmer and its moisture-holding capacity increases. This air draws moisture out of plants and the soil rather than giving it up as rain. Thus, a **rain shadow** exists on the leeward side of the mountains, where the lack of abundant rainfall creates arid or semiarid conditions. ("Leeward" means the direction facing the wind; "windward" means the direction from which the wind is blowing.) Rain shadows also exist in the mountain ranges of Europe, the Himalayas of Asia, and the Andes of South America.

Aquatic regions also are influenced by features that interrupt or channel air and water movements. These movements have profound effects on aquatic ecosystems, in ways that will be described later in the chapter.

THE WORLD'S BIOMES

Ever since global explorations began in the sixteenth century (Chapter Two), biologists have recognized that huge tracts of grasslands, deserts, forests, and tundra exist. There have been numerous attempts to group all such tracts into simple classification schemes. For example, W. Sclater and then Alfred Wallace were the first to propose six **biogeographic realms**, which contain characteristic types of animals and plants. Barriers such as oceans, mountain ranges, and deserts keep the species of each realm more or less isolated from the others by restricting their dispersal. (Thus the kangaroos of the Australian realm could not, on their own, radiate elsewhere.) Figure 46.8 depicts this biogeographic scheme, which is still widely accepted.

Each biogeographic realm is actually a mosaic created by variations in climate, topography, and the composition of regional soils. Thus there are distinct vegetational subdivisions in which the climax communities are dominated by plant species adapted to a particular set of conditions. These broad, vegetational subdivisions, together with the animals and other organisms living there, are known as the **biomes** of each realm. As Figure 46.9 indicates, biomes include different types of deserts, shrublands, grasslands, forests, and tundra. Figures 46.10 and 46.11 together indicate the types of soils that are characteristc of some major biomes.

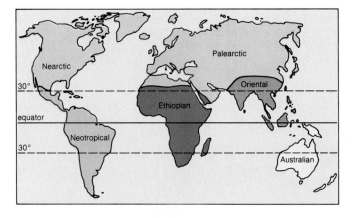

Figure 46.8 Major biogeographic realms, as first proposed by W. Sclater and Alfred Wallace in the 1800s. The scheme is still widely accepted.

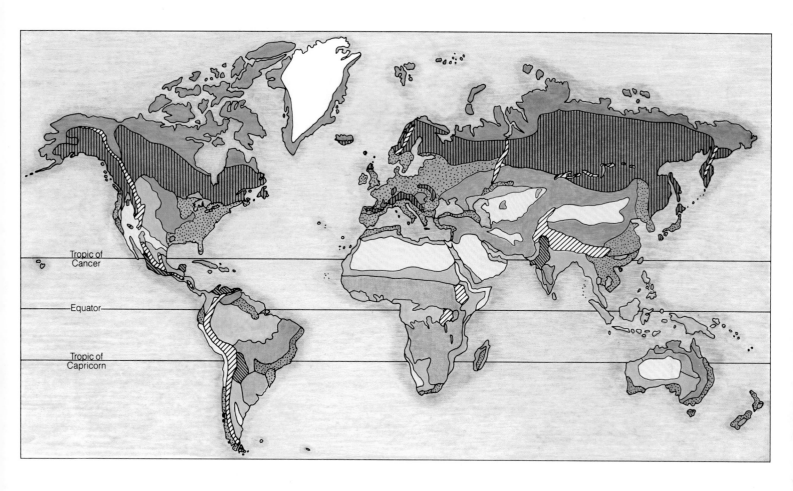

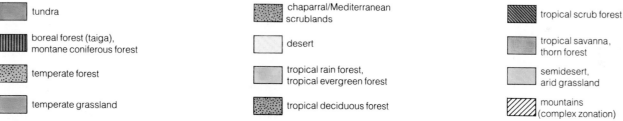

	tundra		chaparral/Mediterranean scrublands		tropical scrub forest
	boreal forest (taiga), montane coniferous forest		desert		tropical savanna, thorn forest
	temperate forest		tropical rain forest, tropical evergreen forest		semidesert, arid grassland
	temperate grassland		tropical deciduous forest		mountains (complex zonation)

Figure 46.9 Simplified picture of the world's major biomes. Arctic tundra and boreal forests are continuous over vast regions and contain ecologically equivalent and taxonomically related species. Other biomes that are isolated but of the same type generally have ecologically equivalent but often taxonomically unrelated species. The overall pattern of biome distribution roughly corresponds with patterns of soil type distributions and climate.

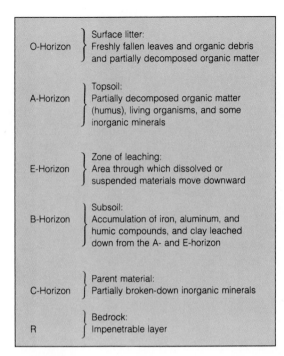

O-Horizon	Surface litter: Freshly fallen leaves and organic debris and partially decomposed organic matter
A-Horizon	Topsoil: Partially decomposed organic matter (humus), living organisms, and some inorganic minerals
E-Horizon	Zone of leaching: Area through which dissolved or suspended materials move downward
B-Horizon	Subsoil: Accumulation of iron, aluminum, and humic compounds, and clay leached down from the A- and E-horizon
C-Horizon	Parent material: Partially broken-down inorganic minerals
R	Bedrock: Impenetrable layer

Figure 46.10 Categories of horizontal layers of soils, or soil horizons.

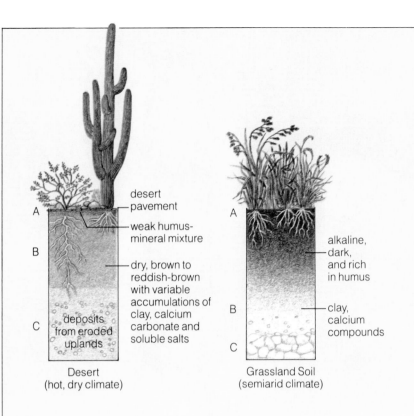

Figure 46.11 Characteristics of soil, together with soil profiles from a few different biomes.

The distribution of different ecosystems on land depends on the composition of regional soils as well as on climate and topography. *Soil* is a mixture of rock, mineral ions, and organic matter in some state of physical and chemical decomposition. Depending on the degree of weathering, the rock component ranges from coarse-grained gravel to sand, silt, and fine-grained clay. Partially decomposed organic matter, called *humus*, helps retain water-soluble ions (such as potassium and ammonium ions) that are released during the activities of soil bacteria and other decomposers.

The most productive soil is loam, a mixture of sand, silt, clay, and humus that holds adequate nutrients and mineral ions and that is well aerated. (The roots of most plants, recall, do not do well in poorly aerated, poorly draining soils.) Water percolates too rapidly through soils with a high gravel and sand content;

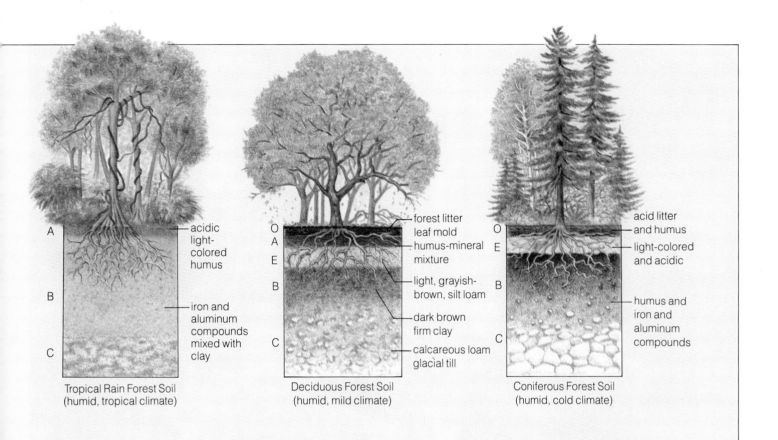

Tropical Rain Forest Soil
(humid, tropical climate)

A
B
C

acidic
light-
colored
humus

iron and
aluminum
compounds
mixed with
clay

Deciduous Forest Soil
(humid, mild climate)

O
A
E
B
C

forest litter
leaf mold
humus-mineral
mixture
light, grayish-
brown, silt loam
dark brown
firm clay
calcareous loam
glacial till

Coniferous Forest Soil
(humid, cold climate)

O
E
B
C

acid litter
and humus
light-colored
and acidic

humus and
iron and
aluminum
compounds

hence these soils are quickly depleted of mineral ions. This process is called *leaching*. Clay soils with small, closely packed particles are poorly aerated and do not drain well; few plants can grow in such soils when they are waterlogged.

Topsoil, the soil layer below any surface litter, ranges in thickness from less than a centimeter on steep slopes to more than a meter in grasslands. Topsoil is the most fertile of all soil layers, the so-called A-horizon. Biomes vary considerably in the depth of this layer. Most of the world's crops are grown on grassland soils.

When tropical rain forests are cleared (as for agriculture), the torrential, seasonal rains leach most of the nutrients and remaining minerals from the A-horizon of the exposed soil. When exposed to the air and sun, the iron oxide that remains can form red rock called *laterite* or *ironstone*—a rock so hard it is used for paving highways. Once laterite forms, the land can be lost to cultivation permanently.

About one-third of the area covered by desert soils is not useful for growing crops because of too little rainfall. With proper soil management and extensive irrigation, soils in certain desert areas (such as the Imperial Valley of California) can produce a variety of crops. Such soils, however, often become unproductive from waterlogging and salt buildup when they are irrigated without sufficient drainage.

Figure 46.12 Warm desert near Tucson, Arizona. The vegetation includes creosote bushes, ocotillo, saguaro cacti, and prickly pear cacti.

Figure 46.13 Chaparral-covered hills east of San Diego, California.

Deserts

Where the potential for evaporation greatly exceeds rainfall, there you will find deserts. These climatic conditions are pronounced at latitudes 30° from the equator, where rainfall is scarce because of the high subtropical air pressure. The immense Sahara and Australian deserts are at these latitudes; so are the deserts of the American Southwest (Figure 46.12). Other deserts, such as the deserts of eastern Oregon, have formed through rain shadow effects, as described earlier.

The vegetational cover is minimal in deserts, compared with other biomes. When rain does fall, it falls in heavy, brief pulses. The unprotected soil erodes rapidly during violent windstorms and thunderstorms. Because of the low humidity, sunlight penetration through the atmosphere is high and the ground heats up rapidly. At night, the heat quickly radiates back to the air. Page 455

Figure 46.14 Rolling shortgrass prairie to the east of the Rocky Mountains.

described some of the behavioral responses that desert animals make to the temperature extremes.

Today, more than a third of the world's total land area is arid or semiarid. Prolonged periods of drought, overgrazing, and other factors together are leading to *desertification* (conversion of grasslands and other regions to deserts) on a grand scale. During the past fifty years, about 650,000 square kilometers (250,000 square miles) of grazing land along the southern margins of the Sahara have become desert.

Shrublands

Shrublands occur in semiarid regions with cool, moist winters and long, hot, dry summers. Most of the rains fall during the winter, and temperatures typically fall below 15°C for one of the winter months. There is a burst

Figure 46.15 A patch of natural tallgrass prairie.

of plant growth at the end of the wet season. The dominant plant species are generally well-branched, short, and woody.

Successional shrublands occur in transitional zones between grasslands and forests. If dense enough, they can persist for long periods. Several regions of *Mediterranean-type shrublands* are concentrated mostly between 30° and 40° latitudes. The chaparral shrublands of California, covering about 5 to 6 million acres, fall in this broad zone (Figure 46.13). There is no understory or much ground litter in these shrublands. Because of the aromatic oils in their leaves, many of the plants are highly flammable. Many species actually depend on heat and scarring to prepare their seeds for germination. For centuries, Mediterranean-type shrublands have been maintained by episodes of fire, which clears away old growth and helps recycle nutrients.

Grasslands

The world's great grasslands are in South Africa, South America, and the midcontinental regions of North America and the USSR. At one time, grasslands covered more than forty percent of the land; today, most have been cleared for agriculture. All grassland biomes share several characteristics. The land is flat or rolling; the climate is dry, with high rates of evaporation. Rainfall averages between ten and thirty inches a year—enough to keep the grasslands from turning into deserts, but not enough to support heavy forest growth. Grazing and burrowing species are the dominant forms of animal life. In many grasslands, grazing and episodes of fire help maintain the climax communities by creating conditions necessary for renewed growth.

There are different types of grassland biomes. For example, *shortgrass prairie* occurs where winds are strong, rainfall light and infrequent, and evaporation rapid (Figure 46.14). Plant roots soak up the brief, seasonal rainfall on the surface, above the permanently dry subsoil. Much of the shortgrass prairie of the American Midwest was overgrazed and plowed under for wheat, which requires more moisture than the region provides, without irrigation. During the 1930s, removal of the tight vegetation cover, drought, and strong prevailing winds combined to turn much of the prairie into a Dust Bowl. John Steinbeck's *Grapes of Wrath* and James Michener's *Centennial* are two historical novels that speak eloquently of this disruption and its consequences.

Tallgrass prairie once extended westward from the temperate deciduous forests of North America (Figure 46.15). In some tallgrass regions, species of composites such as daisies actually outnumbered the species of grasses. Legumes were also abundant; these nitrogen-fixing plants increased the net primary productivity for the region. Most tallgrass prairie has been converted to agriculture.

Tropical grasslands include the broad belts of African savanna, described in Chapter One. In savanna regions of low rainfall, the main species are rapid-growing, tufted grasses. Where rainfall increases slightly, shrubs such as acacia grow in scattered patches. Savanna regions of high rainfall are mosaics of tall, coarse grasses, shrubs, and low trees, even of humid forests. Tropical grasslands also extend into Southeast Asia, where they are known as monsoon grasslands.

Forests

There are three major types of forests: coniferous, deciduous, and rain forests. The forests of all three biomes are stratified, which means that the amounts of light, moisture, and temperature and vegetation types change at different levels between the tree canopy and the forest floor. In all three biomes, gross primary productivity is high. However, much of the production is allocated to maintaining the existing forest structure. Nutrients accumulate in woody biomass and are locked away from short-term cycling. Minerals become available more from decaying roots and litter than from the soil (Figure 46.10).

Tropical Rain Forests. In most equatorial regions, rainfall is heavy during at least one season, humidity is high, and the annual mean temperature is about 28°C. In the tropical rain forests of these regions (Figure 46.16), vines clamber up toward the sun. Mosses, orchids, lichens, and bromeliads (plants related to the pineapple) grow on tree branches, absorbing minerals dissolved in tiny pockets of water. (The minerals are released during the decay of bits of leaves, insects, and litter.) Entire communities of insects, spiders, and amphibians live, breed, and die in the small pools of water that collect in the leaves of these aerial plants. Many insects, birds, and monkeys spend most of their lives at a single level in the stratified forest. In some cases, the number of species living in or on a single tree exceeds the number living in an entire forest to the north.

With all of its diversity, a tropical rain forest is one of the worst places to grow crops. There is practically no soil organic matter, and most nutrients are tied up in the standing biomass. Decomposition is rapid in the hot, humid climate. Minerals released during decomposition are rapidly picked up by roots and mycorrhizae concentrated in the top layers of soil; there is very little nutrient storage in the subsoil. Thus, when such forests

bromeliad

Figure 46.16 Tropical rain forest, showing aerial plants (including bromeliads) and vines in the canopy. Where light breaks through the canopy, luxuriant new growth occurs (see, for example, Figure 44.1).

are cleared for agriculture, most of the nutrients are permanently cleared away (Figure 46.11).

With *slash-and-burn agriculture*, the forest biomass is reduced to nutrient-rich ashes, which are tilled into the soil. Even then, heavy rains wash away most of the nutrients from the exposed clay soils. After a few years, the cleared fields become infertile and usually are abandoned. Because nutrients are so depleted, successional replacement is extremely slow.

Deciduous Forests. Deciduous forests dominate in regions where a growing season characterized by moderate rainfall and mild temperatures alternates with a pronounced dry or cold season. Mineral cycling is moderate in these forests. Conditions are favorable for decomposition, and the seasonal leaf drop assures abundant litter on the forest floor. Yet the trees go dormant during the dry or cold season, reducing the nutrient losses (through leaching) and thereby preserving soil fertility.

In many tropical regions, including parts of Southeast Asia, temperatures remain mild but rainfall dwindles for

Spring

Autumn

Figure 46.17 The changing character of a temperate deciduous forest in spring, summer, autumn, and winter.

part of the year. These regions support *tropical seasonal forests*, which are dominated by deciduous and semi-deciduous species. Regions of North America, Europe, and eastern Asia support temperate deciduous forests. In the deciduous forests of the eastern United States, the dormant season coincides with cold winter conditions (Figure 46.17).

Coniferous Forests. There are two major types of coniferous forests. One, the *montane coniferous forest*, occurs in the Sierra Nevada, Cascades, and Rocky Mountains of the United States, in Europe, and in parts of Central America. At high elevations, winters are long and snow is heavy, yet summers are warm enough to promote a dense growth of coniferous trees, such as spruces, firs,

and pines. At lower elevations, Douglas fir and ponderosa pine predominate. Figure 46.18 shows a montane coniferous forest.

South of the tree line in northern regions of the world is an extensive zone known as the *boreal forest*, or *taiga* (Figure 46.19). For the most part, these coniferous forests occupy glaciated regions punctuated with many cold lakes, streams, bogs, and marshes ("taiga" is Russian for swamp forest). Winters are prolonged and extremely cold, with a fairly constant snow cover. As with the montane forests, however, the summer is warm enough to promote dense growth.

Figure 46.18 The montane coniferous forest of Yosemite beneath the first snows of winter.

Ansel Adams

Tundra

Tundra is a word derived from the Finnish *tuntura*, meaning a treeless plain. *Arctic tundra* lies to the north, between the huge belts of coniferous forests in North America, Europe, and Asia and the polar ice cap. In Alaska, much of the tundra is flat, windswept, and desolate (Figure 46.20). Temperatures average 5°C (41°F) in midsummer, and −32°C (−26°F) in midwinter. The air is too cold to permit formation of much water vapor, so rainfall is sparse. Yet for three summer months, sunlight is nearly continuous. Then, plant growth is profuse, with flowering and seed ripening completed quickly.

Although the tundra is not completely covered with snow all year long, the brief summer thaw is not enough to warm much more than the surface soil. Just beneath the surface is the **permafrost**, a permanently frozen layer more than 500 meters thick in some regions. Permafrost

forms an impenetrable basement beneath the flat terrain; hence drainage is inhibited. In combination with the low temperatures, permafrost has a major effect on nutrient cycling. Organic matter cannot completely decompose here. It is gradually becoming locked up in soggy masses (peat). At present, more than ninety-five percent of the carbon in the arctic is inaccessible to the tundra organisms. Less than two percent of the total carbon, nitrogen, and phosphorus is found in living plants, mostly concentrated in underground plant parts.

An *alpine tundra* occurs in the high mountain ranges at northern latitudes. Here also, we find low temperatures, limited rainfall, and short growing seasons with correspondingly low growth rates for the plants. However, alpine tundra has no permafrost layer below the soil surface. The plant species often form cushions and mats that can withstand buffeting from the strong winds that sweep through the mountains.

Figure 46.19 Example of a boreal forest, or taiga.

Figure 46.20 (Below) A view of the arctic tundra, looking toward the Brooks Range.

Figure 46.21 (a) A lake ecosystem in Canada. (b) Lake zonation. The littoral includes all areas where light penetrates to the lake bottom. The profundal includes areas below the depth of light penetration. Above the profundal zone are the open, sunlit waters of the limnetic zone.

a

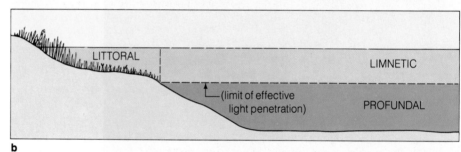

b

THE WATER PROVINCES

As vast as the biogeographic realms are, they cover less than thirty percent of the earth's surface; the water provinces are far more extensive and they, too, encompass diverse ecosystems. Inland are the *lentic* ecosystems (bodies of standing fresh water such as lakes), the *lotic* ecosystems (running fresh water such as rivers), and brackish seas. Associated with the oceans and seas are estuaries, tidal marshes, rocky and sandy coasts, coral reefs, and other diverse ecosystems.

There is no such thing as a "typical" aquatic ecosystem. Lentic ecosystems alone vary in size from small ponds to lakes that cover many thousands of square kilometers. Some are shallow enough to wade across; one is more than 1,700 meters deep. Most have gradients in light penetration, temperature, and dissolved gases, but the gradients vary from one ecosystem to another, both daily and seasonally. All we can do here is provide a sampling of the diversity that exists.

Lake Ecosystems

The topography, climate, and geologic history of a lake influence the kinds and numbers of organisms found

there, how they are distributed, and how nutrients are cycled among them. All lakes form in land basins (Figure 46.21), hence all are doomed to become filled with sediments over time. Succession has been observed in some smaller ones, with the encroachment of marshes and then swampy meadows eventually giving way to forests. Long-term changes in climate or geologic events can bring to an end even the vast, deep lakes. Human sewage, mining wastes, and agricultural runoff also influence the character of a lake.

Lake Zones. In terms of primary productivity, a lake has three zones: the littoral, profundal, and limnetic zones.

Waves, changing temperatures, and (in cold regions) grinding ice act in the shallows of a lake and lead to the formation of coarse sediments. Typically, light penetration is good in these shallow waters, and numerous aquatic plants take hold here, as do numerous aquatic animals and decomposers. The **littoral zone** includes all those areas where light penetrates to the lake bottom. It extends from the lake shore to the depth where the rooted plants disappear (Figure 46.21b).

The **profundal zone** includes those areas below the depth of effective light penetration (that is, the wave-

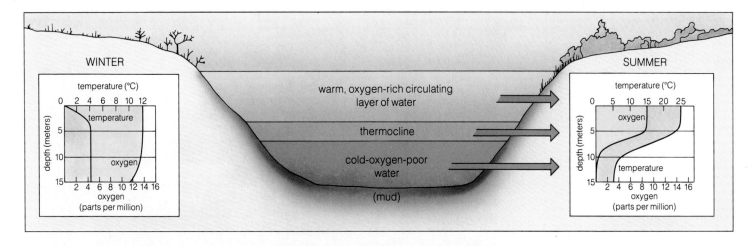

Figure 46.22 Thermal stratification in a north temperate lake in Connecticut. The temperatures and oxygen levels in winter are shown to the left; they are shown to the right for summer. Notice that in summer, the cold, oxygen-poor water below is separated from the upper layer of warm, oxygen-rich water by a broad zone, the *thermocline*, in which temperature and oxygen levels change rapidly with increasing depth.

lengths of light necessary for photosynthesis do not reach these areas). Bacterial decomposers dominate in the profundal zone. Above this zone are the open, sunlit waters of the **limnetic zone**, where tiny free-floating or swimming organisms, collectively called **plankton**, abound. Photosynthetic algae are the major primary producers of the limnetic zone.

Spring and Fall Overturns. Let's take a look at how nutrients are cycled in lakes that show thermal stratification. During winter, the lake waters are cold, low in oxygen, and high in carbon dioxide. With the lengthening of daylight in spring, the surface waters become warmer. The strong spring winds and destratification set up currents that mix the water from the surface with water from the bottom of the lake basin. This *spring overturn* mixes bottom nutrients, oxygen, and plankton throughout the lake and is accompanied by a burst in primary productivity.

As the spring sun warms the surface waters, they become less dense than the underlying waters. In summer, differences in temperature and density are greatest. These differences oppose the force of winds, and waters resist vertical mixing. Below the surface layer (where waters freely circulate), temperatures drop rapidly (about 1°C for each meter of depth). This midlayer of steeply declining temperature and noncirculating waters is the **thermocline** (Figure 46.22). The bottom layer is deep and cold, with gradual drops in temperature.

Figure 46.23 Field experiment demonstrating the effect of nutrient enrichment on a lake. A plastic curtain was stretched across a narrow channel between two basins of the same lake. Phosphorus, carbon, and nitrogen were added to the basin at the left; only carbon and nitrogen were added to the basin in the right foreground. Within two months, the phosphorus-enriched basin showed accelerated eutrophication, with a dense algal bloom.

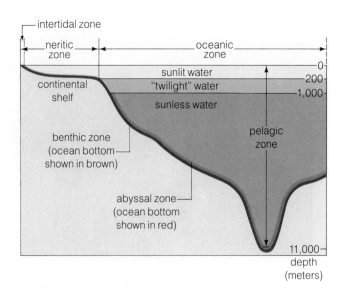

Figure 46.24 Oceanic province. Regions of the ocean are defined in terms of their location relative to the continents, to the ocean bottom, and to its surface.

In autumn, air temperature drops and the surface layer becomes cooler. Then the surface waters mix with those of the thermocline and temperature becomes much the same through the lake. Winds then create a *fall overturn*: nutrients, oxygen, and plankton are mixed once more throughout the lake basin, and there may be another burst in productivity before the coming of winter.

Types of Lakes. Someone once identified seventy-five types of lakes, but here we will concentrate on two categories based on productivity. The primary productivity is influenced by lake depth; it is influenced also by the amount of nutrients draining into the lake from the surrounding region. If the supply of nutrients is high, the lake is **eutrophic**; if nutrients are in short supply, the lake is **oligotrophic**. Typically, eutrophic lakes are productive, relatively shallow, and subject to fluctuations in oxygen levels in the surface and profundal zones. The water is not very transparent and looks green or yellowish. Oligotrophic lakes are often deep and steeply banked, with oxygen abundant at all levels throughout the year. These lakes show lower productivity, and their waters are blue or green, and very transparent. Many lakes are undergoing human-caused eutrophication, with nutrients such as nitrogen and phosphates being added rapidly from agricultural runoff, sewage, and industrial wastes. Figure 46.23 shows the results of one experiment designed to test the effect of nutrient enrichment in an otherwise oligotrophic lake.

Figure 46.25 Angler fish from the ocean depths, sporting a prey-attracting luminescent lure on its head.

Ecosystems of the Oceans

Figure 46.24 shows the major zones of the oceans. The shallow regions above the continental shelves are the **neritic zone**; the **oceanic zone** encompasses the deeper ocean basins. Regardless of depth, the bottom of the ocean is called the **benthic zone** and the water above it, the **pelagic zone**. The part of the benthic zone where no light penetrates is called the **abyssal zone**.

Photosynthetic activity is restricted to the upper surface waters of open oceans, where phytoplankton often drift with ocean currents. Zooplankton (including copepods, planktonic arthropods, and shrimplike krill) graze on the suspended marine pastures. In turn, strong-swimming forms (small fishes, squids, and other carnivores) feed on the herbivores.

There are scattered communities in the abyssal zone; Figure 46.25 shows a fierce-looking representative of an array of deep-sea fishes. A remarkable type of ecosystem, one *not* based on photosynthesis, occurs around hydrothermal vents in this zone (Figure 46.26).

Upwellings and Coastal Waters. In all oceans, the remains and wastes of marine organisms sink downward, below a permanent thermocline. Decomposition releases nutrients from this organic matter, which then moves up again to the surface waters through upwelling along the western coasts of the continents and around islands.

Upwelling is an upward movement of deep water that carries nutrients to the surface zone. The movement is associated with reversals in prevailing coastal winds.

a

b

Figure 46.26 Hydrothermal vent ecosystems. In 1977, biologists discovered a distinct type of ecosystem deep in the Pacific Ocean, where sunlight never penetrates. John Corliss and his coworkers were exploring the Galápagos Rift, a volcanically active boundary between two of the earth's crustal plates. There, on the ocean floor, the near-freezing seawater seeps into fissures, becomes heated, and is spewed out through vents at temperatures exceeding 700°F.

This hydrothermal effluent deposits zinc, iron, and copper sulfides as well as calcium and magnesium sulfates, all leached from rocks as pressure forces the heated water upward. In marked contrast to most of the deep ocean floor, these nutrient-rich, warm "oases" support diverse marine communities. Chemosynthetic bacteria and other microbes use the inorganic deposits as energy sources. They are the primary producers in a food web that includes tube worms, clams, sea anemones, crabs, and fishes (**a,b**).

So far, other hydrothermal vent ecosystems have been discovered near Easter Island, off the northwestern United States, in the Gulf of California, and about 150 miles south of the tip of Baja California, Mexico.

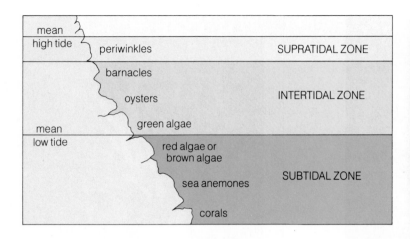

Figure 46.27 Organisms adapted to the tides along rocky coasts. In general, periwinkles live in the high-tide area, above the rocks where barnacles and mussels predominate. Algae grow profusely in and below the low-tide area. Plenty of other organisms are present, but the ones mentioned here are the dominant types.

These reversals occur primarily along the western coasts of Peru, southern California, northern and southwestern Africa, and the Antarctic. In winter, winds slam into the coastal regions, and the wind-driven water piles up. In summer, the coastal winds flow north to south, toward the equator. The force of the winds causes surface waters that are piled up along the coast to move away from the shore. When the surface water moves out, deep water moves in vertically to replace it.

For example, upwelling near Peru is coupled with the northward flow of the cold Humboldt Current (Figure 46.4). Together, these movements bring up tremendous amounts of nitrate and phosphates from below; they are the foundation for one of the world's richest fisheries. Periodically, prevailing wind direction shifts and warm equatorial waters displace the Humboldt Current. This condition, known as El Niño, prevents upwelling and productivity plummets.

Life Along the Coasts.   Like the littoral zone of lakes, the littoral zone along the coasts of oceans and seas is rich in species diversity. Along rocky coasts, stratified communities include plants and animals equipped to hold on or burrow in and thereby resist being yanked from the substrates by action of surf and tides (Figure 46.27). Competition for living space is intense here, for the waters bring in abundant food to the glued-in-place organisms.

Estuaries.   In estuaries, nutrient-rich fresh water from rivers and streams mixes with seawater. The water becomes enriched with organic matter from surrounding tidal marshes of the sort shown in Figure 46.28. These marshes are flooded daily by incoming tides. Litter from the marsh plants, such as *Spartina*, is carried in with the water and enriches the estuarine nutrient base.

Estuaries have both grazing food webs (planktonic) and detrital food webs. In the water, dinoflagellates and diatoms are major producers in these webs; aquatic plants are also important. Estuaries serve as nurseries, where many marine animals spend the early part of their life cycle (Figure 46.28).

SUMMARY

1. The biosphere is the narrow zone of water, the lower region of the atmosphere, and the fraction of the earth's crust in which organisms live.

2. The biosphere is composed of ecosystems, each of which is influenced by the flow of energy and the movement of materials on a global scale.

3. The main determinant of the distribution of ecosystems is climate (prevailing weather conditions, including temperature, humidity, wind velocity, degree of cloud cover, and rainfall).

4. Climate is influenced by variations in the amount of incoming solar radiation, the earth's daily rotation and its annual path around the sun, the distribution of continents and oceans, and the elevation of land masses. All such factors interact to produce prevailing winds and ocean currents which together influence global weather patterns. The weather in turn affects the composition of soils and sediments as well as water availability, which

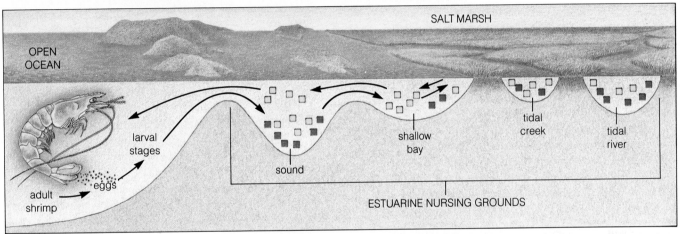

Figure 46.28 Life history of a shrimp that uses estuaries as nursery grounds. Adult shrimp spawn offshore and the larval stages move shoreward. In the semi-enclosed estuaries they find the food and protection they need for rapid growth during the juvenile stage (red boxes) and adolescent stages (yellow boxes). As they mature, the shrimp move back into deeper waters of the sounds, then of the open ocean. The photograph shows a New England salt marsh, with *Spartina* predominating. This salt marsh grass is the major producer, with its microbial-enriched litter providing the food for consumers in the creeks and sounds.

influences the growth and distribution of primary producers (mostly photosynthetic plants)—hence the nature of the ecosystems.

5. A rainshadow forms when a mountain range modifies the pattern of rainfall. When prevailing winds reach the range, the air rises, cools, and loses moisture. The air flowing over the crest descends, becoming warmer and able to hold more moisture, which evaporates from plants and soil. Arid or semiarid conditions result.

6. The land masses can be divided into six biogeographic realms, each with characteristic types of animals and plants and each more or less isolated by oceans, mountain ranges, and desert barriers.

7. Biomes are distinct vegetational subdivisions of the six major realms, created by variations in climate, topography, and the composition of regional soils. Each is dominated by plant species adapted to a particular set of conditions.

a. Deserts: evaporation greatly exceeds rainfall, minimal vegetational cover, drought-tolerant plants. Many deserts occur at latitudes 30° from the equator; others have formed through rain shadow effects.

b. Shrublands: semiarid regions with cool, moist winters and long, hot, dry summers. Well-branched, short, woody plants dominate.

c. Grasslands: flat or rolling land; dry climate with high rate of evaporation but ten to thirty inches of rainfall annually. Depending on moisture level, shortgrass or tallgrass species dominate. Grazing and episodes of fire help maintain the climax community, creating conditions necessary for renewed growth.

d. Forests: In *tropical rain forests* of equatorial regions, rainfall is heavy at least one season, humidity is high, and the annual mean temperature is mild; plant species are quite diverse. In *deciduous forests*, rainfall is moderate and temperatures are mild during the growing season, which alternates with a pronounced dry or cold season. In *coniferous forests* (montane or boreal), winters are long with heavy snowfall, but warm summers promote dense growth.

e. Tundra: low temperatures, limited rainfall, and a short growing season. *Arctic tundra* lies to the south of the polar ice cap and north of the tree line of coniferous forests in the Northern Hemisphere; it is characterized by permafrost (a permanently frozen layer, sometimes more than 500 meters thick, just beneath the soil surface). Grasses, sedges, dwarf woody plants, and lichens dominate. *Alpine tundra* occurs in high mountain ranges at northern latitudes. Growth of grasses and lichens is scantier than in arctic tundra.

8. The water provinces, which cover more than seventy percent of the earth's surface, include:

a. Lentic ecosystems (standing fresh water such as ponds and lakes), lotic ecosystems (running fresh water such as streams and rivers), and brackish inland seas.

b. Marine ecosystems, including oceans and seas, estuaries, tidal marshes, rocky and sandy coasts, and coral reefs.

9. All aquatic ecosystems have gradients in light penetration, temperature, and dissolved gases, features that vary daily and seasonally from one type of ecosystem to another.

10. In terms of primary productivity, a lake has three zones: the littoral (all areas where light penetrates to the lake bottom and where aquatic plants live); the profundal (areas too deep to be penetrated by wavelengths most useful for photosynthesis, hence where primary productivity is low); and the limnetic (all open, sunlit waters above the profundal, and extending outward from the littoral, where phytoplankton dominate).

11. *Eutrophic* lakes (in which nutrient levels and hence primary productivity are high) typically are shallow and subject to shifts in oxygen levels; transparency is reduced in the greenish or yellowish water. *Oligotrophic* lakes (in which nutrient levels and hence primary productivity are low) often are deep and characterized by abundant oxygen throughout the year; the water is blue or green and transparent.

12. The oceanic province is divided into the following major zones: neritic (shallow regions above continental shelves), oceanic (deeper ocean basins), benthic (ocean bottom, regardless of depth; no light reaches its abyssal portions), and pelagic (waters above the benthic zone).

13. Competition is intense in the littoral zone along coasts of oceans and seas. Along rocky coasts, highly stratified communities consist of organisms able to cling to or burrow into the rocks (and resist being swept away by surf and tides).

14. In estuaries, nutrient-rich fresh water from rivers and streams mixes with seawater. Estuaries are nursery grounds for shrimp and many other marine animals.

Review Questions

1. Describe the three main physical components of the biosphere.

2. Define climate. What four interacting factors influence climate? What does climate in turn influence?

3. How do prevailing air and ocean currents help dictate the distribution of different types of ecosystems?

4. Define rain shadow. How does a rain shadow affect ecosystems on both sides of a mountain range?

5. What is the difference between the littoral and profundal zones of a lake ecosystem? What kinds of lakes are eutrophic? Oligotrophic?

6. What are the three zones of the open ocean, and which shows the greatest productivity?

7. What is a thermocline? Are thermoclines most pronounced in temperate or tropical oceans? In shallow or deep lakes?

8. Define upwelling, and give an example of a region where upwelling has a profound effect on primary productivity.

9. How does the composition of regional soils affect ecosystem distribution?

10. Distinguish between biogeographic realm and biome. In what type of biome region would you say you live?

11. How do climatic conditions affect the character of the following biomes: desert, shrublands, tropical rain forests, deciduous forests, coniferous forests, and tundra?

Readings

Cole, G. 1983. *Textbook of Limnology*. Third edition. St. Louis: Mosby. Nice introduction to the study of lakes.

Miller, G. T., Jr. 1986. *Environmental Science: An Introduction*. Belmont, California: Wadsworth.

Odum, E. 1983. *Basic Ecology*. Philadelphia: Saunders.

Smith, R. L. 1980. *Ecology and Field Biology*. Third edition. New York: Harper & Row. Good introduction to the structure and functioning of the biosphere.

West, S. 1980. "Smokers, Red Worms, and Deep Sea-Plumbing." *Science News* 117(2):28–30. An account of deep-sea explorations and discoveries in the Galápagos Rift.

The ecosystems described in the preceding chapter have had a long history of development. Their arrays of species have been locked for so long in coevolutionary relationships that energy use and nutrient cycling through these systems are remarkably efficient over the long term. Sometimes resources dwindle severely, as they do when snowshoe hares overbrowse in Canadian forests (Figure 44.8). Yet the plants fight back, so to speak, and the forest ecosystem more or less bounces back over the years. Sometimes toxins build up, as happened when oxygen first started accumulating in the atmosphere of the primordial earth. Yet gradually, some organisms were able to adapt to this photosynthetically produced "toxin" by using it as an electron acceptor in aerobic respiration. Sometimes, too, species introductions can lead to wholesale displacements of other species, as happened when water hyacinths underwent explosive growth in waterways of the American South.

Natural ecosystems, then, are not static—but they tend to recover from disruptions or adjust to them. If this is true, why is there such widespread concern over the disruptions that human populations are creating in ecosystems throughout the biosphere? *The answer is that natural disruptions are not of the same magnitude in time and space.* The changes we are introducing are global in dimension and they are occurring at an accelerated pace. As one indication of our effects, species are becoming extinct fifty times faster then they were even a century ago.

The demands of the human population are increasing at a phenomenal rate, one that parallels our J-shaped curve of population growth. At the same time, resource

47

HUMAN IMPACT ON THE BIOSPHERE

Figure 47.1 The city as ecosystem.

utilization is not exactly what you would call efficient. Energy resources are not being conserved. Also, the products of our existence are accumulating to levels harmful enough to be called pollutants. **Pollutants** can be *any* substances with which ecosystems have had no prior evolutionary experience, in terms of kinds or amounts, and therefore have no established mechanisms for dealing with those substances.

We can reduce, collect, concentrate, bury, and burn waste products or spread them out in other ecosystems. Yet we are never completely rid of the pollutants we are generating. Over the past four decades, 350 million gallons of sewage from metropolitan New York and New Jersey have been dumped every day off the shore of Long Island. They are out of sight, but it is not easy to put them out of mind. In the summer of 1970, a dead sea of sewage began moving back toward land, bringing with it the specter of hepatitis, encephalitis, and other terrible diseases.

What options are available to us? The following case studies will give you an idea of how limited the current options are.

AN EXCESS OF OUTPUTS

Case Study: Solid Wastes

On the outskirts of Del Mar, a small coastal town in southern California, a plot of land is set aside as a recycling center for newspapers, glass, and aluminum cans. Centers of this kind have popped up all over the United States. When one of our daughters entered college, she decided to take part in the recycling program. There was a brief period of adjustment: family members had to begin stacking newspapers in the garage, putting glass jars and bottles in one container, and putting aluminum cans in another. By the weekend, we were surprised at how much had accumulated; we had assumed it would take months to stack up. It took only a few minutes for our daughter to take the wastes to the recycling center. However, along the way she became aware, for the first time, of how many trashcans lined the streets, waiting to be emptied. Two or three trashcans sat in front of each house. At the time there were about 5,000 homes in Del Mar. That meant about 12,000 trashcans a week had to be emptied somewhere. In the surrounding metropolitan area, there were about 200,000 homes—meaning about 400,000 to 600,000 trashcans *each week.*

After our daughter deposited the wastes in collecting bins at the recycling center, she drove out to the county "sanitary landfill station" where the nonrecycled solid wastes end up. There, one trash-filled truck after another rumbled through the gate, antlike in their line to the dumping ground. Mounds of refuse were being bull-dozed down the sides of what had once been a chaparral-covered canyon. There are only so many canyons; what happens when they are all filled?

This same thing had to be happening all over the nation. What could one isolated commitment possibly mean? How could one small action help turn such a tide of waste? For a while our daughter thought about starting a campaign for public awareness, but with college starting soon, how could she get involved? Somebody else would have to do it.

Which "somebody" is going to tackle the billions of metric tons of solid wastes that are dumped, burned, and buried each year in the United States? Who is going to decide where the landfills go next? Conversely, is recycling itself a workable alternative? Recycling is part of the answer, but it cannot be a hit-or-miss effort. For instance, each drive to a recycling center uses up some gasoline—a nonrenewable energy source that ultimately must be figured as part of the energy cost of the program. Although the energy cost is lower than it would be to extract and use new raw materials, it is still significant—particularly when you multiply the energy cost by all the individuals driving separately to the center.

What it will take, in the long run, is a change in basic living habits. We have a "throwaway" mentality: use something once, discard it, and buy another. For instance, about half the volume of urban solid wastes is composed of paper products—of which only one-fourth is now being recycled. If we recycled half the paper being thrown away each year, we would do more than conserve trees. The energy it takes to produce an equivalent amount of new paper could be diverted to provide electricity to about 10 million homes each year. (It takes 150 acres of forest to produce the paper in each Sunday edition of the *New York Times* alone.) Or consider that about 60 billion beverage containers are sold annually in the United States. About 50 billion are nonreturnable cans and bottles, many of which are discarded in public places. These containers account for three-fourths of the litter picked up along highways—a time-consuming, energy-draining activity that costs thousands of barrels of petroleum (not to mention hundreds of millions of tax dollars) every year.

A transition from a throwaway life-style to one based on conservation and reuse is economically feasible, and we have most of the technology needed to implement the change. The question becomes one of commitment. We can put pressure on manufacturers by refusing to buy goods that are lavishly wrapped, excessively boxed, and designed for one-time use. Individuals can ask the local post office to turn off their daily flow of junk mail, which wastes an astounding amount of paper, time, and energy—and means higher postage rates for everyone. We can urge our local city and county governments to develop well-designed, large-scale resource recovery

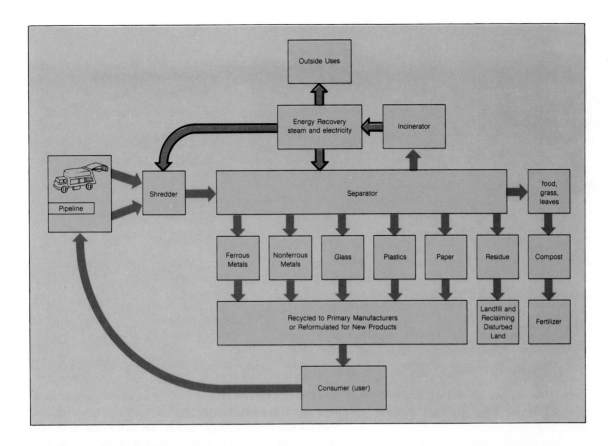

centers, of the sort depicted in Figure 47.2. Such a plant has been operating in Saugus, Massachusetts. In such centers, existing dumps and landfills would be urban "mines" from which we might recover nearly one-quarter of our past and present solid wastes.

Case Study: Air Pollution

If you were to compare the earth with an apple from the grocery store, the atmosphere would be no thicker than the layer of shiny wax applied to it. Yet this thin, finite layer of air receives more than 700,000 metric tons of pollutants each day in the United States alone.

Air pollutants do more than make the air smell, and cut down visibility, and discolor buildings. They *corrode* buildings. They ruin oranges, wilt lettuce, stunt the growth of peaches and corn, and damage leaves on conifers hundreds of kilometers away. They can cause humans to suffer headaches, burning eyes, bronchitis, emphysema, and lung cancer. Even in rush-hour traffic, prolonged exposure to the fumes may cause headaches, nausea, stomach cramps, and impaired coordination and vision. Air pollutants come from many sources and contain diverse chemicals. Here we will focus on two of the most serious kinds: acid deposition and smog.

Figure 47.2 Generalized resource recovery system. Solid wastes are treated as urban "ore."

Bulk items are sorted out, and the remainder is shredded and separated. Electromagnets could be used to extract steel and iron; air blowers could send plastic and paper to different recovery chambers. Mechanical screening, flotation, and centrifugal hurling could sort out metals, glass, and garbage.

Once sorted, metals could be returned to mills, smelters, and foundries; glass to various glass processing plants. Organic matter could go to compost centers, later to be used as fertilizers or soil conditioners, perhaps as fuel or animal feed. Wood could be used as fuel. Paper could be recycled or used as fuel.

Even residues from incinerated materials (including particles removed from smoke to limit air pollution) could be processed into road or building material.

Figure 47.3 Effect of calcium depletion in a lake subject to acid deposition. Calcium reservoirs in the bone tissue of adult fish are depleted, causing the fish to become humpbacked, dwarfed, or otherwise deformed. The muscles remain strong, but without a skeleton to act against, they slowly are pulled out of shape.

Acid Deposition. Ozone and oxides of sulfur and nitrogen are among the most dangerous air pollutants. Coal-burning power plants, factories, and metal smelters are the main sources of sulfur-dioxide emissions. Vehicles and fossil fuel power plants are the main sources of nitrogen oxides. Ozone forms during complex reactions of nitrogen oxides in the atmosphere.

Depending on climatic conditions, some of these emissions may remain airborne for a time as tiny particles. When they fall to earth and come in contact with objects in the environment, they are said to be **dry acid deposition**. Between seventy and ninety percent of the sulfur and nitrogen dioxides dissolve in atmospheric water to form a weak solution of sulfuric acid and nitric acid. Winds distribute these acids over great distances before they fall to earth, as **wet acid deposition**, in rain and snow. Normal rainwater has a pH of about 5.6; acid rain is between four and forty times more acidic, sometimes as much as lemon juice.

Acid deposition and ozone originating in industrial regions of England and West Germany are destroying large tracts of forests in northern Europe. In Canada and the United States, acid deposition damages forests and crops. Acid deposition attacks marble, metals, mortar, rubber, plastic, even nylon stockings. Nutrients and toxic metals (such as aluminum) leached from soils as a result of acid deposition enter nearby streams, lakes, and rivers.

Because the soils and vegetation vary in different watersheds (page 703), acid deposition does not have the same effect on all aquatic ecosystems. Highly alkaline soils neutralize the acids before runoff carries them into the system. Water with high concentrations of carbonates also will neutralize the acids. However, in watersheds throughout much of northern Europe, southeastern Canada, and in scattered regions of the United States, thin soils overlie solid granite—and such soils do not provide any buffer against the acids. Excess acidity is already affecting aquatic life in thousands of lakes and thousands of kilometers of streams in the United States. Some Canadian biologists predict that 48,000 lakes in Ontario will be devoid of higher aquatic life within the next two decades (see, for example, Figures 47.3 and 47.4).

In 1983, a United States government task force finally confirmed that power plants, factories, and vehicles are indeed the main sources of acid depositions, and that these airborne pollutants are indeed damaging the environment; as of 1986, not much had been done about it. Ironically, local air pollution standards have in some cases contributed to the problem by calling for remarkably tall smokestacks on power plants and smelting plants. The idea is to dump the acid-laden smoke high enough in the atmosphere for winds to distribute it elsewhere—which winds readily do. The world's tallest smokestack, in Sudbury, Ontario, accounts all by itself for about one percent of the annual worldwide emissions of sulfur dioxide. Canada, however, cannot be singled out in this issue. Canada presently receives more acid depositions from industrialized regions of the northeastern United States than it sends across its southern border (Figure 47.4). Prevailing winds do not stop at national boundaries; the problem is of global concern.

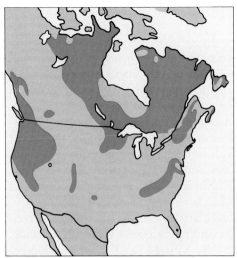

(a) Generalized diagram of areas in North America containing lakes sensitive to acid precipitation.

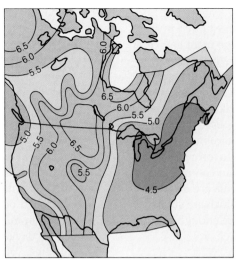

(b) Regions receiving acid rain.

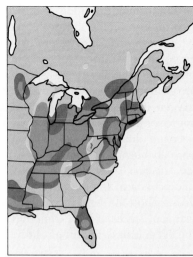

(c) Distribution of emissions of sulfur dioxide (yellow) and nitrous oxides (red).

Figure 47.4 Acid deposition patterns and sensitive regions in the continental United States and Canada.

Industrial and Photochemical Smog. Climate, topography, and air pollutants interact in ways that intensify the effects of local air pollution. For example, in a **thermal inversion**, a layer of dense, cool air becomes trapped beneath a layer of warm air (Figure 47.5). When that happens, pollutants cannot diffuse into the higher atmosphere or be dispersed by winds; they accumulate to dangerous levels right above their sources. By intensifying a phenomenon known as smog, thermal inversions have contributed to some of the worst air pollution disasters.

There are two types of smog, industrial and photochemical, both of which occur in major cities. **Industrial smog** is gray air. It predominates in industrialized cities that have cold, wet winters, such as London, New York, Pittsburgh, and Chicago. Such cities burn considerable amounts of fossil fuel for heating, manufacturing, and producing electric power. The burning fuel releases two major classes of pollutants:

particulates   *airborne solid particles or liquid droplets (including dust, smoke, ashes, soot, asbestos, oil, and bits of heavy metals such as lead)*

oxides of sulfur   *in particular, sulfur dioxide and sulfur trioxide*

These substances may reach lethal concentrations when winds and rain do not disperse them. For example, industrial smog was the source of London's 1952 air pollution disaster, in which 4,000 people died.

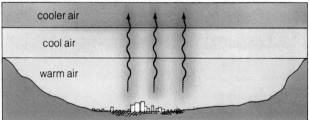

a Normal pattern

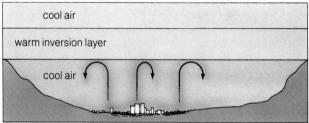

b Thermal inversion

Figure 47.5 Trapping of airborne pollutants by a thermal inversion layer.

Photochemical smog is brown, it smells, and it is hazardous to living things. It occurs most often in cities with warm, dry climates and where the main source of air pollution is nitric oxide, produced in the internal combustion engine. Los Angeles, Denver, and Salt Lake City are examples of brown air cities. Nitric oxide reacts with oxygen in the air to form nitrogen dioxide, the main source of the brownish haze. When exposed to sunlight, nitrogen dioxide can react with hydrocarbons (spilled or partially burned gasoline, most often) to form *photochemical oxidants*. These are among the main components of photochemical smog. Other components are ozone and PANs (short for *peroxyacylnitrates*), which are similar to tear gas. Traces of PAN compounds are enough to sting the eyes, irritate the lungs, and damage crops.

Case Study: Water Pollution

There is a tremendous amount of water in the world, yet three of every four humans do not have enough water or, if they do, it is contaminated. Most water is saline (too salty) and cannot be used for human consumption or agriculture. In fact, for every 1 million liters of water, only about 6 liters are in a readily usable form.

Yet we have tapped into the hydrologic cycle and are using it, directly or indirectly , as a dumping ground for the by-products of human existence. Water becomes unfit to drink (even to swim in) once it contains human sewage and animal wastes. Agricultural runoff becomes polluted with sediments, insecticides, herbicides, and plant nutrients. Often the nutrients cause explosive growth of cyanobacteria in lakes and slow-moving rivers. When those bacteria die, the water can become putrid. Industrial activity and power-generating plants pollute water with chemicals, radioactive materials, and excess heat (thermal pollution).

Table 47.1 lists some water pollutants and highlights some suggestions for their control. To get an idea of how difficult the clean-up task will be, consider the problems inherent in existing waste-water treatment methods. About 15,000 municipal plants treat the liquid wastes from about seventy percent of the population and from about 87,000 industries. The wastes of the remaining population (mostly suburban and rural) are treated in lagoons or septic tanks, or they are discharged untreated directly into waterways.

In a mechanical process called *primary treatment*, screens and settling tanks remove coarse suspended solids (called sludge) from the waste water. Sludge must be disposed of by burning, in landfills, or by further treatment. Although chlorine is often used to kill pathogenic microbes in the waste water, it does not kill them all. Moreover, chlorine may react with certain industrial chemicals to produce chlorinated organic chemicals, some of which are suspected carcinogens (cancer-causing agents).

In *secondary treatment*, various microbes are used to degrade the organic matter. After primary treatment (but before chlorination), the waste water is either (1) sprayed and trickled through large beds of exposed gravel in which monerans and protistans live or (2) aerated in tanks and "seeded" with microbes to promote degradation. The microbes used in this biologically based treatment are vulnerable to poisoning by toxic substances in the waste water, and every so often the treatment plants are shut down to replace the destroyed microbial populations.

Even after secondary treatment, the water still contains oxygen-demanding wastes, suspended solids, nitrates, phosphates, viruses, and toxic substances, including heavy metals, pesticides, and industrial chemicals. Perhaps the waters are chlorinated, but then they are usually discharged into the waterways.

Tertiary treatment adequately reduces the pollutant levels in waste water. Among the processes involved are advanced methods of precipitating suspended solids and phosphate compounds, adsorption of dissolved organic compounds, reverse osmosis, stripping nitrogen from ammonia, and disinfecting the water through chlorination or ultrasonic energy vibrations. Tertiary treatment, however, is used on only five percent of the waste water. It is largely in the experimental stage and is expensive.

What all this means is that most waste water is not being properly treated. A typical pattern is repeated thousands of times along our waterways. Water for drinking is removed *upstream* from a city, and wastes from industry and sewage treatment are discharged *downstream*. It takes no great leap of the imagination to see that pollution intensifies as rivers flow toward the oceans. In Louisiana, where the waters drained from the central states flow toward the Gulf of Mexico, pollution levels are high enough to be a real threat to public health. Water destined for drinking does get treated to remove pathogens—but the treatment does not remove poisonous heavy metals (such as mercury) being dumped into the waterways by numerous factories upstream. *You may find it illuminating to investigate where your own city's supply of water comes from and where it has been.*

REPERCUSSIONS OF MODERN AGRICULTURE

Through the efforts of the environmental scientist Tyler Miller and others, there is increasing awareness that the agricultural practices necessary to sustain the burgeoning human population collectively represent one of the most important trends affecting life on earth. Here we

Table 47.1 Major Water Pollutants: Sources, Effects, and Possible Controls

Pollutant	Main Sources	Effects	Possible Controls
Organic oxygen-demanding wastes	Human sewage, animal wastes, decaying plant life, industrial wastes	*Overload depletes dissolved oxygen in water; animal life destroyed or migrates away; plant life destroyed*	Provide secondary and tertiary waste-water treatment; minimize agricultural runoff
Plant nutrients	Agricultural runoff, detergents, industrial wastes, inadequate waste-water treatment	*Algal blooms and excessive aquatic plant growth upset ecological balances; eutrophication*	Agricultural runoff too widespread, diffuse for adequate control
Pathogenic bacteria and viruses	Presence of sewage and animal wastes in water	*Outbreaks of such diseases as typhoid, infectious hepatitis*	Provide secondary and tertiary waste-water treatment; minimize agricultural runoff
Inorganic chemicals and minerals	Mining, manufacturing, irrigation, oil fields	*Alter acidity, basicity, or salinity; also render water toxic*	Disinfect during waste-water treatment; stop pollutants at source
Synthetic organic chemicals (plastics, pesticides, etc.)	At least 10,000 agricultural, manufacturing, and consumer uses	*Many are not biodegradable; chemical interactions in environment are poorly understood. Many poisonous*	Push for biodegradable materials; prevent entry into water supply at source; enforce existing laws
Fossil fuels (oil particularly)	Two-thirds from machinery, automobile wastes; pipeline breaks; offshore blowouts and seepage, supertanker accidents, spills, and wrecks; heating; transportation; industry; agriculture	*Vary with location, duration, and type of fossil fuel; potential disruption of ecosystems; economic, recreational, and aesthetic damage to coasts*	Strictly regulate oil drilling, transportation, storage; collect and reprocess engine oil and grease; develop means to contain spills
Sediments	Natural erosion, poor soil conservation practices in agriculture, mining, construction	*Fill in waterways, reduce shellfish and fish populations*	Put already existing soil conservation practices to use

will focus on only two of the consequences of the following practices (Table 47.2 lists others):

1. Expansion of agriculture into marginal lands, including the clearing of forests for grazing, farming, lumber, and fuel.

2. Increased reliance on irrigation.

Conversion of Marginal Lands

Almost 3.5 billion acres are now under cultivation. It has been proposed that another 7 or 8 billion acres be converted to agriculture. Aside from the environmental impact of such expansion, the best land available for agriculture is already being used for agriculture (Figure 47.6). What remains as arable land (able to support cultivation) is less desirable. In some heavily populated regions such as Asia, food problems are severe, yet more than eighty percent of the arable land is being intensively cultivated already.

There have been valiant efforts to improve crop production on existing land. Under the banner of the so-called **green revolution**, considerable research has been directed toward (1) improving the varieties of crop plants for higher yields and (2) exporting modern agricultural practices and equipment to developing countries. Those countries rely on subsistence agriculture (with energy inputs from sunlight and from human labor) and animal-assisted agriculture (with energy inputs from human labor and the work of draft animals, such as oxen).

By contrast, modern agriculture is based on massive inputs of fertilizers, pesticides, and ample irrigation to sustain high-yield crops. It is based also on fossil fuel energy to drive the farming machines. Although crop yields are four times as high, the modern practices use up a hundred times more energy and mineral resources.

Besides, the plain truth is that developing countries depend on subsistence farmers who cannot afford to take widespread advantage of the new crop strains. The ones who can afford to make the investment come to depend on industrialized producers of fertilizers and machinery. Of necessity, the costs of fertilizers and machinery are

Table 47.2 Some Environmental Consequences of Large-Scale Food Production and Proposed Solutions

Effect	Proposed Solutions
Soil erosion and loss of soil fertility	Conservation-tillage cultivation, crop rotation, and other well-known conservation practices; land-use control
Salination and waterlogging of irrigated soils	Better irrigation and drainage systems; higher water prices
Pollution from pesticides	Increased use of biological control and integrated pest management
Water pollution from runoff of fertilizer and animal wastes	Preventing soil erosion and recycling animal wastes to land
Overgrazing	Laws and land-use control; less dependence on meat in diets
Deforestation	Replanting forests; land-use controls
Endangered wildlife from habitat loss	Laws for wildlife protection; land-use controls
Loss of genetic diversity	Genetic storage banks for plants; wildlife protection
Overfishing	Fish farming; limitations on catches

Adapted from G. T. Miller, Jr. 1986.

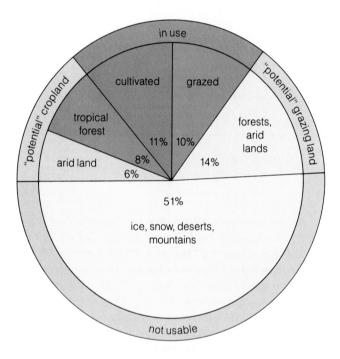

Figure 47.6 Classification of the earth's land. Theoretically, the world's cropland could be doubled in size by clearing tropical forests and irrigating arid lands. But converting this marginal land to cropland would destroy valuable forest resources, cause serious environmental problems, and possibly cost more than it is worth.

reflected in market food prices—which are too high for much of the country's own population.

Pressures are most intense in food-deficit areas of Asia, the Middle East, Africa, and the Andean countries. The livestock population often grows at a pace similar to the human population's in order to expand food supplies, draft animal power, or family wealth and security. As the herds increase, they overgraze more and more of the countryside and open the land to severe soil erosion.

Each year, the world's forests shrink by nearly one percent as tracts are cleared for farming and grazing or overcut for lumber and firewood. About one-third of the original expanse of the world's tropical forests has already been cleared or seriously degraded (Figure 47.7). Deforestation, especially on steep slopes, leads to loss of the fragile soil layer and disrupts the watershed. In the tropics, soil loss means long-term fertility loss as nutrients are quickly washed out of the system, leaving nutrient-poor soil behind (page 719).

Clearing of the tropical forests, with their incredibly diverse organisms, probably means extinction for thousands of species by the close of this century. It will be our loss as well as theirs. Consider that a very small number of crop and livestock species form the basis of most of the world's agriculture. However, this base of food production can be broadened and made less vulnerable if we develop new crops or hybrids. Recently a remarkable variety of wild corn was discovered. Unlike cultivated corn, which must be planted each year, the wild variety is perennial—and it is more resistant to disease. If, by genetic engineering, the attributes of this wild variety can be transferred to cultivated strains, the agricultural benefits would be enormous. Only a few

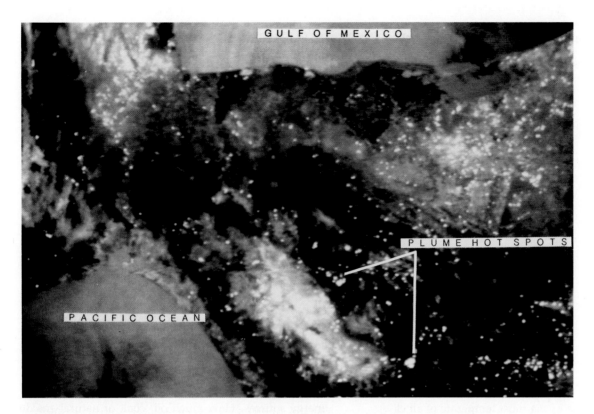

Figure 47.7 Satellite photograph of southern Mexico and northern Guatemala taken on April 18, 1984. The well-defined white spots are fires, most of which are associated with land clearing for agriculture. The white area near the center of the photograph (the Grijalva Basin) is a major agricultural region; most of the burning here is for clearing previously cultivated land. The area in the upper left (around Veracruz) and in the lower right (around the Guatemala border) are primarily virgin tropical forest, being cleared for agriculture at great ecological cost. (Satellite sensory devices penetrated the smoke to reveal the underlying fire activity.)

thousand plants of the wild corn are known to exist; they are extremely vulnerable to extinction.

Irrigation

Almost half the world's food is already produced on irrigated land. The supplies are drawn from surface waters (such as lakes, rivers, and reservoirs) or groundwater, but most of these sources are not located where they are needed to support irrigation.

Irrigation has serious environmental consequences. As irrigation water flows over and through the soil, water evaporation and salt buildup in the soil (salination) can stunt growth, decrease yields, and eventually kill crop plants. Improperly drained irrigated lands can become waterlogged. Water accumulating underground gradually raises the water table close to the soil surface, saturating the plant roots in toxic saline water. Salinity and waterlogging can be corrected with proper management of the water-soil system, but the economic cost is high.

Groundwater is used for many purposes, but irrigation is often paramount. Consider what is happening to the Ogallala aquifer (Figure 47.8) in the United States. (An *aquifer* is a water-bearing stratum of permeable rock, gravel, or sand.) Farmers withdraw so much water from the Ogallala aquifer that the annual overdraft (the amount of water not replenished) is nearly equal to the annual flow of the Colorado River! As a result, the already low water tables in much of the region are falling rapidly, and stream and underground spring flows are dwindling. Where will the water come from when the aquifer is depleted?

A QUESTION OF ENERGY INPUTS

Paralleling the J-shaped curve of human population growth is a steep rise in energy consumption. The rise is due not only to increased numbers of energy users, but also to extravagant consumption and waste.

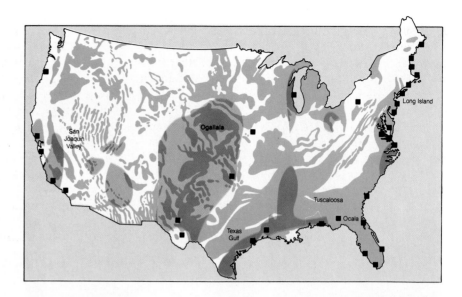

Figure 47.8 Major underground aquifers containing 95 percent of all fresh water in the United States are being depleted in many areas and contaminated elsewhere through pollution and saltwater intrusion. Blue areas indicate major aquifers; gold, the areas of groundwater depletion; and the black boxes indicate areas of saltwater intrusion.

For example, in one of the most temperate of all climates, a major university constructed seven- and eight-story buildings with narrow, sealed windows. The windows cannot be opened to catch the prevailing ocean breezes; the windows and the buildings themselves were not designed or aligned to take advantage of the abundant sunlight for passive solar heating and breezes for passive solar cooling. Massive energy-demanding cooling and heating systems are used instead.

Inefficiencies of this sort may be curtailed sooner than might be expected, for current energy supplies are limited. When you hear talk of abundant energy supplies, keep in mind there is an enormous difference between the total supply and the net amount available. *Net energy* is the energy left over after subtracting the energy used to locate, extract, transport, store, and deliver energy to consumers. Some sources of energy, such as direct solar energy, are renewable; others, such as coal and petroleum, are not. Currently, eighty-two percent of the energy stores being tapped fall in the second category (Figure 47.9).

Fossil Fuels

Fossil fuels are the legacies of primary producers that lived hundreds of millions of years ago. The carbon-containing remains of plants were buried and compressed in sediments and gradually transformed into coal, petroleum (oil), and natural gas. Often you will see references to our annual "production" rates for these energy sources. How much oil, coal, or natural gas do we really produce each year? None. We simply *extract* them from the earth.

Even with stringent conservation efforts, known petroleum and perhaps natural gas reserves may be depleted during the next century. As petroleum and natural gas deposits become depleted in easily accessible areas, we begin to seek new sources, often in wilderness areas such as Alaska and in other fragile environments such as the continental shelves. The *net* energy decreases as costs of extraction and transportation increase; the environmental costs of extraction and transportation escalate.

Colorado, Utah, and Wyoming probably have more potential oil than the entire Middle East. These states have vast deposits of buried rock containing the hydrocarbon *kerogen*, which can be converted to a heavy oil called shale oil. However, collecting, concentrating, heating, and converting kerogen into shale oil may cost so much that the net energy yield would be negligible. The extraction process would disfigure the land, increase water and air pollution, and tax existing water supplies in regions already facing water shortages. Moreover, the extraction process produces benzopyrene (a known carcinogen) by the ton, even though there is no known way to use it or to dispose of it safely.

What about coal? The United States has one-fourth of the world's known coal reserves. In principle, world reserves can meet the energy needs of the entire human population for at least several centuries. But coal-burning has been the largest single source of air pollution, for most coal reserves contain low-quality, high-sulfur material. Unless the sulfur is removed before burning or after (from smokestack gases), quantities of sulfur oxides enter the air, leading to acid deposition. Fossil fuel burning also adds large amounts of carbon dioxide to the atmosphere and may be contributing to global warming due to the greenhouse effect (page 705).

Pressure is on to permit widespread strip-mining of coal reserves close to the earth's surface. How many millions of acres should be opened to mining companies? Strip-mining greatly limits the usefulness of the land for agriculture, grazing, and wildlife habitat—unless it is restored. Restoration is difficult and expensive in arid and semiarid lands, where much of strip-mining is now attempted. Who will pay for restoration, and will restoration be adequate?

Solar Energy

"Solar energy" is not merely a direct source of energy. As we have seen, radiant energy from the sun drives the hydrologic cycle and the winds—both of which can be tapped for energy production. Direct solar energy and wind are the most abundant, clean, and safe energy sources now being considered as alternatives to fossil fuels. Moreover, solar energy captured by way of photosynthesis yields biomass (page 699). Biomass in the form of wood, cattle dung, and crop residues represents about thirteen percent of the world's energy consumption.

Hydroelectric power is produced as flowing water passes through and spins turbines, producing electricity. In 1984, hydroelectric power supplied about one-fourth of the world's electricity and about five percent of the world's total primary energy. Most environmental impacts come about because of the placement of large dams on rivers to form reservoirs and to house the turbines. These large water projects often destroy scenic landscapes and wilderness areas, flood good agricultural lands, displace people, and disrupt natural aquatic and terrestrial wildlife habitat. The current trend in the United States seems to be toward a large number of small-scale power plants on smaller rivers instead of massive dams of the sort built on the Columbia, Colorado, and Tennessee rivers in North America.

In many regions, winds are strong and predictable enough to provide a source of energy. For example, high prevailing winds sweep through the American Midwest, from the Dakotas to Texas, and through the canyons of California. Wind energy seems to be an attractive energy option. It would be an unlimited, nonpolluting source of energy. It has been proposed that perhaps 300,000 turbine towers harnessing the winds of the Great Plains would provide fully half of the present energy needs of the United States. California now leads the world in the use of wind energy.

Modern wind turbines are of a simple design that disfigures the landscape far less than many other energy alternatives. Turbine towers would take up less land than solar-energy, nuclear, and fossil fuel generating plants. They would be far less disruptive than strip-mining for coal. Possibly, wind turbines constructed on existing electrical transmission towers could feed electrical output directly into regional utility lines. The possibility of harnessing the winds is economically and technologically feasible now.

So far, the most important use of direct solar energy is to collect and concentrate it for heating water and buildings. Millions of homeowners throughout the world already use solar energy for these purposes. On

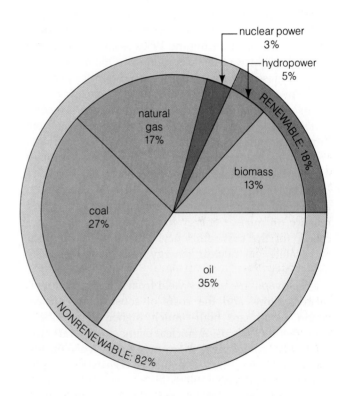

Figure 47.9 World consumption of primary nonrenewable and renewable energy sources in 1984.

a lifetime cost basis, this is the most economical way to heat a house and to get hot water. Although large-scale solar power plants that produce electricity have been built for demonstration purposes, they are expensive to operate and have a low net energy yield. Thus, the use of solar energy to produce temperatures high enough to generate electricity will not become cheap enough to compete with other energy alternatives, at least for the foreseeable future. However, the development of solar photovoltaic cells that can convert sunlight directly to electricity is rapidly progressing and may be a competitive electricity source by the mid-1990s.

Nuclear Energy

As Hiroshima burned in 1945, the world recoiled in horror from the destructive potential of nuclear energy. Optimism soon replaced horror as nuclear energy became publicized during the 1950s as an instrument of progress. That was the beginning of Operation Plowshare, a massive effort to harness the atom for peaceful use. Now, more than four decades later, nuclear power plants dot the landscape. By 1985, more than 300 commercial nuclear reactors in twenty-five countries were providing nine percent of the world's electricity, amounting to about three percent of the world's primary energy. Industrialized nations such as France, which are poor in energy resources, now depend heavily on nuclear power. Yet in most countries, plans to extend reliance on nuclear energy have been delayed or cancelled, for the cost, efficiency, environmental impact, and safety of nuclear energy are being seriously questioned.

The overall net energy yield from nuclear reactors is relatively low, and the costs of constructing nuclear power plants are high—much higher than initially expected. By 1990, using nuclear energy to generate electricity may cost slightly more than using coal—even if the coal-burning plants are equipped with expensive pollution control devices.

What about safety? The radioactivity escaping from a nuclear plant during normal operation is actually less than the amount released from a coal-burning plant of the same capacity. Also, nuclear plants do not add carbon dioxide to the atmosphere, as coal-burning plants do. However, there is the potential danger of a *meltdown*—a massive internal explosion. As nuclear fuel breaks down, it releases considerable heat. Typically, water circulates under high pressure over the fuel and absorbs the heat. The heated water produces the steam that drives electricity-generating turbines. Should a leak develop in the circulating water system, water levels around the fuel might plummet. The nuclear fuel would heat rapidly, past its melting point. The melting fuel would pour onto the floor of the generator, where it would come into contact with the remaining water and instantly convert it to steam. Formation of enough steam, along with other chemical reactions, could blow the system apart, releasing radioactive material into the surroundings. In addition, the overheated reactor core could melt through its thick concrete containment slab and into the earth, causing groundwater contamination.

In 1986, the potential dangers of nuclear power were brought home to people throughout the world. The Chernobyl power station in the Soviet Union experienced an explosion and a major graphite fire. Large amounts of radiation were released into the atmosphere as the plant's containment structures were breached. Within a few weeks, thirty-one people died from exposure to large amounts of radiation, and it has been projected that tens of thousands may die prematurely from radiation-induced cancers and other disorders. No one knows how long the environment around Chernobyl will remain contaminated.

The Chernobyl incident clearly underscores the consequences of nuclear accidents. What about routine nuclear wastes? Nuclear fuel cannot be burned to harmless ashes, like coal. After about three years, the fuel elements of a reactor are spent. They still contain about a third of the useful uranium fuel, but they also contain hundreds of new radioactive isotopes produced during the reactor operation. Altogether, these wastes are an enormously radioactive, extremely dangerous collection of materials. As they undergo radioactive decay, they produce tremendous heat. They are immediately plunged into water-filled pools and stored for several months (sometimes years) at the power plant. The water cools the wastes and keeps radioactive material from escaping. At the end of the holding period, however, the remaining isotopes are still lethal. But the decay rates of some remaining isotopes mean that they must be kept out of the environment for thousands of years. If one isotope of plutonium (Pu-239) is not removed, the wastes must be stored for *a quarter of a million years!*

There are plans to seal radioactive wastes in ceramic material, place them in steel cylinders, then bury them deep underground in supposedly stable rock formations that are free from exposure to water. No radioactive wastes have yet been put into permanent underground storage in the United States. Such facilities are not expected to be available until after the turn of the century, and some geologists contend that none of the proposed storage sites is acceptable.

The development of another type of reactor is being considered. In theory, breeder reactors would consume a rare isotope of uranium and, in the process, convert a much greater amount of a common form of uranium

(U-238) to an isotope of plutonium that can be used as nuclear fuel. However, breeder reactors cannot be water cooled. The design most actively considered uses liquid metallic sodium, a highly dangerous and corrosive substance, as the coolant. There are major problems in containing the molten sodium and in maintaining the reactors in a functional state. Also, even though a conventional reactor cannot explode like an atomic bomb, a breeder reactor potentially could undergo a very small nuclear explosion.

A third type of nuclear power source on the drawing boards is fusion power, in which hydrogen atoms are fused to form helium atoms with considerable release of energy. The process is analogous to the reactions creating the heat energy of the sun. The scientific, technological, and economic problems associated with developing fusion power are so great that, without a major breakthrough, fusion power is not expected to be available to produce electricity on a commercial basis until the last half of the next century, if ever.

Nuclear Winter

One more point should be made here. Earlier, we considered the catastrophic extinctions that marked the boundaries of the great eras of geologic time. We can only speculate on their causes—collisions between asteroids and the earth being at least one of them. Quite probably, nuclear energy has placed in human hands the means of equally catastrophic extinctions. We will not know this for sure until it ever happens, but several predictions have been made on the basis of detailed computer simulations and analyses.

According to one scenario, a nuclear exchange involving about one-third of the existing American and Soviet arsenals would probably kill between forty and sixty-five percent of the human population (along with a good portion of most other forms of life). Those escaping rapid death would have to remain in shelters for a week to three months or more to avoid exposure to dangerous radiation levels. The nuclear detonations would inject a huge, dark cloud of soot and smoke over most of the earth, especially the Northern Hemisphere, and would block out the sun. Consequently, much of the earth would experience darkness and below-freezing temperatures for months—an effect called **nuclear winter**. Cold temperatures and darkness would be well beyond the tolerance limits of many plant and animal species, including some crops and livestock needed by humans, as well as most humans themselves. Will we be the "catastrophe" that brings the Cenozoic Era to a close? And what, we can only wonder, will we usher in as a result?

PERSPECTIVE

Molecules, cells, tissues, organs, organ systems, multi-celled organisms, populations, communities, ecosystems, the biosphere. These are the architectural systems of life, assembled in increasingly complex ways over the past $3\frac{1}{2}$ billion years. We are latecomers to this immense biological building program. Yet, during the relatively short span of 50,000 years, we have been restructuring the stuff of life at all levels—from recombining DNA of different species to changing the nature of the thin, life-giving atmosphere.

It would be presumptuous to think we are the only organisms that have ever changed the nature of living systems. Even during the Proterozoic Era, photosynthetic organisms were irrevocably changing the course of biological evolution by gradually enriching the atmosphere with oxygen. In the present as well as the past, competitive adaptations have assured the rise of some groups, whose dominance has assured the decline of others. Thus change is nothing new to this biological building program. What *is* new is the accelerated, potentially cataclysmic change being brought on by the human population. We now have the population size, the technology, and the cultural inclination to use up energy and modify the environment at frightening rates.

Where will rampant, accelerated change lead us? Will feedback controls begin to operate as they do, for example, when population growth exceeds the carrying capacity of the environment? In other words, will negative feedback controls come into play and keep things from getting too far out of hand?

Feedback control will not be enough, for it operates only when deviation already exists. Our explosive population growth and patterns of resource consumption are founded on an illusion of unlimited resources and a forgiving environment. A prolonged, global shortage of food or the passing of a critical threshold for some toxic pollutant in the atmosphere can come too fast to be corrected. At some point, such deviations may have too great an impact to be reversed.

What about feedforward mechanisms? Many organisms have early warning systems. For example, skin receptors sense a drop in outside air temperature. Each sends messages to the nervous system, which responds by triggering mechanisms that raise internal body temperature before the body itself becomes dangerously chilled. With feedforward control, corrective measures can begin before change in the external environment significantly alters the system.

Even feedforward controls are not enough for us, for they go into operation only when change is under way. Consider, by analogy, the DEW line—the Distant Early Warning system. This system is like a sensory receptor, one that detects intercontinental ballistic missiles that may be launched against North America. By the time this system detects what it is designed to detect, it may be far too late, not only for North America but for the entire biosphere.

It would be naive to assume we can ever reverse who we are at this point in evolutionary time, to de-evolve ourselves culturally and biologically into becoming less complex in the hope of averting disaster. However, there is no reason to assume that we cannot avert disaster by using a third kind of control mechanism, one that is uniquely our own. We have the capacity to anticipate events *before* they happen. We are not locked into responding only after irreversible change has begun. We have the capacity to anticipate the future—it is the essence of our visions of utopia or of nightmarish hell. Thus we all have the capacity for adapting to a future which we can partly shape. We can, for example, learn to live with less. Far from being a return to primitive simplicity, it would be one of the most complex and intelligent behaviors of which we are capable.

Having that capacity and using it are not the same thing. We have already put the world of life on dangerous grounds because we have not yet mobilized ourselves as a species to work toward self-control. Our survival depends on predicting possible futures. It depends on designing and constructing ecosystems that are in harmony not only with what we define as basic human values but also with the biological models available to us. Human values can change; our expectations can and must be adapted to biological reality. *For the principles of energy flow and resource utilization, which govern the survival of all systems of life, do not change.* It is our biological and cultural imperative that we come to terms at last with these principles, and with what will be the long-term contribution of the human species to the unity and diversity of life.

Review Questions

1. Under which conditions would an ecologically based system for handling solid wastes benefit the culture more than a recycling system?

2. What is the basic plan of a generalized resource recovery system?

3. What is meant by primary, secondary, and tertiary waste-water treatment?

4. Where do organic oxygen-demanding wastes come from, how do they affect organisms, and how can this type of pollutant be controlled?

5. Categorize the principal air pollutants, their sources, and the methods of controlling each.

6. How can energy overconsumption and waste be reduced by the average U.S. citizen?

7. What is a meltdown? Can a nuclear power plant explode like an atom bomb?

Readings

Audubon (bimonthly). National Audubon Society. 950 Third Avenue, New York, New York 10022.

BioScience (monthly). American Institute of Biological Sciences. 1401 Wilson Blvd., Arlington, Virginia 22209. Official publication of AIBS; major coverage of environmental concerns.

Brown, L. R., et al. 1984, 1985, 1986. *State of the World.* New York: Norton. Excellent summaries of world environmental issues.

Environment (monthly). Heldref Publication, 4000 Albemarle St., NW, Washington, D.C. 20016. Excellent in-depth articles of key environmental issues.

Miller, G. T., Jr. 1985. *Environmental Science: An Introduction.* Belmont, California: Wadsworth. In our view, the best introduction to environmental science available.

Myers, N. 1983. *A Wealth of Wild Species: Storehouse for Human Welfare.* Boulder, Colorado: Westview Press. Good discussion of the value of preserving endangered species.

Myers, N. 1985. *The Primary Source: Tropical Forests and Our Future.* New York: Norton. Excellent information on the status of tropical forest destruction and why these forests should be protected.

Scientific American (monthly). 415 Madison Avenue, New York, New York 10017.

With the arrival of spring, a male white-throated sparrow living in a patch of swampy Canadian forest whistles a song that sounds something like "Sam Peabody, Peabody, Peabody." He repeats the Sam-Peabody song thousands of times and with such clarity and consistency, we might well wonder how he does it, and what good it does him. We might also wonder why the male swamp sparrows and white-crowned sparrows living in the same patch of forest eschew the Sam-Peabody song and belt out distinctive songs of their own. Moreover, do all those male birds automatically "know" what they are supposed to sing the first time they do it, or does each one learn something from the environment that influences the way they sing?

Questions of this sort lead us into the world of animal behavior studies. Earlier, we considered various aspects of the external environment and the internal state that stimulate an animal's sensory receptors (page 356). We saw how the nervous system and endocrine system process information about such stimuli and then mobilize the body's effectors (muscles and glands) to bring about appropriate, coordinated responses to the stimulation. Here we will consider specific examples of the internal mechanisms underlying predictable patterns of behavior, as well as examples of learning experiences by which some predictable responses can be modified in novel ways. The following premises provide a framework for the chapter:

1. *Animal behavior* refers to the coordinated neuromotor responses an animal makes to external and internal stimuli.

2. Behavioral responses are outcomes of the integration of sensory, neural, endocrine, and effector components, all of which have a genetic basis and therefore are subject to evolution by natural selection.

3. Heritable, genetically based neural programs provide each new individual with a means of responding to situations that members of its species are likely to encounter in their environment (page 355).

4. Interactions between the animal and its environment can lead to modifications in the neural mechanisms underlying behavior; this can happen while the animal embryo is developing as well as later, when juvenile and adult animals undergo learning experiences.

5. Behavior is adaptive, to the extent that it contributes to the reproductive success of the individual.

48

ANIMAL BEHAVIOR

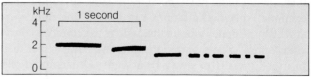

Figure 48.1 A male white-throated sparrow and a sound spectrogram of his territorial song.

MECHANISMS OF BEHAVIOR

Genetic Foundations of Behavior

The motor responses an animal can make to its environment are dictated by the gridwork of neurons in its nervous system and the patterns of activity among them (page 355). Thus, for example, our capacity to control our lips and tongue when speaking resides in the specific design features of our brain (Figure 25.11), just as a white-throated sparrow's ability to sing depends on the coordinated action of millions of neurons in his brain. Genetic instructions are necessary for the growth and "design" of the brain; hence genes contribute in an important way to the development of all behavior.

For example, suppose a researcher takes eggs from the nests of white-throated sparrows and white-crowned sparrows. She places the eggs in an incubator and hand-rears the young after they hatch (which, by the way, is a monumentally difficult task that gives one great respect for the parental abilities of birds). The baby white-throats and white-crowns are both reared under the exact same conditions, fed exactly the same foods, and in all ways treated similarly. As the birds grow up, they are permitted to hear the tape-recorded songs of adult white-throats and white-crowns. Before a year passes, the males will be singing. The white-throats will sing "Sam Peabody, Peabody, Peabody." And the white-crowns will sing a different song, one with all the whistles and buzzy trills characteristic of their species.

This experiment does not demonstrate that bird song is "genetically determined" in the sense that the environment plays no role whatsoever in its development. However, since the white-throated and white-crowned birds were exposed to identical environmental conditions, the *difference* in their singing behavior must be the result of genetic differences between them. The number of genes involved to produce the difference is not important to this argument. Even a one-gene difference could produce this result. One gene-specified enzyme acting at an early stage of embryonic development can have large effects on the sequence of events by which part of the nervous system develops. By analogy, suppose you substitute garlic powder for sugar when following a recipe for chocolate cake. Even if all the other ingredients and all subsequent cake-making steps are identical, that one substitution will have a dramatic effect on the actual cake.

The effects of single genes on behavioral development have been studied in great detail. Studies of fruit flies (*Drosophila*) have been especially revealing. For example, specific mutations can change the frequency at which the males vibrate their wings to produce a courtship "song." Some mutations alter the tendency of flies to move toward or away from light. Other mutations

affect the capacity of flies to remember to avoid entering tubes with distinctive odors and wired to give the entrant an electric shock.

Keep in mind that it is incorrect to say there is a "gene for courtship song" in fruit flies. Rather, gene mutation can affect the development of this and other forms of behavior, as can happen when the gene codes for a key enzyme in a developmental pathway that shapes the part of the nervous system concerned with courtship song.

Hormonal Effects on Behavior

Among the proteins specified by genes are *hormones*, the signaling molecules produced by cells in one body region and carried by the bloodstream to target cells elsewhere in the body. The cellular responses to a hormonal message may ultimately have profound behavioral consequences.

Melatonin, recall, is a hormone produced by cells in the pineal gland of many vertebrates (page 352). This hormone serves as a *behavioral primer*; changes in its concentration in the blood prepare the nervous system for changes in reproductive activity by way of its regulatory effect over the gonads (the testes and ovaries). The pineal gland secretes melatonin mostly at night; photoreceptors activated during the day cause a decrease in its secretion. Thus the glandular output varies with the changing day-length of different times of year; it shows **seasonal photoperiodicity**.

In birds, light penetrating directly through the head feathers and the skull influences melatonin production. Imagine what goes on in the brain of a white-throated sparrow during the lengthening days of spring. The pineal puts out less and less melatonin, the gonads are freed from the hormone's inhibitory effects, and the gonads increase in size compared to what they were in the fall. As the gonads grow, they release hormones of their own, thereby setting in motion a sequence of behavioral responses that will include migration, singing and other forms of territorial behavior, and mating. All of these forms of behavior will occur in a predictable sequence, at a time of year that will most favor reproductive success.

But now a new question arises: Why do the males and *not* the females of so many songbirds sing? Anatomical studies of some songbirds revealed differences between the sexes in the structure and size of brain regions that connect with pathways leading to the syrinx, or vocal organ. (The syrinx is a modified region of the bird windpipe, bronchial tubes, or both.) We know now that hormones influence the development of this entire network, the so-called **song system**.

In very young *male* but not female songbirds, estrogen secretions act on embryonic brain cells, initiating the masculinization of the brain and development of the song system. (Interestingly, when researchers implanted tiny pellets of estrogen in nestling *female* zebra finches, the females developed a brain with a malelike song system.)

A masculinized brain is necessary for singing behavior, but in itself it is not enough to cause a bird to sing. At the start of the breeding season in temperate zones, testosterone is manufactured in the enlarged testes of the male bird. Cells of the song system have receptors that can bind this hormone. Binding triggers changes in metabolic activity that enable the bird to sing after it has staked out a territory and must repel other males and attract a female. As you might predict, if female zebra finches receive testosterone implants when they are adults, they will not sing—*unless* they also have been exposed to estrogen secretions early in life (Figure 48.2).

Thus, estrogen *organizes* the development of the song system; then testosterone *activates* the song system and prepares the bird to sing when properly stimulated. Coordinated interactions between nerve cells and hormones are necessary if a white-throated sparrow is ever to sing so much as a single "Sam Peabody, Peabody, Peabody."

INSTINCTIVE BEHAVIOR

When presented with essence of slug, a newly born garter snake flicks its tongue, bringing the chemical scent into its mouth. Even when a cotton swab doused with slug extract is the snake's first encounter with potential food, it will complete this functional response; sometimes the snake goes so far as to strike at the swab. In nature, this is how the garter snake detects and captures small slugs for food without ever having seen another snake do the same thing.

The tongue-flicking response is an example of a behavior that could be called instinctive. In the past, "instinctive" or "innate" behavior was said to be genetically determined, the mistaken idea being that some forms of behavior can arise without environmental influences. However, *no aspect of phenotype—including behavior—can develop without both genetic and environmental contributions*. The tongue-flicking response requires a tongue, chemical receptors, and a nervous system, all of which developed in the snake embryo as a result of complex interactions between genes and the embryonic environment.

Nevertheless, the term **instinct** can be salvaged if we limit its meaning to the capacity of an animal to complete a fairly complex, stereotyped response to a first-time encounter with a key stimulus—in other words, without having had prior experience with the stimulus.

a

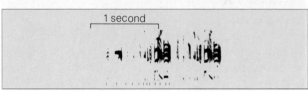

b

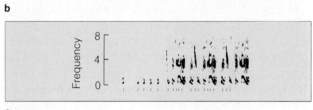

c

Figure 48.2 (**a**) Zebra finch, with sound spectrograms of the full song of (**b**) the male and of the song of (**c**) a female that had been exposed to estrogen as a nestling, then given a testosterone implant as an adult.

What is striking about instinctive behavior is that it seems to be automatically triggered by rather limited environmental cues. When you shake their nests, the baby birds of many species respond with gaping behavior, which is exactly what they do when the (food-carrying) parent lands on the nest rim (see, for example, Figure 34.2d). Male stickleback fish aggressively chase and strike almost any red object placed near their nest even if they have never before seen a red object of any sort. Male sticklebacks happen to have red bellies and they happen to guard nests containing the eggs that they themselves have fertilized, although they will eat the eggs of other sticklebacks in other nests. Under natural

a

b

Figure 48.3 A complex, innate behavioral response by a newly hatched cuckoo to a sign stimulus—round objects in its "foster parent's" nest. The European cuckoo lays its eggs in the nests of other species. Even before a newly hatched cuckoo can open its eyes, it responds to the shape of the host's eggs and shoves them out of the nest (**a**). The foster parents continue to feed the usurper, even when it has grown larger than they are (**b**)

conditions, these fish chase away rival red-bellied males. Very young human infants can be induced to smile simply by showing them a face-sized flat mask with two dark spots corresponding to where eyes would be on a face (one "eye" will not do the trick).

Many years ago, Konrad Lorenz, Niko Tinbergen, and others noted that part of a complex object triggers a complete instinctive response in a host of animals. They named the triggering component of the stimulus the **sign stimulus**. They also proposed that the nervous system of the responding animal was organized into neural units, called *innate releasing mechanisms*, each of which controlled a specific response to a specific sign stimulus. Although these ideas have been modified over the years, it is still worth recognizing that the neural organization of many animals promotes effective first-time responses to sign stimuli in their environment (Figure 48.3).

LEARNING

Many animals respond innately to certain objects yet also store information about the *connections* between various experiences and the consequences of their actions. This information storage leads to **learning**, the adaptive modification of behavior in response to specific experiences during the individual's life. Thus, although the initial feeding response of a newly hatched bird is innate, later on it learns from personal experience where it can find food most profitably—in other words, which food items occur in which areas, and even the best time of day to search for a particular kind of food.

Often, learning has been said to be "environmentally determined" as if it depends solely on experience. Yet the ability to learn resides within the nervous system, where information is stored and the neural wiring is modified in ways that lead to altered behavior. Nervous systems depend on genetic instructions for their development, and the neural wiring that results greatly influences what, exactly, its owner can learn. Thus it is just as inaccurate to ignore the genetic component of learning as it is to consider instincts purely genetic.

Categories of Learning

The conditions under which learning takes place are varied. Here we will briefly describe a few of the different categories of learning as reference points for later discussions.

Associative Learning: In this form of behavior, an animal has a capacity to make a connection between a new stimulus and a familiar one. Ivan Pavlov, a physiologist who was interested in digestive juice secretion, provided

a

b

Figure 48.4 (**a**) Human imprinting objects. No one can tell these goslings that Konrad Lorenz is not Mother Goose. (**b**) An imprinted rooster wading out to meet the objects of his affections. During a critical period of the rooster's life, he was exposed to a mallard duck. Although sexual behavior patterns were not yet developing during that period, the imprinting object became fixed in the rooster's mind for life. Then, with the maturation of sexual behavior, the rooster sought out ducks, forsaking birds of his own kind, and lending further support to the finding that imprinting may be one of the reasons why birds of a feather do flock together.

a classic, controlled study of associative learning. Pavlov observed that his laboratory dogs salivated just after he placed a meat extract on their tongues. He interpreted this to be a simple reflex response. Then he found that if he rang a bell just before giving the dogs the extract, the dogs began salivating at the sound of the bell alone. Pavlov called this new response a **conditioned reflex**, for the dogs had come to associate the sound of the bell (a conditioned stimulus) with food (a reinforcing stimulus).

Instrumental conditioning is another form of associative learning. Here, a reinforcing stimulus (either reward or punishment) appears after a particular behavior is performed by chance. The animal learns by trial and error. For example, earthworms can do this in simple T-mazes (which have a base and two arms shaped like a "T"). They enter the maze at the base, and if they turn down one of its arms, they encounter an irritating stimulus such as an electric shock. If they turn down the other arm, they encounter a moist, darkened chamber that approximates the earthworm habitat. After many trials and enough shocks, the "right" response becomes more frequent.

Extinction: In the forms of learning just defined, the behavior persists for as long as the reinforcement persists. However, if the reinforcing stimulus is withdrawn so the animal does not encounter it again, the learned behavior may soon become extinguished. This, too, is a learning process; it is called extinction.

Latent Learning: This term refers to an ability to store information about features of the environment, information that is later used in guiding the animal through its habitat. For example, a rat will learn to find its way through a complicated maze with fewer food-reinforced learning trials *if* it is first given a chance to explore the maze, without any reward when it happens to find the way out.

Insight Learning: It is only among some primates that insight learning has been adequately demonstrated. With this behavior, novel problems can be solved without trial-and-error practice. In a sense, insight learning is a trial-and-error process that goes on in the brain. It is a synthesis of accumulated experiences that can suggest what responses might be appropriate in new situations. For example, some captive chimpanzees will study a banana dangling from the high ceiling of their enclosure, then they will pile up boxes scattered around the enclosure, having perceived that they can climb the boxes to reach the out-of-reach banana.

Imprinting

Among at least some animal species, the capacity to learn specific kinds of information may be especially pronounced during certain early stages of development. This form of learning, which seems to be time dependent, is called **imprinting**.

For example, some birds are able to walk a few hours after hatching and will traipse after their mother as she moves away from the nest. In one of his studies, Lorenz separated newly hatched goslings from their mother and discovered they followed *any* moving object, even a human being (Figure 48.4). The young birds must learn something when they walk after an object, because they form an attachment to their guide and hurry after it whenever it moves off. However, if young chicks or goslings are not offered any object to follow within a couple of days after hatching, they lose their readiness to imprint. *Thus it appears that the young animals are primed during a short sensitive period early in life to form a learned attachment to a moving object, normally their mother.*

Imprinting has significant long-term consequences. When male goslings or ducklings have matured, they direct their sexual behavior toward members of whatever species they followed as hatchlings. Again, in nature this normally would be a female member of their own species. However, sexually mature male goslings that had imprinted on Lorenz early in life preferred to court humans!

Imprinting and Migration

One of the most intriguing examples of imprinting comes from studies of bird migration. Many animals, including a large number of birds, have the capacity to travel from a summer breeding area to a winter refuge of some sort. The length of the trip varies from species to species. The arctic tern makes round-trip journeys of 22,000 miles per year, a truly remarkable achievement. To migrate across unfamiliar terrain, an animal must possess a **compass sense** (an ability to travel in a constant direction) and a **navigational sense** (a sense of its destination).

Stephen Emlen's work with a small songbird, the indigo bunting, revealed that imprinting plays a role in the development of its compass sense. This bird breeds in the northeastern United States but spends its winters in Mexico south to Panama. The young depart on their own for the wintering grounds when they are just a few months old, and they fly at night. The buntings, Emlen suspected, must use the position of the stars as an orientation guide for their nocturnal travels. Emlen took young, hand-reared birds into a planetarium where he could control the position of the apparent stars. Each bird had its own cage with an ink pad on the floor and sloping paper-covered walls (Figure 48.5). In the fall, the buntings became restless at night and attempted to fly. As they jumped from the floor of their cages they landed on the paper walls and left a mark of their chosen direction. When the planetarium sky mimicked the natural night sky of the fall, the marks were clumped at the south end of the cage; clearly the birds were using the position of the stars as a compass. When the planetarium star pattern was shifted 90° from the actual fall pattern, the birds also shifted the orientation of their nightly leaps by 90°.

Interestingly, when the North Star and the stars of the Big Dipper were eliminated from the artificial night sky in the planetarium, the buntings became disoriented. The North Star, which seems to remain stationary all night in the natural sky, was acting as a fixed compass cue.

In nature, young indigo buntings imprint on the visual image of the night sky, and what they learn is essential if they are to migrate successfully at night later on. When they are hand-reared and kept in a completely dark room for several months, or when they grow up in a planetarium under an artificial sky from which the North Star and the Big Dipper are missing, then they cannot orient themselves in the proper migratory direction. *As with sexual imprinting, there is a sensitive period for the selective acquisition of information that has long-term effects on critical aspects of an animal's behavior.*

Song Learning

What does all of this mean for our singing sparrows? Apparently song behavior is learned—but the song that *is* learned requires species-specific wiring in the bird brain. For example, some male white-crowned sparrows were hand-reared in chambers where they were isolated from sounds produced by birds of any species, including their own. They sang at maturity, but the song was only vaguely reminiscent of typical white-crowned sparrow song. However, when a similarly isolated sparrow heard tape recordings of its species song from ten to fifty days after hatching—and long before it would start to sing itself—the bird remembered what it heard. When it started to sing, it gradually came closer and closer to matching its output with its memory of the adult song. Eventually the sparrow produced a nearly perfect copy of the taped song that it had heard months before.

THE ADAPTIVE VALUE OF BEHAVIOR

Foundations of Behavioral Evolution

Now that we have considered some of the mechanisms underlying the behavior of an individual animal, let's turn to evolutionary mechanisms by which diverse forms of behavior come about. We can begin by defining a few terms currently used in animal behavior studies:

1. *Reproductive success:* survival and production of offspring.

2. *Adaptive behavior:* forms of behavior that promote reproductive success and that thereby tend to occur at increased frequency in successive generations.

3. *Selfish behavior:* forms of behavior by which an individual protects or increases its own chance of producing offspring, regardless of the consequences for the group to which it belongs.

4. *Altruistic behavior:* self-sacrificing behavior; the individual behaves in a way that helps others but, in so doing, decreases its own chance to produce offspring.

5. *Natural selection:* differential reproduction among individual members of a group that vary in heritable traits—including many behavioral traits—that promote survival and reproduction.

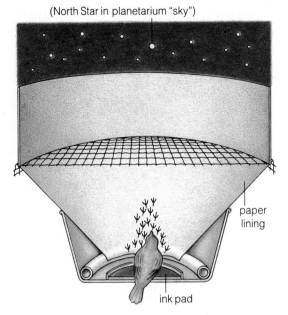

(North Star in planetarium "sky")

paper lining

ink pad

Figure 48.5 Cross-section through the cage used to test the ability of indigo buntings to use the position of the stars as an orientation guide. The bird stands on an ink pad while looking at the night sky through the wire-mesh cage cover. When it jumps up in its attempt to fly, it lands on the paper funnel lining the cage, leaving an ink mark that shows which direction it selected.

When a behavioral biologist talks about a "selfish" animal or "altruistic" animal, there is no implication that the animal is consciously aware of what it is doing or that it knows its behavior is related ultimately to reproductive success. Talking about animals as if they knew they were trying to reproduce as much as possible is evolutionary shorthand for what would be a cumbersome mouthful—namely, that the evolution of animal nervous systems (the physical foundation for behavior) is correlated with the production of surviving descendants. A lion does not have to know that eating zebras is good for its survival and reproductive success in order to want to eat zebras.

Before 1966, the idea of "group selection" dominated the thinking of many behavioral researchers. Implicit in this idea is the assumption that individuals of a group (or species) stand ready to sacrifice their own chance of producing offspring if their self-sacrificing behavior will promote the welfare of the group as a whole.

Then the evolutionary biologist George Williams asked a key question. Assuming a population was composed of self-sacrificing types, what would happen if mutation gave rise to a "selfish" individual, able to increase its own reproductive success without regard to the welfare of other members of the group? If the selfish individual out-reproduced the "good-of-the-species" types, then its distinctive genetic makeup would become somewhat more common in the next generation. If the process continued, the good-of-the-species types would inevitably be completely replaced because they were not as successful at leaving surviving descendants as the selfish types over the generations.

A simple example will illustrate Williams's point. Norwegian lemmings disperse when population densities become extremely high, and many perish by acci-

dental drowning, starvation, and predatory attack during their travels. The dispersing animals were once thought to be sacrificing themselves in order to prevent overpopulation. After all, an over-large population would destroy its environment, and this could lead to extinction of the species as a whole. The idea was given popular impetus by a nature film that showed masses of lemmings marching over a cliff and drowning in the water below. Lemmings do disperse from areas with dense populations, and some die as a result. But the question is, do they die to help their species?

There is a wonderfully instructive cartoon by Gary Larson in which a legion of lemmings is shown plunging over a cliff into the water, presumably in the act of suicide. However, one member of the group has come equipped with an inflated inner tube about its waist. The point is, if lemming populations really were composed of animals programmed to commit suicide so a few unrelated survivors could carry on the species, then individual selection would favor mutants able to avoid the ultimate sacrifice. Those mutants would be among the few left to exploit remaining resources, produce offspring, and leave copies of their "selfish" genes in surviving "selfish" descendants. Over time, nonsuicidal types obviously would completely replace those with suicidal tendencies.

Figure 48.6 Beneath the hot African sun, a black heron holds its wings over its head, like an umbrella. Minnows drawn to the shade of the umbrella are within range of the heron's bill.

In sum, the current consensus among behavioral researchers is this: *In developing a working hypothesis to explain some behavioral trait, it will almost always be more profitable to use individual selection (not good-of-the-group selection).* A few examples will illustrate how hypotheses consistent with this concept of individual selection can be developed and tested. We will draw these examples from three categories of behavior: feeding, avoidance of predators, and reproductive behavior.

Adaptation and Feeding Behavior

Figure 48.6 shows a black heron holding its wings like an umbrella, creating a patch of shade that may attract minnows nearby in the water. Minnows require shaded cover from bird predators—and the black heron exploits their requirement with often fatal results for the minnows. In terms of individual selection, the bird is behaving in a way that will enhance its own survival and reproduction by providing it with something to eat.

An elegant example of natural selection for feeding behavior is Stevan Arnold's study of garter snakes from two populations in California. Garter snakes living in moist habitats along the coast search for and eat banana slugs, exceptionally slimy animals almost the size of a mouse (Figure 48.7). Inland, the populations of garter snakes will have nothing to do with slugs and besides, the slugs don't live in those relatively dry regions. There, the snakes feed primarily on tadpoles and small fish.

Even newborn garter snakes from coastal and inland habitats react differently to slugs. For their first meal, Arnold offered recently born and isolated garter snakes a (previously frozen) chunk of thawed banana slug.

Coastal newborns almost always approached and ate the slug, but inland baby snakes rarely did. Thus the difference between adults from the two regions could not stem from the effects of feeding experience; different feeding preferences had to be present right from birth.

Could the different reactions to slug cubes be attributable to differences in their responsiveness to slug odor? Arnold found that newborn snakes from inland populations tongue-flicked at cotton swabs drenched in essence of tadpole but were essentially unresponsive to swabs drenched in essence of slug. In contrast, coastal snakes actively tongue-flicked at the slug-treated swabs (Figure 48.7c). Clearly the two populations differed in their sensitivity to the chemicals inherent in banana slugs.

When males and females from the two populations were mated in the laboratory, their offspring were intermediate in responsiveness to slug extracts. Overall, the "hybrid" offspring were much more variable in slug acceptance than either parental population. By maintaining similar environments for the young animals, Arnold was able to show that the behavioral differences among the offspring were largely due to genetic differences. Members of the two parental populations had to differ with respect to genes regulating chemical responsiveness to slug odors, and when they interbred, the outcome was a variety of genotypes (and phenotypes). The chemoreceptor system of the young snakes had been affected during development, hence the wide range of degrees of slug acceptance or slug rejection (Figure 48.8).

In Arnold's view, the genetic basis of slug acceptance by inland populations has been selected against over evolutionary time. Although there are no slugs inland, there are leeches. Slug acceptors also attack and try to

a

b

c

Figure 48.7 (**a**) Banana slug of the Pacific coastal regions of California, food for garter snakes (**b**). (**c**) Newborn garter snake from a coastal population tongue-flicking at a cube of banana slug.

eat leeches, because leeches and slugs share some chemical similarity. But a leech is hard to digest. Some leeches even remain alive in the digestive tracts of their consumers, perhaps feeding upon or damaging the internal organs. If the dangers associated with ingesting leeches are as great as they seem to be, then individual inland snakes that happened to have a genetic predisposition for slug acceptance would leave fewer progeny, on the average, than those endowed with different genes. Similarly, on the California coast, where edible slugs abound and potentially dangerous leeches are absent, those individuals with a preference for slug odors would gain access to a rich food supply, leave more descendants, and help spread the genetic basis for slug acceptance.

Anti-Predator Behavior

It does an animal no good to gather food with great efficiency if it is likely to be attacked and killed in the process. By compromising their foraging efficiency to some extent, most animals improve their chances of escaping detection by predators or of avoiding them even if detected.

The hoary marmot of Alaska is a good example of this point (Figure 48.9). This chunky "woodchuck" feeds on grasses and low-lying vegetation. If maximizing energy intake were its only concern, it should forage in those grassy patches of meadow with the most dense plant growth. But a foraging marmot is at constant risk from golden eagles and wolves. Given a choice between food patches of equal quality, a marmot always selects the one growing closest to safe burrow retreats (often located under rocky slopes on mountainsides). More-

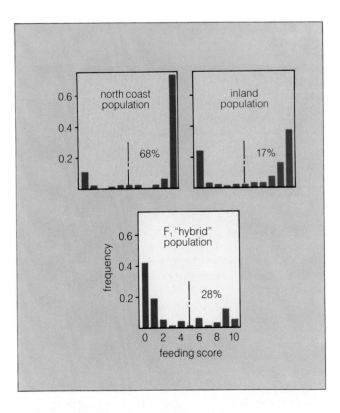

Figure 48.8 Variation in the rate of acceptance of slug cubes of garter snakes that were produced by crossing adult snakes from coastal and inland populations. (Adults from inland populations avoid slugs, for the most part; adults from the coastal regions of California almost always eat them.)

Figure 48.9 Hoary marmot, an animal that compromises foraging efficiency but, in so doing, increases its chances of surviving a predator attack.

Figure 48.10 Predator avoidance behavior by a snake caterpillar. This insect altered the shape of its body in response to being poked. The anterior segments of the caterpillar let go of the vine and puff up like a snake head, which "strikes" at whatever touches it.

over, when it eats, it constantly interrupts the meal and looks around. Although vigilance reduces the rate of food consumption, it prolongs survival; if a marmot can spot an eagle in time, it can race back to its burrow.

Caterpillars of very large tropical moths employ a special behavioral response to predation. When fully grown, these caterpillars are nearly fifteen centimeters long. During the day, they remain motionless on vines. When poked sharply (as by a bird beak), the caterpillar drops part way from the vine and puffs up its anterior segments, so that it looks like a snake (Figure 48.10). The pseudosnake even strikes at whatever touches it. A small bird typically hesitates before continuing an attack. A reluctance to deal with snakes generally prolongs the bird's life, a response that the caterpillar exploits to its own advantage.

Reproductive Behavior

After the rediscovery of the importance of individual selection in the 1960s, it became widely recognized that, *among sexually reproducing animals, the other members of one's species create obstacles to reproductive success.* The intraspecific pressures created as a result of competition for mates and discrimination among potential mates constitutes **sexual selection**, a special reproductive-related category of individual selection. Consider a male white-throated sparrow singing his "Sam-Peabody" song in a Canadian forest. He faces constant challenges from other males that may force him away from the breeding site he has selected. Moreoever, even if he succeeds in holding a small patch of the forest as his own territory, females of his species may refuse to settle there, leaving him with no descendants to show for all his effort. This is the "discriminating mate" aspect of sexual selection. The sparrow is not unique, for male-male competition and female selectivity are extremely widespread. Why?

Among sexually reproducing animals, the male produces smaller gametes (sperm) and vastly more of them compared to the oocytes, or "eggs," produced by the female. It takes a much greater energy investment to produce the larger and far more complex eggs. The disparity of investment in each potential offspring is even more pronounced when the females (but not the males) of the species care for their young. Said another way, females (and their eggs) are a limited resource for the males of most species.

If we measure a male's reproductive success in terms of the number of his descendants, then it will usually be the case that the more females he mates with, the more eggs can be fertilized, and the more successful he will be. In contrast, a female's reproductive success generally will be dictated largely by how many eggs she can produce or how many offspring she can care for. Thus for females, the quality of a mate, not the quantity of

Figure 48.11 Contest between male bighorn sheep for the possession of a cluster of females.

matings, should usually be the prime factor influencing her sexual behavior.

With this simple theoretical background, we can interpret a great deal of male and female reproductive behavior, including the adaptive value of singing by male songbirds. To leave offspring, a male bird must "convince" females that he rather than some other individual should fertilize her (very large and energetically expensive) eggs. The female songbird's reproductive success will depend on how well her offspring are fed.

Nestling white-throated sparrows eat astonishing quantities of insects. Thus the males compete with one another to monopolize an insect-rich patch of forest, which will attract females. In part, they compete by vocally advertising their control over a resource-based

territory. (A "territory" is an area in which an animal or group of animals establishes residency and forages and which may be defended against others of the species.) Through many studies, we know that simply hearing an established resident is enough to discourage intruders from settling in an area, for it means they will have to fight for its possession when there may be unoccupied areas elsewhere.

Females also represent a selective pressure by favoring vocal males, which "announce" potentially suitable habitats for them. However, although male song functions as an attractant, the females go on to assess the quality of the territory for themselves.

The point is, *environmental factors influencing the distribution of females are central to male competition for access*

Figure 48.12 A male hangingfly (*Harpobittacus*) holding a prey insect that he will offer to a female if one comes to visit him. He will mate with the female if she accepts his food present.

to mates. If the resources sought by females are clumped in space, then **resource-defense** behavior should evolve, with the winners getting mates and the losers kept from reproducing (at least temporarily).

Resource-defense behavior has evolved independently in many groups besides birds. For example, the females of some dragonflies lay their eggs only in particular habitats on ponds or streams. Males of their species attempt to defend this limited resource, and the winners mate with females that come to the spot suitable for laying eggs.

As another example, sometimes females are not concentrated at discrete patches of a useful resource but they live together, often for defense against predators. In such cases, males usually show **female-defense behavior**.

Male bison, lions, elk, and bighorn sheep compete fiercely as a consequence of the female clustering behavior. The competition favors large males with formidable combat abilities, obvious in the strength of male lions, the antlers of elk, and the readiness of bison and bighorn sheep to use their horns on each other (Figure 48.11). The return can be great; by winning possession of females, the males of these species have a ready-made harem.

What happens when resources are widely distributed and females do *not* cluster? In such cases, identifying what the males are actually competing for is difficult. For example, the territories defended by male sage grouse of the central United States do not have concentrated resources, nor are the territories centered in places where females live in groups. During the breeding season, males gather in groups on small patches of prairie. Each individual defends his own tiny territory, where he prances about in a truly astonishing display. With tail feathers erect and splayed, head held forward, and neck pouches inflated (Figure 36.9b), a male projects booming calls over the prairie, all the while stamping about like a wind-up toy on its display ground. Females come to the lek, observe the males, and eventually select a partner from among the many. After mating, a female returns to her nest site and rears the young by herself. Many females choose the same male from those at the lek, conferring exceptional reproductive success on the preferred male. What is the basis of this peculiar reproductive strategy? No one knows, although it may be that the males form leks only as a last resort to advertise their qualities, there being no other "workable" and productive mode of mate selection.

So far, we have focused on the advantages that certain forms of behavior bestow on males. Now we conclude with a look at how females may enhance their reproductive success through the judicious selection of a mate. *Two main types of benefits may accrue as a result of her choice: superior genes or superior material benefits.* Consider the male sage grouse, which provide females with the genes in their gametes, nothing more. Do the females "assess" a male's genetic quality by observing his display? Perhaps the nature of his calls, the vigor of his visual display, or his endurance provide cues about his survival ability or foraging skills. To the extent that these behavioral attributes are heritable, a male that puts on a superior show could endow his offspring with useful abilities; and females that could identify such a male and be inseminated by him would enjoy heightened reproductive success. This possibility, like the related issue of the evolution of leks, remains an open question for further research.

One example in which female choice clearly benefits the female is provided by the hangingfly (*Harpobittacus apicalis*). In this species, males capture and kill a fly or a moth and hold it for a female, which they attract by releasing a sex pheromone (Figure 48.12). The size of a "nuptial gift" can vary. If females exercise mate choice in terms of material benefits, they should prefer males that can feed them more, and they do in fact choose males on the criterion of prey size. When a female approaches a male, he offers her his prey; and if she begins to feed on it, she allows mating to begin. However, she will not allow him to transfer *any sperm* until she has eaten for about five minutes. Thereafter, she accepts a steady flow of sperm into her reproductive tract, *but only as long as the food holds out.* At any point up to twenty minutes, she can break off the mating and leave without a full complement of sperm. If so, she will mate again and accept another male's gametes, diluting or replacing her first partner's sperm.

Because the females manufacture a limited resource (eggs), they dictate the rules of male competition, with males attempting to fertilize as many eggs as possible. The improved understanding that has been gained by testing hypotheses on the value of reproductive traits to *individuals,* not to a species as a whole, represents a major triumph of the individual selection approach.

SUMMARY

1. Genetic differences among individuals can lead to differences in the development of the endocrine and nervous systems that control an animal's behavior.

2. Both instinctive and learned behavior are based on genetic information and environmental inputs. Instincts are not purely genetic; learned responses are not purely environmental, as clearly seen in the various forms of imprinting and imprinting-like behavior.

3. The view that behavior is adaptive in advancing individual reproductive success, not group survival, stems from the logic of natural selection and is the basis for most current research on the ultimate causes of behavior.

4. Hypotheses about the adaptive (reproductive) value of a trait can be tested in several ways. For example, there are ways to determine what the "best" response of animals should be to a certain set of obstacles to reproductive success. The animal can then be observed to see if its behavior matches the prediction. Alternatively, one can determine whether different species confronted with the same ecological problem have developed similar ways of dealing with that problem.

5. Both approaches have been used in many studies of feeding, anti-predator, and reproductive behavior. Major advances in the study of sexual behavior have come from the recognition that members of a species create problems for each other as they compete to reproduce.

6. In particular, males are predicted to compete (often aggressively) for access to receptive females because winners of this competition will tend to have many offspring. Females, on the other hand, produce a limited resource (their large and valuable eggs) that the males need, and therefore they should not often compete for males but instead exercise mate choice to accept the best possible partner from among the many suitors available to them.

Review Questions

1. Rephrase the statement "There is a gene for sexual imprinting by greylag geese." The reworded statement should avoid the implication that genes "make" behavioral traits in a one-to-one relationship.

2. What role does the environment play in the development of an instinct?

3. What contributions have behavioral geneticists made to an understanding of the evolutionary basis of behavior?

4. Why is song learning by white-throated sparrows a good illustration of the principle that genetic mechanisms contribute to the development of a learned response?

5. When an adult female sparrow receives a testosterone implant, why won't she sing the territorial song of males of her species?

6. Give a group selectionist and individual selectionist explanation for territorial behavior in a species, and then criticize the logic of the group selectionist hypothesis.

7. You find an insect that looks and behaves like a piece of bark. How could you compare a number of insect species to test the hypothesis that the behavior of the animal is an adaptation to avoid being eaten by a predator?

8. If you were presented with a species that was totally unfamiliar to you, how would you identify males and females with complete reliability?

9. Develop a hypothesis for the observation that male lions kill the offspring of females they acquire after they chase away the males that had been the previous pride holders and mates of these females. How would you test your hypothesis?

Readings

Alcock, J. 1984. *Animal Behavior: An Evolutionary Approach.* Third edition. Sunderland, Massachusetts: Sinauer Associates. A broad-based survey of the many topics that make up the study of behavior, with a strong emphasis on evolutionary theory.

Dawkins, R. 1976. *The Selfish Gene.* New York: Oxford University Press. An entertainingly written book that explains the basis for the argument that natural selection favors traits that are associated with individual success in propagating genes.

Krebs, J., and N. Davies. 1984. *Behavioral Ecology: An Evolutionary Approach.* Sunderland, Massachusetts: Sinauer Associates. A more advanced text consisting of chapters written by experts on many of the issues covered in this chapter.

Williams, G. C. 1966. *Adaptation and Natural Selection.* Princeton, New Jersey: Princeton University Press. The classic book that brought about the revolution in thinking about animal behavior in terms of individual selection.

49

SOCIAL BEHAVIOR

COMMUNICATION AND SOCIAL BEHAVIOR

Consider the Termite

In a forest in southern Queensland, Australia, a long tube about a centimeter across runs up the white trunk of a dead eucalyptus tree. Here, millions of glued-together pellets form a continuous tunnel above the wood's surface. When a fragment of the firm but brittle tube chips away and exposes the interior, several small, nearly white insects scurry away from the light, then a larger number of brown insects quickly line the breach. Those companions of the pale insects look like something out of science fiction; their swollen, eyeless heads taper into long, pointed "noses" (Figure 49.1). When you blow a puff of air at them, each shoots a thin, silvery filament out of its nose!

These insects, one of many species of nasute termites, live in various tropical and subtropical areas around the world. The tunnel is really a covered highway running many meters from an underground nest to areas where the pale workers gather dead wood, which they carry in their mouth to underground gardens. After processing the wood (they cannot digest its tough cellulose fibers), they add it to their gardens, where they are growing a special fungus that *can* digest the cellulose. The pale insects harvest "the mushrooms" (basidiocarps) of the fungus and use them as their food source. The entire colony is defended by the snout-shooters—soldiers that fire off silky strands of glue that entangle and immobilize ants attempting to invade the tunnel.

Perhaps a million or more workers and soldiers in one nasute colony cooperate to build tunnels, create and tend gardens, and defend the colony against predators—and this remarkable cooperation is based on their ability to *communicate*. Despite their pinhead-sized brain, termites use a complex array of communication signals to exchange information in ways that keep the colony running smoothly.

Not one of those million workers and soldiers will ever reproduce. Within the underground bunker beneath the dead eucalyptus are one king and one queen termite, the original fertile founders of the colony, parents to all one million sterile progeny.

With this example of life among the termites, we turn our attention to **social behavior**, the tendency of individual animals to enter into cooperative and interdependent relationships with others of their kind. Because communication signals are the means by which animals manage to be social, we will first explore what "communication" means and how animals use different channels of communication. Then we will turn to the diversity of animal societies, which range from simple aggregations to complex societies.

Communication Defined

When a termite tunnel is breached, the pale workers bang their head against the ceiling and floor of the tunnel, creating vibrations that alert nearby soldiers and cause them to rush to the point of disturbance. The soldiers point their nose in the direction of potential danger, as announced by the scent of ants or a sudden jostling of the tunnel wall. When they shoot out their sticky strands, chemical odors released from the glue attract more soldiers to the site; the odors serve as an alarm scent (a type of pheromone).

The alarm scent of a soldier is clearly a communication signal. Is the scent of the ant one also? In both cases, airborne chemicals associated with one animal change the behavior of another. However, to most biologists, the term **communication signal** means a stimulus produced by one animal that changes the behavior of another individual, in ways that benefit both the signaler and the receiver.

Natural selection has favored the evolution of information exchanges between some signalers and some receivers. When one soldier termite attracts another with an alarm scent, they both benefit through improved defense of their colony. When the original signaler appeared in a population of termites, its scent proved beneficial for itself and its nestmates. Had the scent *harmed* the signaler, then natural selection would have favored nonsignalers, eventually eliminating the "signal" from the population. Similarly, if the respondents had been harmed by reacting to the signal, then selection would have favored types that ignored the signal.

Thus, a communication signal evolves as a benefit for *both* the signaler and the receiver. Such benefits do not accrue in ant-termite interactions. The ant odors did not evolve so that ants could announce their impending attacks on termites; they evolved because they serve useful functions within *ant* colonies.

Communication signals evolved in one context are sometimes exploited by animals of other species. Thus,

Figure 49.1 Nasute termites—highly social insects with a soldier caste that defends the colony by shooting strands of sticky silk from the noselike extension of their individual heads.

from the perspective of the raiding ants, the soldier termites are "illegitimate receivers" of one of their communication signals. Nature also has its share of "illegitimate signalers," which fool the normal receivers. An assassin bug covers its body with some nest material of its termite prey. Thus the bug adopts an accepted odor and is tolerated by living members of the colony, even as it goes about consuming them, one by one.

CHANNELS OF COMMUNICATION

Communication signals are based on diverse sensory modes. For example, nasute termites make use of tactile senses (when they transmit alarm vibrations through wood) and olfactory senses (in the release of alarm pheromones). Humans are more familiar with acoustical signals (as in speech) and visual displays (as in facial expressions like smiles and frowns). Let's consider a few examples of signals in these major sensory modes.

Figure 49.2 Example of the "yawns" by which male baboons threaten each other. Such facial displays often resolve a conflict without actual fighting.

Visual Signals

Visual signals are nearly universal among species of animals that are active by day. Often male baboons "yawn" at each other when they are both interested in a receptive female or some other resource. This facial movement is a visual display, designed to threaten a rival (Figure 49.2), and often precedes an attack on the rival.

How can it be advantageous for a baboon to signal an intention to attack or for a receiver to "accept" this information? Here, the signaler benefits if the receiver backs down after perceiving the threat signal, rather than physically attacking and possibly injuring the signaler. The receiver also benefits if, after mulling over the threat signal, he can accurately judge that his opponent can inflict a real beating unless he gives ground.

Visual communication is essential in the courtship of many animal species. For example, male and female albatross have a battery of visual displays, each with its own message (Figure 49.3). As is true for many birds, the albatross adopt exaggerated and sometimes contorted postures that convey readiness to mate, to take over incubation duties from the partner sitting on the eggs, and so on. Possibly, these highly distinctive poses represent unambiguous messages for the intended receiver and minimize the chance of confusion.

Visual communication is limited at night, but a number of organisms use *bioluminescent* signals (based on special metabolic reactions that are accompanied by a release of light energy). For example, fireflies (beetles, really) have light-generating organs at the tip of the abdomen. Each male produces a signal characteristic of his species and one that a receptive female may answer. After seeing his signal, she waits a standard time and replies with a single flash. A male approaches an answering female, engaging her in a visual "dialogue" until he precisely locates and mates with her.

Females of a predatory species of firefly have "broken the code" of their prey. When they detect a signaling male, they wait the appropriate interval and then give the "come-hither" answering flash. The male approaches and is embraced by the predator, which consumes her victim.

Chemical Signals

Chemical signals abound among the insects, which use them as sex attractants, trail markers, alarm calls, and so on. Male hangingflies lure females to themselves with sex pheromones (page 760). More often, a receptive female insect releases sex pheromones and waits for the male. The male insects have large antennae, packed with specialized receptors. After detecting the pheromone, males fly upwind, tracking the odor in a race to be first to reach the chemically beckoning female (page 359).

Figure 49.3 Courtship behavior of various species of albatross. (**a**) The male spreads his wings as part of a courtship ritual that also includes pointing his head harmlessly at the sky. These birds are at a future nest site in the male's territory (**b–d**). After pair formation is well advanced, the birds begin to touch bills. This contact display precedes copulation.

Chemical signaling is also notable among some of the primates. A female baboon uses chemical and visual signals· to announce her readiness to mate. Her external genitalia become swollen and she produces distinctive sex pheromones that elicit mating behavior from the males. In humans, male pheromones may influence the physiological health of the female's reproductive system.

Tactile Signals

Many animals communicate over short ranges, where tactile signals (distinct patterns of touch) become important.

When a foraging honeybee returns from several successful trips to a rich nectar or pollen source, she may begin to "dance" on the vertical surface of the comb in the darkness of the beehive. The forager moves in circles, jostling her way through the crowded mass of workers on the comb. Other individuals may follow her, keeping in physical contact with the dancer and, in doing so, acquiring information about the distance and direction of the food source.

Karl von Frisch was the first to discern the message encoded in this tactile signaling. In one set of experiments, he trained marked foragers to come to stations baited with concentrated sugar water. He then set out a

a

b

Figure 49.4 (Above) (**a**) The round dance of the honeybee. When foragers perform these movements upon their return to the hive, other bees are stimulated to fly out fairly close to the hive in search of the food that the round-dancing bee had found. (**b**) The waggle dance of the honeybee. Bees that have found a rich source more than eighty meters from the hive perform a waggle dance to recruit other bees in the hive.

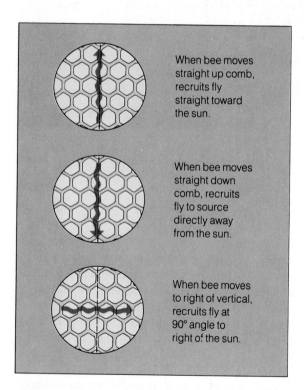

When bee moves straight up comb, recruits fly straight toward the sun.

When bee moves straight down comb, recruits fly to source directly away from the sun.

When bee moves to right of vertical, recruits fly at 90° angle to right of the sun.

Figure 49.5 The manner in which the angle of the run of the waggle dance conveys information about the angle between the food source, the hive, and the sun.

series of stations with identical baits at different distances or directions from the hive. Generally, the station closest to the original training site attracted by far the greatest number of new (unmarked) bees in a short time. The trained bees apparently were acting as recruiters, directing new bees to the general area where they had found so much food.

By observing marked bees and knowing where they had been trained, von Frisch was able to break the code of the dancing bees. He saw that bees trained at stations close to a hive performed a round dance (Figure 49.4a). Worker bees trailing the round-dancing bees were stimulated to fly out of the hive in search of nearby food but they did not learn which direction to fly. If, however, the station was more than eighty meters distant, the foragers trained at this station performed waggle dances (Figure 49.6b). The frequency with which they waggled their abdomen during the straight runs of the dance and the speed with which they completed each loop of the dance conveyed information about the distance to the food. *The more waggles and the faster the dance, the closer the food source* (down to about eighty meters, at which point bees switch to round dancing).

The angle of the straight run with respect to vertical provided information about the direction of the food source relative to the sun and the hive. If a dancer went straight up the comb surface on her straight runs, followers learned that food could be found by flying from the hive in the direction of the sun. If, however, a dancer made her straight run 90° to the right of vertical, she was trained at a food station located at 90° to the right of a line drawn between the hive and the sun (Figure 49.5). Thus dancing bees transpose information about the position of the sun, hive, and food source into a code, and they convey the symbolic message via tactile stimulation to their hivemates.

Figure 49.6 Frog-eating bat (*Trachops cirrhosus*) catching a singing tropical frog (*Physalaemus*). The bat can distinguish between poisonous and edible frogs, as well as locate them, by the frog's calls. The frogs call to attract mates but face the dilemma of how not to attract the bat instead. Bat predation obviously has major influence on frog courtship behavior.

Acoustical Signals

Both chemical and visual signals travel long distances in air (and water), and the same is true for acoustical messages (distinctive sounds). Bird song, described in the preceding chapter, is a familiar example of acoustical signaling. Male frogs also use acoustical signals to communicate with rival males and receptive females. As is true for bird song, calling is primarily a male activity, and each frog or toad species has its own distinctive call.

Frog calling is complicated for many species because (1) some predators use the calls of their amphibian prey to locate victims, and (2) many rival males may be concentrated in the same relatively small area, creating intense competition for space on the "air waves." For example, a tropical frog (*Physalaemus pustulosus*) is a favorite prey of a bat that tracks down calling individuals and snatches them from the water with its formidable jaws (Figure 49.6). The frog call has two parts, a whine followed by a "chuck." A frog calling all by itself often drops the "chuck" which, because of its acoustical properties, gives better information about his location than the initial whine. Although the male thereby makes himself less attractive to females, he is more likely to live to

call another day. He gives the complete call when part of a male chorus; being one of many that could be grabbed that evening by a frog-eating bat, he may live *and* get to mate.

COSTS AND BENEFITS OF SOCIAL LIFE

The forms of communication just described are the foundation for diverse social units. Although there are certain advantages to being highly social (in the sense of living together), some animals live in small family groups, others as loose aggregations of unrelated individuals, and still others in solitude. The degrees of sociality (or its absence) represent different adaptations to different environmental challenges.

The reason for diversity in social life becomes clearer if we consider the risks as well as benefits of social life *in terms of individual reproductive success*. Consider the vast rookeries of egrets, gulls, or terns, in which hundreds of individuals form pairs and nest in apparent harmony

Figure 49.7 Porpoises engaging in social pursuit of schooling fish. Sociality sometimes exists because individuals can forage more effectively for prey when they do so as a group.

a short distance from one another. The appearance of harmony is superficial. A colonial nesting gull does not hesitate to eat its neighbor's eggs or chicks if they are left unguarded for a moment. Moreover, many birds deplete the local food supplies more rapidly than would happen if they were dispersed. Given the crowded conditions, they are all probably more vulnerable than solitary birds would be to contagious diseases.

What does a social animal gain that more than compensates for the price of associating with others? No one overriding benefit applies to all social species. However, lions and porpoises living in groups are able to capture more prey than if they operated alone. Through cooperative hunting, a pride of lions can bring down a giraffe or a Cape buffalo that no one lion would dare attack. When porpoises collectively go after a school of fish, the members of the school break apart in a wild dash to safety. But many escape from one porpoise only to run into the mouth of another (Figure 49.7).

PREDATION AND SOCIALITY

Predation is by far the dominant selection pressure favoring the evolution of social behavior. Here we will consider some of the ways in which group-living animals minimize (at least in theory) the risk of predation.

Dilution Effect

Simply by being part of a large group, a prey animal "dilutes" the risk of being the one animal that predators will select out of the many. For example, the diminutive fairy penguins of Australia become social in the evening when they return to nest burrows on land. The birds tend to come ashore in groups. The surging waves deposit them upon the beach, where they stand about

uncertainly before beginning to waddle together inland. Penguins on land are not noted for agility or speed; they are vulnerable to large hawks and eagles patrolling the beaches for prey. But by making their move across the beach only when large numbers of their fellow penguins have assembled, each animal improves its odds of surviving. While an eagle is dispatching one victim, the others can waddle along to safety.

Improved Vigilance

In effect, animals living in groups have many eyes available to spot predators. When the first animal to detect danger flees for cover, others can follow its lead even if they themselves have detected nothing.

In a study of captive starlings, one researcher showed that isolated birds reacted less quickly to an artificial hawk model released over their cages than did a flock of ten starlings. Moreover, the flock flew up sooner after the release of the "hawk," even though any one bird in the group had been looking up *less often* than the isolated starling. By virtue of being a flock member, a starling is not only safer, it has more time to feed because less time is needed for vigilance.

Group Defense

Often, social animals can better repel a predator through coordinated group defense. Nasute termites work as a team to alert soldiers, which then combine forces to spray intruders with their sticky glue. Members of a herd of musk oxen also employ group defense when threatened by attacking wolves (Figure 49.8). Musk oxen have powerful hooked horns, but a solitary animal cannot protect its flanks from a wolf pack. The array of Cape buffalo shown in Figure 1.9 would be formidable even to a pride of lions.

Figure 49.8 Defensive formation of musk oxen. Predation pressure can favor the evolution of social life when members of the group are safer than are solitary animals.

The Selfish Herd

Some animals may live in groups simply to "use" other individuals as living shields against predators. Such groups have been labeled "selfish herds." When hyenas or hunting dogs attack a herd of wildebeest on an African savanna, the prey animals cluster into a tight mass that dashes across the plain as if it were a single animal. Although their escape behavior seems to be a cooperative undertaking, it almost certainly results from the efforts of individuals not to be isolated from the group. Instead they maneuver themselves as closely as they can to the center of the herd, where it is safest. Hyenas and hunting dogs separate an easy victim (from the periphery of the herd) rather than plunge into a mass of flying hooves.

Given the survival advantages, we can predict that members of a selfish herd will compete for the safe central positions. Such competition is evident in nesting colonies of freshwater bluegill fish. The many predators of these small fish include largemouth bass (which can inhale an entire adult) to snails that slip into a nest and eat the fish eggs. Male bluegill fish build their nests together, with each male constructing a depression in the lake bottom to receive the eggs that his mate(s) will deposit there. The largest, most powerful males claim the center of the nesting area, forcing other, smaller males to assemble around them. Thus the smaller males form a living shield against bass and snails, which attack from the edge of the colony inward.

SOCIAL LIFE AND SELF-SACRIFICE

From the preceding discussion, it should be clear that the individuals of many animal societies do not make any personal sacrifice for other members of their group. However, the degree to which individuals tolerate and interact with others is influenced by natural selection. If the environment is such that social members of a species leave fewer descendants on the average than the more solitary types, what will the species become—social or solitary?

Yet if it is true that natural selection influences the evolution of social behavior, then there would never be any cases of self-sacrificing behavior—*unless such behavior enhances the continuity of the genetic line of the altruistic individual.* Let's look at a few examples that will clarify this point.

Parental Behavior

Parenting is the most familiar form of self-sacrificing behavior. A breeding pair of Caspian terns (Figure 49.9) incubate their eggs, shelter the nestlings, bring them food, defend them from predators, and accompany them for some time even after they have begun to fly and feed partly on their own. Such parental activities use up time and energy the adults might spend in ways that would improve their own chance of living to reproduce another time.

Figure 49.9 Male and female Caspian terns, which cooperate in the care of their young. Male parental behavior occurs in most bird species, but overall is rare in the animal kingdom.

However, if a parent can significantly help some of its offspring live to reproductive age, then its sacrifice could increase the frequency of the parental genes in the population. *Because of the genetic similarity between a parent and its offspring, which share fifty percent of their genes, it is possible for an individual to give up some of its own chances for survival and future reproduction—and still have its genes spread through the population.*

The critical point is this: the parent must sacrifice for its *own* offspring, not for those of another. Typically, parents are adept at recognizing their own offspring, which they help, while ignoring or even attacking the offspring of others. Again, many terns and gulls nest at the same time in extraordinarily dense colonies. After eggs have hatched and the young have grown a bit, there may be hundreds or thousands of dependent chicks, all about the same age, in a relatively small area. Yet the parents never feed the chicks of other parents, and they may even eat the "foreign" chicks if they catch them unprotected.

Cooperative Societies

Despite the existence of selfish herds, discriminating parents, and cannibalistic neighbors, there are also numerous examples of apparently self-sacrificing behavior, in which friendly, helpful behavior is not directed exclusively to offspring.

For example, membership in a troop of baboons is stable, all the animals recognize each other as individuals, and they actively seek out the company of their fellows, spending hours grooming a favored companion to remove parasites and burrs. At some risk to themselves, the adult males may join forces and confront an attacking leopard while the smaller females and juveniles flee to safety.

Similarly, the members of a wolf pack do more than merely tolerate each other. They have active greeting ceremonies after they have been separated for a time and they spend long periods in very close association (Figure 49.10). They coordinate their hunting activities and share food in a kill; returning hunters often regurgitate food to pack members that have remained at the den to protect the pups.

This last example raises an interesting question. When a wolf brings back food to pack members that are *not* its offspring, it seems to be sacrificing something (time, food, or both) that could be used to further its own reproductive success. How can we account for such behavior in evolutionary terms? Let's take a look.

Dominance Hierarchies

To gain insight into self-sacrificing behavior, we must first realize that the apparent harmony within even the most cooperative societies is, to some extent, superficial.

a   b

Figure 49.10 A dominant male and female member of a wolf pack (**a**). Typically, the dominant male is the only pack member likely to breed successfully. (**b**) Subordinate members of the pack greet a dominant male.

Some "sacrifices" are made simply because some members are less dominant than others, and their aggression and other expressions of individuality are held in check. Thus we see a **dominance hierarchy** in some societies, with some members subordinate to less subordinate ones, which are dominated by other members, and so on up the ladder of who socks it to whom.

The top male in a baboon troop is the so-called alpha member of the hierarchy. Merely by walking toward another baboon, an alpha male can cause it to move quickly to the side, leaving him in charge of a safe sleeping place, choice bits of food, or a receptive female. The second-ranking male can preempt all but the alpha individual. The third-ranking male dominates all but the alpha and beta baboons, and so on down the line.

Because the group members accept their status, aggression is minimized. Subordinates almost never actively challenge more dominant individuals over limited resources. They are often forced to show appeasement gestures if they are to remain in the company of a more dominant group member (Figure 49.11).

What is notable about this behavior is that subordinates almost never reproduce as much as the dominant members do. How, then, can subordinate behavior evolve by way of natural selection—in other words, through pressures that favor individual reproductive success? According to one view, *even dominant members die or become injured, old, or feeble—and a patient subordinate may then move up to a higher position in the social hierarchy, and ultimately reproduce.*

A young, small, or weak animal has three options. First, it could pull out all the stops and compete for dominance—but a challenge to a much stronger individual might lead to injury or death, so any future chance at reproduction would be lost. Second, the subordinate animal could strike out on its own, find a mate, and form a new breeding group. This, too, usually is a prescription for injury or early death. (A wandering, solitary baboon no doubt quickens the pulse of the first leopard to see it.) Third, the subordinate animal could *accept* its low status and thereby secure better protection from predators and more safety when foraging. It may even end up producing offspring if chance events favor its upward movement in the dominance hierarchy.

EVOLUTION OF ALTRUISM

So far, we have seen that often it is not necessary to invoke anything other than individual selection to account for the evolution of helpful behavior. However, a special evolutionary process could, in theory, promote the evolution of altruistic behavior toward individuals other than offspring. To explain this process, we will be using some terms used by behavioral biologists:

Figure 49.11 Appeasement behavior among baboons. Notice the assured position of the dominant animal—and the abject stare and groveling posture of the subordinate one, who is intent on making little conciliatory smacking noises with its lips.

1. *Altruistic behavior:* helpful actions that reduce the individual's production of surviving offspring while increasing the reproductive success of the helped individual.

2. *Cooperation:* helpful behavior that raises the helper's output of its own surviving offspring as well as the number produced by the helped individual.

3. *Self-sacrificing behavior:* helpful actions that reduce the probability of survival of an individual but can, under some circumstances, increase the reproductive success of the individual.

4. *Kin selection:* selection for aid to relatives, which increases an individual's transmission of genes to subsequent generations. The *direct* form of kin selection favors aid given by adults to their offspring when such aid increases the number of their surviving offspring. The *indirect* form favors aid given to relatives (and others with similar genotypes)—not the helper's offspring but surviving relatives that preserve the gene lineage they share with the helpful individual.

Kin Selection

Even if a subordinate never actually succeeds in reproducing, it may still help pass on copies of its genes. *True altruism, even if it involves reproductive self-sacrifice, can be genetically advantageous if the beneficiaries of the sacrifice are relatives of the altruist.* Individuals share a certain proportion of genes not only with their offspring but also with their relatives. Two siblings will have about half the same genotype as the other, because they have inherited some of the same genes from the same parents. Due to common descent, about one-fourth of the genotype of an uncle and nephew is the same. Similarly, about one-eighth of the genotype of cousins is the same. Thus an individual that bears genes leading to the development of helpful behavior can *indirectly* propagate those genes by helping preserve and produce more relatives.

In many animal societies, altruistic behavior is directed strictly to close relatives. For example, a communal flock of scrubjays is composed of a breeding pair and as many as six helpers-at-the-nest. The helpers almost always are older offspring of the pair. They help, feed, and protect younger brothers and sisters. They may never reproduce, yet the sacrificing individuals help perpetuate some "shared" genes when their siblings reproduce.

Suicide, Sterility, and Social Insects

What about animals that are permanently incapable of reproduction? What genetic advantage could be attributed to their altruistic behavior in a group? Consider sterile helpers or workers, which are found among social insects such as honeybees. A honeybee colony is characterized by great sacrifices of the sterile workers (Figure 49.12). A single queen bee is the sole reproductively active female among tens of thousands of bees. Workers force-feed her with food that they collect, groom her, and distribute her socially binding pheromones throughout the colony. They also feed one another, build honeycomb, attend to the queen's eggs, feed larvae, remove dead or diseased pupae, and guard the entrance to the hive.

When a guard bee detects a nest parasite or predator (such as a raccoon), it attacks and stings the intruder. It also releases a chemical attractant that recruits still more defenders, which also attack and sting. Because the loss of a stinger and associated poison glands is fatal, these workers are animals that commit suicide! Suicidal behavior is regularly exhibited by worker ants, wasps, and soldier termites, as well as by worker bees.

How can suicidal behavior and sterility evolve, given that individuals having these traits fail to reproduce? Here, too, close relatives play a role. For example, at some stage a queen bee departs with about half of the work force and starts a new colony. Before she does, drone sons and future queen daughters are produced. All are fertile. One of these future queens inherits the remaining half of the work force, and she stays at the hive. The drones leave the colony and may mate with new queens produced at other colonies. Thus worker bees help the queen mother rear one reproducing sister, and many workers stay with this new queen and help her after their mother leaves. They also help produce a number of potentially reproducing brothers. *When workers sacrifice their own reproductive chances, they increase the number of potentially reproducing siblings—which have a substantial proportion of particular genes in common with them.* This is true of social insects generally.

Sterility, then, can be explained in evolutionary terms as a mechanism that increases the representation of certain genes in certain populations. Modern analysts of worker castes now suggest that worker members do not exist to benefit the species as a whole. Rather, the focus is on how even extreme cooperation in social animals may have evolved through competition at the genetic level.

Human Social Behavior

So far, we have considered examples of how the concept of individual selection can be used to explain animal behavior. Can the concept also be applied to analysis of human behavior? The question is controversial. Since Darwin's time, strong evidence has accumulated in support of the principle of biological evolution, and today there is widespread belief that we share an evolutionary heritage with all other forms of life. Yet there is still a prevailing belief that humans are so evolutionarily advanced that they are totally unique, even compared with our closest relatives in the animal kingdom.

To be sure, our cultural evolution has been so extraordinary that we are unique in many ways. But surely, beneath the elaborate and diverse layerings of culture, there still resides a biological core for human behavior.

Think about some forms of behavior that occur in all human societies. For example, smiling has an innate basis. It starts among newborns, even those born prematurely. Apparently, smiling helps establish emotional ties between infants and the adults who help assure their survival. As another example, beginning at four or five weeks after birth, all infants make strong visual contact with their mothers—even infants who were born blind. Here, too, the behavior apparently helps strengthen social bonds between the protector and the protected. Universally, humans express pleasure, anger, distress, surprise, and rage with the same kinds of facial movements. The meaning of all such behavioral expressions is universally recognized; they are all evolved mechanisms that help promote survival and reproductive success in human social groups.

Some contend that if you were to ask people why they behaved in a particular way, no one would ever say "My genes made me do it." They contend that any behavior in any environmental context is an entirely learned response. Yet conscious awareness of the genetic contribution to some behavior is not a requirement for its expression. If it could talk, a baby egret battering its younger sibling to death (as baby egrets are prone to do) wouldn't say "My genes made me do it" either; it simply has an inherited capacity to behave in ways that help assure its own survival and, ultimately, its reproduction.

Notice the word *capacity*. Having a genetic basis for the development of some behavior does not mean that the trait is biologically determined, in the sense of being impossible to alter. There are no "genes for behavioral traits." There are only genes that code for enzymes and other products—and the synthesis and activity of those products are profoundly influenced by the environment. As pointed out repeatedly in this book, *change the environment, and the course of development can change.* Thus, whatever the extent of its genetic foundation, expression of different forms of behavior that are considered to be unacceptable in a social environment can also be modified.

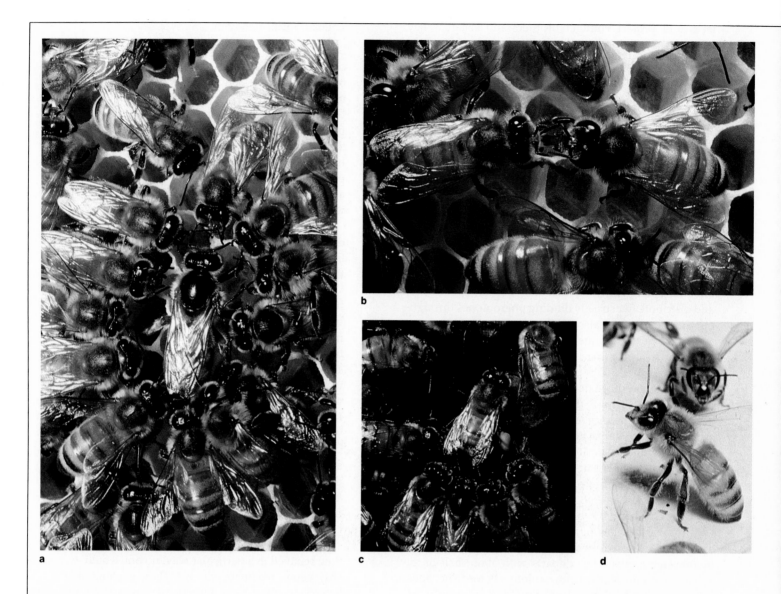

Figure 49.12 Life in a honeybee colony. (**a**) The only fertile female in the hive is the queen bee, shown here surrounded by her court of sterile worker daughters. These individuals feed her and relay her pheromone throughout the hive, regulating the activity of all of its members.

(**b**) Transfer of food from bee to bee, one of the helpful actions within honeybee society.

(**c**) Bee dance. The central bee is in the midst of a complex dance maneuver, which contains information about the direction and distance to a food source. Floral odors on the dancer's body are also useful to the recruits, which follow her movements before flying out in search of the food.

(**d**) Guard bees. Worker females assume a typical stance at the colony entrance. They are quick to repel intruders from the hive.

(**e**) The queen is much larger than the workers, in part because her ovaries are fully developed (unlike those of her sterile daughters).

(**f**) Stingless drones are produced at certain times of the year. They do not work for the colony but instead attempt to mate with queens of other hives.

(**g**) Worker bees forage, feed larvae, guard the colony, construct honeycomb, and clean and maintain the nest. Between 30,000 and 50,000 are present in a colony. They live

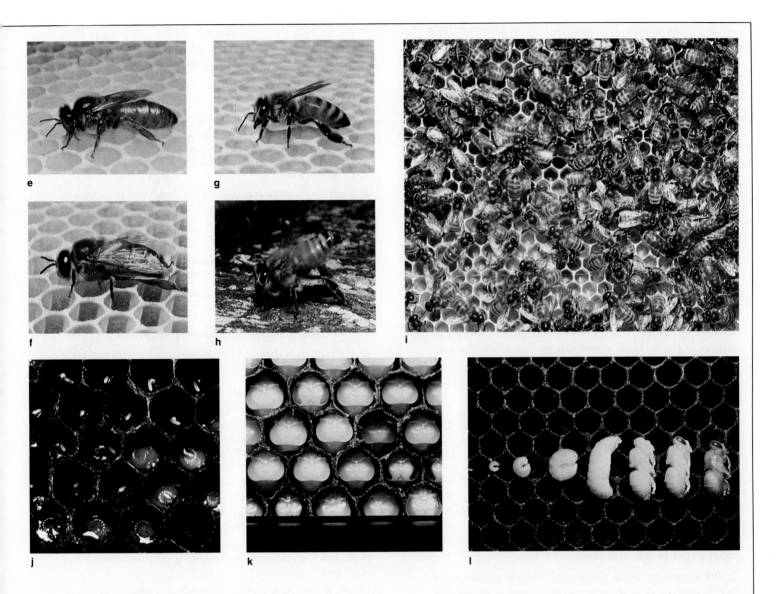

about six weeks in the spring and summer, and can survive about four months in an overwintering colony.

(**h**) Scent-fanning, another cooperative action by the worker. As air is fanned, it passes over the exposed scent gland of the bee. The pheromones released from the gland help other bees orient to the colony entrance.

(**i**) Worker bees constructing new honeycomb from wax secretions. Here, honey or pollen may be stored or new generations may be cared for from the egg stage to the emergence of the adult.

(**j**) The initial stages in the life cycle of a bee. The brood cell caps have been removed, revealing eggs and larvae of various ages. Larvae are fed by young worker bees.

(**k**) Worker pupae. Once again the cell caps have been removed, exposing pupae that will metamorphose into future workers.

(**l**) Complete sequence of developmental stages of a worker bee, from the egg (far left) to a six-day-old larva (fourth from left), to a twenty-one-day-old pupa about to become an adult (far right).

An evolutionary approach to human behavior may be under most attack by individuals who fear that what it reveals will be misused. They may think that to suggest a trait such as aggression is adaptive (genetically advantageous) is to imply that it is desirable in moral or social terms. Yet there is a clear difference between trying to explain something (by reference to its possible evolutionary history) and trying to justify it. "Adaptive" does not mean the same thing as "moral." It means *valuable in the transmission of an individual's genes.*

Research into the relationship between evolution, genes, and our behavioral capacity is in its infancy. Yet improved knowledge of ourselves could be used to help resolve many critical issues related to our biology as a species, including overpopulation, exploitation of the environment, and aggression among modern nations. It remains to be seen how quickly this knowledge will be gained, and whether it will actually be employed to alleviate negative aspects of the human condition. In the meantime, perhaps some readers will find that an appreciation of the natural world can be deepened through application of an evolutionary approach to one of its most diverse expressions—human behavior.

SUMMARY

1. Social behavior depends on communication among animals living together. Communication signals are a form of cooperation in which both a signaler and a receiver benefit from the transfer of information between them. Signaling animals employ many different sensory modes as channels of communication.

2. Social species, those that live in groups, are not more highly evolved than solitary species whose members live largely in isolation from each other. Sociality carries with it higher risks of communicable diseases, exploitation by others, and increased competition for scarce resources. But particularly if predation pressure is severe, the various benefits that arise from improved defense against predation can outweigh the costs of social life.

3. Most animal societies are composed of individuals that do not sacrifice their reproductive success to benefit others in their group. In some species, parents gain by living with and helping their offspring. In other species, weaker subordinate animals gain by deferring reproduction temporarily in order to take advantage of the benefits, especially improved safety, that come from belonging to a group.

4. It is theoretically possible for individuals to reduce their net lifetime production of offspring in order to help other kinds of relatives reproduce. Kin selection can lead to the extreme altruism of workers of some social insects;

even though the workers are sterile, they help their reproducing siblings survive, passing on "by proxy" the genes underlying the development of altruism.

Review Questions

1. A hyena places scent marks on vegetation in its territory by releasing specific chemicals from certain glands. What evidence would you need to demonstrate that this action is an evolved communication signal?

2. Explain how a threat display can be considered an example of cooperation.

3. Why don't the members of a "selfish herd" live apart if each member is "trying to take advantage" of the others?

4. How can an animal be cooperative and still pass on more of its genes than a noncooperative individual?

5. Is parental behavior always adaptive?

6. What is true altruism?

7. How can an individual propagate its genes even if it is sterile?

8. In evolutionary terms, when is solitary living a "superior" mode of existence compared to complex social life?

Readings

In addition to the readings outlined at the end of the preceding chapter, all of which deal with animal behavior, there are a number of excellent texts that focus primarily on material covered in this chapter, including:

Daly, M., and M. Wilson. 1983. *Sex, Evolution, and Behavior.* Second edition. Boston: Willard Grant Press. This book has two particularly good chapters on the evolutionary analysis of human behavior.

Frisch, K. von. 1961. *The Dancing Bees.* New York: Harcourt Brace Jovanovich. A classic on the natural history and behavior of honeybees.

Trivers, R. 1985. *Social Evolution.* Menlo Park, California: Benjamin/Cummings. A complete and readable account of all the topics that make up an evolutionary approach to social behavior.

Wilson, E. O. 1975. *Sociobiology: The New Synthesis.* Cambridge, Massachusetts: Harvard University Press. The pivotal book in presenting a new approach to the study of social behavior.

Wittenberger, J. F. 1981. *Animal Social Behavior.* Boston: Willard Grant Press. A book that covers much the same ground as *Social Evolution* by Robert Trivers but at a somewhat more advanced level. The book does an excellent job of laying out competing hypotheses on aspects of social behavior and evaluating the evidence for them.

APPENDIX I
BRIEF SYSTEM OF CLASSIFICATION

This classification scheme is a composite of several used in microbiology, botany, and zoology.* There is no universally accepted scheme. Although major groupings are fairly well understood, what to call them and, sometimes, where to place them are not. Our scheme is not all-encompassing; it simply covers organisms mentioned in the text. Numerals refer to pages on which a few representatives are illustrated or described.

KINGDOM MONERA Prokaryotes (single-celled, with no nucleus or other organelles; autotrophic and heterotrophic forms) — 32, 33, 569–574

SUBKINGDOM ARCHAEBACTERIA. Methanogens, halophiles, thermoacidophiles — 554, 554, 572

SUBKINGDOM EUBACTERIA. Gram-positive eubacteria (e.g., *Clostridium*); purple bacteria, pseudomonas (e.g., *Escherichia coli*); spirochetes; gram-negative anaerobic rods, cyanobacteria (e.g., *Anabaena*), green sulfur bacteria, green nonsulfur bacteria; sulfur reducers, myxobacteria (e.g., *Bdellovibrios*); radiation-resistant micrococci; planctomyces group — 219, 232, 556, 571, 572

KINGDOM PROTISTA Single-celled eukaryotes (with nucleus and other organelles; predatory and parasitic heterotrophs; some photosynthetic autotrophs) — 32, 33, 575

PHYLUM EUGLENOPHYTA.** Euglenids — 575, 575
PHYLUM CHRYSOPHYTA. Golden algae, diatoms — 576, 576
PHYLUM PYRROPHYTA. Dinoflagellates — 576, 577
PHYLUM PROTOZOA. Protozoans — 576ff.
  Subphylum Mastigophora. Flagellate protozoans — 577, 578
  Subphylum Rhizopoda. Amoeboid protozoans — 578–579, 578
  Subphylum Sporozoa. Sporozoans — 528, 579
  Subphylum Ciliophora. Cilates — 580, 581

KINGDOM FUNGI Multicelled eukaryotes (heterotrophs; some parasites, most saprobes; all rely on extracellular digestion, absorption of nutrients)

DIVISION GYMNOMYCOTA.*** Cellular slime molds (e.g., *Dictyostelium*); true slime molds (e.g., *Physarum, Stemonitis*) — 223, 590
DIVISION OOMYCOTA. *Egg-forming fungi* — 585–586
DIVISION ZYGOMYCOTA. *Zygospore-forming fungi* — 584, 586
DIVISION ASCOMYCOTA. Sac fungi — 586–587, 586
DIVISION BASIDIOMYCOTA. Club fungi — 587–588, 587, 588
DIVISION DEUTEROMYCOTA. Imperfect fungi — 588, 589

KINGDOM PLANTAE Mostly multicelled eukaryotes (photosynthetic autotrophs) — 33, 33, 240–241, 590ff.

DIVISION RHODOPHYTA. Red algae — 66, 594, 594
DIVISION PHAEOPHYTA. Brown Algae — 594, 595
DIVISION CHLOROPHYTA. Green algae — 10, 592, 594, 595

DIVISION BRYOPHYTA. Mosses, liverworts — 596–597, 596
DIVISION RHYNIOPHYTA. Group of extinct land plants — 559, 559
DIVISION LYCOPHYTA. Club mosses — 597, 597
DIVISION SPHENOPHYTA. Horsetails — 597, 600
DIVISION PTEROPHYTA. Ferns — 600–601, 600
DIVISION CYCADOPHYTA. Cycads — 601, 601
DIVISION GINKGOPHYTA. Ginkgo — 601, 601
DIVISION GNETOPHYTA. Gnetophytes — 601–602, 602
DIVISION CONIFEROPHYTA. Gymnosperms — 241, 602–603, 603

DIVISION ANTHOPHYTA. Angiosperms (flowering plants) — 241ff., 270ff.
  Class Monocotyledonae. Monocots (grasses, orchids, lilies, etc.) — 241, 598, 603, 604

  Class Dicotyledonae. Dicots (most of the common flowering plants not mentioned under monocots) — 241, 598, 603, 604

KINGDOM ANIMALIA Multicelled eukaryotes (heterotrophs of varied sorts; herbivores, carnivores, parasites, decomposers, detritivores) — 33, 33, 606ff., 607

PHYLUM PLACAZOA. *Trichoplax adhaerens* — 582, 582
PHYLUM PORIFERA. Sponges — 610–612, 611
PHYLUM CNIDARIA. Jellyfishes, hydras, corals — 10, 12, 326, 612–613
PHYLUM CTENOPHORA. Comb jellies — 614, 615
PHYLUM PLATYHELMINTHES. Flatworms, flukes, tapeworms — 326, 614–617, 616
PHYLUM NERMERTEA. Ribbon worms — 605
PHYLUM NEMATODA. Roundworms — 617
PHYLUM ROTIFERA. Rotifers — 585
PHYLUM ANNELIDA. Oligochaetes (e.g., earthworms, leeches), Polychaetes (marine worms) — 385, 433, 618–621, 619, 620

PHYLUM ARTHROPODA
  Subphylum Trilobita. Trilobites (extinct) — 557, 557, 564, 621

  Subphylum Chelicerata. Chelicerates (horseshoe crabs, spiders, scorpions, ticks) — 524, 373, 622, 680

  Subphylum Crustacea. Crustaceans (crabs, lobsters, etc.) — 327, 623–624, 624

  Subphylum Urinamia. Centipedes, millipedes, insects — 7, 484, 486, 624–625, 624, 625, 682

---

*The most influential references used in its compilation: Stanier et al, *The Microbial World* (5th, 1986); Bold et al, *Morphology of Plants and Fungi* (1980); Romer and Parsons, *The Vertebrate Body* (6th, 1986).

**Sometimes called *divisions* instead of *phyla*; the terms are equivalent.

***Undetermined affiliation; sometimes placed with protistans.

APPENDIX II
SOLUTIONS TO GENETICS PROBLEMS

Chapter Twelve

1. a. *AB*
   b. *AB* and *aB*
   c. *Ab* and *ab*
   d. *Ab, aB, Ab,* and *ab*

2. a. *AaBB* will occur in all the offspring.
   b. 25% *AABB*; 25% *AaBB*; 25% *AABb*; 25% *AaBb*.
   c. 25% *AaBb*; 25% *Aabb*; 25% *aaBb*; 25% *aabb*.
   d. $\frac{1}{16}$ *AABB* (6.25%)
      $\frac{1}{8}$ *AaBB* (12.5%)
      $\frac{1}{16}$ *aaBB* (6.25%)
      $\frac{1}{8}$ *AABb* (12.5%)
      $\frac{1}{4}$ *AaBb* (25 %)
      $\frac{1}{8}$ *aaBb* (12.5%)
      $\frac{1}{16}$ *AAbb* (6.25%)
      $\frac{1}{8}$ *Aabb* (12.5%)
      $\frac{1}{16}$ *aabb* (6.25%)

3. a. Yellow is recessive. Because the first-generation plants must be heterozygous and had a green phenotype, green must be dominant over the recessive yellow.
   b. $(\frac{3}{4})(135) = 101.25$ green in second generation.
      $(\frac{1}{4})(135) = 33.75$ yellow in second generation.

      Obviously one cannot have fractional offspring. However, in actual situations, the expected number will often turn out to be in fractional form. From a practical standpoint, geneticists would probably respond by asserting that about 101 green second-generation plants would be expected and about 34 yellow, although slight variation from these predictions would be common and expected.

4. a. Mother must be heterozygous for both genes; father is homozygous recessive for both genes. The first child is also homozygous recessive for both genes.
   b. The probability that the second child will not be able to roll the tongue and will have free earlobes is $\frac{1}{4}$ (25%).

5. a. *ABC*
   b. *ABc* and *aBc*
   c. *ABC, aBC, ABc,* and *aBc*
   d. *ABC, aBC, AbC, abC, ABc, aBc, Abc,* and *abc*

6. Because the man can only produce one type of allele for each of his ten genes, he can only produce one type of sperm. The woman, on the other hand, can produce two types of alleles for each of her two heterozygous genes; she can produce 2×2 or 4 different kinds of eggs. As can be observed, as the number of heterozygous genes increases, more and more different types of gametes can be produced. For a population or species, this might be advantageous, as it would introduce more genetic variability into the population. Such variability might prove useful, in that the variant individuals might have better chances of surviving under a variety of environmental conditions.

7. The first-generation plants must all be double heterozygotes. When these plants are self-pollinated, $\frac{1}{4}$ (25%) of the second-generation plants will be doubly heterozygous.

8. The most direct way to accomplish this would be to allow a true-breeding mouse having yellow fur to mate with a true-breeding mouse having brown fur. Such true-breeding strains could be obtained by repeated inbreeding (mating of related individuals; for example, a male and a female of the same litter) of yellow and brown strains. In this way, it should be possible to obtain homozygous yellow and homozygous brown mice.

When true-breeding yellow and true-breeding brown mice are crossed, the progeny should all be heterozygous. If the progeny phenotype is either yellow or brown, then the dominance is simple or complete, and the phenotype reflects the dominant allele. If the phenotype is intermediate between yellow and brown, there is incomplete dominance. If the phenotype shows both yellow and brown, there is codominance.

9. a. The mother must be heterozygous ($I^A i$). The man having type B blood could have fathered the child if he were also heterozygous ($I^B i$).
   b. If the man is heterozygous, then he *could be* the father. However, because any other type B heterozygous male also could be the father, one cannot say that this particular man absolutely must be. Actually, any male who could contribute an O allele (i) could have fathered the child. This would include males with type O blood (ii) or type A blood who are heterozygous ($I^A i$).

10. a. F_1 genotypes and phenotypes: 100% *Bb Cc*, brown progeny.
    F_2 phenotypes: $\frac{9}{16}$ brown $+ \frac{3}{16}$ tan $+ \frac{4}{16}$ albino.
    F_2 genotypes: $\begin{cases} \frac{1}{16} BB\ CC + \frac{2}{16} BB\ Cc + \frac{2}{16} Bb\ CC \\ + \frac{4}{16} Bb\ Cc; (\frac{9}{16}\ \text{brown}) \\ \frac{1}{16} bb\ CC + \frac{2}{16} bb\ Cc; (\frac{3}{16}\ \text{tan}) \\ \frac{1}{16} BB\ cc + \frac{2}{16} Bb\ cc + \frac{1}{16} bb\ cc; (\frac{4}{16}\ \text{albino}) \end{cases}$

    b. Backcross phenotypes: $\frac{1}{4}$ brown $+ \frac{1}{4}$ tan $+ \frac{2}{4}$ albino.
    Backcross genotypes: $\begin{cases} \frac{1}{4} Bb\ Cc; (\frac{1}{4}\ \text{brown}) \\ \frac{1}{4} bb\ Cc; (\frac{1}{4}\ \text{tan}) \\ \frac{1}{4} Bb\ cc + \frac{1}{4} bb\ cc; (\frac{1}{2}\ \text{albino}) \end{cases}$

    Parental cross (diagram): $BB\ CC \times bb\ cc$
    P_1 gametes: (BC) (bc)
    F_1 progeny: 100% *Bb Cc* (black)
    $F_2 \times F_1$ cross: $Bb\ Cc \times Bb\ Cc$
    F_1 gametes: (BC) (Bc) (bC) (bc) × (BC) (Bc) (bC) (bc)

    F_2 genotypes:

BB CC	BB Cc	Bb CC	Bb Cc
BB Cc	BB cc	Bb Cc	Bb cc
Bb CC	Bb Cc	bb CC	bb Cc
Bb Cc	Bb cc	bb Cc	bb cc

11. Parents: *Ff Ss* × *Ff Ss*
    Progeny: $\begin{cases} \frac{1}{16} FF\ SS + \frac{2}{16} FF\ Ss + \frac{2}{16} Ff\ SS + \\ \frac{4}{16} Ff\ Ss + \frac{1}{16} FF\ ss + \frac{2}{16} Ff\ ss + \\ \frac{1}{16} ff\ SS + \frac{2}{16} ff\ Ss; (\frac{15}{16}\ \text{scaly shanks}) \\ \frac{1}{16} ff\ ss; (\frac{1}{16}\ \text{feathered shanks}) \end{cases}$

12. The mating is *Ll* × *Ll*.
    Progeny genotypes: $\frac{1}{4} LL + \frac{1}{2} Ll + \frac{1}{4} ll$
    Phenotypes: $\begin{cases} \frac{1}{4}\ \text{homozygous survivors } (LL) + \\ \frac{1}{2}\ \text{heterozygous survivors } (Ll) + \\ \frac{1}{4}\ \text{lethal } (ll)\ \text{nonsurvivors.} \end{cases}$
    Thus, among the survivors, $\frac{2}{3}$ will be heterozygous.

13. To work this problem, consider the effect of each pair of genes separately.

   1. $Aa \times aa$: The resulting progeny genotypes are $\frac{1}{2}$ Aa and $\frac{1}{2}$ aa. Thus, half of the progeny will receive an A allele, which permits kernel pigmentation.
   2. $cc \times Cc$: Here, too, half of the progeny will receive a dominant C allele, which permits kernel pigmentation.
   3. $Rr \times Rr$: In this case, $\frac{3}{4}$ of the progeny will receive at least one R allele, which permits kernel pigmentation.

   Remember that in order to have pigmented kernels, at least one dominant allele of each and every one of these three gene loci must be simultaneously present. Since the A, C, and R loci assort independently, to find the overall fraction of kernels that are pigmented, you must multiply together the fraction of pigmented kernels produced separately by the A, C, and R loci, and that is: $\frac{1}{2} \times \frac{1}{2} \times \frac{3}{4} = \frac{3}{16}$.

Chapter Thirteen

1. a. Males inherit their X chromosome from their mothers.
   b. A male can produce two types of gametes with respect to an X-linked gene. One type will lack this gene and possess a Y chromosome. The other will have an X chromosome and the linked gene.
   c. A female homozygous for an X-linked gene will produce just one type of gamete containing an X chromosome with the gene.
   d. A female heterozygous for an X-linked gene will produce two types of gametes. One will contain an X chromosome with the dominant allele, and the other type will contain an X chromosome with the recessive allele.

2. a. Because this gene is only carried on Y chromosomes, females would not be expected to have hairy pinnae because they normally do not have Y chromosomes.
   b. Because sons always inherit a Y chromosome from their fathers and because daughters never do, a man having hairy pinnae will always transmit this trait to his sons and never to his daughters.

3. a. A 0% crossover frequency means that 50% of the gametes will be AB and 50% will be ab.
   b. Of the gametes, 10% will be recombinant (5% Ab and 5% aB), and 90% will be non-recombinant (45% AB and 45% ab).
   c. Of the gametes, 25% will be recombinant (12.5% Ab and 12.5% aB), and 75% will be non-recombinant (37.5% AB and 37.5% ab).

4. Ten map units means that 10% of the gametes produced by the double-heterozygote parent will be recombinant according to the following Punnett square:

	AB (45%)	ab (45%)	Ab (5%)	aB (5%)
ab	$AaBb$ 45% of the progeny	$aabb$ 45% of the progeny	$Aabb$ 5% of the progeny	$aaBb$ 5% of the progeny

5. The first-generation females must be heterozygous for both genes. The 42 red-eyed, vestigial-winged and the 30 purple-eyed, long-winged progeny represent recombinant gametes from these females. Because the first-generation females must have produced 600 gametes to give these 600 progeny, and because 42 + 30 of these were recombinant, the percentage of recombinant gametes is 72/600, or 12%, which implies that 12 map units separate the two genes.

6. The rare vestigial-winged flies could be explained by a deletion of the dominant allele from one of the chromosomes, due to the action of the x-rays.

7. If the longer-than-normal chromosome 14 represented the translocation of most of chromosome 21 to the end of a normal chromosome 14, then this individual would be afflicted with Down's syndrome due to the presence of this attached chromosome 21 as well as two normal chromosomes 21. The total chromosome number, however, would be 46.

8. The initial cells of the hybrid from a cross between *T. turgidum* and *T. tauschii* did have twenty-one chromosomes. These chromosomes were induced to double in the absence of cell division by application of colchicine; the result was the hexaploid, with forty-two chromosomes. Gametes from *T. aestivum* should have twenty-one chromosomes (ABD); when these combine with gametes from *T. turgidum* (fourteen chromosomes, AB), offspring with 21 + 14 = 35 chromosomes (AABBD) would arise. Because there is only one set of D chromosomes, normal meiotic pairing should not occur and these hybrids would be sterile.

9. Using *c* as the symbol for colorblindness and *C* for normal vision, then the cross can be diagrammed as follows:

   C(Y) female × Cc male

   In mugwumps, a son receives one sex-linked allele from each of his parents, but a daughter inherits her unpaired sex-linked allele solely from her father. In this cross, half of the sons will be CC and half Cc, but none will be colorblind. Of the daughters, half will be C(Y) and half c(Y). Thus, there is a 50% chance that a daughter will be colorblind. (Note: this answer is backward from the way it would be in humans; but it is correct not only for mugwumps but also for all birds and Lepodoptera (moths and butterflies), as well as a few other forms. It may be easier for you to work sex-linked problems involving such organisms if you first work them exactly as you would a human or fruit fly problem; but at the end, simply change all progeny called male to female and all female to male. Remember it this way: "Birds is backwards!")

10. In order to produce a female suffering from childhood muscular dystrophy, not only must her mother be a carrier of the disease, but her father must have it. Few, if any, such males who survive to adulthood are capable of having children. His bride, should he marry, would surely think twice before bearing a child.

Chapter Fourteen

1. The gene for hemophilia occurs on the X but not the Y chromosome. A male has only one X chromosome. Therefore, it would be impossible for a male simply to be a carrier; the allele associated with hemophilia would always be expressed. For each child produced by Victoria and Leopold, there would be a probability of $\frac{1}{2}$ (50%) that it would be a nonhemophilic son and $\frac{1}{2}$ (50%) that it would be a carrier daughter.

2. Assuming the mother is heterozygous (most individuals with Huntington's chorea are), the woman has a $\frac{1}{2}$ (50%) chance of being heterozygous and therefore of later developing the disorder. Also, if this woman married a normal male, they would have a 50% chance of having a child with the disorder. Thus the *total* probability of their having a child with Huntington's chorea is $(0.5)(0.5) = 0.25$, or $\frac{1}{4}$ (25%).

3. The first child can only be color blind if it is a boy. Why? The probability of this happening is 25%. Similarly, their second child also has a 25% chance of being color blind. The probability that both will be color blind is $(0.25)(0.25) = 0.0625$, or 6.25%.

4. This indicates that genetic information other than that necessary for sex determination must reside on the X chromosome. Such information is necessary for survival regardless of whether one is male or female. Obviously, this is not true for the Y chromosome, in that individuals (females) survive quite nicely in its absence. Perhaps one could simply imagine that the major function of the Y chromosome is to change what would have been a female individual into a male.

5. The only parent from whom this child could have received an X chromosome that bears a nonhemophilia allele is the mother. Therefore, nondisjunction must have occurred in the male.

6. A child with Klinefelter's syndrome could be produced if a Y-bearing sperm fertilized an egg having two X chromosomes (as a result of nondisjunction during egg development). Such a child could also be produced if a normal egg (with one X chromosome) were fertilized by a sperm having an X and Y chromosome (as a result of nondisjunction during sperm development).

7. The Punnett square for this situation would be as follows:

	X-bearing sperm	Y-bearing sperm
XX-bearing egg	trisomic XXX	Klinefelter's XXY
no X in egg	Turner's XO	dies before birth; has only Y

8. a. An unaffected male selected at random has 1 chance in 50 (2%) of being heterozygous.

b. An unaffected female selected at random also has 1 chance in 50 (2%) of being heterozygous.

c. If you selected an unaffected male and unaffected female at random, the probability that both will be heterozygous is $(0.02)(0.02) = 0.0004$, or 0.04%.

d. The probability that a pair of unaffected individuals selected at random could have a child afflicted with PKU is given by $(0.02)(0.02)(0.25) = 0.0001$, or 0.01%. This is the same thing as 1/10,000, which suggests that about one birth in every 10,000 will be a child with PKU, assuming that only heterozygous individuals have such children (an assumption which is not completely true).

9. There would be 1 chance in 50 (2%) that a randomly selected mate would also be a PKU carrier. There would be a much greater chance that your first cousin would be a PKU carrier, because both of you would share a common ancestor and thus, possibly, a common source for the PKU allele. In fact, your first cousin would have 1 chance in 8 (12.5%) of also being a PKU carrier.

CREDITS

Chapter 1

1.1 James A. Kern / **1.2** (a) Robert Stern; (b) W. Hargreaves and D. Deamer; (c) Centers for Disease Control; (d) G. Cohen-Bazire; (e) Michael A. Walsh; (f) M.P.L. Fogden/Bruce Coleman Ltd. / **1.4** Jack de Coningh / **1.5** Jim Doran / **1.6** Douglas Faulkner/ Sally Faulkner Collection / **1.7** Douglas Faulkner/ Sally Faulkner Collection / **1.8** Roger K. Burnard / **1.9** Roger K. Burnard / **1.10** (a) Norman Meyers/ Bruce Coleman Inc.; (b) Timothy Ransom / **1.11** Roger K. Burnard

Chapter 2

2.1 The Bettmann Archive / **2.4** Art by D. & V. Hennings / **2.5** Courtesy George P. Darwin / **2.6** (above) Bianca Lavies, © 1982 National Geographic Society; (below) Charles R. Knight, © Field Museum of Natural History / **2.7** (a) The Bettmann Archive / **2.8** (b) Alan Root/Bruce Coleman Ltd. / **2.9** Encyclopedia of Pigeon Breeds / **2.9** Down House and The Royal College of Surgeons of England / **2.11** (a) Neg. # 325288 Courtesy Department Library Services, American Museum of Natural History; (b) John D. Cunningham/Visuals Unlimited

Chapter 3

3.9 Robert D. Schuster / **3.10** William H. Amos / **3.11** Roger K. Burnard

Chapter 4

4.1 Computer Graphics Laboratory, University of California, San Francisco, © Regents, University of California / **4.6** Micrograph E. Frei and R.D. Preston / **4.10** Roland R. Dute / **4.15–4.16** Arthur M. Lesk, *Trends in Biochemical Sciences*, June 1984 / **4.17** Micrograph Jerome Gross; sketch after D. Eyre, *Science*, 207, 21 March 1980, copyright 1980 AAAS

Chapter 5

5.1 Biophoto Associates / **5.3** (a) William H. Amos; (b) F. Sauer/Bruce Coleman Ltd.; (c),(d) Gary W. Grimes and Stephen W. L'Hernault / **5.5** (a) Kim Taylor/Bruce Coleman Ltd.; (b) Hervé Chaumeton/ Agence Nature; (c) McCutcheon/ZEFA / **5.6** Micrograph G. Cohen-Bazire / **5.9** Micrograph M.C. Ledbetter, Brookhaven National Laboratory; art by D. & V. Hennings / **5.10** Micrograph G.L. Decker; art by D. & V. Hennings / **5.11** D. Fawcett, *The Cell*, Philadelphia: W.B. Saunders Co., 1966; art by D. & V. Hennings / **5.12** A.C. Faberge, *Cell and Tissue Research*, 151: 403–415, 1974; (b) E.G. Pollock / **5.13** (a) After W. Bloom and D. Fawcett, *A Textbook of Histology*, Philadelphia: W.B. Saunders Co., 1975; (b),(c) Daniel S. Friend, M.D. / **5.15** (a) Gary W. Grimes; (b) Hilton H. Mollenhauer / **5.16** Gary W. Grimes / **5.17** Micrograph Eugene S. Vigil / **5.18** Micrograph Keith R. Porter / **5.19** Micrograph L.K. Shumway / **5.20** After K. Porter and J. Tucker, *Scientific American*, March 1981, copyright © by Scientific American, Inc., all rights reserved / **5.21** E. Lazarides / **5.22** Y. Kersey and N. Wessels, *Journal of Cell Biology*, 68:264, 1976, by copyright permission of The Rockefeller University Press / **5.23** (a) R. Sloboda, *American Scientist*, 68:290, 1980; (b) R. Barton / **5.24** Sketch by D. & V. Hennings after P. Raven et al., *Biology of Plants*, Third edition, Worth Publishers, 1981; micrograph P.A. Roelofsen / **5.25** Susumu Ito / **5.26** Gary Gaard and Arthur Kelman / **5.27** D.H. Matulionis / **5.29** (a),(b) I.R. Gibbons; (c) U.W. Goodenough and J.E. Heuser / **5.30**

Micrograph S. Wolfe, *Biology of the Cell*, Second edition, Wadsworth, 1981

Chapter 6

6.2 Micrograph J.D. Robertson; art by D. & V. Hennings / **6.3** Micrograph P. Pinto da Silva and D. Branton, *Journal Cell Biology*, 45:598, 1970, by copyright permission of The Rockefeller University Press / **6.5** Micrograph M. Sheetz, R. Painter, and S. Singer, *Journal Cell Biology*, 70:193, 1976, by copyright permission of The Rockefeller University Press / **6.6** (a),(b) Bruce Russell/BioMedia Associates; (c),(d) Thomas Eisner, Cornell University / **6.7** Frank B. Salisbury / **6.11** Micrographs K.W. Jeon / **6.12** M.M. Perry and A.B. Gilbert

Chapter 7

7.1 NASA / **7.2** Manfred Kage/Peter Arnold / **7.4** W.S. Bennett and T.A. Steitz / **7.7** Art by Victor Royer / **7.8** Michael Connolly, *Science*, 19 August 1983, 221:709–713, copyright 1983 AAAS

Chapter 8

8.2 (a) Photograph Klaus Hackenberg/ZEFA; (b),(c) Harry T. Horner; art by Joel Ito / **8.3** Art by Victor Royer / **8.9** Micrograph M.D. Hatch

Chapter 9

9.1 (a) Michael Freeman/Bruce Coleman Ltd.; (b) Adrian Warren/Ardea, London; (c) Tom McHugh / **9.6** (a) Keith R. Porter; art by Joel Ito / **9.11** Roger K. Burnard

Chapter 10

10.1 Micrograph A. Ryter / **10.2** Micrograph E.J. Dupraw, *DNA and Chromosomes*, Holt, Rinehart and Winston, 1970 / **10.3** S. Brecher / **10.4** Andrew S. Bajer, University of Oregon / **10.6** Andrew S. Bajer, University of Oregon / **10.9** B.R. Brinkley / **10.10** H. Beams and R.G. Kessel, *American Scientist*, 64:279–290, 1976 / **10.11** Micrograph F.D. Hess

Chapter 11

11.1 Micrograph E.J. Dupraw, *DNA and Chromosomes*, Holt, Rinehart and Winston, 1970 / **11.4** Micrograph B. John / **11.6** (b) © 1986 David M. Phillips/Visuals Unlimited

Chapter 12

12.1 The Granger Collection, New York / **12.2** Jean M. Labat/Ardea, London / **12.10** Jack Carey / **12.11** F. Claussen, *Zietschr. Abstgs. und Vererbgsl.*, 76:30, 1939 in U.W. Goodenough, *Genetics*, Third edition, Saunders College Publishing, 1984 / **12.13** Tedd Somes / **12.14** David Hosking / **12.15** (a) Richard Kolberg; (b) Neal Nichols / **12.16** Douglas Faulkner/ Sally Faulkner Collection / **12.17** After John G. Torrey, *Development in Flowering Plants*, by permission of Macmillan Publishing Company, copyright © 1967 by John G. Torrey / **12.18** (a) F. Blakeslee, *Journal of Heredity*, 1914

Chapter 13

13.8 Chuck Brown / **13.9** A.D. Stock, City of Hope National Medical Center / **13.11** (a) H.P. Olmo, University of California, Davis; (b) Allison L.

Hansen / **13.12** After W. Jensen and F.B. Salisbury, *Botany: An Ecological Approach*, Wadsworth, 1972

Chapter 14

14.2 After V. McKusick, *Human Genetics*, Second edition, copyright 1969. Reprinted by permission of Prentice-Hall, Inc., Englewood Cliffs, N.J. / **14.4** (a) Cytogenetics Laboratory, University of California, San Francisco; (b) After Collmann and Stoller, *American Journal of Public Health*, 52, 1962

Chapter 15

15.2 (a) Lee D. Simon, Waksman Institute of Microbiology, Rutgers University; (b) A.K. Kleinschmidt / **15.6** Photograph Computer Graphics Laboratory, University of California, San Francisco, © Regents, University of California; art by D. & V. Hennings / **15.7** A.C. Barrington Brown, © 1968 J.D. Watson

Chapter 16

16.2 After S. Wolfe, *Biology of the Cell*, Second edition, Wadsworth, 1981 / **16.7** Micrograph O.L. Miller, Jr. and B.R. Beatty, *Science*, 164:955–957, copyright 1969 AAAS / **16.9** Sung Hou Kim / **16.10** (b) O.L. Miller, Jr. and B. Hamkalo, *Int. Rev. Cytol.* 33:7, 1972 / **16.11** Art by L. Calver / **16.12** Art by L. Calver

Chapter 17

17.3 I. Paulsen and U. Laemmli, *Cell*, 12:827–828, 1977 / **17.4** Micrograph O.L. Miller, Jr. and Steve McKnight / **17.5** (a) E.J. Dupraw; (b) J.G. Gall; (c) B. Hamkalo; (d) V. Foe / **17.6** (a) John T. Bonner; (b) London Scientific Films Ltd.; (c–e) Carolina Biological Supply Company / **17.7** Micrograph J.G. Gall / **17.8** W. Beermann / **17.9** Murray L. Barr / **17.10** Jack Carey / **17.11** Photograph A. Kerstitch/ Tom Stack & Associates

Chapter 18

18.3 Micrograph Robley C. Williams / **18.4** Peter Starlinger / **18.5** Dr. Huntington Potter and Dr. David Dressler, Harvard Medical School / **18.6** C.C. Brinton, Jr. and J. Carnahan / **18.9** John D. Cunningham/Visuals Unlimited / **18.10** R.L. Brinster

Chapter 19

19.3–19.5 Photographs Biophoto Associates / **19.6** Thomas Eisner, Cornell University / **19.7–19.8** Micrographs H.A. Core, W.A. Coté, and A.C. Day, *Wood Structure and Identification*, Second edition, Syracuse University Press, 1979 / **19.10** (b) G. Shih and R.G. Kessel, *Living Images*, Jones and Bartlett, Publishers, Inc., Boston, ©1982 / **19.11** Micrograph Chuck Brown; art by D. & V. Hennings / **19.12** (a) Robert and Linda Mitchell; (b),(c) Roland R. Dute / **19.13** E.R. Degginger / **19.14** (a) Ray F. Evert; (b) Gene Cox/Bruce Coleman Ltd. / **19.17** (a) John N.A. Lott, *Scanning Electron Microscope Study of Green Plants*, St. Louis: C.V. Mosby Company, 1976; (b–d) Robert and Linda Mitchell / **19.18** J. Troughton and L.A. Donaldson; art by D. & V. Hennings / **19.19** G. Shih and R.G. Kessel, *Living Images*, Jones and Bartlett, Publishers, Inc., Boston, © 1982 / **19.20** W. Thomson, *American Journal of Botany*, 57(3):316, 1970 / **19.21** John Hodgin / **19.22** Sketch Marian Reeve; photograph E.R. Degginger / **19.23** (a),(b)

Chuck Brown; (c),(d) after T. Weier et al., *Botany: An Introduction to Plant Biology*, Sixth edition, Wiley, 1982 / **19.24** Thomas Eisner, Cornell University / **19.25** Ripon Microslides, Inc. / **19.26** (b),(c) Ray F. Evert / **19.27** After Marian Reeve / **19.29** H.A. Core, W.A. Coté, and A.C. Day, *Wood Structure and Identification*, Second edition, Syracuse University Press, 1979 / **19.30** Photograph Biophoto Associates

Chapter 20

20.1 Dennis Brokaw / **20.2** Adrian P. Davies/Bruce Coleman Ltd. / **20.3** J. Mexal, C. Reid, and E. Burke, *Botanical Gazette*, 140: No. 3, University of Chicago Press, 1979 / **20.4** F.B. Reeves / **20.5** Jean Paul Revel / **20.7** (a),(b) John Troughton and L.A. Donaldson / **20.8** T.A. Mansfield / **20.10** Martin Zimmerman, *Science*, 133:73–79, © AAAS 1961

Chapter 21

21.1 Wardene Wiesser/Ardea, London / **21.5** F.D. Hess / **21.6** Art by D. & V. Hennings / **21.7** (a) Ted Schwartz; (b) R. Taggart; (c) Harlo H. Hadow / **21.8** Thomas Eisner, Cornell University / **21.9** Edward S. Ross / **21.10** (a) R. Taggart; (b) Edward S. Ross; (c) Ted Schwartz / **21.11** (a) Patricia Schulz; (b),(c) Ray F. Evert; (d),(e) Ripon Microslides, Inc. / **21.12** (a) Kjell B. Sandved; (b) Edward S. Ross / **21.13** B. Bracegirdle and P. Miles, *An Atlas of Plant Structure*, Heinemann Educational Books, 1977 / **21.14** Janet Jones / **21.15** (a) B.J. Miller, Fairfax, VA/BPS; (b) R. Carr/Bruce Coleman Ltd.; (c) Richard H. Gross, Motlow State Community College

Chapter 22

22.1 (a) Carolina Biological Supply Company; (b) Hervé Chaumeton/Agence Nature / **22.6** (a) H.E. Thompson / **22.7** Frank B. Salisbury / **22.8** John Digby and Richard Firn / **22.9** B.E. Juniper / **22.10** Cary Mitchell / **22.13** Frank B. Salisbury / **22.15** Jan Zeevaart / **22.16** N.R. Lersten, Iowa State University / **22.17** A.C. Leopold et al., *Plant Physiology*, 34:570, 1958 / **22.18** A.C. Leopold and M. Kawase, *American Journal of Botany*, 51:294–298, 1964 / **22.19** R.J. Downs in T.T. Kozlowski, ed., *Tree Growth*, The Ronald Press, 1962 / **22.20** (left) Edward S. Ross; (right) Dennis Brokaw

Chapter 23

23.1 Art by D. & V. Hennings / **23.2** (a),(b) Manfred Kage/Bruce Coleman Ltd.; (c) Ed Reschke / **23.4** (a) Richard Kolberg; (b) Chuck Brown / **23.5** Micrograph W.K. Elwood / **23.6** Art by Joel Ito; micrograph Bruce Russell/BioMedia Associates / **23.8** Art by Joel Ito / **23.9** Photograph Chaumeton-Lanceau/Agence Nature; sketch after T. Storer et al., *General Zoology*, Sixth edition, McGraw-Hill, 1979

Chapter 24

24.3 After J.E. Crouch, *Functional Human Anatomy*, Fourth edition, Lea & Febiger, Philadelphia, 1985 / **24.4** Art by D. & V. Hennings / **24.6** (b) A.L. Hodgkin, *Journal of Physiology*, vol. 131, 1956 / **24.11** (a) Carolina Biological Supply Company / **24.12** Art by D. & V. Hennings; (d) J.E. Heuser and T.S. Reese / **24.14** Art by Joel Ito

Chapter 25

25.2 Art by D. & V. Hennings / **25.4** Art by Joel Ito / **25.7** Art by Joel Ito / **25.9** Art by D. & V. Hennings after Romer and others / **25.10** C. Yokochi and J. Rohen, *Photographic Anatomy of the Human Body*, Second edition, Igaku-Shoin Ltd., 1979 / **25.11** Art by Joel Ito / **25.12** After Penfield and Rasmussen, *The Cerebral Cortex of Man*, copyright © 1950 Macmillan Publishing Company, Inc. Renewed 1978 by Theodore Rasmussen / **25.13** (a) Based on

A. Vander et al., *Human Physiology*, Third edition, © 1980 by McGraw-Hill / **25.14** After H. Jasper, 1941

Chapter 26

26.1 Art by D. & V. Hennings / **26.2** Art by Joel Ito / **26.3** Art by Joel Ito / **26.4** Art by Joel Ito / **26.7** The Bettmann Archive / **26.9** Art by Joel Ito

Chapter 27

27.1 Eric A. Newman / **27.2** From Hensel and Bowman, *Journal of Physiology*, 23:564–568, 1960 / **27.3** After Penfield and Rasmussen, *The Cerebral Cortex of Man*, copyright © 1950 Macmillan Publishing Company, Inc. Renewed 1978 by Theodore Rasmussen / **27.4** (a) After R. Murray and A. Murray, *Taste and Smell in Vertebrates*, Churchill, 1970; (b) Ed Reschke / **27.5** Art by Ron Ervin / **27.6** Art by D. & V. Hennings / **27.8** Photograph Douglas Faulkner/Sally Faulkner Collection / **27.10** (a) Dr. Goran Bredburg, Uppsala, Sweden; (b),(c) Robert E. Preston, courtesy Joseph E. Hawkins, Kresge Hearing Research Institute, University of Michigan Medical School / **27.11** Merlin D. Tuttle, Bat Conservation International / **27.12** (a) After M. Gardiner, *The Biology of Vertebrates*, McGraw-Hill, 1972 / **27.13** (a) Keith Gillett/Tom Stack and Associates; (b) J. Grossauer/ZEFA / **27.14** (a) E.R. Degginger; (b) L.M. Beidler / **27.15** G.A. Mazohkin-Porshnykov (1958). Reprinted with permission from *Insect Vision*, © 1969 Plenum Press / **27.18** Micrograph E.R. Lewis, F.S. Werblin, and Y.Y. Zeevi / **27.19** Sketches after S. Kuffler and J. Nicholls, *From Neuron to Brain*, Sinauer, 1977

Chapter 28

28.2 D.A. Parry, *Journal of Experimental Biology*, 36:654, 1959 / **28.3** Art by D. & V. Hennings / **Page 377** Photograph C. Yokochi and J. Rohen, *Photographic Anatomy of the Human Body*, Second edition, 1979 / **28.11** Micrograph D. Fawcett, *The Cell*, Philadelphia: W.B. Saunders Co., 1966 / **28.13** Inset after L.D. Peachey, *Journal of Cell Biology*, 25:222, 1965 / **28.14** Adapted from R. Eckert and D. Randall, *Animal Physiology: Mechanisms and Adaptations*, Second edition, W.H. Freeman and Co., 1983

Chapter 29

29.1 After M. Labarbera and S. Vogel, *American Scientist*, 70:54–60, 1982 / **29.3** (a) Michael H. Ross; (b) Tony Brain/Science Photo Library Ltd./Photo Researchers / **29.4** Art by Kathleen Talaro / **29.6** Art by Joel Ito / **29.7** Photograph C. Yokochi and J. Rohen, *Photographic Anatomy of the Human Body*, Second edition, Igaku-Shoin Ltd., 1979; art by Joel Ito / **29.12** Art by L. Calver based on A. Spence, *Basic Human Anatomy*, Benjamin-Cummings, 1982 / **29.13** Art by Victor Royer based on A. Vander, J. Sherman, and D. Luciano, *Human Physiology*, Fourth edition, McGraw-Hill, 1985 / **29.14** Data from A. Vander, J. Sherman, and D. Luciano, *Human Physiology*, Fourth edition, McGraw-Hill, 1985 / **Page 397** (a),(b) Robert LaPorta; (c) Ripon Microslides, Inc. / **29.16** T. Hovig, Oslo University / **29.17** Micrograph Emil Bernstein, *Science*, cover 27 August 1971, copyright 1971 AAAS / **29.18** (a) After F. Ayala and J. Kiger, *Modern Genetics*, © 1980 Benjamin-Cummings; (b),(c) Lester V. Bergman, NY / **29.19** Art by D. & V. Hennings

Chapter 30

30.1 The Granger Collection, New York / **30.3** Photograph Arthur J. Olson, © 1985 Scripps Clinic Research Foundation / **30.4** Art by L. Calver after S. Tonegawa, *Scientific American*, October 1985 / **30.5** After B. Alberts et al., *Molecular Biology of the Cell*, Garland Publishing Co., 1983 / **30.7** Jeffrey Laurence, M.D., The New York Hospital, Cornell

Medical Center / **Page 415** Photographs R. Albrecht from J. Krahenbuhl and J. Remington, *Human Nature*, 1(1):52–59, 1978

Chapter 31

31.1 Art by D. & V. Hennings; (a–c) after C.P. Hickman et al., *Integrated Principles of Zoology*, Sixth edition, St. Louis: C.V. Mosby Co., 1979 / **31.2** Ed Reschke / **31.3** Ed Reschke / **31.4** After C.P. Hickman et al., *Integrated Principles of Zoology*, Sixth edition, St. Louis: C.V. Mosby Co., 1979 / **31.5** Art by Victor Royer; micrograph H.R. Duncker, Justus-Liebig University, Giessen, West Germany; diagram after W.L. Bretz and K. Schmidt-Nielsen, *Journal of Experimental Biology*, 56:57–65, 1972 / **31.6** Art by L. Calver / **31.7** After J. Sobotta, *Atlas of Human Anatomy*, Urban & Schwarzenberg, 1978 / **31.8** After A. Vander, J. Sherman, and D. Luciano, *Human Physiology*, Third edition, McGraw-Hill, 1980 / **31.11** After J.M. Kinney, *Anesthesiology*, 21:615, 1980 / **31.12** David Steinberg / **31.13** R.G. Kessel and R.H. Kardon, *A Text-Atlas of Scanning Electron Microscopy*, W.H. Freeman and Co., copyright © 1979 / **31.14** Gerard D. McLane

Chapter 32

32.1 Martin W. Grosnick/Ardea, London / **32.3** (a) Kim Taylor/Bruce Coleman Ltd.; (b) P. Morris/Ardea, London; (c) Wardene Weisser/Ardea, London / **32.6** Art by L. Calver; photograph C. Yokochi and J. Rohen, *Photographic Anatomy of the Human Body*, Igaku-Shoin Ltd., 1979 / **32.7** After A. Vander, J. Sherman, and D. Luciano, *Human Physiology*, Fourth edition, McGraw-Hill, 1985 / **32.9** (a–c) R.G. Kessel and R.H. Kardon, *Tissues and Organs: A Text-Atlas of Scanning Electron Microscopy*, W.H. Freeman and Co., copyright © 1979; (d) J.D. Hoskins, W.G. Henk, and Y.Z. Abdelbaki, *American Journal of Veterinary Research*, 43:10, 1982 / **32.10** Photograph P. Morris/Ardea, London / **32.12** Modified from A. Vander, J. Sherman, and D. Luciano, *Human Physiology*, Fourth edition, McGraw-Hill, 1985

Chapter 33

33.1 David Mcdonald / **33.2** After A. Vander, J. Sherman, and D. Luciano, *Human Physiology*, Fourth edition, McGraw-Hill, 1985 / **33.3** Terry Vaughan / **33.4** Art by Ron Ervin / **33.5** Art by D. & V. Hennings / **33.10** Art by Joel Ito / **33.11** Tom McHugh/Photo Researchers

Chapter 34

34.1 (a) Hans Pfletschinger; (b–e) John H. Gerard / **34.2** (a) Leonard Lee Rue III; (b) Frieder Sauer/Bruce Coleman Ltd.; (c) Warren Garst/Tom Stack & Associates; (d) Wisniewski/ZEFA; (e) Jack Dermid; (f) Jean-Paul Ferrero/Ardea, London; (g) Alan Root/Bruce Coleman Ltd. / **34.4** Adapted from R.G. Ham and M.J. Veomett, *Mechanisms of Development*, St. Louis: C.V. Mosby Co., 1980 / **34.6** R.G. Kessel and C.Y. Shih, *Scanning Electron Microscopy in Biology*, Springer-Verlag, 1976 / **34.7** Photograph Jeff Hardin / **34.8** Photographs Carolina Biological Supply Company; sketches after M.B. Patten, *Early Embryology of the Chick*, Fifth edition, McGraw-Hill, 1971 / **34.9** From L.B. Arey, *Developmental Anatomy*, Philadelphia: W.B. Saunders Co., 1965 / **34.10** Sketches after B. Burnside, *Developmental Biology*, 26:416–441, 1971: micrograph K.W. Tosney / **34.11** K.W. Tosney / **34.12** (a–c) After John W. Saunders, Jr., *Patterns and Principles of Animal Development*, copyright © by John W. Saunders, Jr., Macmillan Publishing Company; (d) S.R. Hilfer and J.W. Yang, *The Anatomical Record*, 197:423–433, 1980 / **34.13** After J.W. Fristrom et al., in E.W. Hanly, ed., *Problems in Biology: RNA in Development*, University of Utah Press / **34.14–34.15** Sketches after Willier, Weiss, and Hamburger, *Analysis of Development*,

Philadelphia: W.B. Saunders Co., 1955; photographs Roger K. Burnard / **34.16** F. Seidel, *Roux' Archives of Developmental Biology* 152:43–130, 1960

Chapter 35

35.1 Art by Ron Ervin / **35.2** Art by Ron Ervin; micrograph R.G. Kessel and R.H. Kardon, *Tissues and Organs: A Text-Atlas of Scanning Electron Microscopy*, W.H. Freeman and Co., copyright © 1979 / **35.3** Art by Ron Ervin / **35.4** Art by Ron Ervin / **35.6** Art by Ron Ervin / **35.10** (a) From Lennart Nilsson, *A Child Is Born*, Dell Publishing Company, New York, 1977; (b) From Lennart Nilsson, *Behold Man*, © 1974 by Albert Bonniers Forlag and Little, Brown and Co. (Canada) Ltd. / **35.11** The Carnegie Institution of Washington and Ronan O'Rahilly, M.D., Director / **35.12** Art by L. Calver / **35.13** Art by L. Calver; (c) after A.S. Romer and T.S. Parsons, *The Vertebrate Body*, Sixth edition, Saunders College Publishing, © 1986 CBS College Publishing / **35.14–35.17** From Lennart Nilsson, *A Child Is Born*, Delacourt Press, New York, 1977

Chapter 36

36.1 (a) Hans Reinhard/Bruce Coleman Ltd.; (b) Gary Zahm; (c) G. Alan Solem / **36.3** After D. Futuyma, *Evolutionary Biology*, Sinauer, 1979 / **36.5** After M. Karns and L. Penrose, *Annals of Eugenics*, 15:206–233, 1951 / **36.6** (a) Peabody Museum of Natural History, Yale University; (b) Leonard Lee Rue III / **36.7** J.A. Bishop and L.M. Cook / **36.9** (a) Charles W. Fowler/National Marine Fisheries; (b) C.G. Summers/Tom Stack & Associates; (c) G. Sirena/ZEFA / **36.10** After F. Ayala and J. Valentine, *Evolving*, Benjamin-Cummings, 1979 / **36.11** After V. Grant, *Organismic Evolution*, W.H. Freeman and Co., 1977

Chapter 37

37.3 After S.M. Stanley, *Macroevolution: Pattern and Process*, W.H. Freeman and Co., 1979 / **37.4** (a) A. Feduccia, *The Age of Birds*, Harvard University Press, 1980; (b) Donald Baird, Princeton Museum of Natural History / **37.5** Art by Joel Ito / **37.6** (top) D.P. Wilson/Eric & David Hosking; (center) J. Kirschner, Courtesy Department Library Services, American Museum of Natural History; (bottom) E.R. Degginger / **37.7** L.R. Maxson and A.C. Wilson, *Systematic Zoology*, 24(1):1–15, 1975 / **37.8** After C.G. Sibley and J.E. Ahlquist, *Scientific American*, February 1986, copyright © by Scientific American, Inc., all rights reserved / **37.11** After E.O. Dodson and P. Dodson, *Evolution: Process and Product*, Third edition, Prindle, Weber & Schmidt, 1985 / **37.12** Art by Victor Royer; (a) after D.M. Raup and J.J. Sepkoski, Jr., *Science*, 215:1501–1503, 19 March 1982; (b) after J.J. Sepkoski, Jr., *Paleobiology*, 7(1):36–53, 1981

Chapter 38

38.1 Chesley Bonestell / **38.4** (a) Sidney W. Fox; (b) W. Hargreaves and D. Deamer / **38.5** (a) Stanley M. Awramik; (b) J.W. Schopf; (c) M.R. Walter / **38.6** J.W. Schopf / **38.7** Art by Victor Royer / **38.8** David Chase / **38.9** (a) H. Stolp; (b) L. Margulis / **38.10** (a),(b) G.R. Adlington; (c) National Museum of Natural History / **38.11** National Museum of Natural History / **38.13** Jack Carey / **38.14** (a) D.S. Edwards; (b) T.N. Taylor / **38.16** Negative #322871, Courtesy Department Library Services, American Museum of Natural History / **38.17** (a) R.T. Bakker; (b) Negative #109403, Courtesy Department Library Services, American Museum of Natural History; (c) Charles R. Knight, Field Museum of Natural History; (d) Peabody Museum of Natural History, Yale University / **38.18** Art by D. & V. Hennings / **38.19** Data from J.J. Sepkoski, Jr., *Paleobiology* 7(1):36–53 and J.J. Sepkoski, Jr. and M.L. Hulver in Valentine, ed., *Phanerozoic Diversity Patterns: Profiles in*

Macroevolution, Princeton University Press, 1985 / **38.20** Smithsonian Museum

Chapter 39

39.1 Tony Brain/Science Photo Library/Photo Researchers / **39.2** Richard Blakemore / **39.3** David Vitale and George B. Chapman, Georgetown University / **39.4** J.R. Waaland, University of Washington/BPS / **39.6** Robley C. Williams / **39.7** C. McLaren and F. Siegel/Burroughs Wellcome Co. / **39.8** (a) P.L. Walne and J.H. Arnott, *Planta*, 77:325–354, 1967; (b) G. Shih and R.G. Kessel, *Living Images*, Jones and Bartlett Publishers, Inc., Boston, © 1982 / **39.9** (a) Hans Paerl; (b) G. Shih and R.G. Kessel, *Living Images*, Jones and Bartlett Publishers, Inc., Boston, © 1982 / **39.10** Edward Gabrielle / **39.11** (a) Florida Department of Natural Resources, Bureau of Marine Research; (b) C.C. Lockwood / **39.12** From Stanier, Adelberg, and Ingraham, *The Microbial World*, Fourth edition, © 1976 Prentice-Hall Inc., Englewood Cliffs, N.J. / **39.13** John Clegg/Ardea, London / **39.14** Negative #318863 Courtesy Department Library Services, American Museum of Natural History / **39.15** Micrograph Steven W. L'Hernault; illustration based on correspondence with Raul J. Cano / **39.16** Micrograph Gary W. Grimes and Steven W. L'Hernault / **39.17** (a) Thomas Eisner, Cornell University; (b) Gary W. Grimes and Steven W. L'Hernault / **39.18** Richard W. Greene / **39.19** Laszlo Meszoly in L. Margulis, *Early Life*, Jones and Bartlett Publishers, Inc., Boston, © 1982

Chapter 40

40.1 John E. Hodgin / **40.3** Sketch after T. Rost et al., *Botany*, Wiley, 1979; Photograph Victor Duran / **40.4** Victor Duran / **40.5** (a) Roger K. Burnard; (b),(c) Victor Duran / **40.7** Photograph Victor Duran; micrographs G. Shih and R.G. Kessel, *Living Images*, Jones and Bartlett Publishers, Inc., Boston, © 1982 / **40.8** N. Allin and G.L. Barron, University of Guelph / **40.9** After Raven, Evert, and Eichhorn, *Biology of Plants*, Fourth edition, Worth Publishers, New York, 1986 / **40.10** (a) Edward S. Ross; (b) Ken Davis/Tom Stack & Associates / **40.11** (a) Edward S. Ross; (b) John Shaw/Bruce Coleman Ltd. / **40.12** From Mary Barkworth, Utah State University / **40.13** (a: left) Douglas Faulkner/Sally Faulkner Collection; (a: right) D.P. Wilson/Eric & David Hosking; (c) Dennis Brokaw; (d) Hervé Chaumeton/Agence Nature / **40.14** Photograph Hervé Chaumeton/Agence Nature / **40.15** (a) Jane Burton/Bruce Coleman Ltd.; (b) Anthony & Elizabeth Bomford/Ardea, London / **40.16** Field Museum of Natural History / **40.17** (a) Edward S. Ross; (b) W.H. Hodge; (c) Kratz/ZEFA / **40.18** Jean-Paul Ferrero/Ardea, London / **40.19** Photograph Anthony & Elizabeth Bomford/Ardea, London / **40.20** (a),(c) Edward S. Ross; (b),(d) John H. Gerard / **40.21** (a) Edward S. Ross; (b) F.J. Odendaal, Duke University/BPS / **40.22** Photograph Edward S. Ross / **40.23** Edward S. Ross

Chapter 41

41.1 Kjell B. Sandved / **41.4** David C. Haas/Tom Stack & Associates / **41.5** H.M. Reiswig / **41.7** Kim Taylor/Bruce Coleman Ltd. / **41.8** Frieder Sauer/Bruce Coleman Ltd. / **41.9** After T. Storer et al., *General Zoology*, Sixth edition, © 1979 McGraw-Hill / **41.11** Photograph Kim Taylor/Bruce Coleman Ltd. / **41.12** After T. Storer et al., *General Zoology*, Sixth edition, © 1979 McGraw-Hill / **41.14** (a) C.P. Hickman; (b) Eugene Kozloff / **41.15** After C.P. Hickman, Jr. and L.S. Roberts, *Integrated Principles of Zoology*, Seventh edition, St. Louis: Times Mirror/Mosby College Publishing, 1984 / **41.16** (a) Lorus J. and Margery Milne; (b) Ed Reschke / **41.18** (a) Hervé Chaumeton/Agence Nature; (b) Jon Kenfield/Bruce Coleman Ltd. / **41.19** Art by Ron Ervin / **41.20** After

C.P. Hickman et al., *Integrated Principles of Zoology*, Sixth edition, St. Louis: C.V. Mosby Co., 1979 / **41.21** Edward S. Ross / **41.23** John H. Gerard / **41.24** Photograph Hervé Chaumeton/Agence Nature; art by Laszlo Meszoly / **41.25** Tom McHugh / **41.26** Kjell B. Sandved / **41.27** Agence Nature / **41.28** Steve Martin/Tom Stack & Associates / **41.29** C.P. Hickman, Jr. / **41.30** Art by D. & V. Hennings / **41.31** Art by Joel Ito / **41.32** (a) Anthony & Elizabeth Bomford/Ardea, London; (b) Kjell B. Sandved / **41.33** (top) Eugene Kozloff; (bottom) Rick M. Harbo / **41.34** Art by Laszlo Meszoly and D. & V. Hennings / **41.35** Hervé Chaumeton/Agence Nature / **41.36** Art by Laszlo Meszoly and D. & V. Hennings / **41.37** Douglas Faulkner/Sally Faulkner Collection / **41.38** (a–c),(e) Douglas Faulkner/Sally Faulkner Collection; (d) Kjell B. Sandved / **41.41** (a),(b) After W.D. Russell-Hunter, *A Biology of Higher Vertebrates*, copyright © 1969 W.D. Russell-Hunter, Macmillan Publishing Company / **41.42** Photograph Hervé Chaumeton/Agence Nature; art by Laszlo Meszoly and D. & V. Hennings / **41.43** After A.S. Romer and T.S. Parsons, *The Vertebrate Body*, Sixth edition, Saunders College Publishing, © 1986, CBS College Publishing; art by Laszlo Meszoly and D. & V. Hennings / **41.44** After C.P. Hickman, Jr. and L.S. Roberts, *Integrated Principles of Zoology*, Seventh edition, St. Louis: Times Mirror/Mosby College Publishing, 1984 / **41.46** Art by Laszlo Meszoly and D. & V. Hennings / **41.47** (page 638) Top: (left) Bill Wood/Bruce Coleman Ltd.; (right) Allan Power/Bruce Coleman Ltd.; Bottom: (left) E.R. Degginger; (right) Reinhard/ZEFA; (page 639) Top: (left) Erwin Christian/ZEFA; (right) C.B. & D.W. Firth/Bruce Coleman Ltd.; Bottom: (left) Christopher Crowley

Chapter 42

42.1 Noel Badrian / **42.2** (a) Bruce Coleman Ltd.; (b) Larry Burrows/Aspect Picture Library, London / **42.3** Owen Newmann/Bruce Coleman Ltd. / **42.6** After A.S. Romer and T.S. Parsons, *The Vertebrate Body*, Sixth edition, Saunders College Publishing, © 1986 CBS College Publishing / **42.7** After J. Cronin et al., *Nature*, 292:113–122, 1981 / **42.8** Dr. Donald Johanson, Institute of Human Origins / **42.9** Louise M. Robbins / **42.10** R.I.M. Campbell/Bruce Coleman Ltd.

Chapter 43

43.1 (left) Robert K. Colwell; (right) William J. Weber/Visuals Unlimited / **43.3** E.R. Degginger / **43.6** Data from V. Scheffer, *Science Monthly*, 73:356–362, 1951; photograph E. Vetter/ZEFA / **43.7** Data from A.J. Nicholson, *Cold Spring Harbor Symposium on Quantitative Biology*, 22:153–173, 1957 / **43.8** (d) Eric Crichton/Bruce Coleman Ltd. / **43.10** John Endler / **43.13** After G.T. Miller, *Living in the Environment*, Wadsworth, 1982 / **43.14** Data from Population Reference Bureau

Chapter 44

44.1 (left) Dona Hutchins; (right) Edward S. Ross / **44.2** (a),(c) Harlo H. Hadow; (b) Bob and Miriam Francis/Tom Stack & Associates / **44.3** After G. Gause, 1934 / **44.4** Photograph Clara Calhoun/Bruce Coleman Ltd. / **44.5** John Dominis, Life Magazine, © 1965, Time Inc. / **44.7** Photograph Ed Cesar/NAS/Photo Researchers / **44.8** W.E. Ruth / **44.9** W.M. Laetsch / **44.10** (a–c) Edward S. Ross; (d) James H. Carmichael, Jr. / **44.11** (a) Roger T. Peterson/NAS/Photo Researchers; (b),(c) Thomas Eisner, Cornell University / **44.12** Edward S. Ross / **44.13** Data from P. Price and H. Tripp, *Canadian Entomology*, 104:1003–1016, 1972 / **44.14** After N. Weland and F. Bazzaz, *Ecology*, 56:681–688, © 1975 Ecological Society of America / **44.15** (a–f), (i) Roger K. Burnard; (g),(h) E.R. Degginger / **44.16** (top) Dr. Harold Simon/Tom Stack & Associates; (bottom) After S. Fridriksson, *Evolution of Life on a Volcanic Island*, Butterworth: London, 1975 / **44.17** After J.M.

Diamond, *Proceedings of the National Academy of Sciences*, 69:3199–3201, 1972 / **44.18** After M.H. Williamson, *Island Populations*, Oxford University Press, 1981 / **44.19** (a) After F.G. Stehli et al., *Geological Society of America Bulletin*, 78:455–466, 1967; (b) After M. Kusenov, *Evolution*, 11:298–299, 1957; (c) After T. Dobzhansky, *American Scientist*, 38:209–221, 1950

Chapter 45

45.1 Roger K. Burnard / **45.2** Photograph Sharon R. Chester / **45.3** Photograph Alan D. Briere/Tom Stack & Associates / **45.5** After H.T. Odum, "Trophic Structure and Productivity of Silver Springs," *Ecological Monographs*, 27:55–112, copyright © 1957 Ecological Society of America / **45.8** USDA, Forest Service / **45.9** Art by Raychel Ciemma / **45.10** G. Likens et al., *Biogeochemistry of a Forest System*, Springer-Verlag, 1977

Chapter 46

46.1 NASA / **46.2** Art by Victor Royer / **46.4** After the U.S. Navy Oceanographic Department / **46.5** Edward S. Ross / **46.7** Art by Raychel Ciemma / **46.11** After G.T. Miller, Jr., *Environmental Science: An Introduction*, Wadsworth, 1986 / **46.12** Harlo H. Hadow / **46.13** Dennis Brokaw / **46.14** Kenneth W. Fink/Ardea, London / **46.15** Ray Wagner/Save the Tallgrass Prairie, Inc. / **46.16** Thase Daniel / **46.17** Thomas E. Hemmerly / **46.18** Ansel Adams / **46.19** Dennis Brokaw / **46.20** Lynn Eckmann, University of Washington/BPS / **46.21** D.W. MacManiman / **46.22** Art by Victor Royer; after Edward S. Deevy, Jr., *Scientific American*, October 1951, copyright © 1951 Scientific American Inc., all rights reserved / **46.23** D.W. Schindler, *Science*, 184:897–899, 1974 / **46.24** Photograph Dennis Brokaw / **46.25** William H. Amos / **46.26** (a) Robert Hessler; (b) Fred Grassle, Woods Hole Institution of Oceanography / **46.27** Photograph Chuck Niklin / **46.28** Photograph E.R. Degginger; art by D. & V. Hennings

Chapter 47

47.1 © 1983 Billy Grimes / **47.2** After G.T. Miller, Jr., *Living in the Environment*, Fourth edition, Wadsworth, 1985 / **47.3** From "Downwind: The Acid Rain Story," Environment Canada, 1981 / **47.4** After "Downwind: The Acid Rain Story," Environment Canada, 1981 / **47.6** After G.T. Miller, Jr., *Environmental Science: An Introduction*, Wadsworth, 1986 / **47.7** National Oceanic and Atmospheric Administration/NESDIS / **47.8** From Water Resources Council, 1979 / **47.9** After G.T. Miller, Jr., *Environmental Science: An Introduction*, Wadsworth, 1986

Chapter 48

48.1 Photograph John S. Dunning/Ardea, London; sonogram J. Bruce Falls and Tom Dickinson, University of Toronto / **48.2** (a) Hans Reinhard/ Bruce Coleman Ltd.; (b),(c) G. Pohl-Apel and R. Sussinka, *Journal für Ornithologie*, 123:211–214 / **48.3** (a) Eric & David Hosking; (b) Stephen Dalton/Photo Researchers / **48.4** (a) Nina Leen in *Animal Behavior*, Life Nature Library (b) F. Schulz / **48.5** After J. Alcock, *Animal Behavior*, Second edition, Sinauer, 1979 / **48.6** Eric Hosking / **48.7** (a) Eugene Kozloff; (b),(c) Stevan Arnold / **48.9** John Alcock / **48.10** Lincoln P. Brower / **48.11** Pat and Tom Leeson/ Photo Researchers / **48.12** John Alcock

Chapter 49

49.1 John Alcock / **49.2** Edward S. Ross / **49.3** (a) E. Mickleburgh/Ardea, London; (b–d) G. Ziesler/ ZEFA / **49.4** Art by D. & V. Hennings / **49.6** Merlin D. Tuttle, Bat Conservation International / **49.7** Scott Preiss / **49.8** Fred Bruemmer / **49.9** Frank Lane Agency/Bruce Coleman Inc. / **49.10** Patricia Caulfield / **49.11** Timothy Ransom / **49.12** Kenneth Lorenzen

GLOSSARY

abortion. Spontaneous or induced expulsion of the embryo or fetus from the uterus. Spontaneous abortions are also called miscarriages.

abscission (ab-SIH-zhun) [L. *abscissus*, to cut off]. Leaf (or fruit or flower) drop after hormonal action causes a corky cell layer to form where a leaf stalk joins a stem; nutrient and water flow is thereby shut off.

acid [L. *acidus*, sour]. A substance that releases a hydrogen ion (H⁺) in solution.

acoelomate (ay-SEE-la-mate). Type of animal that has no fluid-filled cavity between the gut and body wall.

actin (AK-tin). A protein that functions in contraction; together with myosin, a component of the threadlike myofibrils of muscle cells.

action potential. Nerve impulse; a sudden, dramatic reversal of the polarity of charge across the plasma membrane of neurons and some other cells.

activation energy. The minimum amount of collision energy needed to boost reactant molecules to the point at which a reaction will proceed spontaneously.

active site. A crevice in the surface region of an enzyme in which a particular reaction is catalyzed.

active transport. Movement of ions and molecules across a cell membrane, against a concentration gradient, by ATP expenditure. The ion or molecule is moved in a direction other than the one in which simple diffusion would take it.

adaptation [L. *adaptare*, to fit]. An existing structural, physiological, or behavioral trait of an individual that promotes survival and reproduction under prevailing conditions.

adaptive radiation. A burst of evolutionary activity in geologic time, with lineages branching away from one another as they partition the existing environment or invade new ones.

adaptive zone. A way of life, such as "catching insects in the air at night." A lineage must have physical, ecological, and evolutionary access to an adaptive zone to become a successful occupant of it.

adenine (AH-de-neen). A purine; a nitrogen-containing base found in nucleotides.

adenosine diphosphate (ah-DEN-uh-seen die-FOSS-fate). ADP, a molecule involved in cellular energy transfers; typically formed by hydrolysis of ATP.

adenosine triphosphate. ATP, a molecule that is a major carrier of energy (by way of its phosphate groups) from one reaction site to another in all living cells.

aerobic cell (air-OH-bik) [Gk. *aer*, air, + *bios*, life]. A cell that is able to use free oxygen as a final electron acceptor in carbohydrate metabolism.

aerobic respiration. Pathway of carbohydrate metabolism, including glycolysis, the Krebs cycle, and electron transport phosphorylation. The "spent" electrons are transferred finally to oxygen. Much greater net energy yield than from anaerobic pathways.

afferent (AFF-uh-rent) [L. *affere*, to bring to]. Conducting inward to a body part or region, as from sensory receptor endings toward the spinal cord or brain.

allantois (ah-LAN-twahz) [Gk. *allas*, sausage]. Vascularized extraembryonic membrane of reptiles, birds, and mammals that develops as a bladderlike pouch from the primitive gut. Functions in excretion and respiration in reptiles and birds; functions in oxygen transport by way of the umbilical cord in placental mammals.

allele (uh-LEEL). One of two or more alternative forms of a gene at a given gene locus.

allele frequency. The relative abundance of different alleles carried by the individuals of a population. Also called gene frequency, which is something of a non sequitur.

allopatric speciation [Gk. *allos*, different, + *patria*, native land]. Speciation that occurs when geographic separation prevents gene flow and has assured reproductive isolation of different parts of a population or of different populations of the same species.

alternation of generations. In many plant life cycles, the alternation of diploid multicelled bodies with haploid multicelled bodies.

altruistic behavior. Helpful actions that reduce the individual's production of surviving offspring while increasing the reproductive success of the helped individual.

alveolus, plural **alveoli** (ahl-VEE-uh-luss) [L. *alveus*, small cavity]. One of many small, thin-walled pouches in the lungs; sites of gas exchange between air in the lungs and the bloodstream.

amino acid (uh-MEE-no). A molecule having an amino group (NH₂) and an acid group (—COOH); a subunit for protein synthesis.

ammonification (uh-moan-ih-fih-KAY-shun). Decomposition of nitrogenous wastes and remains of organisms by certain bacteria and fungi.

amnion (AM-nee-on). In reptiles, birds, and mammals, an extraembryonic membrane that arises from the inner cell mass of a blastocyst; becomes a fluid-filled sac in which the embryo develops freely.

amyloplast (AM-uh-low-plast) [L. *amylum*, starch, + Gk. *plastos*, formed]. A plastid having no pigments; functions in starch storage.

anaerobic cell (an-uh-ROW-bik) [Gk. *an*, without, + *aer*, air). A cell that either cannot use free oxygen as a final electron acceptor in carbohydrate metabolism or dies upon exposure to it.

anaerobic electron transport. Pathway of carbohydrate metabolism that begins with glycolysis and ends with the transfer of "spent" electrons to inorganic compounds in the environment.

anaphase (AN-uh-faze). In mitosis, the stage when the two sister chromatids of each chromosome are separated and moved to opposite poles of the microtubular spindle.

angiosperm (AN-gee-oh-sperm) [Gk. *angeion*, vessel, + *sperma*, seed]. Flowering plant.

animal. Multicelled heterotroph; except for a few parasites, most ingest food, which is then digested and absorbed.

annual plant. Vascular plant that completes its life cycle in one growing season.

anther [Gk. *anthos*, flower]. In flowering plants, the pollen-bearing part of the male reproductive structure (stamen).

antheridium (an-thuh-RID-ee-um). Protective layer of sterile cells surrounding the haploid sperm in a male gametophyte.

antibody [Gk. *anti*, against]. Any of various Y-shaped proteins, of the immunoglobulin class, produced by B cells. Antibodies bind specific foreign agents invading the body, thus tagging them for destruction by phagocytes or by activating the complement system. Some are bound to the plasma membrane of B cells; others circulate freely in blood and lymph vessels.

antigen [Gk. *anti*, against, + *genos*, race, kind; against "self"]. Any of various specific molecular patterns that are on the surface of foreign agents invading the body and that trigger defense responses.

anus. In some invertebrates and all vertebrates, the terminal opening of the gut through which solid residues of digestion are eliminated.

aorta (ay-OR-tah) [Gk. *airein*, to lift, heave]. Main artery of systemic circulation; carries oxygenated blood away from the heart to all regions except the lungs.

apical meristem (AY-pih-kul MARE-ih-stem) [L. *apex*, top, + Gk. *meristos*, divisible]. In most plants, a mass of self-perpetuating cells at a root or shoot tip that is responsible for primary growth, or elongation, of plant parts.

archegonium (ar-kih-GO-nee-um) [Gk. *archegonos*, first of a kind]. Protective layer of sterile cells surrounding the haploid egg in a female gametophyte.

asexual reproduction. Production of new individuals by any process that does not involve gametes.

atom [Gk. *atomos*, indivisible]. Smallest unit of an element that still retains the properties of that element.

atomic number. A relative number assigned to each kind of element based on the number of protons in one of its atoms.

atomic weight. The weight of an atom of any element relative to the weight of the most abundant isotope of carbon (which is set at 12).

australopithecine (ohss-trah-low-PITH-uh-seen) [L. *australis*, southern, + Gk. *pithekos*, ape]. Any of the earliest known species of hominids; that is, the first species on the human evolutionary branch.

autonomic nervous system (auto-NOM-ik). Those efferent nerves leading from the central nervous system to cardiac cells, muscle cells, smooth muscle cells (such as those of the stomach), and glands—that is, the visceral portion of the body; generally not under conscious control.

autosome. Any of those chromosomes that are of the same number and kind in both males and females of the species.

autotroph (AH-toe-trofe) [Gk. *autos*, self, + *trophos*, feeder]. An organism able to build all the organic molecules it requires using carbon dioxide (present in air and in water) and energy from the physical environment. Photosynthetic autotrophs use sunlight energy; chemosynthetic autotrophs extract energy from chemical reactions involving inorganic substances. Compare *heterotroph*.

axon. Nerve cell process serving as a through-conducting pathway for action potentials, which are messages that travel rapidly, without alteration, from one region to another.

bacillus, plural **bacilli** (bah-SILL-us, bah-SILL-eye) [L. *baculus*, small staff, rod]. Rodlike form of bacterium.

bacteriophage (bak-TEER-ee-oh-fahj) [Gk. *baktērion*, small staff, rod, + *phagein*, to eat]. Category of viruses that infect and destroy certain bacterial cells.

basal body. A centriole that has given rise to the microtubular system of a cilium or flagellum and that remains attached at the base of the motile structure, just beneath the plasma membrane.

base. Any substance that combines with a hydrogen ion (H$^+$) in solution.

behavior. Any coordinated neuromotor response that an animal makes to external and internal stimuli. The responses are outcomes of the integration of sensory, neural, endocrine, and effector components, all of which have a genetic basis (hence are subject to natural selection); and they may be modified by learning processes.

biennial (by-EN-ee-ull). Flowering plant that lives two growing seasons.

bilateral symmetry. Body plan by which an animal has an anterior and a posterior end, a left and right side, and a dorsal and ventral surface. The only plane in which the body can be divided into two equivalent halves extends from dorsal to ventral along the midline.

binary fission. Asexual reproduction by division of a body (or cell) into two equivalent parts.

biogeographic realm [Gk. *bios*, life, + *geographein*, to describe the surface of the earth]. In one scheme, one of six major regions having a characteristic array of species that are generally isolated from the other realms by physical barriers that restrict dispersal.

biological clock. Internal timing mechanism that allows organisms to anticipate and adjust to environmental change. In plants, phytochromes may figure in the timing mechanism; in some vertebrates, the pineal gland seems to.

biological magnification. Increasing concentration of a relatively nondegradable substance in body tissues, beginning at low trophic levels and moving up through those organisms that are diners, then are dined upon in food webs.

biomass. The total dry weight of all organisms at a given trophic level of an ecosystem.

biome. A broad, vegetational subdivision of some biogeographic realm, shaped by climate, topography, and the composition of regional soils.

biosphere [Gk. *bios*, life, + *sphaira*, globe]. Narrow zone that harbors life, limited to the waters of the earth, a fraction of its crust, and the lower region of the surrounding air.

biosynthesis [Gk. *bios*, life, + *synthesis*, a putting together]. Assembly of the lipids, carbohydrates, proteins, and nucleic acids that make up a cell.

blastocyst (BLASS-tuh-sist) [Gk. *blastos*, sprout, + *kystis*, pouch]. In mammalian development, a modified blastula stage consisting of a hollow ball of surface cells (trophoblast) having inner cells massed at one end.

blastula (BLASS-chew-lah). In many animal species, an embryonic stage consisting of a hollow, fluid-filled ball of cells one layer thick.

blood pressure. Fluid pressure, generated by heart contractions, that keeps blood circulating. Generally measured at large arteries of systemic circulation.

bronchus, plural **bronchi** (BRONG-cuss, BRONG-kee) [Gk. *bronchos*, windpipe]. Tubelike branchings of the trachea (windpipe) that lead to the lungs.

budding. Asexual reproduction in which some cells differentiate and grow outward from the parent body, then the bud breaks away to form a new individual.

buffer. In living cells, a substance that combines with and/or releases hydrogen ions as a function of pH.

bulk flow. In response to a pressure gradient, a movement of more than one kind of molecule in the same direction in the same medium (gas or liquid).

calorie (KAL-uh-ree) [L. *calor*, heat]. The amount of heat needed to raise the temperature of one gram of water by 1°C. Nutritionists sometimes use "calorie" to mean kilocalorie (1,000 calories), which is a source of much confusion.

Calvin-Benson cycle. Stage of light-independent reactions of photosynthesis in which carbon-containing compounds are used to form carbohydrates (such as glucose) and to regenerate a sugar phosphate (RuBP) required in carbon dioxide fixation. The first product is a three-carbon compound (PGA).

cambium, plural **cambia** (KAM-bee-um). In vascular plants, one of two types of embryonic tissue masses that are responsible for secondary growth (increase in stem or root diameter). Vascular cambium gives rise to secondary xylem and phloem; cork cambium gives rise to periderm.

camouflage. Adaptation in form, coloration, and/or behavior that enables an organism to blend with its background, the advantage being to escape detection. Also known as crypsis.

cancer. Malignancy arising from cells that are characterized by profound abnormalities in the plasma membrane and in the cytoplasm, abnormal growth and division, and diminished capacity for adhesion to substrates.

capillary [L. *capillus*, hair]. Small blood vessel whose thin walls are permeable to many materials; exchange point between blood and interstitial fluid.

carbohydrate [L. *carbo*, charcoal, + *hydro*, water]. Monomer or polymer of a sugar, which is a compound of carbon, hydrogen, and oxygen present in about a 1:2:1 ratio.

carbohydrate metabolism [Gk. *metabolē*, change]. The release of chemical bond energy from carbohydrates by means of phosphorylation and oxidation-reduction reactions.

carbon dioxide fixation. First stage of light-independent reactions of photosynthesis; carbon dioxide from the air is combined with a sugar phosphate to form an intermediate necessary in the synthesis of glucose and other carbon compounds.

carcinogen (CAR-sin-oh-jen). Any agent capable of promoting cancer.

cardiac cycle [Gk. *kardia*, heart, + *kyklos*, circle]. The sequence of muscle contractions and relaxation constituting one heartbeat.

cardiovascular system. Of animals, an organ system composed of blood, one or more hearts, and blood vessels.

carnivore [L. *caro*, *carnis*, flesh, + *vovare*, to devour]. An animal that eats other animals; a type of heterotroph.

carpel. The central whorl of modified leaves that represents the female reproductive structure of a flower. Typically consists of a stigma, style, and ovary. There may be more than one carpel per flower.

carrying capacity. The equilibrium size at which a particular population in a particular environment will stabilize when its supply of resources (including nutrients, energy, and living space) remains constant.

cell [L. *cella*, small room]. The basic *living* unit. There are large, complex organic molecules below this level of organization, but such molecules by themselves are nonliving.

cell plate. In plant cell division, a partition that forms from vesicles at the equator of the mitotic spindle, between the two newly forming cells.

cell wall. One or more layers of surface deposits outside the plasma membrane that generally provide support and resist mechanical pressure. Cell walls occur among bacteria, protistans, fungi, and plants.

central nervous system. Brain and spinal cord of vertebrates.

centriole. One of a pair of short, barrel-shaped structures that gives rise to the microtubule system of a cilium or a flagellum.

centromere (SEN-troh-meer) [Gk. *kentron*, center, + *meros*, a part]. A special region of the chromosome serving as the attachment site for spindle microtubules during nuclear division.

cephalization (sef-ah-lah-ZAY-shun) [Gk. *kephalikos*, head]. Differentiation of one end of the animal body into a head in which nervous tissue and sensory organs are especially concentrated.

cerebellum (ser-ah-BELL-um) [L. diminutive of *cerebrum*, brain]. Hindbrain region that coordinates motor activity for refined limb movements, maintaining posture, and spatial orientation.

cerebrum (suh-REE-bruhm). In vertebrate forebrain, paired masses of gray matter; cerebral hemispheres overlying thalamus, hypothalamus, and pituitary. Includes primary receiving centers for receptors at body periphery, association centers for coordinating and processing sensory input, and motor centers for coordinating motor responses.

chemical bond. A union between the electron structures of two or more atoms or ions.

chemiosmotic theory (kem-ee-oz-MOT-ik). Concept that an electrochemical gradient across a cell membrane drives ATP synthesis. Operation of electron transport systems builds up the hydrogen ion concentration on one side of the membrane. Then the electrical and chemical force of the H^+ flow down the gradient is linked to enzyme machinery that combines ADP with inorganic phosphate to form ATP.

chemoreceptor (KEE-moe-ree-SEP-tur). Sensory cell or cell part that directly or indirectly transforms chemical stimuli into action potentials, the means of communication in nervous systems.

chemosynthetic autotroph (KEE-moe-sin-THET-ik). One of a few kinds of bacteria able to build all the organic molecules it requires using carbon dioxide as the carbon source and certain inorganic substances (such as sulfur) as the energy source.

chiasma, plural **chiasmata** (kai-AZ-mah, kai-az-MAH-tah) [Gk. *chiasma*, cross]. A crossing between two nonsister chromatids during prophase I of meiosis; evidence that breakage and exchange (a crossover) occurred earlier between them.

chlorophyll (KLOR-uh-fill) [Gk. *chloros*, green, + *phyllon*, leaf]. Light-trapping pigment molecule that acts as an electron donor in photosynthesis.

chloroplast. Eukaryotic organelle that houses membranes, pigments, and enzymes of photosynthesis.

chorion (CORE-ee-on). In reptiles, birds, and mammals, outermost membrane around embryo. Its vascularized villi form a nutritional link with the mother; its hormonal secretion (chorionic gonadotropin) helps maintain the corpus luteum (hence the uterine lining) following implantation.

chromatid (CROW-mah-tid) [Gk. *chroma*, color]. One of the two threadlike forms of a replicated chromosome, for as long as they remain attached at the centromere prior to and during nuclear division. Two sister chromatids joined at the centromere represent duplicate sets of hereditary information.

chromosome (CROW-muh-sohme) [Gk. *chrōma*, color, + *soma*, body]. A DNA molecule and the proteins intimately associated with it; the vehicle by which hereditary information is transmitted from one generation to the next. For eukaryotes, the word can refer to a single or a duplicated chromosome (with its two sister chromatids).

cilium, plural **cilia** (SILL-ee-um) [L. *cilium*, eyelid]. Short, hairlike process extending from the plasma membrane and containing a regular array of microtubules. Some function as motile structures, others in creating currents of fluids; modified cilia are components of diverse sensory structures.

circulatory system. An organ system consisting of a muscular pump, blood vessels, and blood itself; the means by which materials are transported to and from cells; in many animals, also helps stabilize body temperature and pH.

cleavage (KLEE-vidj). Rapid, successive divisions in an animal zygote that lead to increase in cell number but not in cell size to produce the early embryo.

coccus, plural **cocci** (COCK-us, COCK-eye). Spherical form of bacterium.

codominance. The discernible expression of both alleles of a pair in heterozygotes.

coelum (SEE-lum) [Gk. *koilos*, hollow]. A type of body cavity defined by the peritoneum (a continuous lining of epithelial cells).

coenzyme. A large, nonprotein organic molecule that serves as a carrier of electrons or atoms in metabolic reactions and that is necessary for proper functioning of many enzymes. NAD^+ and coenzyme A are examples.

cofactor. A metal ion or coenzyme that helps make substrates bind to an active site of an enzyme or that makes them more reactive. Some are bound tightly to the enzyme, others diffuse freely to and away from it.

cohesion. Condition in which molecular bonds resist rupturing when under tension.

commensalism [L. *com*, together, + *mensa*, table]. Two-species interaction in which one species benefits significantly while the other is neither helped nor harmed to any great degree.

communication signal. A stimulus produced by one animal that changes the behavior of another animal of the same species, in ways that benefit both the signaler and the receiver.

community. The populations of all species that occupy a habitat. Also used to refer to certain groups of organisms (such as the bird community) in a habitat.

competition, interspecific. Two-species interaction in which both species can be harmed as a result of their overlapping niches (that is, they have some requirements or activity in common).

competition, intraspecific. Interaction among individuals of the same species that are competing for the same resources.

competitive exclusion. The concept that if the resources required by two competing species are similar enough, the population growth rate of the one better able to exploit those resources (because of structural, physiological, or behavioral differences) will depress the growth rate of the other and lead to its exclusion from the habitat.

complement system. About twenty proteins, circulating in inactive form in the bloodstream until contact is made with certain bacterial or fungal invaders, whereupon they become activated and contribute to the inflammatory response. They are activated during general defense responses *and* specific immune responses.

concentration gradient. For a given substance, a greater concentration of its molecules in one region of a system than in another. A concentration difference between extracellular fluid and cytoplasm is an example.

condensation. Covalent linkage of small molecules in an enzyme-mediated reaction that can also involve formation of water.

conditioning. Form of learning in which a behavioral response becomes associated (by means of a reinforcing stimulus) with a new stimulus that was not previously associated with the response.

conjugation [L. *conjugatio*, a joining]. In some bacteria, transfer of DNA between two different mating strains that have made cell-to-cell contact.

consumers [L. *consumere*, to take completely]. Any organism that is not self-feeding (it is heterotrophic) and that ingests other (usually) living organisms in whole or in part to obtain organic nutrients. Herbivores, carnivores, omnivores, and parasites are consumers.

continuous variation. Small degrees of phenotypic variation that occur over a more or less continuous range in a population.

contractile vacuole (kun-TRAK-till VAK-you-ohl) [L. *contractus*, to draw together]. In some single-celled organisms, a membranous chamber that takes up excess water in the cell body, then contracts, expelling the water through a pore to the outside.

convergence. An outcome of natural selection whereby morphologically dissimilar and only distantly related lineages adopt a similar mode of life and come to resemble one another rather closely.

corpus luteum (CORE-pus LOO-tee-um). A glandular structure that develops from cells of a ruptured ovarian follicle and that secretes progesterone and some estrogen, both of which maintain the endometrium (the lining of the uterus).

cortex [L. *cortex*, bark]. In general, a rindlike layer; the kidney cortex is an example. In vascular plants, ground tissue that makes up most of the primary plant body, supports plant parts, and stores food.

cotyledon (cot-ill-EE-don). "Seed leaf"; often contains stored nutrients that are used in early growth when a seed germinates.

covalent bond (koe-VAY-lunt) [L. *con*, together, + *valere*, to be strong]. A sharing of one or more electrons between atoms or groups of atoms. When electrons are shared equally, it is a nonpolar covalent bond. When they are shared unequally, it is a polar covalent bond.

creatine phosphate (KREE-uh-teen FOSS-fate). Compound that readily gives up phosphate to ADP; important storage form of phosphate that is used in regenerating ATP for muscle contraction.

crossing over. During prophase I of meiosis, the breakage and exchange of corresponding segments of nonsister chromatids (of homologous chromosomes) at one or more sites along their length, resulting in genetic recombination.

cyclic adenosine monophosphate; cyclic AMP (SIK-lik ah-DEN-uh-seen mon-oh-FOSS-fate). A nucleotide that serves as an intracellular mediator of the cellular response to hormonal signals; a type of second messenger.

cyclic photophosphorylation (SIK-lik foe-toe-FOSS-for-ih-LAY-shun). Photosynthetic pathway in which electrons excited by sunlight energy move from a photosystem to a transport chain, then back to the photosystem. Energy released in the transport chain is coupled to ATP formation.

cytochrome (SIGH-toe-krome) [Gk. *kytos*, hollow vessel, + *chrōma*, color]. Iron-containing protein molecule that occurs in electron transport systems used in photosynthesis and aerobic respiration.

cytokinesis (SIGH-toe-kih-NEE-sis) [Gk. *kinesis*, motion]. Cytoplasmic division accompanying or following nuclear division.

cytomembrane system. Those organelles concerned with modifying newly formed proteins for use in the cell or for secretion, protein transport, lipid synthesis, and adding or replacing the protein and lipid components of cell membranes. Includes endoplasmic reticulum, Golgi bodies, lysosomes, microbodies, and transport vesicles.

cytoplasm (SIGH-toe-plaz-um) [Gk. *plassein*, to mold]. In a cell, everything but the plasma membrane and the nucleus (or, in bacteria, the nucleoid). Includes internal membranes and other structures that function in metabolism, biosynthesis, and cell movements. The structures are bathed in cytosol, a semifluid substance.

cytosine (SIGH-toe-seen). A pyrimidine; one of the nitrogen-containing bases in nucleotides.

cytoskeleton. In the cytoplasm of eukaryotic cells, an internal framework of microtubules, microfilaments, and other fine strands by which organelles and other structures are anchored, organized, and moved about.

cytosol. The continuous aqueous portion of cytoplasm, with its dissolved solutes.

decomposers [L. *de*, down, away, + *companere*, to put together]. Mostly heterotrophic bacteria and fungi that obtain organic nutrients by breaking down the remains or products of other organisms; their activities help cycle simple compounds back to autotrophs.

denaturation (deh-NAY-chur-AY-shun). Disruption of bonds holding a protein in its three-dimensional form, such that its polypeptide chain(s) unfolds partially or completely.

dendrite (DEN-drite) [Gk. *dendron*, tree]. Nerve cell process, typically short and slender, which together with the cell body receives and integrates most incoming signals.

denitrification. Reduction of nitrate or nitrite to gaseous nitrogen (N_2) and a small amount of nitrous oxide (N_2O) by soil bacteria.

deoxyribonucleic acid (dee-ox-ee-rye-bow-new-CLAY-ik). DNA; double-stranded helically coiled nucleic acid, in which the hydrogen bonds between strands can be unzipped and DNA's chemical messages exposed to agents of protein synthesis or DNA replication. Overall, a stable molecule in which genetic information is stored, yet which is subject to change in some of its structural details.

detritivore (dih-TRY-tih-vore) [L. *detritus*; after *deterere*, to wear down]. An earthworm, crab, nematode, or other heterotroph that feeds on particles of organic matter, such as would be produced by the partial decomposition of plant and animal tissues.

deuterostome (DYEW-ter-oh-stome) [Gk. *deuteros*, second, + *stoma*, mouth]. Any of those bilateral animals, including echinoderms and chordates, in which the first opening that appears in the gastrula stage of embryonic development becomes the anus and the second opening, the mouth. Also characterized by radial cleavage and formation of the coelom from pouches in the embryonic gut wall.

diaphragm (DIE-uh-fram) [Gk. *diaphragma*, to partition]. Muscular partition between the thoracic and abdominal cavities, the contraction and relaxation of which contribute to breathing. Also, a contraceptive device used temporarily to close off and thus prevent sperm from entering the uterus during sexual intercourse.

dicot (DIE-kot) [Gk. *di*, two, + *kotylēdōn*, cup-shaped vessel]. Short for dicotyledon; class of flowering plants characterized primarily by seeds having embryos with two cotyledons (seed leaves), generally net-veined leaves, and floral parts generally arranged in fours, fives, or multiples of these.

differential reproduction. The tendency of bearers of adaptive traits to reproduce successfully more than bearers of less adaptive traits. Because their offspring tend to make up an increasingly greater proportion of the reproductive base for each new generation, the adaptive traits increase in frequency also.

differentiation. Processes by which eukaryotic cells of identical genetic makeup become structurally and functionally different from one another, according to the genetically controlled developmental program of the species.

diffusion. Tendency of like molecules to move from their region of greater concentration to a region where they are less concentrated; occurs through random energetic movements of individual molecules, which tend to become dispersed uniformly in a given system.

digestive system. Some form of body cavity or tube by which food is ingested and prepared for absorption into the internal environment and from which the residues are eliminated.

diploid (DIP-loyd). State in which a somatic cell contains two sets of chromosomes. Each chromosome of one set has a partner (homologue) from the other set with which it pairs at meiosis. Except for sex chromosomes, the homologues resemble their partner in length, shape, and which genes they carry. Compare *haploid*.

directional selection. Mode of natural selection that moves the frequency distribution of alleles in a steady, consistent direction, such that the phenotypic character of a population shifts as a whole.

disaccharide (die-SAK-uh-ride) [Gk. *di*, two, + *sakcharon*, sugar]. A carbohydrate; two monosaccharides covalently bonded.

disruptive selection. Mode of natural selection that increases the frequency of two or more alleles that give rise to extreme forms of a trait, such that intermediate forms are selected against.

divergence [L. *dis*, apart, + *vergere*, to incline]. A buildup of differences in allele frequencies between reproductively isolated populations of a species or local breeding units of the same population. In *morphological* divergence, selection leads to departures from the ancestral species form.

diversity, organismic. Sum total of variations in form, functioning, and behavior that have accumulated in different lineages. Those variations generally are adaptive to prevailing conditions or were once adaptive to conditions that existed in the past.

dominance hierarchy. Social ranking of members of a group.

dominant allele. In a diploid cell, an allele whose expression masks the expression of its partner at the same gene locus on the homologous chromosome.

dormancy [L. *dormire*, to sleep]. Cessation of growth under physical conditions that could be quite suitable for growth.

ecology [Gk. *oikos*, home, + *logos*, reason]. Study of the interactions of organisms with one another and with the physical and chemical environment.

ecosystem. A community and its physical and chemical environment. Its biotic (living) component includes producers, consumers, decomposers, and detritivores. Its abiotic (nonliving) component includes soils, temperature, and rainfall.

ectoderm [Gk. *ecto*, outside, + *derma*, skin]. In an animal embryo, an outermost cell layer that gives rise to the outer layer of skin and to tissues of the nervous system.

effector. A muscle (or gland) that responds to nerve signals by producing movement (or chemical change) that helps adjust the body to changes in internal and/or external conditions.

efferent (EFF-uh-rent) [L. *effere*, to carry outward]. Conducting away from a body part or region, as from the brain or spinal cord to muscles or glands by way of motor neurons.

electric charge. A property of matter that enables ions, atoms, and molecules to attract or repel one another.

electron. Negatively charged particle that orbits the nucleus of an atom.

electron transport system. In a cell membrane, electron carriers and enzymes positioned in an organized array that enhances oxidation-reduction reactions. Such systems function in the release of energy that is used in ATP formation and other reactions.

element. Any substance that cannot be decomposed into substances with different properties.

embryo (EM-bree-oh) [Gk. *en*, in, + probably *bryein*, to swell]. In animals generally, the early stages of development (including cleavage, gastrulation, organogenesis, and morphogenesis) after fertilization. In most plants, the young sporophyte, from the first cell divisions after fertilization until germination.

endergonic reaction (en-dur-GONE-ik). Chemical reaction showing a net gain in energy.

endocrine element (EN-doe-krin) [Gk. *endon*, within, + *krinein*, to separate]. Cell or gland that produces and/or secretes hormones.

endocrine system. System of cells, tissues, and organs functionally linked to the nervous system and whose chemical secretions (hormones) help control body functioning.

endocytosis (EN-doe-sigh-TOE-sis). The process by which a region of the plasma membrane encloses substances (or cells, in the case of phagocytes) at or near the cell surface, then pinches off to form a vesicle that transports the substances into the cytoplasm.

endoderm [Gk. *endon*, within, + *derma*, skin]. In an animal embryo, the innermost cell layer, which differentiates into the inner lining of the gut and organs derived from it.

endometrium (EN-doh-MEET-ree-um) [Gk. *metrios*, of the womb]. Inner lining of the uterus, consisting of connective tissues, glands, and blood vessels.

endoplasmic reticulum, or **ER** (EN-doe-PLAZ-mik reh-TIK-you-lum). A system of membranous tubes, channels, and flattened sacs that form compartments within the cytoplasm of eukaryotic cells and that function in the processing of proteins destined for secretion from the cell, and in the manufacture of the protein and lipid components of most organelles. *Rough* ER has ribosomes attached to the side of the membrane facing the cytoplasm; *smooth* ER does not.

endoskeleton [Gk. *endon*, within, + *skleros*, hard, stiff]. In chordates, the internal framework of bone, cartilage, or both. Together with skeletal muscle, supports and protects other body parts, helps maintain posture, and moves the body.

endosperm [Gk. *endon*, within, + *sperma*, seed]. Mass of tissue that surrounds embryo in a seed; in monocots, storage site for nutrients needed after seed germination.

energy. Capacity to make things happen, to cause change, to do work.

entropy (EN-trohp-ee). A measure of the degree of disorganization of a system—that is, how much energy in a system has become so dispersed (usually as evenly distributed, low-quality heat) that it is no longer available to do work.

enzyme. Any of a class of proteins that catalyze reactions (they greatly enhance the rate at which a reaction approaches equilibrium) by lowering the required activation energy.

epidermis. Outermost tissue layer of a multicelled animal or plant.

epithelium (EP-ih-THEE-lee-um). Sheet of cells, one or more layers thick, lining internal or external surfaces of the multicelled animal body.

equilibrium, dynamic [Gk. *aequus*, equal, + *libra*, balance]. The point at which a chemical reaction runs forward as fast as it runs in reverse, so that there is no net change in the concentrations of products or reactants.

erythrocyte (eh-RITH-row-site) [Gk. *erythros*, red, + *kytos*, vessel]. Red blood cell.

estrus (ESS-truss) [Gk. *oistrus*, frenzy]. For mammals generally, the cyclic period of a female's sexual receptivity to the male.

estuary (ESS-chew-airy). A region where fresh water from a river or stream mixes with salt water from the sea.

eukaryote (yoo-CARRY-oht) [Gk. *eu*, good, + *karyon*, nut]. Having a true nucleus; a cell that has membranous organelles, most notably the nucleus.

evaporation [L. *e-*, out, + *vapor*, steam]. The changes by which a substance is converted from a liquid state into (and carried off in) vapor.

evolution [L. *evolutio*, act of unrolling]. In biology, successive changes in allele frequencies in a population, as brought about by such occurrences as mutation, genetic drift, gene flow, and selection pressure.

excretion. Elimination of excess water and excess or harmful solutes from the internal environment, as by kidneys.

exergonic reaction (EX-ur-GONE-ik). A chemical reaction that shows a net loss in energy.

exocrine gland (EK-suh-krin) [Gk. *ex*, out of, + *krinein*, to separate]. Secretory structure whose products travel through ducts that empty at a free epithelial surface.

exocytosis (EK-so-sigh-TOE-sis). The process by which substances are moved out of a cell. The substances are transported in cytoplasmic vesicles, the surrounding membrane of which merges with the plasma membrane in such a way that the substances are dumped outside.

exoskeleton [Gk. *exō*, out, + *skleros*, hard, stiff]. An external skeleton, as in arthropods.

exponential growth (EX-poe-NEN-shul). Pattern of population growth in which the number of individuals increases in doubling increments (2, 4, 8, 16, 32 . . .). Occurs when the birth rate remains even slightly above the death rate.

extinction, background. The steady rate of species turnover that characterizes lineages through most of their histories.

extinction, mass. An abrupt increase in the rate at which higher taxa disappear, with several higher taxa being affected simultaneously.

extracellular fluid. In animals generally, the medium through which substances are continuously exchanged between cells and between body regions. In vertebrates most is interstitial fluid; the rest is blood plasma.

facilitated diffusion. The movement of specific solutes across a plasma membrane in the direction that diffusion would take them but with the passive assistance of carrier proteins that span the lipid bilayer of the membrane.

fat. A lipid with one, two, or three fatty acid tails attached to a glycerol backbone.

fatty acid. A long, unbranched hydrocarbon with a —COOH group at the end.

feedback inhibition. Control mechanism whereby an increase in some substance or activity inhibits the very process leading to (or allowing) the increase.

fermentation [L. *fermentum*, yeast]. Pathway of carbohydrate metabolism that begins with glycolysis and ends with the "spent" electrons being transferred back to one of the carbohydrate breakdown products or intermediates.

fertilization [L. *fertilis*, to carry, to bear]. Fusion of sperm nucleus with egg nucleus. In flowering plants, an additional sperm nucleus fuses with two other nuclei present in the ovule, forming a single triploid nucleus that will divide and give rise to endosperm; this is *double fertilization*.

first law of thermodynamics [Gk. *thermē*, heat, + *dynamikos*, powerful]. The total amount of energy in the universe remains constant; more energy cannot be created and existing energy cannot be destroyed. What already exists can only undergo conversion from one form to another.

flagellum, plural **flagella** (fluh-JELL-um) [L. *flagellum*, whip]. A motile structure that some cells use to move rapidly through the environment. Longer and less numerous than cilia; contains system of microtubules.

fluid mosaic membrane structure. Current model of membrane structure, in which diverse proteins are embedded in a lipid bilayer or attached to one of its two surfaces. The lipids give the membrane its basic structure and its relative impermeability to water-soluble molecules; packing variations and movements of lipids impart fluidity to the membrane. The proteins carry out most membrane functions, such as transport, enzyme action, and reception of chemical signals or substances.

follicle (FOLL-ih-kul). In a mammalian ovary, one of the spherical chambers containing an oocyte on the way to becoming a mature ovum (egg).

food chain. Linear sequence of who eats whom in an ecosystem.

food web. Network of many interlinked food chains, encompassing primary producers, consumers, decomposers, and detritivores.

fruit [L. after *frui*, to enjoy]. In flowering plants, the ripened ovary of one or more carpels, sometimes with accessory structures incorporated.

functional group. Atom or groups of atoms bonded covalently to the carbon backbone of an organic molecule and contributing to its characteristic structure and properties.

fungus [probably modification of Gk. *spongos*, sponge]. A heterotrophic organism the cells of which secrete enzymes that promote digestion of large organic molecules *outside* the cell, which then absorbs the breakdown products.

gamete (GAM-eet) [Gk. *gametēs*, husband, and *gametē*, wife]. Mature haploid cell (sperm or egg) that functions in sexual reproduction.

gametogenesis (gam-EET-oh-JEN-ih-sis). Formation of gametes by way of meiosis.

gametophyte (gam-EET-oh-fight) [Gk. *phyton*, plant]. Haploid, multicelled, gamete-producing phase in the life cycle of most plants.

ganglion (GANG-lee-un) [Gk. *ganglion*, a swelling]. A clustering of cell bodies of neurons into a distinct structure in body regions other than the brain or spinal cord. (Such clusterings in the brain or spinal cord are called nuclei.)

gastrula (GAS-truh-lah). Stage of animal development in which elaborate patterns of cell migrations in the embryo bring about the formation of two or three embryonic tissue layers (which, in most animals, are the endoderm, mesoderm, and ectoderm).

gene (jeen) [short for German *pangen*, after Gk. *pan*, all, + *-genēs*, to be born]. The basic unit of inheritance; a specific region of DNA coding for a tRNA, rRNA, or mRNA molecule, with the translation product of the mRNA being a polypeptide chain (the basic structural unit of proteins).

gene flow. Change in allele frequencies due to immigration (new individuals join the population), emigration (some individuals leave), or both.

gene frequency. More precisely, allele frequency: the relative abundance of different alleles carried by the individuals of a population.

gene locus. Particular location on a chromosome for a given gene.

gene pair. In diploid cells, the two alleles at a given gene locus on homologous chromosomes.

gene pool. Sum total of all genotypes in a given population. More accurately, allele pool.

genetic code [after L. *genesis*, to be born]. Basic language of protein synthesis, by which nucleotide triplets in DNA (and then mRNA) call for specific amino acids used in protein synthesis.

genetic drift. Random fluctuation in allele frequencies over time, due to chance occurrence alone.

genetic engineering. Altering the information content of DNA through use of recombinant DNA technology.

genetic equilibrium. Stability of allelic and genotypic ratios in a population over succeeding generations. Reference point, signifying zero evolution, for measuring rates of evolutionary change.

genetic recombination. Presence of new combinations of alleles in a DNA molecule as a result of crossing over at meiosis, chromosomal aberration, or gene mutation.

genotype (JEEN-oh-type). Genetic constitution of an individual. Can mean a single gene pair or the sum total of the individual's genes. Compare *phenotype*.

genus, plural **genera** (JEEN-us, JEN-er-ah) [L. *genus*, race, origin]. A taxon (that is, a category of relationship based on phenotypic similarities, descent, or both) in which all species sharing (or exhibiting) certain characteristics are grouped.

germ cell. Animal cell that may develop into gametes. Compare *somatic cell*.

gill. A respiratory organ, typically with a moist, thin, vascularized layer of epidermis that functions in gas exchange.

glomerulus (glow-MARE-yoo-luss) [L. *glomus*, ball]. Cluster of capillaries in Bowman's cap sule of the nephron, the functional unit of the kidney.

glucagon (GLUE-kuh-gone). Animal hormone secreted by pancreatic cells and essential in breakdown of glycogen (a polysaccharide) to glucose subunits during the post-absorptive state of organic metabolism.

glycerol (GLISS-er-ohl) [Gk. *glykys*, sweet, + L. *oleum*, oil]. Three-carbon molecule with three hydroxyl groups attached; combines with fatty acids to form fat or oil.

glycogen (GLY-kuh-jen). In animals, a starch that is a main food reserve; can be readily broken down into glucose subunits during the post-absorptive state of organic metabolism.

glycolysis (gly-CALL-ih-sis) [Gk. *glykys*, sweet, + *lysis*, loosening or breaking apart]. In carbohydrate metabolism, the initial breaking apart of sugar molecules such as glucose, with the release of energy. Glycolysis may proceed under aerobic as well as anaerobic conditions; but it does not require oxygen to do so.

gonad (GO-nad). Primary reproductive organ in which gametes are produced.

graded potential. At chemical synapses and receptors, a change in membrane permeability characteristics that can vary in magnitude, depending on the stimulus. Many graded potentials acting at the same time can so change the voltage difference across the membrane that they initiate an action potential. Compare *action potential*.

granum, plural **grana**. In chloroplasts, stacked membrane system where sunlight energy is actually trapped and where ATP is formed.

gravitropism (GRAV-ih-TROPE-izm) [L. *gravis*, heavy, + Gk. *trepein*, to turn]. Directional growth of a coleoptile, root, or shoot in response to the earth's gravitational force. Also called *geotropism*.

ground meristem (MARE-ih-stem) [Gk. *meristos*, divisible]. A primary meristem that produces ground tissue (hence the bulk of the plant body).

gymnosperm [Gk. *gymnos*, naked, + *sperma*, seed]. One of two divisions of seed plants, characterized by having their seeds borne on surfaces of reproductive structures, without protective tissue layers. Conifers such as pines are examples.

habitat [L. *habitare*, to live in]. Place where an individual or population of a given species lives; its "mailing address."

haploid (HAP-loyd). State in which a nucleus contains half the number of chromosome sets characteristic of the somatic cells of a species; brought about by meiosis, which is necessary for gamete formation. For example, meiosis reduces a diploid nucleus (with two chromosome sets) to the haploid state (one chromosome set).

heart. Muscular pump that keeps blood circulating through the animal body.

hemoglobin (HEEM-oh-glow-bin) [Gk. *haima*, blood, + L. *globus*, ball]. Iron-containing protein that gives red blood cells their color; functions in oxygen transport.

hemorrhage. Bulk flow of blood from damaged vessels.

herbivore [L. *herba*, grass, + *vovare*, to devour]. Plant-eating animal.

heterotroph (HET-er-oh-trofe) [Gk. *heteros*, other, + *trophos*, feeder]. Organisms that obtain carbon and all metabolic energy from organic molecules that have already been assembled by autotrophs. Animals, fungi, many protistans, and most bacteria are heterotrophs.

heterozygote (HET-er-oh-ZYE-gote) [Gk. *zygoun*, join together]. Individual having nonidentical alleles at a given gene locus.

histone. Any of a class of structural proteins complexed with DNA in the eukaryotic chromosome.

homeostasis (HOE-me-oh-STAY-sis) [Gk. *homo*, same, + *stasis*, standing]. For multicelled organisms, maintaining the internal environment (that is, the extracellular fluid environment) within some tolerable range throughout the life cycle even when conditions change.

homeostasis, dynamic. Maintaining the living state by adjusting the organism's form and behavior over time, as a function of the genetically prescribed developmental program of the species.

hominid [L. *homo*, man]. All species on the human evolutionary branch. *Homo sapiens* is the only living representative.

hominoid [after Gk. *eidos*; resembling the form of]. Ape or human species.

homologous chromosome (huh-MOLL-uh-gus) [Gk. *homologia*, agreement; correspondence]. In the nucleus of a somatic cell, one of a pair of chromosomes that resemble each other in length, shape, and which genes they carry and that synapse at meiosis. Sex chromosomes, which differ morphologically in males and females, also function as homologues. (Typically the two chromosomes of a homologous pair are derived from two different parents, but exceptions do occur, as in the case of self-fertilizing plants.)

homologous recombination. Crossing over and the exchange of segments between homologous chromosomes during meiosis I. An exchange can occur almost anywhere along the chromosomes, the exchange is reciprocal, and a fairly long, staggered joint forms between the interacting DNA strands. Compare *site-specific recombination*.

homozygote [Gk. *homos*, same, + *zygoun*, join together]. Individual having two identical alleles at a given gene locus.

hormone [Gk. *hormōn*, to stir up, set in motion]. A secretion from an endocrine cell or gland, transported by the bloodstream to nonadjacent target cells, these being any cells having receptors to which the hormone can bind.

hydrogen bond. Type of chemical bond in which an electronegative atom interacts weakly with a hydrogen atom that is already participating in a polar covalent bond.

hydrolysis (high-DRAWL-ih-sis) [L. *hydro*, water, + Gk. *lysis*, loosening or breaking apart]. Reaction in which covalent bonds between parts of molecules are broken and an H^+ ion and an OH group derived from water become attached to the fragments.

hydrophilic [Gk. *philos*, loving]. Having an attraction for water molecules; refers to a polar substance that readily dissolves in water.

hydrophobic [Gk. *phobos*, dreading]. Repelled by water molecules; refers to a nonpolar substance that does not readily dissolve in water.

hypha, plural **hyphae** [Gk. *hyphē*, web]. A filament with a thin, transparent wall, usually reinforced with chitin, that is the structural component of the meshlike mycelium of fungal life cycles.

hypothalamus [Gk. *hypo*, under, + *thalamos*, inner chamber or possibly *tholos*, rotunda]. Region of vertebrate forebrain concerned with neural-endocrine control of visceral activities (e.g., salt-water balance, temperature control, reproduction).

immune system, vertebrate. Three types of white blood cells (phagocytes, T lympho-cytes, and B lymphocytes) and their products, all of which make a *specific* response to a particular invader of the body (as opposed to a general attack response) and which are characterized by *memory* (an ability to mount a rapid attack when the same type of invader returns). The complement system and other components of the general, inflammatory response are also activated during immune responses.

independent assortment. Mendelian principle that the alleles of two (or more) gene pairs located on nonhomologous chromosomes tend to be assorted independently of one another into gametes.

inflammation. Nonspecific defense response involving mobilization of phagocytic cells and the complement system; a series of homeostatic events that restore damaged tissues and intercellular conditions.

instinctive behavior. The capacity of an animal to complete a fairly complex, stereotyped response to a first-time encounter with a key stimulus (without having had prior experience with that stimulus).

integration, neural [L. *integrare*, to coordinate]. Moment-by-moment summation of all excitatory and inhibitory synapses acting on a neuron; occurs at each level of synapsing in a nervous system.

interneuron. Main component of integrating centers such as the brain and spinal cord; integrates information arriving from sensory neurons, then influences other neurons in turn.

interphase. Time interval (variable among species) in which a cell increases its mass, approximately doubles the number of its structures and organelles, and finally replicates its DNA, prior to nuclear division.

interstitial fluid (IN-ter-STISH-ul) [L. *interstitus*, to stand in the middle of something]. In vertebrates, that portion of the extracellular fluid occupying spaces between cells and tissues. (The remaining portion is blood plasma.)

ion, negatively charged. An atom or a compound that has gained one or more electrons, hence has acquired an overall negative charge.

ion, positively charged. An atom or a compound that has lost one or more electrons, hence has acquired an overall positive charge.

ionic bond. An association between ions of opposite charge.

isogamy (EYE-soh-gam-ee) [Gk. *isos*, equal, + *gametēs*, husband, and *gametē*, wife]. In some sexual reproductive modes, gametes that are all identical (no differentiation into sperm and eggs).

isolating mechanism. Some aspect of structure, functioning, or behavior that prevents interbreeding between populations that are undergoing or have undergone speciation. Also applies to local breeding units within a population.

isotope (EYE-so-tope). Individual atom that contains the same number of protons as other atoms of a given element, but that has a different number of neutrons.

kidney. Organ of salt and water regulation; its nephron/capillary units are concerned with filtration of water and other noncellular components of blood, selective reabsorption of solutes and most of the water, and tubular secretion (through active transport) of certain substances from the capillaries.

Krebs cycle. Stage of aerobic respiration in which pyruvate fragments are completely broken down into carbon dioxide; molecules reduced in the process can be used in forming many electron carriers for use in the last stage of the aerobic pathway.

larva, plural **larvae.** A sexually immature, free-living and free-feeding animal that grows and develops into the sexually mature adult form.

larynx (LARE-inks). Tube that leads to the lungs. In humans, contains vocal cords, the production site of sound waves used in speech.

lateral meristem. Either vascular cambium or cork cambium, the meristems responsible for secondary growth (increases in diameter) in most plants.

learning. Modification of behavior, arising from specific experiences during the lifetime of an animal.

leucoplast (LEW-kuh-plast). In some plant cells, colorless plastid in which starch grains and other substances may be stored.

life cycle. For any species, the genetically programmed sequence of events by which individuals are produced, grow, develop, and themselves reproduce.

light-dependent reactions. First stage of photosynthesis, concerned with harnessing sunlight and using it as an energy source for synthesizing ATP alone (the cyclic pathway) or ATP and NADPH (the noncyclic pathway).

light-independent reactions. Second stage of photosynthesis, in which sugars and other organic compounds are assembled with the ATP and NADPH produced during the first stage.

linkage. The tendency of genes physically located on the same chromosome to be inherited together instead of undergoing independent assortment.

lipid [Gk. *lipos*, fat]. A hydrocarbon (mostly) that generally does not dissolve in water but will dissolve in nonpolar substances. Some lipids have fatty acid components (e.g., oils, waxes), others do not (e.g., steroids).

lymph vascular system [L. *lympha*, water, + *vasculum*, a small vessel]. Network of vessels that supplements the blood circulation system; reclaims water that has entered interstitial regions from the bloodstream; also transports fats from small intestine to bloodstream. Fluid in its vessels is called *lymph*.

lymphocyte. Any of various white blood cells that take part in vertebrate immune responses.

lymphoid organs. Those organs (and some tissue regions) that function as blood cell production centers and as sites for some defense responses; bone marrow, thymus, lymph nodes, spleen, appendix, tonsils, adenoids, and patches of small intestine.

lymphokine. Any of a class of proteins by which the cells of the vertebrate immune system communicate with one another.

lysosome. Membrane-bound organelle containing hydrolytic enzymes that can break down all polysaccharides, nucleic acids, and proteins as well as some lipids. Central in the cell's materials recycling and biosynthesis programs.

macroevolution. Large-scale rates, trends, and patterns of change among groups of species since the beginning of life.

mantle [L. *mantellum*, a loose external garment]. In mollusks, a body wall surrounding internal parts; secretes substances that form the molluscan shell.

mass number. Total number of protons and neutrons in the nucleus of atoms of a given element.

mechanoreceptor. Sensory cell or cell part that detects mechanical energy associated with changes in pressure, position, or acceleration.

medusa (meh-DOO-sah) [Gk. *Medousa*, one of three sisters in Greek mythology having snake-entwined hair; this image probably evoked by the tentacles and oral arms extending from the medusa]. Free-swimming, bell-shaped stage in cnidarian life cycles.

megaspore. In seed plants, a meiospore that develops into a female gametophyte.

meiosis (my-OH-sis) [Gk. *meioun*, to diminish]. Two-stage nuclear division process in which the number of chromosome sets in each daughter nucleus becomes haploid (half of what it was in the parent nucleus). Basis of gamete and meiospore formation. Compare *mitosis*.

meiospore. Haploid cell that divides by mitosis and ˙differentiates into multicelled haploid bodies (gametophytes).

menopause (MEN-uh-pozz) [L. *mensis*, month, + *pausa*, stop]. End of the period of a human female's reproductive potential.

menstrual cycle. The cyclic reproductive capacity of female humans and other primates.

menstruation. Periodic sloughing of the blood-enriched lining of the uterus when pregnancy does not occur.

meristem (MARE-ih-stem) [Gk. *meristos*, divisible]. In most plants, a mass of self-perpetuating cells not yet committed to developing into a specialized cell type. Compare *apical*, *lateral*, and *primary meristems*.

mesoderm (MEH-so-derm) [Gk. *mesos*, middle, + *derm*, skin]. In most animal embryos, a tissue layer between ectoderm and endoderm; gives rise to muscle, the organs of circulation, reproduction, and excretion, most of the internal skeleton (when present), and connective tissue layers of the gut and body covering.

metabolic pathway. In a cell, breakdown or synthesis reactions that occur in sequential, stepwise fashion.

metabolic reaction. Some form of internal energy change in a cell.

metabolism (meh-TAB-oh-lizm) [Gk. *metabolē*, change]. All activities by which organisms extract and transform energy from their environment, and use it in manipulating materials in ways that assure maintenance, growth, and reproduction.

metamorphosis (met-ah-more-FOE-sis) [Gk. *meta-*, change, + *morphē*, form]. For animals that undergo indirect development, the reactivation of development from the larval stage to the adult form.

metaphase. In mitosis, stage when microtubules increase in number and become organized into mitotic spindle, which is responsible for separating the sister chromatids of chromosomes from each other prior to cell division.

metazoan (MET-ah-ZOE-un). Multicelled animal.

microfilament [Gk. *mikros*, small, + L. *filum*, thread]. Component of the cytoskeleton; involved in cell shape, motion, and growth.

microspore. In seed plants, a meiospore that develops into a male gametophyte.

microtubular spindle. An array of microtubules that helps establish the polarity necessary for chromosome movements during nuclear division.

microtubule. Hollow cylinder of (mostly) tubulin subunits; involved in cell shape, motion, and growth; functional unit of cilia and flagella.

microvillus (MY-crow-VILL-us) [L. *villus*, shaggy hair]. A slender, cylindrical extension of the animal cell surface that functions in absorption or secretion.

migration. A cyclic movement between two distant regions at times of year corresponding to seasonal change.

mimicry (MIM-ik-ree). Situation in which one species (the mimic) bears deceptive resemblance in color, form, and/or behavior to another species (the model) that enjoys some survival advantage.

mitochondrion, plural **mitochondria** (MY-toe-KON-dree-on). Eukaryotic organelle that specializes in aerobic respiration.

mitosis (my-TOE-sis) [Gk. *mitos*, thread]. Nuclear division in which the number of chromosome sets is maintained from one cell generation to the next. Basis of reproduction of single-celled eukaryotes; basis of physical growth (through cell divisions) in multicelled eukaryotes.

molecule [diminutive of L. *moles*, mass]. A unit of two or more atoms of the same or different elements bonded together.

monocot. Short for monocotyledon; a flowering plant in which seeds have only one cotyledon, whose floral parts generally occur in threes (or multiples of threes), and whose leaves typically are parallel-veined. Compare *dicot*.

monomer (MON-oh-mur). Any of numerous individual subunits that become incorporated into polymers.

monosaccharide (MON-oh-SAK-ah-ride) [Gk. *mono*, alone, single, + *sakcharon*, sugar]. A sugar monomer, typically with a backbone of three to seven carbon atoms, an aldehyde or ketone group, and two or more hydroxyl groups.

morphogenesis (MORE-foe-GEN-ih-sis) [Gk. *morphē*, form, + *genesis*, origin]. The growth, shaping, and arrangement of body parts according to genetically predefined patterns. The extent, direction, and rate of morphogenesis depend on genetic controls and environmental factors.

motor neuron. Type of neuron that relays information away from integrating centers (such as the brain and spinal cord) to the body's effectors (muscles and glands).

multiple allele system. More than two forms of alleles that can occur at a given gene locus.

muscle fiber. Contractile cell of skeletal, smooth, or cardiac muscle tissue.

mutation [L. *mutatus*, a change, + *-ion*, result of a process or an act]. A heritable change in the kind, structure, sequence, or number of the component parts of a DNA molecule.

mutualism [L. *mutuus*, reciprocal]. A major type of community interaction from which both members of a pair of species benefit clearly and directly. When such a mutually beneficial relationship involves continuous, intimate contact, it is called symbiosis.

mycelium, plural **mycelia** (my-SEE-lee-um) [Gk. *mykēs*, fungus, mushroom, + *hēlos*, callus]. A multicelled structure, in the form of a mesh of branching, microscopic filaments, that forms during the life cycle of most fungi.

NAD⁺. Nicotinamide adenine dinucleotide, oxidized form.

NADH. Nicotinamide adenine dinucleotide, reduced form.

NADP⁺. Nicotinamide adenine dinucleotide phosphate, oxidized form.

NADPH. Nicotinamide adenine dinucleotide phosphate, reduced form.

natural selection. The result of differential reproduction of genotypes within a population, this being one of the most important mechanisms bringing about evolutionary change. See also *differential reproduction*.

negative feedback mechanism. Homeostatic control whereby an increase in some substance or activity sooner or later inhibits the very processes leading to (or allowing) the increase.

nematocyst (NEM-add-uh-sist) [Gk. *nēma*, thread, + *kystis*, bladder, pouch]. A stinging capsule that assists in capturing prey and that may serve in protection; a distinguishing feature of cnidarians such as jellyfishes.

nephridium, plural **nephridia** (neh-FRID-ee-um). In invertebrates such as earthworms, a system for regulating water and solute levels.

nephron (NEH-frohn) [Gk. *nephros*, kidney]. One of more than a million long, slender tubules in the kidney in which urine is formed by processes of filtration, reabsorption, and secretion.

nerve. Cordlike communication line of nervous systems, composed of axons of sensory or motor neurons (or both) packed tightly in bundles within connective tissue. In the brain and spinal cord, such bundles are called nerve pathways or tracts.

nerve impulse. Action potential.

nerve net. Cnidarian nervous system consisting of nerve cells associated with the gastrodermis and epidermis and concerned primarily with feeding behavior.

nervous system. Constellations of neurons oriented relative to one another in precise message-conducting and information-processing pathways.

neuroglia (NUR-oh-GLEE-uh). Cells intimately associated with neurons and functioning in their structural and metabolic support, maintenance, and in some cases as axonal sheaths. In vertebrates they represent at least half of the volume of the nervous system.

neuromuscular junction. The synapses between the splayed-out axon terminals of a motor neuron and a muscle cell. The terminals are positioned in troughs (in the muscle cell membrane) called the motor end plate.

neuron. In most animals, a cell that responds to specific chemical, electrical, or mechanical stimuli in three ways: it can *integrate* different incoming signals, *propagate* excitation as a pulse of information along its plasma membrane, and *transmit* information about change to other neurons, muscles, or glands.

neutron. Subatomic particle of about the same size and mass as a proton but having no electric charge.

niche (nitch) [L. *nidus*, nest]. The full range of abiotic and biotic conditions under which a particular species can live and reproduce.

nicotinamide adenine dinucleotide, or **NAD⁺**. A local electron carrier that transfers hydrogen atoms and electrons *within* metabolic pathways. NAD⁺ is a free-moving carrier, not membrane-bound in a transport system.

nicotinamide adenine dinucleotide phosphate. NADPH. Together with ATP, a major coupling agent *between* degradative and biosynthetic pathways.

nitrification. Process by which certain soil bacteria strip ammonia or ammonium of electrons, and nitrite (NO_2^-) is released as a reaction product, then other soil bacteria use nitrite for energy metabolism, yielding nitrate (NO_3^-).

nitrogen fixation. Among some bacteria, assimilation of gaseous nitrogen (N_2) from the air; through reduction reactions, electrons (and associated H^+) become attached to the nitrogen, thereby forming ammonia (NH_3) or ammonium (NH_4^+).

noncyclic photophosphorylation (non-SIK-lik foe-toe-FOSS-for-ih-LAY-shun) [L. *non*, not, + Gk. *kylos*, circle]. Photosynthetic pathway in which new electrons derived from water molecules flow through two photosystems and two transport chains, the result being formation of ATP and NADPH.

notochord (NO-toe-kord) [Gk. *nōtos*, back, + L. *chorda*, cord]. A stiffened but flexible supporting structure, just beneath the nerve cord, that is present during at least some stages of the life cycle of all chordates. During embryonic development of complex vertebrates, it is replaced by a vertebral column.

nuclear envelope. Double membrane forming the surface boundary of a eukaryotic nucleus.

nucleic acid (new-CLAY-ik). Long-chain, single- or double-stranded nucleotide; DNA and RNA are examples.

nucleoid. In bacterial cells, the irregularly shaped region in which DNA is concentrated but not bound by a membrane.

nucleolus (new-KLEE-oh-lus) [L. *nucleolus*, a little kernel]. Within the nucleus of a non-dividing cell, a mass of proteins, RNA, and other material used in ribosome synthesis.

nucleotide (NEW-klee-oh-tide). Molecule containing a five-carbon sugar (ribose), a nitrogen-containing base (either purine or pyrimidine), and a phosphate group. Structural unit of adenosine phosphates, nucleotide coenzymes, and nucleic acids.

nucleus (NEW-klee-us) [L. *nucleus*, a kernel]. In atoms, the central core of one or more positively charged protons and (in all but hydrogen) electrically neutral neutrons. In eukaryotic cells, the membranous organelle that houses the DNA.

omnivore [L. *omnis*, all, + *vovare*, to devour]. An organism able to obtain energy from more than one source rather than being limited to one trophic level.

oogamy (oo-AH-gam-ee) [Gk. *ōion*, egg, + *gamos*, marriage]. In some sexual reproductive modes, gametes that differ in size and motility. One gamete typically is small and motile (a sperm, for example); the other is larger and nonmotile (the egg).

oogenesis (oo-oh-JEN-uh-sis). Formation of a female gamete, from a germ cell (oogonium) to a mature haploid ovum.

operon. A set of related protein-coding genes transcribed as a coordinated unit under the direction of a single set of controls built into the DNA.

organ. A structure of definite form and function that is composed of more than one tissue and the character of which is influenced by the type, combination, and arrangement of those tissues.

organelle. Any of various membranous sacs, envelopes, and other compartmented portions of cytoplasm that separate different, often incompatible metabolic reactions in the space of the cytoplasm and in time (through specific reaction sequences).

osmosis (oss-MOE-sis) [Gk. *ōsmos*, act of pushing]. Movement of water molecules across a differentially permeable membrane in response to a concentration and/or pressure gradient.

ovary. The primary female reproductive organ in which oogenesis occurs.

oviduct (OH-vih-dukt). Passageway through which ova travel from the ovary to the uterus.

ovule (OH-vewl) [L. *ovum*, egg]. In seed-bearing plants, the structure destined to become the seed; includes the nucellus, integuments, and the stalk attaching them to the ovarian wall. Cell divisions in the nucellus give rise to the female gametophyte, with its egg cell and endosperm mother cell.

oxidation. The loss of one or more electrons from an atom or molecule.

oxidation-reduction reaction. An electron transfer from one atom or molecule to another. Often hydrogen is also transferred along with the electron or electrons.

oxidative phosphorylation. Use of electron energy being released during oxidation reactions to phosphorylate (tack a phosphate group onto) a molecule such as ADP (which yields energy-rich ATP).

parapatric speciation [Gk. *para*, alongside, + *patria*, native land]. Speciation that occurs when adjoining populations undergo divergence despite some gene flow.

parasite [Gk. *para*, alongside, + *sitos*, food]. A type of heterotroph that lives on or in a living host organism during some part of its life cycle, obtains nutrients from the host's tissues, and may or may not end up killing the host as a consequence of the association (but usually produces an intermediate level of negative effects).

parasitoid. A type of larva that kills a host insect by completely consuming its soft tissues before the host metamorphoses into an adult.

parasympathetic nerves. A division of the autonomic nervous system, consisting of those efferent nerves concerned more with slowing down overall body activity and diverting energy to basic housekeeping tasks. Compare *sympathetic nerves*.

passive transport. Movement of a substance across a cell membrane without any direct energy outlay by the cell. Diffusion is an example.

pathogen (PATH-oh-jen) [Gk. *pathos*, suffering, + *-genēs*, origin]. Disease-causing organism.

penis. Component of the male reproductive system of many species; the copulatory organ by which sperm are deposited into a specialized duct of the female reproductive system.

perennial [L. *per-*, throughout, + *annus*, year]. A plant that lives year after year.

pericycle (PARE-ih-sigh-kul) [Gk. *peri-*, around, + *kyklos*, circle]. One or more layers of parenchyma cells, just inside the endodermis of the root vascular column, that maintain the potential for meristematic activity. Gives rise to lateral roots and, in species showing secondary growth, contributes to the formation of vascular cambium and cork cambium.

periderm. In stems and roots of gymnosperms and flowering plants, a protective covering that replaces epidermis during secondary growth.

peripheral nervous system (per-IF-ur-uhl) [Gk. *peripherein*, to carry around]. In vertebrates, the nerves leading into and out from the spinal cord and brain and the ganglia along those communication lines.

pH. Whole number referring to the number of hydrogen ions present in a liter of a given fluid.

phagocytosis (FAG-uh-sigh-TOE-sis) [Gk. *phagein*, to eat, + *kytos*, hollow vessel]. Engulfment of foreign cells or substances by amoebas and some white blood cells, by means of endocytosis.

pharynx (FAR-inks). A muscular tube by which food is taken into the gut. In humans, the gateway to the digestive tract and to the windpipe (trachea).

phenotype (FEE-no-type) [Gk. *phainein*, to show, + *typos*, image]. Observable trait or traits of an individual; arises from interactions between genes, and between genes and the environment.

pheromone (FARE-oh-moan) [Gk. *phero*, to carry, + *-mone*, as in hormone]. A chemical secreted by an exocrine gland that serves as a communication signal between individuals of the same species.

phloem (FLOW-um). The food-conducting tissue of vascular plants, the main components of which are sieve tube members and companion cells (in flowering plants) or sieve cells (in gymnosperms).

phospholipid. A key component of cell membranes in plants and animals; a molecule with a glycerol backbone, two fatty acid tails, and a phosphate group to which an alcohol is attached.

phosphorylation (FOSS-for-ih-LAY-shun). Addition of one or more phosphate groups to a molecule.

photolysis (foe-TALL-ih-sis) [Gk. *photos*, light, + *-lysis*, breaking apart]. First step in noncyclic photophosphorylation, when water is split into oxygen, hydrogen, and associated electrons; photon energy indirectly drives the reaction.

photoreceptor. Light-sensitive sensory cell.

photosynthesis. The trapping of solar energy and its conversion to chemical energy (ATP, NADPH, or both), which is used in manufacturing food molecules from carbon dioxide and water.

photosynthetic autotroph. An organism able to build all of the organic molecules it requires using carbon dioxide as the carbon source and sunlight as the energy source. All plants, some protistans, and a few bacteria are photosynthetic autotrophs.

photosystem. Functional light-trapping unit in photosynthetic membranes; contains pigment molecules and enzymes.

phototropism [Gk. *photos*, light, + *trope*, turning, direction]. Movement or growth curvature toward light.

phytochrome. Light-sensitive pigment molecule whose activation and inactivation trigger hormone activities governing leaf expansion, stem branching, stem length, and, in many plants, seed germination and flowering.

phytoplankton (FIE-toe-PLANK-tun) [Gk. *phyton*, plant, + *planktos*, wandering]. Community of photosynthetic microorganisms in freshwater or saltwater environments.

pinocytosis (PIN-oh-sigh-TOE-sis) [Gk. *pinein*, to drink, + *kytos*, vessel]. "Cell-drinking"; engulfment of liquid droplets.

placenta (play-SEN-tuh). In the uterus, an organ made of extensions of extraembryonic membranes (the chorion especially) and the endometrium. Through this composite of embryonic and maternal tissues and vessels, nutrients reach the embryo and wastes are carried away.

plant. Generally, multicelled autotroph able to build its own food molecules through photosynthesis.

plasma. Liquid component of blood in which numerous plasma proteins, ions, simple sugars, amino acids, vitamins, hormones, and gases are dissolved.

plasma membrane. Outermost membrane of a cell. Its surface has molecular regions that detect changes in external conditions. Spanning the membrane are passageways through which substances move inward and outward in controlled ways. Transport vesicles form from or fuse with its lipid bilayer.

plasmid. In some bacteria, a small circle of DNA in addition to the main DNA molecule.

plasmodesma (PLAZ-moe-DEZ-muh). In a multicelled plant, a junction between the linked walls of adjacent cells through which nutrients and other substances are transported.

plastid. In some plant cells, a storage organelle; some plastids also function in photosynthesis.

platelet. Component of blood that functions in clotting.

pleiotropism (PLEE-oh-TROE-pizm) [Gk. *pleōn*, more, + *trope*, direction]. Multiple phenotypic effect of a single gene; the action of the gene affects many developmental or maintenance activities.

pollen grain [L. *pollen*, fine dust]. In gymnosperms and flowering plants, the immature male gametophyte (gamete-producing body).

pollination. The transfer of pollen grains to the female gametophyte.

pollutant. Any substance with which an ecosystem has had no prior evolutionary experience, in terms of kinds or amounts, and that can accumulate to disruptive or harmful levels. Can be naturally occurring or synthetic.

polymer (POH-lih-mur) [Gk. *polus*, many, + *meris*, part]. A molecule composed of from three to millions of subunits of relatively low molecular weight that may or may not be identical.

polymorphism (poly-MORE-fizz-um) [Gk. *polus*, many, + *morphe*, form]. In a population, the persistence of two or more forms of a trait, at a frequency that is greater than can be maintained by newly arising mutations alone; and that frequency, if changed, will return to its former value over several generations.

polyp (POH-lip). Vase-shaped, sedentary stage of cnidarian life cycles.

polypeptide. Chain of amino acids linked by peptide bonds, which form through condensation reactions.

polyploid. Having three or more sets of chromosomes in the somatic cells of a species.

polyribosome. During protein synthesis, a clustering of ribosomes engaged in translation of a messenger RNA molecule.

polysaccharide [Gk. *polus*, many, + *sakcharon*, sugar]. Three or more monosaccharides bonded together covalently.

population. Group of individuals of the same species occupying a given area.

predator [L. *prehendere*, to grasp, seize]. Any of various organisms that get food from other living organisms (their prey), but they do not live on or in the prey and they may or may not kill it. Compare *parasite*.

primary growth. Following seed germination, the cell divisions, elongation, and differentiation that produce the primary plant body.

primary productivity, gross. The total rate at which the autotrophs of an ecosystem fix energy in organic compounds (as plants do in photosynthesis).

primary productivity, net. The rate of energy storage in the tissues of autotrophs in excess of the rate of respiration during a measured time interval.

prion (PRY-on). Infectious agent apparently consisting of protein only and typically causing slow but fatal diseases of the central nervous system.

procambium (pro-KAM-bee-um). A primary meristem (formed from apical meristem) that gives rise to the vascular tissues of the primary plant body.

producer. An autotrophic organism; able to build its own complex organic molecules from simple inorganic substances in the environment.

prokaryote (pro-CARRY-oht) [L. *pro*, before, + Gk. *karyon*, kernel]. Single-celled organism that has no membrane-bound nucleus or other internal organelles; all bacteria are prokaryotes.

prokaryotic fission. Form of bacterial cell reproduction; membrane growth divides the replicated DNA and the cytoplasm into daughter cells.

prophase. In mitosis, the stage when chromatin coils into compact chromosome bodies. In meiosis I, the stage when crossing over occurs.

protein. Molecule composed of one or more chains of amino acids (polypeptide chains).

protistan (pro-TISS-tun) [Gk. *prōtistos*, primal, very first]. Single-celled eukaryote.

proton. Positively charged unit of energy that is found in the atomic nucleus.

protostome (PRO-toe-stome) [Gk. *proto*, first, + *stoma*, mouth]. Any of those bilateral animals, including annelids, arthropods, and mollusks, in which the first opening that appears in the gastrula stage of development becomes the mouth and the second opening, the anus. Also characterized by spiral cleavage and formation of the coelom by the splitting of two masses of mesoderm on either side of the gut.

pseudocoelomate (SOO-doe-SEE-la-mate) [Gk. *pseudos*, false, + *koilos*, a hollow]. Type of invertebrate in which the body lacks a continuous peritoneal lining (which occurs in animals having a coelom).

pseudopod (SOO-doe-pod). "False foot"; a nonpermanent cytoplasmic extension of the cell body.

pulmonary circulation. Pathways of blood flow leading to and from the lungs.

purine. Nucleotide base having a double ring structure. Examples are adenine and guanine.

pyrimidine (pih-RIM-ih-deen). Nucleotide base having a single ring structure. Cytosine and thymine are examples.

pyruvate (PIE-roo-vate). Three-carbon compound produced by the initial breakdown of a glucose molecule during glycolysis.

radial symmetry. Body plan in which the body can be divided into four or more pieces that are equal with respect to the structures they contain.

radicle. In plant seeds, the embryonic root; typically, longitudinal growth of its cells gives rise to a slender primary root.

receptor. Sensory cell or cell part that may be activated by a specific stimulus in the internal or external environment.

recessive allele [L. *recedere*, to recede]. In the heterozygous state, an allele whose expression is fully or partially masked by expression of its partner; recessive alleles can be fully expressed in the homozygous state.

recombinant DNA. Whole molecules or fragments that incorporate parts of different parent DNA molecules, as formed by natural recombination mechanisms or by recombinant DNA technology.

reduction. The gaining of one or more electrons by an atom or molecule.

reflex [L. *reflectere*, to bend back]. A simple, stereotyped, and repeatable motor action that is elicited by a sensory stimulus.

reproduction, asexual. Production of new individuals by any process that does not involve gametes.

reproduction, sexual. Process of reproduction that begins with meiosis, proceeds through gamete formation, and ends at fertilization.

reproductive isolating mechanism. Any aspect of structure, functioning, or behavior that prevents successful interbreeding (hence gene flow) between populations or between local breeding units within a population.

respiration [L. *respirare*, to breathe]. In most animals, the overall exchange of oxygen from the environment and carbon dioxide wastes from cells by way of circulating blood. Compare *aerobic respiration*.

resting membrane potential. Steady voltage difference that exists across the plasma membrane of a neuron (or some other excitable cell) that is not being stimulated.

rhizoid (RYE-zoid) [Gk. *rhiza*, root]. Long, single cell or filament that anchors some gametophytes to the ground.

ribonucleic acid (RYE-bow-new-CLAY-ik). RNA; a category of nucleotides used in translating the genetic message of DNA into actual protein structure.

ribosome. In both prokaryotic and eukaryotic cells, a structure made of RNA and proteins and the site of protein synthesis.

salt. Ionic compound (such as NaCl) that forms by the reaction between an acid and a base; usually dissociates into positive and negative ions in water.

saprobe. Heterotroph that obtains nutrients from nonliving organic matter. Most fungi are saprobes.

sarcolemma (SAR-koe-LEM-uh) [Gk. *sarx*, the flesh, + *lemma*, husk]. The plasma membrane surrounding a muscle fiber.

sarcomere (SAR-koe-meer). Fundamental unit of contraction in skeletal muscle; repeating bands of actin and myosin that appear between two Z-lines.

sarcoplasmic reticulum (SAR-koe-PLAZ-mik reh-TIK-you-lum). A continuous system of membrane-bound chambers that surrounds myofibrils within a muscle fiber and that stores calcium ions necessary for the mechanism of muscle contraction.

second law of thermodynamics. When left to itself, any system along with its surroundings undergoes energy conversions, spontaneously, to less organized forms. When that happens, some energy gets randomly dispersed in a form (often evenly distributed, low-grade heat) that is not as readily available to do work.

secondary growth. In vascular plants, an increase in stem and root diameter, made possible by the activity of two types of lateral meristems (vascular cambium and cork cambium).

seed. In gymnosperms and flowering plants, a fully mature ovule (contains the plant embryo), with its integuments forming the seed coat.

segmentation. In many animal species, a series of body units that may be externally similar to or quite different from one another.

segregation, allelic [L. *se-*, apart, + *grex*, herd]. Mendelian principle that two units of heredity (alleles) exist for a trait, and that during gamete formation, the two units of each pair are separated from each other and end up in different gametes.

semen (SEE-mun) [L. *serere*, to sow]. Sperm-bearing fluid expelled from the penis during male orgasm.

semiconservative replication [Gk. *hēmi*, half, + L. *conservare*, to keep]. Manner in which a DNA molecule is reproduced; formation of a complementary strand on each of the unzipping strands of a DNA double helix, the outcome being two "half-old, half-new" molecules.

senescence (seh-NESS-sens) [L. *senescere*, to grow old]. Sum total of processes leading to death of a plant or any of its organs; cause appears to be built into the life cycle of the species.

sensory neuron. Type of neuron that carries signals about changing conditions into integrating centers (such as the central nervous system).

sex chromosomes. In most animals and some plants, chromosomes that differ in number or kind between males and females but that still function as homologues during meiosis. All other chromosomes are called autosomes.

sex-linked gene. A gene located only on a female X chromosome; has no allelic partner on the male Y chromosome.

sexual selection. Mode of natural selection based on any trait that gives the individual a preferential advantage in mating and in producing offspring.

site-specific recombination. Genetic recombination in which the exchange is limited to a short, specific base sequence that is identical on both of two interacting DNA molecules; a single enzyme mediates the exchange. Compare *homologous recombination*.

sodium-potassium pump. A carrier protein spanning the lipid bilayer of the plasma membrane. When the protein receives an energy boost from ATP, its shape changes in such a way that it selectively transports sodium ions *out* of the cell and potassium ions *in*. This active transport process maintains the ion distributions (hence the voltage difference) across the membrane.

solute (SOL-yoot) [L. *solvere*, to loosen]. Any substance dissolved in some solution. In water, this means its individual molecules are surrounded by spheres of hydration that keep their charged parts from interacting, so the molecules remain dispersed in the water.

solvent. Fluid in which one or more substances is dissolved.

somatic cell (so-MAT-ik) [Gk. *sōma*, body]. Any cell of the animal body that is not a germ cell (which develops by meiosis into sperm or eggs).

somatic nervous system. Those efferent nerves leading from the central nervous system to skeletal muscles.

species (SPEE-sheez) [L. *species*, a kind]. For sexually reproducing organisms, one or more populations whose members interbreed under natural conditions and produce fertile offspring, and who are reproductively isolated from other such groups.

sperm [Gk. *sperma*, seed]. Mature male gamete.

spermatogenesis (sperm-AT-oh-JEN-ih-sis). Formation of a mature sperm from a germ cell (spermatogonium).

sphere of hydration. Through positive or negative interactions, a clustering of water molecules around the individual molecules of a substance placed in water. Compare *solute*.

sphincter (SFINK-tur). Ring of muscle that serves as a gate between regions of a tubelike system (as between the stomach and small intestine).

sporangium, plural **sporangia** (spore-AN-gee-um) [Gk. *spora*, seed]. Protective layer of sterile cells surrounding haploid spores in a sporophyte.

spore. In plants, a meiospore that gives rise to one or more haploid gametophytes. Among fungi, an asexual reproductive cell that gives rise to new hyphae of the fungal mat (mycelium).

sporophyte [Gk. *phyton*, plant]. Diploid, spore-producing stage of plant life cycles.

sporozoite. Infective, sporelike stage of sporozoan life cycles.

stabilizing selection. Mode of natural selection that decreases the frequency of alleles giving rise to extreme forms of a trait, such that intermediate forms already well adapted to prevailing conditions are favored.

stamen (STAY-mun). In flowering plants, the male reproductive structure; commonly consists of pollen-bearing structures (anthers) positioned on single stalks (filaments).

steroid (STAIR-oid). A lipid with a backbone of four carbon rings. Steroids differ in the number and location of double bonds in the backbone and in their number, position, and type of functional groups.

stimulus [L. *stimulus*, goad]. Any form of energy that the body is able to detect by means of its receptors.

stoma, plural **stomata** (STOW-muh) [Gk. *stoma*, mouth]. An opening, defined by two guard cells, across the epidermis of a leaf or stem, through which water vapor moves out of the plant and carbon dioxide moves in, in amounts governed by controls over stomatal widening and closure.

stroma [Gk. *strōma*, bed]. In chloroplasts, the semifluid matrix, surrounding the grana, where complex organic molecules are assembled.

substrate. Molecule or molecules of a reactant on which an enzyme acts.

succession, primary (suk-SESH-un) [L. *succedere*, to follow after]. The orderly changes in species composition from the pioneer species that inhabit an area previously devoid of life to the more or less constant array of species that constitutes the climax community.

succession, secondary. Reestablishment of a climax community that has been disrupted in whole or in part.

surface-to-volume ratio. In cells, a physical constraint on increased size: as the cell's linear dimensions grow, its surface area does not increase at the same rate as its volume (hence each unit of plasma membrane would be called upon to serve increasing amounts of cytoplasm).

symbiosis (sim-by-OH-sis) [Gk. *syn*, together, + *bios*, life, mode of life]. A mutually beneficial relationship involving continuous, intimate contact between species. Compare *mutualism*.

sympathetic nerves. A division of the autonomic nervous system, consisting of those efferent nerves concerned with slowing down the body's housekeeping tasks and with increasing overall body activity during times of stress, danger, excitement, and heightened awareness. Compare *parasympathetic nerves*.

sympatric speciation (sim-PAT-rik) [Gk. *syn*, together, + *patria*, native land]. The origin of a species as a result of ecological, behavioral, or genetic barriers that arise *within* the boundaries of a single population. The emergence of a new polyploid species of plant is an example.

synapse, chemical (SIN-aps) [Gk. *synapsis*, union]. A junction between two neurons, or between a neuron and a muscle or gland cell, that are separated by a small gap. At an *excitatory* synapse, a transmitter substance released from the first neuron produces changes in the receiving cell that bring its membrane closer to threshold. At an *inhibitory* synapse, a transmitter substance released from the first neuron produces changes in the receiving cell that drive membrane potential away from threshold.

synapsis (sin-AP-sis). At prophase I of meiosis, the point-by-point alignment of the two sister chromatids of each chromosome with the two sister chromatids of its homologue.

systemic circulation. Pathways of blood flow leading to and from all body parts except the lungs.

telophase (TEE-low-faze). Final stage of mitosis, during which the separated chromosomes decondense and become enclosed within newly forming daughter nuclei.

testis, plural **testes**. Male gonad; primary reproductive organ in which male gametes and sex hormones are produced.

threshold value. The minimum voltage change across the plasma membrane necessary to produce an action potential in neurons and some other excitable cells.

thymine. Nitrogen-containing base found in some nucleotides.

tissue. A group of cells and intercellular substances functioning together in a specialized activity.

tonicity. The relative concentrations of solutes in the fluid inside and outside the cell. When solute concentrations are *isotonic* (equal in both fluids), water shows no net osmotic movement in either direction. When one of the fluids is *hypotonic* (has less solutes than the other), the other is *hypertonic* (has more solutes) and is the direction in which water tends to move.

trachea, plural **tracheae** (TRAY-kee-uh). A tube for breathing; in land vertebrates, the windpipe that carries air between the larynx and bronchi.

tracheid (TRAY-kid). Typically elongated cell, dead at maturity, that passively conducts water and solutes in xylem.

transcription [L. *trans*, across, + *scribere*, to write]. The assembly of an RNA strand on one of the two strands of a DNA double helix; the resulting transcript has a nucleotide sequence that is complementary to the DNA region on which it is assembled.

translation. The interaction of rRNA, tRNA, and mRNA in converting the DNA instructions encoded in the mRNA molecule into a particular sequence of amino acids to form a polypeptide chain.

translocation. In vascular plants, the transport of soluble food molecules (mostly sucrose) from one plant organ to another by way of the phloem tissue.

transmitter substance. Chemical secretion, released in tiny amounts from a neuron, that triggers change in the membrane potential of an adjacent cell.

transpiration. Evaporative water loss from stems and leaves.

transposition. Genetic recombination involving nonhomologous base sequences in which transposable elements move from one site in the DNA to another. Transposable elements are also called jumping genes.

trophic level (TROE-fik) [Gk. *trophos*, feeder]. All organisms that are the same number of energy transfers away from the original source of energy (e.g., sunlight) that enters an ecosystem.

turgor pressure (TUR-gore) [L. *turgere*, to swell]. Internal pressure on a cell wall caused by osmotic movement of water into the cell body.

urinary system. An organ system concerned with regulating water and solute levels in the body.

uterus (YOU-tur-us) [L. *uterus*, womb]. A chamber in which the developing embryo is contained and nurtured during pregnancy.

vacuole, central (VAK-you-ohle). In plant cells, a membrane-bound, fluid-filled sac that may take up most of the cell interior; main function is to increase cell size and surface area, thereby enhancing absorption of relatively dilute concentrations of nutrients from the external environment.

vagina. Part of the female reproductive system that receives sperm from the male penis; forms part of the birth canal, and acts as a channel to the exterior for menstrual flow.

vascular cambium. In vascular plants, one of the lateral meristems that increase stem or root diameter.

vernalization [L. *ver*, spring]. Cold-temperature stimulation of the flowering process.

vertebra, plural **vertebrae**. One of a series of hard bones that form the backbone in most chordates.

vertebrate. Animal having a backbone made of bony segments called vertebrae.

vesicle (VESS-ih-kul) [L. *vesicula*, little bladder]. In cytoplasm, a small, membrane-bound sac in which various substances may be transported or stored.

vessel element. Typically elongated cell, dead at maturity, that passively conducts water and solutes in xylem.

viroid. An infectious nucleic acid that has no protein coat; a tiny rod or circle of single-stranded RNA.

virus. Infectious agent consisting of nucleic acid encased in protein; incapable of metabolism or reproduction without a host cell, hence is often not considered alive.

vision. Precise light focusing onto a layer of photoreceptive cells that is dense enough to sample details concerning a given light stimulus, followed by image formation in the brain.

water potential. The sum of two opposing forces (osmosis and turgor pressure) that can cause the directional movement of water into or out of a walled cell.

white matter. Mainly axons of interneurons, so named because of the glistening myelin sheaths around them.

wild-type allele. The normal or most common allele at a given gene locus.

xylem (ZYE-lum) [Gk. *xylon*, wood]. In vascular plants, a tissue that transports water and solutes through the plant body.

zygote (ZYE-gote). In most multicelled eukaryotes, the first diploid cell formed after fertilization (fusion of nuclei from a male and a female gamete).

INDEX

Italic numerals refer to illustrations.

E